Pour Marion

Contents

Introduction

This initiation to the theory of probability and random processes has three objectives. The first is to provide the *theoretical background* necessary for the student to feel comfortable and secure in the utilization of probabilistic models, particularly those involving stochastic processes. The second is to introduce the *specific stochastic processes* most often encountered in applications (operations research, insurance, finance, biology, physics, computer and communications sciences and signal processing), that is, Markov chains in discrete and continuous time, Poisson processes in time and space, renewal and regenerative processes, queues and their networks, wide-sense stationary processes, and Brownian motion. In addition to the above topics, which form the indispensable cultural background of an applied probabilist, this text gives—and this is the third objective—the *advanced tools* such as martingales, ergodic theory, Palm theory and stochastic integration, useful for the analysis of the more complex stochastic models.

The only prerequisite is an elementary knowledge of calculus and of linear algebra usually acquired at the undergraduate or beginning graduate level, the occasional material beyond this being given in the Appendix or in the main text.

The above cartoon symbolizes the feeling of awe and distress that may invade the potential reader's mind when considering the size of the book. The fact is that this text has been written for an ideal reader who has had no previous exposure to probability theory and who is willing to learn in a *self-teaching* mode the basics of the theory of stochastic processes that will constitute a solid platform both for applications and for more specialized studies. This implies that the material should be presented in a *detailed, progressive* and *self-contained* manner. However, there is sufficient modularity to devise a table of contents adapted to one's appetite and interest.

The material of this book has been divided into three parts roughly corresponding to the three objectives listed at the beginning of this introduction.

PART ONE: Probability theory

This part stands alone as a short course in probability at the intermediate level. It contains all the results referenced in the rest of the book, and the initiate can therefore skip it and use it as an appendix.

The first chapter introduces the basic notions in the elementary framework of discrete random variables and gives a few tricks that permit us to obtain at an early stage non-trivial results such as the strong law of large numbers for coin tossing or the extinction probability of a branching process. One of its purposes is to persuade the neophyte of the power of a formal approach to probability while introducing the main concepts of expectation, independence and conditional expectation. Having acquired familiarity with the vocabulary and the spirit of probability theory, the reader will be ready for the development of this discipline in the framework of integration theory, of which the second chapter provides a detailed account. This theory requires more concentration from the beginner but the effort is worthwhile as it will provide her/him with enough confidence in the manipulation of stochastic models. Probability theory is usually developed in the more theoretical texts without a preliminary exposition of integration theory, whose results are then presented as need arises. But, as far as stochastic processes are concerned, the theory of integration with respect to a finite measure (such as a probability) is not sufficient. The third chapter translates the previous one into the probabilistic language and formalizes the concepts of distribution, independence and conditional expectation. The fourth chapter features the various notions of convergence of a sequence of random variables (almost-sure, in distribution, in variation, in probability and in the mean square) and their interconnections, and closes the first part on the probabilistic background directly useful in the rest of the book.

PART TWO: Standard stochastic processes

This part forms a basic course on stochastic processes.

It begins with the pivotal Chapter 5, devoted to general issues such as trajectory continuity, measurability and stopping times. Chapter 6 and Chapter 7 introduce the stochastic processes that are the most popular and that can be treated at an elementary level, namely discrete-time homogeneous Markov chains, homogeneous Poisson process on the line and continuous-time homogeneous Markov chains. Chapter 8 is devoted to Poisson processes in space, with or without marks, a versatile source of spatial models. Chapter 10 gives the essentials of renewal theory and its application to regenerative processes. Chapter 9 gives a panoramic view of the classical queues and their networks at the elementary level. Chapter 11 features the Brownian motion and the Doob–Wiener stochastic integral. The latter is, together with Bochner's representation of characteristic functions, one of the foundations of the theory of wide-sense stationary stochastic processes of Chapter 12.

PART THREE: Advanced topics

This part complements the previous one in that it is not exclusively devoted to specific random processes, but rather to general classes of such processes. Only a taste of the topics treated here will be given since each one of them requires and deserves considerably more space.

Chapter 13 gives the basic theory of martingales, one of the most important all-purpose tools of probability theory. Chapter 14 is a short introduction to the Brownian motion stochastic calculus based on the Itô integral. It is a natural continuation of the chapters on Brownian motion and on martingales. Chapter 15, a novel item in the table of contents of a textbook on stochastic processes, introduces point processes on the line admitting a stochastic intensity and the associated stochastic calculus. Chapter 16 gives the essentials of ergodic theory. The presentation is not the classical one and gives the opportunity to extend the elementary results of Chapter 9 on queueing. Another novel item of the table of contents for a book at this level is Chapter 17 on Palm probability, a natural complement to the theory of renewal point processes.

Practical issues

The index gives the page(s) where a particular notation or abbreviation is used. These items will appear at the beginning of the list corresponding to their first letter. The special numbering of equations, such as ($\star$), ($\dagger$) and the like, is used only locally, inside proofs.

Just before the Exercises section of each chapter there is a subsubsection entitled "Complementary reading" pointing at books (only books) where additional material connected with the current topic can be found. The selection is mainly based on two criteria: accessibility by a reader of this book and relevance to applications (with, of course, a natural bias towards the author's own interests). No attempt at exhaustivity or proper crediting has been made, the reader being directed to the Bibliography or to the sporadic footnotes for this. In these subsubsections, only the year of the last edition is given. The full history is given in the Bibliography.

Acknowledgements

I wish to acknowledge the precious help of Eva Löcherbach (University of Paris I Sorbonne), Anne Bouillard (Nokia France), Paolo Baldi (University Roma II Torre Vergata) and Léo Miolane (Inria Paris). I warmly thank them and also, last but not least, Marina Reizakis, the patient and diligent editor of this book.

Pierre Brémaud

Paris, July 14, 2019

I: PROBABILITY THEORY

Chapter 1

Warming Up

We apparently live in a random world. There are events that we are unable to predict with absolute certainty. Is this world inherently random or does randomness just refer to our incapacity to solve the highly complex deterministic equations that rule the universe? In fact, probabilists do not attempt to take sides in this debate and just observe that there are some phenomena that *look random* and yet seem to exhibit *some kind of regularity*. The canonical example is a coin tossed over and over by a non-mischievous person: the result is an erratic sequence of heads and tails, yet there seems to be a balance between heads and tails. This regularity takes the form of the law of large numbers: the long run proportion of heads is $\frac{1}{2}$, that is, as the number of tosses tends to infinity, the frequency of heads approaches $\frac{1}{2}$. Is this a physical law, or is it a mathematical theorem? At first sight, it is a physical law, and indeed it has to do with a complex process, but so complex that it is preferable to view the law of large numbers as a theorem resulting from a mathematical model.

It took some time for the corresponding mathematical theory to emerge. The modern era, announced by the proof of the strong law of large numbers for coin tossing by Émile Borel in 1909, really started with Andreï Nikolaïevitch Kolmogorov who axiomatized probability in 1933 in terms of the theory of measure and integration, which is presented in the next chapter. Before this, however, it is wise to introduce the terminology and the probabilistic concepts (expectation, independence, conditional expectation) in the elementary framework of discrete random variables. This is done in the current chapter, which contains the proofs of two results that demonstrate the power of probabilistic reasoning, already at an elementary level: Borel's strong law of large numbers and the computation of the extinction probability of a branching process.

1.1 Sample Space, Events and Probability

The study of random phenomena requires a clear and precise language. That of probability theory features familiar mathematical objects such as points, sets and functions, which, however, receive a particular interpretation: points are *outcomes* (of an experiment), sets are *events*, functions are *random numbers*. The meaning of these terms will be given just after we recall the notation concerning the elementary operations on sets: *union, intersection* and *complementation*.

© Springer Nature Switzerland AG 2020
P. Brémaud, *Probability Theory and Stochastic Processes*, Universitext,
https://doi.org/10.1007/978-3-030-40183-2_1

1.1.1   Events

If A and B are subsets of some set Ω, $A \cup B$ denotes their union and $A \cap B$ their intersection. In this book, $\overline{A}$ (rather than $\complement A$ or A^c) denotes the complement of A in Ω. The notation $A + B$ (the *sum* of A and B) implies *by convention* that A and B are *disjoint*, in which case it represents the union $A \cup B$. Similarly, the notation $\sum_{k=1}^{\infty} A_k$ is used for $\cup_{k=1}^{\infty} A_k$ only when the A_k's are pairwise disjoint. The notation $A - B$ is used only if $B \subseteq A$, and it stands for $A \cap \overline{B}$. In particular, if $B \subseteq A$, then $A = B + (A - B)$. The *symmetric difference* of A and B, that is, the set $(A \cup B) - (A \cap B)$, is denoted by $A \triangle B$.

The *indicator function* of the subset $A \subseteq \Omega$ is the function $1_A : \Omega \to \{0, 1\}$ defined by

$$1_A(\omega) = \begin{cases} 1 & \text{if } \omega \in A, \\ 0 & \text{if } \omega \notin A. \end{cases}$$

Random phenomena are observed by means of experiments (performed either by man or nature). Each experiment results in an *outcome*. The collection of all possible outcomes ω is called the *sample space* Ω. Any subset A of the sample space Ω will be regarded fro the time being[1] as a representation of some *event*.

EXAMPLE 1.1.1: TOSSING A DIE, TAKE 1. The experiment consists in tossing a die once. The possible outcomes are $\omega = 1, 2, \ldots, 6$ and the sample space is the set $\Omega = \{1, 2, 3, 4, 5, 6\}$. The subset $A = \{1, 3, 5\}$ is the event "the result is odd."

EXAMPLE 1.1.2: THROWING A DART. The experiment consists in throwing a dart at a wall. The sample space can be chosen to be the plane $\mathbb{R}^2$. An outcome is the position $\omega = (x, y) \in \mathbb{R}^2$ hit by the dart. The subset $A = \{(x, y); x^2 + y^2 > 1\}$ is an event that could be named "you missed the dartboard" if the dartboard is a closed disk of radius 1 centered at 0.

EXAMPLE 1.1.3: HEADS OR TAILS, TAKE 1. The experiment is an infinite succession of coin tosses. One can take for the sample space the collection of all sequences $\omega = \{x_n\}_{n \geq 1}$, where $x_n = 1$ or 0, depending on whether the n-th toss results in heads or tails. The subset $A = \{\omega; x_k = 1 \text{ for } k = 1 \text{ to } 1000\}$ is a lucky event for anyone betting on heads!

The Language of Probabilists

Probabilists have their own dialect. They say that outcome ω *realizes* event A if $\omega \in A$. For instance, in the die model of Example 1.1.1, the outcome $\omega = 1$ realizes the event "result is odd", since $1 \in A = \{1, 3, 5\}$. Obviously, if ω does not realize A, it realizes $\overline{A}$. Event $A \cap B$ is realized by outcome ω if and only if ω realizes both A and B. Similarly, $A \cup B$ is realized by ω if and only if *at least* one event among A and B is realized (both can be realized). Two events A and B are called *incompatible* when $A \cap B = \varnothing$. In other words, event $A \cap B$ is impossible: no outcome ω can realize both A and B. For this reason one refers to the empty set $\varnothing$ as the *impossible* event. Naturally, Ω is called the *certain* event.

---

[1]See Definition 1.1.4.

Recall now that the notation $\sum_{k=1}^{\infty} A_k$ is used for $\cup_{k=1}^{\infty} A_k$ only when the subsets A_k are pairwise disjoint. In the terminology of sets, the sets $A_1, A_2, \ldots$ form a *partition* of Ω if

$$\sum_{k=0}^{\infty} A_k = \Omega.$$

The probabilists then say that the events $A_1, A_2, \ldots$ are *mutually exclusive* and *exhaustive*. They are exhaustive in the sense that any outcome ω realizes at least one among them. They are mutually exclusive in the sense that any two distinct events among them are incompatible. Therefore, any ω realizes *one and only one* of the events A_n $(n \geq 1)$.

If $B \subseteq A$, event B is said to *imply* event A, since ω realizes A when it realizes B.

The σ-field of Events

Probability theory assigns to each event a number, the *probability* of the said event. The collection $\mathcal{F}$ of events to which a probability is assigned is not always identical to the collection of all subsets of Ω. The requirement on $\mathcal{F}$ is that it should be a σ-field:

Definition 1.1.4 *Let $\mathcal{F}$ be a collection of subsets of Ω, such that*

(i) Ω is in $\mathcal{F}$,

(ii) if A belongs to $\mathcal{F}$, then so does its complement $\overline{A}$, and

(iii) if $A_1, A_2, \ldots$ belong to $\mathcal{F}$, then so does their union $\cup_{k=1}^{\infty} A_k$.

One then calls $\mathcal{F}$ a σ-field on Ω (here the σ-field of events).

Note that the impossible event $\varnothing$, being the complement of the certain event Ω, is in $\mathcal{F}$. Note also that if $A_1, A_2, \ldots$ belong to $\mathcal{F}$, then so does their intersection $\cap_{k=1}^{\infty} A_k$ (Exercise 1.6.5).

EXAMPLE 1.1.5: TRIVIAL σ-FIELD, GROSS σ-FIELD. These are respectively the collection $\mathcal{P}(\Omega)$ of all subsets of Ω and the σ-field with only two sets: $\{\Omega, \varnothing\}$. If the sample space Ω is finite or countable, one usually (but not always and not necessarily) considers any subset of Ω to be an event, that is, $\mathcal{F} = \mathcal{P}(\Omega)$.

EXAMPLE 1.1.6: BOREL σ-FIELD. The Borel σ-field on the n-dimensional euclidean space $\mathbb{R}^n$, denoted $\mathcal{B}(\mathbb{R}^n)$ and called the *Borel σ-field* on $\mathbb{R}^n$, is, by definition, the smallest σ-field on $\mathbb{R}^n$ that contains all rectangles, that is, all sets of the form $\prod_{j=1}^{n} I_j$, where the I_j's are intervals of $\mathbb{R}$. This definition is not constructive and therefore one may wonder if there exists sets that are not Borel sets (that is, not sets in $\mathcal{B}(\mathbb{R}^n)$). The theory tells us that there are indeed such sets, but they are in a sense "pathological": the proof of existence of non-Borel sets is not constructive, in the sense that it involves the axiom of choice. In any case, all the sets whose n-volume you have ever been able to compute in your early life are Borel sets. More about this in Chapter 2.

EXAMPLE 1.1.7: HEADS OR TAILS, TAKE 2. Take $\mathcal{F}$ to be the smallest σ-field that contains all the sets of the form $\{\omega\,;\,x_k = 1\}$ $(k \geq 1)$. This σ-field also contains all the sets of the form $\{\omega\,;\,x_k = 0\}$ $(k \geq 1)$ (pass to the complements) and therefore (take intersections) all the sets of the form $\{\omega\,;\,x_1 = a_1, \ldots, x_n = a_n\}$ for all $n \geq 1$ and all $a_1, \ldots, a_n \in \{0,1\}$.

1.1.2   Probability of Events

The probability $P(A)$ of an event $A \in \mathcal{F}$ measures the likeliness of its occurrence. As a function defined on $\mathcal{F}$, the probability P is required to satisfy a few properties, the *axioms of probability*.

Definition 1.1.8 *A* probability *on* $(\Omega, \mathcal{F})$ *is a mapping* $P : \mathcal{F} \to \mathbb{R}$ *such that*

(i) $0 \leq P(A) \leq 1$ *for all* $A \in \mathcal{F}$,

(ii) $P(\Omega) = 1$, *and*

(iii) $P(\sum_{k=1}^{\infty} A_k) = \sum_{k=1}^{\infty} P(A_k)$ *for all sequences* $\{A_k\}_{k \geq 1}$ *of pairwise disjoint events in* $\mathcal{F}$.

Property (iii) is called σ-*additivity*. The triple $(\Omega, \mathcal{F}, P)$ is called a *probability space*, or *probability model*.

EXAMPLE 1.1.9: TOSSING A DIE, TAKE 2. An event A is a subset of $\Omega = \{1, 2, 3, 4, 5, 6\}$. The formula

$$P(A) = \frac{|A|}{6},$$

where $|A|$ is the *cardinality* of A (the number of elements in A), defines a probability P.

EXAMPLE 1.1.10: HEADS OR TAILS, TAKE 3. Choose a probability P such that for any event of the form $A = \{x_1 = a_1, \ldots, x_n = a_n\}$, where $a_1, \ldots, a_n$ are in $\{0, 1\}$,

$$P(A) = \frac{1}{2^n}.$$

Note that this does not define the probability of all events of $\mathcal{F}$. But the theory (wait until Chapter 3) tells us that there exists such a probability satisfying the above requirement and that this probability is unique.

EXAMPLE 1.1.11: RANDOM POINT IN A SQUARE, TAKE 1. The following is a possible model of a random point inside the unit square $[0, 1]^2 = [0, 1] \times [0, 1]$: $\Omega = [0, 1]^2$, $\mathcal{F}$ is the collection of sets in the Borel σ-field $\mathcal{B}(\mathbb{R}^2)$ that are contained in $[0, 1]^2$. The theory tells us that there does indeed exist one and only one probability P satisfying the above requirement, called the *Lebesgue measure* on $[0, 1]^2$, which formalizes the intuitive notion of "area". (More about this in Chapters 2 and 3.)

The probability of Example 1.1.9 suggests an unbiased die, where each outcome 1, 2, 3, 4, 5, or 6 has the same probability. As we shall soon see, probability P of Example

1.1.10 implies an *unbiased* coin and *independent* tosses (the emphasized terms will be defined later).

The axioms of probability are motivated by the heuristic interpretation of $P(A)$ as the *empirical frequency* of occurrence of event A. If n "independent" experiments are performed, among which n_A result in the realization of A, then the empirical frequency

$$F(A) = \frac{n_A}{n}$$

should be close to $P(A)$ if n is "sufficiently large". (This statement has to be made precise. It is in fact a loose expression of the law of large numbers that will be given later on.) Clearly, the empirical frequency function F satisfies the axioms of probability.

Basic Formulas

We shall now list the properties of probability that follow directly from the axioms:

Theorem 1.1.12 *For any event A*

$$P(\overline{A}) = 1 - P(A), \tag{1.1}$$

and

$$P(\varnothing) = 0. \tag{1.2}$$

Proof. For a proof of (1.1), use additivity:

$$1 = P(\Omega) = P(A + \overline{A}) = P(A) + P(\overline{A}).$$

Applying (1.1) with $A = \Omega$ gives (1.2). $\square$

Theorem 1.1.13 *Probability is monotone, that is, for any events A and B,*

$$A \subseteq B \Longrightarrow P(A) \leq P(B). \tag{1.3}$$

Proof. For a proof, observe that when $A \subseteq B$, $B = A + (B - A)$, and therefore

$$P(B) = P(A) + P(B - A) \geq P(A).$$

$\square$

Theorem 1.1.14 *Probability is* sub-σ-additive: *for any sequence $A_1, A_2, \ldots$ of events,*

$$P(\cup_{k=1}^{\infty} A_k) \leq \sum_{k=1}^{\infty} P(A_k). \tag{1.4}$$

Proof. Observe that

$$\cup_{k=1}^{\infty} A_k = \sum_{k=1}^{\infty} A'_k,$$

where $A'_1 := A_1$ and

$$A'_k := A_k \cap \left\{ \overline{\cup_{i=1}^{k-1} A_i} \right\} \quad (k \geq 2).$$

Therefore,

$$P\left(\cup_{k=1}^{\infty}A_k\right) = P\left(\sum_{k=1}^{\infty}A'_k\right) = \sum_{k=1}^{\infty}P(A'_k).$$

But $A'_k \subseteq A_k$, and therefore $P(A'_k) \leq P(A_k)$. $\square$

The next (very important) property is the *sequential continuity* of probability:

Theorem 1.1.15 *Let* $\{A_n\}_{n\geq 1}$ *be a* non-decreasing *sequence of events, that is,* $A_{n+1} \supseteq A_n$ *for all* $n \geq 1$. *Then*

$$P(\cup_{n=1}^{\infty}A_n) = \lim_{n\uparrow\infty}P(A_n).$$ (1.5)

Proof. Write

$$A_n = A_1 + (A_2 - A_1) + \cdots + (A_n - A_{n-1})$$

and

$$\cup_{k=1}^{\infty}A_k = A_1 + (A_2 - A_1) + (A_3 - A_2) + \cdots.$$

Therefore,

$$P(\cup_{k=1}^{\infty}A_k) = P(A_1) + \sum_{j=2}^{\infty}P(A_j - A_{j-1})$$

$$= \lim_{n\uparrow\infty}\left\{P(A_1) + \sum_{j=2}^{n}P(A_j - A_{j-1})\right\} = \lim_{n\uparrow\infty}P(A_n).$$

$\square$

Corollary 1.1.16 *Let* $\{B_n\}_{n\geq 1}$ *be a* non-increasing *sequence of events, that is,* $B_{n+1} \subseteq B_n$ *for all* $n \geq 1$. *Then*

$$P(\cap_{n=1}^{\infty}B_n) = \lim_{n\uparrow\infty}P(B_n).$$ (1.6)

Proof. To obtain (1.6), write (using De Morgan's identity; see Exercise 1.6.1):

$$P(\cap_{n=1}^{\infty}B_n) = 1 - P\left(\overline{\cap_{n=1}^{\infty}B_n}\right) = 1 - P(\cup_{n=1}^{\infty}\overline{B}_n),$$

and apply (1.5) with $A_n = \overline{B}_n$:

$$1 - P(\cup_{n=1}^{\infty}\overline{B}_n) = 1 - \lim_{n\uparrow\infty}P(\overline{B}_n) = \lim_{n\uparrow\infty}(1 - P(\overline{B}_n)) = \lim_{n\uparrow\infty}P(B_n).$$

$\square$

Negligible Sets

A central notion of probability is that of a *negligible set*.

Definition 1.1.17 *A set* $N \subset \Omega$ *is called* P-negligible *if it is contained in an event* $A \in \mathcal{F}$ *of probability* $P(A) = 0$.

Theorem 1.1.18 *A countable union of negligible sets is a negligible set.*

Proof. Let N_k $(k \geq 1)$ be P-negligible sets. By definition there exists a sequence A_k $(k \geq 1)$ of events of null probability such that $N_k \subseteq A_k$ $(k \geq 1)$. We have

$$N := \cup_{k \geq 1} N_k \subseteq A := \cup_{k \geq 1} A_k,$$

and by the sub-σ-additivity property of probability, $P(A) = 0$. □

EXAMPLE 1.1.19: RANDOM POINT IN A SQUARE, 2. (Example 1.1.11 continued) Recall the model of a random point inside the unit square $[0,1]^2 = [0,1] \times [0,1]$. Each rational point therein has a null area and therefore a null probability. Therefore, the (countable) set of rational points of the square has null probability. In other words, the probability of drawing a rational point is, in this particular model, null.

1.2 Independence and Conditioning

In the frequency interpretation of probability, a situation where $n_{A \cap B}/n \approx (n_A/n) \times (n_B/n)$, or

$$\frac{n_{A \cap B}}{n_B} \approx \frac{n_A}{n}$$

(here $\approx$ is a non-mathematical symbol meaning "approximately equal") suggests some kind of "independence" of A and B, in the sense that statistics relative to A do not vary when passing from a neutral sample of population to a selected sample characterized by the property B. For example, the proportion of people with a family name beginning with H is "approximately" the same among a large population with the usual mix of men and women as it would be among a "large" all-male population. Therefore, one's gender is "independent" of the fact that one's name begins with an H.[2].

1.2.1 Independent Events

The above discussion prompts us to give the following formal definition of independence, the single most important concept of probability theory.

Definition 1.2.1 *Two events A and B are called* independent *if and only if*

$$P(A \cap B) = P(A)P(B). \tag{1.7}$$

Remark 1.2.2 One should be aware that incompatibility is different from independence. As a matter of fact, two incompatible events A and B are independent if and only if at least one of them has null probability. Indeed, if A and B are incompatible, $P(A \cap B) = P(\varnothing) = 0$, and therefore (1.7) holds if and only if $P(A)P(B) = 0$.

The notion of independence carries over to families of events in the following manner.

Definition 1.2.3 *A family $\{A_n\}_{n \in \mathbb{N}}$ of events is called* independent *if for any finite set of indices $i_1 < \ldots < i_r$ where $i_j \in \mathbb{N}$ $(1 \leq j \leq r)$,*

$$P\left(A_{i_1} \cap A_{i_2} \cap \cdots \cap A_{i_r}\right) = P(A_{i_1}) \times P(A_{i_2}) \times \cdots \times P(A_{i_r}).$$

One also says that the A_n's $(n \in \mathbb{N})$ are jointly independent.

---

[2]As far as we know...

EXAMPLE 1.2.4: THE SWITCHES. Two locations A and B in a communications network are connected by three different paths, and each path contains a number of links that can fail. These are represented symbolically in the figure below by switches that are in the lifted position if the link is unable to operate. The number associated with a switch is the probability that the switch is lifted. The switches are lifted independently. What is the probability that A is accessible from B, that is, that there exists at least one available path for communications?

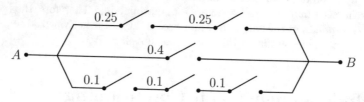

Let U_1 be the event "no switch lifted in the upper path". Defining U_2 and U_3 similarly, we see that the probability to be computed is that of $U_1 \cup U_2 \cup U_3$, or by de Morgan's law, that of the complement of $\overline{U}_1 \cap \overline{U}_2 \cap \overline{U}_3$:

$$1 - P(\overline{U}_1 \cap \overline{U}_2 \cap \overline{U}_3) = 1 - P(\overline{U}_1)P(\overline{U}_2)(P\overline{U}_3),$$

where the last equality follows from the independence assumption concerning the switches. Letting now $U_1^1 = $ "switch 1 (first from left) in the upper path is not lifted" and $U_1^2 = $ "switch 2 in the upper path is not lifted", we have $U_1 = U_1^1 \cap U_1^2$, therefore, in view of the independence assumption,

$$P(\overline{U}_1) = 1 - P(U_1) = 1 - P(U_1^1)P(U_1^2).$$

We must now use the data $P(U_1^1) = 1 - 0.25, P(U_1^2) = 1 - 0.25$ to obtain $P(\overline{U}_1) = 1 - (0.75)^2$. Similarly $P(\overline{U}_2) = 1 - 0.6$ and $P(\overline{U}_3) = 1 - (0.9)^3$. The final result (of rather limited interest) is $1 - (0.4375)(0.4)(0.271) = 0.952575$.

EXAMPLE 1.2.5: IS THIS NUMBER THE LARGER ONE? Let a and b be two numbers in $\{1, 2, \ldots, 10\,000\}$. Nothing is known about these numbers, except that they are not equal, say $a > b$. Only one of these numbers is shown to you, secretly chosen at random and equiprobably. Call this random number X. Is there a good strategy for guessing if the number shown to you is the larger one? Of course, one would like to have a probability of success strictly larger than $\frac{1}{2}$.

Perhaps surprisingly, there is such a strategy, that we now describe. Select at random, uniformly on $\{1, 2, \ldots, 10\,000\}$, a number Y. If $X \geq Y$, say that X is the largest $(= a)$, otherwise say that it is the smallest.

Let us compute the probability P_E of a wrong guess. An error occurs when either (i) $X \geq Y$ and $X = b$, or (ii) $X < Y$ and $X = a$. These events are exclusive of one another, and therefore

$$\begin{aligned}
P_E &= P(X \geq Y, X = b) + P(X < Y, X = a) \\
&= P(b \geq Y, X = b) + P(a < Y, X = a) \\
&= P(b \geq Y)P(X = b) + P(a < Y)P(X = a) \\
&= P(b \geq Y)\frac{1}{2} + P(a < Y)\frac{1}{2} = \frac{1}{2}(P(b \geq Y) + P(a < Y)) \\
&= \frac{1}{2}(1 - P(Y \in [b+1, a])) = \frac{1}{2}\left(1 - \frac{a-b}{10\,000}\right) < \frac{1}{2}.
\end{aligned}$$

1.2.2 Bayes' Calculus

We continue the heuristic discussion of Subsection 1.2.1 in terms of empirical frequencies. Dependence between A and B occurs when $P(A \cap B) \neq P(A)P(B)$. In this case the relative frequency $n_{A \cap B}/n_B \approx P(A \cap B)/P(B)$ is different from the frequency n_A/n. This suggests the following definition.

Definition 1.2.6 *The* conditional probability *of A given B is the number*

$$P(A \mid B) := \frac{P(A \cap B)}{P(B)}, \tag{1.8}$$

defined when $P(B) > 0$.

Remark 1.2.7 The quantity $P(A \mid B)$ represents our expectation of A being realized when the only available information is that B is realized. Indeed, this expectation is based on the relative frequency $n_{A \cap B}/n_B$ alone. Of course, if A and B are independent, then $P(A \mid B) = P(A)$.

Probability theory is primarily concerned with the computation of probabilities of complex events. The following formulas, called *Bayes' rules*, are useful for that purpose.

Theorem 1.2.8 *With $P(A) > 0$, we have the Bayes rule of* retrodiction*:*

$$P(B \mid A) = \frac{P(A \mid B)P(B)}{P(A)}. \tag{1.9}$$

Proof. Rewrite Definition 1.8 symmetrically in A and B:

$$P(A \cap B) = P(A \mid B)P(B) = P(B \mid A)P(A).$$

$\square$

Theorem 1.2.9 *Let $B_1, B_2, \ldots$ be events forming partition of Ω. Then for any event A, we have the Bayes rule of* total causes*:*

$$P(A) = \sum_{i=1}^{\infty} P(A \mid B_i)P(B_i). \tag{1.10}$$

Proof. Decompose A as follows:

$$A = A \cap \Omega = A \cap \left(\sum_{i=1}^{\infty} B_i \right) = \sum_{i=1}^{\infty} (A \cap B_i).$$

Therefore (by σ-additivity and the definition of conditional probability):

$$P(A) = P\left(\sum_{i=1}^{\infty} (A \cap B_i) \right) = \sum_{i=1}^{\infty} P(A \cap B_i) = \sum_{i=1}^{\infty} P(A \mid B_i) P(B_i).$$

$\square$

Theorem 1.2.10 *For any sequence of events* $A_1, \ldots, A_n$, *we have the Bayes sequential formula:*

$$P\left(\cap_{i=1}^{k} A_i \right) = P(A_1) P(A_2 \mid A_1) P(A_3 \mid A_1 \cap A_2) \cdots P\left(A_k \mid \cap_{i=1}^{k-1} A_i \right). \qquad (1.11)$$

Proof. By induction. First observe that (1.11) is true for $k = 2$ by definition of conditional probability. Suppose that (1.11) is true for k. Write

$$P\left(\cap_{i=1}^{k+1} A_i \right) = P\left(\left(\cap_{i=1}^{k} A_i \right) \cap A_{k+1} \right) = P\left(A_{k+1} \mid \cap_{i=1}^{k} A_i \right) P\left(\cap_{i=1}^{k} A_i \right),$$

and replace $P\left(\cap_{i=1}^{k} A_i \right)$ by the assumed equality (1.11) to obtain the same equality with $k + 1$ replacing k.

$\square$

EXAMPLE 1.2.11: SHOULD WE ALWAYS BELIEVE DOCTORS? Doctors apply a test that gives a positive result in 99% of the cases where the patient is affected by the disease. However it happens in 2% of the cases that a healthy patient has a positive test. Statistical data show that one individual out of 1000 has the disease. What is the probability that a patient with a positive test is affected by the disease?

Solution: Let M be the event "patient is ill," and let $+$ and $-$ be the events "test is positive" and "test is negative" respectively. We have the data

$$P(M) = 0.001, \ P(+ \mid M) = 0.99, \ P(+ \mid \overline{M}) = 0.02,$$

and we must compute $P(M \mid +)$. By the Bayes retrodiction formula,

$$P(M \mid +) = \frac{P(+ \mid M) P(M)}{P(+)}.$$

By the Bayes formula of total causes,

$$P(+) = P(+ \mid M) P(M) + P(+ \mid \overline{M}) P(\overline{M}).$$

Therefore,

$$P(M \mid +) = \frac{(0.99)(0.001)}{(0.99)(0.001) + (0.02)(0.999)},$$

that is, approximately 0.005.

Remark 1.2.12 The quantitative result of the above example may be disquieting. In fact, this may happen with grouped blood tests (maybe in a prison or in the army) to detect, say AIDS. A single individual will provoke a positive test alert for all his mates. Of course, the doctor in charge will then proceed to individual tests. See Exercise 1.6.19.

EXAMPLE 1.2.13: THE BALLOT PROBLEM. In an election, candidates I and II have obtained a and b votes, respectively. Candidate I won, that is, $a > b$. We seek to compute the probability that in the course of the vote counting procedure, candidate I has always had the lead.

Let $p_{a,b}$ be the probability that A is always ahead. We have by the formula of total causes, conditioning on the last vote:

$$p_{a,b} = P(A \text{ always ahead} \,|\, A \text{ gets the last vote})P(A \text{ gets the last vote})$$
$$+ P(A \text{ always ahead} \,|\, B \text{ gets the last vote})P(B \text{ gets the last vote})$$

$$= p_{a-1,b}\frac{a}{a+b} + p_{a,b-1}\frac{b}{a+b},$$

with the convention that for $a = b+1$, $p_{a-1,b} = p_{b,b} = 0$. The result follows by induction on the total number of votes $a + b$:

$$p_{a,b} = \frac{a-b}{a+b}.$$

1.2.3 Conditional Independence

Definition 1.2.14 *Let A, B and C be events, where $P(C) > 0$. One says that A and B are* conditionally independent *given C if*

$$P(A \cap B \,|\, C) = P(A \,|\, C)P(B \,|\, C). \tag{1.12}$$

In other words, A and B are independent with respect to the probability P_C defined by $P_C(A) = P(A\,|\,C)$ (see Exercise 1.6.11).

EXAMPLE 1.2.15: CHEAP WATCHES. Two factories A and B manufacture watches. Factory A produces on average one defective item out of 100, and B produces on average one bad watch out of 200. A retailer receives a container of watches from one of the two above factories, but he does not know which. He checks the first watch. It works!

(a) What is the probability that the second watch he will check is good?

(b) Are the states of the first two watches independent?

You will need to invent reasonable hypotheses when needed.

Solution: (a) Let X_n be the state of the n-th watch in the container, with $X_n = 1$ if it works and $X_n = 0$ if it does not. Let Y be the factory of origin. We express our *a priori* ignorance of where the case comes from by

$$P(Y = A) = P(Y = B) = \frac{1}{2}.$$

(Note that this is a hypothesis.) Also, we assume that given $Y = A$ (resp., $Y = B$), the states of the successive watches are independent. For instance,

$$P(X_1 = 1, X_2 = 0 \mid Y = A) = P(X_1 = 1 \mid Y = A)P(X_2 = 0 \mid Y = A).$$

We have the data

$$P(X_n = 0 \mid Y = A) = 0.01 \qquad P(X_n = 0 \mid Y = B) = 0.005.$$

We are required to compute

$$P(X_2 = 1 \mid X_1 = 1) = \frac{P(X_1 = 1, X_2 = 1)}{P(X_1 = 1)}.$$

By the Bayes formula of total causes, the numerator of this fraction equals

$$P(X_1 = 1, X_2 = 1 \mid Y = A)P(Y = A) + P(X_1 = 1, X_2 = 1 \mid Y = B)P(Y = B),$$

that is, $(0.5)(0.99)^2 + (0.5)(0.995)^2$, and the denominator is

$$P(X_1 = 1 \mid Y = A)P(Y = A) + P(X_1 = 1 \mid Y = B)P(Y = B),$$

that is, $(0.5)(0.99) + (0.5)(0.995)$. Therefore,

$$P(X_2 = 1 \mid X_1 = 1) = \frac{(0.99)^2 + (0.995)^2}{0.99 + 0.995}.$$

(b) The states of the two watches are not independent. Indeed, if they were, then

$$P(X_2 = 1 \mid X_1 = 1) = P(X_2 = 1) = (0.5)\,(0.99 + 0.995),$$

a result different from what we obtained.

Remark 1.2.16 The above example shows that two events A and B that are conditionally independent given some event C and at the same time conditionally independent given $\overline{C}$, may yet *not* be independent.

1.3   Discrete Random Variables

The number of heads in a sequence of 1000 coin tosses, the number of days it takes until the next rain and the size of a genealogical tree are random numbers. All are functions of the outcome of a random experiment performed either by man or nature, and these outcomes take discrete values, that is, values in a countable set. These values are integers in the above examples, but they could be more complex mathematical objects. This section gives the basic theory of *discrete random variables*.

1.3.1   Probability Distributions and Expectation

Definition 1.3.1 *Let E be a countable set. A function $X : \Omega \to E$ such that for all $x \in E$*

$$\{\omega;\, X(\omega) = x\} \in \mathcal{F}$$

is called a discrete random variable. *(Being in $\mathcal{F}$, the event $\{X = x\}$ can be assigned a probability.)*

Remark 1.3.2 Calling an integer-valued random variable X a *random number* is an innocuous habit as long as one is aware that it is *not the function* X that is random, but the outcome ω. This in turn makes the *number* $X(\omega)$ random.

EXAMPLE 1.3.3: TOSSING A DIE, TAKE 3. The sample space is the set $\Omega = \{1, 2, 3, 4, 5, 6\}$. Take for X the identity: $X(\omega) = \omega$. Therefore X is a random number obtained by tossing a die.

EXAMPLE 1.3.4: HEADS OR TAILS, TAKE 4. (Example 1.1.10 continued.) The sample space Ω is the collection of all sequences $\omega = \{x_n\}_{n \geq 1}$, where $x_n = 1$ or 0. Define a random variable X_n by $X_n(\omega) = x_n$. It is the random number obtained at the n-th toss. It is indeed a random variable since for all $a_n \in \{0, 1\}$, $\{\omega\,;\, X_n(\omega) = a_n\} = \{\omega\,;\, x_n = a_n\} \in \mathcal{F}$, by definition of $\mathcal{F}$.

The following are elementary remarks.

Let E and F be countable sets. Let X be a random variable with values in E, and let $f : E \to F$ be a function. Then $Y := f(X)$ is a random variable.

Proof. Let $y \in F$. The set $\{\omega\,;\, Y(\omega) = y\}$ is in $\mathcal{F}$ since it is a countable union of sets in $\mathcal{F}$, namely:

$$\{Y = y\} = \sum_{x \in E;\, f(x) = y} \{X = x\}.$$

$\square$

Let E_1 and E_2 be countable sets. Let X_1 and X_2 be random variable with values in E_1 and E_2, respectively. Then $Y := (X_1, X_2)$ is a random variable with values in $E = E_1 \times E_2$.

Proof. Let $x = (x_1, x_2) \in E$. The set $\{\omega\,;\, X(\omega) = x\}$ is in $\mathcal{F}$ since it is the intersection of sets in $\mathcal{F}$, namely:

$$\{X = x\} = \{X_1 = x_1\} \cap \{X_2 = x_2\}.$$

$\square$

Definition 1.3.5 *Let X be a discrete random variable taking its values in E. Its probability distribution function is the function $\pi : E \to [0, 1]$, where*

$$\pi(x) := P(X = x) \quad (x \in E).$$

EXAMPLE 1.3.6: THE GAMBLER'S FORTUNE. This is a continuation of the coin tosses example (Example 1.1.10). The number of occurrences of heads in n tosses is $S_n = X_1 + \cdots + X_n$. This random variable is the fortune at time n of a gambler systematically betting on heads. It takes integer values from 0 to n. We have

$$P(S_n = k) = \binom{n}{k}\frac{1}{2^n}.$$

Proof. The event $\{S_n = k\}$ is "k among $X_1, \ldots, X_n$ are equal to 1". There are $\binom{n}{k}$ distinct ways of assigning k values of 1 and $n - k$ values of 0 to $X_1, \ldots, X_n$, and all have the same probability 2^{-n}.

$\square$

Remark 1.3.7 One may have to prove that a random variable X, taking its values in $\overline{\mathbb{N}}$ (and therefore for which the value ∞ is *a priori* possible) is in fact almost surely finite, that is, to prove that $P(X = \infty) = 0$ or, equivalently, that $P(X < \infty) = 1$. Since

$$\{X < \infty\} = \sum_{n=0}^{\infty}\{X = n\},$$

we have

$$P(X < \infty) = \sum_{n=0}^{\infty} P(X = n).$$

(This remark provides an opportunity to recall that in an expression such as $\sum_{n=0}^{\infty}$, the sum is over $\mathbb{N}$ and does not include ∞ as the notation seems to suggest. A less ambiguous notation would be $\sum_{n\in\mathbb{N}}$. If we want to sum over all integers plus ∞, we shall *always* use the notation $\sum_{n\in\overline{\mathbb{N}}}$.)

Expectation for Discrete Random Variables

Definition 1.3.8 *Let X be a discrete random variable taking its values in a countable set E and let the function $g : E \to \mathbb{R}$ be either non-negative or such that it satisfies the absolute summability condition*

$$\sum_{x\in E} |g(x)|P(X = x) < \infty.$$

Then one defines $\mathrm{E}[g(X)]$, *the expectation of $g(X)$, by the formula*

$$\mathrm{E}[g(X)] := \sum_{x\in E} g(x)P(X = x).$$

If the absolute summability condition is satisfied, the random variable $g(X)$ is called *integrable*, and in this case the expectation $\mathrm{E}[g(X)]$ is a *finite* number. If it is only assumed that g is non-negative, the expectation may well be infinite.

EXAMPLE 1.3.9: THE GAMBLER'S FORTUNE. This is a continuation of Example 1.3.6. Consider the random variable $S_n = X_1 + \cdots + X_n$ taking its values in $\{0, 1, \ldots, n\}$. Its expectation is $\mathrm{E}[S_n] = n/2$, as the following straightforward computation shows:

$$\mathrm{E}[S_n] = \sum_{k=0}^{n} kP(S_n = k)$$

$$= \frac{1}{2^n} \sum_{k=1}^{n} k\frac{n!}{k!(n - k)!}$$

$$= \frac{n}{2^n} \sum_{k=1}^{n} \frac{(n - 1)!}{(k - 1)!((n - 1) - (k - 1))!}$$

$$= \frac{n}{2^n} \sum_{j=0}^{n-1} \frac{(n - 1)!}{j!(n - 1 - j)!} = \frac{n}{2^n} 2^{n-1} = \frac{n}{2}.$$

EXAMPLE 1.3.10: FINITE RANDOM VARIABLES WITH INFINITE EXPECTATIONS. One should be aware that a discrete random variable taking *finite values* may have an *infinite expectation*. The canonical example is the random variable X taking its values in $E = \overline{\mathbb{N}}$ and with probability distribution

$$P(X = n) = \frac{1}{cn^2},$$

where the constant c is chosen such that

$$P(X < \infty) = \sum_{n=1}^{\infty} P(X = n) = \sum_{n=1}^{\infty} \frac{1}{cn^2} = 1$$

(that is, $c = \sum_{n=1}^{\infty} \frac{1}{n^2} = \frac{\pi^2}{6}$). Indeed, the expectation of X is

$$E[X] = \sum_{n=1}^{\infty} nP(X = n) = \sum_{n=1}^{\infty} n\frac{1}{cn^2} = \sum_{n=1}^{\infty} \frac{1}{cn} = \infty.$$

Remark 1.3.11 The above example seems artificial. It is however *not pathological*, and there are a lot more natural occurrences of the phenomenon. Consider for instance Example 1.3.9, and let T be the first integer n (necessarily even) such that $2S_n - n = 0$. (The quantity $2S_n - n$ is the fortune at time n of a gambler systematically betting on heads.) Then as it turns out and as we shall prove later (in Example 6.3.6), T is a *finite* random variable with *infinite expectation*.

The *telescope formula* below gives an alternative way of computing the expectation of an integer-valued random variable.

Theorem 1.3.12 *For a random variable X taking its values in $\mathbb{N}$,*

$$E[X] = \sum_{n=1}^{\infty} P(X \geq n).$$

Proof.

$$E[X] = P(X = 1)+2P(X = 2) + 3P(X = 3) + \cdots$$
$$= P(X = 1) \ +P(X = 2) + P(X = 3) + \cdots$$
$$+P(X = 2) + P(X = 3) + \cdots$$
$$+ P(X = 3) + \cdots$$

$\square$

Basic Properties of Expectation

Let A be some event. The expectation of the indicator random variable $X = 1_A$ is

$$E[1_A] = P(A).$$

(We call this the expectation formula for indicator functions.)

Proof. The random variable $X = 1_A$ takes the value 1 with probability $P(X = 1) = P(A)$ and the value 0 with probability $P(X = 0) = P(\overline{A}) = 1 - P(A)$. Therefore,

$$E[X] = 0 \times P(X = 0) + 1 \times P(X = 1) = P(X = 1) = P(A).$$

$\square$

Theorem 1.3.13 *Let g_1 and g_2 be functions from E to $\overline{\mathbb{R}}$ such that $g_1(X)$ and $g_2(X)$ are integrable (resp., non-negative), and let $\lambda_1, \lambda_2 \in \mathbb{R}$ (resp., $\in \mathbb{R}_+$). Then*

$$E[\lambda_1 g_1(X) + \lambda_2 g_2(X)] = \lambda_1 E[g_1(X)] + \lambda_2 E[g_2(X)]$$

(linearity *of expectation). Also, if $g_1(x) \le g_2(x)$ for all $x \in E$,*

$$E[g_1(X)] \le E[g_2(X)]$$

(monotonicity *of expectation). Finally, we have the* triangle inequality

$$|E[g(X)]| \le E[|g(X)|].$$

Proof. These properties follow directly from the corresponding properties of series. $\square$

EXAMPLE 1.3.14:   THE MATCHING PARADOX. There are n boxes $B_1, \cdots, B_n$ and n objects $O_1, \cdots, O_n$. These objects are placed "at random" in the boxes, one and only one per box. What is the average number of matchings, that is, of boxes that receive an object with the same index? The problem will be stated mathematically in a way that gives meaning to the phrase "at random". Let Π_n be the set of permutations of $\{1, 2, \ldots, n\}$. Let σ a random permutation, that is, $P(\sigma = \sigma^{(0)}) = \frac{1}{n!}$ for all $\sigma^{(0)} \in \Pi_n$. The random placement is assimilated to such a random permutation, and a matching at position (box) i is said to occur if $\sigma_i = i$. Let $X_i = 1$ if a matching occurs at position i, and $X_i = 0$ otherwise. The total number of matches is therefore $Z_n := \sum_{i=1}^{n} X_i$, so that the average number of matches is

$$E[Z_n] = \sum_{i=1}^{n} E[X_i] = \sum_{i=1}^{n} P(X_i = 1).$$

By symmetry, $P(X_i = 1) = P(X_1 = 1)$ $(1 \le i \le n)$, so that

$$E[Z_n] = nP(X_1 = 1).$$

But there are $(n-1)!$ permutations $\sigma^{(0)}$ such that $\sigma_1^{(0)} = 1$, each one occurring with probability $\frac{1}{n!}$, so that

$$P(X_1 = 1) = \frac{(n-1)!}{n!} = \frac{1}{n}.$$

Therefore

$$E[Z_n] = n \times \frac{1}{n} = 1.$$

This number remains constant and does not increase with n as one (maybe) expects!

Mean and Variance

Definition 1.3.15 *Let X be a random variable such that $E[|X|] < \infty$ (X is integrable). In this case (and only in this case) the* mean μ *of X is defined by*

$$\mu := E[X] = \sum_{n=0}^{+\infty} nP(X = n).$$

From the inequality $|a| \leq 1 + a^2$, true for all $a \in \mathbb{R}$, we have that $|X| \leq 1 + X^2$, and therefore, by the monotonicity and linearity properties of expectation, $E[|X|] \leq 1 + E[X^2]$ (we also used the fact that $E[1] = 1$). Therefore if $E[X^2] < \infty$ (in which case we say that X is *square-integrable*) then X is integrable. The following definition then makes sense.

Definition 1.3.16 *Let X be a square-integrable random variable. Its* variance *is, by definition, the quantity*

$$\sigma^2 := E[(X - \mu)^2] = \sum_{n=0}^{+\infty} (n - \mu)^2 P(X = n).$$

The variance is also denoted by $\mathrm{Var}\,(X)$. From the linearity of expectation, it follows that $E[(X - m)^2] = E[X^2] - 2mE[X] + m^2$, that is,

$$\mathrm{Var}\,(X) = E[X^2] - m^2.$$

The mean is the "center of inertia" of a random variable. More precisely,

Theorem 1.3.17 *Let X be a real random variable with mean μ and finite variance σ^2. Then, for all $a \in \mathbb{R}$, $a \neq \mu$,*

$$E[(X - a)^2] > E[(X - \mu)^2] = \sigma^2.$$

Proof.

$$\begin{aligned}
E\left[(X - a)^2\right] &= E\left[((X - \mu) + (\mu - a))^2\right] \\
&= E\left[(X - \mu)^2\right] + (\mu - a)^2 + 2(\mu - a)E\left[(X - \mu)\right] \\
&= E\left[(X - \mu)^2\right] + (\mu - a)^2 > E\left[(X - \mu)^2\right].
\end{aligned}$$

$\square$

Independent Variables

Definition 1.3.18 *Two discrete random variables X and Y are called* independent *if*

$$P(X = i, Y = j) = P(X = i)P(Y = j) \quad (i, j \in E).$$

Remark 1.3.19 The left-hand side of the last display is $P(\{X = i\} \cap \{Y = j\})$. This is a general feature of the notational system: commas replace intersection signs. For instance, $P(A, B)$ is the probability that both events A and B occur.

Definition 1.3.18 extends to a finite number of random variables:

Definition 1.3.20 *The discrete random variables $X_1, \ldots, X_k$ taking their values in $E_1, \ldots, E_k$ respectively are said to be* independent *if for all $i_1 \in E_1, \ldots, i_k \in E_k$,*

$$P(X_1 = i_1, \ldots, X_k = i_k) = P(X_1 = i_1) \cdots P(X_k = i_k).$$

Definition 1.3.21 *A sequence $\{X_n\}_{n \geq 1}$ of discrete random variables taking their values in the sets $\{E_n\}_{n \geq 1}$ respectively is called* independent *if any finite collection of distinct random variables $X_{i_1}, \ldots, X_{i_r}$ extracted from this sequence is independent. It is said to be* IID *(independent and identically distributed) if $E_n \equiv E$ for all $n \geq 1$, if it is independent and if the probability distribution function of X_n does not depend on n.*

EXAMPLE 1.3.22: HEADS OR TAILS, TAKE 5. (Example 1.3.4 continued) We are going to show that the sequence $\{X_n\}_{n \geq 1}$ is IID. Therefore, we have a model for *independent* tosses of an *unbiased* coin.

Proof. Event $\{X_k = a_k\}$ is the direct sum of events $\{X_1 = a_1, \ldots, X_{k-1} = a_{k-1}, X_k = a_k\}$ for all possible values of $(a_1, \ldots, a_{k-1})$. Since there are 2^{k-1} such values and each one has probability 2^{-k}, we have $P(X_k = a_k) = 2^{k-1} 2^{-k}$, that is,

$$P(X_k = 1) = P(X_k = 0) = \frac{1}{2}.$$

Therefore,

$$P(X_1 = a_1, \ldots, X_k = a_k) = P(X_1 = a_1) \cdots P(X_k = a_k)$$

for all $a_1, \ldots, a_k \in \{0, 1\}$, from which it follows by definition that $X_1, \ldots, X_k$ are independent random variables, and more generally that $\{X_n\}_{n \geq 1}$ is a family of independent random variables. $\square$

Definition 1.3.23 *Let $\{X_n\}_{n \geq 1}$ and $\{Y_n\}_{n \geq 1}$ be sequences of discrete random variables taking their values in the sets $\{E_n\}_{n \geq 1}$ and $\{F_n\}_{n \geq 1}$, respectively. They are said to be* independent *if for any finite collection of random variables $X_{i_1}, \ldots, X_{i_r}$ and $Y_{j_1}, \ldots, Y_{j_s}$ extracted from their respective sequences, the discrete random variables $(X_{i_1}, \ldots, X_{i_r})$ and $(Y_{j_1}, \ldots, Y_{j_s})$ are independent.*

(This means that for all $a_1 \in E_1, \ldots, a_r \in E_r$, $b_1 \in F_1, \ldots, b_s \in F_s$,

$$P\left((\cap_{\ell=1}^r \{X_{i_\ell} = a_\ell\}) \cap (\cap_{m=1}^s \{Y_{j_m} = b_m\})\right)$$
$$= P\left(\cap_{\ell=1}^r \{X_{i_\ell} = a_\ell\}\right) P\left(\cap_{m=1}^s \{Y_{j_m} = b_m\}\right).)$$

The notion of conditional independence for events (Definition 1.2.14) extends naturally to discrete random variables.

Definition 1.3.24 *Let X, Y, Z be random variables taking their values in the countable sets E, F, G, respectively. One says that X and Y are* conditionally independent *given Z if for all x, y, z in E, F, G, respectively, the events $\{X = x\}$ and $\{Y = y\}$ are conditionally independent given $\{Z = z\}$.*

Recall that the events $\{X = x\}$ and $\{Y = y\}$ are said to be conditionally independent given $\{Z = z\}$ if

$$P(X = x, Y = y \mid Z = z) = P(X = x \mid Z = z)P(Y = y \mid Z = z).$$

The Product Formula for Expectations

Theorem 1.3.25 *Let Y and Z be two discrete random variables with values in the countable sets F and G, respectively, and let $v : F \to \overline{\mathbb{R}}$, $w : G \to \overline{\mathbb{R}}$ be functions that are either non-negative or such that $v(Y)$ and $w(Z)$ are both integrable. Then*

$$E[v(Y)w(Z)] = E[v(Y)]E[w(Z)].$$

Proof. Consider the discrete random variable X with values in $E = F \times G$ defined by $X = (Y, Z)$, and consider the function $g : E \to \overline{\mathbb{R}}$ defined by $g(x) = v(y)w(z)$, where $x = (y, z)$. We have, under the above stated conditions

$$\begin{aligned}
E[v(Y)w(Z)] = E[g(X)] &= \sum_{x \in E} g(x) P(X = x) \\
&= \sum_{y \in F} \sum_{z \in F} v(y) w(z) P(Y = y, Z = z) \\
&= \sum_{y \in F} \sum_{z \in F} v(y) w(z) P(Y = y) P(Z = z) \\
&= \left(\sum_{y \in F} v(y) P(Y = y) \right) \left(\sum_{z \in F} w(z) P(Z = z) \right) \\
&= E[v(Y)]E[w(Z)].
\end{aligned}$$

$\square$

For *independent* random variables, "variances add up":

Corollary 1.3.26 *Let $X_1, \ldots, X_n$ be independent integrable random variables with values in $\mathbb{N}$. Then*

$$\sigma^2_{X_1 + \cdots + X_n} = \sigma^2_{X_1} + \cdots + \sigma^2_{X_n}. \tag{1.13}$$

Proof. Let $\mu_1, \ldots, \mu_n$ be the respective means of $X_1, \ldots, X_n$. The mean of the sum $X := X_1 + \cdots + X_n$ is $\mu := \mu_1 + \cdots + \mu_n$. By the product formula for expectations, if $i \neq k$,

$$E\left[(X_i - \mu_i)(X_k - \mu_k)\right] = E\left[(X_i - \mu_i)\right] E\left[(X_k - \mu_k)\right] = 0.$$

Therefore

$$\begin{aligned}
\operatorname{Var}(X) = E\left[(X - \mu)^2\right] \\
= E\left[\left(\sum_{i=1}^{n}(X_i - \mu_i)\right)^2\right] = E\left[\sum_{i=1}^{n}\sum_{k=1}^{n}(X_i - \mu_i)(X_k - \mu_k)\right] \\
= \sum_{i=1}^{n}\sum_{k=1}^{n} E\left[(X_i - \mu_i)(X_k - \mu_k)\right] \\
= \sum_{i=1}^{n} E\left[(X_i - \mu_i)^2\right] = \sum_{i=1}^{n} \operatorname{Var}(X_i).
\end{aligned}$$

$\square$

Remark 1.3.27 Note that means always add up, even when the random variables are not independent.

Let X be an integrable random variable. Then, clearly, for any $a \in \mathbb{R}$, aX is integrable and its variance is given by the formula

$$\text{Var}\,(aX) = a^2\,\text{Var}\,(X).$$

EXAMPLE 1.3.28: VARIANCE OF THE EMPIRICAL MEAN. From this remark and Corollary 1.3.26, we immediately obtain that if $X_1, \ldots, X_n$ are independent and identically distributed *integrable* random variables with values in $\mathbb{N}$ with common variance σ^2, then

$$\text{Var}\,\left(\frac{X_1 + \cdots + X_n}{n}\right) = \frac{\sigma^2}{n}.$$

1.3.2 Famous Discrete Probability Distributions

The Binomial Distribution

Consider an IID sequence $\{X_n\}_{n \geq 1}$ of random variables taking their values in the set $\{0, 1\}$ and with a common distribution given by

$$P(X_n = 1) = p \quad (p \in (0, 1)).$$

This may be taken as a model for a game of heads and tails with a possibly biased coin (when $p \neq \frac{1}{2}$). Since $P(X_j = a_j) = p$ or $1 - p$ depending on whether $a_i = 1$ or 0, and since there are exactly $h(a) := \sum_{j=1}^{k} a_j$ coordinates of $a = (a_1, \ldots, a_k)$ that are equal to 1,

$$P(X_1 = a_1, \ldots, X_k = a_k) = p^{h(a)} q^{k - h(a)}, \tag{1.14}$$

where $q := 1 - p$. (The integer $h(a)$ is called the *Hamming weight* of the binary vector a.)

The heads and tails framework shelters two important discrete random variables: the binomial random variable and the geometric random variable.

The Binomial Distribution

Definition 1.3.29 *A random variable X taking its values in the set $E = \{0, 1, \ldots, n\}$ and with the probability distribution*

$$P(X = i) = \binom{n}{i} p^i (1-p)^{n-i} \quad (0 \leq i \leq n)$$

is called a binomial random variable *of size n and parameter $p \in (0, 1)$. This is denoted by $X \sim \mathcal{B}(n, p)$.*

EXAMPLE 1.3.30: NUMBER OF HEADS IN COIN TOSSING. Define

$$S_n = X_1 + \cdots + X_n \,.$$

This random variable takes the values $0, 1, \ldots, n$. To obtain $S_n = i$, where $0 \le i \le n$, one must have $X_1 = a_1, \ldots, X_n = a_n$ with $\sum_{j=1}^n a_j = i$. There are $\binom{n}{i}$ distinct ways of having this, and each occurs with probability $p^i(1-p)^{n-i}$. Therefore, for $0 \le i \le n$,

$$P(S_n = i) = \binom{n}{i} p^i(1-p)^{n-i} \,.$$

Theorem 1.3.31 *The mean and the variance of a binomial random variable X of size n and parameter p are respectively*

$$\mathrm{E}[X] = np$$

and

$$\mathrm{Var}\,(X) = np(1-p) \,.$$

Proof. Consider the random variable S_n of Example 1.3.30, which is a binomial random variable. We have, since expectations add up,

$$\mathrm{E}\,[S_n] = \sum_{i=1}^n \mathrm{E}\,[X_i] = n\mathrm{E}\,[X_1] \,,$$

and since the X_i's are IID (and therefore in this case variances add up),

$$\mathrm{Var}\,(S_n) = \sum_{i=1}^n \mathrm{Var}\,(X_i) = nV(X_1) \,.$$

Now,

$$\mathrm{E}\,[X_1] = 0 \times P(X_1 = 0) + 1 \times P(X_1 = 1) = P(X_1 = 1) = p \,,$$

and since $X_1^2 = X_1$,

$$\mathrm{E}\,[X_1^2] = \mathrm{E}\,[X_1] = p \,.$$

Therefore

$$\mathrm{Var}\,(X_1) = \mathrm{E}\,[X_1^2] - \mathrm{E}\,[X_1]^2 = p - p^2 = p(1-p). $$

$\square$

The Geometric Distribution

Definition 1.3.32 *A random variable X taking its values in $\mathbb{N}_+ := \{1, 2, \ldots\}$ and with the distribution*

$$P(X = k) = (1-p)^{k-1}p \quad (k \ge 1),$$

where $0 < p < 1$, is called a geometric random variable *with parameter p. This is denoted $X \sim \mathcal{G}eo(p)$.*

EXAMPLE 1.3.33: FIRST "HEADS" IN THE SEQUENCE. Let $\{X_n\}_{n\geq 1}$ be an IID sequence of random variables taking their values in the set $\{0,1\}$ with common distribution given by $P(X_n = 1) = p \in (0,1)$. Define the random variable T to be the first time of occurrence of 1 in this sequence, that is,

$$T = \inf\{n \geq 1; X_n = 1\},$$

with the convention that if $X_n = 0$ for all $n \geq 1$, then $T = \infty$. The event $\{T = k\}$ is exactly $\{X_1 = 0, \ldots, X_{k-1} = 0, X_k = 1\}$, and therefore,

$$P(T = k) = P(X_1 = 0) \cdots P(X_{k-1} = 0)P(X_k = 1),$$

that is,

$$P(T = k) = (1 - p)^{k-1}p.$$

Theorem 1.3.34 *The mean of a geometric random variable X with parameter $p > 0$ is*

$$\mathrm{E}[X] = \frac{1}{p}.$$

Proof.

$$\mathrm{E}[X] = \sum_{k=1}^{\infty} k(1-p)^{k-1}p.$$

But for $\alpha \in (0,1)$,

$$\sum_{k=1}^{\infty} k\alpha^{k-1} = \frac{d}{d\alpha}\left(\sum_{k=1}^{\infty} \alpha^k\right) = \frac{d}{d\alpha}\left(\frac{1}{1-\alpha} - 1\right) = \frac{1}{(1-\alpha)^2}.$$

Therefore, with $\alpha = 1 - p$,

$$\mathrm{E}[X] = \frac{1}{p^2} \times p = \frac{1}{p}.$$

$\square$

Theorem 1.3.35 *A geometric random variable T with parameter $p \in (0,1)$ is memoryless in the sense that for any integer $k_0 \geq 1$,*

$$P(T = k + k_0 \mid T > k_0) = P(T = k) \quad (k \geq 1).$$

Proof. We first compute

$$P(T > k_0) = \sum_{k=k_0+1}^{\infty} (1-p)^{k-1}p$$

$$= p(1-p)^{k_0} \sum_{n=0}^{\infty} (1-p)^n = \frac{p(1-p)^{k_0}}{1-(1-p)} = (1-p)^{k_0}.$$

Therefore,

$$P\left(T = k_0 + k | T > k_0\right) = \frac{P\left(T = k_0 + k, T > k_0\right)}{P\left(T > k_0\right)} = \frac{P\left(T = k_0 + k\right)}{P\left(T > k_0\right)}$$

$$= \frac{p\left(1 - p\right)^{k+k_0-1}}{\left(1 - p\right)^{k_0}} = p\left(1 - p\right)^{k-1} = P\left(T = k\right).$$

$\square$

EXAMPLE 1.3.36: THE COUPON COLLECTOR. Each chocolate tablet of a certain brand contains a coupon, randomly and independently chosen among n types. A prize may be claimed once the chocolate amateur has gathered a collection containing a subset with all the types of coupons. What is the average value of the number X of chocolate tablets bought when this happens for the first time?

Solution: Let X_i ($0 \le i \le n-1$) be the number of tablets bought during the time where there are exactly i different types of coupons in the collector's box, so that

$$X = \sum_{i=0}^{n-1} X_i.$$

Each X_i is a geometric random variable with parameter $p_i = 1 - \frac{i}{n}$. In particular, $\mathrm{E}\left[X_i\right] = \frac{1}{p_i} = \frac{n}{n-i}$, and therefore

$$\mathrm{E}\left[X\right] = \sum_{i=0}^{n-1} \mathrm{E}\left[X_i\right] = n \sum_{i=1}^{n} \frac{1}{i}.$$

We can have a more precise idea of how far away from its mean the random variable X can be. Observing that $\left|\sum_{i=1}^{n} 1/i - \ln n\right| \le 1$, we have that $\left|\mathrm{E}\left[X\right] - n \ln n\right| \le n$. We shall now prove that for all $c > 0$,

$$P\left(X > \lceil n \ln n + cn \rceil\right) \le \mathrm{e}^{-c}. \tag{$\star$}$$

For this, define A_α to be the event that no coupon of type α shows up in the first $\lceil n \ln n + cn \rceil$ tablets. Then (by sub-additivity)

$$P\left(X > \lceil n \ln n + cn \rceil\right) = P\left(\cup_{\alpha=1}^{n} A_\alpha\right) \le \sum_{\alpha=1}^{n} P\left(A_\alpha\right)$$

$$= \sum_{\alpha=1}^{n} \left(1 - \frac{1}{n}\right)^{\lceil n \ln n + cn \rceil} = n \left(1 - \frac{1}{n}\right)^{\lceil n \ln n + cn \rceil},$$

and therefore, since $1 + x \le \mathrm{e}^x$ for all $x \in \mathbb{R}$,

$$P\left(X > \lceil n \ln n + cn \rceil\right) \le n \left(\mathrm{e}^{-\frac{1}{n}}\right)^{n \ln n + cn} = n \mathrm{e}^{-\ln n - cn} = \mathrm{e}^{-c}.$$

Remark 1.3.37 An inequality such as ($\star$) is called a *concentration inequality*. It reads

$$P\left(\frac{X - \mathrm{E}[X]}{\mathrm{E}[X]} > \frac{c}{\ln n}\right) \le \mathrm{e}^{-c},$$

which explains the terminology, since $\frac{X-\mathrm{E}[X]}{\mathrm{E}[X]}$ measures the relative dispersion of X around its mean.

The Poisson Distribution

Definition 1.3.38 *A random variable X taking its values in $\mathbb{N}$ and such that for all $k \geq 0$,*

$$P(X = k) = \mathrm{e}^{-\theta}\frac{\theta^k}{k!}, \qquad (1.15)$$

is called a Poisson random variable *with parameter $\theta > 0$. This is denoted by $X \sim \mathcal{P}oi(\theta)$.*

EXAMPLE 1.3.39: POISSON'S LAW OF RARE EVENTS, TAKE 1. A veterinary surgeon of the Prussian army collecting data relative to accidents due to horse kicks found that the yearly number of such casualties was approximately following a Poisson distribution. Here is an explanation of his findings.

Suppose that you play "heads or tails" for a large number n of (independent) tosses of a coin such that

$$P(X_i = 1) = \frac{\alpha}{n}.$$

In the example, n is the (large) number of soldiers, $X_i = 1$ if the i-th soldier was hurt and $X_i = 0$ otherwise. Let S_n be the total number of *heads* (wounded soldiers) and let $p_n(k) := P(S_n = k)$. It turns out that

$$\lim_{n \uparrow \infty} p_n(k) = \mathrm{e}^{-\alpha}\frac{\alpha^k}{k!} \qquad (\star)$$

(with the convention $0! = 1$). (The average number of heads is α and the choice $P(X_i = 1) = \frac{\alpha}{n}$ guarantees this. Letting $n \uparrow \infty$ accounts for n being large but unknown.)

Here is the proof of this result, which is known as *Poisson's law of rare events*. As we know, the random variable S_n follows a binomial law:

$$P(S_n = k) = \binom{n}{k} \left(\frac{\alpha}{n}\right)^k \left(1 - \frac{\alpha}{n}\right)^{n-k}$$

of mean $n \times \frac{\alpha}{n} = \alpha$. Denoting by $p_n(k) = P(S_n = k)$, we see that $p_n(0) = \left(1 - \frac{\alpha}{n}\right)^n \to \mathrm{e}^{-\alpha}$ as $n \uparrow \infty$. Also,

$$\frac{p_n(k+1)}{p_n(k)} = \frac{\frac{n-k}{k+1}\frac{\alpha}{n}}{1 - \frac{\alpha}{n}} \to \frac{\alpha}{k+1}$$

as $n \uparrow \infty$. Therefore, $(\star)$ holds true for all $k \geq 0$. The limit distribution is therefore a Poisson distribution of mean α.

Theorem 1.3.40 *The mean of a Poisson random variable with parameter $\theta > 0$ is given by*

$$\mathrm{E}[X] = \theta,$$

and its variance is

$$\mathrm{Var}(X) = \theta.$$

Proof. We have

$$\mathrm{E}[X] = \mathrm{e}^{-\theta}\sum_{k=1}^{\infty}\frac{k\theta^k}{k!} = \mathrm{e}^{-\theta}\theta\sum_{k=1}^{\infty}\frac{\theta^{k-1}}{(k-1)!}$$

$$= \mathrm{e}^{-\theta}\theta\sum_{j=0}^{\infty}\frac{\theta^j}{j!} = \mathrm{e}^{-\theta}\theta\mathrm{e}^{\theta} = \theta.$$

Also:

$$\mathrm{E}\left[X^2 - X\right] = \mathrm{e}^{-\theta} \sum_{k=0}^{\infty} \left(k^2 - k\right) \frac{\theta^k}{k!} = \mathrm{e}^{-\theta} \sum_{k=2}^{\infty} k\left(k-1\right) \frac{\theta^k}{k!}$$

$$= \mathrm{e}^{-\theta} \theta^2 \sum_{k=2}^{\infty} \frac{\theta^{k-2}}{(k-2)!} = \mathrm{e}^{-\theta} \theta^2 \sum_{j=0}^{\infty} \frac{\theta^j}{j!} = \mathrm{e}^{-\theta} \theta^2 \mathrm{e}^{\theta} = \theta^2 \,.$$

Therefore,

$$\begin{aligned}
\mathrm{Var}\,(X) &= \mathrm{E}\left[X^2\right] - \mathrm{E}\left[X\right]^2 \\
&= \mathrm{E}\left[X^2 - X\right] + \mathrm{E}\left[X\right] - \mathrm{E}\left[X\right]^2 \\
&= \theta^2 + \theta - \theta^2 = \theta.
\end{aligned}$$

$\square$

Theorem 1.3.41 *Let X_1 and X_2 be two independent Poisson random variables with means $\theta_1 > 0$ and $\theta_2 > 0$, respectively. Then $X = X_1 + X_2$ is a Poisson random variable with mean $\theta = \theta_1 + \theta_2$.*

Proof. For $k \geq 0$,

$$\begin{aligned}
P(X = k) &= P(X_1 + X_2 = k) \\
&= P\left(\cup_{i=0}^{k}\{X_1 = i, X_2 = k - i\}\right) \\
&= \sum_{i=0}^{k} P(X_1 = i, X_2 = k - i) \\
&= \sum_{i=0}^{k} P(X_1 = i)P(X_2 = k - i) \\
&= \sum_{i=0}^{k} \mathrm{e}^{-\theta_1} \frac{\theta_1^i}{i!} \, \mathrm{e}^{-\theta_2} \frac{\theta_2^{k-i}}{(k-i)!} \\
&= \frac{\mathrm{e}^{-(\theta_1+\theta_2)}}{k!} \sum_{i=0}^{k} \frac{k!}{i!(k-i)!} \theta_1^i \, \theta_2^{k-i} \\
&= \mathrm{e}^{-(\theta_1+\theta_2)} \frac{(\theta_1 + \theta_2)^k}{k!}.
\end{aligned}$$

$\square$

Remark 1.3.42 See Example 1.4.8 for an alternative shorter proof using generating functions (defined in Subsection 1.4.1).

The Multinomial Distribution

Consider the random vector $X = (X_1, \ldots, X_n)$ where all the random variables X_i take their values in the *same* countable space E (this restriction is not essential, but it simplifies the notation). Let $\pi : E^n \to \mathbb{R}_+$ be a function such that

$$\sum_{x \in \mathrm{E}^n} \pi(x) = 1 \,.$$

The discrete random vector X above is said to admit the probability distribution π if

$$P(X = x) = \pi(x) \quad (x \in \mathrm{E}^n) \,.$$

In fact, there is nothing new here with respect to previous definitions, since X is a discrete random variable taking its values in the countable set $\mathcal{X} := \mathrm{E}^n$.

EXAMPLE 1.3.43: MULTINOMIAL RANDOM VECTOR. We place, independently of one another, k balls in n boxes $B_1, \ldots, B_n$, with probability p_i for a given ball to be assigned to box B_i. Of course,

$$\sum_{i=1}^n p_i = 1 \,.$$

After placing all the balls in the boxes, there are X_i balls in box B_i, where

$$\sum_{i=1}^n X_i = k \,.$$

The random vector $X = (X_1, \ldots, X_n)$ is a *multinomial* vector of size (n, k) and parameters $p_1, \ldots, p_n$, that is, its probability distribution is

$$P(X_1 = m_1, \ldots, X_n = m_n) = \frac{k!}{\prod_{i=1}^n (m_i)!} \prod_{i=1}^n p_i^{m_i} \,,$$

where $m_1 + \cdots + m_n = k$.

Proof. Observe that (α): there are $k! / \prod_{i=1}^n (m_i)!$ distinct ways of placing k balls in n boxes in such a manner that m_1 balls are in box B_1, m_2 are in B_2, etc., and (β): each of these distinct ways occurs with the same probability $\prod_{i=1}^n p_i^{m_i}$. $\square$

Random Graphs

A graph is a discrete object and therefore random graphs are, from a purely formal point of view, discrete random variables. The random graphs considered below are in fact described by a finite collection of IID $\{0, 1\}$-valued random variables.

A (finite) graph $G = (V, \mathcal{E})$ consists of a finite collection V of vertices v and of a collection $\mathcal{E}$ of unordered pairs of distinct vertices, $\langle u, v \rangle$, called the edges. If $\langle u, v \rangle \in \mathcal{E}$, then u and v are called neighbors, and this is also denoted by $u \sim v$. The degree of vertex $v \in V$ is the number of edges stemming from it.

Definition 1.3.44 (Gilbert, 1959) *Let n be a fixed positive integer and let $V = \{1, 2, \ldots, n\}$ be a finite set of vertices. To each unordered pair of distinct vertices $\langle u, v \rangle$, associate a random variable $X_{\langle u, v \rangle}$ taking its values in $\{0, 1\}$ and suppose that all such variables are IID with probability $p \in (0, 1)$ for the value 1. This defines a random graph denoted by $\mathcal{G}(n, p)$, a random element taking its values in the (finite) set of all graphs with vertices $\{1, 2, \ldots, n\}$ and admitting for an edge the unordered pair of vertices $\langle u, v \rangle$ if and only if $X_{\langle u, v \rangle} = 1$.*

Note that $\mathcal{G}(n, p)$ is indeed a discrete random variable (taking its values in the finite set consisting of the collection of graphs with vertex set $V = \{1, 2, \ldots, n\}$). Similarly, the set $\mathcal{E}_{n,p}$ of edges of $\mathcal{G}(n, p)$ is also a discrete random variable. If we call any unordered pair of vertices $\langle u, v \rangle$ a potential edge (there are $\binom{n}{2}$ such edges forming the set $\mathbf{E_n}$), $\mathcal{G}(n, p)$ is constructed by accepting a potential edge as one of its edges with probability p, independently of all other potential edges. The probability of occurrence of a graph G with exactly m edges is then

$$P(\mathcal{G}(n, p) = G) = P(|\mathcal{E}_{n,p}| = m) = p^m (1 - p)^{\binom{n}{2} - m}.$$

Note that the degree of a given vertex, that is, the number of edges stemming from it, is a binomial random variable $\mathcal{B}(n - 1, p)$. In particular, the average degree is $d = (n - 1)p$.

Another type of random graph is the Erdös–Rényi random graph (Definition 1.3.45 below). It is closely related to the Gilbert graph (Exercise 1.6.27).

Definition 1.3.45 (Erdös and Rényi, 1959) *Consider the collection $\mathbf{G_m}$ of graphs $G = (V, \mathcal{E})$ where $V = \{1, 2, \ldots, n\}$ with exactly m edges ($|\mathcal{E}| = m$). There are $\binom{\binom{n}{2}}{m}$ such graphs. The Erdös–Rényi random graph $\mathcal{G}_{n,m}$ is a random graph uniformly distributed on $\mathbf{G_m}$.*

1.3.3 Conditional Expectation

This subsection introduces the concept of conditional expectation given a random element (variable or vector).[3]

Let Z be a discrete random variable with values in E, and let $f : E \to \mathbb{R}$ be a non-negative function. Let A be some event of positive probability. The conditional expectation of $f(Z)$ given A, denoted by $\mathrm{E}[f(Z) \,|\, A]$, is by definition the expectation when the distribution of Z is replaced by its conditional distribution given A:

$$\mathrm{E}[f(Z) \,|\, A] := \sum_z f(z) P(Z = z \,|\, A).$$

Let $\{A_i\}_{i \in \mathbb{N}}$ be a partition of the sample space. The following formula is then a direct consequence of Bayes' formula of total causes:

$$\mathrm{E}[f(Z)] = \sum_{i \in \mathbb{N}} \mathrm{E}[f(Z) \,|\, A_i] \, P(A_i).$$

EXAMPLE 1.3.46: THE POISSON AND MULTINOMIAL DISTRIBUTIONS. Suppose we have N bins in which we place balls in the following manner. The number of balls in any given bin is a Poisson variable of mean $\frac{m}{N}$, and is independent of numbers in the other bins. In particular, the total number of balls $Y_1 + \cdots + Y_N$ is, as the sum of independent Poisson random variables, a Poisson random variable whose mean is the sum of the means of the coordinates, that is, m.

For a given integer k, we will compute the conditional probability that there are k_1 balls in bin 1, k_2 balls in bin 2, etc, given that the total number of balls is $k_1 + \cdots + k_N = k$:

---

[3]The general theory of conditional expectation will be given in Section 3.3.

$$P\left(Y_1 = k_1, \ldots, Y_N = k_N \mid Y_1 + \cdots + Y_N = k\right)$$

$$= \frac{P\left(Y_1 = k_1, \ldots, Y_N = k_N, Y_1 + \cdots + Y_N = k\right)}{P\left(Y_1 + \cdots + Y_N = k\right)}$$

$$= \frac{P\left(Y_1 = k_1, \ldots, Y_N = k_N\right)}{P\left(Y_1 + \cdots + Y_N = k\right)}.$$

By independence of the Y_i's, and since they are Poisson variables with mean $\frac{m}{N}$,

$$P\left(Y_1 = k_1, \ldots, Y_N = k_N\right) = \prod_{i=1}^{N} \left(e^{-\frac{m}{N}} \frac{\left(\frac{m}{N}\right)^{k_i}}{k_i!}\right).$$

Also,

$$P\left(Y_1 + \cdots + Y_N = k\right) = e^{-m} \frac{m^k}{k!}.$$

Therefore

$$P\left(Y_1 = k_1, \ldots, Y_N = k_N \mid Y_1 + \cdots + Y_N = k\right) = \frac{k!}{k_1! \cdots k_N!} \left(\frac{1}{N}\right)^N.$$

But this is equal to $P(Z_1 = k_1, \ldots, Z_N = k_N)$, where Z_i is the number of balls in bin i when $k = k_1 + \cdots + k_N$ balls are placed independently and at random in the N bins. Note that the above equality is independent of m.

_____

The conditional expectation of some discrete random variable Z given some other discrete random variable Y is the expectation of Z using the probability measure modified by the observation of Y. For instance, if $Y = y$, instead of the original probability assigning the mass $P(A)$ to the event A, we use the conditional probability given $Y = y$ assigning the mass $P(A|Y = y)$ to this event.

Definition 1.3.47 *Let X and Y be two discrete random variables taking their values in the countable sets F and G, respectively, and let $g : F \times G \to \mathbb{R}_+$ be either non-negative, or such that $\mathrm{E}[|g(X,Y)|] < \infty$. Define for each $y \in G$ such that $P(Y = y) > 0$,*

$$\psi(y) = \sum_{x \in F} g(x, y) P(X = x \mid Y = y), \tag{1.16}$$

and if $(P(Y = y) = 0)$, let $\psi(y) = 0$. This quantity is called the conditional expectation *of $g(X,Y)$ given $Y = y$, and is denoted by $\mathrm{E}^{Y=y}[g(X,Y)]$, or $\mathrm{E}[g(X,Y) \mid Y = y]$. The random variable $\psi(Y)$ is called the conditional expectation of $g(X,Y)$ given Y, and is denoted by $\mathrm{E}^Y[g(X,Y)]$ or $\mathrm{E}[g(X,Y) \mid Y]$.*

The sum in (1.16) is well defined (possibly infinite however) when g is non-negative. Note that in the non-negative case, we have that

$$\sum_{y \in G} \psi(y) P(Y = y) = \sum_{y \in G} \sum_{x \in F} g(x, y) P(X = x \mid Y = y) P(Y = y)$$

$$= \sum_{x} \sum_{y} g(x, y) P(X = x, Y = y) = \mathrm{E}[g(X,Y)].$$

In particular, if $\mathrm{E}[g(X,Y)] < \infty$, then

$$\sum_{y \in G} \psi(y) P(Y = y) < \infty,$$

which implies that $\psi(y) < \infty$ for all $y \in G$ such that $P(Y = y) > 0$. We observe (for reference in a few lines) that in this case, $\psi(Y) < \infty$ almost surely, that is to say $P(\psi(Y) < \infty) = 1$ (in fact, $P(\psi(Y) = \infty) = \sum_{y; \psi(y) = \infty} P(Y = y) = 0$).

Let now $g : F \times G \to \mathbb{R}$ be a function of arbitrary sign such that $\mathrm{E}[|g(X, Y)|] < \infty$, and in particular $\mathrm{E}[g^{\pm}(X, Y)] < \infty$. Denote by $\psi^{\pm}$ the functions associated to $g^{\pm}$ as in (1.16). As we just saw, for all $y \in G$, $\psi^{\pm}(y) < \infty$, and therefore $\psi(y) = \psi^+(y) - \psi^-(y)$ is well defined (not an indeterminate $\infty - \infty$ form). Thus, the conditional expectation is also well defined in the integrable case. From the observation made a few lines above, in this case,

$$|\mathrm{E}^Y[g(X, Y)]| < \infty.$$

EXAMPLE 1.3.48: BINOMIAL EXAMPLE. Let X_1 and X_2 be independent binomial random variables of the same size N and same parameter p. We are going to show that

$$\mathrm{E}^{X_1 + X_2}[X_1] = h(X_1 + X_2) = \frac{X_1 + X_2}{2}.$$

We have

$$P(X_1 = k | X_1 + X_2 = n) = \frac{P(X_1 = k) P(X_2 = n - k)}{P(X_1 + X_2 = n)}$$

$$= \frac{\binom{N}{k} p^k (1-p)^{N-k} \binom{N}{n-k} p^{n-k} (1-p)^{N-n+k}}{\binom{2N}{n} p^n (1-p)^{N-n}} = \frac{\binom{N}{k} \binom{N}{n-k}}{\binom{2N}{n}},$$

where we have used the fact that the sum of two independent binomial random variables with size N and parameter p is a binomial random variable with size $2N$ and parameter p. This is the *hypergeometric distribution*. The right-hand side of the last display is the probability of obtaining k black balls when a sample of n balls is randomly selected from an urn containing N black balls and N red balls. The mean of such a distribution is (by reason of symmetry) $\frac{n}{2}$, therefore

$$\mathrm{E}^{X_1 + X_2 = n}[X_1] = \frac{n}{2} = h(n),$$

and this gives the announced result.

EXAMPLE 1.3.49: POISSON EXAMPLE. Let X_1 and X_2 be two independent Poisson random variables with respective means $\theta_1 > 0$ and $\theta_2 > 0$. We seek to compute $\mathrm{E}^{X_1 + X_2}[X_1]$, that is $\mathrm{E}^Y[X]$, where $X = X_1$, $Y = X_1 + X_2$. Following the instructions of Definition 1.3.47, we must first compute (only for $y \geq x$, why?)

$$P(X = x \mid Y = y) = \frac{P(X = x, Y = y)}{P(Y = y)} = \frac{P(X_1 = x, X_1 + X_2 = y)}{P(X_1 + X_2 = y)}$$

$$= \frac{P(X_1 = x, X_2 = y - x)}{P(X_1 + X_2 = y)} = \frac{P(X_1 = x) P(X_2 = y - x)}{P(X_1 + X_2 = y)}$$

$$= \frac{e^{-\theta_1} \frac{\theta_1^x}{x!} e^{-\theta_2} \frac{\theta_2^{y-x}}{(y-x)!}}{e^{-(\theta_1 + \theta_2)} \frac{(\theta_1 + \theta_2)^y}{y!}} = \binom{y}{x} \left(\frac{\theta_1}{\theta_1 + \theta_2}\right)^x \left(\frac{\theta_2}{\theta_1 + \theta_2}\right)^{y-x}.$$

Therefore, letting $\alpha = \frac{\theta_1}{\theta_1 + \theta_2}$,

$$\psi(y) = \mathrm{E}^{Y=y}[X] = \sum_{x=0}^{y} x \binom{y}{x} \alpha^x (1-\alpha)^{y-x} = \alpha y.$$

Finally, $\mathrm{E}^Y[X] = \psi(Y) = \alpha Y$, that is,

$$\mathrm{E}^{X_1 + X_2}[X_1] = \frac{\theta_1}{\theta_1 + \theta_2}(X_1 + X_2).$$

1.4   The Branching Process

The branching process is also known as the *Galton–Watson process*. Sir Francis Galton, a cousin of Darwin, was interested in the survival probability of a given line of English peerage. He posed the problem in the *Educational Times* in 1873. In the same year and the same journal, Reverend Watson proposed the method of solution that has become a textbook classic, and thereby initiated an important branch of probability.

The elementary theory of branching processes of Subsection 1.4.2 provides the opportunity to introduce the tool of generating functions.

1.4.1   Generating Functions

The computation of probabilities in discrete probability models often requires an enumeration of all the possible outcomes realizing this particular event. Generating functions are very useful for this task, and more generally, for obtaining distribution functions of integer-valued random variables. In order to introduce this versatile tool, we shall need to define the expectation of a complex-valued function of an integer-valued variable.

Let X be a discrete random variable with values in $\mathbb{N}$, and let $\varphi : \mathbb{N} \to \mathbb{C}$ be a complex function with real and imaginary parts φ_R and φ_I, respectively. The expectation $\mathrm{E}[\varphi(X)]$ is naturally defined by

$$\mathrm{E}[\varphi(X)] = \mathrm{E}[\varphi_R(X)] + i\mathrm{E}[\varphi_I(X)],$$

provided that the expectations on the right-hand side are well defined and finite.

Definition 1.4.1 *Let X be an $\mathbb{N}$-valued random variable. Its* generating function (GF) *is the function $g : \overline{D}(0;1) := \{z \in \mathbb{C}; |z| \leq 1\} \to \mathbb{C}$ defined by*

$$g(z) = \mathrm{E}[z^X] = \sum_{k=0}^{\infty} P(X=k)z^k. \tag{1.17}$$

The power series associated with the sequence $\{P(X = n)\}_{n \geq 0}$ has a radius of convergence $R \geq 1$, since $\sum_{n=0}^{\infty} P(X = n) = 1 < \infty$. The domain of definition of g could be, in specific cases, larger than the closed unit disk centered at the origin. In the next two examples below, the domain of absolute convergence is the whole complex plane.

EXAMPLE 1.4.2: THE GF OF THE BINOMIAL VARIABLE. For the *binomial random variable* of size n and parameter p,

$$\sum_{k=0}^{n} P(X = k)z^k = \sum_{k=0}^{n} \binom{n}{k}(zp)^k(1-p)^{n-k},$$

and therefore

$$g(z) = (1 - p + pz)^n. \tag{1.18}$$

EXAMPLE 1.4.3: THE GF OF THE POISSON VARIABLE. For the *Poisson random variable* of mean θ,

$$\sum_{k=0}^{\infty} P(X = k)z^k = e^{-\theta} \sum_{k=0}^{\infty} \frac{(\theta z)^k}{k!},$$

and therefore

$$g(z) = e^{\theta(z-1)}. \tag{1.19}$$

Here is an example where the radius of convergence is finite.

EXAMPLE 1.4.4: THE GF OF THE GEOMETRIC VARIABLE. For the *geometric random variable* of Definition 1.3.32,

$$\sum_{k=1}^{\infty} P(X = k)z^k = \sum_{k=0}^{\infty} p(1-p)^{k-1} z^k,$$

and therefore, with $q = 1 - p$,

$$g(z) = \frac{pz}{1-qz}.$$

The radius of convergence of this generating function power series is $\frac{1}{q}$.

Moments from the Generating Function

Generating functions are powerful computational tools. First of all, they can be used to obtain moments of a discrete random variable.

Theorem 1.4.5 *We have*

$$g'(1) = E[X] \tag{1.20}$$

and

$$g''(1) = E[X(X - 1)]. \tag{1.21}$$

Proof. Inside the open disk centered at the origin and of radius R, the power series defining the generating function g is continuous, and differentiable at any order term by term. In particular, differentiating both sides of (1.17) twice inside the open disk $D(0; R)$ gives

$$g'(z) = \sum_{n=1}^{\infty} nP(X = n)z^{n-1}, \tag{1.22}$$

and

$$g''(z) = \sum_{n=2}^{\infty} n(n-1)P(X = n)z^{n-2}. \tag{1.23}$$

When the radius of convergence R is *strictly larger* than 1, we obtained the announced results by letting $z = 1$ in the previous identities.

If $R = 1$, the same is basically true but the mathematical argument is more subtle. The difficulty is not with the right-hand side of (1.22), which is always well defined at $z = 1$, being equal to $\sum_{n=1}^{\infty} nP(X = n)$, a non-negative and possibly infinite quantity. The difficulty is that g may not be differentiable at $z = 1$, a border point of the disk (here of radius 1) on which it is defined. However, by *Abel's theorem* (Theorem B.2.3), the limit as the *real* variable x increases to 1 of $\sum_{n=1}^{\infty} nP(X = n)x^{n-1}$ is $\sum_{n=1}^{\infty} nP(X = n)$. Therefore g', as a function on the real interval $[0, 1)$, can be extended to $[0, 1]$ by (1.20), and this extension preserves continuity. With this *definition* of $g'(1)$, Formula (1.20) holds true. Similarly, when $R = 1$, the function g'' defined on $[0, 1)$ by (1.23) is extended to a continuous function on $[0, 1]$ by *defining $g''(1)$* by (1.21). $\qquad\square$

Theorem 1.4.6 *The generating function characterizes the distribution of a random variable.*

This means the following. Suppose that, without knowing the distribution of X, you have been able to compute its generating function g, and that, moreover, you are able to give its power series expansion in a neighborhood of the origin:[4]

$$g(z) = \sum_{n=0}^{\infty} a_n z^n.$$

Since $g(z)$ is the generating function of X,

$$g(z) = \sum_{n=0}^{\infty} P(X = n)z^n,$$

and since the power series expansion around the origin is unique, the distribution of X is identified as

$$P(X = n) = a_n$$

for all $n \geq 0$. Similarly, if two $\mathbb{N}$-valued random variables X and Y have the same generating function, they have the same distribution. Indeed, the identity in a neighborhood of the origin of the power series:

$$\sum_{n=0}^{\infty} P(X = n)z^n = \sum_{n=0}^{\infty} P(Y = n)z^n$$

implies the identity of their coefficients.

Theorem 1.4.7 *Let X and Y be two independent integer-valued random variables with respective generating functions g_X and g_Y. Then the sum $X + Y$ has the GF*

$$g_{X+Y}(z) = g_X(z) \times g_Y(z). \tag{1.24}$$

---

[4]This is a common situation; see Theorem 1.4.10 for instance.

Proof. Use the product formula for expectations:

$$g_{X+Y}(z) = \mathrm{E}\left[z^{X+Y}\right]$$
$$= \mathrm{E}\left[z^X z^Y\right] = \mathrm{E}\left[z^X\right]\mathrm{E}\left[z^Y\right].$$

$\square$

EXAMPLE 1.4.8: SUMS OF INDEPENDENT POISSON VARIABLES. Let X and Y be two *independent* Poisson random variables of means α and β respectively. We shall prove that the sum $X + Y$ is a Poisson random variable with mean $\alpha + \beta$.

Indeed, according to (1.24) and (1.19),

$$g_{X+Y}(z) = g_X(z) \times g_Y(z)$$
$$= e^{\alpha(z-1)} e^{\beta(z-1)} = e^{(\alpha+\beta)(z-1)},$$

and the assertion follows directly from Theorem 1.4.6 since g_{X+Y} is the GF of a Poisson random variable with mean $\alpha + \beta$.

Counting with Generating Functions

The following example is typical of the use of generating functions in combinatorics (the art of counting).

EXAMPLE 1.4.9: THE LOTTERY. Let X_1, X_2, X_3, X_4, X_5, and X_6 be independent random variables uniformly distributed over $\{0, 1, \ldots, 9\}$. We shall compute the generating function of $Y = 27 + X_1 + X_2 + X_3 - X_4 - X_5 - X_6$ and use the result to obtain the probability that in a 6-digit lottery the sum of the first three digits equals the sum of the last three digits. We have

$$\mathrm{E}[z^{X_i}] = \frac{1}{10}(1 + z + \cdots + z^9) = \frac{1}{10}\frac{1 - z^{10}}{1 - z},$$

$$\mathrm{E}[z^{-X_i}] = \frac{1}{10}\left(1 + \frac{1}{z} + \cdots + \frac{1}{z^9}\right) = \frac{1}{10}\frac{1 - z^{-10}}{1 - z^{-1}} = \frac{1}{10}\frac{1}{z^9}\frac{1 - z^{10}}{1 - z},$$

and

$$\mathrm{E}[z^Y] = \mathrm{E}\left[z^{27 + \sum_{i=1}^{3} X_i - \sum_{i=4}^{6} X_i}\right]$$
$$= \mathrm{E}\left[z^{27}\prod_{i=1}^{3} z^{X_i}\prod_{i=4}^{6} z^{-X_i}\right] = z^{27}\prod_{i=1}^{3}\mathrm{E}[z^{X_i}]\prod_{i=4}^{6}\mathrm{E}[z^{-X_i}].$$

Therefore,

$$g_Y(z) = \frac{1}{10^6}\frac{\left(1 - z^{10}\right)^6}{(1 - z)^6}.$$

But $P(X_1 + X_2 + X_3 = X_4 + X_5 + X_6) = P(Y = 27)$ is the factor of z^{27} in the power series expansion of $g_Y(z)$. Since

$$(1 - z^{10})^6 = 1 - \binom{6}{1}z^{10} + \binom{6}{2}z^{20} + \cdots$$

and

$$(1 - z)^{-6} = 1 + \binom{6}{5}z + \binom{7}{5}z^2 + \binom{8}{5}z^3 + \cdots$$

(negative binomial formula), we find that

$$P(Y = 27) = \frac{1}{10^6}\left(\binom{32}{5} - \binom{6}{1}\binom{22}{5} + \binom{6}{2}\binom{12}{5}\right).$$

Random Sums

How to compute the distribution of *random sums*? Here again, generating functions help.

Theorem 1.4.10 *Let* $\{Y_n\}_{n\geq 1}$ *be an* IID *sequence of integer-valued random variables with the common generating function* g_Y. *Let* T *be another random variable, integer-valued, independent of the sequence* $\{Y_n\}_{n\geq 1}$, *and let* g_T *be its generating function. The generating function of*

$$X = \sum_{n=1}^{T} Y_n,$$

where by convention $\sum_{n=1}^{0} = 0$, *is*

$$g_X(z) = g_T(g_Y(z)).\tag{1.25}$$

Proof. Since $\{T = k\}_{k\geq 0}$ is a sequence forming a partition of Ω, we have (Exercise 1.6.3) $1 = \sum_{k=0}^{\infty} 1_{\{T=k\}}$. Therefore

$$z^X = z^{\sum_{n=1}^{T} Y_n} = \left(\sum_{k=0}^{\infty} 1_{\{T=k\}}\right) z^{\sum_{n=1}^{T} Y_n}$$

$$= \sum_{k=0}^{\infty}\left\{\left(z^{\sum_{n=1}^{T} Y_n}\right)1_{\{T=k\}}\right\} = \sum_{k=0}^{\infty}\left(z^{\sum_{n=1}^{k} Y_n}\right)1_{\{T=k\}}.$$

Taking expectations,

$$E[z^X] = \sum_{k=0}^{\infty} E\left[1_{\{T=k\}}\left(z^{\sum_{n=1}^{k} Y_n}\right)\right]$$

$$= \sum_{k=0}^{\infty} E[1_{\{T=k\}}]E[z^{\sum_{n=1}^{k} Y_n}],$$

where we have used independence of T and $\{Y_n\}_{n\geq 1}$. Now, $E[1_{\{T=k\}}] = P(T = k)$, and $E[z^{\sum_{n=1}^{k} Y_n}] = g_Y(x)^k$, and therefore

$$E[z^X] = \sum_{k=0}^{\infty} P(T = k)g_Y(z)^k = g_T(g_Y(z)).$$

$\square$

Another useful result is *Wald's identity* below (Formula (13.2.10)), which gives the expectation of a random sum of independent and identically distributed integer-valued variables.

By taking derivatives in (1.25) of Theorem 1.4.10,

$$E[X] = g'_X(1) = g'_Y(1)g'_T(g_Y(1)) = E[Y_1]E[T].$$

A stronger version of this result is often needed:

Theorem 1.4.11 *Let* $\{Y_n\}_{n\geq 1}$ *be a sequence of integer-valued integrable random variables such that* $E[Y_n] = E[Y_1]$ *for all* $n \geq 1$. *Let* T *be an integer-valued random variable such that for all* $n \geq 1$, *the event* $\{T \geq n\}$ *is independent of* Y_n. *Define*

$$X = \sum_{n=1}^{T} Y_n.$$

Then

$$E[X] = E[Y_1]E[T]. \tag{1.26}$$

Proof. We have

$$E[X] = E\left[\sum_{n=1}^{\infty} Y_n 1_{\{n \leq T\}}\right] = \sum_{n=1}^{\infty} E[Y_n 1_{\{n \leq T\}}].$$

But

$$E[Y_n 1_{\{n \leq T\}}] = E[Y_n]E[1_{\{n \leq T\}}] = E[Y_1]P(n \leq T).$$

The result then follows from the telescope formula. □

The following technical result will be needed in the next subsection on branching processes. It gives details concerning the shape of the generating function restricted to the interval $[0, 1]$.

Theorem 1.4.12

(α) *Let* $g : [0,1] \to \mathbb{R}$ *be defined by* $g(x) = E[x^X]$, *where* X *is a non-negative integer-valued random variable. Then* g *is non-decreasing and convex. Moreover, if* $P(X = 0) < 1$, *then* g *is strictly increasing, and if* $P(X \leq 1) < 1$, *it is strictly convex.*

(β) *Suppose* $P(X \leq 1) < 1$. *If* $E[X] \leq 1$, *the equation* $x = g(x)$ *has a unique solution* $x \in [0,1]$, *namely* $x = 1$. *If* $E[X] > 1$, *it has two solutions in* $[0,1]$, $x = 1$ *and* $x = x_0 \in (0,1)$.

Proof. Just observe that for $x \in [0,1]$,

$$g'(x) = \sum_{n=1}^{\infty} nP(X = n)x^{n-1} \geq 0,$$

and therefore g is non-decreasing, and

$$g''(x) = \sum_{n=2}^{\infty} n(n-1)P(X-n)x^{n-2} \geq 0,$$

and therefore g is convex. For $g'(x)$ to be null for some $x \in (0,1)$, it is necessary to have $P(X = n) = 0$ for all $n \geq 1$, and therefore $P(X = 0) = 1$. For $g''(x)$ to be null for some $x \in (0,1)$, one must have $P(X = n) = 0$ for all $n \geq 2$, and therefore $P(X = 0) + P(X = 1) = 1$.

The graph of $g : [0,1] \to \mathbb{R}$ has, in the strictly increasing strictly convex case, $P(X = 0) + P(X = 1) < 1$, the general shape shown in the figure, where we distinguish two cases: $E[X] = g'(1) \leq 1$, and $E[X] = g'(1) > 1$. The rest of the proof is then easy. □

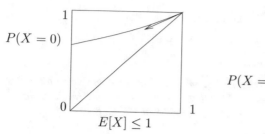
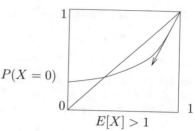

Two aspects of the generating function

1.4.2   Probability of Extinction

We shall first formally define the branching process. Let $Z_n = (Z_n^{(1)}, Z_n^{(2)}, \ldots)$, where the random variables $\{Z_n^{(j)}\}_{n\geq 1, j\geq 1}$ are IID and integer-valued. The recurrence equation

$$X_{n+1} = \sum_{k=1}^{X_n} Z_{n+1}^{(k)} \tag{1.27}$$

($X_{n+1} = 0$ if $X_n = 0$) may be interpreted as follows: X_n is the number of individuals in the nth generation of a given population (humans, particles, etc.). Individual number k of the nth generation gives birth to $Z_{n+1}^{(k)}$ descendants, and this accounts for Eqn. (1.27).

The number X_0 of ancestors is assumed to be independent of $\{Z_n\}_{n\geq 1}$. The sequence of random variables $\{X_n\}_{n\geq 0}$ is called a branching process because of the genealogical tree that it generates (see the figure below).

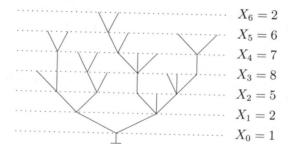

$X_6 = 2$
$X_5 = 6$
$X_4 = 7$
$X_3 = 8$
$X_2 = 5$
$X_1 = 2$
$X_0 = 1$

Sample tree of a branching process (one ancestor)

The event $\mathcal{E} =$ "an extinction occurs" is just "at least one generation is empty", that is,

$$\mathcal{E} = \cup_{n=1}^{\infty}\{X_n = 0\}.$$

We now proceed to the computation of the extinction probability when there is one ancestor. We discard trivialities by supposing that $P(Z \leq 1) < 1$. Let g be the common generating function of the variables $Z_n^{(k)}$. The generating function of the number of individuals in the nth generation is denoted

$$\psi_n(z) = \mathrm{E}[z^{X_n}].$$

We prove successively that

(a) $P(X_{n+1} = 0) = g(P(X_n = 0))$,

(b) $P(\mathcal{E}) = g(P(\mathcal{E}))$, and

(c) if $E[Z_1] < 1$, the probability of extinction is 1; and if $E[Z_1] > 1$, the probability of extinction is < 1 but nonzero.

Proof.

(a) In (1.27), X^n is independent of the $Z_{n+1}^{(k)}$'s. Therefore, by Theorem 1.4.10,

$$\psi_{n+1}(z) = \psi_n(g(z)).$$

Iterating this equality, we obtain

$$\psi_{n+1}(z) = \psi_0(g^{(n+1)}(z)),$$

where $g^{(n)}$ is the nth iterate of g. Since there is only *one ancestor*, $\psi_0(z) = z$, and therefore $\psi_{n+1}(z) = g^{(n+1)}(z) = g(g^{(n)}(z))$, that is,

$$\psi_{n+1}(z) = g(\psi_n(z)).$$

In particular, since $\psi_n(0) = P(X_n = 0)$, (a) is proved.

(b) Since $X_n = 0$ implies $X_{n+1} = 0$, the sequence $\{X_n = 0\}$, $n \geq 1$, is non-decreasing, and therefore, by monotone sequential continuity,

$$P(\mathcal{E}) = \lim_{n\uparrow\infty} P(X_n = 0).$$

The generating function g is continuous, and therefore from (a) and the last equation, the probability of extinction satisfies (b).

(c) Let Z be any of the random variables $Z_n^{(k)}$. Since the trivial cases where $P(Z = 0) = 1$ or $P(Z \geq 2) = 0$ have been eliminated, by Theorem 1.4.12:

(α) If $E[Z] \leq 1$, the only solution of $x = g(x)$ in $[0, 1]$ is 1, and therefore $P(\mathcal{E}) = 1$. The branching process eventually becomes extinct.

(β) If $E[Z] > 1$, there are two solutions of $x = g(x)$ in $[0, 1]$, 1 and x_0 such that $0 < x_0 < 1$. From the strict convexity of $f : [0, 1] \rightarrow [0, 1]$, it follows that the sequence $y_n = P(X_n = 0)$ that satisfies $y_0 = 0$ and $y_{n+1} = g(y_n)$ converges to x_0. Therefore, when the mean number of descendants $E[Z]$ is strictly larger than 1, $P(\mathcal{E}) \in (0, 1)$.

$\square$

1.5   Borel's Strong Law of Large Numbers

The empirical frequency of heads in a sequence of independent tosses of a fair coin is $\frac{1}{2}$. This is a special case of Borel's strong law of large numbers:

Theorem 1.5.1 *Let $\{X_n\}_{n\geq 1}$ be an* IID *sequence of $\{0,1\}$-valued random variables taking the value 1 with probability $p \in [0,1]$. Then*

$$P\left(\lim_{n\uparrow\infty}\frac{1}{n}\sum_{k=1}^{n}X_k = p\right) = 1.$$

We then say: the sequence $\{\frac{1}{n}\sum_{k=1}^{n}X_k\}_{n\geq 1}$ converges almost surely to p.

For the proof, some preliminaries are in order.

1.5.1   The Borel–Cantelli Lemma

Consider a sequence of events $\{A_n\}_{n\geq 1}$. Let

$$\{A_n \, i.o.\} := \{\omega; \omega \in A_n \text{ for an infinity of indices } n\}.$$

Here *i.o.* means *infinitely often*. We have the *Borel–Cantelli lemma*:

Theorem 1.5.2

$$\sum_{n=1}^{\infty} P(A_n) < \infty \implies P(A_n \, i.o.) = 0.$$

Proof. First observe that

$$\{A_n \text{ i.o.}\} = \bigcap_{n=1}^{\infty}\bigcup_{k\geq n} A_k.$$

(Indeed, if ω belongs to the set on the right-hand side, then for *all* $n \geq 1$, ω belongs to at least one among $A_n, A_{n+1}, \ldots$, which implies that ω is in A_n for an infinite number of indices n. Conversely, if ω is in A_n for an infinite number of indices n, it is for *all* $n \geq 1$ in at least one of the sets $A_n, A_{n+1}, \ldots.$)

The set $\cup_{k\geq n} A_k$ decreases as n increases, so that by the sequential continuity property of probability,

$$P(A_n \text{ i.o.}) = \lim_{n\uparrow\infty} P\left(\bigcup_{k\geq n} A_k\right). \tag{1.28}$$

But by sub-σ-additivity,

$$P\left(\bigcup_{k\geq n} A_k\right) \leq \sum_{k\geq n} P(A_k),$$

and by the summability assumption, the right-hand side of this inequality goes to 0 as $n \uparrow \infty$. $\qquad\square$

For the *converse Borel–Cantelli lemma* below, an additional assumption of independence is needed.

Theorem 1.5.3 *Let $\{A_n\}_{n\geq 1}$ be a sequence of* independent *events. Then,*

$$\sum_{n=1}^{\infty} P(A_n) = \infty \implies P(A_n \ i.o.) = 1.$$

Proof. We may without loss of generality assume that $P(A_n) > 0$ for all $n \geq 1$ (why?). The divergence hypothesis implies that for all $n \geq 1$,

$$\prod_{k\geq n}(1 - P(A_k)) = 0.$$

This infinite product equals, in view of the independence assumption,

$$\prod_{k\geq n} P\left(\overline{A_k}\right) = P\left(\bigcap_{k=n}^{\infty} \overline{A_k}\right).$$

Passing to the complement and using De Morgan's identity,

$$P\left(\bigcup_{k\geq n} A_k\right) = 1.$$

Therefore, by (1.28),

$$P(A_n \ i.o.) = \lim_{n\uparrow\infty} P\left(\bigcup_{k\geq n} A_k\right) = 1.$$

$\square$

1.5.2 Markov's Inequality

Theorem 1.5.4 *Let Z be a non-negative real random variable and let $a > 0$. Then,*

$$P(Z \geq a) \leq \frac{E[Z]}{a}. \tag{1.29}$$

Proof. From the inequality

$$Z \geq a\mathbf{1}_{\{Z\geq a\}},$$

it follows by taking expectations that

$$E[Z] \geq aE[\mathbf{1}_{\{Z\geq a\}}] = aP(Z \geq a).$$

$\square$

EXAMPLE 1.5.5: CHEBYSHEV'S INEQUALITY. Let X be a real (discrete) random variable. Specializing the Markov inequality of Theorem 1.5.4 to $Z = (X - \mu)^2$, $a = \varepsilon^2 > 0$, we obtain *Chebyshev's inequality*: For all $\varepsilon > 0$,

$$P(|X - \mu| \geq \varepsilon) \leq \frac{\sigma^2}{\varepsilon^2}.$$

EXAMPLE 1.5.6: THE WEAK LAW OF LARGE NUMBERS. Let $\{X_n\}_{n\geq 1}$ be an IID sequence of real square-integrable random variables with common mean μ and common variance $\sigma^2 < \infty$. Since the variance of the empirical mean $\frac{S_n}{n} := \frac{X_1 + \cdots + X_n}{n}$ is equal to $\frac{\sigma^2}{n}$, we have by Chebyshev's inequality, for all $\varepsilon > 0$,

$$P\left(\left|\frac{S_n}{n} - \mu\right| \geq \varepsilon\right) = P\left(\left|\frac{\sum_{i=1}^{n}(X_i - \mu)}{n}\right| \geq \varepsilon\right) \leq \frac{\sigma^2}{n^2 \varepsilon}.$$

In other words, the empirical mean $\frac{S_n}{n}$ converges to the mean μ in probability, which means exactly (by definition of the convergence in probability) that, for all $\varepsilon > 0$,

$$\lim_{n\uparrow\infty} P\left(\left|\frac{S_n}{n} - \mu\right| \geq \varepsilon\right) = 0.$$

This specific result is called the *weak law of large numbers*.

1.5.3   Proof of Borel's Strong Law

Let $S_n := \sum_{k=1}^{n} X_k$ and $Z_n := \frac{1}{n} S_n$.

Lemma 1.5.7 *If*

$$\sum_{n\geq 1} P(|Z_n - p| \geq \varepsilon_n) < \infty \tag{1.30}$$

for some sequence of positive numbers $\{\varepsilon_n\}_{n\geq 1}$ converging to 0, then the sequence $\{Z_n\}_{n\geq 1}$ converges P-a.s. to p.

Proof. If, for a given ω, $|Z_n(\omega) - p| \geq \varepsilon_n$ finitely often (or *f.o.*; that is, for all but a finite number of indices n), then $\lim_{n\uparrow\infty} |Z_n(\omega) - p| \leq \lim_{n\uparrow\infty} \varepsilon_n = 0$. Therefore

$$P(\lim_{n\uparrow\infty} Z_n = p) \geq P(|Z_n - p| \geq \varepsilon_n \quad f.o.).$$

On the other hand,

$$\{|Z_n - p| \geq \varepsilon_n \quad f.o.\} = \overline{\{|Z_n - p| \geq \varepsilon_n \quad i.o.\}}.$$

Therefore

$$P(|Z_n - p| \geq \varepsilon_n \quad f.o.) = 1 - P(|Z_n - p| \geq \varepsilon_n \quad i.o.).$$

Hypothesis (1.30) implies (Borel–Cantelli lemma) that

$$P(|Z_n - p| \geq \varepsilon_n \quad i.o.) = 0.$$

By linking the above facts, we obtain

$$P(\lim_{n\uparrow\infty} Z_n = p) \geq 1,$$

and of course, the only possibility is $= 1$. $\square$

In order to prove almost-sure convergence using Lemma 1.5.7, we must find some adequate upper bound for the general term of the series occurring in the left-hand side of (1.30). The basic tool for this is the *Markov inequality*:

$$P\left(\left|\frac{S_n}{n} - p\right| \geq \varepsilon\right) = P\left(\left(\frac{S_n}{n} - p\right)^4 \geq \varepsilon^4\right)$$

$$\leq \frac{E\left[\left(\frac{S_n}{n} - p\right)^4\right]}{\varepsilon^4} \leq \frac{E\left[\left(\sum_{i=1}^n Y_i\right)^4\right]}{n^4 \varepsilon^4},$$

where $Y_i := X_i - p$. In view of the independence hypothesis,

$$E[Y_1 Y_2 Y_3 Y_4] = E[Y_1]E[Y_2]E[Y_3]E[Y_4] = 0,$$
$$E[Y_1 Y_2^3] = E[Y_1]E[Y_2^3] = 0,$$

and the like. Finally, in the expansion

$$E\left[\left(\sum_{i=1}^n Y_i\right)^4\right] = \sum_{i,j,k,\ell=1}^n E[Y_i Y_j Y_k Y_\ell],$$

only the terms of the form $E[Y_i^4]$ and $E[Y_i^2 Y_j^2]$, $i \neq j$, remain. There are n terms of the first type and $3n(n-1)$ terms of the second type. Therefore, $nE[Y_1^4] + 3n(n-1)E[Y_1^2 Y_2^2]$ remains, which is less than Kn^2 for some finite K. Therefore

$$P\left(\left|\frac{S_n}{n} - p\right| \geq \varepsilon\right) \leq \frac{K}{n^2 \varepsilon^4},$$

and in particular, with $\varepsilon = n^{-\frac{1}{8}}$,

$$P\left(\left|\frac{S_n}{n} - p\right| \geq n^{-\frac{1}{8}}\right) \leq \frac{K}{n^{\frac{3}{2}}},$$

from which it follows that

$$\sum_{n=1}^\infty P\left(\left|\frac{S_n}{n} - p\right| \geq n^{-\frac{1}{8}}\right) < \infty.$$

Therefore, by Lemma 1.5.7, $\left|\frac{S_n}{n} - p\right|$ converges almost surely to 0.

Complementary reading

[Brémaud, 2017] is entirely devoted to the main discrete probability models and methods, using only the elementary tools presented in this chapter.

1.6 Exercises

Exercise 1.6.1. DE MORGAN'S RULES
Let $\{A_n\}_{n \geq 1}$ be a sequence of subsets of Ω. Prove *De Morgan's identities*:

$$\overline{\left(\bigcap_{n=1}^\infty A_n\right)} = \bigcup_{n=1}^\infty \overline{A}_n \text{ and } \overline{\left(\bigcup_{n=1}^\infty A_n\right)} = \bigcap_{n=1}^\infty \overline{A}_n.$$

Exercise 1.6.2. FINITELY OFTEN

Let $\{A_n\}_{n\geq 1}$ be a sequence of subsets of Ω. Show that $\omega \in B := \bigcup_{n=1}^{\infty}\bigcap_{k=n}^{\infty} \overline{A_k}$ if and only if there exists at most a *finite* number (depending on ω) of indices k such that $\omega \in A_k$. (The event B is therefore the event that events A_n occur finitely often.)

Exercise 1.6.3. INDICATOR FUNCTIONS
Show that for all subsets $A, B \subset \Omega$, where Ω is an arbitrary set,

$$1_{A\cap B} = 1_A \times 1_B \text{ and } 1_{\overline{A}} = 1 - 1_A.$$

Show that if $\{A_n\}_{n\geq 1}$ is a partition of Ω,

$$1 = \sum_{n\geq 1} 1_{A_n}.$$

Exercise 1.6.4. SMALL σ-FIELDS
Is there a σ-field on Ω with 6 elements (including of course Ω and $\varnothing$)?

Exercise 1.6.5. COMPOSED EVENTS
Let $\mathcal{F}$ be a σ-field on some set Ω.

(1) Show that if $A_1, A_2, \ldots$ are in $\mathcal{F}$, then so is $\bigcap_{k=1}^{\infty} A_k$.

(2) Show that if A_1, A_2 are in $\mathcal{F}$, then so is their *symmetric difference* $A_1 \triangle A_2 := A_1 \cup A_2 - A_1 \cap A_2$.

Exercise 1.6.6. SET INVERSE FUNCTIONS
Let $f : U \to E$ be a function, where U and E are arbitrary sets. For any subset $A \subseteq E$, define

$$f^{-1}(A) = \{u \in U \,;\, f(u) \in A\}.$$

(i) Show that for all $u \in U$,

$$1_A(f(u)) = 1_{f^{-1}(A)}(u).$$

(ii) Prove that if $\mathcal{E}$ is a σ-field on E, then the collection of subsets of U

$$f^{-1}(\mathcal{E}) := \{f^{-1}(A) \,;\, A \in \mathcal{E}\}$$

is a σ-field on U.

Exercise 1.6.7. IDENTITIES
Prove the set identities

$$P(A \cup B) = 1 - P(\overline{A} \cap \overline{B})$$

and

$$P(A \cup B) = P(A) + P(B) - P(A \cap B).$$

Exercise 1.6.8. URNS

1. An urn contains 17 red balls and 19 white balls. Balls are drawn in succession at random and without replacement. What is the probability that the first 2 balls are red?

2. An urn contains N balls numbered from 1 to N. Someone draws n balls ($1 \leq n \leq N$) simultaneously from the urn. What is the probability that the lowest number drawn is k?

Exercise 1.6.9. INDEPENDENCE OF A FAMILY OF EVENTS

1. Give a simple example of a probability space $(\Omega, \mathcal{F}, P)$ with three events A_1, A_2, A_3 that are pairwise independent but *not* globally independent (the family $\{A_1, A_2, A_3\}$ is not independent).

2. If $\{A_i\}_{i \in \mathbb{N}}$ is an independent family of events, is it true that $\{\widetilde{A}_i\}_{i \in \mathbb{N}}$ is also an independent family of events, where for each $i \in \mathbb{N}$, $\widetilde{A}_i = A_i$ or $\overline{A}_i$ (your choice, for instance, $\widetilde{A}_0 = A_0, \widetilde{A}_1 = \overline{A}_1, \widetilde{A}_3 = A_3, \ldots$)?

Exercise 1.6.10. EXTENSION OF THE PRODUCT FORMULA FOR INDEPENDENT EVENTS
Let $\{C_n\}_{n \geq 1}$ be a sequence of *independent* events. Then

$$P(\cap_{n=1}^{\infty} C_n) = \Pi_{n=1}^{\infty} P(C_n).$$

(This extends formula (1.7) to a countable number of sets.)

Exercise 1.6.11. CONDITIONAL INDEPENDENCE AND THE MARKOV PROPERTY

1. Let $(\Omega, \mathcal{F}, P)$ be a probability space. Define for a fixed event C of positive probability, $P_C(A) := P(A \mid C)$. Show that P_C is a probability on $(\Omega, \mathcal{F})$. (Note that A and B are independent with respect to this probability if and only if they are conditionally independent given C.)

2. Let A_1, A_2, A_3 be three events of positive probability. Show that events A_1 and A_3 are conditionally independent given A_2 if and only if the "Markov property" holds, that is, $P(A_3 \mid A_1 \cap A_2) = P(A_3 \mid A_2)$.

Exercise 1.6.12. ROLL IT!
You roll fairly and simultaneously three unbiased dice.

(i) What is the probability that you obtain the (unordered) outcome $\{1, 2, 4\}$?

(ii) What is the probability that *some* die shows 2, given that the sum of the 3 values equals 5?

Exercise 1.6.13. HEADS OR TAILS AS USUAL
A person, A, tossing an *unbiased* coin N times obtains T_A tails. Another person, B, tossing her own unbiased coin $N + 1$ times has T_B tails. What is the probability that $T_A \geq T_B$? *Hint:* Introduce H_A and H_B, the number of heads obtained by A and B respectively, and use a symmetry argument.

Exercise 1.6.14. APARTHEID UNIVERSITY
In the renowned Social Apartheid University, students have been separated into three social groups for pedagogical purposes. In group A, one finds students who individually

have a probability of passing equal to 0.95. In group B this probability is 0.75, and in group C only 0.65. The three groups are of equal size. What is the probability that a student passing the course comes from group A? B? C?

Exercise 1.6.15. A WISE BET
There are three cards. The first one has both faces red, the second one has both faces white, and the third one is white on one face, red on the other. A card is drawn at random, and the color of a randomly selected face of this card is shown to you (the other remains hidden). What is the winning strategy if you must bet on the color of the hidden face?

Exercise 1.6.16. A SEQUENCE OF LIARS
Consider a sequence of n "liars" $L_1, \ldots, L_n$. The first liar L_1 receives information about the occurrence of some event in the form "yes or no", and transmits it to L_2, who transmits it to L_3, etc... Each liar transmits what he hears with probability $p \in (0, 1)$, and the contrary with probability $q = 1 - p$. The decision of lying or not is made independently by each liar. What is the probability x_n of obtaining the correct information from L_n? What is the limit of x_n as n increases to infinity?

Exercise 1.6.17. THE CAMPUS LIBRARY COMPLAINT
You are looking for a book in the campus libraries. Each library has it with probability 0.60 but the book of each given library may have been stolen with probability 0.25. If there are three libraries, what are your chances of obtaining the book?

Exercise 1.6.18. PROFESSOR NEBULOUS
Professor Nebulous travels from Los Angeles to Paris with stopovers in New York and London. At each stop his luggage is transferred from one plane to another. In each airport, including Los Angeles, the chances are that with probability p his luggage is not placed in the right plane. Professor Nebulous finds that his suitcase has not reached Paris. What are the chances that the mishap took place in Los Angeles, New York, and London, respectively?

Exercise 1.6.19. BLOOD TEST
Give a mathematical model and invent data to corroborate the informal discussion of Remark 1.2.12.

Exercise 1.6.20. SAFARI BUTCHERS
Three tourists participate in a safari in Africa. Here comes an elephant, unaware of the rules of the game. The innocent beast is killed, having received two out of the three bullets simultaneously shot by the tourists. The tourist's hit probabilities are: Tourist A: $\frac{1}{4}$, Tourist B: $\frac{1}{2}$, Tourist C: $\frac{3}{4}$. Give for each tourist the probability that he was the one who missed.

Exercise 1.6.21. ONE IS THE SUM OF THE OTHER TWO
You perform three independent tosses of an unbiased die. What is the probability that one of these tosses results in a number that is the sum of the other two numbers?

Exercise 1.6.22. POINCARÉ'S FORMULA
Let $A_1, \ldots, A_n$ be events and let $X_1, \ldots, X_n$ be their indicator functions. By expanding the expression $E\left[\Pi_{i=1}^n (1 - X_i)\right]$, deduce *Poincaré's formula*:

$$P(\cup_{i=1}^{n} A_i) = \sum_{i=1}^{n} P(A_i) \quad - \sum_{i=1,j=1;i\neq j}^{n} P(A_i \cap A_j)$$

$$+ \sum_{i=1,j=1,k=1;i\neq j\neq k}^{n} P(A_i \cap A_j \cap A_k) - \cdots$$

Exercise 1.6.23. NO NAME
Let X be a discrete random variable taking its values in E, with probability distribution $p(x)$ ($x \in E$). Let $A := \{\omega;\, p(X(\omega)) = 0\}$. What is the probability of this event?

Exercise 1.6.24. NULL VARIANCE
Prove that a null variance implies that the random variable is almost surely constant.

Exercise 1.6.25. MOMENT INEQUALITIES

(a) Prove that for any integer-valued random variable X,

$$P(X \neq 0) \leq \mathrm{E}[X].$$

(b) Prove that for any square-integrable real-valued discrete random variable X,

$$P(X = 0) \leq \frac{\mathrm{Var}(X)}{\mathrm{E}[X]^2}.$$

Exercise 1.6.26. CHECKING CONDITIONAL INDEPENDENCE
Let X, Y and Z be three discrete random variables with values in E, F, and G, respectively. Prove the following: If for some function $g : E \times F \to [0,1]$, $P(X = x \mid Y = y, Z = z) = g(x,y)$ for all x, y, z, then $P(X = x \mid Y = y) = g(x,y)$ for all x, y, and X and Z are conditionally independent given Y.

Exercise 1.6.27. $\mathcal{G}(n,p)$ WITH A GIVEN NUMBER OF EDGES
Prove that the conditional distribution of $\mathcal{G}(n,p)$ given that the number of edges is $m \leq \binom{n}{2}$ is uniform on the set $\mathbf{G_m}$ of graphs $G = (V, \mathcal{E})$, where $V = \{1, 2, \ldots, n\}$ with exactly m edges.

Exercise 1.6.28. SUM OF GEOMETRIC VARIABLES
Let T_1 and T_2 be two independent geometric random variables with the same parameter $p \in (0,1)$. Give the probability distribution of their sum $X = T_1 + T_2$.

Exercise 1.6.29. THE COUPON COLLECTOR, TAKE 2
In the coupon collector problem of Example 1.3.36, show that the number X of chocolate tablets bought when all the n coupons have been collected for the first time satisfies the inequality

$$|\mathrm{E}[X] - n \ln n| \leq n.$$

Exercise 1.6.30. THE COUPON COLLECTOR, TAKE 3

In the coupon collector problem of Example 1.3.36, compute the variance σ_X^2 of X (the number of chocolate tablets needed to complete the collection of the n different coupons) and show that $\frac{\sigma_X^2}{n^2}$ has a limit (to be identified) as n grows indefinitely.

Exercise 1.6.31. THE COUPON COLLECTOR, TAKE 4
In the coupon collector problem of Example 1.3.36, prove that for all $c > 0$,

$$P\left(X > \lceil n \ln n + cn \rceil\right) \le e^{-c}.$$

Hint: you might find it useful to define A_i to be the event that the Type i coupon has not shown up in the first $\lceil n \ln n + cn \rceil$ tablets.

Exercise 1.6.32. FACTORIAL OF POISSON

1. Let X be a Poisson random variable with mean $\theta > 0$. Compute the mean of the random variable $X!$ (factorial, not exclamation mark!).

2. Compute $E\left[\theta^X\right]$.

Exercise 1.6.33. EVEN AND ODD POISSON
Let X be a Poisson random variable with mean $\theta > 0$. What is the probability that X is odd? even?

Exercise 1.6.34. A RANDOM SUM
Let $\{X_n\}_{n \ge 1}$ be independent random variables taking the values 0 and 1 with probability $q = 1 - p$ and p, respectively, where $p \in (0, 1)$. Let T be a Poisson random variable with mean $\theta > 0$, independent of $\{X_n\}_{n \ge 1}$. Define $S = X_1 + \cdots + X_T$. Show that S is a Poisson random variable with mean $p\theta$.

Exercise 1.6.35. MULTIPLICATIVE BERNOULLI
Let $X_1, \ldots, X_{2n}$ be independent random variables taking the values 0 or 1, and such that for all i ($1 \le i \le 2n$) $P(X_i = 1) = p \in [0, 1]$. Define $Z = \sum_{i=1}^{n} X_i X_{n+i}$. Compute $P(Z = k)$ ($1 \le k \le n$).

Exercise 1.6.36. THE MATCHBOX
A smoker has one matchbox with n matches in each pocket. He reaches at random for one box or the other. What is the probability that, having eventually found an empty matchbox, there will be k matches left in the other box?

Exercise 1.6.37. MEANS AND VARIANCES VIA GENERATING FUNCTIONS

(a) Compute the mean and variance of the binomial random variable B of size n and parameter p from its generating function. Do the same for the Poisson random variable P of mean θ.

(b) What is the generating function g_T of the geometric random variable T with parameter $p \in (0, 1)$ (recall $P(T = n) = (1 - p)^{n-1}p$, $n \ge 1$). Compute its first two derivatives and deduce from the result the variance of T.

Exercise 1.6.38. FACTORIAL MOMENT OF POISSON
What is the n-th factorial moment $(\mathrm{E}\left[X(X-1)\cdots(X-n+1)\right])$ of a Poisson random variable X of mean $\theta > 0$?

Exercise 1.6.39. FROM GENERATING FUNCTION TO PROBABILITY DISTRIBUTION
What is the probability distribution of the integer-valued random variable X with generating function $g(z) = \frac{1}{(2-z)^2}$? Compute its variance.

Exercise 1.6.40. NEGATIVE BINOMIAL FORMULA
Prove that for all $z \in \mathbb{C}$, $|z| \le 1$,

$$(1-z)^{-p} = 1 + \binom{p}{p-1}z + \binom{p+1}{p-1}z^2 + \binom{p+2}{p-1}z^3 + \cdots .$$

Exercise 1.6.41. THROW A DIE
You perform three independent tosses of an unbiased die. What is the probability that one of these tosses results in a number that is the sum of the other two numbers? (You are required to find a solution using generating functions.)

Exercise 1.6.42. RESIDUAL TIME
Let X be a random variable with values in $\mathbb{N}$ and with finite mean m. We know (telescope formula; Theorem 1.3.12) that $p_n = \frac{1}{m}P(X > n)$, $n \in \mathbb{N}$, defines a probability distribution on $\mathbb{N}$. Compute its generating function.

Exercise 1.6.43. THE BLUE PINKO
The blue pinko (an extravagant Australian bird) lays T eggs, each egg blue or pink, with probability p that a given egg is blue. The colors of the successive eggs are independent, and independent of the number of eggs laid. Exercise 1.6.34 shows that if the number of eggs is Poisson with mean θ, then the number of blue eggs is Poisson with mean θp and the number of pink eggs is Poisson with mean θq. Show that the number of blue eggs and the number of pink eggs are independent random variables.

Exercise 1.6.44. THE ENTOMOLOGIST
Each individual of a specific breed of insects has, independently of the others, the probability θ of being a male. An entomologist seeks to collect exactly $M > 1$ males, and therefore stops hunting as soon as she captures M males. She has to capture an insect in order to determine its gender. What is the distribution of X, the number of insects she must catch to collect *exactly* M males?

Exercise 1.6.45. THE RETURN OF THE ENTOMOLOGIST
The situation is as in Exercise 1.6.44. What is the distribution of X, the smallest number of insects that the entomologist must catch to collect *at least* M males and N females?

Exercise 1.6.46. THE ENTOMOLOGIST STRIKES AGAIN
This continues Exercise 1.6.44. What is the expectation of X? (In Exercise 1.6.44, you computed the distribution of X, from which you can of course compute the mean. However you can give the solution directly, and this is what is required in the present exercise.)

Exercise 1.6.47. A RECURRENCE EQUATION
Recall the notation $a^+ = \max(a, 0)$. Consider the recurrence equation

$$X_{n+1} = (X_n - 1)^+ + Z_{n+1} \quad (n \geq 0),$$

where X_0 is a random variable taking its values in $\mathbb{N}$, and $\{Z_n\}_{n\geq 1}$ is a sequence of independent random variables taking their values in $\mathbb{N}$, and independent of X_0. Express the generating function ψ_{n+1} of X_{n+1} in terms of the generating function φ of Z_1.

Exercise 1.6.48. EXTINCTION OF A BRANCHING PROCESS
Compute the probability of extinction of a branching process with one ancestor when the probabilities of having $0, 1$, or 2 sons are respectively $\frac{1}{4}$, $\frac{1}{4}$, and $\frac{1}{2}$.

Exercise 1.6.49. SEVERAL ANCESTORS

(a) Give the survival probability in the model of Section 1.4.2 with k ancestors, $k > 1$.

(b) Give the mean and variance of X_n in the model of Section 1.4.2 with one ancestor.

Exercise 1.6.50. SIZE OF THE BRANCHING TREE
When the probability of extinction is 1 ($m < 1$), call Y the size of the branching tree ($Y = \sum_{n\geq 0} X_n$). Prove that $g_Y(z) = z\, g_Z(g_Y(z))$.

Chapter 2

Integration

From a technical point of view, a probability is a particular kind of measure and expectation is integration with respect to that measure. However we shall see in the next chapter that this point of view is short-sighted because the probabilistic notions of independence and conditioning, which are absent from the theory of measure and integration, play a fundamental role in probability. Nevertheless, the foundations of probability theory rest on Lebesgue's integration theory, of which the present chapter gives a detailed outline and the main results.

The reader is assumed to have a working knowledge of the Riemann integral. This type of integral is sufficient for many purposes, but it has a few weak points when compared to the Lebesgue integral. For instance:

(1) The class of Riemann-integrable functions is too narrow. As a matter of fact, there are functions that have a Lebesgue integral and yet are not Riemann integrable (see Example 2.2.14).

(2) The stability properties under the limit operation of the functions that admit a Riemann integral are too weak. Indeed, it often happens that such limits do not have a Riemann integral whereas the limit, for instance, of non-negative functions for which the Lebesgue integral is well defined also admits a well-defined Lebesgue integral.

(3) The Riemann integral is defined with respect to the Lebesgue measure (length, area, volume, etc.) whereas Lebesgue's integral can be defined with respect to a general *abstract measure*, a probability for instance.

This last advantage makes it worthwhile to invest a little time in order to understand the fundamental results of the Lebesgue integration theory, because the return is considerable. In fact, the Lebesgue integral of a function f with respect to an abstract measure μ contains a variety of mathematical objects besides the usual Lebesgue integral on $\mathbb{R}^d$

$$\int_{\mathbb{R}^d} f(x_1, \cdots, x_d)\, \mathrm{d}x_1 \cdots \mathrm{d}x_d \, .$$

In fact, an infinite sum

$$\sum_{n \in \mathbb{Z}} f(n)$$

can also be regarded as a Lebesgue integral with respect to the counting measure on $\mathbb{Z}$. The Stieltjes–Lebesgue integral

© Springer Nature Switzerland AG 2020
P. Brémaud, *Probability Theory and Stochastic Processes*, Universitext,
https://doi.org/10.1007/978-3-030-40183-2_2

$$\int_{\mathbb{R}} f(x)\, \mathrm{d}F(x)$$

with respect to a right-continuous non-decreasing function F is again a special case of the Lebesgue integral. Most importantly, the expectation

$$\mathrm{E}[Z]$$

of a random variable Z will be recognized as an abstract integral. It is the latter type of integral that is of interest in this book and the purpose of the present chapter is to define the Lebesgue integral and to give its properties directly useful to probability theory.

2.1   Measurability and Measure

This section describes the functions that Lebesgue's integration theory admits for integrands and gives the formal definition and properties of the measure with respect to which such functions are integrated.

2.1.1   Measurable Functions

Remember the definition of a σ-field:

Definition 2.1.1 *Denote by $\mathcal{P}(X)$ the collection of all subsets of a given set X. A collection of subsets $\mathcal{X} \subseteq \mathcal{P}(X)$ is called a σ-field on X if:*

(α) $X \in \mathcal{X}$,

(β) $A \in \mathcal{X} \implies \overline{A} \in \mathcal{X}$ and

(γ) $A_n \in \mathcal{X}$ for all $n \in \mathbb{N} \implies \cup_{n=0}^{\infty} A_n \in \mathcal{X}$.

One then says that $(X, \mathcal{X})$ is a *measurable space*. A set $A \in \mathcal{X}$ is called a *measurable set*.

Observe that (See Exercise 1.6.5):

(γ') $A_n \in \mathcal{X}$ for all $n \in \mathbb{N} \implies \cap_{n=0}^{\infty} A_n \in \mathcal{X}$.

In fact, given the properties (α) and (β), properties (γ) and (γ') are equivalent. Note also that $\varnothing \in \mathcal{X}$, being the complement of X. Therefore, a σ-field on X is a collection of subsets of X that contains X and $\varnothing$, and is closed under countable unions, countable intersections and complementation.

The two simplest examples of σ-fields on X are the *gross* σ-field $\mathcal{X} = \{\varnothing, X\}$ and the *trivial* σ-field $\mathcal{X} = \mathcal{P}(X)$.

Definition 2.1.2 *The σ-field generated by a non-empty collection of subsets $\mathcal{C} \subseteq \mathcal{P}(X)$ is, by definition, the smallest σ-field on X containing all the sets in $\mathcal{C}$ (see Exercise 2.4.1). It is denoted by $\sigma(\mathcal{C})$.*

Of course, if $\mathcal{G}$ is a σ-field, $\sigma(\mathcal{G}) = \mathcal{G}$, a fact that will often be used.

We now review some basic notions of topology.

Definition 2.1.3 *A* topology *on a set X is a collection $\mathcal{O}$ of subsets of X satisfying the following properties:*

(i) X and $\varnothing$ belong to $\mathcal{O}$;

(ii) the union of an arbitrary *collection of sets in $\mathcal{O}$ is a set in $\mathcal{O}$; and*

(iii) the intersection of a finite *number of sets in $\mathcal{O}$ is a set in $\mathcal{O}$.*

The elements $O \in \mathcal{O}$ are called open sets *and the pair $(X, \mathcal{O})$ is called a* topological space. *(Usually, when the context clearly defines the open sets, we say: "the topological space X".)*

EXAMPLE 2.1.4: METRIC SPACES. Let X be a set. A function $d : X \times X \to \overline{\mathbb{R}}_+$ such that for all $x, y, z \in X$,

(i) $d(x, y) = 0 \Rightarrow x = y$,

(ii) $d(x, y) = d(y, x)$ and

(iii) $d(x, z) \leq d(x, y) + d(y, z)$

is called a *metric* on X. The pair (X, d) is called a *metric space*.

Any metric induces a topology as follows. A set $O \in X$ is called open if for all $x \in O$ there is an open ball $B(x, a) := \{y \in X; d(y, x) < a\}$ contained in O. The collection of such sets is indeed a topology as can be straightforwardly checked.

EXAMPLE 2.1.5: THE EUCLIDEAN TOPOLOGY. In $\mathbb{R}^n$, the usual euclidean distance defines a topology called the *euclidean topology*.

Topology concerns continuity:

Definition 2.1.6 *Let $(X, \mathcal{O})$ and $(E, \mathcal{V})$ be two topological spaces. A function $f : X \to E$ is said to be* continuous *(with respect to these topologies) if*

$$V \in \mathcal{V} \Rightarrow f^{-1}(V) \in \mathcal{O}.$$

In other terms, $f^{-1}(\mathcal{V}) \subseteq \mathcal{O}$. This is a rather abstract definition (however it will turn out to be quite convenient, as we shall soon see). For metric spaces, continuity is more explicit, and is usually defined in terms of metrics:

Theorem 2.1.7 *Let (X, d) and (E, ρ) be two metric spaces. A necessary and sufficient condition for a mapping $f : X \to E$ to be continuous according to the above abstract definition (and with respect to the topologies induced by the corresponding metrics) is that for any $\varepsilon > 0$ and any $x \in X$, there exists a $\delta > 0$ such that $d(y, x) \leq \delta$ implies $\rho(f(y), f(x)) \leq \varepsilon$.*

The proof is given in Section B.7 of the appendix.

We shall now be more "concrete" about σ-fields.

Definition 2.1.8 Let $(X, \mathcal{O})$ be a topological space. The σ-field $\sigma(\mathcal{O})$ will be denoted by $\mathcal{B}(X)$ and called the Borel σ-field on X (associated with topology $\mathcal{O}$). A set $B \in \mathcal{B}(X)$ is called a Borel set of X (with respect to the topology $\mathcal{O}$).

When $X = \mathbb{R}^n$ is endowed with the euclidean topology, the Borel σ-field is denoted by $\mathcal{B}(\mathbb{R}^n)$.

The next result gives a sometimes more convenient characterization of $\mathcal{B}(\mathbb{R}^n)$.

Theorem 2.1.9 $\mathcal{B}(\mathbb{R}^n)$ is generated by the collection $\mathcal{C}$ of all rectangles of the type $\prod_{i=1}^{n}(-\infty, a_i]$, where $a_i \in \mathbb{Q}$ for all $i \in \{1, \ldots, n\}$.

Proof. It suffices to show that $\mathcal{B}(\mathbb{R}^n)$ is generated by the collection $\mathcal{C}'$ of all rectangles $\prod_{i=1}^{n}(a_i, b_i)$ with rational endpoints (that is, such that $a_i, b_i \in \mathbb{Q}$ for all $i \in \{1, \ldots, n\}$). Note that $\mathcal{C}'$ is a countable collection and that all its elements are open sets for the euclidean topology (the latter we denote by $\mathcal{O}$). It follows that $\mathcal{C}' \subseteq \mathcal{O}$ and therefore $\sigma(\mathcal{C}') \subseteq \sigma(\mathcal{O}) = \mathcal{B}(\mathbb{R}^n)$.

It remains to show that $\mathcal{O} \subseteq \sigma(\mathcal{C}')$, since this implies that $\sigma(\mathcal{O}) \subseteq \sigma(\mathcal{C}')$. For this it suffices to show that any set $O \in \mathcal{O}$ is a countable union of elements in $\mathcal{C}'$. Take $x \in O$. By definition of the euclidean topology, there exists a non-empty open ball $B(x, r)$ centered at x and contained in O. Now we can always choose a rational rectangle $R_x \in \mathcal{C}'$ that contains x and that is contained in $B(x, r)$. Clearly $\cup_{x \in O} R_x = O$. Since the R_x are chosen in a countable family of sets, the union $\cup_{x \in O} R_x$ is in fact countable. As a countable union of sets in $\mathcal{C}'$ it is in $\sigma(\mathcal{C}')$. Therefore $O \in \sigma(\mathcal{C}')$. $\qquad \square$

Definition 2.1.10 $\mathcal{B}(\overline{\mathbb{R}})$ is the σ-field on $\overline{\mathbb{R}}$ generated by the intervals of type $[-\infty, a]$, $a \in \mathbb{R}$.

If $I = \prod_{j=1}^{n} I_j$, where I_j is an interval of $\mathbb{R}$, the Borel σ-field $\mathcal{B}(I)$ on I is the collection of all the Borel sets contained in I.

Remark 2.1.11 A question naturally arises at this point. How can we be sure that the Borel σ-field (of $\mathbb{R}$ for instance) is not just the trivial σ-field? In other terms, does there exist at least one subset of $\mathbb{R}$ that is not a Borel set? The answer is yes: there exist such "pathological" subsets, but we shall not prove this here.

The central concept of Lebesgue's integration theory is that of a measurable function.

Definition 2.1.12 Let $(X, \mathcal{X})$ and $(E, \mathcal{E})$ be two measurable spaces. A function $f : X \to E$ is said to be a measurable function with respect to $\mathcal{X}$ and $\mathcal{E}$ if $f^{-1}(C) \in \mathcal{X}$ for all $C \in \mathcal{E}$.

In other terms, $f^{-1}(\mathcal{E}) \subseteq \mathcal{X}$. This will be denoted by:

$$f : (X, \mathcal{X}) \to (E, \mathcal{E}) \text{ or } f \in \mathcal{E}/\mathcal{X}.$$

Let $(X, \mathcal{X})$ be a measurable space. A function $f : (X, \mathcal{X}) \to (\mathbb{R}^k, \mathcal{B}(\mathbb{R}^k))$ is called a *Borel function* from X to $\mathbb{R}^k$. A function $f : (X, \mathcal{X}) \to (\overline{\mathbb{R}}, \mathcal{B}(\overline{\mathbb{R}}))$ is called an *extended Borel function*, or simply a *Borel function*. As for functions $f : (X, \mathcal{X}) \to (\mathbb{R}, \mathcal{B}(\mathbb{R}))$, they are called *real* Borel functions. In general, in a sentence such as "f is a Borel function defined on X", the σ-field $\mathcal{X}$ is assumed to be the obvious one in the given context.

The key to the definition of the Lebesgue integral is the *theorem of approximation* of non-negative measurable functions by simple Borel functions.

Definition 2.1.13 *A function $f : X \to \mathbb{R}$ of the form*

$$f(x) = \sum_{i=1}^{k} a_i \, 1_{A_i}(x),$$

where $k \in \mathbb{N}_+$, $a_1, \ldots, a_k \in \mathbb{R}$ and $A_1, \ldots, A_k$ are sets in $\mathcal{X}$, is called a simple Borel function *(defined on X).*

Theorem 2.1.14 *Let $f : (X, \mathcal{X}) \to (\overline{\mathbb{R}}, \mathcal{B}(\overline{\mathbb{R}}))$ be a non-negative Borel function. There exists a non-decreasing sequence $\{f_n\}_{n \geq 1}$ of non-negative simple Borel functions converging pointwise to the function f.*

Proof. Let

$$f_n(x) := \sum_{k=0}^{n2^{-n}-1} k2^{-n} \, 1_{A_{k,n}}(x) + n1_{A_n}(x),$$

where

$$A_{k,n} := \{x \in X : k2^{-n} < f(x) \leq (k+1)2^{-n}\}, \ A_n = \{x \in X : f(x) > n\} \, .$$

This sequence of functions has the announced properties. In fact, for any $x \in X$ such that $f(x) < \infty$ and for n large enough,

$$|f(x) - f_n(x)| \leq 2^{-n} \, ,$$

and for any $x \in X$ such that $f(x) = \infty$, $f_n(x) = n$ indeed converges to $f(x) = +\infty$. $\square$

It seems difficult to prove measurability since σ-fields are often not defined explicitly (see the definition of $\mathcal{B}(\mathbb{R}^n)$, for instance). However, the following result renders the task feasible. The corollaries will provide examples of application.

Theorem 2.1.15 *Let $(X, \mathcal{X})$ and $(E, \mathcal{E})$ be two measurable spaces, where $\mathcal{E} = \sigma(\mathcal{C})$ for some collection $\mathcal{C}$ of subsets of E. Let $f : X \to E$ be some function. Then $f : (X, \mathcal{X}) \to (E, \mathcal{E})$ if and only if $f^{-1}(C) \in \mathcal{X}$ for all $C \in \mathcal{C}$.*

Proof. Only sufficiency requires a proof. We start with the following trivial observations. Let X and E be sets, let $\mathcal{G}$ be a σ-field on E, and let $\mathcal{C}_1, \mathcal{C}_2$ be non-empty collections of subsets of E. Then

(i) $\sigma(\mathcal{G}) = \mathcal{G}$, and

(ii) $\mathcal{C}_1 \subseteq \mathcal{C}_2 \Rightarrow \sigma(\mathcal{C}_1) \subseteq \sigma(\mathcal{C}_2)$.

Let now $f : X \to E$ be a function from X to E, and define

$$\mathcal{G} := \{C \subseteq E; f^{-1}(C) \in \mathcal{X}\}.$$

One checks that $\mathcal{G}$ is a σ-field. But by hypothesis, $\mathcal{C} \subseteq \mathcal{G}$. Therefore, by (ii) and (i), $\mathcal{X} = \sigma(\mathcal{C}) \subseteq \sigma(\mathcal{G}) = \mathcal{G}$. $\square$

Stability Properties of Measurable Functions

Measurability is a stable property, in the sense that all the usual operations on measurable functions preserve measurability. (This is not the case for continuity, which is in general not stable with respect to limits.) Also, the class of measurable functions is a rich one. In particular, "continuous functions are measurable". More precisely:

Corollary 2.1.16 *Let X and E be two topological spaces with respective Borel σ-fields $\mathcal{B}(X)$ and $\mathcal{B}(E)$. Any* continuous *function $f : X \to E$ is measurable with respect to $\mathcal{B}(X)$ and $\mathcal{B}(E)$.*

Proof. By definition of continuity, the inverse image of an open set of E is an open set of X and is therefore in $\mathcal{B}(X)$. By Theorem 2.1.15, since the open sets of E generate $\mathcal{B}(E)$, the function f is measurable with respect to $\mathcal{B}(X)$ and $\mathcal{B}(E)$. $\square$

Corollary 2.1.17 *Let $(X, \mathcal{X})$ be a measurable space and let $n \geq 1$ be an integer. Then $f = (f_1, \ldots, f_n) : (X, \mathcal{X}) \to (\mathbb{R}^n, \mathcal{B}(\mathbb{R}^n))$ if and only if for all $1 \leq i \leq n$, $\{f_i \leq a_i\} \in \mathcal{X}$ for all $a_i \in \mathbb{Q}$ (the rational numbers).*

Proof. Since by Theorem 2.1.9, $\mathcal{B}(\mathbb{R}^n)$ is generated by the sets $\prod_{i=1}^{n}(-\infty, a_i]$, where $a_i \in \mathbb{Q}$ for all $i \in \{1, \ldots, n\}$, it suffices by Theorem 2.1.15 to show that for all $a \in \mathbb{Q}^n$, $\{f \leq a\} \in \mathcal{X}$. This is indeed the case since

$$\{f \leq a\} = \cap_{i=1}^{n}\{f_i \leq a_i\},$$

and therefore $\{f \leq a\} \in \mathcal{X}$, being the intersection of a countable (actually: finite) number of sets in $\mathcal{X}$. $\square$

Measurability is *closed under composition*:

Theorem 2.1.18 *Let $(X, \mathcal{X})$, $(Y, \mathcal{Y})$ and $(E, \mathcal{E})$ be measurable spaces and let $\varphi : (X, \mathcal{X}) \to (Y, \mathcal{Y})$, $g : (Y, \mathcal{Y}) \to (E, \mathcal{E})$. Then $g \circ \varphi : (X, \mathcal{X}) \to (E, \mathcal{E})$.*

Proof. Let $f := g \circ \varphi$ (meaning: $f(x) = g(\varphi(x))$ for all $x \in X$). For all $C \in \mathcal{E}$,

$$f^{-1}(C) = \varphi^{-1}(g^{-1}(C)) = \varphi^{-1}(D) \in \mathcal{X},$$

because $D = g^{-1}(C)$ is a set in $\mathcal{Y}$ since $g \in \mathcal{E}/\mathcal{Y}$, and therefore $\varphi^{-1}(D) \in \mathcal{X}$ since $\varphi \in \mathcal{Y}/\mathcal{X}$. $\square$

Corollary 2.1.19 *Let $\varphi = (\varphi_1, \ldots, \varphi_n)$ be a measurable function from $(X, \mathcal{X})$ to $(\mathbb{R}^n, \mathcal{B}(\mathbb{R}^n))$ and let $g : \mathbb{R}^n \to \mathbb{R}$ be a continuous function. Then $g \circ \varphi : (X, \mathcal{X}) \to (\mathbb{R}, \mathcal{B}(\mathbb{R}))$.*

Proof. Follows directly from Theorem 2.1.18 and Corollary 2.1.16. $\square$

This corollary in turn allows us to show that addition, multiplication and quotients preserve measurability.

Corollary 2.1.20 *Let* $\varphi_1, \varphi_2 : (X, \mathcal{X}) \to (\mathbb{R}, \mathcal{B}(\mathbb{R}))$ *and let* $\lambda \in \mathbb{R}$. *Then* $\varphi_1 \times \varphi_2$, $\varphi_1 + \varphi_2$, $\lambda\varphi_1$, $(\varphi_1/\varphi_2)1_{\{\varphi_2 \neq 0\}}$ *are real Borel functions. Moreover, the set* $\{\varphi_1 = \varphi_2\}$ *is a measurable set.*

Proof. For the first three functions, take in Corollary 2.1.19 $g(x_1, x_2) = x_1 \times x_2, = x_1 + x_2, = \lambda x_1$ successively. For $(\varphi_1/\varphi_2)1_{\{\varphi_2 \neq 0\}}$, let $\psi_2 := \frac{1_{\{\varphi_2 \neq 0\}}}{\varphi_2}$, check that the latter function is measurable, and use the just proved fact that the product $\varphi_1 \psi_2$ is then measurable. Finally, $\{\varphi_1 = \varphi_2\} = \{\varphi_1 - \varphi_2 = 0\} = (\varphi_1 - \varphi_2)^{-1}(\{0\})$ is a measurable set since $\varphi_1 - \varphi_2$ is a measurable function and $\{0\}$ is a measurable set (any singleton is in $\mathcal{B}(\mathbb{R})$; exercise). $\square$

Finally, taking the limit preserves measurability, as will now be proved. Without otherwise explicitly mentioned, the limits of functions must be understood as *pointwise limits*.

Theorem 2.1.21 *Let* $f_n : (X, \mathcal{X}) \to (\overline{\mathbb{R}}, \mathcal{B}(\overline{\mathbb{R}}))$ *($n \in \mathbb{N}$). Then* $\liminf_{n\uparrow\infty} f_n$ *and* $\limsup_{n\uparrow\infty} f_n$ *are Borel functions, and the set*

$$\{\limsup_{n\uparrow\infty} f_n = \liminf_{n\uparrow\infty} f_n\} = \{\exists \lim_{n\uparrow\infty} f_n\}$$

belongs to $\mathcal{X}$. *In particular, if* $\{\exists \lim_{n\uparrow\infty} f_n\} = X$, *the function* $\lim_{n\uparrow\infty} f_n$ *is a Borel function.*

Proof. We first prove the result in the particular case when the sequence of functions is non-decreasing. Denote by f the limit of this sequence. By Theorem 2.1.15 it suffices to show that for all $a \in \mathbb{R}$, $\{f \leq a\} \in \mathcal{X}$. But since the sequence $\{f_n\}_{n \geq 1}$ is non-decreasing, we have that $\{f \leq a\} = \cap_{n=1}^{\infty} \{f_n \leq a\}$, which is indeed in $\mathcal{X}$, as a countable intersection of sets in $\mathcal{X}$. Now recall that by definition,

$$\liminf_{n\uparrow\infty} f_n := \lim_{n\uparrow\infty} g_n,$$

where

$$g_n := \inf_{k \geq n} f_k.$$

The function g_n is measurable since for all $a \in \mathbb{R}$, $\{\inf_{k \geq n} f_k \leq a\}$ is a measurable set, being the complement of $\{\inf_{k \geq n} f_k > a\} = \cap_{k \geq n}\{f_k > a\}$, a measurable set (as the countable intersection of measurable sets). Since the sequence $\{g_n\}_{n \geq 1}$ is non-decreasing, the measurability of $\liminf_{n\uparrow\infty} f_n$ follows from the particular case of non-decreasing functions. Similarly, $\limsup_{n\uparrow\infty} f_n = -\liminf_{n\uparrow\infty}(-f_n)$ is measurable.

The set $\{\limsup_{n\uparrow\infty} f_n = \liminf_{n\uparrow\infty} f_n\}$ is the set on which two measurable functions are equal, and therefore, by the last assertion of Corollary 2.1.20, it is a measurable set.

Finally, if $\lim_{n\uparrow\infty} f_n$ exists, it is equal to $\limsup_{n\uparrow\infty} f_n$, which is, as we just proved, a measurable function. $\square$

Dynkin's Systems

Proving that a given property is common to all measurable functions, or to all measurable sets, may at times appear difficult because there is usually no constructive definition of the σ-fields involved (these are often defined as "the smallest σ-field containing a certain class of subsets"). There is however a technical tool that allows us to do this, the *Dynkin theorem*(s).

The central notions in this respect are those of a π-system and of a d-system.

Definition 2.1.22 *Let X be a set. The collection $\mathcal{S} \subseteq \mathcal{P}(X)$ is called a π-system of X if it is closed under finite intersections.*

For instance, the collection of finite unions of intervals of the type $(a, b]$ is a π-system of $\mathbb{R}$.

Definition 2.1.23 *Let X be a set. A non-empty collection of sets $\mathcal{S} \in \mathcal{P}(X)$ is called a d-system, or Dynkin system, of sets if*

(a) $X, \varnothing \in \mathcal{S}$.

(b) $\mathcal{S}$ is closed under strict difference (that is, if $A, B \in \mathcal{S}$ and $A \subseteq B$, then $B - A \in \mathcal{S}$).

(c) $\mathcal{S}$ is closed under sequential non-decreasing limits (that is, the limit of a non-decreasing sequence of sets in $\mathcal{S}$ is in $\mathcal{S}$).

Theorem 2.1.24 *Let X be a set. If the collection $\mathcal{S} \in \mathcal{P}(X)$ is a π-system and a d-system, it is a σ-field.*

Proof.

(i) $\mathcal{S}$ contains X and $\varnothing$, by definition of a d-system.

(ii) $\mathcal{S}$ is closed under complementation. (Apply (b) in Definition 2.1.23 with $B = X$ and $A \in \mathcal{S}$, to obtain that $\overline{A} \in \mathcal{S}$.)

(iii) $\mathcal{S}$ is closed under countable unions. To prove this, we first show that it is closed under finite unions. Indeed, if $A, B \in \mathcal{S}$, then by (ii), $\overline{A}, \overline{B} \in \mathcal{S}$, and therefore, since $\mathcal{S}$ is a π-system, $\overline{A} \cap \overline{B} \in \mathcal{S}$. Taking the complement we obtain by (ii) that $A \cup B \in \mathcal{S}$. Now for countable unions, consider a sequence $\{A_n\}_{n \geq 1}$ of elements of $\mathcal{S}$. The union $\cup_{n \geq 1} A_n$ can be written as $\cup_{n \geq 1} \left(\cup_{k=1}^n A_k \right)$, which is a countable union of non-decreasing sets in $\mathcal{S}$, and therefore it is in $\mathcal{S}$, by (c) of Definition 2.1.23. $\qquad \square$

The smallest d-system containing a non-empty collection of sets $\mathcal{C} \subseteq \mathcal{P}(X)$ is denoted by $d(\mathcal{C})$. Observe that since a σ-field $\mathcal{G}$ is already a d-system, $d(\mathcal{G}) = \mathcal{G}$. In particular, for any collection of sets $\mathcal{C} \subseteq \mathcal{P}(X)$, $d(\mathcal{C}) \subseteq \sigma(\mathcal{C})$. Also note that if $\mathcal{C}_1 \subseteq \mathcal{C}_2$, then $d(\mathcal{C}_1) \subseteq d(\mathcal{C}_2)$.

We now state Dynkin's theorem (sometimes called the monotone class theorem).

Theorem 2.1.25 *Let S be a π-system defined on X. Then $d(S) = \sigma(S)$.*

Proof. Any σ-field containing S will contain $d(S)$. Therefore $\sigma(d(S)) = \sigma(S)$. If we can show that $d(S)$ is a σ-field, then $\sigma(d(S)) = d(S)$, so that $d(S) = \sigma(S)$.

To prove that $d(S)$ is a σ-field, it suffices by Theorem 2.1.24 to show that it is a π-system. For this, define

$$\mathcal{D}_1 := \{A \in d(S); A \cap C \in d(S) \text{ for all } C \in S\}.$$

One checks that $\mathcal{D}_1$ is a d-system and that $S \subseteq \mathcal{D}_1$ since S is a π-system by assumption. Therefore $d(S) \subseteq \mathcal{D}_1$. By definition of $\mathcal{D}_1$, $\mathcal{D}_1 \subseteq d(S)$. Therefore $\mathcal{D}_1 = d(S)$. Define now

$$\mathcal{D}_2 := \{A \in d(S); A \cap C \in d(S) \text{ for all } C \in d(S)\}.$$

This is a d-system. Also, if $C \in S$, then $A \cap C \in d(S)$ for all $A \in \mathcal{D}_1 = d(S)$, and therefore $S \subseteq \mathcal{D}_2$. Therefore $d(S) \subseteq d(\mathcal{D}_2) = \mathcal{D}_2$. Also by definition of $\mathcal{D}_2$, $\mathcal{D}_2 \subseteq d(S)$, so that finally $\mathcal{D}_2 = d(S)$. In particular, $d(S)$ is a π-system. $\square$

There is a *functional form of Dynkin's theorem*. We first define a d-system of functions.

Definition 2.1.26 *Let $\mathcal{H}$ be a collection of non-negative functions $f : X \to \overline{\mathbb{R}}_+$ such that*

(α) $1 \in \mathcal{H}$,

(β) $\mathcal{H}$ *is closed under monotone non-decreasing sequential limits (that is, if $\{f_n\}_{n\geq 1}$ is a non-decreasing sequence of functions in $\mathcal{H}$, then $\lim_{n\uparrow\infty} f_n \in \mathcal{H}$), and*

(γ) *if $f_1, f_2 \in \mathcal{H}$, then $\lambda_1 f_1 + \lambda_2 f_2 \in \mathcal{H}$ for all $\lambda_1, \lambda_2 \in \mathbb{R}$ such that $\lambda_1 f_1 + \lambda_2 f_2$ is a non-negative function.*

Then $\mathcal{H}$ is called a d-system, or Dynkin system, of functions.

Theorem 2.1.27 *Let $\mathcal{H}$ be a family of non-negative functions $f : X \to \overline{\mathbb{R}}_+$ that form a d-system. Let S be π-system on X such that $1_C \in \mathcal{H}$ for all $C \in S$. Then $\mathcal{H}$ contains all non-negative functions $f : (X, \sigma(S)) \to (\mathbb{R}, \mathcal{B}(\mathbb{R}))$.*

Proof. The collection

$$\mathcal{D} := \{A \subseteq X; 1_A \in \mathcal{H}\}$$

is a d-system containing S by hypothesis. Therefore $d(S) \subseteq d(\mathcal{D}) = \mathcal{D}$. Since $d(S) = \sigma(S)$ by Dynkin's theorem, $\sigma(S) \subseteq \mathcal{D}$. This means that $\mathcal{H}$ contains the indicator functions of all the sets in $\sigma(S)$. Being a d-system of functions, it contains all the non-negative simple $\sigma(S)$-measurable functions. The rest of the proof follows from the theorem of approximation of non-negative measurable functions by non-negative simple functions (Theorem 2.1.14) and property (γ) of Definition 2.1.26. $\square$

2.1.2   Measure

The next most important notion of integration theory after that of measurable sets and measurable functions is that of *measure*.

Definition 2.1.28 *Let $(X, \mathcal{X})$ be a measurable space. A set function $\mu : \mathcal{X} \to [0, \infty]$ is called a* measure *on $(X, \mathcal{X})$ if $\mu(\varnothing) = 0$ and if for any countable sequence $\{A_n\}_{n \geq 0}$ of pairwise disjoint sets in $\mathcal{X}$, the following property (σ-additivity) is satisfied*

$$\mu\left(\sum_{n=0}^{\infty} A_n\right) = \sum_{n=0}^{\infty} \mu(A_n).$$

The triple $(X, \mathcal{X}, \mu)$ is then called a measure space.

The next two properties have been proved in the first chapter. We repeat the proofs for self-containedness. First, the *monotonicity* property:

$$A \subseteq B \text{ and } A, B \in \mathcal{X} \implies \mu(A) \leq \mu(B).$$

Indeed, $B = A + (B - A)$ and therefore, $\mu(B) = \mu(A) + \mu(B - A) \geq \mu(A)$. The *sub-$\sigma$-additivity* property:

$$A_n \in \mathcal{X} \text{ for all } n \in \mathbb{N} \implies \mu\left(\bigcup_{n=0}^{\infty} A_n\right) \leq \sum_{n=0}^{\infty} \mu(A_n),$$

is obtained by writing

$$\mu\left(\bigcup_{n=0}^{\infty} A_n\right) = \mu\left(\sum_{n=0}^{\infty} A'_n\right),$$

where $A'_0 = A_0$ and for $n \geq 1$,

$$A'_n = A_n \cap \left(\overline{\cup_{j=1}^{n-1} A_j}\right) \subseteq A_n,$$

so that $\mu(A'_n) \leq \mu(A_n)$ by the monotonicity property.

EXAMPLE 2.1.29:   THE DIRAC MEASURE. Let $a \in X$. The measure ϵ_a defined by $\epsilon_a(C) = 1_C(a)$ is called the *Dirac measure* at $a \in X$. The set function $\mu : \mathcal{X} \to [0, \infty]$ defined by

$$\mu(C) = \sum_{i=0}^{\infty} \alpha_i 1_{a_i}(C),$$

where $a_i \in X$, $\alpha_i \in \mathbb{R}_+$ for all $i \in \mathbb{N}$, is a measure denoted $\mu = \sum_{i=0}^{\infty} \alpha_i \epsilon_{a_i}$.

EXAMPLE 2.1.30:   WEIGHTED COUNTING MEASURE. Let $\{\alpha_n\}_{n \in \mathbb{Z}}$ be a sequence of $\mathbb{R}_+$. The set function $\mu : \mathcal{P}(\mathbb{Z}) \to [0, \infty]$ defined by $\mu(C) = \sum_{n \in C} \alpha_n$ is a measure on $(\mathbb{Z}, \mathcal{P}(\mathbb{Z}))$. In the case $\alpha_n \equiv 1$, it is called the *counting measure* on $\mathbb{Z}$ (then $\mu(C) = \operatorname{card}(C)$).

EXAMPLE 2.1.31: THE LEBESGUE MEASURE. There exists one and only one measure ℓ on $(\mathbb{R}, \mathcal{B}(\mathbb{R}))$ such that $\ell((a, b]) = b - a$. This measure is called the *Lebesgue measure* on $\mathbb{R}$. (Beware. The statement of this example is in fact a theorem, which is part of a more general result, Theorem 2.1.53 below.) More generally, the Lebesgue measure on $\mathbb{R}^n$, denoted by ℓ^n, is the unique measure on $(\mathbb{R}^n, \mathcal{B}(\mathbb{R}^n))$ such that

$$\ell^n \left(\prod_{i=1}^{n} (a_i, b_i] \right) = \prod_{i=1}^{n} (b_i - a_i).$$

The proof of existence of ℓ^n for $n \geq 2$, assuming the existence of ℓ, is a consequence of the forthcoming Theorem 2.3.7.

Definition 2.1.32 *Let μ be a measure on $(X, \mathcal{X})$. If $\mu(X) < \infty$ the measure μ is called a finite measure. If there exists a sequence $\{K_n\}_{n \geq 1}$ of $\mathcal{X}$ such that $\mu(K_n) < \infty$ for all $n \geq 1$ and $\cup_{n=1}^{\infty} K_n = X$, the measure μ is called a σ-finite measure. A measure μ on $(\mathbb{R}^m, \mathcal{B}(\mathbb{R}^m))$ such that $\mu(C) < \infty$ for all bounded Borel sets C is called a locally finite measure. It is called non-atomic or diffuse if $\mu(\{a\}) = 0$ for all $a \in \mathbb{R}^m$.*

Remark 2.1.33 Non-atomicity does not imply that μ is the null measure. Indeed, the "proof" that it is null, namely

$$\mu(C) = \sum_{a \in C} \mu(\{a\}) = 0,$$

is not valid because C is not countable.

We shall single out the case of a probability measure:

Definition 2.1.34 *A measure P on a measurable space $(\Omega, \mathcal{F})$ such that $P(\Omega) = 1$ is called a probability measure (for short: a probability).*

EXAMPLE 2.1.35: A FEW EXAMPLES. The Dirac measure ϵ_a is a probability measure. The counting measure ν on $\mathbb{Z}$ is a σ-finite measure. Any locally finite measure on $(\mathbb{R}^n, \mathcal{B}(\mathbb{R}^n))$ is σ-finite. Lebesgue measure is a locally finite measure.

Theorem 2.1.36 *Let $(X, \mathcal{X}, \mu)$ be a measure space. Let $\{A_n\}_{n \geq 1}$ be a sequence of $\mathcal{X}$, non-decreasing (that is, $A_n \subseteq A_{n+1}$ for all $n \geq 1$). Then*

$$\mu(\cup_{n=1}^{\infty} A_n) = \lim_{n \uparrow \infty} \mu(A_n). \tag{2.1}$$

Proof. Since $A_n = A_0 + \cup_{k=1}^{n} (A_k - A_{k-1})$ and $\cup_{n \geq 0} A_n = A_0 + \cup_{k \geq 1} (A_k - A_{k-1})$, by σ additivity

$$\mu(A_n) = \mu(A_0) + \sum_{k=1}^{n} \mu(A_k - A_{k-1})$$

and

$$\mu(\cup_{n \geq 0} A_n) = \mu(A_0) + \sum_{k \geq 1} \mu(A_k - A_{k-1}),$$

from which the result follows. $\square$

(A word of caution: see Exercise 2.4.8.)

Definition 2.1.37 *Let μ be a measure on the measurable topological space $(X, \mathcal{B}(X))$. Its* support $\mathrm{supp}(\mu)$ *is, by definition, the closure of the subset of X consisting of all the points x such that for all open neighborhoods N_x of x, $\mu(N_x) > 0$.*

Negligible Sets

Definition 2.1.38 *Let $(X, \mathcal{X}, \mu)$ be a measure space. A μ-negligible set is a set contained in a measurable set $N \in \mathcal{X}$ such that $\mu(N) = 0$. One says that some property $\mathcal{P}$ relative to the elements $x \in X$ holds μ-almost everywhere (μ-a.e.) if the set $\{x \in X : x$ does not satisfy $\mathcal{P}\}$ is a μ-negligible set.*

For instance, if f and g are two Borel functions defined on X, the expression $f \leq g$ μ-a.e. means that $\mu(\{x : f(x) > g(x)\}) = 0$.

Theorem 2.1.39 *A countable union of μ-negligible sets is a μ-negligible set.*

Proof. Let A_n, $n \geq 1$, be a sequence of μ-negligible sets, and let N_n, $n \geq 1$, be a sequence of measurable sets such that $\mu(N_n) = 0$ and $A_n \subseteq N_n$. Then $N = \cup_{n \geq 1} N_n$ is a measurable set containing $\cup_{n \geq 1} A_n$, and N is of μ-measure 0, by the sub-σ-additivity property.   □

EXAMPLE 2.1.40: THE RATIONALS ARE LEBESGUE-NEGLIGIBLE. Any singleton $\{a\}$, $a \in \mathbb{R}$, is a Borel set of Lebesgue measure 0. The set of rationals $\mathbb{Q}$ is a Borel set of Lebesgue measure 0.

Proof. The Borel σ-field $\mathcal{B}(\mathbb{R})$ is generated by the intervals $I_a = (-\infty, a]$, $a \in \mathbb{R}$ (Theorem 2.1.9), and therefore $\{a\} = \cap_{n \geq 1}(I_a - I_{a-1/n})$ is also in $\mathcal{B}$. Denoting by ℓ the Lebesgue measure, $\ell(I_a - I_{a-1/n}) = 1/n$, and therefore $\ell(\{a\}) = \lim_{n \geq 1} \ell(I_a - I_{a-1/n}) = 0$. $\mathbb{Q}$ is a countable union of sets in $\mathcal{B}$ (singletons) and is therefore in $\mathcal{B}$. It has Lebesgue measure 0 as a countable union of sets of Lebesgue measure 0.   □

Definition 2.1.41 *The measure space $(X, \mathcal{X}, \mu)$ is called complete if $\mathcal{X}$ contains all the μ-negligible subsets of X.*

Given a measure space $(X, \mathcal{X}, \mu)$ that is not necessarily complete, denote by $\mathcal{N}$ the collection of μ-negligible subsets of X (note that it contains the empty set). Let $\overline{\mathcal{X}}$ be formed by the sets $A \cup N$ ($A \in \mathcal{X}$, $N \in \mathcal{N}$). This is obviously a σ-field. Let $\overline{\mu}$ be the function from $\overline{\mathcal{X}}$ to $\mathbb{R}_+$ defined by

$$\overline{\mu}(A \cup N) = \mu(A) \quad (A \in \mathcal{X}, N \in \mathcal{N}).$$

Then $(X, \overline{\mathcal{X}}, \overline{\mu})$ is a complete measure space, called the completion of $(X, \mathcal{X}, \mu)$. It is an extension of μ in the sense that $\overline{\mu}(A) = \mu(A)$ for all $A \in \mathcal{F}$.

Equality of Measures

The forthcoming results, which are consequences of Dynkin's theorem, help to prove that two measures are identical.

Let $\mathcal{P}$ be a collection of subsets of X, and let $\mu : \mathcal{P} \to [0, \infty]$ be σ-additive, that is, for any countable family $\{A_n\}_{n \geq 1}$ of mutually disjoint sets in $\mathcal{P}$ such that $\cup_{n \geq 1} A_n \in \mathcal{P}$, we have $\mu(\cup_{n \geq 1} A_n) = \sum_{n \geq 1} \mu(A_n)$. Then μ is called a measure on $\mathcal{P}$. Let $\mathcal{C} \subseteq \mathcal{P}$ be a collection of subsets of X. A mapping $\mu : \mathcal{P} \to [0, \infty]$ is called σ-finite on $\mathcal{C}$ if there exists a countable family $\{C_n\}_{n \geq 1}$ of sets in $\mathcal{C}$ such that $\cup_{n \geq 1} C_n = X$ and $\mu(C_n) < \infty$ for all $n \geq 1$.

Theorem 2.1.42 *Let μ_1 and μ_2 be two measures on $(X, \mathcal{X})$ and let $\mathcal{S}$ be a π-system of measurable sets generating $\mathcal{X}$. Suppose that μ_1 and μ_2 are σ-finite on $\mathcal{S}$. If μ_1 and μ_2 agree on $\mathcal{S}$ (that is, $\mu_1(C) = \mu_2(C)$ for all $C \in \mathcal{S}$), then they are identical.*

Proof. Let $C \in \mathcal{S}$ be such that $\mu_1(C)(= \mu_2(C)) < \infty$. Consider the collection

$$\mathcal{D}_C = \{A \in \mathcal{X} \,;\, \mu_1(C \cap A) = \mu_2(C \cap A)\}.$$

One verifies that $\mathcal{D}_C$ is a d-system (the finiteness of $\mu_1(C)$ and $\mu_2(C)$ is needed here). Moreover, by hypothesis, $\mathcal{S} \subseteq \mathcal{D}_C$, and therefore $d(\mathcal{S}) \subseteq d(\mathcal{D}_C)$. But $d(\mathcal{D}_C) = \mathcal{D}_C$ and by Dynkin's Theorem 2.1.25, $d(\mathcal{S}) = \sigma(\mathcal{S}) = \mathcal{X}$. Therefore for all $C \in \mathcal{S}$ of finite measure, all $A \in \mathcal{X}$, we have $\mu_1(C \cap A) = \mu_2(C \cap A)$. By the assumed σ-finiteness of μ_1 and μ_2 on $\mathcal{S}$ there exists a countable family $\{C_n\}_{n \geq 1}$ of sets in $\mathcal{S}$ such that $\cup_{n \geq 1} C_n = X$ and $\mu_1(C_n) = \mu_2(C_n) < \infty$ for all $n \geq 1$. For $\alpha = 1, 2$ and all $n \geq 1$ we have that

$$\mu_\alpha(\cup_{i=1}^n (C_i \cap B)) = \sum_{1 \leq i \leq n} \mu_\alpha(C_i \cap B) - \sum_{1 \leq i < j \leq n} \mu_\alpha(C_i \cap C_j \cap B) + \cdots$$

Since $\mathcal{S}$ is a π-system, $C_i \cap C_j, C_i \cap C_j \cap C_k, \ldots$ are in $\mathcal{S}$ and therefore all the quantities in the right-hand side do not depend on $\alpha = 1, 2$. Therefore $\mu_\alpha(B) = \lim_{n \uparrow \infty} \mu_\alpha(\cup_{i=1}^n (C_i \cap B))$ is also independent of $\alpha = 1, 2$. $\square$

Corollary 2.1.43 *Let μ_1 and μ_2 be two finite measures on $(X, \mathcal{X})$ and let $\mathcal{S}$ be a π-system of measurable sets generating $\mathcal{X}$. Suppose that X is a countable union of sets in $\mathcal{S}$. If μ_1 and μ_2 agree on $\mathcal{S}$ (that is, $\mu_1(C) = \mu_2(C)$ for all $C \in \mathcal{S}$), then they are identical.*

Proof. There exists a countable family $\{C_n\}_{n \geq 1}$ of sets in $\mathcal{S}$ such that $\cup_{n \geq 1} C_n = X$. For $\alpha = 1, 2$ we have $\mu_\alpha(C_n) \leq \mu_\alpha(X) < \infty$, and therefore μ_α is σ-finite on $\mathcal{S}$, and Theorem 2.1.42 applies. $\square$

Remark 2.1.44 Consider the π-system $\mathcal{S} = \{\varnothing\}$ and let $\mathcal{X} = \{\varnothing, X\}$ be the gross σ-field. Clearly any two finite measures agree on $\mathcal{S}$ but need not agree on $\mathcal{X}$. In this case, X is *not* a union of sets in $\mathcal{S}$ and therefore Theorem 2.1.43 does not apply. Theorem 2.1.42 does not apply either because no measure on $\mathcal{X}$ is σ-finite on $\mathcal{S}$.

"The only measure on $\mathbb{R}^n$ that is translation-invariant is the Lebesgue measure." We shall be more precise in a few lines. Recall the following notation: For any set $A \subseteq \mathbb{R}^n$ and $h \in \mathbb{R}^n$, $A + h := \{a + h \,;\, a \in \mathbb{R}^n\}$.

Theorem 2.1.45

(a) *The Lebesgue measure ℓ^n on $(\mathbb{R}^n, \mathcal{B}(\mathbb{R}^n))$ is translation-invariant, that is, for all $h \in \mathbb{R}^n$ and all $A \in \mathcal{B}(\mathbb{R}^n)$, $\ell^n(A + h) = \ell^n(A)$.*

(b) *If μ is a locally finite measure on $(\mathbb{R}^n, \mathcal{B}(\mathbb{R}^n))$ that is translation-invariant, there exists a finite constant $c \geq 0$ such that $\mu = c\ell^n$.*

Proof. (a) Apply Theorem 2.1.42 with $\mu_1 = \ell^n$, μ_2 given by $\mu_2(A) = \ell^n(A + h)$ and the π-system $\mathcal{S}$ consisting of the rectangles $\prod_{i=1}^n (a_i, b_i]$.

(b) Let $c := \mu((0, 1]^n)$. For any integer $k \geq 1$, the n-cube $(0, 1]^n$ is the disjoint union of k^n n-cubes, all obtained from one another by translation of $(0, \frac{1}{k}]^n$. Therefore

$$k^n \mu \left(\left(0, \tfrac{1}{k} \right]^n \right) = c.$$

For all $d_1, \ldots, d_n \in \mathbb{R}_+$,

$$\prod_{i=1}^n \left(0, \tfrac{\lfloor kd_i \rfloor}{k} \right] \subseteq \prod_{i=1}^n (0, d_i] \subseteq \prod_{i=1}^n \left(0, \tfrac{\lfloor kd_i \rfloor + 1}{k} \right]$$

and therefore

$$\left(\prod_{i=1}^n \lfloor kd_i \rfloor \right) \frac{c}{k^n} = \mu \left(\prod_{i=1}^n \left(0, \tfrac{\lfloor kd_i \rfloor}{k} \right] \right) \leq \mu \left(\prod_{i=1}^n (0, d_i] \right)$$

$$\leq \mu \left(\prod_{i=1}^n \left(0, \tfrac{\lfloor kd_i \rfloor + 1}{k} \right] \right) \leq \left(\prod_{i=1}^n \lfloor kd_i \rfloor + 1 \right) \frac{c}{k^n}.$$

Letting $k \uparrow \infty$, we obtain

$$\mu \left(\prod_{i=1}^n (0, d_i] \right) = c \prod_{i=1}^n d_i = c\ell^n \left(\prod_{i=1}^n (0, d_i] \right).$$

Since both μ and ℓ^n are translation-invariant, the above equality extends to arbitrary rectangles $\prod_{i=1}^n (a_i, b_i]$. Theorem 2.1.42 concludes the proof. $\qquad\square$

Existence of Measures

We now give the relevant definitions for *Carathéodory's extension theorem*, one of the fundamental tools of measure theory.[1]

Definition 2.1.46 *Let X be a set. The collection $\mathcal{A} \subseteq \mathcal{P}(X)$ is called an* algebra *if*

(α) $X \in \mathcal{A}$;

(β) $A, B \in \mathcal{A} \implies A \cup B \in \mathcal{A}$;

(γ) $A \in \mathcal{A} \implies \overline{A} \in \mathcal{A}$.

---

[1]The proofs concerning the existence and unicity of measures are omitted. See the subsubsection "Complementary reading".

The only difference with a σ-field is that we require it to be closed under *finite* (instead countable) unions. (This is why a σ-field is also called a σ-*algebra*.) Note that, similarly to the σ-field case, $\varnothing \in \mathcal{A}$ and $\mathcal{A}$ is closed under finite intersections.

EXAMPLE 2.1.47: FINITE UNIONS OF DISJOINT INTERVALS. On $\mathbb{R}$, the collection of finite sums of disjoint intervals is an algebra. (By interval, we mean any type of interval: open, closed, semi-open, semi-closed, infinite, etc., in other words a connected subset of $\mathbb{R}$.[2])

Definition 2.1.48 *Let X be a set. The collection $\mathcal{C} \subseteq \mathcal{P}(X)$ is called a* semi-algebra *if*

(α) $X \in \mathcal{C}$,

(β) $A, B \in \mathcal{C} \implies A \cup B \in \mathcal{C}$, *and*

(γ) *if $A \in \mathcal{C}$ then $\overline{A}$ can be expressed as a finite union of disjoint sets of $\mathcal{C}$.*

EXAMPLE 2.1.49: THE COLLECTION OF INTERVALS. On $\mathbb{R}$, the collection of intervals is a semi-algebra.

Theorem 2.1.50 *Let $\mathcal{C}$ be either an algebra or a semi-algebra defined on X. Let μ be a σ-finite measure on $(X, \mathcal{C})$. Then there exists a unique extension of μ to $(X, \sigma(\mathcal{C}))$ that is a measure.*

This result pertains to measure theory and its proof is therefore omitted.

Definition 2.1.51 *A function $F : \mathbb{R} \to \mathbb{R}$ is called a* cumulative distribution function (CDF) *if the following properties are satisfied:*

1. *F is non-decreasing;*

2. *F is right-continuous;*

3. *F admits a left-hand limit, denoted $F(x-)$, at all $x \in \mathbb{R}$.*

EXAMPLE 2.1.52: CUMULATIVE DISTRIBUTION FUNCTION. Let μ be a locally finite measure on $(\mathbb{R}, \mathcal{B})$ and define

$$F_\mu(t) = \begin{cases} +\mu((0,t]) & \text{if } t \geq 0, \\ -\mu((t,0]) & \text{if } t < 0. \end{cases} \tag{2.2}$$

This is clearly a cumulative distribution function, and moreover,

$$F_\mu(b) - F_\mu(a) = \mu((a,b]),$$
$$F_\mu(a) - F_\mu(a-) = \mu(\{a\}).$$

The function F_μ is called the CDF of μ.

Theorem 2.1.53 *Let $F : \mathbb{R} \to \mathbb{R}$ be a CDF. There exists a unique locally finite measure μ on $(\mathbb{R}, \mathcal{B}(\mathbb{R}))$ such that $F_\mu = F$.*

---

[2]A subset C of $\mathbb{R}$ is called connected if for all $a, b \in C$, the segment $[a, b] \subseteq C$.

2.2 The Lebesgue Integral

We are now in a position to define (when it exists) the Lebesgue integral of a measurable function $f : (X, \mathcal{X}) \to (\mathbb{R}, \mathcal{B}(\mathbb{R}))$ with respect to a measure μ, denoted by

$$\int_X f \, d\mu, \quad \text{or} \quad \int_X f(x) \, \mu(dx), \quad \text{or} \quad \mu(f).$$

2.2.1 Construction of the Integral

Denote by $\mathcal{S}^+(X)$ (or $\mathcal{S}^+$ if the context is clear) the set of *non-negative* simple functions $f : (X, \mathcal{X}) \to (\mathbb{R}, \mathcal{B}(\mathbb{R}))$, and by $\mathcal{M}^+(X)$ (or $\mathcal{M}^+$) the set of *non-negative* functions $f : (X, \mathcal{X}) \to (\overline{\mathbb{R}}, \mathcal{B}(\overline{\mathbb{R}}))$.

The integral is defined in three steps:

STEP 1. One first defines it for functions in $\mathcal{S}^+$. Let $f : X \to \mathbb{R}$ be a non-negative simple Borel function as in Definition 2.1.13. The integral of f with respect to μ is defined by

$$\int_X f \, d\mu := \sum_{i=1}^{k} a_i \, \mu(A_i). \tag{2.3}$$

This definition apparently depends on the representation of f. One has to show that if f admits another representation

$$f(x) = \sum_{j=1}^{m} b_j \, 1_{B_j}(x),$$

where $m \in \mathbb{N}_+$, $b_1, \ldots, b_m \in \mathbb{R}$, and $B_1, \ldots, B_m$ are sets in $\mathcal{X}$, then

$$\sum_{j=1}^{m} b_j \, \mu(B_j) = \sum_{i=1}^{k} a_i \, \mu(A_i).$$

This follows from the following chain of equalities, where it is noted that if $A_i \cap B_j \neq \varnothing$, then $a_i = b_j$:

$$\sum_{j=1}^{m} b_j \, \mu(B_j) = \sum_{j=1}^{m} b_j \left(\sum_{i=1}^{k} \mu(B_j \cap A_i) \right)$$

$$= \sum_{i=1}^{k} \sum_{j=1}^{m} b_j \, \mu(B_j \cap A_i)$$

$$= \sum_{i=1}^{k} \sum_{j=1}^{m} a_i \, \mu(B_j \cap A_i)$$

$$= \sum_{i=1}^{k} a_i \left(\sum_{j=1}^{m} \mu(B_j \cap A_i) \right) = \sum_{i=1}^{k} a_i \, \mu(A_i).$$

The next lemma collects a few intermediary results.

Lemma 2.2.1 *Let $f, f_1, f_2, \dots$ be in $\mathcal{S}^+$. Then*

(a) for all $\lambda \geq 0$, $\lambda f \in \mathcal{S}^+$ and $\int_X (\lambda f)\, d\mu = \lambda \int_X f\, d\mu$,

(b) $f_1 + f_2 \in \mathcal{S}^+$ and $\int_X (f_1 + f_2)\, d\mu = \int_X f_1\, d\mu + \int_X f_2\, d\mu$,

(c) $f_1 \leq f_2$ implies $\int_X f_1\, d\mu \leq \int_X f_2\, d\mu$,

(d) $f_1 \wedge f_2$ and $f_1 \vee f_2$ are in $\mathcal{S}^+$, and

(e) if $f_n \leq f_{n+1} \leq f$ for all $n \geq 1$ and $\lim_{n \uparrow \infty} f_n = f$, then $\lim_{n \uparrow \infty} \int_X f_n\, d\mu = \int_X f\, d\mu$.

Proof. Properties (a)–(d) are immediate. For (e), first consider the case $f = 1_A$. Fix $m \geq 1$. For all $n \geq 1$, define $A_{n,m} = \{x : f_n(x) \geq 1 - \frac{1}{m}\}$. Since $\{f_n\}_{n \geq 1}$ is non-decreasing, we have that $A_{n,m} \subseteq A_{n+1,m}$; and since $f_n \uparrow 1_A$, we have that $\cup_{n=1}^{\infty} A_{n,m} = A$. Note that

$$\left(1 - \frac{1}{m}\right) 1_{A_{n,m}} \leq f_n \leq 1_A,$$

and therefore

$$\left(1 - \frac{1}{m}\right) \mu(A_{n,m}) \leq \int_X f_n\, d\mu \leq \mu(A),$$

from which the announced result follows in this particular case by first letting $n \uparrow +\infty$ and then $m \uparrow +\infty$.

Consider now the general case where $f = \sum_{i=1}^k a_i 1_{A_i}$. We may suppose that $\{A_i\}_{i=1}^k$ is a partition of X, so that $f_n = \sum_{i=1}^k f_n 1_{A_i}$ and therefore, by (a),

$$\int_X f_n\, d\mu = \sum_{i=1}^k \int_X f_n 1_{A_i}\, d\mu = \sum_{i=1}^k a_i \int_X \frac{f_n}{a_i} 1_{A_i}\, d\mu.$$

Passing to the limit $n \uparrow \infty$ yields the desired result, since $\frac{f_n}{a_i} 1_{A_i} \uparrow 1_{A_i}$. $\qquad \square$

STEP 2. The integral will now be defined for integrands $f \in \mathcal{M}^+$. For such f, let

$$\mu(f) := \sup \left\{ \int_X \varphi\, d\mu \, ; \, \varphi \leq f, \, \varphi \in \mathcal{S}^+ \right\}.$$

The function f is called μ-integrable if $\mu(f) < \infty$.

We first check that if $f \in \mathcal{S}^+$, $\mu(f) = \int_X f\, d\mu$. For this, let

$$A_f := \left\{ \int_X \varphi\, d\mu \, ; \, \varphi \leq f, \, \varphi \in \mathcal{S}^+ \right\}.$$

Since $f \in \mathcal{S}^+$, $\int_X f\, d\mu \in A_f$ and therefore $\mu(f) \geq \int_X f\, d\mu$. On the other hand, for all $\varphi \leq f$, $\int_X \varphi\, d\mu \leq \int_X f\, d\mu$ and therefore $\mu(f) \leq \int_X f\, d\mu$. Therefore $\mu(f) = \int_X f\, d\mu$.

Having checked this point, it is now safe to call $\mu(f)$ the integral of f with respect to μ, and denote it also by $\int_X f\, d\mu$. Indeed the two ways of defining $\int_X f\, d\mu$ for $f \in \mathcal{S}^+$ (as in Step 1 and Step 2) give the same result.

The next result is due to *Beppo Levi.*

Theorem 2.2.2 (Monotone Convergence Theorem) *Let* $\{f_n\}_{n\geq 1}$ *be a non-decreasing sequence of non-negative measurable functions from X to $\overline{\mathbb{R}}$. Then*

$$\lim_{n\uparrow\infty} \int_X f_n \, d\mu = \int_X (\lim_{n\uparrow\infty} f_n) \, d\mu \,.$$

Proof. We shall need the following monotonicity property. Let f_1, f_2 in $\mathcal{M}^+$ be such that $f_1 \leq f_2$. Then $\int_X f_1 \, d\mu \leq \int_X f_2 \, d\mu$. In fact, $A_{f_1} \subseteq A_{f_2}$, and therefore

$$\mu(f_1) = \sup A_{f_1} \leq \sup A_{f_2} = \mu(f_2) \,.$$

We now turn to the proof of the theorem. Denote $\lim_{n\uparrow\infty} f_n$ by f. By the just proved monotonicity property of the integral of functions in $\mathcal{M}^+$,

$$\int_X f_n \, d\mu \leq \int_X f_{n+1} \, d\mu \leq \int_X f \, d\mu \,.$$

Being a non-decreasing sequence, $\{\int_X f_n \, d\mu\}_{n\geq 1}$ has a limit, and by the previous inequality

$$\lim_{n\uparrow\infty} \int_X f_n \, d\mu \leq \int_X f \, d\mu \,.$$

It remains to prove the converse inequality. For $\varphi \in \mathcal{S}^+$ such that $\varphi \leq f$, $\lambda \in (0,1)$ and $n \geq 1$, define the (measurable) set $E_n = \{f_n \geq \lambda\varphi\}$. We have that $E_n \subseteq E_{n+1}$. Moreover $\cup_{n\geq 1} E_n = X$. Since $\lambda\varphi 1_{E_n} \leq f_n$,

$$\int_X \lambda\varphi 1_{E_n} \, d\mu \leq \int_X f_n \, d\mu \leq \lim_{k\uparrow\infty} \int_X f_k \, d\mu \,.$$

On the other hand, since $E_n \subseteq E_{n+1}$ and $\cup_{n\geq 1} E_n = X$, we have that $1_{E_n} \uparrow 1$ and in particular $1_{E_n}\varphi \uparrow \varphi$. Therefore by (e) of Lemma 2.2.1, $\lim_{n\uparrow\infty} \int_X \lambda\varphi 1_{E_n} \, d\mu = \int_X \lambda\varphi \, d\mu$. Passing to the limit $n \uparrow \infty$ in the last displayed inequalities, we have that for all $\lambda \in (0,1)$,

$$\lambda \int_X \varphi \, d\mu \leq \lim_{n\uparrow\infty} \int_X f_n \, d\mu \,.$$

This equality remains true at the limit $\lambda = 1$. This being true of all $\varphi \in \mathcal{S}^+$ such that $\varphi \leq f$, we have

$$\int_X f \, d\mu \leq \lim_{n\uparrow\infty} \int_X f_n \, d\mu \,.$$

$\square$

Here is another collection of intermediary results for later use that we group in a lemma:

Lemma 2.2.3 *Let f, f_1, f_2 be in $\mathcal{M}^+$. Then*

(i) *for all $\lambda \geq 0$, $\int_X (\lambda f) \, d\mu = \lambda \int_X f \, d\mu$,*

(ii) *$\int_X (f_1 + f_2) \, d\mu = \int_X f_1 \, d\mu + \int_X f_2 \, d\mu$, and*

(iii) *if $f_1 \leq f_2$, then $\int_X f_1 \, d\mu \leq \int_X f_2 \, d\mu$.*

Proof. (iii) was obtained in the proof of Theorem 2.2.2. (i) and (ii) are true for functions in $\mathcal{S}^+$ (Lemma 2.2.1). Using non-decreasing sequences of functions in $\mathcal{S}^+$, $\{f_{1,n}\}_{n\geq 1}$ and $\{f_{2,n}\}_{n\geq 1}$, converging respectively to f_1 and f_2, we have that for all $n \geq 1$,

$$\int_X (\lambda f_{1,n})\,\mathrm{d}\mu = \lambda \int_X f_{1,n}\,\mathrm{d}\mu$$

and

$$\int_X (f_{1,n} + f_{2,n})\,\mathrm{d}\mu = \int_X f_{1,n}\,\mathrm{d}\mu + \int_X f_{2,n}\,\mathrm{d}\mu.$$

Letting $n \uparrow \infty$, the Monotone Convergence Theorem 2.2.2 yields (i) and (ii). □

The next result is a fundamental technical tool, called *Fatou's lemma*:

Theorem 2.2.4 *Let* $\{f_n\}_{n\geq 1}$ *be a sequence of non-negative measurable functions (from* X *to* $\overline{\mathbb{R}}_+$*). Then*

$$\int_X (\liminf_{n\uparrow\infty} f_n)\,\mathrm{d}\mu \leq \liminf_{n\uparrow\infty} \int_X f_n\,\mathrm{d}\mu.$$

Proof. Define $f := \liminf_{n\uparrow\infty} f_n := \lim_{n\uparrow\infty} \inf_{k\geq n} f_k$. Using the Monotone Convergence Theorem 2.2.2 for the second equality, we obtain

$$\int_X f\,\mathrm{d}\mu = \int_X (\liminf_{n\uparrow\infty}\inf_{k\geq n} f_k)\,\mathrm{d}\mu = \lim_{n\uparrow\infty}\int_X (\inf_{k\geq n} f_k)\,\mathrm{d}\mu.$$

On the other hand, since for all $i \geq n$, $\int_X (\inf_{k\geq n} f_k)\,\mathrm{d}\mu \leq \int_X f_i\,\mathrm{d}\mu$, we have that $\int_X (\inf_{k\geq n} f_k)\,\mathrm{d}\mu \leq \inf_{i\geq n}(\int_X f_i\,\mathrm{d}\mu)$. Therefore

$$\int_X f\,\mathrm{d}\mu \leq \lim_{n\uparrow\infty}\inf_{i\geq n}\left(\int_X f_i\,\mathrm{d}\mu\right) = \liminf_{n\uparrow\infty}\int_X f_n\,\mathrm{d}\mu.$$

□

STEP 3. Integrals of functions of arbitrary sign.

Definition 2.2.5 *A measurable function* $f : (X,\mathcal{X}) \to (\overline{\mathbb{R}}, \mathcal{B}(\overline{\mathbb{R}}))$ *satisfying*

$$\int_X |f|\,\mathrm{d}\mu < \infty,$$

is called a μ-*integrable function.*

Define $f^+ := \max(f,0)$ and $f^- := \max(-f,0)$. In particular, $f = f^+ - f^-$ and $f^\pm \leq |f|$. Therefore, by monotonicity (Property (iii) of Lemma 13.3.35),

$$\int_X f^\pm\,\mathrm{d}\mu \leq \int_X |f|\,\mathrm{d}\mu.$$

Thus, if f is integrable, the right-hand side of

$$\int_X f\,\mathrm{d}\mu := \int_X f^+\,\mathrm{d}\mu - \int_X f^-\,\mathrm{d}\mu \tag{2.4}$$

is meaningful and defines the integral of the left-hand side. Moreover, the integral of f with respect to μ defined in this way is finite.

The integral can be defined for some *non-integrable* functions, for instance for all non-negative functions. More generally, if $f : (X, \mathcal{X}) \to (\overline{\mathbb{R}}, \mathcal{B}(\overline{\mathbb{R}}))$ is such that at least one of the integrals $\int_X f^+ \, d\mu$ or $\int_X f^- \, d\mu$ is finite, one defines the integral as in (2.4). This leads to one of the forms "finite minus finite", "finite minus infinite", and "infinite minus finite". The case which is rigorously excluded is that in which $\mu(f^+) = \mu(f^-) = +\infty$.

The following equality is a definition of the left-hand side:

$$\int_A f(x)\,\mu(dx) := \int_X 1_A(x)\,f(x)\,\mu(dx).$$

For a complex Borel function $f : X \to \mathbb{C}$ (that is, $f = f_1 + if_2$, where $f_1, f_2 : (X, \mathcal{X}) \to (\mathbb{R}, \mathcal{B}(\mathbb{R}))$) such that $\mu(|f|) < \infty$, let

$$\int_X f\,d\mu := \int_X f_1\,d\mu + i \int_X f_2\,d\mu.$$

The examples below show how the construction of the integral works in very simple cases.

EXAMPLE 2.2.6: INTEGRAL WITH RESPECT TO THE WEIGHTED COUNTING MEASURE. Clearly, any function $f : \mathbb{Z} \to \mathbb{R}$ is measurable with respect to $\mathcal{P}(\mathbb{Z})$ and $\mathcal{B}(\mathbb{R})$. With the measure μ defined in Example 2.1.30, and with $f \geq 0$ for instance,

$$\mu(f) = \sum_{n \in \mathbb{Z}} \alpha_n f(n).$$

Proof. It suffices to consider the approximating sequence of simple functions

$$f_n(k) = \sum_{j=-n}^{+n} f(j) 1_{\{j\}}(k)$$

whose integral is

$$\mu(f_n) = \sum_{j=-n}^{+n} f(j)\mu(\{j\}) = \sum_{j=-n}^{+n} f(j)\alpha_j$$

and to let n tend to ∞.

When $\alpha_n \equiv 1$, the integral reduces to the sum of a series:

$$\mu(f) = \sum_{n \in \mathbb{Z}} f(n).$$

In this case, integrability means that the series is absolutely convergent. $\square$

The next trivial example should demystify the notion of Dirac measure.

EXAMPLE 2.2.7: INTEGRAL WITH RESPECT TO THE DIRAC MEASURE. Let ϵ_a be the Dirac measure at point $a \in X$. Then any $f : (X, \mathcal{X}) \to (\mathbb{R}, \mathcal{B}(\mathbb{R}))$ is ϵ_a-integrable and

$$\epsilon_a(f) = f(a).$$

Proof. For a simple function f as in Definition 2.1.13,

$$\epsilon_a(f) = \sum_{i=1}^{k} a_i \, \epsilon_a(A_i) = \sum_{i=1}^{k} a_i \, 1_{A_i}(a) = f(a).$$

For a non-negative function f and any non-decreasing sequence of simple non-negative Borel functions $\{f_n\}_{n \geq 1}$ converging to f,

$$\epsilon_a(f) = \lim_{n \uparrow \infty} \epsilon_a(f_n) = \lim_{n \uparrow \infty} f_n(a) = f(a).$$

Finally, for any $f : (X, \mathcal{X}) \to (\mathbb{R}, \mathcal{B}(\mathbb{R}))$

$$\epsilon_a(f) = \epsilon_a(f^+) - \epsilon_a(f^-) = f^+(a) - f^-(a) = f(a)$$

is a well-defined quantity. $\qquad\square$

Remark 2.2.8 In the applied literature, especially the engineering literature, the "Dirac distribution" is often used when the Dirac measure is all that is needed.

The Stieltjes–Lebesgue Integral

Definition 2.2.9 *Let F be a* CDF *on $\mathbb{R}$. By definition, the* Lebesgue–Stieltjes integral *of g with respect to F, denoted by*

$$\int_{\mathbb{R}} g(x) \, \mathrm{d}F(x),$$

is the Lebesgue integral $\int_{\mathbb{R}} g(x) \, \mu_F(\mathrm{d}x)$, where μ_F is the locally finite measure on $(\mathbb{R}, \mathcal{B}(\mathbb{R}))$ whose CDF *is F (Theorem 2.1.53).*

2.2.2 Elementary Properties of the Integral

We are now ready to state and prove one of the most important results of integration theory, the *Lebesgue theorem*, also called the *dominated convergence theorem*.

Theorem 2.2.10 *Let $\{f_n\}_{n \geq 1}$ be a sequence of measurable functions from $(X, \mathcal{X})$ to $(\mathbb{R}, \mathcal{B}(\mathbb{R}))$ that converges to a (necessarily) measurable function f. Suppose moreover that for all $n \geq 1$, $|f_n| \leq g$, where g is integrable. Then*

$$\lim_{n \uparrow \infty} \int_X f_n \, \mathrm{d}\mu = \int_X (\lim_{n \uparrow \infty} f_n) \, \mathrm{d}\mu.$$

Proof. By Fatou's lemma applied to the sequence of non-negative functions $\{g+f_n\}_{n\geq1}$,

$$\int_X (g+f)\,\mathrm{d}\mu = \int_X \lim_{n\uparrow\infty}(g+f_n)\,\mathrm{d}\mu$$

$$\leq \liminf_{n\uparrow\infty} \int_X (g+f_n)\,\mathrm{d}\mu = \int_X g\,\mathrm{d}\mu + \liminf_{n\uparrow\infty} \int_X f_n\,\mathrm{d}\mu.$$

Therefore,

$$\int_X f\,\mathrm{d}\mu \leq \liminf_{n\uparrow\infty} \int_X f_n\,\mathrm{d}\mu.$$

Similarly, replacing f and f_n by $-f$ and $-f_n$ respectively,

$$\int_X f\,\mathrm{d}\mu \geq \limsup_{n\uparrow\infty} \int_X f_n\,\mathrm{d}\mu.$$

In particular, $\lim_{n\uparrow\infty} \int_X f_n\,\mathrm{d}\mu$ exists and is equal to $\int_X f\,\mathrm{d}\mu$. $\square$

EXAMPLE 2.2.11: THE CLASSICAL COUNTEREXAMPLE. Let $(X,\mathcal{X},\mu) = (\mathbb{R},\mathcal{B}(\mathbb{R}),\ell)$, and let

$$f_n(x) := n\,1_{(0,\frac{1}{n}]}(x).$$

One has $\lim_{n\uparrow\infty} f_n = 0$. Therefore $\mu(\lim_{n\uparrow\infty} f_n) = 0$. However, $\mu(f_n) = 1$ for all $n \geq 1$.

More Elementary Properties

Recall that for all $A \in \mathcal{X}$,

$$\int_X 1_A\,\mathrm{d}\mu = \mu(A) \tag{2.5}$$

by definition and that the notation $\int_A f\,\mathrm{d}\mu$ stands for $\int_X 1_A f\,\mathrm{d}\mu$.

Theorem 2.2.12 *Let $f,g : (X,\mathcal{X}) \to (\overline{\mathbb{R}},\mathcal{B}(\overline{\mathbb{R}}))$ be μ-integrable functions, and let $a,b \in \mathbb{R}$. Then*

(a) $af + bg$ is μ-integrable and $\mu(af+bg) = a\mu(f) + b\mu(g)$,

(b) if $f = 0$ μ-a.e., then $\mu(f) = 0$; if $f = g$ μ-a.e., then $\mu(f) = \mu(g)$, and

(c) if $f \leq g$ μ-a.e., then $\mu(f) \leq \mu(g)$.

Proof. We omit the (easy) proofs. $\square$

We continue to list the basic properties of the integral.

Theorem 2.2.13 *Moreover:*

(d) $|\mu(f)| \leq \mu(|f|)$,

(e) if $f \geq 0$ μ-a.e. and $\mu(f) = 0$, then $f = 0$ μ-a.e.,

(f) if $\mu(1_A f) = 0$ for all $A \in \mathcal{X}$, then $f = 0$ μ-a.e., and

(g) if f is μ-integrable, then $|f| < \infty$ μ-a.e.

Proof.

(d) $\mu(f) = \mu(f_+) - \mu(f_-)$. Therefore $|\mu(f)| \leq \mu(f_+) + \mu(f_-) = \mu(f_+ + f_-) = \mu(|f|)$.

(e) Define $A_n = \{f \geq \frac{1}{n}\}$. Since f is non-negative, $f \geq \frac{1}{n}1_{A_n}$, and therefore,

$$\mu(f) \geq \frac{1}{n}\mu(A_n),$$

from which it follows that, since $\mu(f) = 0$, $\mu(A_n) = 0$, and $\lim_{n\uparrow\infty} \mu(A_n) = 0$. But the sequence of sets $\{A_n\}_{n\geq 1}$ increases to $\{f > 0\}$, and therefore by sequential continuity, $\mu(\{f > 0\}) = 0$, that is, $f \leq 0, \mu$-a.e. On the other hand, by hypothesis, $f \geq 0, \mu$-a.e. Therefore $f = 0, \mu$-a.e.

(f) With $A = \{f > 0\}$, $1_A f$ is a non-negative measurable function. By (e), $1_A f = 0, \mu$-a.e. This implies that $1_A = 0, \mu$-a.e., that is to say $f \leq 0, \mu$-a.e. Similarly, $f \geq 0, \mu$-a.e. Therefore, $f = 0, \mu$-a.e.

(g) It is enough to consider the case $f \geq 0$. Since $f \geq n1_{\{f=\infty\}}$ for all $n \geq 1$, we have

$$\infty > \mu(f) \geq n\mu(\{f = \infty\}),$$

and therefore $n\mu(\{f = \infty\}) < \infty$. This cannot be true for all $n \geq 1$ unless $\mu(\{f = \infty\}) = 0$. $\qquad\square$

The extension to complex Borel functions of the properties (a), (b), (d) and (f) is immediate.

EXAMPLE 2.2.14: LEBESGUE-INTEGRABLE BUT NOT RIEMANN-INTEGRABLE. The function f defined by $f := 1_\mathbb{Q}$ ($\mathbb{Q}$ is the set of rational numbers) is a Borel function and it is Lebesgue integrable with its integral equal to zero because $\{f \neq 0\}$ is the set of rational numbers, which has null Lebesgue measure. However, f is not Riemann integrable.

2.2.3   Beppo Levi, Fatou and Lebesgue

The following versions of the theorems of Beppo Levi, Fatou and Lebesgue differ from the previous ones by the introduction of "μ-almost everywhere" in the statements of the conditions. No other proofs are needed since integrals of almost everywhere equal functions are equal and countable unions of negligible sets are negligible. Only a convention must be stated: if the limit of a sequence of real measurable functions exists μ-almost everywhere, that is, outside a μ-negligible set, then the limit is typically assigned some arbitrary value on this μ-negligible set; for example, many people set the limit to be 0.

Remember that we are looking for conditions guaranteeing that

$$\int_X \lim_{n\uparrow\infty} f_n \, d\mu = \lim_{n\uparrow\infty} \int_X f_n \, d\mu. \tag{2.6}$$

We start by restating the *Beppo Levi* or *monotone convergence* theorem.

Theorem 2.2.15 *Let* $f_n : (X, \mathcal{X}) \to (\overline{\mathbb{R}}, \mathcal{B}(\overline{\mathbb{R}}))$ $(n \geq 1)$ *be such that*

(i) $f_n \geq 0$ μ-a.e., and

(ii) $f_{n+1} \geq f_n$ μ-a.e.

Then, there exists a non-negative function $f : (X, \mathcal{X}) \to (\overline{\mathbb{R}}, \mathcal{B}(\overline{\mathbb{R}}))$ *such that*

$$\lim_{n \uparrow \infty} f_n = f \quad \mu\text{-a.e.},$$

and (2.6) holds true.

Next, we restate *Fatou's lemma.*

Theorem 2.2.16 *Let* $f_n : (X, \mathcal{X}) \to (\overline{\mathbb{R}}, \mathcal{B}(\overline{\mathbb{R}}))$ $(n \geq 1)$ *be such that* $f_n \geq 0$ μ-a.e. $(n \geq 1)$. *Then*

$$\int_X (\liminf_{n \uparrow \infty} f_n) \, d\mu \leq \liminf_{n \uparrow \infty} \left(\int_X f_n \, d\mu \right). \tag{2.7}$$

Finally, we restate the *Lebesgue* or *dominated convergence* theorem.

Theorem 2.2.17 *Let* $f_n : (X, \mathcal{X}) \to (\overline{\mathbb{R}}, \mathcal{B}(\overline{\mathbb{R}}))$ $(n \geq 1)$ *be such that, for some function* $f : (X, \mathcal{X}) \to (\overline{\mathbb{R}}, \mathcal{B}(\overline{\mathbb{R}}))$ *and some* μ-integrable function $g : (X, \mathcal{X}) \to (\overline{\mathbb{R}}, \mathcal{B}(\overline{\mathbb{R}}))$:

(i) $\lim_{n \uparrow \infty} f_n = f$, μ-a.e., and

(ii) $|f_n| \leq |g|$ μ-a.e. for all $n \geq 1$.

Then, (2.6) holds true.

Differentiation under the integral sign

Let $(X, \mathcal{X}, \mu)$ be a measure space and let $(a, b) \subseteq \mathbb{R}$. Let $f : (a, b) \times X \to \mathbb{R}$ and for all $t \in (a, b)$, define $f_t : X \to \mathbb{R}$ by $f_t(x) := f(t, x)$. Suppose that for all $t \in (a, b)$, f_t is measurable with respect to $\mathcal{X}$, and define, when possible, the function $I : (a, b) \to \mathbb{R}$ by the formula

$$I(t) = \int_X f(t, x) \, \mu(dx). \tag{2.8}$$

Assume that for μ-almost all x the function $t \mapsto f(t, x)$ is continuous at $t_0 \in (a, b)$ and that there exists a μ-integrable function $g : (X, \mathcal{X}) \to (\overline{\mathbb{R}}, \mathcal{B}(\overline{\mathbb{R}}))$ such that $|f(t, x)| \leq |g(x)|$ μ-a.e. for all t in a neighborhood V of t_0. Then I is well defined and is continuous at t_0.

Proof. Let $\{t_n\}_{n \geq 1}$ be a sequence in $V \setminus \{t_0\}$ such that $\lim_{n \uparrow \infty} t_n = t_0$, and define $f_n(x) = f(t_n, x)$, $\bar{f}(x) = f(t_0, x)$. By dominated convergence,

$$\lim_{n \uparrow \infty} I(t_n) = \lim_{n \uparrow \infty} \mu(f_n) = \mu(f) = I(t_0).$$

$\square$

If we furthermore assume that

(α) $t \to f(t, x)$ is continuously differentiable on V for μ-almost all x, and

(β) for some μ-integrable function $h : (X, \mathcal{X}) \to (\overline{\mathbb{R}}, \mathcal{B}(\overline{\mathbb{R}}))$ and all $t \in V$,

$$|(df/dt)(t, x)| \leq |h(x)| \quad \mu\text{-a.e.},$$

then I is differentiable at t_0 and

$$I'(t_0) = \int_X (df/dt)(t_0, x)\, \mu(dx). \tag{2.9}$$

Proof. Let $\{t_n\}_{n\geq 1}$ be a sequence in $V \setminus \{t_0\}$ such that $\lim_{n\uparrow\infty} t_n = t_0$, and define $f_n(x) = f(t_n, x)$, $f(x) = f(t_0, x)$. By dominated convergence,

$$\lim_{n\uparrow\infty} I(t_n) = \lim_{n\uparrow\infty} \mu(f_n) = \mu(f) = I(t_0).$$

Also

$$\frac{I(t_n) - I(t_0)}{t_n - t_0} = \int_X \frac{f(t_n, x) - f(t_0, x)}{t_n - t_0}\, \mu(dx),$$

and for some $\theta \in (0, 1)$, possibly depending upon n,

$$\left| \frac{f(t_n, x) - f(t_0, x)}{t_n - t_0} \right| \leq |(df/dt)(t_0 + \theta(t_n - t_0), x)|.$$

The latter quantity is bounded by $|h(x)|$. Therefore, by dominated convergence,

$$\lim_{n\uparrow\infty} \frac{I(t_n) - I(t_0)}{t_n - t_0} = \int_X \left(\lim_{n\uparrow\infty} \frac{f(t_n, x) - f(t_0)}{t_n - t_0} \right) \mu(dx)$$
$$= \int_X (df/dt)(t_0, x)\, \mu(dx).$$

$\square$

2.3 The Other Big Theorems

Besides Beppo Levi, Fatou and Lebesgue's theorems, the four main results of integration theory are

(i) the image measure theorem,

(ii) the Fubini–Tonelli theorem relative to the product measures (to be defined in a few lines), which gives conditions allowing one to "choose the order of integration in multiple integrals",

(iii) the Riesz–Fischer theorem relative to the Hilbert space structure of the space of square-integrable functions, and

(iv) the Radon–Nikodým theorem relative to the product of a measure by a function, more precisely a converse of Theorem 2.3.28.

2.3.1   The Image Measure Theorem

Definition 2.3.1 *Let $(X, \mathcal{X})$ and $(E, \mathcal{E})$ be two measurable spaces, let $h : (X, \mathcal{X}) \to (E, \mathcal{E})$ be a measurable function and let μ be a measure on $(X, \mathcal{X})$. Define the set function $\mu \circ h^{-1} : \mathcal{E} \to [0, \infty]$ by*

$$(\mu \circ h^{-1})(C) := \mu(h^{-1}(C)) \quad (C \in \mathcal{E}). \tag{2.10}$$

Then, as one easily checks, $\mu \circ h^{-1}$ is a measure on $(E, \mathcal{E})$ called the image of μ by h.

Integrals can be computed in the original domain or in the image domain. More precisely:

Theorem 2.3.2 *For a non-negative $f : (X, \mathcal{X}) \to (\overline{\mathbb{R}}, \mathcal{B}(\overline{\mathbb{R}}))$*

$$\int_X (f \circ h)(x)\, \mu(\mathrm{d}x) = \int_E f(y)(\mu \circ h^{-1})\,(\mathrm{d}y). \tag{2.11}$$

For functions $f : (X, \mathcal{X}) \to (\overline{\mathbb{R}}, \mathcal{B}(\overline{\mathbb{R}}))$ of arbitrary sign either one of the conditions

(a) $f \circ h$ is μ-integrable,

(b) f is $\mu \circ h^{-1}$-integrable,

implies the other, and equality (2.11) then holds.

Proof. The equality (2.11) is readily verified when f is a non-negative simple Borel function. In the general case one approximates f by a non-decreasing sequence of non-negative simple Borel functions $\{f_n\}_{n \geq 1}$ and (2.11) then follows from the same equality written with $f = f_n$, by letting $n \uparrow \infty$ and using the monotone convergence theorem. For the case of functions of arbitrary sign, apply (2.11) with f^+ and f^-. $\square$

2.3.2   The Fubini–Tonelli Theorem

The first task is to define products of measurable spaces and of measures.

Definition 2.3.3 *Let $(X_i, \mathcal{X}_i)$ $(1 \leq i \leq n)$ be measurable spaces and let $X := \prod_{i=1}^n X_i$. Define the* product σ-field $\mathcal{X} := \otimes_{i=1}^n \mathcal{X}_i$ *to be the smallest σ-field on X containing all the so-called* generalized measurable rectangles $\prod_{i=1}^n A_i$, *where $A_i \in \mathcal{X}_i$ $(1 \leq i \leq n)$.*

If $(X_i, \mathcal{X}_i) = (Y, \mathcal{Y})$ $(1 \leq i \leq n)$, $\prod_{i=1}^n X_i$ is denoted by Y^n and $\otimes_{i=1}^n \mathcal{X}_i$ by $\mathcal{Y}^{\otimes n}$. For the product of σ-fields, we have the rule of associativity. For instance (exercise):

$$\mathcal{X}_1 \otimes \mathcal{X}_2 \otimes \mathcal{X}_3 = (\mathcal{X}_1 \otimes \mathcal{X}_2) \otimes \mathcal{X}_3 = \mathcal{X}_1 \otimes (\mathcal{X}_2 \otimes \mathcal{X}_3).$$

The Borel σ-field $\mathcal{B}(\mathbb{R})^{\otimes n}$ (denoted by $\mathcal{B}(\mathbb{R})^n$ for short) is the σ-field on $\mathbb{R}^n$ generated by the generalized measurable rectangles of $\mathbb{R}^n$. The question is: Is this σ-field identical to $\mathcal{B}(\mathbb{R}^n)$, the σ-field generated by the rectangles of the type $\prod_{i=1}^n (a_i, b_i]$, where $-\infty < a_i \leq b_i < +\infty$ $(1 \leq i \leq n)$? The answer is positive and a consequence of the rule of associativity of the product of σ-fields and of the result below.

Theorem 2.3.4 *Let E and F be two separable metric spaces with respective Borel σ-fields $\mathcal{B}(E)$ and $\mathcal{B}(F)$. Then*

$$\mathcal{B}(E \times F) = \mathcal{B}(E) \otimes \mathcal{B}(F).$$

Proof. (i) For the proof that $\mathcal{B}(E \times F) \supseteq \mathcal{B}(E) \otimes \mathcal{B}(F)$, separability is not needed. We just have to observe that the projections π_1 and π_2 from $E \times F$ to E and F, respectively, are continuous, and therefore measurable functions from $(E \times F, \mathcal{B}(E \times F))$ to $(E, \mathcal{B}(E))$ and $(F, \mathcal{B}(F))$, respectively. In particular, if $C \in \mathcal{B}(E)$, then $C \times F = \pi_1^{-1}(C) \in \mathcal{B}(E \times F)$ and similarly, if $D \in \mathcal{B}(F)$, then $E \times D \in \mathcal{B}(E \times F)$. Therefore $C \times D = (C \times F) \cap (E \times D) \in \mathcal{B}(E \times F)$.

(ii) We now prove that $\mathcal{B}(E \times F) \subseteq \mathcal{B}(E) \otimes \mathcal{B}(F)$. By the separability assumption, there exists a dense countable subset of E. Consider the collection $\mathcal{U}$ consisting of all open balls with rational radius centered at some point of this dense set. It forms a base for the topology in the sense that any open set of E is the union of sets in $\mathcal{U}$. Let $\mathcal{V}$ be a similar base for F. To any $(x, y) \in O$, an open set of $E \times F$, one can associate an open set $U \times V$, where $U \in \mathcal{U}$ and $V \in \mathcal{V}$, that contains (x, y) and is contained in O. Therefore O is the union (at most countable) of sets of the form $U \times V$, where $U \in \mathcal{U}$ and $V \in \mathcal{V}$. In particular, every open set of $E \times F$ is measurable with respect to $\mathcal{B}(E) \otimes \mathcal{B}(F)$. $\quad\square$

Lemma 2.3.5 *Let $X = X_1 \times X_2$ and $\mathcal{X} = \mathcal{X}_1 \otimes \mathcal{X}_2$. Let $f : (X, \mathcal{X}) \to (\overline{\mathbb{R}}, \mathcal{B}(\overline{\mathbb{R}}))$. For fixed $x_1 \in X_1$, let the function $f_{x_1} : X \to \mathbb{R}$ be defined by $f_{x_1}(x_2) := f(x_1, x_2)$. Then f_{x_1} is a measurable function from $(X_2, \mathcal{X}_2)$ to $(\mathbb{R}, \mathcal{B}(\mathbb{R}))$.*

Proof.

STEP 1. We first prove the result for the special case $f = 1_F$, where $F \in \mathcal{X}$. Let for fixed $x_1 \in X_1$

$$F_{x_1} := \{x_2 \in X_2;\ (x_1, x_2) \in F\}.$$

(F_{x_1} is called the *section* of F at x_1.) We have $f_{x_1} = 1_{F_{x_1}}$. Therefore we have to prove that $F_{x_1} \in \mathcal{X}_2$. For this define

$$\mathcal{C}_{x_1} = \{F \subseteq X;\ F_{x_1} \in \mathcal{X}_2\}.$$

We want to show that $\mathcal{C}_{x_1} \supseteq \mathcal{X}$. For this, we first observe that $\mathcal{C}_{x_1}$ is a σ-field since it contains Ω and $\varnothing$, and

(i) if $F \in \mathcal{C}_{x_1}$, then $\overline{F} \in \mathcal{C}_{x_1}$. Indeed, since $F_{x_1} \in \mathcal{X}_2$, we have that $\overline{(F_{x_1})} \in \mathcal{X}_2$. But $\overline{(F_{x_1})} = (\overline{F})_{x_1}$. Therefore $(\overline{F})_{x_1} \in \mathcal{X}_2$.

(ii) if $F_n \in \mathcal{C}_{x_1}$ $(n \geq 1)$, then $\cup_{n \geq 1} F_n \in \mathcal{C}_{x_1}$. Indeed, $(F_n)_{x_1} \in \mathcal{X}_2$ and therefore $\cup_{n \geq 1} (F_n)_{x_1} \in \mathcal{X}_2$. But $\cup_{n \geq 1} (F_n)_{x_1} = (\cup_{n \geq 1} F_n)_{x_1}$. Therefore $(\cup_{n \geq 1} F_n)_{x_1} \in \mathcal{X}_2$.

The σ-field $\mathcal{C}_{x_1}$ contains all the rectangles $A \times B$, where $A \in \mathcal{X}_1$, $B \in \mathcal{X}_2$. Indeed, $(A \times B)_{x_1} = B$ if $x_1 \in A$, $= \varnothing$ if $x_1 \notin A$. We may now conclude that, since $\mathcal{C}_{x_1}$ is a σ-field containing the generators of $\mathcal{X}$, it contains $\mathcal{X}$.

STEP 2. Let now $f : (X, \mathcal{X}) \to (\overline{\mathbb{R}}_+, \mathcal{B}(\overline{\mathbb{R}}_+))$. It is the limit of some non-decreasing sequence $\{f_n\}_{n \geq 1}$ of non-negative simple functions. In particular,

$$f_{x_1} = \lim_{n \uparrow \infty} (f_n)_{x_1}.$$

It therefore suffices to prove that for any non-negative simple function $g = \sum_{i=1}^{k} a_i 1_{A_i}$, the function g_{x_1} is $\mathcal{X}_2$-measurable. This is true since

$$g_{x_1} = \sum_{i=1}^{k} a_i 1_{(A_i)_{x_1}}$$

and since the $(A_i)_{x_1} \in \mathcal{X}_2$ by the result in Step 1.

STEP 3. We now consider a general $f : (X, \mathcal{X}) \to (\overline{\mathbb{R}}, \mathcal{B}(\overline{\mathbb{R}}))$. We have, with the usual notation, $f = f^+ - f^-$, and therefore $f_{x_1} = (f^+)_{x_1} - (f^-)_{x_1}$. By Step 2, $(f^+)_{x_1}$ and $(f^-)_{x_1}$ are $\mathcal{X}_2$-measurable, and therefore so is f_{x_1}. $\square$

Lemma 2.3.6 Let $f : (X, \mathcal{X}) \to (\overline{\mathbb{R}}, \mathcal{B}(\overline{\mathbb{R}}))$ be a non-negative function. Then, if μ_2 is σ-finite, the function $x_1 \to \int_{X_2} f_{x_1}(x_2)\, \mu_2(dx_2)$ is measurable from $(X_1, \mathcal{X}_1)$ to $(\overline{\mathbb{R}}_+, \mathcal{B}(\overline{\mathbb{R}}_+))$.

Proof. For fixed $x_1 \in \mathcal{X}_1$, the integral $\int_{X_2} f_{x_1}\, d\mu_2$ is well defined since f_{x_1} is measurable (by the previous lemma) and non-negative. Observe that the conclusion of the lemma is true for $f = 1_{A \times B}$, where $A \in \mathcal{X}_1$ and $B \in \mathcal{X}_2$. Indeed, in this case,

$$\int_{X_2} f_{x_1}\, d\mu_2 = \int_{X_2} 1_A(x_1) 1_B(x_2)\, \mu_2(dx_2) = 1_A(x_1)\mu_2(B).$$

We now prove that the lemma is true for $f = 1_F$ when $F \in \mathcal{X}$. Let

$$g_F(x_1) := \int_{X_2} 1_F(x_1, x_2)\, \mu_2(dx_2) = \mu_2(F_{x_1}).$$

Consider the collection of sets

$$\mathcal{C} := \{F \in \mathcal{X} \,; g_F \text{ is } \mathcal{X}_1\text{-measurable}\}.$$

First suppose that μ_2 is a finite measure. In this case, $\mathcal{C}$ is a Dynkin system, since

(i) if A and B are in $\mathcal{C}$ and $A \subseteq B$, then $\mu_2((B - A)_{x_1}) = \mu_2(B_{x_1}) - \mu_2(A_{x_1})$ (this is where we need finiteness of μ_2) is $\mathcal{X}_1$-measurable, and

(ii) if $\{C_n\}_{n \geq 1}$ is a non-decreasing sequence,

$$\mu_2((\cup_{n \geq 1} C_n)_{x_1}) = \lim_{n \uparrow \infty} \mu_2((C_n)_{x_1})$$

is $\mathcal{X}_1$-measurable.

Since $\mathcal{C}$ contains the measurable rectangles $A \times B$, it contains $\mathcal{X}$, by Dynkin's theorem (Theorem 2.1.25).

If μ_2 is not finite, but only σ-finite, there exists a sequence $\{K_n\}_{n \geq 1}$ of elements of $\mathcal{X}_2$ increasing to X_2 and such that the measure $\mu_{2,n}$ defined by $\mu_{2,n}(A) = \mu_2(A \cup K_n)$ is finite. Then $\mu_2(F_{x_1}) = \lim_{n \geq 1} \mu_{2,n}(F_{x_1})$ is $\mathcal{X}_1$-measurable.

Finally, we pass from indicator functions of measurable sets to non-negative measurable functions by the usual monotone convergence argument. $\square$

Theorem 2.3.7 *Suppose μ_1 and μ_2 σ-finite. Then there exists a unique measure μ on $(X_1 \times X_2, \mathcal{X}_1 \times \mathcal{X}_2)$ such that*

$$\mu(A_1 \times A_2) = \mu_1(A_1)\mu_2(A_2) \tag{2.12}$$

for all $A_1 \in \mathcal{X}_1$, $A_2 \in \mathcal{X}_2$.

This measure, denoted by $\mu_1 \otimes \mu_2$, or $\mu_1 \times \mu_2$, is called the *product measure* of μ_1 by μ_2.

Proof. Existence. Consider the set function $\mu : \mathcal{X} \to [0, \infty]$ defined by

$$\mu(F) = \int_{X_1} \left(\int_{X_2} 1_F(x_1, x_2)\, \mu_2(dx_2) \right) \mu_1(dx_1).$$

It is a measure on $(X, \mathcal{X})$ (the monotone convergence theorem proves σ-additivity) that is obviously σ-finite and satisfies (2.12).

Uniqueness. Let $\mathcal{A}$ be the algebra consisting of the finite sums of disjoint measurable rectangles. Define (uniquely) the measure μ_0 on $(X, \mathcal{A})$ by $\mu_0(A_1 \times A_2) = \mu_1(A_1)\mu_2(A_2)$. By Carathéodory's theorem (Theorem 2.1.50), there exists a unique extension of μ_0 to $(X, \mathcal{X})$. $\qquad\square$

The above result extends in an obvious manner to a finite number of σ-finite measures.

EXAMPLE 2.3.8: LEBESGUE MEASURE ON $\mathbb{R}^n$. The typical example of a product measure is the Lebesgue measure on the space $(\mathbb{R}^n, \mathcal{B}(\mathbb{R}^n))$: It is the unique measure ℓ^n on that space that is such that

$$\ell^n(\Pi_{i=1}^n A_i) = \Pi_{i=1}^n \ell(A_i) \quad \text{for all } A_1, \dots, A_n \in \mathcal{B}(\mathbb{R}).$$

Going back to the situation with two measure spaces (the case of a finite number of measure spaces is similar) we have the following result:

Theorem 2.3.9 *Let $(X_1, \mathcal{X}_1, \mu_1)$ and $(X_1, \mathcal{X}_2, \mu_2)$ be two measure spaces in which μ_1 and μ_2 are σ-finite. Let $(X, \mathcal{X}, \mu) = (X_1 \times X_2, \mathcal{X}_1 \otimes \mathcal{X}_2, \mu_1 \otimes \mu_2)$.*

(A) Tonelli. If $f : (X, \mathcal{X}) \to (\overline{\mathbb{R}}, \mathcal{B}(\overline{\mathbb{R}}))$ is non-negative, then, for μ_1-almost all x_1, the function $x_2 \to f(x_1, x_2)$ is measurable with respect to $\mathcal{X}_2$, and

$$x_1 \to \int_{X_2} f(x_1, x_2)\, \mu_2(dx_2)$$

is a measurable function with respect to $\mathcal{X}_1$. Furthermore,

$$\int_X f \, d\mu = \int_{X_1} \left[\int_{X_2} f(x_1, x_2)\, \mu_2(dx_2) \right] \mu_1(dx_1)$$
$$= \int_{X_2} \left[\int_{X_1} f(x_1, x_2)\, \mu_2(dx_1) \right] \mu_1(dx_2). \tag{2.14}$$

(B) *Fubini. If $f : (X, \mathcal{X}) \to (\overline{\mathbb{R}}, \mathcal{B}(\overline{\mathbb{R}}))$ is μ-integrable, then, (a): for μ_1-almost all x_1, the function $x_2 \to f(x_1, x_2)$ is μ_2-integrable, and (b): $x_1 \to \int_{X_2} f(x_1, x_2)\,\mu_2(\mathrm{d}x_2)$ is μ_1-integrable, and (2.14) is true.*

The global result is referred to as the *Fubini–Tonelli* theorem.

Remark 2.3.10 Part A (Tonelli) says that one can integrate a non-negative $\mathcal{X}$-measurable function in any order of its variables. Part B (Fubini) says that the same is true of any $\mathcal{X}$-measurable function provided that function is μ-integrable. In general, in order to apply Part (B) one must use Part (A) with $f = |f|$ to ascertain whether or not $\int_X |f|\,\mathrm{d}\mu < \infty$.

Proof.

(A) The σ-finite measures

$$\nu(F) = \int_X 1_F\,\mathrm{d}\mu$$

and

$$\mu(F) = \int_{X_1} \left(\int_{X_2} 1_F(x_1, x_2)\,\mu_2(\mathrm{d}x_2) \right) \mu_1(\mathrm{d}x_1)$$

coincide on the algebra $\mathcal{A}$ consisting of the finite sums of disjoint generalized measurable rectangles. They are therefore identical, by Theorem 2.1.42. Therefore we have proved the theorem for f of the form 1_F, $F \in \mathcal{X}$. The general case of a non-negative measurable function is obtained by the usual monotone convergence argument.

(B) Since f is μ-integrable, by Tonelli's theorem,

$$\int_{X_1} \left[\int_{X_2} |f(x_1, x_2)|\,\mu_2(\mathrm{d}x_2) \right] \mu_1(\mathrm{d}x_1) < \infty$$

and in particular

$$\int_{X_2} |f(x_1, x_2)|\,\mu_2(\mathrm{d}x_2) < \infty\,, \mu_1\text{-a.e.}$$

Therefore, outside a μ_1-negligible set N_1,

$$\int_{X_2} f^{\pm}(x_1, x_2)\,\mu_2(\mathrm{d}x_2) < \infty\,.$$

We may suppose that the above inequalities are true everywhere because we may replace f, without changing its integral with respect to μ, by a function μ-almost everywhere equal to f. By Tonelli,

$$\int_X f^{\pm}\,\mathrm{d}\mu = \int_{X_1} \left[\int_{X_2} f^{\pm}(x_1, x_2)\,\mu_2(\mathrm{d}x_2) \right] \mu_1(\mathrm{d}x_1)$$

and therefore

$$\begin{aligned}
\int_X f\,\mathrm{d}\mu &= \int_X f^+\,\mathrm{d}\mu - \int_X f^-\,\mathrm{d}\mu \\
&= \int_{X_1} \left[\int_{X_2} f^+(x_1, x_2)\,\mu_2(\mathrm{d}x_2) \right] \mu_1(\mathrm{d}x_1) - \int_{X_1} \left[\int_{X_2} f^-(x_1, x_2)\,\mu_2(\mathrm{d}x_2) \right] \mu_1(\mathrm{d}x_1) \\
&= \int_{X_1} \left[\left(\int_{X_2} f^+(x_1, x_2)\,\mu_2(\mathrm{d}x_2) - \int_{X_2} f^-(x_1, x_2)\,\mu_2(\mathrm{d}x_2) \right) \right] \mu_1(\mathrm{d}x_1) \\
&= \int_{X_1} \left[\int_{X_2} f(x_1, x_2)\,\mu_2(\mathrm{d}x_2) \right] \mu_1(\mathrm{d}x_1)\,.
\end{aligned}$$

(All this fuss guarantees that at every step we do not encounter $\infty - \infty$ forms.) $\quad\square$

Remark 2.3.11 Is the hypothesis of σ-finiteness superfluous? In fact it is not, as the following counterexample shows. Take $(X_i, \mathcal{X}_i) = (\mathbb{R}, \mathcal{B}(\mathbb{R}))$ $(i = 1, 2)$, let $\mu_1 = \ell$, the Lebesgue measure, and let $\mu_2 = \nu$, the measure that associates with a measurable set its cardinality (only finite sets have a finite measure, and therefore, since there is no sequence of finite sets increasing to $\mathbb{R}$, this measure is not σ-finite). Now, let $C = \{(x, x); x \in \mathbb{R}\}$ (the diagonal of $\mathbb{R}^2$). Clearly $C_{x_1} = x_1$ and $C_{x_2} = x_2$, so that $\mu_2(C_{x_1}) = 1$ and therefore

$$\int_{X_1} \mu_2(C_{x_1})\,\mu_1(\mathrm{d}x_1) = \int_{\mathbb{R}} 1\,\ell(\mathrm{d}x) = \infty.$$

On the other hand, $\mu_1(C_{x_2}) = 0$, and therefore,

$$\int_{X_2} \mu_1(C_{x_2})\,\mu_2(\mathrm{d}x_2) = \int_{\mathbb{R}} 0\,\nu(\mathrm{d}x) = 0.$$

Integration by Parts Formula

Theorem 2.3.12 *Let μ_1 and μ_2 be two σ-finite measures on $(\mathbb{R}, \mathcal{B}(\mathbb{R}))$. For any interval $(a, b) \subseteq \mathbb{R}$*

$$\mu_1((a, b])\mu_2((a, b]) = \int_{(a,b]} \mu_1((a, t])\,\mu_2(\mathrm{d}t) + \int_{(a,b]} \mu_2((a, t))\,\mu_1(\mathrm{d}t). \tag{2.15}$$

Observe that the first integral features the interval $(a, t]$ (closed on the right), whereas in the second integral, the interval is of the type (a, t) (open on the right).

Proof. The proof consists in computing the $\mu_1 \times \mu_2$-measure of the square $D := (a, b] \times (a, b]$ in two ways. The first one is obvious and gives the left-hand side of (2.15). The second one consists in observing that $\mu(D) = \mu(D_1) + \mu(D_2)$, where $D_1 = \{(x, y); a < y \leq b, a < x \leq y\}$ and $D_2 = \{(a, b] \times (a, b]\} \cap \overline{D_1}$. Then $\mu(D_1)$ and $\mu(D_2)$ are computed using Tonelli's theorem. For instance,

$$\mu(D_1) = \int_{\mathbb{R}} \left(\int_{\mathbb{R}} 1_{D_1}(x, y)\mu_1(\mathrm{d}x) \right) \mu_2(\mathrm{d}y)$$

$$\int_{\mathbb{R}} \left(\int_{\mathbb{R}} 1_{\{a < x \leq y\}}\mu_1(\mathrm{d}x) \right) \mu_2(\mathrm{d}y) = \int_{\mathbb{R}} \mu_1((a, y])\,\mu_2(\mathrm{d}y).$$

$\square$

In terms of Lebesgue–Stieltjes integrals,

$$F_1(b)F_2(b) - F_1(a)F_1(a) = \int_{(a,b]} F_1(x)\,\mathrm{d}F_2(x) + \int_{(a,b]} F_2(x-)\,\mathrm{d}F_1(x),$$

where F_1 and F_2 are CDFs on $\mathbb{R}$. This is the Lebesgue–Stieltjes version of the integration by parts formula of Calculus.

Here are some simple examples involving counting measures on $\mathbb{Z}$.

EXAMPLE 2.3.13: FUBINI FOR THE COUNTING MEASURE. Applied to the product of two counting measures on $\mathbb{Z}$, Fubini's theorem deals with the problem of interchanging the order of summations. It says (observing that almost everywhere relatively to the counting measure in fact means everywhere since for such measure, the only set of measure 0 is the empty set): Let $\{a_{k,n}\}_{k,n \in \mathbb{Z}}$ be a doubly indexed sequence of real numbers. Then if this sequence is absolutely summable, that is, if

$$\sum_{k,n \in \mathbb{Z}} |a_{k,n}| < \infty,$$

then (a): the sum $\sum_{k,n \in \mathbb{Z}} a_{k,n}$ is well defined, (b): for each $n \in \mathbb{Z}$,

$$\sum_{k \in \mathbb{Z}} |a_{k,n}| < \infty$$

and (c):

$$\sum_{k,n \in \mathbb{Z}} a_{k,n} = \sum_{n \in \mathbb{Z}} \left(\sum_{k \in \mathbb{Z}} a_{k,n} \right) = \sum_{k \in \mathbb{Z}} \left(\sum_{n \in \mathbb{Z}} a_{k,n} \right).$$

If the terms of the doubly indexed sequence are non-negative, the latter equality holds without the absolute summability condition. Of course, these results can be obtained without recourse to integration theory.

EXAMPLE 2.3.14: INTEGRAL OF SUMS OF FUNCTIONS. Let $\{f_n\}_{n \in \mathbb{N}}$ be a sequence of measurable functions $f_n : \mathbb{R} \to \mathbb{C}$. Applying Fubini's theorem with the product of the Lebesgue measure on $\mathbb{R}$ by the counting measure on $\mathbb{Z}$ yields: Under the condition

$$\int_{\mathbb{R}} \left(\sum_{n \in \mathbb{Z}} |f_n(t)| \right) dt < \infty, \tag{2.16}$$

for almost all $t \in \mathbb{R}$ (with respect to the Lebesgue measure),

$$\sum_{n \in \mathbb{Z}} |f_n(t)| \, dt < \infty$$

and

$$\int_{\mathbb{R}} \left(\sum_{n \in \mathbb{Z}} f_n(t) \right) dt = \sum_{n \in \mathbb{Z}} \left(\int_{\mathbb{R}} f_n(t) \, dt \right).$$

If the f_n's are non-negative, the latter equality holds without condition (2.16) of the Fubini–Tonelli theorem.

Remark 2.3.15 Note that the result of Example 2.3.14 can be obtained from the monotone and the dominated convergence theorems, and that these theorems were in fact used in the proof of the Fubini–Tonelli theorem.

2.3.3   The Riesz–Fischer Theorem

For a given integer $p \geq 1$, $L^p_{\mathbb{C}}(\mu)$ is, roughly speaking (see the details below), the collection of complex-valued measurable functions f defined on X such that $\int_X |f|^p \, d\mu < \infty$. We shall see that it is a complete normed vector space over $\mathbb{C}$, that is, a Banach space.

Let $(X, \mathcal{X}, \mu)$ be a measure space and let f, g be two complex-valued measurable functions defined on X. The relation $\mathcal{R}$ defined by

$$f \mathcal{R} g \text{ if and only if } f = g \ \mu\text{-a.e.}$$

is an equivalence relation. Denote the equivalence class of f by $\{f\}$. Note that for any $p > 0$ (using property (b) of Theorem 2.2.12),

$$f \mathcal{R} g \implies \int_X |f|^p \, d\mu = \int_X |g|^p \, d\mu .$$

The operations $+$, $\times$, * and multiplication by a scalar $\alpha \in \mathbb{C}$ are defined on the equivalence class by

$$\{f\} + \{g\} = \{f + g\}, \ \ \{f\}\{g\} = \{fg\}, \ \ \{f\}^* = \{f^*\}, \ \ \alpha \{f\} = \{\alpha f\} .$$

The first equality means that $\{f\} + \{g\}$ is, by definition, the equivalence class consisting of the functions $f + g$, where f and g are members of $\{f\}$ and $\{g\}$, respectively. Similar interpretations hold for the other equalities.

By definition, for a given $p \geq 1$, $L^p_{\mathbb{C}}(\mu)$ is the collection of equivalence classes $\{f\}$ such that $\int_X |f|^p \, d\mu < \infty$. Clearly it is a vector space over $\mathbb{C}$ (for the proof recall that

$$\left(\frac{|f| + |g|}{2} \right)^p \leq \tfrac{1}{2} |f|^p + \tfrac{1}{2} |g|^p$$

since $t \to t^p$ is a convex function when $p \geq 1$). In order to avoid cumbersome notation, in this section and in general whenever we consider L^p-spaces, we shall write f for $\{f\}$. This abuse of notation is harmless since two members of the same equivalence class have the same integral if that integral is defined. Therefore, using this loose notation, we may write

$$L^p_{\mathbb{C}}(\mu) = \left\{ f : \int_X |f|^p \, d\mu < \infty \right\} . \tag{2.17}$$

When the measure is the counting measure on the set $\mathbb{Z}$ of relative integers, the traditional notation is $\ell^p_{\mathbb{C}}(\mathbb{Z})$. This is the space of random complex sequences $\{x_n\}_{n \in \mathbb{Z}}$ such that

$$\sum_{n \in \mathbb{Z}} |x_n|^p < \infty.$$

The following is a simple and often used observation.

Theorem 2.3.16 *Let p and q be positive real numbers such that $p > q$. If the measure μ on $(X, \mathcal{X}, \mu)$ is finite, then $L^p_{\mathbb{C}}(\mu) \subseteq L^q_{\mathbb{C}}(\mu)$. In particular, $L^2_{\mathbb{C}}(\mu) \subseteq L^1_{\mathbb{C}}(\mu)$.*

Proof. From the inequality $|a|^q \leq 1 + |a|^p$, true for all $a \in \mathbb{C}$, it follows that $\mu(|f|^q) \leq \mu(1) + \mu(|f|^p)$. Since $\mu(1) = \mu(\mathbb{R}) < \infty$, $\mu(|f|^q) < \infty$ whenever $\mu(|f|^p) < \infty$. $\qquad \square$

Remark 2.3.17 This inclusion is not true in general if μ is not a finite measure, for instance consider the Lebesgue measure ℓ on $\mathbb{R}$: there exist functions in $L^1_{\mathbb{C}}(\ell)$ that are not in $L^2_{\mathbb{C}}(\ell)$ and vice versa (Exercise 2.4.18).

In the case of the (not finite) counting measure on $\mathbb{Z}$, the order of inclusion is the reverse of the one concerning finite measures:

Theorem 2.3.18 $\ell^p_{\mathbb{C}}$ *inclusions. If* $p > q$, $\ell^q_{\mathbb{C}}(\mathbb{Z}) \subset \ell^p_{\mathbb{C}}(\mathbb{Z})$. *In particular,* $\ell^1_{\mathbb{C}}(\mathbb{Z}) \subset \ell^2_{\mathbb{C}}(\mathbb{Z})$.

Proof. Exercise 2.4.19

$\square$

Hölder's Inequality

Theorem 2.3.19 *Let p and q be positive real numbers in $(0, 1)$ such that*

$$\frac{1}{p} + \frac{1}{q} = 1$$

(p and q are then said to be conjugate) and let $f, g : (X, \mathcal{X}) \mapsto (\mathbb{R}, \mathcal{B}(\mathbb{R}))$ be non-negative real functions. Then,

$$\int_X fg\,d\mu \leq \left[\int_X f^p\,d\mu\right]^{1/p} \left[\int_X g^q\,d\mu\right]^{1/q}. \tag{2.18}$$

In particular, if $f, g \in L^2_{\mathbb{C}}(\mathbb{R})$, then $fg \in L^1_{\mathbb{C}}(\mathbb{R})$.

Proof. Let

$$A = \left(\int_X f^p\,d\mu\right)^{1/p}, \quad B = \left(\int_X g^q\,d\mu\right)^{1/q}.$$

It may be assumed that $0 < A, B < \infty$, because otherwise Hölder's inequality is trivially satisfied. Let $F := f/A$, $G := g/B$, so that

$$\int_X F^p\,d\mu = \int_X G^q\,d\mu = 1.$$

Suppose that we have been able to prove that

$$F(x)G(x) \leq \frac{1}{p}F(x)^p + \frac{1}{q}G(x)^q. \tag{2.19}$$

Integrating this inequality yields

$$\int_X (FG)\,d\mu \leq \frac{1}{p} + \frac{1}{q} = 1,$$

and this is just (2.18).

Inequality (2.19) is trivially satisfied if x is such that $F \equiv 0$ or $G \equiv 0$. It is also satisfied in the case when F and G are not μ-almost everywhere null. Indeed, letting

$$s(x) := p\ln(F(x)), \qquad t(x) := q\ln(G(x)),$$

from the convexity of the exponential function and the assumption that $1/p + 1/q = 1$,

$$e^{s(x)/p + t(x)/q} \leq \frac{1}{p}e^{s(x)} + \frac{1}{q}e^{t(x)},$$

and this is precisely inequality (2.19).

For the last assertion of the theorem, take $p = q = 2$.

$\square$

Minkowski's Inequality

Theorem 2.3.20 *Let $p \geq 1$ and let $f, g : (X, \mathcal{X}) \mapsto (\mathbb{R}, \mathcal{B}(\mathbb{R}))$ be non-negative functions in $L^p_{\mathbb{C}}(\mu)$. Then,*

$$\left[\int_X (f + g)^p\right]^{1/p} \leq \left[\int_X f^p \, d\mu\right]^{1/p} + \left[\int_X g^p \, d\mu\right]^{1/p} . \tag{2.20}$$

Proof. For $p = 1$ the inequality (in fact an equality) is obvious. Therefore, assume $p > 1$. From Hölder's inequality

$$\int_X f(f + g)^{p-1} \, d\mu \leq \left[\int_X f^p \, d\mu\right]^{1/p} \left[\int_X (f + g)^{(p-1)q}\right]^{1/q}$$

and

$$\int_X g(f + g)^{p-1} \, d\mu \leq \left[\int_X g^p \, d\mu\right]^{1/p} \left[\int_X (f + g)^{(p-1)q}\right]^{1/q} .$$

Adding up the above two inequalities and observing that $(p - 1)q = p$, we obtain

$$\int_X (f + g)^p \, d\mu \leq \left(\left[\int_X f^p \, d\mu\right]^{1/p} + \left[\int_X g^p \, d\mu\right]^{1/p}\right) \left[\int_x (f + g)^p\right]^{1/q} .$$

One may assume that the right-hand side of (2.20) is finite and that the left-hand side is positive (otherwise the inequality is trivial). Therefore $\int_X (f + g)^p \, d\mu \in (0, \infty)$ and we may therefore divide both sides of the last display by $\left[\int_X (f + g)^p \, d\mu\right]^{1/q}$. Observing that $1 - 1/q = 1/p$ yields the announced inequality (2.20). $\qquad \square$

Theorem 2.3.21 *Let $p \geq 1$. The mapping $\nu_p : L^p_{\mathbb{C}}(\mu) \mapsto [0, \infty)$ defined by*

$$\nu_p(f) := \left(\int_X |f|^p \, d\mu\right)^{1/p} \tag{2.21}$$

is a norm on $L^p_{\mathbb{C}}(\mu)$.

Proof. Clearly, $\nu_p(\alpha f) = |\alpha| \nu_p(f)$ for all $\alpha \in \mathbb{C}$, $f \in L^p_{\mathbb{C}}(\mu)$. Also, $\nu_p(f) = 0$ if and only if $\left(\int_X |f|^p \, d\mu\right)^{1/p} = 0$, which in turn is equivalent to $f = 0, \mu$-a.e. Finally, $\nu_p(f + g) \leq \nu_p(f) + \nu_p(g)$ for all $f, g \in L^p_{\mathbb{C}}(\mu)$, by Minkowski's inequality. $\qquad \square$

The Riesz–Fischer Theorem

Denoting $\nu_p(f)$ by $\|f\|_p$, $L^p_{\mathbb{C}}(\mu)$ is a normed vector space over $\mathbb{C}$, with the norm $\| \cdot \|_p$ and the induced metric $d_p(f, g) := \|f - g\|_p$.

Theorem 2.3.22 *Let $p \geq 1$. The metric d_p makes of $L^p_{\mathbb{C}}(\mu)$ a complete normed vector space.*

In other words, $L^p_{\mathbb{C}}(\mu)$ is a *Banach space* for the norm $\| \cdot \|_p$.

Proof. To show completeness one must prove that for any sequence $\{f_n\}_{n \geq 1}$ of $L^p_{\mathbb{C}}(\mu)$ that is a Cauchy sequence (that is, such that $\lim_{m,n \uparrow \infty} d_p(f_n, f_m) = 0$), there exists an $f \in L^p_{\mathbb{C}}(\mu)$ such that $\lim_{n \uparrow \infty} d_p(f_n, f) = 0$.

Since $\{f_n\}_{n \geq 1}$ is a Cauchy sequence, one can select a subsequence $\{f_{n_i}\}_{i \geq 1}$ such that

$$d_p(f_{n_{i+1}} - f_{n_i}) \leq 2^{-i}. \tag{2.22}$$

Let

$$g_k = \sum_{i=1}^{k} |f_{n_{i+1}} - f_{n_i}|, \, g = \sum_{i=1}^{\infty} |f_{n_{i+1}} - f_{n_i}|.$$

By (2.22) and Minkowski's inequality, $\|g_k\|_p \leq 1$. Fatou's lemma applied to the sequence $\{g_k^p\}_{k \geq 1}$ gives $\|g\|_p \leq 1$. In particular, any member of the equivalence class of g is μ-almost everywhere finite and therefore

$$f_{n_1}(x) + \sum_{i=1}^{\infty} \left(f_{n_{i+1}}(x) - f_{n_i}(x) \right)$$

converges absolutely for μ-almost all x. Call the corresponding limit $f(x)$ (set $f(x) = 0$ when this limit does not exist). Since

$$f_{n_1} + \sum_{i=1}^{k-1} \left(f_{n_{i+1}} - f_{n_i} \right) = f_{n_k}$$

we see that

$$f = \lim_{k \uparrow \infty} f_{n_k} \quad \mu\text{-a.e.}$$

One must show that f is the limit in $L_{\mathbb{C}}^p(\mu)$ of $\{f_{n_k}\}_{k \geq 1}$. Let $\epsilon > 0$. There exists an integer $N = N(\epsilon)$ such that $\|f_n - f_m\|_p \leq \epsilon$ whenever $m, n \geq N$. For all $m > N$, by Fatou's lemma we have

$$\int_X |f - f_m|^p \, d\mu \leq \liminf_{i \to \infty} \int_x |f_{n_i} - f_m|^p \, d\mu \leq \epsilon^p.$$

Therefore $f - f_m \in L_{\mathbb{C}}^p(\mu)$ and consequently $f \in L_{\mathbb{C}}^p(\mu)$. It also follows from the last inequality that

$$\lim_{m \to \infty} \|f - f_m\|_p = 0.$$

$\square$

The next result is a by-product of the proofs of Theorems 2.3.22.

Theorem 2.3.23 *Let $p \geq 1$ and let $\{f_n\}_{n \geq 1}$ be a convergent sequence in $L_{\mathbb{C}}^p(\mu)$. Let f be the corresponding limit in $L_{\mathbb{C}}^p(\mu)$. Then, there exists a subsequence $\{f_{n_i}\}_{i \geq 1}$ such that*

$$\lim_{i \uparrow \infty} f_{n_i} = f \quad \mu\text{-a.e.} \tag{2.23}$$

Note that the statement in (2.23) is about functions and not about equivalence classes. The functions thereof are *any* members of the corresponding equivalence class. In particular, when a given sequence of functions converges μ-a.e. to two functions, these two functions are necessarily equal μ-a.e. Therefore,

Theorem 2.3.24 *If $\{f_n\}_{n \geq 1}$ converges both to f in $L_{\mathbb{C}}^p(\mu)$ and to g μ-a.e., then $f = g$ μ-a.e.*

Of special interest for applications is the space $L^2_{\mathbb{C}}(\mu)$ of complex measurable functions $f : X \to \mathbb{R}$ such that

$$\int_X |f(x)|^2 \, \mu(\mathrm{d}x) < \infty,$$

where two functions f and f' such that $f(x) = f'(x)$, μ-a.e. are not distinguished. We have by the *Riesz–Fischer theorem*:

Theorem 2.3.25 $L^2_{\mathbb{C}}(\mu)$ *is a vector space with scalar field* $\mathbb{C}$, *and when endowed with the inner product*

$$\langle f, g \rangle := \int_X f(x)g(x)^* \, \mu(\mathrm{d}x), \tag{2.24}$$

it is a Hilbert space.

The norm of a function $f \in L^2_{\mathbb{C}}(\mu)$ is

$$\|f\| = \left(\int_X |f(x)|^2 \, \mu(\mathrm{d}x) \right)^{\frac{1}{2}}$$

and the distance between two functions f and g in $L^2_{\mathbb{C}}(\mu)$ is

$$d(f, g) = \left(\int_X |f(x) - g(x)|^2 \, \mu(\mathrm{d}x) \right)^{\frac{1}{2}}.$$

The completeness property of $L^2_{\mathbb{C}}(\mu)$ reads in this case as follows. If $\{f_n\}_{n \geq 1}$ is a sequence of functions in $L^2_{\mathbb{C}}(\mu)$ such that

$$\lim_{m,n \uparrow \infty} \int_X |f_n(x) - f_m(x)|^2 \, \mu(\mathrm{d}x) = 0,$$

then, there exists a function $f \in L^2_{\mathbb{C}}(\mu)$ such that

$$\lim_{n \uparrow \infty} \int_X |f_n(x) - f(x)|^2 \, \mu(\mathrm{d}x) = 0.$$

In $L^2_{\mathbb{C}}(\mu)$, Schwarz's inequality reads as follows:

$$\left| \int_X f(x)g(x)^* \, \mu(\mathrm{d}x) \right| \leq \left(\int_X |f(x)|^2 \, \mu(\mathrm{d}x) \right)^{\frac{1}{2}} \left(\int_X |g(x)|^2 \, \mu(\mathrm{d}x) \right)^{\frac{1}{2}}.$$

EXAMPLE 2.3.26: COMPLEX SEQUENCES. The set of complex sequences $a = \{a_n\}_{n \in \mathbb{Z}}$ such that

$$\sum_{n \in \mathbb{Z}} |a_n|^2 < \infty$$

is, when endowed with the inner product

$$\langle a, b \rangle = \sum_{n \in \mathbb{Z}} a_n b_n^*,$$

a Hilbert space, denoted by $\ell^2_{\mathbb{C}}(\mathbb{Z})$. This is indeed a particular case of a Hilbert space $L^2_{\mathbb{C}}(\mu)$, where $X = \mathbb{Z}$ and μ is the counting measure. In this example, Schwarz's inequality takes the form

$$\left| \sum_{n \in \mathbb{Z}} a_n b_n^* \right| \leq \left(\sum_{n \in \mathbb{Z}} |a_n|^2 \right)^{\frac{1}{2}} \times \left(\sum_{n \in \mathbb{Z}} |b_n|^2 \right)^{\frac{1}{2}}.$$

2.3.4   The Radon–Nikodým Theorem

The Product of a Measure by a Function

Definition 2.3.27 *Let $(X, \mathcal{X}, \mu)$ be a measure space and let $h : (X, \mathcal{X}) \to (\overline{\mathbb{R}}, \mathcal{B}(\overline{\mathbb{R}}))$ be a non-negative measurable function. Define the set function $\nu : \mathcal{X} \to [0, \infty]$ by*

$$\nu(C) = \int_C h(x)\, \mu(\mathrm{d}x)\,.$$

Then ν is a measure on $(X, \mathcal{X})$ called the product of μ by the function h. This is denoted by $\mathrm{d}\nu = h\, \mathrm{d}\mu$.

That ν is a measure is easily checked. First of all, it is obvious that $\nu(\varnothing) = 0$. As for the σ-additivity property, write for any sequence of mutually disjoint measurable sets $\{A_n\}_{n \geq 1}$,

$$
\begin{aligned}
\nu(\cup_{n \geq 1} A_n) &= \int_{\cup_{n \geq 1} A_n} h\, \mathrm{d}\mu = \int_X 1_{\cup_{n \geq 1} A_n} h\, \mathrm{d}\mu \\
&= \int_X \left(\sum_{n \geq 1} 1_{A_n} \right) h\, \mathrm{d}\mu = \int_X \left(\lim_{k \uparrow \infty} \sum_{n=1}^k 1_{A_n} \right) h\, \mathrm{d}\mu \\
&= \lim_{k \uparrow \infty} \int_X \left(\sum_{n=1}^k 1_{A_n} \right) h\, \mathrm{d}\mu = \lim_{k \uparrow \infty} \sum_{n=1}^k \int_X 1_{A_n} h\, \mathrm{d}\mu \\
&= \lim_{k \uparrow \infty} \sum_{n=1}^k \nu(A_n) = \sum_{n \geq 1} \nu(A_n)\,,
\end{aligned}
$$

where the fifth equality is by monotone convergence.

Theorem 2.3.28 *Let μ, h and ν be as in Definition 2.3.27.*

(i) For non-negative $f : (X, \mathcal{X}) \to (\overline{\mathbb{R}}, \mathcal{B}(\overline{\mathbb{R}}))$,

$$\int_X f(x)\, \nu(\mathrm{d}x) = \int_X f(x) h(x)\, \mu(\mathrm{d}x)\,. \tag{2.25}$$

(ii) If $f : (X, \mathcal{X}) \to (\overline{\mathbb{R}}, \mathcal{B}(\overline{\mathbb{R}}))$ has arbitrary sign, then either one of the following conditions

(a) f is ν-integrable,

(b) fh is μ-integrable,

implies the other, and the equality (2.25) then holds.

Proof. Verify (2.25) for elementary non-negative functions and, approximating f by a non-decreasing sequence of such functions, use the monotone convergence theorem as in the proof of (2.11). For the case of functions of arbitrary sign, apply (2.25) with $f = f^+$ and $f = f^-$. $\qquad \square$

Observe that in the situation of Theorem 2.3.28, for all $C \in \mathcal{X}$,

$$\mu(C) = 0 \implies \nu(C) = 0\,. \tag{2.26}$$

Definition 2.3.29 *Let μ and ν be two measures on $(X, \mathcal{X})$.*

(A) If (2.26) holds for all $C \in \mathcal{X}$, ν is said to be absolutely continuous *with respect to μ. This is denoted by $\nu \ll \mu$.*

(B) The measures μ and ν on $(X, \mathcal{X})$ are said to be mutually singular *if there exists a set $A \in \mathcal{X}$ such $\nu(A) = \mu(\overline{A}) = 0$. This is denoted by $\mu \perp \nu$.*

Lebesgue's decomposition

Theorem 2.3.30 *Let μ and ν be two σ-finite measures on $(X, \mathcal{X})$. There exists a unique decomposition (called the* Lebesgue decomposition*)*

$$\nu = \nu_a + \nu_s$$

such that

$$\nu_a \ll \mu \text{ and } \nu_a \perp \mu ,$$

and a non-negative measurable function $g : X \to \mathbb{R}$ such that

$$d\nu_a = g \, d\mu ,$$

this function being μ-essentially unique. [3]

Proof. STEP 1. We first assume that μ and ν are finite and that $\nu \leq \mu$, that is, $\nu(A) \leq \mu(A)$ for all $A \in \mathcal{X}$. Define a mapping $\varphi : L^2_\mathbb{R}(\mu) \to \mathbb{R}$ by

$$\varphi(f) = \int_X f \, d\nu .$$

The latter integral is well defined since the hypothesis of finiteness of μ implies that $L^2_\mathbb{R}(\mu) \subseteq L^1_\mathbb{R}(\mu)$, and hypothesis $\nu \leq \mu$ implies that $\int_\mathbb{R} |f| \, d\nu \leq \int_\mathbb{R} |f| \, d\mu$. Also $\varphi(f)$ does not depend on the function chosen in the equivalence class of $L^2_\mathbb{R}(\mu)$. In fact, letting f' be another such function, $f = f'$ μ-a.e. implies that $f = f'$ ν-a.e. and then $\int_X f \, d\nu = \int_X f' \, d\nu$.

By Schwarz's inequality,

$$|\varphi(f)| \leq \left(\int_X f^2 \, d\nu \right)^{\frac{1}{2}} \nu(X)^{\frac{1}{2}}$$

$$\leq \left(\int_X f^2 \, d\mu \right)^{\frac{1}{2}} \nu(X)^{\frac{1}{2}} = \nu(X)^{\frac{1}{2}} \|f\|_{L^2_\mathbb{R}(\mu)} .$$

Therefore, the (linear) functional φ from the Hilbert space $L^2_\mathbb{R}(\mu)$ to $\mathbb{R}$ is continuous. By Riesz's theorem on the representation of linear functionals on $L^2_\mathbb{R}(\mu)$ (Theorem C.5.2), there exists a $g \in L^2_\mathbb{R}(\mu)$ such that

$$\varphi(f) = \langle f, g \rangle_{L^2_\mathbb{R}(\mu)} := \int_X f g \, d\mu ,$$

that is,

$$\int_X f \, d\nu = \int_X f g \, d\mu .$$

---

[3]If g' is such that $d\nu_a = g' \, d\mu$, then $g(x) = g'(x)$ μ-a.e.

In particular, $\nu(A) = \int_A g \, d\mu$ for all $A \in \mathcal{X}$. With $A = \{g \geq 1 + \varepsilon\}$ where $\varepsilon > 0$,

$$\mu(\{g \geq 1 + \varepsilon\}) \geq \nu(\{g \geq 1 + \varepsilon\})$$
$$= \int_{\{g \geq 1 + \varepsilon\}} g \, d\mu \geq (1 + \varepsilon)\mu(\{g \geq 1 + \varepsilon\}),$$

and therefore $\mu(\{g \geq 1 + \varepsilon\}) = 0$. Since ε is an arbitrary positive number, this implies that $g \leq 1$ μ-a.e. A similar argument shows that $g \geq 0$ μ-a.e. We may in fact suppose that $0 \leq g(x) \leq 1$ for *all* $x \in X$ by replacing if necessary g by $g \, 1_{\{0 \leq g \leq 1\}}$.

STEP 2. We still assume that μ and ν are finite, but not that $\nu \leq \mu$. However, since $\nu \leq \mu + \nu$, we may apply the above results to ν and $\mu + \nu$, to obtain the existence of a measurable function g such that $0 \leq g \leq 1$ and such that for all $f \in L^2_{\mathbb{R}}(\mu + \nu)$,

$$\int_X f \, d\nu = \int_X f g \, d(\mu + \nu).$$

In particular, for any bounded measurable function $f : X \to \mathbb{R}$,

$$\int_X f (1 - g) \, d\nu = \int_X f g \, d\mu. \qquad (\star)$$

By monotone convergence, this inequality extends to all non-negative measurable functions f.

STEP 3. With $f = 1_N$ in $(\star)$, where $N := \{g = 1\}$, we have that $\int_N f \, d\mu = 0$ for all non-negative measurable f, and therefore $\mu(N) = 0$. The measures $\nu_s := 1_{\overline{N}} \nu$ and μ are therefore mutually singular.

STEP 4. Replacing in $(\star)$ f by $\frac{f}{1-g} 1_{\overline{N}}$ gives that for all non-negative measurable functions f,

$$\int_{\overline{N}} f \, d\nu = \int_X f h \, d\mu,$$

where

$$h := 1_{\overline{N}} \frac{g}{1 - g}.$$

It remains to define ν_a by

$$d\nu_a := 1_{\overline{N}} \, d\nu = h \, d\mu$$

to conclude the existence part of the theorem, under the assumption that μ and ν are finite.

STEP 5. To prove the uniqueness of the pair (ν_a, ν_s), consider another such pair $(\widetilde{\nu}_a, \widetilde{\nu}_s)$. For all $A \in \mathcal{X}$,

$$\nu_a(A) - \widetilde{\nu}_a(A) = -\nu_s(A) + \widetilde{\nu}_s(A). \qquad (\dagger)$$

Since the supports N and $\widetilde{N}$ of ν_s and $\widetilde{\nu}_s$ respectively have a null μ-measure, and since $\nu_a \ll \mu$ and $\widetilde{\nu}_a \ll \mu$,

$$\nu_s(A) - \widetilde{\nu}_s(A) = \nu_s(A \cap (N \cup \widetilde{N})) - \widetilde{\nu}_s(A \cap (N \cup \widetilde{N}))$$
$$= -\nu_a(A \cap (N \cup \widetilde{N})) + \widetilde{\nu}_a(A \cap (N \cup \widetilde{N})) = 0.$$

Therefore $\nu_a \equiv \widetilde{\nu}_a$, and consequently, from $(\dagger)$, $\nu_s \equiv \widetilde{\nu}_s$.

STEP 6. To prove the uniqueness of h, just observe that if another measurable non-negative function $\widetilde{h}$ satisfies

$$\nu_a(A) = \int_A \widetilde{h}\,\mathrm{d}\mu = \int_A h\,\mathrm{d}\mu$$

for all $A \in \mathcal{X}$, then necessarily $\widetilde{h} = h$ μ-a.e.

STEP 7. We get rid of the finiteness hypothesis for μ and ν, only assuming that these measures are σ-finite. Therefore, there exists a measurable partition $\{K_n\}_{n\geq 1}$ of X such that $\mu_n := 1_{K_n}\mu$ and $\nu_n := 1_{K_n}\nu$ are finite measures. Applying the above results to μ_n and ν_n, and calling $\nu_{a,n}$, $\nu_{s,n}$ and h_n the corresponding items of the decomposition, define

$$\nu_a = \sum_{n\geq 1}\nu_{a,n}, \quad \nu_s = \sum_{n\geq 1}\nu_{s,n}, \quad h = \sum_{n\geq 1}h_n\,.$$

The verification that ν_a, ν_s and h satisfy the requirement of the theorem is straightforward. $\qquad\square$

The Radon–Nikodým Derivative

Corollary 2.3.31 *Let μ and ν be two σ-finite measures on $(X,\mathcal{X})$ such that $\nu \ll \mu$. Then there exists a non-negative function $h : (X,\mathcal{X}) \to (\overline{\mathbb{R}}, \mathcal{B}(\overline{\mathbb{R}}))$ such that*

$$\nu(dx) = h(x)\,\mu(dx)\,.$$

Proof. From the uniqueness of the Lebesgue decomposition of Theorem 2.3.30 and the hypothesis $\nu \ll \mu$, it follows that $\nu_a = \nu$ and therefore $\nu_s \equiv 0$. $\qquad\square$

The function h is called the *Radon–Nikodým derivative* of ν with respect to μ and is denoted $\mathrm{d}\nu/\mathrm{d}\mu$. With such a notation, we have that

$$\int_X f(x)\,\nu(dx) = \int_X f(x)\,\tfrac{\mathrm{d}\nu}{\mathrm{d}\mu}(x)\,\mu(dx)$$

for all non-negative $f = (X,\mathcal{X}) \to (\overline{\mathbb{R}}, \mathcal{B}(\overline{\mathbb{R}}))$.

Complementary reading

For the omitted proofs of existence and unicity of measures, see for instance [Royden, 1988].

2.4 Exercises

Exercise 2.4.1. THE σ-FIELD GENERATED BY A COLLECTION OF SETS

(1) Let $\{\mathcal{F}_i\}_{i\in I}$ be a non-empty family of σ-fields on some set Ω (the non-empty index set I is arbitrary). Show that the family $\mathcal{F} = \cap_{i\in I}\mathcal{F}_i$ is a σ-field ($A \in \mathcal{F}$ if and only if $A \in \mathcal{F}_i$ for all $i \in I$).

(2) Let $\mathcal{C}$ be a family of subsets of some set Ω. Show the existence of a smallest σ-field $\mathcal{F}$ containing $\mathcal{C}$. (This means, by definition, that $\mathcal{F}$ is a σ-field on Ω containing $\mathcal{C}$, such that if $\mathcal{F}'$ is a σ-field on Ω containing $\mathcal{C}$, then $\mathcal{F} \subseteq \mathcal{F}'$.)

Exercise 2.4.2. SIMPLE FUNCTIONS

(1) Show that a Borel function $f : (X, \mathcal{X}) \to (\mathbb{R}, \mathcal{B})$ taking a finite number of values is a simple function.

(2) Show that a function measurable with respect to the gross σ-field is a constant.

Exercise 2.4.3. $\mathcal{B}(\overline{\mathbb{R}})$
Recall that $\mathcal{B}(\overline{\mathbb{R}})$ is the σ-field on $\overline{\mathbb{R}}$ generated by the intervals of type $(-\infty, a]$ $(a \in \mathbb{R})$. Describe $\mathcal{B}(\overline{\mathbb{R}})$ in terms of $\mathcal{B}(\mathbb{R})$.

Exercise 2.4.4. $\sigma(f^{-1}(\mathcal{C})) = f^{-1}(\sigma(\mathcal{C}))$
Let X and E be sets, $f : X \to E$ a function from X to E and $\mathcal{C}$ a collection of subsets of E. Prove that

$$\sigma(f^{-1}(\mathcal{C})) = f^{-1}(\sigma(\mathcal{C})).$$

Exercise 2.4.5. THE SMALLEST σ-FIELD GUARANTEEING MEASURABILITY
Let $f : X \to E$ be a function. Let $\mathcal{E}$ be a given σ-field on E. What is the smallest σ-field on X such that f is measurable with respect to $\mathcal{X}$ and $\mathcal{E}$?

Exercise 2.4.6. THE MODULUS OF A FUNCTION AND MEASURABILITY
Let $f : X \to E$ be a function. Is it true that if $|f|$ is measurable with respect to $\mathcal{X}$ and $\mathcal{E}$, then so is f itself?

Exercise 2.4.7. MEASURABILITY WITH RESPECT TO THE GROSS σ-FIELD
Prove that a function $f : E \to \mathbb{R}$ measurable with respect to the gross σ-field on E and the Borel σ-field on $\mathbb{R}$ is a constant.

Exercise 2.4.8. DECREASING SEQUENCES OF MEASURABLE SETS
Let $(X, \mathcal{X})$ be a measurable space and $\{B_n\}_{n \geq 1}$ a non-increasing sequence of $\mathcal{X}$ such that $\mu(B_{n_0}) < \infty$ for some $n_0 \in \mathbb{N}_+$. Show that

$$\mu\left(\bigcap_{n=1}^{\infty} B_n\right) = \lim_{n \downarrow \infty} \downarrow \mu(B_n).$$

Give a counterexample for the necessity of condition $\mu(B_{n_0}) < \infty$ for some n_0.

Exercise 2.4.9. ALMOST-EVERYWHERE EQUAL CONTINUOUS FUNCTIONS
Prove that if two continuous functions $f, g : \mathbb{R} \to \mathbb{R}$ are ℓ-a.e. equal, they are everywhere equal.

Exercise 2.4.10. $\frac{\sin x}{x}$
Let $f(x) := \frac{\sin x}{x}$. Prove that the limit $\lim_{t \uparrow \infty} \int_0^t \frac{\sin x}{x} \, dx$ exists. Is f integrable on $\mathbb{R}_+$ with respect to the Lebesgue measure?

Exercise 2.4.11. FROM INTEGRAL TO SERIES
Prove that for all $a, b \in \mathbb{R}$,

$$\int_{\mathbb{R}_+} \frac{t \, e^{-at}}{1 - e^{-bt}} \, dt = \sum_{n=0}^{+\infty} \frac{1}{(a + nb)^2}.$$

Exercise 2.4.12. FUN FUBINI
A bounded rectangle of $\mathbb{R}^2$ is said to have Property (A) if *at least one* of its sides "is an integer" (meaning: its length is an integer). Let Δ be a finite rectangle that is the union of a finite number of disjoint rectangles with Property (A). Show that Δ itself must have Property (A). Hint: $\int_I e^{2i\pi x} dx \ldots$

Exercise 2.4.13. FOURIER TRANSFORM
Let $f : (\mathbb{R}, \mathcal{B}(\mathbb{R})) \to (\mathbb{R}, \mathcal{B}(\mathbb{R}))$ be integrable with respect to the Lebesgue measure. Show that for any $\nu \in \mathbb{R}$,

$$\hat{f}(\nu) = \int_{\mathbb{R}} f(t) \, e^{-2i\pi\nu t} dt$$

is well defined and that the function $\hat{f}$ is continuous and bounded. ($\hat{f}$ is called the *Fourier transform* of f.)

Exercise 2.4.14. CONVOLUTION
Let $f, g : (\mathbb{R}, \mathcal{B}(\mathbb{R})) \to (\mathbb{R}, \mathcal{B}(\mathbb{R}))$ be integrable with respect to the Lebesgue measure ℓ and let $\hat{f}, \hat{g}$ be their respective Fourier transforms (See Exercise 2.4.13).

(1) Show that

$$\int_{\mathbb{R}} \int_{\mathbb{R}} |f(t-s)g(s)| \, dt \, ds < \infty.$$

(2) Deduce from this that for almost all $t \in \mathbb{R}$, the function $s \mapsto f(t-s)g(s)$ is ℓ-integrable, and therefore that the convolution $f * g$, where

$$(f * g)(t) = \int_{\mathbb{R}} f(t-s)g(s) \, ds,$$

is almost everywhere well defined.

(3) For all t such that the last integral is not defined, set $(f * g)(t) = 0$. Show that $f * g$ is ℓ-integrable and that its Fourier transform is $\widehat{f * g} = \hat{f}\hat{g}$.

Exercise 2.4.15. A FUBINI COUNTEREXAMPLE
Let $(X_i, \mathcal{X}_i, \mu_i)$ $(i = 1, 2)$ be two versions of the measure space $(X, \mathcal{X}, \mu)$, where $X = \{1, 2, \ldots\}$, $\mathcal{X} = \mathcal{P}(X)$ and μ is the counting measure. Consider the function $f : (X_1 \times X_2) \to \mathbb{Z}$ whose non null values are $f(m, m) = +1$ and $f(m+1, m) = -1$ $(m \geq 1)$. Show that

$$\sum_m \left(\sum_n f(m, n) \right) = 1 \text{ and } \sum_n \left(\sum_m f(m, n) \right) = 0.$$

Why don't we obtain the same values for both sums?

Exercise 2.4.16. ANOTHER FUBINI COUNTEREXAMPLE
Define $f : [0, 1]^2 \to \mathbb{R}$ by

$$f(x, y) = \frac{x^2 - y^2}{(x^2 + y^2)^2} \, 1_{\{(x,y) \neq (0,0)\}} \, .$$

Compute $\int_{[0,1]} \left(\int_{[0,1]} f(x, y) \, dx \right) dy$ and $\int_{[0,1]} \left(\int_{[0,1]} f(x, y) \, dy \right) dx$. Is f Lebesgue integrable on $[0, 1]^2$?

Exercise 2.4.17. CONVOLUTION OF MEASURES

The convolution product of two finite measures μ_1 and μ_2 on $\mathbb{R}^d$ is the measure ν on $\mathbb{R}^d$ that is the image of the product measure $\mu := \mu_1 \times \mu_2$ on $\mathbb{R}^d \times \mathbb{R}^d$ under the mapping $(x_1, x_2) \to x_1 + x_2$. This measure will be denoted by $\mu_1 * \mu_2$.

(i) Show that for any non-negative measurable function $f : \mathbb{R}^d \to \mathbb{R}$

$$\int_{\mathbb{R}^d} f(x)\,\nu(\mathrm{d}x) = \int_{\mathbb{R}^d} \left(\int_{\mathbb{R}^d} f(x_1 + x_2)\mu_1(\mathrm{d}x_1) \right) \mu_2(\mathrm{d}x_2).$$

(ii) Let μ be a finite measure on $\mathbb{R}^d$ and let ε_a be the Dirac measure (on $\mathbb{R}^d$) at point $a \in \mathbb{R}^d$. What is the convolution product $\mu * \varepsilon_a$?

Exercise 2.4.18. $L_{\mathbb{C}}^1(\ell)$ AND $L_{\mathbb{C}}^2(\ell)$

Show that there exist functions in $L_{\mathbb{C}}^1(\ell)$ that are not in $L_{\mathbb{C}}^2(\ell)$ and vice versa.

Exercise 2.4.19. $\ell_{\mathbb{C}}^p$

Show that if $p > q$, then $\ell_{\mathbb{C}}^q(\mathbb{Z}) \subset \ell_{\mathbb{C}}^p(\mathbb{Z})$.

Exercise 2.4.20. THE LEBESGUE DECOMPOSITION

Let μ and ν be measures on the measurable space $(X, \mathcal{X})$. Describe the Lebesgue decomposition in the following cases:

A. $(X, \mathcal{X}) = (\mathbb{Z}, \mathcal{P}(\mathbb{Z}))$.

B. $(X, \mathcal{X}) = (\mathbb{R}, \mathcal{B}(\mathbb{R}))$, $\mu(\mathrm{d}x) = f(x)\,\mathrm{d}x$ and $\nu(\mathrm{d}x) = g(x)\,\mathrm{d}x$.

Chapter 3
Probability and Expectation

Although from a formal point of view a probability is just a measure with total mass equal to one, and expectation is nothing more than an integral with respect to this measure,[1] probability theory has two ingredients that make the difference: the notion of independence and that of conditional expectation.

Probability theory has a specific terminology adapted to its goals, and therefore we begin with the "translation" of the theory of measure and integration into the theory of probability and expectation.

3.1 From Integral to Expectation

3.1.1 Translation

Recall that abstract (or axiomatic) probability theory features a "sample" space Ω and a collection $\mathcal{F}$ of its subsets that forms a σ-field, the *σ-field of events*. An element $A \in \mathcal{F}$ is called an *event*. A *probability* P on $(\Omega, \mathcal{F})$ is a measure on this measurable space with total mass 1. The results obtained in the previous chapter will now be recast in this specific framework.

The probabilistic version of Theorem 2.1.42 is given below for future reference.

Theorem 3.1.1 *Let P_1 and P_2 be two probability measures on $(\Omega, \mathcal{F})$ and let $\mathcal{S}$ be a π-system of measurable sets generating $\mathcal{F}$. If P_1 and P_2 agree on $\mathcal{S}$, they are identical.*

Let $(E, \mathcal{E})$ be some measurable space. A measurable mapping (or function)

$$X : (\Omega, \mathcal{F}) \to (E, \mathcal{E})$$

is called a *random element* with values in E. If $E = \overline{\mathbb{R}}$ and $\mathcal{F} = \mathcal{B}(\overline{\mathbb{R}})$, it is called a *random variable*. If $E = \mathbb{R}^m$ and $\mathcal{F} = \mathcal{B}(\mathbb{R}^m)$, it is called a *random vector*.

In view of Theorem 2.1.18, for a mapping $X = (X_1, \ldots, X_m)$ to be a random vector in $\mathbb{R}^m$, it suffices that $\{X_i \leq a\} \in \mathcal{F}$ ($1 \leq i \leq m$, $a \in \mathbb{R}$). From Corollary 2.1.17, we have that if X is a random element with values in the measurable space $(E, \mathcal{E})$ and if g is a measurable function from $(E, \mathcal{E})$ to another measurable space $(G, \mathcal{G})$, then $g(X)$ is a random element with values in the measurable space $(G, \mathcal{G})$.

---

[1]But as the wise man said: "He who does not know measure from probability does not know sake from rice."

© Springer Nature Switzerland AG 2020
P. Brémaud, *Probability Theory and Stochastic Processes*, Universitext,
https://doi.org/10.1007/978-3-030-40183-2_3

Corollary 2.1.20 and Theorem 2.1.21 tell us that all ordinary operations on random variables (addition, multiplication, and quotient—if well defined—) and the limit operations (limsup, liminf, and lim—if well defined—) preserve the status of random variable.

Since a random variable X is a measurable function, we can define, under certain circumstances, its integral with respect to the probability measure P, called the *expectation* of X. Therefore

$$\mathrm{E}[X] = \int_\Omega X(\omega) P(\mathrm{d}\omega).$$

The main steps in the definition of the integral (here the expectation) are summarized below in the specific notation of probability theory. If $A \in \mathcal{F}$,

$$\mathrm{E}[1_A] = P(A),$$

and more generally, if X is a simple random variable, that is, $X(\omega) = \sum_{i=1}^{N} \alpha_i 1_{A_i}(\omega)$, where $\alpha_i \in \mathbb{R}$, $A_i \in \mathcal{F}$ and $N < \infty$, then

$$\mathrm{E}[X] = \sum_{i=1}^{N} \alpha_i P(A_i).$$

For a non-negative random variable X, the expectation is always defined by

$$\mathrm{E}[X] = \lim_{n \uparrow \infty} \mathrm{E}[X_n],$$

where $\{X_n\}_{n \geq 1}$ is a non-decreasing sequence of non-negative simple random variables that converges to X. This definition is consistent, that is, it does not depend on the approximating sequence of non-negative simple random variables as long as it is non-decreasing and has X for limit. In particular, with the following special choice of the approximating sequence:

$$X_n = \sum_{k=0}^{n2^n - 1} \frac{k}{2^n} 1_{A_{k,n}} + n 1_{\{X \geq n\}},$$

where $A_{k,n} := \{k \times 2^{-n} \leq X < (k+1) \times 2^{-n}\}$, we have for any non-negative random variable X, the "horizontal slice formula":

$$\mathrm{E}[X] = \lim_{n \uparrow \infty} \sum_{k=0}^{n2^n - 1} \frac{k}{2^n} P(A_{k,n}) + n P(X \geq n).$$

If X is of arbitrary sign, the expectation is defined by $\mathrm{E}[X] = \mathrm{E}[X^+] - \mathrm{E}[X^-]$ if $\mathrm{E}[X^+]$ and $\mathrm{E}[X^-]$ are not both infinite. If $\mathrm{E}[X^+]$ *and* $\mathrm{E}[X^-]$ are infinite, the expectation is *not* defined. If $\mathrm{E}[|X|] < \infty$, X is said to be *integrable* and $\mathrm{E}[X]$ is then a finite number.

The basic properties of the expectation are *linearity* and *monotonicity*. If X_1 and X_2 are integrable (*resp.* non-negative) random variables, then (linearity): for all $\lambda_1, \lambda_2 \in \mathbb{R}$ (*resp.* $\in \mathbb{R}_+$),

$$\mathrm{E}[\lambda_1 X_1 + \lambda_2 X_2] = \lambda_1 \mathrm{E}[X_1] + \lambda_2 \mathrm{E}[X_2], \tag{3.1}$$

and (monotonicity):

$$X_1 \leq X_2 \implies \mathrm{E}[X_1] \leq \mathrm{E}[X_2]. \tag{3.2}$$

It follows from monotonicity that if $\mathrm{E}[X]$ is well defined,

$$|\mathrm{E}[X]| \leq \mathrm{E}[|X|]. \tag{3.3}$$

Mean and Variance

The definitions of the mean m_X and the variance σ_X^2 of a real-valued random variable X are given, when the corresponding expectations are meaningful, by

$$m_X := \mathrm{E}[X], \qquad \sigma_X^2 = \mathrm{E}[(X - m_X)^2] := \mathrm{E}[X^2] - m_X^2.$$

Markov's Inequality

This inequality was given and proved in the specific framework of discrete random variables (Theorem 1.5.4).

Theorem 3.1.2 *Let Z be a non-negative real random variable, and let $a > 0$. We then have*

$$P(Z \geq a) \leq \frac{\mathrm{E}[Z]}{a}.$$

Proof. Reproduce *verbatim* the proof in the special case of discrete variables (Theorem 3.1.2). $\square$

Specializing the Markov inequality of Theorem 3.1.2 to $Z = (X - m_X)^2$ and $a = \varepsilon^2 > 0$, we obtain as in the first chapter *Chebyshev's inequality*: For all $\varepsilon > 0$,

$$P(|X - m_X| \geq \varepsilon) \leq \frac{\sigma_X^2}{\varepsilon^2}.$$

Jensen's Inequality

Jensen's inequality is also a simple consequence of the monotonicity of expectation and of the expectation formula for indicator functions.

Theorem 3.1.3 *Let I be a general interval of $\mathbb{R}$ (closed, open, semi-closed, infinite, etc.) and let (a, b) be its interior, assumed non-empty. Let $\varphi : I \to \mathbb{R}$ be a convex function. Let X be an integrable real-valued random variable such that $P(X \in I) = 1$. Assume moreover that either φ is non-negative, or that $\varphi(X)$ is integrable. Then*

$$\mathrm{E}\left[\varphi(X)\right] \geq \varphi(\mathrm{E}\left[X\right]). \tag{3.4}$$

Proof. Reproduce *verbatim* the proof in the special case of discrete random variables (Theorem 3.1.3). $\square$

3.1.2 Probability Distributions

Definition 3.1.4 *The distribution of a random element X with values in $(E, \mathcal{E})$ is, by definition, the probability measure Q_X on $(E, \mathcal{E})$, the image of the probability measure P by the mapping X from $(\Omega, \mathcal{F})$ to $(E, \mathcal{E})$ (that is, for all $C \in \mathcal{E}$, $Q_X(C) = P(X \in C)$).*

The next result is a rephrasing of Theorem 2.3.2 in the context of probability.

Theorem 3.1.5 *If g is a measurable function from $(E, \mathcal{E})$ to $(\overline{\mathbb{R}}, \mathcal{B}(\overline{\mathbb{R}}))$ that is non-negative, then*

$$\mathrm{E}\left[g(X)\right] = \int_E g(x)\, Q_X(\mathrm{d}x). \tag{3.5}$$

If g is of arbitrary sign, and if one of the following two conditions is satisfied:

(a) $g(X)$ is P-integrable, or

(b) g is Q_X-integrable,

then the other one is also satisfied and equality (3.5) holds true.

Definition 3.1.6 *If X is a random vector $((E, \mathcal{E}) = (\mathbb{R}^m, \mathcal{B}(\mathbb{R}^m)))$ whose probability distribution Q_X is the product of a measurable function f_X by the Lebesgue measure ℓ^n, one calls f_X the probability density function (PDF) of X.*

Remark 3.1.7 The PDF is unique, in the sense that any other PDF f_X' is such that $f_X'(x) = f_X(x)$ Lebesgue-almost everywhere. See Exercise 3.4.1.

Remark 3.1.8 The following is an "obvious" result (a proof is however required in Exercise 3.4.6):

$$P(f_X(X) = 0) = 0.$$

EXAMPLE 3.1.9: THE CASE OF A REAL RANDOM VARIABLE. In the particular case where $(E, \mathcal{E}) = (\mathbb{R}, \mathcal{B}(\mathbb{R}))$, taking $C = (-\infty, x]$, we have

$$Q_X((-\infty, x]) = P(X \le x) = F_X(x),$$

where F_X is the *cumulative distribution function* (CDF) of X, and therefore

$$\mathrm{E}[g(X)] = \int_{\mathbb{R}} g(x)\, \mathrm{d}F(x),$$

by definition of the Stieltjes–Lebesgue integral (Definition 2.2.9).

EXAMPLE 3.1.10: THE CASE OF A DISCRETE RANDOM VARIABLE. In the particular case where $(E, \mathcal{E}) = (\mathbb{N}, \mathcal{P}(\mathbb{N}))$, $Q_X(\{n\}) = P(X = n)$ and

$$\mathrm{E}[g(X)] = \sum_{\mathbb{N}} g(n)P(X = n).$$

EXAMPLE 3.1.11: THE CASE OF A RANDOM VECTOR WITH A PROBABILITY DENSITY. If X is a random vector admitting a probability density f_X, then, by Theorem 2.3.28,

$$\mathrm{E}[g(X)] = \int_{\mathbb{R}^n} g(x) f_X(x)\, \mathrm{d}x.$$

The cumulative distribution of a real random variable X has the following properties:

(i) $F : \mathbb{R} \to [0, 1]$.

(ii) F is non-decreasing.

(iii) F is right-continuous.

(iv) For each $x \in \mathbb{R}$ there exists $F(x-) := \lim_{h \downarrow 0} F(x - h)$.

(v) $F(+\infty) := \lim_{a \uparrow \infty} F(a) = P(X < \infty)$.

(vi) $F(-\infty) := \lim_{a \downarrow -\infty} F(a) = P(X = -\infty)$.

(vii) $P(X = a) = F(a) - F(a-)$ for all $a \in \mathbb{R}$.

Proof. (i) is obvious; (ii) If $a \leq b$, then $\{X \leq a\} \subseteq \{X \leq b\}$, and therefore $P(X \leq a) \leq P(X \leq b)$; (iii) Let $B_n = \{X \leq a + \frac{1}{n}\}$. Since $\cap_{n \geq 1} \{X \leq a + \frac{1}{n}\} = \{X \leq a\}$, we have, by sequential continuity,

$$\lim_{n \uparrow \infty} P\left(X \leq a + \frac{1}{n}\right) = P(X \leq a).$$

(iv) We know from Analysis that a non-decreasing function from $\mathbb{R}$ to $\mathbb{R}$ has at any point a limit to the left; (v) Let $A_n = \{X \leq n\}$ and observe that $\cup_{n=1}^{\infty}\{X \leq n\} = \{X < \infty\}$. The result again follows by sequential continuity; (vi) Apply (1.6) with $B_n = \{X \leq -n\}$ and observe that $\cap_{n=1}^{\infty}\{X \leq -n\} = \{X = -\infty\}$. The result follows by sequential continuity. (vii) The sequence $B_n = \{a - \frac{1}{n} < X \leq a\}$ is decreasing, and $\cap_{n=1}^{\infty} B_n = \{X = a\}$. Therefore, by sequential continuity, $P(X = a) = \lim_{n \uparrow \infty} P\left(a - \frac{1}{n} < X \leq a\right) = \lim_{n \uparrow \infty}\left(F(a) - F\left(a - \frac{1}{n}\right)\right)$, that is to say, $P(X = a) = F(a) - F(a-)$. $\square$

From (vii), we see that the CDF is continuous at $a \in \mathbb{R}$ if and only if $P(X = a) = 0$.

Being a non-decreasing right-continuous function, F has at most a countable set of discontinuity points $\{d_n, n \geq 1\}$. Define the discontinuous part of F by

$$F_d(x) = \sum_{n \geq 1} (F(d_n) - F(d_n-)) 1_{\{d_n \leq x\}}$$

$$= \sum_{n \geq 1} P(X = d_n) 1_{\{d_n \leq x\}}.$$

In particular, when a random variable takes its values in a countable set, its CDF reduces to the discontinuous part F_d. For such (discrete) random variables, the *probability distribution* $\{p(d_n)\}_{n \geq 1}$, where

$$p(d_n) = P(X = d_n),$$

suffices to describe the probabilistic behavior of X.

Famous Continuous Random Variables

An *(absolutely) continuous* random variable is by definition a *real* (no infinite values) random variable with a probability density, that is,

$$P(X \leq x) = \int_{-\infty}^{x} f(x) \mathrm{d}x,$$

where $f(x) \geq 0$, and since X is real, $P(X < \infty) = 1$, that is,

$$\int_{-\infty}^{+\infty} f(x) \mathrm{d}x = 1.$$

Definition 3.1.12 *Let a and b be real numbers. A real random variable X with proba-bility density function*

$$f(x) = \frac{1}{b-a} 1_{[a,b]} \tag{3.6}$$

is called a uniform *random variable on* $[a, b]$.

This is denoted by $X \sim \mathcal{U}([a, b])$.

Theorem 3.1.13 *The mean and the variance of a uniform random variable on* $[a, b]$ *are given by*

$$\mathrm{E}[X] = \frac{a+b}{2}, \;\; \mathrm{Var}\,(X) = \frac{(b-a)^2}{12} \,. \tag{3.7}$$

Proof. Direct computation.

□

Theorem 3.1.14 *Let for* $u \in (0, 1)$

$$F^{\leftarrow}(u) := \inf\{x \,;\, F(x) > u\} \,.$$

If U is a uniform random variable on $(0, 1)$, then $F^{\leftarrow}(U)$ has the same probability dis-tribution as X.

Proof. First note that for all $u \in (0, 1)$, $F^{\leftarrow}(u) \leq t$ implies $F(t) \geq u$. Indeed, in this case, for all $s > t$ there exists an $x < s$ such that $F(x) > u$ and therefore $F(s) > u$; and consequently, by right-continuity of F, $F(t) \geq u$. Conversely, $F(t) \geq u$ implies that $t \in \{x \,;\, F(x) \geq u\}$ and therefore $F^{\leftarrow}(u) \leq t$. Taking all this into account,

$$F(t) = P(U < F(t)) \leq P(F^{\leftarrow}(U) \leq t)$$
$$\leq P(F(t) \geq U) = F(t) \,.$$

This forces $P(F^{\leftarrow}(U) \leq t)$ to equal $F(t)$.

□

Remark 3.1.15 Of course, if F is continuous, $F^{\leftarrow}$ is the inverse F^{-1} in the usual sense.

Definition 3.1.16 *A real random variable X with* PDF

$$f(x) = \frac{1}{\sigma\sqrt{2\pi}} e^{-\frac{1}{2}\frac{(x-m)^2}{\sigma^2}} \,, \tag{3.8}$$

where $m \in \mathbb{R}$ and $\sigma \in \mathbb{R}_+$, is called a Gaussian *random variable.*

This is denoted by $X \sim \mathcal{N}(m, \sigma^2)$. One can check that $\mathrm{E}[X] = m$ and $\mathrm{Var}\,(X) = \sigma^2$ (Exercise 3.4.18).

Definition 3.1.17 *The* tail distribution *of a random variable X is, by definition, the quantity $P(X > x)$.*

The following bounds for the tail distribution of a standard Gaussian variable are useful:

$$\frac{x}{1+x^2}\frac{1}{\sqrt{2\pi}}e^{-\frac{1}{2}x^2} \leq \frac{1}{\sqrt{2\pi}}\int_x^\infty e^{-\frac{1}{2}y^2}\,dy \leq \frac{1}{x}\frac{1}{\sqrt{2\pi}}\int_x^\infty e^{-\frac{1}{2}y^2}\,dy\,.$$

Proof. We have

$$\frac{1}{x^2}\int_x^\infty e^{-\frac{1}{2}y^2}\,dy > \int_x^\infty \frac{1}{y^2}e^{-\frac{1}{2}y^2}\,dy$$

$$= \frac{1}{x}e^{-\frac{1}{2}x^2} - \int_x^\infty e^{-\frac{1}{2}y^2}\,dy,$$

where the equality is obtained by integration by parts. This gives the inequality on the right. Integration by parts again:

$$\int_x^\infty \frac{1}{y^2}e^{-\frac{1}{2}y^2}\,dy = \frac{1}{x}e^{-\frac{1}{2}x^2} - \int_x^\infty e^{-\frac{1}{2}y^2}\,dy\,,$$

and therefore

$$\frac{1}{x}e^{-\frac{1}{2}x^2} = \int_x^\infty \left(1 + \frac{1}{y^2}\right)e^{-\frac{1}{2}y^2}\,dy \geq \int_x^\infty e^{-\frac{1}{2}y^2}\,dy\,,$$

and this is the inequality on the left. $\square$

Definition 3.1.18 *A random variable X with* PDF

$$f(x) = \lambda e^{-\lambda x}1_{\{x\geq 0\}} \tag{3.9}$$

is called an exponential *random variable with parameter λ.*

This is denoted by $X \sim \mathcal{E}(\lambda)$. The CDF of the exponential random variable is

$$F(x) = (1 - e^{-\lambda x})1_{\{x\geq 0\}}\,.$$

Theorem 3.1.19 *The mean of an exponential random variable with parameter λ is*

$$E[X] = \lambda^{-1}\,. \tag{3.10}$$

Proof. Direct computation, or see the Gamma distribution below. $\square$

The exponential distribution lacks memory, in the following sense:

Theorem 3.1.20 *Let $X \sim \mathcal{E}(\lambda)$. For all $t, t_0 \in \mathbb{R}_+$, we have*

$$P(X \geq t_0 + t \mid X \geq t_0) = P(X \geq t).$$

Proof.

$$P(X \geq t_0 + t \mid X \geq t_0) = \frac{P(X \geq t_0 + t,\, X \geq t_0)}{P(X \geq t_0)}$$

$$= \frac{P(X \geq t_0 + t)}{P(X \geq t_0)} = \frac{e^{-\lambda(t_0+t)}}{e^{-\lambda(t_0)}} = e^{-\lambda t} = P(X \geq t).$$

$\square$

Recall the definition of the *gamma function* Γ:

$$\Gamma(\alpha) := \int_0^\infty x^{\alpha-1} e^{-x} \, dx \,.$$

Integration by parts gives, for $\alpha > 0$,

$$0 = u^\alpha e^{-u} \Big|_0^\infty = \int_0^\infty \alpha u^{\alpha-1} e^{-u} du - \int_0^\infty e^{-u} u^\alpha du$$
$$= \alpha \Gamma(\alpha) - \Gamma(\alpha+1).$$

Therefore

$$\Gamma(\alpha+1) = \alpha \, \Gamma(\alpha),$$

from which it follows in particular, since $\Gamma(1) = \int_0^\infty e^{-x} dx = 1$, that for all integers $n \geq 1$,

$$\Gamma(n) = (n-1)!$$

Definition 3.1.21 *Let α and β be two strictly positive real numbers. A non-negative random variable X with the PDF*

$$f(x) = \frac{\beta^\alpha}{\Gamma(\alpha)} x^{\alpha-1} e^{-\beta x} 1_{\{x>0\}} \tag{3.11}$$

is called a Gamma random variable *with parameters α and β.*

This is denoted by $X \sim \gamma(\alpha, \beta)$.

We must check that (3.11) defines a probability density (that is, the integral of f is 1). In fact:

$$\int_{-\infty}^{+\infty} f(x) dx = \frac{\beta^\alpha}{\Gamma(\alpha)} \int_0^\infty x^{\alpha-1} e^{-\beta x} dx$$
$$= \frac{1}{\Gamma(\alpha)} \int_0^\infty y^{\alpha-1} e^{-y} dy = \frac{\Gamma(\alpha)}{\Gamma(\alpha)} = 1,$$

where the second equality has been obtained via the change of variable $y = \beta x$.

Theorem 3.1.22 *If $X \sim \gamma(\alpha, \beta)$, then*

$$E[X] = \tfrac{\alpha}{\beta} \text{ and } \text{Var}(X) = \tfrac{\alpha}{\beta^2} \,. \tag{3.12}$$

Proof.

$$E[X] = \int_0^\infty \frac{\beta^\alpha}{\Gamma(\alpha)} x \, x^{\alpha-1} e^{-\beta x} \, dx$$
$$= \frac{\beta^\alpha}{\Gamma(\alpha)} \int_0^\infty x^\alpha e^{-\beta x} \, dx = \frac{\Gamma(\alpha+1)}{\Gamma(\alpha)} \frac{1}{\beta} = \frac{\alpha}{\beta}.$$

Similarly,

$$E[X^2] = \frac{\Gamma(\alpha+2)}{\Gamma(\alpha)} \frac{1}{\beta^2} = \frac{\alpha(\alpha+1)}{\beta^2}.$$

Therefore

$$\text{Var}(X) = E[X^2] - E[X]^2 = \frac{\alpha(\alpha+1)}{\beta^2} - \left(\frac{\alpha}{\beta}\right)^2 = \frac{\alpha}{\beta^2}.$$

$\square$

The exponential distribution is a particular case of the Gamma distribution. In fact, $\gamma(1, \lambda) \equiv \mathcal{E}(\lambda)$. The so-called *chi-square* distribution *with n degrees of freedom*, denoted by χ_n^2, is just the $\gamma(\frac{n}{2}, \frac{1}{2})$ distribution. It therefore has the PDF

$$f(x) = \frac{1}{2^{\frac{n}{2}} \Gamma(\frac{n}{2})} x^{\frac{n}{2}-1} e^{-\frac{1}{2}x} 1_{\{x>0\}}. \tag{3.13}$$

This is denoted by $X \sim \chi_n^2$.

Definition 3.1.23 *A random variable X with* PDF

$$f(x) = \frac{1}{\pi(1+x^2)} \tag{3.14}$$

is called a Cauchy *random variable.*

It is important to observe that the mean of X is not defined since

$$\int_{\mathbb{R}} \frac{|x|}{\pi(1+x^2)} \, dx = +\infty.$$

Of course, *a fortiori*, its variance is not defined.

Change of Variables

Let $X = (X_1, \ldots, X_n)$ be a random vector with the probability density function f_X, and define the random vector $Y = g(X)$, where $g : \mathbb{R}^n \to \mathbb{R}^n$. More explicitly,

$$\begin{cases} Y_1 = g_1(X_1, \ldots, X_n), \\ \vdots \\ Y_n = g_n(X_1, \ldots, X_n). \end{cases}$$

Under smoothness assumptions on g, the random vector Y is absolutely continuous, and its probability density function can be explicitly computed from g and the probability density function f_X. The conditions allowing this are the following:

A_1: The function g from U to $\mathbb{R}^n$, where U is an open subset of $\mathbb{R}^n$, is one-to-one (injective).

A_2: The coordinate functions g_i $(1 \le i \le n)$ are continuously differentiable.

A_2: Moreover, the Jacobian matrix of the function g,

$$J_g(x) := J_g(x_1, \ldots, x_n) := \left\{ \frac{\partial g_i}{\partial x_j}(x_1, \ldots, x_n) \right\}_{1 \le i,j \le n},$$

satisfies the positivity condition

$$|\det J_g(x)| > 0 \qquad (x \in U).$$

A standard result of Analysis says that $V = g(U)$ is an open subset of $\mathbb{R}^n$, and that the invertible function $g : U \to V$ has an inverse $g^{-1} : V \to U$ with the same properties as the direct function g. In particular, on V,

$$|\det J_{g^{-1}}(y)| > 0.$$

Moreover,

$$J_{g^{-1}}(y) = J_g(g^{-1}(y))^{-1}.$$

Also, under the conditions $A_1 - A_3$, for any function $u : \mathbb{R}^n \to \mathbb{R}^n$,

$$\int_U u(x)\mathrm{d}x = \int_{g(U)} u(g^{-1}(y))|\det J_{g^{-1}}(y)|\mathrm{d}y.$$

Theorem 3.1.24 *Under the conditions just stated for X, g, and U, and if moreover $P(X \in U) = 1$, then Y admits the probability density*

$$f_Y(y) = f_X(g^{-1}(y))|\det J_g(g^{-1}(y))|^{-1}1_V(y). \tag{3.15}$$

Proof. The proof consists in checking that for any bounded function $h : \mathbb{R} \to \mathbb{R}$,

$$E[h(Y)] = \int_{\mathbb{R}^n} h(y)\psi(y)\mathrm{d}y, \tag{3.16}$$

where ψ is the function on the right-hand side of (3.15). Indeed, taking $h(y) = 1_{y \le a} = 1_{y_1 \le a_1} \cdots 1_{y_n \le a_n}$, (3.16) reads

$$P(Y_1 \le a_1, \ldots, Y_n \le a_n) = \int_{-\infty}^{a_1} \cdots \int_{-\infty}^{a_n} \psi(y_1, \ldots, y_n)\mathrm{d}y_1 \cdots \mathrm{d}y_n.$$

To prove that (3.16) holds with the appropriate ψ, one just uses the basic rule of change of variables:

$$E[h(Y)] = E[h(g(X))]$$
$$= \int_U h(g(x))f_X(x)\mathrm{d}x = \int_V h(y)f_X(g^{-1}(y))|\det J_{g-1}(y)|\mathrm{d}y.$$

$\square$

Theorem 3.1.25 *Let X be an n-dimensional random vector with probability density f_X. Let A be an invertible $n \times n$ real matrix and let b be an n-dimensional real vector. Then, the random vector $Y = AX + b$ admits the density*

$$f_Y(y) = f_X(A^{-1}(y - b))\frac{1}{|\det A|}. \tag{3.17}$$

Proof. Here $U = \mathbb{R}^n$, $g(x) = Ax + b$ and $|\det J_{g-1}(y)| = \frac{1}{|\det A|}$. $\square$

EXAMPLE 3.1.26: POLAR COORDINATES. Let (X_1, X_2) be a two-dimensional random vector with probability density $f_{X_1,X_2}(x_1, x_2)$ and let (R, Θ) be its polar coordinates. The probability density of (R, Θ) is given by the formula

$$f_{R,\Theta}(r, \theta) = f_{X_1,X_2}(r\cos\theta, r\sin\theta)\, r.$$

Proof. Here g is the bijective function from the open set U consisting of $\mathbb{R}^2$ without the half-line $\{(x_1, 0)\,;\; x_1 \geq 0\}$ to the open set $V = (0, \infty) \times (0, 2\pi)$. The inverse function is

$$x = r\cos\theta, \qquad y = r\sin\theta.$$

The Jacobian of g^{-1} is

$$J_{g^{-1}}(r, \theta) = \begin{pmatrix} \cos\theta & -r\sin\theta \\ \sin\theta & r\cos\theta \end{pmatrix}$$

of determinant $\det J_{g^{-1}}(r, \theta) = r$. Apply formula (3.15) to obtain the announced result. □

Covariance Matrices

Recall that $L^2_{\mathbb{C}}(P)$ (resp., $L^2_{\mathbb{R}}(P)$) is the set of square-integrable complex (resp., real) random variables, where two variables X and X' such that $P(X = X') = 1$ are not distinguished. Define for X, Y in $L^2_{\mathbb{C}}(P)$ or $L^2_{\mathbb{R}}(P)$

$$\langle X, Y \rangle = \mathrm{E}[XY^*]. \tag{3.18}$$

$L^2_{\mathbb{C}}(P)$ and $L^2_{\mathbb{R}}(P)$ are Hilbert subspaces with scalar field $\mathbb{C}$ and $\mathbb{R}$ respectively (the Riesz–Fischer Theorem 2.3.25).

Definition 3.1.27 *Two complex square-integrable random variables are said to be orthogonal if* $\mathrm{E}[XY^*] = 0$. *They are said to be* uncorrelated *if* $\mathrm{E}[(X - m_X)(Y - m_Y)^*] = 0$.

Recall Schwarz's inequality for square-integrable random variables:

$$|\mathrm{E}[XY]| \leq \mathrm{E}[|XY|] \leq \mathrm{E}[|Y|^2]^{\frac{1}{2}} \times \mathrm{E}[|X|^2]^{\frac{1}{2}}. \tag{3.19}$$

In particular, with $Y = 1$,

$$\mathrm{E}[|X|] \leq \mathrm{E}[|X|^2]^{\frac{1}{2}} < \infty. \tag{3.20}$$

Correlation Coefficient

Definition 3.1.28 *The* cross-variance *of the two complex square integrable variables X and Y is, by definition, the complex number* $\mathrm{E}[(X - m_X)(Y - m_Y)^*]$, *denoted by* σ_{XY}.

Definition 3.1.29 *Let X and Y be square-integrable real random variables with respective means m_X and m_Y, and respective variances $\sigma_X^2 > 0$ and σ_Y^2. Their* correlation coefficient *is the quantity*

$$\rho_{XY} := \frac{\sigma_{XY}}{\sigma_X \sigma_Y},$$

where σ_{XY} is the cross-variance.

By Schwarz's inequality, $|\sigma_{XY}| \leq \sigma_X \sigma_y$, and therefore

$$|\rho_{XY}| \leq 1,$$

with equality if and only if X and Y are colinear.

When $\rho_{XY} = 0$, X and Y are said to be uncorrelated. If $\rho_{XY} > 0$, they are said to be *positively correlated*, whereas if $\rho_{XY} < 0$, they are said to be *negatively correlated*.

Theorem 3.1.30 *Let X be a square-integrable real random variable. Among all variables $Z = aX + b$, where a and b are real numbers, the one that minimizes the error $E[(Z - Y)^2]$ is*

$$\hat{Y} = m_Y + \frac{\sigma_{XY}}{\sigma_X^2}(X - m_X),$$

and the error is then

$$E[(\hat{Y} - Y)^2] = \sigma_Y^2(1 - \rho_{XY}^2).$$

(The proof is left as Exercise 3.4.43.)

Remark 3.1.31 We see that if the variables are not correlated, then the best prediction is the trivial one $\hat{Y} = m_Y$ and the (maximal) error is then σ_Y^2. In imprecise but suggestive terms, high correlation implies high predictability.

Notation: For vectors and matrices, an asterisk superscript (*) denotes complex conjugates, a T superscript (^{T}) is for vector transposition, and the dagger superscript (†) is for conjugation-transposition. When x is a vector of $\mathbb{R}^n$, we always assume in the notation that it is a *column* vector, and therefore x^T will be the corresponding *row* vector.

Definition 3.1.32 *A random vector $X = (X_1, \ldots, X_n)^T$ such that X_1, ..., X_n are square-integrable complex random variables is called a* square-integrable complex vector.

In particular, for all $1 \leq i, j \leq n$, by (3.20), $E[|X_i|] < \infty$, and by Schwarz's inequality (3.19), $E[|X_i X_j|] < \infty$. This allows us to define the *mean of X*

$$m_X := E[X] = (E[X_1], \ldots, E[X_n])^T$$

and the *covariance matrix* of X

$$\begin{aligned}
\Gamma_X &:= E[(X - m_X)(X - m_X)^\dagger] \\
&= \left\{ E\left[(X_i - m_{X_i})(X_j - m_{X_j})^* \right] \right\}_{1 \leq i,j \leq n} \\
&= \left\{ \sigma_{X_i, X_j} \right\}_{1 \leq i,j \leq n}.
\end{aligned}$$

Theorem 3.1.33 *The matrix Γ_X is symmetric Hermitian, that is,*

$$\Gamma_X^\dagger = \Gamma_X, \tag{3.21}$$

and it is non-negative definite (denoted $\Gamma_X \geq 0$), that is,

$$\alpha^\dagger \Gamma_X \alpha \geq 0 \quad (\alpha \in \mathbb{C}^n). \tag{3.22}$$

Proof.

$$\begin{aligned}
\alpha^\dagger \Gamma \alpha = \alpha^T \Gamma \alpha^* &= \sum_{i=1}^n \sum_{j=1}^n \alpha_i \alpha_j^* E[(X_i - E[X_i])(X_j - E[X_j])^*] \\
&= E\left[\sum_{i=1}^n \sum_{j=1}^n \alpha_i \alpha_j^* (X_i - E[X_i])(X_j - E[X_j])^* \right] \\
&= E\left[\left(\sum_{i=1}^n \alpha_i(X_i - E[X_i]) \right) \left(\sum_{j=1}^n \alpha_j(X_j - E[X_j]) \right)^* \right] \\
&= E[|\alpha^T(X - E[X])|^2] \geq 0.
\end{aligned}$$

$\square$

Theorem 3.1.34 *Let X be a square-integrable real random vector with a covariance matrix Γ_X which is degenerate, that is,*

$$\alpha^T \Gamma_X \alpha = 0$$

for some $\alpha \neq 0$. Then X lies almost surely in a given hyperplane of $\mathbb{R}^n$ of dimension strictly less than n.

Proof. For such α, $\mathrm{E}[|\alpha^T(X - \mathrm{E}[X])|^2] = \alpha^T \Gamma_X \alpha = 0$, and therefore

$$\alpha^T(X - \mathrm{E}[X]) = 0$$

almost surely.
□

Remark 3.1.35 Since X lies almost surely in a strict hyperplane of $\mathbb{R}^n$, it cannot have a probability density.

A vector X with degenerate covariance matrix is also called *degenerate*. If Γ_X is non-degenerate, we write $\Gamma_X > 0$.

We now examine the effects of an affine transformation of a random vector on its covariance matrix. Let X be a square-integrable n-dimensional complex random vector, with mean m_X and covariance matrix Γ_X. Let A be an $(n \times k)$-dimensional complex matrix, and b a k-dimensional complex vector.

Theorem 3.1.36 *Then the k-dimensional complex vector $Z = AX + b$ has mean*

$$m_Z = A\, m_X + b,$$

and covariance matrix

$$\Gamma_Z = A\,\Gamma_X A^\dagger.$$

Proof. The formula giving the mean is immediate. As for the other one, it suffices to observe that $(Z - m_Z) = A(X - m_X)$ and to write

$$
\begin{aligned}
\Gamma_Z &= \mathrm{E}\left[(Z - m_Z)(Z - m_Z)^\dagger\right] \\
&= \mathrm{E}\left[A(X - m_X)(A(X - m_X))^\dagger\right] \\
&= \mathrm{E}\left[A(X - m_X)(X - m_X)^\dagger A^\dagger\right] \\
&= A\mathrm{E}\left[(X - m_X)(X - m_X)^T\right] A^\dagger \\
&= A\Gamma_X A^\dagger.
\end{aligned}
$$

□

Let X and Y be square-integrable complex random vectors of respective dimensions n and q. We define the *inter-covariance matrix* of X and Y—in this order—by

$$\Gamma_{XY} = \mathrm{E}[(X - m_X)(Y - m_Y)^\dagger].$$

Note that

$$\Gamma_{YX} = \Gamma_{XY}^\dagger.$$

Also if we define the $(n+q)$-dimensional vector Z by

$$Z = (X_1, \ldots, X_n, Y_1, \ldots, Y_q)^T$$

then its covariance takes the block form

$$\Gamma_Z = \begin{pmatrix} \Gamma_X & \Gamma_{XY} \\ \Gamma_{YX} & \Gamma_Y \end{pmatrix}.$$

3.1.3 Independence and the Product Formula

Recall the definition of independence for events. Two events A and B are said to be independent if

$$P(A \cap B) = P(A)P(B). \tag{3.23}$$

More generally, a family $\{A_i\}_{i \in I}$ of events, where I is an arbitrary index, is called *independent* if for every *finite* subset $J \in I$,

$$P\left(\bigcap_{j \in J} A_j\right) = \prod_{j \in J} P(A_j).$$

Definition 3.1.37 *Two random elements* $X : (\Omega, \mathcal{F}) \to (E, \mathcal{E})$ *and* $Y : (\Omega, \mathcal{F}) \to (G, \mathcal{G})$ *are called independent if for all* $C \in \mathcal{E}$, $D \in \mathcal{G}$,

$$P(\{X \in C\} \cap \{Y \in D\}) = P(X \in C)P(Y \in D). \tag{3.24}$$

More generally, let I *be an arbitrary index. The family of random elements* $\{X_i\}_{i \in I}$, *where* $X_i : (\Omega, \mathcal{F}) \to (E_i, \mathcal{E}_i)$ *$(i \in I)$, is called* independent *if for every finite subset* $J \in I$,

$$P\left(\bigcap_{j \in J} \{X_j \in C_j\}\right) = \prod_{j \in J} P(X_j \in C_j)$$

for all $C_j \in \mathcal{E}_j$ $(j \in J)$.

Theorem 3.1.38 *If the random elements* X *and* Y *taking their values in* $(E, \mathcal{E})$ *and* $(G, \mathcal{G})$ *respectively are independent , then so are the random elements* $\varphi(X)$ *and* $\psi(Y)$, *where* $\varphi : (E, \mathcal{E}) \to (E', \mathcal{E}')$, $\psi : (G, \mathcal{G}) \to (G', \mathcal{G}')$.

Proof. For all $C' \in \mathcal{E}'$, $D' \in \mathcal{G}'$, the sets $C = \varphi^{-1}(C')$ and $D = \psi^{-1}(D')$ are in $\mathcal{E}$ and $\mathcal{G}$ respectively, since φ and ψ are measurable. We have

$$\begin{aligned} P\left(\varphi(X) \in C', \psi(Y) \in D'\right) &= P\left(X \in C, Y \in D\right) \\ &= P\left(X \in C\right) P\left(Y \in D\right) \\ &= P\left(\varphi(X) \in C'\right) P\left(\psi(Y) \in D'\right). \end{aligned}$$

$\square$

The above result is stated for two random variables for simplicity, and it extends in the obvious way to a finite number of independent random variables.

The next result simplifies the task of proving that two σ-fields are independent.

Theorem 3.1.39 *Let $(\Omega, \mathcal{F}, P)$ be a probability space and let $\mathcal{S}_1$ and $\mathcal{S}_2$ be two π-systems of sets in $\mathcal{F}$. If $\mathcal{S}_1$ and $\mathcal{S}_2$ are independent, then so are $\sigma(\mathcal{S}_1)$ and $\sigma(\mathcal{S}_2)$.*

Proof. Fix $A \in \mathcal{S}_1$. Let

$$\mathcal{V}_2 = \{B \subseteq X; \ P(A \cap B) = P(A)P(B)\}.$$

This is a d-system (easy to check) and $\mathcal{S}_2 \subseteq \mathcal{V}_2$. Therefore $d(\mathcal{S}_2) \subseteq d(\mathcal{V}_2) = \mathcal{V}_2$. On the other hand, by Dynkin's theorem, $d(\mathcal{S}_2) = \sigma(\mathcal{S}_2)$. We have therefore proved that $\mathcal{S}_1$ and $\sigma(\mathcal{S}_2)$ are independent.

Now fix $B \in \sigma(\mathcal{S}_2)$. Let

$$\mathcal{V}_1 = \{A \subseteq X; \ P(A \cap B) = P(A)P(\cap B)\}.$$

This is a d-system and $\mathcal{S}_1 \subseteq \mathcal{V}_1$. Therefore $d(\mathcal{S}_1) \subseteq d(\mathcal{V}_1) = \mathcal{V}_1$. On the other hand, by Dynkin's theorem, $d(\mathcal{S}_1) = \sigma(\mathcal{S}_1)$. We therefore have proved that $\sigma(\mathcal{S}_1)$ and $\sigma(\mathcal{S}_2)$ are independent. $\square$

Corollary 3.1.40 *Let $(\Omega, \mathcal{F}, P)$ be a probability space on which are given two real random variables X and Y. For these two random variables to be independent, it is necessary and sufficient that for all $a, b \in \mathbb{R}$, $P(X \leq a, Y \leq b) = P(X \leq a)P(Y \leq b)$.*

Proof. This follows from Theorem 3.1.39, remembering that the collection $\{(-\infty, a]; \ a \in \mathbb{R}\}$ is a π-system generating $\mathcal{B}(\mathbb{R})$. $\square$

The independence of two random elements X and Y is equivalent to the factorization of their joint distribution:

$$Q_{(X,Y)} = Q_X \times Q_Y,$$

where $Q_{(X,Y)}$, Q_X, and Q_Y are the distributions of, respectively, (X, Y), X and Y. Indeed, for all sets of the form $C \times D$, where $C \in \mathcal{E}$ and $D \in \mathcal{G}$,

$$\begin{aligned}
Q_{(X,Y)}(C \times D) &= P((X, Y) \in C \times D) \\
&= P(X \in C, Y \in D) \\
&= P(X \in C)P(Y \in D) = Q_X(C)Q_Y(D),
\end{aligned}$$

and therefore (Theorem 2.3.7) $Q_{(X,Y)}$ is the product measure of Q_X and Q_Y.

In particular, the Fubini–Tonelli theorem immediately gives a result that we have already seen in the particular case of discrete random variables: the *product formula* for expectations (Formula (3.25) below).

Theorem 3.1.41 *Let the random variables X and Y taking their values in $(E, \mathcal{E})$ and $(G, \mathcal{G})$ respectively be independent, and let $g : (E, \mathcal{E}) \to (\mathbb{R}, \mathcal{B})$, $h : (G, \mathcal{G}) \to (\mathbb{R}, \mathcal{B})$ such that either one of the following two conditions is satisfied:*

(i) $\mathrm{E}\,[|g(X)|] < \infty$ and $\mathrm{E}\,[|h(Y)|] < \infty$, and

(ii) $g \geq 0$ and $h \geq 0$.

Then

$$\mathrm{E}\,[g(X)h(Y)] = \mathrm{E}\,[g(X)]\,\mathrm{E}\,[h(Y)]. \tag{3.25}$$

Proof. It suffices to give the proof in the non-negative case. We have:

$$
\begin{aligned}
\mathrm{E}\left[g(X)h(Y)\right] &= \int_E \int_G g(x)h(y)Q_{(X,Y)}(\mathrm{d}x \times \mathrm{d}y) \\
&= \int_E \int_G g(x)h(y)Q_X(\mathrm{d}x)Q_Y(\mathrm{d}y) \\
&= \int_E g(x)h(y)Q_X(\mathrm{d}x) \int_G h(y)Q_Y(\mathrm{d}y) \\
&= \mathrm{E}\left[g(X)\right]\mathrm{E}\left[h(Y)\right].
\end{aligned}
$$

$\square$

Theorem 3.1.42 *Let X be a random vector of $\mathbb{R}^n$ admitting the PDF f_X. The (measurable) set of samples ω such that there exists i, j ($i \neq j$) such that $X_i(\omega) = X_j(\omega)$ has a null probability.*

Proof. Let A be this set and let

$$
C := \{x_1, \ldots, x_n \, ; \, x_i = x_j \text{ for some } i \neq j\}.
$$

The set C has null Lebesgue measure, and therefore, since $1_A(\omega) \equiv 1_C(X(\omega))$,

$$
P(A) = \mathrm{E}\left[1_C(X(\omega))\right] = \int_{\mathbb{R}^n} 1_C(x)f_X(x)\,\mathrm{d}x = 0.
$$

$\square$

Order Statistics

Let $X_1, \ldots, X_n$ be independent random variables with the same PDF f. By Theorem 3.1.42, the probability that two or more among $X_1, \ldots, X_n$ take the same value is null. Therefore one can define unambiguously the random variables $Z_1, \ldots, Z_n$ obtained by arranging $X_1, \ldots, X_n$ in increasing order:

$$
\begin{cases}
Z_i \in \{X_1, \ldots, X_n\}, \\
Z_1 < Z_2 < \cdots < Z_n.
\end{cases}
$$

In particular, $Z_1 = \min(X_1, \ldots, X_n)$ and $Z_n = \max(X_1, \ldots, X_n)$.

Theorem 3.1.43 *The probability density of the reordered vector $Z = (Z_1, \ldots, Z_n)$ (defined above) is*

$$
f_Z(z_1, \ldots, z_n) = n! \left\{ \prod_{j=1}^n f(z_j) \right\} 1_C(z_1, \ldots, z_n), \tag{3.26}
$$

where

$$
C = \{(z_1, \ldots, z_n) \in \mathbb{R}^n \, ; \, z_1 < z_2 < \cdots < z_n\}.
$$

Proof. Let σ be the permutation of $\{1, \ldots, n\}$ that orders $X_1, \ldots, X_n$ in ascending order, that is,

$$
X_{\sigma(i)} = Z_i
$$

(note that σ is a *random* permutation). For any set $A \subseteq \mathbb{R}^n$,

$$P(Z \in A) = P(Z \in A \cap C)$$
$$= P(X_\sigma \in A \cap C) = \sum_{\sigma_o} P(X_{\sigma_o} \in A \cap C, \sigma = \sigma_o),$$

where the sum is over all permutations of $\{1, \ldots, n\}$. Observing that $X_{\sigma_o} \in A \cap C$ implies $\sigma = \sigma_o$,

$$P(X_{\sigma_o} \in A \cap C, \sigma = \sigma_o) = P(X_{\sigma_o} \in A \cap C)$$

and therefore since the probability distribution of X_{σ_o} does not depend upon a fixed permutation σ_o (here we need the independence and equidistribution assumptions for the X_i's),

$$P(X_{\sigma_o} \in A \cap C) = P(X \in A \cap C).$$

Therefore,

$$P(Z \in A) = \sum_{\sigma_o} P(X \in A \cap C) = n! P(X \in A \cap C)$$
$$= n! \int_{A \cap C} f_X(x) \mathrm{d}x = \int_A n! f_X(x) 1_C(x) \mathrm{d}x.$$

$\square$

EXAMPLE 3.1.44: VOLUME OF THE RIGHT-ANGLED PYRAMID. We shall apply the above result to prove the formula

$$\int_a^b \cdots \int_a^b 1_C(z_1, \ldots, z_n) \mathrm{d}z_1 \cdots \mathrm{d}z_n = \frac{(b-a)^n}{n!}. \tag{3.27}$$

Indeed, when the X_i's are uniformly distributed over $[a, b]$,

$$f_Z(z_1, \ldots, z_n) = \frac{n!}{(b-a)^n} 1_{[a,b]^n}(z_1, \ldots, z_n) 1_C(z_1, \ldots, z_n). \tag{3.28}$$

The result follows since $\int_{\mathbb{R}^n} f_Z(z) \mathrm{d}z = 1$.

Sampling from a Distribution

The problem that we address now (which arises in the simulation of stochastic systems) is to generate a random variable with prescribed CDF or, in other terms, to *sample* the said CDF. For this, one is allowed to use a random generator that produces a sequence $U_1, U_2, \ldots$ of independent real random variables, uniformly distributed on $[0, 1]$. In practice, the numbers that such random generators produce are not quite random, but they look as if they are (the generators are called *pseudo-random generators*). The topic of how to devise a good pseudo-random generator is out of our scope, and we shall admit that we can trust our favourite computer to provide us with an IID sequence of random variables uniformly distributed on $[0, 1]$ (from now on we call them *random numbers*).

Given such a sequence, we are going to describe methods for constructing a random variable Z with CDF

$$F(z) = P(Z \le z).$$

In the case where Z is a discrete random variable with distribution $P(Z = a_i) = p_i$ $(0 \le i \le K)$, the basic principle of the sampling algorithm is the following:

Draw $U \sim \mathcal{U}([0,1])$.

Set $Z = a_\ell$ if $p_0 + p_1 + \ldots + p_{\ell-1} < U \leq p_0 + p_1 + \ldots + p_\ell$.

This method is called the *method of the inverse*.

A crude generation algorithm would successively perform the tests $U \leq p_0$?, $U \leq p_0 + p_1$?, $\ldots$, until the answer is positive. The average number of iterations required would therefore be $\sum_{i \geq 0}(i+1)p_i = 1 + \mathrm{E}[Z]$. This number may be too large, but there are ways of improving it, as the example below will show for the Poisson random variable.

For absolutely continuous variables, the inverse method takes the following form. Draw a random number $U \sim \mathcal{U}([0,1])$ and set

$$Z = F^{-1}(U),$$

where F^{-1} is the inverse of F. Indeed,

$$P(Z \leq z) = P(F^{-1}(U) \leq z)$$
$$= P(U \leq F(z)) = F(z).$$

EXAMPLE 3.1.45: EXPONENTIAL DISTRIBUTION. We want to sample from $\mathcal{E}(\lambda)$. The corresponding CDF is

$$F(z) = 1 - \mathrm{e}^{-\lambda z} \quad (z \geq 0).$$

The solution of $y = 1 - \mathrm{e}^{-\lambda z}$ is $z = -\frac{1}{\lambda}\ln(1-y) = F^{-1}(y)$, and therefore, $Z = -\frac{1}{\lambda}\ln(1-U)$ will do, or since U and $1 - U$ have the same distribution,

$$Z = -\frac{1}{\lambda}\ln U.$$

Remark 3.1.46 Both the discrete case and the absolutely continuous cases are particular cases of the more general result of Theorem 3.1.14.

EXAMPLE 3.1.47: SYMMETRIC EXPONENTIAL DISTRIBUTION. This example features a simple trick. We want to sample from the symmetric exponential distribution with PDF

$$f(x) = \frac{1}{2}\mathrm{e}^{-|x|}.$$

One way is to generate two independent random variables Y and Z, where $Z \sim \mathcal{E}(1)$ and $P(Y = +1) = P(Y = -1) = \frac{1}{2}$. Taking $X = YZ$ we have that

$$P(X \leq x) = P(U = +1, Z \leq x) + P(U = -1, Z \geq -x) = \frac{1}{2}\left(F_Z(x) + 1 - F_Z(-x)\right),$$

and therefore, by differentiation,

$$f_X(x) = \frac{1}{2}\left(f_Z(x) + f_Z(-x)\right) = \frac{1}{2}f_Z(|x|).$$

It is not always easy to compute the inverse of the cumulative distribution function of the random variable to be generated. An alternative method is the *method of acceptance-rejection* below.

Let $\{Y_n\}_{n \geq 1}$ be a sequence of IID random variables with the probability density $g(x)$ satisfying for all $x \in \mathbb{R}$

$$\frac{f(x)}{g(x)} \leq c \tag{3.29}$$

for some finite constant c (necessarily larger or equal to 1). Let $\{U_n\}_{n \geq 1}$ be a sequence of IID random variables uniformly distributed on $[0, 1]$.

Theorem 3.1.48 *Let τ be the first index $n \geq 1$ for which*

$$U_n \leq \frac{f(Y_n)}{cg(Y_n)}$$

and let $Z = Y_\tau$. Then

(a) Z admits the probability density function f, and

(b) $\mathrm{E}[\tau] = c$.

Proof. We have

$$P(Z \leq x) = P(Y_\tau \leq x) = \sum_{n \geq 1} P(\tau = n, Y_n \leq x).$$

Denote by A_k the event $\left\{ U_k > \frac{f(Y_k)}{cg(Y_k)} \right\}$. Then

$$\begin{aligned}
P(\tau = n, Y_n \leq x) &= P(A_1, \ldots, A_{n-1}, \overline{A_n}, Y_n \leq x) \\
&= P(A_1) \cdots P(A_{n-1}) P(\overline{A_n}, Y_n \leq x).
\end{aligned}$$

$$\begin{aligned}
P\left(\overline{A_k}\right) &= \int_{\mathbb{R}} P\left(U_k \leq \frac{f(y)}{cg(y)} \right) g(y)\, dy \\
&= \int_{\mathbb{R}} \frac{f(y)}{cg(y)} g(y)\, dy = \int_{\mathbb{R}} \frac{f(y)}{c}\, dy = \frac{1}{c}.
\end{aligned}$$

$$\begin{aligned}
P\left(\overline{A_k}, Y_k \leq x\right) &= \int_{\mathbb{R}} P\left(U_k \leq \frac{f(y)}{cg(y)} \right) 1_{y \leq x} g(y)\, dy \\
&= \int_{-\infty}^{x} \frac{f(y)}{cg(y)} g(y)\, dy = \int_{-\infty}^{x} \frac{f(y)}{c}\, dy = \frac{1}{c} \int_{-\infty}^{x} f(y)\, dy.
\end{aligned}$$

Therefore

$$P(Z \leq x) = \sum_{n \geq 1} \left(1 - \frac{1}{c} \right)^{n-1} \frac{1}{c} \int_{-\infty}^{x} f(y)\, dy = \int_{-\infty}^{x} f(y)\, dy.$$

Also, using the above calculations,

$$\begin{aligned}
P(\tau = n) &= P\left(A_1, \ldots, A_{n-1}, \overline{A_n} \right) \\
&= P(A_1) \cdots P(A_{n-1}) P\left(\overline{A_n}\right) = \left(1 - \frac{1}{c} \right)^{n-1} \frac{1}{c},
\end{aligned}$$

from which it follows that $\mathrm{E}[\tau] = c$. $\qquad \square$

The method depends on one's ability to easily generate random vectors with the probability density g. Such a PDF must satisfy (3.29) and c should be as small as possible under this constraint.

3.1.4   Characteristic Functions

Recall that for a complex-valued random variable $X = X_R + iX_I$, where X_R and X_I are real-valued integrable random variables, $\mathrm{E}[X] = \mathrm{E}[X_R] + i\mathrm{E}[X_I]$ defines the expectation of X. The characteristic function φ_X of a real-valued random variable X is defined by

$$\varphi_X(u) = \mathrm{E}\left[e^{iuX}\right].$$

Similarly, the characteristic function $\varphi_X : \mathbb{R}^d \to \mathbb{C}$ of a real random vector $X \in \mathbb{R}^d$ is defined by

$$\varphi_X(u) = \mathrm{E}\left[e^{iu^T X}\right].$$

Theorem 3.1.49 *Let $X \in \mathbb{R}^d$ be a random vector with characteristic function φ. Then for all $1 \leq j \leq d$, all $a_j, b_j \in \mathbb{R}^d$ such that $a_j < b_j$,*

$$\lim_{c\uparrow+\infty} \frac{1}{(2\pi)^d} \int_{-c}^{+c} \cdots \int_{-c}^{+c} \left(\prod_{j=1}^{d} \frac{e^{-iu_j a_j} - e^{-iu_j b_j}}{iu_j} \right) \varphi(u_1, \ldots, u_d)\, du_1 \cdots du_d$$

$$= \mathrm{E}\left[\prod_{j=1}^{d} \left(\frac{1}{2} 1_{\{X_j = a_j \ or \ b_j\}} + 1_{\{a_j < X_j < b_j\}} \right) \right].$$

Proof. We prove the theorem in the univariate case for the sake of notational ease. The multivariate case is a straightforward adaptation of it. Let X be a real-valued random variable with cumulative distribution function F and characteristic function φ. We have to show that for any pair of points a, b with $(a < b)$,

$$\lim_{c\uparrow+\infty} \frac{1}{2\pi} \int_{-c}^{+c} \frac{e^{-iua} - e^{-iub}}{iu} \varphi(u)\, du = \mathrm{E}\left[\left(\frac{1}{2} 1_{\{X = a \ or \ b\}} + 1_{\{a < X < b\}} \right) \right]. \qquad (\star)$$

For this, write

$$\Phi_c := \frac{1}{2\pi} \int_{-c}^{+c} \frac{e^{-iua} - e^{-iub}}{iu} \varphi(u)\, du$$

$$= \frac{1}{2\pi} \int_{-c}^{+c} \frac{e^{-iua} - e^{-iub}}{iu} \left(\int_{-\infty}^{+\infty} e^{iux}\, dF(x) \right) du$$

$$= \frac{1}{2\pi} \int_{-\infty}^{+\infty} \left(\int_{-c}^{+c} \frac{e^{-iua} - e^{-iub}}{iu} e^{iux}\, du \right) dF(x) = \int_{-\infty}^{+\infty} \Psi_c(x)\, dF(x),$$

where

$$\Psi_c(x) := \frac{1}{2\pi} \int_{-c}^{+c} \frac{e^{-iua} - e^{-iub}}{iu} e^{+iux}\, du.$$

The above computations make use of Fubini's theorem. This is allowed since, observing that

$$\left| \frac{e^{-iua} - e^{-iub}}{iu} \right| = \left| \int_{a}^{b} e^{-iux}\, dx \right| \leq (b - a),$$

we have

$$\int_{-c}^{+c} \int_{-\infty}^{+\infty} \left| \frac{e^{-iua} - e^{-iub}}{iu} e^{iux} \right| dF(x)\, du$$

$$= \int_{-c}^{+c} \int_{-\infty}^{+\infty} \left| \frac{e^{-iua} - e^{-iub}}{iu} \right| dF(x)\, du$$

$$\leq \int_{-c}^{+c} \int_{-\infty}^{+\infty} (b - a)\, dF(x)\, du = 2c(b - a) < \infty.$$

Since the function $u \to \frac{\cos(au)}{u}$ is antisymmetric, $\int_{-c}^{+c} \frac{\cos(au)}{u} \, du = 0$, and therefore

$$\Psi_c(x) = \frac{1}{2\pi} \int_{-c}^{+c} \frac{\sin u(x-a) - \sin u(x-b)}{u} \, du$$

$$= \frac{1}{2\pi} \int_{-c(x-a)}^{+c(x-a)} \frac{\sin u}{u} \, du - \frac{1}{2\pi} \int_{-c(x-b)}^{+c(x-b)} \frac{\sin u}{u} \, du \, .$$

The function $c \to \int_0^c \frac{\sin u}{u} \, du = \int_{-c}^0 \frac{\sin u}{u}$ is uniformly continuous in c and tends to $\int_0^{+\infty} \frac{\sin u}{u} \, du = \frac{1}{2}\pi$ as $c \uparrow +\infty$. Therefore the function $(c,x) \to \Psi_c(x)$ is uniformly bounded. Moreover, in view of the above expression for Ψ_c,

$$\lim_{c \uparrow \infty} \Psi_c(x) := \Psi(x) = \begin{cases} 0 & \text{if} \quad x < a \text{ or } x > b \\ \frac{1}{2} & \text{if} \quad x = a \text{ or } x = b \\ 1 & \text{if} \quad a < x < b. \end{cases}$$

Therefore, by dominated convergence,

$$\lim_{c \uparrow \infty} \Phi_c = \int_{-\infty}^{+\infty} \lim_{c \uparrow \infty} \Psi_c(x) \, dF(x)$$

$$= \int_{-\infty}^{+\infty} \Psi(x) \, dF(x) = E\left[\left(\frac{1}{2} 1_{\{X=a \text{ or } b\}} + 1_{\{a<X<b\}} \right) \right].$$

$\square$

Note that, in the univariate case, denoting by F the cumulative distribution function of the random variable X,

$$E\left[\left(\frac{1}{2} 1_{\{X=a \text{ or } b\}} + 1_{\{a<X<b\}} \right) \right] = \frac{F(b) + F(b-)}{2} - \frac{F(a) + F(a-)}{2},$$

so that formula $(\star)$ takes the perhaps more familiar form

$$\frac{F(b) + F(b-)}{2} - \frac{F(a) + F(a-)}{2} = \lim_{c \uparrow +\infty} \frac{1}{2\pi} \int_{-c}^{+c} \frac{e^{-iua} - e^{-iub}}{iu} \varphi(u) \, du \, .$$

This is Paul Lévy's inversion formula.

Corollary 3.1.50 *If the random variable X admits a probability density f and if moreover its characteristic function φ is integrable, then*

$$f(x) = \frac{1}{2\pi} \int_{-\infty}^{+\infty} \varphi(u) e^{-iux} \, du \, . \tag{3.30}$$

Proof. With f defined as in (3.30), we have, by Fubini,

$$\int_a^b f(x) \, dx = \int_a^b \frac{1}{2\pi} \int_{-\infty}^{+\infty} \varphi(u) e^{-iux} \, du \, dx$$

$$= \frac{1}{2\pi} \int_{-\infty}^{+\infty} \varphi(u) \left(\int_a^b e^{-iux} \, du \right) dx$$

$$= \lim_{c \uparrow \infty} \frac{1}{2\pi} \int_{-c}^{+c} \varphi(u) \left(\int_a^b e^{-iux} \, du \right) dx$$

$$= \lim_{c \uparrow \infty} \frac{1}{2\pi} \int_{-c}^{+c} \varphi(u) \frac{e^{-iua} - e^{-iub}}{iu} \, dx$$

$$= F(b) - F(a),$$

by Paul Lévy's inversion formula. This proves that f is a PDF of X.

$\square$

Corollary 3.1.51 *The distribution of a random vector of $\mathbb{R}^d$ is uniquely determined by its characteristic function.*

Proof. Let X and Y be two vectors of $\mathbb{R}^d$ with the same characteristic function φ. Lévy's inversion formula shows that the distributions of X and Y agree on $\mathcal{A}$, the class of rectangles $A = \prod_{j=1}^d (a_j, b_j]$ whose boundary has a null measure with respect to the distributions of both X and Y. Since there are at most a countable number of rectangles whose boundary has positive measure with respect to the distributions of both X and Y, $\mathcal{A}$ generates $\mathcal{B}(\mathbb{R}^d)$. Moreover, $\mathcal{A}$ is a π-system and therefore, by Theorem 2.1.43, the two distributions coincide. □

Theorem 3.1.52 *For the random variables $X_1, \ldots, X_d$ to be independent, a necessary and sufficient condition is that the characteristic function φ_X of $X = (X_1, \ldots, X_d)$ factorizes as*

$$\varphi_X(u_1, \ldots, u_d) = \prod_{j=1}^d \varphi_j(u_j),$$

where for all $1 \le j \le d$, φ_j is a characteristic function. In this case, for all $1 \le j \le d$, $\varphi_j = \varphi_{X_j}$, the characteristic function of X_j.

Proof. Necessity. Write

$$\varphi_X(u) = \mathrm{E}\left[e^{i\sum_{j=1}^d u_j X_j}\right]$$

$$= \mathrm{E}\left[\prod_{j=1}^d e^{iu_j X_j}\right] = \prod_{j=1}^d \mathrm{E}\left[e^{iu_j X_j}\right] = \prod_{j=1}^d \varphi_{X_j}(u_j),$$

by the product formula for expectations (Theorem 3.1.41).

Sufficiency. Let $X' := (X_1', \ldots, X_d') \in \mathbb{R}^d$ be a random vector whose *independent* coordinate random variables $X_1', \ldots, X_d'$ have the respective characteristic functions $\varphi_1, \ldots, \varphi_d$. The characteristic function of X' is $\prod_{j=1}^d \varphi_j(u_j)$ and therefore X and X' have the same distribution. In particular, $X_1, \ldots, X_d$ are independent random variables with respective characteristic functions $\varphi_1, \ldots, \varphi_d$. □

Ladder Random Variables

Definition 3.1.53 *A real random variable X is called a* ladder random variable *if there exists a and h in $\mathbb{R}$ such that*

$$\sum_{n \in \mathbb{Z}} P(X = a + nh) = 1.$$

It is called degenerate *if $P(X = a) = 1$ for some $a \in \mathbb{R}$.*

Theorem 3.1.54 *Let φ be the characteristic function of a real random variable X. If $|\varphi(t_0)| = 1$ for some $t_0 \in \mathbb{R}$, $t_0 \neq 1$, then X is a ladder random variable.*

Proof. The hypothesis implies that there exists an $a \in \mathbb{R}$ such that

$$e^{iat_0} = \mathrm{E}\left[e^{it_0 X}\right].$$

This implies in particular (considering the real parts) that

$$1 - \mathrm{E}\left[\cos(t_0(X - a))\right] = \mathrm{E}\left[1 - \cos(t_0(X - a))\right] = 0.$$

Since $1 - \cos(t_0(X - a)) \geq 0$, this implies that, P-a.s.,

$$1 = \cos(t_0(X - a)),$$

which implies the announced result. □

Random Sums and Wald's Identity

The next two results concerning random sums were proved in the discrete case (Theorems 1.4.10 and 1.4.11). The proofs are similar.

Theorem 3.1.55 *Let $\{Y_n\}_{n \geq 1}$ be an IID sequence of random variables with the common characteristic function φ_Y. Let T be a random variable, integer-valued, independent of the sequence $\{Y_n\}_{n \geq 1}$, and let g_T be its generating function. The characteristic function of the random variable*

$$X = \sum_{n=1}^{T} Y_n,$$

where by convention $\sum_{n=1}^{0} = 0$, is

$$\varphi_X(u) = g_T(\varphi_Y(u)). \tag{3.31}$$

Proof. We just adapt the proof of Theorem 1.4.10.

$$e^{iuX} = e^{iu\sum_{n=1}^{T} Y_n} = \left(\sum_{k=0}^{\infty} 1_{\{T=k\}}\right) e^{iu\sum_{n=1}^{T} Y_n}$$

$$= \sum_{k=0}^{\infty} \left\{\left(e^{iu\sum_{n=1}^{T} Y_n}\right) 1_{\{T=k\}}\right\} = \sum_{k=0}^{\infty} \left(e^{iu\sum_{n=1}^{k} Y_n}\right) 1_{\{T=k\}}.$$

Therefore,

$$\mathrm{E}[e^{iuX}] = \sum_{k=0}^{\infty} \mathrm{E}\left[1_{\{T=k\}}\left(e^{iu\sum_{n=1}^{k} Y_n}\right)\right]$$

$$= \sum_{k=0}^{\infty} \mathrm{E}[1_{\{T=k\}}]\mathrm{E}[e^{iu\sum_{n=1}^{k} Y_n}],$$

where we have used the independence of T and $\{Y_n\}_{n \geq 1}$. Now, $\mathrm{E}[1_{\{T=k\}}] = P(T = k)$, and $\mathrm{E}[e^{iu\sum_{n=1}^{k} Y_n}] = \varphi_Y(u)^k$, and therefore

$$\mathrm{E}[e^{iuX}] = \sum_{k=0}^{\infty} P(T = k)\,\varphi_Y(u)^k = g_T(\varphi_Y(u)).$$

□

Theorem 3.1.56 *Let* $\{Y_n\}_{n\geq 1}$ *be a sequence of integrable random variables such that* $E[Y_n] = E[Y_1]$ $(n \geq 1)$. *Let* T *be an integer-valued random variable such that for all* $n \geq 1$, *the event* $\{T \geq n\}$ *is independent of* Y_n. *Let*

$$X := \sum_{n=1}^{T} Y_n \,.$$

Then

$$\mathrm{E}\left[X\right] = \mathrm{E}[Y_1]\mathrm{E}[T] \,. \tag{3.32}$$

Proof. Same as that of Theorem 1.4.11. □

3.1.5   Laplace Transforms

In some circumstances, it is preferable to work with Laplace transforms rather than with characteristic functions.

Definition 3.1.57 *The* Laplace transform *of a non-negative random variable* X *(resp., of a cumulative probability distribution function* F *on* $\mathbb{R}_+$*) is the function* $t \in \mathbb{R}_+ \mapsto \mathrm{e}^{-tX}$ *(resp.* $t \in \mathbb{R}_+ \mapsto \int_{\mathbb{R}_+} \mathrm{e}^{-tx}\,\mathrm{d}F(x)$*).*

It characterizes the distribution of a non-negative random variable, in the sense that

Theorem 3.1.58 *Two non-negative random variables* X *and* Y *with the same Laplace transforms have the same distribution.*

The proof will be based on the following lemma of intrinsic interest.

Lemma 3.1.59 *Two bounded random variables* X *and* Y *such that* $\mathrm{E}\left[X^n\right] = \mathrm{E}\left[Y^n\right]$ *for all* $n = 0, 1, \ldots$ *have the same distribution.*

Proof. Let $M < \infty$ be a common bound of these variables. The hypothesis implies that for any polynomial P, $\mathrm{E}\left[P(X)\right] = \mathrm{E}\left[P(Y)\right]$. By the Weierstrass approximation theorem,[2] if $h : [0, M] \to \mathbb{R}$ is a continuous function, there exists for any $\varepsilon > 0$ a polynomial P_ε such that $\sup_{0 \leq x \leq M} |h(x) - P_\varepsilon(x)| \leq \varepsilon$. Therefore

$$\mathrm{E}\left[|h(X) - P_\varepsilon(X)|\right] \leq \varepsilon \text{ and } \mathrm{E}\left[|h(Y) - P_\varepsilon(Y)|\right] \leq \varepsilon$$

and

$$\begin{aligned}
\mathrm{E}&\left[|h(X) - h(Y)|\right] \\
&\leq \mathrm{E}\left[|h(X) - P_\varepsilon(X)|\right] + \mathrm{E}\left[|P_\varepsilon(X) - P_\varepsilon(Y)|\right] + \mathrm{E}\left[|h(Y) - P_\varepsilon(Y)|\right] \\
&= \mathrm{E}\left[|h(X) - P_\varepsilon(X)|\right] + \mathrm{E}\left[|h(Y) - P_\varepsilon(Y)|\right] \leq 2\varepsilon \,.
\end{aligned}$$

Since ε can be chosen arbitrarily small, $\mathrm{E}\left[h(X)\right] = \mathrm{E}\left[h(Y)\right]$. By uniformly approximating the indicator function of any interval $(a, b] \subseteq [0, M]$ by a continuous function, we deduce that

$$P(X \in (a, b]) = \mathrm{E}\left[1_{(a,b]}(X)\right] = \mathrm{E}\left[1_{(a,b]}(Y)\right] = P(Y \in (a, b]) \,,$$

that is, $X \overset{\mathcal{D}}{=} Y$. □

---

[2]A refinement of this fundamental result, due to Bernstein, is given in Example 4.2.2.

Remark 3.1.60 Lemma 3.1.59 is a particular instance where the moments $\mu_n :=$ $\int_{-\infty}^{+\infty} |x|^n \, dF(x)$ $(n = 1, 2, \ldots)$ characterize the distribution F of a real random variable. This is not the case in general. A necessary and sufficient condition for this is that $\limsup_{n \uparrow \infty} \frac{\mu_n^{1/n}}{n} < \infty.$[3]

We now prove Theorem 3.1.58.

Proof. The variables $U := \mathrm{e}^{-X}$ and $V := \mathrm{e}^{-Y}$ are bounded and such that for all $n = 0, 1, \ldots,$ $\mathrm{E}\,[U^n] = \mathrm{E}\,[V^n]$, so that $U \stackrel{\mathcal{D}}{=} V$, and therefore $X = -\log U$ and $Y = -\log V$ have the same distribution. $\qquad\square$

A proof similar to the above gives the corresponding result for vectors of non-negative random variables.

Definition 3.1.61 *The* Laplace transform *of a vector of non-negative random variables* $(X_1, \ldots, X_m)$ *is the function* $(t_1, \ldots, t_m) \in (\mathbb{R}_+)^m \mapsto \mathrm{e}^{-(t_1 X_1 + t_m X_m)} \in [0, 1]$.

Theorem 3.1.62 *Two non-negative random vectors* $(X_1, \ldots, X_m)$ *and* $(Y_1, \ldots, Y_m)$ *with the same Laplace transforms have the same distribution.*

3.2 Gaussian vectors

The importance of Gaussian vectors is due to their mathematical tractability, their stability with respect to linear transformations and the fact that their distribution is entirely characterized by their mean and their covariance matrix.

3.2.1 Two Equivalent Definitions

A slight extension of the definition of a Gaussian variable will be needed.

Definition 3.2.1 *An* extended Gaussian variable X *is a real random variable with a characteristic function of the form*

$$\varphi_X(u) = \exp\{imu - \tfrac{1}{2}\sigma^2 u^2\}, \tag{3.33}$$

where $m \in \mathbb{R}$ *and* $\sigma^2 \in \mathbb{R}_+$.

Remark 3.2.2 Note that $\sigma = 0$ is allowed, and in this case, the random variable X is the constant m.

Definition 3.2.3 *An* n-*dimensional real random vector* X *is called a* Gaussian random vector *if the random variable* $\alpha^T X$ *is an extended Gaussian random variable for all* $\alpha \in \mathbb{R}^n$.

We now make the connection with the more classical definition of Gaussian vectors, in terms of their characteristic function.

---

[3]See, for instance, [Shiryaev, 1984], Theorem 7, p. 293.

Theorem 3.2.4 *For a real n-dimensional random vector X to be a Gaussian vector it is necessary and sufficient that its characteristic function φ_X be of the following form:*

$$\varphi_X(u) = \exp\{iu^T m_X - \tfrac{1}{2}u^T \Gamma_X u\}, \tag{3.34}$$

where $m_X \in \mathbb{R}^n$ and where Γ_X is an $n \times n$ matrix that is symmetric and non-negative definite. In this case, the parameters m_X and Γ_X are respectively the mean and the covariance of X.

Proof. Necessary condition. The characteristic function of a Gaussian vector as defined in Definition 3.2.3 is

$$\mathrm{E}[e^{iu^T X}] = \varphi_Z(1),$$

where φ_Z is the CF of $Z = u^T X$. The random variable Z being an extended Gaussian variable,

$$\varphi_Z(1) = \exp\{im_Z - \frac{1}{2}\sigma_Z^2\},$$

where

$$m_Z := \mathrm{E}[Z] = u^T \mathrm{E}[X] = u^T m_X$$

and

$$\sigma_Z^2 := \mathrm{E}[(u^T(X - m_X))(u^T(X - m_X))^T]$$
$$= u^T \mathrm{E}[(X - m_X)(X - m_X)^T]u = u^T \Gamma_X u.$$

Therefore, finally,

$$\varphi_X(u) = \exp\{iu^T m_X - \frac{1}{2}u^T \Gamma_X u\}.$$

Sufficient condition. Let X be a random vector with characteristic function given by (3.34). Let $Z = \alpha^T X$, where $\alpha \in \mathbb{C}^n$. The characteristic function of the random variable Z is

$$\varphi_Z(v) = \mathrm{E}[\exp\{ivZ\}] = \mathrm{E}[\exp\{iv\alpha^T X\}]$$
$$= \exp\{iv(\alpha^T m_X) - \frac{1}{2}v^2(\alpha^T \Gamma_X \alpha)\}.$$

Therefore Z is an extended Gaussian random variable. □

It remains to prove the existence of a random vector with the characteristic function (3.34):

Theorem 3.2.5 *Let Γ be a non-negative definite $d \times d$ matrix and let $m \in \mathbb{R}^d$. There exists a vector $X \in \mathbb{R}^d$ with characteristic functions (3.34).*

Proof. Since Γ is non-negative definite, there exists a matrix A of the same dimension and such that $\Gamma = AA^T$. Let $X = m + AZ$ where Z is a vector of independent standard Gaussian variables $Z_1, \ldots, Z_d$. Then X has the required characteristic functions. In fact, the characteristic function of Z is

$$\varphi_Z(u_1, \ldots, u_d) = \mathrm{E}\left[e^{i\sum_{j=1}^d u_j Z_j}\right] = \prod_{j=1}^d \mathrm{E}\left[e^{iu_j Z_j}\right]$$

$$= \prod_{j=1}^d \varphi_{Z_j}(u_j) = e^{-\frac{1}{2}\sum_{j=1}^d u_j^2} = e^{-\frac{1}{2}\|u\|^2},$$

and therefore

$$\varphi_X(u) = \mathrm{E}\left[e^{i\langle u, m+AZ\rangle} \right] = e^{i\langle u, m\rangle} \varphi_Z(u^T A)$$
$$= e^{i\langle u, m\rangle} e^{-\frac{1}{2}\|u^T A\|^2} = e^{i\langle u, m\rangle - \frac{1}{2}\langle u, \Gamma u\rangle}.$$

$\square$

It is clear from the above definition of X that the parameters m and Γ in (3.34) are respectively its mean vector and covariance matrix.

We shall now give useful formulas concerning the moments of a centered n-dimensional Gaussian vector $X = (X_1, \ldots, X_n)^T$ with the covariance matrix $\Gamma = \{\sigma_{ij}\}$. First, we have

$$\mathrm{E}[X_{i_1} X_{i_2}, \ldots, X_{i_{2k}}] = \sum_{\substack{(j_1, \ldots, j_{2k}) \\ j_1 < j_2, \ldots, j_{2k-1} < j_{2k}}} \sigma_{j_1 j_2} \sigma_{j_3 j_4} \cdots \sigma_{j_{2k-1} j_{2k}}, \qquad (3.35)$$

where the summation extends over all permutations $(j_1, \ldots, j_{2k})$ of $\{i_1, \ldots, i_{2k}\}$ such that $j_1 < j_2, \ldots, j_{2k-1} < j_{2k}$. There are $1 \cdot 3 \cdot 5 \ldots (2k-1)$ terms in the right-hand side of Eq. (3.35). The indices $i_1, \ldots, i_{2k}$ are in $\{1, \ldots, n\}$ and they may occur with repetitions. For instance

$$\mathrm{E}[X_1 X_2 X_3 X_4] = \sigma_{12}\sigma_{34} + \sigma_{13}\sigma_{24} + \sigma_{14}\sigma_{23}$$
$$\mathrm{E}[X_1^2 X_2^2] = \sigma_{11}\sigma_{22} + \sigma_{12}\sigma_{12} + \sigma_{12}\sigma_{12} = \sigma_1^2 \sigma_2^2 - 2\sigma_{12}^2$$
$$\mathrm{E}[X_1^4] = 3\sigma_{11}^2 = 3\sigma_1^4$$
$$\mathrm{E}[X_1^{2k}] = 1 \cdot 3 \ldots (2k-1)\sigma_1^{2k}.$$

Also the odd moments of a centered Gaussian vector are null:

$$\mathrm{E}[X_{i_1} \ldots X_{i_{2k+1}}] = 0, \qquad (3.36)$$

for all $(i_1, \ldots, i_{2k+1}) \in \{1, 2, \ldots, n\}^{2k+1}$. The proof is by differentiation of the characteristic function (Theorem 4.4.12). The details are left to the reader (Exercise 3.4.21).

3.2.2 Independence and Non-correlation

In general, non-correlation does not imply independence. However, this is nearly (see Example 3.2.8 below) true in the case of Gaussian vectors. We start with a definition in view of correctly stating the announced result.

Definition 3.2.6 *Two random real vectors X and Y of respective dimensions n and q are said to be* jointly Gaussian *if the vector Z defined by*

$$Z^T = (X^T, Y^T) = (X_1, \ldots, X_n, Y_1, \ldots, Y_q)$$

is a Gaussian vector.

Theorem 3.2.7 *Two jointly Gaussian random vectors* X *and* Y *of respective dimensions* n *and* q *are independent if and only if they are uncorrelated (that is,* $\Gamma_{XY} = 0$*).*

Proof. Necessity: If X and Y are independent then, by the product formula for expectations,

$$\mathrm{E}[(X - m_X)(Y - m_Y)^T] = \mathrm{E}[X - m_X]\mathrm{E}[Y - m_Y]^T = 0 \,.$$

Sufficiency: If X and Y are uncorrelated the vector Z has for covariance matrix

$$\Gamma_Z = \begin{pmatrix} \Gamma_X & 0 \\ 0 & \Gamma_Y \end{pmatrix},$$

and mean

$$m_Z = \begin{pmatrix} m_X \\ m_Y \end{pmatrix}.$$

It is a Gaussian vector by hypothesis and therefore, with $w = (u_1, \ldots, u_n, v_1, \ldots, v_q)^T$,

$$\mathrm{E}[\exp\{i(u^T X + v^T Y)\}] = \mathrm{E}[\exp\{iw^T Z\}]$$
$$= \exp\{iw^T m_Z - \frac{1}{2}w^T \Gamma_Z w\}$$
$$= \exp\{i(u^T m_X + v^T m_Y) - \frac{1}{2}u^T \Gamma_X u - \frac{1}{2}v^T \Gamma_Y v\}$$
$$= \mathrm{E}[\exp\{iu^T X\}]\mathrm{E}[\exp\{iv^T Y\}]\,,$$

and the conclusion follows from the factorization theorem of characteristic functions (Theorem 3.1.52). $\qquad\square$

EXAMPLE 3.2.8: GAUSSIAN, UNCORRELATED, NOT JOINTLY GAUSSIAN. Let X and U be two independent random variables, where $X \sim \mathcal{N}(0,1)$ and $U \in \{-1, 1\}$, $P(U = \pm 1) = \frac{1}{2}$. We show that

$$Y = UX \sim \mathcal{N}(0,1)$$

and therefore X and Y are *separately* Gaussian. However, we also show that they are *not* jointly Gaussian, that they are uncorrelated and yet not independent. The proof of the above statements is as follows.

$$P(Y \le x) = P(UY \le x) = P(U = 1, X \le x) + P(U = -1, X \ge -x)$$
$$= P(U = 1)P(X \le x) + P(U = -1)P(X \ge -x)$$
$$= \frac{1}{2}P(X \le x) + \frac{1}{2}P(X \ge -x) = P(X \le x)\,.$$

Also, $\mathrm{E}[YX] = \mathrm{E}[UX^2] = \mathrm{E}[U]\mathrm{E}[X^2] = 0$, that is, Y and X are uncorrelated. We show that they are *not* independent. We have

$$P(X^2 = Y^2) = 1\,.$$

If X and Y were independent, since they are absolutely continuous, (X, Y) would admit a probability density, say, $f_{X,Y}(x, y)$. Then

$$P(X^2 = Y^2) = \int_{\mathbb{R}} \int_{\mathbb{R}} 1_{\{x^2 = y^2\}}\, f_{X,Y}(x, y)\mathrm{d}x\, \mathrm{d}y = 0,$$

since the set $\{(x, y); x^2 = y^2\}$ has null Lebesgue measure. Hence a contradiction.

The reason why Theorem 3.2.7 cannot applied is that (X, Y) is not a Gaussian vector. If it were, then $X - Y$ would be an extended Gaussian random variable. Obviously $X - Y$ is not a constant. The only case remaining is that in which $X - Y$ has a probability density, and therefore $P(X - Y = 0) = 0$. But this is incompatible with

$$P(X - Y = 0) = P(U = +1) = \frac{1}{2}.$$

3.2.3 The PDF of a Non-degenerate Gaussian Vector

If Γ_X is degenerate, X *cannot* have a probability density (see Theorem 3.1.34). However:

Theorem 3.2.9 *Let X be an n-dimensional Gaussian vector with mean m and covariance matrix Γ, and assume that it is non-degenerate, that is, $(u^T \Gamma u = 0)$ implies $u = 0$. Then X admits the probability density*

$$f_X(x) = \frac{1}{(2\pi)^{n/2}(\det \Gamma)^{1/2}} \exp\left\{-\tfrac{1}{2}(x - m)^T \Gamma^{-1}(x - m)\right\}. \tag{3.37}$$

Proof. Since $\Gamma > 0$ there exists a non-singular matrix A of the same dimension as Γ and such that $\Gamma = AA^T$. Define the n-vector $Z = A^{-1}(X - m)$. By Definition 3.2.3 it is a Gaussian vector, and furthermore $E[Z] = 0$ and

$$\Gamma_Z = A^{-1}\Gamma(A^T)^{-1} = A^{-1}AA^T(A^T)^{-1} = I.$$

Therefore

$$\sigma_Z(u) = E[\exp\{iu^T Z\}] = \exp\left\{-\frac{1}{2}\sum_{i=1}^{n} u_i^2\right\}.$$

Since this is the characteristic function of a Gaussian vector of zero mean and having independent coordinates, we can assert that $Z_1, \ldots, Z_n$ are independent standard random variables. In particular, Z admits the probability density

$$f_Z(z) = \prod_{i=1}^{n} \frac{1}{\sqrt{2\pi}} e^{-\frac{1}{2}z_i^2/2} = \frac{1}{(2\pi)^{n/2}} \exp\left\{-\frac{1}{2}\|z\|^2\right\}.$$

Now, $X = AZ + m$, and therefore by the formula for the smooth change of variables

$$f_X(x) = \frac{1}{|\det A|} f_Y(A^{-1}(x - m))$$

$$= \frac{1}{(\det \Gamma)^{1/2}} \frac{1}{(2\pi)^{n/2}} \exp\left\{-\frac{1}{2}\|A^{-1}(x - m)\|^2\right\},$$

and this is precisely (3.37) since

$$\|A^{-1}(x - m)\|^2 = \left(A^{-1}(x - m)\right)^T \left(A^{-1}(x - m)\right)$$
$$= (x - m)^T A^{-T} A^{-1}(x - m)$$
$$= (x - m)^T \Gamma^{-1}(x - m).$$

$\square$

EXAMPLE 3.2.10: PROBABILITY OF THE QUADRANT. Let (X, Y) be a 2-dimensional Gaussian vector with probability density

$$f(x, y) = \frac{1}{2\pi(1-\rho^2)^{1/2}} \exp\left\{-\frac{1}{2(1-\rho^2)}\left(x^2 - 2\rho xy + y^2\right)\right\},$$

where $|\rho| < 1$. We shall prove that X and $(Y - \rho X)/(1-\rho^2)^{1/2}$ are independent Gaussian random variables with mean 0 and variance 1, and deduce from this that

$$P(X > 0, Y > 0) = \frac{1}{4} + \frac{1}{2\pi}\sin^{-1}(\rho).$$

Proof. The random variables X, Y are Gaussian, of mean 0, variance 1, and $E[XY] = \text{cov}(X, Y) = \rho$. Therefore $Z = Y - \rho X/(1-\rho^2)^{1/2}$ is a Gaussian random variable, of mean 0, and variance

$$\frac{1}{1-\rho^2}\left(\sigma_Y^2 + \rho^2\sigma_X^2 - 2\rho\,\text{cov}(X, Y)\right) = \frac{1}{1-\rho^2}\left(1 + \rho^2 - 2\rho^2\right) = \frac{1-\rho^2}{1-\rho^2} = 1.$$

Moreover,

$$\text{cov}(X, Z) = E[XZ] = E\left[X\frac{Y - \rho X}{(1-\rho^2)^{1/2}}\right]$$

$$= \frac{1}{(1-\rho^2)^{1/2}}\left(E[XY] - \rho E[X^2]\right) = \frac{1}{(1-\rho^2)^{1/2}}(\rho - \rho) = 0.$$

Thus X, Z are uncorrelated, and since they are jointly Gaussian (as linear combinations of the same Gaussian vectors), they are independent.

The probability density of (X, Y) is therefore

$$g(x, z) = \frac{1}{2\pi}\exp\left\{-\frac{1}{2}\left(x^2 + z^2\right)\right\}.$$

On the other hand

$$P(X > 0, Y > 0) = P\left(X > 0, \rho X + Z(1-\rho^2)^{1/2} > 0\right)$$

$$= P\left(X > 0, Z > -\frac{\rho}{(1-\rho^2)^{1/2}}X\right).$$

Let $\alpha := -\tan^{-1}\left(\rho/(1-\rho^2)^{1/2}\right) = (\sin^{-1}\rho)$. Passing to polar coordinates, we have

$$P(X > 0, Y > 0) = \int_0^\infty \int_{-\frac{\rho}{(1-\rho^2)^{1/2}}x}^\infty g(x, z)\,dxdz$$

$$= \int_\alpha^{\frac{\pi}{2}} \int_0^\infty \frac{1}{2\pi}\exp\left\{-\frac{1}{2}r^2\right\} r\,drd\theta$$

$$= \int_\alpha^{\frac{\pi}{2}} \frac{1}{2\pi}d\theta = \frac{1}{4} - \frac{\alpha}{2\pi}.$$

$\square$

3.3 Conditional Expectation

This central notion of probability theory was introduced in the elementary discrete variable setting of the first chapter (Section 1.3.3). We shall now see the intermediate theory. However, the impatient reader or the initiate can skip to Subsection 3.3.2.

3.3.1 The Intermediate Theory

Let X and Y be two discrete random variables with values in E and F, respectively. Let the function $g : E \times F \to \mathbb{R}_+$ be either non-negative or such that $g(X, Y)$ is integrable. For any non-negative bounded function $\varphi : \mathbb{R}^n \to \mathbb{R}$, it is readily checked that

$$\mathrm{E}\left[\mathrm{E}^Y\left[g(X,Y)\right]\varphi(Y)\right] = \mathrm{E}\left[g(X,Y)\varphi(Y)\right]. \tag{3.38}$$

This suggests to take (3.38) as a definition of conditional expectation, where X and Y are random elements taking their values in spaces E and F, respectively, which can be either discrete or some $\mathbb{R}^k$. This would include mixed cases of the type discrete/absolutely continuous, but also more complex situations, such as the following: $E = \mathbb{R}^p$ and $F = \mathbb{R}^n$, X and Y admit a PDF, but the pair (X, Y) does not admit a PDF (for instance, in the univariate case, if $Y = X^2$).

We shall now give more generality to the study (but only in appearance) and take for the conditioned variable a real random variable Z (previously, in Subsection 1.2.3, we treated the case $Z = g(X, Y)$). In the sequel, we call Y-*measurable* a random variable U of the form $U = \varphi(Y)$, where $\varphi : \mathbb{R}^n \to \mathbb{R}$.

Definition 3.3.1 *Let Z and Y be as above, and suppose that Z is either non-negative or integrable. The* conditional expectation $\mathrm{E}^Y[Z]$ *is by definition the essentially unique variable of the form $\psi(Y)$ such that the equality*

$$\mathrm{E}\left[\psi(Y)U\right] = \mathrm{E}\left[ZU\right] \tag{3.39}$$

holds for any non-negative bounded Y-measurable real random variable U.

By "essentially unique" the following is meant. If there are two functions ψ_1 and ψ_2 that meet the requirement, then $\psi_1(Y) = \psi_2(Y)$ almost surely, that is, $P(\psi_1(Y) = \psi_2(Y)) = 1$.

Theorem 3.3.2 *In the situation described in the above definition, the conditional expectation exists and is essentially unique.*

Proof. The proof of existence is postponed to the next subsection. In practice, we shall always be able to find "a" function ψ by construction. The uniqueness part that we now agree to prove will guarantee that it is (essentially) "the" function ψ.

For uniqueness, suppose that ψ_1 and ψ_2 meet the requirement. In particular, $\mathrm{E}\left[\psi_1(Y)\varphi(Y)\right] = \mathrm{E}\left[\psi_2(Y)\varphi(Y)\right]$ ($= \mathrm{E}\left[Z\varphi(Y)\right]$), or $\mathrm{E}\left[(\psi_1(Y) - \psi_2(Y))\varphi(Y)\right] = 0$, for any non-negative bounded function $\varphi : \mathbb{R}^n \to \mathbb{R}$. Choose $\varphi(Y) = 1_{\{\psi_1(Y)-\psi_2(Y)>0\}}$ to obtain

$$\mathrm{E}\left[(\psi_1(Y) - \psi_2(Y))1_{\{\psi_1(Y)-\psi_2(Y)>0\}}\right] = 0.$$

Since the random variable $(\psi_1(Y) - \psi_2(Y))1_{\{\psi_1(Y)-\psi_2(Y)>0\}}$ is non-negative and has a null expectation, it must be almost surely null. In other words, $\psi_1(Y) - \psi_2(Y) \leq 0$ almost surely. Exchanging the roles of ψ_1 and ψ_2, we have that $\psi_1(Y) - \psi_2(Y) \geq 0$ almost surely. Therefore $\psi_1(Y) - \psi_2(Y) = 0$ almost surely. $\square$

EXAMPLE 3.3.3: THE DISCRETE CASE REVISITED. If Y is a positive integer-valued random variable, then

$$E^Y[Z] = \sum_{n=1}^{\infty} \frac{E[Z 1_{\{Y=n\}}]}{P(Y=n)} 1_{\{Y=n\}},$$

where, by convention, $\frac{E[Z 1_{\{Y=n\}}]}{P(Y=n)} = 0$ when $P(Y=n) = 0$ (in other terms, the sum is over all n such that $P(Y=n) > 0$).

Proof. We must verify (3.39) for all bounded measurable $\varphi : \mathbb{R} \to \mathbb{R}$. The right-hand side is equal to

$$E\left[\left(\sum_{n\geq 1} \frac{E[Z 1_{\{Y=n\}}]}{P(Y=n)} 1_{\{Y=n\}}\right)\left(\sum_{k\geq 1} \varphi(k) 1_{\{Y=k\}}\right)\right]$$

$$= E\left[\sum_{n\geq 1} \frac{E[Z 1_{\{Y=n\}}]}{P(Y=n)} \varphi(n) 1_{\{Y=n\}}\right] = \sum_{n\geq 1} \frac{E[Z 1_{\{Y=n\}}]}{P(Y=n)} \varphi(n) E[1_{\{Y=n\}}]$$

$$= \sum_{n\geq 1} \frac{E[Z 1_{\{Y=n\}}]}{P(Y=n)} \varphi(n) P(Y=n) = \sum_{n\geq 1} E[Z 1_{\{Y=n\}}] \varphi(n)$$

$$= \sum_{n\geq 1} E[Z 1_{\{Y=n\}} \varphi(n)] = E[Z(\sum_{n\geq 1} \varphi(n) 1_{\{Y=n\}})] = E[Z\varphi(Y)] \, .$$

$\square$

EXAMPLE 3.3.4: THE ABSOLUTELY CONTINUOUS CASE. Let X and Y be the random vectors of dimensions p and n, respectively, with the joint probability density $f_{X,Y}(x, y)$. Let $g : \mathbb{R}^{p+n} \to \mathbb{R}$ be a measurable function, and suppose that the random variable $Z = g(X, Y)$ is integrable. The conditional expectation of Z given Y is the random variable

$$\psi(Y) = \int_{\mathbb{R}^P} g(x, Y) f(x \mid Y) \mathrm{d}x \, ,$$

where

$$f(x \mid Y) := \frac{f_{X,Y}(x, Y)}{f_Y(Y)} \, .$$

(Recall that $P(f_Y(Y) = 0) = 0$; see Remark 3.1.8.)

Proof. We first check that $\psi(Y)$ is integrable. We have

$$|\psi(Y)| \leq \int_{\mathbb{R}^P} |g(x, Y)| f(x \mid Y) \, \mathrm{d}x$$

and therefore

$$E[|\psi(Y)|] = \int_{\mathbb{R}^n} |\psi(y)| f_Y(y) \, \mathrm{d}y \leq \int_{\mathbb{R}^n} \left(\int_{\mathbb{R}^P} |g(x,y)| f(x \mid y) \mathrm{d}x\right) f_Y(y) \, \mathrm{d}y$$

$$= \int_{\mathbb{R}^P} \int_{\mathbb{R}^n} |g(x,y)| f_{X,Y}(x, y) \, \mathrm{d}x \, \mathrm{d}y = E[|g(X, Y)|] = E[|Z|] < \infty.$$

We must now check that (3.39) is true with $U = \varphi(Y)$ bounded. The right-hand side is

$$
\begin{aligned}
E[\psi(Y)\varphi(Y)] &= \int_{\mathbb{R}^n} \psi(y)\varphi(y)f_Y(y)\,\mathrm{d}y \\
&= \int_{\mathbb{R}^n} \left(\int_{\mathbb{R}^p} g(x,y)f(x\mid y)\mathrm{d}x \right) \varphi(y)f_Y(y)\,\mathrm{d}y \\
&= \int_{\mathbb{R}^p} \int_{\mathbb{R}^n} g(x,y)\varphi(y)f_{X,Y}(x,y)\,\mathrm{d}x\,\mathrm{d}y \\
&= E[g(X,Y)\varphi(Y)] = E[Z\varphi(Y)].
\end{aligned}
$$

$\square$

EXAMPLE 3.3.5: 2-D GAUSSIAN VECTORS. Let X_1 and X_2 be two random variables with the joint probability density

$$
f_{X_1,X_2}(x_1,x_2) = \frac{1}{2\pi\sigma_1\sigma_2\sqrt{1-\rho^2}} \exp\left\{ -\frac{1}{2(1-\rho^2)} \left(\frac{x_1^2}{\sigma_1^2} - 2\rho\frac{x_1 x_2}{\sigma_1\sigma_2} + \frac{x_2^2}{\sigma_2^2} \right) \right\}.
$$

The random variable X_2 is Gaussian random with mean 0 and variance σ_2^2, that is,

$$
f_{X_2}(x_2) = \frac{1}{\sqrt{2\pi}\sigma_2} \exp\left\{ -\frac{1}{2}\frac{x_2^2}{\sigma_2^2} \right\}.
$$

We then find that

$$
f_{X_1}^{X_2=x_2}(x_1) = \frac{1}{\sqrt{2\pi}\sigma_1\sqrt{1-\rho^2}} \exp\left\{ -\frac{1}{2\sigma_1^2(1-\rho^2)} \left(x_1 - \rho\frac{\sigma_1}{\sigma_2}x_2 \right) \right\}.
$$

Note that this is the probability density (in x_1) of a Gaussian random variable with mean $\rho\frac{\sigma_1}{\sigma_2}x_2$ and variance $\sigma_1^2(1-\rho^2)$.

EXAMPLE 3.3.6: THE MIXED CASE, I. Consider the situation, often encountered in practice, where X is a random vector of dimension p and Y takes its values in $\mathbb{N}_+$. We suppose that for all $k \geq 1$, there is a PDF f_k such that

$$
P(X \in A\mid Y = k) = \int_A f_k(x)\mathrm{d}x
$$

for all $A \in \mathcal{B}(\mathbb{R}^p)$. Then, for any function $g: \mathbb{R}^p \times \mathbb{N}_+ \to \mathbb{R}$ that is non-negative or such that $g(X,Y)$ is integrable, we have

$$
E^X[g(X,Y)] = \psi(Y),
$$

where

$$
\psi(k) = \int_{\mathbb{R}^p} g(x,k)f_k(x)\,\mathrm{d}x.
$$

The proof is left to the reader, and is similar to the proof when (X,Y) has a joint probability distribution.

EXAMPLE 3.3.7: THE MIXED CASE, II. Y is as above and let now X be of the form

$$X = Y + \xi$$

where ξ is a random variable independent of Y with probability density f_ξ. Let $h : \mathbb{R} \to \mathbb{R}$ be such that $E[|h(X)|] < \infty$. We shall compute $E^Y[h(X)] = \psi(Y)$.

We have

$$\psi(k) = \int_\mathbb{R} h(X) f_k(x) \mathrm{d}x$$

and f_k is defined by

$$\int_A f_k(x) \mathrm{d}x = P(X \in A | Y = k),$$

that is,

$$\int_A f_k(x) \mathrm{d}x = \frac{P(X \in A, Y = k)}{P(Y = k)}$$

$$= \frac{P(k + \xi \in A, Y = k)}{P(Y = k)} = \frac{P(k + \xi \in A) P(Y = k)}{P(Y = k)}$$

$$= P(k + \xi \in A) = P(\xi \in A - k) = \int_A f_\xi(x + k) \mathrm{d}x.$$

Therefore

$$f_k(x) = f_\xi(x + k)$$

and

$$\psi(k) = \int_\mathbb{R} h(x) f_\xi(x + k) \mathrm{d}x = \int_\mathbb{R} h(x + k) f_\xi(x) \mathrm{d}x,$$

that is,

$$\psi(k) = E[h(\xi + k)].$$

EXAMPLE 3.3.8: THE MIXED CASE, III. We now treat the second type of mixed case, where the conditioning variable X is a random vector of dimension n, Y is an $\mathbb{N}_+$-valued random variable, with the joint distribution of (X, Y) given by

$$P(Y = i) = \pi(i)$$

for all $i \geq 1$, and

$$P(X \in A | Y = i) = \int_A f_i(x) \, \mathrm{d}x$$

for all $k \geq 1$, and all $A \in \mathbb{R}^n$. For all $i \geq 1, x \in \mathbb{R}^n$, we define

$$\pi_{Y|X}(i|x) = \frac{\pi(i) f_i(y)}{f_X(x)}$$

if $f_X(x) = \sum_{i \geq 1} \pi(i) f_i(x) > 0$, and $\pi_{Y|X}(i|x) = 0$ otherwise. We verify that for all $g : \mathbb{N} \times \mathbb{R}^n \to \mathbb{R}$ such that $E[|g(Y, X)|] < \infty$,

$$E^X[g(Y, X)] = \psi(X),$$

where

$$\psi(x) = \sum_{i \geq 1} g(i, x) \pi_{Y|X}(i|x).$$

Bayesian Tests of Hypotheses

The above example gives an opportunity to present an elementary version of the theory of *hypothesis testing*. We now interpret the set $\{1, 2, \ldots, k\}$ as the possible states of Nature. The variable Y is the actual random state of nature that is not directly available to the observer, who can however see the vector X. The observer devises a guessing strategy in terms of this observation. More precisely, this strategy consists in choosing a partition $\{A_1, A_2, \ldots, A_k\}$ of the range space $\mathbb{R}^n$ of X, the observer deciding that the state of Nature is $\widehat{Y} = i$ if $X \in A_i$ $(1 \geq i \geq k)$. Of course there is a possibility that $\widehat{Y} \neq Y$, in which case one says that an error has occurred. The goal is to find the strategy (the partition) that minimizes this probability of error P_E or, equivalently, maximizes $1 - P_E$. The latter is given by a straightforward application of Bayes' rules:

$$
\begin{aligned}
1 - P_E &= P(\widehat{Y} = Y) \\
&= \sum_{i=1}^{k} P(\widehat{Y} = i \mid Y = i) P(Y = i) \\
&= \sum_{i=1}^{k} P(X \in A_i \mid Y = i) P(Y = i) \\
&= \sum_{i=1}^{k} \int_{A_i} f_i(x) \, \mathrm{d}x \times \pi_i \\
&= \int_{\mathbb{R}^n} \sum_{i=1}^{k} 1_{A_i}(x) \pi_i f_i(x) \, \mathrm{d}x \,,
\end{aligned}
$$

It is clear from this expression that any partition such that

$$
x \in A_i \Longleftarrow \pi_i f_i(x) \geq \max_{k} \pi_k f_k(x)
$$

is optimal.

EXAMPLE 3.3.9: TESTING TWO GAUSSIAN HYPOTHESES. In this example, X is a non-degenerate Gaussian vector with covariance matrix Γ, with mean m_1 under hypothesis 1 and m_2 under hypothesis 2. It is moreover assumed that the two hypotheses are equiprobable. In this case

$$
f_i(x) \sim \exp\{-\frac{1}{2}(x - m_i)^T \Gamma^{-1}(x - m_i)\}
$$

and an optimal strategy is (passing to logarithms)

$$
x \in A_i \Longleftarrow (x - m_i)^T \Gamma^{-1}(x - m_i) = \min_{i} .
$$

For further simplification, suppose that the covariance matrix is the unit matrix, in which case, an optimal strategy is

$$
x \in A_i \Longleftarrow ||x - m_i|| = \min_{i} .
$$

In other words, choose hypothesis i corresponding to the mean vector m_i closest to the observed X.

We shall now give, in the setting of the intermediate theory, the main rules that are useful in computing conditional expectations.

In the following, Y is a random variable, Z, Z_1, Z_2 are integrable (resp. non-negative finite) random variables and $\lambda_1, \lambda_2 \in \mathbb{R}$ (resp. $\in \mathbb{R}_+$).

Theorem 3.3.10 Rule 1. *(linearity)*

$$E^Y[\lambda_1 Z_1 + \lambda_2 Z_2] = \lambda_1 E^Y[Z_1] + \lambda_2 E^Y[Z_2].$$

Proof. We consider the integrable case. The non-negative case follows *mutatis mutandis*. We must check that $\lambda_1 E^Y[Z_1] + \lambda_2 E^Y[Z_2]$ is Y-measurable (which is part of the definition of a conditional expectation with respect to Y) and that for every bounded Y-measurable random variable U

$$E[(\lambda_1 E^Y[X_1] + \lambda_2 E^Y[X_1])U] = E[(\lambda_1 Z_1 + \lambda_2 Z_2)U].$$

This follows immediately from the definition of $E^Y[Z_i]$, which says that $E[E^Y[Z_i]U] = E[Z_i U], i = 1, 2.$ □

Theorem 3.3.11 Rule 2. *If Z is independent of Y, then*

$$E^Y[Z] = E[Z].$$

Proof. (Non-negative case.) The constant $E[Z]$ (as any constant) is Y-measurable. Moreover, for all bounded Y-measurable random variables U, $E[E[Z]U] = E[ZU]$. In fact, Z and U are independent and therefore $E[ZU] = E[Z]E[U]$. □

Theorem 3.3.12 Rule 3. *If Z is Y-measurable, then*

$$E^Y[Z] = Z.$$

Proof. (Nonnegative case.) In fact, Z is Y-measurable and $E[ZU] = E[ZU]$. □

Theorem 3.3.13 Rule 4. *If $Z_1 \leq Z_2$ P-a.s., then*

$$E^Y[Z_1] \leq E^Y[Z_2], P\text{-a.s.}$$

In particular, if Z is a non-negative random variable, then $E^Y[Z] \geq 0, P$-a.s.

Proof. We consider the integrable case. For any bounded Y-measurable random variable U,

$$E[E^Y[Z_1]U] = E[Z_1 U] \leq E[Z_2 U] = E[E^Y[Z_2]U].$$

Therefore

$$E[(E^Y[Z_2] - E^Y[Z_1])U] \geq 0.$$

Set $U = 1_{\{E^Y[Z_2] < E^Y[Z_2]\}}$ to obtain the announced inequality. □

Theorem 3.3.14 Rule 5. *Let Y_1 and Y_2 be random variables, and let Z be either integrable or non-negative. Then*

$$E^{Y_2}[E^{Y_1,Y_2}[Z]] = E^{Y_2}[Z].$$

Proof. We just have to check that $E^{Y_2}[E^{Y_1,Y_2}[Z]]$ is a version of $E^{Y_2}[Z]$. Since it is a Y_2-measurable variable it remains to show that

$$E[E^{Y_2}[E^{Y_1,Y_2}[Z]]U] = E[ZU],$$

for any bounded (resp. bounded non-negative) Y_2-measurable variable U. Since such a variable is *a fortiori* (Y_1, Y_2)-measurable,

$$E[[E^{Y_1,Y_2}[Z]U] = E[ZU].$$

Moreover,

$$E[E^{Y_2}[E^{Y_1,Y_2}[Z]]U] = E[[E^{Y_1,Y_2}[Z]U],$$

by definition of $E^{Y_2}[E^{Y_1,Y_2}[Z]]$. □

Theorem 3.3.15 *Let Y be a random vector and let Z be of the form $Z = VZ'$, where V is a Y-measurable bounded (resp. non-negative finite) random variable, and Z' is an integrable (resp. non-negative finite) random variable. Then*

$$E^Y[VZ'] = VE^Y[Z'].$$

Proof. We consider the integrable case. We observe that $VE^Y[Z]$ is Y-measurable, and it therefore remains to prove that for all bounded Y-measurable random variables U,

$$E[VZ'U] = E[VE^Y[Z']U].$$

But, since VU is bounded, by definition of $E^Y[Z']$,

$$E[VE^Y[Z']U] = E[VZ'U].$$

□

3.3.2　The General Theory

Previously, the conditioning was with respect to random variables or vectors. We now condition with respect to σ-fields.

Definition 3.3.16 *Let Y be an integrable (resp. finite non-negative) random variable, and let $\mathcal{G}$ be a sub-σ-field of $\mathcal{F}$. A version of the conditional expectation of Y given $\mathcal{G}$ is any integrable (resp. finite non-negative) $\mathcal{G}$-measurable random variable Z such that*

$$E[YU] = E[ZU] \tag{3.40}$$

for all bounded (resp. bounded non-negative) $\mathcal{G}$-measurable random variables U.

Theorem 3.3.17 *Let Y and $\mathcal{G}$ be as above. There exists at least one version of the conditional expectation of Y given $\mathcal{G}$, and it is essentially unique, that is, if Z' is another version of the conditional expectation of Y given $\mathcal{G}$, then $Z = Z'$, P-a.s.*

There will be no problem in representing two versions of this conditional expectation by the same symbol, since, as we just saw, they are P-almost surely equal. We choose the symbol $\mathrm{E}[Y|\mathcal{G}]$ or $\mathrm{E}^{\mathcal{G}}[Y]$ indifferently. From now on we say: $\mathrm{E}^{\mathcal{G}}[Y]$ (or $\mathrm{E}[Y|\mathcal{G}]$) is the conditional expectation of Y given $\mathcal{G}$. The defining equality (3.40) reads

$$\mathrm{E}[YU] = \mathrm{E}[\mathrm{E}^{\mathcal{G}}[Y]U]$$

for all bounded (resp. bounded non-negative) $\mathcal{G}$-measurable random variables U.

Proof. Uniqueness. The integrable case will be treated, the other case being similar. First observe that

$$0 = \mathrm{E}[ZU] - \mathrm{E}[Z'U] = \mathrm{E}[(Z - Z')U]$$

for all bounded non-negative $\mathcal{G}$-measurable U. In particular, with $U = 1_{\{Z>Z'\}}$,

$$\mathrm{E}[(Z - Z')1_{\{Z>Z'\}}] = 0.$$

Since the random variable in the expectation is non-negative, it can have a null expectation only if it is P-a.s. null, that is if P-a.s., $Z \le Z'$. By symmetry, P-a.s., $Z \ge Z'$, and therefore, as announced, $Z = Z'$, P-a.s.

Existence. We do this for the non-negative integrable case, the general case following easily from this special case. Consider the measure ν on $(\Omega, \mathcal{G})$ defined by

$$\nu(A) = \int_A Y \, dP \quad (A \in \mathcal{G}).$$

It is finite (resp. σ-finite) since Y is assumed integrable (resp. finite non-negative). Moreover, if $P(A) = 0$ then $\nu(A) = 0$. Therefore the measure μ on $(\Omega, \mathcal{G})$ that is the restriction of P to $(\Omega, \mathcal{G})$ is absolutely continuous with respect to ν, so that, by the Radon–Nikodým theorem, there exists an integrable (resp. finite non-negative) random variable of $(\Omega, \mathcal{G})$, that is, an integrable (resp. finite non-negative) $\mathcal{G}$-measurable random variable Z of $(\Omega, \mathcal{F})$, such that

$$\nu(A) = \int_A Z \, dP \quad (A \in \mathcal{G}).$$

In particular,

$$\int_\Omega U Y \, dP = \int_\Omega U Z \, dP$$

for all bounded (resp. non-negative bounded) $\mathcal{G}$-measurable random variables U. $\qquad\square$

Connection with the Intermediate Theory

A special case is when

$$\mathcal{G} = \sigma(X). \tag{3.41}$$

Here $X = (X_1, \dots, X_N)$ is an arbitrary random vector defined on $(\Omega, \mathcal{F})$ and $\sigma(X)$ is, by definition, the smallest σ-field that contains all the sets of the form $\{X \in C\}$ where $C \in \mathcal{B}(\mathbb{R}^N)$. In this situation, we adopt the notation $\mathrm{E}^X[Y]$ for $\mathrm{E}^{\mathcal{G}}[Y]$ (or for $\mathrm{E}[Y|\mathcal{G}]$), and we call this (equivalence class of) random variable(s), the conditional expectation of Y given X.

Theorem 3.3.18 *Let X be a random vector with values in the measurable space $(\mathbb{R}^k, \mathcal{B}(\mathbb{R}^k))$. A random variable $Z : (\Omega, \mathcal{F}) \to (\overline{\mathbb{R}}, \mathcal{B}(\overline{\mathbb{R}}))$ is $\sigma(X)$-measurable if and only if there exists a measurable function $g : (\mathbb{R}^k, \mathcal{B}(\mathbb{R}^k)) \to (\overline{\mathbb{R}}, \mathcal{B}(\overline{\mathbb{R}}))$ such that $Z = g(X)$.*

Proof. The "if" part is just the stability of measurability under composition (Theorem 2.1.18). For the necessity, first observe that this is true of simple random variables. It therefore remains to show that it is true for a non-negative random variable Z (from which the general case straightforwardly follows). Such a random variable is the limit of a non-decreasing sequence $\{Z_n\}_{n \geq 1}$ of non-negative simple random variables of the form $g_n(X)$ for some measurable function $g_n : (\mathbb{R}^k, \mathcal{B}(\mathbb{R}^k)) \to (\overline{\mathbb{R}}, \mathcal{B}(\overline{\mathbb{R}}))$. Let M be the (measurable) set on which the sequence $\{g_n\}_{n \geq 1}$ admits a limit. Define $g(x) = \lim g_n(x) 1_M(x)$ (a measurable function). For each ω, $Z(\omega) = \lim g_n(X(\omega))$, which implies that $Z(\omega) \in M$ and that $Z(\omega) = \lim g_n(X(\omega)) = g(X(\omega))$. $\square$

Properties of the Conditional Expectation

The main rules that are useful in computing conditional expectations will be given once more, but this time in the general abstract framework.

Let $\mathcal{G}$ be a sub-σ-field of $\mathcal{F}$, and let Y, Y_1, Y_2 be integrable (resp. non-negative finite) random variables, $\lambda_1, \lambda_2 \in \mathbb{R}$ (resp. $\in \mathbb{R}_+$).

Theorem 3.3.19 Rule 1. *(linearity)*

$$\mathrm{E}^{\mathcal{G}}[\lambda_1 Y_1 + \lambda_2 Y_2] = \lambda_1 \mathrm{E}^{\mathcal{G}}[Y_1] + \lambda_2 \mathrm{E}^{\mathcal{G}}[Y_2].$$

Proof. We consider the integrable case. We must check that $\lambda_1 \mathrm{E}\mathcal{G}[X_1] + \lambda_2 \mathrm{E}\mathcal{G}[X_2]$ is $\mathcal{G}$-measurable (which is part of the definition of a conditional expectation with respect to $\mathcal{G}$) and that for all bounded $\mathcal{G}$-measurable random variable U

$$\mathrm{E}[(\lambda_1 \mathrm{E}\mathcal{G}[X_1] + \lambda_2 \mathrm{E}\mathcal{G}[X_1])U] = \mathrm{E}[(\lambda_1 Y_1 + \lambda_2 Y_2)U].$$

This follows immediately from the definition of $\mathrm{E}\mathcal{G}[X_i]$, which says that $\mathrm{E}[\mathrm{E}\mathcal{G}[X_i]U] = \mathrm{E}[Y_i U]$ $(i = 1, 2)$. $\square$

Theorem 3.3.20 Rule 2. *If Y is independent of $\mathcal{G}$, then*

$$\mathrm{E}^{\mathcal{G}}[Y] = \mathrm{E}[Y].$$

Proof. We consider the integrable case. First recall that the constant $\mathrm{E}[Y]$ is $\mathcal{G}$-measurable. It remains to prove that for all bounded $\mathcal{G}$-measurable random variables U, $\mathrm{E}[\mathrm{E}[Y]U] = \mathrm{E}[YU]$. This is the case since Y and U are independent and therefore $\mathrm{E}[YU] = \mathrm{E}[Y]\mathrm{E}[U]$. $\square$

Theorem 3.3.21 Rule 3. *If Y is $\mathcal{G}$-measurable,*

$$\mathrm{E}^{\mathcal{G}}[Y] = Y.$$

Proof. We consider the integrable case. We must check that Y is $\mathcal{G}$-measurable and that $\mathrm{E}[YU] = \mathrm{E}[YU]$. $\square$

Theorem 3.3.22 Rule 4. *If $Y_1 \leq Y_2$ P-a.s., then*

$$\mathrm{E}^{\mathcal{G}}[Y_1] \leq \mathrm{E}^{\mathcal{G}}[Y_2] \quad P\text{-a.s.} \tag{3.42}$$

In particular, if Y is a non-negative random variable, then $\mathrm{E}^{\mathcal{G}}[Y] \geq 0, P$-a.s.

Proof. We consider the integrable case. The non-negative case follows *mutatis mutandis*. For any bounded $\mathcal{G}$-measurable random variable $U \geq 0$,

$$E[E^{\mathcal{G}}[Y_1]U] = E[Y_1 U] \leq E[Y_2 U] = E[E^{\mathcal{G}}[Y_2]U].$$

Therefore

$$E[(E^{\mathcal{G}}[Y_2] - E^{\mathcal{G}}[Y_1])U] \geq 0.$$

Taking $U = 1_{\{E^{\mathcal{G}}[Y_2] < E^{\mathcal{G}}[Y_2]\}}$, we obtain (3.42). $\qquad\square$

Theorem 3.3.23 Rule 5. *(successive conditioning). Let $\mathcal{H}$ be a sub-σ-field of $\mathcal{F}$ such that $\mathcal{H} \subseteq \mathcal{G}$. Then*

$$E^{\mathcal{H}}[E^{\mathcal{G}}[Y]] = E^{\mathcal{H}}[Y].$$

Proof. We just have to check that $E^{\mathcal{H}}[E^{\mathcal{G}}[Y]]$ is a version of $E^{\mathcal{H}}[Y]$. Since it is an $\mathcal{H}$-measurable variable, it remains to show that it satisfies the equality

$$E[E^{\mathcal{H}}[E^{\mathcal{G}}[Y]]U] = E[YU],$$

for any bounded (resp. bounded non-negative) $\mathcal{H}$-measurable variable U. Since such a variable is *a fortiori* $\mathcal{G}$-measurable,

$$E[[E^{\mathcal{G}}[Y]]U] = E[YU].$$

Moreover,

$$E[E^{\mathcal{H}}[E^{\mathcal{G}}[Y]]U] = E[[E^{\mathcal{G}}[Y]]U],$$

by definition of $E^{\mathcal{H}}[E^{\mathcal{G}}[Y]]$. $\qquad\square$

Theorem 3.3.24 *Let Y be of the form $Y = VZ$, where V is a $\mathcal{G}$-measurable bounded (resp. non-negative finite) random variable, and Z is an integrable (resp. non-negative finite) random variable. Then*

$$E^{\mathcal{G}}[VZ] = VE^{\mathcal{G}}[Z].$$

Proof. We consider the integrable case. We observe that $VE^{\mathcal{G}}[Z]$ is $\mathcal{G}$-measurable, and it remains to prove that for all bounded $\mathcal{G}$-measurable random variables U,

$$E[VZU] = E[VE^{\mathcal{G}}[Z]U].$$

But, since VU is bounded, by definition of $E^{\mathcal{G}}[Z]$,

$$E[VE^{\mathcal{G}}[Z]U] = E[VZU].$$

$\qquad\square$

3.3.3 The Doubly Stochastic Framework

In imprecise but suggestive terms, the expression "doubly stochastic" refers to the situation where the samples are of the form $\omega = (\omega_1, \omega_2) \in \Omega_1 \times \Omega_2$ and are constructed in two steps. One draws ω_1 according to a probability P_1 on some measurable space $(\Omega_1, \mathcal{F}_1)$, and then draws ω_2 in another measurable space $(\Omega_2, \mathcal{F}_2)$ according to a probability "that depends on ω_1". In order to formalize this notion, an extension of the Fubini–Tonelli theorem is needed, whose proof is a straightforward adaptation of the proof of the original result in Subsection 2.3.2.

Let $(\Omega_1, \mathcal{F}_1)$ and $(\Omega_2, \mathcal{F}_2)$ be two measurable spaces, let P_1 be a probability measure on $(\Omega_1, \mathcal{F}_1)$ and let $P_2 : \Omega_1 \times \mathcal{F}_2 \to [0, 1]$ be a probability kernel from $(\Omega_1, \mathcal{F}_1)$ to $(\Omega_2, \mathcal{F}_2)$. By this we mean that

(i) for all $A_2 \in \mathcal{F}_2$, the mapping from Ω_1 to $[0, 1]$ defined by $\omega_1 \to P_2(\omega_1, A_2)$ is measurable with respect to $\mathcal{F}_1$ and $\mathcal{B}([0, 1])$, and

(ii) for all $\omega_1 \in \Omega_1$, $P_2(\omega_1, \cdot)$ is a probability measure on $(\Omega_2, \mathcal{F}_2)$.

Let $(\Omega, \mathcal{F}) := (\Omega_1 \times \Omega_2, \mathcal{F}_1 \otimes \mathcal{F}_2)$. Then, for any non-negative function $X : (\Omega, \mathcal{F}) \to \mathbb{R}$, the mapping $\omega_2 \to X(\omega_1, \omega_2)$ is $\mathcal{F}_2$-measurable for all $\omega_1 \in \Omega_1$ (this is Lemma 2.3.5) and the mapping $\omega_1 \to \int_{\Omega_2} X(\omega_1, \omega_2) P_2(\omega_1, d\omega_2)$ is $\mathcal{F}_1$-measurable (the proof is an immediate adaptation of that of Lemma 2.3.6).

Also (this is the analog of Theorem 2.3.7) there exists a unique probability measure P on $(\Omega, \mathcal{F})$ such that for all $A_1 \in \mathcal{F}_1$, $A_2 \in \mathcal{F}_2$,

$$P(A_1 \times A_2) := \int_{A_1} P_2(\omega_1, A_2) \, P_1(d\omega_1) \,.$$

Finally (this is the analog of Theorem 2.3.9), if X is a non-negative random variable defined on $(\Omega, \mathcal{F})$, then (Tonelli)

$$\int_{\Omega} X(\omega) \, P(d\omega) = \int_{\Omega_1} \left[\int_{\Omega_2} X(\omega_1, \omega_2) P_2(\omega_1, d\omega_2) \right] P_1(d\omega_1) \,.$$

The same is true for a random variable X of arbitrary sign, provided X is P-integrable (Fubini). (The integrability condition is usually checked by applying Tonelli's theorem to the non-negative variable $|X|$.)

We shall slightly abuse notation by denoting the σ-field $\mathcal{F}_1 \otimes \{\Omega_2, \varnothing\}$ on $(\Omega, \mathcal{F})$ by $\mathcal{F}_1$.

Theorem 3.3.25 *Let X be integrable. A version of the conditional expectation of X given $\mathcal{F}_1$ is*

$$\mathrm{E}\left[X \mid \mathcal{F}_1\right](\omega) := \int_{\Omega_2} X(\omega_1, \omega_2) \, P_2(\omega_1, d\omega_2) \,. \tag{3.43}$$

Proof. The right-hand side of (3.43) is well defined and $\mathcal{F}_1$-measurable. Moreover, for all $A_1 \in \mathcal{F}_1$,

$$\int_{A_1 \times \Omega_2} \mathrm{E}\left[X \mid \mathcal{F}_1\right](\omega_1, \omega_2) P(d(\omega_1, \omega_2)) = \int_{A_1} \left[\int_{\Omega_2} X(\omega_1, \omega_2) P_2(\omega_1, d\omega_2) \right] P_1(d\omega_1) \,,$$

that is, with the abuse of notation mentioned above, for all $A_1 \in \mathcal{F}_1$,

$$\int_{A_1} \mathrm{E}\left[X \mid \mathcal{F}_1\right](\omega) P(d\omega) = \int_{A_1} \left[\int_{\Omega_2} X(\omega_1, \omega_2) P_2(\omega_1, d\omega_2) \right] P_1(d\omega_1) \,.$$

$\square$

3.3.4   The L^2-theory of Conditional Expectation

Conditional expectation may be defined for square-integrable random variables in terms of projection from a Hilbert space onto a Hilbert subspace of the latter. More precisely, let $(\Omega, \mathcal{F}, P)$ be a probability space and let $\mathcal{G}$ be a sub-σ-field of $\mathcal{F}$. Denote by $L^2_{\mathbb{R}}(\mathcal{F}, P)$ and $L^2_{\mathbb{R}}(\mathcal{G}, P)$ the Hilbert spaces of $\mathcal{F}$-measurable (resp. $\mathcal{G}$-measurable) square-integrable real random variables. Clearly, $L^2_{\mathbb{R}}(\mathcal{G}, P)$ is a Hilbert subspace of $L^2_{\mathbb{R}}(\mathcal{F}, P)$, and therefore, one can define the projection of an $\mathcal{F}$-measurable variable X on $L^2_{\mathbb{R}}(\mathcal{G}, P)$, denoted by $P^{\mathcal{G}}(X)$. From the general theory of projection (see Theorems C.4.6 and C.4.5), this random variable is the unique (in the L^2-sense) variable Y such that

(i) $Y \in L^2_{\mathbb{R}}(\mathcal{G}, P)$, and

(ii) $\langle U, Y \rangle_{L^2_{\mathbb{R}}(\mathcal{F}, P)} = \langle U, X \rangle_{L^2_{\mathbb{R}}(\mathcal{F}, P)}$ for all $U \in L^2_{\mathbb{R}}(\mathcal{G}, P)$.

In other terms, $P^{\mathcal{G}}(X)$ is the unique (in the L^2-sense) square-integrable random variable Y such that

(a) Y is $\mathcal{G}$-measurable, and

(b) $\mathrm{E}\,[UY] = \mathrm{E}\,[UX]$ for all square-integrable $\mathcal{G}$-measurable variables U.

This shows that $P^{\mathcal{G}}(X) = \mathrm{E}\,[X \mid \mathcal{G}]$.

Starting from there, a proof of Theorem 3.3.17 is easy and left as an exercise.

Complementary reading

[Billingsley, 1979, 1992] and [Kallenberg, 2002].

3.4   Exercises

Exercise 3.4.1. THE PDF IS UNIQUE
Let f_X and f'_X be two PDFs of the same random vector X. Show that $f'_X(x) = f_X(x)$ Lebesgue-almost everywhere.

Exercise 3.4.2. THE TELESCOPE FORMULA, TAKE 1
Prove that for any *non-negative* random variable X,

$$\mathrm{E}[X] = \int_0^\infty [1 - F(x)]\mathrm{d}x\,,$$

by means of the Tonnelli theorem applied to the product measure $\ell \times P$.

Exercise 3.4.3. RESIDUAL TIME
Let $G : \mathbb{R}_+ \to \mathbb{C}$ be the primitive function of the locally integrable function $g : \mathbb{R}_+ \to \mathbb{C}$, that is, for all $x \geq 0$,

$$G(x) = G(0) + \int_0^x g(u)\,\mathrm{d}u\,.$$

(a) Let X be a non-negative random variable with finite mean μ and such that $\mathrm{E}\,[G(X)] < \infty$. Show that

$$\mathrm{E}\,[G(X)] = G(0) + \int_0^\infty g(x)P(X \geq x)\,\mathrm{d}x\,.$$

(b) Let X be as in (a) and let $\widetilde{X}$ be a non-negative random variable with the probability density
$$\mu^{-1} P(X \geq x).$$
Check that this is indeed a probability density[4] and show that its characteristic distribution is
$$E\left[e^{iu\widetilde{X}}\right] = \frac{E\left[e^{iuX}\right] - 1}{i\mu u}.$$

Exercise 3.4.4. DEGENERATE RANDOM VARIABLES
Let φ be the characteristic function of a real random variable X.

(i) Prove that if $|\varphi(t)| = 1$ for all $t \in \mathbb{R}$, then X is degenerate (that is, there exists an $a \in \mathbb{R}$ such that $P(X = a) = 1$).

(ii) Prove that if $|\varphi(t)| = |\varphi(\alpha t)| = 1$ for two different real numbers t and αt in $\mathbb{R}$, where α is irrational, then X is degenerate.

Exercise 3.4.5. THE LAPLACE TRANSFORM
Let X be a non-negative random variable. Prove that
$$\lim_{\theta \uparrow \infty; \theta > 0} E\left[e^{-\theta X}\right] = P(X = 0).$$

Exercise 3.4.6. $P(f(X) = 0) = 0$
Let X be a random vector of $\mathbb{R}^d$ admitting the PDF f. Show that $P(f(X) = 0) = 0$.

Exercise 3.4.7. A RADON–NIKODÝM DERIVATIVE
Let X be a random vector of $\mathbb{R}^d$ defined on a measurable space $(\Omega, \mathcal{F})$, and admitting the PDF f_1 with respect to some probability P_1 on $(\Omega, \mathcal{F})$. Let $f_2 : \mathbb{R}^d \to \mathbb{R}$ be a non-negative measurable function such that $\int_{\mathbb{R}^d} f_2(x)\,dx = 1$ (f_2 is a PDF). Under what condition can we construct a probability P_2 on $(\Omega, \mathcal{F})$ that is absolutely continuous with respect to P_1 and such that X admits the PDF f_2 with respect to P_2. Give the corresponding Radon–Nikodým derivative $L = \frac{dP_2}{dP_1}$.

Exercise 3.4.8. SUMS OF IID UNIFORM VARIABLES
A point inside the unit square $[0, 1]^2 = [0, 1] \times [0, 1]$ is chosen at random according to the following model: $\Omega = [0, 1]^2$, $P(A) = \ell^2 A$. Let X and Y be the coordinate random variables.

1. Compute the PDF of $Z = X + Y$.

2. Compute $E[Z^2]$.

Exercise 3.4.9. THE SQUARE ROOT OF A RANDOM VARIABLE
Let X be a non-negative real random variable with the probability density function f_X. What is the probability density function of the non-negative square root Z of X?

---

[4]The corresponding distribution is called the residual time distribution of X.

Exercise 3.4.10. UNIFORM PDF IN THE DISK
Let $X = (X_1, X_2)$ be an absolutely continuous 2-dimensional random vector uniformly distributed on the disk $D := \{(x_1, x_2); x_1^2 + x_2^2 \leq 1\}$. Its PDF is

$$f_{X_1, X_2}(x_1, x_2) = \frac{1}{\pi} 1_D(x_1, x_2).$$

A. Prove that X_1 and X_2 are not independent random variables.

B. Letting (Z, Θ) be the polar coordinates of (X_1, X_2), that is, $\Theta \in (0, 2\pi]$, $Z \in [0, 1]$ and $X_1 + iX_2 = Ze^{i\Theta}$, show that Z and Θ are independent and give their respective probability distributions.

Exercise 3.4.11. $Z = XY$

(1) Let X be a real-valued random variable with the PDF f_X. Let Y be a positive integer-valued random variable with the distribution $P(Y = k) = p_k$ $(k \geq 1)$. Suppose that X and Y are independent. Show that the random variable $Z = XY$ is absolutely continuous and give its PDF f_Z.

(2) Same setting as in (1) except that Y may take the value 0, with positive probability p_0. What is the CDF of Z?

Exercise 3.4.12. CAUCHY VIA GAUSS
Let X and Y be two independent standard Gaussian variables. Prove that the ratio $\frac{X}{|Y|}$ is a Cauchy random variable.

Exercise 3.4.13. THE HAZARD RATE
Let F be the CDF of a non-negative random variable with a PDF f. Let $I = [-\infty, t_0)$ (t_0 possibly infinite) be the set of $t \in \mathbb{R}_+$ such that $F(t) < 1$. Define the hazard rate function

$$\lambda(t) := \frac{f(t)}{1 - F(t)} \quad (t \in I).$$

(1) Show that for $t \in I$

$$f(t) = \lambda(t) e^{-\int_0^t \lambda(s) ds}.$$

(2) Compute the hazard rate of the exponential variable.

Exercise 3.4.14. MORE HAZARD RATE

(1) Let T_1 and T_2 be two non-negative random variables with hazard rate (see Exercise 3.4.13) $\lambda_1(t)$ and $\lambda_2(t)$. Show that the property $P(T_2 > t) = P(T_1 > t)^\alpha$ (for some $\alpha > 0$) is equivalent to: $\lambda_2(t) = \alpha \lambda_1(t)$.

(2) Let T_1 and T_2 be two non-negative random variables with hazard rate $\lambda_1(t)$ and $\lambda_2(t)$. What is the hazard rate of $T = \min(T_1, T_2)$?

Exercise 3.4.15. $X = \cos(\Phi)$
Let Φ be a random variable uniformly distributed on the interval $[0, 2\pi]$. Compute the mean and the variance of $X := \cos(\Phi)$.

3.4. EXERCISES

Exercise 3.4.16. $Z = UV$
Let U and V be two independent random variables uniformly distributed on $[0, 1]$. Show that the variable $Z = UV$ has a probability density and compute it.

Exercise 3.4.17. THE MAXIMUM OF IID VARIABLES
Let $X_1, X_2, \ldots, X_n$ be independent random variables uniformly distributed on $[0, 1]$. Compute the expectation of $Z = \max(X_1, \ldots, X_n)$.

Exercise 3.4.18. GAUSSIAN VARIABLES
Let $\sigma, m \in \mathbb{R}$, $\sigma > 0$.

(i) Prove that $\int_{\mathbb{R}} e^{-\frac{1}{2}x^2} dx = 2\pi$, and deduce that $f(x) = \frac{1}{\sqrt{2\pi\sigma^2}} e^{-\frac{x^2}{2\sigma^2}}$ is a probability density function on $\mathbb{R}$.

(ii) Prove that $\frac{1}{\sqrt{2\pi\sigma^2}} \int_{\mathbb{R}} x e^{-\frac{1}{2}\left(\frac{x-m}{\sigma}\right)^2} dx = m$.

(iii) Prove that $\frac{1}{\sqrt{2\pi\sigma^2}} \int_{\mathbb{R}} (x - m)^2 e^{-\frac{1}{2}\left(\frac{x-m}{\sigma}\right)^2} dx = \sigma^2$.

Exercise 3.4.19. THE SQUARE OF A GAUSSIAN VARIABLE
Let X be a real random variable with the probability density function $f_X(x) = \frac{1}{(2\pi\sigma^2)^{\frac{1}{2}}} e^{-\frac{x^2}{2\sigma^2}}$. Compute the probability density function of $Y = X^2$.

Exercise 3.4.20. TWO INDEPENDENT RANDOM POINTS ON $[0, 1]$
Two numbers are drawn independently and completely at random on $[0, 1]$. The smaller is larger than $\frac{1}{3}$. Given this information, what is the probability that the larger number exceeds $\frac{3}{4}$.

Exercise 3.4.21. GAUSSIAN MOMENTS
Let $X = (X_1, \ldots, X_n)^T$ be a centered n-dimensional Gaussian vector with covariance matrix $\Gamma = \{\sigma_{ij}\}$. Prove that

$$E[X_{i_1} X_{i_2}, \ldots, X_{i_{2k}}] = \sum_{\substack{(j_1, \ldots, j_{2k}) \\ j_1 < j_2, \ldots, j_{2k-1} < j_{2k}}} \sigma_{j_1 j_2} \sigma_{j_3 j_4} \cdots \sigma_{j_{2k-1} j_{2k}}, \qquad (3.44)$$

where the summation extends over all permutations $(j_1, \ldots, j_{2k})$ of $\{i_1, \ldots, i_{2k}\}$ such that $j_1 < j_2, \ldots, j_{2k-1} < j_{2k}$. (There are $1 \cdot 3 \cdot 5 \ldots (2k - 1)$ terms in the right-hand side of Eq. (3.35). The indices $i_1, \ldots, i_{2k}$ are in $\{1, \ldots, n\}$ and they may occur with repetitions.) Deduce from this the following specific Gaussian moments:

$$E[X_1 X_2 X_3 X_4] = \sigma_{12}\sigma_{34} + \sigma_{13}\sigma_{24} + \sigma_{14}\sigma_{23}$$
$$E[X_1^2 X_2^2] = \sigma_{11}\sigma_{22} + \sigma_{12}\sigma_{12} + \sigma_{12}\sigma_{12} = \sigma_1^2\sigma_2^2 - 2\sigma_{12}^2$$
$$E[X_1^4] = 3\sigma_{11}^2 = 3\sigma_1^4$$
$$E[X_1^{2k}] = 1 \cdot 3 \ldots (2k - 1)\sigma_1^{2k}.$$

Also show that the odd moments of a centered gaussian vector are null:

$$E[X_{i_1} \ldots X_{i_{2k+1}}] = 0, \qquad (3.45)$$

for all $(i_1, \ldots, i_{2k+1}) \in \{1, 2, \ldots, n\}^{2k+1}$.

Exercise 3.4.22. SUMS OF EXPONENTIALS
Compute the PDF of the sum Z of n IID exponential random variables of parameter θ.

Exercise 3.4.23. GEOMETRIC SUMS OF EXPONENTIALS
Let $\{X_n\}_{n \geq 1}$ be a sequence of IID exponential random variables with common mean $\lambda^{-1} > 0$, and let T be a geometric random variable with mean $p^{-1} > 0$ and independent of the above sequence. Show that $Z := X_1 + \ldots + X_T$ admits a PDF. Which one?

Exercise 3.4.24. QUOTIENTS OF UNIFORMS
Let X_1 and X_2 be two independent random variables uniformly distributed over $[0,1]$. Find the PDF of X_1/X_2.

Exercise 3.4.25. QUOTIENTS OF EXPONENTIALS
Let X_1 and X_2 be two independent random variables with a common exponential distribution of mean θ^{-1}. Give the PDF of the vector $(X_1/X_2, X_2)$ and of the variable X_1/X_2.

Exercise 3.4.26. RANDOM ROOTS
The numbers A and B are selected independently and uniformly on the segment $[-1, +1]$. Find the probability that the roots of the equation $x^2 + 2Ax + B$ are real.

Exercise 3.4.27. ISN'T THIS PUZZLING?
Some guy uses his wild imagination to preselect two different numbers, and he does not tell you which ones. Then he chooses one of the two preselected numbers at random (probability $\frac{1}{2}, \frac{1}{2}$). He shows this number to you and asks you to guess if it is the largest of the two numbers he preselected. Are you interested in playing? (meaning: do you think that you have a better guess than the random guess "yes-no probability $\frac{1}{2}, \frac{1}{2}$")? Hint: you might fix for yourself a "reference number", and answer "yes you showed me the largest number" if the guy's number is larger than your reference number, and otherwise answer "no".

Exercise 3.4.28. $X_1 - X_2$
Let X_1 and X_2 be two independent random variables admitting the PDFs f_1 and f_2 respectively. What is the PDF of $X_1 - X_2$?

Exercise 3.4.29. DISCRETE PLUS CONTINUOUS
Let X be a real-valued random variable with a PDF f_X and let Y be an integer-valued random variable with the distribution $P(Y = k) = p_k$, $k \geq 0$. Suppose that X and Y are independent. Show that the sum $Z = X + Y$ is an absolutely continuous random variable and give its PDF.

Exercise 3.4.30. AN AUTOREGRESSIVE GAUSSIAN MODEL
Consider the stochastic sequence $\{X_n\}_{n \geq 0}$ defined by

$$X_{n+1} = aX_n + \epsilon_{n+1} \quad (n \geq 0),$$

where X_0 is a Gaussian random variable of mean 0 and variance c^2, and where $\{\epsilon_n\}_{n \geq 0}$ is an IID sequence of Gaussian variables of mean 0 and variance σ^2, and independent of X_0.

1. Show that for all $n \geq 1$, the vector $(X_0, \ldots, X_n)$ is a Gaussian vector.

2. Express X_n in terms of $X_0, \epsilon_1, \ldots, \epsilon_n$ (and a). Give the mean and variance of X_n.

Exercise 3.4.31. THE FIRST BOX
Let $X = (X_1, \ldots, X_K)$ be a multinomial vector of size (n, K) and parameters $p_1, \ldots, p_K$. Show that X_1 is a binomial random variable of size n and parameter p_1.

Exercise 3.4.32. SUMS OF MULTINOMIAL RANDOM VECTORS
Let $X = (X_1, \ldots, X_k)$ and $Y = (Y_1, \ldots, Y_k)$ be two independent multinomial random vectors of sizes (n, K) and (m, K), respectively, and with the same parameters $p_1, \ldots, p_K$. What is the distribution of $Z = X + Y$?

Exercise 3.4.33. THE INFIMUM OF INDEPENDENT EXPONENTIALS
Let $X_1, \ldots, X_n$ be independent exponential random variables with the respective parameters λ_i, $i \in [0, n]$. Define $Z = \inf(X_1, \ldots, X_n)$ and let J be the (random) index such that $X_J = Z$ (J is for almost all $\omega \in \Omega$ unambiguously defined because, P-almost surely, $X_1, \ldots, X_n$ take different values). Show that Z and J are independent, and give their respective distributions.

Exercise 3.4.34. SUMS OF IID EXPONENTIALS
Let $\{X_n\}_{n \geq 1}$ be an IID sequence of exponential random variables with mean $1/\theta$, where $\theta \in (0, \infty)$. What is the distribution of $Z = X_1 + \cdots + X_n$?

Exercise 3.4.35. PRODUCTS OF INDEPENDENT UNIFORMS
Let $U_1, \ldots, U_n$ be independent uniform random variables on $[0, 1]$. Give the cdf of the random variable $U_1 \times U_2 \times \cdots \times U_n$. (Hint: logarithms and Exercise 3.4.34.)

Exercise 3.4.36. SUMS OF IID EXPONENTIALS
Let $X_1, \ldots, X_n$ be IID exponential random variables with mean λ^{-1}. Give the characteristic function of $X_1 + \cdots + X_n$, and deduce from the result its PDF.

Exercise 3.4.37. $\mathrm{E}[X] - \mathrm{E}[Y]$

1. Let X and Y be real integrable random variables. Prove the following:

$$\mathrm{E}[X] - \mathrm{E}[Y] = \int_{\mathbb{R}} (P(X < t \leq Y) - P(Y < t \leq X)) \, dt.$$

2. Let X and Y be real integrable random variables such that $P(X < Y) = 1$. Show that the expected length of the interval $(X, Y]$ is the Lebesgue integral with respect to t of the probability that this interval covers t.

Exercise 3.4.38. THE COVARIANCE MATRIX OF $AX + b$
Let ψ_X be the characteristic function of the random vector X. What is the characteristic function of the random vector $Y = AX + b$, where A is a matrix and b is a vector of appropriate dimensions?

Exercise 3.4.39. INDEPENDENCE OF AN EVENT AND OF A RANDOM VARIABLE
Let A be some event of positive probability, and let P_A denote the probability P conditioned by A, that is,

$$P_A(\cdot) = P(\cdot \mid A).$$

The random variables X and Y are said to be conditionally independent given A if they are independent with respect to probability P_A. Prove that this is the case if and only if for all $u, v \in \mathbb{R}$,

$$P(A)\mathrm{E}[e^{iuX}e^{ivY}1_A] = \mathrm{E}[e^{iuX}1_A]\mathrm{E}[e^{ivY}1_A].$$

Exercise 3.4.40. INDEPENDENT CAUCHY RANDOM VARIABLES
Let $\{X_n\}_{n\geq 1}$ be a sequence of independent Cauchy random variables. Let T be a positive integer-valued random variable, independent of this sequence. Define $Y = \sum_{n=1}^{T} X_n$. What is the probability distribution of $Z = \frac{Y}{T}$?

Exercise 3.4.41. RANDOM SPLITTING OF A POISSON RANDOM VARIABLE
Let $\{X_n\}_{n\geq 1}$ be independent random variables taking the values 0 and 1 with probability $q = 1 - p$ and p, respectively, where $p \in (0,1)$. Let T be a Poisson random variable with mean $\theta > 0$, independent of $\{X_n\}_{n\geq 1}$. Define

$$S = X_1 + \cdots + X_T.$$

Compute the characteristic function of the vector $(S, T - S)$. Deduce from this that S and $T - S$ are independent Poisson random variable with respective means $p\theta$ and $q\theta$.

Exercise 3.4.42. THE CHARACTERISTIC FUNCTION OF A MULTINOMIAL VECTOR
Let $(X_1, \ldots, X_k)$ be a multinomial random vector of size k and parameters $p_1, \ldots, p_k$ ($p_i > 0$, $p_1 + \ldots + p_k = 1$). Compute the characteristic function of $(X_1, \ldots, X_{k-1})$.

Exercise 3.4.43. A LEAST-SQUARE ESTIMATE
Let X be a square-integrable real random variable. Prove that among all variables $Z = aX + b$, where a and b are real numbers, the one that minimizes the error $\mathrm{E}[(Z - Y)^2]$ is

$$\hat{Y} = m_Y + \frac{\sigma_{XY}}{\sigma_X^2}(X - m_X)$$

and the error is then

$$\mathrm{E}[(\hat{Y} - Y)^2] = \sigma_Y^2(1 - \rho_{XY}^2).$$

Exercise 3.4.44. POISSON AND THE COVARIANCE MATRIX
Let $Z_1, Z_2, \ldots, Z_n$ be independent Poisson random variables with respective means $\theta_1, \theta_2, \ldots, \theta_n$. Define for $1 \leq i \leq n$,

$$X_i = Z_1 + \cdots + Z_i.$$

Compute the covariance matrix of $X = (X_1, \ldots, X_n)^T$.

Exercise 3.4.45. CORRELATION COEFFICIENTS
Let X and Y be square-integrable random variables. Let a, b, c, d be real numbers, $a \neq 0$, $c \neq 0$. Give the correlation coefficient of $aX + b$ and $cY + d$ in terms of the correlation coefficient ρ_{XY} of X and Y.

Exercise 3.4.46. A FUNCTION OF A GAUSSIAN VECTOR
Let X and Y be two independent Gaussian random variables with mean 0 and variance
1. Show that the random variables $\frac{X+Y}{\sqrt{2}}$ and $\frac{X-Y}{\sqrt{2}}$ are independent Gaussian random
variables, and give their means and variances.

Exercise 3.4.47. DEGENERATE RANDOM VECTORS HAVE NO PROBABILITY DENSITY
Show that when the covariance matrix Γ_X of a random vector X is degenerate, this
random vector *cannot* have a probability density.

Exercise 3.4.48. THE I-TH SMALLEST UNIFORM
Find the probability distribution of the random variable Z_i, the i-th smallest among
$X_1, \ldots, X_n$, when the X_i's are independent $[0, 1]$-uniform random variables.

Exercise 3.4.49. THE COVARIANCE MATRIX OF A MULTINOMIAL VECTOR
Compute the covariance matrix of a multinomial random vector of size k and with
parameters $p_1, \ldots, p_k$.

Exercise 3.4.50. CONDITIONAL JENSEN'S INEQUALITY
Let I be a general interval of $\mathbb{R}$ (closed, open, semi-closed, infinite, etc.) and let (a, b)
be its interior, assumed non-empty. Let $\varphi : I \to \mathbb{R}$ be a convex function. Let X be an
integrable real-valued random variable such that $P(X \in I) = 1$. Assume moreover that
either φ is non-negative, or that $\varphi(X)$ is integrable. Prove that for any subσ-field $\mathcal{G} \subseteq \mathcal{F}$

$$E\left[\varphi(X) \,|\, \mathcal{G}\right] \geq \varphi(E\left[X \,|\, \mathcal{G}\right]).$$

Exercise 3.4.51. HYPOTHESIS TESTING
In Example 3.3.9 compute the probability of error for the case where $\Gamma = I$, the identity
matrix.

Exercise 3.4.52. $E[h(X)|X^2]$
Let X be a real random variable with probability density f_X. Let $h : \mathbb{R} \to \mathbb{R}$ be a
function such that $h(X)$ is integrable. Compute $E[h(X)|X^2]$.

Chapter 4

Convergences

Order hidden in chaos: an erratic sequence of coin tosses exhibits a remarkable balance between heads and tails, at least "when the coin is fair and fairly tossed". This phenomenon is captured by a theorem, the *strong law of large numbers*, the heart of probability theory. The relevant mathematical notion is that of *almost-sure convergence* of a sequence of random variables, with close relations with two other types of convergence: in probability and in the quadratic mean.

The most important notion of convergence together with almost-sure convergence is *convergence in distribution* and the main result there is the *central limit theorem*, the core of Statistics, which is the art of assessing the correctness of a proposed probability model (are these variables really independent?) or to estimate a parameter of such a model (is this coin fair?).

4.1 Almost-sure Convergence

4.1.1 A Sufficient Condition and a Criterion

The definition of almost-sure convergence is deceptively simple:

Definition 4.1.1 *A sequence* $\{Z_n\}_{n\geq 1}$ *of random variables with values in* $\mathbb{C}$ *(resp. in* $\overline{\mathbb{R}}$*) is said to* converge P-almost surely *to the random variable* Z *with values in* $\mathbb{C}$ *(resp. in* $\overline{\mathbb{R}}$*) if*

$$P(\lim_{n\uparrow\infty} Z_n = Z) = 1. \tag{4.1}$$

For instance, Theorem 1.1.6 states that the empirical frequency of heads in a sequence of independent tosses of a fair coin converges almost surely to $\frac{1}{2}$.

In the case where the sequence takes values in $\overline{\mathbb{R}}$, the limit may be infinite. Otherwise, when $P(Z < \infty) = 1$, we shall add the precision: "converges to a *finite* limit".

The most frequently used tool for asserting almost sure convergence of a sequence of random variables is via the following result, a corollary of the Borel–Cantelli lemma.

© Springer Nature Switzerland AG 2020
P. Brémaud, *Probability Theory and Stochastic Processes*, Universitext,
https://doi.org/10.1007/978-3-030-40183-2_4

Theorem 4.1.2 *Let $\{Z_n\}_{n\geq 1}$ and Z be complex random variables. If*

$$\sum_{n\geq 1} P(|Z_n - Z| \geq \varepsilon_n) < \infty \qquad (4.2)$$

for some sequence of positive numbers $\{\varepsilon_n\}_{n\geq 1}$ converging to 0, then the sequence $\{Z_n\}_{n\geq 1}$ converges P-a.s. to Z.

Proof. If for a given ω, $|Z_n(\omega) - Z(\omega)| \geq \varepsilon_n$ finitely often (or *f.o.*; that is, for no more that a finite number of indices n), then $\lim_{n\uparrow\infty} |Z_n(\omega) - Z(\omega)| \leq \lim_{n\uparrow\infty} \varepsilon_n = 0$. Therefore

$$P(\lim_{n\uparrow\infty} Z_n = Z) \geq P(|Z_n - Z| \geq \varepsilon_n \quad f.o.).$$

On the other hand,

$$\{|Z_n - Z| \geq \varepsilon_n \quad f.o.\} = \overline{\{|Z_n - Z| \geq \varepsilon_n \quad i.o.\}}.$$

Therefore

$$P(|Z_n - Z| \geq \varepsilon_n \quad f.o.) = 1 - P(|Z_n - Z| \geq \varepsilon_n \quad i.o.).$$

Hypothesis (4.2) implies (Borel–Cantelli lemma) that

$$P(|Z_n - Z| \geq \varepsilon_n \quad i.o.) = 0.$$

By linking the above facts, we obtain

$$P(\lim_{n\uparrow\infty} Z_n = Z) \geq 1,$$

and of course, the only possibility is $= 1$. $\qquad\square$

In order to prove almost-sure convergence with the help of Theorem 4.1.2, some adequate upper bound for the general term of the series occurring in the left-hand side of (4.2) is needed. See the proof of Borel's strong law of large numbers in Section 1.5. The proof of Kolmogorov's strong law of large numbers will also make use of Theorem 4.1.2, however in a more sophisticated manner.

A Criterion

The following *criterion* (necessary and sufficient condition) of almost-sure convergence will be useful for comparing convergence almost-sure to convergence in probability (to be defined in Section 4.2.1).

Theorem 4.1.3 *The sequence $\{Z_n\}_{n\geq 1}$ of complex random variables converges P-a.s. to the complex random variable Z if and only if for all $\epsilon > 0$,*

$$P(|Z_n - Z| \geq \epsilon \ i.o.) = 0. \qquad (4.3)$$

Proof. For the necessity, observe that

$$\{|Z_n - Z| \geq \epsilon \ i.o.\} \subseteq \overline{\{\omega; \lim_{n\uparrow\infty} Z_n(\omega) = Z(\omega)\}},$$

and therefore

$$P(|Z_n - Z| \geq \epsilon \text{ i.o.}) \leq 1 - P(\lim_{n\uparrow\infty} Z_n = Z) = 0.$$

For the sufficiency, let N_k be the last index n such that $|Z_n - Z| \geq \frac{1}{k}$ (let $N_k = \infty$ if $|Z_n - Z| \geq \frac{1}{k}$ for infinitely many indices $n \geq 1$). By (4.3) with $\epsilon = \frac{1}{k}$, we have $P(N_k = \infty) = 0$. By sub-σ-additivity, $P(\cup_{k\geq1}\{N_k = \infty\}) = 0$. Equivalently, $P(N_k < \infty$, for all $k \geq 1) = 1$, which implies $P(\lim_{n\uparrow\infty} Z_n = Z) = 1$. □

Corollary 4.1.4

A. The sequence $\{Z_n\}_{n\geq1}$ of complex random variables converges P-a.s. to the complex random variable Z if and only if for all $\epsilon > 0$,

$$P\left(\sup_{k\geq n}|Z_k - Z| \geq \epsilon\right) = 0. \tag{4.4}$$

B. The sequence $\{Z_n\}_{n\geq1}$ of complex random variables converges P-a.s. if and only if for all $\epsilon > 0$,

$$P\left(\sup_{k\geq n, \ell\geq n}|Z_k - Z_\ell| \geq \epsilon\right) = 0, \tag{4.5}$$

or, equivalently

$$P\left(\sup_{k\geq 0}|Z_{n+k} - Z_k| \geq \epsilon\right) = 0, \tag{4.6}$$

Proof. A. Since

$$\{|Z_n - Z| \geq \epsilon \text{ i.o.}\} \equiv \bigcap_n \bigcup_{k\geq n}\{|Z_k - Z| \geq \epsilon\},$$

by the sequential continuity property of probability

$$P(|Z_n - Z| \geq \epsilon \text{ i.o.}) = \lim_{n\uparrow\infty} P\left(\bigcup_{k\geq n}\{|Z_k - Z| \geq \epsilon\}\right) = \lim_{n\uparrow\infty} P\left(\sup_{k\geq n}|Z_k - Z| \geq \epsilon\right).$$

Therefore (4.4) and (4.3) are equivalent statements.

B. From the set identity

$$P\left(\{Z_n\}_{n\geq1} \text{ is not a Cauchy sequence}\right) = P\left(\bigcup_\epsilon C_\epsilon\right),$$

where

$$C_\epsilon := \bigcap_{n\geq1} \bigcup_{k,\ell\geq n}\{|Z_k - Z_\ell| \geq \epsilon\}.$$

As in A, it is shown that

$$P\left(\{Z_n\}_{n\geq1} \text{ is not a Cauchy sequence}\right) = 0$$

and (4.3) are equivalent statements. As for the equivalence of (4.5) and (4.6), it follows from the observation that

$$\sup_{k\geq 0}|Z_{n+k} - Z_k| \leq \sup_{k,\ell\geq n}|Z_k - Z_\ell| \leq 2\sup_{k\geq 0}|Z_{n+k} - Z_k|.$$

□

4.1.2 Beppo Levi, Fatou and Lebesgue

The theorems of Beppo Levi and of Lebesgue will now be given in terms of expectations. Remember that we seek conditions guaranteeing that

$$\mathrm{E}\left[\lim_{n\uparrow\infty} X_n\right] = \lim_{n\uparrow\infty} \mathrm{E}\left[X_n\right] . \tag{4.7}$$

We begin with the (Beppo Levi) *monotone convergence* theorem (Theorem 2.2.15) in the context of expectations.

Theorem 4.1.5 *Let $\{X_n\}_{n\geq 1}$ be a sequence of random variables such that for all $n \geq 1$,*

$$0 \leq X_n \leq X_{n+1}, \ P\text{-a.s.}$$

Then (4.7) holds.

Next, we restate in terms of expectations the (Lebesgue) *dominated convergence* theorem (Theorem 2.2.17).

Theorem 4.1.6 *Let $\{X_n\}_{n\geq 1}$ be a sequence of random variables such that for all ω outside a set $\mathcal{N}$ of null probability there exists $\lim_{n\uparrow\infty} X_n(\omega)$ and such that for all $n \geq 1$*

$$|X_n| \leq Y, \ P\text{-a.s.},$$

where Y is some integrable random variable. Then (4.7) holds.

Finally, we give the probabilistic version of *Fatou's lemma*.

Theorem 4.1.7 *Let $\{X_n\}_{n\geq 1}$ be a sequence of P-almost surely non-negative random variables. Then*

$$\mathrm{E}\left[\liminf_{n\uparrow\infty} X_n\right] \leq \liminf_{n\uparrow\infty} \mathrm{E}\left[X_n\right] .$$

The above results have conditioned versions:

Theorem 4.1.8 *Let $\mathcal{G}$ be a sub-σ-field of $\mathcal{F}$, and let $\{Y_n\}_{n\geq 1}$ be a P-a.s. non-decreasing sequence of non-negative random variables converging P-a.s. to the random variable Y. Then $\{\mathrm{E}^{\mathcal{G}}[Y_n]\}_{n\geq 1}$ is a P-a.s. non-decreasing sequence of random variables converging P-a.s. to $\mathrm{E}^{\mathcal{G}}[Y]$.*

Proof. By monotonicity of conditional expectation, $\{\mathrm{E}^{\mathcal{G}}[Y_n]\}_{n\geq 1}$ is a P-a.s. non-decreasing sequence of $\mathcal{G}$-measurable random variables. In particular, there exists a P-a.s. limit W, $\mathcal{G}$-measurable, of this sequence. By monotone convergence, for any bounded non-negative $\mathcal{G}$-measurable random variable U

$$\lim_{n\uparrow\infty} \mathrm{E}[Y_n U] = \mathrm{E}[YU],$$

and

$$\lim_{n\uparrow\infty} \mathrm{E}[\mathrm{E}^{\mathcal{G}}[Y_n]U] = \mathrm{E}[WU].$$

Therefore, since $\mathrm{E}[Y_n U] = \mathrm{E}[\mathrm{E}^{\mathcal{G}}[Y_n]U]$ for all $n \geq 1$, $\mathrm{E}[YU] = \mathrm{E}[WU]$. This being true for all bounded $\mathcal{G}$-measurable random variables U, $W = \mathrm{E}[Y \mid \mathcal{G}]$. $\qquad\square$

Theorem 4.1.9 *Let $\mathcal{G}$ be a sub-σ-field of $\mathcal{F}$ and let $\{Y_n\}_{n\geq 1}$ be a sequence of random variables converging P-a.s. to the random variable Y, and such that $|Y_n| \leq Z$ for some integrable random variable Z. Then $\{E^{\mathcal{G}}[Y_n]\}_{n\geq 1}$ converges P-a.s. to $E^{\mathcal{G}}[Y]$.*

Proof. Define $W_n := \sup_{m\geq n} |Y_m - Y|$. The sequence $\{W_n\}_{n\geq 1}$ decreases P-a.s. to 0. We have

$$|E^{\mathcal{G}}[Y_n] - E^{\mathcal{G}}[Y]| = |E^{\mathcal{G}}[Y_n - Y]|$$
$$\leq E^{\mathcal{G}}[|Y_n - Y|] \leq E^{\mathcal{G}}[W_n].$$

The non-negative sequence $\{E^{\mathcal{G}}[W_n]\}_{n\geq 1}$ decreases P-a.s. Let $H \geq 0$ be its limit. Then

$$0 \leq |E[H]| \leq E\left[E^{\mathcal{G}}[W_n]\right] = E[W_n],$$

where the latter quantity tends to 0 by dominated convergence (because $0 \leq W_n \leq 2Z$). Therefore $E[H] = 0$, which implies that $P(H = 0) = 1$ since H is P-a.s non-negative. $\square$

4.1.3 The Strong Law of Large Numbers

Émile Borel proved his theorem in 1909. In 1933, Kolmogorov gave the general form of the theorem.

Theorem 4.1.10 *Let $\{X_n\}_{n\geq 1}$ be an IID sequence of random variables such that*

$$E[|X_1|] < \infty. \tag{4.8}$$

Then, with $S_n := X_1 + \cdots + X_n$,

$$P\left(\lim_{n\uparrow\infty} \frac{S_n}{n} = E[X_1]\right) = 1. \tag{4.9}$$

Remark 4.1.11 Kolmogorov's version does not restrict the X_n's to be discrete random variables. Borel's law is a special case, with X_n taking the values 0 or 1. Note that if you suppose that the X_n's are not discrete but only bounded, then the proof of Borel's law can be used almost as it is for this more general case.

Proof. Assume (without loss of generality) that $E[X_1] = 0$. The main ingredient of the proof is *Kolmogorov's inequality*:

Lemma 4.1.12 *Let $X_1, \ldots, X_n$ be independent random variables such that $E[|X_i|^2] < \infty$ and $E[X_i] = 0$ $(1 \leq i \leq n)$. Let $S_k := X_1 + \ldots + X_k$. Then for all $\lambda > 0$,*

$$P\left(\max_{1\leq k\leq n} |S_k| \geq \lambda\right) \leq \frac{E[S_n^2]}{\lambda^2}. \tag{4.10}$$

Proof. Let T be the first (random) index $k \leq n$ such that $|S_k| \geq \lambda$, with $T = \infty$ if $\max_{1\leq k\leq n} |S_k| < \lambda$. For $k \leq n$,

$$E[1_{\{T=k\}}S_n^2] = E\left[1_{\{T=k\}}\{(S_n - S_k)^2 + 2S_k(S_n - S_k) + S_k^2\}\right]$$
$$= E\left[1_{\{T=k\}}\{(S_n - S_k)^2 + S_k^2\}\right] \geq E[1_{\{T=k\}}S_k^2],$$

where we used the fact that $1_{\{T=k\}}S_k$ is $\sigma(X_1, \ldots, X_k)$-measurable and therefore independent of $S_n - S_k$. By the product rule for expectations, $E[1_{\{T=k\}}S_k(S_n - S_k)] = E[1_{\{T=k\}}S_k]E[S_n - S_k] = 0$. Therefore,

$$E[|S_n|^2] \geq \sum_{k=1}^{n} E[1_{\{T=k\}}S_k^2] \geq \sum_{k=1}^{n} E[1_{\{T=k\}}\lambda^2]$$

$$= \lambda^2 \sum_{k=1}^{n} P(T = k) = \lambda^2 P(T \leq n) = \lambda^2 P\left(\max_{1\leq k\leq n} |S_k| \geq \lambda\right).$$

$\square$

Lemma 4.1.13 *Let $\{X_n\}_{n\geq1}$ be a sequence of independent random variables such that for all $n \geq 1$, $E[|X_n|^2] < \infty$ and $E[X_n] = 0$. If moreover*

$$\sum_{n\geq1} \frac{E[X_n^2]}{n^2} < \infty,$$ (4.11)

then

$$\frac{1}{n} \sum_{k=1}^{n} X_k \to 0, \quad P\text{-}a.s.$$ (4.12)

(Note that this lemma is already the SLLN under the additional assumption $E[|X_1|^2] < \infty$.)

Proof. If $2^{k-1} \leq n \leq 2^k$, then $|S_n| \geq n\varepsilon$ implies $|S_n| \geq 2^{k-1}\varepsilon$. Therefore, for all $\varepsilon > 0$, and all $k \geq 1$,

$$P\left(\frac{|S_n|}{n} \geq \varepsilon\right) \leq P\left(|S_n| \geq \varepsilon 2^{k-1} \text{ for some } n \in [2^{k-1}, 2^k]\right)$$

$$\leq P\left(|S_n| \geq \varepsilon 2^{k-1} \text{ for some } n \in [1, 2^k]\right)$$

$$= P\left(\max_{1\leq n\leq 2^k} |S_n| \geq \varepsilon 2^{k-1}\right)$$

$$\leq \frac{4}{\varepsilon^2} \frac{1}{(2^k)^2} \sum_{n=1}^{2^k} E[X_n^2],$$

the last inequality by Kolmogorov's inequality. Since

$$\sum_{k=1}^{\infty} \frac{1}{(2^k)^2} \sum_{n=1}^{2^k} E[X_n^2] = \sum_{n=1}^{\infty} E[X_n^2] \left(\sum_{k=1}^{\infty} 1_{\{2^k\geq n\}} \frac{1}{(2^k)^2}\right) \leq \sum_{n=1}^{\infty} E[X_n^2] \frac{K}{n^2}$$

for some finite K, by (4.11),

$$\sum_{k=1}^{\infty} P\left(\frac{|S_n|}{n} \geq \varepsilon \text{ for some } n \in [2^{k-1}, 2^k]\right) < \infty,$$

and by the Borel–Cantelli lemma,

$$P\left(\frac{|S_n|}{n} \geq \varepsilon \text{ i.o.}\right) = 0.$$

The result then follows from Theorem 4.1.3. □

We are now ready for the proof of Theorem 4.1.10. Having proved the above lemma, it remains to get rid of the assumption $E[|X_n|^2] < \infty$, and the natural technique for this is truncation. Define

$$\widetilde{X}_n = \begin{cases} X_n & \text{if } |X_n| \leq n, \\ 0 & \text{otherwise.} \end{cases}$$

A. We first show that

$$\lim_{n\uparrow\infty} \frac{1}{n} \sum_{k=1}^{n} (\widetilde{X}_k - E[\widetilde{X}_k]) = 0.$$

In view of Lemma 4.1.13, it suffices to prove that

$$\sum_{n=1}^{\infty} \frac{E[(\widetilde{X}_n - E[\widetilde{X}_n])^2]}{n^2} < \infty.$$

But

$$E[(\widetilde{X}_n - E[\widetilde{X}_n])^2] \leq E[\widetilde{X}_n^2] = E[X_1^2 1_{\{|X_1|\leq n\}}].$$

It is therefore enough to show that

$$\sum_{n=1}^{\infty} \frac{E[X_1^2 1_{\{|X_1|\leq n\}}]}{n^2} < \infty.$$

The left-hand side of the above inequality is equal to

$$\sum_{n=1}^{\infty} \frac{1}{n^2} \sum_{k=1}^{n} E[X_1^2 1_{\{k-1<|X_1|\leq k\}}] = \sum_{k=1}^{\infty} \sum_{n=k}^{\infty} \frac{1}{n^2} E[X_1^2 1_{\{k-1<|X_1|\leq k\}}].$$

Using the fact that

$$\sum_{n=k}^{\infty} \frac{1}{n^2} \leq \frac{1}{k^2} + \int_k^{\infty} \frac{1}{x^2} dx = \frac{1}{k^2} + \frac{1}{k} \leq \frac{2}{k}$$

(draw the graph of $x \to x^{-2}$), this is less than or equal to

$$\sum_{k=1}^{\infty} \frac{2}{k} E[X_1^2 1_{\{k-1<|X_1|\leq k\}}] = 2\sum_{k=1}^{\infty} E\left[\frac{X_1^2}{k} 1_{\{k-1<|X_1|\leq k\}}\right]$$

$$\leq 2\sum_{k=1}^{\infty} E[|X_1| 1_{\{k-1<|X_1|\leq k\}}]$$

$$= 2E[|X_1|] < \infty.$$

B. Since $E[|X_1|] < \infty$, we have by dominated convergence that

$$\lim_{n\uparrow\infty} E[X_1 1_{\{|X_1|\leq n\}}] = E[X_1] = 0.$$

Since X_n has the same distribution as X_1,

$$\lim_{n\uparrow\infty} E[\widetilde{X}_n] = \lim_{n\uparrow\infty} E[X_n 1_{\{|X_n|\leq n\}}] = \lim_{n\uparrow\infty} E[X_1 1_{\{|X_1|\leq n\}}] = E[X_1] = 0.$$

In particular (Cesàro's lemma, see the Appendix),

$$\lim_{n\uparrow\infty} \frac{1}{n} \sum_{k=1}^{n} E[\widetilde{X}_k] = 0.$$

C. We have

$$\sum_{n=1}^{\infty} P(|X_n| > n) = \sum_{n=1}^{\infty} P(|X_1| > n) \le E[|X_1|] < \infty,$$

and therefore, by the Borel–Cantelli lemma,

$$P(\widetilde{X}_n \ne X_n \text{ i.o.}) = P(X_n > n \text{ i.o.}) = 0,$$

which implies that

$$\lim_{n\uparrow\infty} \frac{\widetilde{S}_n}{n} = \lim_{n\uparrow\infty} \frac{S_n}{n}.$$

$\square$

The integrability condition in the strong law of large numbers is almost necessary. More precisely:

Theorem 4.1.14 *Let $\{X_n\}_{n\ge 1}$ be a sequence of* IID *random variables such that*

$$\frac{S_n}{n} \to C < \infty \quad P-a.s.,$$

where $S_n := X_1 + \cdots X_n$. Then $E\,[|X_1|] < \infty$ and $C = E\,[X_1]$.

Proof. Under these circumstances,

$$\frac{X_n}{n} = \frac{S_n}{n} - \frac{n-1}{n}\frac{S_{n-1}}{n-1} \to 0$$

and therefore, $P(|X_n| > n\,i.o.) = 0$. By the converse Borel–Cantelli lemma,

$$\sum_{n=1}^{\infty} P(|X_n| > n) < \infty$$

or, since the distribution of any X_n is that of X_1

$$\sum_{n=1}^{\infty} P(|X_1| > n) < \infty.$$

But, by the following inequalities concerning any non-negative random variable X (Exercise 4.6.1) and $|X_1|$ in particular,

$$\sum_{n=1}^{\infty} P(X \ge n) \le E\,[X] \le 1 + \sum_{n=1}^{\infty} P(X \ge n),$$

we have that $E\,[|X_1|] < \infty$. The identification of C and $E\,[|X_1|]$ is then just the strong law of large numbers.

$\square$

Theorem 4.1.15 *If a sequence $\{\xi_n\}_{n\ge 1}$ of independent centered random variables is such that*

$$\sum_{n\ge 1} E\,[\xi_n^2] < \infty,$$

then $S_n := \sum_{n\ge 1} \xi_n \to$ almost surely as $n \uparrow \infty$.

Proof. According to Corollary 4.1.4, it suffices to show that for all $\varepsilon > 0$,

$$\lim_{n\uparrow\infty} P\left(\sup_{k\geq 1}|S_{n+k} - S_n| \geq \varepsilon\right) = 0.$$

But, by Kolmogorov's inequality,

$$\lim_{n\uparrow\infty} P\left(\sup_{1\leq k\leq N}|S_{n+k} - S_n| \geq \varepsilon\right) = \lim_{N\uparrow\infty} P\left(\max_{k\geq 1}|S_{n+k} - S_n| \geq \varepsilon\right)$$

$$\leq \lim_{N\uparrow\infty} \frac{\sum_{k=n}^{n+N} E\left[\xi_k^2\right]}{\varepsilon^2} = \frac{\sum_{k=n}^{\infty} E\left[\xi_k^2\right]}{\varepsilon^2} \to 0.$$

$\square$

Large Deviations

The large deviations theory of random variables produces estimates for the deviation of such variables from their means. When applied to sums of IID variables $S_n = X_1 + \cdots X_n$, they complement the strong law of large numbers. The type of result produced by this theory is, in the case where the X_i's are integrable of common mean m,

$$\lim_{n\uparrow\infty} \frac{1}{n} \log P\left(\left|\frac{S_n}{n} - m\right| \geq a\right) = -h(a), \tag{$\star$}$$

where $a > 0$ and $h(a) > 0$. Such bounds have important theoretical implications, but they are somewhat imprecise in that the meaning of $(\star)$ is

$$P\left(\left|\frac{S_n}{n} - m\right| \geq a\right) = g(n)e^{-n(h(a))},$$

where $n^{-1}\log g(n)$ tends to 0 as $n \uparrow \infty$, but perhaps too slowly and in an uncontrolled manner. To obtain practical bounds, it is often useful to look at specific cases, using the Chernoff bound below at the origin of the general abstract theory. These powerful bounds are easy consequences of the elementary Markov inequality.

Theorem 4.1.16 *Let X be a real-valued random variable and let $a \in \mathbb{R}$. Then* (Chernoff's bound)

$$P(X \geq a) \leq \min_{t>0} \frac{E\left[e^{tX}\right]}{e^{ta}}, \tag{4.13}$$

and

$$P(X \leq a) \leq \min_{t<0} \frac{E\left[e^{tX}\right]}{e^{ta}}. \tag{4.14}$$

Proof. By the $\uparrow$-monotony of $x \to e^x$ and Markov's inequality: For $t > 0$,

$$P(X \geq a) = P(e^{tX} \geq e^{ta}) \leq \frac{E\left[e^{tX}\right]}{e^{ta}},$$

and for $t < 0$,

$$P(X \leq a) = P(e^{tX} \geq e^{ta}) \leq \frac{E\left[e^{tX}\right]}{e^{ta}}.$$

The announced result follows by minimizing the right-hand sides with respect to $t > 0$ and $t < 0$, respectively. □

EXAMPLE 4.1.17: LARGE DEVIATIONS FOR THE POISSON VARIABLE. Let X be a Poisson variable with mean θ and therefore $\mathrm{E}\left[e^{tX}\right] = e^{\theta(e^t - 1)}$. We prove that for $c \geq 0$

$$P\left(X \geq \theta + c\right) \leq \exp\left\{-\frac{1}{e}\left(\frac{\theta + c}{e\theta}\right)^{\theta + c}\right\}.$$

With $a = \theta + c$ in (4.13):

$$P\left(X \geq \theta + c\right) \leq \min_{t > 0} \frac{e^{\theta(e^t - 1)}}{e^{t(\theta + c)}}$$
$$= e^{-\max_{t > 0}\{t(\theta + c) - \theta(e^t - 1)\}}.$$

The derivative of the function $f : t \to t(\theta + c) - \theta(e^t - 1)$ at $t \geq 0$ is $\theta + c - \theta e^t$, and it is null for $e^t = \frac{\theta + c}{\theta}$ or equivalently $t = \ln(\theta + c) - \ln(\theta)$, and this corresponds to a maximum since the second derivative $-e^t$ is negative. Therefore

$$\max_{t > 0}\left\{t(\theta + c) - \theta(e^t - 1)\right\} = \frac{1}{e}\left(\frac{\theta + c}{e\theta}\right)^{\theta + c}$$

and finally

$$P\left(X \geq \theta + c\right) \leq \exp\left\{-\frac{1}{e}\left(\frac{\theta + c}{e\theta}\right)^{\theta + c}\right\}.$$

Theorem 4.1.18 *Let $X_1, \ldots, X_n$ be IID real-valued random variables and let $a \in \mathbb{R}$. Then,*

$$P\left(\sum_{i=1}^{n} X_i \geq na\right) \leq e^{-nh^+(a)},$$

where

$$h^+(a) = \sup_{t \geq 0}\{at - \ln \mathrm{E}\left[e^{tX_1}\right]\}. \tag{4.15}$$

Proof. For all $t \geq 0$, Markov's inequality gives

$$P\left(\sum_{i=1}^{n} X_i \geq na\right) = P\left(\exp\left\{t\sum_{i=1}^{n} X_i\right\} \geq \exp\{nta\}\right)$$
$$\leq \mathrm{E}\left[\exp\left\{t\sum_{i=1}^{n} X_i\right\}\right] \times e^{-nta}$$
$$\leq \exp\{-n\left(at - \ln \mathrm{E}\left[e^{tX_1}\right]\right)\},$$

from which the result follows by optimizing this bound over $t \geq 0$. □

Suppose that $\mathrm{E}\left[e^{tX_1}\right] < \infty$ for all $t \geq 0$. Differentiating $t \to at - \ln \mathrm{E}\left[e^{tX_1}\right]$ yields $a - \frac{\mathrm{E}\left[X_1 e^{tX_1}\right]}{\mathrm{E}\left[e^{tX_1}\right]}$, and therefore the function $t \to at - \ln \mathrm{E}\left[e^{tX_1}\right]$ is finite and differentiable on $\mathbb{R}$, with derivative at 0_+ equal to $a - \mathrm{E}[X_1]$, which implies that when $a > \mathrm{E}[X_1]$, $h^+(a)$ is positive.

Similarly to (4.15), we obtain that

$$P\left(\sum_{i=1}^{n} X_i \leq na\right) \leq \mathrm{e}^{-nh^-(a)},$$

where

$$h^-(a) = \sup_{t \leq 0}\{at - \ln \mathrm{E}\left[\mathrm{e}^{tX_1}\right]\}.$$

Moreover, if $a < \mathrm{E}[X_1]$, $h^-(a)$ is positive.

The Chernoff bound can be interpreted in terms of *large deviations* from the law of large numbers. Denote by μ the common mean of the X_n's, and define for $\varepsilon > 0$ the (positive) quantities

$$H^+(\varepsilon) = \sup_{t \geq 0}\left\{\varepsilon t - \ln \mathrm{E}\left[\mathrm{e}^{t(X_1-\mu)}\right]\right\},$$

$$H^-(\varepsilon) = \sup_{t \leq 0}\left\{\varepsilon t - \ln \mathrm{E}\left[\mathrm{e}^{t(X_1-\mu)}\right]\right\}.$$

Then

$$P\left(\left|\frac{1}{n}\sum_{i=1}^{n} X_i\right| \geq +\varepsilon\right) \leq \mathrm{e}^{-nH^+(\varepsilon)} + \mathrm{e}^{-nH^-(\varepsilon)}.$$

Remark 4.1.19 The computation of the supremum in (4.15) may be fastidious, There are shortcuts leading to practical bounds that are not as good but nevertheless satisfactory for certain applications.

EXAMPLE 4.1.20: Suppose for instance that $\{X_n\}_{n \geq 1}$ is IID, the X_n's taking the values -1 and $+1$ equiprobably so that $\mathrm{E}\left[\mathrm{e}^{tX}\right] = \frac{1}{2}\mathrm{e}^{+t} + \frac{1}{2}\mathrm{e}^{-t}$. We do not keep this expression as such but instead replace it by an upper bound, namely $\mathrm{e}^{\frac{t^2}{2}}$, and therefore, for $a > 0$,

$$P\left(\sum_{i=1}^{n} X_i \geq na\right) \leq \mathrm{e}^{-n\left(at - \ln \mathrm{E}[\mathrm{e}^{tX_1}]\right)}$$

$$\leq \mathrm{e}^{-n\left(at - \frac{1}{2}t^2\right)},$$

so that, with $t = a$,

$$P\left(\sum_{i=1}^{n} X_i \geq na\right) \leq \mathrm{e}^{-n\frac{1}{2}a^2}.$$

By symmetry of the distribution of $\sum_{i=1}^{n} X_i$, one would obtain for $a > 0$

$$P\left(\sum_{i=1}^{n} X_i \leq -na\right) = P\left(\sum_{i=1}^{n} X_i \geq na\right) \leq \mathrm{e}^{-n\frac{1}{2}a^2},$$

and therefore combining the two bounds,

$$P\left(\left|\sum_{i=1}^{n} X_i\right| \geq na\right) \leq 2\mathrm{e}^{-n\frac{1}{2}a^2}.$$

4.2 Two Other Types of Convergence

These are

(i) convergence *in probability*, the "parent pauvre" of almost-sure convergence, and

(ii) convergence *in the quadratic mean*, that is, convergence in $L^2_{\mathbb{C}}(P)$.

(Convergences in distribution and in variation will be treated in the next section.)

4.2.1 Convergence in Probability

Recall the definition already given for discrete random variables in the first chapter:

Definition 4.2.1 *A sequence $\{Z_n\}_{n\geq 1}$ of variables is said to* converge in probability *to the random variable Z if, for all $\varepsilon > 0$,*

$$\lim_{n\uparrow\infty} P(|Z_n - Z| \geq \varepsilon) = 0. \tag{4.16}$$

EXAMPLE 4.2.2: BERNSTEIN'S POLYNOMIAL APPROXIMATION. This example is a particular instance of the fruitful interaction between probability and analysis. Here, we shall give a probabilistic proof of the fact that a continuous function f from $[0,1]$ into $\mathbb{R}$ can be uniformly approximated by a polynomial. More precisely, for all $x \in [0,1]$,

$$f(x) = \lim_{n\uparrow\infty} P_n(x), \tag{$\star$}$$

where

$$P_n(x) = \sum_{k=0}^{n} f\left(\frac{k}{n}\right) \frac{n!}{k!(n-k)!} x^k(1-x)^{n-k},$$

and the convergence of the series in the right-hand side is *uniform* in $[0,1]$. A proof of this classical theorem of analysis using probabilistic arguments and in particular the notion of convergence in probability is as follows. Since $S_n \sim B(n,p)$,

$$E\left[f\left(\frac{S_n}{n}\right)\right] = \sum_{k=0}^{n} f\left(\frac{k}{n}\right) P(S_n = k) = \sum_{k=0}^{n} f\left(\frac{k}{n}\right) \frac{n!}{k!(n-k)!} x^k(1-x)^{n-k}.$$

The function f is continuous on the *bounded* $[0,1]$ and therefore *uniformly* continuous on this interval. Therefore to any $\varepsilon > 0$, one can associate a number $\delta(\varepsilon)$ such that if $|y - x| \leq \delta(\varepsilon)$, then $|f(x) - f(y)| \leq \varepsilon$. Being continuous on $[0,1]$, f is bounded on $[0,1]$ by some finite number, say M. Now

$$|P_n(x) - f(x)| = \left|E\left[f\left(\frac{S_n}{n}\right) - f(x)\right]\right| \leq E\left[\left|f\left(\frac{S_n}{n}\right) - f(x)\right|\right]$$

$$= E\left[\left|\left[f\left(\frac{S_n}{n}\right) - f(x)\right]1_A\right|\right] + E\left[\left|f\left(\frac{S_n}{n}\right) - f(x)\right|1_{\overline{A}}\right],$$

where $A := \{|S_n(\omega)/n| - x| \leq \delta(\varepsilon)\}$. Since $|f(S_n/n) - f(x)|1_{\overline{A}} \leq 2M1_{\overline{A}}$, we have

$$E\left[\left|f\left(\frac{S_n}{n}\right) - f(x)\right|1_{\overline{A}}\right] \leq 2MP(\overline{A}) = 2MP\left(\left|\frac{S_n}{n} - x\right| \geq \delta(\varepsilon)\right).$$

Also, by definition A and $\delta(\varepsilon)$,

$$\mathrm{E}\left[\left|f\left(\frac{S_n}{n}\right)-f(x)\right|1_A\right]\le\varepsilon.$$

Therefore

$$|P_n(x)-f(x)|\le\varepsilon+2MP\left(\left|\frac{S_n}{n}-x\right|\ge\delta(\varepsilon)\right).$$

But x is the mean of S_n/n, and the variance of S_n/n is $nx(1-x)\le n/4$. Therefore, by Tchebyshev's inequality,

$$P\left(\left|\frac{S_n}{n}-x\right|\ge\delta(\varepsilon)\right)\le\frac{4}{n[\delta(\varepsilon)]^2}.$$

Finally

$$|f(x)-P_n(x)|\le\varepsilon+\frac{4}{n[\delta(\varepsilon)]^2}.$$

Since $\varepsilon>0$ is otherwise arbitrary, this suffices to prove the convergence in $(\star)$. The convergence is *uniform* since the right-hand side of the latter inequality does not depend on $x\in[0,1]$.

———————

There is a *Cauchy-type criterion* for convergence in probability.

Theorem 4.2.3 *For a sequence $\{Z_n\}_{n\ge1}$ of random variables to converge in probability to some random variable, it is necessary and sufficient that for all $\varepsilon>0$,*

$$\lim_{m,n\uparrow\infty}P(|Z_m-Z_n|\ge\varepsilon)=0.$$

Proof. Necessity. We have the inclusion

$$\{|Z_m-Z_n|\ge\varepsilon\}\subseteq\{|Z_m-Z|\ge\frac{1}{2}\varepsilon\}\cup\{|Z_m-Z|\ge\frac{1}{2}\varepsilon\}$$

and therefore

$$P(|Z_m-Z_n|\ge\varepsilon)\le P(|Z_m-Z|\ge\frac{1}{2}\varepsilon)+P(|Z_m-Z|\ge\frac{1}{2}\varepsilon).$$

Sufficiency. Let $n_1:=+1$ and let for $j\ge2$,

$$n_j=\inf\{N>n_{j-1};\,P(|Z_r-Z_s|>\frac{1}{2^j})<\frac{1}{3^j}\text{ if }r,s>N\}.$$

Then

$$\sum_j P(|Z_{n_j}-Z_{n_{j-1}}|>\frac{1}{2^{j-1}}<\infty,$$

and therefore, there exists a random variable Z such that $Z_{n_j}\overset{a.s.}{\to}Z$ as $j\uparrow\infty$. Now:

$$P(|Z-Z_n|\ge\varepsilon)\le P(|Z_n-Z_{n_j}|\ge\frac{1}{2}\varepsilon)+P(|Z_{n_j}-Z|\ge\frac{1}{2}\varepsilon)$$

can be made arbitrarily close to 0 as $n\uparrow\infty$, by definition of the n_j's and the fact that almost sure convergence implies convergence in probability, as we shall see next, in Theorem 4.5.1. $\square$

In fact, there exists a distance between random variables that metrizes convergence in probability, namely

$$d(X,Y):=\mathrm{E}\left[|X-Y|\wedge1\right].$$

The verification that d is indeed a metric is left as an exercise.

Theorem 4.2.4 *The sequence* $\{X_n\}_{n\geq 1}$ *converges in probability to the variable* X *if and only if*

$$\lim_{n\uparrow\infty} d(X_n, X) = 0.$$

Proof. If: By Markov's inequality, for $\varepsilon \in (0, 1]$,

$$P(|X_n - X| \geq \varepsilon) = P(|X_n - X| \wedge 1 \geq \varepsilon) \leq \frac{d(X_n, X)}{\varepsilon}.$$

Only if: For all $\varepsilon > 0$,

$$\begin{aligned}
d(X_n, X) &= \int_{\{|X_n - X| \geq \varepsilon\}} (|X_n - X| \wedge 1)\, dP + \int_{\{|X_n - X| < \varepsilon\}} (|X_n - X| \wedge 1)\, dP \\
&\leq P(|X_n - X| \geq \varepsilon) + \varepsilon.
\end{aligned}$$

If the sequence converges in probability, there exists an n_0 such that for $n \geq n_0$, $P(|X_n - X| \geq \varepsilon) \leq \varepsilon$ and therefore $d(X_n, X) \leq 2\varepsilon$. Since $\varepsilon > 0$ is arbitrary, we have shown that $\lim_{n\uparrow\infty} d(X_n, X) = 0$. $\qquad\square$

4.2.2 Convergence in L^p

Definition 4.2.5 *Let* p *be a positive integer. A sequence* $\{Z_n\}_{n\geq 1}$ *of complex random variables of* $L^p_{\mathbb{C}}(P)$ *is said to* converge in L^p *to the complex random variable* $Z \in L^p_{\mathbb{C}}(P)$ *if*

$$\lim_{n\uparrow\infty} \mathrm{E}[|Z_n - Z|^p] = 0. \tag{4.17}$$

In the case $p = 2$, the sequence $\{Z_n\}_{n\geq 1}$ of square-integrable complex random variables is said to converge *in the quadratic mean* to Z.

By the Riesz–Fischer theorem (Theorem 2.3.22):

Theorem 4.2.6 *For the sequence* $\{Z_n\}_{n\geq 1}$ *of square-integrable complex random variables to converge in* L^p *to some random variable* $Z \in L^p_{\mathbb{C}}(P)$, *it is necessary and sufficient that*

$$\lim_{n,m\uparrow\infty} \mathrm{E}[|Z_n - Z_m|^p] = 0. \tag{4.18}$$

Recall that $L^2_{\mathbb{C}}(P)$ is a Hilbert space with inner product $\langle X, Y \rangle = \mathrm{E}[XY^*]$ with the following property of continuity.

Theorem 4.2.7 *Let* $\{X_n\}_{n\geq 1}$ *and* $\{Y_n\}_{n\geq 1}$ *be two sequences of square-integrable complex random variables that converge in quadratic mean to the square-integrable complex random variables* X *and* Y, *respectively. Then,*

$$\lim_{n,m\uparrow\infty} \mathrm{E}[X_n Y_m^*] = \mathrm{E}[XY^*]. \tag{4.19}$$

Proof. We have

$$|E[X_n Y_m^*] - E[XY^*]|$$
$$= |E[(X_n - X)(Y_m - Y)^*] + E[(X_n - X)Y^*] + E[X(Y_m - Y)^*]|$$
$$\leq |E[(X_n - X)(Y_m - Y)^*]| + |E[(X_n - X)Y^*]| + |E[X(Y_m - Y)^*]|$$

and the right-hand side of this inequality is, by Schwarz's inequality, less than

$$\left(E[|X_n - X|^2]\right)^{\frac{1}{2}} \left(E[|Y_m - Y|^2]\right)^{\frac{1}{2}}$$
$$+ \left(E[|X_n - X|^2]\right)^{\frac{1}{2}} \left(E[|Y|^2]\right)^{\frac{1}{2}}$$
$$+ \left(E[|X|^2]\right)^{\frac{1}{2}} \left(E[|Y_m - Y|^2]\right)^{\frac{1}{2}},$$

which tends to 0 as $n, m \uparrow \infty$. $\square$

EXAMPLE 4.2.8: L^2-CONVERGENCE OF SERIES. Let $\{A_n\}_{n\in\mathbb{Z}}$ and $\{B_n\}_{n\in\mathbb{Z}}$ be two sequences of centered square-integrable complex random variables such that

$$\sum_{j\in\mathbb{Z}} E[|A_j|^2] < \infty, \ \sum_{j\in\mathbb{Z}} E[|B_j|^2] < \infty.$$

Suppose, moreover, that for all $i \neq j$,

$$E[A_i A_j^*] = E[B_i B_j^*] = E[A_i B_j^*] = 0 \text{ for all } i \neq j.$$

Define

$$U_n = \sum_{j=-n}^{n} A_j, \ V_n = \sum_{j=-n}^{n} B_j.$$

Then $\{U_n\}_{n\geq 1}$ (resp., $\{V_n\}_{n\geq 1}$) converges in quadratic mean to some square-integrable random variable U (resp., V) and

$$E[U] = E[V] = 0 \text{ and } E[UV^*] = \sum_{j\in\mathbb{Z}} E[A_j B_j^*].$$

Proof. We have

$$E[|U_n - U_m|^2] = E\left[\left|\sum_{j=n+1}^{m} A_j\right|^2\right] = \sum_{j=n+1}^{m}\sum_{i=n+1}^{m} E[A_j A_i^*] = \sum_{j=n+1}^{m} E[|A_j|^2]$$

since when $i \neq j$, $E[A_j A_i^*] = 0$. The conclusion then follows from the Cauchy criterion for convergence in quadratic mean, since

$$\lim_{m,n\uparrow\infty} E[|U_n - U_m|^2] = \lim_{m,n\uparrow\infty} \sum_{j=n+1}^{m} E[|A_j|^2] = 0,$$

in view of hypothesis $\sum_{j\in\mathbb{Z}} E[|A_j|^2] < \infty$. By continuity of the inner product in $L^2_{\mathbb{C}}(P)$,

$$E[UV^*] = \lim_{n\uparrow\infty} E[U_n V_n^*] = \lim_{n\uparrow\infty} \sum_{j=1}^{n}\sum_{\ell=1}^{n} E[A_j B_\ell^*]$$

$$= \lim_{n\uparrow\infty} \sum_{j=1}^{n} E[A_j B_j^*] = \sum_{j\in\mathbb{Z}} E[A_j B_j^*].$$

$\square$

4.2.3   Uniform Integrability

The monotone and dominated convergence theorems are not all the tools that we have at
our disposition giving conditions under which it is possible to exchange limits and expectations. Uniform integrability, which will be introduced now, is another such sufficient
condition.

Definition 4.2.9 *A collection $\{X_i\}_{i \in I}$ (where I is an arbitrary index) of integrable
random variables is called uniformly integrable if*

$$\lim_{c \uparrow \infty} \int_{\{|X_i| > c\}} |X_i| \, dP = 0 \text{ uniformly in } i \in I \,.$$

EXAMPLE 4.2.10:   COLLECTION DOMINATED BY AN INTEGRABLE VARIABLE. If, for
some integrable random variable, $P(|X_i| \le X) = 1$ for all $i \in I$, then $\{X_i\}_{i \in I}$ is uniformly
integrable. Indeed, in this case,

$$\int_{\{|X_i| > c\}} |X_i| \, dP \le \int_{\{X > c\}} X \, dP$$

and by monotone convergence the right-hand side of the above inequality tends to 0 as
$c \uparrow \infty$.

Remark 4.2.11 Clearly, if one adds a finite number of integrable variables to a uniformly integrable collection, the augmented collection will also be uniformly integrable.

Theorem 4.2.12 *The collection $\{X_i\}_{i \in I}$ of integrable random variables is uniformly
integrable if and only if*

(a) $\sup_i \mathrm{E}\left[|X_i|\right] < \infty$, *and*

(b) for every $\varepsilon > 0$, there exists a $\delta(\varepsilon) > 0$ such that

$$\sup_n \int_A |X_i| \, dP \le \varepsilon \text{ whenever } P(A) \le \delta(\varepsilon) \,.$$

(In other words, $\int_A |X_i| \, dP \to 0$ uniformly in i as $P(A) \to 0$.)

Proof. Assume uniform integrability. For any $\varepsilon > 0$, there exists a c such that
$\int_{\{X_i > c\}} |X_i| \, dP \le \varepsilon$ for all $i \in I$. For all $A \in \mathcal{F}$, all $i \in I$,

$$\int_A |X_i| \, dP \le cP(A) + \int_{\{|X_i| > c\}} |X_i| \, dP \le cP(A) + \frac{1}{2}\varepsilon \,.$$

Therefore we have (b) by taking $\delta(\varepsilon) = \frac{\varepsilon}{2c}$ and (a) with $A = \Omega$.

Conversely, let $M := \sup_i \mathrm{E}\left[|X_i|\right] < \infty$. Let ε and $\delta(\varepsilon)$ be as in (b). Let $c_0 := \frac{M}{\delta(\varepsilon)}$.
For all $c \ge c_0$ and all $i \in I$, $P(|X_i| > c) \le \delta_\varepsilon$ (Markov's inequality). Apply (b) with
$A = \{|X_c| > c\}$ to obtain that $\sup_n \int_{\{|X_c| > c\}} |X_i| \, dP \le \varepsilon$.   □

Since the "collection" consisting of a single integrable variable X is uniformly integrable, condition (b) of the theorem above reads

$$\sup_{A \,;\, P(A) < \delta} \mathrm{E}\left[|X| 1_A\right] \to 0 \text{ as } \delta \to 0 \,. \qquad (4.20)$$

This simple observation will be used in the proof of the next result.

Theorem 4.2.13 *Let Y be an integrable random variable and let $\{\mathcal{F}_i\}_{i \in I}$ be a collection of sub-σ fields of $\mathcal{F}$. The collection $X_i := \mathrm{E}\left[Y \mid \mathcal{F}_i\right]$ $(i \in I)$ is uniformly integrable.*

Proof. By Jensen's inequality,

$$|X_i| = |\mathrm{E}\left[Y \mid \mathcal{F}_i\right]| \leq \mathrm{E}\left[|Y| \mid \mathcal{F}_i\right]$$

and therefore, for all $a > 0$,

$$\mathrm{E}\left[|X_i| 1_{\{|X_i| \geq a\}}\right] \leq \mathrm{E}\left[Z_i\right] 1_{\{Z_i \geq a\}},$$

where $Z_i := \mathrm{E}\left[|Y| \mid \mathcal{F}_i\right]$. By definition of conditional expectation, since $\{Z_i \geq a\} \in \mathcal{F}_i$,

$$\mathrm{E}\left[(|Y| - Z_i) 1_{\{Z_i \geq a\}}\right] = 0$$

and therefore

$$\mathrm{E}\left[|X_i| 1_{\{|X_i| \geq a\}}\right] \leq \mathrm{E}\left[|Y| 1_{\{Z_i \geq a\}}\right]. \qquad (\star)$$

By Markov's inequality,

$$P(Z_i \geq a) \leq \frac{\mathrm{E}\left[Z_i\right]}{a} = \frac{\mathrm{E}\left[|Y|\right]}{a},$$

and therefore $P(Z_i \geq a) \to 0$ as $a \to \infty$ uniformly in i. Use (15.42) to obtain that $\mathrm{E}\left[|Y| 1_{\{Z_i \geq a\}}\right] \to 0$ as $a \to \infty$ uniformly in i. Conclude with $(\star)$. $\qquad \square$

Theorem 4.2.14 *A sufficient condition for the collection $\{X_i\}_{i \in I}$ of integrable random variables to be uniformly integrable is the existence of a non-negative non-decreasing function $G : \mathbb{R} \to \mathbb{R}$ such that*

$$\lim_{t \uparrow \infty} \frac{G(t)}{t} = +\infty$$

and

$$\sup_i \mathrm{E}\left[G(|X_i|)\right] < \infty.$$

Proof. Fix $\varepsilon > 0$ and let $a = \frac{M}{\varepsilon}$ where $M := \sup_n (\mathrm{E}\left[G(|X_i|)\right])$. Take c large enough so that $G(t)/t \geq a$ for $t \geq c$. In particular, $|X_i| \leq \frac{G(|X_i|)}{a}$ on $\{|X_i| > c\}$ and therefore

$$\int_{\{|X_i| > c\}} |X_i| \, dP \leq \frac{1}{a} \mathrm{E}\left[G(|X_i|) 1_{\{|X_i| > c\}}\right] \leq \frac{M}{a} = \varepsilon$$

uniformly in i. $\qquad \square$

EXAMPLE 4.2.15: TWO SUFFICIENT CONDITIONS FOR UNIFORM INTEGRABILITY. Two frequently used sufficient conditions guaranteeing uniform integrability are

$$\sup_i \mathrm{E}\left[|X_i|^{1+\alpha}\right] < \infty \quad (\alpha > 1)$$

and

$$\sup_i \mathrm{E}\left[|X_i| \log^+ |X_i|\right] < \infty.$$

Almost-sure convergence of a sequence of integrable random variables to an integrable random variable does not necessarily imply convergence in L^1. However:

Theorem 4.2.16 *Let $\{X_n\}_{n\geq 1}$ be a sequence of integrable random variables and let X be some random variable. The following are equivalent:*

(a) $\{X_n\}_{n\geq 1}$ *is uniformly integrable and* $X_n \overset{Pr.}{\to} X$ *as* $n \to \infty$.

(b) X *is integrable and* $X_n \overset{L^1}{\to} X$ *as* $n \to \infty$.

Proof. (a) implies (b): Since $X_n \overset{Pr.}{\to} X$, there exists a subsequence $\{X_{n_k}\}_{k\geq 1}$ such that $X_{n_k} \overset{a.s.}{\to} X$. By Fatou's lemma,

$$\mathrm{E}\left[|X|\right] \leq \liminf_k \mathrm{E}\left[|X_{n_k}|\right] \leq \sup_{n_k} \mathrm{E}\left[|X_{n_k}|\right] \leq \sup_n \mathrm{E}\left[|X_n|\right] < \infty.$$

Therefore $X \in L^1$. Also for fixed $\varepsilon > 0$,

$$\mathrm{E}\left[|X_n - X|\right] \leq \int_{\{|X_n - X| < \varepsilon\}} |X_n - X|\, \mathrm{d}P + \int_{\{|X_n - X| \geq \varepsilon\}} |X_n|\, \mathrm{d}P + \int_{\{|X_n - X| \geq \varepsilon\}} |X|\, \mathrm{d}P$$

$$\leq \varepsilon + \int_{\{|X_n - X| \geq \varepsilon\}} |X_n|\, \mathrm{d}P + \int_{\{|X_n - X| \geq \varepsilon\}} |X|\, \mathrm{d}P.$$

Recall (Remark 4.2.11) that adding an integrable random variable to a uniformly integrable collection retains uniformly integrability. Apply (b) of Theorem 4.2.12 to the uniformly integrable family $\{X_n\}_{n\geq 0}$ where $X_0 := X$, denoting by δ' the corresponding δ. By hypothesis, $P(|X_n - X| \geq \varepsilon) \leq \delta'$ for large enough n. By (b) of Theorem 4.2.12 with $A := \{|X_n - X| \geq \varepsilon\}$, for large enough n, $\int_{\{|X_n - X| \geq \varepsilon\}} |X_n|\, \mathrm{d}P \leq \varepsilon$ and $\int_{\{|X_n - X| \geq \varepsilon\}} |X|\, \mathrm{d}P \leq \varepsilon$. Therefore, $\mathrm{E}\left[|X_n - X|\right] \leq 3\varepsilon$ for large enough n, thus proving convergence in L^1.

(b) implies (a): Let $\varepsilon > 0$ be given and let n_0 be such that $\mathrm{E}\left[|X_n - X|\right] \leq \varepsilon$ for all $n \geq n_0$. The random variables $X, X_1, \ldots, X_{n_0}$ being integrable, there exists a $\delta > 0$ such that if $P(A) \leq \delta$, $\int_A |X|\, \mathrm{d}P \leq \frac{\varepsilon}{2}$ and $\int_A |X_n|\, \mathrm{d}P \leq \frac{\varepsilon}{2}$ for $n \leq n_0$. If $n \geq n_0$, by the triangle inequality,

$$\int_A |X_n|\, \mathrm{d}P \leq \int_A |X|\, \mathrm{d}P + \int_A |X_n - X|\, \mathrm{d}P \leq 2\varepsilon,$$

and therefore (b) of Theorem 4.2.12 is satisfied. Whereas (a) of Theorem 4.2.12 is satisfied since $\mathrm{E}\left[|X_n|\right] \leq \mathrm{E}\left[|X_n - X|\right] + \mathrm{E}\left[|X|\right]$. $\qquad\square$

4.3   Zero-one Laws

4.3.1   Kolmogorov's Zero-one Law

Definition 4.3.1 *Let $\{X_n\}_{n\geq 1}$ be a sequence of random variables and let $\mathcal{F}_n^X := \sigma(X_1, \ldots, X_n)$. The σ-field $\mathcal{T}^X := \cap_{n\geq 1}\sigma(X_n, X_{n+1}, \ldots)$ is called the* tail σ-field *of this sequence.*

EXAMPLE 4.3.2: For any $a \in \mathbb{R}$, the event $\{\lim_{n\uparrow\infty} \frac{X_1+\cdots+X_n}{n} \leq a\}$ belongs to the tail σ-field, since the existence and the value of the limit of $\frac{X_1+\cdots+X_n}{n}$ does not depend on any fixed finite number of terms of the sequence. More generally, any event concerning $\lim_{n\uparrow\infty} \frac{X_1+\cdots+X_n}{n}$ such as, for instance, the event that such limit exists, is in the tail σ-field.

Recall the notation $\mathcal{F}_\infty^X := \vee_{n\geq 1}\mathcal{F}_n^X$.

Theorem 4.3.3 *The tail σ-field of a sequence $\{X_n\}_{n\geq 1}$ of independent random variables is trivial, that is, if $A \in \mathcal{T}^X$, then $P(A) = 0$ or 1.*

Proof. The σ-fields $\mathcal{F}_n^X$ and $\sigma(X_{n+k}, X_{n+k+1}, \ldots)$ are independent for all $k \geq 1$ and therefore, since $\mathcal{T}^X = \cap_{k\geq 1}\sigma(X_{n+k}, X_{n+k+1})$, the σ-fields $\mathcal{F}_n^X$ and $\mathcal{T}^X$ are independent. Therefore the *algebra* $\cup_{n\geq 1}\mathcal{F}_n^X$ and $\mathcal{T}^X$ are independent, and consequently (Theorem 3.1.39) $\mathcal{F}_\infty^X$ and $\mathcal{T}^X$ are independent. But $\mathcal{F}_\infty^X \supseteq \mathcal{T}^X$, so that $\mathcal{T}^X$ is independent of itself. In particular, for all $A \in \mathcal{T}^X$, $P(A\cap A) = PA)P(A)$, that is $P(A) = P(A)^2$, which implies that $P(A) = 0$ or 1. $\qquad\square$

4.3.2 The Hewitt–Savage Zero-one Law

Let $(S, \mathcal{S})$ be some measurable space and let μ be a probability measure on it. We shall work on the canonical measurable space $(\Omega, \mathcal{F}) := (S^\mathbb{N}, \mathcal{S}^{\otimes\mathbb{N}})$ of S-valued random sequences, endowed with the probability measure $P := \mu^{\otimes\infty}$. In particular, an element of Ω has the form $\omega := x := (x_1, x_2, \ldots) \in S^\mathbb{N}$ and moreover, the sequence $\{X_n\}_{n\geq 1}$ defined by

$$X_n(\omega) := x_n \quad (\omega \in \Omega, n \geq 1)$$

is IID with common probability distribution μ.

Definition 4.3.4

(a) *A finite permutation of $\mathbb{N}$ is a permutation π such that $\pi(i) = i$ for all but a finite number of indices $i \geq 1$.*

(b) *An event $A \in \mathcal{F}$ such $\pi^{-1}A = A$ for all finite permutations π is called exchangeable.*

(c) *The sub-σ-field $\mathcal{E}$ consisting of the collection of exchangeable events is called the exchangeable σ-field.*

(Note that $X_n(\pi\omega) = X_{\pi(n)}(\omega)$ for any permutation π on Ω.)

EXAMPLE 4.3.5: TAIL EVENTS ARE EXCHANGEABLE. All the events of the tail σ-field $\mathcal{T}$ are exchangeable. Indeed, for all $n \geq 1$, an event $B \in \sigma(X_{n+1}, X_{n+2}, \ldots)$ is unaltered by a permutation bearing on only the first n coordinates. Therefore any event $B \in \cap_{n\geq 1}\sigma(X_{n+1}, X_{n+2}, \ldots)$ is unaltered by any finite permutation.

EXAMPLE 4.3.6: THERE EXIST EXCHANGEABLE EVENTS THAT ARE NOT TAIL EVENTS.
In Example 4.3.5, we have seen that $\mathcal{T} \subset \mathcal{E}$. The current example shows that we do not
have the reverse inclusion. Indeed, the event $A := X_1 + \cdots + X_n \in C\,i.o.\}$ is exchangeable
(if the finite permutation π bears on only the first K integers, then for all $n \geq K+1$,
$X_1 + \cdots + X_n = X_{\pi(1)} + \cdots + X_{\pi(n)}$). However, it is not a tail event.

Theorem 4.3.7 *The events of the exchangeable σ-field are trivial, that is for any $A \in \mathcal{E}$,*
$P(A) = 0$ *or* 1.

The proof depends on the following lemma of approximation of an element of $\mathcal{F}$
by an element of an algebra $\mathcal{A}$ generating $\mathcal{F}$. More precisely (recalling the notation
$A \Delta B := (A - A \cap B) \cup (B - A \cap B)$):

Lemma 4.3.8 *Let $\mathcal{A}$ be an algebra generating the σ-field $\mathcal{F}$ and let P be a probability
on $\mathcal{F}$. With any event $B \in \mathcal{F}$ and any $\varepsilon > 0$, one can associate an event $A \in \mathcal{A}$ such
that $P(A \triangle B) \leq \varepsilon$.*

Proof. The collection of sets

$$\mathcal{G} := \{B \in \mathcal{F};\ \forall \varepsilon > 0, \exists A \in \mathcal{A} \text{ with } P(A \triangle B) \leq \varepsilon\}$$

contains $\mathcal{A}$. It is moreover a σ-field, as we now show. First, $\Omega \in \mathcal{A} \subseteq \mathcal{G}$ and the stability
of $\mathcal{G}$ under complementation is clear. For the stability of $\mathcal{G}$ under countable unions, let
B_n ($n \geq 1$) be in $\mathcal{G}$ and let $\varepsilon > 0$ be given. Also, by definition of $\mathcal{G}$, there exist A_n's in
$\mathcal{A}$ such that $P(A_n \triangle B_n) \leq 2^{-n-1}\varepsilon$. Therefore, for all $K \geq 1$,

$$P((\cup_{n=1}^{K} A_n) \triangle (\cup_{n=1}^{K} B_n)) \leq \sum_{n=1}^{K} 2^{-n-1}\varepsilon \leq \sum_{n \geq 1} 2^{-n-1}\varepsilon = 2^{-1}\varepsilon.$$

By the sequential continuity property of probability, there exists an integer $K = K(\varepsilon)$
such that $P(\cup_{n \geq 1} B_n - \cup_{n=1}^{K} B_n) \leq 2^{-1}\varepsilon$. Therefore, for such an integer,

$$P((\cup_{n=1}^{K} A_n) \triangle (\cup_{n \geq 1} B_n)) \leq \varepsilon.$$

The proof of stability of $\mathcal{G}$ under countable unions is completed since $\mathcal{A}$ is an algebra and
therefore $\cup_{n=1}^{K} A_n \in \mathcal{A}$. Therefore $\mathcal{G}$ is a σ-field containing $\mathcal{A}$ and consequently contains
the σ-field $\mathcal{F}$ generated by $\mathcal{A}$. $\square$

We now proceed to the proof of Theorem 4.3.7.

Proof. Let $A \in \mathcal{E}$. Lemma 4.3.8 guarantees that for any $n \geq 1$, there exists an $A_n \in$
$\sigma(X_1, \ldots, X_n)$ such that

$$P(A_n \Delta A) \to 0.$$

Note that for all $n \geq 1$,

$$A_n = \{\omega\,;\ (x_1, \ldots, x_n) \in B_n\}$$

for some $B_n \in \mathcal{S}^{\otimes \mathbb{N}}$. Define the finite permutation $\pi_n = \pi$ by

$$\pi(j) = j + n \text{ if } 1 \le j \le n$$
$$= j - n \text{ if } n + 1 \le j \le 2n$$
$$= j \text{ if } j \ge 2n + 1.$$

Note that $\pi^2 \equiv \pi$ and in particular $\pi = \pi^{-1}$, and that by the IID assumption, the sequence obtained by finite permutation of an IID sequence is IID. Therefore

$$P(\omega ; \omega \in A_n \Delta A) = P(\omega ; \pi\omega \in A_n \Delta A). \qquad (\star)$$

Now $\{\omega ; \pi\omega \in A\} = \{\omega ; \omega \in A\}$ by exchangeability of A, and

$$\{\omega ; \pi\omega \in A_n\} = \{\omega ; (x_{n+1}, \dots, x_{2n}) \in B_n\}.$$

Therefore denoting by A'_n the event in the right-hand side of the above equality,

$$\{\omega ; \pi\omega \in A_n \Delta A\} = \{\omega ; \omega \in A'_n \Delta A\}. \qquad (\star\star)$$

Combining $(\star)$ and $(\star\star)$:
$$P(A_n \Delta A) = P(A'_n \Delta A). \qquad (\dagger)$$

From the set inclusion $A \Delta C \subseteq (A \Delta B) \cup (B \Delta C)$, $(\dagger)$ and $P(A_n \Delta A) \to 0$,

$$P(A_n \Delta A'_n) + P(A \Delta A'_n) \to 0. \qquad (\dagger\dagger)$$

Therefore

$$0 \le P(A_n) - P(A_n \cap A'_n)$$
$$\le P(A_n \cup A'_n) - P(A_n \cap A'_n) = P(A_n \Delta A'_n) \to 0.$$

Therefore $P(A_n \cap A'_n) \to P(A)$. Since A_n and A'_n are independent (and recalling that $P(A'_n) = P(A_n)$)

$$P(A_n \cap A'_n) = P(A_n)P(A'_n) = P(A_n)^2 \to P(A)^2.$$

Comparing with $(\dagger\dagger)$, we see that $P(A)^2 = P(A)$, which implies that $P(A) = 0$ or $P(A) = 1$. $\qquad \square$

The Hewitt–Savage zero-one law will now applied to the asymptotic behavior of random walks. By definition, a random walk on $\mathbb{R}$ is a sequence $\{S_n\}_{n \ge 1}$ of real-valued random variables of the form

$$S_n = X_1 + \cdots + X_n,$$

where $\{X_n\}_{n \ge 1}$ is an IID sequence of real-valued random variables.

Theorem 4.3.9 *Discarding the trivial case where $P(X_1 = 0) = 1$, with probability one, one and only one of the following occurs:*

(a) $\lim_n S_n = +\infty$,

(b) $\lim_n S_n = -\infty$,

(c) $-\infty = \liminf_n S_n < \limsup_n S_n = +\infty$.

If, moreover, the distribution of X_1 is symmetric around 0, (c) occurs with probability 1.

Proof. The random variable $\limsup_n S_n$ is exchangeable (its value is independent of any finite permutation of the X_i's) and therefore is a constant c, possibly $+\infty$ or $-\infty$. Since

$$S_n = X_1 + S'_n\,,$$

where $S'_n = X_2 + \cdots + X_n$, we have that

$$\limsup_n S_n = X_1 + \limsup_n S'_n,$$

where $\limsup_n S'_n$ has the same distribution as $\limsup_n S_n$ and is independent of X_1, and therefore

$$c = X_1 + c\,.$$

Since $P(X_1 \neq 0) > 0$, this implies that $\limsup_n S_n$ cannot be finite, and is therefore either $+\infty$ or $-\infty$. Similarly for $\liminf_n S_n$, which is therefore either $+\infty$ or $-\infty$. Since we cannot have simultaneously $\liminf_n S_n = +\infty$ and $\limsup_n S_n = -\infty$, the first part of the theorem in proved.

In the symmetric case, only (c) is possible because one of the events (a) or (b) entails the other. $\square$

4.4   Convergence in Distribution and in Variation

4.4.1   The Role of Characteristic Functions

Let $\{X_n\}_{n\geq 1}$ and X be real-valued random variables with respective cumulative distribution functions $\{F_n\}_{n\geq 1}$ and F. A natural definition of convergence in distribution of $\{X_n\}_{n\geq 1}$ to X is the following:

$$\lim_{n\uparrow\infty} F_n(x) = F(x)\,. \tag{$\star$}$$

We have not specified for what $x \in \mathbb{R}$ ($\star$) is required. If we want this to hold for *all* x, then we could not say that the "random" (actually deterministic) sequence of random variables $X_n \equiv a + \frac{1}{n}$ where $a \in \mathbb{R}$ converges to $X \equiv a$. In fact, ($\star$) holds in this case for all points of continuity of the cumulative distribution of X, here $F(x) = 1_{x\geq a}$.

It turns out that a "good" definition would be precisely that ($\star$) should hold for all continuity points of the target CDF F.

EXAMPLE 4.4.1:   MAGNIFIED MINIMUM. Let $\{Y_n\}_{n\geq 1}$ be a sequence of IID random variables uniformly distributed on $[0, 1]$. Then

$$X_n = n\min(Y_1, \ldots, Y_n) \overset{\mathcal{D}}{\to} \mathcal{E}(1)\,,$$

(the exponential distribution with mean 1). In fact, for all $x \in [0, n]$,

$$P(X_n > x) = P\left(\min(Y_1, \ldots, Y_n) > \frac{x}{n}\right) = \prod_{i=1}^{n} P\left(Y_i > \frac{x}{n}\right) = \left(1 - \frac{x}{n}\right)^n,$$

and therefore $\lim_{n\uparrow\infty} P(X_n > x) = e^{-x} 1_{\mathbb{R}_+}(x)$.

For technical reasons, our starting point will differ from the definition $(\star)$, properly modified.

Denote by $M^+(\mathbb{R}^d)$ the collection of finite measures on $(\mathbb{R}^d, \mathcal{B}(\mathbb{R}^d))$ and by $C_b(\mathbb{R}^d)$ the collection of continuous bounded functions $f : \mathbb{R}^d \to \mathbb{R}$.

Definition 4.4.2

(a) *The sequence $\{\mu_n\}_{n\geq 1}$ in $M^+(\mathbb{R}^d)$ is said to converge weakly to μ if $\lim_{n\uparrow\infty} \int_{\mathbb{R}^d} f \, d\mu_n = \int_{\mathbb{R}^d} f \, d\mu$ for all $f \in C_b(\mathbb{R}^d)$. This is denoted by*

$$\mu_n \overset{w}{\to} \mu.$$

(b) *The sequence of random vectors $\{X_n\}_{n\geq 1}$ of $\mathbb{R}^d$ with respective probability distributions $\{Q_{X_n}\}_{n\geq 1}$ is said to converge in distribution to the random vector $X \in \mathbb{R}^d$ with distribution Q_X if $Q_{X_n} \overset{w}{\to} Q_X$. (In other words, for all continuous and bounded functions $f : \mathbb{R}^d \to \mathbb{R}$, $\lim_{n\uparrow\infty} E[f(X_n)] = E[f(X)]$.) This is denoted by*

$$X_n \overset{\mathcal{D}}{\to} X.$$

Remark 4.4.3 Observe that X and the X_n's need not be defined on the same probability space. Convergence in distribution concerns only probability distributions. As a matter of fact, very often the X_n's are defined on the same probability space but there is no "visible" (that is, defined on the same probability space) limit random vector X. Therefore one sometimes denotes convergence in distribution as follows: $X_n \overset{\mathcal{D}}{\to} Q$, where Q is a probability distribution on $\mathbb{R}^d$. If Q is a "famous" probability distribution, for instance a standard Gaussian variable, one then says that $\{X_n\}_{n\geq 1}$ "converges in distribution to a standard Gaussian distribution". This is also denoted by $X_n \overset{\mathcal{D}}{\to} \mathcal{N}(0,1)$.

Let B° and B^c be respectively the interior and the closure of the set $B \in \mathbb{R}^d$ and let ∂B be its boundary $(:= B^c \backslash B^\circ)$. The following theorem is a major tool of the theory of convergence in distribution:

Theorem 4.4.4 *Let $\{\mu_n\}_{n\geq 1}$ and μ be probability distributions on $\mathbb{R}^d$. The following conditions are equivalent:*

(i) $\mu_n \overset{w}{\to} \mu$.

(ii) *For any open set $G \subseteq \mathbb{R}^d$, $\liminf_n \mu_n(G) \geq \mu(G)$.*

(iii) *For any closed set $F \subseteq \mathbb{R}^d$, $\limsup_n \mu_n(F) \leq \mu(F)$.*

(iv) *For any measurable set $B \subseteq \mathbb{R}^d$ such that $\mu(\partial B) = 0$, $\lim_n \mu_n(B) = \mu(B)$.*

Proof. (i) $\Rightarrow$ (ii). For any open set $G \in \mathbb{R}^d$ there exists a non-decreasing sequence $\{\varphi_k\}_{k\geq 1}$ of non-negative functions of $C_b(\mathbb{R}^d)$ such that $0 \leq \varphi_k \leq 1$ and $\varphi_k \uparrow 1_G$ (for instance, $\varphi_k(x) = 1 - e^{-kd(x,\overline{G})}$). Since $\int 1_G \, d\mu_n \geq \int \varphi_k \, d\mu_n$ $(k \geq 1)$,

$$\liminf_n \mu_n(G) = \liminf_n \int 1_G \, d\mu_n \geq \liminf_n \int \varphi_k \, d\mu_n.$$

This being true for all $k \geq 1$,

$$\liminf_n \mu_n(G) \geq \sup_k \left(\liminf_n \int \varphi_k \, d\mu_n \right)$$
$$= \sup_k \left(\lim_n \int \varphi_k \, d\mu_n \right) = \sup_k \int \varphi_k \, d\mu = \mu(G).$$

(ii) $\Leftrightarrow$ (iii). Take complements.

(ii) + (iii) $\Rightarrow$ (iv). Indeed, by (ii) and (iii),

$$\limsup_n \mu_n(B) \leq \limsup_n \mu_n(\bar{B}) \leq \mu(\bar{B})$$

and

$$\liminf_n \mu_n(B) \geq \liminf_n \mu_n(B^o) \geq \mu(B^o).$$

But since $\mu(\partial B) = 0$, $\mu(B^o) = \mu(\bar{B}) = \mu(B)$, and therefore (iv) is verified.

(iv) $\Rightarrow$ (i). Let $f \in C_b(\mathbb{R}^d)$. We must show that $\lim_{n\uparrow\infty} \int_{\mathbb{R}^d} f \, d\mu_n = \int_{\mathbb{R}^d} f \, d\mu$. It is enough to show this for $f \geq 0$. Let $K < \infty$ be a bound of f. By Fubini,

$$\int_{\mathbb{R}^d} f(x) \, d\mu(x) = \int_{\mathbb{R}^d} \left(\int_0^K 1_{\{t \leq f(x)\}} \, dt \right) d\mu(x)$$
$$= \int_0^K \mu(\{x \,;\, t \leq f(x)\}) \, dt = \int_0^K \mu(D_t^f) \, dt,$$

where $D_t^f := \{x \,;\, t \leq f(x)\}$. Observe that $\partial D_t^f \subseteq \{x \,;\, t = f(x)\}$ and that the collection of positive t such that $\mu(\{x \,;\, t = f(x)\}) > 0$ is at most countable (for each positive integer k there are at most k values of t such that $\mu(\{x \,;\, t = f(x)\}) \geq \frac{1}{k}$). Therefore, by (iv), for almost all t (with respect to the Lebesgue measure),

$$\lim_n \mu_n(D_t^f) = \mu(D_t^f) \quad (\ell\text{-almost everywhere})$$

and by dominated convergence,

$$\lim_n \int f \, d\mu_n = \lim_n \int_0^K \mu_n(D_t^f) \, dt = \int_0^K \mu(D_t^f) \, dt = \int f \, d\mu.$$

$\square$

Paul Lévy's Characterization

Theorem 4.4.5 *A necessary and sufficient condition for the sequence $\{X_n\}_{n\geq 1}$ of random vectors of $\mathbb{R}^d$ to converge in distribution is that the sequence of their characteristic functions $\{\varphi_n\}_{n\geq 1}$ converges to some function φ that is continuous at 0. In such a case, φ is the characteristic function of the limit probability distribution.*

The (technical) proof is postponed to Section 4.4.4.

Corollary 4.4.6 *Let $\{X_n\}_{n\geq 1}$ and X be random vectors of $\mathbb{R}^d$ with respective characteristic functions $\{\varphi_n\}_{n\geq 1}$ and φ. The following two statements are equivalent.*

(A) $X_n \overset{\mathcal{D}}{\to} X$.

(B) $\lim_{n\uparrow\infty} \varphi_n = \varphi$.

Corollary 4.4.7 *In the univariate case, denote by F_n and F the cumulative distribution functions of X_n and X, respectively. Call a point $x \in \mathbb{R}$ a continuity point of F if $F(x) = F(x_-)$. Then $X_n \overset{\mathcal{D}}{\to} X$ if and only*

$$\lim_n F_n(x) = F(x) \text{ for all continuity points } x \text{ of } F.$$

Proof. Necessity. Let Q_X be the probability distribution of X. If x is a continuity point of F, the boundary of $C := (-\infty, x]$ is $\{x\}$ of null Q_X-measure. Therefore by (iv) of Theorem 4.4.4, $\lim_n Q_{X_n}((-\infty, x]) = Q_X((\infty, x])$, that is, $\lim_n F_n(x) = F(x)$.

Sufficiency. Let $f \in C_b(\mathbb{R})$ and let $M < \infty$ be an upper bound of f. For $\varepsilon > 0$, there exists a subdivision $-\infty < a = x_0 < x_1 < \cdots < x_k = b < +\infty$ formed by continuity points of F, such that $F(a) < \varepsilon$, $F(b) > 1 - \varepsilon$ and $|f(x) - f(x_i)| < \varepsilon$ on $[x_{i-1}, x_i]$. By hypothesis,

$$S_n := \sum_{i=1}^{k} f(x_i)(F_n(x_i) - F_n(x_{i-1})) \to S := \sum_{i=1}^{k} f(x_i)(F(x_i) - F(x_{i-1})).$$

Also

$$|\mathrm{E}\,[f(X)] - S| \leq \varepsilon + MF(a) + M(1 - F(b)) \leq (2M + 1)\varepsilon$$

and

$$|\mathrm{E}\,[f(X_n)] - S_n|$$
$$\leq \varepsilon + MF_n(a) + M(1 - F_n(b)) \to \varepsilon + MF(a) + M(1 - F(b))$$
$$\leq (2M + 1)\varepsilon.$$

Therefore,

$$\limsup_n |\mathrm{E}\,[f(X_n)] - \mathrm{E}\,[f(X)]|$$
$$\leq \limsup_n |\mathrm{E}\,[f(X_n)] - S_n| + \limsup_n |S_n - S| + |\mathrm{E}\,[f(X)] - S|$$
$$\leq (4M + 2)\varepsilon.$$

Since ε is arbitrary, $\lim_n |\mathrm{E}\,[f(X_n)] - \mathrm{E}\,[f(X)]| = 0$. $\qquad\square$

Theorem 4.4.8 *Let $\{X_n\}_{n\geq 1}$ and $\{Y_n\}_{n\geq 1}$ be sequences of random vectors of $\mathbb{R}^d$ such that $X_n \overset{\mathcal{D}}{\to} X$ and $d(X_n, Y_n) \overset{Pr.}{\to} 0$, where d denotes the euclidean distance. Then $Y_n \overset{\mathcal{D}}{\to} X$.*

Proof. By (iii) of Theorem 4.4.4, it suffices to show that for all closed sets F, $\limsup_n P(Y_n \in F) \leq P(X \in F)$. For all $\varepsilon > 0$, define the closed set $F_\varepsilon = \{x \in \mathbb{R}^d \,;\, d(x, F) \leq \varepsilon\}$. Then

$$P(Y_n \in F) \leq P(d(X_n, F) \geq \varepsilon) + P(X_n \in F_\varepsilon),$$

and therefore

$$\limsup_n P(Y_n \in F) \leq \limsup_n P(d(X_n, F) \geq \varepsilon) + \limsup_n P(X_n \in F_\varepsilon)$$
$$= \limsup_n P(X_n \in F_\varepsilon) \leq P(X \in F_\varepsilon).$$

Since $\varepsilon > 0$ is arbitrary and $\lim_{\varepsilon \downarrow 0} P(X \in F_\varepsilon) = P(X \in F)$, $\limsup_n P(Y_n \in F) \leq P(X \in F)$. $\qquad\square$

Corollary 4.4.9 *Let $\{X_n\}_{n \geq 1}$ be a sequence of random vectors of $\mathbb{R}^d$ such that $X_n \overset{\mathcal{D}}{\to} X$. If the sequence of real numbers $\{a_n\}_{n \geq 1}$ converges to the real number a, then $a_n X_n \overset{\mathcal{D}}{\to} aX$.*

Bochner's Theorem

Bochner's theorem will play a central role in the theory of wide-sense stationary processes of Chapter 12.

The characteristic function φ of a real random variable X has the following properties:

A. it is hermitian symmetric (that is, $\varphi(-u) = \varphi(u)^*$) and uniformly bounded (in fact, $|\varphi(u)| \leq \varphi(0)$);

B. it is uniformly continuous on $\mathbb{R}$; and

C. it is definite non-negative, in the sense that for all integers n, all $u_1, \ldots, u_n \in \mathbb{R}$, and all $z_1, \ldots, z_n \in \mathbb{C}$,

$$\sum_{j=1}^n \sum_{k=1}^n \varphi(u_j - u_k) z_j z_k^* \geq 0$$

(just observe that the left-hand side equals $\mathrm{E}\left[\left|\sum_{j=1}^n z_j e^{iu_j X}\right|^2\right]$).

It turns out that Properties A, B and C characterize characteristic functions (up to a multiplicative constant). This is *Bochner's theorem*:

Theorem 4.4.10 *Let $\varphi : \mathbb{R} \to \mathbb{C}$ be a function satisfying properties A, B and C. Then there exists a constant $0 \leq \beta < \infty$ and a real random variable X such that for all $u \in \mathbb{R}$,*

$$\varphi(u) = \beta \mathrm{E}\left[e^{iuX}\right].$$

Proof. We henceforth eliminate the trivial case where $\varphi(0) = 0$ (implying, in view of condition A, that φ is the null function). For any continuous function $z : \mathbb{R} \to \mathbb{C}$ and any $A \geq 0$,

$$\int_0^A \int_0^A \varphi(u - v) z(u) z^*(v) \, du \, dv \geq 0. \tag{$\star$}$$

Indeed, since the integrand is continuous, the integral is the limit as $n \uparrow \infty$ of

$$\frac{A^2}{4^n} \sum_{j=1}^{2^n} \sum_{k=1}^{2^n} \varphi\left(\frac{A(j-k)}{2^n}\right) z\left(\frac{Aj}{2^n}\right) z\left(\frac{Ak}{2^n}\right)^*,$$

a non-negative quantity by condition C. From $(\star)$ with $z(u) := e^{-ixu}$, we have that

$$g(x, A) := \frac{1}{2\pi A} \int_0^A \int_0^A \varphi(u - v) e^{-ix(u-v)} \, du \, dv \geq 0 \,.$$

Changing variables, we obtain for $g(x, A)$ the alternative expression

$$g(x, A) := \frac{1}{2\pi} \int_{-A}^A \left(1 - \frac{|u|}{A}\right) \varphi(u) e^{-iux} \, du$$

$$= \frac{1}{2\pi} \int_{-\infty}^{+\infty} h\left(\frac{u}{A}\right) \varphi(u) e^{-iux} \, du \,,$$

where $h(u) = (1 - |u|) \, 1_{\{|u| \leq 1\}}$. Let $M > 0$. We have

$$\int_{-\infty}^{+\infty} h\left(\frac{x}{2M}\right) g(x, A) \, dx$$

$$= \frac{1}{2\pi} \int_{-\infty}^{+\infty} h\left(\frac{u}{A}\right) \varphi(u) \left(\int_{-\infty}^{+\infty} h\left(\frac{x}{2M}\right) e^{-iux} \, dx\right) du$$

$$= \frac{1}{\pi} M \int_{-\infty}^{+\infty} h\left(\frac{u}{A}\right) \varphi(u) \left(\frac{\sin Mu}{Mu}\right)^2 du \,.$$

Therefore

$$\int_{-\infty}^{+\infty} h\left(\frac{x}{2M}\right) g(x, A) \, dx \leq \frac{1}{\pi} M \int_{-\infty}^{+\infty} h\left(\frac{u}{A}\right) |\varphi(u)| \left(\frac{\sin Mu}{Mu}\right)^2 du$$

$$\leq \frac{1}{\pi} \varphi(0) \int_{-\infty}^{+\infty} \left(\frac{\sin u}{u}\right)^2 du = \varphi(0) \,.$$

By monotone convergence,

$$\lim_{M \uparrow \infty} \int_{-\infty}^{+\infty} h\left(\frac{x}{2M}\right) g(x, A) \, dx = \int_{-\infty}^{+\infty} g(x, A) \, dx \,,$$

and therefore

$$\int_{-\infty}^{+\infty} g(x, A) \, dx \leq \varphi(0) \,.$$

The function $x \to g(x, A)$ is therefore integrable and it is the Fourier transform of the integrable and continuous function $u \to h\left(\frac{u}{A}\right) \varphi(u)$. Therefore, by the Fourier inversion formula:

$$h\left(\frac{u}{A}\right) \varphi(u) = \int_{-\infty}^{+\infty} g(x, A) e^{iux} \, dx \,.$$

In particular, with $u = 0$, $\int_{-\infty}^{+\infty} g(x, A) \, dx = \varphi(0)$. Therefore, $f(x, A) := \frac{g(x,A)}{\varphi(0)}$ is the probability density of some real random variable with characteristic function $h\left(\frac{u}{A}\right) \frac{\varphi(u)}{\varphi(0)}$. But

$$\lim_{A \uparrow \infty} h\left(\frac{u}{A}\right) \frac{\varphi(u)}{\varphi(0)} = \frac{\varphi(u)}{\varphi(0)} \,.$$

This limit of a sequence of characteristic functions is continuous at 0 and is therefore a characteristic function (Paul Lévy's criterion, Theorem 4.4.5). $\qquad \square$

4.4.2   The Central Limit Theorem

The emblematic theorem of Statistics is the so-called *central limit theorem (CLT)*.

Theorem 4.4.11 *Let $\{X_n\}_{n\geq 1}$ be an IID sequence of real random variables such that*

$$E[X_1^2] < \infty. \tag{4.21}$$

(In particular, $E[|X_1|] < \infty$.) Then, for all $x \in \mathbb{R}$,

$$\lim_{n\uparrow\infty} P\left(\left(\frac{S_n - nE[X_1]}{\sigma_{X_1}\sqrt{n}}\right) \leq x\right) = P(\mathcal{N}(0;1) \leq x), \tag{4.22}$$

where $\mathcal{N}(0;1)$ is a Gaussian variable with mean 0 and variance 1.

The proof depends in part on the following theorem, which says in particular that under certain conditions, the moments of a random variable can be extracted from its characteristic function:

Theorem 4.4.12 *Let X be a real random variable with characteristic function ψ, and suppose that $E[|X|^n] < \infty$ for some integer $n \geq 1$. Then for all integers $r \leq n$, the r-th derivative $\psi^{(r)}$ of ψ exists and is given by*

$$\psi^{(r)}(u) = i^r E\left[X^r e^{iuX}\right], \tag{4.23}$$

and in particular $E[X^r] = \frac{\psi^{(r)}(0)}{i^r}$. Moreover,

$$\psi(u) = \sum_{r=0}^{n} \frac{(iu)^r}{r!} E[X^r] + \frac{(iu)^n}{n!}\varepsilon_n(u), \tag{4.24}$$

where $\lim_{n\uparrow\infty} \varepsilon_n(u) = 0$ and $|\varepsilon_n(u)| \leq 3E[|X|^n]$.

Proof. First we observe that for any non-negative real number a, and all integers $r \leq n$, $a^r \leq 1 + a^n$ (Indeed, if $a \leq 1$, then $a^r \leq 1$, and if $a \geq 1$, then $a^r \leq a^n$). In particular,

$$E[|X|^r] \leq E[1 + |X|^n] = 1 + E[|X|^n] < \infty.$$

Suppose that for some $r < n$,

$$\psi^{(r)}(u) = i^r E\left[X^r e^{iuX}\right].$$

In

$$\frac{\psi^{(r)}(u+h) - \psi^{(r)}(u)}{h} = i^r E\left[X^r \frac{e^{i(u+h)X} - e^{iuX}}{h}\right]$$

$$= i^r E\left[X^r e^{iuX} \frac{e^{ihX} - 1}{h}\right],$$

the quantity under the expectation sign tends to $X^{r+1}e^{iuX}$ as $h \to 0$, and moreover, it is bounded in absolute value by an integrable function since

$$\left|X^r e^{iuX} \frac{e^{ihX} - 1}{h}\right| \leq \left|X^r \frac{e^{ihX} - 1}{h}\right| \leq |X|^{r+1}.$$

(For the last inequality, use the fact that $\left|e^{ia} - 1\right|^2 = 2(1 - \cos a) \le a^2$.) Therefore, by dominated convergence,

$$\psi^{(r+1)}(u) = \lim_{h \to 0} \frac{\psi(u + h) - \psi(u)}{h}$$

$$= i^r E\left[\lim_{h \to 0} X^r e^{iuX} \frac{e^{ihX} - 1}{h}\right] = i^r E\left[X^{r+1} e^{iuX}\right].$$

Equality (4.23) follows since the induction hypothesis is trivially true for $r = 0$.

We now prove (4.24). By Taylor's formula, for $y \in \mathbb{R}$,

$$e^{iy} = \cos y + i \sin y = \sum_{k=0}^{n-1} \frac{(iy)^k}{k!} + \frac{(iy)^n}{n!} \left(\cos(\theta_1 y) + i \sin(\theta_2 y)\right)$$

for some $\theta_1, \theta_2 \in [-1, +1]$. Therefore

$$e^{iuX} = \sum_{k=0}^{n-1} \frac{(iuX)^k}{k!} + \frac{(iuX)^n}{n!} \left(\cos(\theta_1 uX) + i \sin(\theta_2 uX)\right),$$

where $\theta_1 = \theta_1(\omega), \theta_2 = \theta_2(\omega) \in [-1, +1]$, and

$$E\left[e^{iuX}\right] = \sum_{k=0}^{n-1} \frac{(iu)^k}{k!} E[X^k] + \frac{(iu)^n}{n!} \left(E[X^n] + \varepsilon_n(u)\right),$$

where

$$\varepsilon_n(u) = E\left[X^n \left(\cos \theta_1 uX + i \sin \theta_2 uX - 1\right)\right].$$

Clearly $|\varepsilon_n(u)| \le 3E\left[|X|^n\right]$. Also, since the random variable

$$X^n \left(\cos \theta_1 uX + i \sin \theta_2 uX - 1\right)$$

is bounded in absolute value by the integrable random variable $3|X|^n$ and tends to 0 as $u \to 0$, we have by dominated convergence $\lim_{u \to 0} \varepsilon_n(u) = 0$. $\qquad\square$

We now proceed to the proof of Theorem 4.22.

Proof. Assume without loss of generality that $E[X_1] = 0$. Then call σ^2 the variance of X_1. By the characteristic function criterion for convergence in distribution, it suffices to show that

$$\lim_{n \uparrow \infty} \varphi_n(u) = e^{-\sigma^2 u^2 / 2},$$

where

$$\varphi_n(u) = E\left[\exp\left\{iu \frac{\sum_{j=1}^n X_j}{\sqrt{n}}\right\}\right]$$

$$= \prod_{j=1}^n E\left[\exp\left\{i \frac{u}{\sqrt{n}} X_j\right\}\right] = \psi\left(\frac{u}{\sqrt{n}}\right)^n,$$

where ψ is the characteristic function of X_1. From the Taylor expansion of ψ about zero,

$$\psi(u) = 1 + \frac{\psi''(0)}{2!} u^2 + o(u^2),$$

we have, for fixed $u \in \mathbb{R}$,

$$\psi\left(\frac{u}{\sqrt{n}}\right) = 1 - \frac{1}{n}\frac{\sigma^2 u^2}{2} + o\left(\frac{1}{n}\right),$$

and therefore

$$\lim_{n\uparrow\infty} \ln\{\varphi_n(u)\} = \lim_{n\uparrow\infty} n\left(\ln\left\{1 - \frac{\sigma^2 u^2}{2n} + o\left(\frac{1}{n}\right)\right\}\right) = -\frac{1}{2}\sigma^2 u^2.$$

The result follows by Theorem 4.4.6.

$\square$

Remark 4.4.13 The random variable

$$\frac{S_n - n\mathrm{E}[X_1]}{\sigma\sqrt{n}}$$

is obtained by *centering* the sum S_n (subtracting its mean $n\mathrm{E}[X_1]$) and then *normalizing* it (dividing by the square root of its variance to make the resulting variance equal to 1).

Theorem 4.4.14 *Let $\{X_n\}_{n\geq 1}$ be a sequence of independent random vectors of dimension d, and let $\{a_n\}_{n\geq 1}$ be a sequence of real numbers such that $\lim_{n\uparrow\infty} a_n = \infty$. Suppose that*

$$X_n \overset{p.s.}{\to} m$$

and

$$\sqrt{a_n}(X_n - m) \overset{D}{\to} \mathcal{N}(0, \Gamma).$$

Let $g : \mathbb{R}^d \to \mathbb{R}^q$ be twice continuously differentiable in a neighborhood U of m. Then

$$g(X_n) \overset{p.s.}{\to} g(m)$$

and

$$\sqrt{a_n}(g(X_n) - g(m)) \overset{D}{\to} \mathcal{N}\left(0, J_g(m)^T \, \Gamma \, J_g(m)\right)$$

where $J_g(m)$ is the Jacobian matrix of g evaluated at m.

Proof. U can be chosen convex and compact. Let g_j denote the j-th coordinate of g, and let $D^2 g_j$ denote the second differential matrix of g_j. By Taylor's formula,

$$g_j(x) - g_j(m) = (x - m)^T (\mathrm{grad}\, g_j(m)) + \frac{1}{2}(x - m)^T D^2 g_j(m^*)(x - m)$$

for some m^* in the closed segment linking m to x, denoted $[m, x]$. Therefore, if $X_n \in U$

$$\sqrt{a_n}(g_j(X_n) - g_j(m)) = \sqrt{a_n}(X_n - m)^T(\mathrm{grad}\, g_j(m))$$
$$+ \frac{1}{2}a_n(X_n - m)^T \frac{1}{\sqrt{a_n}}D^2 g_j(m_n^*)(X_n - m),$$

where $m_n^* \in [m, X_n]$.

Suppose $X_n \in U$. Since U is convex and $m \in U$, also $m_n^* \in U$. Now since U is compact, the continuous function $D^2 g_j$ is bounded in U. Therefore, since $a_n \uparrow \infty$, $\frac{1}{\sqrt{a_n}}D^2 g_j(m_n^*)1_U(X_n) \to 0$. Since $X_n \overset{a.s.}{\to} m$, we deduce from the above remarks that

$$\sqrt{a_n}(g_j(X_n) - g_j(m)) - \sqrt{a_n}(X_n - m)^T(\mathrm{grad}\, g_j(m)) \overset{a.s.}{\to} 0,$$

and therefore

$$\sqrt{a_n}(g(X_n) - g(m)) - J_g(m)\sqrt{a_n}(X_n - m) \overset{p.s.}{\to} 0.$$

But $\sqrt{a_n}(X_n - m) \overset{D}{\to} \mathcal{N}(0, \Gamma)$ and therefore

$$\sqrt{a_n}(g(X_n) - g(m)) \overset{D}{\to} J_g(m)\mathcal{N}(0, \Gamma) = \mathcal{N}\left(0, J_g(m)^T \Gamma J_g(m)\right).$$

□

Statistical Applications

A basic methodology of Statistics is based on the notion of *confidence interval* and on the central limit theorem.

Theorem (4.22) implies that for $x \geq 0$

$$\lim_{n\uparrow\infty} P\left(\mathrm{E}[X_1] - \frac{\sigma}{\sqrt{n}}x \leq \frac{S_n}{n} \leq \mathrm{E}[X_1] + \frac{\sigma}{\sqrt{n}}x\right) = P(|\mathcal{N}(0; 1)| \leq x).$$

Under the condition $\mathrm{E}\left[|X_1|^3\right] < \infty$, this limit is uniform in $x \in \mathbb{R}$ [1] and therefore, with $\frac{\sigma}{\sqrt{n}}x = a$,

$$\lim_{n\uparrow\infty}\left[P\left(\mathrm{E}[X_1] - a \leq \frac{S_n}{n} \leq \mathrm{E}[X_1] + a\right) - P\left(|\mathcal{N}(0; 1)| \leq \frac{a\sqrt{n}}{\sigma}\right)\right] = 0.$$

That is, for large n,

$$P\left(\mathrm{E}[X_1] - a \leq \frac{S_n}{n} \leq \mathrm{E}[X_1] + a\right) \simeq P\left(|\mathcal{N}(0; 1)| \leq \frac{a\sqrt{n}}{\sigma}\right).$$

In other words, for large n, the SLLN estimate of $\mathrm{E}[X_1]$, that is $\frac{S_n}{n}$, lies within distance a of $\mathrm{E}[X_1]$ with probability $P\left(|\mathcal{N}(0; 1)| \leq \frac{a\sqrt{n}}{\sigma}\right)$.

In statistical practice, this result is used in two manners.

(1) One wishes to know the number n experiments that guarantee that with probability, say 0.99, the estimation error is less than a. Choose n such that

$$P\left(|\mathcal{N}(0; 1)| \leq \frac{a\sqrt{n}}{\sigma}\right) = 0.99.$$

Since

$$P(|\mathcal{N}(0; 1)| \leq 2.58) = 0.99,$$

we have

$$2.58 = \frac{a\sqrt{n}}{\sigma}, \tag{4.25}$$

and therefore

$$n = \left(\frac{2.58a}{\sigma}\right)^2.$$

---

[1] This is the content of the Berry–Essen theorem. The proof is omitted.

(2) The (usually large) number n of experiments is fixed. We want to determine the interval $\left[\frac{S_n}{n} - a, \frac{S_n}{n} + a\right]$ within which the mean $E[X_1]$ lies with probability at least 0.99. From (4.25):

$$a = \frac{2.58\sigma}{\sqrt{n}}.$$

If the standard deviation σ is unknown, it may be replaced by an SLLN estimate of it (but then of course...), or the conservative method can be used, which consists of replacing σ by an upper bound.

EXAMPLE 4.4.15: TESTING A COIN. Consider the problem of estimating the bias p of a coin. Here, X_n takes two values, 1 and 0 with probability p and $1-p$ respectively, and in particular $E[X_1] = p$, $\text{Var}(X_1) = \sigma^2 = p(1-p)$. Clearly, since we are trying to estimate p, the standard deviation σ is unknown. Here the upper bound of σ is the maximum of $\sqrt{p(1-p)}$ for $p \in [0,1]$, which is attained for $p = \frac{1}{2}$. Thus $\sigma \le \frac{1}{2}$.

Suppose the coin was tossed $10,000$ times and that the experiment produced the estimate $\frac{S_n}{n} = 0.4925$. Can we "believe 99 percent" that the coin is unbiased. For this we would check that the corresponding confidence interval contains the value $\frac{1}{2}$. Using the conservative method (not a big problem since obviously the actual bias is not far from $\frac{1}{2}$), we have

$$a = \frac{\sigma 2.58}{\sqrt{n}} = 0.0129.$$

and indeed $\frac{1}{2} \in [0.4925 - 0.0129, 0.4925 - 0.0129]$, so that we are at least 99 percent confident that the coin is unbiased.

4.4.3   Convergence in Variation

The Variation Distance

Convergence in variation is convergence with respect to the variation distance, a notion that is now first introduced in the discrete case.[2]

Definition 4.4.16 *Let E be a countable space. The distance in variation between two probability distributions α and β on E is the quantity*

$$d_V(\alpha, \beta) := \frac{1}{2} \sum_{i \in E} |\alpha(i) - \beta(i)|. \tag{4.26}$$

That d_V is indeed a metric is clear.

Lemma 4.4.17 *Let α and β be two probability distributions on the same countable space E. Then*

$$d_V(\alpha, \beta) = \sup_{A \subseteq E} \{|\alpha(A) - \beta(A)|\}$$

$$= \sup_{A \subseteq E} \{\alpha(A) - \beta(A)\}.$$

---

[2]Only the discrete case will be used in this book, in Chapter 6 on discrete-time Markov chains.

Proof. For the second equality observe that for each subset A there is a subset B such that $|\alpha(A) - \beta(A)| = \alpha(B) - \beta(B)$ (take $B = A$ or $\bar{A}$). For the first equality, write

$$\alpha(A) - \beta(A) = \sum_{i \in E} 1_A(i)\{\alpha(i) - \beta(i)\}$$

and observe that the right-hand side is maximal for

$$A = \{i \in E; \ \alpha(i) > \beta(i)\}.$$

Therefore, with $g(i) = \alpha(i) - \beta(i)$,

$$\sup_{A \subseteq E} \{\alpha(A) - \beta(A)\} = \sum_{i \in E} g^+(i) = \frac{1}{2} \sum_{i \in E} |g(i)|$$

since $\sum_{i \in E} g(i) = 0$. $\qquad\square$

The distance in variation *between two random variables* X and Y with values in E is the distance in variation between their probability distributions, and it is denoted (with a slight abuse of notation) by $d_V(X, Y)$. Therefore

$$d_V(X, Y) := \frac{1}{2} \sum_{i \in E} |P(X = i) - P(Y = i)|.$$

The distance in variation *between a random variable X with values in E and a probability distribution* α *on* E, denoted (again with a slight abuse of notation) by $d_V(X, \alpha)$, is defined by

$$d_V(X, \alpha) := \frac{1}{2} \sum_{i \in E} |P(X = i) - \alpha(i)|.$$

The Coupling Inequality

Coupling of two discrete probability distributions π' on E' and π'' on E'' consists, by definition, of the construction of a probability distribution π on $E := E' \times E''$ such that the marginal distributions of π on E' and E'', respectively, are π' and π'', that is,

$$\sum_{j \in E''} \pi(i, j) = \pi'(i) \text{ and } \sum_{i \in E'} \pi(i, j) = \pi''(j).$$

For two probability distributions α and β on the countable set E, let $\mathcal{D}(\alpha, \beta)$ be the collection of random vectors (X, Y) taking their values in $E \times E$ and with given marginal distributions α and β, that is,

$$P(X = i) = \alpha(i), \ P(Y = i) = \beta(i). \tag{4.27}$$

Theorem 4.4.18 *For any pair $(X, Y) \in \mathcal{D}(\alpha, \beta)$, we have the fundamental coupling inequality*

$$d_V(\alpha, \beta) \leq P(X \neq Y),$$

and equality is attained by some pair $(X, Y) \in \mathcal{D}(\alpha, \beta)$, which is then said to realize maximal coincidence.

Proof. For $A \subset E$,

$$
\begin{aligned}
P(X \neq Y) &\geq P(X \in A, Y \in \bar{A}) \\
&= P(X \in A) - P(X \in A, Y \in A) \\
&\geq P(X \in A) - P(Y \in A),
\end{aligned}
$$

and therefore

$$
P(X \neq Y) \geq \sup_{A \subset E} \{P(X \in A) - P(Y \in A)\} = d_V(\alpha, \beta).
$$

We now construct $(X, Y) \in \mathcal{D}(\alpha, \beta)$ realizing equality. Let $U, Z, V,$ and W be independent random variables; U takes its values in $\{0, 1\}$, and Z, V, W take their values in E. The distributions of these random variables are given by

$$
\begin{aligned}
P(U = 1) &= 1 - d_V(\alpha, \beta), \\
P(Z = i) &= \alpha(i) \wedge \beta(i) / (1 - d_V(\alpha, \beta)), \\
P(V = i) &= (\alpha(i) - \beta(i))^+ / d_V(\alpha, \beta), \\
P(W = i) &= (\beta(i) - \alpha(i))^+ / d_V(\alpha, \beta).
\end{aligned}
$$

Observe that $P(V = W) = 0$. Defining

$$
(X, Y) = \begin{cases} (Z, Z) & \text{if} \quad U = 1 \\ (V, W) & \text{if} \quad U = 0 \end{cases}
$$

we have

$$
\begin{aligned}
P(X = i) &= P(U = 1, Z = i) + P(U = 0, V = i) \\
&= P(U = 1)P(Z = i) + P(U = 0)P(V = i) \\
&= \alpha(i) \wedge \beta(i) + (\alpha(i) - \beta(i))^+ = \alpha(i),
\end{aligned}
$$

and similarly, $P(Y = i) = \beta(i)$. Therefore, $(X, Y) \in \mathcal{D}(\alpha, \beta)$. Also, $P(X = Y) = P(U = 1) = 1 - d_V(\alpha, \beta)$. $\square$

EXAMPLE 4.4.19: POISSON'S LAW OF RARE EVENTS, TAKE 2. Let $Y_1, \ldots, Y_n$ be independent random variables taking their values in $\{0, 1\}$, with $P(Y_i = 1) = \pi_i$, $1 \leq i \leq n$. Let $X := \sum_{i=1}^{n} Y_i$ and $\lambda := \sum_{i=1}^{n} \pi_i$. Let p_λ be the Poisson distribution with mean λ. We wish to bound the variation distance between the distribution q of X and p_λ. For this we construct a coupling of the two distributions as follows. First we generate independent pairs $(Y_1, Y_1'), \ldots, (Y_n, Y_n')$ such that

$$
P(Y_i = j, Y_i' = k) = \begin{cases} 1 - \pi_i & \text{if } j = 0, k = 0, \\ e^{-\pi_i} \dfrac{\pi_i^k}{k!} & \text{if } j = 1, k \geq 1, \\ e^{-\pi_i} - (1 - \pi_i) & \text{if } j = 1, k = 0. \end{cases}
$$

One verifies that for all $1 \leq i \leq n$, $P(Y_i = 1) = \pi_i$ and $Y_i' \sim \mathrm{Poi}(\pi_i)$. In particular, $X' := \sum_{i=1}^{n} Y_i'$ is a Poisson variable with mean λ. Now

$$
P(X \neq X') = P\left(\sum_{i=1}^{n} Y_i \neq \sum_{i=1}^{n} Y_i'\right)
$$

$$
\leq P\left(Y_i \neq Y_i' \text{ for some } i\right) \leq \sum_{i=1}^{n} P\left(Y_i \neq Y_i'\right).
$$

But

$$P\left(Y_i \neq Y_i'\right) = e^{-\pi_i} - (1 - \pi_i) + P(Y_1' > 1)$$
$$= \pi_i \left(1 - e^{-\pi_i}\right) \leq \pi_i^2.$$

Therefore $P(X \neq X') \leq \sum_{i=1}^{n} \pi_i^2$, and by the coupling inequality

$$d_V(q, p_\lambda) \leq \sum_{i=1}^{n} \pi_i^2.$$

For instance, with $\pi_i = p := \frac{\lambda}{n}$, we have

$$d_V(q, p_\lambda) \leq \frac{\lambda^2}{n}.$$

In other terms, the binomial distribution of size n and mean λ differs in variation of less than $\frac{\lambda^2}{n}$ from a Poisson variable with the same mean. This is obviously a refinement of the Poisson approximation theorem since it gives exploitable estimates for finite n.

___

A More General Definition

The extension of the notions of the previous subsection to probability distributions on more general spaces is conceptually straightforward and necessitates only obvious adaptations.

Definition 4.4.20 *Let P_1 and P_2 be two probability measures on the same measurable space $(X, \mathcal{X})$. The quantity*

$$d_V(P_1, P_2) := \sup_{A \in \mathcal{X}} |P_1(A) - P_2(A)|$$

is called the distance in variation *between P_1 and P_2.*

Let Q be a probability measure such that $P_i \ll Q$ $(i = 1, 2)$, for instance $Q = \frac{P+Q}{2}$. Therefore there exist (Radon–Nikodým theorem) two non-negative measurable real functions f_i $(i = 1, 2)$ such that

$$P_i(A) = \int_A f_i \, dQ \qquad (A \in \mathcal{X}).$$

Theorem 4.4.21 *We have that*

$$d_V(P_1, P_2) = \frac{1}{2} \int_X |f_1 - f_2| \, dQ. \tag{4.28}$$

Proof. We first observe that

$$\sup_{A \in \mathcal{X}} |P_1(A) - P_2(A)| = \sup_{A \in \mathcal{X}} (P_1(A) - P_2(A))$$

since for any $A \in \mathcal{X}$ there exists a $B \in \mathcal{X}$ such that $P_1(A) - P_2(A) = -(P_1(B) - P_2(B))$ (take $B = \overline{A}$). Therefore

$$d_V(P_1, P_2) = \sup_{A \in \mathcal{X}} (P_1(A) - P_2(A)) = \sup_{A \in \mathcal{X}} \int_A (f_1 - f_2) \, dQ \, .$$

The supremum is attained for $A = \{f_1 - f_2 \geq 0\}$:

$$d_V(P_1, P_2) = \int_{\{f_1 - f_2 \geq 0\}} (f_1 - f_2) \, dQ \, .$$

Since $\int_X (f_1 - f_2) \, dQ = 0$, we have that $\int_{\{f_1 - f_2 \geq 0\}} (f_1 - f_2) \, dQ = \frac{1}{2} \int_X |f_1 - f_2| \, dQ$. $\quad\square$

It follows from the expression (4.28) that d_V is indeed a metric.

A Bayesian Interpretation

Let $X \in \mathbb{R}^d$ be a random vector called the *observation* and $H \in \{1, 2\}$ be a random variable called the *hypothesis*. The joint law of (X, H) is described as follows:

$$P(X \in C \mid H_i) = \int_C f_i(x) \, dx \quad (i \in \{1, 2\}, C \in \mathcal{B}(\mathbb{R}^d))$$

and

$$P(H = i) = \frac{1}{2} \quad (i \in \{1, 2\}).$$

We seek to devise a test based on the observation of X alone that will help us to decide which is the value of H. In other words, we must select a measurable partition $\{A_1, A_2\}$ of $\mathbb{R}^d$ and decide $H = i$ if $X \in A_i$ $(i = 1, 2)$. This partition is called a *test*. A *probability of error* (wrong guess) is associated with this test:

$$P_E = P(H = 1)P(X \in A_2 \mid H = 1) + P(H = 2)P(X \in A_2 \mid H = 2) \, ,$$

that is,

$$
\begin{aligned}
P_E &= \frac{1}{2} \int_{A_2} f_1(x) \, dx + \frac{1}{2} \int_{A_1} f_2(x) \, dx \\
&= \frac{1}{2} \int_{A_2} f_1(x) \, dx + \frac{1}{2} \left(1 - \int_{A_2} f_2(x) \, dx \right) \\
&= \frac{1}{2} + \frac{1}{2} \int_{A_2} (f_1(x) - f_2(x)) \, dx \, .
\end{aligned}
$$

We seek to minimize this quantity (with respect to A_2) or, equivalently, to maximize the quantity

$$\int_{A_2} (f_2(x) - f_1(x)) \, dx = P_2(A_2) - P_1(A_2) \, ,$$

where $P_i(C) := \int_C f_i(x) \, dx$ $(i = 1, 2)$. This is done by the choice $A_2 := \{x \, ; \, f_2(x) \geq f_1(x)\}$ and the resulting (minimal) probability of error is then

$$P_E^* = \frac{1}{2}(1 - d_V(P_1, P_2)) \, ,$$

where $P_i(\cdot) := P(\cdot \mid H_i)$ $(i = 1, 2)$.

Convergence in Variation

Definition 4.4.22 *The sequence $\{P_n\}_{n\geq 1}$ of probability measures on $(X, \mathcal{X})$ is said to converge in variation to the probability P on $(X, \mathcal{X})$ if*

$$\lim_{n\uparrow\infty} d_V(P_n, P) = 0.$$

This is denoted $P_n \overset{\text{Var.}}{\to} P$.

Let Q be a probability measure such that $P_n \ll Q$ ($n \geq 1$), for instance

$$Q = \sum_{n\geq 1} \frac{1}{2^n} P_n$$

defines a probability measure Q such that for all $n \geq 1$, $P_n \ll Q$. Denote by f_n (resp., f) the Radon–Nikodým derivative of P_n (resp., P) with respect to Q. By Theorem 4.4.21, $P_n \overset{\text{Var.}}{\to} P$ if and only if $f_n \overset{L^1}{\to} f$, where $L^1 = L^1_{\mathbb{C}}(Q)$. Note also that if $\varphi : X \to \mathbb{C}$ is a bounded function, then

$$\int_X \varphi \, dP_n \to \int_X \varphi \, dP,$$

as follows from the fact that

$$\int_X \varphi \, dP_n - \int_X \varphi \, dP = \int_X \varphi \times (f_n - f) \, dQ$$

and dominated convergence.

Theorem 4.4.23 *Let P_n, Q and f_n be defined as above. Suppose that there exists a non-negative measurable function f from $(X, \mathcal{X})$ to $(\mathbb{R}, \mathcal{B}(\mathbb{R}))$ such that Q-a.e, $f_n \to f$. Then $P_n \overset{\text{Var.}}{\to} P$ where P is the probability defined by $P(A) = \int_A f \, dQ$, $A \in \mathcal{X}$.*

The proof is a direct consequence of *Scheffé's lemma*:

Lemma 4.4.24 *Let f and f_n ($n \geq 1$) be Q-integrable non-negative real functions from $(X, \mathcal{X})$ to $(\mathbb{R}, \mathcal{B}(\mathbb{R}))$, with $\lim_{n\uparrow\infty} f_n = f$ Q-a.e. and $\lim_{n\uparrow\infty} \int_X f_n \, dQ = \int_X f \, dQ$. Then $\lim_{n\uparrow\infty} \int_X |f_n - f| \, dQ = 0$.*

Proof. The function $\inf(f_n, f)$ is bounded by the (integrable) function f (this is where the non-negativeness assumption is used). Moreover, it converges to f. Therefore, by dominated convergence, $\lim_{n\uparrow\infty} \int_X \inf(f_n, f) \, dQ = \int_X f \, dQ$. The rest of the proof follows from

$$\int_X |f_n - f| \, dQ = \int_X f_n \, dQ + \int_X f \, dQ - \int_X \inf(f_n, f) \, dQ.$$

$\square$

Definition 4.4.25

A. *Let X_1, X_2 be random elements with values in the measurable space $(E, \mathcal{E})$, with respective distributions α_1, α_2. The distance in variation between X_1, and X_2 is, by definition the quantity*

$$d_V(X_1, X_2) := d_V(\alpha_1, \alpha_2).$$

B. Let X and $\{X_n\}_{n\geq 1}$ be random elements with values in the measurable space $(E, \mathcal{E})$, with respective distributions α, $\{\alpha_n\}_{n\geq 1}$. The sequence $\{X_n\}_{n\geq 1}$ is said to converge in variation to X if $\lim_{n\uparrow\infty} d_V(X_n, X) = 0$. This is denoted by $X_n \overset{\text{Var.}}{\rightarrow} X$.

Let $\{X_n\}_{n\geq 1}$ be random elements with values in some measurable space $(E, \mathcal{E})$ and let α be some probability distribution on $(E, \mathcal{E})$. The notation $X_n \overset{\text{Var.}}{\rightarrow} \alpha$ means, by convention, that $\alpha_n \overset{\text{Var.}}{\rightarrow} \alpha$, where α_n is the distribution of X_n. (This convention is similar to the one introduced above in the context of convergence in distribution.)

4.4.4   Proof of Paul Lévy's Criterion

Radon Linear Forms

Denote by $C_0(\mathbb{R}^d)$ the set of continuous functions from $\varphi : \mathbb{R}^d \to \mathbb{R}$ that vanish at infinity ($\lim_{|x|\uparrow\infty} \varphi(x) = 0$), endowed with the norm of uniform convergence $||\varphi|| :=$ $\sup_{x\in\mathbb{R}^d} ||\varphi(x)||$. Let $C_c(\mathbb{R}^d)$ be the set of continuous functions from $\mathbb{R}^d$ to $\mathbb{R}$ with compact support. In particular, $C_c(\mathbb{R}^d) \subset C_0(\mathbb{R}^d)$.

Definition 4.4.26 A linear form $L : C_c(\mathbb{R}^d) \to \mathbb{R}$ such that $L(f) \geq 0$ whenever $f \geq 0$ is called a positive Radon linear form.

We quote without proof the following fundamental result of Riesz:[3]

Theorem 4.4.27 Let $L : C_c(\mathbb{R}^d) \to \mathbb{R}$ be a positive Radon linear form. There exists a unique locally finite measure μ on $(\mathbb{R}^d, \mathcal{B}(\mathbb{R}^d))$ such that for all $f \in C_c(\mathbb{R}^d)$,

$$L(f) = \int_{\mathbb{R}^d} f \, d\mu.$$

We shall need a slight extension of Riesz's theorem (Theorem 4.4.27), Part (ii) of the following:

Theorem 4.4.28

(i) Let $\mu \in M^+(\mathbb{R}^d)$. The linear form $L : C_0(\mathbb{R}^d) \to \mathbb{R}$ defined by $L(f) := \int_{\mathbb{R}^d} f \, d\mu$ is positive ($L(f) \geq 0$ whenever $f \geq 0$) and continuous, and its norm is $\mu(\mathbb{R}^d)$.

(ii) Let $L : C_0(\mathbb{R}^d) \to \mathbb{R}$ be a positive continuous linear form. There exists a unique measure $\mu \in M^+(\mathbb{R}^d)$ such that for all $f \in C_0(\mathbb{R}^d)$,

$$L(f) = \int_{\mathbb{R}^d} f \, d\mu.$$

Proof. Part (i) is left as an exercise. We turn to the proof of (ii). The restriction of L to C_c is a positive Radon linear form, and therefore, according to Riesz's Theorem 4.4.27, there exists a locally finite μ on $(\mathbb{R}^d, \mathcal{B}(\mathbb{R}^d))$ such that for all $f \in C_c(\mathbb{R}^d)$, $L(f) = \int_{\mathbb{R}^d} f \, d\mu$.

The measure μ is a *finite* (not just locally finite) measure. If not, there would exist a sequence $\{K_m\}_{m\geq 1}$ of compact subsets of $\mathbb{R}^d$ such that $\mu(K_m) \geq 3^m$ for all $m \geq 1$. Let

---

[3] See for instance [Rudin, 1986], Theorem 2.14.

then $\{\varphi_m\}_{m\geq 1}$ be a sequence of non-negative functions in $C_c(\mathbb{R}^d)$ with values in $[0,1]$ and such that for all $m \geq 1$, $\varphi_m(x) = 1$ for all $x \in K_m$. In particular, the function $\varphi := \sum_{m\geq 1} 2^{-m}\varphi_m$ is in $C_0(\mathbb{R}^d)$ and

$$L(\varphi) \geq L\left(\sum_{m=1}^{k} 2^{-m}\varphi_m\right) = \sum_{m=1}^{k} 2^{-m}L(\varphi_m)$$

$$= \sum_{m=1}^{k} 2^{-m}\int_{\mathbb{R}^d} \varphi_m \, d\mu \geq \sum_{m=1}^{k} 2^{-m}\mu(K_m) \geq \left(\frac{3}{2}\right)^k.$$

Letting $k \uparrow \infty$ leads to $L(\varphi) = \infty$, a contradiction.

The mapping L is continuous. Suppose it is not. One could then find a sequence $\{\varphi_m\}_{m\geq 1}$ of functions in $C_c(\mathbb{R}^d)$ such that $|\varphi_m| \leq 1$ and $L(\varphi_m) \geq 3^m$. The function $\varphi := \sum_{m\geq 1} 2^{-m}\varphi_m$ is in $C_0(\mathbb{R}^d)$ and

$$L(\varphi) \geq L(\sum_{m=1}^{k} 2^{-m}\varphi_m) = \sum_{m=1}^{k} 2^{-m}L(\varphi_m)$$

$$= \sum_{m=1}^{k} 2^{-m}\int_{\mathbb{R}^d} \varphi_m \, d\mu \geq \left(\frac{3}{2}\right)^k.$$

Letting $k \uparrow \infty$ again leads to $L(\varphi) = \infty$, a contradiction.

It remains to show that $L(f) = \int_{\mathbb{R}^d} f \, d\mu$ for all $f \in C_0(\mathbb{R}^d)$. For this, consider a sequence $\{f_m\}_{m\geq 1}$ of functions in $C_c(\mathbb{R}^d)$ converging uniformly to $f \in C_0(\mathbb{R}^d)$. We have, since L is continuous, $\lim_{m\uparrow\infty} L(f_m) = L(f)$ and, since μ is finite, $\lim_{m\uparrow\infty} \int f_m \, d\mu = \int f \, d\mu$ by dominated convergence. Therefore, since $L(f_m) = \int f_m \, d\mu$ for all $m \geq 1$, $L(f) = \int f \, d\mu$. $\qquad\square$

Vague Convergence

Definition 4.4.29 *The sequence $\{\mu_n\}_{n\geq 1}$ in $M^+(\mathbb{R}^d)$ is said to converge* vaguely *(resp., weakly) to μ if, for all $f \in C_0(\mathbb{R}^d)$ (resp. $f \in C_b(\mathbb{R}^d)$), $\lim_{n\uparrow\infty} \int_{\mathbb{R}^d} f \, d\mu_n = \int_{\mathbb{R}^d} f \, d\mu$.*

Remark 4.4.30 When applied to probability measure this notion is weaker than weak convergence, because a continuous function vanishing at infinity is a particular case of a bounded continuous function.

Theorem 4.4.31 *The sequence $\{\mu_n\}_{n\geq 1}$ in $M^+(\mathbb{R}^d)$ converges vaguely if and only if*

(a) $\sup_n \mu_n(\mathbb{R}^d) < \infty$, *and*

(b) *there exists a dense subset $\mathcal{E}$ in $C_0(\mathbb{R}_d)$ such that for all $f \in \mathcal{E}$, there exists $\lim_{n\uparrow\infty} \int_{\mathbb{R}^d} f \, d\mu_n$.*

Proof. *Necessity.* If the sequence converges vaguely, it obviously satisfies (b). As for (a), it is a consequence of the Banach–Steinhaus theorem.[4] Indeed, $\mu_n(\mathbb{R}^d)$ is the norm of L_n,

---

[4]Let E be a Banach space and F be a normed vector space. Let $\{L_i\}_{i\in I}$ be a family of continuous linear mappings from E to F such that $\sup_{i\in I} \|L_i(x)\| < \infty$ for all $x \in E$. Then $\sup_{i\in I} \|L_i\| < \infty$. See for instance [Rudin, 1986], Thm. 5.8.

where L_n is the continuous linear form $f \to \int_{\mathbb{R}^d} f \, d\mu_n$ from the Banach space $C_0(\mathbb{R}^d)$ (with the sup norm) to $\mathbb{R}$, and for all $f \in C_0(\mathbb{R}^d)$, $\sup_n |\int_{\mathbb{R}^d} f \, d\mu_n| < \infty$.

Sufficiency. Suppose the sequence satisfies (a) and (b). Let $f \in C_0(\mathbb{R}^d)$. For all $\varphi \in \mathcal{E}$,

$$\left| \int f \, d\mu_m - \int f \, d\mu_n \right|$$

$$\leq \left| \int \varphi \, d\mu_m - \int \varphi \, d\mu_n \right| + \left| \int f \, d\mu_m - \int \varphi \, d\mu_m \right| + \left| \int f \, d\mu_n - \int \varphi \, d\mu_n \right|$$

$$\leq \left| \int \varphi \, d\mu_m - \int \varphi \, d\mu_n \right| + \sup_{x \in \mathbb{R}^d} |f(x) - \varphi(x)| \times \sup_n \mu_n(\mathbb{R}^d).$$

Since $\sup_{x \in \mathbb{R}^d} |f(x) - \varphi(x)|$ can be made arbitrarily small by a proper choice of φ, this shows that the sequence $\{\int f \, d\mu_n\}_{n \geq 1}$ is a Cauchy sequence. It therefore converges to some $L(f)$, and L so defined is a positive linear form on $C_0(\mathbb{R}^d)$. Therefore, there exists a $\mu \in M^+(\mathbb{R}^d)$ such that $L(f) = \int_{\mathbb{R}^d} f \, d\mu$ and $\{\mu_n\}_{n \geq 1}$ converges vaguely to μ. □

Helly's Theorem

Theorem 4.4.32 *From any bounded sequence of $M^+(\mathbb{R}^d)$, one can extract a vaguely convergent subsequence.*

Proof. Let $\{\mu_n\}_{n \geq 1}$ be a bounded sequence of $M^+(\mathbb{R}^d)$. Let $\{f_n\}_{n \geq 1}$ be a dense sequence of $C_0(\mathbb{R}^d)$.

Since the sequence $\{\int f_1 \, d\mu_n\}_{n \geq 1}$ is bounded, one can extract from it a convergent subsequence $\{\int f_1 \, d\mu_{1,n}\}_{n \geq 1}$. Since the sequence $\{\int f_2 \, d\mu_{1,n}\}_{n \geq 1}$ is bounded, one can extract from it a convergent subsequence $\{\int f_2 \, d\mu_{2,n}\}_{n \geq 1}$. This diagonal selection process is continued. At step k, since the sequence $\{\int f_{k+1} \, d\mu_{k,n}\}_{n \geq 1}$ is bounded, one can extract from it a convergent subsequence $\{\int f_{k+1} \, d\mu_{k+1,n}\}_{n \geq 1}$. The sequence $\{\nu_k\}_{k \geq 1}$ where $\nu_k = \mu_{k,k}$ (the "diagonal" sequence) is extracted from the original sequence and for all f_n, the sequence $\{\int f_n \, d\nu_k\}_{k \geq 1}$ converges. The conclusion follows from Theorem 4.4.31. □

Fourier Transforms of Finite Measures

Definition 4.4.33 *The* Fourier transform *of a measure $\mu \in M^+(\mathbb{R}^d)$ is the function $\widehat{\mu} : \mathbb{R}^d \to \mathbb{C}$ defined by*

$$\widehat{\mu}(\nu) = \int_{\mathbb{R}^d} e^{-2i\pi\langle \nu, x \rangle} \, \mu(dx),$$

where $\langle \nu, x \rangle := \sum_{j=1}^d \nu_j x_j$.

Theorem 4.4.34 *The Fourier transform of a measure $\mu \in M^+(\mathbb{R}^d)$ is bounded and uniformly continuous.*

Proof. From the definition, we have that

$$|\widehat{\mu}(\nu)| \leq \int_{\mathbb{R}^d} |e^{-2i\pi\langle \nu, x \rangle}| \, \mu(dx)$$

$$= \int_{\mathbb{R}^d} \mu(dx) = \mu(\mathbb{R}^d),$$

where the last term does not depend on ν and is finite. Also, for all $h \in \mathbb{R}^d$,

$$|\widehat{\mu}(\nu + h) - \widehat{\mu}(\nu)| \leq \int_{\mathbb{R}^d} \left| e^{-2i\pi\langle \nu, x+h \rangle} - e^{-2i\pi\langle \nu, x \rangle} \right| \mu(\mathrm{d}x)$$

$$= \int_{\mathbb{R}^d} \left| e^{-2i\pi\langle h, x \rangle} - 1 \right| \mu(\mathrm{d}x).$$

The last term is independent of ν and tends to 0 as $h \to 0$ by dominated convergence (recall that μ is finite). $\qquad\square$

Theorem 4.4.35 *Let $\mu \in M^+(\mathbb{R}^d)$ and let $\widehat{f}$ be the Fourier transform of $f \in L^1_{\mathbb{C}}(\mathbb{R}^d)$. Then*

$$\int_{\mathbb{R}^d} \widehat{f} \, \mathrm{d}\mu = \int_{\mathbb{R}^d} f \widehat{\mu} \, \mathrm{d}x.$$

Proof. This follows from Fubini's theorem. In fact,

$$\int_{\mathbb{R}^d} \widehat{f} \, \mathrm{d}\mu = \int_{\mathbb{R}^d} \left(\int_{\mathbb{R}^d} f(x) e^{-2i\pi\langle \nu, x \rangle} \mathrm{d}x \right) \mu(\mathrm{d}\nu)$$

$$= \int_{\mathbb{R}^d} f(x) \left(\int_{\mathbb{R}^d} e^{-2i\pi\langle \nu, x \rangle} \mu(\mathrm{d}\nu) \right) \mathrm{d}x.$$

(Interversion of the order of integration is justified by the fact that the function $(x, \nu) \to \left| f(x) e^{-2i\pi\langle \nu, x \rangle} \right| = |f(x)|$ is integrable with respect to the product measure $\mathrm{d}x \times \mu(\mathrm{d}\nu)$. Recall that μ is finite.) $\qquad\square$

For the next definition, recall that $C_b(\mathbb{R}^d)$ denotes the collection of uniformly bounded and continuous functions from $\mathbb{R}^d$ to $\mathbb{R}$.

Theorem 4.4.36 *The sequence $\{\mu_n\}_{n \geq 1}$ in $M^+(\mathbb{R}^d)$ converges weakly to μ if and only if*

(i) It converges vaguely to μ, and

(ii) $\lim_{n \uparrow \infty} \mu_n(\mathbb{R}^d) = \mu(\mathbb{R}^d)$.

Proof. The necessity of (i) immediately follows from the observation that $C_0(\mathbb{R}^d) \subset C_b(\mathbb{R}^d)$. The necessity of (ii) follows from the fact that the function that is the constant 1 is in $C_b(\mathbb{R}^d)$ and therefore $\int 1 \, \mathrm{d}\mu_n = \mu_n(\mathbb{R}^d)$ tends to $\int 1 \, \mathrm{d}\mu = \mu(\mathbb{R}^d)$ as $n \uparrow \infty$.

Sufficiency. Suppose that (i) and (ii) are satisfied. To prove weak convergence, it suffices to prove that $\lim_{n \uparrow \infty} \int_{\mathbb{R}^d} f \, \mathrm{d}\mu_n = \int_{\mathbb{R}^d} f \, \mathrm{d}\mu$ for any non-negative function $f \in C_b(\mathbb{R}^d)$.

Since the measure μ is of finite total mass, for any $\varepsilon > 0$ one can find a compact set $K_\varepsilon = K$ such that $\mu(\overline{K}) \leq \varepsilon$. Choose a continuous function with compact support φ with values in $[0, 1]$ and such that $\varphi \geq 1_K$. Since $|f - f\varphi| \leq \|f\|(1 - \varphi)$ (where $\|f\| = \sup_{x \in \mathbb{R}^d} |f(x)|$),

$$\limsup_{n \uparrow \infty} \left| \int f \, \mathrm{d}\mu_n - \int f\varphi \, \mathrm{d}\mu_n \right| \leq \limsup_{n \uparrow \infty} \|f\| \int (1 - \varphi) \, \mathrm{d}\mu_n$$

$$= \|f\| \left(\lim_{n \uparrow \infty} \int \mathrm{d}\mu_n - \lim_{n \uparrow \infty} \int \varphi \, \mathrm{d}\mu_n \right)$$

$$= \|f\| \int (1 - \varphi) \, \mathrm{d}\mu \leq \varepsilon \|f\|.$$

Similarly, $|\int f\,\mathrm{d}\mu - \int f\varphi\,\mathrm{d}\mu| \leq \varepsilon\|f\|$. Therefore, for all $\varepsilon > 0$,

$$\limsup_{n\uparrow\infty}\left|\int f\,\mathrm{d}\mu_n - \int f\,\mathrm{d}\mu\right| \leq 2\varepsilon\|f\|,$$

and this completes the proof. $\square$

The Proof of Paul Lévy's criterion

We shall in fact prove a slightly more general result:

Theorem 4.4.37 *Let $\{\mu_n\}_{n\geq 1}$ be a sequence of $M^+(\mathbb{R}^d)$ such that for all $\nu \in \mathbb{R}^d$, there exists $\lim_{n\uparrow\infty}\widehat{\mu}_n(\nu) = \varphi(\nu)$ for some function φ that is continuous at 0. Then $\{\mu_n\}_{n\geq 1}$ converges weakly to a finite measure μ whose Fourier transform is φ.*

We will be ready for the generalization of Paul Lévy's criterion of convergence in distribution after a few preliminaries.

Definition 4.4.38 *A family $\{\alpha_t\}_{t>0}$ of functions $\alpha_t : \mathbb{R}^d \to \mathbb{C}$ in $L^1_{\mathbb{C}}(\mathbb{R}^d)$ is called an* approximation of the Dirac distribution *in $\mathbb{R}^d$ if it satisfies the following three conditions:*

(i) $\int_{\mathbb{R}^d} \alpha_t(x)\,\mathrm{d}x = 1$;

(ii) $\sup_{t>0} \int_{\mathbb{R}^d} |\alpha_t(x)|\,\mathrm{d}x := M < \infty$; *and*

(iii) *for any compact neighborhood V of $0 \in \mathbb{R}^d$, $\lim_{t\downarrow 0}\int_V |\alpha_t(x)|\,\mathrm{d}x = 0$.*

Lemma 4.4.39 *Let $\{\alpha_t\}_{t>0}$ be an approximation of the Dirac distribution in $\mathbb{R}^d$. Let $f : \mathbb{R}^d \to \mathbb{C}$ be a bounded function continuous at all points of a compact $K \subset \mathbb{R}^d$. Then $\lim_{t\downarrow 0} f * \alpha_t$ uniformly in K.*

Proof. We will show later that

$$\lim_{y\to 0}\ \sup_{x\in K} |f(x-y) - f(x)| \to 0. \qquad (\star)$$

V being a compact neighborhood of 0, we have that

$$\sup_{x\in K} |f(x) - (f * \alpha_t)(x)| \leq M \sup_{y\in V}\sup_{x\in K} |f(x-y) - f(x)| + 2|\sup_{x\in\mathbb{R}^d} |f(x)| \int_V |\alpha_t(y)|\,\mathrm{d}y.$$

This quantity can be made smaller than any $\varepsilon > 0$ by choosing V such that the first term is $< \frac{1}{2}\varepsilon$ (uniform continuity of f on compact sets) and the second term can then be made $< \frac{1}{2}\varepsilon$ by letting $t \downarrow$ (condition (iii) of Definition 4.4.38).

Proof of $(\star)$. Let $\varepsilon > 0$ be given. For all $x \in K$, there exists an open and symmetric neighborhood V_x of 0 such that for all $y \in V_x$, $f(x-y)f(x)| \leq \frac{1}{2}\varepsilon$. Also, one can find an open and symmetric neighborhood W_x of 0 such that $W_x + W_x \subset V_x$. The union of open sets $\cup_{x\in K}\{x + W_x\}$ obviously covers K, and since the latter is a compact set, one can extract a finite covering of K: $\cup_{j=1}^m (x_j + W_{x_j})$. Define $W = \cap_{j=1}^m W_{x_j}$, an open neighborhood of 0.

Let $y \in W$. Any $x \in K$ belongs to some $x_j + W_{x_j}$, and for such j,

$$|f(x-y) - f(x)| \leq |f(x_j) - f(x_j - (x_j - x))|$$
$$+ |f(x_j) - f(x_j - (x_j - x + y))|.$$

But $x_j - x \in W_{x_j}$ and $x_j - x + y \in W_{x_j} + W \subset V_{x_j}$. Therefore

$$|f(x-y) - f(x)| \leq \frac{1}{2}\varepsilon + \frac{1}{2}\varepsilon = \varepsilon.$$

$\square$

We can now prove Theorem 4.4.37.

Proof. The sequence $\{\mu_n\}_{n \geq 1}$ is bounded (that is, $\sup_n \mu_n(\mathbb{R}^d) < \infty$) since $\mu_n(\mathbb{R}^d) = \widehat{\mu}_n(1)$ has a limit as $n \uparrow \infty$. In particular,

$$\widehat{\mu}_n(\nu) \leq \mu_n(\mathbb{R}^d) \leq \sup_n \mu_n(\mathbb{R}^d) < \infty. \tag{$\dagger$}$$

If $f \in L^1_{\mathbb{R}}(\mathbb{R}^d)$, then by Theorem 4.4.35, $\int \widehat{f} \, d\mu_n = \int f \widehat{\mu}_n \, dx$. By dominated convergence (using ($\dagger$)), $\lim_{n \uparrow \infty} \int f \widehat{\mu}_n \, dx = \int f\varphi \, dx$. Therefore

$$\lim_{n \uparrow \infty} \int \widehat{f} \, d\mu_n = \int f\varphi \, dx.$$

One can replace in the above equality $\widehat{f}$ by any function in $\mathcal{D}(\mathbb{R}^d)$, since such a function is always the Fourier transform of some integrable function. Therefore, by Theorem 4.4.31, $\{\mu_n\}_{n \geq 1}$ converges vaguely to some finite measure μ.

We now show that it converges *weakly* to μ. Let f be an integrable function with integral 1 such that $f(x) = f(-x)$ and $\widehat{f} \in \mathcal{D}(\mathbb{R}^d)$. For $t > 0$, define $f_t(x) := t^{-d} f(t^{-1}x))$. Using Theorem 4.4.35, we have

$$\int \widehat{f}(tx) \, \mu_n(dx) = \int f_t(x) \, \widehat{\mu}_n(x) \, dx = (f_t * \widehat{\mu}_n)(0).$$

By dominated convergence,

$$\lim_{n \uparrow \infty} \int \widehat{f}(tx) \, \mu_n(dx) = (f_t * \varphi)(0),$$

and by vague convergence,

$$\lim_{n \uparrow \infty} \int \widehat{f}(tx) \, \mu_n(dx) = \int \widehat{f}(tx) \, \mu(dx).$$

Therefore, for all $t > 0$, $\int \widehat{f}(tx) \, \mu(dx) = (f_t * \varphi)(0)$.

Since the function φ is bounded and continuous at the origin, by Lemma 4.4.39, $\lim_{t \downarrow 0} (f_t * \varphi)(0) = \varphi(0)$. Also, by dominated convergence $\lim_{t \downarrow 0} \int \widehat{f}(tx) \, \mu(dx) = \mu(\mathbb{R}^d)$. Therefore,

$$\mu(\mathbb{R}^d) = \varphi(0) = \lim_{n \uparrow \infty} \mu_n(\mathbb{R}^d).$$

Therefore, by Theorem 4.4.36, $\{\mu_n\}_{n \geq 1}$ converges *weakly* to μ. Since the function $x \to e^{-2i\pi\langle\nu,x\rangle}$ is continuous and bounded,

$$\widehat{\mu}(\nu) = \int e^{-2i\pi\langle\nu,x\rangle} \mu(dx) = \lim_{n \uparrow \infty} \int e^{-2i\pi\langle\nu,x\rangle} \mu_n(dx) = \varphi(\nu).$$

$\square$

4.5   The Hierarchy of Convergences

4.5.1   Almost-sure *vs* in Probability

Theorem 4.5.1

A. If the sequence $\{Z_n\}_{n\geq 1}$ of complex random variables converges almost surely to some complex random variable Z, it also converges in probability to the same random variable Z.

B. If the sequence of complex random variables $\{X_n\}_{n\geq 1}$ converges in probability to the complex random variable X, one can find a sequence of integers $\{n_k\}_{k\geq 1}$, strictly increasing, such that $\{X_{n_k}\}_{k\geq 1}$ converges almost surely to X.

(B says, in other words: From a sequence converging in probability, one can extract a subsequence converging almost surely.)

Proof. A. Suppose almost-sure convergence. By Theorem 4.1.3 , for all $\varepsilon > 0$,

$$P(|Z_n - Z| \geq \varepsilon \ i.o.) = 0,$$

that is

$$P(\cap_{n\geq 1} \cup_{k=n}^\infty (|Z_k - Z| \geq \varepsilon)) = 0,$$

or (sequential continuity of probability)

$$\lim_{n\uparrow\infty} P(\cup_{k=n}^\infty (|Z_k - Z| \geq \varepsilon)) = 0,$$

which in turn implies that

$$\lim_{n\uparrow\infty} P(|Z_n - Z| \geq \varepsilon) = 0.$$

B. By definition of convergence in probability, for all $\varepsilon > 0$,

$$\lim_{n\uparrow\infty} P\left(|X_n - X| \geq \varepsilon\right) = 0.$$

Therefore one can find n_1 such that $P\left(|X_{n_1} - X| \geq \frac{1}{1}\right) \leq \left(\frac{1}{2}\right)$. Then, one can find $n_2 > n_1$ such that $P\left(|X_{n_2} - X| \geq \frac{1}{2}\right) \leq \left(\frac{1}{2}\right)^2$, and so on, until we have a strictly increasing sequence of integers n_k $(k \geq 1)$ such that

$$P\left(|X_{n_k} - X| \geq \frac{1}{k}\right) \leq \left(\frac{1}{2}\right)^k.$$

It then follows from Theorem 4.1.2 that

$$\lim_{k\uparrow\infty} X_{n_k} = X \quad \text{a.s.}$$

$\square$

Remark 4.5.2 Exercise 4.6.7 gives an example of a sequence converging in probability, but not almost surely. Thus, convergence in probability is a notion strictly weaker than almost-sure convergence.

Theorem 4.5.3 *If the sequence $\{Z_n\}_{n \geq 1}$ of square-integrable complex random variables converges in quadratic mean to the complex random variable Z, it also converges in probability to the same random variable.*

Proof. It suffices to observe that, by Markov's inequality, for all $\varepsilon > 0$,

$$P(|Z_n - Z| \geq \varepsilon) \leq \frac{1}{\varepsilon^2} \mathrm{E}[|Z_n - Z|^2].$$

□

4.5.2 The Rank of Convergence in Distribution

We now compare convergence in distribution to the other types of convergence. Convergence in distribution is weaker than almost-sure convergence:

Theorem 4.5.4 *If the sequence $\{X_n\}_{n \geq 1}$ of random vectors of $\mathbb{R}^d$ converges almost surely to some random vector X, it also converges in distribution to the same vector X.*

Proof. By dominated convergence, for all $u \in \mathbb{R}$,

$$\lim_{n \uparrow \infty} \mathrm{E}\left[e^{i\langle u, X_n \rangle}\right] = \mathrm{E}\left[e^{i\langle u, X \rangle}\right]$$

which implies, by Theorem 4.4.6 that $\{X_n\}_{n \geq 1}$ converges in distribution to X. □

In fact, convergence in distribution is even weaker than convergence in probability.

Theorem 4.5.5 *If the sequence $\{X_n\}_{n \geq 1}$ of random vectors of $\mathbb{R}^d$ converges in probability to some random vector X, it also converges in distribution to X.*

Proof. If this were not the case, one could find a function $f \in C_b(\mathbb{R}^d)$ such that $\mathrm{E}[f(X_n)]$ does not converge to $\mathrm{E}[f(X)]$. In particular, there would exist a subsequence n_k and some $\varepsilon > 0$ such that $|\mathrm{E}[f(X_{n_k})] - \mathrm{E}[f(X)]| \geq \varepsilon$ for all k. As $\{X_{n_k}\}_{k \geq 1}$ converges in probability to X, one can extract from it a subsequence $\{X_{n_{k_\ell}}\}_{\ell \geq 1}$ converging almost surely to X. In particular, since f is bounded and continuous, $\lim_\ell \mathrm{E}[f(X_{n_{k_\ell}})] = \mathrm{E}[f(X)]$ by dominated convergence, a contradiction. □

Combining Theorems 4.5.3 and 4.5.5, we have that convergence in distribution is weaker than convergence in the quadratic mean:

Theorem 4.5.6 *If the sequence of real random variables $\{Z_n\}_{n \geq 1}$ converges in quadratic mean to some random variable Z, it also converges in distribution to the same random variable Z.*

Theorem 4.5.6 can be refined in the Gaussian case, where the distribution of the limit can be proved to be Gaussian.

A Stability Property of Gaussian Vectors

Theorem 4.5.7 *If* $\{Z_n\}_{n\geq 1}$, *where* $Z_n = (Z_n^{(1)}, \ldots, Z_n^{(m)})$, *is a sequence of Gaussian random vectors of fixed dimension* m *that converges componentwise in quadratic mean to some vector* $Z = (Z^{(1)}, \ldots, Z^{(m)})$, *the latter vector is Gaussian.*

Proof. In fact, by continuity of the inner product in $L^2_{\mathbb{R}}(P)$, for all $1 \leq i, j \leq m$, $\lim_{n\uparrow\infty} \mathrm{E}[Z_n^{(i)} Z_n^{(j)}] = \mathrm{E}[Z^{(i)} Z^{(j)}]$ and $\lim_{n\uparrow\infty} \mathrm{E}[Z_n^{(i)}] = \mathrm{E}[Z^{(i)}]$, that is,

$$\lim_{n\uparrow\infty} m_{Z_n} = m_Z, \qquad \lim_{n\uparrow\infty} \Gamma_{Z_n} = \Gamma_Z,$$

and in particular, for all $u \in \mathbb{R}^m$,

$$\lim_{n\uparrow\infty} \mathrm{E}\left[e^{iu^T Z_n}\right] = \lim_{n\uparrow\infty} e^{iu^T \mu_{Z_n} - \frac{1}{2}u^T \Gamma_{Z_n} u}$$
$$= e^{iu^T \mu_Z - \frac{1}{2}u^T \Gamma_Z u}.$$

The sequence $\{u^T Z_n\}_{n\geq 1}$, converging in quadratic mean to $u^T Z$, also converges in distribution to $u^T Z$. Therefore, $\lim_{n\uparrow\infty} \mathrm{E}\left[e^{iu^T Z_n}\right] = \mathrm{E}[e^{iu^T Z}]$, and finally

$$\mathrm{E}[e^{iu^T Z}] = e^{iu^T \mu_Z - \frac{i}{2}u^T \Gamma_Z u}$$

for all $u \in \mathbb{R}^m$. This shows that Z is a Gaussian vector.

Therefore, limits in the quadratic mean preserve the Gaussian nature of random vectors. This is the stability property referred to in the title of this example. Note that the Gaussian nature of random vectors is also preserved by linear transformations, as we already know. □

Convergence in distribution is weaker that convergence in variation:

Theorem 4.5.8 *If the sequence of real random variables* $\{X_n\}_{n\geq 1}$ *converges in variation to* X, *it converges in distribution to the same random variable.*

Proof. Indeed, for all x (not just the continuity points of the distribution of X),

$$|P(X_n \leq x) - P(X \leq x)| \leq d_V(X_n, X) \to 0.$$

□

Complementary reading

[Billingsley, 1979, 1992], [Kallenberg, 2002].

4.6 Exercises

Exercise 4.6.1. THE TELESCOPE FORMULA, TAKE 2
Prove the following inequalities concerning a non-negative random variable X:

$$\sum_{n=1}^{\infty} P(X \geq n) \leq \mathrm{E}[X] \leq 1 + \sum_{n=1}^{\infty} P(X \geq n).$$

Exercise 4.6.2. A RECURRENCE EQUATION
Consider the recurrence equation

$$X_{n+1} = (X_n - 1)^+ + Z_{n+1}, \ n \geq 0$$

$(a^+ := \sup(a, 0))$, where $X_0 = 0$ and where $\{Z_n\}_{n \geq 1}$ is an IID sequence of random variables with values in $\mathbb{N}$. Denote by T_0 the first index $n \geq 1$ such that $X_n = 0$ ($T_0 = \infty$ if such index does not exist)

a) Show that if $E[Z_1] < 1$, then $P(T_0 < \infty) = 1$.

b) Show that if $E[Z_1] > 1$, there exists a (random) index n_0 such that $X_n > 0$ for all $n \geq n_0$.

Exercise 4.6.3. POISSON ASYMPTOTICS
Let $\{S_n\}_{n \geq 1}$ be an IID sequence of real random variables such that $P(S_1 \in (0, \infty)) = 1$ and $E[S_1] < \infty$, and let for each $t \geq 0$, $N(t) = \sum_{n \geq 1} 1_{(0,t]}(T_n)$, where $T_n = S_1 + \cdots + S_n$. Prove that $\lim_{t \to \infty} \frac{N(t)}{t} = \frac{1}{E[S_1]}$.

Exercise 4.6.4. SLLN AND INFINITE EXPECTATION
Let $\{Z_n\}, n \geq 1$, be an IID sequence of non-negative random variables such that $E[Z_1] = \infty$. Show that

$$\lim_{n \uparrow \infty} \frac{Z_1 + \ldots + Z_n}{n} = \infty \quad (= E[Z_1]).$$

Exercise 4.6.5. EXCHANGING THE ORDER OF EXPECTATION AND SUMMATION

(a) Let $\{S_n\}_{n \geq 1}$ be a sequence of non-negative random variables. Show that

$$E\left[\sum_{n=1}^{\infty} S_n\right] = \sum_{n=1}^{\infty} E[S_n]. \tag{$\star$}$$

(b) Let $\{S_n\}_{n \geq 1}$ be a sequence of real random variables such that $\sum_{n \geq 1} E[|S_n|] < \infty$. Show that $(\star)$ holds as well.

Exercise 4.6.6. A SUFFICIENT CONDITION FOR ALMOST-SURE CONVERGENCE
Show that

$$\sum_{n \geq 1} P(|Z_n - Z| \geq \epsilon) < \infty$$

for all $\varepsilon > 0$ is a sufficient condition for the sequence of random variables $\{Z_n\}_{n \geq 1}$ to converge to Z.

Exercise 4.6.7. CONVERGENCE ALMOST SURE vs CONVERGENCE IN PROBABILITY
Let $\{X_n\}_{n \geq 1}$ be a sequence of independent random variables taking only 2 values, 0 and 1.

(A) Show that a necessary and sufficient condition of almost-sure convergence to 0 is that

$$\sum_{n \geq 1} P(X_n = 1) < \infty.$$

(B) Show that a necessary and sufficient condition of convergence in probability to 0 is that

$$\lim_{n\uparrow\infty} P(X_n = 1) = 0.$$

(C) Deduce from the above that convergence in probability does not imply almost-sure convergence.

Exercise 4.6.8. CONVERGENCE IN PROBABILITY AND IN THE QUADRATIC MEAN
Let $\alpha > 0$, and let $\{Z_n\}_{n\geq 1}$ be a sequence of random variables such that

$$P(Z_n = 1) = 1 - \frac{\alpha}{n}, \ P(Z_n = n) = \frac{\alpha}{n},$$

where $\alpha < 1$.

Show that $\{Z_n\}_{n\geq 1}$ converges in *probability* to some variable Z.

For what values of α does $\{Z_n\}_{n\geq 1}$ converge to Z *in quadratic mean*?

Exercise 4.6.9. CONVERGENCE IN PROBABILITY
Suppose the sequence of random variables $\{Z_n\}_{n\geq 1}$ converges to a in probability. Let $g : \mathbb{R} \to \mathbb{R}$ be a *continuous* function. Show that $\{g(Z_n)\}_{n\geq 1}$ converges to $g(a)$ in probability.

Exercise 4.6.10. INNER PRODUCT IN $L^2(P)$
Prove the following: If the sequence $\{Z_n\}_{n\geq 1}$ of square-integrable complex random variables converges in quadratic mean to the complex random variable Z, then

$$\lim_{n\uparrow\infty} \mathrm{E}\,[Z_n] = \mathrm{E}\,[Z] \ \text{ and } \ \lim_{n\uparrow\infty} \mathrm{E}\,\big[|Z_n|^2\big] = \mathrm{E}\,\big[|Z|^2\big]\,.$$

Exercise 4.6.11. CONVERGENCE IN PROBABILITY BUT NOT ALMOST-SURE
Let $\{Z_n\}_{n\geq 1}$ be an independent sequence of random variables such that

$$P(Z_n = \pm 1) = \frac{1}{2n\log n}, \ P(Z_n = 0) = 1 - \frac{1}{n\log n}.$$

Let $S_n = \sum_{j=1}^{n} Z_j$. Show the limit in probability of $\frac{S_n}{n}$ exists, but not the almost-sure limit.

Exercise 4.6.12. CONVERGENCE IN DISTRIBUTION BUT NOT IN PROBABILITY
Let Z be a random variable with a symmetric distribution (that is, Z and $-Z$ have the same distribution). Define the sequence $\{Z_n\}_{n\geq 1}$ as follows: $Z_n = Z$ if n is odd, $Z_n = -Z$ if n is even. In particular, $\{Z_n\}_{n\geq 1}$ converges in *distribution* to Z. Show that if Z is not degenerate, then $\{Z_n\}_{n\geq 1}$ does NOT converge to Z *in probability*.

Exercise 4.6.13. THE UNLIMITED GAMBLER
This exercise anticipates the gambling situation described in Example 13.1.4 to which the reader is referred for the notation. Suppose that the stakes are bounded, say by M, and that the initial fortune of the gambler is a. The gambler can borrow whatever amount is needed, so that his "fortune" Y_n at any time n can take arbitrary values.

Prove that

$$P(|Y_n - a| \geq \lambda) \leq 2 \exp\left(-\frac{\lambda^2}{2nM^2}\right).$$

Exercise 4.6.14. FAIR COIN TOSSES
Consider a Bernoulli sequence of parameter $\frac{1}{2}$ representing a fair game of HEADS or TAILS. Let X be the number of HEADS after n tosses. Use Hoeffding's inequality to prove that

$$P(|X - E[X]| \geq \lambda) \leq 2 \exp\left(-\frac{\lambda^2}{n}\right).$$

Exercise 4.6.15. EMPTY BINS
Consider the usual "balls and bins" setting with n bins and m balls (the multinomial distribution). Let X be the number of empty bins. Prove that

$$P(|X - E[X]| \geq \lambda) \leq 2 \exp\left(-\frac{\lambda^2}{m}\right).$$

Exercise 4.6.16. BERNOULLI IS BOREL
Let the sequence $\{X_n\}_{n \geq 1}$ be as in Theorem 1.1.6 (it is sometimes called a *Bernoulli sequence*). Prove that

$$P(\{X_n\}_{n \geq 1} \text{ is a Borel sequence}) = 1.$$

A sequence $\{x_n\}_{n \geq 1}$ taking its values in the set $\{0, 1\}$ is called a Borel sequence if for all $k \geq 1$, all $a_1, \ldots, a_k$ in $\{0, 1\}$,

$$\lim_{n \uparrow \infty} \frac{1}{n} \sum_{j=1}^n 1_{\{x_{j+1} = a_1, \ldots, x_{j+k} = a_k\}} = \frac{1}{2}^k.$$

Exercise 4.6.17. METRIZATION OF CONVERGENCE IN PROBABILITY
Define for any two random variables X and Y,

$$d(X, Y) := E\left[\frac{|X - Y|}{1 + |X - Y|}\right].$$

Prove that d so defined is a metric. Prove the following variant of Theorem 4.2.4: The sequence $\{X_n\}_{n \geq 1}$ converges in probability to the variable X if and only if

$$\lim_{n \uparrow \infty} d(X_n, X) = 0.$$

Exercise 4.6.18. POISSON'S LAW OF RARE EVENTS IN THE PLANE
Let $Z_1, \ldots, Z_M$ be M bidimensional IID random vectors uniformly distributed on the square $[0, A] \times [0, A] = \Gamma_A$. For any measurable set $C \subseteq \Gamma_A$, define $N(C)$ to be the number of random vectors Z_i that fall in C. Let $C_1, \ldots, C_K$ be measurable *disjoint* subsets of Γ_A.

i) Give the characteristic function of the vectors $(N(C_1), \ldots, N(C_K))$.

ii) We now let M be a function of A such that

$$\frac{M(A)}{A^2} = \lambda > 0.$$

Show that, as $A \uparrow \infty$, $(N(C_1), \ldots, N(C_K))$ converges in distribution. Identify the limit distribution.

Exercise 4.6.19. A CHARACTERIZATION OF THE GAUSSIAN DISTRIBUTION
Let G be a cumulative distribution function on $\mathbb{R}$ with

$$\int_{\mathbb{R}} x dG(x) = 0, \quad \int_{\mathbb{R}} x^2 dG(x) = \sigma^2 < \infty.$$

In addition, suppose that G has the following property: If X_1 and X_2 are independent random variables with the CDF G, then $\frac{X_1 + X_2}{\sqrt{2}}$ also admits G as CDF.
Prove that G is the CDF of a Gaussian variable with mean 0 and variance σ^2.

Exercise 4.6.20. THE CENTRAL LIMIT THEOREM
Prove, using the central limit theorem, that

$$\lim_{n \to \infty} \sum_{k=1}^{n} e^{-n} \frac{n^k}{k!} = \frac{1}{2}.$$

Exercise 4.6.21. A CONFIDENCE INTERVAL
In Example 4.4.15 how would you find the best statement of the kind: "This coin is x percent guaranteed unbiased"? In other words, how would you obtain the largest x in this claim? (You are not required to give the actual value, just the method for obtaining it.)

Exercise 4.6.22. MAXIMAL COINCIDENCE OF BIASED COINS
Find a pair of $\{0,1\}$-valued random variables with prescribed marginals

$$P(X = 1) = a, \ P(Y = 1) = b,$$

where $a, b \in (0,1)$, and such that $P(X = Y)$ is maximal.

Exercise 4.6.23. FUNCTIONS OF RANDOM VARIABLES AND DISTANCE IN VARIATION
Let $(E, \mathcal{E})$ and $(G, \mathcal{G})$ be measurable spaces and let $f : (E, \mathcal{E}) \mapsto (G, \mathcal{G})$ be some measurable function. For α and β, probability distributions on $(E, \mathcal{E})$, define the probability distribution αf^{-1} on $(G, \mathcal{G})$ by $\alpha f^{-1}(B) = \alpha(f^{-1}(B))$, and define likewise βf^{-1}. Prove that

$$d_V(\alpha, \beta) \geq d_V(\alpha f^{-1}, \beta f^{-1}).$$

Exercise 4.6.24. THE VARIATION DISTANCE OF TWO POISSON VARIABLES
Let p_λ denote the Poisson distribution with mean λ. Let $\mu > 0$. Prove that

$$d_V(p_\lambda, p_\mu) \leq 1 - e^{-|\mu - \lambda|}.$$

Exercise 4.6.25. CONVEXITY OF THE DISTANCE IN VARIATION
Let α_i and β_i, $1 \leq i \leq K$, be probability distributions on the countable space E. Show that if $\lambda_i \in [0,1]$ and $\sum_{i=1}^{K} \lambda_i = 1$, then

$$d_V\left(\sum_{i=1}^{K} \lambda_i \alpha_i, \sum_{i=1}^{K} \lambda_i \beta_i\right) \leq \sum_{i=1}^{K} \lambda_i d_V(\alpha_i, \beta_i).$$

State and prove the analogous result in the general case of an arbitrary measurable space $(E, \mathcal{E})$.

Exercise 4.6.26. AN ALTERNATIVE EXPRESSION OF THE DISTANCE IN VARIATION
Let α and β be two probability distributions on some countable space E. Show that

$$d_V(\alpha, \beta) = \frac{1}{2} \sup_{|f| \leq 1} \left(\sum_i f(i)\alpha(i) - \sum_i f(i)\beta(i)\right),$$

where $|f| := \sup_{i \in E} |f(i)|$.

State and prove the analogous result in the general case of an arbitrary measurable space $(E, \mathcal{E})$.

Exercise 4.6.27. ANOTHER EXPRESSION OF THE DISTANCE IN VARIATION
Let α and β be two probability distributions on the same countable space E. Prove the following alternative expressions of the distance in variation:

$$d_V(\alpha, \beta) = 1 - \sum_{i \in E} \alpha(i) \wedge \beta(i)$$

$$= \sum_{i \in E} (\alpha(i) - \beta(i))^+ = \sum_{i \in E} (\beta(i) - \alpha(i))^+.$$

State and prove the analogous result in the general case of an arbitrary measurable space $(E, \mathcal{E})$.

Exercise 4.6.28. CONVERGENCE IN PROBABILITY AND CONVERGENCE IN VARIATION
Let $\{Z_n\}_{n \geq 0}$ be a sequence of $\{0, 1\}$-valued random variables. Show that it converges in variation to 0 if and only if it converges in probability to 0. Deduce from this that there exist sequences of random variables that converge in distribution but not in variation.

Exercise 4.6.29. TRICKY CAUCHY
Let $\{X_n\}_{n \geq 1}$ be a sequence of IID Cauchy random variables.

(a) What is the limit in distribution of $\frac{X_1 + \cdots + X_n}{n}$?

(b) Does $\frac{X_1 + \cdots + X_n}{n^2}$ converge in distribution?

(c) Does $\frac{X_1 + \cdots + X_n}{n}$ converge almost surely to a (non-random) constant?

II: STANDARD STOCHASTIC PROCESSES

Chapter 5

Generalities on Random Processes

A random process (or stochastic process) is a collection of random variables indexed by time, which may record the evolution of some phenomenon. This definition is generalized to accommodate models where space, and not just time, plays a role. This chapter addresses the basic issues concerning the distribution of such processes, their sample path properties and the various notions of measurability.

5.1 The Distribution of a Random Process

5.1.1 Kolmogorov's Theorem on Distributions

Let $\mathbf{T}$ be an arbitrary index. In general in this book, it will be one of the following: $\mathbb{N}$ (the natural numbers), $\mathbb{Z}$ (the integers), $\mathbb{R}$ (the real numbers) or $\mathbb{R}_+$ (the non-negative real numbers) (see, however, Example 5.1.24).

Random Processes as Collections of Random Variables

Definition 5.1.1 *A stochastic process (or* random process*) is a family $\{X(t)\}_{t \in \mathbf{T}}$ of random elements defined on the same probability space $(\Omega, \mathcal{F}, P)$ and taking their values in some given measurable space $(E, \mathcal{E})$.*

It is called a *real* (resp., *complex*) stochastic process if it takes real (resp., complex) values. It is called a *continuous-time* stochastic process when the index set is $\mathbb{R}$ or $\mathbb{R}_+$, and a *discrete-time* stochastic process when it is $\mathbb{N}$ or $\mathbb{Z}$. When the index set is $\mathbb{N}$ or $\mathbb{Z}$, we also use the notation n instead of t for the time index, and write X_n instead of $X(t)$.

EXAMPLE 5.1.2: RANDOM SINUSOID. Let A be some real non-negative random variable, let $\nu_0 \in \mathbb{R}$ be a positive constant and let Φ be a random variable with values in $[0, 2\pi]$. The formula

$$X(t) = A \sin(2\pi\nu_0 t + \Phi)$$

defines a stochastic process. For each sample $\omega \in \Omega$, the function $t \mapsto X(t, \omega)$ is a sinusoid with frequency ν_0, random amplitude $A(\omega)$ and random phase $\Phi(\omega)$.

© Springer Nature Switzerland AG 2020
P. Brémaud, *Probability Theory and Stochastic Processes*, Universitext,
https://doi.org/10.1007/978-3-030-40183-2_5

EXAMPLE 5.1.3: COUNTING PROCESSES. A counting process $\{N(t)\}_{t\geq 0}$ is, by definition, an integer-valued stochastic process such that the functions $t \mapsto N(t,\omega)$ are almost-
surely integer-valued, non-decreasing, right-continuous with left-hand limits, such that
$N(t) - N(t-) \leq 1$ $(t \geq 0)$ and $N(0) = 0$. For instance, $N(t)$ could be the number of
arrivals of cars at a highway toll in the interval $(0, t]$.

Finite-dimensional Distributions

One way of describing the probabilistic behavior of a stochastic process is by means of
its finite-dimensional distribution.

Definition 5.1.4 *By definition, the* finite-dimensional (fidi) *distribution of a stochastic
process* $\{X(t)\}_{t\in\mathbf{T}}$ *is the collection of probability distributions of the random vectors*

$$(X(t_1),\dots,X(t_k))$$

for all $k \geq 1$ *and all* $t_1,\dots,t_k \in \mathbf{T}$.

EXAMPLE 5.1.5: A PARTICULAR COUNTING PROCESS. Let $\{N(t)\}_{t\geq 0}$ be a counting
process. Suppose that for all $k \geq 1$, all $0 = t_0 \leq t_1 < \cdots < t_k$, and all integers $m_1 \dots, m_k$,

$$P(\cap_{j=1}^k \{N(t_j) - N(t_{j-1}) = m_j\}) = \prod_{j=1}^k e^{\lambda(t_j - t_{j-1})} \frac{(\lambda(t_j - t_{j-1}))^{m_j}}{m_j!}.$$

The fact that this completely describes the fidi distribution of $\{N(t)\}_{t\geq 0}$ is obvious. The
existence of such a process is at this point not proved but will be guaranteed by Theorem
5.1.7 below. We shall see later on that this is the counting process of a homogeneous
Poisson process on the positive half-line of intensity λ.

There are two pending issues. Firstly, is there a stochastic process having a prescribed
finite distribution? Secondly, is it unique? The answer is rather simple and it is best
answered in the setting of canonical measurable spaces of functions.

Let $E^{\mathbf{T}}$ be the set of functions $x : \mathbf{T} \to E$. An element $x \in E^{\mathbf{T}}$ is therefore a function
from $\mathbf{T}$ to E:

$$x := (x(t), t \in \mathbf{T}),$$

where $x(t) \in E$. Let $\mathcal{E}^{\mathbf{T}}$ [1] be the smallest σ-field containing all the sets of the form

$$\{x \in E^{\mathbf{T}} ; x(t) \in C\},$$

where t ranges over $\mathbf{T}$ and C ranges over $\mathcal{E}$. The measurable space $(E^{\mathbf{T}}, \mathcal{E}^{\mathbf{T}})$ so defined is
called the *canonical (measurable) space* of stochastic processes indexed by $\mathbf{T}$ with values
in $(E, \mathcal{E})$ (we say: "with values in E" if the choice of the σ-field $\mathcal{E}$ is clear in the given
context).

---

[1]The notation $\mathcal{E}^{\otimes\mathbf{T}}$ is also used in order to distinguish this mathematical object from a
collection of $\mathcal{E}$-valued functions.

Denote by π_t the *coordinate map* at $t \in \mathbf{T}$, that is, the mapping from $E^{\mathbf{T}}$ to E defined by

$$\pi_t(x) := x(t) \quad (x \in E).$$

This is a random variable since when $C \in \mathcal{B}(\mathbb{R})$, the set $\{x \, ; \, \pi_t(x) \in C\} = \{x \, ; \, x(t) \in C\}$, and therefore belongs to $\mathcal{E}^{\mathbf{T}}$, by definition of the latter. The family $\{\pi_t\}_{t \in \mathbf{T}}$ is called the *coordinate process* on the (canonical) measurable space $(E^{\mathbf{T}}, \mathcal{E}^{\mathbf{T}})$.

The probability distribution of the vector $(X(t_1), \ldots, X(t_k))$ is a probability measure $Q_{(t_1, \ldots, t_k)}$ on $(E^k, \mathcal{E}^k)$. It satisfies the following obvious properties, called the *compatibility conditions*:

C_1. For all $(t_1, \ldots, t_k) \in \mathbf{T}^k$, and any permutation σ on $\mathbf{T}^k$,

$$Q_{\sigma(t_1, \ldots, t_k)} = Q_{t_1, \ldots, t_k} \circ \sigma^{-1}. \tag{5.1}$$

C_2. For all $(t_1, \ldots, t_k, t_{k+1}) \in \mathbf{T}^{k+1}$ and all $A \in \mathcal{E}^k$

$$Q_{(t_1, \ldots, t_k)}(A) = Q_{(t_1, \ldots, t_{k+1})}(A \times E). \tag{5.2}$$

Remark 5.1.6 Conditions C_1 and C_2 just acknowledge the obvious facts of the type

$$P(X(t_1) \in A_1, X(t_2) \in A_2) = P(X(t_2) \in A_2, X(t_1) \in A_1) \text{ and}$$
$$P(X(t_1) \in A_1, X(t_2) \in E) = P(X(t_1) \in A_1).$$

Recall the definition of a *Polish space*.[2] It is a topological space whose topology is metrizable (generated by some metric), complete (with respect to this metric) and separable (there exists a countable dense subset).

Theorem 5.1.7 *Let E be a Polish space and let $\mathcal{E} := \mathcal{B}(E)$ be its Borel σ-field. Let $\mathcal{Q} = \{Q_{(t_1, \ldots, t_k)}; k \geq 1, (t_1, \ldots, t_k) \in \mathbf{T}^k\}$ be a family of probability distributions on $(E^k, \mathcal{E}^k)$ satisfying the compatibility conditions C1 and C2. Then there exists a unique probability P on the canonical measurable space $(E^{\mathbf{T}}, \mathcal{E}^{\mathbf{T}})$ such that the coordinate process $\{\pi_t\}_{t \in \mathbf{T}}$ admits the finite distribution $\mathcal{Q}$.*

This is the *Kolmogorov existence and uniqueness theorem*.[3]

EXAMPLE 5.1.8: IID SEQUENCES. Take $E = \mathbb{R}$, $\mathbf{T} = \mathbb{Z}$, and let the *fidi* distributions $\mathcal{Q}_{t_1, \ldots, t_k}$ be of the form

$$\mathcal{Q}_{(t_1, \ldots, t_k)} = Q_{t_1} \times \cdots \times Q_{t_k},$$

where for each $t \in \mathbf{T}$, Q_t is a probability distribution on $(E, \mathcal{E})$. This collection of finite-dimensional distributions obviously satisfies the compatibility conditions, and the resulting coordinate process is an independent random sequence indexed by the relative integers. It is an IID (that is: independent and identically distributed) sequence if $Q_t = Q$ for all $t \in \mathbf{T}$. Note that the restriction to a Polish space E is superfluous (Exercise 5.4.1).

---

[2]For most applications in this book, the Polish space in question will be some $\mathbb{R}^m$ with the euclidean topology.

[3]For a proof of Kolmogorov's distribution theorem, see for instance [Shiryaev, 1996].

Independence

Let $\{X(t)\}_{t\in\mathbf{T}}$ be a stochastic process. The σ-field

$$\mathcal{F}^X := \sigma(X(t); t \in \mathbf{T})$$

is called the *global history* of this process.

Definition 5.1.9 *Two stochastic processes $\{X(t)\}_{t\in\mathbf{T}}$ and $\{Y(t)\}_{t\in\mathbf{T}'}$ defined on the same probability space, with values in $(E,\mathcal{E})$ and $(E',\mathcal{E}')$ respectively, are called independent if the σ-fields $\mathcal{F}^X$ and $\mathcal{F}^Y$ are independent.*

The verification of independence is simplified by the following result.

Theorem 5.1.10 *For the stochastic processes $\{X(t)\}_{t\in\mathbf{T}}$ and $\{Y(t)\}_{t\in\mathbf{T}'}$, with values in $(E,\mathcal{E})$ and $(E',\mathcal{E}')$ respectively, to be independent, it suffices that for all $t_1,\ldots,t_k \in \mathbf{T}$ and all $s_1,\ldots,s_\ell \in \mathbf{T}'$, the vectors $(X(t_1),\ldots,X(t_k))$ and $(Y(s_1),\ldots,Y(s_\ell))$ be independent.*

Proof. The collection of events of the type $\{X(t_1) \in C_1,\ldots,X(t_k) \in C_k\}$, where the C_i's belong to $\mathcal{E}$, is a π-system generating $\mathcal{F}^X$, with a similar observation for $\mathcal{F}^Y$. The result then follows from these observations and Theorem 3.1.39.   □

Transfer to Canonical Spaces

Let a stochastic process $\{X(t)\}_{t\in\mathbf{T}}$ be given with values in a Polish space E and defined on the probability space $(\Omega, \mathcal{F}, P)$. Define the mapping $h : (\Omega, \mathcal{F}) \to (E^\mathbf{T}, \mathcal{E}^\mathbf{T})$ by

$$h(\omega) = (X(t,\omega), t \in \mathbf{T}).$$

This mapping is measurable. To show this, it is enough to verify that $h^{-1}(C) \in \mathcal{F}$ for all $C \in \mathcal{C}$, where $\mathcal{C}$ is a collection of subsets of $E^\mathbf{T}$ that generates $\mathcal{E}^\mathbf{T}$. Here, we choose $\mathcal{C} = (\{x; x(t) \in A\}, t \in \mathbf{T}, A \in \mathcal{E})$. But $h^{-1}(\{x; x(t) \in A\}) = (\{\omega; X(t,\omega) \in A\}) \in \mathcal{F}$ since $X(t)$ is a random variable. Now, denote by $\mathcal{P}_X$ the image of P by h. The fidi distribution of the coordinate process of the canonical measurable space is the same as that of the original stochastic process.

Definition 5.1.11 *The probability $\mathcal{P}_X$ on $(E^\mathbf{T}, \mathcal{E}^\mathbf{T})$ is called the* distribution *of $\{X(t)\}_{t\in\mathbf{T}}$.*

An immediate consequence of Kolmogorov's (existence and) uniqueness theorem is:

Corollary 5.1.12 *Two stochastic processes with the same fidi distribution have the same distribution.*

Stationarity

In this subsubsection, the index set is one of the following: $\mathbb{N}$, $\mathbb{Z}$, $\mathbb{R}_+$, $\mathbb{R}$.

Definition 5.1.13 *A stochastic process* $\{X(t)\}_{t \in \mathbf{T}}$ *is called (strictly) stationary if for all* $k \geq 1$, *all* $(t_1, \ldots, t_k) \in \mathbf{T}^k$, *the probability distribution of the random vector*

$$(X(t_1 + h), \ldots, X(t_k + h))$$

is independent of $h \in \mathbf{T}$ *such that* $t_1 + h, \ldots, t_k + h \in \mathbf{T}$.

A stochastic process with index set $\mathbb{R}_+$ (resp., $\mathbb{N}$) that is stationary can always be uniquely extended to $\mathbb{R}$ (resp., $\mathbb{Z}$) in such a way that stationarity is preserved. More precisely:

Theorem 5.1.14 *Consider the canonical space* $(E^{\mathbf{T}}, \mathcal{E}^{\mathbf{T}})$ *of stochastic processes with values in the Polish space* E *and with index set* $\mathbf{T} = \mathbb{R}_+$ *(resp.,* $\mathbf{T} = \mathbb{N}$). *Let* $\mathcal{P}_+$ *be a probability measure on this canonical space that makes the canonical process stationary. Then there exists a (unique) probability measure* $\mathcal{P}$ *on the canonical space of stochastic processes with values in* E *with index set* $\mathbb{R}$ *(resp.,* $\mathbb{Z}$) *such that the restriction of* $\mathcal{P}$ *to* $(E^{\mathbb{R}_+}, \mathcal{E}^{\mathbb{R}_+})$ *(resp.,* $(E^{\mathbb{N}}, \mathcal{E}^{\mathbb{N}})$) *is* $\mathcal{P}_+$.

Proof. Let $\{Q^+_{(t_1, \ldots, t_k)}; k \geq 1, (t_1, \ldots, t_k) \in \mathbf{T}^k\}$ be the finite-dimensional distributions relative to $\mathcal{P}_+$. Define the collection of finite-dimensional distributions

$$\{Q_{(t_1, \ldots, t_k)}; k \geq 1, (t_1, \ldots, t_k) \in \mathbf{T}^k\},$$

where the index set is now $\mathbb{R}$ (resp., $\mathbb{Z}$), by

$$Q_{(t_1, \ldots, t_k)} := Q^+_{(t_1 + h, \ldots, t_k + h)},$$

where h is an element of $\mathbb{R}_+$ (resp., $\mathbb{N}$) such that $t_1 + h, \ldots, t_k + h \in \mathbb{R}_+$ (resp., $\in \mathbb{N}$). Observe that this family satisfies the compatibility conditions (5.1) and (5.1). The result then follows from Kolmogorov's existence and uniqueness theorem. $\qquad\square$

5.1.2 Second-order Stochastic Processes

In this subsection, $\mathbf{T}$ represents any of the following index sets: $\mathbb{R}$, $\mathbb{R}_+$, $\mathbb{Z}$ and $\mathbb{N}$.

Definition 5.1.15 *A measurable complex stochastic process* $\{X(t)\}_{t \in \mathbf{T}}$ *satisfying the condition*

$$\mathrm{E}[|X(t)|^2] < \infty \quad (t \in \mathbf{T})$$

is called a second-order *stochastic process.*

In other words, for all $t \in \mathbf{T}$, the complex random variable $X(t) \in L^2_{\mathbb{C}}(P)$. This implies that the *mean* function $m : \mathbf{T} \to \mathbb{C}$ and the *covariance* function $\Gamma : \mathbf{T} \times \mathbf{T} \to \mathbb{C}$ are well defined by

$$m(t) := \mathrm{E}[X(t)]$$

and

$$\Gamma(t, s) := \mathrm{cov}(X(t), X(s)) = \mathrm{E}[X(t)X(s)^*] - m(t)m(s)^*.$$

When the mean function is the null function, the stochastic process is said to be *centered*.

Theorem 5.1.16 *Let $\{X(t)\}_{t \in \mathbf{T}}$ be a second-order stochastic process with mean function m and covariance function Γ. Then, for all $s, t \in \mathbf{T}$,*

$$\mathrm{E}\left[|X(t) - m(t)|\right] \leq \Gamma(t, t)^{\frac{1}{2}}$$

and

$$|\Gamma(t, s)| \leq \Gamma(t, t)^{\frac{1}{2}} \Gamma(s, s)^{\frac{1}{2}}.$$

Proof. Apply Schwarz's inequality

$$\mathrm{E}\left[|X|\,|Y|\right] \leq \mathrm{E}\left[|X|^2\right]^{\frac{1}{2}} \mathrm{E}\left[|Y|^2\right]^{\frac{1}{2}}$$

with $X := X(t) - m(t)$ and $Y := 1$ for the first inequality, and with $X := X(t) - m(t)$ and $Y := X(s) - m(s)$ for the second one. $\qquad \square$

For a stationary second-order stochastic process, for all $s, t \in \mathbf{T}$,

$$m(t) \equiv m, \tag{5.3}$$

where $m \in \mathbb{C}$ and

$$\Gamma(t, s) = C(t - s) \tag{5.4}$$

for some function $C : \mathbf{T} \to \mathbb{C}$, also called the *covariance function* of the process. The complex number m is called the *mean* of the process.

Wide-sense Stationarity

A notion weaker than strict stationarity concerns second-order processes with values in $E = \mathbb{C}$:

Definition 5.1.17 *If conditions (5.3) and (5.4) are satisfied for all $s, t \in \mathbf{T}$, the complex second-order stochastic process $\{X(t)\}_{t \in \mathbf{T}}$ is called* wide-sense stationary.

Remark 5.1.18 There exist stochastic processes that are wide-sense stationary but not strictly stationary (Exercise 5.4.3).

In continuous time ($\mathbf{T} = \mathbb{R}$ or $\mathbb{R}_+$) this appellation will be reserved in this book to wide-sense stationary processes that have in addition a continuous covariance function. For this condition to be satisfied, it suffices that the covariance function be continuous at the origin. This is in turn equivalent to continuity in the quadratic mean of the stochastic process, that is: For all $t \in \mathbf{T}$,

$$\lim_{h \to 0} \mathrm{E}\left[|X(t + h) - X(t)|^2\right] = 0.$$

In fact, the covariance function is then uniformly continuous on $\mathbb{R}$.

Proof.

$$\mathrm{E}\left[|X(t + h) - X(t)|^2\right] = \mathrm{E}\left[|X(t + h)|^2\right] + \mathrm{E}\left[|X(t)|^2\right]$$
$$- \mathrm{E}\left[X(t)X(t + h)^*\right] - \mathrm{E}\left[X(t)^*X(t + h)\right]$$
$$= 2C(0) - C(h) - C(h)^*,$$

and therefore, uniform continuity in quadratic mean follows from the continuity at the origin of the autocovariance function. On the other hand,

$$
\begin{aligned}
|C(\tau + h) - C(\tau)| &= |\mathrm{E}\left[X(\tau + h)X(0)^*\right] - \mathrm{E}\left[X(\tau)X(0)^*\right]| \\
&= |\mathrm{E}\left[(X(\tau + h) - X(\tau))X(0)^*\right]| \\
&\leq \mathrm{E}\left[|X(0)|^2\right]^{\frac{1}{2}} \times \mathrm{E}\left[|X(\tau + h) - X(\tau)|^2\right]^{\frac{1}{2}} \\
&= \mathrm{E}\left[|X(0)|^2\right]^{\frac{1}{2}} \times \mathrm{E}\left[|X(h) - X(0)|^2\right]^{\frac{1}{2}},
\end{aligned}
$$

and therefore, uniform continuity of the autocovariance function follows from the continuity in quadratic mean at the origin. $\qquad\square$

Note that $C(0) = \sigma_X^2$, the variance of any of the random variables $X(t)$.

As an immediate corollary of Theorem 5.1.16, we have:

Corollary 5.1.19 *Let* $\{X(t)\}_{t \in \mathrm{T}}$ *be a wide-sense stationary stochastic process with mean m and covariance function C. Then*

$$
\mathrm{E}\left[|X(t) - m|\right] \leq C(0)^{\frac{1}{2}}
$$

and

$$
|C(\tau)| \leq C(0).
$$

EXAMPLE 5.1.20: HARMONIC PROCESSES. Let $\{U_k\}_{k \geq 1}$ be centered random variables of $L^2_{\mathbb{C}}(P)$ that are mutually uncorrelated. Let $\{\Phi_k\}_{k \geq 1}$ be completely random phases, that is, real random variables uniformly distributed on $[0, 2\pi]$. Suppose moreover that the U variables are independent of the Φ variables. Finally, suppose that $\sum_{k=1}^{\infty} \mathrm{E}[|U_k|^2] < \infty$. Then (Exercise 5.4.5): For all $t \in \mathbb{R}$, the series in the right-hand side of

$$
X(t) = \sum_{k=1}^{\infty} U_k \cos(2\pi\nu_k t + \Phi_k),
$$

where the ν_k's are real numbers (frequencies), is convergent in $L^2_{\mathbb{C}}(P)$ and defines a centered WSS stochastic process (called a *harmonic process*) with covariance function

$$
C(\tau) = \sum_{k=1}^{\infty} \frac{1}{2} \mathrm{E}[|U_k|^2] \cos(2\pi\nu_k \tau).
$$

Recall the definition of the *correlation coefficient* ρ between two non-trivial real square-integrable random variables X and Y with respective means m_X and m_Y and respective variances σ_X^2 and σ_Y^2:

$$
\rho = \frac{\mathrm{cov}\,(X, Y)}{\sigma_X \sigma_Y}.
$$

The variable $aX + b$ that minimizes the function $F(a, b) := \mathrm{E}\left[(Y - aX - b)^2\right]$ is

$$\widehat{Y} = m_Y + \frac{\operatorname{cov}(X, Y)}{\sigma_X^2}(X - m_X)$$

and moreover

$$\mathrm{E}\left[\left(\widehat{Y} - Y\right)^2\right] = \left(1 - \rho^2\right)\sigma_Y^2$$

(Exercise 3.4.43). This variable is called the *best linear-quadratic estimate* of Y given X, or the *linear regression* of Y on X.

For a WSS stochastic process with covariance function C, the function

$$\rho(\tau) = \frac{C(\tau)}{C(0)}$$

is called the *autocorrelation function*. It is in fact, for any t, the correlation coefficient between $X(t)$ and $X(t+\tau)$. In particular, the best linear-quadratic estimate of $X(t+\tau)$ given $X(t)$ is

$$\widehat{X}(t+\tau \mid t) := m + \rho(\tau)(X(t) - m).$$

The estimation error is then, according to the above,

$$\mathrm{E}\left[\left(\widehat{X}(t+\tau \mid t) - X(t+\tau)\right)^2\right] = \sigma_X^2\left(1 - \rho(\tau)^2\right).$$

Remark 5.1.21 In the continuous time case, this shows that if the support of the covariance function is concentrated around $\tau = 0$, the process tends to be "unpredictable". We shall come back to this when we introduce the notion of white noise.

5.1.3 Gaussian Processes

This particular type of stochastic process is an important one for many reasons, for instance:

(1) because of its mathematical tractability due in particular to the stability of the Gaussianity of stochastic processes by linear transformations and limits,

(2) because of its ubiquity due to the many forms of the central limit theorem for stochastic processes,

(3) because the most famous example of a Gaussian process, Brownian motion (Chapter 11), is the basis of a very productive stochastic calculus (Chapter 14).

Gaussian processes will at this point serve to substantiate the definitions and theoretical results of this chapter.

Definition 5.1.22 *Let* $\mathbf{T}$ *be an arbitrary index. The real-valued stochastic process* $\{X(t)\}_{t \in \mathbf{T}}$ *is called a* Gaussian process *if for all* $n \geq 1$ *and for all* $t_1, \ldots, t_n \in \mathbf{T}$, *the random vector* $(X(t_1), \ldots, X(t_n))$ *is Gaussian.*

In particular, its characteristic function is given by the formula

$$\mathrm{E}\left[\exp\left\{i\sum_{j=1}^n u_j X(t_j)\right\}\right] = \exp\left\{i\sum_{j=1}^n u_j m(t_j) - \frac{1}{2}\sum_{j=1}^n\sum_{k=1}^n u_j u_k \Gamma(t_j, t_k)\right\}, \quad (5.5)$$

where $u_1, \ldots, u_n \in \mathbb{R}$, $m(t) := \mathrm{E}[X(t)]$ and $\Gamma(t, s) := \mathrm{E}[(X(t) - m(t))(X(s) - m(s))]$.

The next result is an existence theorem.

Theorem 5.1.23 *Let* $\Gamma : \mathbf{T}^2 \to \mathbb{R}$ *be a non-negative definite function, that is, such that for all* $t_1, \ldots, t_k \in \mathbf{T}$ *and all* $u_1, \ldots, u_k \in \mathbb{R}$,

$$\sum_{i=1}^{k}\sum_{j=1}^{k} u_i u_j \Gamma(t_i, t_j) \geq 0. \tag{5.6}$$

Then, there exists a centered Gaussian process with covariance Γ.

Proof. By Theorem 3.2.5, for any $k \in \mathbb{N}_+$, any $t_1, \ldots, t_k \in \mathbf{T}$, there exists a centered Gaussian vector with covariance matrix $\{\Gamma(t_i, t_j)\}_{1 \leq i,j \leq k}$. Let $Q_{t_1,\ldots,t_k}$ be the probability distribution of this vector. The family $\{Q_{t_1,\ldots,t_k}; t_1, \ldots, t_k \in \mathbf{T}\}$ is obviously compatible and therefore, by Kolmogorov's theorem (Theorem 5.1.7), a centered Gaussian process with covariance Γ exists and is unique in distribution. $\square$

EXAMPLE 5.1.24: A GAUSSIAN FIELD ON $\mathbb{R}^d$. Let μ be locally finite measure on $\mathbb{R}^d$ and let $\mathcal{B}_b(\mathbb{R}^d)$ be the collection of bounded Borelian sets of $\mathbb{R}^d$. There exists a unique (distributionwise) centered Gaussian process $\{X(A)\}_{A \in \mathcal{B}_b(\mathbb{R}^d)}$ with autocorrelation function

$$\Gamma(A, B) = \mu(A \cap B) \quad (A, B \in \mathcal{B}_b(\mathbb{R}^d)).$$

To prove this it suffices to verify condition (5.6). This is done by observing that

$$\sum_{i=1}^{k}\sum_{j=1}^{k} u_i u_j \Gamma(A_i, A_j) = \sum_{i=1}^{k}\sum_{j=1}^{k} u_i u_j \mu(A_i \cap A_j)$$

$$= \int_{\mathbb{R}^d} \left(\sum_{j=1}^{k} u_j 1_{A_j}(x)\right)^2 \mu(\mathrm{d}x) \geq 0.$$

Theorem 5.1.25 *For a Gaussian process with index set* $\mathbf{T} = \mathbb{R}^d$ *or* $\mathbb{Z}^d$ *(*$d \in \mathbb{N}_+$*) to be stationary, it is necessary and sufficient that for some real number* m *and some function* $C : \mathbf{T} \to \mathbb{R}$, $m(t) = m$ *and* $\Gamma(t, s) = C(t - s)$ *for all* $s, t \in \mathbf{T}$.

Proof. The necessity is obvious, whereas the sufficiency is proved by replacing the t_ℓ's in (5.5) by $t_\ell + h$ to obtain the characteristic function of $(X(t_1 + h), \ldots, X(t_n + h))$, namely,

$$\exp\left\{i\sum_{j=1}^{n} u_j m - \frac{1}{2}\sum_{j=1}^{n}\sum_{k=1}^{n} u_j u_k C(t_j - t_k)\right\},$$

and then observing that this quantity is independent of h. $\square$

Gaussian Subspaces

With any second-order stochastic process is associated a Hilbert subspace of the Hilbert space of square-integrable random variables. More precisely: let $\{X_i\}_{i \in I}$, where I is an arbitrary index set, be a collection of complex (resp., real) random variables in $L^2_{\mathbb{C}}(P)$ (resp., $L^2_{\mathbb{R}}(P)$). The Hilbert subspace of $L^2_{\mathbb{C}}(P)$ (resp., $L^2_{\mathbb{R}}(P)$) consisting of the closure in

$L^2_{\mathbb{C}}(P)$ (resp., $L^2_{\mathbb{R}}(P)$) of the vector space of finite linear complex (resp., real) combinations of elements of $\{X_i\}_{i\in I}$ is called the complex (resp., real) *Hilbert subspace generated by* $\{X_i\}_{i\in I}$ and is denoted by $H_{\mathbb{C}}(X_i, i \in I)$ (resp., $H_{\mathbb{R}}(X_i, i \in I)$).

A collection $\{X_i\}_{i\in I}$ of real random variables defined on the same probability space, where I is an arbitrary index set, is called a *Gaussian family* if for all finite set of indices $i_1, \ldots, i_k \in I$, the random vector $(X_{i_1}, \ldots, X_{i_k})$ is Gaussian.

Definition 5.1.26 *A Hilbert subspace G of the real Hilbert space $L^2_{\mathbb{R}}(P)$ is called a* Gaussian (Hilbert) subspace *if it is a Gaussian family.*

Theorem 5.1.27 *Let $\{X_i\}_{i\in I}$, where I is an arbitrary index set, be a Gaussian family of random variables of $L^2_{\mathbb{R}}(P)$. Then the Hilbert subspace $H_{\mathbb{R}}(X_i, i \in I)$ generated by $\{X_i\}_{i\in I}$ is a Gaussian subspace of $L^2_{\mathbb{R}}(P)$.*

Proof. By definition, the Hilbert subspace $H_{\mathbb{R}}(X_i, i \in I)$ consists of all the random variables in $L^2_{\mathbb{R}}(P)$ that are limits in quadratic mean of finite linear combinations of elements of the family $\{X_i\}_{i\in I}$. To prove the announced result, it suffices to show that if $\{Z_n\}_{n\geq 1}$, where $Z_n = (Z_n^{(1)}, \ldots, Z_n^{(m)})$, is a sequence of Gaussian random vectors of fixed dimension m that converges componentwise in quadratic mean to some vector $Z = (Z^{(1)}, \ldots, Z^{(m)})$ of $H_{\mathbb{R}}(X_i, i \in I)$, then the latter vector is Gaussian. By continuity of the inner product in $L^2_{\mathbb{R}}(P)$, for all $1 \leq i, j \leq m$, $\lim_{n\uparrow\infty} \mathrm{E}[Z_n^{(i)}Z_n^{(j)}] = \mathrm{E}[Z^{(i)}Z^{(j)}]$ and $\lim_{n\uparrow\infty} \mathrm{E}[Z_n^{(i)}] = \mathrm{E}[Z^{(i)}]$, that is

$$\lim_{n\uparrow\infty} m_{Z_n} = m_Z, \qquad \lim_{n\uparrow\infty} \Gamma_{Z_n} = \Gamma_Z$$

and in particular, for all $u \in \mathbb{R}^m$,

$$\lim_{n\uparrow\infty} \mathrm{E}\left[e^{iu^T Z_n}\right] = \lim_{n\uparrow\infty} e^{iu^T \mu_{Z_n} - \frac{i}{2}u^T \Gamma_{Z_n} u}$$
$$= e^{iu^T \mu_Z - \frac{i}{2}u^T \Gamma_Z u}.$$

The sequence $\{u^T Z_n\}_{n\geq 1}$ converges to $u^T Z$ in quadratic mean, and therefore in distribution (Theorem 4.5.6). In particular, $\lim_{n\uparrow\infty} \mathrm{E}\left[e^{iu^T Z_n}\right] = \mathrm{E}[e^{iu^T Z}]$, and finally

$$\mathrm{E}[e^{iu^T Z}] = e^{iu^T \mu_Z - \frac{i}{2}u^T \Gamma_Z u} \quad (u \in \mathbb{R}^m),$$

which shows that Z is Gaussian. $\square$

5.2   Random Processes as Random Functions

5.2.1   Versions and Modifications

For each $\omega \in \Omega$, the function $t \in \mathbf{T} \mapsto X(t, \omega) \in E$ is called the ω-*trajectory*, or ω-*sample path*, of the stochastic process $\{X(t)\}_{t\in\mathbf{T}}$.

A stochastic process can be viewed as a *random function*, associating to each $\omega \in \Omega$ the trajectory $t \mapsto X(t, \omega)$. When the state space E is some $\mathbb{R}^m$ and the index set is $\mathbb{R}$, for fixed $\omega \in \Omega$, we can then discuss the continuity properties of the associated sample path. For example, if for all $\omega \in \Omega$ the ω-sample path is right-continuous, we call this stochastic process right-continuous. It is called P-a.s. right-continuous if the ω-sample paths are right-continuous for all $\omega \in \Omega$, except perhaps for $\omega \in \mathcal{N}$, where $\mathcal{N}$ is a P-negligible set. One defines similarly (P-a.s.) left-continuity, (P-a.s.) continuity, etc.

Definition 5.2.1 *Two stochastic processes $\{X(t)\}_{t\in \mathbf{T}}$ and $\{Y(t)\}_{t\in \mathbf{T}}$ defined on the same probability space $(\Omega, \mathcal{F}, P)$ are said to be* versions of one another *if*

$$P(\{\omega ;\, X(t,\omega) \neq Y(t,\omega)\}) = 0 \ for \ all \ t \in \mathbf{T}.$$

They are said to be undistinguishable *if*

$$P(\{\omega ;\, X(t,\omega) = Y(t,\omega) \ for \ all \ t \in \mathbf{T}\}) = 1,$$

that is, if they have identical trajectories except on a P-null set.

Clearly two undistinguishable processes are versions of one another.

EXAMPLE 5.2.2: TWO DISTINGUISHABLE VERSIONS. The two processes

$$X(t) = 1 \quad (t \in [0,1])$$

and

$$X(t) = 1_{U=t} \quad (t \in [0,1]),$$

where U is a random variable uniformly distributed on $[0,1]$, have the same distributions, and therefore are versions of one another, but they are not undistinguishable.

It is useful to find conditions bearing only on the finite-dimensional distributions and guaranteeing that the sample paths have certain desired properties. This is not always feasible but, in certain cases, there is a version possessing the desired properties. One such result is Kolmogorov's continuity theorem below.

5.2.2   Kolmogorov's Continuity Condition

Theorem 5.2.3 *Let (E, d) be a complete metric space. Let $\{X(t)\}_{t\in [0,1]}$ be a stochastic process with values in E. Suppose that for some positive real numbers α, β, K,*

$$\mathrm{E}\left[d(X(t), X(s))^\alpha\right] \leq K|t - s|^{1+\beta},$$

for all $s, t \in [0,1)$. Then there exists a version of this stochastic process whose sample paths are almost surely continuous.

Proof. We will encounter in the proof the following subsets of $[0,1]$: the set D of dyadic rationals of $[0,1]$ and for each $n \geq 1$ the set $D_n := \{k2^{-n} ;\, k = 0, \cdots, n\}$. The countable set D is dense in $[0,1]$.

STEP 1. The original process is continuous in probability, that is, for all $\varepsilon > 0$,

$$\lim_{t_n \to t} P(d(X(t_n) - X(t)) \geq \varepsilon) = 0. \qquad (\star)$$

This follows from Markov's inequality:

$$P(d(X(t_n), X(t)) \geq \varepsilon) = P(d(X(t_n), X(t))^\alpha > \varepsilon^\alpha)$$
$$\leq \frac{\mathrm{E}\left[d(X(t_n), X(t))^\alpha\right]}{\varepsilon^\alpha} \leq \frac{K|t_n - t|^{1+\beta}}{\varepsilon^\alpha}.$$

STEP 2. The original process is uniformly continuous on D. To see this, let $\gamma \in (0, \beta/\alpha)$ and obtain from $(\star)$ that

$$P(d(X(k2^{-n}), X((k+1)2^{-n})) \geq 2^{-\gamma n}) \leq K2^{-n(1+\beta-\alpha\gamma)}.$$

In particular, with

$$A_n := \left\{ \max_{1 \leq k \leq 2^n} \|X(k2^{-n}) - X((k-1)2^{-n})\| \geq 2^{-\gamma n} \right\}$$

and by sub-σ-additivity,

$$P(A_n) \leq \sum_{k=1}^{2^n} P(\|X(k2^{-n}) - X((k-1)2^{-n})\| \geq 2^{-\gamma n})$$

$$\leq 2^n K2^{-n(1+\beta-\alpha\gamma)} = K2^{-n(\beta-\alpha\gamma)},$$

and therefore, since $\beta - \alpha\gamma > 0$,

$$\sum_n P(A_n) < \infty.$$

By the Borel–Cantelli lemma, there exists a P-negligible subset $\mathcal{N}$ and an almost surely finite random integer N such that outside $\mathcal{N}$, for $n \geq N$ and $k = 1, \ldots, 2^n$,

$$d(X(k2^{-n}), X((k-1)2^{-n})) < 2^{-\gamma n}. \tag{†}$$

In particular,

$$K_\gamma := \sup_{n \geq 1} \left(\sup_{1 \leq k \leq 2^n} \frac{d(X((k-1)2^{-n}), X(k2^{-n}))}{2^{-\gamma n}} \right)$$

is an almost surely finite random variable. It will follow from this (Lemma 5.2.4 below) that for all $s, t \in D$, almost surely

$$d(X(t), X(s)) \leq C_\gamma |t - s|^\gamma, \tag{5.7}$$

where

$$C_\gamma := \frac{2}{1 - 2^{-\gamma}} K_\gamma.$$

Therefore, almost surely, the mapping $t \to X(t)$ is continuous, and therefore uniformly continuous on D.

Step 3. Let $Y(t) = 0$ on $\mathcal{N}$ and if $t \notin \mathcal{N}$, $Y(t) = X(t)$ for $t \in D$, and

$$Y(t) = \lim_{t_n \to t, \, t_n \in D} X(t_n) \qquad (t \in \overline{D}).$$

Outside $\mathcal{N}$, the function $t \in [0, 1) \mapsto Y(t)$ is a continuous extension of the uniformly continuous function $t \in D \mapsto X(t)$ to $[0, 1)$. Indeed, for any $s, t \in [0, 1)$, there exist a sequence in D, $t_k \to t$, and a sequence in D, $s_k \to s$, such that $Y(t) = \lim_{t_k \to t} X(t_k)$ and $Y(s) = \lim_{s_k \to t} X(s_k)$, so that

$$\|Y(t) - Y(s)\| \leq \|Y(t) - X(t_k)\| + \|X(t_k) - X(s_k)\| + \|X(s_k) - Y(s)\|.$$

One can choose the t_k's and the s_k's inside the interval $[s, t]$. With any $\varepsilon > 0$ one can then associate δ such that if $|t - s| \leq \delta$, the middle term of the right-hand side is less than $\varepsilon/3$ whatever k (by the uniform continuity of $t \in D \to X(t)$) and such that the extreme terms are less than $\varepsilon/3$ (by an appropriate choice of s_k and t_k, by construction of $t \in [0, 1) \to Y(t)$) and therefore finally $\|Y(t) - Y(s)\| \leq \varepsilon$. Therefore, outside $\mathcal{N}$, $t \in [0, 1) \to Y(t)$ is a continuous function.

Step 4. We now show that $\{Y(t)\}_{t \in [0,1)}$ is a version of $\{X(t)\}_{t \in [0,1)}$, that is, $P(X(t) = Y(t)) = 1$ for all $t \in [0, 1)$. This follows from the fact that $\{X(t)\}_{t \in [0,1)}$ is continuous in probability and that limits in probability and almost-sure limits coincide when both exist.

Step 5. It remains to prove the inequality (5.7). This follows from Lemma 5.2.4 below. □

Lemma 5.2.4 *Let f be a mapping from D to the metric space (E, d). Suppose that there exists a finite constant K such that for $n \geq N$ and $k = 1, \ldots, 2^n$,*

$$d(f(k2^{-n}), f((k-1)2^{-n})) < K2^{-\gamma n}.$$

Then for all $s, t \in D$,

$$d(f(t), f(s)) \leq \frac{2}{1 - 2^{-\gamma}} K |t - s|^{\gamma}.$$

Proof. Let $s, t \in D$, $s < t$. Let p be the smallest integer such that $2^{-p} \leq t - s$. Let k be the smallest integer such that $k2^{-p} \geq s$. Then it is possible to write

$$s = k2^{-p} - \varepsilon_1 2^{-p-1} - \cdots - \varepsilon_\ell 2^{-p-\ell},$$
$$t = k2^{-p} + \varepsilon_1' 2^{-p-1} + \cdots + \varepsilon_m' 2^{-p-m}$$

for some non-negative integers ℓ, m and ε's and ε''s taking the values 0 or 1. Define

$$s_i = k2^{-p} - \varepsilon_1 2^{-p-1} - \cdots - \varepsilon_\ell 2^{-p-i} \quad (0 \leq i \leq \ell),$$
$$t_j = k2^{-p} + \varepsilon_1' 2^{-p-1} + \cdots + \varepsilon_m' 2^{-p-j} \quad (0 \leq j \leq m).$$

Then, observing that $s = s_\ell$ and $t = t_m$,

$$d(f(s), f(t)) = d(f(s_\ell), f(t_m))$$

$$= d(f(s_0), f(t_0)) + \sum_{i=1}^{\ell} d(f(s_{i-1}), f(s_i)) + \sum_{j=1}^{m} d(f(t_{j-1}), f(t_j))$$

$$\leq K2^{-p\gamma} + \sum_{i=1}^{\ell} K2^{-(p+i)\gamma} + \sum_{j=1}^{m} K2^{-(p+j)\gamma}$$

$$\leq 2K(1 - 2^{-\gamma})^{-1} 2^{-p\gamma} \leq 2K(1 - 2^{-\gamma})^{-1}(t - s)^{\gamma}.$$

□

Remark 5.2.5 Going back to Example 5.1.5, Kolmogorov's theorem guarantees the existence of an integer-valued process $\{N(t)\}_{t \geq 0}$ with the fidi distribution described in this example. It does not say, however, that it is a counting process in the sense of Example 5.1.3. It turns out that such a version exists, as we shall see in Chapter 8, where a completely different definition is given.

5.3   Measurability Issues

5.3.1   Measurable Processes and their Integrals

Viewing a stochastic process as a *mapping* $X : \mathbf{T} \times \Omega \to E$, defined by $(t, \omega) \mapsto X(t, \omega)$, opens the way to various measurability concepts.

Definition 5.3.1 *The stochastic process* $\{X(t)\}_{t \in \mathbb{R}}$ *is said to be* measurable *iff the mapping from* $\mathbb{R} \times \Omega$ *into* E *defined by* $(t, \omega) \mapsto X(t, \omega)$ *is measurable with respect to* $\mathcal{B} \otimes \mathcal{F}$ *and* $\mathcal{E}$.

In particular, for any $\omega \in \Omega$ the mapping $t \to X(t, \omega)$ is measurable with respect to the σ-fields $\mathcal{B}(\mathbb{R})$ and $\mathcal{E}$ (Lemma 2.3.6). Also, if $E = \mathbb{R}$ and if $X(t)$ is non-negative, one can define the Lebesgue integral

$$\int_{\mathbb{R}} X(t, \omega) dt$$

for each $\omega \in \Omega$, and also apply Tonelli's theorem (Theorem 2.3.9) to obtain

$$\mathrm{E}\left[\int_{\mathbb{R}} X(t) dt\right] = \int_{\mathbb{R}} \mathrm{E}[X(t)] \, dt.$$

By Fubini's theorem, the last equality also holds true for measurable stochastic processes of arbitrary sign such that $\int_{\mathbb{R}} \mathrm{E}[|X(t)|] \, dt < \infty$.

Theorem 5.3.2 *Let* $\{X(t)\}_{t \in \mathbb{R}}$ *be a second-order complex-valued measurable stochastic process with mean function* m *and covariance function* Γ. *Let* $f : \mathbb{R} \to \mathbb{C}$ *be an integrable function such that*

$$\int_{\mathbb{R}} |f(t)| \mathrm{E}[|X(t)|] \, dt < \infty. \tag{5.8}$$

Then the integral $\int_{\mathbb{R}} f(t) X(t) \, dt$ *is almost surely well defined and*

$$\mathrm{E}\left[\int_{\mathbb{R}} f(t) X(t) \, dt\right] = \int_{\mathbb{R}} f(t) m(t) \, dt.$$

Suppose in addition that f *satisfies the condition*

$$\int_{\mathbb{R}} |f(t)| |\Gamma(t, t)|^{\frac{1}{2}} \, dt < \infty \tag{5.9}$$

and let $g : \mathbb{R} \to \mathbb{C}$ *be a function with the same properties as* f. *Then* $\int_{\mathbb{R}} f(t) X(t) \, dt$ *is square-integrable and*

$$\mathrm{cov}\left(\int_{\mathbb{R}} f(t) X(t) \, dt, \int_{\mathbb{R}} g(t) X(t) \, dt\right) = \int_{\mathbb{R}} \int_{\mathbb{R}} f(t) g^*(s) \Gamma(t, s) \, dt \, ds.$$

Proof. By Tonelli's theorem

$$\mathrm{E}\left[\int_{\mathbb{R}} |f(t)| |X(t)| \, dt\right] = \int_{\mathbb{R}} |f(t)| \mathrm{E}[|X(t)|] \, dt < \infty$$

and therefore, almost surely $\int_{\mathbb{R}} |f(t)| |X(t)| \, dt < \infty$, so that almost surely the integral $\int_{\mathbb{R}} f(t) X(t) \, dt$ is well defined and finite. Also (Fubini)

$$\mathrm{E}\left[\int_{\mathbb{R}} f(t)X(t)\,\mathrm{d}t\right] = \int_{\mathbb{R}} \mathrm{E}\left[f(t)X(t)\right]\,\mathrm{d}t = \int_{\mathbb{R}} f(t)\mathrm{E}\left[X(t)\right]\,\mathrm{d}t.$$

Suppose now (without loss of generality) that the process is centered. By Tonelli's theorem

$$\mathrm{E}\left[\left(\int_{\mathbb{R}} |f(t)||X(t)|\,\mathrm{d}t\right)\left(\int_{\mathbb{R}} |g(t)||X(t)|\,\mathrm{d}t\right)\right]$$
$$= \int_{\mathbb{R}}\int_{\mathbb{R}} |f(t)||g(s)|\mathrm{E}\left[|X(t)||X(s)|\right]\,\mathrm{d}t\,\mathrm{d}s.$$

But (Schwarz's inequality) $\mathrm{E}\left[|X(t)||X(s)|\right] \leq \Gamma(t,t)^{\frac{1}{2}}\Gamma(s,s)^{\frac{1}{2}}$, and therefore the right-hand side of the last equality is bounded by

$$\left(\int_{\mathbb{R}} |f(t)|\Gamma(t,t)^{\frac{1}{2}}\,\mathrm{d}t\right)\left(\int_{\mathbb{R}} |g(s)|\Gamma(s,s)^{\frac{1}{2}}\,\mathrm{d}s\right) < \infty.$$

One may therefore apply Fubini's theorem to obtain

$$\mathrm{E}\left[\left(\int_{\mathbb{R}} f(t)X(t)\,\mathrm{d}t\right)\left(\int_{\mathbb{R}} g(t)X(t)\,\mathrm{d}t\right)\right] = \int_{\mathbb{R}}\int_{\mathbb{R}} f(t)g^*(s)\mathrm{E}\left[X(t)X(s)\right]\,\mathrm{d}t\,\mathrm{d}s.$$

$\square$

Remark 5.3.3 Since $\mathrm{E}[|X(t)|] \leq \mathrm{E}[1 + |X(t)|^2] = 1 + \Gamma(t,t)$, condition (5.8) is satisfied if f is an integrable function such that $\int_{\mathbb{R}} |f(t)|\Gamma(t,t)\,\mathrm{d}t < \infty$.

5.3.2 Histories and Stopping Times

In the following, the index set $\mathbf{T}$ is any of the following: $\mathbb{R}, \mathbb{R}_+, \mathbb{N}, \mathbb{Z}$.

Definition 5.3.4 *Let* $(\Omega, \mathcal{F})$ *be a measurable space. The family* $\{\mathcal{F}_t\}_{t\in\mathbf{T}}$ *of sub-σ-fields of $\mathcal{F}$ is called a* history *(or* filtration*) on* $(\Omega, \mathcal{F})$ *if for all* $s,t \in \mathbf{T}$ *such that* $s \leq t$,

$$\mathcal{F}_s \subseteq \mathcal{F}_t.$$

In other words, a history is a non-decreasing family of sub-σ-fields of $\mathcal{F}$ indexed by $\mathbf{T}$. In applications $\mathcal{F}_t$ often represents the information available at time t to an observer.

The σ-field $\mathcal{F}_\infty := \vee_{t\in\mathbf{T}}\mathcal{F}_t$ is, by definition, the smallest σ-field that contains $\mathcal{F}_t$ for all $t \in \mathbf{T}$.

Definition 5.3.5 *Let* $\{X(t)\}_{t\in\mathbf{T}}$ *be stochastic process defined on* $(\Omega, \mathcal{F})$. *The history* $\{\mathcal{F}_t^X\}_{t\in\mathbf{T}}$ *defined by*

$$\mathcal{F}_t^X = \sigma(X(s)\,;s \leq t)$$

is called the internal history *of* $\{X(t)\}_{t\in\mathbf{T}}$. *Any history* $\{\mathcal{F}_t\}_{t\in\mathbf{T}}$ *such that*

$$\mathcal{F}_t \supseteq \mathcal{F}_t^X \quad (t \in \mathbf{T})$$

is called a history *of* $\{X(t)\}_{t\in\mathbf{T}}$. *The stochastic process* $\{X(t)\}_{t\in\mathbf{T}}$ *is then said to be* adapted *to the history* $\{\mathcal{F}_t\}_{t\in\mathbf{T}}$, *or $\mathcal{F}_t$-adapted.*

Definition 5.3.6 *Let* $\mathbf{T} = \mathbb{R}$ *or* $\mathbb{R}_+$. *Define for all* $t \in \mathbf{T}$

$$\mathcal{F}_{t+} := \cap_{s>t}\mathcal{F}_s.$$

The history $\{\mathcal{F}_t\}_{t\in\mathbf{T}}$ *is called* right-continuous *if for all* $t \in \mathbf{T}$, $\mathcal{F}_t = \mathcal{F}_{t+}$.

Progressive Measurability

Definition 5.3.7 *A stochastic process* $\{X(t)\}_{t \in \mathbb{R}_+}$ *taking its values in the measurable space* $(E, \mathcal{E})$ *is said to be* $\mathcal{F}_t$-*progressively measurable if for all* $t \in \mathbb{R}_+$ *the mapping* $(s, \omega) \to X(s, \omega)$ *from* $[0, t] \times \Omega$ *into* E *is measurable with respect to the* σ-*fields* $\mathcal{B}([0, t]) \otimes \mathcal{F}_t$ *and* $\mathcal{E}$.

A $\mathcal{F}_t$-progressively measurable $\{X(t)\}_{t \in \mathbb{R}_+}$ is then $\mathcal{F}_t$-adapted and measurable.

Theorem 5.3.8 *Let* $\{X(t)\}_{t \in \mathbb{R}_+}$ *be a stochastic process, taking its values in a topological space* E *endowed with its Borel* σ-*field* $\mathcal{E} = \mathcal{B}(E)$, *adapted to* $\{\mathcal{F}_t\}_{t \in \mathbb{R}_+}$ *and right-continuous (resp., left-continuous). Then* $\{X(t)\}_{t \in \mathbb{R}_+}$ *is* $\mathcal{F}_t$-*progressively measurable.*

Proof. Let t be a non-negative real number. For all $n \geq 0$ and all $s \in [0, t)$, let

$$X_n(s) := \sum_{k=0}^{2^n - 1} X((k+1)t/2^{-n}) 1_{\{[k2^{-n}t, (k+1)2^{-n}t)\}}(s),$$

and let $X_n(t) := X(t)$. This defines a function $[0, t] \times \Omega \to E$ which is measurable with respect to $\mathcal{B}([0, t]) \otimes \mathcal{F}_t$ and $\mathcal{E}$. If $t \mapsto X(t, \omega)$ is right-continuous, $X(s, \omega)$ is the limit of $X_n(s, \omega)$ for all $(s, \omega) \in [0, t] \times \Omega$, and therefore $(s, \omega) \to X(s, \omega)$ is measurable with respect to $\mathcal{B}([0, t]) \otimes \mathcal{F}_t$ and $\mathcal{E}$ as a function of $[0, t] \times \Omega$ into E. The case of a left-continuous process is treated in a similar way. $\square$

Theorem 5.3.9 *If the non-negative stochastic process* $\{X(t)\}_{t \in \mathbb{R}_+}$ *is* $\mathcal{F}_t$-*progressively measurable, then, for each* $t \in \mathbb{R}_+$, *the random variable* $\int_{(0, t]} X(s, \omega) \, ds$ *is* $\mathcal{F}_t$-*measurable.*

The proof is left as Exercise 5.4.2.

Stopping Times

A principal notion in the theory of stochastic processes is that of stopping time. In this subsection, the index set is $\mathbf{T} = \mathbb{N}$ or $\mathbb{R}_+$, and $\overline{\mathbf{T}} := \mathbf{T} \cup \{+\infty\}$.

Definition 5.3.10 *Let* $\{\mathcal{F}_t\}_{t \in \mathbf{T}}$ *be a history. A* $\overline{\mathbf{T}}$-*valued random variable* τ *is called an* $\mathcal{F}_t$-*stopping time iff for all* $t \in \mathbf{T}$,

$$\{\tau \leq t\} \subset \mathcal{F}_t.$$

Remark 5.3.11 In the case $\mathbf{T} = \mathbb{R}_+$, the condition

$$\{\tau < t\} \subset \mathcal{F}_t \quad (t \geq 0)$$

does not guarantee that T is an $\mathcal{F}_t$-stopping time. But since

$$\{\tau \leq t\} = \cap_n \{\tau < t + \frac{1}{n}\} \in \cap_{s > t} \mathcal{F}_s = \mathcal{F}_{t+},$$

we have that T is an $\mathcal{F}_{t+}$-stopping time, and therefore an $\mathcal{F}_t$-stopping time *if the history is right-continuous.*

EXAMPLE 5.3.12: COUNTEREXAMPLE. Define a (not right-continuous) history by $\mathcal{F}_t = \{\varnothing, \Omega\}$ if $t \le 1$, and $\mathcal{F}_t = \sigma(A)$ if $t > 1$, where $A \notin \{\varnothing, \Omega\}$. The random variable $\tau := 1 + 1_A$ is such that $\{\tau < t\} \subset \mathcal{F}_t$ for all $t \ge 0$, but it is not an $\mathcal{F}_t$-stopping time because $\{\tau \le 1\}$ is not in $\mathcal{F}_1$.

The following approximation of a stopping time by simple random variables will be of frequent use in the sequel.

Theorem 5.3.13 *Let $\{\mathcal{F}_t\}_{t\in\mathbb{R}_+}$ be a history, and let for all $n \ge 1$,*

$$\tau(n,\omega) := \begin{cases} 0 & \text{if } \tau(\omega) = 0 \\ \frac{k+1}{2^n} & \text{if } \frac{k}{2^n} < \tau(\omega) \le \frac{k+1}{2^n} \\ +\infty & \text{if } \tau(\omega) = \infty. \end{cases}$$

Then $\tau(n)$ is an $\mathcal{F}_t$-stopping-time decreasing to τ as $n \uparrow \infty$.

Proof. In fact, for all $t \ge 0$,

$$\{\tau(n) \le t\} = \cup_{k\,;\,(k+1)2^{-n}\le t}\{\tau_n = (k+1)2^{-n}\}$$
$$= \{\tau = 0\} \bigcup \left(\cup_{k\,;\,(k+1)2^{-n}\le t}\{k2^{-n} < \tau \le (k+1)2^{-n}\} \right) \in \mathcal{F}_t.$$

The decreasing convergence to τ is obvious. $\qquad\square$

Let $\{X(t)\}_{t\in\mathbb{R}_+}$ be a stochastic process with values in E. For any set $C \subset E$, let

$$\tau(C) := \inf\{t \ge 0\,;\, X(t) \in C\}.$$

Theorem 5.3.14 *Let $\{X(t)\}_{t\in\mathbb{R}_+}$ be a right-continuous stochastic process with values in a metric space E and adapted to the history $\{\mathcal{F}_t\}_{t\ge 0}$.*

  A. Let G be an open set of E. The random time $\tau(G)$ is an $\mathcal{F}_{t+}$-stopping time.

  B. Suppose moreover that $\{X(t)\}_{t\in\mathbb{R}_+}$ has left limits for all $t > 0$. Let Γ be a closed set of E. Then, the random time $\tau(\Gamma)$ is an $\mathcal{F}_t$-stopping time.

Proof.

  A. This comes from the identity

$$\{\tau(G) < t\} = \cap_{r\in\mathbf{Q},r<t}\{X(r) \in G\} \in \mathcal{F}_t$$

and Remark 5.3.11.

  B. The closed set Γ is identical to the intersection of the open sets $G_n := \{x \in E\,;\, d(x,\Gamma) < \frac{1}{n}\}$. Therefore

$$\{\tau(\Gamma) \le t\} = \cap_n\{\tau(G_n) < t\} \in \mathcal{F}_t.$$

$\qquad\square$

Remark 5.3.15 Part A will be used in this book only for right-continuous histories, and therefore in this case the entrance time to an open set of a continuous stochastic process is an $\mathcal{F}_t$-stopping time.

Theorem 5.3.16 *Let* $\{X(t)\}_{t\in\mathbb{R}_+}$ *be a right-continuous (resp., left-continuous) real-valued process adapted to* $\{\mathcal{F}_t\}_{t\geq 0}$. *For any real number* c, *the random time* $\tau := \inf\{t \mid X(t) \geq c\}$ *is an* $\mathcal{F}_t$-*stopping time. In the left-continuous case,* $X(\tau) \leq c$.

Proof. Right-continuous case: The event $\{\tau > t\} = \{X(s) < c \text{ for all } s \in [0,t]\}$ is identical to

$$\bigcap_{n\geq 1}\{X(kt/2^n) < c \quad (k = 0, 1, \ldots, 2^n)\},$$

which is in $\mathcal{F}_t$. The left-continuous case is similar. In the left-continuous case, suppose that for a given ω, $X(\tau(\omega),\omega) = c + \varepsilon > c$. Then there exists a $\delta > 0$ such that for all $t \in [\tau(\omega) - \delta, \tau(\omega))$, $X(t,\omega) \geq c + \frac{1}{2}\varepsilon > c$, and this is in contradiction with the definition of τ.

$\square$

Remark 5.3.17 The situation depicted in the left-continuous time guarantees that the entrance time τ_n in $[n,\infty)$ is such that the stopped process $\{X(t\wedge\tau_n)\}_{t\in\mathbb{R}_+}$ is bounded (by n). This remark will be of frequent use in the sequel.

Theorem 5.3.18 *Let* $\mathbf{T} = \mathbb{N}$ *or* $\mathbb{R}_+$. *Let* $\{\mathcal{F}_t\}_{t\in\mathbf{T}}$ *be a history. Let* τ *be an* $\mathcal{F}_t$-*stopping time. The collection of events*

$$\mathcal{F}_\tau = \{A \in \mathcal{F}_\infty \mid A \cap \{\tau \leq t\} \in \mathcal{F}_t, \text{ for all } t \in \mathbf{T}\}$$

is a σ-*field, and* τ *is* $\mathcal{F}_\tau$-*measurable. Let* $\{X(t)\}_{t\in\mathbf{T}}$ *be an* E-*valued* $\mathcal{F}_t$-*adapted stochastic process, and let* τ *be a finite* $\mathcal{F}_t$-*stopping time. Define the random variable* $X(\tau)$ *by* $X(\tau)(\omega) := X(\tau(\omega),\omega)$. *Then, if* $\mathbf{T} = \mathbb{N}$ *(resp.,* $\mathbf{T} = \mathbb{R}_+$ *and* $\{X(t)\}_{t\in\mathbb{R}_+}$ *is* $\mathcal{F}_t$-*progressively measurable)* $X(\tau)$ *is* $\mathcal{F}_\tau$-*measurable.*

Proof. The verification that $\mathcal{F}_\tau$ is a σ-field is straightforward. In order to show that τ is $\mathcal{F}_\tau$-measurable, it is enough to show that for all $c \geq 0, \{\tau \leq c\} \in \mathcal{F}_\tau$, that is, for all $t \in \mathbf{T}$, $\{\tau \leq c\}\cap\{\tau \leq t\} \in \mathcal{F}_t$. But this last event is just $\{\tau \leq c\wedge t\} \in \mathcal{F}_{c\wedge t}$, by definition of an $\mathcal{F}_t$-stopping time, and $\mathcal{F}_{c\wedge t} \subseteq \mathcal{F}_t$.

We treat the case $\mathbf{T} = \mathbb{R}_+$. Let $A \in \mathcal{E}$ and $a \geq 0$. The set $\{X(\tau) \in A\}\cap\{\tau \leq a\}$ is identical to $(\{X(S) \in A\}\cap\{S < a\})\cup(\{X(a) \in A\}\cap\{\tau = a\})$, where $S = \tau\wedge a$. Therefore, it suffices to prove that $\{X(S) \in A\}$ is in $\mathcal{F}_a$, as we now show. Indeed, the random variable S is an $\mathcal{F}_t$-stopping time and it is also an $\mathcal{F}_a$-measurable random variable. The $\mathcal{F}_a$-measurability of $X(S)$ follows from the fact that it is obtained by composition of $\omega \to (S(\omega),\omega)$ from $(\Omega,\mathcal{F}_a)$ into $([0,a]\times\Omega, \mathcal{B}([0,a])\otimes\mathcal{F}_a)$, and $(s,\omega) \to X(s,\omega)$ from $([0,a]\times\Omega, \mathcal{B}([0,a])\otimes\mathcal{F}_a)$ into $(E,\mathcal{E})$, which are measurable: the first by definition of $\mathcal{F}_t$-stopping times and the second by definition of $\mathcal{F}_t$-progressiveness. $\square$

Theorem 5.3.19

(i) If S and T are $\mathcal{F}_t$-stopping times, then so are $S \wedge T$ and $S \vee T$.

(ii) An $\mathcal{F}_t$-stopping time T is $\mathcal{F}_T$-measurable.

(iii) If T is an $\mathcal{F}_t$-stopping time and S is $\mathcal{F}_T$-measurable and such that $S \geq T$, then S is an $\mathcal{F}_t$-stopping time.

(iv) If S and T are $\mathcal{F}_t$-stopping times and $A \in \mathcal{F}_S$, then $A \cap \{S \leq T\} \in \mathcal{F}_T$.

(v) If S and T are $\mathcal{F}_t$-stopping times such that $S \leq T$, then $\mathcal{F}_S \subseteq \mathcal{F}_T$.

(vi) If $\{T_n\}_{n \geq 1}$ is a sequence of $\mathcal{F}_t$-stopping times, then $\sup_n T_n$ is an $\mathcal{F}_t$-stopping time.

Proof. (i) and (ii) are left as exercises (Exercise 5.4.10).

(iii) By hypothesis $\{S \leq t\} \in \mathcal{F}_T$ and therefore, by definition of $\mathcal{F}_T$,

$$\{S \leq t\} \cap \{T \leq t\} \in \mathcal{F}_t.$$

But the last intersection is just $\{S \leq t\}$.

(iv) By definition of $\mathcal{F}_T$, we must check that

$$[A \cap \{S \leq T\}] \cap \{T \leq t\} \in \mathcal{F}_t \quad (t \geq 0).$$

But this intersection is equal to

$$[A \cap \{S \leq t\}] \cap \{T \leq t\} \cap \{S \wedge t \leq T \wedge t\},$$

and all the three sets therein are in $\mathcal{F}_t$, the first one because $A \in \mathcal{F}_S$, the second one because T is an $\mathcal{F}_t$-stopping time and the last one because $S \wedge t$ and $S \vee t$ are $\mathcal{F}_t$-measurable.

(v) Let $A \in \mathcal{F}_S$. According to (iv), $A = A \cap \{S \leq T\} \in \mathcal{F}_T$.

(vi) Just observe that

$$\{\sup_n T_n \leq t\} = \cap_n \{T_n \leq t\} \in \mathcal{F}_t.$$

$\square$

Remark 5.3.20 It is not true in general that $\inf_n T_n$ is an $\mathcal{F}_t$-stopping time. In fact, it is not always true that $\{\inf_n T_n \leq t\} = \cup_n \{T_n \leq t\}$. However,

$$\{\inf_n T_n < t\} = \cup_n \{T_n < t\} \in \mathcal{F}_t \quad (t \geq 0),$$

and therefore, if $\{\mathcal{F}_t\}_{t \geq 0}$ is a right-continuous history, $\inf_n T_n$ is an $\mathcal{F}_t$-stopping time (Remark 5.3.11).

Theorem 5.3.21 Let $\{\mathcal{F}_t\}_{t\geq 0}$ be a right-continuous history and $\{T_n\}_{n\geq 1}$ be a sequence of $\mathcal{F}_t$-stopping times. Then

(i) $\liminf_n T_n$ and $\limsup_n T_n$ are $\mathcal{F}_t$-stopping times.

(ii) If $\{T_n\}_{n\geq 1}$ is decreasing, with limit T, then $\mathcal{F}_T = \cap_n \mathcal{F}_{T_n}$.

Proof. The proof of (i) is left as an exercise. It ensures that T of (ii) is indeed an $\mathcal{F}_t$-stopping time. We have from (v) of Theorem 5.3.19 that $\mathcal{F}_T \subseteq \cap_n \mathcal{F}_{T_n}$. Now, let $A \in \cap_n \mathcal{F}_{T_n}$. Then

$$A \cap \{T_n < t\} \in \mathcal{F}_t \quad (t \geq 0)$$

and therefore

$$\cup_n (A \cap \{T_n < t\}) = A \cap \{T < t\} \in \mathcal{F}_t \quad (t \geq 0),$$

which guarantees, because of the right-continuity hypothesis for the history, that T is a stopping time for this history (Remark 5.3.11). $\square$

Complementary reading

[Meyer, 1975] (the first chapters) and [Durrett, 1996] are advanced references.

5.4   Exercises

Exercise 5.4.1. THE CASE OF IID SEQUENCES
Prove directly (without referring to Theorem 5.1.7) the last statement in Example 5.1.8.

Exercise 5.4.2. WHY PROGRESSIVE MEASURABILITY?
Prove that if the non-negative stochastic process $\{X(t)\}_{t\in\mathbb{R}_+}$ is $\mathcal{F}_t$-progressively measurable, then, for each $t \in \mathbb{R}_+$, the random variable $\int_{(0,t]} X(s)\,\mathrm{d}s$ is $\mathcal{F}_t$-measurable.

Exercise 5.4.3. WIDE-SENSE STATIONARY BUT NOT STATIONARY
Give a simple example of a discrete-time stochastic process that is wide-sense stationary, but not strictly stationary. Give a similar example in continuous time.

Exercise 5.4.4. STATIONARIZATION
Let $\{Y(t)\}_{t\in\mathbb{R}}$ be the stochastic process taking its values in $\{-1, +1\}$ defined by

$$Y(t) = Z \times (-1)^n \text{ on } (nT, (n+1)T],$$

where T is a positive real number and Z is a random variable equidistributed on $\{-1, +1\}$.

(1) Show that $\{Y(t)\}_{t\in\mathbb{R}}$ is not a stationary (strictly or in the wide sense) stochastic process.

(2) Let now U be a random variable uniformly distributed on $[0, T]$ and independent of Z. Define for all $t \in \mathbb{R}$, $X(t) = Y(t - U)$. Show that $\{X(t)\}_{t\in\mathbb{R}}$ is a wide-sense stationary stochastic process and compute its covariance function.

Exercise 5.4.5. A HARMONIC PROCESS
Let $\{U_k\}_{k\geq 1}$ be centered random variables of $L^2_{\mathbb{C}}(P)$ that are mutually uncorrelated. Let $\{\Phi_k\}_{k\geq 1}$ be completely random phases, that is, real random variables uniformly distributed on $[0, 2\pi]$. Suppose, moreover, that the U variables are independent of the Φ variables. Finally, suppose that $\sum_{k=1}^{\infty} \mathrm{E}[|U_k|^2] < \infty$. Prove that for all $t \in \mathbb{R}$, the series in the right-hand side of

$$X(t) = \sum_{k=1}^{\infty} U_k \cos(2\pi\nu_k t + \Phi_k),$$

where the ν_k's are real numbers (frequencies), is convergent in $L^2_{\mathbb{C}}(P)$ and defines a WSS stochastic process. Give its covariance function.

Exercise 5.4.6. JUST A JOKE
Let $\{X(t)\}_{t\in\mathbb{R}}$ be a centered Gaussian process, and let $t_1, t_2 \in \mathbb{R}$ be fixed times. Compute the probability that $X(t_1) > X(t_2)$.

Exercise 5.4.7. A CLIPPED GAUSSIAN PROCESS
Let $\{X(t)\}_{t\in\mathbb{R}}$ be a centered stationary Gaussian process with covariance function C_X. Define the *clipped* (or *hard-limited*) process

$$Y(t) = \mathrm{sign}\, X(t),$$

with the convention sign $X(t) = 0$ if $X(t) = 0$ (note however that this occurs with null probability if $C_X(0) = \sigma_X^2 > 0$, which we assume to hold). Clearly this stochastic process is centered. Moreover, it is unchanged when $\{X(t)\}_{t\in\mathbb{R}}$ is multiplied by a positive constant. In particular, we may assume that the variance $C_X(0)$ equals 1, so that the covariance matrix of the vector $(X(0), X(\tau))^T$ is

$$\Gamma(\tau) = \begin{pmatrix} 1 & \rho_X(\tau) \\ \rho_X(\tau) & 1 \end{pmatrix},$$

where $\rho_X(\tau)$ is the correlation coefficient of $X(0)$ and $X(\tau)$. We assume that $\Gamma(\tau)$ is invertible, that is, $|\rho_X(\tau)| < 1$.

Prove the following formula:

$$C_Y(\tau) = \frac{2}{\pi} \sin^{-1}\left(\frac{C_X(\tau)}{C_X(0)}\right).$$

Exercise 5.4.8. THE BLACK AND SCHOLES FORMULA
This formula concerns a certain type of financial product called the *European call option*. The value of a stock at time $t \geq 0$ is

$$V(t) = V(0) \exp\{\theta t + \sigma W(t)\}.$$

In particular,

$$\mathrm{E}[V(t)] = \mathrm{E}[V(0)] \exp\left\{\left(\theta + \frac{1}{2}\sigma^2\right)t\right\},$$

that is, one euro invested in this stock at time 0 will yield $\exp\left\{\left(\theta + \frac{1}{2}\sigma^2\right)t\right\}$ at time t. On the other hand, an investment of 1 euro in a risk-free instrument (bonds, saving account) returns e^{rt} euros at time t, where r is the fixed return rate of the risk-free investment. In a competitive market, the return rates are the same:[4]

$$\theta + \frac{1}{2}\sigma^2 = r.$$

Therefore

$$V(t) = V(0)\exp\left\{\left(r - \frac{1}{2}\sigma^2\right)t + \sigma W(t)\right\}.$$

The investor has the right to exercise the following option (the European call option). At some time T in the future, called the expiration date, he can buy one share of the stock at a fixed price K (the *strike price*) and immediately sell it at price $V(T)$ and therefore make a profit $V(T) - K$. If he does not exercise the option he will do nothing and therefore the profit will be

$$\max(V(T) - K, 0).$$

Of course, the investor must pay an entrance fee C in order to enter the deal. The value of C should be such that the expected return of an investment C in the option should equal the expected return when exercising the option:

$$C = e^{rT} = \mathrm{E}\left[\max(V(T) - K, 0)\right].$$

This is called the "no arbitrage" condition. Give an explicit formula for C in terms of r, σ, $V(0)$ and T.

Exercise 5.4.9. AN ELEMENTARY ERGODIC THEOREM
Let $\{X(t)\}_{t\in\mathbb{R}}$ be a WSS stochastic process with mean m and covariance function $C(\tau)$. Prove that in order that

$$\lim_{T\uparrow\infty} \frac{1}{T}\int_0^T X(s)\,\mathrm{d}s = m_X$$

holds in the quadratic mean, it is necessary and sufficient that

$$\lim_{T\uparrow\infty} \frac{1}{T}\int_0^T \left(1 - \frac{u}{T}\right) C(u)\,\mathrm{d}u = 0. \tag{5.10}$$

Show that this condition is satisfied in particular when the covariance function is integrable.

Exercise 5.4.10. ABOUT STOPPING TIMES
Let S and T be $\mathcal{F}_t$-stopping times. Prove that

(1) the events $\{S < T\}$, $\{S \le T\}$ and $\{S = T\}$ are in $\mathcal{F}_S \cap \mathcal{F}_T$,

(2) $S \wedge T$ and $S \vee T$ are $\mathcal{F}_t$-stopping times, and

(3) T is $\mathcal{F}_T$-measurable.

---

[4]We do not attempt here to define a "competitive market" or to prove the corresponding statement.

Chapter 6

Markov Chains, Discrete Time

A sequence $\{X_n\}_{n\geq 0}$ of random variables with values in a set E is called a *discrete-time stochastic process* with *state space* E. According to such a definition, sequences of independent and identically distributed random variables *are* stochastic processes. However, in order to introduce more variability, one may wish to allow for some dependence on the past in the manner of deterministic recurrence equations. Discrete-time homogeneous Markov chains possess the required feature since they can always be represented (in a sense to be made precise) by a stochastic recurrence equation $X_{n+1} = f(X_n, Z_{n+1})$, where $\{Z_n\}_{n\geq 1}$ is an IID sequence independent of the initial state X_0. The probabilistic dependence on the past is only through the previous state, but this limited amount of memory suffices to produce enough varied and complex behavior to make Markov chains a most important source of models.

6.1 The Markov Property

6.1.1 The Markov Property on the Integers

Let $\{X_n\}_{n\geq 0}$ be a *discrete-time stochastic process* with countable *state space* E. The elements of the state space will be denoted by $i, j, k, \ldots$ If $X_n = i$, the process is said to be in state i at time n, or to visit state i at time n.

Definition 6.1.1 *If for all integers $n \geq 0$ and all states $i_0, i_1, \ldots, i_{n-1}, i, j$,*

$$P(X_{n+1} = j \mid X_n = i, X_{n-1} = i_{n-1}, \ldots, X_0 = i_0) = P(X_{n+1} = j \mid X_n = i), \qquad (6.1)$$

the above stochastic process is called a Markov chain, *and a* homogeneous *Markov chain* (HMC) *if, in addition, the right-hand side of (6.1) is independent of n.*

In the homogeneous case, the matrix $\mathbf{P} = \{p_{ij}\}_{i,j\in E}$, where

$$p_{ij} := P(X_{n+1} = j \mid X_n = i),$$

is called the *transition matrix* of the HMC. Since the entries are probabilities and since a transition from any state i must be to some state, it follows that

$$p_{ij} \geq 0 \quad \text{and} \quad \sum_{k\in E} p_{ik} = 1 \qquad (i, j \in E).$$

© Springer Nature Switzerland AG 2020
P. Brémaud, *Probability Theory and Stochastic Processes*, Universitext,
https://doi.org/10.1007/978-3-030-40183-2_6

A matrix $\mathbf{P}$ indexed by E and satisfying the above properties is called a *stochastic matrix*. The state space may be infinite, and therefore such a matrix is in general not of the kind studied in linear algebra. However, the basic operations of addition and multiplication will be defined by the same formal rules. The notation $x = \{x(i)\}_{i \in E}$ formally represents a column vector, and x^T is the corresponding row vector. For instance,

$$(x^T \mathbf{P})(j) = \sum_{k \in E} x(k) p_{kj} \, .$$

A transition matrix $\mathbf{P}$ is sometimes described by its *transition graph* G, that is, a graph having for nodes (or vertices) the states of E, and with an oriented edge from i to j if and only if $p_{ij} > 0$.

The Markov property (6.1) extends to

$$P(A \mid X_n = i, B) = P(A \mid X_n = i) \, ,$$

where

$$A := \{X_{n+1} = j_1, \ldots, X_{n+k} = j_k\}, \; B = \{X_0 = i_0, \ldots, X_{n-1} = i_{n-1}\}$$

(Exercise 6.6.1). This is in turn equivalent to

$$P(A \cap B \mid X_n = i) = P(A \mid X_n = i) P(B \mid X_n = i) \, .$$

In other words, A and B are conditionally independent given $X_n = i$. Therefore, the future at time n and the past at time n are conditionally independent given the present state $X_n = i$. More generally:

Theorem 6.1.2 *For all $n \geq 2$ and all $i \in E$, the σ-fields $\sigma(X_0, \ldots, X_{n-1})$ and $\sigma(X_{n+1}, X_{n+2}, \ldots)$ are independent given $X_n = i$.*

Proof. This is a direct consequence of the above observations and of Theorem 3.1.39. $\square$

Theorem 6.1.2 shows in particular that the Markov property is independent of the direction of time.

Notation. We shall from now on abbreviate $P(C \mid X_0 = i)$ as $P_i(C)$ $(C \in \mathcal{F})$. Also, if μ is a probability distribution on E, then $P_\mu(C)$ is the probability of C given that the initial state X_0 is distributed according to μ:

$$P_\mu(C) = \sum_{i \in E} \mu(i) P(C \mid X_0 = i) = \sum_{i \in E} \mu(i) P_i(C) \, .$$

The distribution at time n of the chain is the vector ν_n indexed by E and defined by

$$\nu_n(i) := P(X_n = i) \qquad (i \in E) \, .$$

From Bayes' rule of total causes,

$$P(X_{n+1} = j) = \sum_{k \in E} P(X_n = k) P(X_{n+1} = j \mid X_n = k),$$

that is, $\nu_{n+1}(j) = \sum_{k\in E} \nu_n(k)p_{kj}$. In matrix form: $\nu_{n+1}^T = \nu_n^T \mathbf{P}$. Iteration of this equality yields

$$\nu_n^T = \nu_0^T \mathbf{P}^n . \tag{6.2}$$

The matrix $\mathbf{P}^m$ is called the *m-step transition matrix* because its general term is

$$p_{ij}(m) := P(X_{n+m} = j \mid X_n = i).$$

Indeed, the Bayes sequential rule and the Markov property give for the right-hand side of the latter equality

$$\sum_{i_1,\ldots,i_{m-1}\in E} p_{ii_1}p_{i_1 i_2}\cdots p_{i_{m-1}j},$$

which is the general term of the *m*-th power of $\mathbf{P}$.

The probability distribution ν_0 of the *initial state* X_0 is called the *initial distribution*. From Bayes' sequential rule, the homogeneous Markov property and the definition of the transition matrix,

$$P(X_0 = i_0, X_1 = i_1,\ldots, X_k = i_k) = \nu_0(i_0)p_{i_0 i_1}\cdots p_{i_{k-1}i_k}. \tag{6.3}$$

Therefore, by Theorem 5.1.7, we have the following

Theorem 6.1.3 *The distribution of a discrete-time* HMC *is uniquely determined by its initial distribution and its transition matrix.*

Many HMCs receive a natural description in terms of a recurrence equation.

Theorem 6.1.4 *Let* $\{Z_n\}_{n\geq 1}$ *be an* IID *sequence of random variables with values in some measurable space* $(G, \mathcal{G})$. *Let E be a countable space and let* $f : (E\times G, \mathcal{P}(E)\otimes\mathcal{G}) \to (E, \mathcal{P}(E))$ *be some measurable function. Let* X_0 *be a random variable with values in E, independent of* $\{Z_n\}_{n\geq 1}$. *The recurrence equation*

$$X_{n+1} = f(X_n, Z_{n+1}) \tag{6.4}$$

then defines an HMC.

Proof. Iteration of recurrence (6.4) shows that for all $n \geq 1$, there is a function g_n such that $X_n = g_n(X_0, Z_1,\ldots, Z_n)$, and therefore $P(X_{n+1} = j \mid X_n = i, X_{n-1} = i_{n-1},\ldots, X_0 = i_0) = P(f(i, Z_{n+1}) = j \mid X_n = i, X_{n-1} = i_{n-1},\ldots, X_0 = i_0) = P(f(i, Z_{n+1}) = j)$, since the event $\{X_0 = i_0,\ldots, X_{n-1} = i_{n-1}, X_n = i\}$ is expressible in terms of $X_0, Z_1,\ldots, Z_n$ and is therefore independent of Z_{n+1}. Similarly, $P(X_{n+1} = j \mid X_n = i) = P(f(i, Z_{n+1}) = j)$. We therefore have a Markov chain, and it is homogeneous since the right-hand side of the last equality does not depend on n. Explicitly:

$$p_{ij} = P(f(i, Z_1) = j). \tag{6.5}$$

$\square$

Not all homogeneous Markov chains receive a "natural" description of the type featured in Theorem 6.1.4. However, it is always possible to find a "theoretical" description of this kind. More exactly, we have

Theorem 6.1.5 *For any transition matrix* $\mathbf{P}$ *on* E, *there exists a homogeneous Markov chain with this transition matrix and with a representation such as in Theorem 6.1.4.*

Proof. Since E is countable, we may identify it with $\mathbb{N}$, which we do in this proof. Let

$$X_{n+1} = j \text{ if } \sum_{k=0}^{j-1} p_{X_n k} \le Z_{n+1} < \sum_{k=0}^{j} p_{X_n k},$$

where $\{Z_n\}_{n \ge 1}$ is IID, uniform on $[0, 1]$. By application of Theorem 6.1.4 and of formula (6.5), we check that this HMC has the announced transition matrix, the function f of (6.4) being defined by

$$f(i, z) = j \iff \sum_{k=0}^{j-1} p_{ik} \le z < \sum_{k=0}^{j} p_{ik}.$$

$\square$

EXAMPLE 6.1.6: RANDOM WALKS ON $\mathbb{Z}$. Let X_0 be a random variable with values in $\mathbb{Z}$. Let $\{Z_n\}_{n \ge 1}$ be a sequence of IID random variables, independent of X_0, taking the values $+1$ or -1, and with the probability distribution

$$P(Z_n = +1) = p, \ P(Z_n = -1) = 1 - p,$$

where $p \in (0, 1)$. The stochastic process $\{X_n\}_{n \ge 1}$ defined by

$$X_{n+1} := X_n + Z_{n+1}$$

is, in view of Theorem 6.1.4, an HMC, called a *random walk* on $\mathbb{Z}$ (with drift p to the right). If $p = \frac{1}{2}$, it is called the *symmetric random walk* on $\mathbb{Z}$.

EXAMPLE 6.1.7: REPAIR SHOP, TAKE 1. During day n, Z_{n+1} machines break down, and they enter the repair shop on day $n+1$. Every day one machine among those waiting for service is repaired. Therefore, denoting by X_n the number of machines in the shop on day n,

$$X_{n+1} = (X_n - 1)^+ + Z_{n+1}, \tag{6.6}$$

where $a^+ = \max(a, 0)$. In particular, if $\{Z_n\}_{n \ge 1}$ is an IID sequence independent of the initial state X_0, then $\{X_n\}_{n \ge 0}$ is a homogeneous Markov chain. Let

$$a_k := P(Z_1 = k) \qquad (k \ge 0).$$

By formula (6.5),

$$p_{ij} = P((i - 1)^+ + Z_1 = j) = P(Z_1 = j - (i - 1)^+) = a_{j - (i-1)^+}.$$

Therefore

$$\mathbf{P} = \begin{pmatrix} a_0 & a_1 & a_2 & a_3 & \cdots \\ a_0 & a_1 & a_2 & a_3 & \cdots \\ 0 & a_0 & a_1 & a_2 & \cdots \\ 0 & 0 & a_0 & a_1 & \cdots \\ \vdots & \vdots & \vdots & \vdots & \end{pmatrix}.$$

A slight modification of the result of Theorem 6.1.4 considerably enlarges its scope.

Theorem 6.1.8 *Let everything be as in Theorem 6.1.4 except for the joint distribution of $X_0, Z_1, Z_2, \ldots$. Suppose instead that for all $n \geq 0$, Z_{n+1} is conditionally independent of $Z_n, \ldots, Z_1, X_{n-1}, \ldots, X_0$ given X_n, and that for all $i \in E$, the conditional distribution of Z_{n+1} given $X_n = i$ is independent of n. Then $\{X_n\}_{n \geq 0}$ is an HMC, with transition probabilities*

$$p_{ij} = P(f(i, Z_1) = j \mid X_0 = i).$$

Proof. The proof is quite similar to that of Theorem 6.1.4 and is left for the reader. $\square$

EXAMPLE 6.1.9: THE EHRENFEST DIFFUSION MODEL, TAKE 1. This simplified model of diffusion through a porous membrane was proposed in 1907 by the Austrian physicists Tatiana and Paul Ehrenfest in order to describe in terms of statistical mechanics the exchange of heat between two systems at different temperatures. (This model considerably helped in understanding the phenomenon of thermodynamic irreversibility.) There are N particles that can be either in compartment A or in compartment B. Suppose that at time $n \geq 0$, $X_n = i$ particles are in A. One then chooses a particle at random independently of whatever happened before, and this particle is moved at time $n+1$ from where it is to the other compartment. We prove that $\{X_n\}_{n \geq 0}$ is an HMC and compute its transition matrix. If $X_n = i$, the next state X_{n+1} is either $i-1$ (the displaced particle was found in compartment A) with probability $\frac{i}{N}$, or $i+1$ (it was found in B) with probability $\frac{N-i}{N}$. This model pertains to Theorem 6.1.8. For all $n \geq 0$,

$$X_{n+1} = X_n + Z_{n+1},$$

where $Z_n \in \{-1, +1\}$ and $P(Z_{n+1} = -1 \mid X_n = i) = \frac{i}{N}$. The non-null entries of the transition matrix are therefore

$$p_{i,i+1} = \frac{N-i}{N}, \quad p_{i,i-1} = \frac{i}{N}.$$

First-step Analysis

Some functionals of homogeneous Markov chains such as probabilities of absorption by a closed set (A is called closed if $\sum_{j \in A} p_{ij} = 1$ for all $i \in A$) and average times before absorption can be evaluated by a technique called first-step analysis.

EXAMPLE 6.1.10: THE GAMBLER'S RUIN, TAKE 1. Two players A and B play "heads or tails", heads occurring with probability $p \in (0, 1)$ and the successive outcomes forming an independent sequence. Calling X_n the fortune in dollars of player A at time n, then $X_{n+1} = X_n + Z_{n+1}$, where $Z_{n+1} = +1$ (resp., -1) with probability p (resp., $q := 1 - p$) and $\{Z_n\}_{n \geq 1}$ is IID. In other words, A bets \$1 on heads at each toss, and B bets \$1 on tails. The respective initial fortunes of A and B are a and b (positive integers). The game ends when a player is ruined and therefore the process $\{X_n\}_{n \geq 1}$ is a random walk on $\mathbb{Z}$ as described in Example 6.1.6 and restricted to $E = \{0, \ldots, a, a+1, \ldots, a+b = c\}$. The duration of the game is T, that is, the first time n at which $X_n = 0$ or c, and the probability of winning for A is $u(a) = P(X_T = c \mid X_0 = a)$.

Instead of computing $u(a)$ alone, first-step analysis computes

$$u(i) = P(X_T = c \mid X_0 = i)$$

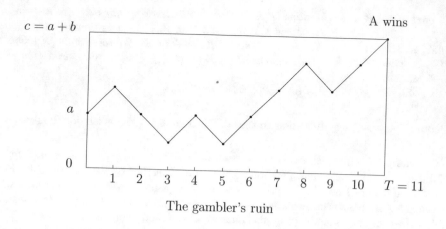

The gambler's ruin

for all states i $(0 \leq i \leq c)$ and for this, it first generates a recurrence equation for $u(i)$ by breaking down event "A wins" according to what can happen after the first step (the first toss) and using the rule of total causes. If $X_0 = i$, $1 \leq i \leq c - 1$, then $X_1 = i + 1$ (resp., $X_1 = i - 1$) with probability p (resp., q), and the probability of winning for A with updated initial fortune $i + 1$ (resp., $i - 1$) is $u(i + 1)$ (resp., $u(i - 1)$). Therefore, for i $(1 \leq i \leq c - 1)$

$$u(i) = pu(i + 1) + qu(i - 1),$$

with the boundary conditions $u(0) = 0$, $u(c) = 1$. The characteristic equation associated with this linear recurrence equation is $pr^2 - r + q = 0$. It has two distinct roots, $r_1 = 1$ and $r_2 = \frac{q}{p}$, if $p \neq \frac{1}{2}$, and a double root, $r_1 = 1$, if $p = \frac{1}{2}$. Therefore, the general solution is $u(i) = \lambda r_1^i + \mu r_2^i = \lambda + \mu \left(\frac{q}{p}\right)^i$ when $p \neq q$, and $u(i) = \lambda r_1^i + \mu i r_1^i = \lambda + \mu i$ when $p = q = \frac{1}{2}$. Taking into account the boundary conditions, one can determine the values of λ and μ. The result is, for $p \neq q$,

$$u(i) = \frac{1 - \left(\frac{q}{p}\right)^i}{1 - \left(\frac{q}{p}\right)^c},$$

and for $p = q = \frac{1}{2}$,

$$u(i) = \frac{i}{c}.$$

In the case $p = q = \frac{1}{2}$, the probability $v(i)$ that B wins when the initial fortune of B is $c - i$ is obtained by replacing i by $c - i$ in the expression for $u(i)$: $v(i) = \frac{c-i}{c} = 1 - \frac{i}{c}$. One checks that $u(i) + v(i) = 1$, which means in particular that the probability that the game lasts forever is null. The reader is invited to check that the same is true in the case $p \neq q$.

First-step analysis can also be used to compute average times before absorption (Exercise 6.6.21).

6.1.2   The Markov Property on a Graph

It can be shown by elementary computations that for an HMC $\{X_n\}_{n \geq 0}$ with countable state space E, X_n is independent of X_k ($k \leq n-2$) and X_m ($m \geq n+2$) given X_{n-1} and X_{n+1} ([1]). This property will be generalized to a large class of stochastic processes indexed by a finite set where the notion of time is absent, the Markov random fields.

The introduction of such processes at this point is motivated, besides the fact that their defining property extends the Markov property to a non-oriented time index, by the fact that they often occur as states of an ordinary HMC, only with a special state space (see the subsection on local characteristics below).

Let $G = (V, \mathcal{E})$ be a finite graph, and let $v_1 \sim v_2$ denote the fact that $\langle v_1, v_2 \rangle$ is an edge of the graph. Such vertices are also called neighbors (one of the other). We shall also refer to vertices of V as *sites*. The boundary with respect to $\sim$ of a set $A \subset V$ is the set

$$\partial A := \{v \in V \backslash A \,;\, v \sim w \text{ for some } w \in A\}\,.$$

Let Λ be a finite set, the *phase space*. A *random field* on V with phases in Λ is a collection $X = \{X(v)\}_{v \in V}$ of random variables with values in Λ. A random field can be regarded as a random variable taking its values in the *configuration space* $E := \Lambda^V$. A configuration $x \in \Lambda^V$ is of the form $x = (x(v), v \in V)$, where $x(v) \in \Lambda$ for all $v \in V$. For a given configuration x and a given subset $A \subseteq V$, let

$$x(A) := (x(v), v \in A)$$

be the restriction of x to A. If $V \backslash A$ denotes the complement of A in V, one writes $x = (x(A), x(V \backslash A))$. In particular, for fixed $v \in V$, $x = (x(v), x(V \backslash v))$, where $V \backslash v$ is an abbreviation for $V \backslash \{v\}$.

Local characteristics

Of special interest are the random fields characterized by *local interactions*. This leads to the notion of a Markov random field. The "locality" is in terms of the neighborhood structure inherited from the graph structure. More precisely, for any $v \in V$, $\mathcal{N}_v := \{w \in V \,;\, w \sim v\}$ is the neighborhood of v. In the following, $\tilde{\mathcal{N}}_v$ denotes the set $\mathcal{N}_v \cup \{v\}$.

Definition 6.1.11 *The random field X is called a Markov random field (MRF) with respect to $\sim$ if for all sites $v \in V$, the random elements $X(v)$ and $X(V \backslash \tilde{\mathcal{N}}_v)$ are independent given $X(\mathcal{N}_v)$.*

In mathematical symbols:

$$P(X(v) = x(v) \mid X(V \backslash v) = x(V \backslash v)) = P(X(v) = x(v) \mid X(\mathcal{N}_v) = x(\mathcal{N}_v)) \qquad (6.7)$$

for all $x \in \Lambda^V$ and all $v \in V$. Property (6.7) is clearly of the Markov type: the distribution of the phase at a given site is directly influenced only by the phases of the neighboring sites.

Remark 6.1.12 Note that any random field is Markovian with respect to the trivial topology, where the neighborhood of any site v is $V \backslash v$. However, the interesting Markov fields (from the point of view of modeling, simulation, and optimization) are those with relatively small neighborhoods.

---

[1] See Example 6.1.22 below for an educated proof.

Definition 6.1.13 *The* local characteristic *of the* MRF *at site* v *is the function* $\pi^v : \Lambda^V \to [0,1]$ *defined by*

$$\pi^v(x) := P(X(v) = x(v) \mid X(\mathcal{N}_v) = x(\mathcal{N}_v)).$$

The family $\{\pi^v\}_{v \in V}$ *is called the* local specification *of the* MRF.

One sometimes writes

$$\pi^v(x) := \pi(x(v) \mid x(\mathcal{N}_v))$$

in order to stress the role of the neighborhoods.

Theorem 6.1.14 *Two positive distributions of a random field with a* finite *configuration space* Λ^V *that have the same local specification are identical.*

Proof. Enumerate V as $\{1, 2, \ldots, K\}$. Therefore a configuration $x \in \Lambda^V$ is represented as $x = (x_1, \ldots, x_{K-1}, x_K)$ where $x_i \in \Lambda$, $1 \le i \le K$. The following identity

$$\pi(z_1, z_2, \ldots, z_K) = \prod_{i=1}^{K} \frac{\pi(z_i \mid z_1, \ldots, z_{i-1}, y_{i+1}, \ldots, y_K)}{\pi(y_i \mid z_1, \ldots, z_{i-1}, y_{i+1}, \ldots, y_K)} \pi(y_1, y_2, \ldots, y_K) \qquad (\star)$$

holds for any $z, y \in \Lambda^K$. For the proof, write

$$\pi(z) = \prod_{i=1}^{K} \frac{\pi(z_1, \ldots, z_{i-1}, z_i, y_{i+1}, \ldots, y_K)}{\pi(z_1, \ldots, z_{i-1}, y_i, y_{i+1}, \ldots, y_K)} \pi(y)$$

and use Bayes' rule to obtain for each i, $1 \le i \le K$:

$$\frac{\pi(z_1, \ldots, z_{i-1}, z_i, y_{i+1}, \ldots, y_K)}{\pi(z_1, \ldots, z_{i-1}, y_i, y_{i+1}, \ldots, y_K)} = \frac{\pi(z_i \mid z_1, \ldots, z_{i-1}, y_{i+1}, \ldots, y_K)}{\pi(y_i \mid z_1, \ldots, z_{i-1}, y_{i+1}, \ldots, y_K)}.$$

Let now π and π' be two positive probability distributions on V with the same local specification. Choose any $y \in \Lambda^V$. Identity $(\star)$ shows that for all $z \in \Lambda^V$,

$$\frac{\pi'(z)}{\pi(z)} = \frac{\pi'(y)}{\pi(y)}.$$

Therefore $\frac{\pi'(z)}{\pi(z)}$ is a constant, necessarily equal to 1 since π and π' are probability distributions. $\square$

Gibbs Distributions

Consider the probability distribution

$$\pi_T(x) = \frac{1}{Z_T} e^{-\frac{1}{T} U(x)} \tag{6.8}$$

on the configuration space Λ^V, where $T > 0$ is the *temperature*, $U(x)$ is the *energy* of configuration x, and Z_T is the normalizing constant, called the *partition function*. Since $\pi_T(x)$ takes its values in $[0,1]$, necessarily $-\infty < U(x) \le +\infty$. Note that $U(x) < +\infty$ if and only if $\pi_T(x) > 0$. One of the challenges associated with Gibbs models is obtaining explicit formulas for averages, considering that it is generally hard to compute the partition function. This is feasible in exceptional cases (see Exercise 6.6.3).

Such distributions are of interest to physicists when the energy is expressed in terms of a potential function describing the local interactions. The notion of clique then plays a central role.

Definition 6.1.15 *Any singleton $\{v\} \subset V$ is a clique. A subset $C \subseteq V$ with more than one element is called a* clique *(with respect to $\sim$) if and only if any two distinct sites of C are mutual neighbors. A clique C is called* maximal *if for any site $v \notin C$, $C \cup \{v\}$ is not a clique.*

The collection of cliques will be denoted by $\mathcal{C}$.

Definition 6.1.16 *A* Gibbs potential *on Λ^V relative to $\sim$ is a collection $\{V_C\}_{C \subseteq V}$ of functions $V_C : \Lambda^V \to \mathbb{R} \cup \{+\infty\}$ such that*

(i) $V_C \equiv 0$ if C is not a clique, and

(ii) for all $x, x' \in \Lambda^V$ and all $C \subseteq V$,

$$x(C) = x'(C) \Rightarrow V_C(x) = V_C(x').$$

The energy function U is said to derive *from the potential $\{V_C\}_{C \subseteq V}$ if*

$$U(x) = \sum_C V_C(x).$$

The function V_C depends only on the phases at the sites inside subset C. One could write more explicitly $V_C(x(C))$ instead of $V_C(x)$, but this notation will not be used.

In this context, the distribution in (6.8) is called a *Gibbs distribution* (w.r.t. $\sim$).

EXAMPLE 6.1.17: THE ISING MODEL, TAKE 1. In statistical physics, the following model is regarded as a qualitatively correct idealization of a piece of ferromagnetic material. Here $V = \mathbb{Z}_m^2 = \{(i,j) \in \mathbb{Z}^2, i, j \in [1,m]\}$ and $\Lambda = \{+1,-1\}$, where ± 1 is the orientation of the magnetic spin at a given site. The figure below depicts two particular neighborhood systems, their respective cliques, and the boundary of a 2×2 square for both cases. The neighborhood system in the original Ising model is as in column (α) of the figure below, and the Gibbs potential is

$$V_{\{v\}}(x) = -\frac{H}{k}x(v),$$

$$V_{\langle v,w \rangle}(x) = -\frac{J}{k}x(v)x(w),$$

where $\langle v, w \rangle$ is the 2-element clique ($v \sim w$). For physicists, k is the Boltzmann constant, H is the external magnetic field, and J is the internal energy of an elementary magnetic dipole. The energy function corresponding to this potential is therefore

$$U(x) = -\frac{J}{k}\sum_{\langle v,w \rangle} x(v)x(w) - \frac{H}{k}\sum_{v \in V} x(v).$$

EXAMPLE 6.1.18: THE AUTOBINOMIAL MODEL[2] For the purpose of image synthesis, one seeks Gibbs distributions describing pictures featuring various textures, lines separating patches with different textures (boundaries), lines per se (roads, rail tracks), randomly located objects (moon craters), etc. The following is an all-purpose texture model that

---

[2][Besag, 1974].

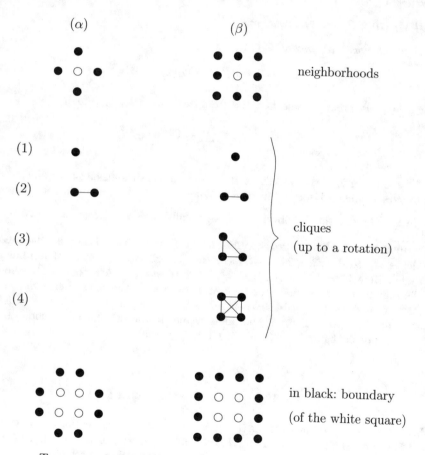

Two examples of neighborhoods, cliques, and boundaries

may be used to describe the texture of various materials. The set of sites is $V = \mathbb{Z}_m^2$, and the phase space is $\Lambda = \{0, 1, \ldots, L\}$. In the context of image processing, a site v is a pixel (PICTure ELement), and a phase $\lambda \in \Lambda$ is a shade of grey, or a color. The neighborhood system is

$$\mathcal{N}_v = \{w \in V; w \neq v; \|w - v\|^2 \leq d\}, \tag{6.9}$$

where d is a fixed positive integer and where $\|w - v\|$ is the euclidean distance between v and w. In this model the only cliques participating in the energy function are singletons and pairs of mutual neighbors. The set of cliques appearing in the energy function is a disjoint sum of collections of cliques

$$\mathcal{C} = \sum_{j=1}^{m(d)} \mathcal{C}_j,$$

where $\mathcal{C}_1$ is the collection of singletons, and all pairs $\{v, w\}$ in $\mathcal{C}_j$, $2 \leq j \leq m(d)$, have the same distance $\|w - v\|$ and the same direction, as shown in the figure below. The potential is given by

$$V_C(x) = \begin{cases} -\log \binom{L}{x(v)} + \alpha_1 x(v) & \text{if } C = \{v\} \in \mathcal{C}_1, \\ \alpha_j x(v) x(w) & \text{if } C = \{v, w\} \in \mathcal{C}_j, \end{cases}$$

where $\alpha_j \in \mathbb{R}$. For any clique C not of type $\mathcal{C}_j$, $V_C \equiv 0$.

The terminology ("autobinomial") is motivated by the fact that the local system has the form

$$\pi^v(x) = \binom{L}{x(v)} \tau^{x(v)} (1 - \tau)^{L - x(v)}, \tag{6.10}$$

where τ is a parameter depending on $x(\mathcal{N}_v)$ as follows:

$$\tau = \tau(\mathcal{N}_v) = \frac{e^{-\langle \alpha, b \rangle}}{1 + e^{-\langle \alpha, b \rangle}}. \tag{6.11}$$

Here $\langle \alpha, b \rangle$ is the scalar product of

$$\alpha = (\alpha_1, \ldots, \alpha_{m(d)}) \text{ and } b = (b_1, \ldots, b_{m(d)}),$$

where $b_1 = 1$, and for all j, $2 \leq j \leq m(d)$,

$$b_j = b_j(x(\mathcal{N}_v)) = x(u) + x(w),$$

where $\{v, u\}$ and $\{v, w\}$ are the two pairs in $\mathcal{C}_j$ containing v.

Proof. From the explicit formula (6.12) giving the local characteristic at site v,

$$\pi^v(x) = \frac{\exp\left\{ \log\binom{L}{x(v)} - \alpha_1 x(v) - \left[\sum_{j=2}^{m(d)} \alpha_j \sum_{v; \{v,w\} \in \mathcal{C}_j} x(w) \right] x(v) \right\}}{\sum_{\lambda \in \Lambda} \exp\left\{ \log\binom{L}{\lambda} - \alpha_1 \lambda - \left[\sum_{j=2}^{m(d)} \alpha_j \sum_{t; \{v,w\} \in \mathcal{C}_j} x(w) \right] \lambda \right\}}.$$

The numerator equals

$$\binom{L}{x(v)} e^{-\langle \alpha, b \rangle x(v)},$$

and the denominator is

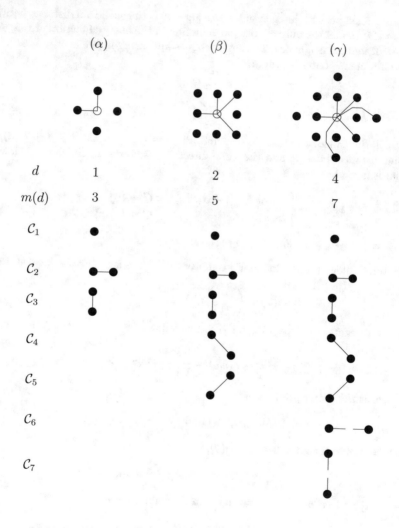

Neighborhoods and cliques of three autobinomial models

$$\sum_{\lambda \in \Lambda} \binom{L}{\lambda} e^{(-\alpha, b)\lambda} = \sum_{\ell=0}^{L} \binom{L}{\ell} \left(e^{-\langle \alpha, b \rangle}\right)^{\ell} = \left(1 + e^{-\langle \alpha, b \rangle}\right)^{L}.$$

$\square$

Equality (6.10) then follows.

Expression (6.10) shows that τ is the average level of grey at site v, given $x(\mathcal{N}_v)$, and expression (6.11) shows that τ is a function of $\langle \alpha, b \rangle$. The parameter α_j controls the bond in the direction and at the distance that characterize $\mathcal{C}_j$.

The Hammersley–Clifford Theorem

Gibbs distributions with an energy deriving from a Gibbs potential relative to a neighborhood system are distributions of Markov fields relative to the same neighborhood system.

Theorem 6.1.19 *If X is a random field with a distribution π of the form $\pi(x) = \frac{1}{Z} e^{-U(x)}$, where the energy function U derives from a Gibbs potential $\{V_C\}_{C \subseteq V}$ relative to $\sim$, then X is a Markov random field with respect to $\sim$. Moreover, its local specification is given by the formula*

$$\pi^v(x) = \frac{e^{-\sum_{C \ni v} V_C(x)}}{\sum_{\lambda \in \Lambda} e^{-\sum_{C \ni v} V_C(\lambda, x(V \setminus v))}}, \tag{6.12}$$

where the notation $\sum_{C \ni v}$ means that the sum extends over the sets C that contain the site v.

Proof. First observe that the right-hand side of (6.12) depends on x only through $x(v)$ and $x(\mathcal{N}_v)$. Indeed, $V_C(x)$ depends only on $(x(w), w \in C)$, and for a clique C, if $w \in C$ and $v \in C$, then either $w = v$ or $w \sim v$. Therefore, if it can be shown that $P(X(v) = x(v) | X(V \setminus v) = x(V \setminus v))$ equals the right-hand side of (6.12), then (see Exercise 6.6.6) the Markov property will be proved. By definition of conditional probability,

$$P(X(v) = x(v) \mid X(V \setminus v) = x(V \setminus v)) = \frac{\pi(x)}{\sum_{\lambda \in \Lambda} \pi(\lambda, x(V \setminus v))}. \tag{\dagger}$$

But

$$\pi(x) = \frac{1}{Z} e^{-\sum_{C \ni v} V_C(x) - \sum_{C \not\ni v} V_C(x)},$$

and similarly,

$$\pi(\lambda, x(V \setminus v)) = \frac{1}{Z} e^{-\sum_{C \ni v} V_C(\lambda, x(V \setminus v)) - \sum_{C \not\ni v} V_C(\lambda, x(V \setminus v))}.$$

If C is a clique and v is not in C, then $V_C(\lambda, x(V \setminus v)) = V_C(x)$ and is therefore independent of $\lambda \in \Lambda$. Therefore, after factoring out $\exp\left\{-\sum_{C \not\ni v} V_C(x)\right\}$, the right-hand side of ($\dagger$) is found to be equal to the right-hand side of (6.12).
$\square$

The *local energy* at site v of configuration x is

$$U_v(x) = \sum_{C \ni v} V_C(x).$$

With this notation, (6.12) becomes

$$\pi^v(x) = \frac{e^{-U_v(x)}}{\sum_{\lambda \in \Lambda} e^{-U_v(\lambda, x(V \setminus v))}}.$$

EXAMPLE 6.1.20:   THE ISING MODEL, TAKE 2. The local characteristics in the Ising model are

$$\pi_T^v(x) = \frac{e^{\frac{1}{kT}\{J\sum_{w;w \sim v} x(w) + H\}x(v)}}{e^{+\frac{1}{kT}\{J\sum_{w;w \sim v} x(w) + H\}} + e^{-\frac{1}{kT}\{J\sum_{w;w \sim v} x(w) + H\}}}.$$

Theorem 6.1.19 above is the direct part of the *Gibbs–Markov equivalence* theorem: A Gibbs distribution relative to a neighborhood system is the distribution of a Markov field with respect to the same neighborhood system. The converse part (Hammersley–Clifford theorem) is important from a theoretical point of view, since together with the direct part it concludes that Gibbs distributions and MRFs are essentially the same objects.

Theorem 6.1.21 *Let $\pi > 0$ be the distribution of a Markov random field with respect to $\sim$. Then*

$$\pi(x) = \frac{1}{Z} e^{-U(x)}$$

for some energy function U deriving from a Gibbs potential $\{V_C\}_{C \subseteq V}$ with respect to $\sim$.

The proof is omitted with little inconvenience since, in practice, the potential as well as the topology of V can be obtained directly from the expression of the energy, as the following example shows.

EXAMPLE 6.1.22:   MARKOV CHAINS AS MARKOV FIELDS. Let $V = \{0, 1, \ldots N\}$ and $\Lambda = E$, a finite space. A random field X on V with phase space Λ is therefore a vector X with values in E^{N+1}. Suppose that $X_0, \ldots, X_N$ is a homogeneous Markov chain with transition matrix $\mathbf{P} = \{p_{ij}\}_{i,j \in E}$ and initial distribution $\nu = \{\nu_i\}_{i \in E}$. In particular, with $x = (x_0, \ldots, x_N)$,

$$\pi(x) = \nu_{x_0} p_{x_0 x_1} \cdots p_{x_{N-1} x_N},$$

that is,

$$\pi(x) = e^{-U(x)},$$

where

$$U(x) = -\log \nu_{x_0} - \sum_{n=0}^{N-1} (\log p_{x_n x_{n+1}}).$$

Clearly, this energy derives from a Gibbs potential associated with the nearest-neighbor topology for which the cliques are, besides the singletons, the pairs of adjacent sites. The potential functions are:

$$V_{\{0\}}(x) = -\log \nu_{x_0}, \quad V_{\{n,n+1\}}(x) = -\log p_{x_n x_{n+1}}.$$

The local characteristic at site n, $2 \leq n \leq N - 1$, can be computed from formula (6.12), which gives

$$\pi^n(x) = \frac{\exp(\log p_{x_{n-1} x_n} + \log p_{x_n x_{n+1}})}{\sum_{y \in E} \exp(\log p_{x_{n-1} y} + \log p_{y x_{n+1}})},$$

that is,

$$\pi^n(x) = \frac{p_{x_{n-1}x_n}p_{x_nx_{n+1}}}{p^{(2)}_{x_{n-1}x_{n+1}}},$$

where $p^{(2)}_{ij}$ is the general term of the two-step transition matrix $\mathbf{P}^2$. Similar computations give $\pi^0(x)$ and $\pi^N(x)$. We note that, in view of the neighborhood structure, for $2 \leq n \leq N-1$, X_n is independent of $X_0, \ldots, X_{n-2}, X_{n+2}, \ldots, X_N$ given X_{n-1} and X_{n+1}.

6.2 The Transition Matrix

6.2.1 Topological Notions

The notions introduced in this subsection (communication and periodicity) are of a *topological* nature, in the sense that they concern only the *naked* transition graph (without the labels).

Communication Classes

Definition 6.2.1 *State j is said to be accessible from state i if there exists an integer $M \geq 0$ such that $p_{ij}(M) > 0$. In particular, a state i is always accessible from itself, since $p_{ii}(0) = 1$. States i and j are said to communicate if i is accessible from j and j is accessible from i, and this is denoted by $i \leftrightarrow j$.*

For $M \geq 1$, $p_{ij}(M) = \sum_{i_1,\ldots,i_{M-1}} p_{ii_1} \cdots p_{i_{M-1}j}$, and therefore $p_{ij}(M) > 0$ if and only if there exists at least one path $i, i_1, \ldots, i_{M-1}, j$ from i to j such that

$$p_{ii_1}p_{i_1i_2} \cdots p_{i_{M-1}j} > 0,$$

or, equivalently, if there is an oriented path from i to j in the transition graph G. Clearly,

$$i \leftrightarrow i \qquad \text{(reflexivity)},$$
$$i \leftrightarrow j \Rightarrow j \leftrightarrow i \qquad \text{(symmetry)},$$
$$i \leftrightarrow j, j \leftrightarrow k \Rightarrow i \leftrightarrow k \qquad \text{(transivity)}.$$

Therefore, the communication relation ($\leftrightarrow$) is an equivalence relation, and it generates a partition of the state space E into disjoint equivalence classes called *communication classes*.

Definition 6.2.2 *When there exists only one communication class, the chain, its transition matrix and its transition graph are said to be* irreducible.

EXAMPLE 6.2.3: REPAIR SHOP, TAKE 2. Example 6.1.7 continued. A necessary and sufficient condition of irreducibility of the repair shop chain of Example 6.1.7 is that $P(Z_1 = 0) > 0$ *and* $P(Z_1 \geq 2) > 0$ (Exercise 6.6.7).

Period

Consider the random walk on $\mathbb{Z}$ (Example 6.1.6). Since $p \in (0,1)$, it is irreducible. Observe that $E = C_0 + C_1$, where C_0 and C_1, the set of even and odd relative integers respectively, have the following property. If you start from $i \in C_0$ (resp., C_1), then in one step you can go only to a state $j \in C_1$ (resp., C_0). The chain $\{X_n\}$ passes alternately from one cyclic class to the other. In this sense, the chain has a periodic behavior, corresponding to the period 2. More generally, for any *irreducible* Markov chain, one can find a *partition* of E into d classes $C_0, C_1, \ldots, C_{d-1}$ such that for all $k, i \in C_k$,

$$\sum_{j \in C_{k+1}} p_{ij} = 1,$$

where by convention $C_d = C_0$. The proof follows directly from Theorem 6.2.7 below.

The number $d \geq 1$ is called the *period* of the chain (resp., of the transition matrix, of the transition graph). The classes $C_0, C_1, \ldots, C_{d-1}$ are called the *cyclic classes*.

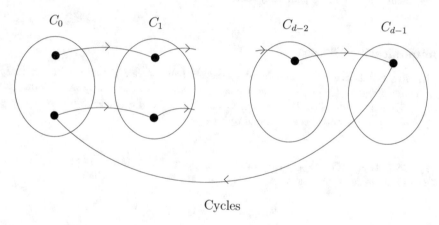

Cycles

The chain therefore moves from one class to the other at each transition, and this cyclically, as shown in the figure.

We shall proceed to substantiate the above description of periodicity starting with the formal definition of period based on the notion of *greatest common divisor* (GCD) of a set of integers.

Definition 6.2.4 *The* period d_i *of state* $i \in E$ *is, by definition,*

$$d_i = \text{GCD}\{n \geq 1 \; ; \; p_{ii}(n) > 0\},$$

with the convention $d_i = +\infty$ *if there is no* $n \geq 1$ *with* $p_{ii}(n) > 0$. *If* $d_i = 1$, *the state* i *is called* aperiodic.

Remark 6.2.5 Very often aperiodicity follows from the following simple observation: An irreducible transition matrix $\mathbf{P}$ with at least one state $i \in E$ such that $p_{ii} > 0$ is aperiodic (in fact, in this case $1 \in \{n \geq 1 \; ; \; p_{ii}(n) > 0\}$ and therefore $d_i = 1$).

Period is a (communication) class property in the following sense:

Theorem 6.2.6 *Two states i and j which communicate have the same period.*

Proof. As i and j communicate, there exist integers N and M such that $p_{ij}(M) > 0$ and $p_{ji}(N) > 0$. For any $k \geq 1$,

$$p_{ii}(M + nk + N) \geq p_{ij}(M)(p_{jj}(k))^n p_{ji}(N)$$

(indeed, the trajectories $X_0, \ldots, X_{M+nk+N}$ such that $X_0 = i, X_M = j, X_{M+k} = j, \ldots,$ $X_{M+nk} = j, X_{M+nk+N} = i$ are a subset of the trajectories starting from i and returning to i in $M + nk + N$ steps). Therefore, for any $k \geq 1$ such that $p_{jj}(k) > 0$, we have that $p_{ii}(M + nk + N) > 0$ for all $n \geq 1$. Therefore, d_i divides $M + nk + N$ for all $n \geq 1$, and in particular, d_i divides k. We have therefore shown that d_i divides all k such that $p_{jj}(k) > 0$, and in particular, d_i divides d_j. By symmetry, d_j divides d_i, and therefore, finally, $d_i = d_j$. □

We may therefore henceforth speak of the period of a communication class or of an irreducible chain.

The important result concerning periodicity is the following.

Theorem 6.2.7 *Let $\mathbf{P}$ be an irreducible stochastic matrix with period d. Then for all states i, j there exist $m \geq 0$ and $n_0 \geq 0$ (m and n_0 possibly depending on i, j) such that*

$$p_{ij}(m + nd) > 0, \quad \text{for all } n \geq n_0.$$

Proof. It suffices to prove the theorem for $i = j$. Indeed, there exists an m such that $p_{ij}(m) > 0$, because j is accessible from i, the chain being irreducible, and therefore, if for some $n_0 \geq 0$ we have $p_{jj}(nd) > 0$ for all $n \geq n_0$, then $p_{ij}(m+nd) \geq p_{ij}(m)p_{jj}(nd) > 0$ for all $n \geq n_0$.

The rest of the proof is an immediate consequence of a classical result of number theory. Indeed, the GCD of the set $A = \{k \geq 1; p_{jj}(k) > 0\}$ is d, and A is closed under addition. The set A therefore contains all but a finite number of the positive multiples of d. In other words, there exists an n_0 such that $n > n_0$ implies $p_{jj}(nd) > 0$. □

6.2.2 Stationary Distributions and Reversibility

We now introduce the central notion of the stability theory of discrete-time HMCs.

Definition 6.2.8 *A probability distribution π satisfying*

$$\pi^T = \pi^T \mathbf{P} \tag{6.13}$$

is called a stationary distribution *(of the transition matrix $\mathbf{P}$ or of the corresponding HMC).*

The so-called *global balance equation* (6.13) says that

$$\pi(i) = \sum_{j \in E} \pi(j) p_{ji} \qquad (i \in E).$$

Iteration of (6.13) gives $\pi^T = \pi^T \mathbf{P}^n$ for all $n \geq 0$, and therefore, in view of (6.2), if the initial distribution $\nu = \pi$, then $\nu_n = \pi$ for all $n \geq 0$. In particular, a chain starting with a stationary distribution keeps the same distribution forever. But there is more, because then,

$$P(X_n = i_0, X_{n+1} = i_1, \ldots, X_{n+k} = i_k) = P(X_n = i_0)p_{i_0 i_1} \ldots p_{i_{k-1} i_k}$$
$$= \pi(i_0)p_{i_0 i_1} \ldots p_{i_{k-1} i_k}$$

does not depend on n. In this sense the chain is *stationary*. One also says that the chain is in a *stationary regime*, or in *equilibrium*, or in *steady state*. In summary:

Theorem 6.2.9 *An* HMC *whose initial distribution is a stationary distribution is stationary.*

Remark 6.2.10 The balance equation $\pi^T\mathbf{P} = \pi^T$, together with the requirement that π is a probability vector, that is, $\pi^T\mathbf{1} = 1$ (where $\mathbf{1}$ is a column vector with all its entries equal to 1), constitute, when E is finite, $|E| + 1$ equations for $|E|$ unknown variables. One of the $|E|$ equations in $\pi^T\mathbf{P} = \pi^T$ is superfluous given the constraint $\pi^T\mathbf{1} = 1$. In fact, summation of all equalities of $\pi^T\mathbf{P} = \pi^T$ yields the equality $\pi^T\mathbf{P}\mathbf{1} = \pi^T\mathbf{1}$, that is, $\pi^T\mathbf{1} = 1$.

EXAMPLE 6.2.11: A TWO-STATE MARKOV CHAIN. The state space $E = \{1, 2\}$ and the transition matrix is

$$\mathbf{P} = \begin{pmatrix} 1 - \alpha & \alpha \\ \beta & 1 - \beta \end{pmatrix},$$

where $\alpha, \beta \in (0, 1)$. The global balance equations are

$$\pi(1) = \pi(1)(1 - \alpha) + \pi(2)\beta, \qquad \pi(2) = \pi(1)\alpha + \pi(2)(1 - \beta).$$

This is a dependent system which reduces to the single equation $\pi(1)\alpha = \pi(2)\beta$, to which must be added the equality $\pi(1) + \pi(2) = 1$ expressing that π is a probability vector. We obtain

$$\pi(1) = \frac{\beta}{\alpha + \beta}, \quad \pi(2) = \frac{\alpha}{\alpha + \beta}.$$

EXAMPLE 6.2.12: THE EHRENFEST DIFFUSION MODEL, TAKE 2. The corresponding HMC was described in Example 6.1.9. The global balance equations are, for $i \in [1, N-1]$,

$$\pi(i) = \pi(i - 1)\left(1 - \frac{i - 1}{N}\right) + \pi(i + 1)\frac{i + 1}{N}$$

and, for the boundary states,

$$\pi(0) = \pi(1)\frac{1}{N}, \quad \pi(N) = \pi(N - 1)\frac{1}{N}.$$

Leaving $\pi(0)$ undetermined, one can solve the balance equations for $i = 0, 1, \ldots, N$ successively, to obtain

$$\pi(i) = \pi(0)\binom{N}{i}.$$

The value of $\pi(0)$ is then determined by writing that π is a probability vector:

$$1 = \sum_{i=0}^{N} \pi(i) = \pi(0) \sum_{i=0}^{N} \binom{N}{i} = \pi(0)2^N.$$

This gives for π the binomial distribution of size N and parameter $\frac{1}{2}$:

$$\pi(i) = \frac{1}{2^N} \binom{N}{i}.$$

This is the distribution one would obtain by placing independently each particle in the compartments, with probability $\frac{1}{2}$ for each compartment.

There may be many stationary distributions. Take the identity as transition matrix. Then any probability distribution on the state space is a stationary distribution. Also there may well not exist any stationary distribution. (See Exercise 6.6.16.)

Remark 6.2.13 An immediate consequence of Theorem 5.1.14 is that if an HMC $\{X_n\}_{n\geq 0}$ is stationary, it may be extended to a stationary HMC $\{X_n\}_{n\in\mathbb{Z}}$ with the same distribution.

Recurrence equations can be used to obtain the stationary distribution when the latter exists and is unique. Generating functions sometimes usefully exploit the dynamics.

EXAMPLE 6.2.14: REPAIR SHOP, TAKE 3. Examples 6.1.7 and 6.2.3 continued. For any complex number z with modulus not larger than 1, it follows from the recurrence equation (6.6) that

$$z^{X_{n+1}+1} = \left(z^{(X_n-1)^+ +1}\right)z^{Z_{n+1}} = \left(z^{X_n}1_{\{X_n>0\}} + z1_{\{X_n=0\}}\right)z^{Z_{n+1}}$$
$$= \left(z^{X_n} - 1_{\{X_n=0\}} + z1_{\{X_n=0\}}\right)z^{Z_{n+1}},$$

and therefore

$$zz^{X_{n+1}} - z^{X_n}z^{Z_{n+1}} = (z-1)1_{\{X_n=0\}}z^{Z_{n+1}}.$$

From the independence of X_n and Z_{n+1}, $E[z^{X_n}z^{Z_{n+1}}] = E[z^{X_n}]g_Z(z)$, where g_Z is the generating function of Z_{n+1}, and $E[1_{\{X_n=0\}}z^{Z_{n+1}}] = \pi(0)g_Z(z)$, where $\pi(0) = P(X_n = 0)$. Therefore,

$$zE[z^{X_{n+1}}] - g_Z(z)E[z^{X_n}] = (z-1)\pi(0)g_Z(z).$$

Suppose that the chain is in steady state, in which case $E[z^{X_{n+1}}] = E[z^{X_n}] = g_X(z)$, and therefore

$$g_X(z)(z - g_Z(z)) = \pi(0)(z-1)g_Z(z). \qquad (\star)$$

This gives the generating function $g_X(z) = \sum_{i=0}^{\infty} \pi(i)z^i$, as long as $\pi(0)$ is available. To obtain $\pi(0)$, differentiate $(\star)$:

$$g'_X(z)(z - g_Z(z)) + g_X(z)(1 - g'_Z(z)) = \pi(0)\left(g_Z(z) + (z-1)g'_Z(z)\right),$$

and let $z = 1$, to obtain, taking into account the equalities $g_X(1) = g_Z(1) = 1$ and $g'_Z(1) = E[Z]$,

$$\pi(0) = 1 - E[Z]. \qquad (\star\star)$$

Since $\pi(0)$ must be non-negative, this immediately gives the necessary condition $E[Z] \leq 1$. Actually, one must have, if the trivial case $Z_1 \equiv 1$ is excluded,

$$E[Z] < 1. \tag{6.14}$$

Indeed, if $E[Z] = 1$, implying $\pi(0) = 0$, it follows from $(\star)$ that $g_X(x)(x - g_Z(x)) = 0$ for all $x \in [0, 1]$. But, if the case $Z_1 \equiv 1$ (that is, $g_Z(x) \equiv x$) is excluded, equation $x - g_Z(x) = 0$ has only $x = 1$ for a solution when $g'_Z(1) = E[Z] \le 1$. Therefore, $g_X(x) \equiv 0$ for $x \in [0, 1)$, and consequently $g_X(z) \equiv 0$ on $\{|z| < 1\}$ (g_Z is analytic inside the open unit disk centered at the origin). This leads to a contradiction, since the generating function of an integer-valued random variable cannot be identically null.

It turns out that $E[Z] < 1$ is also a sufficient condition for the existence of a steady state (Example 6.3.22). For the time being, we have from $(\star)$ and $(\star\star)$ that, if the stationary distribution exists, then its generating function is given by the formula

$$\sum_{i=0}^{\infty} \pi(i)z^i = (1 - E[Z])\frac{(z-1)g_Z(z)}{z - g_Z(z)}.$$

Reversibility

Let $\{X_n\}_{n \in \mathbb{Z}}$ be an HMC with transition matrix $\mathbf{P}$ and admitting a stationary distribution $\pi > 0$ (see Remark 6.2.13). Define the matrix $\mathbf{Q}$, indexed by E, by

$$\pi(i)q_{ij} = \pi(j)p_{ji}. \tag{$\star$}$$

This matrix is stochastic, since

$$\sum_{j \in E} q_{ij} = \sum_{j \in E} \frac{\pi(j)}{\pi(i)}p_{ji} = \frac{1}{\pi(i)}\sum_{j \in E}\pi(j)p_{ji} = \frac{\pi(i)}{\pi(i)} = 1,$$

where the third equality uses the global balance equations. From Bayes' retrodiction formula,

$$P(X_n = j \mid X_{n+1} = i) = \frac{P(X_{n+1} = i \mid X_n = j)P(X_n = j)}{P(X_{n+1} = i)},$$

that is, in view of $(\star)$,

$$P(X_n = j \mid X_{n+1} = i) = q_{ji}. \tag{6.15}$$

Therefore $\mathbf{Q}$ is the transition matrix of the initial chain when time is reversed.

Theorem 6.2.15 *Let $\mathbf{P}$ be a stochastic matrix indexed by a countable set E, and let π be a probability distribution on E. Define the matrix $\mathbf{Q}$ indexed by E by $\pi(i)q_{ij} = \pi(j)p_{ji}$. If $\mathbf{Q}$ is a stochastic matrix, then π is a stationary distribution of $\mathbf{P}$.*

Proof. Just verify that the global balance equation is satisfied.   □

Definition 6.2.16 *One calls* reversible *a stationary Markov chain with initial distribution π (a stationary distribution) if for all $i, j \in E$, we have the so-called* detailed balance equations

$$\pi(i)p_{ij} = \pi(j)p_{ji}.$$

We then say that the pair $(\mathbf{P}, \pi)$ is reversible. In this case, $q_{ij} = p_{ij}$, and therefore the chain and the time-reversed chain are statistically the same, since the distribution of a homogeneous Markov chain is entirely determined by its initial distribution and its transition matrix. The following is an immediate corollary of Theorem 6.2.15.

Theorem 6.2.17 *Let* $\mathbf{P}$ *be a transition matrix on the countable state space* E, *and let* π *be some probability distribution on* E. *If for all* $i, j \in E$, *the detailed balance equations are satisfied, then* π *is a stationary distribution of* $\mathbf{P}$.

EXAMPLE 6.2.18: THE EHRENFEST DIFFUSION MODEL, TAKE 3. This example continues Examples 6.1.9 and 6.2.12. Recall that we obtained the expression

$$\pi(i) = \frac{1}{2^N}\binom{N}{i}$$

for the stationary distribution. We can also find this by checking the detailed balance equations

$$\pi(i)p_{i,i+1} = \pi(i+1)p_{i+1,i}.$$

EXAMPLE 6.2.19: RANDOM WALK ON A GRAPH. Consider a finite non-oriented graph and denote by E the set of vertices, or nodes, of this graph. Let d_i be the *index* of vertex i (the number of edges adjacent to i). Transform this graph into an oriented graph by splitting each edge into two oriented edges of opposite directions, and make it a transition graph by associating to the oriented edge from i to j the transition probability $\frac{1}{d_i}$ (see the figure below). It will be assumed, as is the case in the figure, that $d_i > 0$ for all states i (that is, the graph is *connected*).

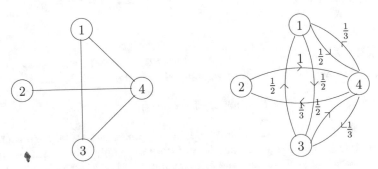

A random walk on a graph

This chain is irreducible due to the connectedness of the graph and the fact that $p_{ij} > 0$ whenever $p_{ji} > 0$. It admits the distribution

$$\pi(i) = \frac{d_i}{2|\text{edges}|} \qquad (i \in E),$$

where $|\text{edges}|$ is the number of edges of the original graph. This follows from Theorem 6.2.17. In fact, if i and j are connected in the graph, $p_{ij} = \frac{1}{d_i}$ and $p_{ji} = \frac{1}{d_j}$, and therefore the detailed balance equation between these two states is

$$\pi(i)\frac{1}{d_i} = \pi(j)\frac{1}{d_j}.$$

This gives

$$\pi(i) = K d_i \quad (i \in E),$$

where K is obtained by normalization: $K = \left(\sum_{j\in E} d_j\right)^{-1}$. But $\sum_{j\in E} d_j = 2|\text{edges}|$.

6.2.3   The Strong Markov Property

The Markov property relative to the past and future at a given instant n can be extended to the situation where this deterministic time is replaced by an $\mathcal{F}_n^X$-stopping time (where $\mathcal{F}_n^X := \sigma(X_0, \cdots, X_n)$), whose definition we recall below.

Definition 6.2.20 *Let $\{\mathcal{F}_n\}_{n\in\mathbb{N}}$ be a non-decreasing sequence of sub-σ-fields of $\mathcal{F}$. A random variable τ taking its values in $\overline{\mathbb{N}}$ and such that, for all $m \in \mathbb{N}$, the event $\{\tau = m\}$ is in $\mathcal{F}_m$ is called an $\mathcal{F}_n$-stopping time.*

In other words, τ is an $\mathcal{F}_n^X$-stopping time if, for all $m \in \mathbb{N}$, the event $\{\tau = m\}$ can be expressed as

$$1_{\{\tau=m\}} = \psi_m(X_0, \ldots, X_m),$$

for some measurable function ψ_m with values in $\{0, 1\}$ (Theorem 3.3.18).

EXAMPLE 6.2.21:   FIXED TIMES AND DELAYED STOPPING TIMES. A constant time is a stopping time. If τ is a stopping time and n_0 a non-negative deterministic time, then $\tau + n_0$ is a stopping time. Indeed, $\{\tau + n_0 = m\} \equiv \{\tau = m - n_0\}$ is expressible in terms of $X_0, X_1, \ldots, X_{m-n_0}$.

EXAMPLE 6.2.22:   RETURN TIMES AND HITTING TIMES. In the theory of Markov chains, a typical and most important stopping time is the *return time* to state $i \in E$,

$$T_i = \inf\{n \geq 1;\ X_n = i\},$$

where $T_i = \infty$ if $X_n \neq i$ for all $n \geq 1$. It is a stopping time, as we shall soon prove. Observe that $T_i \geq 1$, and in particular, $X_0 = i$ does *not* imply $T_i = 0$. This is why T_i is called the *return* time to i, and not the *hitting* time of i. The latter is $S_i = T_i$ if $X_0 \neq i$, and $S_i = 0$ if $X_0 = i$. It is also a stopping time.

More generally, let $\tau_1 = T_i, \tau_2, \ldots$ be the successive return times to state i. If there are only r returns to state i, let $\tau_{r+1} = \tau_{r+2} = \cdots = \infty$. These random times are stopping times with respect to $\{X_n\}_{n\geq 0}$, since for any $m \geq 1$,

$$\{\tau_k = m\} \equiv \left\{\sum_{n=1}^{m} 1_{\{X_n=i\}} = k, X_m = i\right\}$$

is indeed expressible in terms of $X_0, \ldots, X_m$.

Remark 6.2.23 For a given stopping time τ, one can decide whether $\tau = m$ just by observing $X_0, X_1, \ldots, X_m$. This is why stopping times are said to be *nonanticipative*. The random time

$$\tau = \inf\{n \geq 0;\ X_{n+1} = i\},$$

where $\tau = \infty$ if $X_{n+1} \neq i$ for all $n \geq 0$, is anticipative because $\{\tau = m\} = \{X_1 \neq i, \ldots, X_m \neq i, X_{m+1} = i\}$ for all $m \geq 0$. Knowledge of this random time provides information about the value of the process just after it. It is *not* a stopping time.

Let τ be a random time taking its values in $\mathbb{N} \cup \{+\infty\}$, and let $\{X_n\}_{n\geq 0}$ be a stochastic process with values in the countable set E. In order to define X_τ when $\tau = \infty$, one must decide how to define X_∞. This is done by taking some element Δ not in E, and setting

$$X_\infty = \Delta.$$

By definition, the "process $\{X_n\}$ after τ" is the stochastic process

$$\{X_{n+\tau}\}_{n\geq 0}.$$

The "process $\{X_n\}$ before τ" is the process

$$\{X_{n\wedge(\tau-1)}\}_{n\geq 0},$$

where by convention $X_{n\wedge(0-1)} = X_0$.

The main result of the present subsection is the *strong Markov property*. It says that the Markov property, that is, the independence of past and future given the present state, extends to the situation where the present time is a stopping time. More precisely:

Theorem 6.2.24 *Let $\{X_n\}_{n\geq 0}$ be an HMC with countable state space E and transition matrix $\mathbf{P}$. If τ is an $\mathcal{F}_n^X$-stopping time, then given that $X_\tau = i \in E$ (in particular, $\tau < \infty$, since $i \neq \Delta$),*

(α) *the process after τ and the process before τ are independent, and*

(β) *the process after τ is an HMC with transition matrix $\mathbf{P}$.*

Proof. (α) By Theorem 3.1.39 it suffices to show that for all times $k \geq 1, n \geq 0$, and all states $i_0, \ldots, i_n, i, j_1, \ldots, j_k$,

$$P(X_{\tau+1} = j_1, \ldots, X_{\tau+k} = j_k \mid X_\tau = i, X_{(\tau-1)\wedge 0} = i_0, \ldots, X_{(\tau-1)\wedge n} = i_n)$$
$$= P(X_{\tau+1} = j_1, \ldots, X_{\tau+k} = j_k \mid X_\tau = i).$$

We shall prove a simplified version of the above equality, namely

$$P(X_{\tau+k} = j \mid X_\tau = i, X_{(\tau-1)\wedge n} = i_n) = P(X_{\tau+k} = j \mid X_\tau = i) \qquad (\star)$$

(the general case is obtained by the same arguments). The left-hand side of the above equality is equal to

$$\frac{P(X_{\tau+k} = j, X_\tau = i, X_{(\tau-1)\wedge n} = i_n)}{P(X_\tau = i, X_{(\tau-1)\wedge n} = i_n)}.$$

The numerator can be expanded as

$$\sum_{r\geq 0} P(\tau = r, X_{r+k} = j, X_r = i, X_{(r-1)\wedge n} = i_n). \tag{6.16}$$

But

$$P(\tau = r, X_{r+k} = j, X_r = i, X_{(r-1)\wedge n} = i_n)$$
$$= P(X_{r+k} = j \mid X_r = i, X_{(r-1)\wedge n} = i_n, \tau = r)\, P(\tau = r, X_{(r-1)\wedge n} = i_n, X_r = i),$$

and since $(r-1)\wedge n \leq r$ and $\{\tau = r\} \in \mathcal{F}_r^X$, the event $B = \{X_{(r-1)\wedge n} = i_n, \tau = r\}$ is in $\mathcal{F}_r^X$. Therefore, by the Markov property,

$$P(X_{r+k} = j \mid X_r = i, X_{(r-1)\wedge n} = i_n, \tau = r\} = P(X_{r+k} = j \mid X_r = i) = p_{ij}(k).$$

Finally, expression (6.16) reduces to

$$\sum_{r\geq 0} p_{ij}(k) P(\tau = r, X_{(r-1)\wedge n} = i_n, X_r = i) = p_{ij}(k) P(X_\tau = i, X_{\tau \wedge n} = i_n).$$

Therefore, the left-hand side of $(\star)$ is just $p_{ij}(k)$. Similar computations show that the right-hand side of $(\star)$ is also $p_{ij}(k)$, so that (α) is proved.

(β) We must show that for all states $i, j, k, i_{n-1}, \dots, i_1$,

$$P(X_{\tau+n+1} = k \mid X_{\tau+n} = j, X_{\tau+n-1} = i_{n-1}, \dots, X_\tau = i)$$
$$= P(X_{\tau+n+1} = k \mid X_{\tau+n} = j) = p_{jk}.$$

But the first equality follows from the fact proved in (α) that for the stopping time $\tau' = \tau + n$, the processes before and after τ' are independent given $X_{\tau'} = j$. The second equality is obtained by the same calculations as in the proof of (α). □

Regenerative Cycles

Consider a Markov chain with a state conventionally denoted by 0 such that $P_0(T_0 < \infty) = 1$. As a consequence of the strong Markov property, the chain starting from state 0 will return infinitely often to this state. Let $\tau_1 = T_0, \tau_2, \dots$ be the successive return times to 0, and set $\tau_0 \equiv 0$. By the strong Markov property, for any $k \geq 1$, the process after τ_k is independent of the process before τ_k (observe that condition $X_{\tau_k} = 0$ is always satisfied), and the process after τ_k is a Markov chain with the same transition matrix as the original chain, and with initial state 0, by construction. Therefore, the successive times of visit to 0, the pieces of the trajectory

$$\{X_{\tau_k}, X_{\tau_k+1}, \dots, X_{\tau_{k+1}-1}\} \qquad (k \geq 0),$$

are independent and identically distributed. Such pieces are called the *regenerative cycles* of the chain between visits to state 0. Each random time τ_k is a *regeneration time*, in the sense that $\{X_{\tau_k+n}\}_{n\geq 0}$ is independent of the past $X_0, \dots, X_{\tau_k-1}$ and has the same distribution as $\{X_n\}_{n\geq 0}$. In particular, the sequence $\{\tau_k - \tau_{k-1}\}_{k\geq 1}$ is IID.

6.3 Recurrence and Transience

Consider a Markov chain taking its values in $E = \mathbb{N}$. There is a possibility that for any initial state $i \in \mathbb{N}$ the chain will never visit i after some finite random time. This is often an undesirable feature. For example, if the chain counts the number of customers waiting in line at a service counter, such a behavior implies that the waiting line will eventually grow beyond the limits of the waiting room, whatever its size. In a sense, the corresponding system is unstable. The good notion of stability for an irreducible HMC is that of *positive recurrence*, when any given state is visited infinitely often and when, moreover, the average time between two successive visits to this state is finite.

6.3.1 Classification of States

Denote by

$$N_i := \sum_{n \geq 1} 1_{\{X_n = i\}}$$

the number of visits to state i strictly after time 0.

Theorem 6.3.1 *The distribution of N_i given $X_0 = j$ is*

$$P_j(N_i = r) = \begin{cases} f_{ji} f_{ii}^{r-1}(1 - f_{ii}) & \text{for } r \geq 1, \\ 1 - f_{ji} & \text{for } r = 0, \end{cases}$$

where

$$f_{ji} = P_j(T_i < \infty)$$

and T_i is the return time to i.

Proof. An informal proof goes like this: We first go from j to i (probability f_{ji}) and then, $r - 1$ times in succession, from i to i (each time with probability f_{ii}), and the last time, that is the $r + 1$-st time, we leave i never to return to it (probability $1 - f_{ii}$). By the independent cycle property, all these "jumps" are independent, so that the successive probabilities multiply. Here is a formal proof if someone needs it.

For $r = 0$, this is just the definition of f_{ji}. Now let $r \geq 1$, and suppose that $P_j(N_i = k) = f_{ji} f_{ii}^{k-1}(1 - f_{ii})$ is true for k $(1 \leq k \leq r)$. In particular,

$$P_j(N_i > r) = f_{ji} f_{ii}^{r}.$$

Denoting by τ_r the rth return time to state i,

$$\begin{aligned} P_j(N_i = r + 1) &= P_j(N_i = r + 1, X_{\tau_{r+1}} = i) \\ &= P_j(\tau_{r+2} - \tau_{r+1} = \infty, X_{\tau_{r+1}} = i) \\ &= P_j(\tau_{r+2} - \tau_{r+1} = \infty \mid X_{\tau_{r+1}} = i) P_j(X_{\tau_{r+1}} = i). \end{aligned}$$

But

$$P_j(\tau_{r+2} - \tau_{r+1} = \infty \mid X_{\tau_{r+1}} = i) = 1 - f_{ii}$$

by the strong Markov property ($\tau_{r+2} - \tau_{r+1}$ is the return time to i of the process after τ_{r+1}). Also,

$$P_j(X_{\tau_{r+1}} = i) = P_j(N_i > r).$$

Therefore,

$$P_j(N_i = r + 1) = P_i(T_i = \infty)P_j(N_i > r) = (1 - f_{ii})f_{ji}f_{ii}^r.$$

The result then follows by induction. $\square$

The distribution of N_i given $X_0 = i$ is geometric with parameter $1 - f_{ii}$. A geometric random variable with parameter $p = 1$ is in fact equal to infinity, and in particular has an infinite mean. If $p < 1$, however, it is almost surely finite *and* it has a finite mean. From these remarks, we deduce that

$$P_i(T_i < \infty) = 1 \Leftrightarrow P_i(N_i = \infty) = 1,$$

(in words: if starting from i you almost surely return to i, then you will visit i infinitely often) and

$$P_i(T_i < \infty) < 1 \Leftrightarrow E_i[N_i] < \infty.$$

We collect the results just obtained for future reference.

Theorem 6.3.2 *For any state $i \in E$,*

$$P_i(T_i < \infty) = 1 \Longleftrightarrow P_i(N_i = \infty) = 1,$$

and

$$P_i(T_i < \infty) < 1 \Longleftrightarrow P_i(N_i = \infty) = 0 \Longleftrightarrow E_i[N_i] < \infty.$$

In particular, the event $\{N_i = \infty\}$ has P_i-probability 0 or 1.

We are now ready for the basic definitions concerning recurrence. First recall that T_i denotes the *return* time to state i.

Definition 6.3.3 *State $i \in E$ is called* recurrent *if*

$$P_i(T_i < \infty) = 1,$$

and otherwise it is called transient. *A recurrent state $i \in E$ is called* positive *recurrent* if

$$E_i[T_i] < \infty,$$

and otherwise it is called null *recurrent*.

The Potential Matrix Criterion of Recurrence

In general, it is not easy to check whether a given state is transient or recurrent. One of the goals of the theory of Markov chains is to provide criteria of recurrence. Sometimes, one is happy with just a sufficient condition. The problem of finding useful (easy to check) conditions of recurrence is an active area of research. However, the theory has a few conditions that qualify as useful and are applicable to many practical situations. Although the next criterion is of theoretical rather than practical interest, it can be helpful in a few situations, for instance in the study of recurrence of random walks (Example 6.3.5.)

The *potential matrix* $\mathbf{G}$ associated with the transition matrix $\mathbf{P}$ is defined by

$$\mathbf{G} := \sum_{n \geq 0} \mathbf{P}^n.$$

Its general term

$$g_{ij} = \sum_{n=0}^{\infty} p_{ij}(n) = \sum_{n=0}^{\infty} P_i(X_n = j) = \sum_{n=0}^{\infty} E_i[1_{\{X_n=j\}}] = E_i \left[\sum_{n=0}^{\infty} 1_{\{X_n=j\}} \right]$$

is the average number of visits to state j, given that the chain starts from state i.

Theorem 6.3.4 *State $i \in E$ is recurrent if and only if*

$$\sum_{n=0}^{\infty} p_{ii}(n) = \infty.$$

Proof. This merely rephrases Theorem 6.3.2 since

$$\sum_{n \geq 1} p_{ii}(n) = E_i[N_i].$$

In fact,

$$E_i[N_i] = E_i \left[\sum_{n \geq 1} 1_{\{X_n=i\}} \right] = \sum_{n \geq 1} E_i \left[1_{\{X_n=i\}} \right] = \sum_{n \geq 1} P_i(X_n = i) = \sum_{n \geq 1} p_{ii}(n).$$

$\square$

EXAMPLE 6.3.5: RANDOM WALKS ON $\mathbb{Z}$. The corresponding Markov chain was described in Example 6.1.6. The nonzero terms of its transition matrix are

$$p_{i,i+1} = p, \; p_{i,i-1} = 1 - p,$$

where $p \in (0,1)$. We shall study the nature (recurrent or transient) of any one of its states, say, 0. We have $p_{00}(2n + 1) = 0$ and

$$p_{00}(2n) = \frac{(2n)!}{n!n!} p^n (1 - p)^n.$$

By Stirling's equivalence formula $n! \sim (n/e)^n \sqrt{2\pi n}$, the above quantity is equivalent to

$$\frac{[4p(1 - p)]^n}{\sqrt{\pi n}}, \tag{6.17}$$

and the nature of the series $\sum_{n=0}^{\infty} p_{00}(n)$ (convergent or divergent) is that of the series with general term (6.17). If $p \neq \frac{1}{2}$, in which case $4p(1-p) < 1$, the latter series converges. And if $p = \frac{1}{2}$, in which case $4p(1 - p) = 1$, it diverges. In summary, the states of the random walk on $\mathbb{Z}$ are transient if $p \neq \frac{1}{2}$, recurrent if $p = \frac{1}{2}$.

EXAMPLE 6.3.6: RETURNS TO ZERO OF THE SYMMETRIC RANDOM WALK. Consider the *symmetric* ($p = \frac{1}{2}$) 1-D random walk. Let $\tau_1 = T_0, \tau_2, \ldots$ be the successive return times to state 0. We just learnt in the previous example that $P_0(T_0 < \infty) = 1$. We will compute the generating function of T_0 given $X_0 = 0$, and show that the expected return time to 0 is infinite (and therefore the symmetric random walk on $\mathbb{Z}$ is *null* recurrent).

Observe that for $n \geq 1$,

$$P_0(X_{2n} = 0) = \sum_{k \geq 1} P_0(\tau_k = 2n),$$

and therefore, for all $z \in \mathbb{C}$ such that $|z| < 1$,

$$\sum_{n \geq 1} P_0(X_{2n} = 0)z^{2n} = \sum_{k \geq 1} \sum_{n \geq 1} P_0(\tau_k = 2n)z^{2n} = \sum_{k \geq 1} E_0[z^{\tau_k}].$$

But $\tau_k = \tau_1 + (\tau_2 - \tau_1) + \cdots + (\tau_k - \tau_{k-1})$ and therefore, in view of the IID property of the regenerative cycles, and since $\tau_1 = T_0$,

$$E_0[z^{\tau_k}] = (E_0[z^{T_0}])^k.$$

In particular,

$$\sum_{n \geq 0} P_0(X_{2n} = 0)z^{2n} = \frac{1}{1 - E_0[z^{T_0}]}$$

(note that the latter sum includes the term for $n = 0$, that is, 1). Direct evaluation of the left-hand side yields

$$\sum_{n \geq 0} \frac{1}{2^{2n}} \frac{(2n)!}{n!n!} z^{2n} = \frac{1}{\sqrt{1 - z^2}}.$$

Therefore, the generating function of the return time to 0 given $X_0 = 0$ is

$$E_0[z^{T_0}] = 1 - \sqrt{1 - z^2}.$$

Its first derivative $\frac{z}{\sqrt{1-z^2}}$ tends to ∞ as $z \to 1$ from below via real values. Therefore, by Abel's theorem, $E_0[T_0] = \infty$. We see that although the return time to state 0 is almost surely finite, it has an infinite expectation.

EXAMPLE 6.3.7: 3-D SYMMETRIC RANDOM WALK.
The state space of this Markov chain is $E = \mathbb{Z}^3$. Denoting by e_1, e_2, and e_3 the canonical basis vectors of $\mathbb{R}^3$ (respectively $(1, 0, 0)$, $(0, 1, 0)$, and $(0, 0, 1)$), the non-null terms of the transition matrix of the 3-D symmetric random walk are given by

$$p_{x, x \pm e_i} = \frac{1}{6}.$$

We elucidate the nature of state, say, $0 = (0, 0, 0)$. Clearly, $p_{00}(2n + 1) = 0$ for all $n \geq 0$, and (exercise)

$$p_{00}(2n) = \sum_{0 \leq i+j \leq n} \frac{(2n)!}{(i!j!(n-i-j)!)^2} \left(\frac{1}{6}\right)^{2n}.$$

This can be rewritten as

$$p_{00}(2n) = \sum_{0 \leq i+j \leq n} \frac{1}{2^{2n}} \binom{2n}{n} \left(\frac{n!}{i!j!(n-i-j)!}\right)^2 \left(\frac{1}{3}\right)^{2n}.$$

Using the *trinomial formula*

$$\sum_{0 \leq i+j \leq n} \frac{n!}{i!j!(n-i-j)!} \left(\frac{1}{3}\right)^n = 1,$$

we obtain the bound

$$p_{00}(2n) \leq K_n \frac{1}{2^{2n}} \binom{2n}{n} \left(\frac{1}{3}\right)^n,$$

where

$$K_n = \max_{0 \leq i+j \leq n} \frac{n!}{i!j!(n-i-j)!}.$$

For large values of n, K_n is bounded as follows. Let i_0 and j_0 be the values of i, j that maximize $n!/(i!j!(n+j)!)$ in the domain of interest $0 \leq i+j \leq n$. From the definition of i_0 and j_0, the quantities

$$\frac{n!}{(i_0-1)!j_0!(n-i_0-j_0+1)!}$$

$$\frac{n!}{(i_0+1)!j_0!(n-i_0-j_0-1)!}$$

$$\frac{n!}{i_0!(j_0-1)!(n-i_0-j_0+1)!}$$

$$\frac{n!}{i_0!(j_0+1)!(n-i_0-j_0-1)!}$$

are bounded by $\frac{n!}{i_0!j_0!(n-i_0-j_0)!}$. The corresponding inequalities reduce to

$$n-i_0-1 \leq 2j_0 \leq n-i_0+1 \text{ and } n-j_0-1 \leq 2i_0 \leq n-j_0+1,$$

and this shows that for large n, $i_0 \sim n/3$ and $j_0 \sim n/3$. Therefore, for large n,

$$p_{00}(2n) \sim \frac{n!}{(n/3)!(n/3)!2^{2n}e^n} \binom{2n}{n}.$$

By Stirling's equivalence formula, the right-hand side of the latter equivalence is in turn equivalent to $\frac{3\sqrt{3}}{2(\pi n)^{3/2}}$, the general term of a divergent series. State 0 is therefore transient.

A theoretical application of the potential matrix criterion is to the proof that recurrence is a (communication) class property.

Theorem 6.3.8 *If i and j communicate, they are either both recurrent or both transient.*

Proof. States i and j communicate if and only if there exist integers M and N such that $p_{ij}(M) > 0, p_{ji}(N) > 0$. Going from i to j in M steps, then from j to j in n steps, then from j to i in N steps, is just one way of going from i back to i in $M+n+N$ steps. Therefore, $p_{ii}(M+n+N) \geq p_{ij}(M)p_{jj}(n)p_{ji}(N)$. Similarly, $p_{jj}(N+n+M) \geq p_{ji}(N)p_{ii}(n)p_{ij}(M)$. Therefore, writing $\alpha = p_{ij}(M)p_{ji}(N)$ (a strictly positive quantity), we have $p_{ii}(M+N+n) \geq \alpha p_{jj}(n)$ and $p_{jj}(M+N+n) \geq \alpha p_{ii}(n)$. This implies that the series $\sum_{n=0}^{\infty} p_{ii}(n)$ and $\sum_{n=0}^{\infty} p_{jj}(n)$ either both converge or both diverge. Theorem 6.3.4 concludes the proof. $\square$

6.3.2   The Stationary Distribution Criterion

The notion of invariant measure extends the notion of stationary distribution and plays a technical role in the recurrence theory of Markov chains.

Definition 6.3.9 *A nontrivial (that is, non-null) vector $x = \{x_i\}_{i \in E}$ of non-negative real numbers is called an* invariant measure *of the stochastic matrix $\mathbf{P} = \{p_{ij}\}_{i,j \in E}$ if for all $i \in E$,*

$$x_i = \sum_{j \in E} x_j p_{ji}.\tag{6.18}$$

(In abbreviated notation, $0 \leq x < \infty$ and $x^T \mathbf{P} = x^T$.)

Theorem 6.3.10 *Let $\mathbf{P}$ be the transition matrix of an irreducible recurrent* HMC *$\{X_n\}_{n \geq 0}$. Let 0 be a state and let T_0 be the return time to 0. Define for all $i \in E$*

$$x_i = \mathrm{E}_0 \left[\sum_{n \geq 1} 1_{\{X_n = i\}} 1_{\{n \leq T_0\}} \right]\tag{6.19}$$

(for $i \neq 0$, x_i is therefore the expected number of visits to state i before returning to 0). Then, for all $i \in E$,

$$x_i \in (0, \infty),\tag{6.20}$$

and x is an invariant measure of $\mathbf{P}$.

Proof. We make two preliminary observations. First, when $1 \leq n \leq T_0$, $X_n = 0$ if and only if $n = T_0$. Therefore,

$$x_0 = 1.$$

Also,

$$\sum_{i \in E} \sum_{n \geq 1} 1_{\{X_n = i\}} 1_{\{n \leq T_0\}} = \sum_{n \geq 1} \left\{ \sum_{i \in E} 1_{\{X_n = i\}} \right\} 1_{\{n \leq T_0\}}$$
$$= \sum_{n \geq 1} 1_{\{n \leq T_0\}} = T_0,$$

and therefore

$$\sum_{i \in E} x_i = \mathrm{E}_0[T_0].\tag{6.21}$$

We now introduce the so-called *taboo transition probability*

$$_0 p_{0i}(n) := \mathrm{E}_0[1_{\{X_n = i\}} 1_{\{n \leq T_0\}}] = P_0(X_1 \neq 0, \cdots, X_{n-1} \neq 0, X_n = i),$$

the probability, starting from state 0, of visiting i at time n before returning to 0 (the "taboo" state). From the definition of x,

$$x_i = \sum_{n \geq 1} {}_0 p_{0i}(n).\tag{6.22}$$

We first prove (6.18). Observe that

$$_0 p_{0i}(1) = p_{0i}$$

and (first-step analysis) for all $n \geq 2$,

$$_0p_{0i}(n) = \sum_{j \neq 0} {_0p_{0j}(n-1)p_{ji}} . \qquad (6.23)$$

Summing up all the above equalities, and taking (6.22) into account, we obtain

$$x_i = p_{0i} + \sum_{j \neq 0} x_j p_{ji} ,$$

that is, (6.18), since $x_0 = 1$.

Next we show that $x_i > 0$ for all $i \in E$. Indeed, iterating (6.18), we find $x^T = x^T \mathbf{P}^n$, that is, since $x_0 = 1$,

$$x_i = \sum_{j \in E} x_j p_{ji}(n) = p_{0i}(n) + \sum_{j \neq 0} x_j p_{ji}(n) .$$

If x_i were null for some $i \in E$, $i \neq 0$, the latter equality would imply that $p_{0i}(n) = 0$ for all $n \geq 0$, which means that 0 and i do not communicate, in contradiction to the irreducibility assumption.

It remains to show that $x_i < \infty$ for all $i \in E$. As before, we find that

$$1 = x_0 = \sum_{j \in E} x_j p_{j0}(n)$$

for all $n \geq 1$, and therefore if $x_i = \infty$ for some i, necessarily $p_{i0}(n) = 0$ for all $n \geq 1$, and this also contradicts irreducibility. $\qquad \square$

Theorem 6.3.11 *The invariant measure of an irreducible recurrent stochastic matrix is unique up to a multiplicative factor.*

Proof. In the proof of Theorem 6.3.10, we showed that for an invariant measure y of an irreducible chain, $y_i > 0$ for all $i \in E$, and therefore, one can define, for all $i, j \in E$, the matrix $\mathbf{Q}$ by

$$q_{ji} = \frac{y_i}{y_j} p_{ij}. \qquad (6.24)$$

It is a transition matrix, since $\sum_{i \in E} q_{ji} = \frac{1}{y_j} \sum_{i \in E} y_i p_{ij} = \frac{y_j}{y_j} = 1$. The general term of $\mathbf{Q}^n$ is

$$q_{ji}(n) = \frac{y_i}{y_j} p_{ij}(n). \qquad (6.25)$$

Indeed, supposing (6.25) true for n,

$$\begin{aligned} q_{ji}(n+1) &= \sum_{k \in E} q_{jk} q_{ki}(n) = \sum_{k \in E} \frac{y_k}{y_j} p_{kj} \frac{y_i}{y_k} p_{ik}(n) \\ &= \frac{y_i}{y_j} \sum_{k \in E} p_{ik}(n) p_{kj} = \frac{y_i}{y_j} p_{ij}(n+1), \end{aligned}$$

and (6.25) follows by induction.

Clearly, $\mathbf{Q}$ is irreducible, since $\mathbf{P}$ is irreducible (just observe that in view of (6.25) $q_{ji}(n) > 0$ if and only if $p_{ij}(n) > 0$). Also, $p_{ii}(n) = q_{ii}(n)$, and therefore $\sum_{n \geq 0} q_{ii}(n) =$

$\sum_{n\geq 0} p_{ii}(n)$, which ensures that $\mathbf{Q}$ is recurrent (potential matrix criterion). Call $g_{ji}(n)$ the probability, relative to the chain governed by the transition matrix $\mathbf{Q}$, of returning to state i for the first time at step n when starting from j. First-step analysis gives

$$g_{i0}(n+1) = \sum_{j\neq 0} q_{ij} g_{j0}(n),\qquad(6.26)$$

that is, using (6.24),

$$y_i g_{i0}(n+1) = \sum_{j\neq 0} (y_j g_{j0}(n)) p_{ji}.$$

Recall that $_0 p_{0i}(n+1) = \sum_{j\neq 0}\ _0 p_{0j}(n) p_{ji}$, or, equivalently,

$$y_0\ _0 p_{0i}(n+1) = \sum_{j\neq 0} (y_0\ _0 p_{0j}(n)) p_{ji}.$$

We therefore see that the sequences $\{y_0\ _0 p_{0i}(n)\}$ and $\{y_i g_{i0}(n)\}$ satisfy the same recurrence equation. Their first terms $(n = 1)$, respectively $y_0\ _0 p_{0i}(1) = y_0 p_{0i}$ and $y_i g_{i0}(1) = y_i q_{i0}$, are equal in view of (6.24). Therefore, for all $n \geq 1$,

$$_0 p_{0i}(n) = \frac{y_i}{y_0} g_{i0}(n).$$

Summing with respect to $n \geq 1$ and using $\sum_{n\geq 1} g_{i0}(n) = 1$ ($\mathbf{Q}$ is recurrent), we obtain the announced result $x_i = \frac{y_i}{y_0}$.   □

Equality (6.21) and the definition of positive recurrence give the following result:

Theorem 6.3.12 *An irreducible recurrent* HMC *is positive recurrent if and only if its invariant measures x satisfy*

$$\sum_{i\in E} x_i < \infty.\qquad(6.27)$$

Remark 6.3.13 An HMC may well be irreducible and possess an invariant measure, and yet not be recurrent. The simplest example is the one-dimensional non-symmetric random walk, which was shown to be transient and yet admits $x_i \equiv 1$ as an invariant measure. It turns out, however, that the existence of a stationary probability distribution is necessary and sufficient for an irreducible chain (not *a priori* assumed recurrent) to be recurrent positive.

Theorem 6.3.14 *An irreducible homogeneous Markov chain is positive recurrent if and only if there exists a stationary distribution. Moreover, the stationary distribution π is, when it exists, unique, and $\pi > 0$.*

Proof. The direct part follows from Theorems 6.3.10 and 6.3.12. For the converse part, assume the existence of a stationary distribution π. Iterating $\pi^T = \pi^T \mathbf{P}$, we obtain $\pi^T = \pi^T \mathbf{P}^n$, that is, for all $i \in E$,

$$\pi(i) = \sum_{j\in E} \pi(j) p_{ji}(n).$$

If the chain were transient, then, for all states i, j,

$$\lim_{n\uparrow\infty} p_{ji}(n) = 0.$$

Indeed $p_{ji}(n) = E_j[1_{\{X_n=i\}}]$, $\lim_{n\uparrow\infty} 1_{\{X_n=i\}} = 0$ (j is transient), and $1_{\{X_n=i\}} \leq 1$, so that, by dominated convergence $\lim_{n\uparrow\infty} E_j[1_{\{X_n=i\}}] = 0$. Since $p_{ji}(n)$ is bounded by 1 uniformly in j and n, we have by dominated convergence

$$\pi(i) = \lim_{n\uparrow\infty} \sum_{j\in E} \pi(j)p_{ji}(n) = \sum_{j\in E} \pi(j) \left(\lim_{n\uparrow\infty} p_{ji}(n) \right) = 0.$$

This contradicts the assumption that π is a stationary distribution ($\sum_{i\in E} \pi(i) = 1$). The chain must therefore be recurrent, and by Theorem 6.3.12, it is positive recurrent.

The stationary distribution π of an irreducible positive recurrent chain is unique (use Theorem 6.3.11 and the fact that there is no choice for a multiplicative factor but 1). Also recall that $\pi(i) > 0$ for all $i \in E$ (see Theorem 6.3.10). $\square$

Theorem 6.3.15 *Let π be the unique stationary distribution of an irreducible positive recurrent chain, and let T_i be the return time to state i. Then*

$$\pi(i)E_i[T_i] = 1. \tag{6.28}$$

Proof. This equality is a direct consequence of expression (6.19) for the invariant measure. Indeed, π is obtained by normalization of x: for all $i \in E$,

$$\pi(i) = \frac{x_i}{\sum_{j\in E} x_j},$$

and in particular, for $i = 0$, recalling that $x_0 = 1$ and using (6.21),

$$\pi(0) = \frac{x_0}{\sum_{j\in E} x_j} = \frac{1}{E_0[T_0]}.$$

Since state 0 does not play a special role in the analysis, (6.28) is true for all $i \in E$. $\square$

The situation is extremely simple when the state space is finite.

Theorem 6.3.16 *An irreducible* HMC *with finite state space is positive recurrent.*

Proof. We first show recurrence. If the chain were transient, then, for all $i, j \in E$,

$$\lim_{n\uparrow\infty} p_{ij}(n) = 0$$

(see the argument in the proof of Theorem 6.3.14), and therefore, since the state space is finite

$$\lim_{n\uparrow\infty} \sum_{j\in E} p_{ij}(n) = 0.$$

But for all $n \geq 0$,

$$\sum_{j\in E} p_{ij}(n) = 1,$$

a contradiction. Therefore, the chain is recurrent. By Theorem 6.3.10 it has an invariant measure x. Since E is finite, $\sum_{i \in E} x_i < \infty$, and therefore the chain is positive recurrent, by Theorem 6.3.12.

$\square$

EXAMPLE 6.3.17: A RANDOM WALK ON $\mathbb{Z}$ REFLECTED AT 0. This chain has the state space $E = \mathbb{N}$ and the transition graph of the figure below. It is assumed that p_i (and therefore $q_i = 1 - p_i$) are in the open interval $(0, 1)$ for all $i \in E$, so that the chain is irreducible.

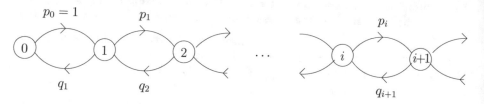

Reflected random walk

The invariant measure equation $x^T = x^T \mathbf{P}$ takes in this case the form

$$
\begin{aligned}
x_0 &= x_1 q_1, \\
x_i &= x_{i-1} p_{i-1} + x_{i+1} q_{i+1}, \ i \geq 1,
\end{aligned}
$$

with $p_0 = 1$. The general solution is, for $i \geq 1$, $x_i = x_0 \frac{p_0 \cdots p_{i-1}}{q_1 \cdots q_i}$. The positive recurrence condition $\sum_{i \in E} x_i < \infty$ is

$$
1 + \sum_{i \geq 1} \frac{p_0 \cdots p_{i-1}}{q_1 \cdots q_i} < \infty,
$$

and if it is satisfied, the stationary distribution π is obtained by normalization of the general solution. This gives

$$
\pi(0) = \left(1 + \sum_{i \geq 1} \frac{p_0 \cdots p_{i-1}}{q_1 \cdots q_i} \right)^{-1},
$$

and for $i \geq 1$,

$$
\pi(i) = \pi(0) \frac{p_0 \cdots p_{i-1}}{q_1 \cdots q_i}.
$$

In the special case where $p_i = p$, $q_i = q = 1 - p$, the positive recurrence condition becomes $1 + \frac{1}{q} \sum_{j \geq 0} \left(\frac{p}{q} \right)^j < \infty$, that is to say $p < q$, or equivalently, $p < \frac{1}{2}$.

Birth-and-death Markov Chains

Birth-and-death process models are omnipresent in operations research and, of course, in biology. We first define the birth-and-death process with a bounded population. The state space of such a chain is $E = \{0, 1, \ldots, N\}$ and its transition matrix is

$$\mathbf{P} = \begin{pmatrix} r_0 & p_0 & & & & & & \\ q_1 & r_1 & p_1 & & & & & \\ & q_2 & r_2 & p_2 & & & & \\ & & \ddots & & & & & \\ & & & q_i & r_i & p_i & & \\ & & & & \ddots & \ddots & \ddots & \\ & & & & & q_{N-1} & r_{N-1} & p_{N-1} \\ & & & & & & p_N & r_N \end{pmatrix},$$

where $p_i > 0$ for all $i \in E \backslash \{N\}$, $q_i > 0$ for all $i \in E \backslash \{0\}$, $r_i \geq 0$ for all $i \in E$, and $p_i + q_i + r_i = 1$ for all $i \in E$. The positivity conditions placed on the p_i's and q_i's guarantee that the chain is irreducible. Since the state space is finite, it is positive recurrent (Theorem 6.3.16), and it has a unique stationary distribution. Motivated by the Ehrenfest HMC, which is reversible in the stationary state, we make the educated guess that the birth-and-death process considered has the same property. This will be the case if and only if there exists a probability distribution π on E satisfying the detailed balance equations, that is, such that $\pi(i-1)p_{i-1} = \pi(i)q_i$ $(1 \leq i \leq N)$. Letting $w_0 = 1$ and

$$w_i = \prod_{k=1}^{i} \frac{p_{k-1}}{q_k} \quad (1 \leq i \leq N),$$

we find that

$$\pi(i) = \frac{w_i}{\sum_{j=0}^{N} w_j} \quad (0 \leq i \leq N) \tag{6.29}$$

indeed satisfies the detailed balance equations and is therefore the (unique) stationary distribution of the chain.

We now treat the unbounded birth-and-death process with state space $E = \mathbb{N}$ and transition matrix as in the previous example (except that the state is now "unbounded on the right"). We assume that the p_i's and q_i's are positive in order to guarantee irreducibility. The same reversibility argument as above applies with a little difference. In fact we can show that the w_i's defined above satisfy the detailed balance equations and therefore the global balance equations. Therefore the vector $\{w_i\}_{i \in E}$ is the unique, up to a multiplicative factor, invariant measure of the chain. It can be normalized to a probability distribution if and only if

$$\sum_{j=0}^{\infty} w_j < \infty.$$

Therefore in this case, and only in this case, there exists a (unique) stationary distribution, also given by (6.29).

Note that the stationary distribution, when it exists, does not depend on the r_i's. The recurrence properties of the above unbounded birth-and-death process are therefore the same as those of the chain below, which is however not aperiodic. For aperiodicity of the original chain, it suffices to assume at least one of the r_i's to be positive (Remark 6.2.5).

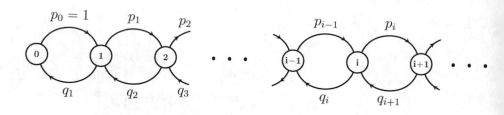

We now compute for the (bounded or unbounded) irreducible birth-and-death process the average time it takes to reach a state b from a state $a < b$. In fact, we shall prove that

$$E_a[T_b] = \sum_{k=a+1}^{b} \frac{1}{q_k w_k} \sum_{j=0}^{k-1} w_j . \qquad (6.30)$$

Since obviously $E_a[T_b] = \sum_{k=a+1}^{b} E_{k-1}[T_k]$, it suffices to prove that

$$E_{k-1}[T_k] = \frac{1}{q_k w_k} \sum_{j=0}^{k-1} w_j . \qquad (\star)$$

For this, consider for any given $k \in \{0, 1, \ldots, N\}$ the truncated chain which moves on the state space $\{0, 1, \ldots, k\}$ as the original chain, except in state k where it moves one step down with probability q_k and stays still with probability $p_k + r_k$. Use $\widetilde{E}$ to symbolize expectations with respect to the modified chain. The unique stationary distribution of this chain is

$$\widetilde{\pi}_\ell = \frac{w_\ell}{\sum_{j=0}^{k} w_\ell} \qquad (0 \le \ell \le k) .$$

First-step analysis yields $\widetilde{E}_k[T_k] = (r_k + p_k) \times 1 + q_k \left(1 + \widetilde{E}_{k-1}[T_k] \right)$, that is,

$$\widetilde{E}_k[T_k] = 1 + q_k \widetilde{E}_{k-1}[T_k] .$$

Also

$$\widetilde{E}_k[T_k] = \frac{1}{\widetilde{\pi}_k} = \frac{1}{w_k} \sum_{j=0}^{k} w_j ,$$

and therefore, since $\widetilde{E}_{k-1}[T_k] = E_{k-1}[T_k]$, we have $(\star)$.

EXAMPLE 6.3.18: SPECIAL CASES. In the special case where $(p_j, q_j, r_j) = (p, q, r)$ for all $j \ne 0, N$, $(p_0, q_0, r_0) = (p, q+r, 0)$ and $(p_N, q_N, r_N) = (0, p+r, q)$, we have $w_i = \left(\frac{p}{q} \right)^i$, and for $1 \le k \le N$,

$$E_{k-1}[T_k] = \frac{1}{q \left(\frac{p}{q} \right)^k} \sum_{j=0}^{k-1} \left(\frac{p}{q} \right)^j = \frac{1}{p - q} \left(1 - \left(\frac{q}{p} \right)^k \right) .$$

6.3.3 Foster's Theorem

The stationary distribution criterion of positive recurrence of an irreducible chain requires solving the balance equation, and this is not always feasible in practice. The following result (*Foster's theorem*) gives a more tractable, and in fact quite powerful sufficient condition.

Theorem 6.3.19 ([3]) *Let the transition matrix* $\mathbf{P}$ *on the countable state space* E *be irreducible and suppose that there exists a function* $h : E \to \mathbb{R}$ *such that* $\inf_i h(i) > -\infty$ *and*

$$\sum_{k \in E} p_{ik} h(k) < \infty \text{ for all } i \in F, \tag{6.31}$$

$$\sum_{k \in E} p_{ik} h(k) \le h(i) - \epsilon \text{ for all } i \notin F, \tag{6.32}$$

for some finite *set* F *and some* $\epsilon > 0$. *Then the corresponding* HMC *is positive recurrent.*

Proof. Since $\inf_i h(i) > -\infty$, one may assume without loss of generality that $h \ge 0$, by adding a constant if necessary. Call τ the return time to F, and define $Y_n = h(X_n) 1_{\{n < \tau\}}$. Equality (6.32) is just $\mathrm{E}[h(X_{n+1}) \mid X_n = i] \le h(i) - \epsilon$ for all $i \notin F$. For $i \notin F$,

$$
\begin{aligned}
\mathrm{E}_i[Y_{n+1} \mid X_0^n] &= \mathrm{E}_i[Y_{n+1} 1_{\{n < \tau\}} \mid X_0^n] + \mathrm{E}_i(Y_{n+1} 1_{\{n \ge \tau\}} \mid X_0^n] \\
&= \mathrm{E}_i[Y_{n+1} 1_{\{n < \tau\}} \mid X_0^n] \le \mathrm{E}_i[h(X_{n+1}) 1_{\{n < \tau\}} \mid X_0^n] \\
&= 1_{\{n < \tau\}} \mathrm{E}_i[h(X_{n+1}) \mid X_0^n] = 1_{\{n < \tau\}} \mathrm{E}_i[h(X_{n+1}) \mid X_n] \\
&\le 1_{\{n < \tau\}} h(X_n) - \epsilon 1_{\{n < \tau\}},
\end{aligned}
$$

where the third *equality* comes from the fact that $1_{\{n < \tau\}}$ is a function of X_0^n, the fourth *equality* is the Markov property, and the last *inequality* is true because P_i-a.s., $X_n \notin F$ on $n < \tau$. Therefore, P_i-a.s.,

$$\mathrm{E}_i[Y_{n+1} \mid X_0^n] \le Y_n - \epsilon 1_{\{n < \tau\}},$$

and taking expectations,

$$\mathrm{E}_i[Y_{n+1}] \le \mathrm{E}_i[Y_n] - \epsilon P_i(\tau > n).$$

Iterating the above equality, and observing that Y_n is non-negative, we obtain

$$0 \le \mathrm{E}_i[Y_{n+1}] \le \mathrm{E}_i[Y_0] - \epsilon \sum_{k=0}^{n} P_i(\tau > k).$$

But $Y_0 = h(i)$, P_i-a.s., and $\sum_{k=0}^{\infty} P_i(\tau > k) = \mathrm{E}_i[\tau]$. Therefore, for all $i \notin F$,

$$\mathrm{E}_i[\tau] \le \epsilon^{-1} h(i).$$

For $j \in F$, first-step analysis yields

$$\mathrm{E}_j[\tau] = 1 + \sum_{i \notin F} p_{ji} \mathrm{E}_i[\tau].$$

Thus $\mathrm{E}_j[\tau] \le 1 + \epsilon^{-1} \sum_{i \notin F} p_{ji} h(i)$, and this quantity is finite in view of assumption (6.31). Therefore, the return time to F starting anywhere in F has finite expectation. Since F is a finite set, this implies positive recurrence in view of the following lemma. $\square$

---

[3][Foster, 1953].

Lemma 6.3.20 Let $\{X_n\}_{n\geq 0}$ be an irreducible HMC, let F be a finite subset of the state space E, and let $\tau(F)$ be the return time to F. If $\mathrm{E}_j[\tau(F)] < \infty$ for all $j \notin F$, the chain is positive recurrent.

Proof. Select $i \in F$, and let T_i be the return time of $\{X_n\}$ to i. Let $\tau_1 = \tau(F), \tau_2, \tau_3, \ldots$ be the successive return times to F. It follows from the strong Markov property that $\{Y_n\}_{n\geq 0}$ defined by $Y_0 = X_0 = i$ and $Y_n = X_{\tau_n}$ for $n \geq 1$ is an HMC with state space F. Since $\{X_n\}$ is irreducible, so is $\{Y_n\}$. Since F is finite, $\{Y_n\}$ is positive recurrent, and in particular, $\mathrm{E}_i[\tilde{T}_i] < \infty$, where $\tilde{T}_i$ is the return time to i of $\{Y_n\}$. Defining $S_0 = \tau_1$ and $S_k = \tau_{k+1} - \tau_k$ for $k \geq 1$, we have

$$T_i = \sum_{k=0}^{\infty} S_k 1_{\{k<\tilde{T}_i\}} \,,$$

and therefore

$$\mathrm{E}_i[T_i] = \sum_{k=0}^{\infty} \mathrm{E}_i[S_k 1_{\{k<\tilde{T}_i\}}].$$

Now,

$$\mathrm{E}_i[S_k 1_{\{k<\tilde{T}_i\}}] = \sum_{\ell\in F} \mathrm{E}_i[S_k 1_{\{k<\tilde{T}_i\}} 1_{\{X_{\tau_k}=\ell\}}] \,,$$

and by the strong Markov property applied to $\{X_n\}$ and the stopping time τ_k, and the fact that the event $\{k < \tilde{T}_i\}$ belongs to the past of $\{X_n\}$ at time τ_k,

$$\begin{aligned}
\mathrm{E}_i[S_k 1_{\{k<\tilde{T}_i\}} 1_{\{X_{\tau_k}=\ell\}}] &= \mathrm{E}_i[S_k \mid k < \tilde{T}_i, X_{\tau_k} = \ell] P_i(k < \tilde{T}_i, X_{\tau_k} = \ell) \\
&= \mathrm{E}_i[S_k \mid X_{\tau_k} = \ell] P_i(k < \tilde{T}_i, X_{\tau_k} = \ell) \,.
\end{aligned}$$

Observing that $\mathrm{E}_i[S_k \mid X_{\tau_k} = \ell] = \mathrm{E}_\ell[\tau(F)]$, we see that the latter expression is bounded by $(\max_{\ell\in F} \mathrm{E}_\ell[\tau(F)]) P_i(k < \tilde{T}_i, X_{\tau_k} = \ell)$, and therefore

$$\mathrm{E}_i[T_i] \leq \left(\max_{\ell\in F} \mathrm{E}_\ell(\tau(F))\right) \sum_{k=0}^{\infty} P_i(\tilde{T}_i > k) \tag{6.33}$$

$$= \left(\max_{\ell\in F} \mathrm{E}_\ell(\tau(F))\right) \mathrm{E}_i[\tilde{T}_i] < \infty. \tag{6.34}$$

$\square$

The function h in Foster's theorem is called a *Lyapunov function* because it plays a role similar to the Lyapunov functions in the stability theory of ordinary differential equations. The corollary below is called *Pakes' lemma*.

Corollary 6.3.21 Let $\{X_n\}_{n\geq 0}$ be an irreducible HMC on $E = \mathbb{N}$ such that for all $n \geq 0$ and all $i \in E$,

$$\mathrm{E}[X_{n+1} \mid X_n = i] < \infty \tag{6.35}$$

and

$$\limsup_{i\uparrow\infty} \mathrm{E}[X_{n+1} - X_n \mid X_n = i] < 0. \tag{6.36}$$

Such an HMC is positive recurrent.

Proof. Let -2ϵ be the left-hand side of (6.36). In particular, $\epsilon > 0$. By (6.36), for i sufficiently large, say $i > i_0$, $\mathrm{E}[X_{n+1} - X_n \mid X_n = i] < -\epsilon$. We are therefore in the conditions of Foster's theorem with $h(i) = i$ and $F = \{i; i \le i_0\}$. □

EXAMPLE 6.3.22: REPAIR SHOP, TAKE 4. Recall the recurrence equation:

$$X_{n+1} = (X_n - 1)^+ + Z_{n+1}.$$

Recall that the condition

$$P(Z_1 = 0) > 0 \text{ and } P(Z_1 \ge 2) > 0$$

is necessary and sufficient for irreducibility (Example 6.6.7). We henceforth assume irreducibility and show that if $\mathrm{E}[Z_1] < 1$, the chain is positive recurrent. This follows immediately from Pakes' lemma, since for $i \ge 1$, $\mathrm{E}[X_{n+1} - X_n \mid X_n = i] = \mathrm{E}[Z] - 1 < 0$.

The following is a Foster-type theorem, only with a negative conclusion.

Theorem 6.3.23 *Let the transition matrix* $\mathbf{P}$ *on the countable state space* E *be irreducible and suppose that there exists a finite set* F *and a function* $h : E \to \mathbb{R}_+$ *such that*

$$\text{there exists a state } j \notin F \text{ such that } h(j) > \max_{i \in F} h(i) \tag{6.37}$$

$$\sup_{i \in E} \sum_{k \in E} p_{ik} |h(k) - h(i)| < \infty, \tag{6.38}$$

$$\sum_{k \in E} p_{ik}(h(k) - h(i)) \ge 0 \text{ for all } i \notin F. \tag{6.39}$$

Then the corresponding HMC *cannot be positive recurrent.*

Proof. Let τ be the return time to F. Observe that

$$h(X_\tau) 1_{\{\tau < \infty\}} = h(X_0) + \sum_{n=0}^{\infty} (h(X_{n+1}) - h(X_n)) 1_{\{\tau > n\}}.$$

Now, with $j \notin F$,

$$\sum_{n=0}^{\infty} \mathrm{E}_j \left[|h(X_{n+1}) - h(X_n)| 1_{\{\tau > n\}} \right]$$

$$= \sum_{n=0}^{\infty} \mathrm{E}_j \left[\mathrm{E}_j \left[|h(X_{n+1}) - h(X_n)| \mid X_0^n \right] 1_{\{\tau > n\}} \right]$$

$$= \sum_{n=0}^{\infty} \mathrm{E}_j \left[\mathrm{E}_j \left[|h(X_{n+1}) - h(X_n)| \mid X_n \right] 1_{\{\tau > n\}} \right] \le K \sum_{n=0}^{\infty} P_j(\tau > n)$$

for some finite positive constant K by (6.38). Therefore, if the chain is positive recurrent, the latter bound is $K \mathrm{E}_j [\tau] < \infty$. Therefore

$$\mathrm{E}_j [h(X_\tau)] = \mathrm{E}_j \left[h(X_\tau) 1_{\{\tau < \infty\}} \right]$$

$$= h(j) + \sum_{n=0}^{\infty} \mathrm{E}_j \left[(h(X_{n+1}) - h(X_n)) 1_{\{\tau > n\}} \right] \ge h(j),$$

by (6.39). In view of assumption (6.37), we have $h(j) > \max_{i \in F} h(i) \ge \mathrm{E}_j [h(X_\tau)]$, hence a contradiction. The chain therefore cannot be positive recurrent. □

6.4　Long-run Behavior

6.4.1　The Markov Chain Ergodic Theorem

The ergodic theorem for Markov chains gives conditions which guarantee that empirical averages of the type

$$\frac{1}{N}\sum_{k=1}^{N} f(X_k,\ldots,X_{k+L})$$

converge to probabilistic averages. As a matter of fact, for an irreducible positive recurrent chain with stationary distribution π, this empirical average converges P_μ-almost surely to $E_\pi[f(X_0,\ldots,X_L)]$ for any initial distribution μ, at least if $E_\pi[|f(X_0,\ldots,X_L)|] < \infty$. We shall obtain this result (Corollary 6.4.3) as a corollary of the following proposition concerning irreducible recurrent (not necessarily positive recurrent) HMCs.

Proposition 6.4.1 *Let $\{X_n\}_{n\geq 0}$ be an irreducible recurrent HMC, and let x denote the canonical invariant measure associated with state $0 \in E$:*

$$x_i = E_0\left[\sum_{n\geq 1} 1_{\{X_n=i\}} 1_{\{n\leq T_0\}}\right] \qquad (i \in E),\tag{6.40}$$

where T_0 is the return time to 0. Define for $n \geq 1$, $\nu(n) := \sum_{k=1}^{n} 1_{\{X_k=0\}}$ and let $f : E \to \mathbb{R}$ be such that

$$\sum_{i\in E} |f(i)| x_i < \infty.\tag{6.41}$$

Then, for any initial distribution μ, P_μ-a.s.,

$$\lim_{N\uparrow\infty} \frac{1}{\nu(N)}\sum_{k=1}^{N} f(X_k) = \sum_{i\in E} f(i) x_i.\tag{6.42}$$

Proof. Let $T_0 = \tau_1, \tau_2, \tau_3, \ldots$ be the successive return times to state 0, and define

$$U_p = \sum_{n=\tau_p+1}^{\tau_{p+1}} f(X_n).$$

By the independence property of the regenerative cycles, $\{U_p\}_{p\geq 1}$ is an IID sequence. Moreover, assuming $f \geq 0$ and using the strong Markov property,

$$E[U_1] = E_0\left[\sum_{n=1}^{T_0} f(X_n)\right] = E_0\left[\sum_{n=1}^{T_0}\sum_{i\in E} f(i) 1_{\{X_n=i\}}\right]$$

$$= \sum_{i\in E} f(i) E_0\left[\sum_{n=1}^{T_0} 1_{\{X_n=i\}}\right] = \sum_{i\in E} f(i) x_i.$$

By hypothesis, this quantity is finite, and therefore the strong law of large numbers applies, to give

$$\lim_{n\uparrow\infty} \frac{1}{n}\sum_{p=1}^{n} U_p = \sum_{i\in E} f(i) x_i,$$

that is,

$$\lim_{n \uparrow \infty} \frac{1}{n} \sum_{k=T_0+1}^{\tau_{n+1}} f(X_k) = \sum_{i \in E} f(i) x_i \,. \tag{6.43}$$

Observing that $\tau_{\nu(n)} \le n < \tau_{\nu(n)+1}$, we have

$$\frac{\sum_{k=1}^{\tau_{\nu(n)}} f(X_k)}{\nu(n)} \le \frac{\sum_{k=1}^{n} f(X_k)}{\nu(n)} \le \frac{\sum_{k=1}^{\tau_{\nu(n)+1}} f(X_i)}{\nu(n)} \,.$$

Since the chain is recurrent, $\lim_{n \uparrow \infty} \nu(n) = \infty$, and therefore, from (6.43), the extreme terms of the above chain of inequality tend to $\sum_{i \in E} f(i) x_i$ as n goes to ∞, which implies (6.42). The case of a function f of arbitrary sign is obtained by considering (6.42) written separately for $f^+ = \max(0, f)$ and $f^- = \max(0, -f)$, and then taking the difference of the two equalities obtained this way. The difference is not an undetermined form $\infty - \infty$ due to hypothesis (6.41). $\qquad\square$

Theorem 6.4.2 *Let $\{X_n\}_{n \ge 0}$ be an irreducible positive recurrent Markov chain with the stationary distribution π, and let $f : E \to \mathbb{R}$ be such that*

$$\sum_{i \in E} |f(i)| \pi(i) < \infty \,. \tag{6.44}$$

Then, for any initial distribution μ, P_μ-a.s.,

$$\lim_{n \uparrow \infty} \frac{1}{N} \sum_{k=1}^{N} f(X_k) = \sum_{i \in E} f(i) \pi(i). \tag{6.45}$$

Proof. Apply Proposition 6.4.1 to $f \equiv 1$. Condition (6.41) is satisfied, since in the positive recurrent case, $\sum_{i \in E} x_i = \mathrm{E}_0[T_0] < \infty$. Therefore, P_μ-a.s.,

$$\lim_{N \uparrow \infty} \frac{N}{\nu(N)} = \sum_{j \in E} x_j \,.$$

Now, f satisfying (6.44) also satisfies (6.41), since x and π are proportional, and therefore, P_μ-a.s.,

$$\lim_{N \uparrow \infty} \frac{1}{\nu(N)} \sum_{k=1}^{N} f(X_k) = \sum_{i \in E} f(i) x_i \,.$$

Combining the above equalities gives, P_μ-a.s.,

$$\lim_{N \to \infty} \frac{1}{N} \sum_{k=1}^{N} f(X_k) = \lim_{N \to \infty} \frac{\nu(N)}{N} \frac{1}{\nu(N)} \sum_{k=1}^{N} f(X_k) = \frac{\sum_{i \in E} f(i) x_i}{\sum_{j \in E} x_j},$$

from which (6.45) follows, since π is obtained by normalization of x. $\qquad\square$

Corollary 6.4.3 *Let $\{X_n\}_{n \ge 1}$ be an irreducible positive recurrent Markov chain with the stationary distribution π and let $g : E^{L+1} \to \mathbb{R}$ be such that*

$$\sum_{i_0, \dots, i_L} |g(i_0, \dots, i_L)| \pi(i_0) p_{i_0 i_1} \cdots p_{i_{L-1} i_L} < \infty \,.$$

Then for all initial distributions μ, P_μ-a.s.

$$\lim \frac{1}{N} \sum_{k=1}^{N} g(X_k, X_{k+1}, \ldots, X_{k+L})$$

$$= \sum_{i_0, i_1, \ldots, i_L} g(i_0, i_1, \ldots, i_L) \pi(i_0) p_{i_0 i_1} \cdots p_{i_{L-1} i_L}.$$

Proof. We introduce the so-called "snake chain". Let for $L \geq 1$

$$Y_n := (X_n, X_{n+1}, \ldots, X_{n+L}).$$

Exercise 6.6.15 requires us to prove the following facts. The process $\{Y_n\}_{n \geq 0}$ taking its values in $F = E^{L+1}$ is an HMC. If the original chain is irreducible, then so is the snake chain if we restrict the state space of the latter to be $F = \{(i_0, \ldots, i_L) \in E^{L+1};\ p_{i_0 i_1} p_{i_1 i_2} \cdots p_{i_{L-1} i_L} > 0\}$. If the original chain is irreducible aperiodic, so is the snake chain. If $\{X_n\}_{n \geq 0}$ has a stationary distribution π, then $\{Y_n\}_{n \geq 0}$ also has a stationary distribution σ given by

$$\sigma(i_0, i_1, \cdots, i_{L-1}, i_L) := \pi(i_0) p_{i_0 i_1} \cdots p_{i_{L-1} i_L}.$$

Given these facts, it suffices to apply Theorem 6.4.2 to the snake chain. $\square$

Note that

$$\sum_{i_0, i_1, \ldots, i_L} g(i_0, i_1, \ldots, i_L) \pi(i_0) p_{i_0 i_1} \cdots p_{i_{L-1} i_L} = \mathrm{E}_\pi[g(X_0, \ldots, X_L)].$$

Remark 6.4.4 The results of this section will be generalized in Chapter 16 (see Remark 16.1.11 therein).

6.4.2   Convergence in Variation to Steady State

Consider an HMC that is irreducible and positive recurrent. In particular, if its initial distribution is the stationary distribution, it keeps the same distribution at all times. The chain is then said to be in the *stationary regime*, or in *equilibrium*, or in *steady state*.

A question arises naturally: What is the long-run behavior of the chain when the initial distribution μ is *arbitrary*? For instance, will it *converge to equilibrium*? The classical form of the result is that for arbitrary states i and j,

$$\lim_{n \uparrow \infty} p_{ij}(n) = \pi(j), \tag{6.46}$$

if the chain is *ergodic*, according to the following definition:

Definition 6.4.5 *An irreducible positive recurrent and aperiodic* HMC *is called* ergodic.

A stronger result will be proved:

Theorem 6.4.6 *Let* $\mathbf{P}$ *be an ergodic transition matrix on the countable state space* E *with stationary distribution* π, *and let* μ *and* ν *be initial distributions. Then*

$$\lim_{n \uparrow \infty} d_V(\mu^T \mathbf{P}^n, \nu^T \mathbf{P}^n) = 0.$$

Statement (6.46) is a particular case, for which $\mu = \pi$ and ν puts mass 1 on i.

Some preliminary results will pave the way to the proof.

Definition 6.4.7 *Two stochastic processes* $\{X'_n\}_{n\geq 0}$ *and* $\{X''_n\}_{n\geq 0}$ *taking their values in the same state space* E *are said to* couple *if there exists an almost surely finite random time* τ *such that*

$$n \geq \tau \Rightarrow X'_n = X''_n. \tag{6.47}$$

The random variable τ *is called a* coupling time *of the two processes.*

Theorem 6.4.8 *For any coupling time* τ *of* $\{X'_n\}_{n\geq 0}$ *and* $\{X''_n\}_{n\geq 0}$, *we have the coupling inequality*

$$d_V(X'_n, X''_n) \leq P(\tau > n). \tag{6.48}$$

Proof. For all $A \subseteq E$,

$$
\begin{aligned}
P(X'_n \in A) - P(X''_n \in A) =& P(X'_n \in A,\ \tau \leq n) + P(X'_n \in A,\ \tau > n) \\
& - P(X''_n \in A,\ \tau \leq n) - P(X''_n \in A, \tau > n) \\
=& P(X'_n \in A,\ \tau > n) - P(X''_n \in A,\ \tau > n) \\
\leq& P(X'_n \in A,\ \tau > n) \leq P(\tau > n).
\end{aligned}
$$

$\square$

Lemma 6.4.9 *Let* $X_0^{(1)}, X_0^{(2)}, Z_n^{(1)}, Z_n^{(2)}$ $(n \geq 1)$ *be independent random variables, and suppose moreover that* $Z_n^{(1)}, Z_n^{(2)}$ $(n \geq 1)$ *are identically distributed. Let* τ *be a nonnegative integer-valued random variable such that for all* $m \in \mathbb{N}$, *the event* $\{\tau = m\}$ *is expressible in terms of* $X_0^{(1)}, X_0^{(2)}, Z_n^{(1)}, Z_n^{(2)}$ $(n \leq m)$, *that is, more formally* $\{\tau = m\} \in \sigma(X_0^{(1)}, X_0^{(2)}, Z_n^{(1)}, Z_n^{(2)}$ $(n \leq m))$. *Define the sequence* $\{Z_n\}_{n\geq 1}$ *by*

$$
Z_n = \begin{cases} Z_n^{(1)} & \text{if } n \leq \tau, \\ Z_n^{(2)} & \text{if } n > \tau. \end{cases}
$$

Then $\{Z_n\}_{n\geq 1}$ *has the same distribution as* $\{Z_n^{(1)}\}_{n\geq 1}$ *and is independent of* $X_0^{(1)}, X_0^{(2)}$.

Proof. The proof is by verification.

$$P(X_0^{(1)} \in C_1, X_0^{(2)} \in C_2, Z_1 \in A_1, \ldots, Z_k \in A_k)$$

$$
\begin{aligned}
=& \sum_{m=0}^{k} P(X_0^{(1)} \in C_1, X_0^{(2)} \in C_2, Z_1 \in A_1, \ldots, Z_k \in A_k, \tau = m) \\
& + P(X_0^{(1)} \in C_1, X_0^{(2)} \in C_2, Z_1 \in A_1, \ldots, Z_k \in A_k, \tau > k)
\end{aligned}
$$

$$
\begin{aligned}
=& \sum_{m=0}^{k} P(X_0^{(1)} \in C_1, X_0^{(2)} \in C_2, Z_1^{(1)} \in A_1, \ldots, Z_m^{(1)} \in A_m, \tau = m, Z_{m+1}^{(2)} \in A_{m+1}, \ldots, Z_k^{(2)} \in A_k) \\
& + P(X_0^{(1)} \in C_1, X_0^{(2)} \in C_2, Z_1^{(1)} \in A_1, \ldots, Z_k^{(1)} \in A_k, \tau > k).
\end{aligned}
$$

Since $\{\tau = m\}$ is independent of $Z_{m+1}^{(2)} \in A_{m+1}, \ldots, Z_k^{(2)} \in A_k$ $(k \geq m)$,

$$= \sum_{m=0}^{k} P(X_0^{(1)} \in C_1, X_0^{(2)} \in C_2, Z_1^{(1)} \in A_1, \ldots, Z_m^{(1)} \in A_m, \tau = m) P(Z_{m+1}^{(2)} \in A_{m+1}, \ldots, Z_k^{(2)} \in A_k)$$

$$+ P(X_0^{(1)} \in C_1, X_0^{(2)} \in C_2, Z_1^{(1)} \in A_1, \ldots, Z_k^{(1)} \in A_k, \tau > k)$$

$$= \sum_{m=0}^{k} P(X_0^{(1)} \in C_1, X_0^{(2)} \in C_2, Z_1^{(1)} \in A_1, \ldots, Z_m^{(1)} \in A_m, \tau = m) P(Z_{m+1}^{(1)} \in A_{m+1}, \ldots, Z_k^{(1)} \in A_k)$$

$$+ P(X_0^{(1)} \in C_1, X_0^{(2)} \in C_2, Z_1^{(1)} \in A_1, \ldots, Z_k^{(1)} \in A_k, \tau > k)$$

$$= \sum_{m=0}^{k} P(X_0^{(1)} \in C_1, X_0^{(2)} \in C_2, Z_1^{(1)} \in A_1, \ldots, Z_m^{(1)} \in A_m, \tau = m), Z_{m+1}^{(1)} \in A_{m+1}, \ldots, Z_k^{(1)} \in A_k)$$

$$+ P(X_0^{(1)} \in C_1, X_0^{(2)} \in C_2, Z_1^{(1)} \in A_1, \ldots, Z_k^{(1)} \in A_k, \tau > k)$$

$$= P(X_0^{(1)} \in C_1, X_0^{(2)} \in C_2, Z_1^{(1)} \in A_1, \ldots, Z_k^{(1)} \in A_k).$$

$\square$

Theorem 6.4.10 *Let $\{X_n^{(1)}\}_{n \geq 0}$ and $\{X_n^{(2)}\}_{n \geq 0}$ be independent* HMCs *with the same transition matrix $\mathbf{P}$ and respective initial distributions μ and ν. Let τ be defined by*

$$\tau = \inf\{n \geq 0 \,;\, X_n^{(1)} = X_n^{(2)}\}$$

with the usual convention $\inf \varnothing = \infty$. Suppose that $P(\tau < \infty) = 1$. Define $\{X_n'\}_{n \geq 1}$, by

$$X_n' = \begin{cases} X_n^{(1)} & \text{if} \quad n \leq \tau, \\ X_n^{(2)} & \text{if} \quad n > \tau. \end{cases}$$

Then $\{X_n'\}_{n \geq 1}$ has the same distribution as $\{X_n^{(1)}\}_{n \geq 1}$. In particular, it is an HMC *with transition matrix $\mathbf{P}$ and initial distribution μ.*

Proof. This concerns only the distributions of $\{X_n^{(1)}\}_{n \geq 0}$ and $\{X_n^{(2)}\}_{n \geq 0}$, and therefore we can assume a representation

$$X_{n+1}^{(\ell)} = f(X_n^{(\ell)}, Z_{n+1}^{(\ell)}) \quad (\ell = 1, 2),$$

where $X_0^{(1)}, X_0^{(2)}, Z_n^{(1)}, Z_n^{(2)}$ $(n \geq 1)$ satisfy the conditions stated in Lemma 6.4.9. One checks that τ of the current theorem satisfies the same condition as τ of Lemma 6.4.9. Defining $\{Z_n\}_{n \geq 1}$ in the same manner as in Lemma 6.4.9, we have

$$X_{n+1}' = f(X_n', Z_{n+1}),$$

which proves the announced result.

$\square$

Theorem 6.4.11 *Let $\{X_n^{(1)}\}_{n \geq 0}$ and $\{X_n^{(2)}\}_{n \geq 0}$ be two independent ergodic* HMCs *with the same transition matrix $\mathbf{P}$ and initial distributions μ and ν, respectively. Let $\tau = \inf\{n \geq 0 \,;\, X_n^{(1)} = X_n^{(2)}\}$. Then $P(\tau < \infty) = 1$.*

Proof. Consider the product HMC $\{(X_n^{(1)}, X_n^{(2)})\}_{n \geq 0}$. It takes values in $E \times E$, and the probability of transition from (i, k) to (j, ℓ) in n steps is $p_{ij}(n) p_{k\ell}(n)$. This chain is irreducible (this is where the aperiodicity hypothesis is needed; Exercise 6.6.29).

Clearly, $\{\pi(i)\pi(j)\}_{(i,j)\in E^2}$ is a stationary distribution for the product chain, where π is the stationary distribution of $\mathbf{P}$. Therefore, by the stationary distribution criterion, the product chain is positive recurrent. In particular, it reaches the diagonal of E^2 in finite time, and consequently, $P(\tau < \infty) = 1$. $\square$

We now give the proof of Theorem 6.4.6:

Proof. Let $\{X_n^{(1)}\}_{n\geq 0}$, $\{X_n^{(2)}\}_{n\geq 0}$ and τ be as in Theorem 6.4.10. Let $\{X_n'\}_{n\geq 0}$ be as in Theorem 6.4.10 and $\{X_n''\}_{n\geq 0} \equiv \{X_n^{(2)}\}_{n\geq 0}$. By Theorem 6.4.11, these couple in finite time τ and therefore, by Theorem 6.4.8,

$$d_V(\mu^T \mathbf{P}^n, \nu^T \mathbf{P}^n) = d_V(X_n', X_n'') \leq P(\tau > n) \to 0.$$

$\square$

6.4.3 Null Recurrent Case: Orey's Theorem

Theorem 6.4.6 concerns the positive recurrent case. In the null recurrent case we have the following result (*Orey's theorem*):

Theorem 6.4.12 *Let $\mathbf{P}$ be an irreducible null recurrent transition matrix on E. Then for all $i, j \in E$,*

$$\lim_{n\uparrow\infty} p_{ij}(n) = 0. \tag{6.49}$$

Proof. The periodic case follows from the aperiodic case by considering the restriction of $\mathbf{P}^d$ to C_0, some cyclic class, and observing that this restriction is also null recurrent. Therefore, $\mathbf{P}$ will be assumed aperiodic.

In this case, we have seen that the product HMC $\{Z_n\}_{n\geq 0} = \{X_n^{(1)}, X_n^{(2)})\}_{n\geq 0}$ is irreducible and aperiodic. However, it cannot be argued that it is recurrent, even if each of its components is recurrent. One must therefore separate the two possible cases.

First, suppose the product chain is transient. Its n-step transition probability from (i, i) to (j, j) is $[p_{ij}(n)]^2$, and it tends to 0 as $n \to \infty$. The result is then proved in this particular case.

Suppose now the product chain is recurrent. The coupling argument used in the aperiodic case applies and yields

$$\lim_{n\uparrow\infty} d_V(\mu^T \mathbf{P}^n - \nu^T \mathbf{P}^n) = 0 \tag{6.50}$$

for initial distributions μ and ν. Suppose in view of contradiction that for some $i, j \in E$, (6.49) is not true. One can then find a sequence $\{n_k\}_{k\geq 0}$ of integers strictly increasing to ∞ such that

$$\lim_{k\uparrow\infty} p_{ij}(n_k) = \alpha > 0.$$

Therefore, by Tychonov's theorem (Theorem B.3.1), there exists a subsequence $\{m_\ell\}_{\ell\geq 0}$ of integers strictly increasing to ∞ and a vector $\{x_s, s \in E\} \in [0,1]^E$ such that for all $s \in E$,

$$\lim_{\ell\uparrow\infty} p_{is}(m_\ell) = x_s.$$

Now, $x_j = \alpha > 0$ and therefore $\{x_s, s \in E\}$ is nontrivial, and since $\sum_{s \in E} p_{is}(m_\ell) = 1$, it follows from Fatou's lemma that

$$\sum_{s \in E} x_s \leq 1.$$

Writing (6.50) with $\mu = \delta_i$ and $\nu = \delta_i \mathbf{P}$, we have that

$$\lim_{n \uparrow \infty} d_V(\delta_i^T \mathbf{P}^n - (\delta_i^T \mathbf{P}) \mathbf{P}^n) = \lim_{n \uparrow \infty} d_V(\delta_i^T \mathbf{P}^n - (\delta_i^T \mathbf{P}^n) \mathbf{P}) = 0,$$

and in particular,

$$\lim_{n \uparrow \infty} \left(p_{is}(n) - \sum_{k \in E} p_{ik}(n) p_{ks} \right) = 0,$$

and then,

$$\lim_{\ell \uparrow \infty} \left(p_{is}(m_\ell) - \sum_{k \in E} p_{ik}(m_\ell) p_{ks} \right) = 0.$$

Therefore,

$$x_s = \lim_{\ell \uparrow \infty} \sum_{k \in E} p_{ik}(m_\ell) p_{ks}.$$

By Fatou's lemma again,

$$\lim_{\ell \uparrow \infty} \sum_{k \in E} p_{ik}(m_\ell) p_{ks} \geq \sum_{k \in E} \lim_{\ell \uparrow \infty} p_{ik}(m_\ell) p_{ks} = \sum_{k \in E} x_k p_{ks}.$$

Therefore

$$x_s \geq \sum_{k \in E} x_k p_{ks}.$$

Summing with respect to s:

$$\sum_{s \in E} x_s \geq \sum_{s \in E} \sum_{k \in E} x_k p_{ks} = \sum_{k \in E} \left(x_k \sum_{s \in E} x_k p_{ks} \right) = \sum_{k \in E} x_k,$$

which shows that the inequality ($\geq$) can only be an equality. In other words, $\{x_s, s \in E\}$ is a non-trivial invariant measure of $\mathbf{P}$ of finite total mass, which implies that $\mathbf{P}$ is positive recurrent, a contradiction. Therefore, (6.49) cannot be contradicted. $\square$

6.4.4 Absorption

The special nature of the branching process allowed for a simple and elegant computation of the probability of absorption into state 0. We now consider the absorption problem for HMCs with no special structure,[4] based only on the transition matrix $\mathbf{P}$, not necessarily assumed irreducible. The state space E is then decomposable as $E = T + \sum_j R_j$, where $R_1, R_2, \ldots$ are the disjoint recurrent classes and T is the collection of transient states. (Note that the number of recurrent classes as well as the number of transient states may be infinite.) The transition matrix can therefore be block-partitioned as

---

[4]Such as those occurring in sociology, for instance in models describing migration (whether geographical or sociological) of populations, for which the transition matrix is obtained empirically.

$$\mathbf{P} = \begin{pmatrix} \mathbf{P}_1 & 0 & \cdots & 0 \\ 0 & \mathbf{P}_2 & \cdots & 0 \\ \vdots & \vdots & \ddots & \vdots \\ B(1) & B(2) & \cdots & \mathbf{Q} \end{pmatrix}$$

or in condensed notation,

$$\mathbf{P} = \begin{pmatrix} D & 0 \\ B & \mathbf{Q} \end{pmatrix}. \tag{6.51}$$

This structure of the transition matrix accounts for the fact that one cannot go from a state in a given recurrent class to any state not belonging to this recurrent class. In other words, a recurrent class is closed.

What is the probability of being absorbed by a given recurrent class when starting from a given transient state? This kind of problem was already addressed when the first-step analysis method was introduced. This method leads to a system of linear equations with boundary conditions, for which the solution is unique, due to the finiteness of the state space. With an infinite state space, the uniqueness issue cannot be overlooked, and the absorption problem will be reconsidered with this in mind, and also with the intention of finding general matrix-algebraic expressions for the solutions. Another phenomenon not manifesting itself in the finite case is the possibility, when the set of transient states is infinite, of never being absorbed by the recurrent set. We shall consider this problem first, and then proceed to derive the distribution of the time to absorption by the recurrent set, and the probability of being absorbed by a given recurrent class.

Before Absorption

Let A be a subset of the state space E (typically the set of transient states, but not necessarily). We aim at computing for any initial state $i \in A$ the probability of remaining forever in A,

$$v(i) = P_i(X_r \in A; \ r \geq 0).$$

Defining $v_n(i) := P_i(X_1 \in A, \ldots, X_n \in A)$, we have, by monotone sequential continuity,

$$\lim_{n \uparrow \infty} \downarrow v_n(i) = v(i).$$

But for $j \in A$, $P_i(X_1 \in A, \ldots, X_{n-1} \in A, X_n = j) = \sum_{i_1 \in A} \cdots \sum_{i_{n-1} \in A} p_{ii_1} \cdots p_{i_{n-1}j}$ is the general term $q_{ij}(n)$ of the n-th iterate of the restriction $\mathbf{Q}$ of $\mathbf{P}$ to the set A. Therefore $v_n(i) = \sum_{j \in A} q_{ij}(n)$, that is, in vector notation,

$$v_n = \mathbf{Q}^n \mathbf{1}_A,$$

where $\mathbf{1}_A$ is the column vector indexed by A with all entries equal to 1. From this equality we obtain

$$v_{n+1} = \mathbf{Q}v_n,$$

and by dominated convergence $v = \mathbf{Q}v$. Moreover, $\mathbf{0}_A \leq v \leq \mathbf{1}_A$, where $\mathbf{0}_A$ is the column vector indexed by A with all entries equal to 0. The above result can be refined as follows:

Theorem 6.4.13 *The vector v is the* maximal *solution of*

$$v = \mathbf{Q}v, \mathbf{0}_A \leq v \leq \mathbf{1}_A.$$

Moreover, either $v = \mathbf{0}_A$ or $\sup_{i \in A} v(i) = 1$. In the case of a finite transient set T, the probability of infinite sojourn in T is null.

Proof. Only maximality and the last statement remain to be proved. To prove maximality consider a vector u indexed by A such that $u = \mathbf{Q}u$ and $\mathbf{0}_A \leq u \leq \mathbf{1}_A$. Iteration of $u = \mathbf{Q}u$ yields $u = \mathbf{Q}^n u$, and $u \leq \mathbf{1}_A$ implies that $\mathbf{Q}^n u \leq \mathbf{Q}^n \mathbf{1}_A = v_n$. Therefore $u \leq v_n$, which gives $u \leq v$ by passage to the limit.

To prove the last statement of the theorem, let $c = \sup_{i \in A} v(i)$. From $v \leq c \mathbf{1}_A$, we obtain $v \leq c v_n$ as above, and therefore, at the limit, $v \leq cv$. This implies either $v = \mathbf{0}_A$ or $c = 1$.

When the set T is *finite*, the probability of infinite sojourn in T is null, because otherwise at least one transient state would be visited infinitely often. $\qquad\square$

Remark 6.4.14 Equation $v = \mathbf{Q}v$ reads

$$v(i) = \sum_{j \in A} p_{ij} v(j) \ (i \in A).$$

First-step analysis gives this equality as a *necessary* condition. However, it does not help to determine which solution to choose, in case there are several.

EXAMPLE 6.4.15: REPAIR SHOP, TAKE 5. We shall prove in a different way a result already obtained in Example 6.3.22, that is: the repair shop chain is recurrent if and only if $\rho \leq 1$. Observe that the restriction of $\mathbf{P}$ to $A_i := \{i+1, i+2, \ldots\}$, namely

$$\mathbf{Q} = \begin{pmatrix} a_1 & a_2 & a_3 & \cdots \\ a_0 & a_1 & a_2 & \cdots \\ & a_0 & a_1 & \cdots \\ & & & \cdots \end{pmatrix},$$

does not depend on $i \geq 0$. In particular, the maximal solution of $v = \mathbf{Q}v$, $\mathbf{0}_A \leq v \leq \mathbf{1}_A$ when $A \equiv A_i$ has, in view of Theorem 6.4.13, the following two interpretations. Firstly, for $i \geq 1$, $1 - v(i)$ is the probability of visiting 0 when starting from $i \geq 1$. Secondly, $(1 - v(1))$ is the probability of visiting $\{0, 1, \ldots, i\}$ when starting from $i+1$. But when starting from $i+1$, the chain visits $\{0, 1, \ldots, i\}$ if and only if it visits i, and therefore $(1 - v(1))$ is also the probability of visiting i when starting from $i+1$. The probability of visiting 0 when starting from $i+1$ is

$$1 - v(i+1) = (1 - v(1))(1 - v(i)),$$

because in order to go from $i+1$ to 0 one must first reach i, and then go to 0. Therefore, for all $i \geq 1$,

$$v(i) = 1 - \beta^i,$$

where $\beta = 1 - v(1)$. To determine β, write the first equality of $v = \mathbf{Q}v$:

$$v(1) = a_1 v(1) + a_2 v(2) + \cdots,$$

that is,

$$(1 - \beta) = a_1(1 - \beta) + a_2(1 - \beta^2) + \cdots.$$

Since $\sum_{i \geq 0} a_i = 1$, this reduces to

$$\beta = g(\beta), \qquad\qquad (\star)$$

where g is the generating function of the probability distribution $(a_k, k \geq 0)$. Also, all other equations of $v = \mathbf{Q}v$ reduce to $(\star)$.

Under the irreducibility assumptions $a_0 > 0$, $a_0 + a_1 < 1$, $(\star)$ has only one solution in $[0, 1]$, namely $\beta = 1$ if $\rho \leq 1$, whereas if $\rho > 1$, it has two solutions in $[0, 1]$, this probability is $\beta = 1$ and $\beta = \beta_0 \in (0, 1)$. We must take the smallest solution. Therefore, if $\rho > 1$, the probability of visiting state 0 when starting from state $i \geq 1$ is $1 - v(i) = \beta_0^i < 1$, and therefore the chain is transient. If $\rho \leq 1$, the latter probability is $1 - v(i) = 1$, and therefore the chain is recurrent.

EXAMPLE 6.4.16: 1-D RANDOM WALK, TAKE 3. The transition matrix of the random walk on $\mathbb{N}$ with a reflecting barrier at 0,

$$\mathbf{P} = \begin{pmatrix} 0 & 1 & & & \\ q & 0 & p & & \\ & q & 0 & p & \\ & & q & 0 & p \\ & & & \ddots & \ddots & \ddots \end{pmatrix},$$

where $p \in (0, 1)$, is clearly irreducible. Intuitively, if $p > q$, there is a drift to the right, and one expects the chain to be transient. This will be proved formally by showing that the probability $v(i)$ of never visiting state 0 when starting from state $i \geq 1$ is strictly positive. In order to apply Theorem 6.4.13 with $A = \mathbb{N} - \{0\}$, we must find the general solution of $u = \mathbf{Q}u$. This equation reads

$$\begin{aligned} u(1) &= pu(2), \\ u(2) &= qu(1) + pu(3), \\ u(3) &= qu(2) + pu(4), \end{aligned}$$

$$\cdots$$

and its general solution is $u(i) = u(1) \sum_{j=0}^{i-1} \left(\frac{q}{p}\right)^j$. The largest value of $u(1)$ respecting the constraint $u(i) \in [0, 1]$ is $u(1) = 1 - \left(\frac{q}{p}\right)$. The solution $v(i)$ is therefore

$$v(i) = 1 - \left(\frac{q}{p}\right)^i.$$

Time to Absorption

We now turn to the determination of the distribution of τ, the time of exit from the transient set T. Theorem 6.4.13 tells that $v = \{v(i)\}_{i \in T}$, where $v(i) = P_i(\tau = \infty)$, is the largest solution of $v = \mathbf{Q}v$ subject to the constraints $\mathbf{0}_T \leq v \leq \mathbf{1}_T$, where $\mathbf{Q}$ is the restriction of $\mathbf{P}$ to the transient set T. The probability distribution of τ when the initial state is $i \in T$ is readily computed starting from the identity

$$P_i(\tau = n) = P_i(\tau \geq n) - P_i(\tau \geq n+1)$$

and the observation that for $n \geq 1$, $\{\tau \geq n\} = \{X_{n-1} \in T\}$, from which we obtain, for $n \geq 1$,

$$P_i(\tau = n) = P_i(X_{n-1} \in T) - P(X_n \in T) = \sum_{j \in T}(p_{ij}(n-1) - p_{ij}(n)).$$

Now, $p_{ij}(n)$ $(i, j \in T)$ is the general term of the matrix $\mathbf{Q}^n$, and therefore:

Theorem 6.4.17
$$P_i(\tau = n) = \{(\mathbf{Q}^{n-1} - \mathbf{Q}^n)\mathbf{1}_T\}_i. \tag{6.52}$$

In particular, if $P_i(\tau = \infty) = 0$,

$$P_i(\tau > n) = \{\mathbf{Q}^n\mathbf{1}_T\}_i.$$

Proof. Only the last statement remains to be proved. From (6.52),

$$P_i(n < \tau \le n + m) = \sum_{j=0}^{m-1}\{(\mathbf{Q}^{n+j} - \mathbf{Q}^{n+j-1})\mathbf{1}_T\}_i$$
$$= \{(\mathbf{Q}^n - \mathbf{Q}^{n+m})\mathbf{1}_T\}_i,$$

and therefore, if $P_i(\tau = \infty) = 0$, we obtain (6.52) by letting $m \uparrow \infty$. $\qquad\square$

Final Destination

We seek to compute the probability of absorption by a given recurrent class when starting from a given transient state. As we shall see later, it suffices for the theory to treat the case where the recurrent classes are singletons. We therefore suppose that the transition matrix has the form

$$\mathbf{P} = \begin{pmatrix} I & 0 \\ B & \mathbf{Q} \end{pmatrix}. \tag{6.53}$$

Let f_{ij} be the probability of absorption by recurrent class $R_j = \{j\}$ when starting from the transient state i. We have

$$\mathbf{P}^n = \begin{pmatrix} I & 0 \\ L_n & \mathbf{Q}^n \end{pmatrix},$$

where $L_n = (I + \mathbf{Q} + \cdots + \mathbf{Q}^n)B$. Therefore, $\lim_{n \uparrow \infty} L_n = SB$. For $i \in T$, the (i, j) term of L_n is

$$L_n(i, j) = P(X_n = j | X_0 = i).$$

Now, if T_{R_j} is the first time of visit to R_j after time 0, then

$$L_n(i, j) = P_i(T_{R_j} \le n),$$

since R_j is a closed state. Letting n go to ∞ gives the following:

Theorem 6.4.18 *For an* HMC *with transition matrix* $\mathbf{P}$ *of the form* (6.53), *the probability of absorption by recurrent class* $R_j = \{j\}$ *starting from transient state* i *is*

$$P_i(T_{R_j} < \infty) = (SB)_{i, R_j}.$$

The general case, where the recurrence classes are not necessarily singletons, can be reduced to the singleton case as follows. Let $\mathbf{P}^*$ be the matrix obtained from the transition matrix $\mathbf{P}$, by grouping for each j the states of recurrent class R_j into a single state $\hat{j}$:

$$\mathbf{P}^* = \begin{pmatrix} 1 & 0 & 0 & 0 \\ 0 & 1 & 0 & 0 \\ 0 & 0 & \ddots & 0 \\ b_{\hat{1}} & b_{\hat{2}} & \cdots & \mathbf{Q} \end{pmatrix} \tag{6.54}$$

where $b_j = B(j)\mathbf{1}_T$ is obtained by summation of the columns of $B(j)$, the matrix consisting of the columns $i \in R_j$ of B. The probability f_{iR_j} of absorption by class R_j when starting from $i \in T$ equals $\hat{f}_{i\hat{j}}$, the probability of ever visiting $\hat{j}$ when starting from i, computed for the chain with transition matrix $\mathbf{P}^*$.

EXAMPLE 6.4.19: SIBMATING. In the reproduction model called *sibmating* (sister-brother mating), two individuals are mated and two individuals from their offspring are chosen at random to be mated, and this incestuous process goes on through the subsequent generations.

Denote by X_n the genetic type of the mating pair at the nth generation. Clearly, $\{X_n\}_{n\geq 0}$ is an HMC with six states representing the different pairs of genotypes $AA \times AA$, $aa \times aa$, $AA \times Aa$, $Aa \times Aa$, $Aa \times aa$, $AA \times aa$, denoted respectively 1, 2, 3, 4, 5, 6. The following table gives the probabilities of occurrence of the three possible genotypes in the descent of a mating pair:

	AA	Aa	aa
$AA\,AA$	1	0	0
$aa\,aa$	0	0	1
$AA\,Aa$	1/2	1/2	0
$Aa\,Aa$	1/4	1/2	1/4
$Aa\,aa$	0	1/2	1/2
$AA\,aa$	0	1	0

$\left.\right\}$ parents' genotype

$\underbrace{}$ descendant's genotype

The transition matrix of $\{X_n\}_{n\geq 0}$ is then easily deduced:

$$\mathbf{P} = \begin{pmatrix} 1 & & & & & \\ & 1 & & & & \\ 1/4 & & 1/2 & 1/4 & & \\ 1/16 & 1/16 & 1/4 & 1/4 & 1/4 & 1/8 \\ & 1/4 & & 1/4 & 1/2 & \\ & & & & & 1 \end{pmatrix}.$$

The set $R = \{1, 2\}$ is absorbing, and the restriction of the transition matrix to the transient set $T = \{3, 4, 5, 6\}$ is

$$\mathbf{Q} = \begin{pmatrix} 1/2 & 1/4 & 0 & 0 \\ 1/4 & 1/4 & 1/4 & 1/8 \\ 0 & 1/4 & 1/2 & 0 \\ 0 & 1 & 0 & 0 \end{pmatrix}.$$

We find

$$S = (1 - \mathbf{Q})^{-1} = \frac{1}{6} \begin{pmatrix} 16 & 8 & 4 & 1 \\ 8 & 16 & 8 & 2 \\ 4 & 8 & 16 & 1 \\ 8 & 16 & 8 & 8 \end{pmatrix},$$

and the absorption probability matrix is

$$SB = S \begin{pmatrix} 1/4 & 0 \\ 1/16 & 1/16 \\ 0 & 1/4 \\ 0 & 0 \end{pmatrix} = \begin{pmatrix} 3/4 & 1/4 \\ 1/2 & 1/2 \\ 1/4 & 3/4 \\ 1/2 & 1/2 \end{pmatrix}.$$

For instance, the $(3, 2)$ entry, $\frac{3}{4}$, is the probability that when starting from a couple of ancestors of type $Aa \times aa$, the race will end up in genotype $aa \times aa$.

6.5 Monte Carlo Markov Chain Simulation

6.5.1 Basic Principle and Algorithms

Recall the method of the inverse in order to generate a discrete random variable Z with distribution $P(Z = a_i) = p_i$ $(0 \le i \le K)$. A crude algorithm based on this method would perform successively the tests $U \le p_0?$, $U \le p_0 + p_1?$, ..., until the answer is positive. Although very simple in principle, the inverse method has the following drawbacks when the size r of the state space E is large.

(a) Problems arise that are due to the small size of the intervals partitioning $[0, 1]$ and to the cost of precision in computing.

(b) In random field simulation, another, maybe more important, reason is the necessity to enumerate the configurations, which implies coding and decoding of a mapping from the integers to the usually very large configuration space.

(c) Another situation is that in which the probability density π is known only up to a normalizing factor, that is, $\pi(i) = K\tilde{\pi}(i)$, and when the sum $\sum_{i \in E} \pi(i) = K^{-1}$ that gives the normalizing factor is prohibitively difficult to compute. In physics, this is a frequent case.

The quest for a random generator without these ailments is at the origin of the Monte Carlo Markov chain (MCMC) sampling methodology.

The basic principle is the following. One constructs an irreducible aperiodic HMC $\{X_n\}_{n \ge 0}$ with state space E and stationary distribution π. Since the state space is finite, the chain is ergodic, and therefore, by Theorem 6.4.6, for any initial distribution μ and all $i \in E$,

$$\lim_{n \to \infty} P_\mu(X_n = i) = \pi(i). \tag{6.55}$$

Therefore, for large n, X_n has a distribution close to π.

The first task is that of designing the MCMC algorithm. One must find an ergodic transition matrix $\mathbf{P}$ on E, with stationary distribution π. In the Monte Carlo context, the transition mechanism of the chain is called a *sampling algorithm*, and the asymptotic distribution π is called the *target distribution*, or *sampled distribution*.

There are infinitely many transition matrices with a given target distribution, and among them there are infinitely many that correspond to a reversible chain, that is, such that

$$\pi(i)p_{ij} = \pi(j)p_{ji}.$$

We seek solutions of the form

$$p_{ij} = q_{ij}\alpha_{ij} \tag{6.56}$$

for $j \neq i$, where $Q = \{q_{ij}\}_{i,j \in E}$ is an irreducible transition matrix on E, called the *candidate-generator* matrix. When the present state is i, the next *tentative* state j is chosen with probability q_{ij}. When $j \neq i$, this new state is accepted with probability α_{ij}. Otherwise, the next state is the same state i. Hence, the resulting probability of moving from i to j when $i \neq j$ is given by (6.56). It remains to select the *acceptance* probabilities α_{ij}.

EXAMPLE 6.5.1: METROPOLIS' ALGORITHM.[5] In this algorithm ,

$$\alpha_{ij} = \min\left(1, \frac{\pi(j)q_{ji}}{\pi(i)q_{ij}}\right).$$

In Physics, one often finds distributions of the form

$$\pi(i) = \frac{e^{-U(i)}}{Z}, \tag{6.57}$$

where $U : E \to \mathbb{R}$ is the "energy function" and Z is the "partition function", the normalizing constant ensuring that π is indeed a probability vector. The acceptance probability of the transition from i to j is then, assuming the candidate-generating matrix to be *symmetric*,

$$\alpha_{ij} = \min\left(1, e^{-(U(j)-U(i))}\right).$$

EXAMPLE 6.5.2: BARKER'S ALGORITHM. ([6]) This algorithm, corresponds to the choice

$$\alpha_{ij} = \frac{\pi(j)q_{ji}}{\pi(j)q_{ji} + \pi(i)q_{ij}}. \tag{6.58}$$

When the distribution π is of the form (6.57), the acceptance probability of the transition from i to j is, assuming the candidate-generating matrix to be *symmetric*,

$$\alpha_{ij} = \frac{e^{-U(i)}}{e^{-U(i)} + e^{-U(j)}}.$$

This corresponds to the basic principle of statistical thermodynamics: when there are two states 1 and 2 with energies E_1 and E_2, Nature chooses 1 with probability $\frac{e^{-E_1}}{e^{-E_1}+e^{-E_2}}$.

---

[5][Metropolis et al., 1953].

[6][Barker, 1965].

Remark 6.5.3 The interest of the above algorithms resides in the fact that their implementation requires the knowledge of the target distribution π only up to a normalizing constant, since it depends only on the ratios $\pi(j)/\pi(i)$ (this in particular avoids the need to compute the normalizing constant Z in (6.57), which is often inaccessible to exact computation). The latter statement is true only as long as the candidate-generating matrix Q is known.

EXAMPLE 6.5.4: THE GIBBS ALGORITHM. Consider a multivariate probability distribution

$$\pi(x(1),\ldots,x(N))$$

on a set $E = \Lambda^N$, where Λ is countable. The basic step of the Gibbs sampler for π consists in selecting a coordinate index i ($1 \leq i \leq N$) at random, and choosing the new value $y(i)$ of the corresponding coordinate, given the present values $x(1),\ldots,$ $x(i-1), x(i+1),\ldots, x(N)$ of the other coordinates, with probability

$$\pi(y(i) \mid x(1),\ldots, x(i-1), x(i+1),\ldots, x(N)).$$

One easily verifies as above that π is the stationary distribution of the corresponding chain.

The next examples feature Markov random fields and graphs.

EXAMPLE 6.5.5: GIBBS SAMPLER. The *Gibbs sampler* uses a strictly positive probability distribution $(q_v, v \in V)$ on V, and the transition from $X_n = x$ to $X_{n+1} = y$ is made according to the following rule. The new state y is obtained from the old state x by changing (or not) the value of the phase at *one site only*. The site v whose phase is to be modified (or not) at time n is chosen independently of the past with probability q_v. When site v has been selected, the current configuration x is changed into y as follows: $y(V\backslash v) = x(V\backslash v)$, and the new phase $y(v)$ at site v is selected with probability $\pi(y(v) \mid x(V\backslash v))$. Thus, configuration x is changed into $y = (y(v), x(V\backslash v))$ with probability $q_v\pi(y(v) \mid x(V\backslash v))$, according to the local specification at site v. This gives for the nonzero entries of the transition matrix

$$P(X_{n+1} = y \mid X_n = x) = q_v\pi(y(v) \mid x(V\backslash v))1_{\{y(V\backslash v)=x(V\backslash v)\}}. \qquad (6.59)$$

Suppose that the corresponding chain is irreducible and aperiodic. To prove that π is the stationary distribution, we check for the detailed balance equations. We must have for all states $x, y \in \Lambda^V$ that differ only the phase at site v,

$$\pi(x)P(X_{n+1} = y \mid X_n = x) = \pi(y)\, P(X_{n+1} = x \mid X_n = y),$$

that is, in view of (6.59), for all $v \in V$,

$$\pi(x)\, q_v\, \pi(y(v) \mid x(V\backslash v)) = \pi(y)\, q_v\, \pi(x(v) \mid x(V\backslash v)).$$

This is indeed so, since the last equality reduces to the identity

$$\pi(x)\, q_v\, \frac{\pi(y(v), x(V\backslash v))}{P(X(V\backslash v) = x(V\backslash v))} = \pi(y(v), x(V\backslash v))\, q_v\, \frac{\pi(x)}{P(X(V\backslash v) = x(V\backslash v))}.$$

EXAMPLE 6.5.6: WHAT MAGNETS DO. In the Ising model, the local characteristic at site v depends only on $x(\mathcal{N}_v)$. The Gibbs sampler is a "natural" sampler, in that it is an idealization of what happens in nature as physicists understand it. In a piece of ferromagnetic material, for instance, the spins are randomly changed according to the local specification. When nature decides to update the orientation of a dipole, it does so according to a law of statistical mechanics. It computes the local energy

$$U(x(v), x(\mathcal{N}_v)) = x(v) \left(\frac{J}{k} \sum_{w \sim v} x(w) + \frac{H}{k} \right)$$

for each possible spin, that is $U_+ = U(+1, x(\mathcal{N}_v))$ and $U_- = U(-1, x(\mathcal{N}_v))$, and takes the corresponding orientation with a probability proportional to e^{-U_+} and e^{-U_-}, respectively, according to the fundamental law of statistical mechanics (the so-called Gibbs principle).

EXAMPLE 6.5.7: PROPERLY COLORED GRAPHS. The phase space Λ consists of a finite number of "colors" labeled from 1 to q. We describe a Markov chain $\{X_n\}_{n \geq 0}$ taking its values in the subset F of $E := \Lambda^V$ consisting of the "properly colored" configurations, that is, configurations x such that $x(v) \neq x(w)$ whenever $v \sim w$. We start from a properly colored configuration X_0. Suppose at time n the state is x. We then choose uniformly at random a site v, and then choose uniformly at random a color in the set of colors allowable at v in configuration x, that is,

$$A_v(x) := \{j \in \{1, 2, \ldots, q\} ; j \neq x(w) \text{ for all } w \text{ such that } w \sim v\}.$$

The new state at time $n + 1$ is then y, which is equal to x except for the new color j at site v. This chain is irreducible if there are at least three colors, which is henceforth assumed. The non null-elements of the transition matrix are

$$p_{xy} = \frac{1}{|V|} \times \frac{1}{|A_v(x)|},$$

where x and y differ only in the color at site v. Note that for such "adjacent" configurations, $A_v(x) = A_v(y)$, and therefore $p_{xy} = p_{yx}$. This implies in particular that the uniform distribution (on F) is the stationary distribution of this chain.

6.5.2 Exact Sampling

The Propp–Wilson Algorithm

The classical Monte Carlo Markov chain method of Section 6.5 provides an approximate sample of a probability distribution π on a finite state space E. The goal is now to construct an *exact* sample of π, that is, a random variable Z such that $P(Z = i) = \pi(i)$ for all $i \in E$. The following algorithm[7] is based on a coupling idea. One starts as usual from an *ergodic* transition matrix $\mathbf{P}$ with stationary distribution π, just as in the classical Monte Carlo Markov chain method.

The algorithm is based on a representation of $\mathbf{P}$ in terms of a recurrence equation, that is, for a given function f and an IID sequence $\{Z_n\}_{n \geq 1}$ independent of the initial state, the chain satisfies the recurrence

---

[7][Propp and Wilson, 1993].

$$X_{n+1} = f(X_n, Z_{n+1}).$$ (6.60)

The algorithm constructs a family of HMCs with transition matrix $\mathbf{P}$ with the help of a unique IID sequence of random vectors $\{Y_n\}_{n \in \mathbb{Z}}$, called the *updating sequence*, where $Y_n = (Z_{n+1}(1), \cdots, Z_{n+1}(r))$ is an r-dimensional random vector, and where the coordinates $Z_{n+1}(i)$ have a common distribution, that of Z_1. For each $N \in \mathbb{Z}$ and each $k \in E$, a process $\{X_n^N(k)\}_{n \geq N}$ is defined recursively by:

$$X_N^N(k) = k,$$

and, for $n \geq N$,

$$X_{n+1}^N(k) = f(X_n^N(k), Z_{n+1}(X_n^N(k))).$$

(If the chain is in state i at time n, it will be at time $n+1$ in state $j = f(i, Z_{n+1}(i))$.) Each one among these processes is therefore an HMC with the transition matrix $\mathbf{P}$. Note that for all $k, \ell \in E$, and all $M, N \in \mathbb{Z}$, the HMCs $\{X_n^N(k)\}_{n \geq N}$ and $\{X_n^M(\ell)\}_{n \geq M}$ use at any time $n \geq \max(M, N)$ the same updating random vector Y_{n+1}.

If, in addition to the independence of $\{Y_n\}_{n \in \mathbb{Z}}$, the components $Z_{n+1}(1)$, $Z_{n+1}(2)$, ..., $Z_{n+1}(r)$ are, for each $n \in \mathbb{Z}$, independent, we say that the updating is *componentwise independent*.

Definition 6.5.8 *The random time*

$$\tau^+ = \inf\{n \geq 0; X_n^0(1) = X_n^0(2) = \cdots = X_n^0(r)\}$$

is called the forward coupling time. *The random time*

$$\tau^- = \inf\{n \geq 1; X_0^{-n}(1) = X_0^{-n}(2) = \cdots = X_0^{-n}(r)\}$$

is called the backward coupling time.

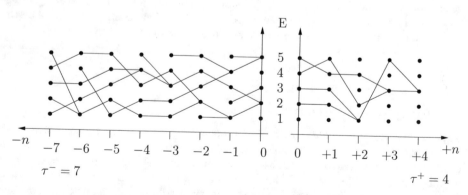

Figure 1. Backward and forward coupling

Thus, τ^+ is the first time at which the chains $\{X_n^0(i)\}_{n \geq 0}$, $1 \leq i \leq r$, coalesce.

Lemma 6.5.9 *When the updating is componentwise independent, the forward coupling time τ^+ is almost surely finite.*

Proof. Consider the (immediate) extension of Theorem 6.4.11 to the case of r independent HMCs with the same transition matrix. It cannot be applied directly to our situation,
because the chains are not independent. However, the probability of coalescence in our
situation is bounded below by the probability of coalescence in the completely independent case. To see this, first construct the independent chains model, using r independent
IID componentwise independent updating sequences. The difference with our model is
that we use too many updatings. In order to construct from this a set of r chains as in
our model, it suffices to use for two chains the same updatings as soon as they meet.
Clearly, the forward coupling time of the so modified model is smaller than or equal to
that of the initial completely independent model. □

Let $\tau := \tau^-$ and $Z := X_0^{-\tau}(i)$. (This random variable is independent of i. In Figure
1, $Z = 2$.) Then, we have the following

Theorem 6.5.10 *With a componentwise independent updating sequence, the backward
coupling time τ is almost surely finite. Also, the random variable Z has the distribution
π.*

Proof. We shall show at the end of the current proof that for all $k \in \mathbb{N}$, $P(\tau \leq k) =
P(\tau^+ \leq k)$, and therefore the finiteness of τ follows from that of τ^+ proved in the last
lemma. Now, since for $n \geq \tau$, $X_0^{-n}(i) = Z$,

$$
\begin{aligned}
P(Z = j) &= P(Z = j, \tau > n) + P(Z = j, \tau \leq n) \\
&= P(Z = j, \tau > n) + P(X_0^{-n}(i) = j, \tau \leq n) \\
&= P(Z = j, \tau > n) - P(X_0^{-n}(i) = j, \tau > n) + P(X_0^{-n}(i) = j) \\
&= P(Z = j, \tau > n) - P(X_0^{-n}(i) = j, \tau > n) + p_{ij}(n) \\
&= A_n - B_n + p_{ij}(n) .
\end{aligned}
$$

But A_n and B_n are bounded above by $P(\tau > n)$, a quantity that tends to 0 as $n \uparrow \infty$
since τ is almost surely finite. Therefore

$$
P(Z = j) = \lim_{n \uparrow \infty} p_{ij}(n) = \pi(j).
$$

It remains to prove the equality of the distributions of the forwards and backwards
coupling time. For this, select an integer $k \in \mathbb{N}$. Consider an updating sequence constructed from a *bona fide* updating sequence $\{Y_n\}_{n \in \mathbb{Z}}$ by replacing $Y_{-k+1}, Y_{-k+2}, \ldots, Y_0$
by $Y_1, Y_2, \ldots, Y_k$. Call τ' the backwards coupling time in the modified model. Clearly τ
and τ' have the same distribution.
Suppose that $\tau^+ \leq k$. Consider in the modified model the chains starting at time $-k$
from states $1, \ldots, r$. They coalesce at time $-k + \tau^+ \leq 0$ (see Figure 2), and consequently
$\tau' \leq k$. Therefore $\tau^+ \leq k$ implies $\tau' \leq k$, so that

$$
P(\tau^+ \leq k) \leq P(\tau' \leq k) = P(\tau \leq k).
$$

Now, suppose that $\tau' \leq k$. Then, in the modified model, the chains starting at time
$k - \tau'$ from states $1, \ldots, r$ must at time $-k + \tau^+ \leq 0$ coalesce at time k. Therefore (see
Figure 3), $\tau^+ \leq k$. Therefore $\tau' \leq k$ implies $\tau^+ \leq k$, so that

$$
P(\tau \leq k) = P(\tau' \leq k) \leq P(\tau^+ \leq k).
$$

□

Remark 6.5.11 The coalesced value at the forward coupling time is in general not a
sample of π (see Exercise 6.6.32).

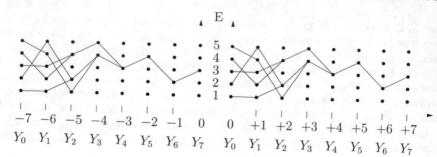

Figure 2. $\tau^+ \leq k$ implies $\tau' \leq k$

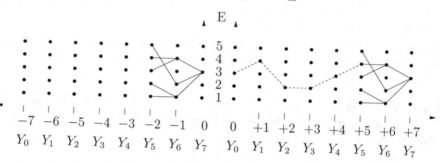

Figure 3. $\tau' \leq k$ implies $\tau^+ \leq k$

Sandwiching

The above exact sampling algorithm is often prohibitively time-consuming when the state space is large. However, if the algorithm required the coalescence of *two*, instead of r processes, then it would take less time. The Propp and Wilson algorithm does this in a special, yet not rare, case.

It is now assumed that there exists a partial order relation on E, denoted by $\preceq$, with a minimal and a maximal element (say, respectively, 1 and r), and that we can perform the updating in such a way that for all $i, j \in E$, all $N \in \mathbb{Z}$, and all $n \geq N$,

$$i \preceq j \Rightarrow X_n^N(i) \preceq X_n^N(j).$$

However, we do not require componentwise independent updating (but the updating vectors sequence remains IID). The corresponding sampling procedure is called the *monotone Propp–Wilson algorithm*.

Define the backwards *monotone* coupling time

$$\tau_m = \inf\{n \geq 1; X_0^{-n}(1) = X_0^{-n}(r)\}.$$

Theorem 6.5.12 *The monotone backwards coupling time τ_m is almost surely finite. Also, the random variable $X_0^{-\tau_m}(1)$ $(= X_0^{-\tau_m}(r))$ has the distribution π.*

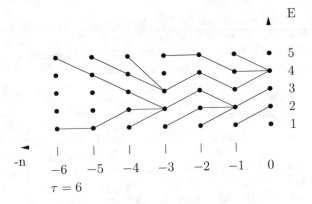

Figure 4. Monotone Propp–Wilson algorithm

Proof. We can use most of the proof of Theorem 6.5.10. We need only to prove independently that τ^+ is finite. This is the case because τ^+ is dominated by the first time $n \geq 0$ such that $X_n^0(r) = 1$, and the latter is finite in view of the recurrence assumption. □

Monotone coupling will occur with representations of the form (6.60) such that for all z,

$$i \preceq j \Rightarrow f(i, z) \preceq f(j, z),$$

and if for all $n \in \mathbb{Z}$ and all $i \in \{1, \ldots, r\}$,

$$Z_{n+1}(i) = Z_{n+1}.$$

EXAMPLE 6.5.13: A DAM MODEL. We consider the following model of a dam reservoir. The corresponding HMC, with values in $E = \{0, 2, \ldots, r\}$, satisfies the recurrence equation

$$X_{n+1} = \min\{r, \max(0, X_n + Z_{n+1})\},$$

where, as usual, $\{Z_n\}_{n \geq 1}$ is IID. In this specific model, X_n is the content at time n of a dam reservoir with maximum capacity r, and $Z_{n+1} = A_{n+1} - c$, where A_{n+1} is the input into the reservoir during the time period from n to $n+1$, and c is the maximum release during the same period. The updating rule is then monotone.

Remark 6.5.14 The average number of trials $E[\tau_-]$ needed for a naive use of the Propp–Wilson algorithm may be forbidding, so that one could be tempted to fix a large value for the number of attempts authorized before giving up and starting a new sequence attempts, and so on until obtaining coalescence within the prescribed limit of time. This will introduce a bias (see Exercise 6.6.33). It is recommended that instead of trying the times $-1, -2$, etc., one uses successive restarting times of the form $\alpha^r T_0$. Let k be the first k for which $\alpha^k T_0 \geq \tau_-$. The number of simulation steps used is $2\left(T_0 + \alpha T_0 + \cdots + \alpha^k T_0\right)$ (the factor 2 accounts for the fact that we are running two chains), that is,

$$2T_0\left(\frac{\alpha^{k+1} - 1}{\alpha - 1}\right) < 2T_0\left(\frac{\alpha^2}{\alpha - 1}\right)\alpha^{k-1} \leq 2\tau_-\frac{\alpha^2}{\alpha - 1}$$

steps, where we have assumed that $T_0 \leq \tau_-$. In the best case, assuming we are informed of the exact value of τ_- by some oracle, the number of steps is $2\tau_-$. The ratio of the worst to best cases is $\frac{\alpha^2}{\alpha-1}$, which is minimized for $\alpha = 2$. This is why it is usually suggested to start the successive attempts of backward coalescence at times of the form $-2^k T_0$ ($k \geq 0$).

Complementary reading

An important issue in the theory of Markov chains is the evaluation of the speed of convergence to stationarity (the algebraic theory is in general of little help since the computation of the second eigenvalue modulus is often impossible). For this, see [Aldous and Fill, 2014] and [Levin, Peres and Wilmer, 2009]. [Brémaud, 1999, 2020] has a chapter on non-homogeneous Markov chains and its application to simulated annealing.

6.6   Exercises

Exercise 6.6.1. Past, present, future.

1. The Markov property does not imply that the past and the future are independent given *any* information concerning the present. Find a simple example of an HMC $\{X_n\}_{n\geq 0}$ with state space $E = \{1, 2, 3, 4, 5, 6\}$ such that

$$P(X_2 = 6 \mid X_1 \in \{3, 4\}, X_0 = 1) \neq P(X_2 = 6 \mid X_1 \in \{3, 4\}).$$

2. For an HMC $\{X_n\}_{n\geq 0}$ with state space E, prove that for all $n \in \mathbb{N}$, and all states $i_0, i_1, \ldots, i_{n-1}, i, j_1, j_2, \ldots, j_k \in E$,

$$P(X_{n+1} = j_1, \ldots, X_{n+k} = j_k \mid X_n = i, X_{n-1} = i_{n-1}, \ldots, X_0 = i_0)$$
$$= P(X_{n+1} = j_1, \ldots, X_{n+k} = j_k \mid X_n = i).$$

3. Let $\{X_n\}_{n\geq 0}$ be an HMC with state space E and transition matrix $\mathbf{P}$. Show that for all $n \geq 1$ and all $k \geq 2$, X_n is conditionally independent of $X_0, \ldots, X_{n-2}, X_{n+2}, \ldots$, X_{n+k} given X_{n-1}, X_{n+1} and compute the conditional distribution of X_n given X_{n-1}, X_{n+1}.

Exercise 6.6.2. A Poissonian version of Besag's model
Consider the model of Example 6.1.18 with the following modifications. Firstly, the phase space is $\Lambda = \mathbb{N}$, and secondly, the potential is now

$$V_C(x) = \begin{cases} -\log(g(x(v))) + \alpha_1 x(v) & \text{if } C = \{v\} \in \mathcal{C}_1, \\ \alpha_j x(v) x(w) & \text{if } C = \{v, w\} \in \mathcal{C}_j, \end{cases}$$

where $\alpha_j \in \mathbb{R}$ and $g : \mathbb{N} \to \mathbb{R}$ is strictly positive. As in the autobinomial model, for any clique C not of the type $\mathcal{C}_j$, $V_C \equiv 0$. For what function g do we have

$$\pi^s(x) = e^{-\rho} \frac{\rho^{x(v)}}{x(v)!},$$

where $\rho = e^{-\langle a, b \rangle}$, and where $\langle a, b \rangle$ is as in Example 6.1.18? (This model is the *auto-Poisson model*.)

6.6. EXERCISES

Exercise 6.6.3. ISING ON A CIRCLE
(Baxter, 1965) Consider the classical Ising model of Example 6.1.17, except that the site space $V = \{1, 2, \ldots, N\}$ consists of N points arranged in this order on a circle. The neighbors of site i are $i + 1$ and $i - 1$, with the convention that site $N + 1$ is site 1. The phase space is $\Lambda = \{+1, -1\}$. Compute the partition function. Hint: express the normalizing constant Z_N in terms of the N-th power of the matrix

$$R = \begin{pmatrix} R(+1, +1) & R(+1, -1) \\ R(-1, +1) & R(-1, -1) \end{pmatrix} = \begin{pmatrix} e^{K+h} & e^{-K} \\ e^{-K} & e^{K-h} \end{pmatrix},$$

where $K := \frac{J}{kT}$ and $h := \frac{H}{kT}$.

Exercise 6.6.4. MARKOV RECURSION
Prove Theorem 6.1.8.

Exercise 6.6.5. RECORDS
Let $\{Z_n\}_{n \geq 1}$ be an IID sequence of geometric random variables (for all $k \geq 0$, $P(Z_n = k) = (1 - p)^k p$, where $p \in (0, 1)$). Let $X_n = \max(Z_1, \ldots, Z_n)$ be the *record value* at time n, and suppose X_0 is an $\mathbb{N}$-valued random variable independent of the sequence $\{Z_n\}_{n \geq 1}$. Show that $\{X_n\}_{n \geq 0}$ is an HMC and give its transition matrix.

Exercise 6.6.6. CHECKING CONDITIONAL INDEPENDENCE
Let X, Y, and Z be three discrete random variables with values in E, F, and G, respectively. Prove the following: If for some function $g : E \times F \to [0, 1]$, $P(X = x \mid Y = y, Z = z) = g(x, y)$ for all x, y, z, then $P(X = x \mid Y = y) = g(x, y)$ for all x, y, and X and Z are conditionally independent given Y.

Exercise 6.6.7. IRREDUCIBILITY OF THE REPAIR SHOP
Prove that a necessary and sufficient condition of irreducibility of the repair shop chain of Example 6.1.7 is that $P(Z_1 = 0) > 0$ *and* $P(Z_1 \geq 2) > 0$.

Exercise 6.6.8. AGGREGATION OF STATES
Let $\{X_n\}_{n \geq 0}$ be an HMC with state space E and transition matrix $\mathbf{P}$, and let $(A_k, k \geq 1)$ be a countable partition of E. Define the process $\{\hat{X}_n\}_{n \geq 0}$ with state space $\hat{E} = \{\hat{1}, \hat{2}, \ldots\}$ by $\hat{X}_n = \hat{k}$ if and only if $X_n \in A_k$. Show that if $\sum_{j \in A_\ell} p_{ij}$ is independent of $i \in A_k$ for all k, ℓ, $\{\hat{X}_n\}_{n \geq 0}$ is an HMC with transition probabilities $\hat{p}_{\hat{k}\hat{\ell}} = \sum_{j \in A_\ell} p_{ij}$ (any $i \in A_k$).

Exercise 6.6.9. STREET GANGS
Three characters, A, B, and C, armed with guns, suddenly meet at the corner of a Washington D.C. street, whereupon they naturally start shooting at one another. Each street-gang kid shoots every tenth second, as long as he is still alive. The probability of a hit for A, B, and C are α, β, and γ respectively. A is the most hated, and therefore, as long as he is alive, B and C ignore each other and shoot at A. For historical reasons not developed here, A cannot stand B, and therefore he shoots only at B while the latter is still alive. Lucky C is shot at if and only if he is in the presence of A alone or B alone. What are the survival probabilities of A, B, and C, respectively?

Exercise 6.6.10. THE GAMBLER'S RUIN
(This exercise continues Example 6.1.10.) Compute the average duration of the game when $p = \frac{1}{2}$.

Exercise 6.6.11. ON THE CIRCLE

Consider the random walk on the circle. More precisely, there are n points labeled $0, 1, 2, \ldots, n-1$ orderly and equidistantly placed on a circle. A particle moves from one point to an adjacent point in the manner of a random walk on $\mathbb{Z}$. This gives rise to an HMC with the transition probabilities $p_{i,i+1} = p \in (0,1)$, $p_{i,i-1} = 1 - p$, where by convention state -1 is $n-1$ and state n is 0. Compute the average time it takes to go back to 0 when initially in 0.

$$S_2 \ \ S_3 \ \ \overset{\curvearrowleft}{S_1} \ \ S_5 \ \ S_4$$

$$\overset{\wedge}{S_2} \ \ S_3 \ \ S_5 \ \ S_1 \ \ S_4$$

$$\overset{\curvearrowleft}{S_2} \ \ S_3 \ \ S_5 \ \ S_1 \ \ S_4$$

$$S_3 \ \ S_2 \ \ S_5 \ \ S_1 \ \ S_4$$

Exercise 6.6.12. TRUNCATED HMCS

Let $\mathbf{P}$ be a transition matrix on the countable state space E, with the positive stationary distribution π. Let A be a subset of the state space, and define the truncation of $\mathbf{P}$ on A to be the transition matrix $\mathbf{Q}$ indexed by A and given by

$$q_{ij} \ = \ p_{ij} \text{ if } i, j \in A, \ i \neq j,$$
$$q_{ii} \ = \ p_{ii} + \sum_{k \in \bar{A}} p_{ik}.$$

Show that if $(\mathbf{P}, \pi)$ is reversible, then so is $(\mathbf{Q}, \frac{\pi}{\pi(A)})$.

Exercise 6.6.13. MOVING STONES

Stones $S_1, \ldots, S_M$ are placed in line. At each time n a stone is selected at random, and this stone and the one ahead of it in the line exchange positions. If the selected stone is at the head of the line, nothing is changed. For instance, with $M = 5$: Let the current configuration be $S_2 S_3 S_1 S_5 S_4$ (S_2 is at the head of the line). If S_5 is selected, the new situation is $S_2 S_3 S_5 S_1 S_4$, whereas if S_2 is selected, the configuration is not altered. At each step, stone S_i is selected with probability $\alpha_i > 0$. Call X_n the situation at time n, for instance $X_n = S_{i_1} \cdots S_{i_M}$, meaning that stone S_{i_j} is in the jth position. Show that $\{X_n\}_{n \geq 0}$ is an irreducible HMC and that it has a stationary distribution given by the formula

$$\pi(S_{i_1} \cdots S_{i_M}) = C \alpha_{i_1}^M \alpha_{i_2}^{M-1} \cdots \alpha_{i_M},$$

for some normalizing constant C.

Exercise 6.6.14. APERIODICITY

(a) Show that an irreducible transition matrix $\mathbf{P}$ with at least one state $i \in E$ such that $p_{ii} > 0$ is aperiodic.

(b) Let $\mathbf{P}$ be an irreducible transition matrix on the *finite* state space E. Show that a necessary and sufficient condition for $\mathbf{P}$ to be aperiodic is the existence of an integer m such that all the entries of $\mathbf{P}^m$ are positive.

Exercise 6.6.15. THE SNAKE CHAIN
Let $\{X_n\}_{n\geq 0}$ be an HMC with state space E and transition matrix $\mathbf{P}$. Define for $L \geq 1, Y_n = (X_n, X_{n+1}, \ldots, X_{n+L})$.

(a) The process $\{Y_n\}_{n\geq 0}$ takes its values in $F = E^{L+1}$. Prove that $\{Y_n\}_{n\geq 0}$ is an HMC and give the general entry of its transition matrix. (The chain $\{Y_n\}_{n\geq 0}$ is called the *snake chain* of length $L+1$ associated with $\{X_n\}_{n\geq 0}$.)

(b) Show that if $\{X_n\}_{n\geq 0}$ is irreducible, then so is $\{Y_n\}_{n\geq 0}$ if we restrict the state space of the latter to be $F = \{(i_0, \ldots, i_L) \in E^{L+1}; p_{i_0 i_1} p_{i_1 i_2} \cdots p_{i_{L-1} i_L} > 0\}$. Show that if the original chain is irreducible aperiodic, so is the snake chain.

(c) Show that if $\{X_n\}_{n\geq 0}$ has a stationary distribution π, then $\{Y_n\}_{n\geq 0}$ also has a stationary distribution. Which one?

Exercise 6.6.16. NO STATIONARY DISTRIBUTION
Show that the symmetric random walk on $\mathbb{Z}$ cannot have a stationary distribution.

Exercise 6.6.17. AN INTERPRETATION OF THE INVARIANT MEASURE
A countable number of particles move independently in the countable space E, each according to a Markov chain with the transition matrix $\mathbf{P}$. Let $A_n(i)$ be the number of particles in state $i \in E$ at time $n \geq 0$, and suppose that the random variables $A_0(i)$ $(i \in E)$ are independent Poisson random variables with respective means $\mu(i)$ $(i \in E)$, where $\mu = \{\mu(i)\}_{i\in E}$ is an invariant measure of $\mathbf{P}$. Show that for all $n \geq 1$, the random variables $A_n(i)$ $(i \in E)$ are independent Poisson random variables with respective means $\mu(i)$ $(i \in E)$.

Exercise 6.6.18. DOUBLY STOCHASTIC TRANSITION MATRIX
A stochastic matrix $\mathbf{P}$ on the state space E is called *doubly stochastic* if for all states i, $\sum_{j\in E} p_{ji} = 1$. Suppose in addition that $\mathbf{P}$ is irreducible, and that E is *infinite*. Find the invariant measure of $\mathbf{P}$. Show that $\mathbf{P}$ cannot be positive recurrent.

Exercise 6.6.19. RETURN TIME TO THE INITIAL STATE
Let τ be the first return time to the initial state of an irreducible positive recurrent HMC $\{X_n\}_{n\geq 0}$, that is,

$$\tau = \inf\{n \geq 1; X_n = X_0\},$$

with $\tau = +\infty$ if $X_n \neq X_0$ for all $n \geq 1$. Compute the expectation of τ when the initial distribution is the stationary distribution π. Conclude that it is finite if and only if E is finite. When E is infinite, is this in contradiction to positive recurrence?

Exercise 6.6.20. THE KNIGHT RETURNS HOME
A knight moves randomly on a chessboard, making each admissible move with equal probability, and starting from a corner. What is the average time he takes to return to the corner he started from?

Exercise 6.6.21. THE GAMBLER'S RUIN
(This exercise continues Example 6.1.10.) Compute the average duration of the game when $p = \frac{1}{2}$.

Exercise 6.6.22. ABBABAA!

A sequence of A's and B's is formed as follows. The first item is chosen at random, $P(A) = P(B) = \frac{1}{2}$, as is the second item, independently of the first one. When the first $n \geq 2$ items have been selected, the $(n+1)$st is chosen, independently of the letters in positions $k \leq n - 2$ conditionally on the pair at position $n - 1$ and n, as follows:

$$P(A \mid AA) = \frac{1}{2}, P(A \mid AB) = \frac{1}{2}, P(A \mid BA) = \frac{1}{4}, P(A \mid BB) = \frac{1}{4}.$$

What is the proportion of A's and B's in a long chain?

Exercise 6.6.23. FIXED-AGE RETIREMENT POLICY

Let $\{U_n\}_{n\geq 1}$ be a sequence of IID random variables taking their values in $\mathbb{N}_+ = \{1, 2, \ldots, \}$. The random variable U_n is interpreted as the lifetime of some equipment, or "machine", the nth one, which is replaced by the $(n+1)$st one upon failure. Thus at time 0, machine 1 is put in service until it breaks down at time U_1, whereupon it is immediately replaced by machine 2, which breaks down at time $U_1 + U_2$, and so on. The time to next failure of the current machine at time n is denoted by X_n. More precisely, the process $\{X_n\}_{n\geq 0}$ takes its values in $E = \mathbb{N}$, equals 0 at time $R_k = \sum_{i=1}^{k} U_i$, equals $U_{k+1} - 1$ at time $R_k + 1$, and then decreases by one unit per unit of time until it reaches the value 0 at time R_{k+1}. It is assumed that for all $k \in \mathbb{N}_+$, $P(U_1 > k) > 0$, so that the state space E is $\mathbb{N}$. Then $\{X_n\}_{n\geq 0}$ is an irreducible HMC called the forward recurrence chain. We assume positive recurrence, that is, $E[U] < \infty$, where $U := U_1$.

A. Show that the chain is irreducible. Give the necessary and sufficient condition for positive recurrence. Assuming positive recurrence, what is the stationary distribution? A visit of the chain to state 0 corresponds to a breakdown of a machine. What is the empirical frequency of breakdowns?

B. Suppose that the cost of a breakdown is so important that it is better to replace a working machine during its lifetime (breakdown implies costly repairs, whereas replacement only implies moderate maintenance costs). The *fixed-age retirement policy* fixes an integer $T \geq 1$ and requires that a machine having reached age T be immediately replaced. What is the empirical frequency of breakdowns (not replacements)?

Exercise 6.6.24. A LAZY RANDOM WALK ON THE CIRCLE

Consider N points on the circle forming the state space $E := \{0, 1, \ldots, N - 1\}$. Two points i, j are said to be neighbors if $j = i \pm 1$ *modulo* n. Consider the Markov chain $\{(X_n, Y_n)\}_{n\geq 0}$ with state space $E \times E$ and representing two particles moving on E as follows. At each time n choose X_n or Y_n with probability $\frac{1}{2}$ and move the corresponding particle to the left or to the right equiprobably while the other particle remains still. The initial positions of the particles are a and b, respectively. Compute the average time it takes until the two particles collide (the average coupling time of two lazy random walks).

Exercise 6.6.25. DIVISIBLE BY k

Let $\{Z_n\}_{n \geq 1}$ be an IID sequence of IID $\{0, 1\}$-valued random variables, $P(Z_n = 1) = p \in (0, 1)$. Show that for all $k \geq 1$,

$$\lim_{n \uparrow \infty} P(Z_1 + Z_2 + \cdots Z_n \text{ is divisible by } k) = 1.$$

Hint: modulo k.

Exercise 6.6.26. CONVERGENCE SPEED VIA COUPLING

Suppose that the coupling time τ in Theorem 6.4.6 satisfies

$$E[\psi(\tau)] < \infty$$

for some non-decreasing function $\psi : \mathbb{N} \to \mathbb{R}_+$ such that $\lim_{n \uparrow \infty} \psi(n) = \infty$. Show that for any initial distributions μ and ν

$$|\mu^T \mathbf{P}^n - \nu^T \mathbf{P}^n| = o\left(\frac{1}{\psi(n)}\right).$$

Exercise 6.6.27. MEAN TIME BETWEEN SUCCESSIVE VISITS OF A SET

Let $\{X_n\}_{n \geq 0}$ be an irreducible positive recurrent HMC with stationary distribution π. Let A be a subset of the state space E and let $\{\tau(k)\}_{k \geq 1}$ be the sequence of return times to A. Show that

$$\lim_{k \uparrow \infty} \frac{\tau(k)}{k} = \frac{1}{\sum_{i \in A} \pi(i)}.$$

(This extends Formula (6.28).)

Exercise 6.6.28. ERGODIC EVALUATION OF THE TRANSITION MATRIX

Let $\{X_n\}_{n \geq 0}$ be a positive recurrent HMC with state space E and transition matrix $\mathbf{P}$. Show that for all $i, j \in E$,

$$\lim_{n \uparrow \infty} \frac{\sum_{k=1}^n 1_{\{X_k = i, X_{k+1} = j\}}}{\sum_{k=1}^n 1_{\{X_k = i\}}} = p_{ij}.$$

Exercise 6.6.29. PRODUCT MARKOV CHAINS

Let $\{X_n^{(1)}\}_{n \geq 0}$ and $\{X_n^{(2)}\}_{n \geq 0}$ be two *independent* irreducible and aperiodic HMCs with the same transition matrix $\mathbf{P}$. Define the *product* HMC $\{Z_n\}_{n \geq 0}$ taking its values in $E \times E$ by $Z_n = (X_n^{(1)}, X_n^{(2)})$. Prove that it is indeed an HMC. What is its n-step transition matrix? Prove that it is irreducible. Give a counterexample if the hypothesis of aperiodicity is omitted.

Exercise 6.6.30. IRREDUCIBILITY OF THE BARKER SAMPLING CHAIN

Show that for both the Metropolis and Barker samplers, if Q is irreducible and U is not a constant, then $\mathbf{P}(T)$ is irreducible and aperiodic for all $T > 0$.

Exercise 6.6.31. THE MODIFIED RANDOM WALK

Consider the usual random walk on a graph. Its stationary distribution is in general non-uniform. We wish to modify it so as to obtain an HMC with uniform stationary distribution. For this, accept a transition from vertex i to vertex j of the original random walk with probability α_{ij}. Find one such acceptance probability depending only on $d(i)$ and $d(j)$ that guarantees that the corresponding Monte Carlo Markov chain admits the uniform distribution as stationary distribution.

Exercise 6.6.32. FORWARD COUPLING DOES NOT YIELD EXACT SAMPLING
Refer to the Propp–Wilson algorithm. Show that the coalesced value at the forwards coupling time is not a sample of π. For a counterexample use the two-state HMC with $E = \{1, 2\}$, $p_{1,2} = 1$, $p_{2,2} = p_{2,1} = 1/2$.

Exercise 6.6.33. THE IMPATIENT SIMULATOR
Find a very simple example showing that use of the Propp–Wilson algorithm by an impatient customer introduces a bias.

Exercise 6.6.34. IID RANDOM FIELDS

A. Let $(Z(v)$ $(v \in V)$ be a family of IID random variables with values in $\{-1, +1\}$ indexed by a finite set V, with $P(Z(v) = -1) = p \in (0, 1)$. Show that

$$P(Z = z) = K e^{\gamma \sum_{v \in V} z(s)},$$

for some constants γ and K to be identified.

B. Do the same when the $Z(v)$s take their values in $\{0, 1\}$, with $P(Z(v) = 0) = p \in (0, 1)$.

Exercise 6.6.35. A TWO-STATE HMC AS A GIBBS FIELD
Consider an HMC $\{X_n\}_{n \geq 0}$ with state space $E = \{-1, 1\}$ and transition matrix

$$\mathbf{P} = \begin{pmatrix} 1 - \alpha & \alpha \\ \beta & 1 - \beta \end{pmatrix} \qquad (\alpha, \beta \in (0, 1))$$

and with the stationary initial distribution

$$(\nu_0, \nu_1) = \frac{1}{\alpha + \beta}(\beta, \alpha).$$

Give a representation of $(X_0, \ldots, X_N)$ as an MRF. What is the normalized potential with respect to phase 1?

Exercise 6.6.36. MONOTONE PROPP–WILSON FOR THE ISING MODEL
Consider the classical Ising model of Example 6.1.17 with energy function $U(x) = \sum_{\langle v, w \rangle} x(v)x(w)$. Define on the state space $E = \{-1, +1\}^S$ the partial order relation $\preceq$ defined as follows: $x = (x(v), v \in V) \preceq y = (y(v), v \in V)$ if and only if for all $v \in V$, $x(v) = +1$ implies $y(v) = +1$. Show that the monotone Propp–Wilson algorithm can be applied.

Exercise 6.6.37. A HARD-CORE MODEL
Consider a random field with finite site space V and phase space $\Lambda := \{0, 1\}$ (with the interpretation that if $x(v) = 1$, the site v is "occupied" and "vacant" otherwise) evolving in time. The resulting sequence $\{X_n\}_{n \geq 0}$ is an HMC with state space F, the subset of $E = \{0, 1\}^V$ consisting of the configurations x such that for all $v \in V$, $x(v) = 1$ implies that $x(w) = 0$ for all $w \sim v$. The updating procedure is the following. If the

current configuration is x, choose a site v uniformly at random, and if no neighbor of v is occupied, make v occupied or vacant equiprobably. Show that the HMC so described is irreducible and that its stationary distribution is the uniform distribution on F.

Exercise 6.6.38. A MONOTONICITY PROPERTY OF THE GIBBS SAMPLER
Let μ be a probability measure on Λ^V and let ν be the probability measure obtained by applying the Gibbs sampler at a site $v \in V$. Show that $d_V(\nu, \pi) \leq d_V(\mu, \pi)$.

Chapter 7

Markov Chains, Continuous Time

There are two complementary points of view in the study of continuous-time HMCs. The traditional approach attempts to mimic the discrete-time theory. It is based on the *transition semigroup*, the continuous-time analogue of the iterates of the transition matrix in discrete time, and the principal mathematical object is then the *infinitesimal generator*. The infinitesimal generator is essentially a continuous-time notion, involving derivatives, and therefore it has no discrete-time analogue. There are technical problems associated with the transition semi-group approach, but these vanish when attention is restricted to HMCs with nice sample paths (namely: regular jump processes, to be defined later), which are the ones arising in applications. In this situation, a continuous-time HMC is just a discrete-time HMC "on an elastic time scale": the times separating the transitions are exponential random variables whose mean depends on the current state of the chain. This is the *regenerative* point of view. Both points of view will be presented.

7.1 Homogeneous Poisson Processes on the Line

This section introduces the simplest non-trivial point process on the line and the stochastic calculus associated with it. The homogeneous Poisson processes on the line are not only the simplest non-trivial continuous-time homogeneous Markov chains, but such chains can be described in terms of and constructed by independent homogeneous Poisson processes.

7.1.1 The Counting Process and the Interval Sequence

A *simple* and *locally finite* stochastic point process on the positive half-line is a sequence $\{T_n\}_{n\geq 1}$ of positive (possibly infinite) random variables such that, almost surely,

(i) $0 < T_n < T_{n+1}$ whenever $T_n < \infty$ (simplicity), and

(ii) $\lim_{n\uparrow\infty} T_n = +\infty$ (local finiteness).

This sequence is the *event time sequence*. Sometimes, one refers to T_n as the n-th "point" of the point process.

© Springer Nature Switzerland AG 2020
P. Brémaud, *Probability Theory and Stochastic Processes*, Universitext,
https://doi.org/10.1007/978-3-030-40183-2_7

The Counting Process

The integer-valued random variable

$$N((a, b]) = \sum_{n \geq 1} 1_{(a,b]}(T_n)$$

counts the event times occurring in the time interval $(a, b] \subset \mathbb{R}_+$. For $t \geq 0$, let

$$N(t) := N((0, t]).$$

In particular, $N(0) = 0$ and $N((a, b]) = N(b) - N(a)$. Since the interval $(0, t]$ is closed on the right, the trajectories (sample paths) $t \mapsto N(t, \omega)$ are right-continuous, non-decreasing, have limits on the left, and jump one unit upwards at each event time of the point process. The stochastic process $\{N(t)\}_{t \geq 0}$ is called the *counting process* of the point process. Since the sequence of event times can be recovered from N, the latter also receives the appellation "point process".

Definition 7.1.1 *A* homogeneous Poisson process *(HPP) or* standard Poisson process *on $\mathbb{R}_+$ with intensity $\lambda > 0$ is a point process on $\mathbb{R}_+$ such that*

(α) *for all $k \in \mathbb{N}_+$, all mutually disjoint intervals $I_j := (a_j, b_j]$ $(1 \leq j \leq k)$, the random variables $N(I_j)$ $(1 \leq j \leq k)$ are independent, and*

(β) *for any interval $(a, b] \subset \mathbb{R}_+$, $N((a, b])$ is a Poisson random variable with mean $\lambda(b - a)$.*

(In particular, $\mathrm{E}[N((a, b])] = \lambda(b - a)$, and therefore, λ is the average density of points.)

Equivalently, for all mutually disjoint intervals $I_j = (a_j, b_j] \subset \mathbb{R}_+$ and all real numbers u_j $(1 \leq j \leq k)$,

$$\mathrm{E}\left[e^{i \sum_{j=1}^k u_j N(I_j)}\right] = \exp\left\{\sum_{j=1}^k \left(e^{iu_j} - 1\right) \lambda(b_j - a_j)\right\}.$$

An HPP is simple and locally finite (see Theorem 8.2.4 below).

EXAMPLE 7.1.2: THE RISK MODEL, TAKE 1. One of the first significant apparition of Poisson point processes occurred in the insurance risk model of Cramér and Lundberg. In the insurance business, one is interested in evaluating the probability of ruin of a given insurance company. The standard model features a point process N on $(0, \infty)$ with event times sequence $\{T_n\}_{n \geq 1}$, and an IID sequence of real-valued marks $\{Z_n\}_{n \geq 1}$, independent of N and having the common cumulative distribution function G, such that $G(0) = 0$, with mean $\mu < \infty$ and variance $\sigma^2 < \infty$. The *risk process* is defined on $[0, \infty)$ by

$$X(t) = ct - \sum_{n=1}^{N(t)} Z_n,$$

where c is a positive real constant. The interpretation is that c is the *gross risk premium* (that is, the rate of income per unit time), T_n is the time of occurrence of a claim and Z_n is the size of the claim. (The process

$$\sum_{n=1}^{N(t)} Z_n \quad (t \geq 0)$$

is called a *compound Poisson process*.)

Superposition of independent HPPs

Theorem 7.1.3 *Let N_i $(i \geq 1)$ be a family of independent HPPs with respective intensities λ_i $(i \geq 1)$. Then,*

(i) two distinct HPPs of this family have no points in common, and

(ii) if $\sum_{i=1}^{\infty} \lambda_i = \lambda < \infty$, the sum $N := \sum_{i=1}^{\infty} N_i$ is an HPP with intensity λ.

Proof. (i) It suffices to prove this for two HPPs, and the result is then a direct consequence of Theorem 8.1.23. For (ii), see Theorem 8.2.3 below. □

The next result is called the *competition theorem* for HPPs.

Theorem 7.1.4 *In the situation of Theorem 7.1.3, where $\sum_{i=1}^{\infty} \lambda_i = \lambda < \infty$, denote by Z the first event time of $N = \sum_{i=1}^{\infty} N_i$ and by J the index of the HPP responsible for it. In particular, Z is the first event of N_J. Then J and Z are independent, $P(J = i) = \frac{\lambda_i}{\lambda}$, and Z is exponential with mean λ^{-1}.*

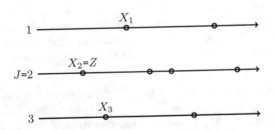

Proof. Call $X_1, X_2, \ldots$ the first event times of $N_1, N_2, \ldots$. These are independent exponential variables with respective parameters $\lambda_1, \lambda_2, \ldots$. In particular, since such variables are absolutely continuous, the index J is almost surely unambiguously defined. We have,

$$P(J = i, Z \geq a) = P(X_i \geq a, X_j \geq X_i \text{ for all } j \neq i)$$

$$= \int_a^{\infty} \lambda_i e^{-\lambda_i s} P(X_j \geq s \text{ for all } j \neq i) \, ds$$

$$= \int_a^{\infty} \lambda_i e^{-\lambda_i s} \Pi_{j \neq i} e^{-\lambda_j s} \, ds = \int_a^{\infty} \lambda_i e^{-\lambda s} \, ds = \frac{\lambda_i}{\lambda} e^{-\lambda a}.$$

Letting $a \to \infty$ yields $P(J = i) = \frac{\lambda_i}{\lambda}$. Summing with respect to i gives $P(Z \geq a) = e^{-\lambda a}$. Therefore $P(J = i, Z \geq a) = P(J = i)P(Z \geq a)$, for all $i \in \mathbb{N}_+$ and all $a \in \mathbb{R}_+$, which implies independence. □

Strong Markov Property

For any random variable τ with values in $[0, +\infty]$, one defines N^τ to be the restriction of N to $\mathbb{R}_+ \cap (0, \tau]$, that is, for all intervals $(a, b] \subset \mathbb{R}_+$,

$$N^\tau((a, b]) := N((a, b] \cap (0, \tau]).$$

The point process N^τ is "N before τ". One defines the point process $S_\tau N$, or "N after τ", by

$$S_\tau N((a, b]) := N((\tau + a, \tau + b])$$

for all intervals $(a, b] \subset \mathbb{R}_+$. In particular, if $\tau = \infty$, $S_\tau N$ is the empty point process.

Recall the definition of an $\mathcal{F}_t$-stopping time, where $\{\mathcal{F}_t\}_{t \geq 0}$ is a family of non-decreasing σ-fields (that is, $\mathcal{F}_s \subseteq \mathcal{F}_t$ whenever $0 \leq s \leq t$). It is a random variable τ with values in $[0, +\infty]$ such that for all $a \in \mathbb{R}_+$, $\{\tau \leq a\} \in \mathcal{F}_a$ (Definition 5.3.10).

Theorem 7.1.5 *Let A be an arbitrary index and let N_α $(\alpha \in A)$ be an independent family of* HPPs *on $\mathbb{R}_+$ with respective intensities λ_α $(\alpha \in A)$. Let $\mathcal{G}$ be a σ-field independent of $\mathcal{F}^N := \vee_{\alpha \in A} \mathcal{F}^{N_\alpha}$. For all $t \geq 0$, define $\mathcal{F}_t^N := \vee_{\alpha \in A} \mathcal{F}_t^{N_\alpha}$. Let τ be an $\mathcal{F}_t^N \vee \mathcal{G}$-stopping-time such that $P(\tau < \infty) > 0$. Then, given $\{\tau < \infty\}$,*

- *(α) $\mathcal{F}^{N^\tau}$ and $\mathcal{G}$ are independent of $\mathcal{F}^{S_\tau N}$, and*

- *(β) the family $S_\tau N_\alpha$ $(\alpha \in A)$ is an independent family of* HPPs *with respective intensities λ_α $(\alpha \in A)$.*

Proof. The proof is given for a single HPP, N. The general case is similar, once it is remembered that the independence statements concerns only finite collections of point processes of the family N_α $(\alpha \in A)$. In the following, the use of Theorems 5.1.7 and 5.1.10 is implicit. (These results allow us to restrict attention to finite-dimensional distributions in order to prove either identity of the distributions of stochastic processes or their independence.)

We first show that it suffices to prove that for all integers k, ℓ, all $u_1, \ldots, u_k, v_1, \ldots, v_\ell$, $w \in \mathbb{R}$, all mutually disjoint intervals $(a_j, b_j]$ $(1 \leq j \leq k)$, all mutually disjoint intervals $[c_j, d_j]$ $(1 \leq j \leq \ell)$, and for any real-valued random variable Z that is $\mathcal{G}$-measurable, it holds that

$$E\left[\left(e^{i\left(\sum_{j=1}^k u_j(S_\tau N)((a_j, b_j]) + \sum_{j=1}^\ell v_j N^\tau((c_j, d_j]) + wZ\right)}\right) 1_{\{\tau < \infty\}}\right]$$
$$= E\left[e^{i\sum_{j=1}^k u_j N((a_j, b_j])}\right] E\left[\left(e^{i\sum_{j=1}^\ell v_j N^\tau((c_j, d_j]) + wZ}\right) 1_{\{\tau < \infty\}}\right]. \qquad (7.1)$$

Indeed, letting $v_1 = \cdots = v_\ell = w = 0$ in the above identity, we obtain

$$E\left[\left(e^{i\sum_{j=1}^k u_j(S_\tau N)((a_j, b_j])}\right) 1_{\{\tau < \infty\}}\right] / P(\tau < \infty)$$

$$= E\left[e^{i\sum_{j=1}^k u_j N((a_j, b_j])}\right] = \exp\left\{\sum_{j=1}^k \left(e^{iu_j} - 1\right) \lambda(b_j - a_j)\right\}. \qquad (7.2)$$

But

$$E\left[\left(e^{i\sum_{j=1}^k u_j(S_\tau N)((a_j, b_j])}\right) 1_{\{\tau < \infty\}}\right] / P(\tau < \infty) = \widetilde{E}\left[e^{i\sum_{j=1}^k u_j(S_\tau N)((a_j, b_j])}\right],$$

where $\widetilde{E}$ denotes expectation with respect to the probability $\widetilde{P}$ defined by

$$\widetilde{P}(A) = P(A \cap \{\tau < \infty\})/P(\tau < \infty)$$

(P conditioned by the event $\{\tau < \infty\}$). Therefore $S_\tau N$ is an HPP of intensity λ under probability $\widetilde{P}$. This proves (β). Next, using (7.2), the identity (7.1) becomes

$$\mathrm{E}\left[\left(e^{i\sum_{j=1}^{k} u_j (S_\tau N)((a_j, b_j]) + i\sum_{j=1}^{\ell} v_j N^\tau((c_j, d_j]) + iwZ}\right) 1_{\{\tau < \infty\}}\right] / P(\tau < \infty)$$

$$= \left(\mathrm{E}\left[\left(e^{i\sum_{j=1}^{k} u_j (S_\tau N)((a_j, b_j])}\right) 1_{\{\tau < \infty\}}\right] / P(\tau < \infty)\right) \cdots$$

$$\times \left(\mathrm{E}\left[\left(e^{i\sum_{j=1}^{\ell} v_j N^\tau((c_j, d_j]) + iwZ}\right) 1_{\{\tau < \infty\}}\right] / P(\tau < \infty)\right),$$

that is,

$$\widetilde{\mathrm{E}}\left[e^{i\sum_{j=1}^{k} u_j (S_\tau N)((a_j, b_j]) + i\sum_{j=1}^{\ell} v_j N^\tau((c_j, d_j]) + iwZ}\right]$$

$$= \widetilde{\mathrm{E}}\left[e^{i\sum_{j=1}^{k} u_j (S_\tau N)((a_j, b_j])}\right] \widetilde{\mathrm{E}}\left[e^{i\sum_{j=1}^{\ell} v_j N^\tau((c_j, d_j]) + iwZ}\right],$$

Therefore, under $\widetilde{P}$, the family of random variables $S_\tau N((a_j, b_j])$ $(1 \le j \le k)$ is independent of Z and $N^\tau([c_j, d_j])$ $(1 \le j \le \ell)$, and this for all integers k, ℓ, all mutually disjoint intervals $(a_j, b_j]$ $(1 \le j \le k)$, all mutually disjoint intervals $[c_j, d_j]$ $(1 \le j \le \ell)$, and for any real-valued random variable Z that is $\mathcal{G}$-measurable. This proves statement (α).

We now proceed to the proof of (7.1), actually a simplified version that contains all the ingredients of the complete proof and saves us from notational hell. We prove that

$$\mathrm{E}\left[\left(e^{iuS_\tau N((a,b]) + ivN^\tau((c,d]) + iwZ}\right) 1_{\{\tau < \infty\}}\right]$$

$$= \mathrm{E}\left[\left(e^{iuN((a,b])}\right) \mathrm{E}[e^{ivN^\tau((c,d]) + iwZ}] 1_{\{\tau < \infty\}}\right]. \tag{7.3}$$

The left-hand side is

$$\mathrm{E}\left[\left(e^{iuN((\tau+a, \tau+b]) + ivN(((0,\tau]\cap(c,d])) + iwZ}\right) 1_{\{\tau < \infty\}}\right],$$

that is, in the special case where τ takes a countable set of values $t_k \in \mathbb{R}_+$ $(k \ge 1)$ and maybe also the value $+\infty$,

$$\sum_{k=1}^{\infty} \mathrm{E}\left[\left(e^{iuN((t_k+a, t_k+b]) + ivN((0,t_k]\cap(c,d]) + iwZ}\right) 1_{\{\tau = t_k\}}\right].$$

But $1_{\{\tau = t_k\}}$ is $\mathcal{G} \vee \mathcal{F}_{t_k}^N$-measurable and therefore the above expression becomes, in view of the independence and homogeneity properties of HPPs,

$$\sum_{k=1}^{\infty} \mathrm{E}\left[e^{iuN((t_k+a, t_k+b])}\right] \mathrm{E}\left[\left(e^{ivN((0,t_k]\cap(c,d]) + iwZ}\right) 1_{\{\tau = t_k\}}\right]$$

$$= \mathrm{E}\left[e^{iuN((a,b])}\right] \left(\sum_{k=1}^{\infty} \mathrm{E}\left[\left(e^{ivN((0,t_k]\cap(c,d]) + iwZ}\right) 1_{\{\tau = t_k\}}\right]\right),$$

and this is the right-hand side of (7.3), as announced.

To pass from the case where the stopping time τ takes a countable number of values to the general case, define for each $n \ge 1$ an approximation $\tau(n)$ of the stopping time τ by:

$$\tau(n, \omega) = \begin{cases} 0 & \text{if } \tau(\omega) = 0 \\ \frac{k+1}{2^n} & \text{if } \frac{k}{2^n} < \tau(\omega) \leq \frac{k+1}{2^n} \\ +\infty & \text{if } \tau(\omega) = \infty. \end{cases}$$

Then (Theorem 5.3.13) $\tau(n)$ is an $\mathcal{F}_t^N \vee \mathcal{G}$-stopping-time with a countable number of values and therefore (7.3) is satisfied for $\tau(n)$. Now, the sequence $\{\tau(n)\}_{n \geq 1}$ decreases to τ, so that the result for τ itself follows by dominated convergence. $\qquad\square$

The Interval Sequence

The sequence $\{S_n\}_{n \geq 1}$ defined by $T_0 = 0$ and

$$S_n := T_n - T_{n-1}$$

($= \infty$ if $T_{n-1} = \infty$) is called the *inter-event sequence* or, in the appropriate context, the *inter-arrival sequence*. The strong Markov property yields the probabilistic description of an HPP in terms of inter-event times.

Theorem 7.1.6 *The inter-event sequence $\{S_n\}_{n \geq 1}$ of an HPP on the positive half-line with intensity $\lambda > 0$ is IID, with exponential distribution of parameter λ, that is, $P(S_n \leq t) = 1 - e^{-\lambda t}$ and in particular $\mathrm{E}[S_n] = \lambda^{-1}$.*

Proof. Apply Theorem 7.1.5 with $\mathcal{G} = (\Omega, \varnothing)$ and $\tau = T_n$ to obtain that the first event time of $S_{T_n} N$, namely S_{n+1}, is independent of N^{T_n} (and in particular of $S_1, \ldots, S_n$) and is an exponential random variable with parameter λ. This being true for all $n \geq 1$, the proof is done.

$\qquad\square$

7.1.2 Stochastic Calculus of HPPs

Let $\{\mathcal{F}_t\}_{t \in \mathbb{R}}$ be a *history*, that is a non-dereasing family of sub-σ-fields of the measurable space $(\Omega, \mathcal{F})$, and let $\mathcal{F}_\infty := \vee_{t \in \mathbb{R}} \mathcal{F}_t$. For a given point process N on $\mathbb{R}_+$ and $t \in \mathbb{R}_+$, let

$$\mathcal{F}_t^N := \sigma(N((a, b]) \, ; \, (a, b] \subseteq (-\infty, t])$$

be the σ-field recording the events of N up to time t. Any history $\{\mathcal{F}_t\}_{t \in \mathbb{R}}$ such that $\mathcal{F}_t^N \subseteq \mathcal{F}_t$ for all $t \in \mathbb{R}$ is called a *history of N*. The history $\{\mathcal{F}_t^N\}_{t \in \mathbb{R}}$ is called the *internal history of N*.

A Smoothing Formula for HPPs

Theorem 7.1.7 *Let $\{\mathcal{F}_t\}_{t \geq 0}$ be a history of N, an HPP on $\mathbb{R}_+$ of intensity λ. Suppose moreover that for all $a > 0$ and all c, d such that $a \leq c \leq d < \infty$, $N(c, d]$ is independent of $\mathcal{F}_a$. Let $\{Z(t)\}_{t \geq 0}$ be a complex left-continuous stochastic process adapted to $\{\mathcal{F}_t\}_{t \geq 0}$. Suppose that at least one of the two conditions below is satisfied.*

(i) $\{Z(t)\}_{t \geq 0}$ is real non-negative.

(ii) $\{Z(t)\}_{t \geq 0}$ is complex-valued and $\mathrm{E}\left[\int_0^\infty |Z(t)| \lambda \, dt\right] < \infty$.

Then,

$$\mathrm{E}\left[\int_{(0,\infty)} Z(s) N(ds)\right] = \mathrm{E}\left[\int_0^\infty Z(s) \lambda \, ds\right]. \tag{7.4}$$

Formula (7.4) is the *smoothing formula* for HPPs. Remember that the Stieltjes–Lebesgue integral in the left-hand side is in this case just the sum $\sum_{n\geq 1} Z(T_n)\,1_{(0,\infty)}(T_n)$.

Proof.

A. We first treat the case where $\{Z(t)\}_{t\geq 0}$ is real, non-negative, and bounded, and prove that

$$\mathrm{E}\left[\int_{(0,T]} Z(s)N(\mathrm{d}s)\right] = \mathrm{E}\left[\int_0^T Z(s)\lambda\,\mathrm{d}s\right], \qquad (\star)$$

where $T < \infty$. Let for all $n \geq 1$, all $\omega \in \Omega$ and all $t \geq 0$,

$$Z_n(t,\omega) := \sum_{k=0}^{2^n-1} Z\left(kT2^{-n},\omega\right) 1_{I(T,n,k)}(t), \qquad (7.5)$$

where $I(T,n,k) := (kT2^{-n}, (k+1)T2^{-n}]$. By the left-continuity hypothesis, for all $\omega \in \Omega$ and all $t \in [0,T]$,

$$\lim_{n\uparrow\infty} Z_n(t,\omega) = Z(t,\omega).$$

We first check that $(\star)$ is true when $\{Z(t)\}_{t\geq 0}$ is replaced by its approximation $\{Z_n(t)\}_{t\geq 0}$. Indeed, the left-hand side of this equality is then

$$\mathrm{E}\left[\int_{(0,T]} Z_n(t)N(\mathrm{d}t)\right] = \mathrm{E}\left[\sum_{k=0}^{2^n-1} Z\left(kT2^{-n}\right) N(I(T,n,k))\right]$$
$$= \sum_{k=0}^{2^n-1} \mathrm{E}\left[Z\left(kT2^{-n}\right) N(I(T,n,k))\right].$$

But $Z\left(kT2^{-n}\right)$ is $\mathcal{F}_{kT2^{-n}}$-measurable, and therefore independent of $N(I(T,n,k))$, so that the last term of the above chain of equalities is equal to

$$\sum_{k=0}^{2^n-1} \mathrm{E}\left[Z\left(kT2^{-n}\right)\right] \mathrm{E}\left[N(I(T,n,k))\right] = \sum_{k=0}^{2^n-1} \mathrm{E}\left[Z\left(kT2^{-n}\right)\right] \lambda T2^{-n}$$
$$= \mathrm{E}\left[\sum_{k=0}^{2^n-1} Z\left(kT2^{-n}\right) \lambda T2^{-n}\right] = \mathrm{E}\left[\int_0^T Z_n(s)\lambda\,\mathrm{d}s\right].$$

Therefore,

$$\mathrm{E}\left[\int_{(0,T]} Z_n(s)N(\mathrm{d}s)\right] = \mathrm{E}\left[\int_0^T Z_n(s)\lambda\,\mathrm{d}s\right]. \qquad (7.6)$$

Denoting by K the upper bound of $Z(t,\omega)$, we have that $\int_0^T Z_n(t)\lambda\,\mathrm{d}t \leq K\lambda T$ and $\int_{(0,T]} Z_n(t)N(\mathrm{d}t) \leq KN(T)$. Also, $\mathrm{E}[N(T)] = \lambda T < \infty$. Therefore, by dominated convergence, letting $n \uparrow \infty$ in both sides of (7.6), we obtain $(\star)$.

B. We now treat the non-negative and unbounded case. Defining

$$g_K(x) = \begin{cases} 1 & \text{if } x \leq K, \\ 0 & \text{if } x \geq K+1, \\ -x+K+1 & \text{if } K \leq x \leq K+1, \end{cases}$$

the process $\{g_K(Z(t))\}_{t\geq 0}$ satisfies the conditions for $(\star)$ and therefore

$$\mathrm{E}\left[\int_{(0,T]} g_K(Z(s))N(\mathrm{d}s)\right] = \mathrm{E}\left[\int_0^T g_K(Z(s)), \lambda\,\mathrm{d}s\right],$$

which gives $(\star)$ as $K \to \infty$, by monotone convergence. It then suffices to let $T \uparrow \infty$ and to invoke the monotone convergence theorem again to obtain (7.4) in the non-negative unbounded case. Note that the quantities in (7.4) can now very well be infinite.

C. The real-valued case follows easily, by first considering separately the positive and negative parts of the integrand. The complex case is a direct consequence of the real case when one considers separately the real and imaginary parts. $\qquad\square$

Watanabe's Characterization

Theorem 7.1.8 *Let N be a simple and locally finite point processes on $\mathbb{R}_+$, and let $\{\mathcal{F}_t\}_{t\geq 0}$ be a history of N. Let λ be a positive real number. Suppose that*

$$\mathrm{E}\left[\int_{(0,T]} Z(t)N(\mathrm{d}t)\right] = \mathrm{E}\left[\int_0^T Z(t)\lambda\,\mathrm{d}t\right] \tag{7.7}$$

holds true for all $T > 0$, and for all non-negative real-valued stochastic processes $\{Z(t)\}_{t\geq 0}$ with left-continuous trajectories and adapted to $\{\mathcal{F}_t\}_{t\geq 0}$. Then, N is a homogeneous Poisson process with intensity λ, and for any interval $(a, b] \in \mathbb{R}_+$, $N((a, b])$ is independent of $\mathcal{F}_a$.

Note that formula (7.7) holds true for all $T > 0$, and for all *complex-valued* stochastic processes $\{Z(t)\}_{t\geq 0}$ with left-continuous trajectories, adapted to $\{\mathcal{F}_t\}_{t\geq 0}$ and such that $\mathrm{E}\left[\int_0^T |Z(t)|\,\mathrm{d}t\right] < \infty$ (the same arguments as in part C of the proof of Theorem 7.1.7).

Proof. We show that it is enough to prove that

$$\mathrm{E}\left[1_A e^{iuN((a,b])}\right] = P(A)\,e^{(e^{iu}-1)\lambda(b-a)}, \tag{$\star$}$$

where $(a, b] \subseteq \mathbb{R}_+$ and $u \in \mathbb{R}$, and A is an event in $\mathcal{F}_a$. In fact, with $A = \Omega$ in $(\star)$, we obtain

$$\mathrm{E}[e^{iuN((a,b])}] = e^{(e^{iu}-1)\lambda(b-a)},$$

which shows that $N((a, b])$ is a Poisson random variable of mean $\lambda(b - a)$. Equality $(\star)$ then reads

$$\mathrm{E}[1_A e^{iuN((a,b])}] = P(A)\mathrm{E}[e^{iuN((a,b])}],$$

from which it follows that $N((a, b])$ and A are independent. Since A is an arbitrary event in $\mathcal{F}_a$, we have proved that $N((a, b])$ is independent of $\mathcal{F}_a$ and that N has independent increments since $\mathcal{F}_a \supseteq \mathcal{F}_a^N$.

For the proof of $(\star)$, consider the process

$$X(t) := 1_A e^{iuN((a,t])},$$

and observe that it is piecewise constant with all discontinuity times located at the event times of N. Therefore,

$$X(b) = X(a) + \sum_{n\geq 1}\{X(T_n) - X(T_n-)\}1_{(a,b]}(T_n).$$

But $X(0) = 1_A$ and, for any $T_n \in (a, b]$ (watch the parentheses and the square brackets),

$$X(T_n) = e^{iuN((a,T_n])} = e^{iu(N((a,T_n))+1)} = e^{iuN((a,T_n))} e^{iu} = X(T_{n-}) e^{iu}.$$

Therefore,

$$X(b) = 1_A + \sum_{n \geq 1} X(T_{n-})(e^{iu} - 1)1_{(a,b]}(T_n)$$

or, in the integral notation,

$$X(b) = 1_A + \int_{(a,b]} X(s-)(e^{iu} - 1)N(\mathrm{d}s). \tag{7.8}$$

Since $Z(t) = X(t-)(e^{iu}-1)$ defines a bounded left-continuous complex-valued stochastic process adapted to $\{\mathcal{F}_t\}_{t\geq0}$, equality (7.7) holds true by hypothesis. Therefore,

$$\mathrm{E}\left[\int_{(a,b]} X(s-)(e^{iu} - 1)N(\mathrm{d}s)\right] = \mathrm{E}\left[\int_a^b X(s-)(e^{iu} - 1)\lambda\,\mathrm{d}s\right].$$

Now, for each $\omega \in \Omega$, $X(t-,\omega) = X(t,\omega)$ except on a countable set, and therefore one may replace $X(t-)$ by $X(t)$ in the right-hand side of the above equality. Taking expectations in (7.8) therefore yields

$$\mathrm{E}[X(b)] = P(A) + \int_a^b \mathrm{E}[X(s)](e^{iu} - 1)\lambda\,\mathrm{d}s.$$

This being true for all $b \geq a$, $(\star)$ follows. $\qquad\square$

The Strong Markov Property via Watanabe's theorem

The strong Markov property of HPPs can be obtained as a consequence of Watanabe's theorem. For instance,

Theorem 7.1.9 *Let N be an HPP on $\mathbb{R}_+$ with intensity λ and let τ be a finite $\mathcal{F}_t^N$-stopping time. Define the point process N^τ on $\mathbb{R}_+$ by*

$$N^\tau(0, t] := N(\tau, \tau + t].$$

Then N^τ is an HPP independent of $\mathcal{F}_\tau^N$.

Proof. Let $(a, b] \subset \mathbb{R}_+$ and $A \in \mathcal{F}_{\tau+a}^N$. The process

$$Z(t) = 1_A 1_{(\tau+a,\tau+b]}(t)$$

is left-continuous and adapted to $\{\mathcal{F}_t^N\}_{t\geq0}$ and therefore

$$\mathrm{E}\left[\int_{\mathbb{R}_+} Z(t)\,N(\mathrm{d}t)\right] = \mathrm{E}\left[\int_{\mathbb{R}_+} Z(t)\,\lambda\,\mathrm{d}t\right],$$

which reads in this particular case as

$$\mathrm{E}\left[1_A N^\tau(a, b]\right] = \mathrm{E}\left[1_A \lambda(b - a)\right].$$

Since this holds for all $(a, b] \subset \mathbb{R}_+$ and all $A \in \mathcal{F}_{\tau+a}^N$, $\{N^\tau(0, t] - \lambda t\}_{t\geq0}$ is an $\mathcal{F}_{\tau+t}^N$-martingale. The conclusion then follows from Corollary 13.4.3. $\qquad\square$

7.2 The Transition Semigroup

7.2.1 The Infinitesimal Generator

Henceforth in this chapter, $t \geq 0$ means $t \in \mathbb{R}_+$, and $n \geq 0$ means $n \in \mathbb{N}$. Let E be a countable set, called the *state space*.

Definition 7.2.1 *The stochastic process $\{X(t)\}_{t \geq 0}$ with values in E is called a* continuous-time *Markov chain if for all $i, j, i_1, \ldots, i_k \in E$, all $t, s \geq 0$ and all $s_1, \ldots, s_k \geq 0$ with $s_\ell < s$ ($1 \leq \ell \leq k$),*

$$P(X(t + s) = j \mid X(s) = i, X(s_1) = i_1, \ldots, X(s_k) = i_k)$$
$$= P(X(t + s) = j \mid X(s) = i). \qquad (7.9)$$

This continuous-time Markov chain is called homogeneous *if the right-hand side of (7.9) is independent of s. We then say: a* continuous-time HMC.

Define the matrix

$$\mathbf{P}(t) := \{p_{ij}(t)\}_{i,j \in E},$$

where

$$p_{ij}(t) := P(X(t + s) = j \mid X(s) = i).$$

The family $\{\mathbf{P}(t)\}_{t \geq 0}$ is called the *transition semi-group* of the continuous-time HMC. A simple application of the Bayes rule of total causes yields the *Chapman–Kolmogorov equation*

$$p_{ij}(t + s) = \sum_{k \in E} p_{ik}(t) p_{kj}(s),$$

that is, in compact form,

$$\mathbf{P}(t + s) = \mathbf{P}(t)\mathbf{P}(s). \qquad (7.10)$$

Also, clearly,

$$\mathbf{P}(0) = I, \qquad (7.11)$$

where I is the identity matrix.

The *distribution at time t* of the chain is the vector $\mu(t) = \{\mu_i(t)\}_{i \in E}$, where $\mu_i(t) = P(X(t) = i)$. It is obtained from the initial distribution by the formula

$$\mu(t)^T = \mu(0)^T \mathbf{P}(t). \qquad (7.12)$$

More generally, for all $0 \leq t_1 \leq t_2 \leq \cdots \leq t_k$ and all states $i_0, i_1, \ldots, i_k$,

$$P\left(\bigcap_{j=1}^{k} \{X(t_j) = i_j\}\right) = \sum_{i_0 \in E} P(X(0) = i_0) \prod_{j=1}^{k} p_{i_{j-1} i_j}(t_j - t_{j-1}).$$

The above formulas are standard applications of the elementary Bayes calculus. The last one shows in particular that (Theorem 5.1.7):

Theorem 7.2.2 *The probability distribution of a continuous-time HMC is entirely determined by its initial distribution and its transition semi-group.*

Formula (7.9) easily extends to

$$P(X(t_1 + s) = j_1, \ldots, X(t_\ell + s) = j_\ell \mid X(s) = i, X(s_1) = i_1, \ldots, X(s_k) = i_k)$$
$$= P(\, X(t_1 + s) = j_1, \ldots, X(t_\ell + s) = j_\ell \mid X(s) = i)\,,$$

for all states $i, i_1, \ldots, i_k, j_1, \ldots, j_\ell$, all $t_1, \ldots, t_\ell, s \geq 0$ and all $s_1, \ldots, s_k \geq 0$ with $s_r < s$ ($1 \leq r \leq k$), or, equivalently,

$$P(A \cap B \mid X(s) = i) = P(A \mid X(s) = i)P(B \mid X(s) = i) \tag{7.13}$$

for all A of the form $\{X(t_1 + s) = j_1, \ldots, X(t_\ell + s) = j_\ell\}$ and all B of the form $\{X(s_1) = i_1, \ldots, X(s_k) = i_k\}$. Therefore, by Theorem 5.1.10, (7.13) is true for all $A \in \sigma(X(u); u \geq s)$ and all $B \in \mathcal{F}_s^X$. This is the content of the following theorem:

Theorem 7.2.3 *The past $\mathcal{F}_s^X$ and the future $\sigma(X(v); v \geq s)$ at time $s \geq 0$ are conditionally independent given the present state $X(s)$.*

In particular, the Markov property is "independent of the arrow of time".

Equality (7.13) can be written as

$$\mathrm{E}[1_A \times 1_B \mid X(s) = i] = \mathrm{E}[1_A \mid X(s) = i]\mathrm{E}[1_B \mid X(s) = i]$$

for all $i \in E$, all $A \in \sigma(X(v); v \geq s)$ and all $B \in \mathcal{F}_s^X$. This extends with the usual argument to

$$\mathrm{E}[Y \times Z \mid X(s) = i] = \mathrm{E}[Y \mid X(s) = i]\mathrm{E}[Z \mid X(s) = i]$$

for all non-negative random variables Y and Z that are respectively $\sigma(X(v); v \geq s)$-measurable and $\mathcal{F}_s^X$-measurable. Since this true for all $i \in E$,

$$\mathrm{E}^{X(s)}[Y \times Z] = \mathrm{E}^{X(s)}[Y]\mathrm{E}^{X(s)}[Z], \tag{7.14}$$

for all non-negative Y and Z as above.

The following notation, similar to the one adopted for discrete-time HMCs, will be used. For all $i \in E$, $P_i(\cdot)$ is an abbreviation for $P(\cdot \mid X(0) = i)$ and for any probability distribution μ on E,

$$P_\mu(\cdot) := \sum_{i \in E} \mu(i)P(\cdot \mid X(0) = i)\,.$$

The symbols E_i and E_μ denote expectation with respect to P_i and P_μ, respectively.

EXAMPLE 7.2.4: HPPS AS CONTINUOUS-TIME HMCS. Let N be an HPP on $\mathbb{R}_+$ with the intensity $\lambda > 0$. The counting process $\{N(t)\}_{t \geq 0}$ is a continuous-time HMC with transition semigroup defined by

$$p_{ij}(t) = 1_{\{j \geq i\}}e^{-\lambda t}\frac{(\lambda t)^{j-i}}{(j-i)!}\,.$$

Proof. With $C := \{N(s_1) = i_1, \ldots, N(s_k) = i_k)\}$, we have, for $i \geq j$,

$$P(N(t + s) = j \mid N(s) = i, C)$$
$$= \frac{P(N(t + s) = j, N(s) = i, C)}{P(N(s) = i, C)}$$
$$= \frac{P(N(s, s + t] = j - i, N(s) = i, C)}{P(N(s) = i, C)}\,.$$

But $N(s, s+t]$ is independent of $N(s)$ and of C, and therefore,

$$P(N(s, s+t] = j - i, N(s) = i, C)$$
$$= P(N(s, s+t] = j - i)P(N(s) = i, C),$$

so that

$$P(N(t+s) = j \mid N(s) = i, C) = P(N(s, s+t] = j - i).$$

Similarly,

$$P(N(t+s) = j \mid N(s) = i) = P(N(s, s+t] = j - i) = e^{-\lambda t}\frac{(\lambda t)^{j-i}}{(j-i)!}.$$

□

EXAMPLE 7.2.5: FLIP-FLOP. Let N be an HPP on $\mathbb{R}_+$ with intensity $\lambda > 0$. Define the *flip-flop process* with state space $\{+1, -1\}$ by

$$X(t) := X(0) \times (-1)^{N(t)},$$

where $X(0)$ is a $\{+1, -1\}$-valued random variable independent of the counting process N. In words: the flip-flop process switches between -1 and $+1$ at each event of N. It is a continuous-time HMC with transition semigroup

$$\mathbf{P}(t) = \frac{1}{2}\begin{pmatrix} 1 + e^{-2\lambda t} & 1 - e^{-2\lambda t} \\ 1 - e^{-2\lambda t} & 1 + e^{-2\lambda t} \end{pmatrix}.$$

Proof. The value $X(t+s)$ depends on $N(s, s+t]$ and $X(s)$. Also, $N(s, s+t]$ is independent of $X(0), N(s_1), \ldots, N(s_k)$ when $s_\ell \le s$ (ℓ, $1 \le \ell \le k$), and the latter random variables determine $X(s_1), \ldots, X(s_k)$. Therefore, $X(t+s)$ is independent of $X(s_1), \ldots, X(s_k)$ given $X(s)$, that is, $\{X(t)\}_{t \ge 0}$ is a Markov chain. Moreover,

$$P(X(t+s) = 1 \mid X(s) = -1) = P(N(s, s+t] = \text{ odd })$$
$$= \sum_{k=0}^{\infty} e^{-\lambda t}\frac{(\lambda t)^{2k+1}}{(2k+1)!} = \frac{1}{2}(1 - e^{-2\lambda t}),$$

that is, $p_{-1,+1}(t) = \frac{1}{2}(1 - e^{-2\lambda t})$. Similar computations give the announced result for $p_{+1,-1}(t)$.

□

The Uniform HMC

Definition 7.2.6 *Let $\{\widehat{X}_n\}_{n \ge 0}$ be a discrete-time HMC with countable state space E and transition matrix $\mathbf{K} = \{k_{ij}\}_{i,j \in E}$ and let N be a HPP on $\mathbb{R}_+$ of intensity $\lambda > 0$ and associated time sequence $\{T_n\}_{n \ge 1}$. Suppose that $\{\widehat{X}_n\}_{n \ge 0}$ and N are independent. The stochastic process*

$$X(t) = \widehat{X}_{N(t)} \quad (t \ge 0)$$

is called a uniform *Markov chain. The Poisson process N is the* clock, *and the chain $\{\widehat{X}_n\}_{n \ge 0}$ is the* subordinated chain.

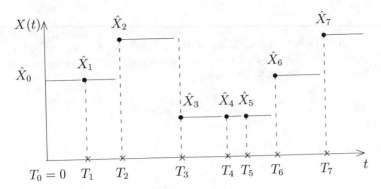

Uniform Markov chain

Remark 7.2.7 Observe that $X(T_n) = \widehat{X}_n$ for all $n \geq 0$. Observe also that the discontinuity times of the uniform chain are all events of N but that not all events of N are discontinuity times, since it may well occur that $\widehat{X}_{n-1} = \widehat{X}_n$ (a "transition" of type $i \to i$ of the subordinated chain).

The process $\{X(t)\}_{t \geq 0}$ is a continuous-time HMC (Exercise 7.5.3). Its transition semigroup is

$$\mathbf{P}(t) = \sum_{n=0}^{\infty} e^{-\lambda t} \frac{(\lambda t)^n}{n!} \mathbf{K}^n, \tag{7.15}$$

that is,

$$p_{ij}(t) = \sum_{n=0}^{\infty} e^{-\lambda t} \frac{(\lambda t)^n}{n!} k_{ij}(n).$$

Indeed,

$$
\begin{aligned}
P_i(X(t) = j) = P_i(\widehat{X}_{N(t)} = j) &= \sum_{n=0}^{\infty} P_i(N(t) = n, \widehat{X}_n = j) \\
&= \sum_{n=0}^{\infty} P_i(N(t) = n) P_i(\widehat{X}_n = j).
\end{aligned}
$$

Definition 7.2.8 *The probability distribution π on E is called a* stationary distribution *of the continuous-time HMC, or of its transition semi-group, if*

$$\pi^T \mathbf{P}(t) = \pi^T \quad (t \geq 0).$$

From (7.12), we see that if the initial distribution of the chain is a stationary distribution π, then the distribution at any time $t \geq 0$ is π, and moreover, the chain is stationary, since for all $k \geq 1$, all $0 \leq t_1 < \ldots < t_k$ and all states $i_1, \ldots, i_k$, the quantity

$$P(X(t_1 + t) = i_1, \ldots, X(t_k + t) = i_k) = \pi(i_1) p_{i_1, i_2}(t_2 - t_1) \cdots p_{i_{k-1}, i_k}(t_k - t_{k-1})$$

does not depend on $t \geq 0$. Therefore

Theorem 7.2.9 *A continuous-time HMC having for initial distribution a stationary distribution of the transition semi-group is stationary.*

EXAMPLE 7.2.10: UNIFORM HMC, TAKE 2. In the case of the uniform HMC of Definition 7.2.6, if π is a stationary distribution of the subordinated chain, $\pi^T \mathbf{K}^n = \pi^T$, and therefore in view of (7.15), $\pi^T \mathbf{P}(t) = \pi^T$. Conversely, if π is a stationary distribution of the continuous-time HMC, then, by (7.15),

$$\pi^T = \sum_{n=0}^{\infty} e^{-\lambda t} \frac{(\lambda t)^n}{n!} \pi^T \mathbf{K}^n,$$

and letting $t \downarrow 0$, we obtain $\pi^T = \pi^T \mathbf{K}$.

7.2.2 The Local Characteristics

Let $\{\mathbf{P}(t)\}_{t \geq 0}$ be a *transition semi-group* on E, that is, for each $t, s \geq 0$,

(a) $\mathbf{P}(t)$ is a stochastic matrix,

(b) $\mathbf{P}(0) = I$,

(c) $\mathbf{P}(t + s) = \mathbf{P}(t)\mathbf{P}(s)$.

Suppose moreover that the semi-group is *continuous at the origin*, that is,

(d) $\lim_{h \downarrow 0} \mathbf{P}(h) = \mathbf{P}(0) = I$, where the convergence therein is pointwise and for each entry.

In Exercise 7.5.4, the reader is invited to prove that continuity at the origin implies continuity at any time, that is, $\lim_{h \to 0} \mathbf{P}(t + h) = \mathbf{P}(t)$ for all $t > 0$.

The result to follow is purely analytical: it does not require $\{\mathbf{P}(t)\}_{t \geq 0}$ to be the transition semigroup of some continuous-time HMC.

Theorem 7.2.11 *Let $\{\mathbf{P}(t)\}_{t \geq 0}$ be a continuous transition semi-group on the countable state space E. For any state i,*

$$\text{there exists } q_i := \lim_{h \downarrow 0} \frac{1 - p_{ii}(h)}{h} \in [0, \infty], \tag{7.16}$$

and for any pair i, j of different states,

$$\text{there exists } q_{ij} := \lim_{h \downarrow 0} \frac{p_{ij}(h)}{h} \in [0, \infty). \tag{7.17}$$

Proof. For all $t \geq 0$ and all $n \geq 1$, we have $\mathbf{P}(t) = [\mathbf{P}\left(\frac{t}{n}\right)]^n$ and therefore $p_{ii}(t) \geq [p_{ii}\left(\frac{t}{n}\right)]^n$ ($i \in E$). Since $\lim_{h \downarrow 0} p_{ii}(h) = 1$, there exists an $\epsilon > 0$ such that $p_{ii}(h) > 0$ for all $h \in [0, \epsilon]$. For n sufficiently large, $\frac{t}{n} \in [0, \epsilon]$. Therefore, for all $t \geq 0$, $p_{ii}(t) > 0$, and the non-negative quantity

$$f_i(t) := -\log p_{ii}(t)$$

is finite. Also, $\lim_{h \downarrow 0} f_i(h) = 0$. Moreover, from $\mathbf{P}(t)\mathbf{P}(s) = \mathbf{P}(t + s)$, we have that $p_{ii}(t + s) \geq p_{ii}(t)p_{ii}(s)$, and therefore, the function f_i is subadditive, that is,

$$f_i(t + s) \leq f_i(t) + f_i(s) \quad (s, t \in \mathbb{R}_+).$$

Define the (possibly infinite) non-negative real number

$$q_i := \sup_{t>0} \frac{f_i(t)}{t} \,.$$

Then (Theorem B.5.1)

$$\lim_{h \downarrow 0} \frac{f_i(h)}{h} = q_i \,.$$

Therefore,

$$\lim_{h \downarrow 0} \frac{1 - p_{ii}(h)}{h} = \lim_{h \downarrow 0} \frac{1 - \mathrm{e}^{-f_i(h)}}{f_i(h)} \frac{f_i(h)}{h} = q_i,$$

and this proves the first equality in (7.16).

It now remains to prove (7.17). For this, take two different states i and j. Since $p_{ii}(t)$ and $p_{jj}(t)$ tend to 1 as $t > 0$ tends to 0, there exists for any $c \in (\frac{1}{2}, 1)$ a number $\delta > 0$ such that for $t \in [0, \delta]$, $p_{ii}(t) > c$ and $p_{jj}(t) > c$. Denote by $\{X_n\}_{n \geq 0}$ the discrete-time HMC defined by $X_n = X(nh)$, with transition matrix $\mathbf{P}(h)$. Let $n > 0$ be an integer and $h > 0$ be such that $0 \leq nh \leq \delta$.

One way to pass from state i at time 0 to state j at time n is to pass through state i at time r for some r, $0 \leq r \leq n-1$, without visiting state j meanwhile, then to pass from i at time r to state j at time $r+1$, and finally to pass from j at time $r+1$ to state j at time n. The paths corresponding to different values of r are different, but they do not exhaust the possibilities of going from $X_0 = i$ to $X_n = j$. Therefore,

$$p_{ij}(nh) \geq \sum_{r=0}^{n-1} P(X_1 \neq j, \ldots, X_{r-1} \neq j, X_r = i \mid X_0 = i) p_{ij}(h) P(X_n = j \mid X_{r+1} = j) \,.$$

The parameters δ, n and h are such that $P(X_n = j \mid X_{r+1} = j) \geq c$. Also

$$P(X_1 \neq j, \ldots, X_{r-1} \neq j, X_r = i \mid X_0 = i)$$
$$= P(X_r = i \mid X_0 = i)$$
$$\qquad - \sum_{k<r} P(X_1 \neq j, \ldots, X_{k-1} \neq j, X_k = j \mid X_0 = i) P(X_r = i \mid X_k = j)$$
$$\geq c - (1-c) \sum_{k<r} P(X_1 \neq j, \ldots, X_{k-1} \neq j, X_k = j \mid X_0 = i)$$
$$\geq c - (1-c) = 2c - 1 \,,$$

where we have observed that for $i \neq j$,

$$P(X_r = i \mid X_k = j) + P(X_r = j \mid X_k = j) \leq 1 \,,$$

and therefore,

$$P(X_r = i \mid X_k = j) \leq 1 - P(X_r = j \mid X_k = j) \leq 1 - c \,.$$

Therefore,

$$p_{ij}(nh) \geq n(2c - 1) p_{ij}(h) c \,.$$

Let now $t < \delta$ and $h < \delta$, and take for $n = n(t, h)$ the integer part of t/h. From the last inequality we obtain

$$\frac{p_{ij}(h)}{h} \leq \frac{1}{c(2c-1)} \frac{p_{ij}(nh)}{nh} \,,$$

and since $\lim_{h\downarrow 0} nh = t$, we see that $\lim_{h\downarrow 0} \frac{p_{ij}(nh)}{nh} = \frac{p_{ij}(t)}{t}$, so that from the last inequality,

$$\limsup_{h\downarrow 0} \frac{p_{ij}(h)}{h} \leq \frac{1}{c(2c-1)} \frac{p_{ij}(t)}{t} < \infty,$$

which in turn gives

$$\limsup_{h\downarrow 0} \frac{p_{ij}(h)}{h} \leq \frac{1}{c(2c-1)} \liminf_{t\downarrow 0} \frac{p_{ij}(t)}{t} < \infty.$$

Since c can be chosen arbitrarily close to 1, we have

$$\limsup_{h\downarrow 0} \frac{p_{ij}(h)}{h} \leq \liminf_{t\downarrow 0} \frac{p_{ij}(t)}{t} < \infty,$$

and this implies the existence of $\lim_{h\downarrow 0} \frac{p_{ij}(h)}{h}$ as well as the finiteness of this limit. □

Definition 7.2.12 *The numbers q_{ij} $(i \neq j)$ and $q_{ii} = -q_i$ are the* local characteristics *of the semi-group, or of the corresponding continuous-time* HMC. *The matrix*

$$\mathbf{A} = \{q_{ij}\}_{i,j \in E}$$

is the infinitesimal generator *of the semi-group (or of the continuous-time* HMC).

In compact notation,

$$\mathbf{A} = \lim_{h\downarrow 0} \frac{\mathbf{P}(h) - \mathbf{P}(0)}{h},$$

where the meaning of this limit is given by (7.16) and (7.17). Thus, in this sense, the infinitesimal generator $\mathbf{A}$ is the derivative at time 0 of the matrix function $t \to \mathbf{P}(t)$.

Definition 7.2.13 *If for all states i,*

$$q_i < \infty,$$

the semigroup $\{\mathbf{P}(t)\}_{t\geq 0}$ is called stable. *If for all states i,*

$$q_i = \sum_{\substack{j \in E \\ j \neq i}} q_{ij},$$

it is called conservative.

Remark 7.2.14 The reason for the last appellation comes from the conservation equality

$$\sum_{j \in E} p_{ij}(h) = 1,$$

or equivalently,

$$\frac{1 - p_{ii}(h)}{h} = \sum_{\substack{j \in E \\ j \neq i}} \frac{p_{ij}(h)}{h},$$

which yields

$$q_i = \lim_{h\downarrow 0} \sum_{\substack{j \in E \\ j \neq i}} \frac{p_{ij}(h)}{h}.$$

And *if* the interchange of summation and limit in the right-hand side is allowed, we obtain the conservation equation. Interchange of sums and limits is always possible if E is finite. In this case the conservation identities hold, and consequently, we also have stability because q_{ij} is finite for all pairs of different sites i, j.

As a matter of fact, a very general class of Markov chains, the so-called *regular jump Markov chains*, are stable and conservative (this is proved in Theorem 7.3.3). Here is the definition:

Definition 7.2.15 *A stochastic process $\{X(t)\}_{t\geq 0}$ taking its values in the (not necessarily countable) state space E is called a* jump process *if for almost all $\omega \in \Omega$, there exists for all $t \geq 0$ a number $\epsilon(t,\omega) > 0$ such that*

$$X(t+s,\omega) = X(t,\omega) \text{ for all } s \in [t, t+\epsilon(t,\omega)).$$

It is called a regular process *if in addition, for almost all $\omega \in \Omega$, the set $A(\omega)$ of discontinuities of the function $t \to X(t,\omega)$ is σ-discrete, that is, for all $c \geq 0$, $A(\omega) \cap [0,c]$ is a finite set. A* regular jump HMC *is by definition a continuous-time HMC that is also a regular jump process.*

Observe that for a jump process (not necessarily regular), there exists a sequence of times $\{\tau_n\}_{n\geq 0}$ where

$$\tau_0 = 0 < \tau_1 < \tau_2 < \tau_3 < \cdots$$

and a sequence $\{X_n\}_{n\geq 0} \in E$ such that

$$X(t) = X_n \text{ if } \tau_n \leq t < \tau_{n+1}.$$

This describes $\{X(t)\}_{t\geq 0}$ on the interval $[0, \tau_\infty)$, where

$$\tau_\infty = \lim_{n \uparrow \infty} \tau_n$$

is the *explosion time*. If, moreover, the process is regular, then $\tau_\infty = \infty$, and $\{X(t)\}_{t\geq 0}$ is right-continuous.

When the chain is stable and conservative, we have from the proof of Theorem 7.2.11 that

$$P(X(t+h) = i \mid X(t) = i) = 1 - q_i h + o(h)$$

and if $i \neq j$,

$$P(X(t+h) = j \mid X(t) = i) = q_{ij} h + o(h)$$

(recall that, by definition of the $o(h)$ symbol, $\lim_{h \to 0} \frac{|o(h)|}{h} = 0$).

Definition 7.2.16 *A continuous-time* birth-and-death process *is a regular jump HMC with state space $\mathbb{N}$, and with an infinitesimal generator of the form*

$$q_{i,i+1} = \lambda_i, \quad q_{i,i-1} = \mu_i 1_{\{i \geq 1\}},$$

and $q_{ij} = 0$ if $j \notin \{i-1, i, i+1\}$.

The parameters λ_i and μ_i are called the *birth and death parameters* respectively, because

$$P(X(t+h) = i+1 \mid X(t) = i) = \lambda_i h + o(h)$$

and

$$P(X(t+h) = i - 1 \mid X(t) = i) = \mu_i 1_{\{i \geq 1\}} h + o(h).$$

Birth-and-death processes are important models in biology (where the terminology comes from, obviously), but also in operations research, and in particular in queueing theory, where they appear as $M/M/1/\infty$, $M/M/K/0$ queues, among many other models of queueing systems (see Chapter 9).

EXAMPLE 7.2.17: UNIFORM HMC, TAKE 3. From expression (7.15) of the transition semigroup of the uniform HMC, we obtain its infinitesimal generator (Exercise 7.5.3)

$$\mathbf{A} = \lambda(\mathbf{K} - I),   \tag{7.18}$$

that is,

$$q_i = \lambda(1 - k_{ii}),$$

and for $i \neq j$,

$$q_{ij} = \lambda k_{ij}.$$

7.2.3   HMCs **from** HPPs

This subsection introduces the notion of stochastic differential equations driven by Poisson processes with the classical example of continuous-time homogeneous Markov chains (HMC).

The problem considered in this subsection is that of *realization* of an infinitesimal generator. It consists in associating to a *generator* $\mathbf{A}$ a continuous-time HMC admitting $\mathbf{A}$ as an *infinitesimal* generator. More precisely, let E be a countable space and let $\mathbf{A} = \{q_{ij}\}_{i,j \in E}$ be a matrix of real numbers indexed by E, such that for all $i, j \in E$ ($i \neq j$),

$$q_i \in [0, \infty], \quad q_{ij} \in [0, \infty).$$

Assume, moreover, that this generator is *stable and conservative*, that is,

$$q_i < \infty, \quad q_i = \sum_{\substack{k \in E \\ k \neq i}} q_{ik},$$

where $q_i = -q_{ii}$. Note also that when speaking about a generator $\mathbf{A}$, one does not refer to a homogeneous continuous-time Markov chain. However, we can construct a continuous-time HMC admitting the given generator as infinitesimal generator.

To do this, start with a family N_{ij} ($i, j \in E, i \neq j$) of independent HPPs with respective intensities q_{ij} ($i, j \in E, i \neq j$) and independent of the initial state $X(0) \in E$. The process is constructed as a jump process:

$$X(t) = X_n \text{ for } t \in [\tau_n, \tau_{n+1}),$$

where the sequence $\{\tau_n, X_n\}_{n \geq 0}$ is defined recursively as follows. First, $\tau_0 = 0$, $X_0 = X(0)$. Before proceeding to the general step, recall that $S_a N$, where $a \in \mathbb{R}_+$, is by definition the point process N on the line shifted by a, that is,

$$(S_a N)(C) := N(C + a) \quad (C \in \mathcal{B}(\mathbb{R})).$$

If $\tau_n < \infty$ and $X_n = X(\tau_n) = i \in E$, then $\tau_{n+1} - \tau_n$ is the first event of the family of HPPs $S_{\tau_n}N_{ij}$ ($j \in E$, $j \neq i$). If $\tau_{n+1} - \tau_n < \infty$, X_{n+1} is the index $k \neq i$ such that $S_{\tau_n}N_{ik}$ is the first among the HPPs $S_{\tau_n}N_{ij}$ ($j \in E$, $j \neq i$) that produces an event (see Theorem 7.1.4). It may occur that $\tau_{n+1} - \tau_n = \infty$. This is the case if and only if $q_i = 0$, and then the construction ends by letting $X_{n+m} = i$ and $\tau_{n+m} = \infty$ for all $m \geq 1$, and $X(t) = X_n$ for all $t \geq \tau_n$.

The process constructed in this way is defined on $[0, \tau_\infty)$, where $\tau_\infty = \lim_{n\uparrow\infty} \tau_n$. If $\tau_\infty = \infty$ P-a.s, then the process is fully defined on $\mathbb{R}_+$ and is therefore a regular jump process on E.

Theorem 7.2.18 *A. For all $n \geq 0$: given X_n, the random variables X_{n+1} and $\tau_{n+1} - \tau_n$ are independent of $X_0, \ldots, X_{n-1}$ and $\tau_1, \ldots, \tau_n$, and for all $i, j \in E$, $i \neq j$, all $a \geq 0$,*

$$P(X_{n+1} = j, \tau_{n+1} - \tau_n > a \mid X_n = i) = e^{-q_i a} p_{ij}, \tag{7.19}$$

where $p_{ij} = \frac{q_{ij}}{q_i}$ if $q_i > 0$, $p_{ij} = 0$ if $q_i = 0$. (In particular, $\{X_n\}_{n\geq 0}$ is an HMC.)

B. If $P(\tau_\infty = \infty) = 1$, the process $\{X(t)\}_{t\geq 0}$ constructed as above is a regular jump HMC with infinitesimal generator $\mathbf{A}$.

Proof. A. The τ_n's form a sequence of $\mathcal{G}_t$-stopping times where

$$\mathcal{G}_t := \sigma(X_0) \vee \left(\vee_{i,j\in E}\mathcal{F}_t^{N_{ij}}\right).$$

The announced result then follows from the strong Markov property for HPPs (Theorem 7.1.5) and the competition theorem (Theorem 7.1.4).

B. By construction, for a given time t, the process after time t depends only upon $X(t)$ and the HPPs $S_t N_{ij}$ ($i, j \in E$, $i \neq j$). The homogeneous Markov property follows immediately from this observation. It remains to show that $\mathbf{A}$ is indeed the infinitesimal generator of the HMC. We first check that, for $i \neq j$,

$$\lim_{t\downarrow 0}\frac{1}{t}P_i(X(t) = j) = q_{ij}.$$

For this, observe that when $X(t) \neq X(0)$, necessarily $\tau_1 < t$ and write

$$\begin{aligned}
P_i(X(t) = j) &= P_i(\tau_2 \leq t, X(t) = j) + P_i(\tau_2 > t, X(t) = j) \\
&= P_i(\tau_2 \leq t, X(t) = j) + P_i(\tau_2 > t, X_1 = j, \tau_1 < t) \\
&= P_i(\tau_2 \leq t, X(t) = j) + P_i(X_1 = j, \tau_1 < t) - P_i(\tau_2 \leq t, X_1 = j, \tau_1 < t).
\end{aligned}$$

By Theorem 7.1.4, $P_i(X_1 = j, \tau_1 < t) = (1 - e^{-q_i t})\frac{q_{ij}}{q_i}$ and then

$$\lim_{t\downarrow 0}\frac{1}{t}P_i(X_1 = j, \tau_1 < t) = q_{ij}.$$

It therefore remains to show that $P_i(\tau_2 \leq t, X(t) = j)$ and $P_i(\tau_2 \leq t, X_1 = j, \tau_1 \leq t)$ are $o(t)$ (obvious if $q_i = 0$, and therefore we suppose $q_i > 0$). Both terms are bounded by $P_i(\tau_2 \leq t)$, and

$$\begin{aligned}
P_i(\tau_2 \leq t) &\leq P_i(\tau_1 \leq t, \tau_2 - \tau_1 \leq t) = \sum_{\substack{k\in E \\ k\neq j}} P_i(\tau_1 \leq t, X_1 = k, \tau_2 - \tau_1 \leq t) \\
&= \sum_{\substack{k\in E \\ k\neq j}}(1 - e^{-q_i t})\frac{q_{ik}}{q_i}(1 - e^{-q_k t}) = (1 - e^{-q_i t})\sum_{\substack{k\in E \\ k\neq i}}\frac{q_{ik}}{q_i}(1 - e^{-q_k t}).
\end{aligned}$$

But $(1 - e^{-q_i t})$ is $O(t)$ (identically null if $q_i = 0$) and $\lim_{t \downarrow 0} \sum_{\substack{k \in E \\ k \neq i}} \frac{q_{ik}}{q_i} (1 - e^{-q_k t}) = 0$ by dominated convergence. Therefore $P_i(\tau_2 \leq t)$ is $o(t)$.

We now check that $\lim_{t \downarrow 0} \frac{1 - p_{ii}(t)}{t} = q_i$. From

$$
\begin{aligned}
1 - p_{ii}(t) &= 1 - P_i(X(t) = i) \\
&= 1 - P_i(X(t) = i, \tau_1 > t) - P_i(X(t) = i, \tau_1 \leq t) \\
&= 1 - P_i(\tau_1 > t) - P_i(X(t) = i, \tau_1 \leq t, \tau_2 \leq t) \\
&= 1 - e^{-q_i t} - P_i(X(t) = i, \tau_1 \leq t, \tau_2 \leq t),
\end{aligned}
$$

and the announced result follows from the fact (proved above) that $P_i(\tau_2 \leq t)$ is $o(t)$. □

Let $Z_i(t) := 1_{\{X(t) = i\}}$. The construction of the state process on $[0, \tau_\infty)$ is summarized by the following equations: for all $i \in E$,

$$
Z_i(t) = Z_i(0) + \sum_{j \in E; j \neq i} \int_{(0,t]} Z_j(s-) \, N_{ji}(\mathrm{d}s) - \sum_{j \in E; j \neq i} \int_{(0,t]} Z_i(s-) \, N_{ij}(\mathrm{d}s) . \tag{7.20}
$$

Equations (7.20) constitute a system of stochastic differential equations driven by the Poisson processes N_{ij} $(i, j \in E, i \neq j)$ for the processes $\{Z_i(t)\}_{t \geq 0}$ $(i \in E)$.

It is sometimes convenient to state conclusion B of Theorem 7.2.18 as

Theorem 7.2.19 *Let $\{X(t)\}_{t \geq 0}$ be a regular jump process with countable state space E, satisfying (7.20) where $Z_i(t) = 1_{\{X(t) = i\}}$ and where $N_{i,j}$ $(i, j \in E, i \neq j)$ is a family of independent HPPs with respective intensities q_{ij} $(i, j \in E, i \neq j)$, and independent of the initial state $X(0)$. Then $\{X(t)\}_{t \geq 0}$ is a regular jump HMC with infinitesimal generator* **A**.

Note that (7.20) is equivalent to the requirement that

$$
f(X(t)) - f(X(0)) = \sum_{\substack{i,j \in E \\ i \neq j}} \{f(j) - f(i)\} \int_{(0,t]} 1_{\{X(s-) = i\}} \mathrm{d}N_{ij}(s) \tag{7.21}
$$

for all non-negative functions $f : E \to \mathbb{R}$. We shall now exploit this *canonical representation*.

Aggregation of States

Consider a regular jump HMC $\{X(t)\}_{t \geq 0}$ with state space E and infinitesimal generator **A**. Let $\tilde{E} = \{\alpha, \beta, \ldots\}$ be a partition of E, and define the process $\{\tilde{X}(t)\}_{t \geq 0}$ taking its values in $\tilde{E}$ by

$$
\tilde{X}(t) = \alpha \iff X(t) \in \alpha . \tag{7.22}
$$

The HMC $\{\tilde{X}(t)\}_{t \geq 0}$ is the aggregated chain of $\{X(t)\}_{t \geq 0}$ (with respect to the partition $\tilde{E}$).

Theorem 7.2.20 *Suppose that for all* $\alpha, \beta \in \tilde{E}$ ($\alpha \neq \beta$),

$$\sum_{j \in \beta} q_{ij} = \tilde{q}_{\alpha\beta} \quad (i \in \alpha). \tag{7.23}$$

(This equality not only defines the quantity in the right-hand side but also states the hypothesis that the left-hand side is independent of $i \in \alpha$.) Then $\{\tilde{X}(t)\}_{t \geq 0}$ is a regular jump HMC with state space $\tilde{E}$ and infinitesimal generator $\tilde{\mathbf{A}}$, with off-diagonal terms given by (7.23).

Proof. This statement concerns the *distribution* of $\{X(t)\}_{t \geq 0}$ and therefore we may suppose that this process is of the form (7.21). Then, for $f : \tilde{E} \to \mathbb{R}$ and $s \leq t$,

$$
\begin{aligned}
f(\tilde{X}(t)) &= f(\tilde{X}(0)) + \sum_{\substack{i,j \in E \\ i \neq j}} \int_{(0,t]} \{f(\tilde{X}(u)) - f(\tilde{X}(u-))\} 1_{\{X(u-)=i\}} \mathrm{d}N_{ij}(u) \\
&= f(\tilde{X}(0)) + \sum_{\substack{\alpha,\beta \in \tilde{E} \\ \alpha \neq \beta}} \{f(\beta) - f(\alpha)\} \sum_{i \in \alpha} \left(\int_{(0,t]} 1_{\{X(u-)=i\}} \left(\sum_{j \in \beta} \mathrm{d}N_{ij}(u) \right) \right).
\end{aligned}
$$

Define for all $\alpha, \beta \in \tilde{E}, \alpha \neq \beta$, the point process $\tilde{N}_{\alpha\beta}$ by

$$
\tilde{N}_{\alpha\beta}(0,t] = \int_{(0,t]} \sum_{i \in \alpha} \left(1_{\{X(s-)=i\}} \left(\sum_{j \in \beta} \mathrm{d}N_{ij}(s) \right) \right) + \int_{(0,t]} 1_{\{\tilde{X}(s-)\neq\alpha\}} \mathrm{d}\hat{N}_{\alpha,\beta}(s),
$$

where the "dummy" point processes $\{\hat{N}_{\alpha\beta}\}_{\substack{\alpha,\beta \in \tilde{E} \\ \alpha \neq \beta}}$ form an independent family of HPPs with intensities $\{\tilde{q}_{\alpha\beta}\}_{\substack{\alpha,\beta \in \tilde{E} \\ \alpha \neq \beta}}$, respectively, and are independent of $X(0)$ and $\{N_{ij}\}_{\substack{i,j \in E \\ i \neq j}}$. Then

$$
f(\tilde{X}(t)) = f(\tilde{X}(0)) + \sum_{\substack{\alpha,\beta \in \tilde{E} \\ \alpha \neq \beta}} (f(\beta) - f(\alpha)) \int_{(0,t]} 1_{\{\tilde{X}(u-)=\alpha\}} \mathrm{d}\tilde{N}_{\alpha\beta}(u). \tag{7.24}
$$

In view of the remark relative to (7.21), it suffices to prove that $\tilde{N}_{\alpha,\beta}$ ($\alpha, \beta \in \tilde{E}$, $\alpha \neq \beta$) is a family of independent HPPs with respective intensities $\tilde{q}_{\alpha\beta}$ ($\alpha, \beta \in \tilde{E}$, $\alpha \neq \beta$). For this, we apply Watanabe's theorem (Theorem 7.1.8). Let $\{Z(t)\}_{t \geq 0}$ be a left-continuous $\mathcal{F}_t$-adapted stochastic process, where

$$
\mathcal{F}_t = \sigma(X(0)) \vee \mathcal{F}_t^{\tilde{N}_{\alpha\beta}} \vee \left(\vee_{u,v \in \tilde{E}; (u,v) \neq (\alpha,\beta)} \mathcal{F}_\infty^{\tilde{N}_{uv}} \right).
$$

We obtain

$$
\begin{aligned}
\mathrm{E} \left[\int_{(0,T]} Z(t) \mathrm{d}\tilde{N}_{\alpha\beta}(t) \right] &= \sum_{i \in \alpha} \sum_{j \in \beta} \mathrm{E} \left[\int_{(0,T]} Z(t) 1_{\{X(t-)=i\}} \mathrm{d}N_{ij}(t) \right] \\
&\quad + \mathrm{E} \left[\int_{(0,T]} Z(t) 1_{\{\tilde{X}(t-)\neq\alpha\}} \mathrm{d}\hat{N}_{\alpha\beta} \right],
\end{aligned}
$$

and this quantity is equal, by the smoothing formula (Theorem 7.1.7), to

$$\sum_{i \in \alpha} \sum_{j \in \beta} \quad E\left[\int_0^T Z(t) 1_{\{X(t-)=i\}} q_{ij} dt\right] + E\left[\int_0^T Z(t) 1_{\{\tilde{X}(t) \neq \alpha\}} \tilde{q}_{\alpha\beta} dt\right]$$

$$= E\left[\int_0^T Z(t)\left[\left(\sum_{j \in \beta} q_{ij}\right) 1_{\{\tilde{X}(t-)=\alpha\}} + \tilde{q}_{\alpha\beta} 1_{\{\tilde{X}(t-) \neq \alpha\}}\right] dt\right]$$

$$= E\left[\int_0^T Z(t) \tilde{q}_{\alpha\beta} dt\right].$$

Therefore, by Theorem 7.1.8, $\tilde{N}_{\alpha\beta}$ is an HPP with intensity $\tilde{q}_{\alpha\beta}$ independent of $\sigma(X(0)) \vee \left(\vee_{u,v \in \tilde{E}; (u,v) \neq (\alpha,\beta)} \mathcal{F}_\infty^{\tilde{N}_{uv}}\right)$. $\qquad\qquad\square$

7.3 Regenerative Structure

7.3.1 The Strong Markov Property

A regular jump HMC has the strong Markov property. More precisely:

Define, similarly to the discrete-time case, the *process after* the random time τ:

$$\{S_\tau X(t)\}_{t \geq 0} = \{X(t + \tau)\}_{t \geq 0}$$

(with the convention $X(\infty) = \Delta$, where Δ is an element not in E) and the *process before* τ:

$$\{X^\tau(t)\}_{t \geq 0} = \{X(t \wedge \tau)\}_{t \geq 0}.$$

Theorem 7.3.1 *Let* $\{X(t)\}_{t \geq 0}$ *be a right-continuous continuous-time HMC with countable state space* E *and transition semigroup* $\{\mathbf{P}(t)\}_{t \geq 0}$, *and let* τ *be a stopping time with respect to* $\{X(t)\}_{t \geq 0}$. *Let* $k \in E$ *be a state. Then,*

(α) *given* $X(\tau) = k$, *the chain after* τ *and the chain before* τ *are independent, and*

(β) *given* $X(\tau) = k$, *the chain after* τ *is a regular jump HMC with transition semigroup* $\{\mathbf{P}(t)\}_{t \geq 0}$.

Proof. Suppose we have proved that for all states k, all positive times $t_1, \ldots, t_n$, $s_1, \ldots, s_p$ and all real numbers $u_1, \ldots, u_n, v_1, \ldots, v_p$,

$$E\left[e^{i \sum_{\ell=1}^n u_\ell X(\tau+t_\ell) + i \sum_{m=1}^p v_m X(\tau \wedge s_m)} 1_{\{X(\tau)=k\}}\right]$$

$$= E\left[e^{i \sum_{\ell=1}^n u_\ell X(t_\ell)} | X(0) = k\right] E\left[e^{i \sum_{m=1}^p v_m X(\tau \wedge s_m)} 1_{\{X(\tau)=k\}}\right]. \qquad (7.25)$$

Then, fixing $v_1 = \cdots = v_p = 0$, we obtain

$$\frac{E\left[e^{i \sum_{\ell=1}^n u_\ell X(\tau+t_\ell)} 1_{\{X(\tau)=k\}}\right]}{P(X(\tau) = k)} = E\left[e^{i \sum_{\ell=1}^n u_\ell X(t_\ell)} | X(0) = k\right],$$

and this shows that given $X(\tau) = k$, $\{X(\tau+t)\}_{t \geq 0}$ had the same distribution as $\{X(t)\}_{t \geq 0}$ given $X(0) = k$. We therefore will have proved (β). For (α), it suffices to rewrite (7.25) as follows, using the previous equality:

$$\mathrm{E}\left[e^{i\sum_{\ell=i}^{n} u_\ell X(\tau+t_\ell)+i\sum_{m=1}^{n} v_m X(\tau\wedge s_m)} \mid X(\tau)=k\right]$$
$$= \mathrm{E}\left[e^{i\sum_{\ell=1}^{n} u_\ell X(\tau+t_\ell)} \mid X(\tau)=k\right] \mathrm{E}\left[e^{i\sum_{n=1}^{n} v_m X(\tau\wedge s_m)} \mid X(\tau)=k\right]. \qquad (7.26)$$

It remains to prove (7.25). For the sake of simplicity, we consider the case where $n = m = 1$, and let $u_1 = u, t_1 = t, v_1 = v, s_1 = s$.

Suppose first that τ takes a countable number of finite values, denoted by a_j, and also, maybe, the value $+\infty$. Note that $X(\tau) = k \in E$ implies $\tau < \infty$. Then

$$\mathrm{E}[e^{iuX(\tau+t)+ivX(\tau\wedge s)}1_{\{X(\tau)=k\}}]$$
$$= \sum_{j\geq 1} \mathrm{E}[e^{iuX(a_j+t)+ivX(a_j\wedge s)}1_{\{X(a_j)=k\}}1_{\{\tau=a_j\}}].$$

For all $j \geq 1$,

$$\mathrm{E}[e^{iuX(a_j+t)+ivX(a_j\wedge s)}1_{\{X(a_j)=k\}}1_{\{\tau=a_j\}}]$$
$$= \mathrm{E}[e^{iuX(a_j+t)}1_{\{X(a_j)=k\}}e^{ivX(a_j\wedge s)}1_{\{\tau=a_j\}}]$$
$$= \mathrm{E}[e^{iuX(a_j+t)} \mid X(a_j) = k]\mathrm{E}[e^{ivX(a_j\wedge s)}1_{\{\tau=a_j\}}1_{\{X(a_j)=k\}}],$$

where for the last equality, we have used the fact that $1_{\{\tau=a_j\}}$ is $\mathcal{F}^X_{a_j}$-measurable and the Markov property at time a_j. Therefore, for all $j \geq 1$,

$$\mathrm{E}[e^{iuX(a_j+t)+ivX(a_j\wedge s)}1_{\{X(a_j)=k\}}1_{\{\tau=a_j\}}]$$
$$= \mathrm{E}[e^{iuX(t)}|X(0) = k]\mathrm{E}[e^{ivX(a_j\wedge s)}1_{\{X(a_j)=k\}}1_{\{\tau=a_j\}}].$$

Summing with respect to j, we obtain the equality corresponding to (7.25).

To pass from the case where the stopping time τ takes a countable number of values to the general case, letting $\tau(n)$ be the approximation of τ of Theorem 5.3.13. The random time $\tau(n)$ is an $\mathcal{F}^X_t$-stopping-time with a countable number of values such that $\lim_{n\uparrow\infty} \downarrow \tau(n,\omega) = \tau(\omega)$. In particular, $\lim_{n\uparrow\infty} X(\tau(n)\wedge a) = X(\tau\wedge a)$, $\lim_{n\uparrow\infty} X(\tau(n)+b) = X(\tau+b)$ and $\lim_{n\uparrow\infty} 1_{\{X(\tau(n))=k\}} = 1_{\{X(\tau)=k\}}$ (use the fact that a regular jump process is right-continuous). Therefore, letting n go to ∞ in (7.25) with τ replaced by $\tau(n)$, we obtain the result for τ itself, by dominated convergence. $\square$

Equality (7.26) says that, given $X(\tau) = i$, the random vectors $(X(\tau \wedge s_1), \ldots, X(\tau \wedge s_m))$ and $(X(\tau + t_1), \ldots, X(\tau + t_n))$ and $(X(\tau + t_1), \ldots, X(\tau + t_n))$ are independent. In particular, the events $\{X(\tau \wedge s_1) = i_1, \ldots, X(\tau \wedge s_m) = i_m\}$ and $\{X(\tau + t_1) = j_1, \ldots, X(\tau + t_n) = j_n\}$ are conditionally independent given $X(\tau) = i$. Since this is true for all $i_1, \ldots, i_m, j_1, \ldots, j_n$, and all $s_1, \ldots, s_m, t_1, \ldots, t_n$, this property extends (see Theorem 5.1.10) to events in A and B respectively in $\mathcal{F}^{S_\tau X}$ and $\mathcal{F}^{X^\tau}$. And then, finally (see the discussion leading to (7.14),

$$\mathrm{E}^{X(\tau)}[Y \times Z] = \mathrm{E}^{X(\tau)}[Y]\mathrm{E}^{X(\tau)}[Z], \qquad (7.27)$$

for all non-negative Y and Z that are respectively $\mathcal{F}^{S_\tau X}$-measurable and $\mathcal{F}^{X^\tau}$-measurable.

7.3.2 Imbedded Chain

Let $\{\tau_n\}_{n\geq 0}$ be the non-decreasing sequence of transition times of a regular jump process $\{X(t)\}_{t\geq 0}$, where $\tau_0 = 0$, and $\tau_n = \infty$ if there are strictly fewer than n transitions in $(0,\infty)$. Note that, for each $n \geq 0$, τ_n is an $\mathcal{F}^X_t$-stopping time.

The discrete-time stochastic process $\{X_n\}_{n\geq 0}$ with values in E defined by

$$X_n := X(\tau_n)$$

if $\tau_n < \infty$ and $X_n = X_{n-1}$ if $\tau_n = \infty$ is called the *imbedded process* of the jump process.

Theorem 7.3.2 *Let $\{X(t)\}_{t\geq 0}$ be a continuous-time regular jump (therefore right-continuous)* HMC*, with infinitesimal generator $\mathbf{A}$, transition times sequence $\{\tau_n\}_{n\geq 0}$, and imbedded process $\{X_n\}_{n\geq 0}$. Then*

(α) $\{X_n\}_{n\geq 0}$ *is a discrete-time* HMC *with transition matrix given by, for $j \neq i$,*

$$p_{ij} = \frac{q_{ij}}{q_i}$$

if $q_i > 0$, and $p_{ij} = 0$ if $q_i = 0$.

(β) *For all $n \geq 0$ and all $a \in \mathbb{R}_+$,*

$$P(\tau_{n+1} - \tau_n \leq a \mid X_0, \ldots, X_n, \tau_1, \ldots, \tau_n) = 1 - e^{-q_{X_n} a}. \tag{7.28}$$

Proof. We begin with the following partial result.

(α') $\{X_n\}_{n\geq 0}$ is a discrete-time HMC on E.

(β') There exists for each $i \in E$ a *finite* real number $\lambda(i) \geq 0$ such that for all $n \geq 0$ and all $a \in \mathbb{R}_+$,

$$P(\tau_{n+1} - \tau_n \leq a \mid X_0, \ldots, X_n, \tau_1, \ldots, \tau_n) = 1 - e^{-\lambda(X_n)a}.$$

It follows from the strong Markov property that given $X(\tau_n) = i \in E$, $\{X(\tau_n+t)\}_{t\geq 0}$ is independent of $\{X(\tau_n \wedge t)\}_{t\geq 0}$ and therefore, given $X_n = i$, the variables $(X_{n+1}, X_{n+2}, \ldots)$ are independent of $(X_0, \ldots, X_n)$, that is, $\{X_n\}_{n\geq 0}$ is a Markov chain. It is clearly homogeneous because the distribution of $\{X(\tau_n + t)\}_{t\geq 0}$ given $X(\tau_n) = i$ is independent of n, being identical with the distribution of $\{X(t)\}_{t\geq 0}$ given $X(0) = i$, again by the strong Markov property. We have therefore proved (α').

Call $p_{ij} = P_i(X(\tau_1) = j)$ the transition probability of $\{X_n\}_{n\geq 1}$ from i to j. To prove (β'), it suffices to show that

$$P_i(X_1 = i, \ldots, X_n = i_n, \tau_1 - \tau_0 > a_1, \ldots, \tau_n - \tau_{n-1} > a_n)$$
$$= e^{-\lambda(i)a_1} p_{ii_1} e^{-\lambda(i_1)a_2} p_{i_1 i_2} \cdots e^{-\lambda(i_{n-1})a_n} p_{i_{n-1} i_n}$$

for all $i, i_1, \ldots, i_n \in E$, all $a_1, \ldots, a_n \in \mathbb{R}_+$ and some function $\lambda : E \to \mathbb{R}_+$. In view of the strong Markov property, it suffices to show that for all $i, j \in E$, $a \in \mathbb{R}_+$, there exists a $\lambda(i) \in [0, \infty)$ such that

$$P_i(X_1 = j, \tau_1 - \tau_0 > a) = P_i(X_1 = j)e^{-\lambda(i)a}. \tag{7.29}$$

Define $g(t) = P_i(\tau_1 > t)$. For $t, s \geq 0$, using the obvious set identities,

$$\begin{aligned}
g(t+s) &= P_i(\tau_1 > t + s) \\
&= P_i(\tau_1 > t + s, \tau_1 > t, X(t) = i) \\
&= P_i(X(t+u) = i \text{ for all } u \in [0, s], \tau_1 > t, X(t) = i).
\end{aligned}$$

The last expression is, in view of the Markov property at time t and using the fact that $\{\tau_1 > t\} \in \mathcal{F}_t^X$,

$$P_i(X(t + u) = i \text{ for all } u \in [0, s] \mid X(t) = i)P_i(\tau_1 > t, X(t) = i)$$
$$= P_i(X(u) = i \text{ for all } u \in [0, s] \mid X(0) = i)P_i(\tau_1 > t)$$
$$= P_i(\tau_1 > s)P_i(\tau_1 > t),$$

where the last two equalities again follow from the obvious set identities. Therefore, for all $s, t \geq 0$,

$$g(t + s) = g(t)g(s).$$

Also, $g(t)$ is non-increasing, and $\lim_{t \downarrow 0} g(t) = 1$ (use the fact that the chain is assumed to be a jump process). It follows that there exists a $\lambda(i) \in [0, \infty)$ such that $g(t) = e^{-\lambda(i)t}$, that is, $P_i(\tau_1 > t) = e^{-\lambda(i)t}$, for all $t \geq 0$. Now, using the Markov property and appropriate set identities,

$$P_i(X_1 = j, \tau_1 > t) = P_i(X(\tau_1) = j, \tau_1 > t, X(t) = i)$$
$$= P_i(\text{first jump of } \{X(t + s)\}_{s \geq 0} \text{ is } j, \tau_1 > t, X(t) = i)$$
$$= P_i(\text{first jump of } \{X(t + s)\}_{s \geq 0} \text{ is } j \mid X(t) = i)P_i(\tau_1 > t, X(t) = i)$$
$$= P_i(\text{first jump of } \{X(s)\}_{s \geq 0} \text{ is } j \mid X(0) = i)P_i(\tau_1 > t)$$
$$= P_i(X(\tau_1) = j)P_i(\tau_1 > t),$$

and this is (7.29).

We have now proved (α) and (β), where q_i is replaced by $\lambda(i) \in \mathbb{R}_+$ (only known to exist but not yet identified with q_i) and where $\frac{q_{ij}}{q_i}$ is replaced by $P_i(X(\tau_1) = j)$ (not yet identified with $\frac{q_{ij}}{q_i}$). We shall now proceed to the required identifications. For this, define the generator $\mathbf{A}'$ on E by

$$q_i' = \lambda(i), \quad q_{ij}' = \lambda(i)P_i(X(\tau_1) = j).$$

This generator is stable and conservative, and we can therefore construct $\{X'(t)\}_{t \geq 0}$, a regular jump HMC associated with $\mathbf{A}'$, via the construction of Section 7.2.3, up to τ_∞'. Then $\{X'(t)\}_{t \geq 0}$ and $\{X(t)\}_{t \geq 0}$ have the same regenerative structure, given by (α') and (β'), and therefore, they have the same distribution (in particular, $\{X'(t)\}_{t \geq 0}$ is regular, since $\{X(t)\}_{t \geq 0}$ is regular, by assumption). Their respective infinitesimal generators $\mathbf{A}$ and $\mathbf{A}'$ are therefore identical. $\qquad \square$

Theorem 7.3.3 *A regular jump HMC is stable and conservative.*

Proof. Indeed, a regular jump HMC is strongly Markovian, and therefore has the regenerative structure of Theorem 7.3.2. In the course of the proof of this theorem, we have identified q_i with a certain *finite* quantity $\lambda(i)$, and therefore $q_i < \infty$. Therefore a regular jump HMC is stable. Also q_{ij} was identified with $\lambda(i)P_i(X(\tau_1) = j)$, that is, $q_iP_i(X(\tau_1) = j)$ and therefore the conservation property is clear. $\qquad \square$

Definition 7.3.4 *A state $i \in E$ such that $q_i = 0$ is called* permanent; *otherwise, it is called* essential.

In view of (7.28), if $X(\tau_n) = i$, a permanent state, then $\tau_{n+1} - \tau_n = \infty$; that is, there is no more transition at finite distance, hence the terminology.

EXAMPLE 7.3.5: UNIFORM HMC, TAKE 3. For the uniform HMC (see Definition 7.2.6), the imbedded process $\{X_n\}_{n \geq 0}$ is an HMC with state space E, and if $i \in E$ is not permanent (that is, in this case, if $k_{ii} < 1$), then for $j \neq i$,

$$p_{ij} = \frac{k_{ij}}{1 - k_{ii}}.$$

Indeed, $\{X_n\}_{n \geq 0}$ is obtained from $\{\hat{X}_n\}_{n \geq 0}$ by considering only the "real" transitions (exercise).

An immediate consequence of Theorem 7.3.2 is:

Corollary 7.3.6 *Two regular jump HMCs with the same infinitesimal generator and the same initial distribution have the same distribution.*

Another way to state this is as follows: Two regular jump HMCs with the same infinitesimal generator have the same transition semi-group.

EXAMPLE 7.3.7: UNIFORMIZATION. A regular jump HMC with infinitesimal generator $\mathbf{A}$ such that $\sup_{i \in E} q_i < \infty$ has the same transition semigroup as a uniform chain. Indeed: select any real number $\lambda > \sup_{i \in E} q_i$, and define the transition matrix $\mathbf{K}$ by (7.18). One checks that it is indeed a stochastic matrix. The uniform chain corresponding to $(\lambda, \mathbf{K})$ has the infinitesimal generator $\mathbf{A}$. Any pair $(\lambda, \mathbf{K})$ as above gives rise to a uniform version of the chain. The *minimal* uniform version is, by definition, that with $\lambda = \sup_{i \in E} q_i$.

Definition 7.3.8 *A continuous time HMC with an infinitesimal generator such that*

$$\sup_{i \in E} q_i < \infty, \tag{7.30}$$

is called uniformizable.

7.3.3 Conditions for Regularity

Theorem 7.3.2 gives a way of constructing a regular jump HMC with values in a countable state space and admitting a *given* generator that is stable and conservative. (We shall also suppose for simplicity that this generator is *essential* ($q_i > 0$ for all $i \in E$).) It suffices to construct a sequence $\tau_0 = 0, X_0, \tau_1 - \tau_0, X_1, \tau_2 - \tau_1, X_2, \ldots$ according to

$$P(X_{n+1} = j, \tau_{n+1} - \tau_n \leq x \mid X_0, \ldots, X_n, \tau_0, \ldots, \tau_n) = q_{X_n j}/q_{X_n}(1 - e^{-q_{X_n} x}), \tag{7.31}$$

the initial state X_0 being chosen at random, with arbitrary distribution. The value of $X(t)$ for $\tau_n \leq t < \tau_{n+1}$ is then X_n. If $\tau_\infty := \lim_{n \uparrow \infty} \uparrow \tau_n = \infty$, we have obtained a regular jump HMC with $\mathbf{A}$ as infinitesimal generator.

Definition 7.3.9 *The generator* $\mathbf{A}$ *is called* non-explosive, *or* regular, *if*

$$P_i(\tau_\infty = \infty) = 1 \quad (i \in E). \tag{7.32}$$

Theorem 7.3.10 *Let* **A** *be a stable and conservative generator on* E. *It is regular if and only if for any real* $\lambda > 0$, *the system of equations*

$$(\lambda + q_i)x_i = \sum_{\substack{j \in E \\ j \neq i}} q_{ij}x_j \quad (i \in E) \tag{7.33}$$

admits no non-negative bounded solution other than the trivial one.

Proof. Let $S_k := \tau_k - \tau_{k-1}$. In particular, $\tau_\infty := \sum_{k=1}^\infty S_k$. The number

$$g_i(\lambda) = \mathrm{E}_i\left[\exp\{-\lambda \sum_{k=1}^\infty S_k\}\right]$$

is uniformly bounded in $\lambda > 0$ and $i \in E$, and if $P_i(\tau_\infty = \infty) < 1$, it is strictly positive. Also, $x_i := g_i(\lambda)$ $(i \in E)$ is a solution of (7.33), as follows from the calculations below:

$$\begin{aligned}
g_i(\lambda) &= \mathrm{E}_i\left[\exp\{-\lambda S_1\}\exp\{-\lambda \sum_{k=2}^\infty S_k\}\right] \\
&= \left(\int_0^\infty e^{-\lambda t}q_i e^{-q_i t}dt\right)\mathrm{E}_i\left[\exp\{-\lambda \sum_{k=2}^\infty S_k\}\right] \\
&= \frac{q_i}{\lambda + q_i}\mathrm{E}_i\left[\exp\{-\lambda \sum_{k=2}^\infty S_k\}\right]
\end{aligned}$$

and, by first-step analysis,

$$\mathrm{E}_i\left[\exp\{-\lambda \sum_{k=2}^\infty S_k\}\right] = \sum_{\substack{j \in E \\ j \neq i}} \frac{q_{ij}}{q_i}\mathrm{E}_j\left[\exp\{-\lambda \sum_{k=2}^\infty S_k\}\right] = \sum_{\substack{j \in E \\ j \neq i}} \frac{q_{ij}}{q_i}g_j(\lambda).$$

Therefore, if **A** is explosive there exists a non-trivial bounded solution of (7.33).

We now prove the converse. Call $\{g_i(\lambda)\}_{i \in E}$ a bounded solution of (7.33) for a fixed real $\lambda > 0$. We have

$$g_i(\lambda) = \mathrm{E}[\exp\{-\lambda S_1\}g_{X_1}(\lambda) \mid X_0 = i], \tag{7.34}$$

since first-step analysis shows that the right-hand side is equal to that of (7.33). We prove by induction that

$$g_i(\lambda) = \mathrm{E}\left[\exp\{-\lambda \sum_{k=1}^n S_k\}g_{X_n}(\lambda) \mid X_0 = i\right]. \tag{7.35}$$

For this, rewrite (7.34) as

$$g_i(\lambda) = \mathrm{E}[\exp\{-\lambda S_{n+1}\}g_{X_{n+1}}(\lambda) \mid X_n = i],$$

that is,

$$g_{X_n}(\lambda) = \mathrm{E}[\exp\{-\lambda S_{n+1}\}g_{X_{n+1}}(\lambda) \mid X_n].$$

Using this expression of $g_{X_n}(\lambda)$ in (7.35), we obtain

$$g_i(\lambda) = \mathrm{E}\left[\exp\{-\lambda \sum_{k=1}^{n+1} S_k\}g_{X_{n+1}}(\lambda) \mid X_0 = i\right]. \tag{7.36}$$

Therefore, (7.35) implies (7.36) (the forward step in the induction argument). Since (7.35) is true for $n = 1$ (Eqn. (7.34)), it is true for all $n \geq 1$ and therefore, since $K := |g_i(\lambda)| < \infty$,

$$|g_i(\lambda)| \leq K \mathrm{E}_i \left[\exp\{-\lambda \sum_{k=1}^{\infty} S_k\} \right].$$

Therefore, if $\{g_i(\lambda)\}_{i \in E}$ is not trivial, it must hold for some $i \in E$ that $P_i(\sum_{k=1}^{\infty} S_k < \infty) > 0$ or, equivalently, $P_i(\tau_\infty = \infty) < 1$.   □

Applied to a birth-and-death process, Theorem 7.3.10 gives *Reuter's criterion*.

Theorem 7.3.11 *Let* **A** *be generator on* $E = \mathbb{N}$ *defined by* $q_{n,n+1} = \lambda_n$ *and* $q_{n,n-1} = \mu_n 1_{n \geq 1}$, *where the birth parameters* λ_n *are strictly positive. A necessary and sufficient condition of non-explosion of this generator is*

$$\sum_{n=1}^{\infty} \left[\frac{1}{\lambda_n} + \frac{\mu_n}{\lambda_n \lambda_{n-1}} + \cdots + \frac{\mu_n \cdots \mu_1}{\lambda_n \cdots \lambda_1 \lambda_0} \right] = \infty. \tag{7.37}$$

Proof. The system of equations (7.33) reads in the particular case of a birth-and-death generator

$$\begin{cases} \lambda x_0 = -\lambda_0 x_0 + \lambda_0 x_1, \\ \lambda x_k = \mu_k x_{k-1} - (\lambda_k + \mu_k) x_k + \lambda_k x_{k+1} \quad (k \geq 1). \end{cases} \tag{7.38}$$

For any fixed x_0, this system admits a unique solution that is identically null if and only if $x_0 = 0$. If $x_0 \neq 0$, the solution is such that x_k/x_0 does not depend on x_0, and therefore, only the case where $x_0 = 1$ needs to be treated.

Writing $y_k = x_{k+1} - x_k$, we obtain from (7.38)

$$y_k = \frac{\lambda}{\lambda_k} x_k + \frac{\mu_k}{\lambda_k} \frac{\lambda}{\lambda_{k-1}} x_{k-1} + \cdots + \frac{\mu_k \cdots \mu_2}{\lambda_k \cdots \lambda_2} \frac{\lambda}{\lambda_1} x_1 + \frac{\mu_k \cdots \mu_1}{\lambda_k \cdots \lambda_1} y_0 \tag{7.39}$$

and $y_0 = \frac{\lambda}{\lambda_0}$. From this we deduce that if $\lambda > 0$, then $y_k > 0$ and therefore $\{x_k\}_{k \geq 0}$ is a strictly increasing sequence.

Therefore, using $y_0 = \frac{\lambda}{\lambda_0}$ in (7.39), we have (since $x_k \geq x_0 = 1$)

$$y_k \geq \lambda \left[\frac{1}{\lambda_k} + \frac{\mu_k}{\lambda_k \lambda_{k-1}} + \cdots + \frac{\mu_k \cdots \mu_1}{\lambda_k \cdots \lambda_1 \lambda_0} \right].$$

Therefore, a necessary condition for $\{x_k\}_{k \geq 0}$ to be bounded is that the left-hand side of (7.37) be finite. This proves the sufficiency of (7.37) for non-explosion.

We now turn to the proof of necessity. For $i \leq k$, bounding in (7.39) x_i by x_k yields the majoration

$$y_k \leq \left[\frac{\lambda}{\lambda_k} + \cdots + \frac{\mu_k \cdots \mu_1 \lambda}{\lambda_k \cdots \lambda_1 \lambda_0} \right] x_k,$$

and therefore, since $y_k = x_{k+1} - x_k$,

$$\begin{aligned} x_{k+1} &\leq \left[1 + \frac{\lambda}{\lambda_k} + \cdots + \frac{\mu_k \cdots \mu_1 \lambda}{\lambda_k \cdots \lambda_1 \lambda_0} \right] x_k \\ &\leq x_k \exp \left\{ \lambda \left[\frac{1}{\lambda_k} + \cdots + \frac{\mu_k \cdots \mu_1}{\lambda_k \cdots \lambda_0} \right] \right\}. \end{aligned}$$

Since $x_0 = 1$, this leads to

$$x_n \leq \exp\left\{\lambda \sum_{k=1}^{n}\left[\frac{1}{\lambda_k} + \cdots + \frac{\mu_k \cdots \mu_1}{\lambda_k \cdots \lambda_0}\right]\right\}.$$

Therefore, a sufficient condition for the solution $\{x_n\}_{n\geq 0}$ to be bounded is that the left-hand side of (7.37) be finite. $\qquad\square$

EXAMPLE 7.3.12: PURE BIRTH. A pure birth generator **A** is a birth-and-death generator with all $\mu_n = 0$. The necessary and sufficient condition of regularity (7.37) reads in this case

$$\sum_{n=0}^{\infty}\frac{1}{\lambda_n} = \infty. \tag{7.40}$$

Remark 7.3.13 There is a large class of HMCs for which the regularity is ensured without recourse to the regularity criterion above (Theorem 7.3.10): see Exercise 7.5.8.

7.4 Long-run Behavior

7.4.1 Recurrence

We shall define irreducibility, recurrence, transience, and positive recurrence for a regular jump HMC.

Definition 7.4.1 *A regular jump* HMC *is called* irreducible *if and only if the imbedded discrete-time* HMC *is irreducible.*

Definition 7.4.2 *A state i is called* recurrent *if and only if it is recurrent for the imbedded chain. Otherwise, it is called* transient.

In order to define positive recurrence, we need the following definitions. The *escape time* from state i is defined by

$$L_i := \inf\{t \geq 0; X(t) \neq i\}$$

($=\infty$ if $X(t) = i$ for all $t \geq 0$). The *return time to i* is

$$R_i := \inf\{t > 0; t > L_i \text{ and } X(t) = i\}$$

($= \infty$ if $E_i = \infty$ or $X(t) \neq i$ if for all $t \geq E_i$). Clearly, E_i and R_i are $\mathcal{F}_t^X$-stopping times (exercise).

Definition 7.4.3 *A recurrent state $i \in E$ is called t-positive recurrent if and only if $E_i[R_i] < \infty$, where R_i is the return time to state i. Otherwise, it is called t-null recurrent.*

Remark 7.4.4 We shall soon see that t-positive recurrence and n-positive recurrence (positive recurrence of the imbedded chain) are not equivalent concepts. Also, observe that recurrence of a given state implies that this state is essential. Finally, in the same vein, note that irreducibility implies that all states are essential.

Remark 7.4.5 Note that there is no notion of periodicity for a continuous-time HMC, for obvious reasons.

Invariant Measures of Recurrent Chains

Definition 7.4.6 *A t-invariant measure is a finite non-trivial vector $\nu = \{\nu(i)\}_{i \in E}$ such that for all $t \geq 0$,*

$$\nu^T \mathbf{P}(t) = \nu^T. \tag{7.41}$$

Of course, an n-invariant measure is, by definition, an invariant measure for the imbedded chain.

Theorem 7.4.7 *Let the regular jump HMC $\{X(t)\}_{t \geq 0}$ with infinitesimal generator $\mathbf{A}$ be irreducible and recurrent. Then there exists a unique (up to a multiplicative factor) t-invariant measure such that $\nu(i) > 0$ for all $i \in E$. Moreover, ν is obtained in one of the following ways:*

(1):

$$\nu(i) = \mathrm{E}_0 \left[\int_0^{R_0} 1_{\{X(s)=i\}} \mathrm{d}s \right], \tag{7.42}$$

where 0 is a state and R_0 is the return time to state 0, or

(2):

$$\nu(i) = \frac{\mu(i)}{q_i} = \frac{\mathrm{E}_0 \left[\sum_{n=1}^{T_0} 1_{\{X_n=i\}} \right]}{q_i}, \tag{7.43}$$

where μ is the canonical invariant measure relative to state 0 of the imbedded chain and T_0 is the return time to 0 of the imbedded chain, or

(3): as a solution of:

$$\nu^T \mathbf{A} = 0. \tag{7.44}$$

Proof.

(α) We first show that (7.42) defines an invariant measure, that is, for all $j \in E$ and all $t \geq 0$,

$$\nu(j) = \sum_{k \in E} \nu(k) p_{kj}(t).$$

The right-hand side of the above equality is equal to

$$A = \sum_{k \in E} \mathrm{E}_0 \left[\int_0^\infty 1_{\{X(s)=k\}} 1_{\{s \le R_0\}} \, \mathrm{d}s \right] p_{kj}(t)$$

$$= \int_0^\infty \sum_{k \in E} P_0(X(t+s) = j \mid X(s) = k) P_0(X(s) = k, s \le R_0) \, \mathrm{d}s$$

$$= \int_0^\infty \sum_{k \in E} P_0(X(t+s) = j \mid X(s) = k) P_0(s \le R_0 \mid X(s) = k) P_0(X(s) = k) \, \mathrm{d}s$$

$$= \int_0^\infty \sum_{k \in E} P_0(X(t+s) = j, s \le R_0 \mid X(s) = k) P_0(X(s) = k) \, \mathrm{d}s$$

$$= \int_0^\infty \sum_{k \in E} P_0(X(t+s) = j, s \le R_0, X(s) = k) \, \mathrm{d}s$$

$$= \int_0^\infty P_0(X(t+s) = j, s \le R_0) \, \mathrm{d}s$$

$$= \int_0^\infty \mathrm{E}_0 \left[1_{\{X(t+s)=j\}} 1_{\{s \le R_0\}} \right] \mathrm{d}s$$

$$= \mathrm{E}_0 \left[\int_0^\infty 1_{\{X(t+s)=j\}} 1_{\{s \le R_0\}} \, \mathrm{d}s \right],$$

where we have used the Markov property for the fourth equality ($\{s \le R_0\} \in \mathcal{F}_s^X$). Therefore,

$$A = \mathrm{E}_0 \left[\int_0^{R_0} 1_{\{X(t+s)=j\}} \mathrm{d}s \right] = \mathrm{E}_0 \left[\int_t^{t+R_0} 1_{\{X(u)=j\}} \mathrm{d}u \right]$$

$$= \mathrm{E}_0 \left[1_{\{t \le R_0\}} \int_t^{R_0} 1_{\{X(u)=j\}} \mathrm{d}u \right] - \mathrm{E}_0 \left[1_{\{t > R_0\}} \int_{R_0}^t 1_{\{X(u)=j\}} \mathrm{d}u \right]$$

$$+ \mathrm{E}_0 \left[\int_{R_0}^{R_0+t} 1_{\{X(u)=j\}} \mathrm{d}u \right].$$

From the strong Markov property applied at R_0,

$$\mathrm{E}_0 \left[\int_{R_0}^{R_0+t} 1_{\{X(u)=j\}} \mathrm{d}u \right] = \mathrm{E}_0 \left[\int_0^t 1_{\{X(u)=j\}} \mathrm{d}u \right].$$

Therefore,

$$A = \mathrm{E}_0 \left[1_{\{t \le R_0\}} \int_t^{R_0} \cdots \right] - \mathrm{E}_0 \left[1_{\{t > R_0\}} \int_{R_0}^t \cdots \right] + \mathrm{E}_0 \left[\int_0^t \cdots \right]$$

$$= \mathrm{E}_0 \left[\int_0^{R_0} 1_{\{X(u)=j\}} \mathrm{d}u \right] = \nu(j).$$

(β) We now show uniqueness. For this consider the *skeleton chain* $\{X(n)\}_{n \ge 0}$. For any state i, consider the sequence $Z_1, Z_2, \ldots$ of successive sojourn times in state i of the state process. This sequence is infinite because the imbedded chain is recurrent, and it is IID with exponential distribution of mean $\frac{1}{q_i}$. In particular, the event $\{Z_n > 1\}$ occurs infinitely often, and this implies that $\{X(n) = i\}$ also occurs infinitely often. This is true for all states. Therefore, the skeleton is irreducible and recurrent. Consequently it has one and only one (up to a multiplicative factor) invariant measure. Since an invariant measure of the continuous-time chain is an invariant measure of the skeleton, the announced uniqueness of the invariant measure of the continuous-time HMC follows.

(γ) Call T_0 the return time to 0 of the imbedded chain. Then

$$\nu(i) = E_0\left[\int_0^{R_0} 1_{\{X(s)=i\}}ds\right] = E_0\left[\sum_{n=0}^{T_0-1} S_{n+1}1_{\{X_n=i\}}\right]$$

$$= E_0\left[\sum_{n=0}^{\infty} S_{n+1}1_{\{X_n=i\}}1_{\{n<T_0\}}\right] = \sum_{n=0}^{\infty} E_0[S_{n+1}1_{\{X_n=i\}}1_{\{n<T_0\}}]$$

$$= \sum_{n=0}^{\infty} E_0[S_{n+1} \mid X_n = i, n < T_0]E_0[1_{\{X_n=i\}}1_{\{n<T_0\}}]$$

$$= \sum_{n=0}^{\infty} E_0[S_{n+1} \mid X_n = i]E_0[1_{\{X_n=i\}}1_{\{n<T_0\}}],$$

where the last equality follows from the strong Markov property at time τ_n. But $E_0[S_{n+1} \mid X_n = i] = \frac{1}{q_i}$. Therefore,

$$\nu(i) = \frac{1}{q_i}\sum_{n=0}^{\infty} E_0[1_{\{X_n=i\}}1_{\{n<T_0\}}] = \frac{1}{q_i}E_0\left[\sum_{n=0}^{T_0-1} 1_{\{X_n=i\}}\right] = \frac{\mu(i)}{q_i}.$$

(δ) The transition matrix $\mathbf{P}$ of the imbedded chain is

$$p_{ii} = 0, \quad p_{ij} = \frac{q_{ij}}{q_i} \text{ if } i \neq j,$$

and the balance equation $\mu^T = \mu^T\mathbf{P}$ reads

$$\frac{\mu(i)}{q_i}q_i = \sum_{\substack{j \in E \\ j \neq i}} \frac{\mu(j)}{q_j}q_{ji},$$

which is just (7.44).   □

Theorem 7.4.8 *An irreducible recurrent regular jump* HMC *with invariant measure* ν *is t-positive recurrent if and only if*

$$\sum_{i \in E} \nu(i) < \infty.$$

In this case the stationary probability π *is related to the mean return times by*

$$q_i\pi_iE_i[R_i] = 1. \tag{7.45}$$

Proof. Observing that

$$\sum_{i \in E} \nu(i) = E_0[R_0],$$

and

$$\nu(0) = E_0[L_0] = \frac{1}{q_0},$$

the proof is then similar to that of the corresponding result in discrete time.   □

Remark 7.4.9 There exists between any t-invariant measure ν and any n-invariant measure μ of the imbedded chain a relationship of the form

$$\mu(i) = Kq_i\nu(i) \quad (i \in E).$$

Therefore all four possibilities concerning the convergence or divergence of the series $\sum_{i \in E} \mu(i)$ and $\sum_{i \in E} \nu(i)$ are open. The mean sojourn times q_i^{-1} make the difference, as is natural since they embody the deformation of the time scale when passing from the discrete time of the imbedded chain to the continuous time of the regular jump HMC.

The Stationary Distribution Criterion of Ergodicity

Definition 7.4.10 *An irreducible regular jump* HMC *is called* ergodic *if it is t-positive recurrent.*

Remark 7.4.11 Note that the notion of periodicity is irrelevant in continuous time. Note also that the imbedded chain of an ergodic regular jump HMC may well be non-ergodic, for two reasons. The first is the possibility that the imbedded chain is null recurrent, as explained in Remark 7.4.9 above, while the second is the possibility that the imbedded chain is periodic.

Theorem 7.4.12 *The irreducible regular jump* HMC *with infinitesimal generator* $\mathbf{A}$ *is ergodic if and only if there exists a probability* π *on E such that*

$$\pi^T \mathbf{A} = 0. \tag{7.46}$$

Proof. In view of Theorem 7.4.7, only sufficiency has to be proved. Therefore, suppose that (7.46) holds for a probability distribution π. We shall prove that $\pi^T \mathbf{P}(t) = \pi^T$ for all $t \geq 0$. For this, define

$$p_{ij}^{(n)}(t) = P_i(X(t) = j, t < \tau_n),$$

where the τ_n are the transition times. A trajectory starting from state i contributes to $p_{ij}^{(n)}(t)$ either if it has no jump before t and $i = j$, or if it has a last jump (say from k to j) at a time $s \leq t$, and therefore at most $n - 1$ jumps before s. Therefore

$$p_{ij}^{(n)}(t) = \delta_{ij}\exp(-q_i t) + \int_0^t \sum_{k \in E} p_{ik}^{(n-1)}(s)q_{kj}\exp(-q_j(t-s))\mathrm{d}s. \tag{7.47}$$

Therefore,

$$\sum_{i \in E} \pi(i)p_{ij}^{(n)}(t) = \pi(j)\exp(-q_j t) \quad + \int_0^t \exp(-q_j(t-s))\sum_{k \in E} q_{kj}\sum_{i \in E}\pi(i)p_{ik}^{(n-1)}(s)\mathrm{d}s.$$

Now,

$$\sum_{i \in E} \pi(i)p_{ik}^{(1)}(s) = \pi(k)\exp(-q_k s) \leq \pi(k),$$

that is, $\pi^T \mathbf{P}^{(1)}(s) \leq \pi$, with the obvious notation. By induction, $\pi^T \mathbf{P}^{(n)}(s) \leq \pi^T$. Indeed, if the latter is true, then

$$\sum_{i \in E} \pi(i) p_{ij}^{(n+1)}(s) \ \leq \ \pi(j) \exp(-q_j t) + \int_0^t \exp(-q_j(t-s)) \sum_{k \in E} q_{kj} \pi(k) ds$$

$$= \ \pi(j) \exp(-q_j t) + \pi(j) q_j \int_0^t \exp(-q_j s) ds = \pi(j).$$

Since the chain is regular,

$$\lim_{n \uparrow \infty} p_{ij}^{(n)}(t) = p_{ij}(t) \text{ and } \sum_{j \in E} p_{ij}(t) = 1.$$

Therefore, by dominated convergence,

$$\sum_{i \in E} \pi(i) p_{ij}(t) \leq \pi(j).$$

Summation with respect to j of both sides of the inequality shows that equality must hold. Therefore, π is a stationary distribution.

If the chain were transient, $\lim_{t \uparrow \infty} 1_{X(t)=j} = 0$, and therefore, by dominated convergence,

$$\lim_{t \to \infty} p_{ij}(t) = 0. \tag{7.48}$$

In particular, by dominated convergence, $(\pi^T \mathbf{P}(t))_j$ would tend to 0 as $t \to \infty$, a contradiction with $\pi^T \mathbf{P}(t) = \pi^T$. The chain is therefore recurrent. It is t-positive recurrent by Theorem 7.4.8.

□

EXAMPLE 7.4.13:   THE BIRTH-AND-DEATH PROCESS The birth-and-death process has an infinitesimal generator of the form

$$\mathbf{A} = \begin{pmatrix} -\lambda_0 & \lambda_0 & 0 & 0 & \cdots \\ \mu_1 & -(\lambda_1 + \mu_1) & \lambda_1 & 0 & \cdots \\ 0 & \mu_2 & -(\lambda_2 + \mu_2) & \lambda_2 & \cdots \\ \vdots & \vdots & \vdots & \vdots & \end{pmatrix}$$

where the state space is $E = \mathbb{N}$, or $E = \{0, 1, \ldots, N\}$ for finite N. We suppose that $\lambda_i > 0$ for all $i \in E$, except $i = N$ when $E = \{0, 1, \ldots, N\}$, and $\mu_i > 0$ for all $i \in E$, except $i = 0$. These conditions obviously guarantee irreducibility. We suppose that it is regular.

The global balance equations are

$$\lambda_0 \pi(0) = \mu_1 \pi(1)$$

and

$$(\lambda_i + \mu_i) \pi(i) = \lambda_{i-1} \pi(i-1) + \mu_{i+1} \pi(i+1)$$

for $i \geq 1$, with the convention $\mu_{N+1} = 0$ if $E = \{0, 1, \ldots, N\}$. Solving this system with initial condition $\pi(o)$ gives

$$\pi(i) = \pi(0) \prod_{n=1}^i \frac{\lambda_{n-1}}{\mu_n}. \tag{7.49}$$

Ergodicity occurs if and only if one can choose $\pi(0) > 0$ such that

$$\pi(0) \left(1 + \sum_{\substack{i \in E \\ i \geq 1}} \prod_{n=1}^{i} \frac{\lambda_{n-1}}{\mu_n} \right)^{-1} = 1. \tag{7.50}$$

Therefore the ergodicity necessary and sufficient condition is

$$\sum_{\substack{i \in E \\ i \geq 1}} \prod_{n=1}^{i} \frac{\lambda_{n-1}}{\mu_n} < \infty. \tag{7.51}$$

The ergodicity condition is, of course, automatically satisfied when the state space is finite.

Reversibility

The concept of *reversibility* in continuous time is basically the same as in discrete time.

Let $\{X(t)\}_{t \geq 0}$ be a regular jump HMC with countable state space E, transition semigroup $\{\mathbf{P}(t)\}_{t \geq 0}$, and infinitesimal generator $\mathbf{A}$. Assume, moreover, that this chain is irreducible and ergodic, and in equilibrium, with the stationary distribution π. For some fixed $T > 0$, consider the process $\{\tilde{X}(t)\}_{t \in [0,T]}$ obtained by time reversal of $\{X(t)\}_{t \in [0,T]}$, that is,

$$\tilde{X}(t) = X(T - t).$$

Then $\{\tilde{X}(t)\}_{t \in [0,T]}$ is an HMC with transition matrix $\tilde{\mathbf{P}}(t)$ given by

$$\pi(i)\tilde{p}_{ij}(t) = \pi(j)p_{ji}(t) \tag{7.52}$$

(the proof is analogous to that of the corresponding result in discrete time), and therefore with infinitesimal generator $\tilde{\mathbf{A}}$, where

$$\pi(i)\tilde{q}_{ij} = \pi(j)\tilde{q}_{ji}, \tag{7.53}$$

as differentiation in (7.52) immediately shows.

Remark 7.4.14 In order not to depend on the choice of T, we can use Theorem 5.1.14 to extend the definition of the chain to negative times. However, this theorem offers no guarantee that the extension retains the right-continuity property of the trajectories. An alternative approach is to make use of a regular jump HMC $\{Y(t)\}_{t \geq 0}$ with transition semigroup $\{\tilde{\mathbf{P}}(t)\}_{t \geq 0}$ and independent of $\{X(t)\}_{t \geq 0}$ given $X(0)$ and with initial state $Y(0)$, and to define for $t \leq 0$, $X(t) = Y(-t)$. In this manner, we obtain a stationary HMC $\{X(t)\}_{t \in \mathbb{R}}$ with transition semigroup $\{\mathbf{P}(t)\}_{t \geq 0}$ and infinitesimal generator $\mathbf{A}$. It will be assumed that the process on the negative times (which is left-continuous) has been modified so as to make it right-continuous. Since the transition times of a continuous-time HMC are absolutely continuous random variables, a modification at these times does not change the distribution of the process. The reversed process $\{X(-t)\}_{t \in \mathbb{R}}$ is therefore a regular jump HMC with transition semigroup $\{\tilde{\mathbf{P}}(t)\}_{t \in \mathbb{R}}$ and infinitesimal generator $\mathbf{A}$ given by (7.52) and (7.53), respectively.

Theorem 7.4.15 *Let $\{X(t)\}_{t \in \mathbb{R}}$ be an irreducible and ergodic stationary regular jump HMC on the countable state space E, with stationary distribution π. If its infinitesimal generator $\mathbf{A}$ satisfies the detailed balance equations*

$$\pi(i)q_{ij} = \pi(j)q_{ji}, \tag{7.54}$$

then the reversed process $\{X(-t)\}_{t \in \mathbb{R}}$ properly modified to be right-continuous is equivalent in distribution to the direct process $\{X(t)\}_{t \in \mathbb{R}}$.

Proof. Comparing (7.53) and (7.54), we see that $\tilde{\mathbf{A}} \equiv \mathbf{A}$, and therefore, the direct and the reversed processes have the same infinitesimal generator. The conclusion then follows from Theorem 7.3.6.

$\square$

The following reversal test, analogous to its discrete-time version, will be needed in Chapter 9 featuring a number of examples of reversible chains concerning queueing networks.

Theorem 7.4.16 Let $\mathbf{A}$ be a stable and conservative generator on the countable state space E, and let π be a strictly positive probability distribution on E. Define $\tilde{\mathbf{A}}$ by (7.53). If for all $i \in E$

$$\sum_{j \in E, j \neq i} \tilde{q}_{ij} = q_i, \tag{7.55}$$

then $\pi^T \mathbf{A} = \pi$.

The proof is a straightforward adaptation of that of the corresponding result in discrete-time.

A regular jump HMC on E $\{X(t)\}_{t \geq 0}$ that is irreducible and stationary, with stationary distribution π and infinitesimal generator $\mathbf{A}$ satisfying the detailed balance equation (7.55) is called *reversible*. Sometimes, the pair $(\mathbf{A}, \pi)$, is called reversible.

7.4.2   Convergence to Equilibrium

The limiting behavior of continuous-time HMCs follows from that of discrete-time HMCs in a usually straightforward manner.

Theorem 7.4.17 Let $\{X(t)\}_{t \geq 0}$ be an ergodic regular jump HMC with state space E and transition semigroup $\{\mathbf{P}(t)\}_{t \geq 0}$. Then, for all $i, j \in E$,

$$\lim_{t \to \infty} p_{ij}(t) = \pi(j), \tag{7.56}$$

where π is the (unique) stationary distribution.

Proof. The *skeleton* $\{X(n)\}_{n \geq 0}$ of this Markov chain is an irreducible HMC (see the proof of Theorem 7.4.7). It is positive recurrent, since it has π for a stationary distribution. We show that it is aperiodic. For this consider the IID sequence of sojourn times in state i, as in part (β) of the proof of Theorem 7.4.7. Since the event $Z_n > 2$ occurs infinitely often, we necessarily have that a one-step transition from i to i has positive probability. And we know that the existence of loops in the transition graph implies aperiodicity.

Two independent continuous-time HMCs with the same transition semigroup $\{\mathbf{P}(t)\}_{t \geq 0}$, but possibly different initial distributions, will meet at a finite integer random time, since their skeletons do. From this observation the result follows by the same coupling argument as in the discrete-time case.

$\square$

Empirical Averages

The next result is the continuous-time version of Corollary 6.4.3 with an analogous proof.

Theorem 7.4.18 Let $\{X(t)\}$ be an ergodic continuous-time HMC with stationary distribution π. Let $g : E^{L+1} \to \mathbb{R}$ be such that

$$\sum_{i_0,\dots,i_L} |g(i_0,\dots,i_L)| \pi(i_0) p(t_-t_0)_{i_0,i_1} \cdots p(t_L - t_{L-1})_{i_{L-1},i_L} < \infty,$$

where $t_0 < t_1 < \cdots < t_{L-1} < t_L$. Then

$$\lim_{t\uparrow\infty} \frac{1}{t} \int_0^t g(X(s+t_0), X(s+t_1),\dots,X(s+t_L)) \mathrm{d}s$$
$$= \sum_{i_0,\dots,i_L} g(i_0,\dots,i_L) \pi(i_0) p(t_1 - t_0)_{i_0,i_1} \cdots p(t_L - t_{L-1})_{i_{L-1},i_L}, \ P_\mu \ a.s.$$

Remark 7.4.19 Remark 16.1.11 can be adapted to the case of a continuous-time irreducible positive HMC to obtain a generalization of Theorem 7.4.18. The statement and the corresponding details are left for the reader.

Complementary reading

[Karlin, 1966] and [Karlin and Taylor, 1975] for examples, especially in biology. [Daley end Gani, 1999] for examples in epidemiology.

7.5 Exercises

Exercise 7.5.1. CHARACTERISTIC FUNCTIONS OF POISSON INTEGRALS
Let N be a Poisson process on $\mathbb{R}^m$ with locally finite intensity measure ν and let $\varphi : \mathbb{R}^m \to \mathbb{R}$ be a function in $L^1_{\mathbb{C}}(\nu)$. Prove the formula

$$\mathrm{E}\left[\mathrm{e}^{iu \int_{\mathbb{R}^m} \varphi(x) N(\mathrm{d}x)}\right] = \exp\left\{ \int_{\mathbb{R}^m} \left(\mathrm{e}^{iu\varphi(x)} - 1\right) \nu(\mathrm{d}x)\right\} \quad (u \in \mathbb{R}).$$

Exercise 7.5.2. CONVERGENCE OF A SHOT NOISE TO A GAUSSIAN PROCESS
Consider the shot noise process $\{X_k(t)\}_{t\in\mathbb{R}}$ given by

$$X_k(t) = \sum_{n\in\mathbb{Z}} h(t - T_{k,n}),$$

where $\{T_{k,n}\}_{n\in\mathbb{Z}}$ is an HPP on $\mathbb{R}$ with intensity $\lambda = k\lambda_0$ and $h(t) = \frac{1}{\sqrt{k}} h_0(t)$ for some integrable function $h_0(t)$ such that $\int_{\mathbb{R}} h_0(t) \, \mathrm{d}t = 0$. Show that the finite distributions of $\{X_k(t)\}_{t\in\mathbb{R}}$ converge as $k \uparrow \infty$ to the finite distributions of a centered Gaussian process $\{Y(t)\}_{t\in\mathbb{R}}$ with covariance function $\lambda_0 \int_{\mathbb{R}} h_0(s+\tau) h_0(t) \, \mathrm{d}t$.

Exercise 7.5.3. A UNIFORM HMC
Consider the uniform HMC of Definition 7.2.6.

1. Prove that it is indeed a continuous-time HMC.

2. Show that its infinitesimal generator is $\mathbf{A} = \lambda(\mathbf{K} - I)$.

Exercise 7.5.4. CONTINUITY OF THE TRANSITION SEMI-GROUP

1. Show that a transition semigroup that is continuous at the origin is continuous at all times $t \geq 0$.

2. Show that the transition semigroup of a right-continuous HMC is continuous.

Exercise 7.5.5. A TWO-STATE HMC

Consider the uniform Markov chain with state space $E = \{0, 1\}$; transition matrix of the subordinated chain

$$\mathbf{K} = \begin{pmatrix} 1 - \alpha & \alpha \\ \beta & 1 - \beta \end{pmatrix},$$

where $\alpha, \beta \in (0, 1)$; and intensity $\lambda > 0$ for the underlying HPP (the clock). Find the transition semigroup $\{\mathbf{P}(t)\}_{t \geq 0}$. Suppose that $X(0) = 0$. Give the joint probability distribution of the sequence $\{\tau_n - \tau_{n-1}\}_{n \geq 1}$, where $\tau_1, \tau_2, \ldots$ are the successive times when $\{X(t)\}_{t \geq 0}$ switches from one value to a *different* value.

Exercise 7.5.6. PATHOLOGY

Let π be a probability distribution on the countable state space E such that $\pi(i) > 0$ for all $i \in E$, and for each $t \in \mathbb{R}_+$, let $X(t)$ be distributed according to π. Also suppose that the family $\{X(t)\}_{t \geq 0}$ is independent. Show that $\{X(t)\}_{t \geq 0}$ is a homogeneous Markov chain. Give its transition semigroup. What about its local characteristics? In particular, is $q_{ij} < \infty$? How does this fit with the main results?

Exercise 7.5.7. A BIRTH-AND-DEATH HMC

Let $\{X(t)\}_{t \geq 0}$ be a birth-and-death process with the parameters $\lambda_n = n\lambda + a$ and $\mu_n = n\mu 1_{n \geq 1}$, where $\lambda > 0, \mu > 0, a > 0$. Suppose that $X(0) = i$ is fixed. Show that $M(t) = \mathrm{E}[X(t)]$ satisfies a differential equation and solve this equation. Do the same for the variance $S(t) = \mathrm{E}[X(t)^2] - M(t)^2$.

Exercise 7.5.8. SUFFICIENT CONDITIONS FOR REGULARITY

(a) Let $\mathbf{A}$ be a stable and conservative generator on E. Define $\mathbf{P} = \{p_{ij}\}_{i,j \in E}$ by $p_{ii} = 0$, $p_{ij} = \frac{q_{ij}}{q_i}$ if $i \neq j$. Show that if the transition matrix $\mathbf{P}$ is irreducible and recurrent, the generator is regular.

(b) Consider the same question, only assuming that the state space is finite ($\mathbf{P}$ is not assumed to be irreducible).

(c) Consider the same question without irreducibility and recurrence assumptions, but assuming $\sup_{i \in E} q_i < \infty$.

Exercise 7.5.9. TRANSIT TIMES OF A BIRTH-AND-DEATH HMC

Let $\{X(t)\}_{t \geq 0}$ be a birth-and-death process with parameters $\lambda_n > 0$ and $\mu_n > 0$ (of course $\mu_0 = 0$). Let w_n be the average time required to pass from state n to state $n+1$. Find a recurrence equation for $\{w_n\}_{n \geq 0}$. Deduce from it the average time required to reach state $n \geq 1$ when starting from state 0. Describe the result when $\lambda_n = \lambda$ and $\mu_n = \mu$.

Exercise 7.5.10. REGULARITY
Show that the birth-and-death generator with birth-and-death parameters $\lambda_0 = \lambda$, $\lambda_n = n\lambda$ for $n \geq 1$, and $n\mu$, respectively, where λ and μ are positive, is regular.

Exercise 7.5.11. THE CIRCLE
The state space is $E = \{0, 1, \ldots, N - 1\}$. The states are equidistant locations on the circle placed clockwise in this order. In particular state 0 and $N - 1$ are adjacent. A particle moves from state to state as follows: if it is in position i, it stays there during an exponential time of mean $m_i \in (0, \infty)$, and then moves to the adjacent set on the right (looking at the center of the circle) with probability $p \in (0, 1)$, on the left with probability $1 - p$. What is the average return time to i?

Exercise 7.5.12. EPIDEMICS
A village of $N + 1$ people suffers an epidemic. Let $X(t)$ be the number of ill people at time t and suppose that $X(0) = 1$ and $\{X(t)\}_{t\geq 0}$ is a pure birth process with birth rates $\lambda_i = \lambda i (N + 1 - i)$. Let T be the amount of time required until every member of the population has succumbed to the illness. Show that

$$E[T] = \frac{2(\log N + \gamma)}{\lambda(N + 1)} + O(N^{-2})$$

where γ is the Euler constant, defined by

$$\sum_{k=1}^{N} \frac{1}{k} = \log N + \gamma + O(N^{-1}).$$

Exercise 7.5.13. RECURRENT BIRTH AND DEATH
Consider an irreducible recurrent birth-and-death process with state space $\mathbb{N}$. Discuss the nature (positive or null) of recurrence of this process and of its imbedded chain in terms of the birth-and-death parameters.

Chapter 8

Spatial Poisson Processes

A point process is a random set of "points" scattered in a "space". Such processes appear in various forms in nature, such as, for instance, the stars in the sky, the small trees in the African savanna, the impacts of lightnings in the Great Plains, the crashes in the stock market or the arrival times of cars at a highway toll. Among the spatial point processes (point processes whose points lie in some Euclidean space), the Poisson processes are the most popular models, in particular because of their mathematical tractability.

8.1 Generalities on Point Processes

This section gives the fundamental results concerning general point processes.

8.1.1 Point Processes as Random Measures

Let E be a locally compact topological space with a denumerable base (for short, $l.c.d.b.$): two distinct points admit disjoint neighborhoods (Hausdorff), each point has a compact neighborhood (local compactness) and there is a countable family of open sets $\{G_n\}_{n\geq 1}$ (the countable base) such that each open set G is the union of open sets of this base. It is in particular a Polish space, that is, a metrizable, complete and separable topological space. The σ-field $\mathcal{B}(E)$ is then the Borel σ-field on this topological space, that is, the σ-field generated by the open sets of the topology. A compact subset of E is a set such that from any cover of it by open sets, one can extract a *finite* subcover. A subset of E is called relatively compact if its closure is compact.

Remark 8.1.1 Those who do not feel comfortable with this generality will be reassured by the fact that in all the applications in this book, E is some $\mathbb{R}^m$. Recall that a subset of $\mathbb{R}^m$ is compact if and only if it is bounded and closed, and that it is relatively compact if and only if it is bounded.

A *locally finite* measure on $(E, \mathcal{B}(E))$ is a measure giving finite mass to relatively compact sets. A locally finite measure on $(E, \mathcal{B}(E))$ is in particular σ-finite, since there exists a sequence of relatively compact sets increasing to E. (This is not true in general if E is only assumed to be a metric space, even a separable one, endowed with its Borel σ-field.) Let $M(E)$ be the set of *locally finite* measures on $(E, \mathcal{B}(E))$ and let $\mathcal{M}(E)$ be the smallest σ-field on $M(E)$ generated by the mappings $p_C : \mu \to \mu(C)$ $(C \in \mathcal{B}(E))$. A measure $\mu \in M(E)$ taking integer (possibly infinite) values is called a *point measure*. Such a point measure is a countable sum of Dirac measures

© Springer Nature Switzerland AG 2020
P. Brémaud, *Probability Theory and Stochastic Processes*, Universitext,
https://doi.org/10.1007/978-3-030-40183-2_8

$$\mu = \sum_{n \in S(\mu)} \varepsilon_{x_n},$$

where $S(\mu)$ is a subset of $\mathbb{N}$ and the x_n's need not be distinct. The subset of $M(E)$ consisting of the locally finite point measures is denoted by $M_p(E)$ and the σ-field $\mathcal{M}_p(E)$ on it is, by definition, the σ-field generated by the collection of sets $\{\mu \in M_p(E); \mu(C) \in F\}$ ($C \in \mathcal{B}(E)$, $F \in \mathcal{B}(\overline{\mathbb{R}}_+)$). The measurable space $(M(E), \mathcal{M}(E))$ is called the *canonical space of locally finite measures* on $(E, \mathcal{B}(E))$, whereas $(M_p(E), \mathcal{M}_p(E))$ is called the *canonical space of locally finite point measures* on $(E, \mathcal{B}(E))$.

Let now $(\Omega, \mathcal{F}, P)$ be some probability space.

Definition 8.1.2 *A* random measure *on* $(E, \mathcal{B}(E))$ *(for short: on E) is a measurable mapping*

$$N : (\Omega, \mathcal{F}) \to (M(E), \mathcal{M}(E)).$$

If, moreover, $N(\omega) \in M_p(E)$ P-a.s., N is called a point process *on* $(E, \mathcal{B}(E))$.

When N is a random measure, $N(C)$ is a random variable for all $C \in \mathcal{B}(E)$, since the mapping $\omega \to N(\omega)(C)$ is the composition of the measurable mappings $\omega \to N(\omega)$ and $\mu \to p_C(\mu) = \mu(C)$.

Definition 8.1.3 *Define $\mathcal{E}_0$ to be a collection of relatively compact subsets of E such that*

(i) *$\mathcal{E}_0$ is a π-system,*

(ii) *$\sigma(\mathcal{E}_0) = \mathcal{B}(E)$, and*

(iii) *there exists a sequence $\{E_n\}_{n \geq 1}$ in $\mathcal{E}_0$ that either (a): forms a partition of E or (b): increases to E.*

For instance, when $E = \mathbb{R}^d$, the class $\mathcal{E}_0$ of sets of the form

$$I = \prod_{j=1}^{d} (a_j, b_j], \text{ where } [a_j, b_j] \subseteq \mathbb{R}.$$

Theorem 8.1.4

A. *For the function $N : \Omega \to M(E)$ to be measurable, it suffices that $N(C) : \Omega \to \overline{\mathbb{R}}_+$ be a random variable for all $C \in \mathcal{E}_0$.*

B. *$\mathcal{M}(E) = \sigma\left(\{\{m\,;\, m(C) \in A\}; C \in \mathcal{E}_0, A \in \mathcal{B}(\overline{\mathbb{R}}_+)\}\right)$. In particular, letting*

$$\mathcal{I} := \{\{m(C_j) \in A_j\,(1 \leq j \leq k)\}(k \in \mathbb{N}_+, C_j \in \mathcal{E}_0, A_j \in \mathbb{N}_+)\}$$

(a π-system), we have that $\sigma(\mathcal{I}) = \mathcal{M}(E)$.

Proof. A. We prove the theorem assuming case (iii(b)) of Definition 8.1.3, the other case being similar. Fix n. The collection

$$\mathcal{G} := \{C \in \mathcal{B}(E)\,;\, \omega \to N(C \cap K_n) \text{ is measurable}\}$$

is a d-system. (The fact that a measure $\mu \in M(E)$ puts finite mass on a relatively compact set, here K_n, plays a role in the proof of stability of $\mathcal{G}$ under proper difference.) Since $\mathcal{G} \supset \mathcal{F}_0$, we have that $\mathcal{G} \supset \sigma(\mathcal{E}_0) = \mathcal{B}(E)$. Therefore, $N(C \cap K_n)$ is a random variable for all $C \in \mathcal{B}(E)$ and all $n \geq 1$. Finally, for all $C \in \mathcal{B}(E)$, $N(C)$ is a random variable, being the limit as $n \uparrow \infty$ of the random variables $N(C \cap K_n)$ $(n \geq 1)$.

B. In A, take $\Omega = M(E)$ and for N the identity mapping on $M(E)$. By definition, $\mathcal{M}(E) = \sigma\{\{m\,;\,m(C) \in A\}, C \in \mathcal{B}(E), A \in \mathcal{B}(\overline{\mathbb{R}}_+)\}$. The conclusion of part A implies that we can replace in the last identity $C \in \mathcal{B}(E)$ by $C \in \mathcal{E}_0$. $\square$

Remark 8.1.5 The same result applies with the same proof when replacing $M(E)$ by $M_p(E)$. In this case, since a point measure takes its values in $\mathbb{N} \cup \{\infty\}$, letting this time

$$\mathcal{I} := \{\{m(C_j) = n_j\,(1 \leq j \leq k)\}; k \in \mathbb{N}_+, C_j \in \mathcal{E}_0, n_j \in \mathbb{N}_+\}$$

(a π-system), we have that $\sigma(\mathcal{I}) = \mathcal{M}_p(E)$.

For the next definition, recall that a *singleton* is a set, denoted by $\{a\}$, consisting of exactly one element $a \in E$.

Definition 8.1.6 *A point measure μ on E is called* simple *if*

$$\mu(\{a\}) = 0 \ or \ 1 \ for \ all \ a \in E.$$

A point process N on E is called simple *if for all $\omega \in \Omega$, $N(\omega)$ is simple.*

The definitions of a P-a.s. σ-finite random measure, of a P-a.s. locally bounded random measure and of a P-a.s. simple point process are the obvious ones. For instance, a point process N on E is called P-a.s. simple if $N(\omega)$ is simple for all ω outside some P-negligible event.

There are situations where for each $\omega \in \Omega$, $N(\omega)$ is only a σ-*finite* measure (not necessarily locally finite). Also, it may be that E is not some $\mathbb{R}^m$, but a space endowed with a σ-field $\mathcal{E}$. We then have to redefine $M(E)$ as the set of measures (of all kinds).

Definition 8.1.7 *A* random measure N *on $(E, \mathcal{E})$ is a mapping $(\omega, C) \to N(\omega, C)$ from $(\Omega \times \mathcal{E})$ into $\overline{\mathbb{R}}_+$, such that*

(i) for each $\omega \in \Omega$, the map $C \in \mathcal{E} \to N(\omega, C)$ is a measure on $(E, \mathcal{E})$, and

(ii) for each $C \in \mathcal{E}$, the map $\omega \in \Omega \to N(\omega, C)$ is measurable.

In the case where there exists a sequence $\{K_n\}_{n \geq 1}$ of measurable sets increasing to E and such that $N(\omega, K_n) < \infty$ for all $\omega \in \Omega$ and all $n \geq 1$, it is called a σ-finite random measure.

Remark 8.1.8 This extension will be useful when we consider marked point processes with marks in a measurable space $(K, \mathcal{K})$ (See Definition 8.1.16). It will sometimes then be convenient to represent these as point processes on $(\mathbb{R}^m \times K, \mathcal{B}(\mathbb{R}^m) \otimes \mathcal{K})$ (see Theorem 8.3.1).

Points

Let μ be a *locally finite* point measure on $(\mathbb{R}, \mathcal{B}(\mathbb{R}))$. Then (exercise):

(i) there exists a non-decreasing sequence $\{t_n\}_{n \in \mathbb{Z}}$ in $\overline{\mathbb{R}}$ such that for all measurable sets C,

$$\mu(C) = \sum_{n \in \mathbb{Z}} \varepsilon_{t_n}(C)$$

(where ε_a is the Dirac measure at a if $a \in \mathbb{R}$, and the null measure if $a = +\infty$ or $-\infty$), and such that $t_0 \leq 0 < t_1$, and

(ii) the mappings $\mu \to t_n = t_n(\mu)$ are measurable from $(M_p(\mathbb{R}), \mathcal{M}_p(\mathbb{R}))$ to $(\overline{\mathbb{R}}, \overline{\mathcal{B}})$, and

(iii) if in addition μ is *simple*, then $|t_n| < \infty \Rightarrow t_n < t_{n+1}$. (In other words, $\{t_n\}_{n \in \mathbb{Z}}$ is strictly increasing on $\mathbb{R}$.)

If N is a locally finite point process on $\mathbb{R}$, define

$$T_n(\omega) := t_n(N(\omega)).$$

In particular, T_n is a random variable since $T_n = t_n \circ N$ is the composition of two measurable functions. If N is *simple*, the sequence of random variables $\{T_n\}_{n \in \mathbb{Z}}$ is *strictly increasing* on $\mathbb{R}$. Recall that $T_0 \leq 0 < T_1$ by convention. The sequence $\{T_n\}_{n \in \mathbb{Z}}$ is the sequence of *event times* or *points* of N. For all $C \in \mathcal{B}(\mathbb{R})$,

$$N(C) = \sum_{n \in \mathbb{Z}} 1_C(T_n).$$

More generally, a locally finite point process on E can be described by a sequence of "points".

Lemma 8.1.9 *Let Δ be an element not in E. A locally finite point process N on E can be represented as*

$$N = \sum_{n \in \mathbb{N}} \varepsilon_{X_n},$$

where $\{X_n\}_{n \in \mathbb{N}}$ is a sequence of random variables with values in $E \cup \{\Delta\}$ endowed with the measurable space generated by $\mathcal{B}(E)$ and $\{\Delta\}$, and ε_a is the Dirac measure at a if $a \in E$, the null measure if $a = \Delta$.

Proof. The proof is left as an exercise (Exercise 8.5.1).   □

Remark 8.1.10 *The Δ element plays the role of ∞. Note that it may occur that for some of the values in the list $\{X_n\}_{n \in \mathbb{N}}$ are the same, representing multiple points.*

EXAMPLE 8.1.11: THE BINOMIAL POINT PROCESS. (In this example, the numbering of points is intrinsic to its description.) This point process on $\mathbb{R}^m$ has a (finite number) of points T, where T is a binomial random variable of size n and parameter $p \in (0, 1)$:

$$P(T = k) = \binom{n}{k} p^k (1 - p)^{n-k} \quad (0 \leq k \leq n).$$

If $T = k$, the k points are located independently of one another on $\mathbb{R}^m$ according to the same probability distribution Q. It is simple if and only if Q is non-atomic.

___

EXAMPLE 8.1.12: THE POISSON PROCESS. Let ν be a σ-finite measure on E. The point process N on E is called a *Poisson process* on E with *intensity measure* ν if

(i) for all finite families of mutually *disjoint* sets $C_1, \ldots, C_K \in \mathcal{B}(E)$, the random variables $N(C_1), \ldots, N(C_K)$ are independent, and

(ii) for any set $C \in \mathcal{B}(E)$ such that $\nu(C) < \infty$,

$$P(N(C) = k) = e^{-\nu(C)} \frac{\nu(C)^k}{k!} \qquad (k \geq 0) .$$

In the case $E = \mathbb{R}^m$, if ν is of the form $\nu(C) = \int_C \lambda(x)\mathrm{d}x$ for some non-negative measurable function $\lambda : \mathbb{R}^m \to \mathbb{R}$, the Poisson process N is said to admit the *intensity function* $\lambda(x)$. If in addition $\lambda(x) \equiv \lambda$, N is called a *homogeneous* Poisson process (HPP) on $\mathbb{R}^m$ with *intensity* or *rate* λ.

___

Let N be a random measure on E. The σ-field $\mathcal{F}^N$ generated by the random variables $N(C)$ $(C \in \mathcal{B}(E))$ is called the σ-field *generated by* N.

Definition 8.1.13 *A family* $\{N_i\}_{i \in I}$ *of random measures defined on the same probability space is said to be* independent *if the family of σ-fields* $\{\mathcal{F}^{N_i}\}_{i \in I}$ *is independent.*

This means that any *finite* subfamily $\{\mathcal{F}^{N_i}\}_{i \in J}$ is independent.

Definition 8.1.14 *A point process N on $\mathbb{R}^m$ is called* stationary *if for all $n \geq 1$, all Borel sets $C_1, \ldots, C_n$ of $\mathbb{R}^m$, the distribution of the random vector* $(N(C_1 + a), \ldots, N(C_n + a))$ *is independent of $a \in \mathbb{R}^m$.*

___

EXAMPLE 8.1.15: A STATIONARY GRID, TAKE 1. A grid on $\mathbb{R}^2$ is a deterministic point process on $\mathbb{R}^2$ whose points are

$$(nT_1, mT_2) \quad (n, m \in \mathbb{Z}) ,$$

where T_1 and T_2 are positive real numbers. It is not a stationary point process. However, the shifted version of it,

$$(nT_1 + V_1, mT_2 + V_2) \quad (n, m \in \mathbb{Z}) ,$$

where V_1 and V_2 are independent random variables uniformly distributed on $[0, T_1)$ and $[0, T_2)$ respectively, is stationary. This can be proved directly, or by using the Laplace functional (Example 8.1.36).

___

Marked Point Processes

Definition 8.1.16 *Let N and $\{X_n\}_{n\in\mathbb{N}}$ be as in Lemma 8.1.9. Let $(K,\mathcal{K})$ be some $(\mathbb{R}^d, \mathcal{B}(\mathbb{R}^d))$ and let $\{Z_n\}_{n\in\mathbb{N}}$ be a random sequence with values in K. The sequence $\{(X_n, Z_n)\}_{n\in\mathbb{N}}$ is called a* marked point process *on E with marks in K, $\{Z_n\}_{n\in\mathbb{N}}$ is the* mark sequence *and N is the* basic point process.

From such a marked point process, define the point process $\overline{N}$ on $E \times K$ by

$$\overline{N}(C \times L) := \sum_{n\in\mathbb{N}} 1_C(X_n)1_L(Z_n) \quad (C \in \mathcal{B}(E), L \in \mathcal{K}). \tag{8.1}$$

Note that since $\Delta \notin C$, the points $X_n \in \{\Delta\}$ do not appear in the sum above ("points at infinity are excluded"). We shall occasionally use the notation N_Z instead of $\overline{N}$. The following phrases are considered equivalent: "the marked point process $\{(X_n, Z_n)\}_{n\in\mathbb{N}}$", "the marked point process $\overline{N}$", "the marked point process N_Z", "the marked point process (N, Z)".

Definition 8.1.17 *If in addition $\{Z_n\}_{n\in\mathbb{N}}$ is* IID *and independent of N, with common probability distribution Q_Z, then $\overline{N}$ is called a* marked point process with independent IID marks.

8.1.2   Point Process Integrals and the Intensity Measure

Point process integrals are integrals with respect to a random point measure. Since this random measure is in this particular case a counting measure, these integrals are in fact sums.

Let N be a locally finite point process on E. Let μ be a measure on $(E, \mathcal{B}(E))$ and let $\varphi : (E, \mathcal{B}(E)) \to (\overline{\mathbb{R}}, \mathcal{B}(\overline{\mathbb{R}}))$ be a measurable function for which the integral $\int_E \varphi\, d\mu$ is well defined. This integral is also denoted by $\mu(\varphi)$. When $\varphi : E \to \overline{\mathbb{R}}$ is a measurable function and N is a point process, the following notations represent the same mathematical object (if it is well defined):

$$\sum_{n\in\mathbb{N}} \varphi(X_n), \qquad \int_E \varphi(x)N(\mathrm{d}x), \qquad N(\varphi).$$

In the first notation, we use the *convention* that the sum extends only to those indices n such that $X_n \in E$, excluding the points "at infinity" (in fact, $\varphi(\Delta)$ is not defined). In the situation of Definition 8.1.16, observe that

$$\int_{E\times K} \varphi(x,z)\overline{N}(\mathrm{d}x \times \mathrm{d}z) = \sum_{n\in\mathbb{N}} \varphi(X_n, Z_n),$$

with the same convention as the one just agreed upon concerning points at infinity.

Definition 8.1.18 *Let N be a locally finite point process on $(E, \mathcal{B}(E))$. The set function*

$$\nu \to \nu(C) := \mathrm{E}\left[N(C)\right] \qquad (C \in \mathcal{B}(E))$$

defines a measure on $(E, \mathcal{B}(E))$ (Exercise 8.5.9), called the mean measure *or the* intensity measure *of N.*

Definition 8.1.19 *The locally finite point process N is called a* first-order point process *if $\mathrm{E}\left[N(C)\right] < \infty$ for all compact sets $C \in \mathcal{E}$.*

Campbell's Formula

Theorem 8.1.20 *Let N be a point process on the measurable space $(E, \mathcal{B}(E))$ with intensity measure ν. Then, for all measurable functions $\varphi : E \to \mathbb{R}$ which are either non-negative or in $L^1_{\mathbb{R}}(\nu)$, the integral $N(\varphi)$ is well defined (possibly infinite when φ is only assumed to be non-negative) and*

$$\mathrm{E}\,[N(\varphi)] = \nu(\varphi)\,. \tag{8.2}$$

In particular, $N(\varphi)$ is a.s. finite if $\varphi \in L^1_{\mathbb{R}}(\nu)$.

Proof. First, suppose that φ is a *simple* non-negative measurable function, that is, of the form

$$\sum_{h=1}^{L} \alpha_h 1_{C_h}\,,$$

where $L \in \mathbb{N}$, $\alpha_h \in \mathbb{R}_+$ and $C_1, \dots, C_L$ are disjoint measurable subsets of E. Then

$$\mathrm{E}[N(\varphi)] = \mathrm{E}\left[\sum_{h=1}^{L} a_h N(C_h)\right] = \sum_{h=1}^{L} a_h \nu(C_h) = \nu(\varphi)\,.$$

Now let φ be a non-negative measurable function and $\{\varphi_n\}_{n\in\mathbb{N}}$ a non-decreasing sequence of simple non-negative measurable functions, with limit φ. Letting $n \uparrow \infty$ in

$$\mathrm{E}[N(\varphi_n)] = \nu(\varphi_n)$$

yields the announced result, by monotone convergence. In the case where $\varphi \in L^1_{\mathbb{R}}(\nu)$, since $\mathrm{E}\,[N(\varphi^{\pm})] = \nu(\varphi^{\pm}) < \infty$, the random variables $N(\varphi^{\pm})$ are P-a.s. finite, and therefore $N(\varphi) = N(\varphi^+) - N(\varphi^-)$ is well defined and finite, and

$$\mathrm{E}[N(\varphi)] = \mathrm{E}[N(\varphi^+)] - \mathrm{E}[N(\varphi^-)] = \nu(\varphi^+) - \nu(\varphi^-) = \nu(\varphi)\,.$$

$\square$

EXAMPLE 8.1.21: CAMPBELL'S FORMULA FOR MARKED POINT PROCESSES WITH IN-DEPENDENT IID MARKS. Let $\overline{N}$ be as in Definition 8.1.16, that is, a marked point process N on E with independent IID marks $\{Z_n\}_{n\in\mathbb{N}}$ taking their values in some measurable space $(K, \mathcal{K})$. Denoting by Q_Z the common distribution of the marks and by ν the intensity measure of the basic point process N, the intensity measure of $\overline{N}$ is the product measure $\overline{\nu}(\mathrm{d}x \times \mathrm{d}z) = \nu(\mathrm{d}x)Q_Z(\mathrm{d}z)$. Campbell's theorem then reads as follows. If the measurable function $\varphi : \mathbb{R}^m \times K \to \mathbb{R}$ is either non-negative or in $L^1_{\mathbb{R}}(\nu \times Q_Z)$, then the sum

$$\sum_{n\in\mathbb{N}} \varphi(X_n, Z_n)$$

is P-a.s. well defined (possibly infinite if φ is only assumed non-negative) and

$$\mathrm{E}\left[\sum_{n\in\mathbb{N}} \varphi(X_n, Z_n)\right] = \int_E \mathrm{E}\,[\varphi(x, Z)]\,\nu(\mathrm{d}x)\,,$$

where Z is a K-valued random variable with distribution Q_Z.

The following avatar of Campbell's formula is useful (see the next theorem for instance):

Corollary 8.1.22 *Let N be a point process on $(E, \mathcal{B}(E))$ with intensity measure ν, defined on the probability space $(\Omega, \mathcal{F}, P)$. Let $\mathcal{G}$ be a sub-σ-field of $\mathcal{F}$, independent of $\mathcal{F}^N$. Then, for all measurable functions $\varphi : (E \times \Omega, \mathcal{B}(E) \otimes \mathcal{G}) \to (\mathbb{R}, \mathcal{B}(\mathbb{R}))$ which are either non-negative or in $L^1_{\mathbb{R}}(\nu \times P)$, the integral*

$$N(\varphi)(\omega) := \int_E \varphi(x, \omega) \, N(\omega, \mathrm{d}x)$$

is a.s. well defined (possibly infinite when φ is only assumed non-negative) and

$$\mathrm{E}\,[N(\varphi)] = \nu(\mathrm{E}\,[\varphi]).$$

In particular, if $\varphi \in L^1_{\mathbb{R}}(\nu \times P)$, $N(\varphi)$ is a.s. finite.

Proof. The assertion is true for

$$\varphi(x, \omega) = u(x)v(\omega),$$

where $u : E \to \mathbb{R}_+$ is $\mathcal{B}(E)$-measurable and $v : \Omega \to \mathbb{R}_+$ is $\mathcal{G}$-measurable, since in this case,

$$\mathrm{E}\,[N(\varphi)] = \mathrm{E}[v]\mathrm{E}\,[N(u)] = \mathrm{E}[v]\nu(u) = \nu(\mathrm{E}[v]u) = \nu(\mathrm{E}\,[\varphi]).$$

The rest is a straightforward application of Dynkin's functional theorem (Theorem 2.1.27). $\qquad\square$

Recall that a measure μ on the measurable space $(E, \mathcal{B}(E))$ is called non-atomic if $\mu(\{a\}) = 0$ for any singleton $\{a\}$.

Theorem 8.1.23 *Two independent point processes N_1 and N_2 on the measurable space $(E, \mathcal{B}(E))$ with respective intensity measures ν_1 and ν_2 have a.s. no points in common if and only if*

$$\int_E \nu_1(\{x\})\, \nu_2(\mathrm{d}x) = \int_E \nu_2(\{x\})\, \nu_1(\mathrm{d}x) = 0. \tag{8.3}$$

This is the case, for instance, if one of them has a non-atomic intensity measure.

Proof. By Corollary 8.1.22,

$$\mathrm{E}\left[\int_E N_1(\{x\})\, N_2(\mathrm{d}x)\right] = \int_E \nu_1(\{x\})\, \nu_2(\mathrm{d}x). \tag{$\star$}$$

Therefore, if $\int_E \nu_1(\{x\})\, \nu_2(\mathrm{d}x) = 0$, $\mathrm{E}\left[\int_E N_1(\{x\})\, N_2(\mathrm{d}x)\right] = 0$ and in particular $\int_E N_1(\{x\})\, N_2(\mathrm{d}x) = 0$, P-a.s., which implies that N_1 and N_2 have P-a.s. no common points. Also, if N_1 and N_2 have a.s. no common points, $P(\int_E N_1(\{x\})\, N_2(\mathrm{d}x) = 0) = 1$ and in particular $\mathrm{E}\left[\int_E N_1(\{x\})\, N_2(\mathrm{d}x)\right] = 0$, that is, $\int_E \nu_1(\{x\})\, \nu_2(\mathrm{d}x) = 0$. $\qquad\square$

Cluster Point Processes

Roughly speaking, a cluster point process consists of a point process (the germ) each point of which is surrounded by a point process (the cluster at this point). More generally, the clusters may be random measures. Formally now:

Let N_0 be a simple locally finite point process on the l.c.d.b. space E with sequence of points $\{X_{0,n}\}_{n \in \mathbb{N}}$ and locally finite intensity measure ν_0. Let $\{(x, \omega, C) \in E \times \Omega \times \mathcal{B}(E) \to$

$Z_n(x, \omega, C)\}_{n \geq 1}$ be mappings such that $Z_n(x, \omega, \cdot) \in \mathcal{M}_p(E)$ for all $(x, \omega) \in E \times \Omega$, and such that for each $C \in \mathcal{B}(E)$ the mapping $(x, \omega) \in E \times \Omega) \to Z_n(x, \omega, C)$ is $\mathcal{B}(E) \otimes \mathcal{F}$-measurable. Write $Z_n(x)(\omega, C) := Z_n(x, \omega, C)$ (therefore $Z_n(x)$ is a point measure on $(E, \mathcal{B}(E))$). Suppose that for each $x \in E$, the sequence of point measures $\{Z_n(x)\}_{n \in \mathbb{Z}}$ is IID, independent of N_0 and such that $E[Z_1(x, C)] = K_Z(x, C)$ for some measurable kernel K from $(E, \mathcal{B}(E))$ to $(E, \mathcal{B}(E))$ such that for any compact set $C \in \mathcal{B}(E)$,

$$\int_E K_Z(x, C - x)\nu_0(dx) < \infty.$$

Definition 8.1.24 *The point process N on E defined by*

$$N(C) := \sum_{n \in \mathbb{N}} Z_n(X_{0,n}, C - X_{0,n}) \tag{8.4}$$

is called a cluster point process *with* germ *(point process) N_0. The point process $Z_n(X_{0,n}, \cdot - X_{0,n})$ is the* cluster at $X_{0,n}$.

A straightforward application of Campbell's formula gives for the intensity measure ν of N

$$\nu(C) = \int_E K_Z(y, C - y)\,\nu_0(dy). \tag{8.5}$$

In fact,

$$E\left[\sum_{n \in \mathbb{N}} Z_n(X_{0,n}, C - X_{0,n})\right] = E\left[E\left[\sum_{n \in \mathbb{N}} Z_n(X_{0,n}, C - X_{0,n}) \,|\, \mathcal{F}^{N_0}\right]\right]$$

$$= E\left[\sum_{n \in \mathbb{N}} E\left[Z_n(X_{0,n}, C - X_{0,n}) \,|\, \mathcal{F}^{N_0}\right]\right]$$

$$= E\left[\sum_{n \in \mathbb{N}} K_Z(X_{0,n}, C - X_{0,n})\right] = \int_E K_Z(y, C - y))\,\nu_0(dy).$$

If this intensity measure is locally finite, the point process considered is a random element of $M_p(E)$. This point process is simple if, for instance, ν_0 is a non-atomic measure.

EXAMPLE 8.1.25: SPACE HOMOGENEOUS CLUSTERS. When Z_1 "does not depend on x", that is, when it is a point process on $(E, \mathcal{B}(E))$ (in which case $K_Z(x, C) = \nu_Z(C)$, the intensity measure of Z_1), we use the notation $N = N_0 * Z$, where Z stands for the "generic" cluster, that is, any point process with the common distribution of the Z_n's. Implicit in this notation is the assumption that the marks Z_n of N_0 are IID and independent of N_0. In this case,

$$\nu = \nu_0 * \nu_Z, \tag{8.6}$$

where ν_Z is the common intensity measure of the Z_n's, and $*$ denotes convolution.

Remark 8.1.26 Note that the Z_n's may have a point at 0 in which case some or all points of the germ point process are part of the cluster point process. When $E = \mathbb{R}^m$, a sufficient condition for the cluster point process to be simple is that its intensity measure be diffuse. This is the case whenever one of the measures of the convolution (8.6) is a multiple of the Lebesgue measure and the other is a finite measure. For instance, if the intensity measure of the germ point process is of the form $\nu_0(dx) = \lambda_0 \ell^m(dx)$, then

$$\nu(C) = \int_{\mathbb{R}^m} \lambda_0 \ell^m(C - x)\nu_Z(dx) = \int_{\mathbb{R}^m} \lambda_0 \ell^m(C)\nu_Z(dx) = \lambda_0 \nu_Z(E)\ell^m(C).$$

EXAMPLE 8.1.27: COX CLUSTER POINT PROCESSES. If for all $x \in E$, $Z_1(x, \cdot)$ is a Poisson process, the cluster point process is called a Cox cluster point process.

EXAMPLE 8.1.28: THINNING AND DISPLACING. If the generic cluster either has only one point at 0 (with probability p) or is empty (with probability $1 - p$), the cluster point process $N = N_0 * Z$ is in fact an independent p-thinning of N_0: each point of N_0 has been erased (with probability $1 - p$) or kept (with probability p) independently of the other points.

If the generic cluster Z has exactly one point at location V, the resulting cluster point process $N = N_0 * Z$ is obtained by independent random displacements, the generic displacement being V. The points of $N = N_0 * Z$ are $\{X_n + V_n\}_{n \in \mathbb{N}}$, where the V_n's form an IID sequence of the same distribution as the generic displacement V.

EXAMPLE 8.1.29: THE NEYMAN–SCOTT CLUSTER POINT PROCESS.[1] In this model, the generic cluster is of the form

$$Z := \{U_1, U_2, \ldots, U_T\},$$

where T is an integer-valued random variable and $\{U_n\}_{n \geq 1}$ is an IID sequence of random vectors of $\mathbb{R}^m$, independent of T.

8.1.3   The Distribution of a Point Process

Finite-dimensional Distributions

Let N be a random measure on $(E, \mathcal{B}(E))$ and let P be a probability measure on $(\Omega, \mathcal{F})$.

Definition 8.1.30 *The probability $\mathcal{P}_N := P \circ N^{-1}$ on $(M(E), \mathcal{M}(E))$ is called the distribution of N.*

Theorem 8.1.31 *Let $\mathcal{E}_0$ be as in Definition 8.1.3. The distribution $\mathcal{P}_N$ of a random measure N on $(E, \mathcal{B}(E))$ is completely characterized by the distributions of the vectors $(N(C_1), \ldots, N(C_m))$ for all integers $m \geq 1$ and all $C_1, \ldots, C_m \in \mathcal{E}_0$.*

Proof. We have seen in Theorem 8.1.4 that the class $\mathcal{I}$ of subsets of $M(E)$ of the form

$$\{m; \, m(C_1) \in A_1, \ldots, m(C_k) \in A_k\},$$

where $C_j \in \mathcal{E}_0$ and $A_j \in \mathcal{B}(E)$ for all j $(1 \leq j \leq k)$, is closed under finite intersection and generates $\mathcal{M}(E)$. Therefore, two probability measures agreeing on $\mathcal{I}$, agree on $\sigma(\mathcal{I})$ (Theorem 2.1.42). □

The *fidi* (finite-dimensional) distribution of a point process N is, by definition, the collection of the probability distributions of the vectors $(N(C_1), \ldots, N(C_m))$, for all $m \in \mathbb{N}_+$, $C_1, \ldots, C_m \in \mathcal{B}(E)$. The above result says in particular that the *fidi* distribution restricted to $\mathcal{I}$ as in the proof of Theorem 8.1.31 characterizes the distribution of a *locally finite* point process on E.

---

[1] [Neyman and Scott, 1958].

The Laplace Functional

This functional plays for point processes a role analogous to that of the usual Laplace transform for random vectors.

Definition 8.1.32 *Let N be a random measure on E. The* Laplace functional *of N is the mapping L_N associating with a non-negative measurable function $\varphi : E \to \mathbb{R}_+$ the non-negative real number*

$$L_N(\varphi) = \mathrm{E}\left[e^{-N(\varphi)}\right].$$

In terms of the distribution $\mathcal{P}_N$ (on $(M(E), \mathcal{M}(E))$) of N,

$$L_N(\varphi) = \int_{M(E)} e^{-\int_E \varphi(x)\, \mu(\mathrm{d}x)}\, \mathcal{P}_N(\mathrm{d}\mu).$$

EXAMPLE 8.1.33: THE LAPLACE FUNCTIONAL OF A POISSON PROCESS. Anticipating a later result (Theorem 8.2.7 thereof), the Laplace functional of a Poisson process on $\mathbb{R}^m$ with intensity measure ν is

$$L_N(\varphi) = \exp\left\{\int_{\mathbb{R}^m} \left(e^{-\varphi(x)} - 1\right) \nu(\mathrm{d}x)\right\}.$$

Theorem 8.1.34 *The Laplace functional of a locally finite random measure N on E characterizes its distribution.*

Proof. It suffices to show that the Laplace functional of a point process N characterizes its finite-dimensional distribution. For this, just observe that for all $K \geq 1$ and all disjoint measurable sets $C_1, \ldots, C_K$ in $\mathcal{B}(E)$, the Laplace transform of the vector $(N(C_1), \ldots, N(C_K))$, that is, the function

$$(t_1, \ldots, t_K) \in \mathbb{R}_+^K \to \mathrm{E}\left[e^{-t_1 N(C_1) - \cdots - t_K N(C_K)}\right],$$

is of the form $\mathrm{E}\left[e^{-N(\varphi)}\right]$, where $\varphi = t_1 1_{C_1} + \cdots + t_K 1_{C_K}$. $\qquad\square$

Corollary 8.1.35 *A point process N on $\mathbb{R}^m$ is stationary if and only if its Laplace transform L_N is such that*

$$L_N(\varphi) = L_N(S_a\varphi)$$

for all non-negative functions φ from $\mathbb{R}^m$ to $\mathbb{R}$ and all $a \in \mathbb{R}^m$, where $S_a\varphi(t) := \varphi(t-a)$.

EXAMPLE 8.1.36: A STATIONARY GRID, TAKE 2. In order to prove the stationarity of the shifted grid of Example 8.1.15, it suffices to show that for any non-negative function φ from $\mathbb{R}^2$ to $\mathbb{R}$, the quantity

$$\mathrm{E}\left[e^{\sum_{n,m\in\mathbb{Z}} \varphi(nT_1 + V_1 + \alpha,\, nT_2 + V_2 + \beta)}\right]$$

is independent of $\alpha, \beta \in \mathbb{R}$. This quantity equals

$$\int_0^{T_1} \left\{ \int_0^{T_2} e^{\sum_{n,m \in \mathbb{Z}} \varphi(nT_1 + v_1 + \alpha, nT_2 + v_2 + \beta)} \, dv_2 \right\} dv_1 .$$

The conclusion follows from the fact that for any non-negative function $\psi : \mathbb{R} \to \mathbb{R}$,

$$\int_0^T \psi(nT + u + \alpha) \, du = \int_0^T \psi(nT + u) \, du$$

for all $\alpha \in \mathbb{R}$, by the shift-invariance of the Lebesgue measure.

The proof of the next result is left as an exercise (Exercise 8.5.4).

Theorem 8.1.37 *Let N_i ($i \in I$) be a collection of point processes on E, where I is an arbitrary index set. If for any finite subset $J \subseteq I$, and any collection φ_i ($i \in J$) of non-negative measurable functions from E to $\mathbb{R}$,*

$$\mathrm{E}\left[e^{-\sum_{i \in J} N_i(\varphi_i)} \right] = \prod_{i \in J} \mathrm{E}\left[e^{-N_i(\varphi_i)} \right], \tag{8.7}$$

then N_i ($i \in I$) is an independent family of point processes.

EXAMPLE 8.1.38: THE LAPLACE FUNCTIONAL OF THINNED POINT PROCESSES. Let N be a simple point process on E with point sequence $\{X_n\}_{n \in \mathbb{N}}$. Let $\{Z_n\}_{n \in \mathbb{N}}$ be an IID sequence of independent marks of N, each Z_n taking its values in $\{0, 1\}$, with probability $p \in (0, 1)$ for the value 1. The point process $N_{thin,p}$ defined by

$$N_{thin,p}(C) := \sum_{n \in \mathbb{N}} 1_C(X_n) Z_n$$

is called the p-thinning of N. Each point of N is retained in $N_{thin,p}$ with probability p, independently of everything else. We compute the Laplace functional of the thinned point process:

$$
\begin{aligned}
L_{N_{thin,p}}(\varphi) &= \mathrm{E}\left[\exp\left\{ -\sum_{n \in \mathbb{Z}} \varphi(X_n) Z_n \right\} \right] \\
&= \mathrm{E}\left[\prod_{n \in \mathbb{Z}} \exp\left(-\varphi(X_n) Z_n \right) \right] \\
&= \mathrm{E}\left[\mathrm{E}\left[\prod_{n \in \mathbb{Z}} \exp\left(-\varphi(X_n) Z_n \right) \mid \mathcal{F}^N \right] \right] \\
&= \mathrm{E}\left[\prod_{n \in \mathbb{Z}} \mathrm{E}\left[\exp\left(-\varphi(X_n) Z_n \right) \mid \mathcal{F}^N \right] \right] \\
&= \mathrm{E}\left[\prod_{n \in \mathbb{Z}} \{ p \exp(-\varphi(X_n)) + (1 - p) \} \right] \\
&= \mathrm{E}\left[\exp\left(\sum_{n \in \mathbb{Z}} \log\left(p \exp(-\varphi(X_n)) + (1 - p) \right) \right) \right] \\
&= L_N\left(-\log\left(p e^{-\varphi(\cdot)} + 1 - p \right) \right) .
\end{aligned}
$$

For future reference, we record the intermediary result obtained in the line before last of the above calculation:

$$L_{N_{thin,p}}(\varphi) = \mathrm{E}\left[\exp\left\{\int_{\mathbb{R}^m} \log\left(1 - p(1 - e^{-\varphi(x)})\right) N(\mathrm{d}x)\right\}\right]. \tag{8.8}$$

If the initial point process is a Poisson process with the locally integrable intensity measure ν,

$$L_{N_{thin,p}}(\varphi) = L_N(\psi) = e^{-\int_E (e^{-\psi(x)}-1)\nu(\mathrm{d}x)},$$

where $\psi(x) := -\log\left(pe^{-\varphi(x)} + 1 - p\right)$. Therefore $e^{-\psi(x)} = pe^{-\varphi(x)} + 1 - p = p(e^{-\varphi(x)} - 1) + 1$ and finally

$$L_{N_{thin,p}}(\varphi) = e^{-\int_E (e^{-\varphi(x)}-1)p\nu(\mathrm{d}x)}.$$

We therefore retrieve the standard result: p-thinning a Poisson process of intensity measure $\nu(\cdot)$ results in a Poisson process of intensity measure $\nu_p(\cdot) = p\nu(\cdot)$.

EXAMPLE 8.1.39: THE LAPLACE FUNCTIONAL OF A CLUSTER POINT PROCESS. Consider the point process N defined by (8.4), assumed simple and with a locally finite intensity measure. Suppose moreover that $Z_n(x, \cdot) = Z_1(\cdot)$ (the general case where the distributions of the clusters depend on their positions can of course be treated without difficulty, only at the price of a more cumbersome notation). Denote by L_{N_0} and L_Z the Laplace transforms of N_0 and any Z_n, respectively. Then,

$$L_N(\varphi) := \mathrm{E}\left[\exp\left\{-\sum_{n\in\mathbb{N}}\left(\int_E \varphi(x + X_{0,n})\, Z_n(\mathrm{d}x)\right)\right\}\right]$$

$$= \mathrm{E}\left[\prod_{n\in\mathbb{Z}} \exp\left(-\int_E \varphi(x + X_{0,n})\, Z_n(\mathrm{d}x)\right)\right]$$

$$= \mathrm{E}\left[\mathrm{E}\left[\prod_{n\in\mathbb{Z}} \exp\left(-\int_E \varphi(x + X_{0,n})\, Z_n(\mathrm{d}x)\right) \mid \mathcal{F}^{N_0}\right]\right]$$

$$= \mathrm{E}\left[\prod_{n\in\mathbb{N}} \mathrm{E}\left[\exp\left(-\int_E \varphi(x + X_{0,n})\, Z_n(\mathrm{d}x)\right) \mid \mathcal{F}^{N_0}\right]\right].$$

Therefore, since

$$\mathrm{E}\left[\exp\left(-\int_E \varphi(x + X_{0,n})\, Z_n(\mathrm{d}x)\right) \mid \mathcal{F}^{N_0}\right] = L_{Z_n}(\varphi(\cdot + X_{0,n})) = L_Z(\varphi(\cdot + X_{0,n})),$$

we obtain

$$L_N(\varphi) = \mathrm{E}\left[\prod_{n\in\mathbb{N}} L_Z(\varphi(\cdot + X_{0,n}))\right]$$

$$= \mathrm{E}\left[\exp\left\{\sum_{n\in\mathbb{N}} \log L_Z(\varphi(\cdot + X_{0,n}))\right\}\right].$$

Finally,

$$L_N(\varphi) = L_{N_0}(-\psi), \tag{8.9}$$

where
$$\psi(x) := \log L_Z(\varphi(\cdot + x)).$$

In the special case where $E = \mathbb{R}^m$ and where the germ point process is a Poisson process of locally finite intensity measure $\widetilde{\mu}$, $L_{N_0}(\varphi) = \exp\left\{\int_{\mathbb{R}^m} \left(e^{-\varphi(x)} - 1\right) \widetilde{\mu}(dx)\right\}$, and therefore

$$L_N(\varphi) = \exp\left\{\int_{\mathbb{R}^m} \left(E\left[e^{-\int_{\mathbb{R}^m} \varphi(y+x)\, Z(dy)}\right] - 1\right) \widetilde{\mu}(dx)\right\}. \tag{8.10}$$

The Avoidance Function

Here is yet another characterization of the distribution of a point process that is sometimes useful. Let N be a point process on the measurable space $(E, \mathcal{E})$. The function $v : \mathcal{E} \to [0, 1]$ defined by

$$v(B) = P(N(B) = 0)$$

is called the *void probability function*, or the *avoidance function* of N.

EXAMPLE 8.1.40: THE AVOIDANCE FUNCTION OF A POISSON PROCESS. For a Poisson process on $\mathbb{R}^m$ with intensity measure ν, $v(B) = \exp(-\lambda\nu(B))$.

Theorem 8.1.41 *Let (E, d) be a complete separable metric space. Let $\mathcal{E}$ be the Borel σ-field on E. The distribution of a simple point process N on $(E, \mathcal{E})$ is characterized by its avoidance function.*

Proof. This result[2] says that if two *simple* point processes have the same avoidance function, they have the same distribution. To show this, it suffices to show that the finite-dimensional distributions of a simple point process N can be obtained from its void probability function alone. For this, it is enough to prove that for all integers $k \geq 1$, all measurable sets $A_1, \ldots, A_k$, and all integers $n_1, \ldots, n_k$,

$$P(N(A_1) \leq n_1, \ldots, N(A_k) \leq n_k)$$

can be expressed in terms of v alone. This is done in four steps.

Step 1. For any Borel sets $A_1, \ldots, A_k, B$, we have

$$P(N(A_1) > 0, \ldots, N(A_k) > 0, N(B) = 0)$$
$$= P(N(A_1) > 0, \ldots, N(A_{k-1}) > 0, N(B) = 0)$$
$$\quad - P(N(A_1) > 0, \ldots, N(A_{k-1}) > 0, N(A_k \cup B) = 0),$$

and for $k = 1$,

$$P(N(A_1) > 0, N(B) = 0) = P(N(B) = 0) - P(N(B \cup A_1) = 0)$$
$$= v(B) - v(B \cup A_1).$$

---

[2] Named after Rényi, who introduced the notion and applied it to Poisson processes [Rényi, 1967].

This shows that $P(N(A_1) > 0, \ldots, N(A_k) > 0, N(B) = 0)$ can be recursively computed from the void probability function v.

Step 2. Suppose that we have a sequence $\mathcal{K}_n = \{K_{n,i}\}_{i=1}^{k_n}$ of nested partitions of E such that for any distinct $x, y \in E$, there exists an n such that x and y belong to two distinct sets of the partition $\mathcal{K}_n$ (in other words, the sequence of partitions $\{\mathcal{K}_n\}_{n \geq 0}$ eventually separates the points of E). Define for $n \geq 1$ and $A \in \mathcal{B}(E)$,

$$H_n(A) = \sum_{i=0}^{k_n} H(A \cap K_{n,i}),$$

where $H(C) := 1_{\{N(C)>0\}}$ $(C \subseteq E)$. Let $\mathcal{K}_n \cap A := \{A \cap K_{n,i}\}_{i=0}^{k_n}$. Since the sequence of partitions $\{\mathcal{K}_n \cap A\}_{n \geq 1}$ of A eventually separates the points of A, and since $H_n(A)$ counts the number of sets of $\mathcal{K}_n \cap A$ that contain at least one point of the point process, we have in view of the assumed simplicity of the point process

$$\lim_{n \uparrow \infty} H_n(A) = N(A), \qquad \text{a.s.}$$

Step 3. The probability $P(H_n(A) = l)$ can be expressed in terms of the void probability function v alone since (with $A_{n,i} = A \cap K_{n,i}$)

$$P(H_n(A) = l) = \sum_{\substack{i_0, \ldots, i_{k_n} \in \{0,1\} \\ \sum_{j=1}^{k_n} i_j = l}} P(H(A_{n,0}) = i_0, \ldots, H(A_{n,k_n}) = i_{k_n})$$

and for $i_0, \ldots, i_{k_n} \in \{0, 1\}$

$$P(H(A_{n,0}) = i_0, \ldots, H(A_{n,k_n}) = i_{k_n})$$
$$= P\left(\cap_{l;i_l=1}\{N(A_{n,l}) > 0\} \cap \{N\left(\cup_{m;i_m=0} A_{n,m}\right) = 0\}\right),$$

a quantity which can be expressed in terms of the void function v alone, as we saw in Step 1. More generally, for all $l_1, \ldots, l_k \in \mathbb{N}$,

$$P(H_n(A_1) = l_1, \ldots, H_n(A_k) = l_k)$$

is expressible in terms of the void probability function, and the same is true of

$$P(H_n(A_1) \leq n_1, \ldots, H_n(A_k) \leq n_k) \qquad (n_1, \ldots, n_k \in \mathbb{N}).$$

Step 4. Finally, observe that

$$\{H_n(A_1) \leq n_1, \ldots, H_n(A_k) \leq n_k\} \downarrow \{N(A_1) \leq n_1, \ldots, N(A_k) \leq n_k\}$$

and therefore

$$\lim_{n \uparrow \infty} P(H_n(A_1) \leq n_1, \ldots, H_n(A_k) \leq n_k) = P(N(A_1) \leq n_1, \ldots, N(A_k) \leq n_k).$$

The proof is now almost done. It just remains to construct the sequence of partitions $\{\mathcal{K}_n\}_{n \geq 1}$. Denote by $\overline{B}(a, r)$ the closed ball of center a and radius r. Since (E, d) is separable, there exists a countable set $\{a_1, a_2, \ldots\}$ that is dense in E. The first partition $\mathcal{K}_1$ consists of two sets

$$K_{11} := \overline{B}(a_1, 1), \qquad K_{10} := E \backslash K_{11}.$$

Suppose that we have constructed $\mathcal{K}_{n-1}$. The next partition $\mathcal{K}_n$ is constructed as follows.

Letting

$$B_{n,i} := \overline{B}\left(a_i, 2^{-(n-i)}\right) \quad (i = 1, \ldots, n) \text{ and } B_{n,0} := E \backslash \bigcup_{i=1}^{n} B_{n,i} \,,$$

define a partition $C_n = \{C_{n,i}\}_{i=0}^{n}$ by

$$C_{n,0} := B_{n,0} ; \quad C_{n,1} := B_{n,1} ; \quad C_{n,i} := B_{n,i} \backslash \bigcup_{j=1}^{i-1} B_{n,j} \quad (i = 2, \ldots, n) \,.$$

In order to obtain the partition $\mathcal{K}_n$ nested in $\mathcal{K}_{n-1}$, we intersect C_n and $\mathcal{K}_{n-1}$, that is,

$$\mathcal{K}_n := \{C_{n,i} \cap K_{n-1,j} \quad (j = 0, \ldots, k_{n-1}, \ i = 0, \ldots, n)\} \,.$$

$\square$

Remark 8.1.42 In the case $E = \mathbb{R}^m$, a simple sequence of nested partitions could be the following, say for $m = 1$ for notational ease:

$$\mathcal{K}_n = \left\{(i2^{-n}, (i+1)2^{-n}] ; i \in \mathbb{Z}\right\} \,.$$

Remark 8.1.43 Note that the assumption of simplicity is necessary in the above result. For instance, doubling the multiplicity of the points of a given point process leaves the avoidance function unchanged.

EXAMPLE 8.1.44: THE AVOIDANCE FUNCTION OF A POISSON CLUSTER PROCESS. Consider the space-homogeneous cluster point process of Example 8.1.25, with the additional specification that the germ N_0 is a Poisson process. In order to compute its void probability function $v_N(C) = P(N(C) = 0)$, we observe that

$$P(N(C) = 0) = \lim_{t \uparrow \infty} E\left[e^{-tN(C)}\right] \,.$$

We have

$$E\left[e^{-tN(C)}\right] = E\left[e^{-t\sum_n Z_n(C-X_n)}\right] = E\left[\prod_n e^{-tZ_n(C-X_n)}\right]$$

$$= E\left[E\left[\prod_n e^{-tZ_n(C-X_n)} \mid \mathcal{F}^{N_0}\right]\right] = E\left[\prod_n E\left[e^{-tZ_n(C-X_n)} \mid \mathcal{F}^{N_0}\right]\right]$$

$$= E\left[\prod_n E\left[e^{-tZ_1(C-X_n)}\right]\right] = E\left[\exp\left(\sum_n \log E\left[e^{-tZ_1(C-X_n)}\right]\right)\right] \,.$$

Since N_0 is a Poisson process with intensity measure ν_0, the last term of the above sequence of equalities is

$$\exp\left(\int_{\mathbb{R}^m} \left(E\left[e^{-tZ_1(C-x)}\right] - 1\right) \nu_0(dx)\right) \,. \tag{$\star$}$$

But

$$\lim_{t \uparrow \infty} E\left[e^{-tZ_1(C-x)} - 1\right] = v_Z(C - x) - 1 \,.$$

Therefore taking the limit as $t \uparrow \infty$ in $(\star)$ yields by dominated convergence:

$$v_N(C) = \exp\left(\int_{\mathbb{R}^m} (v_Z(C - x) - 1) \nu_0(dx)\right) \,.$$

8.2 Unmarked Spatial Poisson Processes

8.2.1 Construction

Recall the definition given in Example 8.1.12.

Definition 8.2.1 *Let ν be a σ-finite measure on E. The point process N on E is called a* Poisson process *on E with intensity measure ν if*

(i) *for all finite families of mutually* disjoint *sets $C_1, \ldots, C_K \in \mathcal{B}(E)$, the random variables $N(C_1), \ldots, N(C_K)$ are independent, and*

(ii) *for any set $C \in \mathcal{B}(E)$ such that $\nu(C) < \infty$,*

$$P(N(C) = k) = e^{-\nu(C)}\frac{\nu(C)^k}{k!} \qquad (k \geq 0).$$

In the case $E = \mathbb{R}^m$, if ν is of the form $\nu(C) = \int_C \lambda(x)\mathrm{d}x$ for some non-negative measurable function $\lambda : \mathbb{R}^m \to \mathbb{R}$, the Poisson process N is said to admit the *intensity function* $\lambda(x)$. If in addition $\lambda(x) \equiv \lambda$, N is called a *homogeneous* Poisson process (HPP) on $\mathbb{R}^m$ with *intensity* or *rate* λ.

We now construct the Poisson process. In other terms, we simulate the distribution of a Poisson process on $\mathbb{R}^m$ of given intensity measure ν. The basic result is the following:

Theorem 8.2.2 *Let T be a Poisson random variable of mean θ. Let $\{Z_n\}_{n\geq 1}$ be an IID sequence of random elements with values in E and common distribution Q. Assume that T is independent of $\{Z_n\}_{n\geq 1}$. The point process N on E defined by*

$$N(C) = \sum_{n=1}^{T} 1_C(Z_n) \quad (C \in \mathcal{B}(E))$$

is a Poisson process with intensity measure $\nu(\cdot) = \theta \times Q(\cdot)$.

Proof. It suffices to show that for any finite family $C_1, \ldots, C_K$ of pairwise disjoint measurable sets of E with finite ν-measure and all non-negative reals $t_1, \ldots, t_K$,

$$\mathrm{E}[e^{-\sum_{j=1}^{K} t_j N(C_j)}] = \Pi_{j=1}^{K} \exp\left\{\nu(C_j)(e^{-t_j} - 1)\right\}.$$

We have

$$\sum_{j=1}^{K} t_j N(C_j) = \sum_{j=1}^{K} t_j \left(\sum_{n=1}^{T} 1_{C_j}(Z_n)\right) = \sum_{n=1}^{T}\left(\sum_{j=1}^{K} t_j 1_{C_j}(Z_n)\right) = \sum_{n=1}^{T} Y_n,$$

where $Y_n = \sum_{j=1}^{K} t_j 1_{C_j}(Z_n)$. By Theorem 3.1.55,

$$\mathrm{E}[e^{-\sum_{n=1}^{T} Y_n}] = g_T(\mathrm{E}[e^{-Y_1}]),$$

where g_T is the generating function of T. Here, since T is Poisson mean θ,

$$g_T(z) = \exp\left\{\theta(z - 1)\right\}.$$

The random variable Y_1 takes the values $t_1, \ldots, t_K$ and 0 with the respective probabilities $Q(C_1), \ldots, Q(C_K)$ and $1 - \sum_{j=1}^{K} Q(C_j)$. Therefore

$$\mathrm{E}[e^{-Y_1}] = \sum_{j=1}^{K} e^{-t_j} Q(C_j) + 1 - \sum_{j=1}^{K} Q(C_j) = 1 + \sum_{j=1}^{K} \left(e^{-t_j} - 1 \right) Q(C_j),$$

from which we get the announced result.

$\square$

The above is a special case of what is to be done, that is, to construct a Poisson process on E with an intensity measure ν that is σ-finite (not just finite). Such a measure can be decomposed as

$$\nu(\cdot) = \sum_{j=1}^{\infty} \theta_j \times Q_j(\cdot),$$

where the θ_j's are positive real numbers and the Q_j's are probability distributions on E. One can construct independent Poisson processes N_j on E with respective intensity measures $\theta_j Q_j(\cdot)$. The announced result then follows from the following theorem:

Theorem 8.2.3 *Let ν be a σ-finite measure on E of the form $\nu = \sum_{i=1}^{\infty} \nu_i$, where the ν_i's $(i \geq 1)$ are σ-finite measures on E. Let N_i $(i \geq 1)$ be a family of independent Poisson processes on E with respective intensity measures ν_i $(i \geq 1)$. Then the point process*

$$N = \sum_{j=1}^{\infty} N_j$$

is a Poisson process with intensity measure ν.

Proof. For mutually disjoint measurable sets $C_1, \ldots, C_K$ of finite ν-measures, and non-negative reals $t_1, \ldots, t_K$,

$$\mathrm{E}\left[e^{-\sum_{\ell=1}^{K} t_\ell N(C_\ell)} \right] = \mathrm{E}\left[e^{-\sum_{\ell=1}^{K} t_\ell \left(\sum_{j=1}^{\infty} N_j(C_\ell) \right)} \right]$$

$$= \mathrm{E}\left[e^{-\lim_{n \uparrow \infty} \sum_{\ell=1}^{K} t_\ell \left(\sum_{j=1}^{n} N_j(C_\ell) \right)} \right]$$

$$= \lim_{n \uparrow \infty} \mathrm{E}\left[e^{-\sum_{\ell=1}^{K} t_\ell \left(\sum_{j=1}^{n} N_j(C_\ell) \right)} \right],$$

by dominated convergence. But

$$\mathrm{E}\left[e^{-\sum_{\ell=1}^{K} t_\ell \left(\sum_{j=1}^{n} N_j(C_\ell) \right)} \right] = \mathrm{E}\left[e^{-\sum_{j=1}^{n} \left(\sum_{\ell=1}^{K} t_\ell N_j(C_\ell) \right)} \right]$$

$$= \prod_{j=1}^{n} \mathrm{E}\left[e^{-\sum_{\ell=1}^{K} t_\ell N_j(C_\ell)} \right] = \prod_{j=1}^{n} \prod_{\ell=1}^{K} e^{-t_\ell N_j(C_\ell)}$$

$$= \prod_{j=1}^{n} \prod_{\ell=1}^{K} \exp\left\{ (e^{-t_\ell} - 1) \nu_j(C_\ell) \right\}$$

$$= \prod_{j=1}^{n} \exp\left\{ \sum_{\ell=1}^{K} \left(e^{-t_\ell} - 1 \right) \nu_j(C_\ell) \right\}$$

$$= \exp\left\{ \sum_{\ell=1}^{K} \left(e^{-t_\ell} - 1 \right) \left(\sum_{j=1}^{n} \nu_j(C_\ell) \right) \right\}.$$

Letting $n \uparrow \infty$ we obtain, by dominated convergence,

$$\mathrm{E}\left[\mathrm{e}^{-\sum_{\ell=1}^{K} t_\ell N(C_\ell)}\right] = \exp\left\{\sum_{\ell=1}^{K} \left(\mathrm{e}^{-t_\ell} - 1\right) \nu(C_\ell)\right\}.$$

Therefore $N(C_1), \ldots, N(C_K)$ are independent Poisson random variables with respective means $\nu(C_1), \ldots, \nu(C_K)$. □

Theorem 8.2.4 *Let N be a Poisson process on $\mathbb{R}^m$ with intensity measure ν.*

 (a) *If ν is locally finite, then N is locally finite.*

 (b) *If ν is locally finite and non-atomic, then N is simple.*

Proof. (a) If C is a bounded measurable set, it is of finite ν-measure, and therefore $\mathrm{E}[N(C)] = \nu(C) < \infty$, which implies that $N(C) < \infty$, P-almost surely.

(b) It suffices to show this for a finite intensity measure $\nu(\cdot) = \theta(\cdot) Q$, where θ is a positive real number and Q is a non-atomic probability measure on $\mathbb{R}^m$, and then use the construction of Theorem 8.2.2. In turn, it suffices to show that for each $n \geq 1$, $P(Z_i = Z_j$ for some pair $(i,j) \, (1 \leq i < j \leq n) \,|\, N(\mathbb{R}^m) = n) = 0$. This is the case because for IID vectors $Z_1, \ldots, Z_n$ with a non-atomic probability distribution, $P(Z_i = Z_j$ for some pair $(i,j) \, (1 \leq i < j \leq n)) = 0$. □

Doubly Stochastic Poisson Processes

Doubly stochastic Poisson processes are also called Cox processes.[3]

Definition 8.2.5 *Let $\mathcal{G}$ be a σ-field containing $\mathcal{F}^\nu$, where ν is a locally finite random measure on $\mathbb{R}^m$. A point process N on $\mathbb{R}^m$ such that given $\mathcal{G}$, N is a Poisson process on $\mathbb{R}^m$ with the intensity measure ν, is called a* doubly stochastic Poisson process *with respect to $\mathcal{G}$ with the (conditional) intensity measure ν.*

If the random measure ν is of the form

$$\nu(dx) = \Lambda \ell^m(dx),$$

where Λ is a non-negative random variable, the corresponding Cox process is also called a *mixed Poisson process*.

8.2.2 Poisson Process Integrals

The Covariance Formula

Let N be a Poisson process on E, with intensity measure ν. Recall Campbell's theorem (Theorem 8.1.20). Let $\varphi : E \to \bar{\mathbb{R}}$ be a ν-integrable measurable function. Then $N(\varphi)$ is a well-defined *integrable* random variable, and

$$\mathrm{E}\left[\int_E \varphi(x) \, N(dx)\right] = \int_E \varphi(x) \, \nu(dx). \tag{8.11}$$

---

[3][Cox, 1955].

Theorem 8.2.6 *Let N be as above. Let $\varphi, \psi : E \to \mathbb{C}$ be two ν-integrable measurable functions such that moreover $|\varphi|^2$ and $|\psi|^2$ are ν-integrable. Then $N(\varphi)$ and $N(\psi)$ are well-defined square-integrable random variables and*

$$\text{cov}\left(\int_E \varphi(x)\,N(\mathrm{d}x), \int_E \psi(x)\,N(\mathrm{d}x)\right) = \int_E \varphi(x)\psi(x)^*\,\nu(\mathrm{d}x)\,. \tag{8.12}$$

Proof. It is enough to consider the case of real functions. First suppose that φ and ψ are *simple* non-negative Borel functions. We can always assume that

$$\varphi := \sum_{h=1}^{K} a_h 1_{C_h}\,, \quad \psi := \sum_{h=1}^{K} b_h 1_{C_h},$$

where $C_1, \ldots, C_K$ are *disjoint* measurable subsets of E. In particular, $\varphi(x)\psi(x) = \sum_{h=1}^{K} a_h b_h 1_{C_h}(x)$. Using the facts that if $i \neq j$, $N(C_i)$ and $N(C_j)$ are independent, and that a Poisson random variable with mean θ has variance θ,

$$
\begin{aligned}
\mathrm{E}[N(\varphi)N(\psi)] &= \sum_{h,l=1}^{K} a_h b_l \mathrm{E}[N(C_h)N(C_l)] \\
&= \sum_{\substack{h,l=1 \\ h \neq l}}^{K} a_h b_l \mathrm{E}[N(C_h)N(C_l)] + \sum_{l=1}^{K} a_l b_l \mathrm{E}[N(C_l)^2] \\
&= \sum_{\substack{h,l=1 \\ h \neq l}}^{K} a_h b_l \mathrm{E}[N(C_h)]\mathrm{E}[N(C_l)] + \sum_{l=1}^{K} a_l b_l \mathrm{E}[N(C_l)^2]\,,
\end{aligned}
$$

and therefore

$$
\begin{aligned}
\mathrm{E}[N(\varphi)N(\psi)] &= \sum_{\substack{h,l=1 \\ h \neq l}}^{K} a_h b_l \nu(C_h)\nu(C_l) + \sum_{l=1}^{k} a_l b_l [\nu(C_l) + \nu(C_l)^2] \\
&= \sum_{h,l=1}^{k} a_h b_l \nu(C_h)\nu(C_l) + \sum_{l=1}^{k} a_l b_l \nu(C_l) \\
&= \nu(\varphi)\nu(\psi) + \nu(\varphi\psi)\,.
\end{aligned}
$$

Let now φ, ψ be non-negative and let $\{\varphi_n\}_{n \geq 1}$, $\{\psi_n\}_{n \geq 1}$ be non-decreasing sequences of simple non-negative functions, with respective limits φ and ψ. Letting n go to ∞ in the equality

$$\mathrm{E}[N(\varphi_n)N(\psi_n)] = \nu(\varphi_n \psi_n) + \nu(\varphi_n)\nu(\psi_n)$$

yields the announced results, by monotone convergence.

We have that for any ν-integrable function $\varphi : E \to \mathbb{C}$

$$\mathrm{E}[N(\varphi)] = \mathrm{E}[N(\varphi^+)] - \mathrm{E}[N(\varphi^-)] = \nu(\varphi^+) - \nu(\varphi^-) = \nu(\varphi)\,.$$

Also by the result in the non-negative case, $\mathrm{E}[N(|\varphi|)^2] = \nu(|\varphi|^2) + \nu(|\varphi|)^2 < \infty$. Therefore, since $|N(\varphi)| \leq N(|\varphi|)$, $N(\varphi)$ is a square-integrable variable, as well as $N(\psi)$ for the same reasons. Therefore, by Schwarz's inequality, $N(\varphi)N(\psi)$ is integrable. We have

$$\begin{aligned}
\mathrm{E}\left[N(\varphi)N(\psi)\right] &= \mathrm{E}\left[\left(N(\varphi^+) - N(\varphi^-)\right)\left(N(\psi^+) - N(\psi^-)\right)\right] \\
&= \mathrm{E}\left[N(\varphi^+)N(\psi^+)\right] + \mathrm{E}\left[N(\varphi^-)N(\psi^-)\right] \\
&\quad - \mathrm{E}\left[N(\varphi^+)N(\psi^-)\right] - \mathrm{E}\left[N(\varphi^-)N(\psi^+)\right] \\
&= \left(\nu(\varphi^+\psi^+) + \nu(\varphi^+)\nu(\psi^+)\right) + \left(\nu(\varphi^-\psi^-) + \nu(\varphi^-)\nu(\psi^-)\right) \\
&\quad - \left(\nu(\varphi^+\psi^-) + \nu(\varphi^+)\nu(\psi^-)\right) - \left(\nu(\varphi^-\psi^+) + \nu(\varphi^-)\nu(\psi^+)\right) \\
&= \nu(\varphi\psi) + \nu(\varphi)\nu(\psi)\,,
\end{aligned}$$

from which (8.12) follows. $\qquad\square$

The Exponential Formula

We now turn to the *exponential formula* for Poisson processes. (It is sometimes called the *second Campbell's formula*. However, in this book, the appellation "Campbell's formula" will be reserved for the first one.)

Theorem 8.2.7 *Let N be a Poisson process on E with intensity measure ν. Let $\varphi : E \to \mathbb{R}$ be a non-negative measurable function. Then,*

$$\mathrm{E}[e^{-\int_E \varphi(x)\,N(\mathrm{d}x))}] = \exp\left\{\int_E (e^{-\varphi(x)} - 1)\,\nu(\mathrm{d}x)\right\}$$

and

$$\mathrm{E}[e^{\int_E \varphi(x)\,N(\mathrm{d}x))}] = \exp\left\{\int_E (e^{\varphi(x)} - 1)\,\nu(\mathrm{d}x)\right\}\,.$$

Proof. We prove the first formula, the proof of the second being similar. Suppose that φ is simple and non-negative: $\varphi = \sum_{h=1}^K a_h 1_{C_h}$ where $C_1, \ldots, C_K$ are mutually disjoint measurable subsets of E. Then

$$\begin{aligned}
\mathrm{E}[e^{-N(\varphi)}] &= \mathrm{E}\left[e^{-\sum_{h=1}^K a_h N(C_h))}\right] = \mathrm{E}\left[\prod_{h=1}^K e^{-a_h N(C_h)}\right] \\
&= \prod_{h=1}^K \mathrm{E}\left[e^{-a_h N(C_h)}\right] = \prod_{h=1}^K \exp\left\{(e^{-a_h} - 1)\nu(C_h)\right\} \\
&= \exp\left\{\sum_{h=1}^K (e^{-a_h} - 1)\nu(C_h)\right\} = \exp\left\{\nu(e^{-\varphi} - 1)\right\}\,.
\end{aligned}$$

The formula is therefore true for non-negative simple functions. Take now a non-decreasing sequence $\{\varphi_n\}_{n\geq 1}$ of such functions converging to φ. For all $n \geq 1$,

$$\mathrm{E}[e^{-N(\varphi_n)}] = \exp\left\{\nu(e^{-\varphi_n} - 1)\right\}\,.$$

By monotone convergence, the limit as n tends to ∞ of $N(\varphi_n)$ is $N(\varphi)$. Consequently, by dominated convergence, the limit of the left-hand side is $\mathrm{E}[e^{-N(\varphi)}]$. The function $g_n = -(e^{-\varphi_n} - 1)$ is a non-negative function increasing to $g = -(e^{-\varphi} - 1)$, and therefore, by monotone convergence, $\nu(e^{-\varphi_n} - 1) = -\nu(g_n)$ converges to $\nu(e^{-\varphi} - 1) = -\nu(g)$, which in turn implies that the right-hand side of the last displayed equality tends to $\exp\left\{\nu(e^{-\varphi} - 1)\right\}$ as n tends to ∞. $\qquad\square$

Remark 8.2.8 The covariance formula can of course be obtained from the exponential formula by differentiation of $t \mapsto \mathrm{E}\left[e^{-tN(\varphi)}\right]$.

EXAMPLE 8.2.9: THE MAXIMUM FORMULA. Let N be a simple Poisson process on E with intensity measure ν and let $\varphi : E \to \mathbb{R}$. Then

$$P(\sup_{n\in\mathbb{N}} \varphi(X_n) \le a) = \exp\left\{-\int_E 1_{\{\varphi(x)>a\}}\nu(\mathrm{d}x)\right\}.$$

A direct proof based on the construction of Poisson processes in Subsection 8.2.1 is possible (Exercise 8.5.21). We take another path and first prove that

$$\lim_{\theta\uparrow\infty} \mathrm{E}\left[e^{-\theta\sum_{n\in\mathbb{N}}1_{\{\varphi(X_n)>a\}}}\right] = P(\sup_{n\in\mathbb{N}} \varphi(X_n) \le a). \qquad (\star)$$

Indeed, the sum $\sum_{n\in\mathbb{N}}1_{\{\varphi(X_n)>a\}}$ is strictly positive, except when $\sup_{n\in\mathbb{N}} \varphi(X_n) \le a$, in which case it is null. Therefore

$$\lim_{\theta\uparrow\infty} e^{-\theta\sum_{n\in\mathbb{N}}1_{\{\varphi(X_n)>a\}}} = 1_{\{\sup_{n\in\mathbb{N}} \varphi(X_n)\le a\}}.$$

Taking expectations yields $(\star)$, by dominated convergence. Now, by Theorem 8.2.7,

$$\mathrm{E}\left[e^{-\theta\sum_{n\in\mathbb{N}}1_{\{\varphi(X_n)>a\}}}\right] = \exp\left\{\int_E \left(e^{-\theta 1_{\{\varphi(x)>a\}}} - 1\right)\nu(\mathrm{d}x)\right\}$$

$$= \exp\left\{\int_E \left(e^{-\theta} - 1\right)1_{\{\varphi(x)>a\}}\nu(\mathrm{d}x)\right\}$$

and the limit of the latter quantity as $\theta \uparrow \infty$ is $\exp\left\{-\int_E 1_{\{\varphi(x)>a\}}\nu(\mathrm{d}x)\right\}$.

EXAMPLE 8.2.10: THE LAPLACE FUNCTIONAL OF A POISSON PROCESS. According to Theorem 8.2.7, the Laplace functional of a Poisson process N on E with intensity measure ν is

$$L_N(\varphi) = \exp\left\{\nu\left(e^{-\varphi} - 1\right)\right\}.$$

Theorem 8.2.11 *Let N_i ($i \in J$) be a finite collection of simple point processes on E. If for any collection $\varphi_i : E \to \mathbb{R}_+$ ($i \in J$) of non-negative measurable functions,*

$$\mathrm{E}\left[e^{-\sum_{i\in J} N_i(\varphi_i)}\right] = \prod_{i\in J}\exp\left\{\int_E \left(e^{-\varphi_i(x)} - 1\right)\nu_i(\mathrm{d}x)\right\}, \qquad (8.13)$$

where ν_i, $i \in J$, is a collection of σ-finite measures on E, then N_i, $i \in J$, is a family of independent Poisson processes with respective intensity measures ν_i, $i \in J$.

Proof. Taking all the φ_i's identically null except the first one, we have

$$\mathrm{E}\left[e^{-N_1(\varphi_1)}\right] = \exp\left\{\int_E \left(e^{-\varphi_1(x)} - 1\right)\nu_1(\mathrm{d}x)\right\},$$

and therefore N_1 is a Poisson process with intensity measure ν_1. Similarly, for any $i \in J$, N_i is a Poisson process with intensity measure ν_i. Independence follows from Theorem 8.1.37. $\square$

8.3 Marked Spatial Poisson Processes

8.3.1 As Unmarked Poisson Processes

Let

(α) N be a simple and locally finite process on E, with point sequence $\{X_n\}_{n\in\mathbb{N}}$, and

(β) $\{Z_n\}_{n\in\mathbb{N}}$ be a sequence of random elements taking their values in the measurable space $(K, \mathcal{K})$.

The sequence $\{X_n, Z_n\}_{n\in\mathbb{N}}$ is a marked point process, with the interpretation that Z_n is the *mark* associated with the *point* X_n. N is the *base* point process of the marked point process, and $\{Z_n\}_{n\in\mathbb{N}}$ is the associated *sequence of marks*. One also calls N a simple and locally finite point process on E with marks $\{Z_n\}_{n\in\mathbb{N}}$ in K. If moreover

(1) N is a Poisson process with intensity measure ν,

(2) $\{Z_n\}_{n\in\mathbb{N}}$ is an IID sequence, and

(3) $\{Z_n\}_{n\in\mathbb{N}}$ and N are independent,

the corresponding marked point process is called a Poisson process on E *with independent* IID *marks*. This model can be slightly generalized by allowing the mark distribution to depend on the location of the marked point. More precisely, we replace (2) and (3) by

(2') $\{Z_n\}_{n\in\mathbb{N}}$ is, conditionally on N, an independent sequence,

(3') given X_n, the random vector Z_n is independent of X_k ($k \in \mathbb{N}, k \neq n$), and

(4') for all $n \in \mathbb{N}$ and all $L \in \mathcal{K}$,

$$P(Z_n \in L \mid X_n = x) = Q(x, L),$$

where $Q(\cdot, \cdot)$ is a stochastic kernel from $(E, \mathcal{B}(E))$ to $(K, \mathcal{K})$, that is, Q is a function from $E \times \mathcal{K}$ to $[0, 1]$ such that for all $L \in \mathcal{K}$ the map $x \to Q(x, L)$ is measurable, and for all $x \in E$, $Q(x, \cdot)$ is a probability measure on $(K, \mathcal{K})$.

Theorem 8.3.1 *Let $\{X_n, Z_n\}_{n\in\mathbb{N}}$ be as in (α) and (β) above, and define the point process $\widetilde{N}$ on $E \times K$ by*

$$\widetilde{N}(A) = \sum_{n\in\mathbb{N}} 1_A(X_n, Z_n) \quad (A \in \mathcal{B}(E) \otimes \mathcal{K}). \tag{8.14}$$

If conditions (1), (2'), (3'), and (4') above are satisfied, then $\widetilde{N}$ is a simple Poisson process with intensity measure $\widetilde{\nu}$ given by

$$\widetilde{\nu}(C \times L) = \int_C Q(x, L)\, \nu(\mathrm{d}x) \quad (C \in \mathcal{B}(E),\, L \in \mathcal{K}).$$

Proof. In view of Theorem 8.1.34, it suffices to show that the Laplace transform of $\widetilde{N}$ has the appropriate form, that is, for any non-negative measurable function $\widetilde{\varphi} : E \times K \to \mathbb{R}$,

$$\mathrm{E}\left[e^{-\widetilde{N}(\widetilde{\varphi})}\right] = \exp\left\{\int_E \int_K \left(e^{-\widetilde{\varphi}(t,z)} - 1\right) \widetilde{\nu}(\mathrm{d}t \times \mathrm{d}z)\right\}.$$

By dominated convergence,

$$\mathrm{E}\left[\mathrm{e}^{-\tilde{N}(\tilde{\varphi})}\right] = \mathrm{E}\left[\mathrm{e}^{-\sum_{n\in\mathbb{N}}\tilde{\varphi}(X_n,Z_n)}\right] = \lim_{L\uparrow\infty}\mathrm{E}\left[\mathrm{e}^{-\sum_{n\leq L}\tilde{\varphi}(X_n,Z_n)}\right].$$

For the time being, fix a positive integer L. Then, taking into account assumptions (2') and (3'),

$$\mathrm{E}\left[\mathrm{e}^{-\sum_{n\leq L}\tilde{\varphi}(X_n,Z_n)}\right] = \mathrm{E}\left[\prod_{n\leq L}\mathrm{e}^{-\tilde{\varphi}(X_n,Z_n)}\right]$$

$$= \mathrm{E}\left[\mathrm{E}\left[\prod_{n\leq L}\mathrm{e}^{-\tilde{\varphi}(X_n,Z_n)}\mid X_j, j\leq L\right]\right]$$

$$= \mathrm{E}\left[\mathrm{e}^{-\sum_{n\leq L}\psi(X_n)}\right],$$

where $\psi(x) := -\log\int_K \mathrm{e}^{-\tilde{\varphi}(x,z)}Q(x,\mathrm{d}z)$, a non-negative function. Letting $L\uparrow\infty$, we have, by dominated convergence,

$$\mathrm{E}\left[\mathrm{e}^{-\tilde{N}(\tilde{\varphi})}\right] = \mathrm{E}\left[\mathrm{e}^{-\sum_{n\in\mathbb{N}}\psi(X_n)}\right] = \mathrm{E}\left[\mathrm{e}^{-N(\psi)}\right]$$

$$= \exp\left\{\int_E\left(\mathrm{e}^{-\psi(x)} - 1\right)\nu(\mathrm{d}x)\right\}$$

$$= \exp\left\{\int_E\left[\int_K \mathrm{e}^{-\tilde{\varphi}(x,z)}Q(x,\mathrm{d}z) - 1\right]\nu(\mathrm{d}x)\right\}$$

$$= \exp\left\{\int_E\left[\int_K\left(\mathrm{e}^{-\tilde{\varphi}(x,z)} - 1\right)Q(x,\mathrm{d}z)\right]\nu(\mathrm{d}x)\right\}$$

$$= \exp\left\{\int_E\int_K\left(\mathrm{e}^{-\tilde{\varphi}(x,z)} - 1\right)\tilde{\nu}(\mathrm{d}x\times\mathrm{d}z)\right\}.$$

$\square$

EXAMPLE 8.3.2: THE M/GI/∞ MODEL, TAKE 1. The model of this example is of interest in queueing theory and in the traffic analysis of communications networks. We adopt the queueing interpretation. Let N be an HPP on $\mathbb{R}$ with intensity λ, and $\{\sigma_n\}_{n\in\mathbb{Z}}$ be a sequence of random vectors taking their values in $\mathbb{R}_+$ with probability distribution Q. Assume moreover that $\{\sigma_n\}_{n\in\mathbb{Z}}$ and N are independent. The n-th event time of N, T_n, is the arrival time of the n-th customer, and σ_n is her service time request. Define the point process $\tilde{N}$ on $\mathbb{R}\times\mathbb{R}_+$ by

$$\tilde{N}(C) = \sum_{n\in\mathbb{Z}}1_C(T_n,\sigma_n)$$

for all $C\in\mathcal{B}(\mathbb{R})\otimes\mathcal{B}(\mathbb{R}_+)$. According to Theorem 8.3.1, $\tilde{N}$ is a simple Poisson process with intensity measure

$$\tilde{\nu}(\mathrm{d}t\times\mathrm{d}z) = \lambda\mathrm{d}t\times Q(\mathrm{d}z).$$

In the M/GI/∞ model,[4] a customer arriving at time T_n is immediately served, and therefore departs from the "system" at time $T_n + \sigma_n$. The number $X(t)$ of customers present in the system at time t is therefore given by the formula

---

[4] "∞" represents the number of servers. This model is sometimes called a "queueing" system, although in reality there is no queueing, since customers are served immediately upon arrival and without interruption. It is in fact a "pure delay" system.

$$X(t) = \sum_{n\in\mathbb{Z}} 1_{(-\infty,t]}(T_n) 1_{(t,\infty)}(T_n + \sigma_n).$$

(The n-th customer is in the system at time t if and only if she arrived at time $T_n \le t$ and departed at time $T_n + \sigma_n > t$.)

Assume that the service times have finite expectation: $\mathrm{E}\,[\sigma_1] < \infty$. Then, for all $t \in \mathbb{R}$, $X(t)$ is a Poisson random variable with mean $\lambda \mathrm{E}\,[\sigma_1]$.

Proof. Observe that
$$X(t) = \widetilde{N}(C(t)),$$
where $C(t) := \{(s,\sigma);\ s \le t,\ s + \sigma > t\} \subset \mathbb{R} \times \mathbb{R}_+$. In particular, $X(t)$ is a Poisson random variable with mean

$$\begin{aligned}
\widetilde{\nu}(C(t)) &= \int_{\mathbb{R}}\int_{\mathbb{R}_+} 1_{\{s+\sigma>t\}}1_{\{s\le t\}}\widetilde{\nu}(\mathrm{d}s \times \mathrm{d}\sigma)\\
&= \int_{\mathbb{R}}\int_{\mathbb{R}_+} 1_{\{s+\sigma>t\}}1_{\{s\le t\}}\lambda\,\mathrm{d}s \times Q(\mathrm{d}\sigma)\\
&= \int_{\mathbb{R}}\left(\int_{\mathbb{R}_+} 1_{\{s+\sigma>t\}}Q(\mathrm{d}\sigma)\right) 1_{\{s\le t\}}\lambda\,\mathrm{d}s\\
&= \lambda\int_{-\infty}^{t} Q((t-s,+\infty))\,\mathrm{d}s\\
&= \lambda\int_{0}^{\infty} Q((s,+\infty))\,\mathrm{d}s = \lambda\int_{0}^{\infty} P(\sigma_1 > s)\mathrm{d}s = \lambda\mathrm{E}[\sigma_1].
\end{aligned}$$

$\square$

It can be shown that the *departure* process D of departure times, defined by
$$D(C) := \sum_{n\in\mathbb{Z}} 1_C(T_n + \sigma_n),$$
is an HPP of intensity λ (Exercise 8.5.13).

Formulas such as Campbell's first formula and the Poisson exponential formula are straightforwardly extended to *marked* point processes. In the situation prevailing in Theorem 8.3.1, consider sums of the type

$$\widetilde{N}(\widetilde{\varphi}) := \sum_{n\in\mathbb{N}} \widetilde{\varphi}(X_n, Z_n), \tag{8.15}$$

for functions $\widetilde{\varphi} : E \times K \to \mathbb{R}$. Note that, denoting by $Z_1(x)$ any random element of K with the distribution $Q(x, \mathrm{d}z)$,

$$\widetilde{\nu}(\widetilde{\varphi}) = \int_{E}\int_{K} \widetilde{\varphi}(x, z)Q(x, \mathrm{d}z)\,\nu(\mathrm{d}x) = \int_{E} \mathrm{E}\,[\widetilde{\varphi}(x, Z_1(x))]\,\nu(\mathrm{d}x),$$

whenever the quantities involved have a meaning. Using this observation, the formulas obtained in the previous subsection can be applied in terms marked point processes. The corollaries below do not require proofs, since they are *reformulations* of previous results, namely Theorem 8.2.6, Theorem 8.2.7 and Exercise 7.5.1.

Let $0 < p < \infty$. Recall that a measurable function $\widetilde{\varphi} : E \times K \to \mathbb{R}$ (resp. $\to \mathbb{C}$) is said to be in $L_{\mathbb{R}}^{p}(\widetilde{\nu})$ (resp. $L_{\mathbb{C}}^{p}(\widetilde{\nu})$) if

$$\int_E \int_K |\widetilde{\varphi}(x,z)|^p \, \nu(\mathrm{d}x) \, Q(x,\mathrm{d}z) < \infty \, .$$

Corollary 8.3.3 *Suppose that* $\widetilde{\varphi} \in L^1_{\mathbb{C}}(\widetilde{\nu})$. *Then the sum (8.15) is well defined, and moreover*

$$\mathrm{E}\left[\sum_{n\in\mathbb{N}} \widetilde{\varphi}(X_n, Z_n)\right] = \int_E \mathrm{E}\left[\widetilde{\varphi}(x, Z_1(x))\right] \nu(\mathrm{d}x) \, .$$

Let $\widetilde{\varphi}, \widetilde{\psi} : \mathbb{R} \times E \to \mathbb{C}$ *be two measurable functions in* $L^1_{\mathbb{C}}(\widetilde{\nu}) \cap L^2_{\mathbb{C}}(\widetilde{\nu})$. *Then*

$$\mathrm{cov}\left(\sum_{n\in\mathbb{N}} \widetilde{\varphi}(X_n, Z_n), \sum_{n\in\mathbb{N}} \widetilde{\psi}(X_n, Z_n)\right)$$
$$= \int_E \mathrm{E}\left[\widetilde{\varphi}(x, Z_1(x))\widetilde{\psi}(x, Z_1(x))^*\right] \nu(\mathrm{d}x) \, .$$

Corollary 8.3.4 *Let* $\widetilde{\varphi}$ *be a non-negative function from* $E \times K$ *to* $\mathbb{R}$. *Then,*

$$\mathrm{E}\left[\mathrm{e}^{-\sum_{n\in\mathbb{N}} \widetilde{\varphi}(X_n, Z_n)}\right] = \exp\left\{\int_E \mathrm{E}\left[\mathrm{e}^{-\widetilde{\varphi}(x, Z_1(x))} - 1\right] \nu(\mathrm{d}x)\right\}$$

8.3.2   Operations on Poisson Processes

Thinning and Coloring

Thinning is the operation of randomly erasing points of a Poisson process. It is a particular case of the independent coloring operation whereby the points of a Poisson process are independently colored with the result of obtaining independent Poisson processes, each one corresponding to a different color.

Theorem 8.3.5 *Consider the situation depicted in Theorem 8.3.1. Let* I *be an arbitrary index set and let* $\{L_i\}_{i\in I}$ *be a family of disjoint measurable sets of* K. *Define for each* $i \in I$ *the simple point process* N_i *on* $\mathbb{R}^m$ *by*

$$N_i(C) = \sum_{n\in\mathbb{N}} 1_C(X_n) 1_{L_i}(Z_n) \, .$$

Then the family N_i *(* $i \in I$ *) is an independent family of Poisson processes with respective intensity measures* ν_i, *$i \in I$, where*

$$\nu_i(\mathrm{d}x) = Q(x, L_i) \, \nu(\mathrm{d}x) \, .$$

Proof. According to the definition of independence, it suffices to consider a *finite* index set I. Define the simple point process $\widetilde{N}$ on $\mathbb{R}^m \times K$ as in (8.14). Then $\widetilde{N}$ is a Poisson process with intensity measure $\widetilde{\nu}(C \times L) = \int_C Q(x, L)\nu(\mathrm{d}x)$. Defining $\widetilde{\varphi}(x, z) = \sum_{i\in I} \varphi_i(x) 1_{L_i}(z)$, we have $\sum_{i\in I} N_i(\varphi_i) = \widetilde{N}(\widetilde{\varphi})$. Therefore

$$E\left[e^{-\sum_{i\in I} N_i(\varphi_i)}\right] = E\left[e^{-\tilde{N}(\tilde{\varphi})}\right]$$

$$= \exp\left\{\int_{\mathbb{R}^m}\int_K \left(e^{-\tilde{\varphi}(x,z)} - 1\right)\tilde{\nu}(\mathrm{d}x \times \mathrm{d}z)\right\}$$

$$= \exp\left\{\int_{\mathbb{R}^m}\int_K \left(e^{-\tilde{\varphi}(x,z)} - 1\right)Q(x,\mathrm{d}z)\nu(\mathrm{d}x)\right\}$$

$$= \exp\left\{\int_{\mathbb{R}^m}\int_K \left(e^{-\sum_{i\in I}\varphi_i(x)1_{L_i}(z)} - 1\right)Q(x,\mathrm{d}z)\nu(\mathrm{d}x)\right\}$$

$$= \exp\left\{\int_{\mathbb{R}^m}\int_K \sum_{i\in I}\left(e^{-\varphi_i(x)} - 1\right)1_{L_i}(z)Q(x,\mathrm{d}z)\nu(\mathrm{d}x)\right\}$$

$$= \exp\left\{\int_{\mathbb{R}^m} \sum_{i\in I}\left(e^{-\varphi_i(x)} - 1\right)Q(x,L_i)\nu(\mathrm{d}x)\right\}$$

$$= \prod_{i\in I}\exp\left\{\int_{\mathbb{R}^m}\left(e^{-\varphi_i(x)} - 1\right)Q(x,L_i)\nu(\mathrm{d}x)\right\}.$$

Therefore,

$$E\left[e^{-\sum_{i\in I} N_i(\varphi_i)}\right] = \prod_{i\in I}\exp\left\{\int_{\mathbb{R}^m}\left(e^{-(\varphi_i)} - 1\right)\nu_i(\mathrm{d}x)\right\}$$

and the result follows from Theorem 8.2.11. □

Transportation

This is the operation of moving the points of a Poisson process. More precisely, consider the situation depicted in Theorem 8.3.1. Form a point process N^* on K by associating to a point $X_n \in \mathbb{R}^m$ a point $Z_n \in K$:

$$N^*(L) := \sum_{n\in\mathbb{N}} 1_L(Z_n),$$

where $L \in \mathcal{B}(\mathbb{R}^m)$. We then say that N^* is obtained by transporting N *via the stochastic kernel* $Q(x,\cdot)$.

Theorem 8.3.6 N^* *is a Poisson process on* K *with intensity measure* ν^* *given by*

$$\nu^*(L) = \int_{\mathbb{R}^m}\nu(\mathrm{d}x)Q(x,L).$$

Proof. Let $\varphi^* : K \to \mathbb{R}$ be a non-negative measurable function. We have

$$E\left[e^{-N^*(\varphi^*)}\right] = E\left[e^{-\sum_{n\in\mathbb{N}}\varphi^*(Z_n)}\right]$$

$$= \exp\left\{\int_{\mathbb{R}^m}\int_K \left(e^{-\varphi^*(z)} - 1\right)\nu(\mathrm{d}x)Q(x,\mathrm{d}z)\right\}$$

$$= \exp\left\{\int_K \left(e^{-\varphi^*(z)} - 1\right)\int_{\mathbb{R}^m}\nu(\mathrm{d}x)Q(x,\mathrm{d}z)\right\}.$$

□

EXAMPLE 8.3.7: TRANSLATION. Let N be a Poisson process on $\mathbb{R}^m$ with intensity measure ν and let $\{V_n\}_{n\in\mathbb{N}}$ be an IID sequence random vectors of $\mathbb{R}^m$ with common distribution Q. Form the point process N^* on $\mathbb{R}^m$ by translating each point X_n of N by V_n. Formally,

$$N^*(C) = \sum_{n\in\mathbb{N}} 1_C(X_n + V_n).$$

We are in the situation of Theorem 8.3.6 with $Z_n = X_n + V_n$. In particular, $Q(x, A) = Q(A - x)$. It follows that N^* is a Poisson process on $\mathbb{R}^m$ with intensity measure

$$\nu^*(L) = \int_{\mathbb{R}^m} Q(L - x)\,\nu(\mathrm{d}x),$$

the convolution of ν and Q.

Poisson Shot Noise

Let N be a simple and locally finite point process on $\mathbb{R}^m$ with point sequence $\{X_n\}_{n\in\mathbb{N}}$ and with marks $\{Z_n\}_{n\in\mathbb{N}}$ in the measurable space $(K, \mathcal{K})$. Let $h : \mathbb{R}^m \times K \to \mathbb{C}$ be a measurable function. The complex-valued spatial stochastic process $\{X(y)\}_{y\in\mathbb{R}^m}$ given by

$$X(y) := \sum_{n\in\mathbb{N}} h(y - X_n, Z_n), \tag{8.16}$$

where the right-hand side is assumed well defined (for instance, when h takes real non-negative values), is called a *spatial shot noise with random impulse response*. If N is a simple and locally finite Poisson process on $\mathbb{R}^m$ with independent IID marks $\{Z_n\}_{n\in\mathbb{N}}$, $\{X(y)\}_{y\in\mathbb{R}^m}$ is called a Poisson spatial shot noise *with random impulse response and independent IID marks*.

The following result is a direct application of Theorems 8.2.6 and 8.3.1.

Theorem 8.3.8 *Consider the above Poisson spatial shot noise with random impulse response and independent IID marks. Suppose that for all $y \in \mathbb{R}^m$,*

$$\int_{\mathbb{R}^m} \mathrm{E}\left[|h(y - x, Z_1)|\right] \nu(\mathrm{d}x) < \infty$$

and

$$\int_{\mathbb{R}^m} \mathrm{E}\left[|h(y - x, Z_1)|^2\right] \nu(\mathrm{d}x) < \infty.$$

Then the complex-valued spatial stochastic process $\{X(y)\}_{y\in\mathbb{R}^m}$ given by (8.16) is well defined, and for any $y, \xi \in \mathbb{R}^m$, we have

$$\mathrm{E}\left[X(y)\right] = \int_{\mathbb{R}^m} \mathrm{E}\left[h(y - x, Z_1)\right] \nu(\mathrm{d}x)$$

and

$$\mathrm{cov}(X(y + \xi), X(y)) = \int_{\mathbb{R}^m} \mathrm{E}\left[h(y - x, Z_1)h^*(y + \xi - x, Z_1)\right] \nu(\mathrm{d}t).$$

In the case where the base point process N is an HPP with intensity λ, we find that

$$\mathrm{E}\left[X(y)\right] = \lambda \int_{\mathbb{R}^m} \mathrm{E}\left[h(x, Z_1)\right] \mathrm{d}x$$

and

$$\mathrm{cov}\left(X(y + \xi), X(y)\right) = \lambda \int_{\mathbb{R}^m} \mathrm{E}\left[h(x, Z_1)h^*(\xi + x, Z_1)\right] \mathrm{d}x\,.$$

Observe that these quantities do not depend on $y \in \mathbb{R}^m$. The process $\{X(y)\}_{y \in \mathbb{R}^m}$ is for that reason called a *wide-sense stationary process* (see Chapter 9).

8.3.3 Change of Probability Measure

Let $(\Omega, \mathcal{F}, P)$ be a probability space on which is given a Poisson process N on $\mathbb{R}^m$ with non-atomic and locally finite intensity measure ν. We shall replace the probability P by another probability $\hat{P}$ in such a way that with respect to this new probability, the *same* point process N is a Poisson process, but with the intensity measure $\hat{\nu}$ given by

$$\hat{\nu}(C) = \int_{\mathbb{R}^m} \mu(x)\,\nu(\mathrm{d}x), \tag{8.17}$$

for some non-negative measurable function $\mu : \mathbb{R}^m \to \mathbb{R}$.

The Case of Finite Intensity Measures

The above program is first carried out under the following hypotheses:

H_1: ν is a *finite* measure, and

H_2: μ is ν-integrable (or equivalently $\hat{\nu}$ is finite).

The change of probability $P \to \hat{P}$ will be an absolutely continuous one, that is, for all $A \in \mathcal{F}$,

$$\hat{P}(A) = \mathrm{E}[L\,\mathbf{1}_A], \tag{8.18}$$

where L is a non-negative random variable such that

$$\mathrm{E}[L] = 1\,, \tag{8.19}$$

called the Radon–Nikodým derivative of $\hat{P}$ with respect to P, and also denoted $\frac{\mathrm{d}\hat{P}}{\mathrm{d}P}$.

Lemma 8.3.9 *Under hypotheses H_1 and H_2, the random variable*

$$L := \left(\prod_{n \in \mathbb{N}} \mu(X_n)\right) \exp\left\{-\int_{\mathbb{R}^m} (\mu(x) - 1)\nu(\mathrm{d}x)\right\} \tag{8.20}$$

satisfies (8.19).

Proof. Let $g(x) = \log(\mu(x))$ and decompose this function into its positive and negative part, $g = g_+ - g_-$. By Theorem 8.2.7 we have that

$$\mathrm{E}\left[\mathrm{e}^{-N(g_-)}\right] = \exp\left\{\int_{\mathbb{R}^m} \left(\mathrm{e}^{-g_-(x)} - 1\right)\nu(\mathrm{d}x)\right\}$$

and

$$\mathrm{E}\left[e^{N(g_+)}\right] = \exp\left\{\int_{\mathbb{R}^m}\left(e^{g_+(x)}-1\right)\nu(\mathrm{d}x)\right\}.$$

Let $B_1 = \{x \in \mathbb{R}^m\,;\, g(x) > 0\}$. By Theorem 8.3.5, the restrictions of N to B_1 and $B_2 = \bar{B}_1$ are independent, and therefore the variables $e^{-N(g_-)}$ and $e^{N(g_+)}$ are independent. In particular, from the two last displays,

$$\mathrm{E}\left[\prod_{n\in\mathbb{N}}\mu(X_n)\right]$$

$$= \mathrm{E}\left[e^{N(\log(\mu))}\right] = \mathrm{E}\left[e^{N(g)}\right] = \mathrm{E}\left[e^{N(g_+)-N(g_-)}\right]$$

$$= \mathrm{E}\left[e^{-N(g_-)}e^{N(g_+)}\right] = \mathrm{E}\left[e^{-N(g_-)}\right]\mathrm{E}\left[e^{N(g_+)}\right]$$

$$= \exp\left\{\int_{\mathbb{R}^m}\left(e^{g_+(x)}-1\right)\nu(\mathrm{d}x)\right\}\exp\left\{\int_{\mathbb{R}^m}\left(e^{-g_-(x)}-1\right)\nu(\mathrm{d}x)\right\}$$

$$= \exp\left\{\int_{\mathbb{R}^m}\left(e^{g(x)}-1\right)1_{\{g(x)>0\}}\nu(\mathrm{d}x)\right\}\exp\left\{\int_{\mathbb{R}^m}\left(e^{g(x)}-1\right)1_{\{g(x)\leq0\}}\nu(\mathrm{d}x)\right\}$$

$$= \exp\left\{\int_{\mathbb{R}^m}\left(e^{g(x)}-1\right)1_{\{g(x)>0\}}\nu(\mathrm{d}x) + \int_{\mathbb{R}^m}\left(e^{g(x)}-1\right)1_{\{g(x)\leq0\}}\nu(\mathrm{d}x)\right\}$$

$$= \exp\left\{\int_{\mathbb{R}^m}\left(e^{g(x)}-1\right)\nu(\mathrm{d}x)\right\} = \exp\left\{\int_{\mathbb{R}^m}(\mu(x)-1)\nu(\mathrm{d}x)\right\}.$$

By assumptions H_1 and H_2 the last quantity is finite and therefore one can divide the first and last terms of the above chain of equalities by it, to obtain (8.19). $\square$

Theorem 8.3.10 *Under the assumptions H_1 and H_2, if we define probability $\hat{P}$ by (8.18) and (8.20), N is under probability $\hat{P}$ a Poisson process with intensity measure $\hat{\nu}$ given by (8.17).*

Proof. Denote expectation with respect to $\hat{P}$ by $\hat{\mathrm{E}}$. It suffices to show that the Laplace transform of N under probability $\hat{P}$ is that of a Poisson process with intensity measure $\hat{\nu}$, that is, for any bounded non-negative measurable function $\varphi : \mathbb{R}^m \to \mathbb{R}$,

$$\hat{\mathrm{E}}\left[e^{-N(\varphi)}\right] = \exp\left[\int_{\mathbb{R}^m}\left(e^{-\varphi(x)}-1\right)\hat{\nu}(\mathrm{d}x)\right].$$

But

$$\hat{\mathrm{E}}\left[e^{-N(\varphi)}\right] = \mathrm{E}\left[L\,e^{-N(\varphi)}\right]$$

$$= \mathrm{E}\left[e^{N(\log(\mu))-\int_{\mathbb{R}^m}(\mu(x)-1)\nu(\mathrm{d}x)}\,e^{-N(\varphi)}\right]$$

$$= \mathrm{E}\left[e^{N(-\varphi+\log(\mu))}\right]e^{-\int_{\mathbb{R}^m}(\mu(x)-1)\nu(\mathrm{d}x)}$$

$$= \exp\left(\int_{\mathbb{R}^m}\left(e^{-\varphi(x)+\log(\mu(x))}-1\right)\nu(\mathrm{d}x)\right)\exp\left(-\int_{\mathbb{R}^m}(\mu(x)-1)\nu(\mathrm{d}x)\right)$$

$$= \exp\left(\int_{\mathbb{R}^m}\left(e^{-\varphi(x)}\mu(x)-1\right)\nu(\mathrm{d}x)\right)\exp\left(-\int_{\mathbb{R}^m}(\mu(x)-1)\nu(\mathrm{d}x)\right)$$

$$= \exp\left(\int_{\mathbb{R}^m}\left(e^{-\varphi(x)}-1\right)\mu(x)\nu(\mathrm{d}x)\right) = \exp\left(\int_{\mathbb{R}^m}\left(e^{-\varphi(x)}-1\right)\hat{\nu}(\mathrm{d}x)\right).$$

$\square$

The Mixed Poisson Case

Let N be a Poisson process on $\mathbb{R}^m$ of finite intensity measure ν and let Λ be a non-negative random variable independent of N. Let

$$L := \Lambda^{N(\mathbb{R}^m)} \exp\{-(\Lambda - 1)N(\mathbb{R}^m)\}\,.$$

The arguments of the proof of Lemma 8.3.9 and Theorem 8.3.10 are immediately adaptable to show that $\mathrm{E}_P[L] = 1$ and that under the probability measure $\hat{P}$ defined by $\frac{d\hat{P}}{dP} = L$, N is a Cox process (here a mixed Poisson process) with $\sigma(\Lambda)$-conditional intensity measure $\Lambda\nu(dx)$.

Theorem 8.3.11 *Under the above conditions, for any non-negative function $g\mathbb{R}_+ \to \mathbb{R}$,*

$$\mathrm{E}_{\hat{P}}\left[g(\Lambda) \,|\, \mathcal{F}^N\right] = \frac{\int g(\lambda)\lambda^{N(\mathbb{R}^m)}\mathrm{e}^{-\lambda N(\mathbb{R}^m)}\,F(d\lambda)}{\int \lambda^{N(\mathbb{R}^m)}\mathrm{e}^{-\lambda N(\mathbb{R}^m)}\,F(d\lambda)}\,. \tag{8.21}$$

Proof. The proof is based on the following fundamental lemma:

Lemma 8.3.12 *Let P and Q be two probability measures on the measurable space $(\Omega, \mathcal{F})$ such that $P \ll Q$ and let $L := \frac{dP}{dQ}$. Let Z be a non-negative random variable. For any sub-σ-field $\mathcal{G}$ of $\mathcal{F}$,*

$$\mathrm{E}_Q[L \,|\, \mathcal{G}]\,\mathrm{E}_P[Z \,|\, \mathcal{G}] = \mathrm{E}_Q[ZL \,|\, \mathcal{G}] \quad Q\text{-a.s.} \tag{8.22}$$

or, equivalently,

$$\mathrm{E}_P[Z \,|\, \mathcal{G}] = \frac{\mathrm{E}_Q[ZL \,|\, \mathcal{G}]}{\mathrm{E}_Q[L \,|\, \mathcal{G}]} \quad P\text{-a.s.} \tag{8.23}$$

Proof. By definition of conditional expectation, for all $A \in \mathcal{G}$

$$\int_A Z\,dP = \int_A \mathrm{E}_P[Z \,|\, \mathcal{G}]\,dP\,.$$

By definition of L and of conditional probability again,

$$\int_A Z\,dP = \int_A ZL\,dQ = \int_A \mathrm{E}_Q[ZL \,|\, \mathcal{G}]\,dQ\,.$$

Also

$$\int_A \mathrm{E}_P[Z \,|\, \mathcal{G}]\,dP = \int_A \mathrm{E}_P[Z \,|\, \mathcal{G}]\,L\,dQ$$
$$= \int_A \mathrm{E}_P[Z \,|\, \mathcal{G}]\,\mathrm{E}_Q[L \,|\, \mathcal{G}]\,dQ\,.$$

Therefore

$$\int_A \mathrm{E}_Q[ZL \,|\, \mathcal{G}]\,dQ = \int_A \mathrm{E}_P[Z \,|\, \mathcal{G}]\,\mathrm{E}_Q[L \,|\, \mathcal{G}]\,dQ\,,$$

which is, since A is arbitrary in $\mathcal{G}$, equivalent to (8.22). Since $P \ll Q$ this equality also holds P-a.s. To obtain (8.23), it remains to show that $P(\mathrm{E}_Q[L \,|\, \mathcal{G}] = 0) = 0$. But

$$P(\mathrm{E}_Q\,[L \mid \mathcal{G}] = 0) = \int 1_{\{\mathrm{E}_Q[L|\mathcal{G}]=0\}}\,\mathrm{d}P$$

$$= \int 1_{\{\mathrm{E}_Q[L|\mathcal{G}]=0\}}L\mathrm{d}Q$$

$$= \int 1_{\{\mathrm{E}_Q[L|\mathcal{G}]=0\}}\mathrm{E}_Q\,[L \mid \mathcal{G}]\,\mathrm{d}Q = 0\,.$$

$\square$

We may now proceed to the proof of Theorem 8.3.11. By Lemma 8.3.12,

$$\mathrm{E}_{\hat{P}}\,\big[g(\Lambda) \mid \mathcal{F}^N\big] = \frac{\mathrm{E}_P\,\big[g(\Lambda)L \mid \mathcal{F}^N\big]}{\mathrm{E}_P\,[L \mid \mathcal{F}^N]}$$

and therefore, since under P, N and Λ are independent,

$$\mathrm{E}_{\hat{P}}\,\big[g(\Lambda) \mid \mathcal{F}^N\big] = \frac{\int_{\mathbb{R}_+} g(\lambda)\lambda^{N(\mathbb{R}^m)}\exp\{-(\lambda-1)N(\mathbb{R}^m)\}\,F(\mathrm{d}\lambda)}{\int_{\mathbb{R}_+} \lambda^{N(\mathbb{R}^m)}\exp\{-(\lambda-1)N(\mathbb{R}^m)\}\,F(\mathrm{d}\lambda)}\,,$$

from which the result follows.

$\square$

8.4 The Boolean Model

Stochastic geometry concerns the study of random shapes. The model considered below pertains to a particular sort of *stochastic geometry*, where the randomness of the shapes is dependent on the positions of the points of an underlying point process.

    Although there exists a sound mathematical theory of random sets, this theory will not be necessary as long as the random sets considered in the applications are "good" sets fully described by a random vector of finite dimension (circle, disk, polygon, line, segment, etc.). We then resort to what can be called the "poor man's random set theory", in which a random set is set of the form $S(Z) \subseteq \mathbb{R}^m$ where $Z \in \mathbb{R}^d$ is a random vector and for each z, $S(z)$ is a measurable set and $S(Z)$ is also a measurable set. These are minimal requirements. We shall consider real-valued functions of S, for instance

$$g(S) = \ell^d(S)\,,\ g(S) = 1_{a \in S}\,,\qquad\qquad(\star)$$

for which the expectation is well defined, as

$$\mathrm{E}\,[g(S)] := \mathrm{E}\,[g(S(Z))] = \int_{\mathbb{R}^d} g(S(z))P(Z \in \mathrm{d}z)\,.\qquad(8.24)$$

We only need to ensure that the function $z \in \mathbb{R}^d \to g(S(z)) \in \mathbb{R}$ is measurable and its integral with respect to the distribution of Z well defined (in the examples, this will generally the case, because g will be a non-negative function, as in $(\star)$). We shall use for (8.24) the abbreviated notation

$$\int_{\mathcal{S}} g(s)\,Q(\mathrm{d}s)\,,$$

thereby pretending that there exists a set $\mathcal{S}$ of shapes with a suitable σ-field $\mathcal{G}$ on it, and an adequate probability distribution Q on $(\mathcal{S}, \mathcal{G})$.

EXAMPLE 8.4.1: RANDOM DISK. In this example, S is the closed disk in $\mathbb{R}^2$ centered on the origin and with radius Z, a non-negative random variable. We have, with $g(S) = \ell^2(S)$

$$\mathrm{E}\left[g(S)\right] = \mathrm{E}\left[\ell^2(S(Z))\right] = \mathrm{E}\left[\pi Z^2\right],$$

and with $g(S) = 1_{a \in S}$,

$$\mathrm{E}\left[g(S)\right] = \mathrm{E}\left[1_{a \in S(Z)}\right] = P(a \in S(Z)) = P(Z \geq a).$$

Definition 8.4.2 *The* capacity functional *of the random set S is the function $K \to T_S(K)$ (K compact) defined by*

$$T_S(K) := P(S \cap K \neq \varnothing).$$

EXAMPLE 8.4.3: THE CAPACITY FUNCTIONAL OF A POINT PROCESS. A simple point process N can also be viewed as a random set $S \equiv N$. In this case the capacity functional is the void probability function:

$$T_N(K) := P(N \cap K \neq \varnothing) = P(N(K) = 0) = v_N(K).$$

We now introduce the Boolean model.[5] Let N be a Poisson process on $\mathbb{R}^m$ with a non-atomic σ-finite intensity measure ν. Denote by $\{X_n\}_{n \in \mathbb{N}}$ its sequence of points. Let now $\{S_n\}_{n \in \mathbb{N}}$ be a sequence of random marks, IID and independent of N. Each S_n is a compact random set. The X_n's are called the *germs* whereas the S_n's are called the *grains*. Recall the following notations: if A and B are subsets of $\mathbb{R}^m$ and $x \in \mathbb{R}^m$, $x + A := \{x + y \,;\, y \in A\}$, and $A \oplus B := \{x + y \,;\, x \in A, \, x \in B\}$.

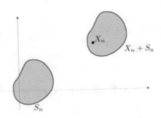

One of the quantities of interest in applications is the probability of intersection of the random set

$$\Sigma = \cup_{n \in \mathbb{N}}(X_n + S_n)$$

with a given compact set $K \subset \mathbb{R}^m$, that is,

$$T_\Sigma(K) := P(\Sigma \cap K \neq \varnothing).$$

In order to compute this quantity, we first observe that

---

[5][Matheron, 1967, 1975], [Serra, 1982].

$$T_\Sigma(K) = P(N(K) \neq 0) \,,$$

where

$$N(K) = \sum_{n \in \mathbb{N}} 1_{(X_n+S_n) \cap K \neq \varnothing} \,.$$

We show that $N(K)$ is a Poisson variable with mean

$$\theta(K) := \int_{\mathbb{R}^m} P((x + S_1) \cap K \neq \varnothing) \, \nu(dx) \qquad\qquad (\star)$$

and therefore

$$T_\Sigma(K) = 1 - \exp\left\{ -\int_{\mathbb{R}^m} P((x + S_1) \cap K \neq \varnothing) \, \nu(dx) \right\} . \qquad (8.25)$$

Proof. The Laplace transform of the distribution of $N(K)$ is given by

$$\mathrm{E}\left[e^{-tN(K)} \right] = \mathrm{E}\left[e^{-t \sum_{n \in \mathbb{N}} 1_{(X_n+S_n) \cap K \neq \varnothing}} \right] = \mathrm{E}\left[e^{-t \sum_{n \in \mathbb{N}} f(X_n, S_n)} \right] ,$$

where $t \geq 0$ and $f(x, s) := 1_{(x+s) \cap K \neq \varnothing}$. To see this, use the formula in Corollary 8.3.4, which gives

$$\mathrm{E}\left[e^{-t \sum_{n \in \mathbb{N}} f(X_n, S_n)} \right] = \exp\left\{ \int_{\mathbb{R}^m} \left(\mathrm{E}\left[e^{-tf(x, S_1)} - 1 \right] \right) \nu(dx) \right\}$$

$$= \exp\left\{ \int_{\mathbb{R}^m} \left(\mathrm{E}\left[e^{-t1_{(x+S_1) \cap K \neq \varnothing}} - 1 \right] \right) \nu(dx) \right\}$$

$$= \exp\left\{ (e^{-t} - 1) \int_{\mathbb{R}^m} P((x + S_1) \cap K \neq \varnothing) \, \nu(dx) \right\} .$$

$\square$

If the germ process is a homogeneous Poisson process of intensity λ, the mean value $\theta(K)$ of $N(K)$ takes the form

$$\theta(K) = \lambda \mathrm{E}\left[\ell^m((-S_1) \oplus K) \right] .$$

Indeed, from $(\star)$,

$$\theta(K) = \lambda \int_{\mathbb{R}^m} P((x + S_1) \cap K \neq \varnothing) \, dx$$

$$= \lambda \mathrm{E}\left[\int_{\mathbb{R}^m} 1_{(x+S_1) \cap K \neq \varnothing} \, dx \right] = \lambda \mathrm{E}\left[\int_{\mathbb{R}^m} 1_{(-S_1) \oplus K}(x) \, dx \right] .$$

Therefore, in the homogeneous case,

$$T_\Sigma(K) = 1 - \exp\left\{ \lambda \mathrm{E}\left[\ell^m((-S_1) \oplus K) \right] \right\} . \qquad (8.26)$$

Another quantity of interest when the germ point process is an HPP with intensity λ is the volume fraction p of the random set Σ, defined by

$$p := \frac{\mathrm{E}\left[\ell^m(\Sigma \cap B) \right]}{\ell^m(B)} .$$

It is independent of B, and by translation invariance of the model, it is equal to

$$p = \frac{\mathrm{E}\left[\int_B 1_{x \in \Sigma}\,\mathrm{d}x\right]}{\ell^m(B)} = P(0 \in \Sigma) = P(\Sigma \cap \{0\} \neq \varnothing)\,.$$

Therefore,

$$p = p(\{0\}) = 1 - \mathrm{e}^{-\int_{\mathbb{R}^m} P((x+S_1)\cap\{0\}\neq\varnothing)\,\lambda\,\mathrm{d}x}$$
$$= 1 - \mathrm{e}^{-\int_{\mathbb{R}^m} P((x+S_1)\cap\{0\}\neq\varnothing)\,\lambda\,\mathrm{d}x} = 1 - \mathrm{e}^{-\int_{\mathbb{R}^m} P(-x\in S_1)\,\lambda\,\mathrm{d}x} = 1 - \mathrm{e}^{-\lambda \mathrm{E}[\ell^m(S_1)]}\,.$$

The *covariance function* $C : \mathbb{R}^m \to \mathbb{R}$ of the random set Σ is defined by

$$C(x) := P(0 \in \Sigma,\, x \in \Sigma)\,.$$

(This is the covariance function, in the usual sense, of the wide-sense stationary stochastic process $\{1_\Sigma(t)\}_{t\in\mathbb{R}^m}$.) In the homogeneous case,

$$C(x) = 2p - 1 - (1-p)^2 \exp\left\{\lambda\mathrm{E}\left[\ell^m(S_1 \cup (S_1 - x))\right]\right\}\,.$$

Proof.

$$
\begin{aligned}
C(x) &= P(0 \in \Sigma \cap (\Sigma - x)) \\
&= P(0 \in \Sigma) + P(x \in \Sigma) - P(0 \in \Sigma \cup (\Sigma - x)) \\
&= 2p - 1 + P(0 \notin \Sigma \cup (\Sigma - x)) \\
&= 2p - 1 + P(\Sigma \cap \{o, x\} = \varnothing) = 2p - 1 + T_\Sigma(\{0, x\})\,.
\end{aligned}
$$

From (8.26),

$$T_\Sigma(\{0, x\}) = 1 - \exp\left\{\lambda\mathrm{E}\left[\ell^m((-S_1) \oplus \{0, x\})\right]\right\}\,.$$

Now

$$
\begin{aligned}
\mathrm{E}\left[\ell^m((-S_1) \oplus \{0, x\})\right] &= \mathrm{E}\left[\ell^m((-S_1) \cup (-S_1 + x))\right] \\
&= \mathrm{E}\left[\ell^m(-S_1)\right] + \mathrm{E}\left[\ell^m(-S_1 + x)\right] - \mathrm{E}\left[\ell^m((-S_1) \cap (-S_1 + x))\right] \\
&= 2\mathrm{E}\left[\ell^m(-S_1)\right] - \mathrm{E}\left[\ell^m((-S_1) \cap (-S_1 + x))\right]\,.
\end{aligned}
$$

Combining the above equalities with the observation that $1 - p = \exp\left\{-\lambda\ell^m(S_1)\right\}$ gives the announced result. $\qquad\square$

EXAMPLE 8.4.4: THE BOUNDARY OF A POISSON CLUSTER OF DISKS. Let N be a homogeneous Poisson process on $\mathbb{R}^2$, of intensity λ. Let $\{X_n\}_{n\in\mathbb{N}}$ be its sequence of points. Draw around each point X_n a closed disk of radius a. The area inside the square $[0, T] \times [0, T]$ that is not covered by a disk is delimited by a curve. We seek to compute its average length, excluding the parts on the boundaries of $[0, T] \times [0, T]$.

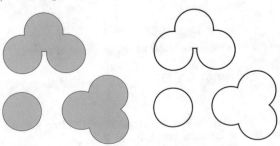

The number of disks covering a given point $y \in \mathbb{R}^2$ is

$$Z(y) = \sum_{n \in \mathbb{N}} 1_{\{\|X_n - y\| < a\}} \, .$$

In particular, $\theta Z(y) = N(\varphi)$ with $\varphi(x) = \theta 1_{\|x-y\| < a}$. Therefore,

$$\mathrm{E}[e^{-\theta Z(y)}] = \mathrm{E}[e^{-N(\varphi)}]$$

$$= \exp\left\{\int_{\mathbb{R}^2} (e^{-\varphi(x)} - 1)\lambda \, \mathrm{d}x\right\} = \exp\left\{\int_{\mathbb{R}^2} (e^{-\theta} - 1)1_{\{\|x-y\| < a\}}\lambda \, \mathrm{d}x\right\}$$

$$= \exp\left\{(e^{-\theta} - 1)\lambda \int_{\mathbb{R}^2} 1_{\|x-y\| < a} \, \mathrm{d}x\right\} = \exp\left\{(e^{-\theta} - 1)\lambda\pi a^2\right\},$$

that is, $Z(y)$ is a Poisson random variable with mean $\lambda\pi a^2$. Let now $S_T = [0, T] \times [0, T]$. The surface not covered by a disk is $\int_{S_T} 1_{\{Z(y)=0\}} \mathrm{d}y$. The mean of this random variable is $\int_{S_T} \mathrm{E}[1_{\{Z(y)=0\}}] \mathrm{d}y = \int_{S_T} P(Z(y) = 0) \, \mathrm{d}y = \int_{S_T} e^{-\lambda\pi a^2} \, \mathrm{d}y = (\ell^2(S_T)) \times e^{-\lambda\pi a^2}$. Call $F(a)$ the average surface computed above. The quantity sought for is $-F'(a) = 2\pi a\lambda e^{-\lambda a^2}\ell^2(S_T)$. (Hint: what does $F(a + \mathrm{d}a) - F(a)$ represent?)

The method of the previous example can be applied to the general Boolean model. Note that the average total area of Σ contained in the bounded set $W \subset \mathbb{R}^m$ is $p\ell^m(W) = (1 - e^{-\lambda\mathrm{E}[\ell^m(S_1)]})\ell^m(W)$. Let $C \subset \mathbb{R}^m$ be a closed convex set with a smooth boundary ∂C. Denote by ∂C_ε the set of points $x \notin C$ such that $d(x, C) \leq \varepsilon$ and let $C_\varepsilon := C \cup \partial C_\varepsilon$. The following informal arguments will lead to the average length L_W of that part of the boundary of Σ which lies in W. With $S \equiv S_1$,

$$L_W = \lim_{\varepsilon \downarrow 0} \frac{1}{\varepsilon}(1 - e^{-\lambda\mathrm{E}[\ell^m(S_\varepsilon)]})\ell^m(W)$$

$$= \ell^m(W) e^{-\lambda\mathrm{E}[\ell^m(S)]} \lim_{\varepsilon \downarrow 0} \frac{1}{\varepsilon}(1 - \mathrm{E}[\ell^m(\partial S_\varepsilon)]) = \ell^m(W) e^{-\lambda\mathrm{E}[\ell^m(S)]}\mathrm{E}[\ell^{m-1}(\partial S)] \, .$$

Therefore $L := \frac{L_W}{\ell^m(W)}$, the average boundary length per unit volume of the Boolean set Σ, is given by the formula

$$L = e^{-\lambda\mathrm{E}[\ell^m(S)]}\mathrm{E}[\ell^{m-1}(\partial S)] \, .$$

The precise conditions that make the above calculations licit are usually satisfied in actual examples, as was the case in Example 8.4.4.

Isolated Points

Consider the following process associated with the Boolean model:

$$Y(y) := \sum_{n \in \mathbb{N}} 1_{X_n + S_n}(y) \quad (y \in \mathbb{R}^d) \, .$$

In rough terms, this quantity counts the number of random shapes S_n centered at a point X_n that cover y. The Laplace transform of $Y(y)$ is

$$E\left[e^{-t\sum_n 1_{y\in X_n+S_n}}\right] = E\left[e^{-t\sum_n 1_{y\in x+s} N(dx\times ds)}\right]$$

$$= \exp\left(\int_{\mathbb{R}^d}\int_S \left(e^{-t1_{y\in x+s}}-1\right)\lambda Q(ds)\,dx\right)$$

$$= \exp\left(\int_{\mathbb{R}^d}\int_S \left(e^{-t1_{x-y\in s}}-1\right)\lambda Q(ds)\,dx\right)$$

$$= \exp\left(\int_{\mathbb{R}^d}\int_S \left(e^{-t1_{x\in s}}-1\right)\lambda Q(ds)\,dx\right)$$

$$= \exp\left(\int_{\mathbb{R}^d}\int_S \left(e^{-t}-1\right)1_{x\in s}\lambda Q(ds)\,dx\right)$$

$$= \exp\left(\int_{\mathbb{R}^d} \left(e^{-t}-1\right)P(x\in -S)\lambda\,dx\right)$$

$$= \exp\left(\left(e^{-t}-1\right)\lambda E\left[\ell^d(S)\right]\right).$$

Therefore $Y(y)$ is a Poisson variable with mean $\lambda E\left[\ell^d(S)\right]$.

A point X_n is called S-isolated if $(X_n+S_n)\cap(X_k+S_k)=\varnothing$ for all $k\neq n$. On a given bounded measurable window $W\subset\mathbb{R}^d$, say a ball of radius a centered on $0\in\mathbb{R}^d$, the average total d-volume of the sets X_n+S_n corresponding to S-isolated points is, ignoring edge effects,

$$\int_W P(Y(y)=1)\,dy = \int_W \lambda E\left[\ell^d(S)\right]e^{-\lambda E[\ell^d(S)]}\,dy = \ell^d(W)\lambda E\left[\ell^d(S)\right]e^{-\lambda E[\ell^d(S)]}.$$

The average of the sum of the volumes of the sets X_n+S_n corresponding to points $X_n\in W$ is $\lambda E\left[\ell^d(S)\right]\ell^d(W)$. Therefore the proportion of S-isolated points is

$$e^{-\lambda E[\ell^d(S)]},$$

and the mean intensity of isolated points is $\lambda e^{-\lambda E[\ell^d(S)]}$.

Note that these quantities depend on the shapes considered only through their average d-volume.

Complementary reading

[Grandell, 1976] for Cox processes. For applications to communications networks, [Franceschetti and Meester, 2007].

Advanced references for the theoretical aspects of point processes: [Neveu, 1976] (in French), [Daley and Vere-Jones, 2003], [Kallenberg, 2017] and [Last and Penrose, 2018] for Poisson processes.

8.5 Exercises

Exercise 8.5.1. THE POINT SEQUENCE
Prove Lemma 8.1.9.

Exercise 8.5.2. RANDOM POINTS UNIFORMLY DISTRIBUTED ON $[0,1]$
Construct a point process N on $\mathbb{R}$ in the following way. First draw a finite integer-valued random variable T, and then an IID sequence $\{U_n\}_{n\geq 1}$ uniformly distributed on $[0,1]$,

independent of T. Define $\alpha_k := P(T = k)$ $(k \geq 0)$. Finally, let $N = \sum_{k=1}^{T} \varepsilon_{U_k}$, where ε_a is the Dirac measure at a, and where $\sum_{k=1}^{0} \varepsilon_{U_k}$ is the null measure by convention. What is the Laplace transform of N? What about the case where T is a Poisson variable of mean θ?

Exercise 8.5.3. THE LAPLACE FUNCTIONAL OF A NEYMAN–SCOTT PROCESS
Compute the Laplace functional of the Neyman–Scott cluster point process of Example 8.1.29 in terms of the generating function of T and of the common probability distribution of the U_n's.

Exercise 8.5.4. AN INDEPENDENT FAMILY OF POINT PROCESSES
Prove the following (Theorem 8.1.37). Let N_i $(i \in I)$ be a collection of point processes on E, where I is an arbitrary index set. If for any finite subset $J \subseteq I$, any collection φ_i $(i \in J)$ of non-negative measurable functions from E to $\mathbb{R}$,

$$\mathrm{E}\left[e^{-\sum_{i \in J} N_i(\varphi_i)}\right] = \prod_{i \in J} \mathrm{E}\left[e^{-N_i(\varphi_i)}\right], \qquad (8.27)$$

then N_i $(i \in I)$ is an independent family of point processes.

Exercise 8.5.5. THE AVOIDANCE FUNCTION OF A MARKED POINT PROCESS
Consider a marked point process as defined in Definition 8.1.16 and assume that the basic point process N is simple. The point process $\overline{N}$ on $E \times K$ being defined as in (8.1), show that the avoidance function $\overline{v}$ of this point process is entirely defined by its values on the subsets of $\mathbb{R}^m \times \mathbb{R}^d$ of the form $C \times L$, where $C \in \mathcal{B}(\mathbb{R}^m)$ and $L \in \mathcal{B}(\mathbb{R}^d)$.

Exercise 8.5.6. THE AVOIDANCE FUNCTION AND THE LAPLACE FUNCTIONAL
Let N be a point process on $\mathbb{R}^m$ with Laplace functional L_N. Show that its void probability function v is given by

$$v(C) = \lim_{t \uparrow +\infty} L_N(t1_C).$$

Exercise 8.5.7. A CHARACTERIZATION OF THE POISSON PROCESS
For a simple point process N on $\mathbb{R}^m$ to be a Poisson process of σ-finite intensity measure ν it is necessary and sufficient that for all Borel sets $C \in \mathbb{R}^m$ of finite ν-measure, $N(C)$ is a Poisson variable of mean $\nu(C)$. (The surprising fact is that one does not need the independence assumption of Definition 8.2.1.)

Exercise 8.5.8. A PROPERTY OF AVOIDANCE FUNCTIONS

(a) Show that if v_1 and v_2 are the void probability functions of some point processes on $\mathbb{R}^m$, so is $v_1 \times v_2$.

(b) Let v be the avoidance function of some point process on $\mathbb{R}^m$ and let g_T be the generating function of some random variable taking its values in $\mathbb{N}$. Show that $w(C) := g_T(v(C))$ is the avoidance function of some point process on $\mathbb{R}^m$.

Exercise 8.5.9. THE INTENSITY MEASURE OF A POINT PROCESS
Prove the statement in Definition 8.1.18.

Exercise 8.5.10. THE LAPLACE FUNCTIONAL OF A CONTRACTED POINT PROCESS
Let N be a simple point process on $\mathbb{R}^m$ with point sequence $\{X_n\}_{n\in\mathbb{N}}$ and let $\alpha > 0$. Define the "contracted"[6] point process $N_{c,\alpha}$ defined by its sequence of points $\{\alpha X_n\}_{n\in\mathbb{N}}$. Prove that its Laplace functional is

$$L_{N_{c,\alpha}}(\varphi) = \mathrm{E}\left[\exp\left(-\sum_{n\in\mathbb{Z}} \varphi(\alpha X_n)\right)\right] = L_N(\varphi(\alpha\cdot)).$$

Exercise 8.5.11. THIRD MOMENTS
Let N be a homogeneous Poisson process on $\mathbb{R}$ with intensity λ. Compute

$$\mathrm{E}\left[\int_{\mathbb{R}} \varphi_1(t)\, N(\mathrm{d}t) \int_{\mathbb{R}} \varphi_2(t)\, N(\mathrm{d}t) \int_{\mathbb{R}} \varphi_3(t)\, N(\mathrm{d}t)\right]$$

for the measurable functions $\varphi_j : \mathbb{R} \to \mathbb{R}$ $(j = 1, 2, 3)$, after giving conditions for this expectation to be well defined and finite.

Exercise 8.5.12. DISTRIBUTION OF THE MAXIMUM INTERFERENCE
Let N be a homogeneous Poisson process on $\mathbb{R}^m$ of positive intensity λ and with point sequence $\{X_n\}_{n\geq 1}$. Let $\{Z_n\}_{n\geq 1}$ be an IID sequence of real non-negative random variables with common distribution Q, and independent of N. Compute the distribution of the random variable

$$\max_{n\geq 1} Z_n e^{-\beta\|X_n\|} \qquad (\beta > 0).$$

(The title of the exercise refers to a situation where Z_n is the noise intensity generated at point X_n, and $e^{-\beta\|X_n\|}$ is an attenuation factor for a receiver located at 0.)

Exercise 8.5.13. THE M/GI/∞ MODEL, TAKE 2
In Example 8.3.2,

  (i) prove that the departure process is a homogeneous Poisson process with intensity λ,

  (ii) compute $\mathrm{cov}(X(t), X(t+\tau))$ for all $t, \tau \in \mathbb{R}$, $\tau \geq 0$, and

  (iii) interpret the process $\{X(t)\}_{t\in\mathbb{R}}$ as a shot noise in order to obtain the results of Example 8.3.2, and of (i) and (ii), from the general results of Section 8.3.

Exercise 8.5.14. LIFTING
Let N be a Poisson process on $\mathbb{R}$ with (locally integrable) intensity function $\lambda : \mathbb{R} \to \mathbb{R}$. Let $\{T_n\}_{n\in\mathbb{Z}}$ be its sequence of points, and let $\{U_n\}_{n\in\mathbb{Z}}$ be an IID sequence of random variables uniformly distributed on $[0, 1]$. Let $\widehat{N}$ be an HPP on $\mathbb{R} \times \mathbb{R}_+$, with intensity 1, independent of N and of $\{U_n\}_{n\in\mathbb{Z}}$. Define a point process $\widetilde{N}$ on $\mathbb{R} \times \mathbb{R}_+$ by

$$\widetilde{N}(C) := \sum_{n\in\mathbb{Z}} 1_C((T_n, U_n\lambda(T_n))) + \widehat{N}(C \cap \bar{H}),$$

where

$$H := \{(t, z) \in \mathbb{R} \times \mathbb{R}_+ \,;\, 0 \leq z \leq \lambda(t)\}.$$

---

[6]Of course, if $\alpha > 1$, it is in fact dilated...

Show that $\widetilde{N}$ is an HPP on $\mathbb{R} \times \mathbb{R}_+$ with intensity 1.

Exercise 8.5.15. WATER BOMBS
You are initially located at the origin $(0,0)$ of the plane at which is centered a disk D of radius R. You run in a straight line from the origin to the "shelter point" $(0, R)$ at constant speed v. The reason why you are running is that water bombs are dropped on the disk D. The times of impact form an HPP of intensity λ, and each impact is located independently of all the rest, uniformly on the disk. You are getting wet if the impact of the bomb is within distance a of your position at the time of impact. Once arrived at the shelter point $(0, R)$, the bombing stops. What are your chances of not getting wet? Given that you did get wet, what is the expected time that you remained dry?

Exercise 8.5.16. HUMANITARIAN AIRDROPS
Packets are dropped on the plane $\mathbb{R}^2$. The impact times $\{T_n\}_{n \in \mathbb{Z}}$ form a simple Poisson process of intensity measure ν, and the impact locations $\{Z_n\}_{n \in \mathbb{Z}}$ are IID and independent of the impact times, with common probability distribution Q. A "shape" moves on $\mathbb{R}^2$ in order to collect the packets as they impact on it. More precisely, there is for each time $t \in \mathbb{R}$ a measurable subset $S(t) \in \mathbb{R}^2$ and the point process $\hat{N}$ counting the packets falling on the shape is defined by

$$\hat{N}(C) = \sum_{n \in \mathbb{Z}} 1_C(T_n) 1_{S(t)}(Z_n).$$

Prove that $\hat{N}$ is a Poisson process and give its intensity measure.

Exercise 8.5.17. SMOKING POT AT SAINT MARY-JANE'S
Smoking pot was recently banned on the Saint Mary-Jane's college campus. The authorities noticed that the violators of the ban make use of a restroom in a secluded wing of the campus. They consequently devised a strategy to send "cops" to capture the culprits. Assume that the schoolboys' arrival times in the restroom premises form a Poisson process with independent IID marks. Let τ_n denote the n-th arrival time of a schoolboy in the pot sanctuary (the restrooms) and let σ_n be the time he spends smoking. Cops also form a Poisson process with independent IID marks. Denote the k-th arrival time of a cop on the potential crime scene by T_k and by S_k the lingering time there of the corresponding representative of the college authority. The probability distribution of σ is Q_s and that of S is Q_c. Assuming the point processes of students and of the cops to be HPPs with respective intensities $\lambda_s > 0$ and $\lambda_c > 0$, compute the average number of students caught per unit of time.

Exercise 8.5.18. THIRD MOMENTS OF A SHOT NOISE
Let N be a homogeneous Poisson process on $\mathbb{R}^d$ with intensity $\lambda > 0$. Let $\{X(t)\}_{t \in \mathbb{R}}$ be the shot noise constructed on it with an impulse function $h : \mathbb{R}^d \to \mathbb{R}$ that is bounded and with compact support (null outside a bounded subset of $\mathbb{R}^d$):

$$X(t) := \int_{\mathbb{R}} h(t - s) \, N(\mathrm{d}s).$$

Let $t_1, t_2, t_3 \in \mathbb{R}^d$.

(a) Compute the characteristic function of the vector $(X(t_1), X(t_2), X(t_3))$.

(b) Compute $\mathrm{E}\,[X(t_1)X(t_2)X(t_3)]$.

Exercise 8.5.19. POINTS IN A RANDOM INTERVAL
Let N be an HPP on $\mathbb{R}_+$ with intensity $\lambda > 0$. Let Z_1 and Z_2 be two non-negative real random variables such that $Z_1 \leq Z_2$. Give the probability distribution of the random variable $X = N\left((Z_1, Z_2]\right)$, assuming that Z_1, Z_2 are independent of N.

Exercise 8.5.20. THE LAPLACE TRANSFORM OF A HOMOGENEOUS COX PROCESS
Let N be a Cox process on $\mathbb{R}^m$ with constant intensity process, that is, $\nu(\mathrm{d}x) := \Lambda \ell^m(\mathrm{d}x)$, where ℓ^m is the Lebesgue measure on $\mathbb{R}^m$ and Λ is a non-negative random variable with Laplace transform $L_\Lambda(t) := \mathrm{E}\left[\mathrm{e}^{-t\Lambda}\right]$. Show that its Laplace transform is

$$L_N(\varphi) = L_\Lambda \left(\int_{\mathbb{R}^m} (1 - \mathrm{e}^{-\varphi(x)})\,\mathrm{d}x \right).$$

Exercise 8.5.21. THE MAXIMUM FORMULA
Give a direct proof of the result of Example 8.2.9 based on the construction of Subsection 8.2.1.

Exercise 8.5.22. SUPERPOSITION VIA THE VOID PROBABILITY DISTRIBUTION
Prove Theorem 8.2.3 via Rényi's theorem.

Exercise 8.5.23. THE VOID PROBABILITY FUNCTION OF A COX PROCESS
Let N be a Cox process on $\mathbb{R}^m$ with random intensity measure ν. Let N' be a Poisson process on $\mathbb{R}^m$ with intensity measure ν' equal to the mean of ν (that is, for all $C \in \mathcal{B}(\mathbb{R}^m)$, $\nu'(C) = \mathrm{E}[\nu(C)]$. For any subset $C \in \mathcal{B}(\mathbb{R}^m)$, show that the probability that there is no point of N in C is greater than or equal to the corresponding probability for N'.

Exercise 8.5.24. RANDOM LINES
Let N be a Poisson process on $\mathbb{R}^2$ with *finite* intensity measure ν and let $\{\Theta_n\}_{n\in\mathbb{N}}$ be an IID sequence random variables uniformly distributed on $[0, \pi)$. A random line process can be generated as follows. Through each point X_n of N let a straight line pass, forming an oriented angle Θ_n with the first coordinate axis of $\mathbb{R}^2$.

Show that the point process on $\mathbb{R}_+ \times [0, 2\pi]$ whose sequence of points is $\{(||X_n||, \Theta_n)\}_{n\in\mathbb{N}}$ is a Poisson process and compute its intensity measure and describe (give the probability distribution of) the point process formed by the intersections of the random lines with the first coordinate axis of $\mathbb{R}^2$.

Exercise 8.5.25. LINE OF SIGHT, TAKE 1
Consider a Poisson process N on $\mathbb{R}^2$ with non-atomic and locally finite intensity measure ν. There is a "random shape" centered around each of its points. Let the generic shape S be "distributed according to some probability distribution Q_S". Now consider two points A, B. We say that A and B can communicate if the line connecting A and B does not intersect any of the existing shapes around the points of the point process (for all $n \geq 1$, the "existing shape around" $X_n \in N$ is $X_n + S_n$, that is, S_n translated by X_n, where S_n is distributed according to Q_S). We assume that $\{S_n\}_{n\in\mathbb{N}}$ is an IID sequence independent of N. Compute the probability that A and B can successfully communicate. Push the computations when N is an HPP of intensity λ, the random shape is a circle of radius a, and the distance from A to B is equal to l.

Exercise 8.5.26. LINE OF SIGHT, TAKE 2
The situation is that of Exercise 8.5.25.

(i) There is a traveler going from B to a third point C. What is the probability that he will have A in sight all the time during his journey?

(ii) It is this time only required that the traveler has B in his line of sight at the beginning and at the end of his journey. What is the probability for this to happen?

Exercise 8.5.27. CELL PHONES AND BASE STATIONS
Let N_1 and N_2 be two independent Poisson processes on $\mathbb{R}^m$ with respective *finite* intensity measures ν_1 and ν_2. Compute the average number of elements in N_1 that see no point of N_2 within distance a. (The title of the example refers to the situation where the points of N_2 are the locations of base stations in a wireless communications network and those of N_1 are the locations of cell phones. Therefore the quantity $(\star)$ is the average number of cell phones that cannot be connected to a base station.)

Chapter 9

Queueing Processes

Queueing theory is of interest in the performance evaluation of systems where the two related phenomena of *congestion* and *waiting* (delay) could be a nuisance. This is the case for communications and computer systems, in addition to the classical systems of operations research featuring ticket booths, freeway tolls and the like. This chapter gives the "elementary theory", that is, the topics that can be treated using the theory of Markov chains. Important aspects of the non-Markovian theory are considered in Sections 16.2 and 17.4.

9.1 Discrete-time Markovian Queues

9.1.1 The Basic Example

This example is just the repair shop model of Example 6.1.7 interpreted in terms of queueing theory. This time it modelizes a communications link operating in slotted time, that is, observations are made and actions are taken at the discrete times $0, 1, \ldots$. A mathematical interval of 1 (separating n from $n + 1$) represents a "slot" of T real time units. During slot n (the interval $(nT, (n + 1)T]$), Z_{n+1} messages are presented to the link for transmission and are stored in a buffer. In slot $n + 1$ one message from the buffer (if there is one there) is transmitted. The others stay in the buffer where new fresh messages will be added. Therefore, denoting by X_n the number of buffered messages in slot n,

$$X_{n+1} = (X_n - 1)^+ + Z_{n+1}, \tag{9.1}$$

where $a^+ = \max(a, 0)$. In particular, if $\{Z_n\}_{n \geq 1}$ is an IID sequence independent of the initial number of messages in the buffer, X_0, then $\{X_n\}_{n \geq 0}$ is a homogeneous Markov chain. Let

$$a_k := P(Z_1 = k) \qquad (k \geq 0).$$

By formula (6.5),

$$p_{ij} = P((i - 1)^+ + Z_1 = j) = P(Z_1 = j - (i - 1)^+) = a_{j-(i-1)^+}.$$

Therefore

$$\mathbf{P} = \begin{pmatrix} a_0 & a_1 & a_2 & a_3 & \cdots \\ a_0 & a_1 & a_2 & a_3 & \cdots \\ 0 & a_0 & a_1 & a_2 & \cdots \\ 0 & 0 & a_0 & a_1 & \cdots \\ \vdots & \vdots & \vdots & \vdots & \end{pmatrix}.$$

© Springer Nature Switzerland AG 2020
P. Brémaud, *Probability Theory and Stochastic Processes*, Universitext,
https://doi.org/10.1007/978-3-030-40183-2_9

We know from Example 6.2.3 that a necessary and sufficient condition of irreducibility of the repair shop chain of Example 6.1.7 is that $P(Z_1 = 0) > 0$ *and* $P(Z_1 \geq 2) > 0$.

In Examples 6.2.14 and 6.3.22, it was proved that $E[Z] < 1$ is a necessary and sufficient condition for the existence of a steady state and that, if the stationary distribution exists, its generating function is given by the formula

$$\sum_{i=0}^{\infty} \pi(i)z^i = (1 - E[Z])\frac{(z-1)g_Z(z)}{z - g_Z(z)}.$$

Therefore, from the point of view of stability, we have all we need and we shall not pursue the study of this type of model, since we shall soon treat its continuous-time version and its extensions.

In this model, there is only one source of messages. The situation becomes more complicated when there are several sources, and this brings us to the next topic.

9.1.2 Multiple Access Communication

A typical situation in a multiple-access satellite communications link is the following. Users—each one identified with a message—contend for access to a single-channel communications link. Two or more messages in the air at the same time jam each other, and are not successfully transmitted. The users are somehow able to detect a collision of this sort and will try to retransmit later the message involved in a collision. The difficulty in such communications systems resides mainly in the absence of cooperation among users, who are all unaware of the intention to transmit of competing users. A first historical attempt was the ALOHA communications system.

ALOHA

The *slotted* ALOHA *protocol* imposes on the users the following rules (see the figure below):

(i) Transmissions and retransmissions of messages can start only at equally spaced times; the interval between two consecutive (re-)transmission times is called a *slot*; the duration of a slot is always larger than that of any message.

(ii) All *backlogged* messages, that is, those messages having already tried unsuccessfully (maybe more than once) to get through the link, require retransmission independently of one another with probability $\nu \in (0, 1)$ at each slot. This is the so-called *Bernoulli retransmission policy*.

(iii) The *fresh messages*—those presenting themselves for the first time—immediately attempt to get through.

Let X_n be the number of backlogged messages at the beginning of slot n. The backlogged messages behave independently, and each one has probability ν of attempting retransmission in slot n. In particular, if there are $X_n = k$ backlogged messages, the probability that i among them attempt to retransmit in slot n is

$$b_i(k) = \binom{k}{i}\nu^i(1 - \nu)^{k-i}.$$

Let A_n be the number of fresh requests for transmission in slot n. The sequence $\{A_n\}_{n \geq 0}$ is assumed IID with the distribution $P(A_n = j) = a_j$. The quantity $\lambda := E[A_n] =$

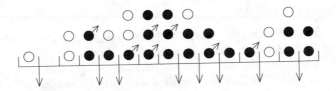

○   fresh message

●   backlogged message, not authorized to attempt retransmission

●↗   backlogged message, authorized to attempt retransmission

↓   successful transmission (or retransmission)

The ALOHA protocol

$\sum_{i=1}^{\infty} i a_i$ is called the *traffic intensity*. We suppose that $a_0 + a_1 \in (0, 1)$, so that $\{X_n\}_{n \geq 0}$ is an irreducible HMC. Its transition matrix is

$$
p_{ij} = \begin{cases}
b_1(i)a_0 & \text{if } j = i - 1, \\
[1 - b_1(i)]a_0 + b_0(i)a_1 & \text{if } j = i, \\
[1 - b_0(i)]a_1 & \text{if } j = i + 1, \\
a_{j-i} & \text{if } j \geq i + 2.
\end{cases}
$$

The proof is by accounting. For instance, the first line corresponds to one among the i backlogged messages having succeeded to retransmit, and for this there should be no fresh arrival (probability a_0) and only one of the i backlogged messages allowed to retransmit (probability $b_1(i)$). The second line corresponds to one of the two events "no fresh arrival and zero or strictly more than two retransmission requests from the backlog" and "zero retransmission request from the backlog and one fresh arrival."

The Instability of ALOHA

It turns out, as will be shown next, that the Bernoulli retransmission policy makes the ALOHA protocol *unstable*, in the sense that the chain $\{X_n\}_{n \geq 0}$ is *not positive recurrent*.

An elementary computation yields, for the ALOHA model,

$$
E[X_{n+1} - X_n \mid X_n = i] = \lambda - b_1(i)a_0 - b_0(i)a_1. \tag{9.2}
$$

Note that $b_1(i)a_0 + b_0(i)a_1$ is the probability of one successful (re-)transmission in a slot given that the backlog at the beginning of the slot is i. Equivalently, since there is at most one successful (re-)transmission in any slot, this is the average number of successful (re-)transmissions in a slot given the backlog i at the start of the slot. An elementary computation shows that $\lim_{i \uparrow \infty} (b_1(i)a_0 + b_0(i)a_1) = 0$. Therefore, outside a finite set F, the conditions of Theorem 6.3.23 are satisfied when we take h to be the identity, and remember the hypothesis that $E[A_1] < \infty$.

Backlog Dependent Policies

Since the ALOHA protocol with a fixed retransmission probability ν is unstable, it seems reasonable to try a retransmission probability $\nu = \nu(k)$ depending on the number k of backlogged messages. In fact, there is a choice of the function $\nu(k)$ that achieves stability of the protocol. The probability that i among the k backlogged messages at the beginning of slot n retransmit in slot n is now

$$b_i(k) = \binom{k}{i}\nu(k)^i(1 - \nu(k))^{k-i}.$$

The transition probabilities are changed accordingly. According to Pakes' lemma and using (9.2), it suffices to find a function $\nu(k)$ guaranteeing that

$$\lambda \leq \lim_{i\uparrow\infty} (b_1(i)a_0 + b_0(i)a_1) - \epsilon, \tag{9.3}$$

for some $\epsilon > 0$. We shall therefore study the function

$$g_k(\nu) = (1 - \nu)^k a_1 + k\nu(1 - \nu)^{k-1} a_0,$$

since condition (9.3) is just $\lambda \leq g_i(\nu(i)) - \epsilon$. The derivative of $g_k(\nu)$ is, for $k \geq 2$,

$$g_k'(\nu) = k(1 - \nu)^{k-2}[(a_0 - a_1) - \nu(ka_0 - a_1)].$$

We first assume that $a_0 > a_1$. In this case, for $k \geq 2$, the derivative is zero for

$$\nu = \nu(k) = \frac{a_0 - a_1}{ka_0 - a_1},$$

and the corresponding value of $g_k(\nu)$ is a maximum equal to

$$g_k(\nu(k)) = a_0 \left(\frac{k - 1}{k - a_1/a_0}\right)^{k-1}.$$

Therefore, $\lim_{k\uparrow\infty} g_k(\nu(k)) = a_0 \exp\left\{\frac{a_1}{a_0} - 1\right\}$, and we see that

$$\lambda < a_0 \exp\left\{\frac{a_1}{a_0} - 1\right\} \tag{9.4}$$

is a sufficient condition for stability of the protocol. For instance, with a Poisson distribution of arrivals

$$a_i = e^{-\lambda}\frac{\lambda^i}{i!},$$

condition (9.4) reads

$$\lambda < e^{-1}$$

(in particular, the condition $a_0 > a_1$ is satisfied a posteriori). If $a_0 \leq a_1$, the protocol can be shown to be unstable, whatever retransmission policy $\nu(k)$ is adopted (the reader is invited to check this).

9.1.3 The Stack Algorithm

At this point, the slotted ALOHA protocol with constant retransmission probability was proved unstable, and it was shown that a backlog-dependent retransmission probability could restore stability. The problem now resides in the necessity for each user to know the size of the backlog in order to implement the retransmission policy. This is not practically feasible, and one must devise policies based on the actual information available by just listening to the link: collision, no transmission, or successful transmission. Such policies, which in a sense estimate the backlog, have been found that yield stability. However, we shall not discuss them here, and instead we shall consider another type of protocol, the so-called *collision-resolution protocol*, or *binary tree protocol*, implemented by the *stack algorithm*.

In this protocol, when a collision occurs, all new requests are buffered until all the messages involved in the collision have found their way through the link. When these messages have resolved their collision problem, the buffered messages then try to retransmit, maybe enter a collision, and then resolve their collision. Time is therefore divided into successive periods, called *collision-resolution intervals* (CRI). Let us examine the fate of the messages arriving in the first slot, which are the messages that arrived during the previous CRI. They all try to retransmit in the first slot of the CRI, and therefore, if there are two or more messages, a collision occurs (in the other case, the CRI has lasted just one slot, and a new CRI begins in the next slot). An unbiased coin is tossed for each colliding message. If it shows heads, the message joins *layer* 0 of a *stack*, whereas if it shows tails, it is placed in layer 1. In the next slot, all messages of layer 0 try the link. If there is no collision (because layer 0 was empty or just contained one message), layer 0 is eliminated, and layer 1 below pops up to become layer 0. If on the contrary there is a collision because layer 0 formed after the first slot contained two or more messages, the colliding messages again flip a coin; those with heads form the new layer 0, those with tails form the new layer 1, and the former layer 1 is pushed bottomwards to form layer 2.

In general, at each step, only layer 0 tries to retransmit. If there is no collision, layer 0 disappears, and the layers $1, 2, 3, \ldots$ become layers $0, 1, 2, \ldots$ If there is a collision, layer 0 splits into layer 0 and layer 1, and layers $1, 2, 3, \ldots$ become layer $2, 3, 4, \ldots$ It should be noted that in this protocol, each user (message) knows at every instant in which layer he or she is, just by listening to the channel that gives the information: collision or no collision. In that sense, the protocol is *distributed*, because there is no central operator broadcasting nonlocally available information, such as the size of the backlog, to all users.

Once a collision is resolved, that is, when all layers have disappeared, a new CRI begins. The number of customers that are starting this CRI are those that have arrived in the CRI that just ended. The figure below gives an example of what happens in a CRI.

Since the fresh requests sequence $\{A_n\}_{n \geq 1}$ is IID, the sequence, $\{X_n\}_{n \geq 0}$, where X_n is the *length* of the nth CRI, forms an irreducible HMC. Stability of the protocol is naturally identified with positive recurrence of this chain, which will now be proved with the help of Foster's theorem.

According to Pakes' lemma, it suffices to show that

$$\limsup_{i \uparrow \infty} \mathrm{E}[X_{n+1} - X_n \mid X_n = i] < 0 \tag{9.5}$$

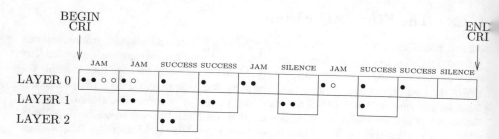

The stack algorithm

and for all i,

$$E[X_{n+1} \mid X_n = i] < \infty. \tag{9.6}$$

For this, let Z_n be the number of fresh arrivals in the nth CRI. We have

$$
\begin{aligned}
E[X_{n+1} \mid X_n = i] &= \sum_{k=0}^{\infty} E[X_{n+1} \mid X_n = i, Z_n = k] P(Z_n = k \mid X_n = i) \\
&= \sum_{k=0}^{\infty} E[X_{n+1} \mid Z_n = k] P(Z_n = k \mid X_n = i).
\end{aligned}
$$

It will be shown that for all $n \geq 0$,

$$E[X_{n+1} \mid Z_n = k] \leq \alpha k + 1, \tag{9.7}$$

where $\alpha = 2.886$, and therefore

$$E[X_{n+1} \mid X_n = i] \leq \sum_{k=0}^{\infty} (\alpha k + 1) P(Z_n = k \mid X_n = i) = \alpha E[Z_n \mid X_n = i] + 1.$$

Using Wald's lemma (Theorem 3.2 of Chapter 1), we have

$$E[Z_n \mid X_n = i] = \lambda i,$$

where λ is the *traffic intensity*, and therefore

$$E[X_{n+1} - X_n \mid X_n = i] \leq 1 + i(\lambda \alpha - 1).$$

We see that condition (9.6) is always satisfied and that (9.5) is satisfied, provided that

$$\lambda < \frac{1}{\alpha} = 0.346. \tag{9.8}$$

It remains to prove (9.7). Let $E[X_{n+1} \mid Z_n = k] = L_k$ (it is indeed a quantity independent of n). Clearly,

$$L_0 = L_1 = 1,$$

since with zero or one packet at the beginning of a CRI, there is no collision. When $k \geq 2$, there is a collision, and the k users toss a coin, and depending on the result they split into two sets, layer 0 and layer 1. Among these k users, i obtain heads with probability

$$q_i(k) = \binom{k}{i} \left(\frac{1}{2}\right)^k.$$

The average length of the CRI given that there are $k \geq 2$ customers at the start, and given that the first layer 0 contains i messages, is

$$L_{k,i} = 1 + L_i + L_{k-i}.$$

Indeed, the first slot saw a collision; the i customers in the first layer 0 will take on average L_i slots to resolve their collision, and L_{k-i} more slots will be needed for the $k-i$ customers in the first-formed layer 1 (these customers are always at the bottom of the stack, in a layer traveling up and down until it becomes layer 0, at which time they start resolving their collision). Since

$$L_k = \sum_{i=0}^{k} q_i(k) L_{k,i},$$

we have

$$L_k = 1 + \sum_{i=0}^{k} q_i(k)(L_i + L_{k-i}).$$

Solving for L_k, we obtain

$$L_k = \frac{1 + \sum_{i=0}^{k-1}[q_i(k) + q_{k-i}(k)]L_i}{1 - q_0(k) - q_k(k)}. \tag{9.9}$$

Suppose that for some $m \geq 2$, and α_m satisfying

$$\alpha_m \geq \sup_{j > m} \frac{\sum_{i=0}^{m-1}(L_i + 1)(q_i(j) + q_{j-i}(j))}{\sum_{i=0}^{m-1} i(q_i(j) + q_{j-i}(j))}, \tag{9.10}$$

it holds that $L_m \leq \alpha_m m - 1$. Then we shall prove that for all $n \geq m$,

$$L_n \leq \alpha_m n - 1. \tag{9.11}$$

We do this by induction, assuming that (9.11) holds true for $n = m, m+1, \ldots, j-1$, and proving that it holds true for $n = j$. Equality (9.9) gives

$$
\begin{aligned}
L_j(1 - q_0(j) - q_j(j)) &= 1 + \sum_{i=0}^{j-1}(q_i(j) + q_{j-i}(j))L_i \\
&= 1 + \sum_{i=0}^{m-1} + \sum_{i=m}^{j-1} \\
&\leq 1 + \sum_{i=0}^{m-1} + \sum_{i=m}^{j-1}(q_i(j) + q_{j-i}(j))(\alpha_m i - 1),
\end{aligned}
$$

where we used the induction hypothesis. The latter term equals

$$1 + \sum_{i=0}^{m-1}(q_i(j) + q_{j-i}(j))(L_i - \alpha_m i + 1) + \sum_{i=0}^{j}(q_i(j)$$
$$+ q_{j-i}(j))(\alpha_m i - 1) - (q_0(j) + q_j(j))(\alpha_m j - 1)$$
$$= 1 + \sum_{i=0}^{m-1}(q_i(j) + q_{j-i}(j))(L_i - \alpha_m i + 1) + \alpha_m j - 2 - (q_0(j) + q_j(j))(\alpha_m j - 1),$$

where we used the identities

$$\sum_{i=0}^{j} q_i(j) = 1, \ \sum_{i=0}^{j} i q_i(j) = jp, \ \sum_{i=0}^{j} i q_{j-i}(j) = j(1-p).$$

Therefore,

$$L_j \le (\alpha_m j - 1) + \frac{\sum_{i=0}^{m-1}(q_i(j) + q_{j-i}(j))(L_i - \alpha_m i + 1)}{1 - q_0(j) - q_j(j)}.$$

Therefore, for $L_j \le \alpha_m j - 1$ to hold, it suffices to have

$$\sum_{i=0}^{m-1}(q_i(j) + q_{j-i}(j))(L_i - \alpha_m i + 1) \le 0.$$

We require this to be true for all $j > m$, and (9.10) guarantees this. It can be checked *numerically* that for $m = 6$ and $\alpha_6 = 2.886$, (9.10) is satisfied, $L_6 \le 2.886 \times 6 - 1$ and equality (9.11) is true for $n = 2, 3, 4, 5$, and this completes the proof.

9.2   Continuous-time Markovian Queues

We now turn to classical queueing theory, based on the theory of continuous-time Markov chains, starting with a sample of the simplest models.

9.2.1   Isolated Markovian Queues

EXAMPLE 9.2.1: M/M/1/∞/FIFO. Consider a ticket booth with a single attendant, or *server*. Customers wait in line in front of the booth, and the facility is so large that no bound is imposed on the number of customers waiting for service. In other words, the *waiting room* has infinite *capacity*. Such a system will be called a $1/\infty$ *service system*, where 1 is for the number of servers, and ∞ is for the capacity of the waiting room.

Customer arrivals are modeled by a homogeneous Poisson process on the positive half-line, $\{T_n\}_{n \ge 1}$, of intensity $\lambda > 0$. Customer n arriving at time T_n brings a service request σ_n, which means that the server will take σ_n units of time for processing this request. The service sequence $\{\sigma_n\}_{n \ge 1}$ is assumed IID, with exponential distribution of mean μ^{-1}. Also, the arrival sequence and the service sequence are assumed to be independent. Such a pattern of arrivals is called an *M/M input process*, introduced by the English probabilist David Kendall. In this notation, M stands for "Markovian". (Indeed, the Poisson process is Markovian, and exponential distributions are intimately connected with the Markov property.)

The server attends one customer at a time and does not remain idle as long as there is at least one customer in the *system* (ticket booth plus waiting room). Once the service of a customer is started it cannot be interrupted before completion.

The above system is called an M/M/1/∞ queue. Its description could be complemented by an indication of the *service discipline* that is enforced: for instance, FIFO (first-in-first-out), where the server, after completion of a service, chooses his next customer at the head of the line.

The stochastic process $\{X(t)\}_{t \ge 0}$, where $X(t)$ is the number of customers present in the system at time t, is called the *congestion process*. It turns out (the proof comes later) that, for a M/M/1/∞/FIFO queue, it is a continuous-time HMC with state space $E = \mathbb{N}$ and infinitesimal generator

$$q_{i,i+1} = \lambda$$
$$q_{i,i-1} = \mu 1_{\{i \geq 1\}}.$$

EXAMPLE 9.2.2: M/M/K/∞/FIFO. This is the M/M/1/∞/FIFO queue, except that there are now K servers. The congestion process is a continuous-time HMC with state space $E = \mathbb{N}$ and infinitesimal generator

$$q_{i,i+1} = \lambda$$
$$q_{i,i-1} = \inf(i, K)\mu.$$

EXAMPLE 9.2.3: M/M/K/0, OR ERLANG QUEUE. This queue is called the *Erlang* queue in honor of the Danish telephone engineer, who used it as a model for a *telephone switch* with K *lines*, or *channels*: a customer finding a free line is connected and holds the line for the time of a conversation. A customer entering the system (that is, arriving and seeing one or more idle servers) will select a free server at random. A customer finding all channels busy is rejected, or in the best case routed to another switch. Therefore the number of customers present in the system at any given time is less than or equal to K. It is a queue with *blocking*. Erlang was able to obtain his famous *blocking formula* giving the probability that, in stationary regime, a given customer finds all lines busy. This is the first formula of queueing theory; see Example 9.2.9 below.

Here, $X(t)$ is the number of busy lines (and therefore the number of customers in the system) at time t. We shall prove later that the congestion process is a continuous-time HMC with state space $E = \{0, 1, \ldots, K\}$ and infinitesimal generator $\mathbf{A}$ given by

$$q_{i,i+1} = \lambda 1_{\{0 \leq i \leq K-1\}}$$
$$q_{i,i-1} = i\mu 1_{\{1 \leq i \leq K\}}.$$

EXAMPLE 9.2.4: M/M/1/∞/LIFO PREEMPTIVE RESUME. This queueing system receives a description similar to that of an M/M/1/∞/FIFO, except for the service discipline, which now becomes LIFO *preemptive resume* (the abbreviation LIFO stands for last in first out). A customer upon arrival goes right to the ticket booth, and the customer presently receiving service is sent back to the waiting room, where he will stand in front of the line (at least until the time when another rude customer shows up, sending the first rude customer to the front of the queue, and so on). This type of discipline is called *preemptive*. The phrase "preemptive resume" means that a preempted customer does not have to start from scratch: when the server sees him next time, he will resume work where it was left.

We shall see that the congestion process of this queue is of exactly the same nature as that of the M/M/1/∞/FIFO queue, and that in particular it has the same infinitesimal generator.

Remark 9.2.5 The type discipline described in the above example is not as unfair as it may appear. First of all, all customers being equally rude, they endure as much as they hurt. A customer with a large service request spends a longer time at the ticket booth and is therefore more exposed than a customer with a modest request, and it is precisely in this sense that the discipline is fair. It makes longer requests who are responsible for congestion wait longer in the system. The precise result is that the expected sojourn time of a customer in the system, given that its service request is x, is equal to $x/(1-\rho)$, where $\rho = \lambda/\mu$.

EXAMPLE 9.2.6: M/M/∞. For simplicity, the second ∞ in M/M/∞/∞ is omitted and we then speak of an M/M/∞ queue. This is not really a queueing system, but a *pure delay* system. The arrival process, the service times sequence, and the waiting room are as in the M/M/1/∞ model, but now there is an infinity of servers, and therefore no queueing, since anyone entering the system finds an idle server just for himself. We shall see that the congestion process is a continuous-time HMC with state space $E = \mathbb{N}$ and infinitesimal generator

$$q_{i,i+1} = \lambda$$
$$q_{i,i-1} = i\mu.$$

EXAMPLE 9.2.7: M/M/1/∞/FIFO QUEUE WITH INSTANTANEOUS FEEDBACK. This is an M/M/1/∞ queue where a customer finishing service either leaves the system with probability $1 - p$ or is immediately recycled with probability p at the end of the waiting line or at the service booth if there is an empty waiting line, with a new independent exponential service request. In this example, the times of service completion of recycled customers do not correspond to a genuine transition of the congestion process.

We shall see that the congestion process of the M/M/1/∞/FIFO queue with instantaneous feedback is a continuous-time HMC with state space $E = \mathbb{N}$ and infinitesimal generator given by

$$q_{i,i+1} = \lambda$$
$$q_{i,i-1} = (1-p)\mu 1_{\{i\geq 1\}}.$$

In the queueing literature, especially when applications to communications and computer networks are considered, a queueing system is sometimes represented by the pictogram of the next figure, where the input arrow represents the arrival stream of *jobs* (customers), the output arrow represents the stream of *completed jobs* (served customers), the circle is a *processor* (the service system), and the stack is a *buffer* (a waiting room, where customers wait for a server to be free).

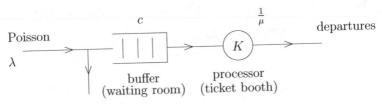

The holy pictogram of queueing theory

This pictogram can be richly adorned. For instance, in the figure, we have a system with K servers, a waiting room of capacity c, λ is the arrival rate of customers, and the derivation shows that the customers finding all K servers busy and a full waiting room are rejected. Also, there is an indication of the average service time, μ^{-1}, and of the fact that the incoming stream is a Poisson process. In this pictogram, one would take it as implicit that the service times sequence is IID and independent of the arrival process, unless otherwise explicitly mentioned. One sometimes also indicates the service discipline (LIFO, FIFO, etc.).

Congestion as a Birth-and-Death Process

As usual in the queueing literature, the rigorous proof that the congestion processes of the queueing systems considered above are continuous-time HMCs with the indicated infinitesimal generators will not be given. We shall be content with a heuristic derivation of the infinitesimal generator.

Consider for instance the queue M/M/K/c, with FIFO discipline. Here c is the capacity of the waiting room, and therefore an arriving customer seeing $K + c$ customers in service is rejected, or routed elsewhere. At a given time t, there are a certain number of exponential random variables "in activity", one for each customer presently receiving service, and one corresponding to the time to the next arrival (which may be refused if the system is full). If $X(t) = i < K$, each one among the i customers is attended by a server, and we have random variables $X_1, \ldots, X_i$ representing the remaining times until completion of service. By the lack of memory property of the exponential random variables, these are IID exponential random variables with mean μ^{-1}. For similar reasons, the time to the next arrival is an exponential random variable X_0 with mean λ^{-1}, independent of $X_1, \ldots, X_i$. Therefore

$$P(X(t + h) = i + 1 \mid X(t) = i) = P(X_0 \leq h, X_1 > h, \ldots, X_i > h) + o(h),$$

taking into account that for infinitesimal h there is at most one of the exponentials $X_0, X_1, \ldots, X_i$ in $(0, h)$. Now, to the first order in h,

$$P(X_0 \leq h, X_1 > h, \ldots, X_i > h) = (1 - e^{\lambda h}) \times e^{\mu h} \times \ldots \times e^{\mu h} \simeq \lambda h.$$

Recall the meaning of $q_{i,i+1}$,

$$P(X(t + h) = i + 1 \mid X(t) = i) = q_{i,i+1}h + o(h).$$

Therefore, $q_{i,i+1} = \lambda$.

Similarly, if $1 \leq i \leq K$,

$$P(X(t + h) = i - 1 \mid X(t) = i) = P(X_0 > h, \inf\{X_1, \ldots, X_i\} \leq h) + o(h).$$

But $\inf \{X_1, \ldots, X_i\}$ is an exponential random variable with mean $(i\mu)^{-1}$, and therefore

$$P(X_0 > h, \inf \{X_1, \ldots, X_i\} \leq h) \simeq i\mu h.$$

Comparing with

$$P(X(t+h) = i - 1 \mid X(t) = i) = q_{i,i-1}h + o(h)$$

we have that $q_{i,i-1} = i\mu$.

If $i \geq K$ there are exactly K customers in service, and the same heuristics as above give $q_{i,i-1} = K\mu$.

In the case $i = 0$, there is of course only one active exponential, namely X_0, the remaining time to the next arrival. We have $q_{i,i-1} = 0$, and $q_{i,i+1} = \lambda$.

In the case $i = c + K$, there are no possible arrivals in the system, since a potential customer will not be accepted. Therefore in this case, $q_{i,i+1} = 0$.

In summary,

$$q_{i,i+1} = \lambda 1_{\{i < c+K\}},$$
$$q_{i,i-1} = \inf(i, K) \mu.$$

Let us now apply similar heuristics to the M/M/1/∞ queue with feedback.

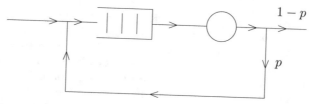

A queue with instantaneous feedback

Suppose $X(t) = i$ and define X_0, X_1 as above for the M/M/K/c queue, taking into account that there is only one server. We have, as the reader may check, $q_{i,i+1} = \lambda$. The treatment differs in the computation of $q_{i,i-1}$ for $i > 0$. We have to introduce a $\{0, 1\}$-valued random variable Z, independent of X_0, X_1, with $P(Z = 1) = p$, indicating whether the next customer to complete service will be recycled ($Z = 1$) or not ($Z = 0$). Here

$$\begin{aligned}
q_{i,i-1}h &\simeq P(X(t+h) = i - 1 \mid X(t) = i) \\
&\simeq P(X_0 > h, X_1 \leq h, Z = 0) \\
&= P(X_0 > h, X_1 \leq h)P(Z = 0) \\
&\simeq \mu h \times (1 - p),
\end{aligned}$$

where the $\simeq$ indicates equality up to the first order in h. Therefore $q_{i,i-1} = \mu(1 - p) 1\{i \geq 1\}$.

The infinitesimal generator of the congestion process of an M/M/1/∞ queue is independent of the service discipline if we consider only service disciplines such that the server works at full speed, equal to 1, whenever there is at least one customer in the system. This general fact is due to the lack of memory of exponential variables, and it can be checked using the same heuristics as in the FIFO case for disciplines without interruption of service, such as LIFO non-preemptive and RANDOM (the next customer is chosen randomly in the waiting room). In the case of interruption of service, with resumed service (that is, the service already provided is not abolished by the preemption) the heuristics will give again the same results, because the remaining time of an interrupted exponential random variable is an exponential random variable with the same mean. If after preemption the service is started from the beginning, again nothing changes as far as the infinitesimal generator is concerned.

The congestion processes (counting the customers in the system) of the queueing systems that we considered above are special cases of birth-and-death processes with an infinitesimal generator of the form

$$
\mathbf{A} = \begin{pmatrix}
-\lambda_0 & \lambda_0 & 0 & 0 & \cdots \\
\mu_1 & -(\lambda_1 + \mu_1) & \lambda_1 & 0 & \cdots \\
0 & \mu_2 & -(\lambda_2 + \mu_2) & \lambda_2 & \cdots \\
\vdots & \vdots & \vdots & \vdots &
\end{pmatrix}
$$

where the state space is $E = \mathbb{N}$, or $E = \{0, 1, \ldots, N\}$ for finite N. For all the queueing processes we consider in the present section, $\lambda_i > 0$ for all $i \in E$, except $i = N$ when $E = \{0, 1, \ldots, N\}$, and $\mu_i > 0$ for all $i \in E$, except $i = 0$. These conditions obviously guarantee irreducibility. We recall the general result obtained in Example 7.4.13, noting first that all the chains arising from the examples above are regular (Exercise 9.4.1). The ergodicity condition is:

$$
\sum_{\substack{i \in E \\ i \geq 1}} \prod_{n=1}^{i} \frac{\lambda_{n-1}}{\mu_n} < \infty, \tag{9.12}
$$

in which case, for $i \geq 1$,

$$
\pi(i) = \pi(0) \prod_{n=1}^{i} \frac{\lambda_{n-1}}{\mu_n}, \tag{9.13}
$$

where $\pi(0)$ is obtained by normalization. The ergodicity condition is, of course, automatically satisfied when the state space is finite.

EXAMPLE 9.2.8: M/M/1/∞. For the M/M/1/∞ queue, $E = \mathbb{N}$, $\lambda_i \equiv \lambda > 0$, and $\mu_i = \mu > 0$ for all $i \geq 1$. The ergodicity condition reads $\sum_{i \geq 1} \left(\frac{\lambda}{\mu}\right)^i < \infty$, that is,

$$
\rho := \frac{\lambda}{\mu} < 1.
$$

This condition is natural, since $\rho = \lambda \mathrm{E}[\sigma_1]$, the *traffic intensity*, is the average rate of work entering the system per unit time, and should not exceed the maximal speed of service, equal to 1. The solution of the balance equation is

$$
\pi(i) = (1 - \rho)\rho^i.
$$

EXAMPLE 9.2.9: M/M/K/0, OR ERLANG QUEUE. For the M/M/K/0 queue (Erlang queue) $E = \{0, \ldots, K\}$ and the solution of the balance equations is

$$\pi(i) = \frac{\rho^i/i!}{\sum_{n=0}^{K} \rho^n/n!}$$

for $0 \le i \le K$. In particular,

$$\pi(K) = \frac{\rho^K/K!}{\sum_{n=0}^{K} \rho^n/n!}$$

is the blocking probability, the probability of finding all channels busy. The corresponding formula is called the *Erlang blocking formula*. The distribution π is called a Poisson distribution truncated at K, since

$$\pi(i) = P(Z = i \mid Z \le K),$$

where Z is a Poisson random variable with mean ρ.

EXAMPLE 9.2.10: M/M/∞. For the M/M/∞ queue, the ergodicity condition (9.12) is $\sum_{i=1}^{\infty} \frac{\rho^i}{i!} < \infty$ and is always satisfied. The stationary distribution is the Poisson distribution with mean equal to the traffic intensity ρ,

$$\pi(i) = e^{-\rho} \frac{\rho^i}{i!}.$$

EXAMPLE 9.2.11: M/M/K/∞. For the M/M/K/∞ queue, the ergodicity condition (9.12) is satisfied only if

$$\rho := \frac{\lambda}{\mu} < K.$$

It says that the average incoming work per unit time cannot exceed the maximal service speed, which is K when all servers are busy. The stationary distribution is then, for $0 \le i \le K$,

$$\pi(i) = \pi(0) \frac{\rho^i}{i!},$$

and when $i \ge K$,

$$\pi(i) = \pi(0) \frac{\rho^K}{s!} \left(\frac{\rho}{K} \right)^{i-K},$$

where

$$\pi(0)^{-1} = \sum_{i=0}^{s-1} \frac{\rho^i}{i!} + \frac{\rho^K}{K!} \frac{1}{1 - \rho/s}.$$

In this system, the probability of waiting is the probability of entering the system when the K servers are busy, that is, $\pi(\ge K) := \sum_{i \ge K} \pi(i)$. One obtains the *Erlang waiting formula*

$$\pi(\ge K) = \frac{\frac{\rho^s}{K!} \frac{1}{1-\rho/K}}{\sum_{i=0}^{K-1} \frac{\rho^i}{i!} + \frac{\rho^K}{K!} \frac{1}{1-\rho/K}}.$$

9.2.2 Markovian Networks

Burke's Output Theorem

A seminal result of Queueing theory says in particular that the point process of the departure times of an $M/M/1/\infty$ queue in equilibrium is a homogeneous Poisson process with the same intensity as the arrival process. In fact, the result extends to more general birth and death processes.

Consider a birth-and-death process on $\mathbb{N}$ with birth parameters of the form

$$\lambda_i \equiv \lambda,$$

and suppose that $\mu_i > 0$ for all $i \geq 1$. The corresponding chain is irreducible, and we shall assume that it is ergodic. At equilibrium, this chain is reversible, since the detailed balance equations $\pi(i+1)q_{i+1,i} = \pi(i)q_{i,i+1}$, that is, in the present case, $\pi(i+1)\mu_{i+1} = \pi(i)\lambda$ are satisfied, as can be readily checked using the explicit form of the stationary distribution.

The upward transitions are due to arriving customers and the downward transitions are due to departing customers. Suppose that the queue is in equilibrium, and therefore the reversed process has the same distribution as the direct process. When time is reversed, the point process of departures becomes the point process of arrivals, and therefore, in view of the reversibility property, the reversed process of departures is an HPP. Now, the probabilistic nature of an HPP does not change when time is reversed. Therefore, the departure process is a Poisson process.

Also, since in direct time, for any time $t \geq 0$, the state $X(t)$ is independent of the future at time t of the arrival process (Markov property of Poisson processes), it follows from reversibility that $X(t)$ is independent of the past at time t of the departure process.

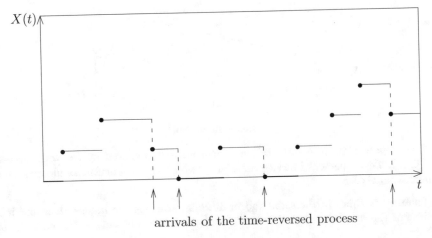

arrivals of the time-reversed process

Proof of Burke's theorem by reversibility

EXAMPLE 9.2.12: TANDEM OF $M/M/1/\infty$ QUEUES. Consider two $M/M/1/\infty$ queues in succession, in the sense that the departure process from the first one is the arrival process into the other. The arrival process, that is, the point process of arrivals into the first queue is an HPP of intensity λ. The service sequences are independent IID exponential sequences with mean $1/\mu_1$ and $1/\mu_2$, respectively. Both service sequences are independent of the arrival process. Let $X_1(t)$ and $X_2(t)$ be the number of customers in the first and the second queueing system, respectively. At time t, $X_2(t)$ depends on the departure process of the first queue before time t and on the second service sequence. Since, by Burke's theorem and the independence property of the service sequence, the latter are independent of $X_1(t)$, it follows that $X_1(t)$ and $X_2(t)$ are independent. Both queues are in isolation $M/M/1/\infty$ with traffic intensities $\rho_1 = \lambda/\mu_1$ and $\rho_2 = \lambda/\mu_2$, and therefore a necessary and sufficient condition of ergodicity of the continuous-time HMC $\{(X_1(t), X_2(t))\}_{t \geq 0}$ is $\lambda < \inf(\mu_1, \mu_2)$, and its stationary distribution has the *product form*

$$\pi(n_1, n_2) = \pi_1(n_1)\pi_2(n_2) = (1 - \rho_1)\rho_1^{n_1}(1 - \rho_2)\rho_1^{n_2}.$$

Jackson Networks

A Jackson network is an open network of interconnected queues. There are K *stations*, and each station has a $1/\infty$ service system, that is, a unique server working at unit speed and an infinite waiting room. There are two types of customers queueing at a given station, (1) those which are fed-back, that is, who have received service in another or the same station and have been rerouted to the given station for more service, and (2) those who are entering the network for the first time (see the figure below).

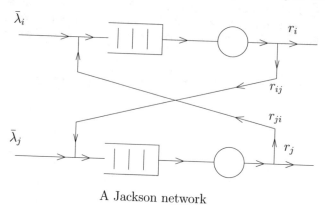

A Jackson network

The *exogenous arrivals* into station i form an HPP, denoted by $\bar{N}_i$, with intensity $\bar{\lambda}_i \in [0, \infty)$. The sequence of service times at station i are exponential random variables of mean $1/\mu_i \in (0, \infty)$.

The service times in the same and in different stations are independent and independent of the exogenous input HPPs $\bar{N}_i$, and the latter HPPs are independent of one another.

The routing is of the Bernoulli type. Each customer just completing service in station i tosses a $(K + 1)$-faced die with probabilities $r_{i,1}, \ldots, r_{i,K}, r_i$ with the effect that the customer is sent to station j with probability r_{ij} or leaves the system with probability $r_i = 1 - \sum_{j=1}^{K} r_{ij}$. The matrix

$$\mathbf{R} = \{r_{ij}\}_{1 \leq i,j \leq K}$$

is the *routing matrix*. The successive tosses of the routing dice of all stations are independent, and independent of the exogenous arrival processes and of all the service times.

This is the original Jackson model, which can be enriched by the introduction of service speeds. If there are n_i customers in station i, the server works at speed $\varphi_i(n_i)$, where $\varphi_i(0) = 0$ and $\varphi_i(n_i) > 0$ for all $n_i \geq 1$.

Let $X_i(t)$ be the number of customers in station i at time t, and define

$$X(t) = (X_1(t), \ldots, X_k(t)).$$

It can be proved that the process $\{X(t)\}_{t \geq 0}$ is a regular jump HMC with state space $E = \mathbb{N}^K$ and with infinitesimal generator $\mathbf{A} = \{q_{n,n'}\}$, where all the non-null off-diagonal terms are

$$
\begin{aligned}
q_{n,n+e_i} &= \bar{\lambda}_i, \\
q_{n,n-e_i} &= \mu_i \varphi_i(n_i) r_i \mathbf{1}_{\{n_i > 0\}}, \\
q_{n,n-e_i+e_j} &= \mu_i \varphi_i(n_i) r_{ij} \mathbf{1}_{\{n_i > 0\}},
\end{aligned}
$$

where $n = (n_1, \ldots, n_K) \in E = \mathbb{N}^K$, and e_i is the ith vector of the canonical basis of $\mathbb{R}^K$.

This generator is independent of the service strategy—LIFO, FIFO, or processor-sharing.

The form of the infinitesimal generator can be obtained from a full description of the network as a Poisson system. However, we shall be content with heuristic arguments. For instance, the expression for $q_{n,n-e_i+e_j}$ follows from the intuitive considerations below. If the state at time t is n, a transfer from station i to station j requires that the exponential random variable with mean $1/\mu_i$ representing the required service of the customer being served at station i at time t be terminated between times t and $t + h$ (probability $\mu_i \varphi_i(n_i)h$ up to the first order in h), and that the corresponding customer be routed to station j (probability r_{ij}).

We shall assume that the chain $\{X(t)\}_{t \geq 0}$ is irreducible. This is the case when

(α) for all $j \in [1, K]$, there exist $i, i_1, \ldots, i_m \in \{1, 2, \ldots, K\}$ such that

$$\bar{\lambda}_i r_{ii_1} r_{i_1 i_2} \cdots r_{i_m j} > 0,$$

and

(β) for all $j \in \{1, 2, \ldots, K\}$, there exist $j_1, j_2, \ldots, j_\ell, k \in \{1, 2, \ldots, K\}$ such that

$$r_{jj_1} r_{j_1 j_2} \cdots r_{j_\ell k} r_k > 0.$$

Recalling that the $\mu_i > 0$ for all stations, condition (α) tells us that any station is *exogenously supplied*, and (β) tells that any station has an *outlet*.

Consider now the $(K+1) \times (K+1)$ matrix

$$\tilde{\mathbf{R}} = \begin{pmatrix} r_{11} & r_{12} & \cdots & r_{1K} & r_1 \\ r_{21} & r_{22} & \cdots & r_{2K} & r_2 \\ \vdots & \vdots & & \vdots & \vdots \\ r_{K1} & r_{K2} & \cdots & r_{KK} & r_K \\ 0 & 0 & \cdots & 0 & 1 \end{pmatrix}.$$

It can be interpreted as the transition matrix of an HMC on the finite state space $\{1, \ldots, K, K+1\}$. Condition (β) implies that state $K+1$ is absorbing and the states $1, \ldots, K$ are transient. In particular, for all i, j in the transient set, $\lim_{n \uparrow \infty} p_{ij}(n) = 0$, that is, $\lim_{n \uparrow \infty} \mathbf{R}^n = 0$. Observing that

$$(I + \mathbf{R} + \cdots + \mathbf{R}^n)(I - \mathbf{R}) = I - \mathbf{R}^{n+1}$$

we see that $(1 - \mathbf{R})^{-1}$ exists and is equal to $(\sum_{n=0}^{\infty} \mathbf{R}^n)$. Therefore the solution of the system of equations

$$\lambda_i = \bar{\lambda}_i + \sum_{j=1}^{K} \lambda_j r_{ji}, \tag{9.14}$$

that is, with obvious notations, $\lambda^T = \bar{\lambda}^T + \lambda^T \mathbf{R}$, has a unique solution

$$\lambda^T = \bar{\lambda}^T (1 - \mathbf{R})^{-1},$$

where the latter expression shows that this solution is non-negative. Equations (9.14) are called the *traffic equations*, because they give a necessary relation between the average numbers of customers λ_i entering station i in steady state if the network is ergodic. Indeed, λ_i is equal to the exogenous rate of arrivals $\bar{\lambda}_i$ plus the sum of all average rates of transfer from other stations. From station j, the corresponding rate is $\alpha_j r_{ji}$, where α_j is the average rate of customers finishing service in station j. But at equilibrium $\alpha_j = \lambda_j$, since the average number of customers in station j remains constant, whence the traffic equations (9.14).

This heuristic argument is not needed, and in particular, one need not attempt to identify λ_i in (9.14) as the average incoming arrival rate in station i (Exercise 15.4.3). This identification is possible if equilibrium is guaranteed. However, even in the non-ergodic cases, the traffic equations have a unique solution.

We now consider the case where the service speeds of all the servers are equal to 1, and refer the reader to Exercise 9.4.5 for the general case.

The global balance equations are

$$\pi(n) \left\{ \sum_{i=1}^{K} (\lambda_i + \mu_i r_i 1_{\{n_i > 0\}}) \right\} = \sum_{i=1}^{K} \pi(n - e_i) \bar{\lambda}_i 1_{\{n_i > 0\}} + \sum_{i=1}^{K} \pi(n + e_i) \mu_i r_i$$
$$+ \sum_{i=1}^{K} \sum_{j=1}^{K} \pi(n + e_i - e_j) \mu_i r_{ij} 1_{\{n_j > 0\}}.$$

It turns out that if the solution of the traffic equation satisfies

$$\rho_i = \frac{\lambda_i}{\mu_i} < 1 \tag{9.15}$$

for all $i \in \{1, 2, \ldots, K\}$, then the network is ergodic, and its stationary distribution is given by

$$\pi(n) = \prod_{i=1}^{K} \pi_i(n_i), \tag{9.16}$$

where π_i is the stationary distribution of an M/M/1/∞ queue with traffic intensity ρ_i,

$$\pi_i(n_i) = \rho_i^{n_i}(1 - \rho_i). \tag{9.17}$$

To prove this, we shall apply the reversal test (Theorem 7.4.16). We define the generator $\tilde{\mathbf{A}}$ on $E = \mathbb{N}^K$ by

$$\pi(n)\tilde{q}_{n,n'} = \pi(n')q_{n',n}$$

and check that

$$\sum \tilde{q}_{n,n'} = q_n.$$

Here

$$q_n = \sum_{i=1}^{K} \lambda_i + \mu_i 1_{\{n_i>0\}}.$$

The generator $\tilde{\mathbf{A}}$ is given by

$$
\begin{aligned}
\pi(n)\tilde{q}_{n,n+e_i} &= \pi(n+e_i)q_{n+e_i,n} = \pi(n+e_i)\mu_i r_i, \\
\pi(n)\tilde{q}_{n,n-e_i} &= \pi(n-e_i)q_{n-e_i,n} = \pi(n-e_i)\bar{\lambda}_i 1_{\{n_i>0\}}, \\
\pi(n)\tilde{q}_{n,n+e_i-e_j} &= \pi(n+e_i-e_j)q_{n+e_i-e_j,n} = \pi(n+e_i-e_j)\mu_i r_{ij}1_{\{n_j>0\}},
\end{aligned}
$$

and therefore, taking into account the specific form of $\pi(n)$ given by (9.16) and (9.17),

$$
\begin{aligned}
\tilde{q}_{n,n+e_i} &= \rho_i\mu_i r_i = \lambda_i r_i, \\
\tilde{q}_{n,n-e_i} &= \frac{\bar{\lambda}_i}{\rho_i}1_{\{n_i>0\}}, \\
\tilde{q}_{n,n+e_i-e_j} &= \frac{\rho_i}{\rho_j}\mu_i r_{ij}1_{\{n_j>0\}} = \frac{1}{\rho_j}\lambda_i r_{ij}1_{\{n_j>0\}}.
\end{aligned}
$$

We have to verify that

$$\sum_{i=1}^{K}(\bar{\lambda}_i + \mu_i 1_{\{r_i>0\}}) = \sum_{i=1}^{K}\left(\lambda_i r_i + \frac{\bar{\lambda}_i}{\rho_i}1_{\{n_i>0\}} + \frac{1}{\rho_i}\left(\sum_{i=1}^{K}\lambda_j r_{ji}\right)1_{\{n_i>0\}}\right).$$

By the traffic equation, $\sum_{i=1}^{K}\lambda_j r_{ji} = \lambda_i - \bar{\lambda}_i$, and therefore the right-hand side of the previous equality is

$$\sum_{i=1}^{K}\left(\lambda_i r_i + \frac{1}{\rho_i}\lambda_i 1_{\{n_i>0\}}\right) = \sum_{i=1}^{K}(\lambda_i r_i + \mu_i 1_{\{n_i>0\}}),$$

and it remains to check that

$$\sum_{i=1}^{K}\bar{\lambda}_i = \sum_{i=1}^{K}\lambda_i r_i.$$

For this we need only sum the traffic equations.

Gordon–Newell Networks

A closed Jackson network, also called a *Gordon–Newell network*, is one for which

$$\bar{\lambda}_i = 0, \; r_i = 0 \tag{9.18}$$

for all $i, 1 \leq i \leq K$. In other words, there is no inlet and no outlet, and therefore the number of customers in the network remains constant, and we shall call it N. The state space is

$$E = \{(n_1, \ldots, n_K) \in \mathbb{N}^K, \sum_{i=1}^K n_i = N\}.$$

The traffic equations are now

$$\lambda^T = \lambda^T \mathbf{R}, \tag{9.19}$$

and since $\mathbf{R}$ is a stochastic matrix in this case, which we shall assume irreducible, it has an infinity of solutions, all multiples of the same vector, which is the stationary distribution of $\mathbf{R}$.

It is true that the vector of average traffics through the stations is a solution of (9.19). We do not know which one, in contrast with the open network, where the solution of the traffic equation is unique. Nevertheless, if we take any positive solution, we see by inspection that for all $n \in E$,

$$\pi(n) = \frac{1}{G(N, K)} \prod_{i=1}^K \rho_i^{n_i}, \tag{9.20}$$

where $\rho_i = \lambda_i / \mu_i$ is a stationary solution. Here $G(N, K)$ is the normalizing factor

$$G(N, K) = \sum_{\substack{n \in \mathbb{N}^K \\ n_1 + \cdots + n_K = N}} \prod_{i=1}^K \rho_i^{n_i}. \tag{9.21}$$

Note that under the irreducibility assumption for the routing matrix $\mathbf{R}$, the chain $\{X(t)\}_{t \geq 0}$ is itself irreducible, and since the state space is finite, it is positive recurrent, with a unique stationary distribution. In particular, for closed networks, there is no ergodicity condition.

The problem with closed Jackson networks resides in the computation of the normalizing constant $G(N, K)$. Brute force summation via (9.21) is practically infeasible for large populations and/or large networks. Instead, we proceed as follows.

Define $G(j, \ell)$ to be the coefficient of z^j in the power series development of

$$g_\ell(z) = \prod_{i=1}^\ell \frac{1}{1 - \rho_i z} = \prod_{i=1}^\ell \left(\sum_{n_i=0}^\infty \rho_i^{n_i} z^{n_i} \right).$$

The normalizing factor is indeed equal to $G(N, K)$. Since

$$g_\ell(z) = g_{\ell-1}(z) + \rho_\ell z g_\ell(z),$$

we find the recurrence equation

$$G(j, \ell) = G(j, \ell - 1) + \rho_\ell G(j - 1, \ell) \tag{9.22}$$

with the initial conditions

$$G(j, 1) = \rho_1^j \ (j \geq 0),$$
$$(9.23)$$

$$G(0, \ell) = 1 \ (\ell \geq 1).$$

This provides an algorithm for computing the normalizing factor.

It is also of interest to be able to compute the *utilization* of server i, defined by

$$U_i(N, K) = P(X_i(t) > 0).$$
$$(9.24)$$

This can be useful, for instance in computing the average *throughput* from station i to station j, that is, the average number of customers transferred from station i to station j in one unit of time:

$$d_{ij}(N, K) = \mu_i U_i(N, K) r_{ij}.$$
$$(9.25)$$

Since

$$U_i(N, K) = \sum_{\substack{n_1 + \cdots + n_K = N \\ n_i > 0}} \pi(n) = \frac{1}{G(N, K)} \sum_{\substack{n_1 + \cdots + n_K = N \\ n_i > 0}} \prod_{j=1}^{K} \rho_j^{n_j},$$

we see that $G(N, K)U_i(N, K)$ is the coefficient of z^N in

$$\tilde{g}_{N,i}(z) = \left(\prod_{\substack{j=1 \\ j \neq i}}^{K} \left(\sum_{n_j=0}^{\infty} \rho_j^{n_j} z^{n_j} \right) \right) \left(\sum_{n_i=1}^{\infty} \rho_i^{n_i} z^{n_i} \right).$$

Now,

$$\tilde{g}_{N,i}(z) = g_N(z) \rho_i z,$$

and therefore $G(N, K)U_i(N, K) = \rho_i G(N-1, K)$, that is,

$$U_i(N, K) = \rho_i \frac{G(N, K)}{G(N-1, K)}.$$
$$(9.26)$$

9.3 Non-exponential Models

Markovian models are easy to handle, but they are not always adequate. The Poissonian assumption for the input is often justified due to some kind of the law of rare events applies. However, the exponential assumption for service times is usually unrealistic. Note however that there are situations where the service time distribution is arbitrary. The typical case is the Erlang queue, in which the substitution of an arbitrary service distribution with the same mean as the original exponential distribution leaves unaltered the stationary distribution of the number of busy lines at an arbitrary time. In particular, Erlang's blocking formula remains valid, and this robustness explains the success of this formula, which was widely applied in the design of early telephone switches. This lucky phenomenon is called *insensitivity*, and is not so rare. For instance, the M/M/∞ queue and the M/M/1/∞/LIFO *preemptive* queue are insensitive. We shall not treat this here, except for two particular cases: the M/GI/∞ delay system of the next subsection, and the M/GI/1/∞ LIFO preemptive queueing system of Example 17.4.3.

9.3.1   M/GI/∞

Let

(a) A be an HPP on $\mathbb{R}$ with intensity λ (the arrival process) and time sequence $\{T_n\}_{n\in\mathbb{Z}}$, with the usual convention $T_0 \leq 0 < T_1$, and let

(b) $\{\sigma_n\}_{n\in\mathbb{Z}}$ be a sequence of random vectors taking their values in $\mathbb{R}_+$ with the probability distribution Q. Suppose, moreover, that

(c) $\{\sigma_n\}_{n\in\mathbb{Z}}$ and N are independent.

Here, T_n is interpreted as the time of arrival of the n-th customer and σ_n as the service time requested by this customer.

Define the point process $\tilde{N}$ on $\mathbb{R} \times \mathbb{R}_+$ by

$$\tilde{N}(C) = \sum_{n\in\mathbb{Z}} 1_C(T_n, \sigma_n)$$

for all $A \in \mathcal{B} \otimes \mathcal{B}_+$. By Theorem 8.3.1, $\tilde{N}$ is a simple Poisson process with mean measure

$$\tilde{\nu}(\mathrm{d}t \times \mathrm{d}z) = \lambda \mathrm{d}t \times Q(\mathrm{d}z).$$

In the M/GI/∞ model,[1] a customer arriving at time T_n is immediately served, and therefore departs from the "system" at time $T_n + \sigma_n$. The number $X(t)$ of customers present in the system at time t is therefore given by the formula

$$X(t) = \sum_{n\in\mathbb{Z}} 1_{(-\infty,t]}(T_n) 1_{(t,\infty)}(T_n + \sigma_n). \tag{9.27}$$

(Indeed, the n-th customer is in the system at time t if and only if she arrived at time $T_n \leq t$ and departed at time $T_n + \sigma_n > t$.)

The *departure* process is, by definition, the point process D of departure times, that is,

$$D(C) = \sum_{n\in\mathbb{Z}} 1_C(T_n + \sigma_n). \tag{9.28}$$

Theorem 9.3.1

A. For all $t \in \mathbb{R}$, $X(t)$ is a Poisson random variable with mean $\lambda \mathrm{E}\,[\sigma_1]$.

B. For all $t, \tau \in \mathbb{R}$, $\tau \geq 0$,

$$\mathrm{cov}\,(X(t), X(t+\tau)) = \lambda \int_\tau^\infty P(\sigma_1 > y)\,\mathrm{d}y.$$

C. The departure process is a homogeneous Poisson process with intensity λ.

---

[1]The ∞ represents the number of servers. This model is sometimes called a "queueing" system, although in reality there is no queueing, since customers are served immediately upon arrival and without interruption. It is in fact a "pure delay" system.

Proof. A. Let

$$C(t) := \{(s,z);\ s \le t,\ s+z > t\} \subset \mathbb{R} \times \mathbb{R}_+ \quad (t \in \mathbb{R}_+).$$

Then, by (9.27),

$$X(t) = \tilde{N}(C(t)).$$

In particular, $X(t)$ is a Poisson random variable whose mean $\tilde{\nu}(C(t))$ is computed as follows, using Fubini's theorem:

$$
\begin{aligned}
\tilde{\nu}(C(t)) &= \int_{\mathbb{R}} \int_{\mathbb{R}_+} 1_{\{s+z>t\}} 1_{\{s\le t\}} \tilde{\nu}(\mathrm{d}s \times \mathrm{d}z) \\
&= \int_{\mathbb{R}} \int_{\mathbb{R}_+} 1_{\{s+z>t\}} 1_{\{s\le t\}} \lambda \, \mathrm{d}s \times Q(\mathrm{d}z) \\
&= \int_{\mathbb{R}} \left(\int_{\mathbb{R}_+} 1_{\{s+z>t\}} Q(\mathrm{d}z) \right) 1_{\{s\le t\}} \lambda \, \mathrm{d}s \\
&= \lambda \int_{-\infty}^{t} Q((t-s,+\infty)) \, \mathrm{d}s \\
&= \lambda \int_{0}^{\infty} Q((s,+\infty)) \, \mathrm{d}s = \lambda \int_{0}^{\infty} P(Z_1 > s) \mathrm{d}s = \lambda \mathrm{E}[Z_1].
\end{aligned}
$$

Therefore $X(t)$ is Poisson, with mean $\lambda \mathrm{E}[Z_1]$.

B. Take $\tau \ge 0$ and define the sets $C = C(t) \cap C(t+\tau)$, $A = C(t) - C$ and $B = C(t+\tau) - C$. In particular, $X(t) = \tilde{N}(A) + \tilde{N}(C)$ and $X(t+\tau) = \tilde{N}(B) + \tilde{N}(C)$. We have therefore

$$
\begin{aligned}
\mathrm{cov}(X(t), X(t+\tau)) &= \mathrm{cov}(\tilde{N}(A) + \tilde{N}(C), \tilde{N}(B) + \tilde{N}(C)) \\
&= \mathrm{cov}(\tilde{N}(A), \tilde{N}(B)) + \mathrm{cov}(\tilde{N}(A), \tilde{N}(C)) \\
&\quad + \mathrm{cov}(\tilde{N}(C), \tilde{N}(B)) + \mathrm{cov}(\tilde{N}(C), \tilde{N}(C)) \\
&= \mathrm{Var}(\tilde{N}(C)) = \mathrm{E}\left[\tilde{N}(C)\right],
\end{aligned}
$$

where we have taken into account that $\tilde{N}(A), \tilde{N}(B), \tilde{N}(C)$ are independent Poisson random variables of variances $\mathrm{E}[\tilde{N}(A)], \mathrm{E}[\tilde{N}(B)], \mathrm{E}[\tilde{N}(C)]$. Therefore

$$\mathrm{cov}\,(X(t), X(t+\tau)) = \tilde{\nu}(C).$$

Now, by Fubini, with computations similar to those performed in detail in Part A above:

$$
\begin{aligned}
\tilde{\nu}(C) &= \lambda \int_{0}^{\infty} P(\sigma_1 > y + \tau) \, \mathrm{d}y \\
&= \lambda \int_{\tau}^{\infty} P(\sigma_1 > y) \, \mathrm{d}y,
\end{aligned}
$$

and therefore finally

$$\mathrm{cov}\,(X(t), X(t+\tau)) = \lambda \int_{\tau}^{\infty} P(\sigma_1 > y) \, \mathrm{d}y.$$

C. From (9.28), we have $D(C) = \tilde{N}(\tilde{C})$, where $\tilde{C} := \{(t,\sigma) \in \mathbb{R} \times \mathbb{R}_+ ;\ t+\sigma \in C\}$, and that if $C_1, \dots, C_K$ are disjoint measurable sets of $\mathbb{R}$, then $\tilde{C}_1, \dots, \tilde{C}_K$ are disjoint measurable sets of $\mathbb{R} \times \mathbb{R}_+$. In particular, $D(C_1) = \tilde{N}(\tilde{C}_1), \dots, D(C_K) = \tilde{N}(\tilde{C}_K)$ are independent Poisson random variables with means $\tilde{\nu}(\tilde{C}_1), \dots, \tilde{\nu}(\tilde{C}_K)$. Since $\tilde{\nu}(\tilde{C}) = \lambda \times \ell(C)$ (the usual Fubini computation), we see that the departure process is a homogeneous Poisson process of intensity λ. $\qquad \square$

9.3.2   M/GI/1/∞/FIFO

This queue is exactly like an M/M/1/∞/FIFO queue, only with a general distribution for the IID service times sequence,

$$P(\sigma_1 \leq x) = G(x).$$

The corresponding congestion process $\{X(t)\}_{t\geq 0}$ is no longer Markovian. Instead, we consider the process $\{X_n\}_{n\geq 0}$, where

$$X_n = X(\tau_n)$$

and τ_n is the nth departure time (see Figure (i)), $\tau_0 = 0$. Since the congestion process is taken as right-continuous, we see that X_n is the number of customers that customer n leaves behind him when he has completed service.

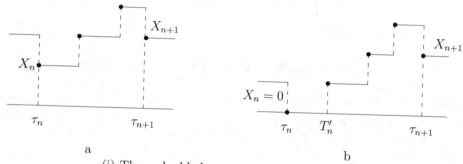

(i) The embedded process at departure times

A glance at Figure (i) reveals that

$$X_{n+1} = (X_n - 1)^+ + Z_{n+1}, \tag{9.29}$$

where Z_{n+1} is the number of customers arriving in the interval $(\alpha_n, \alpha_n + \sigma_{n+1}]$, where $\alpha_n = \tau_n$ if $X_n > 0$ and $\alpha_n = T'_n$ if $X_n = 0$.

The time α_n is an $\mathcal{F}_t^A \vee \mathcal{F}_n^\sigma$-stopping time (where A is the arrival Poisson process, and $\mathcal{F}_n^\sigma$ is the σ-field generated by the first n service requests $\sigma_1, \ldots, \sigma_n$). Also, the $(n+1)$st service request σ_{n+1} is independent of the arrival process before α_n and of the previous requests $(\sigma_1, \ldots, \sigma_n)$. In particular, by the strong Markov property of HPPs,

$$Z_{n+1} = A(\alpha_n, \alpha_n + \sigma_{n+1}]$$

is independent of $X_0, \ldots, X_n$, and $\{Z_n\}_{n\geq 1}$ is an IID sequence distributed as

$$Z = N(0, \sigma],$$

where N is an HPP of intensity λ, independent of the random variable σ with the same distribution as a typical service time, that is, of CDF $G(x)$. Therefore, $\{X_n\}_{n\geq 0}$ is a discrete-time HMC, called the *embedded chain at departure times*. Recall (Example 6.1.7) that its transition matrix is

$$\begin{pmatrix} a_0 & a_1 & a_2 & a_3 & \cdots \\ a_0 & a_1 & a_2 & a_3 & \cdots \\ 0 & a_0 & a_1 & a_2 & \cdots \\ 0 & 0 & a_0 & a_1 & \cdots \\ \vdots & \vdots & \vdots & \vdots & \end{pmatrix},$$

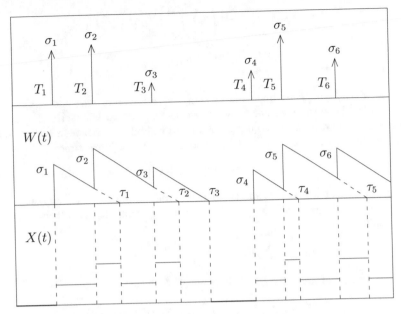

(ii) Arrival process, FIFO workload, and congestion processes

where

$$a_i = P(Z_{n+1} = i)$$
$$= P(Z = i)$$
$$= P(N(0, \sigma] = i)$$
$$= \int_0^\infty P(N((0, t]) = i) \, \mathrm{d}G(t),$$

that is

$$a_i = \int_0^\infty \mathrm{e}^{-\lambda t} \frac{(\lambda t)^i}{i!} \, \mathrm{d}G(t). \tag{9.30}$$

This chain has already been studied. In this particular situation, it is irreducible (see Exercise 9.4.13) and a sufficient condition of positive recurrence is $\mathrm{E}[Z] < 1$, whereas if $\mathrm{E}[Z] > 1$, it is transient. Also, the generating function of the stationary distribution, in the positive recurrent case, is (see Example 6.2.14)

$$\sum_{i=0}^\infty \pi(i) z^i = (1 - \mathrm{E}[Z]) \frac{(z - 1) g_Z(z)}{z - g_Z(z)}. \tag{9.31}$$

In view of (9.30),

$$\mathrm{E}[Z] = \int_0^\infty \left(\sum_{i=0}^\infty i \mathrm{e}^{-\lambda t} \frac{(\lambda t)^i}{i!} \right) \mathrm{d}G(t)$$
$$= \int_0^\infty \lambda t \, \mathrm{d}G(t) = \lambda \mathrm{E}[\sigma]$$

and

$$g_Z(z) = \int_0^\infty e^{-\lambda t} \left(\sum_{i=0}^\infty \int_0^\infty \frac{(\lambda t)^i}{i!} z^i \right) dG(t)$$

$$= \int_0^\infty e^{-\lambda t(1-z)} \, dG(t).$$

In the M/GI/1/∞/FIFO queueing system, the number of customers $X_n = X(\tau_n)$ left behind by the n-th customer when he leaves the system is exactly the number of customers arriving during her sojourn, that is, in the interval between her arrival time T_n and her departure time $\tau_n = T_n + V_n$, where V_n is her sojourn time:

$$X_n = A(T_n, T_n + V_n]. \tag{9.32}$$

Invoking the strong Markov property of HPPs and noting that V_n depends only on $(\sigma_1, \ldots, \sigma_n)$ and the past of the arrival process A at time T_n, it follows from (9.32) that at equilibrium,

$$\sum_{i=0}^\infty \pi(i) z^i = E\left[z^{N((0,V_n])} \right],$$

where N is an HPP of intensity λ independent of V_n. Now,

$$E\left[z^{N((0,V_n])} \right] = \int_0^\infty E\left[z^{N((0,v])} \right] dF_{V_n}(v)$$

$$= \int_0^\infty e^{\lambda v(z-1)} dF_{V_n}(v),$$

where $F_{V_n}(v)$ is the CDF of V_n. This shows in particular that the distribution of V_n is independent of n, and that calling V any random variable with the same distribution as V_n, we have

$$\sum_{i=0}^\infty \pi(i) z^i = \int e^{\lambda s(z-1)} dF_V(s). \tag{9.33}$$

In summary: at equilibrium, X_n is distributed as

$$X = N(V),$$

where N is an HPP of intensity λ, and V is a random variable independent of N with the distribution of the stationary sojourn time.

9.3.3   GI/M/1/∞/FIFO

The GI/M/1/∞/FIFO queue is of the same nature as the M/M/1/∞/FIFO queue, except that now the arrival process is not Poissonian, but renewal. The interarrival times form an IID sequence with cumulative distribution function $F(x)$ and mean λ^{-1}. The service times are exponential with mean μ^{-1}.

As in the M/GI/1/∞/FIFO queue, the congestion process $\{X(t)\}_{t \geq 0}$ is *not* Markovian, but the system is amenable to Markovian analysis. Indeed, if we define

$$X_n = X(T_n-),$$

the number of customers in the system seen upon arrival by the nth customer, we have the relation

$$X_{n+1} = (X_n + 1 - Z_{n+1})^+,$$

where a typical Z_{n+1} is the number of departures in the interval $(T_n, T_{n+1}]$. It can be shown (see Exercise 9.4.11) that $\{Z_n\}_{n\geq 1}$ is IID and independent of X_0, and is distributed as $Z = N(0, \tau]$, where τ and N are independent, N is an HPP with intensity μ and the CDF of τ is F. Therefore $\{X_n\}_{n\geq 0}$ is an HMC, and, writing

$$P(Z_{n+1} = k) = \int_0^\infty e^{-\mu t} \frac{(\mu t)^k}{k!} dF(t) := b_k,$$

the ith row of the transition matrix $\mathbf{P}$ is therefore

$$\left(1 - \sum_{k=0}^i b_k, \ b_i, \ b_{i-1}, \ldots, \ b_0, \ 0, \ 0, \ldots\right).$$

Since $b_k > 0$ for all $k \geq 0$, $\mathbf{P}$ is irreducible and aperiodic. We first observe that

$$\sum_{k=0}^\infty k b_k = \rho^{-1},$$

where

$$\rho = \frac{\lambda}{\mu}$$

is the traffic intensity. Indeed,

$$\sum_{k=0}^\infty k b_k = \sum_{k=0}^\infty k \int_0^\infty e^{-\mu t} \frac{(\mu t)^k}{k!} dF(t)$$

$$= \int_0^\infty \sum_{k=0}^\infty k e^{-\mu t} \frac{(\mu t)^k}{k!} dF(t)$$

$$= \int_0^\infty \mu \, dF(t) = \mu \mathrm{E}[T_{n+1} - T_n] = \frac{\mu}{\lambda}.$$

We show that if $\rho < 1$, the chain is positive recurrent. For this it suffices to prove the existence of a stationary distribution. We make the educated guess that π has the form

$$\pi(i) = \xi^i (1 - \xi) \tag{9.34}$$

for some $\xi \in (0, 1)$. To verify this guess, we must find $\xi \in (0, 1)$ such that for all $i \geq 1$,

$$\sum_{j=i-1}^\infty \xi^{j-i+1} b_{j-i+1} = \xi \tag{9.35}$$

and

$$\sum_{j=0}^\infty \left(\sum_{k=j+1}^\infty b_k\right) \xi^j = 1, \tag{9.36}$$

since these equations are the balance equations when π is given by (9.34).

Equations (9.35) all reduce to

$$\xi = g_Z(\xi), \tag{9.37}$$

where

$$g_Z(\xi) = \sum_{k=0}^\infty b_k \xi^k \tag{9.38}$$

is the generating function of Z_1. Since all the b_k's are positive, there is a unique solution ξ_0 of (9.37) in $(0,1)$ if and only if

$$g_Z'(1) = \sum_{k=1}^{\infty} k b_k > 1,$$

that is, if and only if $\rho < 1$. We must verify (9.36). The left-hand side equals

$$\sum_{k=1}^{\infty} \sum_{j=0}^{k-1} b_k \xi^j = \sum_{k=1}^{\infty} b_k \left(\frac{1-\xi^k}{1-\xi} \right) = \frac{1}{1-\xi} \left(1 - b_0 - \sum_{k=1}^{\infty} b_k \xi^k \right),$$

and in view of (9.37), this equals

$$\frac{1}{1-\xi}(1 - b_0 - (\xi - b_0)) = 1.$$

If $\rho > 1$ the chain is transient (see Exercise 9.4.12).

An arriving customer finds $X_n = X(T_n-)$ customers in front of him, and therefore, in the FIFO discipline, his *waiting* time W_n (the time between his arrival time and the time when he starts receiving service) is the time needed for the server to take care of the X_n customers present at time T_n-. In particular,

$$W_n = \sum_{j=1}^{X_n} Y_j,$$

where the Y_j are IID exponentials of mean μ^{-1} and common characteristic function

$$E[e^{iuY}] = \frac{\mu}{\mu - iu},$$

and are independent of X_n. Also,

$$P(X_n = k) = (1 - \xi_0) \xi_0^k.$$

Therefore, in steady state,

$$
\begin{aligned}
E[e^{iuW_n}] &= E\left[\sum_{k=0}^{\infty} e^{iu \sum_{j=1}^{k} Y_j} 1_{\{X_n=k\}} \right] \\
&= \sum_{k=0}^{\infty} E[e^{iuY}]^k P(X_n = k) \\
&= \sum_{k=0}^{\infty} \left(\frac{\mu}{\mu - iu} \right)^k (1 - \xi_0) \xi_0^k.
\end{aligned}
$$

We find for the characteristic function of the stationary waiting time

$$(1 - \xi_0) + \xi_0 \left(\frac{\mu(1 - \xi_0)}{\mu(1 - \xi_0) + iu} \right).$$

This is the characteristic function of a random variable that is null with probability $1 - \xi_0$ and exponential of mean $[\mu(1 - \xi_0)]^{-1}$ with probability ξ_0.

Complementary reading

[Kelly, 1979] and [Serfozo, 1999] for networks of queues and [Mazumdar, 2013] for the performance evaluation of queueing systems.

9.4 Exercises

Exercise 9.4.1. REGULARITY OF QUEUEING MODELS
Show that all the HMCs arising from the queueing models of Subsections 9.2.1 and 9.2.2 are regular jump processes.

Exercise 9.4.2. M/M/1/∞/PS
In this M/M/1/∞ queue, each customer present at time t receives service at speed $\frac{1}{X(t)}$ (in the context of computer science, "PS" stands for Processor Sharing"). What are the infinitesimal generator, the ergodicity condition, and the stationary distribution of the congestion process?

Exercise 9.4.3. THE M/M/1/∞ QUEUE
Let A and $\overline{D}$ be two independent HPPs with respective intensities λ and μ, and let $X(0)$ be an integer-valued random variable independent of the above point processes. Define for all $t \geq 0$

$$X(t) = X(0) + A(0, t] - \int_{(0,t]} 1_{\{X(s-)>0\}} \, d\overline{D}(s).$$

Show that $\{X(t)\}_{t \geq 0}$ is the congestion process of an M/M/1/∞ queue with arrival rate λ and service times with exponential distribution of mean μ.

Exercise 9.4.4. M/M/1∞ IN HEAVY TRAFFIC
Consider the M/M/1∞ with traffic intensity ρ in equilibrium, and let $X_\rho = X_\rho(0)$ be the congestion process at time 0 (recall that the congestion process at any time $t \geq 0$ has a distribution independent of t since the queue is assumed to be in equilibrium). Show that the random variable $(1 - \rho)X_\rho$ converges in distribution to an exponential random variable of mean 1.

Exercise 9.4.5. A GENERALIZATION OF JACKSON'S NETWORK
The following modification of the basic Jackson network is considered. For all i, $1 \leq i \leq K$, the server at station i has a speed of service $\varphi_i(n_i)$ when there are n_i customers present in station i, where $\varphi_i(k) > 0$ for all $k \geq 1$ and $\varphi_i(0) = 0$. The new infinitesimal generator is obtained from the standard one by replacing μ_i by $\mu_i \varphi(n_i)$. Check this and show that if for all i, $1 \leq i \leq K$,

$$A_i := 1 + \sum_{A_i=1}^{\infty} \left(\frac{\rho_i^{n_i}}{\prod_{k=1}^{n_i} \varphi(k)} \right) < \infty,$$

where $\rho_i = \lambda_i/\mu_i$ and λ_i is the solution of the traffic equation, then the network is ergodic with stationary distribution

$$\pi(n) = \prod_{i=1}^{K} \pi_i(n_i),$$

where

$$\pi_i(n_i) = \frac{1}{A_i} \frac{\rho_i^{n_i}}{\prod_{k=1}^{n_i} \varphi_i(k)}.$$

Exercise 9.4.6. CLOSED JACKSON WITH A SINGLE CUSTOMER
Consider the closed Jackson network of the theory, with $N = 1$ customer. Let $\{Y(t)\}_{t \geq 0}$ be the process giving the position of this customer, that is, $Y(t) = i$ if she is in station i at time t. Show that $\{Y(t)\}_{t \geq 0}$ is a regular jump HMC, irreducible, and give its stationary distribution.

Exercise 9.4.7. ACK
Consider the closed Jackson network below where all service times at different queues have the same (exponential) distribution of mean $\frac{1}{5}$. Compute for $N = 5$ the average time spent by a customer to go from the leftmost point A to the rightmost point B, and the average number of customers passing by A per unit time.

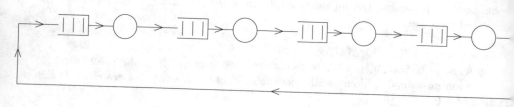

Exercise 9.4.8. SUPPRESSED TRANSITIONS
Let $\{X(t)\}_{t \geq 0}$ be an irreducible positive recurrent HMC with infinitesimal generator $\mathbf{A}$ and stationary distribution π. Suppose that $(\mathbf{A}, \pi)$ is reversible. Let S be a subset of the state space E. Define the infinitesimal generator $\tilde{\mathbf{A}}$ by

$$\tilde{q}_{ij} = \begin{cases} \alpha q_{ij} & \text{if } i \in S, j \in E - S \\ q_{ij} & \text{otherwise} \end{cases}$$

when $i \neq j$. The corresponding HMC is irreducible if $\alpha > 0$. If $\alpha = 0$, the state space will be reduced to S, to maintain irreducibility. Show that the continuous time HMC associated to $\tilde{\mathbf{A}}$ admits the stationary distribution $\tilde{\pi}$ given by

$$\tilde{\pi}_i = \begin{cases} \alpha C \times \pi(i) & \text{if } i \in S, \\ C \times \alpha \pi(i) & \text{if } i \in E - S, \end{cases}$$

with the obvious modification when $\alpha = 0$.

Exercise 9.4.9. LOSS NETWORKS (KELLY'S NETWORKS)
([2]) Consider a telecommunications network with K relays. An incoming call chooses a "route" r among a set $\mathcal{R}$. The network then reserves the set of relays $r_1, \ldots, r_{k(r)}$ corresponding to this route, and processes the call to destination in an exponential time of mean μ_r^{-1}. The incoming calls with route r form a homogeneous Poisson process of intensity λ_r. It is assumed that all the relays are useful, in that they are part of at least one route in $\mathcal{R}$.

---

[2][Kelly, 1979].

(1) The capacity of the system is for the time being assumed to be infinite, that is, the number $X_r(t)$ of calls on route r at time t can take any integer value. All the usual independence hypotheses are made: the processing times and the Poisson processes are independent. Give the stationary distribution of the continuous time HMC $\{X(t)\}_{t \geq 0}$, where $X(t) = (X_r(t), r \in \mathcal{R})$.

(2) The capacity of the system is now restricted as follows. Consider a given pair (a, b) of relays. It represents a "link" in the network. This link has finite capacity C_{ab}. This means that the total number of calls using this link cannot exceed this capacity, with the consequence that an incoming call requiring a route passing through this link will be lost if the link is saturated when it arrives. The process $\{X(t)\}_{t \geq 0}$ therefore has for state space

$$\tilde{E} = \{n = (n_r; r \in \mathcal{R}); \sum_{r \in \mathcal{R}, (a,b) \in r} n_r \leq C_{ab} \text{ for all links } (a, b)\}.$$

What is the stationary distribution of the chain $\{X(t)\}_{t \geq 0}$?

Exercise 9.4.10. M/GI/1/∞/FIFO
Show that the imbedded HMC of an M/GI/1/∞/FIFO is irreducible (as long as the service times are not identically null).

Exercise 9.4.11. $N(0, \tau]$
Prove the statement in Subsection 9.3.3 concerning the sequence $\{Z_n\}_{n \geq 1}$, namely, that it is IID, independent of X_0 and distributed as $Z = N(0, \tau]$, where N is an HPP with intensity μ, τ and N are independent and the CDF of τ is F.

Exercise 9.4.12. GI/M/1/∞
Show that the GI/M/1/∞ queue of Subsection 9.3.3 is transient if $\rho > 1$.

Exercise 9.4.13. THE IMBEDDED HMC OF AN M/GI/1/∞/FIFO QUEUE
Show that the imbedded HMC of an M/GI/1/∞/FIFO is irreducible (as long as the service times are not identically null).

Exercise 9.4.14. CONSTANT SERVICE TIMES MINIMIZE CONGESTION
Show that for a fixed traffic intensity ρ, constant service times minimize average congestion in the $M/GI/1/\infty$ FIFO queue.

Exercise 9.4.15. WORKLOAD OF M/M/1/∞
Show that the stationary distribution of the workload process of an M/M/1/∞ queue with arrival rate λ and mean service time μ^{-1} such that $\rho = \frac{\lambda}{\mu} < 1$ is

$$F_W(x) = 1 - \rho \left(1 - \exp\{-(\mu - \lambda)x\}\right).$$

Chapter 10

Renewal and Regenerative Processes

From the purely analytical point of view, renewal theory is concerned with the renewal equation

$$f(t) = g(t) + \int_{[0,t]} f(t-s)\,\mathrm{d}F(s),$$

where F is the cumulative distribution function of a finite measure on the positive real line. Its main concern is the asymptotic behavior of the solution f (the existence and uniqueness of which is not a real issue under mild conditions, as we shall see). Once embedded in the framework of point processes, renewal theory becomes a fundamental tool of probability theory that is useful in particular in the study of convergence of regenerative processes, a large and important class of stochastic processes that includes the recurrent continuous-time HMCs and the semi-Markov processes.

10.1 Renewal Point processes

10.1.1 The Renewal Measure

Consider an IID sequence $\{S_n\}_{n \geq 1}$ of *non-negative* random variables with common cumulative distribution function

$$F(x) := P(S_n \leq x).$$

This CDF is called *defective* when $F(\infty) := P(S_1 < \infty) < 1$, and *proper* when $F(\infty) = 1$. The uninteresting case where $P(S_1 = 0) = 1$ is henceforth eliminated.

The above sequence is called the *inter-renewal sequence*. The associated *renewal sequence* $\{T_n\}_{n \geq 0}$ is defined by

$$T_n := T_{n-1} + S_n \quad (n \geq 1),$$

where the *initial delay* T_0 is a FINITE non-negative random variable independent of the inter-renewal sequence. When $T_0 = 0$, the renewal sequence is called *undelayed*. Time T_n is called a *renewal time*, or an *event*. The stochastic process

$$N([0,t]) := \sum_{n \geq 0} 1_{\{T_n \leq t\}} \quad (t \geq 0)$$

© Springer Nature Switzerland AG 2020
P. Brémaud, *Probability Theory and Stochastic Processes*, Universitext,
https://doi.org/10.1007/978-3-030-40183-2_10

is the *counting process* of the renewal sequence; $N([0, t])$ counts the number of events in the *closed* interval $[0, t]$. Note that $T_0 \geq 0$ (this convention differs from the usual one, only for this chapter) and that the point at 0 is counted when there is one. Clearly, the random function $t \mapsto N([0, t])$ is almost surely right-continuous and has a limit on the left for each $t > 0$, namely $N[0, t)$.

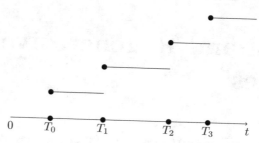

Theorem 10.1.1 *For all $t \geq 0$, $\mathrm{E}[N([0, t])] < \infty$. In particular, almost surely, $N([0, t]) < \infty$ for all $t \geq 0$.*

Proof. It suffices to consider the undelayed case (Exercise 10.5.1). By Markov's inequality,

$$P(T_n \leq t) = P(e^{-T_n} \geq e^{-t}) \leq e^t \mathrm{E}\left[e^{-T_n}\right].$$

But since the renewal sequence is IID,

$$\mathrm{E}\left[e^{-T_n}\right] = \mathrm{E}\left[e^{-\sum_{k=1}^n S_k}\right] = \prod_{k=1}^n \mathrm{E}\left[e^{-S_k}\right] = \alpha^n,$$

with $\alpha = \mathrm{E}\left[e^{-S_1}\right] < 1$ since $P(S_1 = 0) < 1$. Therefore

$$\mathrm{E}\left[N([0, t])\right] - 1 = \mathrm{E}\left[\sum_{n \geq 1} 1_{\{T_n \leq t\}}\right] = \sum_{n \geq 1} \mathrm{E}\left[1_{\{T_n \leq t\}}\right] = \sum_{n \geq 1} P(T_n \leq t) \leq e^t \sum_{n \geq 1} \alpha^n < \infty.$$

$\square$

The *forward recurrence* $\{A(t)\}_{t \geq 0}$ and the *backward recurrence* $\{B(t)\}_{t \geq 0}$ are defined as follows. Both processes are right-continuous with left-hand limits. For $n \geq 0$, they have linear trajectories in (T_n, T_{n+1}) with respective slopes -1 and $+1$, and at a renewal point T_n

$$A(T_n) = T_{n+1} - T_n, \; A(T_{n+1}-) = 0, \; B(T_n) = 0, \; B(T_{n+1}-) = T_{n+1} - T_n.$$

For $0 \leq t < T_0$, $A(t) = T_0 - t$ and $B(t) = t$.

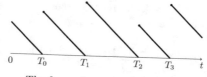

The forward recurrence time process

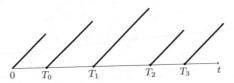

The backward recurrence time process

Definition 10.1.2 *The function $R : \mathbb{R}_+ \to \overline{\mathbb{R}}_+$ defined by*

$$R(t) := \mathrm{E}[N([0,t])],$$

where N is the counting process of the undelayed *renewal sequence, is called the* renewal function.

The renewal function is right-continuous (Exercise 10.5.3) and non-decreasing. Therefore, one can associate with it a unique measure μ_R on $\mathbb{R}_+$ such that $\mu_R([0,a]) = R(a)$. This measure, called the renewal measure, will sometimes be denoted by R, the context avoiding confusion between the measure and its cumulative distribution function. Note that $\mu_R(\{0\}) = R(0) = 1$.

EXAMPLE 10.1.3: THE POISSON PROCESS. Consider the case of exponential inter-event times (that is, $F(t) = 1 - \mathrm{e}^{-\lambda t}$). The undelayed renewal process is then a *homogeneous Poisson process* of intensity λ to which a point at time 0 is added, $R(t) = 1 + \lambda t$.

It will be convenient to express the renewal function in terms of the common cumulative distribution function of the random variables S_n. For this, observe that in the undelayed case $T_n := S_1 + \cdots + S_n$ is the sum of n independent random variables with common cumulative distribution function F and therefore

$$P(T_n \leq t) = F^{*n}(t), \tag{10.1}$$

where F^{*n} is the n-fold convolution of F, defined recursively by

$$F^{*0}(t) = 1_{[0,\infty)}(t), \qquad F^{*n}(t) = \int_{[0,t]} F^{*(n-1)}(t-s)\,\mathrm{d}F(s) \quad (n \geq 1). \tag{10.2}$$

(The role of 0 in the integration over $[0,t]$ is made precise by the following equality:

$$\int_{[0,t]} \varphi(s)\,\mathrm{d}F(s) = \varphi(0)F(0) + \int_{(0,t]} \varphi(s)\,\mathrm{d}F(s).)$$

Writing the renewal function as $\mathrm{E}[N([0,t])] = \mathrm{E}[1 + \sum_{n\geq 1} 1_{\{T_n \leq t\}}] = 1 + \sum_{n\geq 1} P(T_n \leq t)$, we obtain the expression:

$$R(t) = \sum_{n=0}^{\infty} F^{*n}(t). \tag{10.3}$$

Theorem 10.1.4

$$P(S_1 < \infty) < 1 \quad \Longleftrightarrow \quad P(N([0,\infty)) < \infty) = 1 \quad \Longleftrightarrow \quad E[N([0,\infty))] < \infty.$$

Proof. It suffices to prove the theorem in the undelayed case. For all $k \geq 1$,

$$\begin{aligned}
P(N([0,\infty)) = k) &= P(S_1 < \infty, \ldots, S_{k-1} < \infty, S_k = \infty) \\
&= P(S_1 < \infty) \ldots P(S_{k-1} < \infty) P(S_k = \infty) \\
&= F(\infty)^{k-1}(1 - F(\infty)),
\end{aligned}$$

and

$$P(N([0,\infty)) < \infty) = \sum_{k=1}^{\infty} F(\infty)^{k-1}(1 - F(\infty)).$$

In particular, $P(N([0,\infty)) < \infty) = 1$ if $F(\infty) < 1$ and $P(N([0,\infty) < \infty) = 0$ if $F(\infty) = 1$. Also, if $F(\infty) < 1$,

$$E[N([0,\infty))] = \sum_{k=1}^{\infty} k F(\infty)^{k-1}(1 - F(\infty)) = \frac{1}{1 - F(\infty)} < \infty,$$

whereas if $F(\infty) = 1$, $E[N([0,\infty))] = \infty$.

$\square$

A renewal process (delayed or not) is called *recurrent* when $P(S_1 < \infty) = 1$ (F is proper), and *transient* when $P(S_1 < \infty) < 1$ (F is defective).

The following result is called the *elementary renewal theorem*.

Theorem 10.1.5 *We have*

$$\lim_{t\to\infty} \frac{N([0,t])}{t} = \frac{1}{E[S_1]} \quad \text{P-a.s.,} \tag{10.4}$$

and

$$\lim_{t\to\infty} \frac{E[N([0,t])]}{t} = \frac{1}{E[S_1]}. \tag{10.5}$$

Proof. For the proof of (10.4) see Exercise 10.5.2. Proof of (10.5): The transient case follows from the obvious bound $\frac{E[N([0,t])]}{t} \leq \frac{E[N(\infty)]}{t}$, since in this case $E[S_1] = \infty$ and $E[N(\infty)] < \infty$. For the recurrent case, a proof is required (the conditions of the dominated convergence theorem that would guarantee that (10.4) implies (10.5) are not satisfied). However, by Fatou's lemma

$$\liminf_{t\to\infty} E\left[\frac{N([0,t])}{t}\right] \geq E\left[\liminf_{t\to\infty} \frac{N([0,t])}{t}\right] = \frac{1}{E[S_1]}$$

and therefore it suffices to show that $\limsup_{t\to\infty} E[\frac{N([0,t])}{t}] \leq \frac{1}{E[S_1]}$. Define for finite $c > 0$

$$T_0' := T_0, \quad T_1' := T_0' + S_1 \wedge c, \quad T_2' := T_1' + S_2 \wedge c, \quad \ldots$$

where $S_n' := S_n \wedge c$ ($n \geq 1$), and let $N'([0,t]) := \sum_{n \geq 0} 1_{T_n' \leq t}$. Since $N'([0,t]) \geq N([0,t])$ for all $t \geq 0$,

$$\limsup_{t\to\infty} \frac{E[N([0,t])]}{t} \leq \limsup_{t\to\infty} \frac{E[N'([0,t])]}{t}.$$

Observe that $S'_1 + \cdots + S'_{N'([0,t])} \leq t + c$ and therefore, by Wald's lemma (Exercise 10.5.7),

$$E[S'_1]E[N'([0,t])] = E[S'_1 + \cdots + S'_{N'([0,t])}] \leq t + c,$$

so that

$$\limsup_{t \to \infty} \frac{E[N'([0,t])]}{t} \leq \limsup_{t \to \infty} \left(\frac{1}{E[S'_1]} \left(1 + \frac{c}{t} \right) \right) = \frac{1}{E[S'_1]}.$$

Therefore, for all $c > 0$,

$$\limsup_{t \to \infty} \frac{E[N([0,t])]}{t} \leq \frac{1}{E[S_1 \wedge c]}.$$

Since $\lim_{c \uparrow \infty} E[S_1 \wedge c] = E[S_1]$, we finally obtain the desired inequality

$$\limsup_{t \to \infty} \frac{E[N([0,t])]}{t} \leq \frac{1}{E[S_1]}.$$

$\square$

Let $F : \mathbb{R}_+ \to \mathbb{R}_+$ be a *generalized cumulative distribution function* on $\mathbb{R}_+$, that is, $F(x) = c\, G(x)$ where $c > 0$ and G is the cumulative distribution function of a non-negative real random variable that is proper $(G(\infty) = 1)$.

10.1.2 The Renewal Equation

The basic object of renewal theory is the *renewal equation*

$$f = g + f * F,$$

that is, by definition of the convolution symbol $*$,

$$f(t) = g(t) + \int_{[0,t]} f(t-s)\, dF(s) \qquad (t \geq 0), \tag{10.6}$$

where $g : \mathbb{R}_+ \to \mathbb{R}$ is a measurable function called the *data*.

If $F(\infty) = 1$ one refers to the renewal equation as a *proper* renewal equation, or just a renewal equation. The renewal equation is called *defective* if $F(\infty) < 1$ and *excessive* if $F(\infty) > 1$.

The first example features the basic method for obtaining renewal equations.

EXAMPLE 10.1.6: LIFETIME OF A TRANSIENT RENEWAL PROCESS, TAKE 1. The *lifetime* of a renewal sequence is the random variable

$$L := \sup\{T_n \,;\, T_n < \infty\}.$$

Clearly if the renewal process is recurrent, L is almost surely infinite. We therefore consider the transient case, for which L is almost surely finite. In the *undelayed* case, the function $f(t) = P(L > t)$ satisfies the renewal equation

$$f(t) = F(\infty) - F(t) + \int_{[0,t]} f(t-s)\, dF(s).$$

Proof. Define $\widehat{S}_n = S_{n+1}$ $(n \geq 1)$ and let $\{\widehat{T}_n\}_{n \geq 0}$ be the associated undelayed renewal process whose lifetime is denoted by $\widehat{L}$.

Clearly L and $\widehat{L}$ have the same distribution. Also

$$1_{\{L>t\}} = 1_{\{t<T_1\}}1_{\{L>t\}} + 1_{\{t\geq T_1\}}1_{\{L>t\}}.$$

Now, on $\{t \geq T_1\}, \{L > t\} \equiv \{\widehat{L} > t - T_1\}$ and therefore

$$1_{\{L>t\}} = 1_{\{t<T_1\}}1_{\{L>t\}} + 1_{\{t\geq T_1\}}1_{\{\widehat{L}>t-T_1\}}.$$

Taking expectations,

$$P(L > t) = P(L > t, t < T_1) + P(\widehat{L} > t - T_1, T_1 \leq t).$$

Since $\widehat{L}$ and T_1 are independent ($\widehat{L}$ depends only on $S_2, S_3, \ldots$),

$$P(\widehat{L} > t - T_1, T_1 \leq t) = \int_{[0,t]} P(\widehat{L} > t - s)\,\mathrm{d}F(s) = \int_{[0,t]} P(L > t - s)\,\mathrm{d}F(s),$$

where we have used the fact that L and $\widehat{L}$ have the same distribution. Also, $P(L > t, t < T_1) = P(t < T_1, T_1 < \infty) = P(t < T_1 < \infty) = F(\infty) - F(t)$.   □

EXAMPLE 10.1.7: THE RISK MODEL, TAKE 2. This example is a continuation of Example 7.1.2. We find a renewal equation for the *probability of ruin* corresponding to an initial capital u:

$$\Psi(u) := P(u + X(t) < 0 \text{ for some } t > 0).\tag{10.7}$$

This function is non-increasing. It is convenient to work with the non-ruin probability $\Phi(u) := 1 - \Psi(u)$. Of course, $\Phi(u) = 0$ if $u \leq 0$. If the point process N is stationary with average rate λ, the average profit of the insurance company at time t is

$$E[X(t)] = (c - \lambda\mu)t.$$

As expected, insurance companies prefer that $c - \lambda\mu > 0$ or, equivalently, that

$$\rho := \frac{c - \lambda\mu}{\lambda\mu} = \frac{c}{\lambda\mu} - 1 > 0,\tag{10.8}$$

where ρ is the *safety loading*. In fact, by the strong law of large numbers, if the safety loading is negative, the probability of ruin is 1 whatever the initial capital.

If N is a homogeneous Poisson process,

$$\Phi(u) = \Phi(0) + \frac{\lambda}{c}\int_0^u \Phi(u - z)(1 - G(z))\mathrm{d}z.\tag{10.9}$$

Proof. Suppose $u \geq 0$. Since ruin cannot occur at a time $< S_1$,

$$\Phi(u) = \Phi(u + cT_1 - Z_1) = \int_{(0,\infty)} \int_{(0,\infty)} \Phi(u + cs - z)\lambda e^{-\lambda s}\, ds\, dG(z)$$

$$= \int_0^\infty \left(\int_{(0,u+cs]} \Phi(u + cs - z)\, dG(z) \right) \lambda e^{-\lambda s}\, ds$$

$$= \frac{\lambda}{c} e^{\lambda u/c} \int_0^u \left(\int_{(0,x]} \Phi(x - z)\, dG(z) \right) e^{-\lambda x/c}\, dx.$$

The right-hand side is differentiable; differentiation leads to the integro-differential equation

$$\Phi'(u) = \frac{\lambda}{c}\Phi(u) - \frac{\lambda}{c}\int_{(0,u]} \Phi(u - z)\, dG(z). \tag{10.10}$$

Therefore,

$$\Phi(t) - \Phi(0) = \frac{\lambda}{c}\int_0^t \Phi(u)\, du + \frac{\lambda}{c}\int_0^t \int_{(0,u]} \Phi(u - z)\, d(1 - G(z))\, du$$

$$= \frac{\lambda}{c}\int_0^t \Phi(u)\, du + \frac{\lambda}{c}\int_0^t \left[\Phi(0)(1 - G(u)) - \Phi(u) + \int_0^u \Phi'(u - z)(1 - G(z))\, dz \right] du$$

$$= \frac{\lambda}{c}\Phi(0)\int_0^t (1 - G(u))\, du + \frac{\lambda}{c}\int_0^t \left(\int_z^t \Phi'(u - z)\, du \right)(1 - G(z))\, dz$$

$$= \frac{\lambda}{c}\Phi(0)\int_0^t (1 - G(u))\, du + \frac{\lambda}{c}\int_0^t (1 - G(z)(\Phi(t - z) - \Phi(0))\, dz,$$

that is, (10.9). □

Simple arguments (Exercise 10.5.9) show that in the case of positive safety loading, $\Phi(\infty) = 1$. Letting $u \uparrow \infty$ in (10.9), we have by monotone convergence that $\Phi(\infty) = \Phi(0) + \frac{\lambda}{c}\Phi(\infty)$ and therefore the probability of ruin with zero initial capital is

$$\Psi(0) = \frac{\lambda\mu}{c}.$$

From (10.9), when the safety loading is positive,

$$1 - \Psi(u) = 1 - \frac{\lambda\mu}{c} + \frac{\lambda}{c}\int_0^u (1 - \Psi(u - z))(1 - G(z))\, dz$$

$$= 1 - \frac{\lambda}{c}\left(\mu - \int_u^\infty (1 - G(z))dz + \int_0^u \Psi(u - z)(1 - G(z))\, dz \right),$$

that is

$$\Psi(u) = \frac{\lambda}{c}\int_u^\infty (1 - G(z))dz + \frac{\lambda}{c}\int_0^u \Psi(u - z)(1 - G(z))\, dz. \tag{10.11}$$

EXAMPLE 10.1.8: THE LOTKA–VOLTERRA POPULATION MODEL. This model features a population of women. A woman of age a gives birth to girls at the rate $\lambda(a)$ (that is, a woman of age a will have on average $\lambda(a)\, da$ daughters in the infinitesimal time interval $(a, a + da)$ of her lifetime). A woman of age a is alive at time $a + t$ with probability $p(a, t)$. At the origin of time there are $f_0(a)\, da$ women of age between a and $a + da$. The birth rate $f(t)$ at time $t \geq 0$ is the sum of the birth rate $r(t)$ at time t due to women born after time 0 and of the birth rate $g(t)$ due to women born before time 0.

Since women of age a at time 0 are $a + t$ years old at time t,

$$g(t) = \int_0^\infty f_0(a) p(a, t) \lambda(a + t) \, da .$$

Women born at time $t - s \geq 0$ contribute by $f(t - s)p(0, s)\lambda(s)$ to the birth rate at time t and therefore

$$r(t) = \int_0^t f(t - s)p(0, s)\lambda(s) \, ds .$$

Therefore

$$f(t) = g(t) + \int_0^t f(t - s)p(0, s)\lambda(s) \, ds .$$

This is a renewal equation (called *Lotka's equation*) with data g and cumulative distribution function

$$F(t) := \int_0^t p(0, s)\lambda(s) \, ds.$$

Note that

$$F(\infty) = \int_0^\infty p(0, s)\lambda(s) \, ds$$

is the average number of daughters from a given mother's lifetime, that is, the *reproduction rate*.

Theorem 10.1.9 *The renewal function R satisfies the so-called fundamental renewal equation*

$$R = 1 + R * F. \tag{10.12}$$

Proof. By (10.3),

$$R * F = \left(\sum_{n \geq 0} F^{*n} \right) * F = \sum_{n \geq 0} (F^{*n} * F) = \sum_{n \geq 1} F^{*n} = R - F^{*0} = R - 1 .$$

$\square$

The following simple technical result will be needed later on.

Lemma 10.1.10 *For all $t \geq b$, $R([t - b, t]) \leq (1 - F(b))^{-1}$.*

Proof. By (10.12) and for $t \geq b$,

$$1 = R(t) - \int_{[0,t]} F(t - s) \, dR(s) = \int_{[0,t]} (1 - F(t - s)) \, dR(s)$$

$$\geq \int_{[t-b,t]} (1 - F(t - s)) \, dR(s) \geq (1 - F(b)) R([t - b, t]) .$$

$\square$

EXAMPLE 10.1.11: THE ELEMENTARY RENEWAL THEOREM: CORRECTION TERM,
TAKE 1. We have seen that $\lim_{t\uparrow\infty} \frac{R(t)}{t} = \frac{1}{E[S_1]}$. In view of obtaining finer asymptotic results (see Example 10.2.14 below), we study the function

$$f(t) := R(t) - \frac{t}{E[S_1]}.$$

In the case where $E[S_1] < \infty$, f satisfies the renewal equation with data

$$g(t) := \frac{1}{E[S_1]} \int_t^\infty (1 - F(x))\mathrm{d}x.$$

Proof. Let $E[S_1] := m$. By (10.12),

$$
\begin{aligned}
(f * F)(t) &= (R * F)(t) - \frac{1}{m}\int_{[0,t]} (t-s)\,\mathrm{d}F(s) \\
&= R(t) - 1 - \frac{1}{m}\int_{[0,t]} (t-s)\,\mathrm{d}F(s) \\
&= R(t) - \frac{t}{m} - \left\{ \left(1 - \frac{t}{m}\right) + \frac{1}{m}\int_{[0,t]} (t-s)\,\mathrm{d}F(s) \right\} \\
&= f(t) - \frac{1}{m}\int_t^\infty (1 - F(x))\mathrm{d}x,
\end{aligned}
$$

where the last equality is obtained by the following computations. Integration by parts (Theorem 2.3.12) gives

$$t(1 - F(t)) = \int_{[0,t]} (1 - F(s))\,\mathrm{d}s - \int_{[0,t]} s\,\mathrm{d}F(s),$$

and therefore, since $m = \int_0^\infty (1 - F(s))\,\mathrm{d}s$,

$$
\begin{aligned}
\frac{1}{m}\int_t^\infty (1 - F(s))\,\mathrm{d}s &= \frac{1}{m}\int_0^\infty (1 - F(s))\,\mathrm{d}s - \frac{1}{m}\int_0^t (1 - F(s))\,\mathrm{d}s \\
&= 1 - \frac{1}{m}\int_0^t (1 - F(s))\,\mathrm{d}s = 1 - \frac{1}{m}\left(t(1 - F(t)) + \int_{[0,t]} s\,\mathrm{d}F(s) \right) \\
&= 1 - \frac{1}{m}\left(t + \int_{[0,t]} (s-t)\,\mathrm{d}F(s) \right) = 1 - \frac{t}{m} + \frac{1}{m}\int_{[0,t]} (t-s)\,\mathrm{d}F(s).
\end{aligned}
$$

$\square$

Solution of the Renewal Equation

An expression of the solution of the renewal equation in terms of the renewal function is easy to obtain. Recall the following definition: A function $g : \mathbb{R}_+ \to \mathbb{R}$ is called *locally bounded* if for all $a \geq 0$, $\sup_{t\in[0,a]} |g(t)| < \infty$.

Theorem 10.1.12 *If $F(\infty) \leq 1$ and if the measurable data function $g : \mathbb{R}_+ \to \mathbb{R}$ is locally bounded, the renewal equation* (10.6) *admits a unique locally bounded solution $f : \mathbb{R}_+ \to \mathbb{R}$ given by $f = g * R$, that is,*

$$f(t) = \int_{[0,t]} g(t-s)\mathrm{d}R(s). \tag{10.13}$$

Proof. The function $f = g * R$ is indeed locally bounded since g is locally bounded and $R(t)$ is finite for all t. Also

$$f * F = (g * R) * F = g * (R * F) = g * (R - 1) = g * R - g = f - g.$$

Therefore f is indeed a solution of the renewal equation. Let f_1 be another locally bounded solution and let $h := f - f_1$. This is a locally bounded solution which satisfies $h = h * F$. By iteration,

$$h = h * F^{*n}.$$

Therefore, for all $t \geq 0$,

$$|h(t)| \leq \left(\sup_{s \in [0,t]} |h(s)| \right) F^{*n}(t).$$

Since $R(t) = \sum_{n \geq 0} F^{*n}(t) < \infty$, we have $\lim_{n \to \infty} F^{*n}(t) = 0$, which implies in view of the last displayed inequality that $|h(t)| \equiv 0$. $\square$

The first asymptotic result on the solution of the renewal equation concerns the defective case:

Theorem 10.1.13 *If F is defective and if the measurable data function $g : \mathbb{R}_+ \to \mathbb{R}$ is bounded and has a limit $g(\infty) := \lim_{t \to \infty} g(t)$, the unique locally bounded solution of the renewal equation satisfies*

$$\lim_{t \to \infty} f(t) = \frac{g(\infty)}{1 - F(\infty)}.$$

Proof. From previous computations, we have

$$E[N([0, \infty))] = R(\infty) = \frac{1}{1 - F(\infty)},$$

and therefore

$$\frac{g(\infty)}{1 - F(\infty)} = \int_{[0,\infty)} g(\infty) \, dR(s).$$

Also

$$f(t) = \int_{[0,t]} g(t - s) \, dR(s),$$

and therefore

$$f(t) - \frac{g(\infty)}{1 - F(\infty)} = \int_{[0,\infty)} (g(t - s)1_{\{s \leq t\}} - g(\infty)) \, dR(s).$$

The latter integrand is bounded in absolute value by $2 \times \sup |g(t)|$, a finite constant. Considered as a function, a constant is integrable with respect to the renewal measure because, in the defective case, the total mass of the renewal measure is $R(\infty) = E[N([0, \infty))] < \infty$. Now for fixed $s \geq 0$, $\lim_{t \to \infty}(g(t - s)1_{\{s \leq t\}} - g(\infty)) = 0$. Therefore, by dominated convergence, the integral converges to 0 as $t \to \infty$. $\square$

10.1.3   Stationary Renewal Processes

By a proper choice of the initial delay, a renewal process can be made stationary, in a sense to be made precise. Consider a renewal process

$$T_0 = S_0 \,, \ T_1 = S_0 + S_1 \,, \ \dots \,, \ T_n = S_0 + \cdots + S_n$$

where $0 \le S_0 < \infty$. Let G be the cumulative distribution function of the initial delay $S_0 := T_0$ and suppose that $P(S_1 < \infty) = 1$ (the renewal process is proper). As usual, exclude trivialities by imposing the condition $P(S_1 = 0) < 1$. For $t \ge 0$, define

$$S_0(t) := T_{N([0,t])} - t \ \text{and} \ S_n(t) := T_{N([0,t])+n} - T_{N([0,t])+n-1} \ (n \ge 1) . \tag{10.14}$$

In particular, $S_n(0) = S_n$ for all $n \ge 0$. Also observe that $S_0(t) = A(t)$, the forward recurrence time at t.

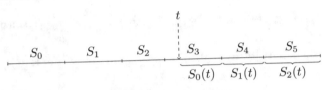

Definition 10.1.14 *The delayed renewal process is called* stationary *if the distribution of the sequence*

$$S_0(t) \,, \ S_1(t) \,, S_2(t) \,, \cdots$$

is independent of time $t \ge 0$.

It turns out that $S_0(t)$ is independent of $\{S_n(t)\}_{n \ge 1}$ and that the latter sequence has the same distribution as $\{S_n\}_{n \ge 1}$ (Exercise 10.5.11). Therefore:

Lemma 10.1.15 *For a delayed renewal process to be stationary it is necessary and sufficient that for all $t \ge 0$, the distribution of $S_0(t) = A(t)$ be the same as that of S_0.*

Lemma 10.1.16 *If the delayed renewal process is stationary, then necessarily $E[S_1] < \infty$ and $E[N([0,t])] = \frac{t}{E[S_1]}$.*

Proof. The measure M on $\mathbb{R}_+$ defined by $M(C) := E[N(C)]$ is translation-invariant and therefore a multiple of the Lebesgue measure (Theorem 2.1.45), that is, $M(C) = K\ell(C)$ for some constant K, which is finite (M is locally finite) and positive (the renewal process is not empty). By the elementary renewal theorem,

$$K = \lim_{t \uparrow \infty} \frac{E[N((0,t])]}{t} = \frac{1}{E[S_1]} \,,$$

and therefore $\frac{1}{E[S_1]} > 0$.   □

Lemma 10.1.17 *If the delayed renewal process is stationary, then necessarily $E[S_1] < \infty$ and the distribution of the initial delay T_0 is*

$$F_0(x) := \frac{1}{E[S_1]} \int_0^x (1 - F(y)) \, dy \,, \tag{10.15}$$

called the stationary forward recurrence time distribution.

Proof. The finiteness of $E[S_1]$ was proved in the previous lemma. For all $u \in \mathbb{R}$, all $t \geq 0$, the following equation

$$e^{iuA(t)} = e^{iuA(0)} + \sum_{n \geq 0} \left(e^{iuA(T_n)} - e^{iuA(T_n-)} \right) 1_{\{T_n \leq t\}} + \int_0^t \frac{d}{ds} e^{iuA(s)} \, ds$$

is obtained by looking at what happens *at* the event times and *between* the event times.[1] Observing that $\frac{d}{ds} e^{iuA(s)} = -iu e^{iuA(s)}$, $A(0) = S_0$, $A(T_n-) = 0$, $A(T_n) = S_{n+1}$, we therefore have

$$e^{iuA(t)} = e^{iuS_0} + \sum_{n \geq 0} \left(e^{iuS_{n+1}} - 1 \right) 1_{\{T_n \leq t\}} - iu \int_0^t e^{iuA(s)} \, ds \,.$$

Therefore, taking into account the independence of S_{n+1} and T_n,

$$E\left[e^{iuA(t)} \right] = E\left[e^{iuS_0} \right] + E\left[e^{iuS_1} - 1 \right] E[N((0, t])] - iu \int_0^t E\left[e^{iuA(s)} \right] \, ds \,.$$

By the assumed stationarity, $E[N((0, t])] = \frac{t}{E[S_1]}$ and $E\left[e^{iuA(t)} \right] = E\left[e^{iuS_0} \right]$, and therefore

$$\frac{E\left[e^{iuS_1} - 1 \right]}{E[S_1]} - iu E\left[e^{iuS_0} \right] = 0 \,,$$

that is,

$$E\left[e^{iuS_0} \right] = \frac{E\left[e^{iuS_1} - 1 \right]}{iu\, E[S_1]} \,.$$

But the right-hand side is the characteristic function of F_0, as the following computation shows:

$$\frac{1}{E[S_1]} \int_0^\infty e^{iux}(1 - F(x)) \, dx = \frac{1}{E[S_1]} \int_0^\infty e^{iux} P(S_1 > x) \, dx$$

$$= \frac{1}{E[S_1]} \int_0^\infty e^{iux} E\left[1_{\{S_1 > x\}} \right] \, dx$$

$$= \frac{1}{E[S_1]} E\left[\int_0^\infty e^{iux} 1_{\{S_1 > x\}} \, dx \right]$$

$$= \frac{1}{E[S_1]} E\left[\int_0^{S_1} e^{iux} \, dx \right] = \frac{1}{E[S_1]} E\left[\frac{e^{iuS_1} - 1}{iu} \right] \,.$$

$\square$

---

[1]Or use the following. Let $f : \mathbb{R}_+ \to \mathbb{R}$ be a right-continuous function with left-hand limits, and with a set of discontinuity times that form a sequence $\{t_n\}_{n \geq 1}$ that is strictly increasing on $\mathbb{R}$. This sequence may be finite, even empty. If $n_0 \in \mathbb{N}$ is the cardinality of this sequence, one conventionally lets $t_n = \infty$ for all $n > n_0$. Suppose in addition that on the intervals $[t_n, t_{n+1})$ that lie in $\mathbb{R}_+$,

$$f(t) = f(t_n) + \int_{t_n}^t f'(s) \, ds \,,$$

for some locally integrable function f' (the derivative). Let now $G : \mathbb{R} \to \mathbb{R}$ be a differentiable function with derivative G'. Then for all $[a, b) \subset \mathbb{R}_+$,

$$G(f(b)) = G(f(a)) + \sum_{t_n \in (a,b)} (G(f(t_n)) - G(f(t_n-)) + \int_a^b f'(s) G'(f(s)) \, ds \,.$$

Theorem 10.1.18 *For a delayed renewal process to be stationary, it is necessary and sufficient that* $\mathrm{E}[S_1] < \infty$ *and that*

$$P(T_0 \leq x) = F_0(x),$$

where F_0 *is the stationary forward recurrence time distribution (10.15).*

Proof. The proof of necessity is contained in Lemmas 10.1.16 and 10.1.17. For sufficiency, we first show that

$$R_{F_0}(t) := \mathrm{E}[N([0,t])] = \frac{t}{\mathrm{E}[S_1]}$$

(the notation emphasizes the role of the initial delay with cumulative distribution function F_0). We have,

$$R_{F_0}(t) = \mathrm{E}\left[\sum_{n \geq 0} 1_{\{T_n \leq t\}}\right] = \sum_{n \geq 0} P(T_n \leq t) = F_0(t) + (F_0 * F)(t) + (F_0 * F^{*2})(t) + \cdots,$$

that is, $R_{F_0} = F_0 + R_{F_0} * F$. Therefore, by Theorem 10.1.12, R_{F_0} is the unique locally bounded solution of the renewal equation $f = F_0 + f * F$. It then suffices to show that $f(t) = \frac{t}{\mathrm{E}[S_1]}$ is indeed a solution. To verify this, observe that for such f,

$$(f * F)(t) = \frac{1}{\mathrm{E}[S_1]} \int_{[0,t]} (t - s)\, \mathrm{d}F(s)$$

and therefore

$$f(t) - (f * F)(t) = \frac{t}{m} - \frac{1}{\mathrm{E}[S_1]} \int_{[0,t]} t\, \mathrm{d}F(s) + \frac{1}{\mathrm{E}[S_1]} \int_{[0,t]} s\, \mathrm{d}F(s)$$

$$= \frac{t}{\mathrm{E}[S_1]}(1 - F(t)) + \frac{1}{\mathrm{E}[S_1]} \int_{[0,t]} s\, \mathrm{d}F(s).$$

It remains to show that the right-hand side of the above equality is $F_0(t)$. Integration by parts does it:

$$F_0(t) = \frac{1}{\mathrm{E}[S_1]} \int_0^t (1 - F(s))\, \mathrm{d}s = \frac{1}{\mathrm{E}[S_1]} \left(t(1 - F(t)) + \int_{[0,t]} s\, \mathrm{d}F(s) \right).$$

Having proved that $\mathrm{E}[N([0,t])] = \frac{t}{\mathrm{E}[S_1]}$, we are almost done. From computations in the proof of Lemma 10.1.17, we extract the identity

$$\mathrm{E}\left[e^{iuA(t)}\right] = \mathrm{E}\left[e^{iuS_0}\right] + \mathrm{E}\left[e^{iuS_1} - 1\right]\frac{t}{\mathrm{E}[S_1]} - iu \int_0^t \mathrm{E}\left[e^{iuA(s)}\right]\, \mathrm{d}s.$$

The function $z(t) := \mathrm{E}\left[e^{iuA(t)}\right]$ is therefore a solution of the ordinary differential equation

$$\frac{\mathrm{d}z}{\mathrm{d}t} = -iuz + \frac{1}{\mathrm{E}[S_1]}\mathrm{E}\left[e^{iuS_1} - 1\right]$$

with initial condition $z(0) = \mathrm{E}\left[e^{iuS_0}\right] = \frac{1}{\mathrm{E}[S_1]}\mathrm{E}\left[e^{iuS_1} - 1\right]$, whose unique solution is

$$\mathrm{E}\left[e^{iuA(t)}\right] = \mathrm{E}\left[e^{iuS_0}\right] = \frac{\mathrm{E}[S_1]}{m}\mathrm{E}\left[e^{iuS_1} - 1\right].$$

Therefore, for all $t \geq 0$, $S_0(t)\ (= A(t))$ has the same distribution as S_0. The conclusion then follows from Lemma 10.1.15. $\qquad\square$

10.2   The Renewal Theorem

Renewal theory deals mainly with the limiting behavior of the solution of the renewal equation. The theory is rather simple when the renewal process is *transient* (Theorem 10.1.13) and becomes more involved in the recurrent case. A very simple example will give the flavor of such a result.

EXAMPLE 10.2.1: THE RENEWAL THEOREM FOR A POISSON PROCESS. If $\{T_n\}_{n \geq 1}$ is a Poisson process of intensity $\lambda > 0$, we know (Theorem 10.1.13) that $R(t) = 1 + \lambda t$. Therefore, the solution of the corresponding renewal equation is, when the data g is non-negative,

$$f(t) = \int_0^t g(t-s)\, R(\mathrm{d}s) = \lambda \int_0^t g(t-s)\mathrm{d}s = \lambda \int_0^t g(s)\mathrm{d}s.$$

Here $\lambda = 1/\mathrm{E}[S_1]$. If we suppose that g is integrable, then

$$\lim_{t \to \infty} f(t) = \frac{\int_0^\infty g(s)\, \mathrm{d}s}{\mathrm{E}[S_1]}. \qquad (\star)$$

Definition 10.2.2 *Let F be the CDF of a non-negative real random variable S_1. Both F and S_1 are called* non-lattice *if there is no strictly positive real a such that $\sum_{k \geq 0} P(S_1 = ka) = 1$.*

10.2.1   The Key Renewal Theorem

It turns out that for non-lattice distributions $(\star)$ is quite general, *modulo* a mild technical assumption on the data g: this function has to be directly Riemann integrable. This is the key renewal theorem (Theorem 10.2.11 below).

Direct Riemann Integrability

Let $g : \mathbb{R}_+ \to \mathbb{R}$ be a *nonnegative* locally bounded function. Define for each $b > 0$ and each $t \geq 0$,

$$\begin{cases} \overline{g}_b(t) & = & \sup\{g(s); nb \leq s < (n+1)b\} \text{ on } [nb, (n+1)b) \\ \underline{g}_b(t) & = & \inf\{g(s); nb \leq s < (n+1)b\} \text{ on } [nb, (n+1)b). \end{cases}$$

The functions $\overline{g}_b$ and $\underline{g}_b$ are finite constants on the intervals $[nb, (n+1)b)$, for all $n \in \mathbb{N}$, and thus Lebesgue integrable[2] on bounded intervals.

Definition 10.2.3 *The function $g \geq 0$ is said to be* Riemann integrable *(Ri) on the bounded interval $[0, a]$ if for some (and then for all) $b > 0$, the integral $\int_0^a \overline{g}_b(t)\mathrm{d}t$ is finite, and*

$$\lim_{b \downarrow 0} \left(\int_0^a \overline{g}_b(t)\mathrm{d}t - \int_0^a \underline{g}_b(t)\mathrm{d}t \right) = 0. \qquad (10.16)$$

---

[2]The theory of the Riemann integral predates that of the Lebesgue integral. We do not follow the historical development in the present treatment of Riemann integrals, and use Lebesgue integration theory – in particular the powerful Lebesgue's dominated convergence theorem– which allows considerably simpler arguments.

(The fact that, when $\overline{g}_b$ is integrable for some $b > 0$, then $\overline{g}_{b'}$ is also integrable for all $b' > 0$, follows from the inequality

$$\overline{g}_{b'}(t) \leq \sum_{k=-n}^{+n} \overline{g}_b(t + kb),$$

where $n = \lceil b'/b \rceil$.)

Theorem 10.2.4 *Let g be a Riemann integrable function on the bounded interval $[0, a]$. Then:*

(i) *g is bounded and almost everywhere continuous on $[0, a]$ and*

(ii) *the limit*

$$\lim_{b \downarrow 0} \int_0^a \overline{g}_b(t) dt$$

exists and is finite. This limit is by definition the R-integral (Riemann integral) of g on $[0, a]$. It is denoted by

$$R\text{-}\int_0^a g(t)\, dt = \lim_{b \downarrow 0} \int_0^a \overline{g}_b(t) dt.$$

It coincides with the Lebesgue integral of g on $[0, a]$.

Proof. (i) Boundedness of g is clear since $\sup_{x \in [0,a]} g(x) \leq b^{-1} \int_0^a \overline{g}_b(t) dt$, which is finite by assumption.

We now show that the set of discontinuity points of g has a null Lebesgue measure. Let $\overline{g}(x) = \limsup_{y \to x} g(y)$, and $\underline{g}(x) = \liminf_{y \to x} g(y)$. Both functions are measurable. In fact, more is true: $\overline{g}$ is upper semi-continuous, that is, for all $A \in \mathbb{R}_+$, the set $\{x : \overline{g}(x) \geq A\}$ is closed, while $\underline{g}$ is lower semi-continuous, that is, for all $A \in \mathbb{R}_+$, the set $\{x : \underline{g}(x) \leq A\}$ is closed. We omit the (easy) proof of these facts.

The set of discontinuity points of g on $[0, a]$, that is, $\{x : \overline{g}(x) > \underline{g}(x)\}$, is therefore measurable. Suppose it is of positive Lebesgue measure. Then, there exists an $\epsilon > 0$ such that the set $\{x : \overline{g}(x) - \underline{g}(x) > \epsilon\}$ is also of positive Lebesgue measure, say δ. Since for all $b > 0$, and since for almost every $t \in [0, a]$

$$\overline{g}_b(t) \geq \overline{g}(t) \geq \underline{g}(t) \geq \underline{g}_b(t),$$

it follows that for all $b > 0$,

$$\int_0^a \overline{g}_b(t) dt - \int_0^a \underline{g}_b(t) dt \geq \epsilon \delta > 0,$$

which contradicts the assumption of Riemann integrability (10.16).

(ii) At a continuity point x of g, it holds that

$$\lim_{b \downarrow 0} \overline{g}_b(x) = \overline{g}(x) = g(x).$$

By (i), this convergence holds almost everywhere. By dominated convergence (with dominating function $\overline{g}_1(x) + \overline{g}_1(x - 1) + \overline{g}_1(x + 1)$),

$$\lim_{b \downarrow 0} \int_0^a \overline{g}_b(t) dt = \int_0^a g(t) dt.$$

$\square$

Definition 10.2.5 *The function $g \geq 0$ is said to be* Riemann integrable *on $[0, \infty)$ if*

$$\lim_{a \uparrow \infty} R\text{-} \int_0^a g(t) \, dt$$

exists and the limit is then, by definition, its Riemann integral on $[0, \infty)$.

Remark 10.2.6 A major result of Riemann's integration theory is that the Riemann integral on the finite interval $[0, a]$ of a function exists if and only if this function is almost everywhere continuous and bounded on this interval, that is, Property (i) in the above proposition is not only necessary, but also sufficient for Riemann-integrability. This result is mainly of theoretical interest and is not needed in this book.

We now turn to the definition of the *direct* Riemann integral. "Direct" means that this integral on $[0, \infty)$ is not defined as a limit of integrals over finite intervals, but "directly" on $[0, \infty)$.

Definition 10.2.7 *The function $g \geq 0$ is said to be* directly Riemann integrable *(dRi) if $\int_0^\infty \bar{g}_b(t) dt < \infty$ for some (and then for all) $b > 0$, and if*

$$\lim_{b \downarrow 0} \left(\int_0^\infty \bar{g}_b(t) dt - \int_0^\infty \underline{g}_b(t) dt \right) = 0.$$

From the definitions, it is clear that for functions vanishing outside a bounded interval, the notions of Riemann integrability and of direct Riemann integrability are the same. Also, the following analog of Theorem 10.2.4 holds for direct Riemann integrability:

Theorem 10.2.8 *Let $g \geq 0$ be dRi. Then:*

(i) g is bounded, and almost everywhere continuous on $\mathbb{R}_+$.

(ii) The limit

$$\lim_{b \downarrow 0} \int_0^\infty \bar{g}_b(t) dt$$

exists and is finite. This limit is, by definition, the dR-integral (direct Riemann integral) of g on $\mathbb{R}_+$, and is denoted by

$$dR\text{-} \int_0^\infty g(t) \, dt = \lim_{b \downarrow 0} \int_0^\infty \bar{g}_b(t) dt.$$

It coincides with the Lebesgue integral of g on $\mathbb{R}_+$.

The proof is identical to that of Theorem 10.2.4.

The following example features a function that is Riemann integrable, but not directly Riemann integrable.

EXAMPLE 10.2.9:   A COUNTEREXAMPLE. Let $\{a_n\}_{n \geq 1}$ and $\{b_n\}_{n \geq 1}$ be sequences of positive real numbers such that $1/2 > a_1 > a_2 > \cdots$, $\sum_{n \geq 1} b_n = \infty$ and $\sum_{n \geq 1} a_n b_n < \infty$. Let g be null outside the union of the intervals $[n - a_n, n + a_n]$, $n \geq 1$, and such that for all $n \geq 1$, $g(n - a_n) = g(n + a_n) = 0$ and $g(n) = b_n$, and g is linear in the intervals $[n - a_n, n]$ and $[n, n + a_n]$. Then, g is Riemann integrable:

$$\text{R-}\int_0^\infty g(t)\,\mathrm{d}t = \sum_{n\geq 1} a_n b_n < \infty.$$

It is however *not* directly Riemann integrable since $\int_0^\infty \bar{g}_b(t)\mathrm{d}t = \infty$ for all $b > 0$.

There exist however a few reassuring results:

Theorem 10.2.10

(a) *If g is directly Riemann integrable, it is Riemann integrable on $[0,\infty)$ and*

$$\text{R-}\int_0^\infty g(t)\mathrm{d}t = \text{dR-}\int_0^\infty g(t)\mathrm{d}t.$$

(b) *Non-negative non-increasing functions are directly Riemann integrable if and only if they are Riemann integrable on $[0,\infty)$.*

(c) *A non-negative function that is Riemann integrable on all finite intervals, and such that $\int_0^\infty \bar{g}_1(t)\mathrm{d}t < \infty$, is directly Riemann integrable. In particular, a non-negative almost everywhere continuous function such that $\int_0^\infty \bar{g}_1(t)\mathrm{d}t < \infty$ is directly Riemann integrable.*

(d) *A non-negative function that is Riemann integrable and bounded above by a directly Riemann-integrable function is directly Riemann integrable.*

Proof.

(a) Since g is directly Riemann integrable,

$$0 = \lim_{b\downarrow 0}\int_0^\infty (\bar{g}_b(t) - \underline{g}_b(t))\mathrm{d}t \geq \lim_{b\downarrow 0}\int_0^a (\bar{g}_b(t) - \underline{g}_b(t))\mathrm{d}t,$$

implying Riemann-integrability on $[0,a]$. For all $a > 0$, recalling that g is (Lebesgue) integrable on $[0,+\infty)$,

$$\left|\text{dR-}\int_0^\infty g(t)\mathrm{d}t - \text{R-}\int_0^a g(t)\mathrm{d}t\right| = \left|\int_0^\infty g(t)\mathrm{d}t - \int_0^a g(t)\mathrm{d}t\right| = \int_a^\infty g(t)\mathrm{d}t.$$

The right-hand side tends to zero as $a \to \infty$ by dominated convergence.

(b) The necessity follows from (a). In view of proving sufficiency, suppose that the non-negative non-increasing function g is Riemann integrable on $[0,+\infty)$. It is in particular (Lebesgue) integrable on $[0,a]$ for all finite $a > 0$, and the integral $\int_0^a g(t)\mathrm{d}t$ admits a finite limit as $a \to \infty$. Therefore, g is (Lebesgue) integrable on $\mathbb{R}_+$, by monotone convergence. Since it is non-increasing, for all $b > 0$

$$\int_0^\infty \bar{g}_b(t)\mathrm{d}t = \sum_{n\geq 0} bg(nb) \leq bg(0) + \int_0^\infty g(t)\mathrm{d}t < \infty.$$

Furthermore,

$$\int_0^\infty \bar{g}_b(t)\mathrm{d}t - \int_0^\infty \underline{g}_b(t)\mathrm{d}t = \sum_{n\geq 0} bg(nb) - \sum_{n>0} bg(nb) = bg(0).$$

The latter term vanishes as $b \to 0$, establishing that g is directly Riemann integrable.

(c) Fix $\varepsilon > 0$ and select $a > 0$ such that $\int_a^\infty \bar{g}_1(t)\mathrm{d}t \le \varepsilon$. For all $b \in (0, 1]$,

$$\underline{g}_b(t) \le \bar{g}_b(t) \le \bar{g}_1(t-1) + \bar{g}_1(t) + \bar{g}_1(t+1).$$

It follows that

$$\int_0^\infty \bar{g}_b(t)\mathrm{d}t - \int_0^\infty \underline{g}_b(t)\mathrm{d}t \le \int_0^{a+1} \left(\bar{g}_b(t)\mathrm{d}t - \underline{g}_b(t) \right) \mathrm{d}t + 3 \int_a^\infty \bar{g}_1(t)\mathrm{d}t$$

$$\le \int_0^{a+1} \left(\bar{g}_b(t)\mathrm{d}t - \underline{g}_b(t) \right) \mathrm{d}t + 3\varepsilon \,.$$

As g is assumed Riemann integrable on $[0, a + 1]$, the rightmost integral tends to zero as $b \to 0$. Since $\varepsilon > 0$ is arbitrary, we conclude that the left-hand side goes to zero as $b \to 0$. Hence, g is directly Riemann integrable.

(d) Follows from (c) since g is Riemann integrable on finite intervals, and calling z the bounding function, we have, since $\bar{g}_1 \le \bar{z}_1$

$$\int_0^\infty \bar{g}_1(t) \, \mathrm{d}t \le \int_0^\infty \bar{z}_1(t) \, \mathrm{d}t,$$

a finite quantity because z is directly Riemann integrable by assumption.   □

The Key Renewal Theorem

The renewal processes considered from now on now are those with a renewal distribution that is non-lattice (Definition 10.2.2).

Theorem 10.2.11 *Let F be a non-lattice distribution function such that $F(\infty) = 1$ (with possibly infinite mean) and let R be the associated renewal function. Then:*

(α) Blackwell's theorem:[3] for all $\tau \ge 0$,

$$\lim_{t \uparrow \infty}\{R(t + \tau) - R(t)\} = \frac{\tau}{\mathrm{E}[S_1]} \,. \tag{10.17}$$

(β) Key renewal theorem: if $g : \mathbb{R}_+ \to \mathbb{R}$ is a non-negative directly Riemann-integrable function,

$$\lim_{t \uparrow \infty}(R * g)(t) = \frac{1}{\mathrm{E}[S_1]} \int_0^\infty g(y) \, \mathrm{d}y \,. \tag{10.18}$$

In fact, (α) and (β) are equivalent.

Remark 10.2.12 Example 10.3.5 below features a spectacular example showing that the direct integrability condition cannot be dispensed with in general. Theorem 10.2.10 above and Theorem 10.3.6 give practical ways to prove direct Riemann integrability.

---

[3][Blackwell, 1948].

Proof.

(α) We shall admit it for the time being. All existing proofs are somewhat technical. A proof, based on the so-called "coupling method", is given later (starting with Theorem 10.2.16) when $E[S_1] < \infty$.

(β) Recall that when g is locally bounded, $f = R * g$ is the *unique* locally bounded solution of the renewal equation

$$f(t) = g(t) + \int_{[0,t]} f(t - s)\, dF(s).$$

STEP 1. Case $g(t) = 1_{[(n-1)b, nb)}(t)$. Then $f(t) = R(t - (n-1)b) - R(t - nb)$, and the result is just Blackwell's theorem.

STEP 2. Case $g(t) = \sum_{n \geq 1} c_n 1_{[(n-1)b, nb)}(t)$, where $c_n \geq 0$, $\sum_{n \geq 1} c_n < \infty$, and b is such that $F(b) < 1$. Then

$$f(t) = \sum_{n \geq 1} c_n (R(t - (n-1)b) - R(t - nb)).$$

By Lemma 10.1.10,

$$\sup_{t \geq 0} (R(t - (n-1)b) - R(t - nb)) \leq (1 - F(b))^{-1} < \infty.$$

In particular, by dominated convergence,

$$\lim_{t \uparrow \infty} \sum_{n \geq 1} c_n (R(t - (n-1)b) - R(t - nb)) = \sum_{n \geq 1} c_n \frac{b}{E[S_1]} = \frac{1}{E[S_1]} \int_0^\infty g(y)\, dy.$$

STEP 3. If g is directly Riemann integrable, the functions $\bar{g}_b$ and $\underline{g}_b$ previously defined are of the type considered in Step 2 since $\int \underline{g}_b \leq \int \bar{g}_b < \infty$. But $\underline{g}_b \leq g \leq \bar{g}_b$ and therefore

$$\frac{1}{E[S_1]} \int \underline{g}_b(s)\, ds = \lim_{t \uparrow \infty} \int \underline{g}_b(t - s)\, R(ds) \leq \liminf_{t \uparrow \infty} \int g(t - s) R(ds)$$

$$\leq \limsup_{t \uparrow \infty} \int g(t - s)\, R(ds)$$

$$\leq \lim_{t \uparrow \infty} \int \bar{g}_b(t - s)\, R(ds) = \frac{1}{E[S_1]} \int \bar{g}_b(s)\, ds.$$

The result follows by letting b tend to 0.

We showed that (α) implies (β). The converse implication follows by choosing $g(t) := 1_{[0,\tau]}(t)$. $\square$

Here is a frequently encountered example of a directly Riemann-integrable function:

EXAMPLE 10.2.13: TAIL DISTRIBUTION OF AN INTEGRABLE VARIABLE. Let F be the CDF of an integrable non-negative random variable S_1. Then $1 - F$ is directly Riemann integrable. This follows from (b) of Theorem 10.2.10 and $\int_0^\infty (1 - F(t))\, dt = E[S_1] < \infty$.

EXAMPLE 10.2.14: ELEMENTARY RENEWAL THEOREM: CORRECTION TERM, TAKE 2. In the recurrent case, we know (Theorem 10.1.5) that

$$\lim_{t\to\infty} \frac{R(t)}{t} = \frac{1}{E[S_1]}.$$

In order to obtain more information on the asymptotic behavior of the renewal function, we shall study the behavior of

$$f(t) := R(t) - \frac{t}{E[S_1]}$$

as t goes to ∞ in the non-lattice case when S_1 has finite first and second moments. Calling σ^2 the common variance of the inter-renewal times and letting $m := E[S_1]$, we have

$$\lim_{t\uparrow\infty} \left\{ R(t) - \frac{t}{E[S_1]} \right\} = \frac{1}{2} \frac{E[S_1]^2 + \sigma^2}{E[S_1]^2}. \tag{10.19}$$

Proof. Recall from Example 10.1.11 that f satisfies the renewal equation $f = g + F * f$ with data

$$g(t) = \frac{1}{E[S_1]} \int_t^\infty (1 - F(x))\,\mathrm{d}x\,.$$

The function g is of the form $1 - F_0$ where F_0 is the CDF of a non-negative variable that is integrable. It is therefore directly Riemann integrable (Example 10.2.13). The key renewal theorem then gives

$$\lim_{t\to\infty} \left(R(t) - \frac{t}{m} \right) = \frac{1}{m} \int_0^\infty g(s)\,\mathrm{d}s.$$

But

$$\frac{1}{m} \int_0^\infty g(s)\,\mathrm{d}s = \frac{1}{m^2} \int_0^\infty \left(\int_s^\infty (1 - F(x))\,\mathrm{d}x \right) \mathrm{d}s$$
$$= \frac{1}{m^2} \int_0^\infty \left(\int_0^x (1 - F(x))\,\mathrm{d}s \right) \mathrm{d}x$$
$$= \frac{1}{m^2} \int_0^\infty x(1 - F(x))\,\mathrm{d}x = \frac{1}{m^2}\frac{1}{2} \int_0^\infty x^2 \mathrm{d}F(x),$$

hence the result. (Proof of the last equality:

$$\frac{1}{2}E\left[X^2\right] = E\left[\int_0^X x\,\mathrm{d}x\right] = E\left[\int_0^\infty 1_{\{x<X\}} x\,\mathrm{d}x\right] = \int_0^\infty E\left[1_{\{x<X\}}\right] x\,\mathrm{d}x\,.)$$

□

Renewal Reward Processes

Consider the undelayed renewal process

$$0 = T_0 < T_1 < \cdots < T_n < \cdots$$

where

$$T_{n+1} = T_n + S_{n+1} \quad (n \geq 0),$$

for an IID sequence $\{S_n\}_{n\geq 1}$ of finite positive random variables. Let $\{Y_n\}_{n\geq 1}$ be an IID sequence of integrable random variables, not necessarily independent of the renewal sequence, where Y_n can be interpreted as a "reward" obtained at time n (it could be negative, hence the quotation marks).

At time $t > 0$ the accumulated rewards amount to

$$C(t) := \sum_{j=1}^{N((0,t])} Y_j \,.$$

This defines the (regenerative) *reward process*. The *reward function* is defined by

$$c(t) := \mathrm{E}\left[C(t)\right] \,.$$

The next result is an extension of Theorem 10.1.5.

Theorem 10.2.15

$$\lim_{t \uparrow \infty} \frac{C(t)}{t} = \frac{\mathrm{E}\left[Y_1\right]}{\mathrm{E}\left[S_1\right]}, \tag{10.20}$$

$$\lim_{t \uparrow \infty} \frac{c(t)}{t} = \frac{\mathrm{E}\left[Y_1\right]}{\mathrm{E}\left[S_1\right]}. \tag{10.21}$$

Proof. Proof of (10.20). By the strong law of large numbers and (10.4),

$$\frac{C(t)}{t} = \frac{\sum_{j=1}^{N((0,t])} Y_j}{N((0,t])} \frac{N((0,t])}{t} \to \frac{\mathrm{E}\left[Y_1\right]}{\mathrm{E}\left[S_1\right]}.$$

Proof of (10.21). This requires a little more work, as was the case for the proof of (10.5). Write (and watch the parentheses and the brackets)

$$c(t) = \mathrm{E}\left[\sum_{j=1}^{N([0,t))} Y_j\right] - \mathrm{E}\left[Y_{N([0,t])}\right] \,.$$

Note that $N([0,t])$ is a stopping time for the sequence $\{(S_n, Y_n)\}_{n\geq 1}$ and therefore, by Wald's lemma,

$$\mathrm{E}\left[\sum_{j=1}^{N([0,t])}\right] = \mathrm{E}\left[N([0,t])\right] \mathrm{E}\left[Y_1\right] \,,$$

so that

$$\frac{c(t)}{t} = \frac{\mathrm{E}\left[N([0,t])\right]}{t} \mathrm{E}\left[Y_1\right] - \frac{\mathrm{E}\left[Y_{N([0,t])}\right]}{t} \,.$$

By (10.4) of Theorem 10.1.5, $\frac{\mathrm{E}[N([0,t])]}{t} \to \frac{1}{\mathrm{E}[S_1]}$, and therefore it remains to prove that

$$\frac{\mathrm{E}\left[Y_{N([0,t])}\right]}{t} \to 0 \,.$$

By conditioning on S_1, we find that $f(t) := \mathrm{E}\left[N([0,t])\right]$ satisfies the renewal equation

$$f(t) = g(t) + \int_{[0,t]} f(t - s) \, \mathrm{d}F(s) \,,$$

where $g(t) = \mathrm{E}\,[Y_1 1_{t<S_1}]$. This function is bounded by $\mathrm{E}\,[|Y_1|]$ and tends to 0 as $t \to \infty$. In particular, for any $\varepsilon > 0$, there exists a T_ε such that $|g(t)| \le \varepsilon$ when $t \ge T_\varepsilon$. The solution of the above renewal equation is

$$f(t) = g(t) + \int_{[0,t]} g(t-s)\,\mathrm{d}R(s)\,,$$

where R is the renewal function. Therefore, for $t \ge T_\varepsilon$,

$$\left|\frac{c(t)}{t}\right| \le \frac{1}{t}\left\{|g(t)| + \int_{[0,t-T_\varepsilon]} |g(t-s)|\,\mathrm{d}R(s) + \int_{(t-T_\varepsilon,t]} |g(t-s)|\,\mathrm{d}R(s)\right\}$$
$$\le \frac{1}{t}\left\{\varepsilon + \varepsilon R(t-T_\varepsilon) + \mathrm{E}\,[Y_1]\,(R(t)-R(t-T_\varepsilon))\right\}\,.$$

Only the middle term of the right-hand side does not tend to 0. It tends to $\frac{\varepsilon}{\mathrm{E}[S_1]}$, but since ε is an arbitrary positive number, we are done. $\qquad\square$

10.2.2 The Coupling Proof of Blackwell's Theorem

We now proceed to the proof of Blackwell's theorem in the non-lattice case and under the additional condition $\mathrm{E}[S_1] < \infty$.

Lemma 10.2.16 *Consider a renewal process with initial delay T_0 (almost surely finite) distributed according to G. Let*

$$F_0(x) = \mu^{-1}\int_0^x (1 - F(y))\,\mathrm{d}y \tag{10.22}$$

and

$$R_G(t) = \mathrm{E}\left[\sum_{n\ge 0} 1_{\{T_n \le t\}}\right]\,.$$

(i) For all $a > 0$, $\lim_{t\uparrow\infty}(R_G(t+a) - R_G(t)) = \mu^{-1}a$ (Blackwell's theorem).

(ii) For all $x \ge 0$, $\lim_{t\uparrow\infty} P(B(t) \le x) = F_0(x)$.

(iii) For all $x \ge 0$, $\lim_{t\uparrow\infty} P(A(t) \le x) = F_0(x)$.

Proof.

(ii) $\Leftrightarrow$ (iii). Just observe that for $t \ge 0$, $x \ge 0$,

$$P(A(t) \le x) = P(N[0, t+x] - N([0,t]) \ge 1)$$
$$= P(B(t+x) \le x).$$

(i) $\Rightarrow$ (ii). When $T_0 \equiv 0$, the function $t \to P(B(t) \le x)$ satisfies a renewal equation with data $g(t) = (1 - F(t))1_{[0,x]}(t)$. This function is directly Riemann integrable, and therefore by the key renewal theorem (a consequence of Blackwell's theorem)

$$\lim_{t\uparrow\infty} P(B(t) \le x) = \mu^{-1}\int_0^\infty g(t)\mathrm{d}t = F_0(x).$$

The case of a non-null and proper initial delay follows by the usual argument (see the proof of Theorem 10.3.4).

(iii) $\Rightarrow$ (i). Define $G_t(x) = P(A(t) \leq x)$ and observe that

$$R_G(t + a) - R_G(t) = (G_t * R)(a)$$
$$= \int_0^a R(a - s)G_t(ds)$$
$$= \int_0^a G_t(a - s)R(ds)$$

and that since by hypothesis $\lim_{t\uparrow\infty} G_t(x) = F_0(x)$, we have by dominated convergence (R gives finite mass to bounded intervals)

$$\lim_{t\uparrow\infty} \int_0^a G_t(a - s)R(ds) = \int_0^a F_0(a - s)R(ds) = \mu^{-1}a.$$

$\square$

In order to prove Blackwell's theorem, it is enough to prove (iii) of Lemma 10.2.16. We do this in the case where $\mu < \infty$.[4] Here is the *coupling argument*.[5]

Consider two independent renewal sequences with the same interarrival distribution F. The first one is undelayed: $S_0 = 0$, S_1, S_2, ...and the second one is stationary: $\tilde{S}_0$, $\tilde{S}_1$, $\tilde{S}_2$. In particular, the distribution of $\tilde{S}_0$ is F_0 given by (10.22). Construct a renewal sequence $\{S_n^*\}_{n\geq 1}$ as follows. Take $S_n^* = S_n$ until the first time where two points of the tilded and untilded processes are ε-close, where ε is fixed. (In this case we say that ε-coupling was successful, which is not granted in general. The technical part of the proof of Blackwell's theorem is to show that ε-coupling is actually realizable with probability 1 when the interval distribution is non-lattice.) Then follow the tilded process. For instance, suppose that T_5 and $\tilde{T}_3$ are at a distance less than ε. Then $S_n^* = S_n$ for $n = 1, 2, 3, 4, 5$, and $S_{5+k}^* = \tilde{S}_{3+k}$ for $k \geq 1$. Denote by $T = T^\varepsilon$ the first point of the tilded process which is ε-close to a point of the untilded process (in the example $T = \tilde{T}_3$).

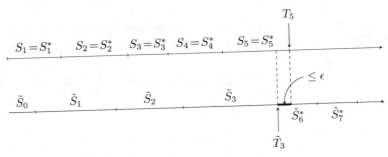

Lemma 10.2.17 *With the assumptions of Theorem 10.2.11, if ε-coupling happens almost surely, that is, if $P(T < \infty) = 1$, then Blackwell's theorem is proved.*

Proof. For simpler notation, let $T := T_\varepsilon$. Let $\{A(t)\}_{t\geq 0}$ and $\{\tilde{A}(t)\}_{t\geq 0}$ be the recurrence times corresponding to the undelayed starred renewal process and the (stationary) tilded

---

[4]For the extension to $\mu = \infty$, see for instance [Lindvall, 1992], p. 76–77.
[5][Lindvall, 1977].

renewal process. For all $t \geq T$, we have $A(t) = \tilde{A}(t + D)$ where $|D| \leq \varepsilon$. Let f be a continuous function bounded by 1, and define

$$M_\varepsilon(t) = \sup_{|s| \leq \varepsilon} \left| f(\tilde{A}(t + s)) - f(\tilde{A}(t)) \right|.$$

Note that $\lim_{\varepsilon \downarrow 0} \mathrm{E}\left[M_\varepsilon(0)\right] = 0$. Now,

$$
\begin{aligned}
\left| \mathrm{E}\left[f(A(t)) - f(\tilde{A}(t)) \right] \right| &\leq \mathrm{E}\left[\left| f(A(t)) - f(\tilde{A}(t)) \right| \right] \\
&= \mathrm{E}\left[\left| (f(A(t)) - f(\tilde{A}(t)) \right| 1\{t < T\} \right] \\
&\quad + \mathrm{E}\left[\left| (f(A(t)) - f(\tilde{A}(t)) \right| 1\{t \geq T\} \right] \\
&\leq 2P(T > t) + \mathrm{E}\left[M_\varepsilon(t)\right] \\
&= 2P(T > t) + \mathrm{E}\left[M_\varepsilon(0)\right],
\end{aligned}
$$

where the last equality follows from stationarity of $\tilde{A}$. Deduce from this that, since $\mathrm{E}\left[f(\tilde{A}(t))\right] = \mathrm{E}\left[f(\tilde{S}_0)\right]$,

$$\lim_{t \uparrow \infty} \mathrm{E}\left[f(A(t))\right] = \mathrm{E}\left[f(\tilde{S}_0)\right].$$

In other words, since f is an arbitrary continuous function bounded by 1, $A(t)$ tends in distribution to $\tilde{S}_0$ as $t \uparrow \infty$. In particular, since the distribution F_0 of $\tilde{S}_0$ is continuous, for all $x \in \mathbb{R}_+$,

$$\lim_{t \uparrow \infty} P(A(t) \leq x) = F_0(x).$$

The conclusion follows from (iii) of Lemma 10.2.16.   □

In order to prove ϵ-coupling, we first examine the role of the non-lattice assumption. Recall that a point x is said to be in the support of the distribution function F if $F(x + \epsilon) - F(x - \epsilon) > 0$ for all $\epsilon > 0$. The set of all such points is called the support of F and is denoted by $\mathrm{supp}(F)$. The key implication of the non-lattice assumption is the following:

Lemma 10.2.18 *Let F be a non-lattice cumulative distribution function. Let $\mathcal{G}$ denote the set of finite linear combinations of elements of $\mathrm{supp}(F)$ with coefficients in $\mathbb{N}$, that is*

$$\mathcal{G} = \bigcup_{n \in \mathbb{N}} \left\{ \sum_{i=1}^{n} g_i \; ; \; g_1, \ldots, g_n \in \mathrm{supp}(F) \right\}. \tag{10.23}$$

Then $\mathcal{G}$ is asymptotically dense in $\mathbb{R}_+$, that is

$$\lim_{x \to \infty} d(x, \mathcal{G}) = 0,$$

where $d(x, \mathcal{G}) = \inf_{g \in \mathcal{G}} |x - g|$.

Observe that the set $\mathcal{G}$ as defined by (10.23) is the union of the supports of the cumulative distribution functions F^{*n} ($n \in \mathbb{N}$) or, equivalently, the support of the renewal function $R = \sum_{n \in \mathbb{N}} F^{*n}$ associated to F.

Proof. Letting

$$\mu := \inf_{g, h \in \mathcal{G}, g > h} \{g - h\}, \tag{10.24}$$

we first prove that $\mu = 0$. Suppose in view of contradiction that $\mu > 0$. The infimum in (10.24) is then necessarily attained, for otherwise there would exist sequences g_n, h_n in $\mathcal{G}$ such that $g_n - h_n > g_{n+1} - h_{n+1}$, and $g_n - h_n \to \mu$ as $n \to \infty$. Then, for n large enough, $g_n - h_n < \mu + \mu/2$. Consequently, letting $g := g_n + h_{n+1}$ and $h = h_n + g_{n+1}$, it holds that

$$g - h = (g_n - h_n) - (g_{n+1} - h_{n+1}) \in (0, \mu/2).$$

This is a contradiction, in view of the definition of μ and the fact that $g, h \in \mathcal{G}$.

There must therefore exist $g, h \in \mathcal{G}$ such that $g - h = \mu$. Since F is non-lattice, there exists $z \in \mathrm{supp}(F)$ such that, for some $k \in \mathbb{N}$,

$$k\mu < z < (k+1)\mu.$$

Define then $g' := z + kh$ and $h' := kg$. Both g' and h' belong to $\mathcal{G}$. Furthermore, $g' - h' = z - k\mu \in (0, m)$, again a contradiction. Necessarily then, $\mu = 0$.

Therefore, for any $\epsilon > 0$, there exist $g, h \in \mathcal{G}$ such that $g - h \in (0, \epsilon)$. Consider the subset $\mathcal{G}'$ of $\mathcal{G}$ consisting of the elements $kg + \ell h$ $(k, \ell \in \mathbb{N})$. We argue that $\lim_{x \to \infty} d(x, \mathcal{G}') \leq \epsilon$. Indeed, let $m = \lceil h/\epsilon \rceil$. Let $x > mh$. Write $x = nh + r$, with $n \in \mathbb{N}$, $n \geq m$, and $r \in [0, h)$. Let $k \in \mathbb{N}$ be such that

$$(n - k)h + kg \leq x < (n - k)h + kg + (g - h).$$

Necessarily $k \leq m$ since $r < h$. The term $(n - k)h + kg$ thus belongs to $\mathcal{G}'$, as $n - k \geq 0$. Furthermore, it is at most ϵ apart from x, by the pair of inequalities displayed above. It follows that

$$\limsup_{x \to \infty} d(x, \mathcal{G}) \leq \limsup_{x \to \infty} d(x, \mathcal{G}') \leq \epsilon.$$

As ϵ is arbitrary, this concludes the proof of the theorem. $\qquad \square$

One says that ϵ-coupling holds for renewal processes with inter-renewal CDF F if for $\epsilon > 0$ and fixed initial delays t_1, t_1', one can construct jointly two renewal processes with the corresponding delays such that, with probability 1, there are indices m, n such that the corresponding renewal times T_m, T_n' are less than ϵ apart.

Lemma 10.2.19 *With the assumptions of Theorem 10.2.11, ε-coupling happens almost surely.*

Proof. Let

$$Z_i := \min\{\tilde{T}_j - T_i \, ; \, \tilde{T}_j - T_i \geq 0\} \quad (i \geq 0).$$

For fixed $\varepsilon > 0$, let

$$A_i := \{Z_j < \varepsilon \text{ for some } j \geq i\}.$$

Then

$$A_0 \supseteq \cdots \supseteq A_\ell \supseteq \cdots \supseteq \cap_{i=0}^\infty A_i = A_\infty := \{Z_i < \varepsilon \, i.o.\}.$$

Since the sequence $\{T_{i+n} - T_i\}_{n \geq 1}$ has a distribution independent of i and is independent of $\tilde{N} \equiv \{\tilde{T}_n\}_{n \geq 0}$, and since $\tilde{N}$ is stationary, the sequence $\{Z_i\}_{i \geq 0}$ is also stationary. Therefore the events A_i $(i \geq 0)$ have the same probability, and in particular

$$P(A_0) = P(A_\infty).$$

Conditionally on $\widetilde{S}_0$, the event A_∞ is an exchangeable event of the symmetric sequence $\{(S_n, \widetilde{S}_n)\}_{n \geq 1}$. Therefore, by the Hewitt–Savage 0-1 law (Theorem 4.3.7), for all $t > 0$

$$P(A_\infty \mid \widetilde{S}_0 = t) = 0 \text{ or } 1. \tag{$\star$}$$

Lemma 10.2.18 guarantees that for sufficiently large u and fixed ε,

$$P(u - t < \widetilde{T}_j - \widetilde{S}_0 < u - t + \varepsilon \text{ for some } j) > 0.$$

Therefore $P(A_0 \mid \widetilde{S}_0 = t) > 0$ for all $t \geq 0$ and in particular

$$P(A_0) = \lambda \int_0^\infty P(A_0 \mid \widetilde{S}_0 = t)(1 - F(t))\, dt > 0,$$

which implies since $P(A_0) = P(A_\infty)$,

$$P(A_\infty) = \lambda \int_0^\infty P(A_\infty \mid \widetilde{S}_0 = t)(1 - F(t))\, dt > 0.$$

In view of $(\star)$, this implies that

$$P(A_\infty \mid \widetilde{S}_0 = t) = 1 \text{ for all } t \text{ such that } F(t) < 1.$$

Therefore $P(A_\infty) = 1 = P(A_0)$, so that $P(Z_i < \varepsilon \text{ for some } i) = 1$. $\qquad \square$

10.2.3   Defective and Excessive Renewal Equations

The key renewal theorem concerns *proper* renewal equations. In a number of situations though, one encounters renewal equations for which $F(\infty) < 1$ or $F(\infty) > 1$. However, when there exists an $\alpha \in \mathbb{R}$ such that

$$\int_{[0,\infty)} e^{\alpha t} dF(t) = 1, \tag{10.25}$$

the asymptotics of the solution of the renewal equation can be obtained from the proper case. In fact, letting

$$\tilde{g}(t) := e^{\alpha t} g(t), \quad \tilde{f}(t) := e^{\alpha t} f(t) \text{ and } \tilde{F}(t) := \int_{[0,t]} e^{\alpha s}\, dF(s),$$

the distribution $\tilde{F}$ is proper, and non-lattice if F itself is non-lattice. One immediately checks that $\tilde{f}$ satisfies the renewal equation $\tilde{f} = \tilde{g} + \tilde{f} * \tilde{F}$. The conclusion of the key renewal theorem is that when $\tilde{g}$ is directly Riemann integrable,

$$\lim_{t \to \infty} e^{\alpha t} f(t) = \frac{\int_0^\infty e^{\alpha t} g(t) dt}{\int_0^\infty t e^{\alpha t} dF(t)}. \tag{10.26}$$

Remark 10.2.20 A number α satisfying (10.25) always exists *in the excessive case*. Indeed, the function $\alpha \to \int_{[0,\infty)} e^{\alpha t} dF(t)$ is continuous on $(-\infty, 0]$ and strictly increases from 0 to $\int_{[0,\infty)} dF(t) > 1$. Therefore there is a unique $\alpha < 0$ satisfying (10.25). In the *defective case*, such α, if it exists, is necessarily positive. But it may not exist. In fact, its existence implies exponential decay of the tail distribution $1 - F$ since by Markov's inequality

$$P(S_1 > t) = P(e^{\alpha S_1} > e^{\alpha t}) \leq e^{-\alpha t} \mathrm{E}[e^{\alpha S_1}].$$

Remark 10.2.21 Clearly, from (10.26), in the non-lattice defective case and assuming the existence of such α, the solution of the renewal equation decays exponentially fast as $t \to \infty$, whereas in the non-lattice excessive case (for which α always exists) the solution of the renewal equation explodes exponentially fast as $t \to \infty$.

EXAMPLE 10.2.22: ASYMPTOTICS IN THE TRANSIENT CASE. Suppose that F is defective ($F(\infty) < 1$) and that there exists α (necessarily > 0) such that

$$\int_0^\infty e^{\alpha t} dF(t) = 1 . \tag{10.27}$$

By Theorem 10.1.13, when the data g is bounded and such that there exists $g(\infty) = \lim_{t \to \infty} g(t)$, the unique solution f of the renewal equation $f = g + f * F$ satisfies

$$\lim_{t \uparrow \infty} f(t) = \frac{g(\infty)}{1 - F(\infty)} . \tag{10.28}$$

With the help of the defective renewal theorem additional information concerning the asymptotic behavior of f can be obtained. In fact, if the function g_1 defined by

$$g_1(t) = g(t) - g(\infty) + g(\infty) \frac{F(t) - F(\infty)}{1 - F(\infty)}$$

is such that the function $t \to \tilde{g}_1(t) = e^{\alpha t} g_1(t)$ is directly Riemann integrable, then

$$\lim_{t \to \infty} e^{\alpha t} \left(f(t) - \frac{g(\infty)}{1 - F(\infty)} \right) = C , \tag{10.29}$$

where

$$C = \frac{\int_0^\infty e^{\alpha t}[g(t) - g(\infty)]dt - \frac{g(\infty)}{\alpha}}{\int_0^\infty t e^{\alpha t} dF(t)} .$$

Proof. Define

$$f_1(t) := f(t) - \frac{g(\infty)}{1 - F(\infty)} .$$

Straightforward computations using the identity $R * F = R - 1$ show that $f_1 = R * g_1$. Therefore f_1 is a solution of the (defective) renewal equation $f_1 = g_1 + f_1 * F$. Since $\tilde{g}_1(t) = e^{\alpha t} g_1(t)$ is assumed to be directly Riemann integrable,

$$\lim_{t \to \infty} e^{\alpha t} f_1(t) = \frac{\int_0^\infty e^{\alpha t} g_1(t)dt}{\int_0^\infty t e^{\alpha t} dF(t)} .$$

But

$$\int_0^\infty e^{\alpha t}(F(\infty) - F(t))dt = \int_0^\infty e^{\alpha t} \left(\int_{(t,\infty)} dF(s) \right) dt$$

$$= \int_0^\infty \left(\int_0^t e^{\alpha s} ds \right) dF(t) = \frac{1}{\alpha}(1 - F(\infty)),$$

from which the above expression for C follows. $\qquad \square$

EXAMPLE 10.2.23: THE RISK MODEL, TAKE 3. This example is a continuation of Example 10.1.7. Recall Eqn. (10.11) in the case where the safety loading is positive, and write this equation in the form

$$\Psi(u) = \frac{\lambda}{c} \int_u^\infty (1 - G(z)) \, dz + \int_0^u \Psi(u - z) \, dF(z),$$

where

$$F(x) := \frac{\lambda}{c} \int_0^x (1 - G(z)) \, dz.$$

This is a renewal equation, and it is defective since $\frac{\lambda}{c} \int_0^\infty (1 - G(z)) \, dz = \frac{\lambda \mu}{c} < 1$ under the positive safety loading condition. *Assume the existence of* $\alpha > 0$ *such that*

$$\frac{\lambda}{c} \int_0^x e^{\alpha z} (1 - G(z)) \, dz = 1.$$

By the defective renewal theorem, if

$$\frac{\lambda}{c} \int_0^\infty e^{\alpha u} \int_u^\infty (1 - G(z)) \, dz \, du < \infty,$$

we have that

$$\lim_{u \uparrow \infty} e^{\alpha u} \Psi(u) = C := \frac{\int_0^\infty e^{\alpha u} \int_u^\infty (1 - G(z)) \, dz \, du}{\int_0^\infty z e^{\alpha z} (1 - G(z)) \, dz}$$

or, equivalently,

$$\Psi(u) = C e^{-\alpha u} + o(u).$$

10.3   Regenerative Processes

10.3.1   Examples

Let $(E, \mathcal{E})$ be a measurable space.

Definition 10.3.1 *Let* $\{X(t)\}_{t \geq 0}$ *be a measurable* E-*valued stochastic process and let* $\{T_n\}_{n \geq 0}$ *be a proper recurrent renewal process, possibly delayed (recall, however, that the initial delay* T_0 *is always assumed finite). The process* $\{X(t)\}_{t \geq 0}$ *is said to be* regenerative *with respect to* $\{T_n\}_{n \geq 0}$ *if for all* $n \geq 0$,

(a) *the distribution of the post-*T_n *process* $S_{n+1}, S_{n+2}, \ldots, \{X(t + T_n)\}_{t \geq 0}$ *is independent of* $n \geq 0$, *and*

(b) *the post-*T_n *process is independent of* $T_0, \ldots, T_n$.

The times T_n *are called* regeneration times *of the regenerative process.*

EXAMPLE 10.3.2: CONTINUOUS-TIME MARKOV CHAINS. Let $\{X(t)\}_{t \geq 0}$ be a recurrent continuous-time homogeneous Markov chain taking its values in the state space $E = \mathbb{N}$. Suppose that it starts from state 0 at time $t = 0$. By the strong Markov property, $\{X(t)\}_{t \geq 0}$ is regenerative with respect to the sequence $\{T_n\}_{n \geq 0}$ where T_n is the n-th time of visit to state 0 of the chain.

Regenerative processes are the main sources of renewal equations.

Theorem 10.3.3 *Let $\{X(t)\}_{t\geq 0}$ and $\{T_n\}_{n\geq 0}$ be as in Definition 10.3.1 except for the additional assumption $T_0 \equiv 0$ (undelayed renewal process) and let $h : E \to \mathbb{R}$ be a non-negative measurable function. The function $f : \mathbb{R}_+ \to \mathbb{R}$ defined by*

$$f(t) := \mathrm{E}\left[h(X(t))\right] \tag{10.30}$$

satisfies the renewal equation with data

$$g(t) = \mathrm{E}\left[h(X(t))1_{\{t<S_1\}}\right]. \tag{10.31}$$

Proof. Following a now familiar line of argument, write:

$$h(X(t)) = h(X(t))1_{\{t<S_1\}} + h(X(t))1_{\{t\geq S_1\}}.$$

But, if $t \geq S_1$, $X(t) = \tilde{X}(t - S_1)$ where $\tilde{X}(t) = X(t - S_1)$, so that

$$h(X(t)) = h(X(t))1_{\{t<S_1\}} + h(\tilde{X}(t - S_1))1_{\{t\geq S_1\}}.$$

Taking expectations and using the hypothesis that $\{\tilde{X}(t)\}_{t\geq 0}$ is independent of S_1 and has the same distribution as $\{X(t)\}_{t\geq 0}$, one obtains the announced result. $\square$

In particular, for $A \in \mathcal{E}$, the function

$$f(t) = P(X(t) \in A) \tag{10.32}$$

satisfies the renewal equation with data

$$g(t) = P(X(t) \in A, t < S_1). \tag{10.33}$$

10.3.2 The Limit Distribution

The following result[6] is called *Smith's regenerative theorem.*

Theorem 10.3.4 *Let $\{X(t)\}_{t\geq 0}$ and $\{T_n\}_{n\geq 0}$ be as in Definition 10.3.1, with $T_0 = 0$, and let A be a measurable subset of the state space of this process. The function $f(t) = P(X(t) \in A)$ satisfies the renewal equation with data $g(t) = P(X(t) \in A, t < S_1)$, and if the renewal distribution is non-lattice and g is directly Riemann integrable,*

$$\lim_{t\to\infty} P(X(t) \in A) = \frac{1}{\mathrm{E}\left[S_1\right]}\mathrm{E}\left[\int_0^{S_1} 1_{X(s)\in A}\mathrm{d}s\right]. \tag{10.34}$$

In the delayed case (with finite delay by definition), $\lim_{t\uparrow\infty} P(X(t) \in A)$ is the same as in the undelayed case, provided the latter exists.

Proof. In the undelayed case, in view of (10.33), this is a direct application of the key renewal theorem. In the delayed case, first observe that, by the usual trick of renewal theory,

$$P(X(t) \in A) = P(X(t) \in A, t < T_0) + \int_{[0,t]} P(X^*(t - s) \in A)\,\mathrm{d}G(s), \tag{10.35}$$

---

[6][Smith, 1955].

where G is the cumulative distribution of the initial delay T_0 and where $X^*(t) = X(t + T_0)$. Since T_0 is almost surely finite, $\lim_{t \to \infty} P(X(t) \in A, t \leq T_0) = 0$. Also, $\{X^*(t)\}_{t \geq 0}$ is an undelayed regenerative process with respect to the sequence

$$T_0' = 0, \ T_1' = S_1, \ T_2' = S_1 + S_2, \ \ldots$$

and therefore $\lim_{t \to \infty} P(X^*(t) \in A)$ exists. By dominated convergence,

$$\lim_{t \to \infty} \int_{[0,t]} P(X^*(t-s) \in A) \, dG(s) = \int_{[0,t]} \lim_{t \to \infty} P(X^*(t-s) \in A) \, dG(s)$$
$$= \int_0^\infty \lim_{t \to \infty} P(X^*(t) \in A) \, dG(s)$$
$$= \lim_{t \to \infty} P(X^*(t) \in A).$$

$\square$

EXAMPLE 10.3.5: THE NEED FOR DIRECT RIEMANN INTEGRABILITY. Suppose that F puts mass 2^{-n} on n^{-1}, for all $n \geq 1$. (This CDF is clearly non-lattice.) Take $X(t) = 1_{\mathbf{Q}}(t)$ where $\mathbf{Q}$ is the set of rationals. This is a regenerative process (with respect to the above renewal sequence) because for all $r \in \mathbf{Q}$, and all $t \geq r$, it holds that $1_{\mathbf{Q}}(t-r) = 1_{\mathbf{Q}}(t)$. But $P(X(t) = 1, t < S_1) = 1_{\mathbf{Q}}(t)(1 - F(t))$ is nowhere continuous and therefore cannot be Riemann integrable or, *a fortiori*, directly Riemann integrable. In this case, we can check directly that the conclusion of the key renewal theorem fails, since $P(X(t) = 1) = 1_{\mathbf{Q}}(t)$ has no limit as $t \uparrow \infty$.

The result below suffices in many cases to guarantee the direct Riemann integrability hypothesis in Theorem 10.3.4. First recall that, by definition, $a \in \mathbb{R}$ is called a fixed discontinuity if

$$P(\{\omega; a \text{ is a discontinuity point of } t \to X(t, \omega)\}) > 0.$$

Theorem 10.3.6 *Let $\{X(t)\}_{t \geq 0}$ be a stochastic process taking its values in a topological space E endowed with its Borel σ-field $\mathcal{E} = \mathcal{B}(E)$ and whose fixed discontinuities form a countable set. Let $A \subseteq E$ be open. The function $t \to K(t, A) = P(X(t) \in A, t < S_1)$ is directly Riemann integrable.*

Proof. The function $K(t, A)$ is bounded by the directly Riemann-integrable function $P(S_1 > t)$. By Theorem 10.2.10, part (b), it is therefore enough to prove that it is Riemann integrable on finite intervals. In fact, as we now show, it is almost everywhere continuous (which implies that it is Riemann integrable on finite intervals since it is bounded). Since

$$|K(t, A) - K(s, A)| \leq P(\{X(t) \in A, t < S_1\} \triangle \{X(s) \in A, s < S_1\}),$$

it suffices to show that the latter tends to 0 as $s \to t \notin D$, where D is the union of the set of fixed discontinuities of the process (assumed countable) and of the (countable) set of discontinuities of F. We therefore have to prove that $P(\{X(t) \in A, t < S_1\} \cap \overline{\{X(s) \in A, s < S_1\}}) \to 0$ and $P(\{X(s) \in A, s < S_1\} \cap \overline{\{X(t) \in A, t < S_1\}}) \to 0$. We take care of the first one, the other one being similar. We have

$$P(\{X(t) \in A, t < S_1\} \cap \overline{\{X(s) \in A, s < S_1\}})$$
$$= P(\{X(t) \in A, t < S_1, S_1 \leq s\} \cup \{X(t) \in A, t < S_1, X(s) \notin A\})$$
$$\leq P(\{X(t) \in A, t < S_1, S_1 \leq s\}) + P(\{X(t) \in A, t < S_1, X(s) \notin A\}).$$

The first term in the right-hand side is bounded above by $F(s \vee t) - F(s \wedge t)$, which tends to 0 since $t \notin D$ is a point of continuity of F. The second term is bounded above by $P(\{X(t) \in A, X(s) \notin A\})$. Since t is not a fixed discontinuity point of the process, $X(s) \to X(t)$. Therefore, since A is open, $\lim_{s \to t} 1_A(X(t)1_{A^c}(X(s)) = 0$, which gives $\lim_{s \to t} P(\{X(t) \in A, X(s) \notin A\}) = 0$. $\qquad \square$

Remark 10.3.7 The hypothesis that $\{X(t)\}_{t \geq 0}$ has fixed discontinuities forming at most a countable set is always satisfied if the trajectories of this process are almost surely in $D([0, \infty))$, the set of functions that are right-continuous on $[0, \infty)$ and have left-hand limits on $(0, \infty)$.

In the following examples, direct Riemann integrability of the data g is a consequence of Theorem 10.3.6.

EXAMPLE 10.3.8: THE FUNDAMENTAL RELIABILITY THEOREM. Consider the situation of a renewal process for which the interrenewal sequence $\{S_n\}_{n \geq 1}$ is of the form $S_n = U_n + V_n$, where $\{U_n\}_{n \geq 1}$ and $\{V_n\}_{n \geq 1}$ are independent IID sequences, and define

$$X(t) = \begin{cases} 1 & \text{if } t \in (T_n, T_n + U_n], \\ 0 & \text{if } t \in (T_n + U_n, T_{n+1} + U_n + V_n]. \end{cases}$$

The interpretation of the process $\{X(t)\}_{t \geq 0}$ in terms of reliability is that $X(t) = 1$ when at time t a given machine is currently in working condition, whereas if $X(t) = 0$ it is in repair. Let a and b be the respective means of U_1 and V_1. In the recurrent non-lattice case, we have that

$$\lim_{t \to \infty} P(X(t) = 1) = \frac{a}{a + b}.$$

It suffices to apply Smith's regenerative formula (10.34) with $A = \{1\}$, and to observe that $E[S_1] = E[U_1] + E[V_1] = a + b$ and $\int_0^\infty P(X(s) = 1, s < S_1) \, ds = \int_0^\infty P(U_1 > s) \, ds = E[U_1] = a$. In a reliability context, $\frac{a}{a+b}$ represents the availability of a given machine with mean lifetime a and mean repair time b.

EXAMPLE 10.3.9: FORWARD AND BACKWARD RECURRENCE TIMES. Let $\{T_n\}_{n \geq 0}$ be an undelayed ($T_0 = 0$) renewal process. Clearly the forward and backward recurrence times are regenerative with respect to the renewal process $\{T_n\}_{n \geq 0}$. From Smith's regenerative formula (10.34), in the non-lattice case

$$\lim_{t \to \infty} P(A(t) > x) = \frac{1}{m} \int_0^\infty P(A(s) > x, s < S_1) \, ds.$$

Since $P(A(s) > x, s < S_1) = P(S_1 > s + x) = 1 - F(s + x)$, we have $\int_0^\infty P(A(s) > x, s < S_1) \, ds = \int_0^\infty (1 - F(s + x)) \, ds = \int_x^\infty (1 - F(s)) \, ds$, and therefore

$$\lim_{t \to \infty} P(A(t) > x) = \frac{1}{m} \int_x^\infty (1 - F(s)) \, ds. \tag{10.36}$$

Similar arguments yield for the backward recurrence time

$$\lim_{t\to\infty} P(B(t) > y) = \frac{1}{m}\int_y^\infty (1 - F(s))\mathrm{d}s. \tag{10.37}$$

(This time, the data function is $(1 - F(t))1_{\{t>y\}}$.) One can prove directly the direct Riemann integrability of the data functions of this example, or use Theorem 10.3.6.

EXAMPLE 10.3.10: THE BUS PARADOX. The sum $A(t) + B(t)$ is the inter-event interval around time t. Interpreting t as the time at which you arrive at a bus stop, and the sequence $\{T_n\}_{n\geq 1}$ as the sequence of times at which buses arrive at (and immediately depart from) the bus stop, $A(t)$ is your waiting time. If t is large enough, one can, in view of (10.36), assume that $A(t)$ is distributed as a random variable A with the distribution

$$P(A > x) = \frac{1}{m}\int_x^\infty (1 - F(s))\,\mathrm{d}s. \tag{10.38}$$

Similarly, the time $B(t)$ by which you missed the previous bus is approximately, when t is large, distributed as a random variable B with the same distribution as A. The bus paradox can be stated in several ways. One of them is: the mean time interval between the bus you missed and the bus you will catch is asymptotically as $t \to \infty$ equal to $E[A + B] = 2E[A]$, and is in general different from the mean of the interval between two successive buses n and $n + 1$, $E[S_1]$.

Let $\{X(t)\}_{t\in\mathbb{R}}$ be a stochastic process taking its values in a metric space E, having right-continuous paths and being regenerative relative to the (possibly delayed) renewal sequence $\{T_n\}_{n\geq 0}$ with non-lattice and finite mean inter-event distribution μ. Let P_0 and E_0 symbolize respectively the probability and the expectation corresponding to the undelayed version of the renewal sequence. One checks easily that

$$P^*(A) := \frac{1}{\mu}E_0\left[\int_0^{S_1} 1_A(X(s))\,\mathrm{d}s\right]$$

defines a probability measure on $(E, \mathcal{B}(E))$.

Theorem 10.3.11 *Under the above conditions, $X(t)$ converges in distribution to P^* as $t\uparrow\infty$.*

Proof. We must show that for all bounded (say, by 1) continuous functions $h : E \to \mathbb{R}$,

$$\lim_{t\uparrow\infty} E\left[h(X(t))\right] = \frac{1}{\mu}E_0\left[\int_0^{S_1} h(X(s))\,\mathrm{d}s\right]. \tag{10.39}$$

By the usual renewal argument (conditioning on T_0, whose cumulative distribution function is denoted by F_{T_0}),

$$E\left[h(X(t))\right] = E\left[h(X(t))1_{\{t<T_0\}}\right] + \int_{(0,t]} f(t - s)\,\mathrm{d}F_{T_0}(s), \tag{$\star$}$$

where

$$f(t) := E_0\left[f(X(t))\right].$$

Since T_0 is finite (*by definition* of a delayed renewal process) and h is bounded, the first term in the right-hand side of $(\star)$ tends to 0 as $t \uparrow \infty$. It remains to prove that the second term converges to $\frac{1}{\mu}\mathrm{E}_0\left[\int_0^{S_1} h(X(s))\,\mathrm{d}s\right]$ as $t \uparrow \infty$. By dominated convergence, this will be true if

$$\lim_{t \uparrow \infty} f(t) = \frac{1}{\mu}\mathrm{E}_0\left[\int_0^{S_1} h(X(s))\,\mathrm{d}s\right].$$

To show this, apply the key renewal theorem to the renewal equation satisfied by

$$f(t) := g(t) + \int_{(0,t]} f(t-s)\,\mathrm{d}F(s),$$

where

$$g(t) = \mathrm{E}_0\left[h(X(t))1_{\{t<S_1\}}\right].$$

One need only verify that g is directly Riemann integrable. This is the case since it is right-continuous and therefore continuous outside a set at most countable[7] and bounded by the directly Riemann-integrable function $1 - F(t)$ (Theorem 10.2.10). $\qquad\square$

10.4 Semi-Markov Processes

A *multivariate renewal equation* is an equation of the type $f = g + F * f$, that is, in developed form

$$f_i(t) = g_i(t) + \sum_{j \in E} \int_{(0,t]} f_j(t-s)\,\mathrm{d}F_{ij}(t) \qquad (i \in E), \qquad (10.40)$$

where E is a countable space, the functions $f_i : \mathbb{R} \to \mathbb{R}$ are the unknowns, the functions $g_i : \mathbb{R} \to \mathbb{R}$ are the data, and the F_{ij}'s are up to a positive multiplicative constant cumulative distribution functions on $\mathbb{R}_+$. The corresponding theory is intimately linked to that of semi-Markov processes.

Let E be a countable space. Construct a stochastic process $\{X(t)\}_{t \geq 0}$ with values in E as follows. Start with a homogeneous Markov chain $\{X_n\}_{n \geq 0}$, called the *leading chain*, with values in E and transition matrix $\mathbf{P} = \{p_{ij}\}_{i,j \in E}$, and then construct a sequence $\{S_n\}_{n \geq 1}$ of positive real random variables which is, conditionally on $\mathcal{F}^X := \sigma(X_0, X_1, \ldots)$, an independent sequence. Suppose, moreover, that for all $n \geq 0$, S_{n+1} is conditionally independent of $X_0, \ldots, X_{n-1}, X_{n+2}, X_{n+3}, \ldots$ given X_n and X_{n+1}. For all $i, j \in E$, let

$$G_{ij}(t) := P(S_{n+1} \leq t \mid X_n = i, X_{n+1} = j).$$

G_{ij} is the cumulative distribution function of some random variable with values in $\mathbb{R}_+$ (in particular $\lim_{t \uparrow \infty} G(t) := G_{ij}(\infty) = 1$) of mean $m_{ij} := \int_{\mathbb{R}} t\,\mathrm{d}G_{ij}(t)$.

Construct an increasing sequence of times $\{T_n\}_{n \geq 0}$ as follows. $T_0 \equiv 0$ and

$$T_n = T_{n-1} + S_n \qquad (n \geq 1).$$

Finally, let

$$X(t) := X_n \text{ for } t \in [T_n, T_{n+1}) \qquad (n \geq 0).$$

---

[7] See Theorem A.3.1 in [Asmussen, 1987].

(The chain $\{X_n\}_{n\geq 0}$ is therefore also called the embedded chain of the semi-Markov process.) In fact, $X(t)$ is defined in this way only for $t < T_\infty := \lim_{n\uparrow\infty} T_n$. In order to guarantee that $X(t)$ is defined for all $t \geq 0$, the following hypothesis is made:

Hypothesis H: The homogeneous Markov chain $\{X_n\}_{n\geq 0}$ is irreducible and recurrent.

To prove that in this case $P(T_\infty = \infty) = 1$, consider two states $i, j \in E$ such that $p_{ij} > 0$. There are an infinite number of occurrences of the event $\{X_n = i, X_{n+1} = j\}$, and therefore T_∞ is, given $\mathcal{F}^X$, larger than an infinite sum of independent positive random variables with the same cumulative distribution G_{ij}, and therefore infinite (law of large numbers).

The stochastic process $\{X(t)\}_{t\geq 0}$ so constructed is called a *semi-Markov process* with *semi-Markov kernel* $F := \{F_{ij}\}_{i,j\in E}$, where the F_{ij}'s are the cumulative distributions of sub-probability measures on $[0, \infty)$ given by

$$F_{ij}(t) := P(X_{n+1} = j, S_{n+1} \leq t \mid X_n = i) = P_i(X_1 = j, S_1 \leq t)$$

(recall the notation $P_i(\cdot) := P(\cdot \mid X_0 = i)$) and in particular

$$F_{ij}(t) = p_{ij}G_{ij}(t).$$

Defining $\|F_{ij}\| := F_{ij}(\infty)$, we see that $\{\|F_{ij}\|\}_{i,j\in E} = \mathbf{P}$, the transition matrix of the leading chain, hence the appellation "stochastic kernel" for F. Let $\mu_{ij} := \int_{\mathbb{R}} t \, dG_{ij}(t) = p_{ij}m_{ij}$.

With each $i \in E$ is associated a renewal point process whose sequence of times are the return times to i ($X(t-) \neq i, X(t) = i$). Since the embedded Markov chain is recurrent, this renewal process is recurrent. The mean inter-arrival time for this renewal process is, denoting by $\nu = \{\nu_i\}_{i\in E}$ the unique (up to a multiplicative constant) positive invariant measure of $\mathbf{P}$,

$$\frac{\mu}{\nu_i} \text{ where } \mu := \sum_{i,j\in E} \nu_i\mu_{ij}.$$

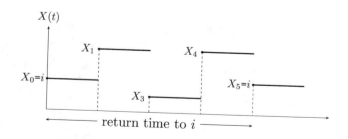

Proof. With $\tau := \inf\{n \geq 1 \,;\, X_n = i\}$,

$$\mathrm{E}_i\left[S_1 + \cdots + S_\tau\right] = \mathrm{E}_i\left[\sum_{n=1}^{\infty} S_n \mathbf{1}_{\{n \leq \tau\}}\right] = \mathrm{E}_i\left[\sum_{n=1}^{\infty} \mathrm{E}_i\left[S_n \mathbf{1}_{\{n \leq \tau\}} \mid \mathcal{F}^X\right]\right]$$

$$= \mathrm{E}_i\left[\sum_{n=1}^{\infty} \mathrm{E}_i\left[S_n \mid \mathcal{F}^X\right] \mathbf{1}_{\{n \leq \tau\}}\right] = \mathrm{E}_i\left[\sum_{n=1}^{\infty} m_{X_{n-1}, X_n} \mathbf{1}_{\{n \leq \tau\}}\right]$$

$$= \sum_{j,k} m_{jk} \mathrm{E}_i\left[\sum_{n=1}^{\infty} \mathbf{1}_{\{X_{n-1}=j, X_n=k, n \leq \tau\}}\right]$$

$$= \sum_{j,k} m_{jk} p_{jk} \mathrm{E}_i\left[\sum_{n=1}^{\infty} \mathbf{1}_{\{X_{n-1}=j, n \leq \tau\}}\right] = \sum_{j,k} \mu_{jk} \frac{\nu_j}{\nu_i} = \frac{\mu}{\nu_i}.$$

$\square$

It is not difficult to show that, if the embedded chain is irreducible recurrent, the interarrival distributions corresponding to the renewal process of returns to i of the above are all non-lattice or all lattice with the same span. The semi-Markov process, or its kernel, is then called non-lattice or lattice respectively.

Define the convolution $H = G * F$ by

$$H_{ij}(t) := \sum_{k \in E} (G_{ik} * F_{kj})(t) := \sum_{k \in E} \int_0^t F_{kj}(t - s)\, \mathrm{d}G_{ik}(t)$$

and the convolution powers F^{*n} in a similar way, starting with $F_{ij}^{*0}(t) := \mathbf{1}_{\{i=j\}}$. In particular,

$$F_{ij}^{*n}(t) = P_i(X_n = j\,,\, S_1 + \cdots + S_n \leq t). \tag{$\star$}$$

Define the Markov renewal kernel $R := \sum_{n \geq 0} F^{*n}$. We have from $(\star)$ that $R_{ij}(t)$ is the expected number of returns to j when starting from i:

$$R_{ij}(t) = \sum_{n \geq 1} \mathrm{E}_i\left[\mathbf{1}_{\{X_{n+1}=j,\, S_1 + \cdots + S_n \leq t\}}\right].$$

In particular, $R_{ij}(t) < \infty$, and for any fixed $a > 0$, we have in the non-lattice case (Blackwell's theorem)

$$\lim_{t \uparrow \infty} \frac{R_{ij}(t + a) - R_{ij}(t)}{a} = \frac{\nu_j}{\mu}. \tag{10.41}$$

The following result is similar to the one in the univariate case:

Theorem 10.4.1 *If the data functions are non-negative and locally bounded, there exists a unique vector $f := \{f_i\}_{i \in E}$ of locally bounded measurable functions from $\mathbb{R}_+ \to \mathbb{R}$ satisfying the multivariate renewal equation (10.40), namely $f = R * g$.*

Proof. The fact that $R * g$ is well defined, locally bounded, and satisfies the renewal equation is proved in the same way as in the univariate case. Let now f and $\tilde{f}$ be two vectors of locally bounded functions satisfying the renewal equation, and let $h := f - \tilde{f}$. Then $h = R * h$, and iteratively, $h = F^{*n} * h$ for all $n \geq 1$, so that $|h| \leq F^{*n} * |h|$. Let $\sup_i |h_i(t)| \leq M(a) < \infty$ on $[0, a]$, say $M(a) = 1$, without loss of generality, so that $|h| \leq F^{*n} * \mathbf{1}$ on $[0, a]$ and therefore, on $[0, a]$,

$$|h_i(t)| \leq \sum_{j \in E} F_{ij}^{*n}(t) = P_i(S_1 + \cdots + S_n \leq t),$$

a quantity that tends to 0 as $n \uparrow \infty$ since $T_\infty = \infty$ under the prevailing conditions (the embedded chain is irreducible recurrent). Therefore $h \equiv 0$ on all $[0, a]$ ($a \in \mathbb{R}_+$) and therefore on $\mathbb{R}_+$.

$\square$

It follows from (10.41) that if g_i is directly Riemann integrable

$$R_{ij} * g_j(t) \to \frac{\nu_j}{\mu} \int_0^\infty g_j(s) \, \mathrm{d}s,$$

and from this, in the case where E is finite,

$$f_i(t) = \sum_{j \in E} R_{ij} * g_j(t) \to \frac{1}{\mu} \sum_{j \in E} \nu_j \int_0^\infty g_j(s) \, \mathrm{d}s.$$

The case when E is infinite requires further conditions and will not be treated here.[8]

Improper Multivariate Renewal Equations

The above results concern the case where $Q := \{\|F_{ij}\|\}_{i,j \in E}$ is a stochastic matrix, that is, the transition matrix of a homogeneous Markov chain (namely, the transition matrix $\mathbf{P}$ of the embedded Markov chain). However the renewal equations (10.40) make sense even if this is not the case. We now give results[9] of the same kind as the ones in the defective or excessive univariate renewal functions when the state space is finite. The matrix Q is no longer a stochastic matrix, but still assumed irreducible. Define for some real β the matrix $\mathbf{A} := \{a_{ij}\}_{i,j \in E}$

$$a_{ij} := \int_0^\infty e^{\beta t} \, \mathrm{d}F_{ij}(t).$$

Assume that β can be chosen such that $\mathbf{A}$ has spectral radius 1. In particular, there exists two positive vectors ν and h such that

$$\nu^T \mathbf{A} = \nu \text{ and } \mathbf{A}h = h.$$

The existence of ν and h is ensured by the Perron–Fröbenius theorem. The following facts are easy. First the matrix

$$\widetilde{Q} := \left\{ \frac{h_j}{h_i} a_{ij} \right\}_{i,j \in E}$$

is an (irreducible) stochastic matrix admitting the invariant measure $\widetilde{\nu}$ given by $\widetilde{\nu}_i = \nu_i h_i$.

Let

$$\widetilde{F}_{ij}(t) := \frac{h_j}{h_i} \int_0^t e^{\beta s} \, \mathrm{d}F_{ij}(s).$$

This defines a semi-Markov kernel for which $\widetilde{Q} = \{\|\widetilde{F}_{ij}\|\}_{i,j \in E}$ is irreducible and recurrent. Defining

$$\widetilde{f}_i(t) := e^{\beta t} f_i(t)/h_i \text{ and } \widetilde{g}_i := e^{\beta t} g_i(t)/h_i,$$

---

[8]See, for instance, [Çinlar, 1975].
[9][Asmussen and Hering, 1977].

we see, analogously to the univariate case, that

$$\widetilde{f} = \widetilde{g} + \widetilde{F} * \widetilde{f}.$$

Therefore, if $\widetilde{F}$ is non-lattice and the $\widetilde{g}_i$'s are locally bounded and integrable,

$$\lim_{t\uparrow\infty} \widetilde{f}_i(t) = \frac{1}{\widetilde{\mu}} \sum_{j\in E} \widetilde{\nu}_j \int_0^\infty \widetilde{g}_j(s)\,\mathrm{d}s\,,$$

that is

$$\lim_{t\uparrow\infty} e^{\beta t} f_i(t) = \frac{h_i \sum_{j\in E} \nu_j \int_0^\infty e^{\beta s} g_j(s)\,\mathrm{d}s}{\sum_{k,j\in E} \nu_k h_j \int_0^\infty s e^{\beta s}\mathrm{d}F_{kj}(s)}\,.$$

The existence of β is guaranteed, for instance,[10] when the spectral radius of $\{\|F_{ij}\|\}_{i,j\in E}$ is strictly less that 1.

Complementary reading

[Asmussen, 2003] for more theory and for applications to random walks and queues.

10.5 Exercises

Exercise 10.5.1. THEOREM 10.1.1 TRUE IN THE UNDELAYED CASE
Prove that as Theorem 10.1.1 is true in the undelayed case, it is then true in the delayed case (recall: with *finite* delay).

Exercise 10.5.2. THE ASYMPTOTIC COUNTING RATE
Prove (10.4), that is,

$$\lim_{t\to\infty} \frac{N([0,t])}{t} = \frac{1}{\mathrm{E}[S_1]}\,,\ \text{P-a.s.}$$

Exercise 10.5.3. RIGHT-CONTINUITY OF THE RENEWAL FUNCTION
Show that the renewal function R is right-continuous.

Exercise 10.5.4. ABOUT THE DISTRIBUTION OF $N([0,t])$
In the undelayed case, compute $P(N([0,t]) = n)$ for $n \geq 1$ in terms of the convolution iterates of the inter-renewal distribution and show that $\mathrm{E}\left[e^{\beta N([0,t])}\right] < \infty$ for all $\beta \in [0,\alpha)$, where $\alpha := \mathrm{E}\left[e^{-S_1}\right]$.

Exercise 10.5.5. FIRST EVENT AFTER A RANDOM TIME
In the undelayed case, let X be a strictly positive random variable, independent of $\{S_n\}_{n\geq 1}$, with cumulative distribution G. Let

$$\widetilde{T} = \inf\{T_n\,;\, T_n > X\}\,.$$

Give an expression of the cumulative distribution of $\widetilde{T}$ in terms of G, the renewal function R and the avoidance function v (defined in Section 8.1.3).

---

[10]See [Asmussen, 1987], Problem 2.3, chap. X.

Exercise 10.5.6. FORWARD RECURRENCE TIME

What is the limit distribution of the forward recurrence time of a renewal process (possibly delayed) when S_1 is deterministic, equal to a?

Exercise 10.5.7. WALD'S LEMMA FOR RENEWAL PROCESSES

Let $\{S_n\}_{n\geq 1}$ be a renewal sequence and let $\{N([0,t])\}_{t\geq 0}$ be the counting process of the corresponding (possibly delayed) renewal process. Assume that $E[S_1] < \infty$. Show that for all $t \geq 0$,

$$E[S_1 + \cdots + S_{N([0,t])}] = E[S_1]E[N([0,t])].$$

Exercise 10.5.8. EXPECTED LIFETIME

In Example 10.1.6, compute the expectation of the lifetime L in the transient case.

Exercise 10.5.9. SAFETY LOAD

This exercise refers to Example 7.1.2. Prove that in case of positive safety loading, $\Phi(\infty) = 1$.

Exercise 10.5.10. THE BUS PARADOX

See Example 10.3.10 for the context. Consider an *undelayed* renewal process with finite mean inter-renewal time $E[S_1]$. For $t \geq 0$, consider the interval between the last renewal time before t and the first renewal time after t. Show that in the Poisson case (the interarrival distribution is exponential) the mean length of this interval is asymptotically $2E[S_1]$. Show that it is equal to $E[S_1]$ if and only S_1 is a constant.

Exercise 10.5.11. FIRST EVENT AFTER A FIXED TIME

Refer to Definition 10.1.14 for the notation. Prove that $S_0(t)$ is independent of $\{S_n(t)\}_{n\geq 1}$ and that the latter sequence has the same distribution as $\{S_n\}_{n\geq 1}$.

Exercise 10.5.12. A LIMIT THEOREM FOR CONTINUOUS-TIME HMCS

Let $\{X(t)\}_{t\geq 0}$, be a positive recurrent continuous-time homogeneous Markov chain taking its values in the state space $E = \mathbb{N}$. Let P_0 and E_0 denote respectively probability and expectation given $X(0) = 0$. Let T_0 be the return time to 0 ($T_0 := \inf\{t > 0\,;\, X(t) = 0, X(t-) \neq 0\}$). Recall that in the positive recurrent case, $E_0[T_0] < \infty$. Show that

$$\lim_{t\uparrow\infty} P(X(t) = i) = \frac{E_0\left[\int_0^{T_0} 1_{\{X(s)=i\}}\,ds\right]}{E_0[T_0]}.$$

Exercise 10.5.13. LOTKA–VOLTERRA ASYMPTOTICS

In the Lotka–Volterra model, give the details concerning the asymptotics of the birth rate f in the cases $F(\infty) = 1$ and $F(\infty) > 1$. What can you say about the defective case $F(\infty) < 1$?

Exercise 10.5.14. BACKWARD AND FORWARD RECURRENCE PROCESSES

Refer to Definition 10.3.9. Compute $\lim_{t\to\infty} P(A(t) > x, B(t) > y)$ for $x, y \geq 0$.

Exercise 10.5.15. ASYMPTOTIC VARIANCE OF $N((0, t])$
For a proper renewal process with an interarrival distribution of finite variance, show that

$$\lim_{t \uparrow \infty} \frac{\text{Var} N((0, t])}{t} = \frac{\text{Var} S_1}{E[S_1]}.$$

Exercise 10.5.16. THE AGE REPLACEMENT POLICY
We interpret the random variables $S_1, S_2, \ldots$ as the lifetimes of machines successively put into service, a new machine immediately replacing a failed one. It will be assumed that $E[S_1] < \infty$, and therefore, by (10.4), $\frac{1}{E[S_1]}$ is the asymptotic failure rate per unit time. In some situations, the inconvenience caused by a failure is too important, and the failure rate must be controlled. The age replacement policy suggests that an engine should be replaced at failure time or at a fixed time $T > 0$, whichever occurs first. What is the asymptotic failure rate? (A replacement is not considered as a failure.)

Exercise 10.5.17. ANOTHER MAINTENANCE POLICY
A given machine can be in either one of three states: G (*good*), M (*in maintenance*), or R (*in repair*). Its successive periods where it is in state G (resp., M, R) form an independent and identically distributed sequence $\{S_n\}_{n \geq 0}$ (resp., $\{U_n\}_{n \geq 0}$, $\{V_n\}_{n \geq 0}$) with finite mean. All these sequences are assumed mutually independent. The maintenance policy uses a number $T > 0$. If the machine has age T and has not failed, it goes to state M. If it fails before it has reached age T, it enters state R. From states M and R, the next state is G. Find the steady state probability that the machine is operational. (Note that "good" does not mean "operational". The machine can be "good" but, due to the operations policy, in maintenance, and therefore not operational. However, after a period of maintenance or of repair, we consider that the machine starts anew, and enters a G period.)

Exercise 10.5.18. A TWO STATE SEMI-MARKOV PROCESS
Let

$$\mathbf{P} = \begin{pmatrix} \alpha & 1 - \alpha \\ 1 - \beta & \beta \end{pmatrix},$$

where $\alpha, \beta \in (0, 1)$, and let G_1 and G_2 be two proper cumulative distribution functions. Let $\{X(t)\}_{t \geq 0}$ be the stochastic process evolving as follows. When in state i $(i = 1, 2)$ it stays there for a random time with distribution G_i $(i = 1, 2)$ after which it moves to state j (possibly the same state) with the probability p_{ij} (the (i, j)-entry of $\mathbf{P}$). The successive sojourn times (in either state) are independent given the knowledge of the state the process is in (the reader will clarify this imprecise sentence). What is the asymptotic distribution of the process, that is, what is $\lim_{t \uparrow \infty} P(X(t) = 1)$?

Chapter 11

Brownian Motion

Brownian motion was originally introduced as an idealized representation of the chaotic motion of an isolated particle in water due to the steady bombardment by neighboring molecules. It plays a fundamental role in the theory of stochastic processes and various domains of application such as mathematical finance and communications theory. This chapter is an introduction to some of its more notable properties and to the Wiener–Doob stochastic integral, which is a fundamental tool in the theory of wide-sense stationary processes (Chapter 12).

11.1 Brownian Motion or Wiener Process

11.1.1 As a Rescaled Random Walk

Recall that two complex random variables X and Y in $L^2_{\mathbb{C}}(P)$ are called orthogonal if $E[XY^*] = 0$.

Definition 11.1.1 *A stochastic process* $\{X(t)\}_{t \in \mathbb{R}}$ *is said to have independent (resp., orthogonal) increments if for all* $n \geq 2$ *and for all mutually disjoint intervals* $(a_1, b_1], \ldots,$ $(a_n, b_n]$ *of* $\mathbb{R}$, *the random variables*

$$X(b_1) - X(a_1), \ldots, X(b_n) - X(a_n)$$

are independent (resp., mutually orthogonal).

Clearly, a *centered* second-order stochastic process with independent increments has *a fortiori* orthogonal increments.

Brownian motion is the fundamental example of a Gaussian process.

Definition 11.1.2 *By definition, a* standard Brownian motion, *or* standard Wiener process, *is a continuous centered Gaussian process* $\{W(t)\}_{t \in \mathbb{R}_+}$ *with independent increments and such that* $W(0) = 0$ *and* $\mathrm{Var}(W(b) - W(a)) = b - a$ $([a, b] \subset \mathbb{R}_+)$.

(The existence of a process with the required distribution is guaranteed by Theorem 5.1.23. The existence of a continuous version is proved in the forthcoming Theorem 11.2.7.)

In particular, the PDFs of the vectors $(W(t_1), \ldots, W(t_k))$ $(0 < t_1 < \ldots < t_k)$ are

443

© Springer Nature Switzerland AG 2020
P. Brémaud, *Probability Theory and Stochastic Processes*, Universitext,
https://doi.org/10.1007/978-3-030-40183-2_11

$$\frac{1}{(\sqrt{2\pi})^k \sqrt{t_1 (t_2 - t_1) \cdots (t_k - t_{k-1})}} e^{-\frac{1}{2}\left(\frac{x_1^2}{t_1} + \frac{(x_1+x_2)^2}{t_2-t_1} + \cdots + \frac{(x_1+\cdots+x_k)^2}{t_k-t_{k-1}}\right)}.$$

Note for future reference that (Exercise 11.5.2) for $s, t \in \mathbb{R}_+$,

$$E[W(t)W(s)] = t \wedge s. \tag{11.1}$$

Definition 11.1.3 *The stochastic process with values in* $\mathbb{R}^k$

$$W(t) := (W_1(t), \ldots, W_k(t)) \quad (t \in \mathbb{R}_+),$$

where $\{W_j(t)\}_{t \in \mathbb{R}_+}$ $(1 \leq j \leq k)$ *are independent (standard) Wiener processes, is called a* k-*dimensional (standard) Wiener process.*

The following extension of the definition of Brownian motion slightly enlarges the scope of the theory by introducing a history that is possibly larger than the internal history.

Definition 11.1.4 *Let* $\{\mathcal{F}_t\}_{t \geq 0}$ *be a filtration. A real stochastic process* $\{W(t)\}_{t \geq 0}$ *is called a* standard $\mathcal{F}_t$-*Brownian motion if*

(i) *for P-almost all* ω, *the trajectories* $t \mapsto W(t, \omega)$ *are continuous,*

(ii) *for all* $t \geq 0$, $W(t)$ *is* $\mathcal{F}_t$-*measurable, and*

(iii) $W(0) = 0$ *and for all* $0 \leq s \leq t$, $W(t) - W(s)$ *is a centered Gaussian random variable independent of* $\mathcal{F}_s$ *and with variance* $t - s$.

Consider a symmetric random walk on $\mathbb{Z}$ with initial state 0. It admits the representation

$$X_n = \sum_{k=1}^n Z_k,$$

where $\{Z_n\}_{n \geq 1}$ is an IID sequence of $\{-1, +1\}$-valued random variables with $P(Z_1 = \pm 1) = \frac{1}{2}$. A time-continuous stochastic process is constructed from this sequence as follows. A time-step equal to 1 in the discrete-time model represents in the continuous-time model Δ units of time. The amplitude scale is also modified, a distance of 1 in the original discrete-time model representing δ state space units in the modified model. We are therefore considering the continuous-time process $\{X(t)\}_{t \geq 0}$ given by

$$X(t) := \delta X_{\lfloor t/\Delta \rfloor} = \delta \sum_{k=1}^{\lfloor t/\Delta \rfloor} Z_k. \tag{11.2}$$

(The dependence on δ and Δ will not be made explicit in the notation for $X(t)$.) Since the Z_k's are centered and of variance 1,

$$E[X(t)] = 0 \text{ and } \operatorname{Var}(X(t)) = \delta^2 \times \lfloor t/\Delta \rfloor.$$

Let now Δ and δ tend to 0 in such a way that the limit in distribution exists and is not trivial. With respect to this goal, the choice $\delta = \Delta$ is not satisfactory since $E[X(t)] = 0$ and $\lim_{\Delta \downarrow 0} \operatorname{Var}(X(t)) = 0$, leading to a null process. With the choice

$\delta^2 = \Delta$, $\mathrm{E}\,[X(t)] = 0$ and $\lim_{\Delta\downarrow0} \mathrm{Var}\,(X(t)) = t$. In this case, since by the central limit theorem

$$\frac{\sum_{k=1}^{n} Z_k}{\sqrt{n}} \xrightarrow{\mathcal{D}} \mathcal{N}(0,1)\,,$$

we have that (using the fact that if a sequence of random variables X_n converges in distribution to some random variable X and if the sequence of real numbers a_n converges to the real number a, then $a_n X_n$ converges in distribution to aX; Theorem 4.4.8)

$$\frac{X(t)}{\sqrt{t}} \;=\; \frac{\sum_{k=1}^{\lfloor t/\Delta\rfloor} Z_k}{\sqrt{\lfloor \frac{t}{\Delta}\rfloor}} \frac{\sqrt{\lfloor \frac{t}{\Delta}\rfloor}}{\sqrt{\frac{t}{\Delta}}} \xrightarrow{\mathcal{D}} \mathcal{N}(0,1)\,.$$

Thus at the limit (in distribution) as $n \uparrow \infty$, $X(t)$ is a centered Gaussian variable with variance t.

In fact, for all $t_1,\ldots,t_k$ in $\mathbb{R}_+$ forming an increasing sequence, the limit distribution of the vector $(X(t_1),\ldots,X(t_k))$ corresponds to a Brownian motion (Exercise 11.5.1).

Behavior at Infinity

In view of the previous approximation of a Wiener process as a symmetric random walk on $\mathbb{Z}$, the following result is expected.

Theorem 11.1.5 *Let $\{W(t)\}_{t\geq0}$ be a standard Brownian motion. Then, P-a.s.,*

$$\limsup_{t\uparrow\infty} W(t) = +\infty\,, \quad \liminf_{t\uparrow+\infty} W(t) = -\infty\,.$$

Proof. This follows from

$$\limsup_{n\uparrow\infty} W(n) = +\infty\,, \quad \liminf_{n\uparrow+\infty} W(n) = -\infty \quad P\text{-a.s.}$$

which in turn is a direct consequence of Theorem 4.3.9 applied to the random walk $S_n = W(n)$ of step $X_n = W(n) - W(n-1)$. $\qquad\square$

Corollary 11.1.6 *For all $a \in \mathbb{R}$, the $\mathcal{F}_t^W$-stopping time $T_a = \inf\{t \geq 0\,;\, W(t) \geq a\}$ is almost surely finite.*

(An explicit expression for the distribution of T_a will be given in Subsection 11.2.1.)

11.1.2 Simple Operations on Brownian motion

These are the operations of

(i) symmetrization:
$$X(t) = -W(t) \quad (t \geq 0)\,,$$

(ii) delay: for $a > 0$,
$$X(t) = W(t+a) - W(t) \quad (t \geq 0)\,,$$

(iii) scaling: for $c > 0$,
$$X(t) = \sqrt{c}W\left(\frac{t}{c}\right) \quad (t \geq 0)\,,$$

(iv) time inversion

$$X(t) = tW\left(\frac{1}{t}\right) \quad (t > 0) \text{ and } X(0) = 0,$$

In each case, the process $\{X(t)\}_{t\geq 0}$ is a standard Brownian motion if $\{W(t)\}_{t\geq 0}$ is a standard Brownian motion. The fact that these processes have the same distribution as the standard Brownian motion is easily checked (Exercise 11.5.3). They are also continuous. This is obvious in all cases except for the continuity at 0 of the process obtained by time inversion. However, almost surely:

$$\lim_{t\downarrow 0} tW\left(\frac{1}{t}\right) = 0.$$

Proof. We prove the equivalent statement

$$\lim_{s\uparrow +\infty} \frac{1}{s}W(s) = 0.$$

First observe that since $W(n)$ is the sum of n IID centered Gaussian variables, $\lim_{n\uparrow\infty}\frac{W(n)}{n} = 0$ by the strong law of large numbers and *a fortiori* $\lim_{n\uparrow\infty}\frac{W(n)}{n^2} = 0$. In the following let $n := n(s)$ be the largest integer less than or equal to s. Therefore, taking into account the first observation, we just have to show that

$$\frac{W(s)}{s} - \frac{W(n)}{n} \to 0 \text{ as } s \to \infty.$$

But

$$\left|\frac{W(s)}{s} - \frac{W(n)}{n}\right| \leq \left|\frac{W(s)}{s} - \frac{W(n)}{s}\right| + \left|\frac{W(n)}{s} - \frac{W(n)}{n}\right|$$

$$\leq \frac{1}{n}\sup_{s\in[n,n+1]}|W(s) - W(n)| + |W(n)|\left|\frac{1}{s} - \frac{1}{n}\right|$$

$$\leq \frac{Z_n}{n} + \frac{W(n)}{n^2},$$

where

$$Z_n := \sup_{s\in[0,1]}|W(s+n) - W(n)|$$

has the same distribution as $\sup_{s\in[0,1]}|W(s)|$ and the sequence $\{Z_n\}_{n\geq 1}$ is IID. Since as observed earlier $\frac{W(n)}{n^2} \to 0$, it remains to show that $\frac{Z_n}{n} \to 0$ or, equivalently (Theorem 4.1.3), that for any $\varepsilon > 0$,

$$P\left(\left|\frac{Z_n}{n}\right| > \varepsilon \text{ i.o.}\right) = 0.$$

By the Borel–Cantelli lemma, it suffices to show that

$$\sum_n P(|Z_n| > n\varepsilon) < \infty$$

or, equivalently, since the Z_n's are identically distributed,

$$\sum_n P(|Z_1| > n\varepsilon) < \infty.$$

But this is true of any integrable random variable Z_1 (see Exercise 4.6.1).   □

The Brownian Bridge

This is the process $\{X(t)\}_{t\in[0,1]}$ obtained from the standard Brownian motion $\{W(t)\}_{t\in[0,1]}$ by

$$X(t) := W(t) - tW(1) \quad (t \in [0,1]).$$

It is a Gaussian process since for all $t_1, \ldots, t_k \in [0,1]$, the random vector $(X(t_1), \ldots, X(t_k))$ is Gaussian, being a linear function of the Gaussian vector $(W(t_1), \ldots, W(t_k), W(1))$. In particular, since it is a centered Gaussian process, its distribution is entirely characterized by its covariance function and a simple calculation (Exercise 11.5.2) gives

$$\text{cov}(X(t), X(s)) = s(1-t) \quad (0 \le s \le t \le 1).$$

In particular, $X(0) = X(1) = 0$.

The Brownian bridge $\{X(t)\}_{t\in[0,1]}$ is distributionwise a Wiener process $\{W(t)\}_{t\in[0,1]}$ conditioned by $W(1) = 0$. This statement is problematic in that the conditioning event has a null probability. However, it is true "at the limit":

Theorem 11.1.7 *Let $f : \mathbb{R}^k \to \mathbb{R}$ be a bounded and continuous function. Then, for any $0 \le t_1 < t_2 < \cdots < t_k \le 1$,*

$$\lim_{\varepsilon\downarrow0} \mathrm{E}\left[f(W(t_1), \ldots, W(t_k)) \mid |W(1)| \le \varepsilon\right] = \mathrm{E}\left[f(X(t_1), \ldots, X(t_k))\right].$$

Proof.

$$\mathrm{E}\left[f(W(t_1), \ldots, W(t_k)) \mid |W(1)| \le \varepsilon\right]$$
$$= \mathrm{E}\left[f(X(t_1) + t_1W(1), \ldots, X(t_k) + t_kW(1)) \mid |W(1)| \le \varepsilon\right]$$
$$= \frac{\mathrm{E}\left[f(X(t_1) + t_1W(1), \ldots, X(t_k) + t_kW(1))\mathbf{1}_{|W(1)|\le\varepsilon}\right]}{P(|W(1)| \le \varepsilon)}.$$

In view of the independence of $\{X(t)\}_{t\in[0,1]}$ and $W(1)$ (Exercise 11.5.13), this last quantity equals

$$\frac{\int_{-\varepsilon}^{+\varepsilon} e^{-\frac{1}{2}x^2}\mathrm{E}\left[f(X(t_1) + t_1x, \ldots, X(t_k) + t_kx)\right]\,\mathrm{d}x}{\int_{-\varepsilon}^{+\varepsilon} e^{-\frac{1}{2}x^2}\,\mathrm{d}x},$$

which tends to $\mathrm{E}\left[f(X(t_1), \ldots, X(t_k))\right]$ as $\varepsilon\downarrow0$. $\qquad\square$

11.1.3 Gauss–Markov Processes

Definition 11.1.8 *Let $\mathbf{T}$ be $\mathbb{R}_+$ or $\mathbb{N}$. A real-valued stochastic process $\{X(t)\}_{t\ge0}$ is called a Markov process if for all $t \ge 0$ and all non-negative (or integrable) random variables Z that are $\sigma(X(s)\,;\,s \ge t)$-measurable*

$$\mathrm{E}\left[Z \mid \mathcal{F}_t^X\right] = \mathrm{E}\left[Z \mid \sigma(X(t))\right]. \tag{11.3}$$

Gaussian processes that are moreover Markovian are called *Gauss–Markov processes*. This class of models receives a simple description in terms of the Brownian motion.

EXAMPLE 11.1.9: THE BROWNIAN MOTION IS GAUSS–MARKOV. The Brownian motion is a Gauss–Markov process (Exercise 11.5.6).

Gauss–Markov processes are characterized among Gaussian processes by a simple property of their covariance function.

Theorem 11.1.10 *Let $\{X(t)\}_{t\geq 0}$ be a centered Gaussian process with continuous co-variance function Γ such that $\Gamma(t,t) > 0$ for all $t \in \mathbb{R}_+$. It is a Markov process if and only if there exist functions f and g such that for all $s,t \in \mathbb{R}_+$*

$$\Gamma(t,s) = f(t \vee s)g(t \wedge s). \tag{11.4}$$

The proof relies on the following lemma.

Lemma 11.1.11 *Let $\{X(t)\}_{t\geq 0}$ be a centered Gaussian process with covariance function Γ such that $\Gamma(t,t) > 0$ for all $t \in \mathbb{R}_+$. If in addition it is a Markov process, then for all $t > s > t_0 \geq 0$,*

$$\Gamma(t,t_0) = \frac{\Gamma(t,s)\Gamma(s,t_0)}{\Gamma(s,s)}. \tag{11.5}$$

Proof. By the Gaussian property, the conditional expectation of $X(t)$ given $X(t_0)$ is equal to the linear regression of $X(t)$ on $X(t_0)$:

$$\mathrm{E}\left[X(t)|X(t_0)\right] = \frac{\Gamma(t,t_0)}{\Gamma(t_0,t_0)}X(t_0). \tag{$\star$}$$

Using this remark and the Markov property,

$$
\begin{aligned}
\mathrm{E}\left[X(t)|X(t_0)\right] &= \mathrm{E}\left[\mathrm{E}\left[X(t)|X(t_0),X(s)\right]|X(t_0)\right] \\
&= \mathrm{E}\left[\mathrm{E}\left[X(t)|X(s)\right]|X(t_0)\right] \\
&= \mathrm{E}\left[\frac{\Gamma(t,s)}{\Gamma(s,s)}X(s)|X(t_0)\right] \\
&= \frac{\Gamma(t,s)}{\Gamma(s,s)}\mathrm{E}\left[X(s)|X(t_0)\right] = \frac{\Gamma(t,s)}{\Gamma(s,s)}\frac{\Gamma(s,t_0)}{\Gamma(t_0,t_0)}X(t_0).
\end{aligned}
$$

Comparing with the right-hand side of $(\star)$, and since $P(X(t_0) \neq 0) > 0$ (in fact $= 1$), we obtain (11.5). $\square$

We now turn to the proof of Theorem 11.1.10.

Proof. Necessity. Suppose the process is Gauss–Markov. Let

$$\rho(t,s) = \frac{\Gamma(t,s)}{\left(\Gamma(t,t)\right)^{\frac{1}{2}}\left(\Gamma(s,s)\right)^{\frac{1}{2}}}$$

be its autocorrelation function. By (11.5), for all $t > s > t_0 \geq 0$,

$$\rho(t,t_0) = \rho(t,s)\rho(s,t_0). \tag{$\star\star$}$$

We show that $\rho(t,s) > 0$ for all $t,s \in \mathbb{R}$. Indeed, assuming $s > t$ and using $(\star\star)$ repeatedly, for all $n \geq 1$,

$$\rho(t,s) = \prod_{k=0}^{n-1}\rho\left(t+\frac{k(s-t)}{n}, t+\frac{(k+1)(s-t)}{n}\right),$$

and therefore, using the facts that $\rho(u,u) = 1$ for all u and that ρ is uniformly continuous on bounded intervals, n can be chosen large enough to guarantee that all the elements in the above product are positive. Therefore, one may divide by $\rho(t, t_0)$ and write $(\star\star)$ as

$$\rho(t, s) = \frac{\rho(t, t_0)}{\rho(s, t_0)}$$

or

$$\Gamma(t, s) = \rho(t, t_0)\Gamma(t, t)^{\frac{1}{2}} \times \frac{\Gamma(s, s)^{\frac{1}{2}}}{\rho(s, t_0)},$$

from which we obtain the desired conclusion (here $s = t \wedge s$ and $t = t \vee s$).

Sufficiency. Suppose that the process is Gaussian and that (11.5) holds true. Assume $t > s$. Therefore $\Gamma(t, s) = f(t)g(s)$. By Schwarz's inequality, $\Gamma(t, s) \leq \Gamma(t, t)^{\frac{1}{2}}\Gamma(s, s)^{\frac{1}{2}}$ or, equivalently, $f(t)g(s) \leq (f(t)g(t)f(s)g(s))^{\frac{1}{2}}$, from which it follows that $f(t)g(s) \leq g(t)f(s)$. Therefore, the function

$$\tau(t) := \frac{g(t)}{f(t)}$$

is monotone non-decreasing. In particular, the centered Gaussian process

$$Y(t) := f(t)W(\tau(t))$$

is a Markov process since the Brownian motion itself is a Markov process. Its covariance function is

$$\begin{aligned} \mathrm{E}\left[Y(t)Y(s)\right] &= f(t)f(s)\mathrm{E}\left[W(\tau(t))W(\tau(s))\right] \\ &= f(t)f(s)(\tau(t) \wedge \tau(s)) \\ &= f(t)f(s)\tau(s) = f(t)g(s). \end{aligned}$$

Since it has the same covariance as $\{X(t)\}_{t\geq 0}$ and since both processes are centered and Gaussian, they have the same distribution. In particular, $\{X(t)\}_{t\geq 0}$ is a Markov process. □

11.2 Properties of Brownian Motion

11.2.1 The Strong Markov Property

Theorem 11.2.1 *Let $\{W(t)\}_{t\geq 0}$ be a standard $\mathcal{F}_t$-Brownian motion and let τ be a finite $\mathcal{F}_t$-stopping time. Then $\{W(\tau + t)\}_{t\geq 0}$ is a standard $\mathcal{F}_t$-Brownian motion independent of $\mathcal{F}_\tau$.*

A proof is given in Subsection 14.3.2 via Itô calculus.

The Reflection Principle

Theorem 11.2.2 *Let $\{W(t)\}_{t\geq 0}$ be a standard $\mathcal{F}_t$-Brownian motion and let T_a be the first time it reaches the value $a > 0$. The stochastic process*

$$Y(t) := W(t)1_{t<T_a} + (2a - W(t))1_{t\geq T_a} \quad (t \geq 0) \tag{11.6}$$

is a standard Brownian motion.

In other words, after time T_a the original trajectory is replaced by its symmetric reflection about the line $y = a$. The operation that changes the Brownian motion $\{W(t)\}_{t \geq 0}$ into the process (11.6) is called "reflection at level a".

Theorem 11.2.2 is an intuitive result in view of the strong Markov property of Brownian motion. See however Exercise 14.4.2 for a proof based on Itô's calculus.

Theorem 11.2.3 Let $M(t) := \sup_{s \in [0,t]} W(s)$. For $a > 0$ and $y \geq 0$,

$$P(W(t) \leq a - y, M(t) \geq a) = P(W(t) \geq a + y). \tag{11.7}$$

Proof. Observing that $\{M(t) \geq a\} \equiv \{T_a \leq t\}$, that $T_a = \inf\{t \geq 0 \,; Y(t) = a\}$ and that $\{W(t) \geq a + y\} \subseteq \{T_a \leq t\}$, and using Theorem 11.2.2,

$$
\begin{aligned}
P(W(t) \leq a - y, M(t) \geq a) &= P(W(t) \leq a - y, T_a \leq t) \\
&= P(Y(t) \leq a - y, T_a \leq t) \\
&= P(2a - W(t) \leq a - y, T_a \leq t) \\
&= P(W(t) \geq a + y, T_a \leq t) = P(W(t) \geq a + y).
\end{aligned}
$$

$\square$

Corollary 11.2.4 For $a > 0$,

$$P(T_a \leq t) = P(M(t) \geq a) = 2P(W(t) > a) = P(|W(t)| > a).$$

Proof. The last equality is by symmetry of Brownian motion. The first equality is a consequence of the identity $\{M(t) \geq a\} \equiv \{T_a \leq t\}$. It remains to prove the second equality. We have

$$
\begin{aligned}
P(M(t) \geq a) &= P(M(t) \geq a, W(t) \leq a) + P(M(t) \geq a, W(t) > a) \\
&= P(M(t) \geq a, W(t) \leq a) + P(W(t) > a) \\
&= P(W(t) > a) + P(W(t) > a),
\end{aligned}
$$

where it was observed that $\{W(t) > a\} \subseteq \{M(t) \geq a\}$ for the second equality, and where (11.7) was applied with $y = 0$ for the third one.

$\square$

The above results immediately yield the distribution of T_a: For $a \geq 0$ and $t \geq 0$,

$$P(T_a \leq t) = \frac{2}{\sqrt{2\pi}} \int_{\frac{a}{\sqrt{t}}}^{\infty} e^{-\frac{y^2}{2}} \, dy.$$

Since the law of T_{-a} is the same as that of T_a, the formula for any a is

$$P(T_a \leq t) = \frac{2}{\sqrt{2\pi}} \int_{\frac{|a|}{\sqrt{t}}}^{\infty} e^{-\frac{y^2}{2}} \, dy.$$

Definition 11.2.5 A real stochastic process $\{X(t)\}_{t \geq 0}$ with stationary increments is called recurrent if for all $x, y \in \mathbb{R}$, this process starting at time 0 from x will almost surely reach y in finite random time. It is called null recurrent if it is recurrent but this random time is not integrable.

Corollary 11.2.6 *Brownian motion is null recurrent.*

Proof. It suffices to verify the conditions of null recurrence of the above definition for $x = 0$. Letting $t \uparrow \infty$ in the expression of the cumulative distribution function of T_y ($y \in \mathbb{R}$), we obtain $P(T_y < \infty) = 1$. On the other hand, with $\alpha := \frac{2}{\sqrt{2\pi}}$ and $\beta := y$,

$$\mathrm{E}\,[T_y] = \int_0^\infty P(T_y \le t)\,\mathrm{d}t = \alpha \int_0^\infty \left(\int_0^{\frac{\beta}{\sqrt{t}}} e^{-\frac{1}{2}u^2}\,\mathrm{d}u \right)\mathrm{d}t$$

$$= \alpha \int_0^\infty \left(\int_0^{\frac{\beta^2}{u^2}} \mathrm{d}t \right) e^{-\frac{1}{2}u^2}\,\mathrm{d}u = \alpha\beta^2 \int_0^\infty \frac{1}{u^2} e^{-\frac{1}{2}u^2}\,\mathrm{d}u = +\infty\,.$$

$\square$

11.2.2 Continuity

Theorem 11.2.7 *Consider a stochastic process such as the Brownian motion of Definition 11.1.2, except that continuity of the trajectories is not assumed. There exists a version of it having almost surely continuous paths.*

Proof. For any $s, t \in \mathbb{R}_+$, $W(t) - W(s)$ has the same distribution as $|t - s|^{\frac{1}{2}} Y$ where Y is a centered Gaussian variable with unit variance. In particular, for any $\alpha > 0$,

$$\mathrm{E}\,[|W(t) - W(s)|^\alpha] = |t - s|^{\frac{\alpha}{2}} E|Y|^\alpha,$$

from which the result immediately follows by application of Theorem 5.2.3: take $\alpha > 2$, $\beta = \frac{1}{2}\alpha - 1$ and $K = E|Y|^\alpha$. $\square$

11.2.3 Non-differentiability

Definition 11.1.2 of the Wiener process does not tell much about the qualitative behavior of its trajectories. Although the trajectories of the (standard) Brownian motion are almost surely continuous functions, their behavior is otherwise rather chaotic. First of all observe that, for fixed $t_0 > 0$, the random variable

$$\frac{W(t_0 + h) - W(t_0)}{h} \overset{\mathcal{D}}{\sim} \mathcal{N}\left(0, h^{-1}\right)$$

does not converge in distribution as $h \downarrow 0$, and *a fortiori* does not converge almost surely. Therefore, for any $t_0 > 0$,

$$P\,(t \mapsto W(t) \text{ is not differentiable at } t_0) = 1\,.$$

But the situation is even more dramatic:

Theorem 11.2.8 *Almost all the paths of the Wiener process are nowhere differentiable.*

Proof. We shall prove that

$$P\left(\limsup_{h \to 0} \frac{W(t + h) - W(t)}{h} = +\infty \text{ for all } t \in [0, 1] \right) = 1\,.$$

Fix $\beta > 0$. If a function $f : (0, 1) \to \mathbb{R}$ has at some point $s \in [0, 1]$ a derivative $f'(s)$ of absolute value smaller than β, then there exists an integer n_0 such that for $n \geq n_0$,

$$|f(t) - f(s)| < 2\beta \, |t - s| \quad \text{if } |t - s| \leq \frac{2}{n}. \tag{$\star$}$$

Let

$$C_n := \{f : [0, 1] \mapsto \mathbb{R} \, ; \quad \text{there exists an } s \in [0, 1] \text{ satisfying } (\star)\}$$

Let

$$A_n := \{\omega \, ; \quad \text{the function } t \in [0, 1] \mapsto W(t, \omega) \text{ is in } C_n\} \, .$$

This event increases with n and its limit A includes all the samples ω corresponding to a trajectory $t \in (0, 1) \to W(t, \omega)$ having at least at one point of $[0, 1]$ a derivative of absolute value smaller than β. Therefore it suffices to show that $P(A) = 0$ for all $\beta > 0$.

If $\omega \in A_n$, letting k be the largest integer such that $\frac{k}{n} \leq s$ (where s is the point in the definition of C_n), and letting

$$Y_k(\omega) := \max_{j=-1,0,+1} \left\{ \left| W\left(\frac{k + j + 1}{n}, \omega\right) - W\left(\frac{k + j}{n}, \omega\right) \right| \right\} \, ,$$

then $Y_k(\omega) \leq \frac{6\beta}{n}$. Therefore,

$$A_n \subseteq B_n := \left\{ \omega \, ; \quad \text{at least one } Y_k(\omega) \leq \frac{6\beta}{n} \right\} \, .$$

In order to prove that $P(A) = 0$, it is then enough to show that $\lim_n P(B_n) = 0$. But

$$B_n = \bigcup_{k=1}^{n-2} \left\{ \omega \, ; \, Y_k(\omega) \leq \frac{6\beta}{n} \right\} \, ,$$

and by sub-σ-additivity,

$$P(B_n) \leq \sum_{k=1}^{n-2} P\left(\max_{j=-1,0,+1} \left\{ \left| W\left(\frac{k + j + 1}{n}\right) - W\left(\frac{k + j}{n}\right) \right| \right\} \leq \frac{6\beta}{n} \right) \, .$$

By the independence property of the increments of a Wiener process and since all the variables involved are $\mathcal{N}\left(0, \frac{1}{n}\right)$,

$$P(B_n) \leq nP\left(\left| \mathcal{N}\left(0, \frac{1}{n}\right) \right| \leq \frac{6\beta}{n} \right)^3$$

$$= n \left(\sqrt{\frac{n}{2\pi}} \int_{-\frac{6\beta}{n}}^{+\frac{6\beta}{n}} e^{-\frac{nx^2}{2}} \, dx \right)^3$$

$$= n \left(\frac{1}{\sqrt{2\pi n}} \int_{-6\beta}^{+6\beta} e^{-\frac{x^2}{2n}} \, dx \right)^3 \to 0 \, .$$

$\square$

If a function $f : [0, 1] \to \mathbb{R}$ is of bounded variation on $[0, 1]$, it has a derivative almost everywhere (with respect to the Lebesgue measure) in $[0, 1]$. Therefore:

Corollary 11.2.9 *Almost every trajectory of a Brownian motion is of unbounded variation on any interval.*

11.2.4 Quadratic Variation

Let

$$\mathcal{D} := \{0 = t_0 \leq t_1 \leq \ldots \leq t_n = t\}$$

be a division of the interval $[0, t]$ with maximum gap

$$\Delta := \max_{1 \leq k \leq n} (t_k - t_{k-1}).$$

Let

$$V_W(t, \mathcal{D}) := \sum_{k=1}^{n} |W(t_k) - W(t_{k-1})| .$$

By Corollary 11.2.9, the variation $\sup_{\mathcal{D}} V_W(t, \mathcal{D})$ of the Wiener process on the interval $[0, t]$ is almost surely infinite. However, the *quadratic variation* of this process on the interval $[0, t]$, defined by

$$Q_W(t, \mathcal{D}) := \sum_{k=1}^{n} (W(t_k) - W(t_{k-1}))^2 ,$$

is such that

$$E[Q_W(t, \mathcal{D})] = \sum_{k=1}^{n} E\left[(W(t_k) - W(t_{k-1}))^2\right] = \sum_{k=1}^{n} (t_k - t_{k-1}) = t.$$

But there is more:

Theorem 11.2.10

A. *As $\Delta \to 0$,*

$$Q_W(t, \mathcal{D}) \to t \text{ in } L^2_{\mathbb{R}}(P).$$

B. *If $\{\mathcal{D}_i\}_{i \geq 0}$ is a sequence of subdivisions of $[0, t]$ such that $\Delta_i = o(i^{-2})$, then*

$$\lim_{i \uparrow \infty} Q_W(t, \mathcal{D}_i) \to t, \quad P\text{-a.s.}$$

Proof. A. Write

$$Q_W(t, \mathcal{D}) - t = \sum_{k=1}^{n} Z_k ,$$

where $Z_k = (W(t_k) - W(t_{k-1}))^2 - (t_k - t_{k-1})$, a centered random variable with variance

$$E[Z_k^2] = 2(t_k - t_{k-1})^2 .$$

Therefore, by independence of the Z_k's,

$$E[(Q_W(t, \mathcal{D}) - t)^2] = \sum_{k=1}^{n} E\left[Z_k^2\right]$$

$$= 2 \sum_{k=1}^{n} (t_k - t_{k-1})^2 \leq 2\Delta \sum_{k=1}^{n} (t_k - t_{k-1}) = 2\Delta t.$$

B. Take $\Delta_i = \varepsilon_i/i^2$ with $\lim_{i\uparrow\infty} \varepsilon_i = 0$. By Markov's inequality,

$$P\left(|Q_W(t,\mathcal{D}_i) - t| > i\sqrt{2\Delta_i}\right) = P\left(|Q_W(t,\mathcal{D}_i) - t|^2 > 2\varepsilon_i\right)$$

$$\leq \frac{\mathrm{E}\left[|Q_W(t,\mathcal{D}_i) - t|^2\right]}{2\varepsilon_i} \leq \frac{2\Delta_i t}{2\varepsilon_i} = \frac{t}{i^2}.$$

The announced result then follows from Theorem 4.1.2.

$\square$

11.3   The Wiener–Doob Integral

11.3.1   Construction

The *Wiener stochastic integral*

$$\int_{\mathbb{R}} f(t)\, \mathrm{d}W(t) \tag{11.8}$$

will be defined for a certain class of measurable functions f. Note, however, that this integral will not be of the usual type. For instance, it cannot be defined pathwise as a Stieltjes–Lebesgue integral since the trajectories of the Brownian motion are of unbounded variation. This integral cannot be interpreted as $\int_{\mathbb{R}} f(t)\dot{W}(t)\,\mathrm{d}t$ either (the dot denotes differentiation) since the Brownian motion does not have a derivative. Therefore, the integral in (11.8) will be defined in a radically different way. In fact, the *Doob–Wiener stochastic integral* is defined with respect to a stochastic process with centered and uncorrelated increments. This generalizes the original Wiener integral (which is defined with respect to the Brownian motion).

Let $\{Z(t)\}_{t\in\mathbb{R}}$ be a complex-valued stochastic process such that

(i) for all intervals $[t_1, t_2] \subset \mathbb{R}$, the increments $Z(t_2) - Z(t_1)$ are centered and in $L^2_{\mathbb{C}}(P)$, and

(ii) there exists a locally finite measure μ on $(\mathbb{R}, \mathcal{B})$ such that

$$\mathrm{E}[(Z(t_2) - Z(t_1))(Z(t_4) - Z(t_3))^*] = \mu((t_1, t_2] \cap (t_3, t_4]) \tag{11.9}$$

for all $[t_1, t_2] \subset \mathbb{R}$ and all $[t_3, t_4] \subset \mathbb{R}$. Note in particular that if $(t_1, t_2] \cap (t_3, t_4] = \varnothing$, $Z(t_2) - Z(t_1)$ and $Z(t_4) - Z(t_3)$ are orthogonal random variables of $L^2_{\mathbb{C}}(P)$.

Definition 11.3.1 *The above stochastic process $\{Z(t)\}_{t\in\mathbb{R}}$ is called a stochastic process with* centered *and* uncorrelated *increments* with *structural measure μ.*

EXAMPLE 11.3.2: THE STRUCTURAL MEASURE OF THE WIENER PROCESS. The Wiener process $\{W(t)\}_{t\in\mathbb{R}}$ is such a process, with structural measure equal to the Lebesgue measure.

EXAMPLE 11.3.3: THE STRUCTURAL MEASURE OF A COMPENSATED HPP. Let N be an HPP on $\mathbb{R}$ with intensity λ. Define $\{Z(t)\}_{t \in \mathbb{R}}$ by $Z(0) = 0$ and, for all $[a, b] \in \mathbb{R}$, $Z(b) - Z(a) = N((a, b]) - \lambda \times (b - a)$. Then $\{Z(t)\}_{t \in \mathbb{R}}$ is a stochastic process with centered and uncorrelated increments whose structural measure is λ times the Lebesgue measure.

The Wiener–Doob integral

$$\int_{\mathbb{R}} f(t) \, dZ(t)$$

is constructed for all $f \in L^2_{\mathbb{C}}(\mu)$ in the following manner. First of all, we define this integral for all $f \in \mathcal{L}$, the vector subspace of $L^2_{\mathbb{C}}(\mu)$ formed by the finite complex linear combinations of interval indicator functions

$$f(t) = \sum_{i=1}^{N} \alpha_i 1_{(a_i, b_i]}(t).$$

For such functions, *by definition,*

$$\int_{\mathbb{R}} f(t) \, dZ(t) := \sum_{i=1}^{N} \alpha_i (Z(b_i) - Z(a_i)). \tag{$\star$}$$

One easily verifies that the linear mapping

$$\varphi : f \in \mathcal{L} \to \int_{\mathbb{R}} f(t) \, dZ(t) \in L^2_{\mathbb{C}}(P)$$

is an isometry, that is, for all $f \in \mathcal{L}$,

$$\int_{\mathbb{R}} |f(t)|^2 \, \mu(dt) = E \left[\left| \int_{\mathbb{R}} f(t) \, dZ(t) \right|^2 \right].$$

Since $\mathcal{L}$ is dense in $L^2_{\mathbb{C}}(\mu)$, φ can be uniquely extended to an isometric linear mapping of $L^2_{\mathbb{C}}(\mu)$ into $L^2_{\mathbb{C}}(P)$. We continue to call this extension φ and then define, for all $f \in L^2_{\mathbb{C}}(\mu)$,

$$\int_{\mathbb{R}} f(t) \, dZ(t) := \varphi(f).$$

The fact that φ is an isometry is expressed by *Doob's isometry formula*:

$$E \left[\left(\int_{\mathbb{R}} f(t) \, dZ(t) \right) \left(\int_{\mathbb{R}} g(t) \, dZ(t) \right)^* \right] = \int_{\mathbb{R}} f(t) g^*(t) \, \mu(dt), \tag{11.10}$$

where f and g are in $L^2_{\mathbb{C}}(\mu)$. Note also that for all $f \in L^2_{\mathbb{C}}(\mu)$,

$$E \left[\int_{\mathbb{R}} f(t) \, dZ(t) \right] = 0, \tag{11.11}$$

since the Doob integral is the limit in $L^2_{\mathbb{C}}(\mu)$ of random variables of the type $\sum_{i=1}^{N} \alpha_i (Z(b_i) - Z(a_i))$ that have mean 0 (use the continuity of the inner product in $L^2_{\mathbb{C}}(P)$).

Remark 11.3.4 In the case where $Z(t) := W(t)$ $(t \in \mathbb{R}_+)$, a Wiener process, the right-hand side of $(\star)$ is in the Gaussian Hilbert subspace $H(W)$, and so are the Wiener integrals, being limits in quadratic mean of elements of $H(W)$.

Series Expansion of Wiener integrals

Let f be a function of $L^2_{\mathbb{R}}([a,b])$ and let $\{W(t)\}_{t\in[a,b]}$ be a standard Wiener process. Let $\{\varphi_n\}_{n\geq 1}$ be an orthonormal basis of the Hilbert space $L^2_{\mathbb{R}}([a,b])$. In particular,

$$f = \sum_{n=1}^{\infty} \langle f, \varphi_n \rangle \varphi_n \,,$$

where the convergence of the series of the right-hand side is in $L^2_{\mathbb{R}}([a,b])$.

Consider now the sequence of random variables

$$Z_n := \int_a^b \varphi_n(t)\, dW(s) \quad (n \geq 1)\,.$$

This is a Gaussian sequence (Remark 11.3.4). Moreover, the Z_n's are uncorrelated since by the isometry formula for the Doob–Wiener integrals, if $n \neq k$,

$$\mathrm{E}\left[Z_n Z_k\right] = \int_a^b \varphi_n(t)\varphi_k(t)\, dt = 0\,.$$

Therefore $\{Z_n\}_{n\geq 1}$ is an IID sequence.

Theorem 11.3.5 *For $f \in L^2_{\mathbb{R}}([a,b])$, we have the expansion*

$$\int_a^b f(t)\, dW(s) = \sum_{n=1}^{\infty} \langle f, \varphi_n \rangle Z_n\,,$$

where the convergence of the series in the right-hand side is in $L^2_{\mathbb{R}}(P)$ and almost surely.

Proof. It is enough to prove convergence in $L^2_{\mathbb{R}}(P)$ since the statement about almost-sure convergence then follows from Theorem 4.1.15 and the fact that when a sequence of random variables converges almost surely and in $L^2_{\mathbb{R}}(P)$, the respective limits are almost surely equal.

For convergence in $L^2_{\mathbb{R}}(P)$:

$$\mathrm{E}\left[\left(\int_a^b f(t)\, dW(t) - \sum_{n=1}^{N}\langle f, \varphi_n\rangle Z_n\right)^2\right]$$

$$= \mathrm{E}\left[\left(\int_a^b f(t)\, dW(t)\right)^2\right] - 2\sum_{n=1}^{N}\langle f, \varphi_n\rangle \mathrm{E}\left[\int_a^b f(t)\, dW(t) Z_n\right]$$

$$+ \sum_{n=1}^{N}\langle f, \varphi_n\rangle^2 \mathrm{E}\left[Z_n^2\right]$$

$$= \int_a^b f(t)^2\, dt - 2\sum_{n=1}^{N}\langle f, \varphi_n\rangle \int_a^b f(t)\varphi_n(t)\, dt + \sum_{n=1}^{N}\langle f, \varphi_n\rangle^2 \mathrm{E}\left[Z_n^2\right]$$

$$= \int_a^b f(t)^2\, dt - 2\sum_{n=1}^{N}\langle f, \varphi_n\rangle^2 + \sum_{n=1}^{N}\langle f, \varphi_n\rangle^2 = \int_a^b f(t)^2\, dt - \sum_{n=1}^{N}\langle f, \varphi_n\rangle^2 \to 0\,.$$

$\square$

Remark 11.3.6 For the purpose of sampling the integral $\int_a^b f(t)\, dW(s)$ (that is, of generating a random variable with the same distribution as $\int_a^b f(t)\, dW(s)$) it is enough to use any sequence $\{Z_n\}_{n\geq 1}$ of independent standard Gaussian random variables.

A Characterization of the Wiener Integral

The following characterisation of the Wiener integral will be useful:

Lemma 11.3.7 *Let $f \in \mathcal{L}_\mathbb{R}^2(\ell)$ and let $\{W(t)\}_{t \in [0,1]}$ be a standard Wiener process. Denote by $H(W)$ the Gaussian Hilbert space generated by this Wiener process. The Wiener integral $Z := \int_{\mathbb{R}_+} f(t)\, dW(t)$ is characterized by the following two properties:*

(a) $Z \in H(W)$, and

(b) $\mathrm{E}[ZW(s)] = \int_0^s f(t)\, dt$ for all $s \geq 0$.

Proof. Necessity: It was already noted that, by construction, $\int_{\mathbb{R}_+} f(t)\, dW(t) \in H(W)$. Also, (b) is just the isometry formula $\mathrm{E}\left[\int_{\mathbb{R}_+} f(t)\, dW(t) \int_{\mathbb{R}_+} 1_{\{s \leq t\}}\, dW(t)\right] = \int_0^s f(t)\, dt$.

Sufficiency: Since $Z - \int_0^t f(s)\, dW(s)$ is in $H(W)$, it suffices to show that this random variable is orthogonal to all the generators $W(s)$ ($s \in \mathbb{R}$) of $H(W)$ to obtain that $Z - \int_0^t f(s)\, dW(s) = 0$ P-a.s. But, by (b) and by the isometry formula,
$$\mathrm{E}\left[\left(Z - \int_0^t f(s)\, dW(s)\right) W(s)\right] = \int_0^s f(t)\, dt - \int_0^s f(t)\, dt = 0. \qquad \square$$

The next lemma features a kind of formula of integration by parts.

Lemma 11.3.8 *Let $\{W(t)\}_{t \in [0,1]}$ be a standard Wiener process. Let T be a positive real number and let $f : [0,T] \to \mathbb{R}$ be a continuously differentiable function (in particular the function $f(t)1_{\{t \leq T\}}$ is in $L_\mathbb{C}^2(\mu)$ and therefore the integral $\int_0^T f(t)\, dW(t)$ is well defined). Then,*
$$\int_0^T f(t)\, dW(t) + \int_0^T f'(t)W(t)\, dt = f(T)W(T). \tag{11.12}$$

Proof. By Lemma 11.3.7, it suffices to prove that for all $s \in \mathbb{R}_+$,
$$\mathrm{E}\left[\left(f(T)W(T) - \int_0^T f'(t)W(t)\, dt\right) W(s)\right] = \int_0^s f(t)1_{\{t \leq T\}}\, dt.$$

Using the equality $\mathrm{E}[W(a)W(b)] = a \wedge b$, the latter reduces to
$$f(T)(T \wedge s) - \mathrm{E}\left[\left(\int_0^T f'(t)W(t)\, dt\right) W(s)\right] = \int_0^s f(t)1_{\{t \leq T\}}\, dt.$$

By Fubini:
$$\mathrm{E}\left[\left(\int_0^T f'(t)W(t)\, dt\right) W(s)\right] = \int_0^T f'(t)\mathrm{E}[W(t)W(s)]\, dt$$
$$= \int_0^T f'(t)(t \wedge s)\, dt.$$

It therefore remains to check that
$$f(T)(T \wedge s) - \int_0^T f'(t)(t \wedge s)\, dt = \int_0^s f(t)1_{\{t \leq T\}}\, dt.$$

When $T \leq s$, this reduces to the identity

$$f(T)T - \int_0^T f'(t)t\,dt = \int_0^T f(t)\,dt,$$

which is verified by integration by parts, and when $T \geq s$, it reduces to

$$f(T)s - \int_0^T f'(t)(s \wedge t)\,dt = \int_0^s f(t)\,dt.$$

This last identity is verified by noting that both sides are null for $s = 0$ and that their derivatives are equal for all $s \leq T$:

$$f(T) - \int_s^T f'(t)\,dt = f(s).$$

$\square$

11.3.2   Langevin's Equation

This is the equation

$$dV(t) + \alpha V(t)\,dt = \sigma dW(t), \tag{11.13}$$

where $\{W(t)\}_{t\in[0,1]}$ is a standard Wiener process and α and σ are positive real numbers, with the following interpretation

$$V(t) - V(0) + \alpha \int_0^t V(s)\,ds = \sigma W(t) \quad (t \geq 0). \tag{11.14}$$

Remark 11.3.9 The motion of a particle of mass m on the line subjected at each instant t to an external force $F(t)$ and to friction is governed by the differential equation

$$mx''(t) = -\alpha x'(t) + F(t),$$

where $\alpha > 0$ is the friction coefficient. (It is assumed that there is no potential energy field.) If the external force is due, as in the Brown experiment, to numerous tiny shocks, one may assume that $\int_a^b F(s)\,ds = \sigma(W(b) - W(a))$ so that, letting $V(t) = x'(t)$ and taking $m = 1$, we obtain equation 11.13.

Theorem 11.3.10 *The unique solution of the Langevin equation (11.15) with initial value $V(0)$ is*

$$V(t) = e^{-\alpha t}V(0) + \int_0^t e^{-\alpha(t-s)}\sigma\,dW(s). \tag{11.15}$$

Proof. Using the integration by parts formula (11.12), (11.15) is found equivalent to

$$V(t) = e^{-\alpha t}V(0) + \sigma W(t) - \int_0^t \alpha e^{-\alpha(t-s)}\sigma W(s)\,ds. \tag{$\star$}$$

Integrating from 0 to u gives

$$\int_0^u V(t)\,dt = \frac{1}{\alpha}(1 - e^{-\alpha u})V(0) + \int_0^u \sigma W(s)\,ds - \int_0^u \left(\int_0^t \alpha e^{-\alpha(t-s)}\sigma W(s)\,ds \right) dt.$$

The last integral is equal to

$$\int_0^\infty \left(\int_0^\infty 1_{s \le t \le u} 1_{s \le u} \alpha e^{-\alpha(t-s)} \sigma W(s) \, ds \right) dt$$

$$= \int_0^\infty \left(\int_0^\infty 1_{s \le t \le u} \alpha e^{-\alpha(t-s)} \, dt \right) 1_{s \le u} \sigma W(s) \, ds$$

$$= \int_0^u \left(\int_s^u \alpha e^{-\alpha(t-s)} \, dt \right) \sigma W(s) \, ds$$

$$= \int_0^u \frac{1}{\alpha} (1 - e^{\alpha(u-s)}) \sigma W(s) \, ds.$$

Replacing this in $(\star)$ gives

$$\alpha \int_0^u V(t) \, dt = (1 - e^{-\alpha u}) V(0) + \int_0^u \alpha e^{-\alpha(u-s)} \sigma W(s) \, ds,$$

and therefore

$$V(u) - V(0) + \alpha \int_0^u V(t) \, dt$$
$$= V(u) - e^{-\alpha u} V(0) + \int_0^u \alpha e^{-\alpha(u-s)} \sigma W(s) \, ds = \sigma W(u).$$

To prove unicity, let V' be another solution of the Langevin equation with the same initial value. Letting $U := V - V'$, we have

$$U(t) = \alpha \int_0^t U(s) \, ds,$$

whose unique solution is the null function (Gronwall's lemma, Theorem B.6.1). $\qquad\square$

11.3.3 The Cameron–Martin Formula

This result is of interest in communications theory. One will recognize the likelihood ratio associated with the hypothesis "signal plus white Gaussian noise" against the hypothesis "white Gaussian noise only".

Theorem 11.3.11 *Let $\{X(t)\}_{t \ge 0}$ be, with respect to probability P, a Wiener process with variance σ^2 and let $\gamma : \mathbb{R} \to \mathbb{R}$ be in $L^2_{\mathbb{R}}(\ell)$. For any $T \in \mathbb{R}_+$, the formula*

$$\frac{dQ}{dP} = e^{\frac{1}{\sigma^2} \{ \int_0^T \gamma(t) dX(t) - \frac{1}{2} \int_0^T \gamma^2(t) dt \}} \tag{11.16}$$

defines a probability measure Q on $(\Omega, \mathcal{F})$ with respect to which

$$X(t) - \int_0^t \gamma(s) \, ds$$

is, on the interval $[0, T]$, a Wiener process with variance σ^2.

The proof of Theorem 11.3.11 is based on the following preliminary result.

Lemma 11.3.12 *Let $\{X(t)\}_{t \ge 0}$ be a Wiener process with variance σ^2 and let $\varphi : \mathbb{R} \to \mathbb{R}$ be in $L^2_{\mathbb{R}}(\ell)$. Then, for any $T \in \mathbb{R}_+$,*

$$E \left[e^{\int_0^T \varphi(t) dX(t)} \right] = e^{\frac{1}{2}\sigma^2 \int_0^T \varphi^2(t) dt}. \tag{11.17}$$

Proof. First consider the case

$$\varphi(t) = \sum_{k=1}^{N} \alpha_k 1_{(a_k, b_k]}(t) \,, \tag{11.18}$$

where $\alpha_k \in \mathbb{R}$ and the intervals $(a_k, b_k]$ are disjoint. For this special case, formula (11.17) reduces to

$$E \left[e^{\sum_{k=1}^{N} \alpha_k (X(b_k) - X(a_k))} \right] = e^{\frac{1}{2}\sigma^2 \sum_{k=1}^{N} \alpha_k^2 (b_k - a_k)} \,,$$

and therefore follows directly from the independence of the increments of a Wiener process and from the Gaussian property of these increments, in particular, the formula giving the Laplace transform of the centered Gaussian variable $X(b) - X(a)$ with variance $\sigma^2(b - a)$:

$$E \left[e^{\alpha(X(b) - X(a))} \right] = e^{\frac{1}{2}\sigma^2 \alpha^2 (b - a)} \,.$$

Let now $\{\varphi_n\}_{n \geq 1}$ be a sequence of functions of type (11.18) converging in $L^2_{\mathbb{R}}(\ell)$ to φ (in particular, $\lim_{n \uparrow \infty} \int_0^T \varphi_n^2(t) \mathrm{d}t = \int_0^T \varphi^2(t) \mathrm{d}t$). Therefore,

$$\lim_{n \uparrow \infty} \int_0^T \varphi_n(t) \mathrm{d}X(t) = \int_0^T \varphi(t) \mathrm{d}X(t),$$

where the latter convergence is in $L^2_{\mathbb{R}}(P)$. This convergence can be assumed to take place almost surely by taking if necessary a subsequence. From the equality

$$E \left[e^{\int_0^T \varphi_n(t) \mathrm{d}X(t)} \right] = e^{\sigma^2 \int_0^T \varphi_n^2(t) \mathrm{d}t}$$

we can then deduce (11.17), at least if the sequence of random variables in the left-hand side is uniformly integrable. This is the case because the quantity

$$E \left[\left| e^{\int_0^T \varphi_n(t) \mathrm{d}X(t)} \right|^2 \right] = E \left[e^{2 \int_0^T \varphi_n(t) \mathrm{d}X(t)} \right] = e^{2\sigma^2 \int_0^T \varphi_n^2(t) \mathrm{d}t}$$

is uniformly bounded, and therefore the uniform integrability claim follows from Theorem 4.2.14, with $G(t) = t^2$. $\qquad\square$

We may now turn to the proof of Theorem 11.3.11.

Proof. The fact that (11.16) properly defines a probability Q, that is, that the expectation of the right-hand side of (11.16) equals 1, follows from Lemma 11.3.12 with $\varphi(t) = \frac{1}{\sigma^2}\gamma(t)$.

Letting

$$Y(t) := X(t) - \int_0^t \gamma(s) \mathrm{d}s \,,$$

we have to prove that this centered stochastic process is Gaussian. To do this, we must show that

$$E_Q \left[e^{\sum_{k=1}^{N} \alpha_k (Y(b_k) - Y(a_k))} \right] = e^{\frac{1}{2}\sigma^2 \sum_{k=1}^{N} \alpha_k^2 (b_k - a_k)} \,,$$

where $\alpha_k \in \mathbb{R}$ and the intervals $(a_k, b_k] \subseteq [0, T]$ are disjoint, that is, letting $\psi(t) = \sum_{k=1}^{N} \alpha_k 1_{(a_k, b_k]}(t)$,

$$E_Q \left[e^{\int_0^T \psi(t) \mathrm{d}Y(t)} \right] = e^{\frac{1}{2}\sigma^2 \int_0^T \psi^2(t) \mathrm{d}t} \,,$$

or equivalently,

$$\mathrm{E}_P\left[\frac{\mathrm{d}Q}{\mathrm{d}P}\mathrm{e}^{\int_0^T \psi(t)(\mathrm{d}X(t)-\gamma(t)\mathrm{d}t)}\right] = \mathrm{e}^{\frac{1}{2}\sigma^2 \int_0^T \psi^2(t)\mathrm{d}t},$$

that is,

$$\mathrm{E}_P\left[\mathrm{e}^{\frac{1}{\sigma^2}\left\{\int_0^T \gamma(t)\mathrm{d}X(t)-\frac{1}{2}\int_0^T \gamma^2(t)\mathrm{d}t\right\}}\mathrm{e}^{\int_0^T \psi(t)(\mathrm{d}X(t)-\gamma(t)\mathrm{d}t)}\right] = \mathrm{e}^{\frac{1}{2}\sigma^2 \int_0^T \psi^2(t)\mathrm{d}t}.$$

Simplifying:

$$\mathrm{E}_P\left[\mathrm{e}^{\int_0^T \left(\psi(t)+\frac{1}{\sigma^2}\gamma(t)\right)\mathrm{d}X(t)-\int_0^T (\gamma(t)\psi(t))\mathrm{d}t-\frac{1}{2}\int_0^T \frac{\gamma^2(t)}{\sigma^2}\mathrm{d}t}\right] = \mathrm{e}^{\frac{1}{2}\sigma^2 \int_0^T \psi^2(t)\mathrm{d}t},$$

and using (11.17) with $\varphi(t) = \psi(t) + \frac{1}{\sigma^2}\gamma(t)$, the left-hand side is equal to

$$\mathrm{E}_P\left[\mathrm{e}^{\frac{1}{2}\sigma^2 \int_0^T \left(\psi(t)+\frac{1}{\sigma^2}\gamma(t)\right)^2 \mathrm{d}t-\int_0^T (\gamma(t)\psi(t))\mathrm{d}t-\frac{1}{2}\int_0^T \frac{\gamma^2(t)}{\sigma^2}\mathrm{d}t}\right].$$

The proof is completed since

$$\frac{1}{2}\sigma^2 \int_0^T \left(\psi(t)+\frac{1}{\sigma^2}\gamma(t)\right)^2 \mathrm{d}t - \int_0^T (\gamma(t)\psi(t))\,\mathrm{d}t - \frac{1}{2}\int_0^T \frac{\gamma^2(t)}{\sigma^2}$$
$$= \frac{1}{2}\sigma^2 \int_0^T \psi^2(t)\mathrm{d}t.$$

$\square$

Remark 11.3.13 A sweeping generalization of the Cameron–Martin theorem, the Girsanov theorem, will be given in Chapter 14 (Theorem 14.3.3). It will require the more advanced tools of the stochastic calculus associated with the Itô integral.

11.4 Fractal Brownian Motion

The Wiener process $\{W(t)\}_{t\geq 0}$ has the following property. If c is a positive constant, the process $\{W_c(t)\}_{t\geq 0} := \{c^{-\frac{1}{2}}W(ct)\}_{t\geq 0}$ is also a Wiener process. It is indeed a centered Gaussian process with independent increments, null at the time origin, and for $0 < a < b$,

$$\mathrm{E}\left[|W_c(b) - W_c(a)|^2\right] = c^{-1}\mathrm{E}\left[|W(cb) - W(ca)|^2\right] = c^{-1}(cb - ca) = b - a.$$

This is a particular instance of a self-similar stochastic process.

Definition 11.4.1 *A real-valued stochastic process* $\{Y(t)\}_{t\geq 0}$ *is called* self-similar *with* (Hurst) self-similarity parameter H *if for any* $c > 0$,

$$\{Y(t)\}_{t\geq 0} \overset{\mathcal{D}}{\sim} \{c^{-H}Y(ct)\}_{t\geq 0}.$$

The Wiener process is therefore self-similar with similarity parameter $H = \frac{1}{2}$.

It follows from the definition that $Y(t) \overset{\mathcal{D}}{\sim} t^H Y(1)$, and therefore, if $P(Y(1) \neq 0) > 0$:

If $H < 0$, $Y(t) \to 0$ in distribution as $t \to \infty$ and $Y(t) \to \infty$ in distribution as $t \to 0$.

If $H > 0$, $Y(t) \to \infty$ in distribution as $t \to 0$ and $Y(t) \to 0$ in distribution as $t \to \infty$.

If $H = 0$, $Y(t)$ has a distribution independent of t.

In particular, when $H \neq 0$, a self-similar process cannot be stationary (strictly or in the wide sense).

We shall be interested in self-similar processes that have *stationary increments*. We must restrict attention to non-negative self-similarity parameters, because of the following negative result:[1] for any strictly negative value of the self-similarity parameter, a self-similar stochastic process with independent increments is not measurable (except of course for the trivial case where the process is identically null).

Theorem 11.4.2 *Let* $\{Y(t)\}_{t \geq 0}$ *be a self-similar stochastic process with stationary increments and self-similarity parameter* $H > 0$ *(in particular,* $Y(0) = 0$*). Its covariance function is given by*

$$\Gamma(s, t) := \operatorname{cov}(Y(s), Y(t)) = \frac{1}{2}\sigma^2 \left[t^{2H} - |t - s|^{2H} + s^{2H} \right],$$

where $\sigma^2 = \operatorname{E}\left[(Y(t+1) - Y(t))^2 \right] = \operatorname{E}\left[Y(1)^2 \right]$.

Proof. Assume without loss of generality that the process is centered. Let $0 \leq s \leq t$. Then

$$\operatorname{E}\left[(Y(t) - Y(s))^2 \right] = \operatorname{E}\left[(Y(t-s) - Y(0))^2 \right]$$
$$= \operatorname{E}\left[(Y(t-s))^2 \right] = \sigma^2 (t - s)^{2H}$$

and

$$2\operatorname{E}\left[Y(t)Y(s) \right] = \operatorname{E}\left[Y(t)^2 \right] + \operatorname{E}\left[Y(s)^2 \right] - \operatorname{E}\left[(Y(t) - Y(s))^2 \right],$$

hence the result.

$\square$

Fractal Brownian motion[2] is a Gaussian process that in a sense generalizes the Wiener process.

Definition 11.4.3 *A fractal Brownian motion on* $\mathbb{R}_+$ *with Hurst parameter* $H \in (0, 1)$ *is a centered Gaussian process* $\{B_H(t)\}_{t \geq 0}$ *with continuous paths such that* $B_H(0) = 0$, *and with covariance function*

$$\operatorname{E}[B_H(t)B_H(s)] = \frac{1}{2} \left(|t|^{2H} + |s|^{2H} - |t - s|^{2H} \right). \tag{11.19}$$

The existence of such process follows from Theorem 5.1.23 as soon as the right-hand side of (11.19) can be shown to be a non-negative definite function. This can be done directly, although we choose another path. We shall prove the existence of the fractal Brownian motion by constructing it as a Doob integral with respect to a Wiener process. More precisely, define for $0 < H < 1$, $w_H(t, s) := 0$ for $t \leq s$,

$$w_H(t, s) := (t - s)^{H - \frac{1}{2}} \text{ for } 0 \leq s \leq t$$

and

---

[1] [Vervaat, 1987].
[2] [Mandelbrot and Van Ness, 1968].

$$w_H(t,s) := (t-s)^{H-\frac{1}{2}} - (-s)^{H-\frac{1}{2}} \text{ for } s < 0.$$

Observe that for any $c > 0$

$$w_H(ct,s) = c^{H-\frac{1}{2}} w_H(t, sc^{-1}).$$

Define

$$B_H(t) := \int_{\mathbb{R}} w_H(t,s) \, dW(s).$$

The Doob integral of the right-hand side is, more explicitly,

$$A - B := \int_0^t (t-s)^{H-\frac{1}{2}} \, dW(s) - \int_{-\infty}^0 \left((t-s)^{H-\frac{1}{2}} - (-s)^{H-\frac{1}{2}} \right) dW(s). \tag{11.20}$$

It is well defined and with the change of variable $u = c^{-1}s$ it becomes

$$c^{H-\frac{1}{2}} \int_{\mathbb{R}} w_H(t,u) \, dW(cu).$$

Using the self-similarity of the Wiener process, the process defined by the last display has the same distribution as the process defined by

$$c^{H-\frac{1}{2}} c^{\frac{1}{2}} \int_{\mathbb{R}} w_H(t,u) \, dW(u).$$

Therefore $\{B_H(t)\}_{t \geq 0}$ is self-similar with similarity parameter H.

The fact that there is a version of this process with continuous paths can be proven using Theorem 5.2.3 along the lines of the proof of Theorem 11.2.7.

It is tempting to rewrite (11.20) as $Z(t) - Z(0)$, where

$$Z(t) = \int_{-\infty}^t (t-s)^{H-\frac{1}{2}} \, dW(s).$$

However this last integral is not well defined as a Doob integral since for all $H > 0$, the function $s \to (t-s)^{H-\frac{1}{2}} 1_{\{s \leq t\}}$ is not in $L^2_{\mathbb{R}}(\mathbb{R})$.

Complementary reading

Chapter 6 of [Resnick, 1992]. [Revuz and Yor, 1999] (more advanced).

11.5 Exercises

Exercise 11.5.1. WIENER AS A LIMIT
Prove that for all $t_1, \ldots, t_n$ in $\mathbb{R}_+$ forming an increasing sequence, the limit distribution of the vector $(X(t_1), \ldots, X(t_n))$, where $X(t)$ is defined by (11.2), is that corresponding to a Wiener process, that is, a centered Gaussian vector such that $X(t_1), X(t_2) - X(t_1), \ldots, X(t_n) - X(t_{n-1})$ are centered Gaussian variables with variances $t_1, t_2 - t_1, \ldots, t_n - t_{n-1}$.

Exercise 11.5.2. A BASIC FORMULA

Let $\{W(t)\}_{t\geq 0}$ be a standard Wiener process. Prove that for $s, t \in \mathbb{R}_+$,

$$E[W(t)W(s)] = t \wedge s.$$

Let $\{Y(t)\}_{t\geq 0}$ be a Brownian bridge. Prove that

$$\text{cov}\,(X(t), X(s)) = s(1 - t) \quad (0 \leq s \leq t \leq 1).$$

Exercise 11.5.3. TRANSFORMING A WIENER PROCESS

Let $\{W(t)\}_{t\geq 0}$ be a standard Wiener process. Prove that the process $\{X(t)\}_{t\in[0,1]}$ is a standard Brownian motion in the following cases:

(i) $X(t) = -W(t)$,

(ii) $X(t) = W(t + a) - W(t)$ $(a > 0)$,

(iii) $X(t) = \sqrt{c}W\left(\frac{t}{c}\right)$ $(t \geq 0)$ $(c > 0)$,

(iv) $X(t) = tW\left(\frac{1}{t}\right)$ $(t > 0)$ and $X(0) = 0$. (Note that the continuity at 0 is already proved in Section 11.1.2.)

Exercise 11.5.4. BROWNIAN BRIDGES

Let $\{W(t)\}_{t\in[0,1]}$ be a Wiener process. Show that the Brownian bridge

$$\{X(t) := W(t) - tW(1)\}_{t\in[0,1]}$$

is a Gaussian process independent of $W(1)$ and compute its autocovariance function. Show that the process $\{X(1 - t)\}_{t\in[0,1]}$ is a Brownian bridge.

Exercise 11.5.5.

Let $\{W(t)\}_{t\in[0,1]}$ be a Wiener process. Let

$$Z(t) := (1 - t)W\left(\frac{t}{1 - t}\right) \quad (0 \leq t < 1)$$

and $Z(1) := 0$. Show that $\{Z(t)\}_{t\in[0,1]}$ is continuous at $t = 1$ and that it has the same distribution as the Brownian bridge.

Exercise 11.5.6. WIENER IS GAUSS–MARKOV

Prove that a Wiener process is a Gauss–Markov process.

Exercise 11.5.7. AN ORNSTEIN–UHLENBECK PROCESS

Let $\{W(t)\}_{t\geq 0}$ be a standard Brownian motion. Show that $\{e^{-\alpha t}W\left(e^{2\alpha t}\right)\}_{t\geq 0}$ is (has the same distribution as) an Ornstein–Uhlenbeck process.

Exercise 11.5.8. EXIT TIME FROM A STRIP

Let $\{W(t)\}_{t\geq 0}$ be a standard Brownian motion, and define for $a > 0$ and $b < 0$ the stopping time

$$T_{a,b} := \inf\{t \geq 0\,;\, W(t) \in \{a, b\}\}.$$

(i) Compute $P(W(T_{a,b}) = a)$.

(ii) Show that $\{W(t)^2 - t\}_{t\geq0}$ is an $\mathcal{F}_t^W$-martingale and deduce from this $\mathrm{E}\,[T_{a,b}]$.

Exercise 11.5.9. THE TRANSIENCE OF BROWNIAN MOTION WITH A POSITIVE DRIFT
Let μ and $\sigma > 0$ be two real numbers. Let

$$X(t) := \sigma W(t) + \mu t \quad (t \geq 0).$$

(i) Show that for all $u \in \mathbb{R}$,

$$Z(t) := \exp\{uX(t) - ut\left(\mu + \frac{u\sigma^2}{2}\right)\} \quad (t \geq 0)$$

is an $\mathcal{F}_t^W$-martingale.

(ii) Take advantage of the choice $u = -\frac{2\mu}{\sigma^2}$ and of Doob's optional sampling theorem (the applicability of which you shall verify) to obtain that the probability r_a that $\{X(t)\}_{t\geq0}$ will reach $a > 0$ before it touches $-b < 0$ is given by

$$r_a = \frac{1 - e^{\frac{2\mu b}{\sigma^2}}}{1 - e^{\frac{2\mu(a+b)}{\sigma^2}}}.$$

(iii) Show that if $\mu > 0$ (or $\mu < 0$), $\{X(t)\}_{t\geq0}$ is transient, that is, for all $a \in \mathbb{R}$, there exists an almost surely finite (random) time after which it does not visit a.

Exercise 11.5.10. INDEPENDENT BROWNIAN MOTIONS WITH A DRIFT
Let for $i = 1, 2$,

$$X_i(t) := x_i + \mu_i t + \sigma_i W_i(t),$$

where $x_i, \mu_i \in \mathbb{R}$, $\sigma_i > 0$, and $\{W_1\}_{t\geq0}$ and $\{W_2\}_{t\geq0}$ are independent standard Brownian motions. Suppose moreover that $x_1 < x_2$. Compute the probability that $\{X_1(t)\}_{t\geq0}$ and $\{X_2(t)\}_{t\geq0}$ never meet.

Exercise 11.5.11. THE LEBESGUE INTEGRAL OF A GAUSSIAN PROCESS
Let $\{X(t)\}_{t\in[0,1)}$ be a continuous Gaussian stochastic process. Prove that the random variable $\int_0^1 X(t)\,dt$ is Gaussian. Compute its mean and variance when $\{X(t)\}_{t\in[0,1)}$ is a Brownian bridge.

Exercise 11.5.12.
Let $\{W(t)\}_{t\geq0}$ be a standard Brownian motion. Show that the stochastic process

$$X(t) := \int_0^{\frac{t}{1-t}} \frac{1}{1-s}\,dW(s) \quad (t \in [0,1))$$

is a Brownian motion.

Exercise 11.5.13. A REPRESENTATION OF THE BROWNIAN BRIDGE
Let $\{W(t)\}_{t\geq0}$ be a standard Brownian motion. Let for $t \in [0,1)$,

$$Y(t) := (1-t) \int_0^t \frac{dW(s)}{1-s}\,ds.$$

(i) Prove that the integral in the right-hand side is well defined on $[0, 1)$ as a Wiener integral.

(ii) Prove that as $t \downarrow 0$, $Y(t) \to 0$ in quadratic mean.

(iii) Define $Y(0) := 0$. Show that $\{Y(t)\}_{t \in [0,1]}$ is a Gaussian process.

(iv) Show that $\{Y(t)\}_{t \in [0,1]}$ is (has the same distribution as) a Brownian bridge.

Exercise 11.5.14. AVERAGE OCCUPATION TIME
Let $\{W(t)\}_{t \geq 0}$ be a standard d-dimensional Brownian motion and let $A \in \mathcal{B}(\mathbb{R}^d)$ be of positive Lebesgue measure. Let, for $\omega \in \Omega$,

$$S_A(\omega) := \{t \geq 0 \,;\, W(t, \omega) \in A\}$$

(the occupation time of A). Prove that $\mathrm{E}\left[\ell^d(S_A)\right] = \infty$ if $d \leq 2$, and that if $d \geq 3$,

$$\mathrm{E}\left[\ell^d(S_A)\right] = \frac{1}{2\pi^{d/2}} \Gamma\left(\frac{d}{2} - 1\right) \int_A \|x\|^{2-d} \, dx \,,$$

where $\Gamma(\alpha) := \int_0^\infty x^{\alpha-1} e^{-x} \, dx$ $(\alpha > 0)$ is the Gamma function.

Exercise 11.5.15. MICROPULSES AND FRACTAL BROWNIAN MOTION
[3] Let $\overline{N}_\varepsilon$ be a Poisson process on $\mathbb{R} \times \mathbb{R}_+$ with intensity measure $\nu(dt \times dz) = \frac{1}{2\varepsilon^2} z^{-1-\theta} \, dt \times dz$ $(0 < \theta < 1, \varepsilon > 0)$. For all $t \geq 0$, let $S_{0,t}^+ = \{(s, z) : 0 < s < t, \, t - s < z\}$ and $S_{0,t}^- = \{(s, z) : -\infty < s < 0, \, -s < z < t - s\}$, and let

$$X_\varepsilon(t) := \varepsilon \left\{ \overline{N}_\varepsilon(S_{0,t}^+) - \overline{N}_\varepsilon(S_{0,t}^-) \right\}.$$

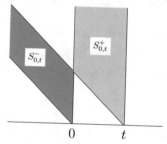

(1) Show that $X_\varepsilon(t)$ is well defined for all $t \geq 0$.

(2) Compute for all $0 \leq t_1 \leq t_2 \ldots \leq t_n$ the characteristic function of

$$(X_\varepsilon(t_1), \ldots, X_\varepsilon(t_n)).$$

(3) Show that for all $0 \leq t_1 \leq t_2 \ldots \leq t_n$, $(X_\varepsilon(t_1), \ldots, X_\varepsilon(t_n))$ converges as $\varepsilon \downarrow 0$ in distribution to $(B_H(t_1), \ldots, B_H(t_n))$, where $\{B_H(t)\}_{t \geq 0}$ is a fractal Brownian motion with Hurst parameter $H = \frac{1-\theta}{2}$ and variance $\mathrm{E}\left[B_H(1)^2\right] = \theta^{-1}(1 - \theta)^{-1}$, that is, $\{B_H(t)\}_{t \geq 0}$ is a centered Gaussian process such that $B_H(0) = 0$ and with covariance function

$$\mathrm{E}[B_H(t) B_H(s)] = \frac{1}{2} \left(|s|^{2H} + |t|^{2H} - |s - t|^{2H}\right) \mathrm{E}\left[B_H(1)^2\right].$$

---

[3][Cioczek-Georges and Mandelbrot, 1995].

Chapter 12

Wide-sense Stationary Stochastic Processes

Wide-sense stationary stochastic processes are of interest in signal analysis and processing, as well as in physics. Their study rests on Bochner's representation of characteristic functions, which immediately leads to the fundamental notion of power spectral measure, and on the Doob–Wiener integral that permits a mathematical definition of white noise as well as the obtention of a spectral decomposition of the trajectories of such stochastic processes, the Cramér–Khinchin decomposition, a fundamental result with importance consequences in signal processing.

12.1 The Power Spectral Measure

12.1.1 Covariance Functions and Characteristic Functions

Recall the simple facts about Fourier theory. Let $f : (\mathbb{R}, \mathcal{B}(\mathbb{R})) \to (\mathbb{R}, \mathcal{B}(\mathbb{R}))$ be integrable with respect to the Lebesgue measure. Then, for any $\nu \in \mathbb{R}$,

$$\widehat{f}(\nu) := \int_{\mathbb{R}} f(t)\,e^{-2i\pi\nu t}\,dt$$

is well defined and the function $\hat{f}$, called the *Fourier transform* of f, is continuous and bounded (Exercise 2.4.13). From classical Fourier analysis, we know that if moreover the function $\hat{f}$ is integrable with respect to the Lebesgue measure, then (*Fourier inversion formula*)

$$f(t) = \int_{\mathbb{R}} \widehat{f}(\nu)\,e^{2i\pi\nu t}\,d\nu\,,$$

where this equality is true almost everywhere, and everywhere if f is continuous (Exercise 2.4.9).

Remark 12.1.1 The notion of Fourier transform does not in general apply as such to the trajectories of a wide-sense stationary stochastic process. Consider, for instance, a square-integrable ergodic process $\{X(t)\}_{t\in\mathbb{R}}$ not identically null. In particular, for $p = 1$ or $p = 2$, P-a.s.,

$$\lim_{t\uparrow\infty} \frac{1}{t} \int_0^t |X(t)|^p\,dt = \mathrm{E}\left[|X(t)|^p\right] > 0\,,$$

from which it follows that almost all trajectories are not in $L^1_{\mathbb{C}}(\mathbb{R})$ nor in $L^2_{\mathbb{C}}(\mathbb{R})$ and therefore do not have a Fourier transform in the usual L^1 or L^2 senses.

467

© Springer Nature Switzerland AG 2020
P. Brémaud, *Probability Theory and Stochastic Processes*, Universitext,
https://doi.org/10.1007/978-3-030-40183-2_12

Nevertheless, there exists a spectral decomposition for the trajectories of a WSS stochastic process, called the *Cramér–Khintchin* decomposition, as we shall see in Section 12.2.

To obtain such a decomposition, the starting point is the Fourier analysis of the covariance function. We begin with a few examples.

Two Particular Cases

EXAMPLE 12.1.2:   ABSOLUTELY CONTINUOUS SPECTRUM. Consider a WSS random process with *integrable and continuous* covariance function C, in which case the Fourier transform f of the latter is well defined by

$$f(\nu) = \int_{\mathbb{R}} e^{-2i\pi\nu\tau} C(\tau) \, d\tau \, .$$

It is called the *power spectral density* (PSD). It turns out, as we shall soon see when we consider the general case, that it is non-negative and integrable. Since it is integrable, the Fourier inversion formula

$$C(\tau) = \int_{\mathbb{R}} e^{2i\pi\nu\tau} f(\nu) \, d\nu \tag{12.1}$$

holds true for all $t \in \mathbb{R}$ since C is continuous. Also f is the unique integrable function such that (12.1) holds. Letting $\tau = 0$ in this formula, we obtain, since $C(0) = \mathrm{Var}(X(t) := \sigma^2$,

$$\sigma^2 = \int_{\mathbb{R}} f(\nu) d\nu \, .$$

EXAMPLE 12.1.3:   THE ORNSTEIN–UHLENBECK PROCESS. The Ornstein–Uhlenbeck process is a centered Gaussian process with covariance function

$$\Gamma(t, s) = C(t - s) = e^{-\alpha|t-s|} \, .$$

The function C is integrable and therefore the power spectral density is the Fourier transform of the covariance function:

$$f(\nu) = \int_{\mathbb{R}} e^{-2i\pi\nu\tau} e^{-\alpha|\tau|} d\tau = \frac{2\alpha}{\alpha^2 + 4\pi^2\nu^2} \, .$$

Not all WSS stochastic processes admit a power spectral density. For instance:

EXAMPLE 12.1.4:   LINE SPECTRUM. Consider a wide-sense stationary process with a covariance function of the form

$$C(\tau) = \sum_{k \in \mathbb{Z}} P_k e^{2i\pi\nu_k\tau} \, ,$$

where

$$P_k \geq 0 \text{ and } \sum_{k \in \mathbb{Z}} P_k < \infty$$

(for instance, the harmonic process of Example 5.1.20). This covariance function is not integrable, and in fact there does not exist a power spectral density. In particular, a representation of the covariance function such as (12.1) is not available, at least if the function f is interpreted in the ordinary sense. However, there is a formula such as (12.1) if we consent, as is usually done in the engineering literature, to define the power spectral density in this case to be the *pseudo-function*

$$f(\nu) = \sum_{k \in \mathbb{Z}} P_k \, \delta(\nu - \nu_k),$$

where $\delta(\nu - a)$ is the delayed Dirac pseudo-function informally defined by

$$\int_{\mathbb{R}} \varphi(\nu) \, \delta(\nu - a) \, d\nu = \varphi(a).$$

Indeed, with such a convention,

$$\int_{\mathbb{R}} f(\nu) e^{2i\pi\nu\tau} f(\nu) \, d\nu = \sum_{k \in \mathbb{Z}} P_k \int_{\mathbb{R}} e^{2i\pi\nu\tau} \delta(\nu - \nu_k) \, d\nu = \sum_{k \in \mathbb{Z}} P_k e^{2i\pi\nu_k\tau}.$$

We can (and perhaps should) however avoid recourse to Dirac pseudo-functions, and the general result to follow (Theorem 12.1.5) will tell us what to do.

In general, it may happen that the covariance function is not integrable and/or that there does not exist a line spectrum. We now turn to the general theory.

The General Case

Remember that the characteristic function φ of a real random variable X has the following properties:

A. it is hermitian symmetric, that is, $\varphi(-u) = \varphi(u)^*$, and it is uniformly bounded: $|\varphi(u)| \le \varphi(0)$,

B. it is uniformly continuous on $\mathbb{R}$, and

C. it is definite non-negative, in the sense that for all integers n, all $u_1, \dots, u_n \in \mathbb{R}$, and all $z_1, \dots, z_n \in \mathbb{C}$,

$$\sum_{j=1}^{n} \sum_{k=1}^{n} \varphi(u_j - u_k) z_j z_k^* \ge 0$$

(just observe that the left-hand side equals $\mathrm{E}\left[\left|\sum_{j=1}^{n} z_j e^{iu_j X}\right|^2\right]$).

It turns out that Properties A , B and C characterize characteristic functions (up to a multiplicative constant). This is *Bochner's theorem* (Theorem 4.4.10), which is now recalled for easier reference:

Let $\varphi : \mathbb{R} \to \mathbb{C}$ be a function satisfying properties A, B and C. Then there exists a constant $0 \le \beta < \infty$ and a real random variable X such that for all $u \in \mathbb{R}$,

$$\varphi(u) = \beta \mathrm{E}\left[e^{iuX}\right].$$

Bochner's theorem is all that is needed to define the power spectral measure of a wide-sense stationary process continuous in the quadratic mean.

Theorem 12.1.5 *Let $\{X(t)\}_{t\in\mathbb{R}}$ be a WSS random process continuous in the quadratic mean, with covariance function C. Then, there exists a unique measure μ on $\mathbb{R}$ such that*

$$C(\tau) = \int_{\mathbb{R}} e^{2i\pi\nu\tau}\mu(d\nu). \tag{12.2}$$

In particular, μ is a *finite* measure:

$$\mu(\mathbb{R}) = C(0) = \mathrm{Var}(X(0)) < \infty. \tag{12.3}$$

Proof. It suffices to observe that the covariance function of a WSS stochastic process that is continuous in the quadratic mean shares the properties A, B and C of the characteristic function of a real random variable. Indeed,

(a) it is hermitian symmetric, and $|C(\tau)| \leq C(0)$ (Schwarz's inequality),

(b) it is uniformly continuous, and

(c) it is definite non-negative, in the sense that for all integers n, all $\tau_1, \ldots, \tau_n \in \mathbb{R}$, and all $z_1, \ldots, z_n \in \mathbb{C}$,

$$\sum_{j=1}^{n}\sum_{k=1}^{n} C(\tau_j - \tau_k)z_j z_k^* \geq 0$$

(just observe that the left-hand side is equal to $\mathrm{E}\left[\left|\sum_{j=1}^{n} z_j X(t_j)\right|^2\right]$). Therefore, by Theorem 4.4.10, the covariance function C is (up to a multiplicative constant) a characteristic function. This is exactly what (12.2) says, since μ thereof is a finite measure, that is, up to a multiplicative constant, a probability distribution.

Uniqueness of the power spectral measure follows from the fact that a finite measure (up to a multiplicative constant: a probability) on $\mathbb{R}^d$ is characterized by its Fourier transform (Theorem 3.1.51).   □

The case of an absolutely continuous spectrum corresponds to the situation where μ admits a density with respect to Lebesgue measure: $\mu(d\nu) = f(\nu)\,d\nu$. We then say that the WSS stochastic process in question admits the *power spectral density* (PSD) f. If such a power spectral density exists, it has the properties mentioned without proof in Example 12.1.2: it is non-negative and it is integrable.

The case of a line spectrum corresponds to a spectral measure that is a weighted sum of Dirac measures:

$$\mu(d\nu) = \sum_{k\in\mathbb{Z}} P_k\,\varepsilon_{\nu_k}(d\nu),$$

where the P_k's are non-negative and have a finite sum, as in Example 12.1.4.

12.1.2 Filtering of WSS Stochastic Processes

We recall a few standard results concerning the (convolutional) filtering of deterministic functions.

Let $f, g : (\mathbb{R}, \mathcal{B}(\mathbb{R})) \to (\mathbb{R}, \mathcal{B}(\mathbb{R}))$ be integrable functions with respective Fourier transforms $\widehat{f}$ and $\widehat{g}$. Then (Exercise 2.4.14),

$$\int_{\mathbb{R}} \int_{\mathbb{R}} |f(t-s)g(s)| \, dt \, ds < \infty,$$

and therefore, for almost all $t \in \mathbb{R}$, the function $s \mapsto f(t-s)g(s)$ is Lebesgue integrable. In particular, the convolution

$$(f * g)(t) := \int_{\mathbb{R}} f(t-s)g(s) \, ds$$

is almost everywhere well defined. For all t such that the last integral is not defined, set $(f * g)(t) = 0$. Then $f * g$ is Lebesgue integrable and its Fourier transform is $\widehat{f * g} = \widehat{f}\widehat{g}$, where $\widehat{f}, \widehat{g}$ are the Fourier transforms of f and g, respectively (Exercise 2.4.14).

Let $h : (\mathbb{R}, \mathcal{B}(\mathbb{R})) \to (\mathbb{R}, \mathcal{B}(\mathbb{R}))$ be an integrable function. The operation that associates to the integrable function $x : (\mathbb{R}, \mathcal{B}(\mathbb{R})) \to (\mathbb{R}, \mathcal{B}(\mathbb{R}))$ the integrable function

$$y(t) := \int_{\mathbb{R}} h(t-s)x(s) \, ds$$

is called a stable convolutional filter. The function h is called the *impulse response* of the filter, x and y are respectively the *input* and the *output* of this filter. The Fourier transform $\widehat{h}$ of the impulse response is the *transmittance* of the filter.

Let now $\{X(t)\}_{t \in \mathbb{R}}$ be a WSS random process with continuous covariance function C_X. We examine the effect of filtering on this process. The output process is the process defined by

$$Y(t) := \int_{\mathbb{R}} h(t-s)X(s) \, ds. \tag{12.4}$$

Note that the integral (12.4) is well defined under the integrability condition for the impulse response h. This follows from Theorem 5.3.2 according to which the integral

$$\int_{\mathbb{R}} f(s)X(s, \omega) \, ds$$

is well defined for P-almost all ω when f is integrable (in the special case of WSS stochastic processes, $m(t) = m$ and $\Gamma(t, t) = C(0) + |m|^2$, and therefore the conditions on f and g thereof reduce to integrability of these functions). Referring to the same theorem, we have

$$\mathrm{E}[\int_{\mathbb{R}} f(t)X(t) \, dt] = \int_{\mathbb{R}} f(t)\mathrm{E}[X(t)] \, dt = m \int_{\mathbb{R}} f(t) \, dt. \tag{12.5}$$

Let now $f, g : \mathbb{R} \to \mathbb{C}$ and be integrable functions. As a special case of Theorem 5.3.2, we have

$$\mathrm{cov}\left(\int_{\mathbb{R}} f(t)X(t) \, dt, \int_{\mathbb{R}} g(s)X(s) \, ds \right) = \int_{\mathbb{R}} \int_{\mathbb{R}} f(t)g^*(s)C(t-s) \, dt \, ds. \tag{12.6}$$

We shall see that, in addition,

$$\text{cov}\left(\int_{\mathbb{R}} f(t)X(t)\,dt, \int_{\mathbb{R}} g(s)X(s)\,ds\right) = \int_{\mathbb{R}}\int_{\mathbb{R}} \widehat{f}(-\nu)\widehat{g}^*(-\nu)\mu(d\nu). \qquad (12.7)$$

Proof. Assume without loss of generality that $m = 0$. From Bochner's representation of the covariance function, we obtain for the last double integral in (12.6)

$$\int_{\mathbb{R}}\int_{\mathbb{R}} f(t)g^*(s)\left(\int_{\mathbb{R}} e^{+2j\pi\nu(t-s)}\mu(d\nu)\right)dt\,ds =$$

$$\int_{\mathbb{R}}\left(\int_{\mathbb{R}} f(t)e^{+2j\pi\nu t}dt\right)\left(\int_{\mathbb{R}} g(s)e^{+2j\pi\nu s}\,ds\right)^*\mu(d\nu).$$

Here again we have to justify the change of order of integration using Fubini's theorem. For this, it suffices to show that the function $(t, s, \nu) \to |f(t)g^*(s)e^{+2j\pi\nu(t-s)}| = |f(t)|\,|g(s)|\,1_{\mathbb{R}}(\nu)$ is integrable with respect to the product measure $\ell \times \ell \times \mu$. This is indeed true, the integral being equal to $(\int_{\mathbb{R}} |f(t)|\,dt) \times (\int_{\mathbb{R}} |g(t)|\,dt) \times \mu(\mathbb{R})$. $\qquad\square$

In view of the above results, the right-hand side of formula (12.4) is well defined. Moreover

Theorem 12.1.6 *When the input process $\{X(t)\}_{t\in\mathbb{R}}$ is a WSS random process with power spectral measure μ_X, the output $\{Y(t)\}_{t\in\mathbb{R}}$ of a stable convolutional filter of transmittance $\widehat{h}$ is a WSS random process with the power spectral measure*

$$\mu_Y(d\nu) = |\widehat{h}(\nu)|^2\mu_X(d\nu). \qquad (12.8)$$

This formula will be referred to as the *fundamental filtering formula* in continuous time.

Proof. Just apply formulas (12.5) and (12.7) with the functions

$$f(u) = h(t - u), \qquad g(v) = h(s - v),$$

to obtain

$$E[Y(t)] = m\int_{\mathbb{R}} h(t)dt,$$

and

$$E[(Y(t) - m)(Y(s) - m)^*] = \int_{\mathbb{R}} |\widehat{h}(\nu)|^2 e^{+2j\pi\nu(t-s)}\mu(d\nu).$$

$\qquad\square$

EXAMPLE 12.1.7: TWO SPECIAL CASES. In particular, if the input process admits a PSD f_X, the output process also admits a PSD given by

$$f_Y(\nu) = |\widehat{h}(\nu)|^2 f_X(\nu)\,d\nu.$$

When the input process has a line spectrum, the power spectral measure of the output process takes the form

$$\mu_Y(d\nu) = \sum_{k=1}^{\infty} P_k|\widehat{h}(\nu_k)|^2 \varepsilon_{\nu_k}(d\nu).$$

12.1.3 White Noise

By analogy with Optics, one calls *white noise* any centered WSS random process $\{B(t)\}_{t \in \mathbb{R}}$ with constant power spectral density $f_B(\nu) = N_0/2$.[1] Such a definition presents a theoretical difficulty, because

$$\int_{-\infty}^{+\infty} f_B(\nu)\, d\nu = +\infty,$$

which contradicts the finite power property of wide-sense stationary processes.

A First Approach

From a pragmatic point of view, one could define a white noise to be a centered WSS stochastic process whose PSD is constant over a "large", yet bounded, range of frequencies $[-A, +A]$. The calculations below show what happens as A tends to infinity. Let therefore $\{X(t)\}_{t \in \mathbb{R}}$ be a centered WSS stochastic process with PSD

$$f(\nu) = \frac{N_0}{2} 1_{[-A,+A]}(\nu)\,.$$

Let $\varphi_1,\ \varphi_2 : \mathbb{R} \to \mathbb{C}$ be two functions in $L^1_{\mathbb{C}}(\mathbb{R}) \cap L^2_{\mathbb{C}}(\mathbb{R})$ with Fourier transforms $\widehat{\varphi}_1$ and $\widehat{\varphi}_2$, respectively. Then

$$\lim_{A \uparrow \infty} \mathrm{E}\left[\left(\int_{\mathbb{R}} \varphi_1(t) X(t)\, dt\right)\left(\int_{\mathbb{R}} \varphi_2(t) X(t)\, dt\right)^*\right] = \frac{N_0}{2} \int_{\mathbb{R}} \varphi_1(t)\varphi_2^*(t)\, dt$$

$$= \frac{N_0}{2} \int_{\mathbb{R}} \widehat{\varphi}_1(\nu)\widehat{\varphi}_2^*(\nu)\, d\nu\,.$$

Proof. We have

$$\mathrm{E}\left[\left(\int_{\mathbb{R}} \varphi_1(t) X(t)\, dt\right)\left(\int_{\mathbb{R}} \varphi_2(t) X(t)\, dt\right)^*\right] = \int_{\mathbb{R}}\int_{\mathbb{R}} \varphi_1(u)\varphi_2(v)^* C_X(u-v)\, du\, dv\,.$$

The latter quantity is equal to

$$\frac{N_0}{2} \int_{-\infty}^{+\infty} \varphi_1(u)\varphi_2(v)^* \left(\int_{-A}^{+A} e^{2i\pi\nu(u-v)}\, d\nu\right) du\, dv$$

$$= \frac{N_0}{2} \int_{-A}^{+A} \left(\int_{-\infty}^{+\infty} \varphi_1(u) e^{2i\pi\nu u}\, du\right)\left(\int_{-\infty}^{+\infty} \varphi_2(v)^* e^{-2i\pi\nu v}\, dv\right) d\nu$$

$$= \frac{N_0}{2} \int_{-A}^{+A} \widehat{\varphi}_1(-\nu)\widehat{\varphi}_2(-\nu)^*\, d\nu,$$

and the limit of this quantity as $A \uparrow \infty$ is:

$$\frac{N_0}{2} \int_{-\infty}^{+\infty} \widehat{\varphi}_1(\nu)\widehat{\varphi}_2^*(\nu)\, d\nu = \frac{N_0}{2} \int_{-\infty}^{+\infty} \varphi_1(t)\varphi_2(t)^*\, dt\,,$$

where the last equality is the Plancherel–Parseval identity. $\qquad\square$

Let now $h : \mathbb{R} \to \mathbb{C}$ be in $L^1_{\mathbb{C}}(\mathbb{R}) \cap L^2_{\mathbb{C}}(\mathbb{R})$, and define

---

[1]The notation $N_0/2$ comes from Physics and is a standard one in communications theory when dealing with the so-called additive white noise channels.

$$Y(t) = \int_{\mathbb{R}} h(t-s)X(s)\,\mathrm{d}s\,.$$

Applying the above result with $\varphi_1(u) = h(t-u)$ and $\varphi_2(v) = h(t+\tau-v)$, we find that the covariance function C_Y of this WSS stochastic process is such that

$$\lim_{A\uparrow\infty} C_Y(\tau) = \int_{\mathbb{R}} e^{2i\pi\nu\tau}|\widehat{h}(\nu)|^2\,\frac{N_0}{2}\,\mathrm{d}\nu\,.$$

The limit is finite since $\widehat{h} \in L^2_{\mathbb{C}}(\mathbb{R})$ and is a covariance function corresponding to a *bona fide* (that is integrable) PDF $f_Y(\nu) = |\widehat{h}(\nu)|^2\,\frac{N_0}{2}$. With $f(\nu) = \frac{N_0}{2}$, we formally retrieve the usual filtering formula,

$$f_Y(\nu) = |\widehat{h}(\nu)|^2 f(\nu)\,.$$

White Noise via the Doob–Wiener Integral

Another approach to white noise, more formal, consists in working right away "at the limit". We do not attempt to define the white noise $\{B(t)\}_{t\in\mathbb{R}}$ directly (for good reasons since it does not exist as a *bona fide* WSS stochastic process, as we noted earlier). Instead, we define directly the symbolic integral $\int_{\mathbb{R}} f(t)B(t)\,\mathrm{d}t$ for integrands f to be described below, by

$$\int_{\mathbb{R}} f(t)B(t)\,\mathrm{d}t := \sqrt{\frac{N_0}{2}}\int_{\mathbb{R}} f(t)\,\mathrm{d}Z(t), \tag{12.9}$$

where $\{Z(t)\}_{t\in\mathbb{R}}$ is a centered stochastic process with uncorrelated increments with unit variance. We say that $\{B(t)\}_{t\in\mathbb{R}}$ is a *white noise* and that $\left\{\sqrt{\frac{N_0}{2}}\,Z(t)\right\}_{t\in\mathbb{R}}$ is an *integrated white noise*. For all $f, g \in L^2_{\mathbb{C}}(\mathbb{R})$, we have that

$$E\left[\int_{\mathbb{R}} f(t)\,B(t)\,\mathrm{d}t\right] = 0\,,$$

and by the isometry formulas for the Doob–Wiener integral,

$$E\left[\left(\int_{\mathbb{R}} f(t)\,B(t)\,\mathrm{d}t\right)\left(\int_{\mathbb{R}} g(t)\,B(t)\,\mathrm{d}t\right)^*\right] = \frac{N_0}{2}\int_{\mathbb{R}} f(t)g(t)^*\,\mathrm{d}t\,,$$

which can be formally rewritten, using the Dirac symbolism:

$$\int_{\mathbb{R}} f(t)g(s)^*\,E[B(t)B^*(s)]\,\mathrm{d}t\,\mathrm{d}s = \int_{\mathbb{R}} f(t)g(s)^*\,\frac{N_0}{2}\delta(t-s)\,\mathrm{d}t\,\mathrm{d}s\,.$$

Hence "the covariance function of the white noise $\{B(t)\}_{t\in\mathbb{R}}$ is a Dirac pseudo-function: $C_B(\tau) = \frac{N_0}{2}\delta(\tau)$".

When $\{Z(t)\}_{t\in\mathbb{R}} \equiv \{W(t)\}_{t\in\mathbb{R}}$, a standard Brownian motion, $\{B(t)\}_{t\in\mathbb{R}}$ is called a *Gaussian white noise*. In this case, the Wiener–Doob integral is certainly not a Stieltjes–Lebesgue integral since the trajectories of the Wiener process are of unbounded variation on any finite interval (Corollary 11.2.9). Also, $B(t)$ cannot be interpreted as the "derivative" $\frac{\mathrm{d}W(t)}{\mathrm{d}t}$ (Theorem 11.2.8).

Let $\{B(t)\}_{t\in\mathbb{R}}$ be a white noise with PSD $N_0/2$. Let $h : \mathbb{R} \to \mathbb{C}$ be in $L^1_{\mathbb{C}} \cap L^2_{\mathbb{C}}$ and define the output of a filter with impulse response h when the white noise $\{B(t)\}_{t\in\mathbb{R}}$ is the input, by

$$Y(t) = \int_{\mathbb{R}} h(t - s)B(t)\,\mathrm{d}s.$$

By the isometry formula for the Wiener–Doob integral,

$$E[Y(t)Y(s)^*] = \frac{N_0}{2} \int_{\mathbb{R}} h(t - s - u)h^*(u)\,\mathrm{d}u,$$

and therefore (Plancherel–Parseval equality)

$$C_Y(\tau) = \int_{\mathbb{R}} e^{2i\pi\nu\tau} |\widehat{h}(\nu)|^2 \frac{N_0}{2}\,\mathrm{d}\nu.$$

The stochastic process $\{Y(t)\}_{t\in\mathbb{R}}$ is therefore centered and WSS, with PSD

$$f_Y(\nu) = |\widehat{h}(\nu)|^2 f_B(\nu),$$

where

$$f_B(\nu) := \frac{N_0}{2}.$$

We therefore once more recover formally the fundamental equation of linear filtering of WSS continuous-time stochastic processes.

The Approximate Derivative Approach

There is a third approach to white noise. The Brownian motion is approximated by the "finitesimal" derivative

$$B_h(t) = \frac{W(t + h) - W(t)}{h}$$

(here we take $N_0/2 = 1$). For fixed $h > 0$ this defines a proper WSS stochastic process centered, with covariance function

$$C_h(\tau) = \frac{(h - |\tau|)^+}{h^2}$$

and power spectral density

$$f_h(\nu) = \left(\frac{\sin \pi\nu h}{\pi\nu h}\right)^2.$$

Note that, as $h \downarrow 0$, the power spectral density tends to the constant function 1, the power spectral density of the "white noise". At the same time, the covariance function "tends to the Dirac function" and the energy $C_h(0) = \frac{1}{h}$ tends to infinity. This is another feature of white noise: unpredictability. Indeed, for $\tau \geq h$, the value $B_h(t+\tau)$ cannot be predicted from the value $B_h(t)$, since both are independent random variables.

The connection with the second approach is the following. For all $f \in L^2_{\mathbb{C}}(\mathbb{R}_+) \cap L^1_{\mathbb{C}}(\mathbb{R}_+)$,

$$\lim_{h\downarrow 0} \int_{\mathbb{R}_+} f(t)B_h(t)\,\mathrm{d}t = \int_{\mathbb{R}_+} f(t)\,\mathrm{d}W(t)$$

in the quadratic mean. The proof is required in Exercise 12.4.4.

12.2   Fourier Analysis of the Trajectories

12.2.1   The Cramér–Khintchin Decomposition

If one seeks a Fourier transform of the trajectories of a non-trivial wide-sense stationary process, there is *a priori* little chance that it will be a classical one, say, in L^1 or L^2 (Remark 12.1.1). However, under quite general conditions there exists a kind of Fourier decomposition of the trajectories of a wide-sense stationary process.

Theorem 12.2.1 *Let* $\{X(t)\}_{t\in\mathbb{R}}$ *be a centered* WSS *stochastic process, continuous in the quadratic mean and with power spectral measure* μ. *There exists a* unique *centered stochastic process* $\{x(\nu)\}_{\nu\in\mathbb{R}}$ *with uncorrelated increments and with structural measure* μ *such that P-a.s.*

$$X(t) = \int_{\mathbb{R}} e^{2i\pi\nu t}\, \mathrm{d}x(\nu) \quad (t \in \mathbb{R}), \tag{12.10}$$

where the integral of the right-hand side is a Doob integral.

Uniqueness is in the following sense: If there exists another centered stochastic process $\{\widetilde{x}(\nu)\}_{\nu\in\mathbb{R}}$ with uncorrelated increments, and with finite structural measure $\widetilde{\mu}$, such that for all $t \in \mathbb{R}$, we have P–a.s., $X(t) = \int_{\mathbb{R}} e^{2i\pi\nu t}\, \mathrm{d}\widetilde{x}(\nu)$, then for all $a, b \in \mathbb{R}$, $a \le b$, $\widetilde{x}(b) - \widetilde{x}(a) = x(b) - x(a)$, P-a.s.

We will occasionally say: "$\mathrm{d}x(\nu)$ is the *Cramér–Khinchin decomposition*" of the WSS stochastic process.

Proof. 1. Denote by $\mathcal{H}(X)$ the vector subspace of $L^2_{\mathbb{C}}(P)$ formed by the finite complex linear combinations of the type

$$Z = \sum_{k=1}^{K} \lambda_k X(t_k),$$

and by φ the mapping of $\mathcal{H}(X)$ into $L^2_{\mathbb{C}}(\mu)$ defined by

$$\varphi : Z \mapsto \sum_{k=1}^{K} \lambda_k e^{2i\pi\nu t_k}.$$

We verify that it is a linear isometry of $\mathcal{H}(X)$ into $L^2_{\mathbb{C}}(\mu)$. In fact,

$$E\left[\left|\sum_{k=1}^{K} \lambda_k X(t_k)\right|^2\right] = \sum_{k=1}^{K}\sum_{\ell=1}^{K} \lambda_k \lambda_\ell^* E\left[X(t_k)X(t_\ell)^*\right]$$

$$= \sum_{k=1}^{K}\sum_{\ell=1}^{K} \lambda_k \lambda_\ell^* C(t_k - t_\ell),$$

and using Bochner's theorem, this quantity is equal to

$$\sum_{k=1}^{K}\sum_{\ell=1}^{K} \lambda_k \lambda_\ell^* \int_R e^{2i\pi\nu(t_k-t_\ell)}\, \mu(\mathrm{d}\nu) = \int_R \left(\sum_{k=1}^{K}\sum_{\ell=1}^{K} \lambda_k \lambda_\ell^* e^{2i\pi\nu(t_k-t_\ell)}\right) \mu(\mathrm{d}\nu)$$

$$= \int_R \left|\sum_{k=1}^{K} \lambda_k e^{2i\pi\nu t_k}\right|^2 \mu(\mathrm{d}\nu).$$

2. This isometric linear mapping can be uniquely extended to an isometric linear mapping (Theorem C.3.2), that we shall continue to call φ, from $H(X)$, the closure of $\mathcal{H}(X)$, into $L^2_{\mathbb{C}}(\mu)$. As the combinations $\sum_{k=1}^K \lambda_k e^{2i\pi\nu t_k}$ are dense in $L^2_{\mathbb{C}}(\mu)$ when μ is a finite measure, φ is *onto*. Therefore, it is a linear isometric bijection between $H(X)$ and $L^2_{\mathbb{C}}(\mu)$.

3. We shall define $x(\nu_0)$ to be the random variable in $H(X)$ that corresponds in this isometry to the function $1_{(-\infty,\nu_0]}(\nu)$ of $L^2_{\mathbb{C}}(\mu)$. First, we observe that

$$E[x(\nu_2) - x(\nu_1)] = 0$$

since $H(X)$ is the closure in $L^2_{\mathbb{C}}(P)$ of a family of centered random variables. Also, by isometry,

$$E[(x(\nu_2) - x(\nu_1))(x(\nu_4) - x(\nu_3))^*] = \int_{\mathbb{R}} 1_{(\nu_1,\nu_2]}(\nu) 1_{(\nu_3,\nu_4]}(\nu) \, \mu(d\nu)$$
$$= \mu((\nu_1,\nu_2] \cap (\nu_3,\nu_4]).$$

We can therefore define the Doob integral $\int_{\mathbb{R}} f(\nu) \, dx(\nu)$ for all $f \in L^2_{\mathbb{C}}(\mu)$.

4. Let now

$$Z_n(t) := \sum_{k\in\mathbb{Z}} e^{2i\pi t(k/2^n)} \left(x\left(\frac{k+1}{2^n}\right) - x\left(\frac{k}{2^n}\right) \right).$$

We have

$$\lim_{n\to\infty} Z_n(t) = \int_{\mathbb{R}} e^{2i\pi\nu t} \, dx(\nu)$$

(limit in $L^2_{\mathbb{C}}(P)$) because

$$Z_n(t) = \int_{\mathbb{R}} f_n(t,\nu) \, dx(\nu),$$

where

$$f_n(t,\nu) = \sum_{k\in\mathbb{Z}} e^{2i\pi t(k/2^n)} 1_{(k/2^n,(k+1)/2^n]}(\nu),$$

and therefore, by isometry,

$$E\left| Z_n(t) - \int_{\mathbb{R}} e^{2i\pi\nu t} \, dx(\nu) \right|^2 = \int_{\mathbb{R}} |e^{2i\pi\nu t} - f_n(t,\nu)|^2 \, \mu(d\nu),$$

a quantity which tends to zero when n tends to infinity (by dominated convergence, using the fact that μ is a bounded measure). On the other hand, by definition of φ,

$$Z_n(t) \overset{\varphi}{\mapsto} f_n(t,\nu).$$

Since, for fixed t, $\lim_{n\to\infty} Z_n(t) = \int_{\mathbb{R}} e^{2i\pi\nu t} \, dx(\nu)$ in $L^2_{\mathbb{C}}(P)$ and $\lim_{n\to\infty} f_n(t,\nu) = e^{2i\pi\nu t}$ in $L^2_{\mathbb{C}}(\mu)$,

$$\int_{\mathbb{R}} e^{2i\pi\nu t} \, dx(\nu) \overset{\varphi}{\mapsto} e^{2i\pi\nu t}.$$

But, by definition of φ,

$$X(t) \overset{\varphi}{\mapsto} e^{2i\pi\nu t}.$$

Therefore $X(t) = \int_{\mathbb{R}} e^{2i\pi\nu t} \, dx(\nu)$.

5. We now prove uniqueness. Suppose that there exists another spectral decomposition $d\widetilde{x}(\nu)$. Denote by $\mathcal{G}$ the set of finite linear combinations of complex exponentials. Since by hypothesis

$$\int_{\mathbb{R}} e^{2i\pi\nu t}\, dx(\nu) = \int_{\mathbb{R}} e^{2i\pi\nu t}\, d\widetilde{x}(\nu) \quad (= X(t))$$

we have

$$\int_{\mathbb{R}} f(\nu)\, dx(\nu) = \int_{\mathbb{R}} f(\nu)\, d\widetilde{x}(\nu)$$

for all $f \in \mathcal{G}$, and therefore, for all $f \in L^2_{\mathbb{C}}(\mu) \cap L^2_{\mathbb{C}}(\widetilde{\mu}) \subseteq L^2_{\mathbb{C}}(\frac{1}{2}(\mu + \widetilde{\mu}))$ because $\mathcal{G}$ is dense in $L^2_{\mathbb{C}}(\frac{1}{2}(\mu + \widetilde{\mu}))$. In particular, with $f = 1_{(a,b]}$,

$$x(b) - x(a) = \widetilde{x}(b) - \widetilde{x}(a).$$

$\square$

More details can be obtained as to the continuity properties (in quadratic mean) of the increments of the spectral decomposition. For instance, it is right-continuous in quadratic mean, and it admits a left-hand limit in quadratic mean at any point $\nu \in \mathbb{R}$. If such limit is denoted by $x(\nu-)$, then, for all $a \in \mathbb{R}$,

$$E[|x(a) - x(a-)|^2] = \mu(\{a\}).$$

Proof. The right-continuity follows from the continuity of the (finite) measure μ:

$$\lim_{h\downarrow 0} E[|x(a+h) - x(a)|^2] = \lim_{h\downarrow 0} \mu((a, a+h]) = \mu(\varnothing) = 0.$$

As for the existence of left-hand limits, it is guaranteed by the Cauchy criterion, since for all $a \in \mathbb{R}$,

$$\lim_{h,h'\downarrow 0, h<h'} E[|x(a-h) - x(a-h')|^2] = \lim_{h,h'\downarrow 0, h<h'} \mu((a-h', a-h]) = 0.$$

Finally,

$$E[|x(a) - x(a-)|^2] = \lim_{h\downarrow 0} E[|x(a) - x(a-h)|^2] = \lim_{h\downarrow 0} \mu((a-h, a]) = \mu(\{a\}).$$

$\square$

Theorem 12.2.2 *Let $\{X(t)\}_{t\in\mathbb{R}}$ be a WSS stochastic process continuous in quadratic mean. It is real if and only if its spectral decomposition is hermitian symmetric, that is, for all $[a, b] \subset \mathbb{R}$,*

$$x(b) - x(a) = (x(-a_-) - x(-b_-))^*.$$

Proof. In the real case,

$$X(t) = \int_{\mathbb{R}} e^{2i\pi\nu t}\, dx(\nu) = \left(\int_{\mathbb{R}} e^{2i\pi\nu t}\, dx(\nu)\right)^*$$

$$= \int_{\mathbb{R}} e^{-2i\pi\nu t}\, dx^*(\nu) = \int_{\mathbb{R}} e^{2i\pi\nu t}\, dx^*(-\nu)$$

and therefore, by uniqueness of the spectral decomposition, $dx(\nu) = dx^*(-\nu)$. Similarly, if $dx(\nu) = dx^*(-\nu)$,

$$X(t) = \int_{\mathbb{R}} e^{2i\pi\nu t}\, dx(\nu)$$

$$= \int_{\mathbb{R}} e^{2i\pi\nu t}\, dx^*(-\nu) = \left(\int_{\mathbb{R}} e^{2i\pi\nu t}\, dx(\nu)\right)^* = X(t)^*$$

and therefore the process is real.

$\square$

Theorem 12.2.3 *Let* $\{X(t)\}_{t\in\mathbb{R}}$ *be a centered* WSS *stochastic process continuous in quadratic mean. Then*

$$H_C(x(\nu); \nu \in \mathbb{R}) = H_C(X(t); t \in \mathbb{R})$$

and both Hilbert subspaces are identical with

$$\left\{ \int_{\mathbb{R}} g(\nu)\, dx(\nu) \,;\, g \in L^2_{\mathbb{C}}(\mu) \right\}.$$

Proof. 1. For all $\nu \in \mathbb{R}$, $x(\nu) \in H_{\mathbb{C}}(X(t); t \in \mathbb{R})$ (by definition of $x(\nu)$; see the proof of Theorem 12.2.1). Therefore,

$$H_C(x(\nu); \nu \in \mathbb{R}) \subseteq H_{\mathbb{C}}(X(t); t \in \mathbb{R}).$$

On the other hand, for all $t \in \mathbb{R}$, $X(t) = \int_{\mathbb{R}} e^{-2i\pi\nu t}\, dx(\nu) \in H_{\mathbb{C}}(x(\nu); \nu \in \mathbb{R})$. Therefore

$$H_{\mathbb{C}}(X(t); t \in \mathbb{R}) \subseteq H_{\mathbb{C}}(x(\nu); \nu \in \mathbb{R}).$$

Combining the last two results gives

$$H_{\mathbb{C}}(X(t); t \in \mathbb{R}) \equiv H_{\mathbb{C}}(x(\nu); \nu \in \mathbb{R}).$$

2. Clearly $H := \{\int_{\mathbb{R}} g(\nu)\, dx(\nu); g \in L^2_{\mathbb{C}}(\mu)\} \subseteq H_C(x(\nu))$. Moreover, $H_{\mathbb{C}}(X(t); t \in \mathbb{R}) \subseteq H$ since H contains all the variables $X(t) = \int_{\mathbb{R}} e^{-2i\pi\nu t}\, dx(\nu)$ $(t \in \mathbb{R})$. Therefore

$$H_{\mathbb{C}}(X(t); t \in \mathbb{R}) \subseteq H \subseteq H_C(x(\nu))$$

and the conclusion then follows from Part 1 of the proof. $\qquad\square$

The Shannon–Nyquist Sampling Theorem

Most signals occurring in communications theory have a power spectral measure of bounded support. We momentarily adopt the usual terminology in this domain of application.

Definition 12.2.4 *A* WSS *stochastic process* $\{X(t)\}_{t\in\mathbb{R}}$ *is said to be* base-band *of bandwidth* $2B$ $(B > 0)$, *or* base-band (B), *if the support of its spectral power measure is contained in the frequency interval* $[-B, +B]$.

The *Shannon–Nyquist sampling theorem* below tells us that such signals can be recovered from their samples if the sampling rate is larger than the so-called *Nyquist rate* $\frac{1}{2B}$.

Theorem 12.2.5 *For a centered* WSS *base-band* (B) *stochastic process* $\{X(t)\}_{t\in\mathbb{R}}$, *for any* $T > 1/(2B)$,

$$X(t) = \lim_{N\uparrow\infty} \sum_{n=-N}^{+N} X(nT)\, \frac{\sin\left(\frac{\pi}{T}(t - nT)\right)}{\frac{\pi}{T}(t - nT)}, \qquad (12.11)$$

where the limit is in $L^2_{\mathbb{C}}(P)$.

Proof. We have

$$X(t) = \int_{[-B,+B]} e^{2i\pi\nu t}\,dx(\nu).$$

Now,

$$e^{2i\pi\nu t} = \lim_{N\uparrow\infty}\sum_{n=-N}^{+N} e^{2i\pi\nu nT}\frac{\sin\left(\frac{\pi}{T}(t-nT)\right)}{\frac{\pi}{T}(t-nT)},$$

where the limit is uniform in $[-B,+B]$ and bounded. Therefore the above limit is also in $L^2_{\mathbb{C}}(\mu)$ because μ is a finite measure. Consequently

$$X(t) = \lim_{N\uparrow\infty}\int_{[-B,+B]}\left\{\sum_{n=-N}^{+N} e^{2i\pi\nu nT}\frac{\sin\left(\frac{\pi}{T}(t-nT)\right)}{\frac{\pi}{T}(t-nT)}\right\}dx(\nu),$$

where the limit is in $L^2_{\mathbb{C}}(P)$. The result then follows by expanding the integral with respect to the sum. $\qquad\square$

12.2.2 A Plancherel–Parseval Formula

The following result is the analog of the Plancherel–Parseval formula of classical Fourier analysis.

Theorem 12.2.6 *Let $f : \mathbb{R} \to \mathbb{C}$ be in $L^1_{\mathbb{C}}(\mathbb{R})$ with Fourier transform $\widehat{f}$. Let $\{X(t)\}_{t\in\mathbb{R}}$ be a centered WSS stochastic process with power spectral measure μ and Cramér–Khintchin spectral decomposition $dx(\nu)$. Then:*

$$\int_{\mathbb{R}}\widehat{f}(\nu)^*\,dx(\nu) = \int_{\mathbb{R}} f(t)^* X(t)\,dt. \tag{12.12}$$

Proof. The function $\widehat{f}$ is bounded and continuous (as the Fourier transform of an integrable function) and μ is a finite measure, so that $\widehat{f} \in L^2_{\mathbb{C}}(\mu)$ and

$$\sum_n \widehat{f}\left(\frac{k}{2^n}\right)1_{\left(\frac{k}{2^n},\frac{k+1}{2^n}\right]} \to \widehat{f} \text{ in } L^2_{\mathbb{C}}(\mu).$$

Therefore (all limits in the following sequence of equalities being in $L^2_{\mathbb{C}}(P)$):

$$\int_{\mathbb{R}}\widehat{f}(\nu)^*\,dx(\nu) = \lim_{n\to\infty}\sum_{-n2^n}^{n2^n-1}\widehat{f}\left(\frac{k}{2^n}\right)^*\left(x\left(\frac{k+1}{2^n}\right)-x\left(\frac{k}{2^n}\right)\right)$$

$$= \lim_{n\to\infty}\sum_{-n2^n}^{n2^n-1}\left(\int_{\mathbb{R}} f^*(t)e^{+2i\pi(k/2^n)t}\,dt\right)\left(x\left(\frac{k+1}{2^n}\right)-x\left(\frac{k}{2^n}\right)\right)$$

$$= \lim_{n\to\infty}\int_{\mathbb{R}} f^*(t)\sum_{-n2^n}^{n2^n-1}\left[e^{+2i\pi(k/2^n)t}\left(x\left(\frac{k+1}{2^n}\right)-x\left(\frac{k}{2^n}\right)\right)\right]dt$$

$$= \lim_{n\to\infty}\int_{\mathbb{R}} f^*(t)X_n(t)\,dt,$$

where

$$X_n(t) = \sum_{-n2^n}^{n2^n-1} e^{+2i\pi(k/2^n)t}\left(x\left(\frac{k+1}{2^n}\right)-x\left(\frac{k}{2^n}\right)\right) \to X(t) \text{ in } L^2_{\mathbb{C}}(P).$$

The announced result will then follow once we prove that

$$\lim_{n \to \infty} \int_{\mathbb{R}} f^*(t) X_n(t) \, dt = \int_{\mathbb{R}} f^*(t) X(t) \, dt,$$

where the limit is in $L_{\mathbb{C}}^2(P)$. In fact, with $Y_n(t) = X(t) - X_n(t)$,

$$\mathrm{E}\left[\left|\int_{\mathbb{R}} f(t) Y_n(t) \, dt\right|^2\right] = \int_{\mathbb{R}} \int_{\mathbb{R}} f(t) f(s)^* \mathrm{E}\left[Y_n(t) Y_n(s)^*\right] \, dt \, ds.$$

But for all $t \in \mathbb{R}$, $\lim_{n \uparrow \infty} Y_n(t) = 0$ (in $L_{\mathbb{C}}^2(P)$) and therefore $\lim_{n \uparrow \infty} \mathrm{E}\left[Y_n(t) Y_n(s)^*\right] = 0$. Moreover, $\mathrm{E}\left[Y_n(t) Y_n(s)^*\right]$ is uniformly bounded in n. Therefore, by dominated convergence,

$$\lim_{n \uparrow \infty} \int_{\mathbb{R}} \int_{\mathbb{R}} f(t) f(s)^* \mathrm{E}\left[Y_n(t) Y_n(s)^*\right] \, dt \, ds = 0.$$

$\square$

EXAMPLE 12.2.7: CONVOLUTIONAL FILTERING. Let $h \in L_{\mathbb{C}}^1(\mathbb{R})$ and let $\widehat{h}$ be its Fourier transform. Then

$$\int_{\mathbb{R}} h(t - s) X(s) \, ds = \int_{\mathbb{R}} \widehat{h}(\nu) e^{2i\pi\nu t} \, dx(\nu). \tag{12.13}$$

Proof. It suffices to apply (12.12) to the function $s \mapsto h^*(t-s)$, whose Fourier transform is $\widehat{h}(\nu)^* e^{-2i\pi\nu t}$.

$\square$

12.2.3 Linear Operations

A function $g : \mathbb{R} \to \mathbb{C}$ in $L_{\mathbb{C}}^2(\mu)$ defines a linear operation on the centered WSS stochastic process $\{X(t)\}_{t \in \mathbb{R}}$ (called the *input*) by associating with it the centered stochastic process (called the *output*)

$$Y(t) = \int_{\mathbb{R}} e^{2i\pi\nu t} g(\nu) \, dx(\nu). \tag{12.14}$$

On the other hand, the calculation of the covariance function

$$C_Y(\tau) = \mathrm{E}[Y(t) Y(t + \tau)^*]$$

of the output gives (isometry formula for Doob's integral),

$$C_Y(\tau) = \int_{\mathbb{R}} e^{2i\pi\nu\tau} |g(\nu)|^2 \, \mu_X(d\nu),$$

where μ_X is the power spectral measure of the input. The power spectral measure of the output process is therefore

$$\mu_Y(d\nu) = |g(\nu)|^2 \, \mu_X(d\nu). \tag{12.15}$$

This is similar to the formula (15.25) obtained for the output of a stable convolutional filter with impulse response. One then says that g is the *transmittance* of the "filter"

(12.14). Note however that this filter is not necessarily of the convolutional type, since g may well not be the Fourier transform of an integrable function (for instance it may be unbounded, as the next example shows).

EXAMPLE 12.2.8: DIFFERENTIATION. Let $\{X(t)\}_{t\in\mathbb{R}}$ be a WSS stochastic processes with spectral measure μ_X such that

$$\int_{\mathbb{R}} |\nu|^2 \mu_X(\mathrm{d}\nu) < \infty. \tag{12.16}$$

Then

$$\lim_{h\to 0} \frac{X(t+h) - X(t)}{h} = \int_{\mathbb{R}} (2i\pi\nu)e^{2i\pi\nu t}\mathrm{d}x(\nu),$$

where the limit is in the quadratic mean. The linear operation corresponding to the transmittance $g(\nu) = 2i\pi\nu$ is therefore the *differentiation in quadratic mean*.

Proof. Let $h \in \mathbb{R}$. From the equality

$$\frac{X(t+h) - X(t)}{h} - \int_{\mathbb{R}} (2i\pi\nu)e^{2i\pi\nu t}\mathrm{d}x(\nu)$$

$$= \int_{\mathbb{R}} e^{2i\pi\nu t}\left(\frac{e^{2i\pi\nu h} - 1}{h} - 2i\pi\nu\right)\mathrm{d}x(\nu)$$

we have, by isometry,

$$\lim_{h\to 0} \mathrm{E}\left[\left|\frac{X(t+h) - X(t)}{h} - \int_{\mathbb{R}} (2i\pi\nu)e^{2i\pi\nu t}\mathrm{d}x(\nu)\right|^2\right]$$

$$= \lim_{h\to 0} \int_{\mathbb{R}} \left|\frac{e^{2i\pi\nu h} - 1}{h} - 2i\pi\nu\right|^2 \mu_X(\mathrm{d}\nu).$$

In view of hypothesis (12.16) and since $\left|\frac{e^{2i\pi\nu h}-1}{h} - 2i\pi\nu\right|^2 \leq 4\pi^2\nu^2$, the latter limit is 0, by dominated convergence.

$\square$

"A line spectrum corresponds to a combination of sinusoids." More precisely:

Theorem 12.2.9 *Let $\{X(t)\}_{t\in\mathbb{R}}$ be a centered* WSS *stochastic processes with spectral measure*

$$\mu_X(\mathrm{d}\nu) = \sum_{k\in\mathbb{Z}} P_k \,\varepsilon_{\nu_k}(\mathrm{d}\nu),$$

where ε_{ν_k} is the Dirac measure at $\nu_k \in \mathbb{R}$, $P_k \in \mathbb{R}_+$ and $\sum_{k\in\mathbb{Z}} P_k < \infty$. Then

$$X(t) = \sum_{k\in\mathbb{Z}} U_k e^{2i\pi\nu_k t}$$

where $\{U_k\}_{k\in\mathbb{Z}}$ is a sequence of centered uncorrelated square-integrable complex variables, and $\mathrm{E}[|U_k|^2] = P_k$.

Proof. The function

$$g(\nu) = \sum_{k\in\mathbb{Z}} 1_{\{\nu_k\}}(\nu)$$

is in $L^2_{\mathbb{C}}(\mu_X)$, as well as the function $1 - g$. Also $\int_{\mathbb{R}} |1 - g(\nu)|^2 \mu_X(\mathrm{d}\nu) = 0$, and in particular $\int_{\mathbb{R}} (1 - g(\nu)) \mathrm{e}^{2i\pi\nu t} \, \mathrm{d}x(\nu) = 0$. Therefore

$$
\begin{aligned}
X(t) &= \int_{\mathbb{R}} g(\nu) \mathrm{e}^{2i\pi\nu t} \, \mathrm{d}x(\nu) \\
&= \sum_{k \in \mathbb{Z}} \mathrm{e}^{2i\pi\nu_k t} (x(\nu_k) - x(\nu_k-)).
\end{aligned}
$$

The conclusion follows by defining $U_k := x(\nu_k) - x(\nu_k-)$. $\qquad\square$

Linear Transformations of Gaussian Processes

Definition 12.2.10 *A linear transformation of a WSS stochastic process $\{X(t)\}_{t \in \mathbb{R}}$ is a transformation of it into the second-order process (not WSS in general)*

$$
Y(t) = \int_{\mathbb{R}} g(\nu, t) \, \mathrm{d}x(\nu), \tag{12.17}
$$

where

$$
\int_{\mathbb{R}} |g(t, \nu)|^2 \, \mu_X(\mathrm{d}\nu) < \infty \quad \text{for all } t \in \mathbb{R}.
$$

Theorem 12.2.11 *A linear transformation of a Gaussian WSS stochastic process yields a Gaussian stochastic process.*

Proof. Let $\{X(t)\}_{t \in \mathbb{R}}$ be a centered Gaussian WSS with Cramér–Khinchin decomposition $\mathrm{d}x(\nu)$. For each $\nu \in \mathbb{R}$, the random variable $x(\nu)$ is in $H_{\mathbb{R}}(X)$, by construction. Now, if $\{X(t)\}_{t \in \mathbb{R}}$ is a Gaussian process, $H_{\mathbb{R}}(X)$ is a Gaussian subspace. But (Theorem 12.2.3) $H_{\mathbb{R}}(X) = H_{\mathbb{R}}(x)$. Therefore the process (12.17) is in $H_{\mathbb{C}}(X)$, hence Gaussian. $\qquad\square$

EXAMPLE 12.2.12: CONVOLUTIONAL FILTERING OF A WSS GAUSSIAN PROCESS. In particular, if $\{X(t)\}_{t \in \mathbb{R}}$ is a Gaussian WSS process with Cramér–Khinchin decomposition $\mathrm{d}x(\nu)$ and if $g \in L^2_{\mathbb{C}}(\mu_X)$, the process

$$
Y(t) = \int_{\mathbb{R}} \mathrm{e}^{2i\pi\nu t} g(\nu) \, \mathrm{d}x(\nu) \quad (t \in \mathbb{R})
$$

is Gaussian. A particular case is when $g = \widehat{h}$, the Fourier transform of a filter with integrable impulse response h. The stochastic process $\{Y(t)\}_{t \in \mathbb{R}}$ is the one obtained by convolutional filtering of $\{X(t)\}_{t \in \mathbb{R}}$ with this filter.

12.3 Multivariate WSS Stochastic Processes

12.3.1 The Power Spectral Matrix

Let

$$
X(t) = (X_1(t), \ldots, X_L(t)) \quad (t \in \mathbb{R})
$$

be a stochastic process with values in $E := \mathbb{C}^L$, where L is an integer greater than or equal to 2. This process is assumed centered and of the second order, that is,

$$E[\|X(t)\|^2] < \infty \quad (t \in \mathbb{R}).$$

Furthermore, it will be assumed that it is wide-sense stationary, in the sense that the mean vector of $X(t)$ and the cross-covariance matrix of the vectors $X(t+\tau)$ and $X(t)$ do not depend upon t. The matrix-valued function C defined by

$$C(\tau) := \text{cov}\,(X(t+\tau), X(t)) \tag{12.18}$$

is called the (matrix) *covariance function* of the stochastic process. Its general entry is

$$C_{ij}(\tau) = \text{cov}(X_i(t), X_j(t+\tau)).$$

The processes $\{X_i(t)\}_{t\in\mathbb{R}}$ $(1 \le i \le L)$ are WSS stochastic processes and in addition, they are *stationarily correlated* or "jointly WSS". Such a vector-valued stochastic process $\{X(t)\}_{t\in\mathbb{R}}$ is called a *multivariate* WSS *stochastic process*.

EXAMPLE 12.3.1: SIGNAL PLUS NOISE. The following model frequently appears in signal processing:

$$Y(t) = S(t) + B(t),$$

where $\{S(t)\}_{t\in\mathbb{R}}$ and $\{B(t)\}_{t\in\mathbb{R}}$ are two *uncorrelated* centered WSS stochastic process with respective covariance functions C_S and C_B. Then, $\{(Y(t), S(t))^T\}_{t\in\mathbb{R}}$ is a bivariate WSS stochastic process. Owing to the assumption of non-correlation,

$$C(\tau) = \begin{pmatrix} C_S(\tau) + C_B(\tau) & C_S(\tau) \\ C_S(\tau) & C_S(\tau) \end{pmatrix}.$$

Theorem 12.3.2 *Let* $\{X(t)\}_{t\in\mathbb{R}}$ *be an L-dimensional multivariate* WSS *stochastic process. For all* r, s $(1 \le r, s \le L)$ *there exists a finite complex measure* μ_{rs} *such that*

$$C_{rs}(\tau) = \int_{\mathbb{R}} e^{2i\pi\nu\tau} \mu_{rs}(d\nu). \tag{12.19}$$

Proof. Say $r = 1$, $s = 2$. Let us consider the stochastic processes

$$Y(t) = X_1(t) + X_2(t), \qquad Z(t) = iX_1(t) + X_2(t).$$

These are WSS stochastic processes with respective covariance functions

$$C_Y(\tau) = C_1(\tau) + C_2(\tau) + C_{12}(\tau) + C_{21}(\tau),$$
$$C_Z(\tau) = -C_1(\tau) + C_2(\tau) + iC_{12}(\tau) - iC_{21}(\tau).$$

From these two equalities we deduce

$$C_{12}(\tau) = \frac{1}{2}\left\{[C_Y(\tau) - C_1(\tau) - C_2(\tau)] - i[C_Z(\tau) - C_1(\tau) + C_2(\tau)]\right\},$$

from which the result follows with

$$\mu_{12} = \frac{1}{2}\left\{[\mu_Y - \mu_1 - \mu_2] - i[\mu_Z - \mu_1 + \mu_2]\right\}.$$

$\square$

The matrix

$$M := \{\mu_{ij}\}_{1 \le i,j \le k}$$

(whose entries are finite complex measures) is the *interspectral power measure matrix* of the multivariate WSS stochastic process $\{X(t)\}_{t \in \mathbb{R}}$. It is clear that for all $z = (z_1, \dots, z_k) \in \mathbb{C}^k$, $U(t) = z^T X(t)$ defines a WSS stochastic process with spectral measure $\mu_U = z\,M\,z^\dagger$ († means transpose conjugate).

The link between the interspectral measure μ_{12} and the Cramér–Khintchine decompositions $dx_1(\nu)$ and $dx_2(\nu)$ is the following:

$$E[x_1(\nu_2) - x_1(\nu_1))(x_2(\nu_4) - x_2(\nu_3))^*] = \mu_{12}((\nu_1, \nu_2] \cup (\nu_3, \nu_4]).$$

This is a particular case of the following result. For all functions $g_i : \mathbb{R} \to \mathbb{C}$, $g_i \in L^2_{\mathbb{C}}(\mu_i)$ $(i = 1, 2)$,

$$E\left[\left(\int_{\mathbb{R}} g_1(\nu)\,dx_1(\nu)\right)\left(\int_{\mathbb{R}} g_2(\nu)\,dx_2(\nu)\right)^*\right] = \int_{\mathbb{R}} g_1(\nu)g_2(\nu)^*\,\mu_{12}(d\nu). \qquad (12.20)$$

Indeed, equality (12.20) is true for $g_1(\nu) = e^{2i\pi t_1 \nu}$ and $g_2(\nu) = e^{2i\pi t_2 \nu}$ since it then reduces to

$$E[X_1(t)X_2(t)^*] = \int_{\pi} e^{2i\pi(t_1 - t_2)\nu}\,\mu_{12}(d\nu).$$

This is therefore verified for $g_1 \in \mathcal{E}$, $g_2 \in \mathcal{E}$, where $\mathcal{E}$ is the set of finite linear combinations of functions of the type $\nu \to e^{2i\pi t\nu}$, $t \in \mathbb{R}$. But $\mathcal{E}$ is dense in $L^2_{\mathbb{C}}(\mu_i)$ $(i = 1, 2)$, and therefore the equality (12.20) is true for all $g_i \in L^2_{\mathbb{C}}(\mu_i)$ $(i = 1, 2)$.

Theorem 12.3.3 *The interspectral measure μ_{12} is absolutely continuous with respect to each spectral measure μ_1 and μ_2.*

Proof. This means that $\mu_{12}(A) = 0$ whenever $\mu_1(A) = 0$ or $\mu_2(A) = 0$. Indeed,

$$\mu_{12}(A) = E\left[\left(\int_A dZ_1\right)\left(\int_A dZ_2\right)^*\right]$$

and $\mu_1(A) = 0$ implies $\int_A dZ_1 = 0$ since

$$E\left[\left|\int_A dZ_1\right|^2\right] = \mu_1(A).$$

$\square$

Therefore, every spectral measure μ_{ij} is absolutely continuous with respect to the trace of the power spectral measure matrix

$$\mathrm{Tr}\,M := \sum_{j=1}^{k} \mu_j.$$

By the Radon–Nikodým theorem there exists a function $g_{ij} : \mathbb{R} \to \mathbb{C}$ such that

$$\mu_{ij}(A) = \int_A g_{ij}(\nu)\,\mathrm{Tr}M(d\nu).$$

The matrix

$$g(\nu) = \{g_{ij}(\nu)\}_{1 \leq i,j \leq k}$$

is called the *canonical spectral density* matrix of $\{X(t)\}_{t \in \mathbb{R}}$. One should insist that it is not required that the stochastic processes $\{X_i(t)\}_{t \in \mathbb{R}}$, $1 \leq i \leq k$, admit power spectral densities.

The correlation matrix $C(\tau)$ has, with the above notation, the representation

$$C(\tau) = \int_{\mathbb{R}} e^{2i\pi\nu\tau} g(\nu) \operatorname{Tr} M(d\nu).$$

If each one among the WSS stochastic processes $\{X_i(t)\}_{t \in \mathbb{R}}$ admits a spectral density, $\{X(t)\}_{t \in \mathbb{R}}$ admits an interspectral density matrix

$$f(\nu) = \{f_{ij}(\nu)\}_{1 \leq i,j \leq k},$$

that is,

$$C_{ij}(\tau) = \operatorname{cov}(X_i(t + \tau), X_j(t)) = \int_{\mathbb{R}} e^{2i\pi\nu\tau} f_{ij}(\nu) \, d\nu.$$

EXAMPLE 12.3.4: INTERFERENCES. Let $\{X(t)\}_{t \in \mathbb{R}}$ be a centered WSS stochastic process with power spectral measure μ_X. Let $h_1, h_2 : \mathbb{R} \to \mathbb{C}$ be integrable functions with respective Fourier transforms $\hat{h}_1$ and $\hat{h}_2$. Define for $i = 1, 2$,

$$Y_i(t) := \int_{\mathbb{R}} h_i(t - s)X(s) \, ds.$$

The WSS stochastic processes $\{Y_1(t)\}_{t \in \mathbb{R}}$ and $\{Y_2(t)\}_{t \in \mathbb{R}}$ are stationarily correlated. In fact (assuming that they are centered, without loss of generality),

$$
\begin{aligned}
E[Y_1(t + \tau)Y_2(t)^*] &= E\left[\left(\int_{\mathbb{R}} h_1(t + \tau - s)X(s) \, ds\right)\left(\int_{\mathbb{R}} h_2(t - s)X(s) \, ds\right)^*\right] \\
&= \int_{\mathbb{R}} \int_{\mathbb{R}} h_1(t + \tau - u)h_2^*(t - v)C_X(u - v) \, du \, dv \\
&= \int_{\mathbb{R}} \int_{\mathbb{R}} h_1(\tau - u)h_2^*(-v)C_X(u - v) \, du \, dv,
\end{aligned}
$$

and this quantity depends only upon τ. Replacing $C_X(u - v)$ by its expression in terms of the spectral measure μ_X, one obtains

$$C_{Y_1Y_2}(\tau) = \int_{\mathbb{R}} e^{2i\pi\nu\tau} T_1(\nu)T_2^*(\nu) \, \mu_X(d\nu).$$

The power spectral matrix of the bivariate process $\{Y_1(t), Y_2(t)\}_{t \in \mathbb{R}}$ is therefore

$$\mu_Y(d\nu) = \begin{pmatrix} |T_1(\nu)|^2 & T_1(\nu)T_2^*(\nu) \\ T_1^*(\nu)T_2(\nu) & |T_2(\nu)|^2 \end{pmatrix} \mu_X(d\nu).$$

12.3.2   Band-pass Stochastic Processes

Let $\{X(t)\}_{t\in\mathbb{R}}$ be a centered WSS stochastic process with power spectral measure μ_X and Cramér–Khinchin decomposition $dx(\nu)$. This process is assumed real, and therefore

$$\mu_X(-d\nu) = \mu_X(d\nu), \qquad dx(-\nu) = dx(\nu)^*.$$

Definition 12.3.5 *The above WSS stochastic process is called* band-pass (ν_0, B), *where* $\nu_0 > B > 0$, *if the support of* μ_X *is contained in the frequency band* $[-\nu_0 - B, -\nu_0 + B] \cup [\nu_0 - B, \nu_0 + B]$.

A band-pass stochastic process admits the following *quadrature decomposition*

$$X(t) = M(t)\cos 2\pi\nu_0 t - N(t)\sin 2\pi\nu_0 t, \tag{12.21}$$

where $\{M(t)\}_{t\in\mathbb{R}}$ and $\{N(t)\}_{t\in\mathbb{R}}$, called the *quadrature components*, are real base-band (B) WSS stochastic process. To prove this, let $G(\nu) := -i\,\text{sign}(\nu)\ (= 0\text{ if }\nu = 0)$. The function G is the so-called *Hilbert filter* transmittance. The *quadrature process* associated with $\{X(t)\}_{t\in\mathbb{R}}$ is defined by

$$Y(t) = \int_{\mathbb{R}} G(\nu)e^{2i\pi\nu t}\,dx(\nu).$$

The right-hand side of the preceding equality is well defined since $\int_{\mathbb{R}} |G(\nu)|^2\,\mu_X(d\nu) = \mu_X(\mathbb{R}) < \infty$. Moreover, this stochastic process is real, since its spectral decomposition is hermitian symmetric. The *analytic process* associated with $\{X(t)\}_{t\in\mathbb{R}}$ is, by definition, the stochastic process

$$Z(t) = X(t) + iY(t) = \int_{\mathbb{R}} (1 + iG(\nu))e^{2i\pi\nu t}\,dx(\nu) = 2\int_{(0,\infty)} e^{2i\pi\nu t}\,dx(\nu).$$

Taking into account that $|G(\nu)|^2 = 1$, the preceding expressions and the Wiener isometry formulas lead to the following properties:

$$\mu_Y(d\nu) = \mu_X(d\nu), \qquad C_Y(\tau) = C_X(\tau), \qquad C_{XY}(\tau) = -C_{YX}(\tau),$$

$$\mu_Z(d\nu) = 4\,\mathbb{1}_{\mathbb{R}_+}(\nu)\,\mu_X(d\nu), \qquad C_Z(\tau) = 2\{C_X(\tau) + iC_{YX}(\tau)\},$$

and

$$E[Z(t+\tau)Z(t)] = 0. \tag{$\star$}$$

Defining the *complex envelope* of $\{X(t)\}_{t\in\mathbb{R}}$ by

$$U(t) = Z(t)e^{-2i\pi\nu_0 t}, \tag{$\star\star$}$$

it follows from this definition that

$$C_U(\tau) = e^{-2i\pi\nu_0\tau}C_Z(\tau), \qquad \mu_U(d\nu) = \mu_Z(d\nu + \nu_0), \tag{$\dagger$}$$

whereas $(\star)$ and $(\star\star)$ give

$$E[U(t+\tau)U(t)] = 0. \tag{$\dagger\dagger$}$$

The quadrature components $\{M(t)\}_{t\in\mathbb{R}}$ and $\{N(t)\}_{t\in\mathbb{R}}$ of $\{X(t)\}_{t\in\mathbb{R}}$ are the *real* WSS stochastic processes defined by

$$U(t) = M(t) + iN(t).$$

Since

$$X(t) = \mathrm{Re}\{Z(t)\} = \mathrm{Re}\{U(t)e^{2i\pi\nu_0 t}\},$$

we have the decomposition (12.21). Taking (††) into account we obtain:

$$C_M(\tau) = C_N(\tau) = \frac{1}{4}\left\{C_U(\tau) + C_U(\tau)^*\right\},$$

and

$$C_{MN}(\tau) = C_{NM}(\tau) = \frac{1}{4i}\left\{C_U(\tau) - C_U(\tau)^*\right\}, \tag{$\diamondsuit$}$$

and the corresponding relations for the spectra

$$\mu_M(d\nu) = \mu_N(d\nu) = \left\{\mu_X(d\nu - \nu_0) + \mu_X(d\nu + \nu_0)\right\}1_{[-B,+B]}(\nu).$$

From ($\diamondsuit$) and the observation that $C_U(0) = C_U(0)^*$ (since $C_U(0) = \mathrm{E}[|U(0)|^2]$ is real), we deduce $C_{MN}(0) = 0$, that is to say,

$$\mathrm{E}[M(t)N(t)] = 0. \tag{12.22}$$

If, furthermore, the original process has a power spectral measure that is symmetric about ν_0 in the band $[\nu_0 - B, \nu_0 + B]$, the same holds for the spectrum of the analytic process and, by (†), the complex envelope has a spectral measure symmetric about 0, which implies $C_U(\tau) = C_U(\tau)^*$ and then, by ($\diamondsuit$),

$$\mathrm{E}[M(t)N(t+\tau)] = 0. \tag{12.23}$$

In summary:

Theorem 12.3.6 *Let $\{X(t)\}_{t\in\mathbb{R}}$ be a centered real band-pass (ν_0, B) WSS stochastic process. The values of its quadrature components at a given time are uncorrelated. Moreover, if the original stochastic process has a power spectral measure symmetric about ν_0, the quadrature component processes are uncorrelated.*

More can be said when the original process is Gaussian. In this case, the quadrature component processes are jointly Gaussian (being obtained from the original Gaussian process by linear operations). In particular, for all $t \in \mathbb{R}$, $M(t)$ and $N(t)$ are jointly Gaussian and uncorrelated, and therefore independent.

If moreover the original process has a spectrum symmetric about ν_0, then, by (12.23), $M(t_1)$ and $N(t_2)$ $(t_1, t_2 \in \mathbb{R})$ are uncorrelated jointly Gaussian variables, and therefore independent. In other words, the quadrature component processes are two independent centered Gaussian WSS stochastic processes.

Complementary reading

[Cramér and Leadbetter, 1967, 1995] is the classical reference. It emphasizes the study of level crossings by wide-sense stationary stochastic processes. [Brémaud, 2014] has a chapter on the spectral measure of point processes.

12.4 Exercises

Exercise 12.4.1. SYMMETRIC POWER SPECTRAL MEASURE
Show that the power spectral measure of a real WSS stochastic process is symmetric.

Exercise 12.4.2. PRODUCTS OF INDEPENDENT WSS STOCHASTIC PROCESSES
Let $\{X(t)\}_{t\in\mathbb{R}}$ and $\{Y(t)\}_{t\in\mathbb{R}}$ be two centered WSS stochastic processes of respective covariance functions $C_X(\tau)$ and $C_Y(\tau)$.

1. Assume the two signals to be independent. Show that $Z(t) := X(t)Y(t)$ $(t \in \mathbb{R})$ is a WSS stochastic process. Give its mean and covariance function.

2. Assume the same hypothesis as in the previous question, but now $\{X(t)\}_{t\in\mathbb{R}}$ is the harmonic process of Example 5.1.20. Suppose that $\{Y(t)\}_{t\in\mathbb{R}}$ admits a power spectral density $f_Y(\nu)$. Give the power spectral density $f_Z(\nu)$ of $\{Z(t)\}_{t\in\mathbb{R}}$.

Exercise 12.4.3. THE APPROXIMATE DERIVATIVE OF A WIENER PROCESS
Let $\{W(t)\}_{t\geq 0}$ be a Wiener process. Show that for $a > 0$, the stochastic process

$$X_a(t) := \frac{W(t+a) - W(t)}{a} \quad (t \in \mathbb{R})$$

is a WSS stochastic process. Compute its mean, its covariance function and its power spectral density.

Exercise 12.4.4. DOOB'S INTEGRAL AND THE FINITESIMAL DERIVATIVE OF BROWNIAN MOTION
Let $\{W(t)\}_{t\geq 0}$ be a standard Brownian motion. Prove the following. For all $f \in L^2_{\mathbb{C}}(\mathbb{R}_+) \cap L^1_{\mathbb{C}}(\mathbb{R}_+)$,

$$\lim_{h\downarrow 0} \int_{\mathbb{R}_+} f(t)B_h(t)\,\mathrm{d}t = \int_{\mathbb{R}_+} f(t)\,\mathrm{d}W(t)$$

in the quadratic mean.

Exercise 12.4.5. THE SQUARE OF A BAND-LIMITED WHITE NOISE
Let $\{X(t)\}_{t\in\mathbb{R}}$ be a wide-sense stationary *centered* Gaussian process with covariance function $C_X(\tau)$ and with the power spectral density

$$f_X(\nu) = \frac{N_0}{2}1_{[-B,+B]}(\nu),$$

where $N_0 > 0$ and $B > 0$.

1. Let $Y(t) = X(t)^2$. Show that $\{Y(t)\}_{t\in\mathbb{R}}$ is a wide-sense stationary process.

2. Give its power spectral density $f_Y(\nu)$.

Exercise 12.4.6. PROJECTION OF WHITE NOISE ONTO AN ORTHONORMAL BASE
Let the set of square-integrable functions $\varphi : [0, T] \to \mathbb{R}$ $(1 \leq i \leq N)$ be such that

$$\int_0^T \varphi_i(t)\varphi_j(t)\,\mathrm{d}t = \delta_{ij} \quad (1 \leq i, j \leq N),$$

and let $\{B(t)\}_{t \in \mathbb{R}}$ be a Gaussian white noise with PSD $N_0/2$. Show that the vector $B = (B_1, \ldots, B_N)^T$ defined by

$$B_i = \int_0^T B(t) \varphi_i(t) \, dt \quad (1 \leq i \leq N)$$

is a centered Gaussian vector with covariance matrix

$$\Gamma_B = \frac{N_0}{2} I. \tag{12.24}$$

(In particular, the components $B_1, \ldots, B_N$ are identically distributed, independent, and centered Gaussian random variables with common variance $N_0/2$.)

Exercise 12.4.7. AN IID SEQUENCE CARRIED BY AN HPP
Let N be a homogeneous Poisson process on $\mathbb{R}_+$ of intensity $\lambda > 0$, and let $\{Z_n\}_{n \geq 0}$ be an IID sequence of integrable real random variables, centered, with finite variance σ^2, and independent of N.

1) Show that $\{Z_{N((0,t])}\}_{t \geq 0}$ is a wide-sense stationary stochastic process and give its covariance function.

2) Give its power spectral density.

3) Compute $P(X(t_1) = X(t_2))$ and $P(X(t_1) > X(t_2))$.

Exercise 12.4.8. POISSON SHOT NOISES
Let N_1, N_2 and N_3 be three independent homogeneous Poisson processes on $\mathbb{R}$ with respective intensities $\theta_1 > 0$, $\theta_2 > 0$ and $\theta_3 > 0$. Let $\{X_1(t)\}_{t \in \mathbb{R}}$ be the shot noise constructed on $N_1 + N_3$ with an impulse function $h : \mathbb{R} \to \mathbb{R}$ that is bounded and with compact support (null outside a finite interval). Let $\{X_2(t)\}_{t \in \mathbb{R}}$ be the shot noise constructed on $N_2 + N_3$ with the same impulse function h.

Compute the power spectral density of the wide-sense stationary process $\{X(t)\}_{t \in \mathbb{R}}$, where $X(t) = X_1(t) + X_2(t)$.

Exercise 12.4.9. FREQUENCY MODULATION
Consider the so-called *frequency modulated (or phase modulated) signal*, a stochastic process $\{X(t)\}_{t \geq 0}$ defined by

$$X(t) = \cos(2\pi(\nu_0 t + \Phi(t) + \alpha)),$$

where

$$\Phi(t) = \int_0^t \nu(s) \, ds,$$

$\{\nu(t)\}_{t \geq 0}$ is a real-valued stochastic process, and α is a real-valued random variable. The following assumptions are made:

(a) $\{\nu(t)\}$ is a strictly stationary process.

(b) α and $\{\nu(t)\}$ are independent.

(c) $\mathbb{E}[e^{2i\pi\alpha}] = \mathbb{E}[e^{4i\pi\alpha}] = 0$.

12.4. EXERCISES

Show that the covariance function of the frequency modulated signal is given by

$$C_X(\tau) = \frac{1}{2}\text{Re}\left\{e^{2i\pi\nu_0\tau}E\left[e^{2i\pi\int_0^\tau \nu(s)\,ds}\right]\right\}.$$

Exercise 12.4.10. GAUSSIAN FREQUENCY MODULATION
This exercise is a continuation of Exercise 12.4.9 to which the reader is referred for the notation and definitions. We now consider a particular case for which the computations are tractable: *Gaussian frequency modulation.* Here $\{\nu(t)\}_{t\geq 0}$ is a stationary Gaussian signal with mean $\bar{\nu}$ and covariance function C_ν. Show that

$$C_X(\tau) = \frac{1}{2}\cos(2\pi(\nu_0 + \bar{\nu})t)e^{-4\pi^2\int_0^\tau C_\nu(s)(\tau-s)\,ds}.$$

Exercise 12.4.11. FLIP-FLOP
Let N be an HPP on $\mathbb{R}_+$ with intensity λ. Define the (*telegraph* or *flip-flop*) process $\{X(t)\}_{t\geq 0}$ with state space $E = \{+1, -1\}$ by

$$X(t) = Z(-1)^{N(t)},$$

where $X(0) = Z$ is an E-valued random variable independent of the counting process N. (Thus the telegraph process switches between -1 and $+1$ at each event of N.) The probability distribution of Z is arbitrary.

1. Compute $P(X(t+s) = j | X(s) = i)$ for all $t,\ s \geq 0$ and all $i,\ j \in E$.

2. Give, for all $i \in E$, the limit of $P(X(t) = i)$ as t tends to ∞.

3. Show that when $P(Z = 1) = \frac{1}{2}$, the process is a stationary process and give its power spectral measure.

Exercise 12.4.12. FLIP-FLOP WITH LIMITED MEMORY
Let N be a HPP on $\mathbb{R}$ with intensity $\lambda > 0$. Define for all $t \in \mathbb{R}$

$$X(t) = (-1)^{N((t,t+a])}.$$

1. Show that $\{X(t)\}_{t\in\mathbb{R}}$ is a WSS stochastic process.

2. Compute its power spectral density.

3. Give the best affine estimate of $X(t+\tau)$ in terms of $X(t)$, that is, find α, β minimizing

$$E\left[|X(t+\tau) - (\alpha + \beta X(t))|^2\right], \quad \text{when } \tau > 0.$$

Exercise 12.4.13. JUMPING PHASE
Define for each $t \in \mathbb{R}$, $t \geq 0$,

$$X(t) = e^{i\Phi_{N(t)}},$$

where $\{N(t)\}_{t\geq 0}$ is the counting process of a homogeneous Poisson process on $\mathbb{R}_+$ with intensity $\lambda > 0$, and $\{\Phi_n\}_{n\geq 0}$ is an IID sequence of random variables uniformly distributed on $[0, 2\pi]$, and independent of the Poisson process.

Show that $\{X(t)\}_{t\geq 0}$ is a wide-sense stationary process, give its covariance function $C_X(\tau)$ and its power spectral measure.

III: ADVANCED TOPICS

Chapter 13

Martingales

A martingale is for the general public a clever way of gambling. In mathematics, it formalizes the notion of fair game and we shall see that martingale theory indeed has something to say about such games. However the interest and scope of martingale theory extends far beyond gambling and has become a fundamental tool of the theory of stochastic processes. The present chapter is an introduction to this topic, featuring the two main pillars on which it rests: the *optional sampling* theorem and the convergence theory of martingales, in discrete as well as in continuous time.

13.1 Martingale Inequalities

13.1.1 The Martingale Property

Let $(\Omega, \mathcal{F}, P)$ be a probability space and let $\{\mathcal{F}_n\}_{n \geq 1}$ be a *history* (or *filtration*) defined on it, that is, a sequence of sub-σ-fields of $\mathcal{F}$ that is non-decreasing: $\mathcal{F}_n \subseteq \mathcal{F}_{n+1}$ $(n \geq 0)$. The *internal history* of a random sequence $\{X_n\}_{n \geq 0}$ is the filtration $\{\mathcal{F}_n^X\}_{n \geq 0}$ defined by $\mathcal{F}_n^X := \sigma(X_0, \ldots, X_n)$.

Definition 13.1.1 *A complex random sequence* $\{Y_n\}_{n \geq 0}$ *such that for all* $n \geq 0$

(i) Y_n *is* $\mathcal{F}_n$*-measurable and*

(ii) $\mathrm{E}[|Y_n|] < \infty$

is called a $(P, \mathcal{F}_n)$*-martingale (resp.,* sub*martingale,* super*martingale) if, in addition, for all* $n \geq 0$,

$$\mathrm{E}[Y_{n+1} \mid \mathcal{F}_n] = Y_n \qquad (resp., \ \geq Y_n, \leq Y_n). \tag{13.1}$$

When the context is clear as to the choice of the underlying probability measure P, we shall abbreviate, saying for instance, "$\mathcal{F}_n$-submartingale" instead of "$(P, \mathcal{F}_n)$-submartingale".

If the history is not mentioned, it is assumed to be the internal history. For instance, the phrase $\{Y_n\}_{n \geq 0}$ is a martingale means that it is an $\mathcal{F}_n^Y$-martingale.

Of course an $\mathcal{F}_n$-martingale is an $\mathcal{F}_n$-submartingale *and* an $\mathcal{F}_n$–supermartingale.

Condition (13.45) implies that for all $k \geq 1$, all $n \geq 0$,

$$\mathrm{E}[Y_{n+k} \mid \mathcal{F}_n] = Y_n \qquad (resp., \ \geq Y_n, \leq Y_n).$$

© Springer Nature Switzerland AG 2020
P. Brémaud, *Probability Theory and Stochastic Processes*, Universitext,
https://doi.org/10.1007/978-3-030-40183-2_13

Proof. In the martingale case, for instance, by the rule of successive conditioning

$$
\begin{aligned}
\mathrm{E}[Y_{n+k} \mid \mathcal{F}_n] &= \mathrm{E}[\mathrm{E}[Y_{n+k} \mid \mathcal{F}_{n+k-1}] \mid \mathcal{F}_n] \\
&= \mathrm{E}[Y_{n+k-1} \mid \mathcal{F}_n] = \mathrm{E}[Y_{n+k-2} \mid \mathcal{F}_n] \\
&= \cdots = \mathrm{E}[Y_n \mid \mathcal{F}_n] = Y_n.
\end{aligned}
$$

$\square$

In particular, taking expectations and letting $n = 0$,

$$
\mathrm{E}[Y_k] = \mathrm{E}[Y_0] \qquad (\text{resp., } \geq \mathrm{E}[Y_0], \, \leq \mathrm{E}[Y_0]).
$$

EXAMPLE 13.1.2: SUMS OF IID RANDOM VARIABLES. Let $\{X_n\}_{n\geq 0}$ be an IID sequence of *centered and integrable* random variables. The random sequence

$$
Y_n := X_0 + X_1 + \cdots + X_n \quad (n \geq 0)
$$

is an $\mathcal{F}_n^X$-martingale. Indeed, for all $n \geq 0$, Y_n is $\mathcal{F}_n^X$-measurable and

$$
\mathrm{E}[Y_{n+1} \mid \mathcal{F}_n^X] = \mathrm{E}[Y_n \mid \mathcal{F}_n] + \mathrm{E}[X_{n+1} \mid \mathcal{F}_n^X] = Y_n + \mathrm{E}[X_{n+1}] = Y_n,
$$

where the second equality is due to the fact that $\mathcal{F}_n^X$ and X_{n+1} are independent (Theorem 3.3.20).

EXAMPLE 13.1.3: PRODUCTS OF IIDS. Let $X = \{X_n\}_{n\geq 0}$ be an IID sequence of integrable random variables with mean 1. The random sequence

$$
Y_n = \prod_{k=0}^{n} X_k \quad (n \geq 0)
$$

is an $\mathcal{F}_n^X$-martingale. Indeed, for all $n \geq 0$, Y_n is $\mathcal{F}_n^X$-measurable and

$$
\begin{aligned}
\mathrm{E}[Y_{n+1} \mid \mathcal{F}_n^X] &= \mathrm{E}\left[X_{n+1} \prod_{k=0}^{n} X_k \mid \mathcal{F}_n^X \right] = \mathrm{E}[X_{n+1} \mid \mathcal{F}_n^X] \prod_{k=0}^{n} X_k \\
&= \mathrm{E}[X_{n+1}] \prod_{k=1}^{n} X_k = 1 \times Y_n = Y_n,
\end{aligned}
$$

where the second equality is due to the fact that $\mathcal{F}_n^X$ and X_{n+1} are independent (Theorem 3.3.20).

EXAMPLE 13.1.4: GAMBLING. Consider the random sequence $\{Y_n\}_{n\geq 0}$ with values in $\mathbb{R}_+$ defined by $Y_0 = a \in \mathbb{R}_+$ and

$$
Y_{n+1} = Y_n + X_{n+1} b_{n+1}(X_0^n) \quad (n \geq 0),
$$

where $X_0^n := (X_0, \ldots, X_n)$, $X_0 = Y_0$, $\{X_n\}_{n\geq 1}$ is an IID sequence of random variables taking the values $+1$ or -1 with equal probability, and the family of functions $b_n : \{0,1\}^n \to \mathbb{N}$ $(n \geq 1)$ is the *betting strategy*, that is, $b_{n+1}(X_0^n)$ is the stake at time $n+1$ of a gambler given the observed history X_0^n of the chance outcomes up to time n. Admissible bets must guarantee that the fortune Y_n remains non-negative at all times

n, that is, $b_{n+1}(X_0^n) \leq Y_n$. The process so defined is an $\mathcal{F}_n^X$-martingale. Indeed, for all $n \geq 0$, Y_n is $\mathcal{F}_n^X$-measurable and

$$
\begin{aligned}
\mathrm{E}\left[Y_{n+1} \mid \mathcal{F}_n^X\right] &= \mathrm{E}\left[Y_n \mid \mathcal{F}_n^X\right] + \mathrm{E}\left[X_{n+1}b_{n+1}(X_0^n) \mid \mathcal{F}_n^X\right] \\
&= Y_n + \mathrm{E}\left[X_{n+1} \mid \mathcal{F}_n^X\right]b_{n+1}(X_0^n) = Y_n\,,
\end{aligned}
$$

where the second equality uses Theorem 3.3.24. The integrability condition should be checked on each application. It is satisfied if the stakes $b_n(X_0^n)$ are uniformly bounded.

EXAMPLE 13.1.5: HARMONIC FUNCTIONS OF AN HMC. Let $\{X_n\}_{n\geq0}$ be an HMC with countable space E and transition matrix $\mathbf{P}$. A function $h : E \to \mathbb{R}$ is called *harmonic* (resp., *subharmonic, superharmonic*) if $\mathbf{P}h$ is well defined and

$$
\mathbf{P}h = h \qquad (\text{resp.},\ \geq h, \leq h)\,, \tag{13.2}
$$

that is,

$$
\sum_{j\in E} p_{ij}h(j) = h(i) \qquad (\text{resp.},\ \geq h(i), \leq h(i)) \quad (i \in E)\,.
$$

Superharmonic functions are also called *excessive* functions.

Equation (13.2) is equivalent, in the harmonic case for instance, to

$$
\mathrm{E}[h(X_{n+1}) \mid X_n = i] = h(i) \quad (i \in E)\,. \tag{$\star$}
$$

In view of the Markov property, the left-hand side of the above equality is also equal to

$$
\mathrm{E}[h(X_{n+1}) \mid X_n = i, X_{n-1} = i_{n-1}, \ldots, X_0 = i_0]\,,
$$

and therefore ($\star$) is equivalent to

$$
\mathrm{E}[h(X_{n+1} \mid \mathcal{F}_n^X] = h(X_n)\,.
$$

Therefore, if $\mathrm{E}[|h(X_n)|] < \infty$ for all $n \geq 0$, the process $\{h(X_n)\}_{n\geq0}$ is an $\mathcal{F}_n^X$-martingale. Similarly, for a subharmonic (*resp.* superharmonic) function h such that $\mathrm{E}[|h(X_n)|] < \infty$ for all $n \geq 0$, the process $\{h(X_n)\}_{n\geq0}$ is an $\mathcal{F}_n^X$-submartingale (*resp.* $\mathcal{F}_n^X$-supermartingale).

EXAMPLE 13.1.6: RADON–NIKODÝM SEQUENCES. Let be given on the measurable space $(\Omega, \mathcal{F})$ two probability measures Q and P and a filtration $\{\mathcal{F}_n\}_{n\geq1}$. Let Q_n and P_n denote the restrictions of Q and P respectively to $(\Omega, \mathcal{F}_n)$. Suppose that for all $n \geq 1$, $Q_n \ll P_n$, in which case we say that Q is locally absolutely continuous along $\{\mathcal{F}_n\}_{n\geq1}$ with respect to P and denote this by $Q \overset{loc.}{\ll} P$. Let $L_n := \frac{\mathrm{d}Q_n}{\mathrm{d}P_n}$. Then $\{L_n\}_{n\geq1}$ is a $(P, \mathcal{F}_n)$-martingale.

Proof. The integrability condition is satisfied since

$$
\mathrm{E}_P[L_n] = \int_\Omega L_n\,\mathrm{d}P_n = \int_\Omega \frac{\mathrm{d}Q_n}{\mathrm{d}P_n}\,\mathrm{d}P_n = \int_\Omega \mathrm{d}Q_n = Q_n(\Omega) = 1\,.
$$

Now, for all $A \in \mathcal{F}_n$ (and *a fortiori* $A \in \mathcal{F}_{n+1}$),

$$\int_A L_{n+1}\,\mathrm{d}P = \int_A L_{n+1}\,\mathrm{d}P_{n+1} = Q_{n+1}(A) = Q_n(A)$$
$$= \int_A L_n\,\mathrm{d}P_n = \int_A L_n\,\mathrm{d}P.$$

$\square$

Definition 13.1.7 *Let $\{\mathcal{F}_n\}_{n\geq 0}$ be some filtration. A complex random sequence $\{X_n\}_{n\geq 0}$ such that for all $n \geq 0$*

(a) *X_n is $\mathcal{F}_n$-measurable,*

(b) *$\mathrm{E}[|X_n|] < \infty$ and $\mathrm{E}[X_n] = 0$, and*

(c) *$\mathrm{E}[X_{n+1} \mid \mathcal{F}_n] = 0$ (resp. ≥ 0, ≤ 0)*

is called a $(P, \mathcal{F}_n)$-martingale difference (resp., submartingale difference, supermartingale difference).

The notion of martingale difference generalizes that of centered IID sequences. Indeed for such IID sequences, X_n is independent of $\mathcal{F}_n^X$, and therefore (Theorem 3.3.20) $\mathrm{E}[X_{n+1} \mid \mathcal{F}_n^X] = 0$.

Convex Functions of Martingales

Theorem 13.1.8 *Let $I \subseteq \mathbb{R}$ be an interval (closed, open, semi-closed, infinite, etc.) and let $\varphi : I \to \mathbb{R}$ be a convex function.*

A. *Let $\{Y_n\}_{n\geq 0}$ be an $\mathcal{F}_n$-martingale such that $P(Y_n \in I) = 1$ for all $n \geq 0$. Assume that $\mathrm{E}[|\varphi(Y_n)|] < \infty$ for all $n \geq 0$. Then, the process $\{\varphi(Y_n)\}_{n\geq 0}$ is an $\mathcal{F}_n$-submartingale.*

B. *Assume moreover that φ is non-decreasing and suppose this time that $\{Y_n\}_{n\geq 0}$ is an $\mathcal{F}_n$-submartingale. Then, the process $\{\varphi(Y_n)\}_{n\geq 0}$ is an $\mathcal{F}_n$-submartingale.*

Proof. By Jensen's inequality for conditional expectations (Exercise 3.4.50),

$$\mathrm{E}\left[\varphi(Y_{n+1})|\mathcal{F}_n\right] \geq \varphi(\mathrm{E}\left[Y_{n+1}|\mathcal{F}_n\right]).$$

Therefore (case A)

$$\mathrm{E}\left[\varphi(Y_{n+1})|\mathcal{F}_n\right] \geq \varphi(\mathrm{E}\left[Y_{n+1}|\mathcal{F}_n\right]) = \varphi(Y_n),$$

and (case B)

$$\mathrm{E}\left[\varphi(Y_{n+1})|\mathcal{F}_n\right] \geq \varphi(\mathrm{E}\left[Y_{n+1}|\mathcal{F}_n\right]) \geq \varphi(Y_n).$$

(For the last inequality, use the submartingale property $\mathrm{E}\left[Y_{n+1}|\mathcal{F}_n\right] \geq Y_n$ and the hypothesis that φ is non-decreasing.)

$\square$

EXAMPLE 13.1.9: Let $\{Y_n\}_{n\geq 0}$ be an $\mathcal{F}_n$-martingale and let $p \geq 1$. As a special case of Theorem 13.1.8 with the convex function $x \to |x|^p$, we have that if $\mathrm{E}[|Y_n|^p] < \infty$, $\{|Y_n|^p\}_{n\geq 0}$ is an $\mathcal{F}_n$-submartingale. Applying Theorem 13.1.8 with the convex function $x \to x^+$, we have that $\{Y_n^+\}_{n\geq 0}$ is an $\mathcal{F}_n$-submartingale.

Martingale Transforms and Stopped Martingales

Let $\{\mathcal{F}_n\}_{n\geq 0}$ be some filtration. The complex stochastic process $\{H_n\}_{n\geq 1}$ is called $\mathcal{F}_n$-*predictable* if

$$H_n \text{ is } \mathcal{F}_{n-1}\text{-measurable} \quad \text{for all } n \geq 1.$$

Let $\{Y_n\}_{n\geq 0}$ be another complex stochastic process. The stochastic process

$$(H \circ Y)_n := \sum_{k=1}^{n} H_k(Y_k - Y_{k-1}) \quad (n \geq 1).$$

is called the *transform* of Y by H.

Theorem 13.1.10

(a) Let $\{Y_n\}_{n\geq 0}$ be an $\mathcal{F}_n$-*submartingale and let* $\{H_n\}_{n\geq 0}$ *be a bounded non-negative* $\mathcal{F}_n$-*predictable process. Then* $\{(H \circ Y)_n\}_{n\geq 0}$ *is an* $\mathcal{F}_n$-*submartingale.*

(b) If $\{Y_n\}_{n\geq 0}$ *is an* $\mathcal{F}_n$-*martingale and if* $\{H_n\}_{n\geq 0}$ *is bounded and* $\mathcal{F}_n$-*predictable, then* $\{(H \circ Y)_n\}_{n\geq 0}$ *is an* $\mathcal{F}_n$-*martingale.*

Proof. Conditions (i) and (ii) of (13.1.1) are obviously satisfied. Moreover,

$$\text{(a)} \quad \mathrm{E}[(H \circ Y)_{n+1} - (H \circ Y)_n \mid \mathcal{F}_n] = \mathrm{E}[H_{n+1}(Y_{n+1} - Y_n) \mid \mathcal{F}_n]$$
$$= H_{n+1}\mathrm{E}[Y_{n+1} - Y_n \mid \mathcal{F}_n] \geq 0,$$

using Theorem 3.3.24 for the second equality.

$$\text{(b)} \quad \mathrm{E}[(H \circ Y)_{n+1} - (H \circ Y)_n \mid \mathcal{F}_n] = H_{n+1}\mathrm{E}[Y_{n+1} - Y_n \mid \mathcal{F}_n] = 0,$$

by the same token. $\qquad\square$

Recall the definition of an $\mathcal{F}_n$-stopping time (Definition 6.2.20): a random variable τ taking its values in $\overline{\mathbb{N}}$ and such that for all $m \in \mathbb{N}$, the event $\{\tau = m\}$ is in $\mathcal{F}_m$.

Theorem 13.1.10 immediately leads to the *stopped martingale* theorem:

Theorem 13.1.11 Let $\{Y_n\}_{n\geq 0}$ *be an* $\mathcal{F}_n$-*submartingale (resp., martingale) and let* τ *be an* $\mathcal{F}_n$-*stopping time. Then* $\{Y_{n\wedge\tau}\}_{n\geq 0}$ *is an* $\mathcal{F}_n$-*submartingale (resp., martingale). In particular,*

$$\mathrm{E}[Y_{n\wedge\tau}] \geq \mathrm{E}[Y_0] \quad (\text{resp., } = \mathrm{E}[Y_0]) \quad (n \geq 0). \tag{13.3}$$

Proof. Let $H_n := 1_{\{n\leq\tau\}}$. The stochastic process H is $\mathcal{F}_n$-predictable since $\{H_n = 0\} = \{\tau \leq n-1\} \in \mathcal{F}_{n-1}$. We have

$$Y_{n\wedge\tau} = Y_0 + \sum_{k=1}^{n\wedge\tau} (Y_k - Y_{k-1})$$

$$= Y_0 + \sum_{k=1}^{n} 1_{\{k\leq\tau\}} (Y_k - Y_{k-1}).$$

The result then follows by Theorem 13.1.10. $\qquad\square$

13.1.2 Kolmogorov's Inequality

It often occurs that a result proved for IID sequences also holds for martingale difference sequences. This is the case for the inequality originally proved in the IID case (Lemma 4.1.12).

Theorem 13.1.12 *Let $\{S_n\}_{n\geq 0}$ be an $\mathcal{F}_n$-submartingale. Then, for all $\lambda \in \mathbb{R}_+$,*

$$\lambda P\left(\max_{0\leq i\leq n} S_i > \lambda\right) \leq \mathrm{E}\left[S_n 1_{\{\max_{0\leq i\leq n} S_i > \lambda\}}\right]. \tag{13.4}$$

Proof. Define the random time

$$\tau = \inf\{n \geq 0\,;\, S_n > \lambda\}.$$

It is an $\mathcal{F}_n$-stopping time since

$$A_i := \{\tau = i\} = \left\{S_i > \lambda,\, \max_{0\leq j\leq i-1} S_j \leq \lambda\right\} \in \mathcal{F}_i.$$

The A_i's so defined are mutually disjoint and

$$A := \left\{\max_{0\leq i\leq n} S_i > \lambda\right\} = \bigcup_{i=1}^{n} A_i.$$

Since $\lambda 1_{A_i} \leq S_i 1_{A_i}$,

$$\lambda P(A) = \lambda \sum_{i=0}^{n} P(A_i) \leq \sum_{i=0}^{n} \mathrm{E}[S_i 1_{A_i}].$$

For all $0 \leq i \leq n$, A_i being $\mathcal{F}_i$-measurable, we have by the submartingale property that $\mathrm{E}[S_n \mid \mathcal{F}_i] \geq S_i$ and therefore $\int_{A_i} S_i \,\mathrm{d}P \leq \int_{A_i} \mathrm{E}[S_n \mid \mathcal{F}_i]\,\mathrm{d}P$. Taking these observations into account,

$$\lambda P(A) \leq \sum_{i=0}^{n} \mathrm{E}[S_i 1_{A_i}]$$

$$\leq \sum_{i=0}^{n} \mathrm{E}\left[\mathrm{E}^{\mathcal{F}_i}[S_n] 1_{A_i}\right]$$

$$= \sum_{i=0}^{n} \mathrm{E}\left[\mathrm{E}^{\mathcal{F}_i}[S_n 1_{A_i}]\right]$$

$$= \sum_{i=0}^{n} \mathrm{E}[S_n 1_{A_i}]$$

$$= \mathrm{E}\left[S_n \sum_{i=0}^{n} 1_{A_i}\right]$$

$$= \mathrm{E}[S_n 1_A].$$

$\square$

Corollary 13.1.13 *Let $\{M_n\}_{n\geq 0}$ be an $\mathcal{F}_n$-martingale. Then, for all $p \geq 1$, all $\lambda \in \mathbb{R}$,*

$$\lambda^p P\left(\max_{0\leq i\leq n} |M_i| > \lambda\right) \leq \mathrm{E}[|M_n|^p]. \tag{13.5}$$

Proof. Let $S_n = |M_n|^p$. This defines an $\mathcal{F}_n$-submartingale (Example 13.1.9) to which one may apply Kolmogorov's inequality with λ replaced by λ^p:

$$\lambda^p P\left(\max_{0\le i\le n}|M_i|^p > \lambda^p\right) \le \mathrm{E}\left[|M_n|^p 1_{\{\max_{0\le i\le n}|M_i|^p>\lambda^p\}}\right] \le \mathrm{E}[|M_n|^p].$$

$\square$

Remark 13.1.14 Note that Kolmogorov's inequality is, as far as martingales are concerned, a considerable improvement with respect to what Markov's inequality would have given: $\lambda^p P\left(|M_i|^p > \lambda^p\right) \le \mathrm{E}[|M_i|^p] \le \mathrm{E}[|M_n|^p]$ $(0 \le i \le n)$.

13.1.3 Doob's Inequality

Recall the notation $\| X \|_p = (\mathrm{E}\left[|X|^p\right])^{1/p}$.

Theorem 13.1.15 *Let $\{M_n\}_{n\ge 0}$ be an $\mathcal{F}_n$-martingale. For all $p > 1$,*

$$\| M_n \|_p \le \| \max_{0\le i\le n}|M_i| \|_p \le q \| M_n \|_p, \tag{13.6}$$

where q (the "conjugate" of p) is defined by $\frac{1}{p} + \frac{1}{q} = 1$.

Proof. The first inequality is trivial. For the second inequality, observe that for all non-negative random variables X, by Fubini's theorem,

$$\mathrm{E}[X^p] = \mathrm{E}\left[\int_0^X px^{p-1}\,\mathrm{d}x\right]$$
$$= \mathrm{E}\left[\int_0^\infty px^{p-1}1_{\{x<X\}}\,\mathrm{d}x\right]$$
$$= p\int_0^\infty x^{p-1}P(X > x)\,\mathrm{d}x.$$

Therefore, applying this and Kolmogorov's inequality (13.4) to the submartingale $S_n = |M_n|$,

$$\mathrm{E}\left[\max_{0\le i\le n}|M_i|^p\right] \le \mathrm{E}\left[\left(\max_{0\le i\le n}|M_i|\right)^p\right]$$
$$= p\int_0^\infty x^{p-1}P\left(\max_{0\le i\le n}|M_i| > x\right)\,\mathrm{d}x$$
$$\le p\int_0^\infty x^{p-2}\mathrm{E}\left[|M_n|1_{\{\max_{0\le i\le n}|M_i|>x\}}\right]\,\mathrm{d}x$$
$$= p\mathrm{E}\left[\int_0^\infty x^{p-2}|M_n|1_{\{\max_{0\le i\le n}|M_i|>x\}}\,\mathrm{d}x\right]$$
$$= p\mathrm{E}\left[|M_n|\int_0^{\max_{0\le i\le n}|M_i|} x^{p-2}\,\mathrm{d}x\right]$$
$$= \frac{p}{p-1}\mathrm{E}\left[|M_n|\left(\max_{0\le i\le n}|M_i|\right)^{p-1}\right]$$
$$= q\mathrm{E}\left[|M_n|\left(\max_{0\le i\le n}|M_i|\right)^{p-1}\right].$$

By Hölder's inequality, and observing that $(p-1)q = p$,

$$E\left[|M_n|\left(\max_{0\le i\le n}|M_i|\right)^{p-1}\right] \le E[|M_n|^p]^{1/p}\, E\left[\left(\max_{0\le i\le n}|M_i|\right)^{(p-1)q}\right]^{1/q}$$

$$= \|\, M_n\,\|_p\, E\left[\left(\max_{0\le i\le n}|M_i|\right)^p\right]^{1/q}.$$

Therefore

$$E\left[\max_{0\le i\le n}|M_i|^p\right] \le q\,\|\, M_n\,\|_p\, E\left[\left(\max_{0\le i\le n}|M_i|\right)^p\right]^{1/q},$$

or (eliminating the trivial case where $E\left[\max_{0\le i\le n}|M_i|^p\right] = \infty$)

$$E\left[\max_{0\le i\le n}|M_i|^p\right]^{1-\frac{1}{q}} \le q\,\|\, M_n\,\|_p,$$

that is, since $1-\frac{1}{q} = \frac{1}{p}$,

$$\|\max_{0\le i\le n}|M_i|\,\|_p \le q\,\|\, M_n\,\|_p.$$

$\square$

13.1.4   Hoeffding's Inequality

Theorem 13.1.16 *Let $\{M_n\}_{n\ge 0}$ be a real $\mathcal{F}_n$-martingale such that, for some sequence $c_1, c_2, \ldots$ of real numbers,*

$$P(|M_n - M_{n-1}| \le c_n) = 1 \quad (n \ge 1). \tag{13.7}$$

Then, for all $x \ge 0$ and all $n \ge 1$,

$$P(|M_n - M_0| \ge x) \le 2\exp\left(-\frac{1}{2}x^2 \Big/ \sum_{i=1}^{n} c_i^2\right). \tag{13.8}$$

Proof. By convexity of $z \mapsto e^{az}$, for $|z| \le 1$ and all $a \in \mathbb{R}$,

$$a^{az} \le \frac{1}{2}(1-z)e^{-a} + \frac{1}{2}(1+z)e^{+a}.$$

In particular, if Z is a centered random variable such that $P(|Z| \le 1) = 1$,

$$E[e^{aZ}] \le \frac{1}{2}(1 - E[Z])e^{-a} + \frac{1}{2}(1 + E[Z])e^{+a}$$

$$= \frac{1}{2}e^{-a} + \frac{1}{2}e^{+a} \le e^{a^2/2}.$$

By similar arguments, for all $a \in \mathbb{R}$,

$$E\left[e^{a\left(\frac{M_n - M_{n-1}}{c_n}\right)}\Big|\mathcal{F}_{n-1}\right]$$

$$\le \frac{1}{2}\left(1 - E\left[\frac{M_n - M_{n-1}}{c_n}\Big|\mathcal{F}_{n-1}\right]\right)e^{-a} + \cdots$$

$$\cdots + \frac{1}{2}\left(1 + E\left[\frac{M_n - M_{n-1}}{c_n}\Big|\mathcal{F}_{n-1}\right]\right)e^{+a} \le e^{a^2/2},$$

and, with a replaced by $c_n a$,

$$\mathrm{E}\left[e^{a(M_n - M_{n-1})} \big| \mathcal{F}_{n-1}\right] \le e^{a^2 c_n^2 / 2} .$$

Therefore,

$$\mathrm{E}\left[e^{a(M_n - M_0)}\right] = \mathrm{E}\left[e^{a(M_{n-1} - M_0)} e^{a(M_n - M_{n-1})}\right]$$

$$= \mathrm{E}\left[e^{a(M_{n-1} - M_0)} \mathrm{E}\left[e^{a(M_n - M_{n-1})} \big| \mathcal{F}_{n-1}\right]\right]$$

$$\le \mathrm{E}\left[e^{a(M_{n-1} - M_0)}\right] \times e^{a^2 c_n^2 / 2} ,$$

and then by recurrence

$$\mathrm{E}\left[e^{a(M_n - M_0)}\right] \le e^{\frac{1}{2} a^2 \sum_{i=1}^n c_i^2} .$$

In particular, with $a > 0$, by Markov's inequality,

$$P(M_n - M_0 \ge x) \le e^{-ax} \mathrm{E}\left[e^{a(M_n - M_0)}\right] \le e^{-ax + \frac{1}{2} a^2 \sum_{i=1}^n c_i^2} .$$

Minimization of the right-hand side with respect to a gives

$$P(M_n - M_0 \ge x) \le e^{-\frac{1}{2} x^2 / \sum_{i=1}^n c_i^2} .$$

The same argument with $M_0 - M_n$ instead of $M_n - M_0$ yields the bound

$$P(-(M_n - M_0) \ge x) \le e^{-\frac{1}{2} x^2 / \sum_{i=1}^n c_i^2} .$$

The announced bound then follows from these two bounds since for any random variable X, and all $x \in \mathbb{R}_+$, $P(|X| \ge x) = P(X \ge x) + P(X \le -x)$. $\qquad \square$

EXAMPLE 13.1.17: THE KNAPSACK. There are n objects, the i-th has a volume V_i and is worth W_i. All these non-negative random variables form an independent family, the V_i's have finite means and the means of the W_i's are bounded by $M < \infty$. You have to choose integers $z_1, \ldots, z_n$ in such a way that the total volume $\sum_{i=1}^n z_i V_i$ does not exceed a given storage capacity c and that the total worth $\sum_{i=1}^n z_i V_i$ is maximized. Call this maximal worth Z. We shall see that

$$P\left(|Z - \mathrm{E}[Z]| \ge x\right) \le 2 \exp\left\{\frac{-x^2}{2nM^2}\right\} \quad (x \ge 0) .$$

For this consider the variables Z_j which are the equivalent of Z when the j-th object has been removed. Let now $M_j := \mathrm{E}[Z \mid \mathcal{F}_j]$, where $\mathcal{F}_j := \sigma\left((V_k, W_k); 1 \le k \le j\right)$. Note that in view of the independence assumptions $\mathrm{E}[Z_j \mid \mathcal{F}_j] = \mathrm{E}[Z_{j-1} \mid \mathcal{F}_j]$. Clearly $Z_j \le Z \le Z_j + M$. Taking conditional expectations given $\mathcal{F}_j$ and then $\mathcal{F}_{j-1}$ in this last chain of inequalities reveals that $|M_j - M_{j-1}| \le M$. The rest is then just Hoeffding's inequality.

A General Framework of Application

Let $\mathcal{X}$ be a finite set, and let $f : \mathcal{X}^N \to \mathbb{R}$ be a given function. We introduce the notation $x = (x_1, \ldots, x_N)$ and $x_1^k = (x_1, \ldots, x_k)$. In particular, $x = x_1^N$. For $x \in \mathcal{X}^N$, $z \in \mathcal{X}$ and $1 \le k \le N$, let

$$f_k(x, z) := f(x_1, \ldots, x_{k-1}, z, x_{k+1}, \ldots, x_N) .$$

The function f is said to satisfy the *Lipschitz condition* with bound c if for all $x \in \mathcal{X}^N$, all $z \in \mathcal{X}$ and all $1 \leq k \leq N$,

$$|f_k(x, z) - f(x)| \leq c.$$

Let $X_1, X_2, \ldots, X_N$ be independent random variables with values in $\mathcal{X}$. Define the martingale

$$M_n = \mathrm{E}\left[f(X) \mid X_1^n\right].$$

By the independence assumption, with obvious notations,

$$\mathrm{E}\left[f(X) \mid X_1^n\right] = \sum_{x_{n+1}^N} f(X_1^{n-1}, X_n, x_{n+1}^N) P(X_{n+1}^N = x_{n+1}^N)$$

and

$$\mathrm{E}\left[f(X) \mid X_1^{n-1}\right] = \sum_{x_{n+1}^N} \sum_{x_n} f(X_1^{n-1}, x_n, x_{n+1}^N) P(X_n = x_n) P(X_{n+1}^N = x_{n+1}^N).$$

Therefore

$$|M_n - M_{n-1}|$$
$$\leq \sum_{x_{n+1}^N} \sum_{x_n} |f(X_1^{n-1}, x_n, x_{n+1}^N) - f(X_1^{n-1}, X_n, x_{n+1}^N)| P(X_n = x_n) P(X_{n+1}^N = x_{n+1}^N) \leq c.$$

EXAMPLE 13.1.18: PATTERN MATCHING. Take $f(x)$ to be the number of occurrences of the fixed pattern $b = (b_1, \ldots, b_k)$ ($k \leq N$) in the sequence $x = (x_1, \ldots, x_N)$, that is

$$f(x) = \sum_{i=1}^{N-k+1} 1_{\{x_i = b_1, \ldots, x_{i+k-1} = b_k\}}.$$

The mean number of matches in an IID sequence $X = (X_1, \ldots, X_N)$ with uniform distribution on $\mathcal{X}$ is therefore

$$\mathrm{E}\left[f(X)\right] = \sum_{i=1}^{N-k+1} \mathrm{E}\left[1_{\{X_i = b_1, \ldots, X_{i+k-1} = b_k\}}\right] = \sum_{i=1}^{N-k+1} \left(\frac{1}{|\mathcal{X}|}\right)^k,$$

that is,

$$\mathrm{E}\left[f(X)\right] = (N - k + 1)\left(\frac{1}{|\mathcal{X}|}\right)^k.$$

The martingale $M_n := \mathrm{E}\left[f(X) \mid X_1^n\right]$ is such that $M_0 = \mathrm{E}\left[f(X)\right]$. Changing the value of one coordinate of $x \in \mathcal{X}^N$ changes $f(x)$ by at most k, we can apply the bound of Theorem 13.8 with $c_i \equiv k$ to obtain the inequality

$$P(|f(X) - \mathrm{E}\left[f(X)\right]| \geq \lambda) \leq 2e^{-\frac{1}{2}\frac{\lambda^2}{Nk^2}}.$$

Exposure Martingales in Erdös–Rényi Graphs

A random graph $\mathcal{G}(n,p)$ (see Definition 1.3.44) with set of vertices V_n of cardinality n may be generated as follows. Enumerate the $N = \binom{n}{2}$ edges of the complete graph on V_n from $i = 1$ to $i = N$. Generate a random vector $X = (X_1, \dots, X_N)$ with independent and identically distributed variables with values in $\{0, 1\}$ and common distribution, $P(X_i = 1) = p$. Then include edge i in $\mathcal{G}(n,p)$ if and only if $X_i = 1$. Any functional of $\mathcal{G}(n,p)$ can always be written as $f(X)$. The *edge exposure martingale* corresponding to this functional is the $\mathcal{F}_n^X$-martingale defined by $M_0 = E\left[f(X)\right]$ and for $i \geq 1$,

$$M_i := E\left[f(X) \mid X_1^i\right].$$

Since the X_i's are independent, the general method of the previous subsection can be applied.

Another type of martingale related to a $\mathcal{G}(n,p)$ graph is useful. Here V_n is identified with $\{1, 2, \dots, n\}$. We denote similarly $\{1, 2, \dots, i\}$ by V_i. For $1 \leq i \leq n$, define the graph G_i to be the restriction of $\mathcal{G}(n,p)$ to V_i. Any functional of $\mathcal{G}(n,p)$ can always be written as $f(G)$, where $G := (G_1, \dots, G_n)$. The *vertex exposure martingale* corresponding to this functional is the G_1^i-martingale defined by $M_0 = E\left[f(G)\right]$ and for $i \geq 1$,

$$M_i := E\left[f(G) \mid G_1^i\right].$$

EXAMPLE 13.1.19: THE CHROMATIC NUMBER OF AN ERDÖS–RÉNYI GRAPH. The chromatic number of a graph G is the minimal number of colors needed to color the vertices in such a way that no adjacent vertices receive the same color. Call $f(G)$ the chromatic number of G. Since the difference between $f(G_0^{i-1}, G_i, g_{i+1}^n)$ and $f(G_0^{i-1}, g_i, g_{i+1}^n)$ for all g_i, g_{i+1}^n is at most one, one can apply Hoeffding's inequality to obtain

$$P\left(|f(G) - E\left[f(G)\right]| \geq \lambda \sqrt{n}\right) \leq \frac{1}{2} e^{-2\lambda^2}. \tag{13.9}$$

But ...the G_i's are not independent! Nevertheless, the general method of the previous subsection can be applied modulo a slight change of point of view. Let X_1 be a constant, and for $2 \leq i \leq n$, let $X_i = \{X_{\langle i,j \rangle}, 1 \leq j \leq i - 1\}$ (recall the definition of $X_{\langle u,v \rangle}$ in Definition 1.3.44). (Here the passage from subgraph G_{i-1} to subgraph G_i is represented by the "difference" X_i between these two subgraphs.) Then $f(G)$ can be rewritten as $h(X) = h(X_1, \dots, X_n)$ and the general method applies since the X_i's are independent.

13.2   Martingales and Stopping Times

13.2.1   Doob's Optional Sampling Theorem

The first pillar of martingale theory is the *optional sampling theorem*. It has many versions and that given next is the most elementary one, sufficient for the elementary examples to be considered now. More general results are given later in this subsection.

Theorem 13.2.1 *Let $\{M_n\}_{n\geq 0}$ be an $\mathcal{F}_n$-martingale, and let τ be an $\mathcal{F}_n$-stopping time (see Definition 6.2.20). Suppose that at least one of the following conditions holds:*

(α) $P(\tau \leq n_0) = 1$ *for some* $n_0 \geq 0$, *or*

(β) $P(\tau < \infty) = 1$ *and* $|M_n| \leq K < \infty$ *when* $n \leq \tau$.

Then

$$E[M_\tau] = E[M_0].$$

(13.10)

Proof. (α) Just apply Theorem 13.1.11 (Formula (13.3) with $n = n_0$).

(β) Apply the result of (α) to the $\mathcal{F}_n$-stopping time $\tau \wedge n_0$ to obtain

$$E[M_{\tau \wedge n_0}] = E[M_0].$$

But, by dominated convergence,

$$\lim_{n_0 \uparrow \infty} E[M_{\tau \wedge n_0}] = E[\lim_{n_0 \uparrow \infty} M_{\tau \wedge n_0}] = E[M_\tau].$$

$\square$

EXAMPLE 13.2.2:   THE RUIN PROBLEM VIA MARTINGALES. The symmetric random walk $\{X_n\}_{n\geq 0}$ on $\mathbb{Z}$ with initial state 0 is an $\mathcal{F}_n^X$-martingale (Example 13.1.2). Let τ be the first time n for which $X_n = -a$ or $+b$, where $a, b > 0$. This is an $\mathcal{F}_n^X$-stopping time and moreover $\tau < \infty$. Part (β) of the above result can be applied with $K = \sup(a, b)$ to obtain $0 = E[X_0] = E[X_\tau]$. Writing $v = P(-a$ is hit before $b)$, we have

$$E[X_\tau] = -av + b(1 - v),$$

and therefore

$$v = \frac{b}{a + b}.$$

EXAMPLE 13.2.3:   A COUNTEREXAMPLE. Consider the symmetric random walk of the previous example, but now define τ to be the hitting time of $b > 0$, an almost surely finite time since the symmetric walk on $\mathbb{Z}$ is recurrent. If the optional sampling theorem applied, one would have

$$0 = E[X_0] = E[X_\tau] = b,$$

an obvious contradiction. Of course, neither condition (α) nor (β) is satisfied.

The following generalization of the elementary result given at the beginning of the present subsection will now be proved after the following theorem.

Theorem 13.2.4 *Let $\{\mathcal{F}_n\}_{n\geq 0}$ be a history and let $\mathcal{F}_\infty := \sigma(\cup_{n\geq 0}\mathcal{F}_n)$. Let τ be an $\mathcal{F}_n$-stopping time. The collection of events*

$$\mathcal{F}_\tau := \{A \in \mathcal{F}_\infty \mid A \cap \{\tau = n\} \in \mathcal{F}_n, \text{ for all } n \geq 1\}$$

is a σ-field, and τ is $\mathcal{F}_\tau$-measurable. Let $\{X_n\}_{n\geq 0}$ be an E-valued $\mathcal{F}_n$-adapted random sequence, and let τ be a finite $\mathcal{F}_n$-stopping time. Then $X(\tau)$ is $\mathcal{F}_\tau$-measurable.

The proof is left as an exercise. A more general result is given in Theorem 13.2.4. If $\{\mathcal{F}_n\}_{n \geq 0}$ is the internal history of some random sequence $\{X_n\}_{n \geq 0}$, that is, if $\mathcal{F}_n = \mathcal{F}_n^X$ ($n \geq 0$), one may interpret $\mathcal{F}_\tau^X$ as the collection of events that are determined by the observation of the random sequence up to time τ (included).

We are now ready for the statement and proof of Doob's optional sampling theorem.

Theorem 13.2.5 *Let $\{Y_n\}_{n \geq 0}$ be an $\mathcal{F}_n$-submartingale (resp., martingale), and let τ_1, τ_2 be finite $\mathcal{F}_n$-stopping times such that $P(\tau_1 \leq \tau_2) = 1$. If for $i = 1, 2$,*

$$\mathrm{E}\left[|Y_{\tau_i}|\right] < \infty, \tag{13.11}$$

and

$$\liminf_{n \uparrow \infty} \mathrm{E}\left[|Y_n| 1_{\{\tau_i > n\}}\right] = 0, \tag{13.12}$$

then, P-a.s.

$$\mathrm{E}[Y_{\tau_2} \mid \mathcal{F}_{\tau_1}] \geq Y_{\tau_1} \ (resp., \ = Y_{\tau_1}). \tag{13.13}$$

Remark 13.2.6 In particular,

$$\mathrm{E}[Y_{\tau_2}] \geq \mathrm{E}[Y_{\tau_1}] \qquad (resp., \ = \mathrm{E}[Y_{\tau_1}]). \tag{13.14}$$

More generally, if $\{\tau_n\}_{n \geq 1}$ is a non-decreasing sequence of finite $\mathcal{F}_n$-stopping times satisfying conditions (13.11) and (13.12), the sequence $\{Y_{\tau_n}\}_{n \geq 1}$ is an $\mathcal{F}_{\tau_n}$-submartingale (resp., martingale).

Proof. It suffices to give the proof for the submartingale case. The meaning of (13.13) is that, for all $A \in \mathcal{F}_{\tau_1}$,

$$\mathrm{E}\left[1_A Y_{\tau_2}\right] \geq \mathrm{E}\left[1_A Y_{\tau_1}\right].$$

It is sufficient to show that for all $n \geq 0$,

$$\mathrm{E}\left[1_{A \cap \{\tau_1 = n\}} Y_{\tau_2}\right] \geq \mathrm{E}\left[1_{A \cap \{\tau_1 = n\}} Y_{\tau_1}\right],$$

or, equivalently since $\tau_1 = n$ implies $\tau_2 \geq n$,

$$\mathrm{E}\left[1_{A \cap \{\tau_1 = n\} \cap \{\tau_2 \geq n\}} Y_{\tau_2}\right] \geq \mathrm{E}\left[1_{A \cap \{\tau_1 = n\} \cap \{\tau_2 \geq n\}} Y_{\tau_1}\right] = \mathrm{E}\left[1_{A \cap \{\tau_1 = n\} \cap \{\tau_2 \geq n\}} Y_n\right].$$

Write this as

$$\mathrm{E}\left[1_{B \cap \{\tau_2 \geq n\}} Y_{\tau_2}\right] \geq \mathrm{E}\left[1_{B \cap \{\tau_2 \geq n\}} Y_n\right], \tag{$\star$}$$

where $B := A \cap \{\tau_1 = n\}$. By definition of $\mathcal{F}_{\tau_1}$, $B \in \mathcal{F}_n$. It is therefore sufficient to show that for all $n \geq 0$, all $B \in \mathcal{F}_n$, $(\star)$ holds. We have

$$\begin{aligned}
\mathrm{E}\left[1_{B \cap \{\tau_2 \geq n\}} Y_n\right] &= \mathrm{E}\left[1_{B \cap \{\tau_2 = n\}} Y_n\right] + \mathrm{E}\left[1_{B \cap \{\tau_2 \geq n+1\}} Y_n\right] \\
&\leq \mathrm{E}\left[1_{B \cap \{\tau_2 = n\}} Y_n\right] + \mathrm{E}\left[1_{B \cap \{\tau_2 \geq n+1\}} \mathrm{E}[Y_{n+1} | \mathcal{F}_n]\right] \\
&= \mathrm{E}\left[1_{B \cap \{\tau_2 = n\}} Y_{\tau_2}\right] + \mathrm{E}\left[1_{B \cap \{\tau_2 \geq n+1\}} Y_{n+1}\right] \\
&\leq \mathrm{E}\left[1_{B \cap \{n \leq \tau_2 \leq n+1\}} Y_{\tau_2}\right] + \mathrm{E}\left[1_{B \cap \{\tau_2 \geq n+2\}} Y_{n+2}\right] \\
& \ \cdots \\
&\leq \mathrm{E}\left[1_{B \cap \{n \leq \tau_2 \leq m\}} Y_{\tau_2}\right] + \mathrm{E}\left[1_{B \cap \{\tau_2 > m\}} Y_m\right],
\end{aligned}$$

that is,

$$\mathrm{E}\left[1_{B \cap \{n \leq \tau_2 \leq m\}} Y_{\tau_2}\right] \geq \mathrm{E}\left[1_{B \cap \{\tau_2 \geq n\}} Y_n\right] - \mathrm{E}\left[1_{B \cap \{\tau_2 > m\}} Y_m\right]$$

for all $m \geq n$. Therefore, by dominated convergence and hypothesis (13.12)

$$\mathrm{E}\big[1_{B \cap \{\tau_2 \geq n\}} Y_{\tau_2}\big] = \mathrm{E}\big[\lim_{m \uparrow \infty} 1_{B \cap \{n \leq \tau_2 \leq m\}} Y_{\tau_2}\big]$$

$$\geq \mathrm{E}\big[1_{B \cap \{\tau_2 \geq n\}} Y_n\big] - \liminf_{m \uparrow \infty} \mathrm{E}\big[1_{B \cap \{\tau_2 > m\}} Y_m\big]$$

$$= \mathrm{E}\big[1_{B \cap \{\tau_2 \geq n\}} Y_n\big].$$

$\square$

Corollary 13.2.7 *Let $\{Y_n\}_{n \geq 0}$ be an $\mathcal{F}_n$-submartingale (resp., martingale). Let τ_1, τ_2 be $\mathcal{F}_n$-stopping times such that $\tau_1 \leq \tau_2 \leq N$ a.s., for some constant $N < \infty$. Then (13.14) holds.*

Proof. This is an immediate consequence of Theorem 13.2.5. $\square$

Corollary 13.2.8 *Let $\{Y_n\}_{n \geq 0}$ be a uniformly integrable $\mathcal{F}_n$-submartingale (resp., martingale). Let τ_1, τ_2 be finite $\mathcal{F}_n$-stopping times. Then (13.13) holds.*

Proof. In order to apply Theorem 13.2.5, we have to show that conditions (13.11) and (13.12) are satisfied when $\{Y_n\}_{n \geq 1}$ is uniformly integrable. Condition (13.12) follows from part (b) of Theorem 4.2.12 since the τ_i's are finite and therefore $P(\tau_i > n) \to 0$ as $n \uparrow \infty$. It remains to show that condition (13.11) is satisfied. Let $N < \infty$ be an integer. By Corollary 13.2.7, if τ is a stopping time (here τ_1 or τ_2),

$$\mathrm{E}[Y_0] \leq \mathrm{E}[Y_{\tau \wedge N}]$$

and therefore

$$\mathrm{E}[|Y_{\tau \wedge N}|] = 2\mathrm{E}[Y_{\tau \wedge N}^+] - \mathrm{E}[Y_{\tau \wedge N}] \leq 2\mathrm{E}[Y_{\tau \wedge N}^+] - \mathrm{E}[Y_0].$$

The submartingale $\{Y_n^+\}_{n \geq 0}$ satisfies

$$\mathrm{E}[Y_{\tau \wedge N}^+] = \sum_{j=0}^{N} \mathrm{E}[1_{\{\tau \wedge N = j\}} Y_j^+] + \mathrm{E}[1_{\{\tau > N\}} Y_N^+]$$

$$\leq \sum_{j=0}^{N} \mathrm{E}[1_{\{\tau \wedge N = j\}} Y_N^+] + \mathrm{E}[1_{\{\tau > N\}} Y_N^+]$$

$$= \mathrm{E}[Y_N^+] \leq \mathrm{E}[|Y_N|].$$

Therefore

$$\mathrm{E}[|Y_{\tau \wedge N}|] \leq 2\mathrm{E}[|Y_N|] + \mathrm{E}[|Y_0|] \leq 3 \sup_N E|Y_N|.$$

Since by Fatou's lemma $\mathrm{E}[|Y_\tau|] \leq \liminf_{N \uparrow \infty} \mathrm{E}[|Y_{\tau \wedge N}|]$, we have

$$\mathrm{E}[|Y_\tau|] \leq 3 \sup_N E[|Y_N|],$$

a finite quantity since $\{Y_n\}_{n \geq 1}$ is uniformly integrable. $\square$

Corollary 13.2.9 *Let $\{Y_n\}_{n \geq 0}$ be an $\mathcal{F}_n$-submartingale (resp., martingale) and let τ be an $\mathcal{F}_n$-stopping time such that*

$$E[\tau] < \infty.$$

Suppose moreover that there exists a constant $c < \infty$ such that, for all $n \geq 0$,

$$E[|Y_{n+1} - Y_n| \,|\, \mathcal{F}_n] \leq c, \quad P\text{-a.s. on } \{\tau \geq n\}.$$

Then $E[|Y_\tau|] < \infty$ and

$$E[Y_\tau] \geq (\text{ resp.,} =) E[Y_0].$$

Proof. In order to apply Theorem 13.2.5 with $\tau_1 = 0$, $\tau_2 = \tau$, one just has to check conditions (13.11) and (13.12) for τ. Let $Z_0 := |Y_0|$. With $Z_n := |Y_n - Y_{n-1}|$ $(n \geq 1)$,

$$E\left[\sum_{j=0}^{\tau} Z_j\right] = \sum_{n=0}^{\infty} E\left[1_{\{\tau=n\}} \sum_{j=0}^{n} Z_j\right]$$

$$= \sum_{n=0}^{\infty} \sum_{j=0}^{n} E\left[1_{\{\tau=n\}} Z_j\right]$$

$$= \sum_{j=0}^{\infty} \sum_{n=j}^{\infty} E\left[1_{\{\tau=n\}} Z_j\right]$$

$$= \sum_{j=0}^{\infty} E\left[1_{\{\tau \geq j\}} Z_j\right].$$

For $j \geq 1$, $\{\tau \geq j\} = \overline{\{\tau < j-1\}} \in \mathcal{F}_{j-1}$ and therefore,

$$E\left[1_{\{\tau \geq j\}} Z_j\right] = E\left[1_{\{\tau \geq j\}} E\left[Z_j \mid \mathcal{F}_{j-1}\right]\right] \leq c P(\tau \geq j), \qquad (\star)$$

and

$$E\left[\sum_{j=0}^{\tau} Z_j\right] \leq E[|Y_0|] + c \sum_{j=1}^{\infty} P(\tau \geq j) = E[|Y_0|] + c E[\tau] < \infty.$$

Therefore condition (13.11) is satisfied since $E\left[|Y_\tau|\right] \leq E\left[\sum_{j=0}^{\tau} Z_j\right]$. Moreover, if $\tau > n$,

$$\sum_{j=0}^{n} Z_j \leq \sum_{j=0}^{\tau} Z_j$$

and therefore

$$E\left[1_{\{\tau>n\}} |Y_n|\right] \leq E\left[1_{\{\tau>n\}} \sum_{j=0}^{\tau} Z_j\right].$$

But, by $(\star)$, $E\left[\sum_{j=0}^{\tau} Z_j\right] < \infty$. Also, $\{\tau > n\} \downarrow \varnothing$ as $n \uparrow \infty$. Therefore, by dominated convergence

$$\liminf_{n \uparrow \infty} E[1_{\{\tau>n\}} |Y_n|] \leq \liminf_{n \uparrow \infty} E\left[1_{\{\tau>n\}} \sum_{j=0}^{\tau} Z_j\right] = 0.$$

This is condition (13.12). $\qquad \square$

13.2.2   Wald's Formulas

Wald's Mean Formula

Theorem 13.2.10 *Let $\{Z_n\}_{n\geq1}$ be an IID sequence of real random variables such that $\mathrm{E}\,[|Z_1|] < \infty$, and let τ be an $\mathcal{F}_n^Z$-stopping time with $\mathrm{E}[\tau] < \infty$. Then*

$$\mathrm{E}\left[\sum_{n=1}^{\tau} Z_n\right] = \mathrm{E}[Z_1]\mathrm{E}[\tau]. \tag{13.15}$$

If, moreover, $\mathrm{E}[Z_1^2] < \infty$,

$$\mathrm{Var}\left(\sum_{n=1}^{\tau} Z_n\right) = \mathrm{Var}\,(Z_1)\mathrm{E}[\tau]. \tag{13.16}$$

Proof. Let $X_0 := 0$, $X_n := (Z_1 + \cdots + Z_n) - n\mathrm{E}[Z_1]$ $(n \geq 1)$. Then $\{X_n\}_{n\geq1}$ is an $\mathcal{F}_n^Z$-martingale such that

$$\mathrm{E}[|X_{n+1} - X_n|\,|\,\mathcal{F}_n^Z] = \mathrm{E}[|Z_{n+1} - \mathrm{E}[Z_1]|\,|\,\mathcal{F}_n^Z]$$
$$= E|Z_n - \mathrm{E}[Z_1]| \leq 2\mathrm{E}\,[|Z_1|] < \infty.$$

Therefore Corollary 13.2.9 can be applied with $Y_n = \sum_{k=1}^{n}(Z_k - \mathrm{E}\,[Z_1])$ to obtain (13.15). For the proof of (13.16), the same kind of argument works, this time with the martingale $Y_n = X_n^2 - n\,\mathrm{Var}\,(Z_1)$. $\square$

Wald's Exponential Formula

Theorem 13.2.11 *Let $\{Z_n\}_{n\geq1}$ be IID real random variables and let $S_n = Z_1 + \cdots + Z_n$. Let $\varphi_Z(t) := \mathrm{E}[e^{tZ_1}]$ and suppose that $\varphi_Z(t_0)$ exists and is greater than or equal to 1 for some $t_0 \neq 0$. Let τ be an $\mathcal{F}_n^Z$-stopping time such that $\mathrm{E}[\tau] < \infty$ and $|S_n| \leq c$ on $\{\tau \geq n\}$ for some constant $c < \infty$. Then*

$$\mathrm{E}\left[\frac{e^{t_0 S_\tau}}{\varphi_Z(t_0)^\tau}\right] = 1. \tag{13.17}$$

Proof. Let $Y_0 := 1$ and for $n \geq 1$,

$$Y_n := \frac{e^{t_0 S_n}}{\varphi_Z(t_0)^n}.$$

By application of the result of Example 13.1.3 with $X_i := \frac{e^{t_0 Z_i}}{\varphi_Z(t_0)}$, we have that the sequence $\{Y_n\}_{n\geq0}$ is an $\mathcal{F}_n^Z$-martingale. Moreover, on $\{\tau \geq n\}$,

$$\mathrm{E}[|Y_{n+1} - Y_n|\,|\,\mathcal{F}_n^Z] = Y_n\mathrm{E}\left[\left|\frac{e^{t_0 Z_{n+1}}}{\varphi_Z(t_0)} - 1\right|\,|\,\mathcal{F}_n^Z\right]$$
$$= \frac{Y_n}{\varphi_Z(t_0)}\mathrm{E}\left[|e^{t_0 Z_1} - \varphi_Z(t_0)|\right] \leq K < \infty$$

since $\varphi_Z(t_0) \geq 1$ and

$$Y_n = \frac{e^{t_0 S_n}}{\varphi_Z(t_0)^n} \leq \frac{e^{|t_0|c}}{\varphi_Z(t_0)^n} \leq e^{|t_0|c}.$$

Therefore, Corollary 13.2.9 applies to give (13.17). $\square$

13.2.3 The Maximum Principle

The general approach to the absorption problem for HMCs of this subsection is in terms of harmonic functions. However, its implementation requires explicit forms of harmonic functions satisfying some boundary conditions, and this is not always easy. In contrast, the purely algebraic method given in the chapter on Markov chains can always be implemented in the finite state space case (of course at the cost of matrix computations).

Let $\{X_n\}_{n\geq 0}$ be an HMC with countable state space E and transition matrix $\mathbf{P}$. Let D be a subset of E, called the domain, and let $\overline{D} := E \backslash D$. Let $c : D \to \mathbb{R}$ and $\varphi : \overline{D} \to \mathbb{R}$ be non-negative functions called the unit time gain function and the final gain function, respectively. Let τ be the hitting time of $\overline{D}$.

For each state $i \in E$, define

$$v(i) = \mathrm{E}_i \left[\sum_{0 \leq k < \tau} c(X_k) + \varphi(X_\tau) 1_{\{\tau < \infty\}} \right]. \tag{13.18}$$

The function $v : E \to \overline{\mathbb{R}}$ so defined is non-negative and possibly infinite. Note that τ is not required to be finite, and that $\overline{D}$ may be empty.

In the context of control theory, v is called the average reward function, since $v(i)$ is the average cost incurred when starting from state i, from the initial time $n = 0$ to the final time $n = \tau$, $c(X_n)$ being the running gain at time n and $\varphi(X_\tau)$ the final reward.

Theorem 13.2.12 *The function $v : E \to \mathbb{R}_+$ defined by (13.18) satisfies the following properties:*

(i) it is non-negative and satisfies

$$v = \begin{cases} \mathbf{P}v + c & \text{on } D, \\ \varphi & \text{on } \overline{D}, \end{cases} \tag{13.19}$$

(ii) it is bounded above by any non-negative function $u : E \to \mathbb{R}$ such that

$$u \geq \begin{cases} \mathbf{P}u + c & \text{on } D, \\ \varphi & \text{on } \overline{D}, \end{cases} \tag{13.20}$$

(iii) and moreover, if for all $i \in E$, $P_i(\tau < \infty) = 1$, then (13.19) has at most one non-negative bounded solution.

Proof. (i) Properties $v \geq 0$ and $v = \varphi$ on $\overline{D}$ are satisfied by definition. For $i \in D$, first-step analysis gives

$$v(i) = c(i) + \sum_{j \in E} p_{ij} v(j). \tag{13.21}$$

(ii) Define for $n \geq 0$ the non-negative function $v_n : E \to \mathbb{R}$ by

$$v_n(i) = \mathrm{E}_i \left[\sum_{k=0}^{n-1} c(X_k) 1_{\{k < \tau\}} + \varphi(X_\tau) 1_{\{\tau < n\}} \right]. \tag{13.22}$$

Observe that $v_0 \equiv 0$ and, by monotone convergence, $\lim_{n \uparrow \infty} \uparrow v_n = v$. Also, with a proof similar to that of (i),

$$v_{n+1} = \begin{cases} \mathbf{P}v_n + c & \text{on } D, \\ \varphi & \text{on } \overline{D}. \end{cases} \qquad (13.23)$$

With u as in (13.20), we have $u \geq v_0$. We show by induction that $u \geq v_n$. This is true for $n = 0$. Suppose it is true for some n. We have $u \geq \mathbf{P}u + c \geq \mathbf{P}v_n + c = v_{n+1}$ on D, and $u \geq \varphi = v_{n+1}$ on $\overline{D}$. Therefore, $u \geq v_{n+1}$. Since $u \geq v_n$ for all $n \geq 0$, $u \geq \lim_{n \to \infty} v_n = v$.

(iii) Suppose that u satisfies

$$u = \begin{cases} \mathbf{P}u + c & \text{on } D, \\ \varphi & \text{on } \overline{D}. \end{cases}$$

Suppose in addition that it is bounded (note that this implies that c and φ are bounded) and non-negative. Then by Exercise 13.5.2,

$$M_n = u(X_n) - u(X_0) - \sum_{k=0}^{n-1} (\mathbf{P} - I)u(X_k) \qquad (13.24)$$

is an $\mathcal{F}_n^X$-martingale. By the optional sampling theorem (Theorem 13.2.1), for all integers K, $\mathrm{E}_i[M_{\tau \wedge K}] = \mathrm{E}_i[M_0] = 0$, and therefore, observing that $(I - \mathbf{P})u = c$ on D,

$$u(i) = \mathrm{E}_i[u(X_{\tau \wedge K})] - \mathrm{E}_i\left[\sum_{k=0}^{\tau \wedge K - 1} (\mathbf{P} - I)u(X_k) \right] = \mathrm{E}_i\left[u(X_{\tau \wedge K}) + \sum_{k=0}^{\tau \wedge K - 1} c(X_k) \right].$$

Since $P_i(\tau < \infty) = 1$, $\lim_{K \uparrow \infty} \mathrm{E}_i[u(X_{\tau \wedge K})] = \mathrm{E}_i[u(X_\tau)]$ by dominated convergence. But $u(X_\tau) = \varphi(X_\tau)$ because $u = \varphi$ on $\overline{D}$. Therefore $\lim_{K \uparrow \infty} \mathrm{E}_i[u(X_{\tau \wedge K})] = \mathrm{E}_i[\varphi(X_\tau)]$. Also, $\lim_{K \uparrow \infty} \mathrm{E}_i[\sum_{k=0}^{\tau \wedge K - 1} c(X_k)] = \mathrm{E}_i[\sum_{k=0}^{\tau - 1} c(X_k)]$ by monotone convergence. Finally,

$$u(i) = \mathrm{E}_i\left[\sum_{k=0}^{\tau - 1} c(X_k) + \varphi(X_\tau) \right] = v(i).$$

$\square$

Theorem 13.2.12 can be rephrased as follows. The function v given by (13.18) is a minorant of all non-negative solutions of (13.20), and for $u = v$, the inequalities in (13.20) become equalities. Moreover, if v is bounded and $P_i(\tau < \infty) = 1$ for all $i \in E$, then v is the *unique* bounded solution of (13.19).

Definition 13.2.13 *If* $\mathbf{P}h = h$ *on* $A \subseteq E$, *we say that* h *is* harmonic on A.

Corollary 13.2.14 *Let* $\varphi : E \to \mathbb{R}$ *be a bounded non-negative function, and let* τ_B *be the hitting time of* $B \subset E$. *Then, if* $P_i(\tau_B < \infty) = 1$ *for all* $i \in E$,

$$v(i) := \mathrm{E}_i[\varphi(X_{\tau_B})]$$

defines the unique bounded non-negative function $v : E \to \mathbb{R}$ *that is harmonic on* $\overline{B}$ *and equal to* φ *on* B.

EXAMPLE 13.2.15: APPLICATION TO THE ABSORPTION PROBLEM. Suppose that the transient set T is finite and that the recurrent classes R_1, R_2, ... are singletons, and therefore absorbing states, denoted by r_1, r_2, ... (As shown before, the general case can always be reduced to this one as far as absorption probabilities are concerned.) In

Corollary 13.2.14, take for B the set of absorbing states, and therefore $\overline{B} = T$. Let $\varphi = 1_{\{r_1\}}$. As T is assumed finite, the time to absorption in one of the absorbing states is finite. The quantity $v(i)$ is just the probability of absorption in r_1. Therefore v is in this case the unique bounded non-negative function $v : E \to \mathbb{R}$ that is harmonic on T and equal to $\varphi = 1_{\{r_1\}}$ on $R := \{r_1, r_2, \ldots\}$.

Suppose that we want to compute the average time to absorption $\mathrm{E}_i[\tau_R]$, $i \in T$. For this, we take in Theorem 13.2.12 $D = R$, $\tau = \tau_R$, $c(i) \equiv 1$, $\varphi \equiv 0$. Then v defined by $v(i) := \mathrm{E}_i[\tau_R]$ is the unique bounded non-negative function such that $v = \mathbf{P}v + 1$ on T and $= 0$ on R.

EXAMPLE 13.2.16: APPLICATION TO OPTIMAL CONTROL. Consider a stochastic process $\{X_n\}_{n\geq0}$ with values in E, that is controlled in the following way. Let $\{\mathbf{P}(a); a \in A\}$, where A is some set, the set of *actions*, be a family of transition matrices on E, with the interpretation that, if at time n the controlled process is in state i, and if the controller takes action a, then at time $n + 1$ the state will be j with probability $p_{ij}(a)$. A *control strategy* u is a (measurable) function $u : E \to A$ which prescribes to take action $u(i)$ when the process is in state i. Therefore, under the strategy u, the controlled process is an HMC with transition matrix $\mathbf{P}^u$, where

$$p_{ij}^u = p_{ij}(u(i)).$$

There is a cost $V^u(i)$ associated with each strategy u and each initial state i, of the form

$$V^u(i) = \mathrm{E}_i^u \left[\sum_{0 \leq k < T} c^u(X_k) + \varphi^u(X_T)1_{\{T<\infty\}} \right],$$

where c^u, φ^u and T are as in Theorem 13.2.12, with D fixed, and moreover, $c^u(i) = c(i, u(i))$ and $\varphi^u(i) = \varphi(i, u(i))$, for appropriate functions c and φ. The problem of *optimal control* is that of finding, if it exists, an *optimal strategy* u^*, such that

$$V^{u^*}(i) \geq V^u(i),$$

for all states i and all strategies u.

We have the following result. Suppose that there exists a function $V : E \to \mathbb{R}$ such that

$$V(i) = \sup_{a \in A} \left\{ \sum_{j \in E} p_{ij}(a)V(j) + c(i, a) \right\} \quad \text{for all } i \in D,$$

and

$$V(i) = \sup_{a \in A} \varphi(i, a) \quad \text{for all } i \in \partial D,$$

and that the suprema above are attained for $a = u^*(i)$, for some (measurable) function $u^* : E \to A$. Then, u^* is an optimal control and $V = V^{u^*}$.

Proof. Since for all controls u,

$$V \geq \mathbf{P}^u V + c^u \quad \text{on } D,$$

and

$$V \geq \varphi^u \text{ on } \partial D$$

(where $\partial D := \{j \notin D$ such that there exists an $i \in D$ such that $p_{ij} > 0\}$), it follows from Theorem 13.2.12 that

$$V \geq V^u$$

for all controls u. Also, $V = V^{u^*}$ and therefore u^* is an optimal control.   □

13.3   Convergence of Martingales

The second pillar of martingale theory is the *martingale convergence theorem*. This result is the probabilistic counterpart of the convergence of a non-negative non-increasing, or bounded non-decreasing, sequence of real numbers to a finite limit. It says in particular (but we shall give a more complete result soon) that a non-negative supermartingale converges almost surely to a finite limit.

13.3.1   The Fundamental Convergence Theorem

The proof of the martingale convergence theorem is based on the *upcrossing inequality*.

Theorem 13.3.1 *Let $\{S_n\}_{n\geq 0}$ be an $\mathcal{F}_n$-submartingale. Let $a, b \in \mathbb{R}$ with $a < b$, and let ν_n be the number of upcrossings[1] of $[a, b]$ before ($\leq$) time n. Then*

$$(b - a)\mathrm{E}[\nu_n] \leq \mathrm{E}[(S_n - a)^+]. \tag{13.25}$$

Proof. Since ν_n is the number of upcrossings of $[0, b - a]$ by the submartingale $\{(S_n - a)^+\}_{n\geq 1}$, we may suppose without loss of generality that $S_n \geq 0$ and take $a = 0$, and then prove that

$$b\mathrm{E}[\nu_n] \leq \mathrm{E}[S_n - S_0], \tag{13.26}$$

where $S_0 = 0$ and $\mathcal{F}_0$ is the gross σ-field. Define a sequence of $\mathcal{F}_n$-stopping times as follows

$$\tau_0 = 0$$
$$\tau_1 = \inf\{n > \tau_0 \, ; \, S_n = 0\}$$
$$\tau_2 = \inf\{n > \tau_1 \, ; \, S_n \geq b\}$$
$$\cdots$$

$$\tau_{2k+1} = \inf\{n > \tau_{2k} \, ; \, S_n = 0\}$$
$$\tau_{2k} = \inf\{n > \tau_{2k+1} \, ; \, S_n \geq b\}$$
$$\cdots$$

For $i \geq 1$, let

$$\varphi_i = 1 \text{ if } \tau_m < i \leq \tau_{m+1} \text{ for some odd } m$$
$$= 0 \text{ if } \tau_m < i \leq \tau_{m+1} \text{ for some even } m.$$

---

[1]By definition, an upcrossing occurs at time ℓ if $S_k \leq a$ and if there exists $\ell > k$ such that $S_j < b$ for $j = 1, \ldots, \ell - 1$ and $S_\ell \geq b$.

Observe that

$$\{\varphi_i = 1\} = \bigcup_{odd\, m} \left(\{\tau_m < i\} \cap \overline{\{\tau_{m+1} < i\}}\right) \in \mathcal{F}_{i-1}$$

and that

$$b\nu_n \leq \sum_{i=1}^{n} \varphi_i(S_i - S_{i-1}).$$

Therefore

$$b\mathrm{E}[\nu_n] \leq \mathrm{E}[\sum_{i=1}^{n} \varphi_i(S_i - S_{i-1})] = \sum_{i=1}^{n} \mathrm{E}[\varphi_i(S_i - S_{i-1})]$$

$$= \sum_{i=1}^{n} \mathrm{E}[\varphi_i \mathrm{E}[(S_i - S_{i-1})|\mathcal{F}_{i-1}]] = \sum_{i=1}^{n} \mathrm{E}[\varphi_i(\mathrm{E}[S_i|\mathcal{F}_{i-1}] - S_{i-1})]$$

$$\leq \sum_{i=1}^{n} \mathrm{E}[(\mathrm{E}[S_i|\mathcal{F}_{i-1}] - S_{i-1})] \leq \sum_{i=1}^{n} (\mathrm{E}[S_i] - \mathrm{E}[S_{i-1}]) = \mathrm{E}[S_n - S_0].$$

$\square$

Theorem 13.3.2 *Let $\{S_n\}_{n\geq 0}$ be an $\mathcal{F}_n$-submartingale. Suppose moreover that it is L^1-bounded, that is,*

$$\sup_{n\geq 0} \mathrm{E}[|S_n|] < \infty. \tag{13.27}$$

Then $\{S_n\}_{n\geq 0}$ converges P-a.s. to an integrable random variable S_∞.

Remark 13.3.3 Condition (13.49) can be replaced by the equivalent condition

$$\sup_{n\geq 0} \mathrm{E}[S_n^+] < \infty.$$

Indeed, if $\{S_n\}_{n\geq 0}$ is an $\mathcal{F}_n$-submartingale,

$$\mathrm{E}\left[S_n^+\right] \leq \mathrm{E}[|S_n|] \leq 2\mathrm{E}\left[S_n^+\right] - \mathrm{E}[S_n] \leq 2\mathrm{E}\left[S_n^+\right] - \mathrm{E}[S_0].$$

Remark 13.3.4 By changing signs, the same hypothesis leads to the same conclusion for a *supermartingale* $\{S_n\}_{n\geq 0}$. Similarly to the previous remark, condition (13.49) can be replaced by the equivalent condition

$$\sup_{n\geq 0} \mathrm{E}[S_n^-] < \infty.$$

Proof. The proof is based on the following observation concerning any deterministic sequence $\{x_n\}_{n\geq 1}$. If this sequence does not converge, then it is possible to find two rational numbers a and b such that

$$\liminf_n x_n < a < b < \limsup_n x_n,$$

which implies that the number of upcrossings of $[a, b]$ by this sequence is infinite. Therefore to prove convergence, it suffices to prove that any interval $[a, b]$ with rational extremities is crossed at most a finite number of times.

Let $\nu_n([a, b])$ be the number of upcrossings of an interval $[a, b]$ prior ($\leq$) to time n and let $\nu_\infty([a, b]) := \lim_{n\uparrow\infty} \nu_n([a, b])$. By (13.25),

$$(b - a)\mathrm{E}[\nu_n([a, b])] \leq \mathrm{E}[(S_n - a)^+]$$
$$\leq \mathrm{E}[S_n^+] + |a|$$
$$\leq \sup_{k\geq 0} \mathrm{E}[S_k^+] + |a|$$
$$\leq \sup_{k\geq 0} \mathrm{E}[|S_k|] + |a| < \infty.$$

Therefore, letting $n \uparrow \infty$,

$$(b - a)\mathrm{E}[\nu_\infty([a, b])] < \infty.$$

In particular, $\nu_\infty([a, b]) < \infty$, P-a.s. Therefore, P-a.s. there is only a finite number of upcrossings of any *rational* interval $[a, b]$. Equivalently, in view of the observation made in the first lines of the proof, $\{S_n\}_{n\geq 0}$ converges P-a.s. to some random variable S_∞. Therefore (by Fatou's lemma for the previous inequality):

$$\mathrm{E}[|S_\infty|] = \mathrm{E}[\lim_{n\uparrow\infty} |S_n|] \leq \liminf_{n\uparrow\infty} E|S_n| \leq \sup_{n\geq 0} E|S_n| < \infty.$$

$\square$

Corollary 13.3.5

(a) *Any non-positive submartingale $\{S_n\}_{n\geq 0}$ almost surely converges to an integrable random variable.*

(b) *Any non-negative supermartingale almost surely converges to an integrable random variable.*

Proof. (b) follows from (a) by changing signs. For (a), we have

$$\mathrm{E}[|S_n|] = -\mathrm{E}[S_n] \leq -\mathrm{E}[S_0] = \mathrm{E}[|S_0|] < \infty.$$

Therefore (13.49) is satisfied and the conclusion then follows from Theorem 13.3.2. $\square$

An immediate application of the martingale convergence theorem is to gambling. The next example teaches us that a gambler in a "fair game" is eventually ruined.

EXAMPLE 13.3.6: FAIR GAME NOT SO FAIR. Consider the situation in Example 13.1.4, assuming that the initial fortune a is a positive integer and that the bets are also positive integers (that is, the functions $b_{n+1}(X_0^n) \in \mathbb{N}_+$ except if $Y_n = 0$, in which case the gambler is not allowed to bet anymore, or equivalently $b_n(X_0^{n-1}*0) := b_n(X_0, X_1, \ldots, X_n, 0) = 0$). In particular, $Y_n \geq 0$ for all $n \geq 0$. Therefore the process $\{Y_n\}_{n\geq 0}$ is a non-negative $\mathcal{F}_n^X$-martingale and by the martingale convergence theorem it almost surely has a finite limit. Since the bets are assumed positive integers when the fortune of the player is positive, this limit cannot be other than 0. Since Y_n is a non-negative integer for all $n \geq 0$, this can happen only if the fortune of the gambler becomes null in finite time.

EXAMPLE 13.3.7: BRANCHING PROCESSES VIA MARTINGALES. The power of the concept of martingale will now be illustrated by revisiting the branching process. It is assumed that $P(Z = 0) < 1$ and $P(Z \geq 2) > 0$ (to get rid of trivialities). The stochastic process

$$Y_n = \frac{X_n}{m^n},$$

where m is the average number of sons of a given individual, is an $\mathcal{F}_n^X$-martingale. Indeed, since each one among the X_n members of the nth generation gives birth on average to m sons and does this independently of the rest of the population, $E[X_{n+1}|X_n] = mX_n$ and

$$E\left[\frac{X_{n+1}}{m^{n+1}}|\mathcal{F}_n^X\right] = E\left[\frac{X_{n+1}}{m^{n+1}}|X_n\right] = \frac{X_n}{m^n}.$$

By the martingale convergence theorem, almost surely

$$\lim_{n\uparrow\infty} \frac{X_n}{m^n} = Y < \infty.$$

In particular, if $m < 1$, then $\lim_{n\uparrow\infty} X_n = 0$ almost surely. Since X_n takes integer values, this implies that the branching process eventually becomes extinct.

If $m = 1$, then $\lim_{n\uparrow\infty} X_n = X_\infty < \infty$ and it is easily argued that this limit must be 0. Therefore, in this case as well the process eventually becomes extinct.

For the case $m > 1$, we consider the unique solution in $(0,1)$ of $x = g(x)$ (g is the generating function of the typical progeny of a member of the population considered). Suppose we can show that $Z_n = x^{X_n}$ is a martingale. Then, by the martingale convergence theorem, Z_n converges to a finite limit and therefore X_n has a limit X_∞, which however can be infinite. One can easily argue that this limit cannot be other than 0 (extinction) or ∞ (non-extinction). Since $\{Z_n\}_{n\geq0}$ is a martingale, $x = E[Z_0] = E[Z_n]$ and therefore, by dominated convergence, $x = E[Z_\infty] = E[x^{X_\infty}] = P(X_\infty = 0)$. Therefore x is the probability of extinction.

It remains to show that $\{Z_n\}_{n\geq0}$ is an $\mathcal{F}_n^X$-martingale. For all $i \in \mathbb{N}$ and all $x \in [0,1]$, $E[x^{X_{n+1}}|X_n = i] = x^i$. This is obvious if $i = 0$. If $i > 0$, X_{n+1} is the sum of i independent random variables with the same generating function g, and therefore, $E[x^{X_{n+1}}|X_n = i] = g(x)^i = x^i$. From this last result and the Markov property,

$$E[x^{X_{n+1}}|\mathcal{F}_n^X] = E[x^{X_{n+1}}|X_n] = x^{X_n}.$$

Theorem 13.3.8 *Let $\{M_n\}_{n\geq0}$ be an $\mathcal{F}_n$-martingale such that for some $p \in (1,\infty)$,*

$$\sup_{n\geq0} E|M_n|^p < \infty. \tag{13.28}$$

Then $\{M_n\}_{n\geq0}$ converges a.s. and in L^p to some finite variable M_∞.

Proof. By hypothesis, the martingale $\{M_n\}_{n\geq0}$ is L^p-bounded and *a fortiori* L^1-bounded since $p > 1$. Therefore it converges almost surely. By Doob's inequality, $E[\max_{0\leq i\leq n}|M_i|^p] \leq q^p E|M_n|^p$ and in particular,

$$E[\max_{0\leq i\leq n}|M_i|^p] \leq q^p \sup_k E|M_k|^p < \infty.$$

Letting $n \uparrow \infty$, we have in view of condition (13.28) that

$$E[\sup_{n\geq 0} |M_n|^p] < \infty. \tag{13.29}$$

Therefore $\{|M_n|^p\}_{n\geq 0}$ is uniformly integrable (Theorem 4.2.14). In particular, since it converges almost surely, it also converges in L^1 (Theorem 4.2.16). In other words, $\{M_n\}_{n\geq 0}$ converges in L^p.

□

Remark 13.3.9 The above result was proved for $p > 1$ (the proof depended on Doob's inequality, which is true for $p > 1$). For $p = 1$, a similar result holds with an additional assumption of uniform integrability. Note however that the next result also applies to submartingales.

Theorem 13.3.10 *A uniformly integrable $\mathcal{F}_n$-submartingale $\{S_n\}_{n\geq 0}$ converges a.s. and in L^1 to an integrable random variable S_∞ and $E[S_\infty \mid \mathcal{F}_n] \geq S_n$.*

Proof. By the uniform integrability hypothesis, $\sup_n E[|S_n|] < \infty$ and therefore, by Theorem 13.3.2, S_n converges almost surely to some integrable random variable S_∞. It also converges to this variable in L^1 since a uniformly integrable sequence that converges almost surely also converges in L^1 (Theorem 4.2.16).

By the submartingale property, for all $A \in \mathcal{F}_n$, all $m \geq n$,

$$E[1_A S_n] \leq E[1_A S_m].$$

Since convergence is in L^1,

$$\lim_{m\uparrow\infty} E[1_A S_m] = E[1_A S_\infty],$$

so that finally $E[1_A S_n] \leq E[1_A S_\infty]$. This being true for all $A \in \mathcal{F}_n$, we have that $E[S_\infty \mid \mathcal{F}_n] \geq S_n$.

□

The following result is *Lévy's continuity theorem for conditional expectations.*

Corollary 13.3.11 *Let $\{\mathcal{F}_n\}_{n\geq 1}$ be a filtration and let ξ be an integrable random variable. Let $\mathcal{F}_\infty := \sigma\left(\cup_{n\geq} \mathcal{F}_n\right)$. Then*

$$\lim_{n\uparrow\infty} E[\xi \mid \mathcal{F}_n] = E[\xi \mid \mathcal{F}_\infty]. \tag{13.30}$$

Proof. It suffices to treat the case where ξ is non-negative. The sequence $\{M_n = E[\xi \mid \mathcal{F}_n]\}_{n\geq 1}$ is a uniformly integrable $\mathcal{F}_n$-martingale (Theorem 4.2.13) and by Theorem 13.3.10, it converges almost surely and in L^1 to some integrable random variable M_∞. We have to show that $M_\infty = E[\xi \mid \mathcal{F}_\infty]$. For $m \geq n$ and $A \in \mathcal{F}_n$,

$$E[1_A M_m] = E[1_A M_n] = E[1_A E[\xi \mid \mathcal{F}_n]] = E[1_A \xi].$$

Since convergence is also in L^1, $\lim_{m\uparrow\infty} E[1_A M_m] = E[1_A M_\infty]$. Therefore

$$E[1_A M_\infty] = E[1_A \xi] \tag{13.31}$$

for all $A \in \mathcal{F}_n$ and therefore for all $A \in \cup_n \mathcal{F}_n$. The σ-finite measures $A \to E[1_A M_\infty]$ and $A \to E[1_A \xi]$ agreeing on the algebra $\cup_n \mathcal{F}_n$ also agree on the smallest σ-algebra containing it, that is $\mathcal{F}_\infty$. Therefore (13.31) holds for all $A \in \mathcal{F}_\infty$ (Theorem 2.1.50) and this implies

$$E[1_A M_\infty] = E[1_A E[\xi \mid \mathcal{F}_\infty]],$$

and finally, since M_∞ is $\mathcal{F}_\infty$-measurable, $M_\infty = E[\xi \mid \mathcal{F}_\infty]$.

□

Kakutani's Theorem

Let $\{X_n\}_{n\geq 1}$ be an independent sequence of non-negative random variables with mean 1. Let $M_0 := 1$ and let

$$M_n := \prod_{i=1}^{n} X_i \quad (n \geq 1).$$

Then (Example 13.1.3) $\{M_n\}_{n\geq 0}$ is a non-negative martingale and (Theorem 13.3.5) it converges almost surely to a finite random variable M_∞. Let $a_n := \mathrm{E}[X_n^{\frac{1}{2}}]$. (Note that $\prod_{n=1}^{\infty} a_n \leq 1$.)

Theorem 13.3.12 *The following conditions are equivalent:*

(i) $\prod_{n=1}^{\infty} a_n > 0$.

(ii) $\mathrm{E}[M_\infty] = 1$.

(iii) $M_n \to M_\infty$ in L^1.

(iv) $\{M_n\}_{n\geq 0}$ is uniformly integrable.

Proof. Note first that (iv) implies (iii) (since $M_n \to M_\infty$ a.s. and by Theorem 13.3.10) which in turn implies (ii) since $1 = \mathrm{E}[M_n] \to \mathrm{E}[M_\infty]$. The announced equivalences will be proved if one can show that (i) implies (iv) and that (ii) implies (i).

A. (i) implies (iv): let $m_0 := 1$ and

$$m_n := \frac{(\prod_{i=1}^{n} X_n)^{\frac{1}{2}}}{\prod_{i=1}^{n} a_n} \quad (n \geq 1).$$

This is a martingale and an L_2-bounded one since

$$\mathrm{E}[m_n^2] = \frac{1}{(a_1 \cdots a_n)^2} \leq \frac{1}{(\prod_{n=1}^{\infty} a_n)^2} < \infty.$$

By Doob's inequality (Theorem 13.1.15) for $p = 2$,

$$\mathrm{E}[\sup_n |m_n|^2] \leq 4 \sup_n \mathrm{E}[m_n^2] < \infty.$$

Also, since $\prod_{n=1}^{\infty} a_n \leq 1$, $M_n \leq m_n^2$ and in particular

$$\mathrm{E}[\sup_n |M_n|] \leq \mathrm{E}[\sup_n |m_n|^2] < \infty.$$

Therefore M_n is uniformly dominated by the *integrable* random variable $\sup_n |M_n|$, which implies that it is uniformly integrable (Example 4.2.10).

B. (ii) implies (i) or, equivalently, if (i) is not true, then (ii) is not true. There-fore suppose in view of contradiction that $\prod a_n = 0$. Being a non-negative martingale, $\{m_n\}_{n\geq 0}$ converges to a finite limit. Since $\prod_{n=1}^{\infty} a_n = 0$, this can happen only if $M_\infty = 0$, a contradiction with (ii). $\qquad\square$

13.3.2   Backwards (or Reverse) Martingales

In the following, pay attention to the indexation: the index set is the set of non-positive relative integers. Let $\{\mathcal{F}_n\}_{n \leq 0}$ be a non-decreasing family of σ-fields, that is, $\mathcal{F}_n \subseteq \mathcal{F}_{n+1}$ for all $n \leq -1$.

There is nothing new in the definition of "backwards" or "reverse" martingales or submartingales, except that the index set is now $\{\ldots, -2, -1, 0\}$. For instance, $\{Y_n\}_{n \leq 0}$ is an $\mathcal{F}_n$-submartingale if $\mathrm{E}\,[Y_n \mid \mathcal{F}_{n-1}] \geq Y_{n-1}$ for all $n \leq 0$. The term "backwards" in fact refers to one of the uses that is made of this notion, that of discussing the limit of Y_n as $n \downarrow -\infty$.

Reverse martingales or submartingales often appear in the following setting. Let $\{Z_k\}_{k \geq 0}$ be a sequence of integrable random variables. Suppose that

$$\mathrm{E}\,[Z_{k-1} \mid Z_k, Z_{k+1}, Z_{k+2}, \ldots] = Z_k \quad (k \geq 0)\,.$$

Clearly, the change of indexation $k \to -n$ gives a "backwards" martingale. The next example concerns that situation.

EXAMPLE 13.3.13:   EMPIRICAL MEAN OF AN IID SEQUENCE. Let $\{X_n\}_{n \geq 1}$ be an IID sequence of integrable random variables and let

$$Z_k := \frac{1}{k} S_k\,,$$

where $S_k := X_1 + \cdots + X_k$. We shall prove that

$$\mathrm{E}\,[Z_{k-1} \mid \mathcal{G}_k] = Z_k\,,$$

where $\mathcal{G}_k = \sigma(Z_k, Z_{k+1}, Z_{k+2}, \ldots)$. It suffices to prove that for all $k \geq 1$,

$$\mathrm{E}\,[Z_1 \mid \mathcal{G}_k] = Z_k\,, \tag{$\star$}$$

since it then follows that for $m \leq k$,

$$\mathrm{E}\,[Z_m \mid \mathcal{G}_k] = \mathrm{E}\,[\mathrm{E}\,[Z_1 \mid \mathcal{G}_m] \mid \mathcal{G}_k] = \mathrm{E}\,[Z_1 \mid \mathcal{G}_k] = Z_k\,.$$

By linearity,

$$S_k = \mathrm{E}\,[S_k \mid \mathcal{G}_k] = \sum_{j=1}^{k} \mathrm{E}\,[X_j \mid \mathcal{G}_k]\,.$$

From the fact that $\mathcal{G}_k = \sigma(Z_k, Z_{k+1}, Z_{k+2}, \ldots) = \sigma(S_k, X_{k+1}, X_{k+2}, \ldots)$ and by the IID assumption for $\{X_n\}_{n \geq 1}$,

$$S_k = \sum_{j=1}^{k} \mathrm{E}\,[X_j \mid S_k, X_{k+1}, X_{k+2}, \ldots] = \sum_{j=1}^{k} \mathrm{E}\,[X_j \mid S_k]\,.$$

But the pairs (X_j, S_k) $(1 \leq j \leq k)$ have the same distribution, and therefore

$$\sum_{j=1}^{k} \mathrm{E}\,[X_j \mid S_k] = k\mathrm{E}\,[X_1 \mid S_k] = k\mathrm{E}\,[X_1 \mid \mathcal{G}_k] = k\mathrm{E}\,[Z_1 \mid \mathcal{G}_k]\,,$$

from which $(\star)$ follows.

Theorem 13.3.14 *Let* $\{\mathcal{F}_n\}_{n\leq 0}$ *be a non-decreasing family of σ-fields. Let* $\{S_n\}_{n\leq 0}$ *be an $\mathcal{F}_n$-submartingale. Then:*

A. S_n *converges P-a.s. and in L^1 as $n \downarrow -\infty$ to an integrable random variable $S_{-\infty}$, and*

B. *with* $\mathcal{F}_{-\infty} := \cap_{n\leq 0}\mathcal{F}_n$,

$$S_{-\infty} \leq \mathrm{E}[S_0 \mid \mathcal{F}_{-\infty}],$$

with equality if $\{S_n\}_{n\leq 0}$ *is an $\mathcal{F}_n$-martingale.*

Proof. First note that by the submartingale property, $S_n \leq \mathrm{E}[S_0 \mid \mathcal{F}_n]$ $(n \leq 0)$. In particular, $\{S_n\}_{n\leq 0}$ is not only L^1-bounded, but also uniformly integrable (Theorem 4.2.13).

A. Denoting by $\nu_m = \nu_m([a,b])$ the number of upcrossings of $[a,b]$ by $\{S_n\}_{n\leq 0}$ in the integer interval $[-m,0]$ and by $\nu = \nu([a,b])$ the total number of upcrossings of $[a,b]$, the upcrossing inequality yields

$$(b-a)\mathrm{E}[\nu_m] \leq \mathrm{E}\left[(S_0 - a)^+\right] < \infty,$$

and letting $m \uparrow \infty$, $\mathrm{E}[\nu] < \infty$. Almost-sure convergence to an integrable random variable $S_{-\infty}$ is then proved as in Theorem 13.3.2. Since $\{S_n\}_{n\leq 0}$ is uniformly integrable, convergence to $S_{-\infty}$ is also in L^1.

B. Clearly, $S_{-\infty}$ is $\mathcal{F}_{-\infty}$-measurable. Also, by the submartingale property, $S_n \leq \mathrm{E}[S_0 \mid \mathcal{F}_n]$ $(n \leq -1)$, that is, for all $n \leq -1$ and all $A \in \mathcal{F}_n$,

$$\int_A S_n \, \mathrm{d}P \leq \int_A S_0 \, \mathrm{d}P.$$

This is true for any $A \in \mathcal{F}_{-\infty}$ because $\mathcal{F}_{-\infty} \subseteq \mathcal{F}_n$ for all $n \leq -1$. Since S_n converges to $S_{-\infty}$ in L^1 as $n \downarrow -\infty$, $\int_A S_n \, \mathrm{d}P \to \int_A S_{-\infty} \, \mathrm{d}P$ and therefore

$$\int_A S_{-\infty} \, \mathrm{d}P \leq \int_A S_0 \, \mathrm{d}P \quad (A \in \mathcal{F}_{-\infty}),$$

which implies that $S_{-\infty} \leq \mathrm{E}[S_0 \mid \mathcal{F}_{-\infty}]$.

The martingale case is obtained using the same proof with each $\leq$ symbol replaced by $=$. $\square$

Remark 13.3.15 Statement B says that $\{S_n\}_{n\in -\mathbb{N}\cup\{-\infty\}}$ is a submartingale relatively to the history $\{\mathcal{F}_n\}_{n\in -\mathbb{N}\cup\{-\infty\}}$.

EXAMPLE 13.3.16: THE STRONG LAW OF LARGE NUMBERS. The situation is that of Example 13.3.13. By Theorem 13.3.14, $S_k/k \to$ converges almost surely. By Kolmogorov's zero-one law (Theorem 4.3.3), $S_k/k \to a$, a deterministic number. It remains to identify a with $\mathrm{E}[X_1]$. We know from the first lines of the proof of Theorem 13.3.14 that $\{S_k/k\}_{k\geq 1}$ is uniformly integrable. Therefore, by Theorem 4.2.16,

$$\lim_{k\uparrow\infty} \mathrm{E}\left[\frac{S_k}{k}\right] = a.$$

But for all $k \geq 1$, $\mathrm{E}\left[S_k/k\right] = \mathrm{E}\left[X_1\right]$.

The uniform integrability of the backwards submartingale in Theorem 13.3.14 followed directly from the submartingale property. This is not the case for a *supermartingale* unless one adds a condition.

Theorem 13.3.17 *Let $\{\mathcal{F}_n\}_{n\leq 0}$ be a filtration and let $\{S_n\}_{n\leq 0}$ be an $\mathcal{F}_n$-supermartingale such that*

$$\sup_{n\leq 0} \mathrm{E}\left[S_n\right] < \infty. \tag{13.32}$$

Then

A. S_n *converges P-a.s. and in L^1 as $n \downarrow -\infty$ to an integrable random variable $S_{-\infty}$, and*

B. *with $\mathcal{F}_{-\infty} := \cap_{n\leq 0}\mathcal{F}_n$,*

$$S_{-\infty} \geq \mathrm{E}\left[S_0 \mid \mathcal{F}_{-\infty}\right] \quad P\text{-}a.s.$$

Proof. (^{2}) It suffices to prove uniform integrability, since the rest of the proof then follows the same lines as in Theorem 13.3.10.

Fix $\varepsilon > 0$ and select $k \leq 0$ such that

$$\lim_{i\downarrow-\infty} \mathrm{E}\left[S_i\right] - \mathrm{E}\left[S_k\right] \leq \varepsilon. \tag{$\star$}$$

Then $0 \leq \mathrm{E}\left[S_n\right] - \mathrm{E}\left[S_k\right] \leq \varepsilon$ for all $n \leq k$. We first show that for sufficiently large $\lambda > 0$,

$$\int_{\{|S_n|>\lambda\}} |S_n|\,\mathrm{d}P \leq \varepsilon.$$

It is enough to prove this for sufficiently large $-n$, here for $-n \geq -k$. The previous integral is equal to

$$-\int_{\{S_n<-\lambda\}} S_n\,\mathrm{d}P + \mathrm{E}\left[S_n\right] - \int_{\{S_n\leq\lambda\}} S_n\,\mathrm{d}P.$$

By the supermartingale hypothesis, this quantity is

$$\leq -\int_{\{S_k<-\lambda\}} S_n\,\mathrm{d}P + \mathrm{E}\left[S_n\right] - \int_{\{S_n\leq\lambda\}} S_k\,\mathrm{d}P.$$

In view of ($\star$), this is less than or equal to

$$-\int_{\{S_n<-\lambda\}} S_k\,\mathrm{d}P + \mathrm{E}\left[S_k\right] - \int_{\{S_n\leq\lambda\}} S_k\,\mathrm{d}P + \varepsilon,$$

which is equal to

$$\int_{\{|S_n|>\lambda\}} |S_k|\,\mathrm{d}P + \varepsilon.$$

Since ε is an arbitrary positive quantity, it remains to show that $\int_{\{|S_n|>\lambda\}} |S_k|\,\mathrm{d}P$ tends to 0 uniformly in $n \leq 0$ as $\lambda \uparrow \infty$. But since $\{S_n^-\}_{n\geq 1}$ is a supermartingale

---

^{2}[Meyer, 1972], VT21.

$$E\left[|S_n|\right] = E\left[S_n\right] + 2E\left[S_n^-\right] \leq \sup_{n \leq 0} E\left[S_n\right] + 2E\left[S_0^-\right].$$

Therefore, in view of hypothesis (13.32),

$$P(|S_n| > \lambda) \leq \frac{E\left[|S_n|\right]}{\lambda} \to 0$$

uniformly in $n \leq 0$, and therefore

$$\int_{\{|S_n| > \lambda\}} |S_k| \, dP \to 0$$

$\square$

uniformly in n.

The following result is the *backwards Lévy's continuity theorem for conditional expectations*.

Corollary 13.3.18 *Let $\{\mathcal{F}_n\}_{n \leq 0}$ be a history and let ξ be an integrable random variable. Then, with $\mathcal{F}_{-\infty} := \cap_{n \leq 0} \mathcal{F}_n$,*

$$\lim_{n \downarrow -\infty} E[\xi \mid \mathcal{F}_n] = E[\xi \mid \mathcal{F}_{-\infty}]. \tag{13.33}$$

Proof. $M_n := E[\xi \mid \mathcal{F}_n]$ $(n \leq 0)$ is an $\mathcal{F}_n$-martingale and therefore by the backwards martingale convergence theorem, it converges as $n \downarrow -\infty$ almost surely and in L^1 to some integrable variable $M_{-\infty}$ and

$$M_{-\infty} = E\left[M_0 \mid \mathcal{F}_{-\infty}\right] = E\left[E\left[\xi \mid \mathcal{F}_0\right] \mid \mathcal{F}_{-\infty}\right] = E\left[\xi \mid \mathcal{F}_{-\infty}\right]$$

$\square$

since $\mathcal{F}_{-\infty} \subseteq \mathcal{F}_0$.

Local Absolute Continuity

Recall the setting of Example 13.1.6. On the measurable space $(\Omega, \mathcal{F})$ are given two probability measures Q and P and a filtration $\{\mathcal{F}_n\}_{n \geq 1}$ such that $\mathcal{F} = \vee_{n \geq 1} \mathcal{F}_n := \mathcal{F}_\infty$. Let Q_n and P_n denote the restrictions to $(\Omega, \mathcal{F}_n)$ of Q and P respectively. Suppose that $Q_n \ll P_n$ $(n \geq 1)$, in which case we say that Q is locally absolutely continuous with respect to P along $\{\mathcal{F}_n\}_{n \geq 1}$ and denote this by $Q \overset{loc.}{\ll} P$. Let then

$$L_n := \frac{dQ_n}{dP_n}$$

denote the corresponding Radon–Nikodým derivative. The question is: under what circumstances can we assert that $Q \ll P$? And what can be said if this is not the case?

That this is not always the case is clear from the following elementary example.

EXAMPLE 13.3.19: INDEPENDENT SEQUENCES OF 0'S AND 1'S. In this example $\mathcal{F}_n = \sigma(X_1, \ldots, X_n)$, where $\{X_n\}_{n \geq 1}$ is an IID sequence of $\{0, 1\}$-valued random variables and

$$Q(X_n = 1) = q_n > 0 \text{ and } P(X_n = 1) = p_n > 0,$$

where $\sum_{n \geq 1} q_n = \infty$ and $\sum_{n \geq 1} p_n < \infty$. Then, by the positivity condition on the p_n's and the q_n's, $Q \overset{loc.}{\ll} P$. However Q and P are mutually singular since $Q(X_n \to 0) = 0$ and $P(X_n \to 0) = 1$ (see Exercise 4.6.7).

Theorem 13.3.20 *The Radon–Nikodým sequence* $\{L_n\}_{n\geq 1}$ *converges Q-almost surely and P-almost surely to some random variable* L_∞ *and*

$$dQ = 1_{\{L_\infty=\infty\}}\, dQ + L_\infty\, dP \qquad (13.34)$$

where $P(L_\infty = \infty) = 0$.

Remark 13.3.21 In particular, the measures $d\lambda := 1_{\{L_\infty=\infty\}}\, dQ$ and $d\mu := L_\infty\, dP$ are mutually singular and λ is absolutely continuous with respect to Q, so that (13.34) is the Lebesgue decomposition of Q with respect to P (Theorem 2.3.30).

Proof. Denote by ν (resp. ν_n) the probability $\frac{1}{2}(P + Q)$ on $(\Omega, \mathcal{F})$ (resp. $\frac{1}{2}(P_n + Q_n)$ on $(\Omega, \mathcal{F}_n)$). Since Q_n and P_n are dominated by ν_n, there exists for each $n \geq 1$ an ($\mathcal{F}_n$-measurable) Radon–Nikodým derivative $U_n := \frac{dQ_n}{d\nu_n}$. The sequence $\{U_n\}_{n\geq 1}$ is a $(\nu, \mathcal{F}_n)$-martingale, since for all $n \geq 1$ and all $A \in \mathcal{F}_n$ (and therefore also in $\mathcal{F}_{n+1}$),

$$\int_A U_{n+1}\, d\nu = Q_{n+1}(A) = Q_n(A) = \int_A U_n\, d\nu.$$

Also, $U_n \leq 2$ because $Q_n \leq 2\frac{P_n+Q_n}{2} = 2\nu_n$. Being a bounded $(\nu, \mathcal{F}_n)$-martingale, $\{U_n\}_{n\geq 1}$ converges ν-a.s. and in $L^1(\nu)$ to some random variable U_∞.

Therefore, for all $k \geq 1$, all $n \geq 0$ and all $A \in \mathcal{F}_k$,

$$\int_A U_\infty\, d\nu = \lim_{n\uparrow\infty} \int_A U_{n+k}\, d\nu$$

and therefore since for all $n \geq 0$ and all $A \in \mathcal{F}_k$, $\int_A U_{n+k}\, d\nu = \int_A U_k\, d\nu = Q(A)$,

$$\int_A U_\infty\, d\nu = Q(A).$$

This being true for all $A \in \mathcal{F}_k$, the probability measures $U_\infty d\nu$ and dQ agree on $\mathcal{F}_k$. This being true for all $k \geq 1$, they agree on the algebra $\cup_{k\geq 1}\mathcal{F}_k$, and therefore on $\mathcal{F}_\infty = \vee_{k\geq 1}\mathcal{F}_k$ (Caratheodory's theorem).

We have just proved that $dQ = U_\infty\, d\frac{P+Q}{2}$, that is,

$$(2 - U_\infty)dQ = U_\infty dP,$$

from which it follows that $P(U_\infty = 2) = 0$ and that if $Q(U_\infty = 2) = 0$, then $Q \ll P$ and $\frac{dQ}{dP} = \frac{U_\infty}{2-U_\infty}$.

Since $L_n = \frac{U_n}{2-U_n}$, $L_n \to L_\infty = \frac{U_\infty}{2-U_\infty}$ $(P + Q)$-a.s. and $P(L_\infty = \infty) = 0$. Now, $dQ_n = L_n\, dP_n$ and $\frac{2}{1+L_\infty}\, dQ = \frac{2L_\infty}{1+L_\infty}\, dP$. Hence the decomposition

$$dQ = 1_{\{L_\infty=\infty\}}\, dQ + L_\infty\, dP$$

where $P(L_\infty = \infty) = 0$ and $Q(L_\infty = 0) = 0$.   □

Theorem 13.3.22

$$Q \ll P \Leftrightarrow \mathrm{E}_P[L_\infty] = 1 \Leftrightarrow Q(L_\infty < \infty) = 1,$$
$$Q \perp P \Leftrightarrow \mathrm{E}_P[L_\infty] = 0 \Leftrightarrow Q(L_\infty = \infty) = 1.$$

Proof. Write (13.34) as

$$Q(A) = \int_A 1_{\{L_\infty = \infty\}} \, dQ + \int_A L_\infty \, dP \quad (A \in \mathcal{F}_\infty).$$

With $A = \Omega$,

$$1 = Q(L_\infty = \infty) + \mathrm{E}_P[L_\infty],$$

and therefore

$$\mathrm{E}_P[L_\infty] = 1 \Leftrightarrow Q(L_\infty < \infty) = 1,$$
$$\mathrm{E}_P[L_\infty] = 0 \Leftrightarrow Q(L_\infty = \infty) = 1.$$

If $Q(L_\infty = \infty) = 0$ it follows by (13.34) that $Q \ll P$. Conversely, if $Q \ll P$, then $Q(L_\infty = \infty) = 0$ since $P(L_\infty = \infty) = 0$.

If $Q \perp P$, there exists a $B \in \mathcal{F}$ such that $Q(B) = 1$ and $P(B) = 0$. In particular, from (13.34), $Q(B \cap \{L_\infty = \infty\}) = 1$ and therefore $Q(L_\infty = \infty) = 1$. Finally, if $Q(L_\infty = \infty) = 1$, $Q \perp P$ since $P(L_\infty = \infty) = 0$. $\square$

Remark 13.3.23 By Theorem 4.2.16, the condition $\mathrm{E}_P[L_\infty] = 1$ is equivalent to the uniform P-integrability of $\{L_n\}_{n \geq 1}$. Therefore, in order to prove that $Q \ll P$, any condition guaranteeing uniform integrability is a sufficient condition for the absolute continuity of Q with respect to P. For instance (Theorem 4.2.14 and Example 4.2.15)

$$\sup_n \mathrm{E}\left[L_n^{1+\alpha}\right] < \infty \quad (\alpha > 1)$$

and

$$\sup_n \mathrm{E}\left[L_n \log^+ L_n\right] < \infty.$$

EXAMPLE 13.3.24: KAKUTANI'S DICHOTOMY THEOREM. Let $\{X_n\}_{n \geq 1}$ be a sequence of random elements with values in the measurable space $(E, \mathcal{E})$. We may suppose that it is the coordinate sequence of the canonical space $(\Omega, \mathcal{F}) := (E^{\mathbb{N}}, \mathcal{E}^{\otimes \mathbb{N}})$. Let Q and P be two probability measures on $(\Omega, \mathcal{F})$ such that the sequence is IID relatively to both. Let Q_{X_n} and P_{X_n} be the restrictions of Q and P respectively to $\sigma(X_n)$ and let Q_n and P_n be the restrictions of Q and P respectively to $\mathcal{F}_n := \sigma(X_1, \ldots, X_n)$. We assume that for all $n \geq 1$, $Q_{X_n} \ll P_{X_n}$ and denote the corresponding Radon–Nikodým derivative $\frac{dQ_{X_n}}{dP_{X_n}}$ by $f_n(X_n)$. Then for all $n \geq 1$, $Q_n \ll P_n$ and $L_n = \frac{dQ_n}{dP_n} = \Pi_{i=1}^n f_i(X_i)$. Since

$$\{L_\infty < \infty\} = \{\log L_\infty < \infty\} = \left\{\sum_{i=1}^n \log f_i(X_i) < \infty\right\}$$

is a tail-event of the sequence, its probability is 0 or 1. Therefore, there are only two possibilities, either $Q \ll P$ or $Q \perp P$.

EXAMPLE 13.3.25: KAKUTANI'S CONDITION. Kakutani's theorem (Theorem 13.3.12) can be applied to the situation (analogous to that of Example 13.3.24 above) where

$$L_n = \prod_{i=1}^{n} Z_i$$

where $\{Z_n\}_{n\geq 1}$ is a sequence of IID non-negative random variables of mean 1. By this theorem, the criterion of absolute continuity of Q with respect to P, $E_P[L_\infty] = 1$, of Theorem 13.3.22 is $\prod_{n=1}^{\infty} E[Z_n^{\frac{1}{2}}] > 0$. By the same argument as in Example 13.3.24, the only alternative to $Q \ll P$ is $Q \perp P$, and therefore a necessary and sufficient condition for the latter is $\prod_{n=1}^{\infty} E[Z_n^{\frac{1}{2}}] = 0$.

Harmonic Functions and Markov Chains

An application to Markov chain theory of the martingale convergence theorem concerns harmonic functions of HMCs and the study of recurrence of HMCs. The basic result is:

Theorem 13.3.26 *An irreducible recurrent HMC $\{X_n\}_{n\geq 0}$ has no non-negative superharmonic or bounded subharmonic functions besides the constants.*

Proof. If h is non-negative superharmonic (resp., bounded subharmonic), the sequence $\{h(X_n)\}_{n\geq 0}$ is a non-negative supermartingale (resp., bounded submartingale) and therefore it converges to a finite limit Y. Since the chain visits any state $i \in E$ infinitely often, one must have $Y = h(i)$ almost surely for all $i \in E$. This can happen only if h is a constant. □

Corollary 13.3.27 *A necessary and sufficient condition for an irreducible HMC to be transient is the existence of some state (henceforth denoted by 0) and of a bounded function $h : E \to \mathbb{R}$, not identically null and satisfying*

$$h(j) = \sum_{k \neq 0} p_{jk} h(k) \quad (j \neq 0). \tag{13.35}$$

Proof. Let T_0 be the return time to state 0. First-step analysis shows that the (bounded) function h defined by

$$h(j) := P_j(T_0 = \infty)$$

satisfies (13.35). If the chain is transient, h is nontrivial (not identically null). This proves necessity.

Conversely, suppose that (13.35) holds for a not identically null bounded function. Define $\tilde{h}$ by $\tilde{h}(0) := 0$ and

$$\tilde{h}(j) := h(j) \quad (j \neq 0),$$

and let $\alpha := \sum_{k \in E} p_{0k} \tilde{h}(k)$. Changing the sign of $\tilde{h}$ if necessary, α can be assumed non-negative. Then $\tilde{h}$ is subharmonic and bounded. If the chain were recurrent, then by Theorem 13.3.26, $\tilde{h}$ would be a constant. This constant would be equal to $\tilde{h}(0) = 0$, and this contradicts the assumed nontriviality of h. □

Here is an application of the martingale convergence theorem in the vein of the previous results and of Foster's theorem (Theorem 6.3.19).

Theorem 13.3.28 *Let the* HMC $\{X_n\}_{n\geq 0}$ *with transition matrix* $\mathbf{P}$ *be irreducible and let* $h : E \to \mathbb{R}$ *be a bounded function such that*

$$\sum_{k\in E} p_{ik}h(k) \leq h(i), \quad \text{for all } i \notin F, \tag{13.36}$$

for some set F (not assumed finite). Suppose, moreover, that there exists an $i \notin F$ such that

$$h(i) < h(j), \quad \text{for all } j \in F. \tag{13.37}$$

Then the chain is transient.

Proof. Let τ be the return time in F and let $i \notin F$ satisfy (13.37). Defining $Y_n = h(X_{n\wedge\tau})$, we have that, under P_i, Y is a (bounded) $\mathcal{F}_n^X$-supermartingale (same proof as in Theorem 6.3.19). By the martingale convergence theorem, the limit Y_∞ of $Y_n = h(X_{n\wedge\tau})$ exists and is finite, P_i-almost surely. By dominated convergence, $\mathrm{E}_i[Y_\infty] = \lim_{n\uparrow\infty} \mathrm{E}_i[Y_n]$, and since $\mathrm{E}_i[Y_n] \leq \mathrm{E}_i[Y_0] = h(i)$ (supermartingale property), we have $\mathrm{E}_i[Y_\infty] \leq h(i)$.

If τ were P_i-a.s. finite, then Y_n would eventually be frozen at a value $h(j)$ for $j \in F$, and therefore by (13.37), $\mathrm{E}_i[Y_\infty] \geq h(i)$, a contradiction with the last inequality.

Therefore, $P_i(\tau < \infty) < 1$, which means that with a strictly positive probability, the chain starting from $i \notin F$ will not return to F. This is incompatible with irreducibility and recurrence. $\qquad\square$

13.3.3 The Robbins–Sigmund Theorem

In applications, one often encounters random sequences that are not quite martingales, submartingales or supermartingales, but "nearly" so, up to "perturbations". The statement of the result below will make this precise.

Theorem 13.3.29 *Let $\{V_n\}_{n\geq 1}$, $\{\beta_n\}_{n\geq 1}$, $\{\gamma_n\}_{n\geq 1}$ and $\{\delta_n\}_{n\geq 1}$ be real non-negative sequences of random variables adapted to some filtration $\{\mathcal{F}_n\}_{n\geq 1}$ and such that*

$$\mathrm{E}[V_{n+1} \mid \mathcal{F}_n] \leq V_n(1 + \beta_n) + \gamma_n - \delta_n \quad (n \geq 1). \tag{13.38}$$

Then, on the set

$$\Gamma = \left\{ \sum_{n\geq 1} \beta_n < \infty \right\} \cap \left\{ \sum_{n\geq 1} \gamma_n < \infty \right\} \tag{13.39}$$

the sequence $\{V_n\}_{n\geq 1}$ converges almost surely to a finite random variable and moreover $\sum_{n\geq 1} \delta_n < \infty$ P-almost surely.

Proof. 1. Let $\alpha_0 := 0$ and

$$\alpha_n := \left(\prod_{k=1}^{n} (1 + \beta_k) \right)^{-1} \quad (n \geq 1),$$

and let

$$V_n' := \alpha_{n-1}V_n, \quad \gamma_n' := \alpha_n\gamma_n, \quad \delta_n' := \alpha_n\delta_n \quad (n \geq 1).$$

Then

$$E[V'_{n+1} \mid \mathcal{F}_n] = \alpha_n E[V_{n+1} \mid \mathcal{F}_n] \leq \alpha_n V_n (1 + \beta_n) + \alpha_n \gamma_n - \alpha_n \delta_n,$$

that is, since $\alpha_n V_n(1 + \beta_n) = \alpha_{n-1} V_n$,

$$E[V'_{n+1} \mid \mathcal{F}_n] \leq V'_n + \gamma'_n - \delta'_n.$$

Therefore, the random sequence $\{Y_n\}_{n \geq 1}$ defined by

$$Y_n := V'_n - \sum_{k=1}^{n-1} (\gamma'_k - \delta'_k)$$

is an $\mathcal{F}_n$-supermartingale.

2. For $a > 0$, let

$$T_a := \inf \left\{ n \geq 1 \, ; \, \sum_{k=1}^{n-1} (\gamma'_k - \delta'_k) \geq a \right\}.$$

The sequence $\{Y_{n \wedge T_a}\}_{n \geq 1}$ is an $\mathcal{F}_n$-supermartingale bounded from below by $-a$. It therefore converges to a finite limit. Therefore, on $\{T_a = \infty\}$, $\{Y_n\}_{n \geq 1}$ converges to a finite limit.

3. On Γ, $\prod_{k=1}^{\infty}(1 + \beta_k)$ converges almost surely to a positive limit and therefore $\lim_{n \uparrow \infty} \alpha_n > 0$. Therefore, condition $\sum_{n \geq 1} \gamma_n < \infty$ implies $\sum_{n \geq 1} \gamma'_n < \infty$.

4. By definition of Y_n,

$$Y_n + \sum_{k=1}^{n-1} \gamma'_k = V'_n + \sum_{k=1}^{n-1} \delta'_k \geq \sum_{k=1}^{n-1} \delta'_k,$$

But on $\Gamma \cap \{T_a = \infty\}$, $\{Y_n\}_{n \geq 1}$ converges to a finite random variable, and therefore $\sum_{n \geq 1} \delta'_n < \infty$.

5. Since on $\Gamma \cap \{T_a = \infty\}$, $\sum_{n \geq 1} \gamma'_n < \infty$, $\sum_{n \geq 1} \delta'_n < \infty$ and $\{Y_n\}_{n \geq 1}$ converges to a finite random variable, it follows that $\{V'_n\}_{n \geq 1}$ converges to a finite limit. Since $\lim_{n \uparrow \infty} \alpha_n > 0$, it follows in turn that $\{V_n\}_{n \geq 1}$ converges to a finite limit and $\sum_{n \geq 1} \delta_n < \infty$ on $\Gamma \cap \{T_a = \infty\}$, and therefore on $\Gamma \cap (\cup_a \{T_a = \infty\}) = \Gamma$. □

Corollary 13.3.30 *Let $\{V_n\}_{n \geq 1}$, $\{\gamma_n\}_{n \geq 1}$ and $\{\delta_n\}_{n \geq 1}$ be real non-negative sequences of random variables adapted to some filtration $\{\mathcal{F}_n\}_{n \geq 1}$. Suppose that for all $n \geq 1$*

$$E[V_{n+1} \mid \mathcal{F}_n] \leq V_n + \gamma_n - \delta_n. \tag{13.40}$$

Let $\{a_n\}_{n \geq 1}$ be a random sequence that is strictly positive and strictly increasing and let

$$\widetilde{\Gamma} := \left\{ \sum_{n \geq 1} \frac{\gamma_n}{a_n} < \infty \right\}. \tag{13.41}$$

Then, almost-surely:

1. *on $\widetilde{\Gamma}$, the series $\sum_{n \geq 1} \frac{V_{n+1} - V_n}{a_n}$ is convergent and $\sum_{n \geq 1} \frac{\delta_n}{a_n} < \infty$,*

2. *on $\widetilde{\Gamma} \cap \{\lim_{n \uparrow \infty} a_n < \infty\}$, $\{V_n\}_{n \geq 1}$ converges almost surely, and*

3. on $\widetilde{\Gamma} \cap \{\lim_{n\uparrow\infty} a_n = \infty\}$, $\lim_{n\uparrow\infty} \frac{V_n}{a_n} = 0$ *and* $\lim_{n\uparrow\infty} \frac{V_{n+1}}{a_n} = 0$.

Proof. 1. Let for $n \geq 1$

$$Z_n := \sum_{k=1}^{n-1} \frac{V_{k+1} - V_k}{a_k} + \frac{V_1}{a_0} = \sum_{k=1}^{n} V_k \left(\frac{1}{a_{k-1}} - \frac{1}{a_k} \right) + \frac{V_n}{a_n}.$$

Since $\frac{1}{a_{k-1}} - \frac{1}{a_k} > 0$, we have that $Z_n \geq 0$ $(n \geq 1)$. Also

$$E[Z_{n+1} \mid \mathcal{F}_n] \leq Z_n + \frac{\gamma_n}{a_n} - \frac{\delta_n}{a_n}.$$

Therefore, by Theorem 13.3.29, on $\widetilde{\Gamma}$, $\{Z_n\}_{n\geq 1}$ converges and $\sum_{n\geq 1} \frac{\delta_n}{a_n} < \infty$. Note that in particular

$$\lim_{n\uparrow\infty} \frac{V_{n+1} - V_n}{a_n} = 0 \text{ on } \widetilde{\Gamma}. \tag{13.42}$$

2. If moreover $\lim_{n\uparrow\infty} a_n = a_\infty < \infty$, the convergence of $\sum_{n\geq 1} \frac{V_{n+1} - V_n}{a_n}$ implies that of $\frac{1}{a_\infty} \sum_{n\geq 1} (V_{n+1} - V_n)$, and therefore $\{V_n\}_{n\geq 1}$ converges.

3. If on the contrary $\lim_{n\uparrow\infty} a_n = \infty$, the convergence of $\sum_{n\geq 1} \frac{V_{n+1} - V_n}{a_n}$ implies that of $\frac{V_{n+1}}{a_n}$ (and therefore that of $\frac{V_n}{a_n}$, by (13.42)) to 0 (recall Kronecker's lemma: if $a_n > 0$ and $a_n \uparrow \infty$, the convergence of $\sum_{n\geq 1} \frac{x_n}{a_n}$ implies that $\lim_{n\uparrow\infty} \frac{1}{a_n} \sum_{k=1}^{n} x_k = 0$). $\quad\square$

13.3.4   Square-integrable Martingales

Doob's decomposition

Let $\{\mathcal{F}_n\}_{n\geq 0}$ be a filtration. Recall that a process $\{H_n\}_{n\geq 0}$ is called $\mathcal{F}_n$-*predictable* if for all $n \geq 1$, H_n is $\mathcal{F}_{n-1}$-measurable.

Theorem 13.3.31 *Let* $\{S_n\}_{n\geq 0}$ *be an* $\mathcal{F}_n$-*submartingale. Then there exists a P-a.s. unique non-decreasing* $\mathcal{F}_n$-*predictable process* $\{A_n\}_{n\geq 0}$ *with* $A_0 \equiv 0$ *and a unique* $\mathcal{F}_n$-*martingale* $\{M_n\}_{n\geq 0}$ *such that for all* $n \geq 0$,

$$S_n = M_n + A_n.$$

Proof. Existence is proved by explicit construction. Let $M_0 := S_0$, $A_0 = 0$ and, for $n \geq 1$,

$$M_n := S_0 + \sum_{j=0}^{n-1} \left\{ S_{j+1} - E[S_{j+1} \mid \mathcal{F}_j] \right\},$$

$$A_n := \sum_{j=0}^{n-1} \left(E[S_{j+1} \mid \mathcal{F}_j] - S_j \right).$$

Clearly, $\{M_n\}_{n\geq 0}$ and $\{A_n\}_{n\geq 0}$ have the announced properties. In order to prove uniqueness, let $\{M'_n\}_{n\geq 0}$ and $\{A'_n\}_{n\geq 0}$ be another such decomposition. In particular, for $n \geq 1$,

$$A'_{n+1} - A'_n = (A_{n+1} - A_n) + (M_{n+1} - M_n) - (M'_{n+1} - M'_n).$$

Therefore

$$E[A'_{n+1} - A'_n \mid \mathcal{F}_n] = E[A_{n+1} - A_n \mid \mathcal{F}_n],$$

and, since $A'_{n+1} - A'_n$ and $A_{n+1} - A_n$ are $\mathcal{F}_n$-measurable,

$$A'_{n+1} - A'_n = A_{n+1} - A_n, \quad \text{P-a.s.} \quad (n \geq 1),$$

from which it follows that $A'_n = A_n$ a.s. for all $n \geq 0$ (recall that $A'_0 = A_0$) and then $M'_n = M_n$ a.s. for all $n \geq 0$.   □

Definition 13.3.32 *The sequence* $\{A_n\}_{n\geq 0}$ *in Theorem 13.3.31 is called the* compensator *of* $\{S_n\}_{n\geq 0}$.

Definition 13.3.33 *Let* $\{M_n\}_{n\geq 0}$ *be a square-integrable* $\mathcal{F}_n$-*martingale (that is,* $E[M_n^2] < \infty$ *for all* $n \geq 0$*). The compensator of the* $\mathcal{F}_n$-*submartingale* $\{M_n^2\}_{n\geq 0}$ *is denoted by* $\{\langle M \rangle_n\}_{n\geq 0}$ *and is called the* bracket process *of* $\{M_n\}_{n\geq 0}$.

By the explicit construction in the proof of Theorem 13.3.31, $\langle M \rangle_0 := 0$ and for $n \geq 1$,

$$\langle M \rangle_n := \sum_{j=0}^{n-1} \left\{ E[M_{j+1}^2 \mid \mathcal{F}_j] - M_j^2 \right\} = \sum_{j=0}^{n-1} \left\{ E[(M_{j+1}^2 - M_j^2) \mid \mathcal{F}_j] \right\}. \tag{13.43}$$

Also, for all $0 \leq k \leq n$,

$$E[(M_n - M_k)^2 \mid \mathcal{F}_k] = E[M_n^2 - M_k^2 \mid \mathcal{F}_k] = E[\langle M \rangle_n - \langle M \rangle_k \mid \mathcal{F}_k].$$

Therefore, $\{M_n^2 - \langle M \rangle_n\}_{n\geq 0}$ is an $\mathcal{F}_n$-martingale. In particular, if $M_0 = 0$, $E[M_n^2] = E[\langle M \rangle_n]$.

EXAMPLE 13.3.34:   Let $\{Z_n\}_{n\geq 0}$ be a sequence of IID centered random variables of finite variance. Let $M_0 := 0$ and $M_n := \sum_{j=1}^n Z_j$ for $n \geq 1$. Then, for $n \geq 1$,

$$\langle M \rangle_n = \sum_{j=1}^n \text{Var}(Z_j).$$

Theorem 13.3.35 *If* $E[\langle M \rangle_\infty] < \infty$, *the square-integrable martingale* $\{M_n\}_{n\geq 0}$ *converges almost surely to a finite limit, and convergence takes place also in* L^2.

Proof. This is Theorem 13.3.8 for the particular case $p = 2$. In fact, condition (13.28) thereof is satisfied since

$$\sup_{n\geq 1} E\left[M_n^2\right] = \sup_{n\geq 1} E[\langle M \rangle_n] = E[\langle M \rangle_\infty] < \infty.$$

□

The Martingale Law of Large Numbers

Theorem 13.3.36 Let $\{M_n\}_{n \geq 0}$ be a square-integrable $\mathcal{F}_n$-martingale. Then:

A. On $\{\langle M \rangle_\infty < \infty\}$, M_n converges to a finite limit.

B. On $\{\langle M \rangle_\infty = \infty\}$, $M_n / \langle M \rangle_n \to 0$.

Proof. A. Let $K > 0$ be fixed, the random time

$$\tau_K := \inf\{n \geq 0: \ \langle M \rangle_{n+1} > K\}$$

is an $\mathcal{F}_n$-stopping time since the bracket process is $\mathcal{F}_n$-predictable. Also $\langle M \rangle_{n \wedge \tau_K} \leq K$ and therefore by Theorem 13.3.35, $\{M_{n \wedge \tau_K}\}_{n \geq 0}$ converges to a finite limit. Therefore $\{M_n\}_{n \geq 0}$ converges to a finite limit on the set $\{\langle M \rangle_\infty < K\}$ contained in $\{\tau_K = \infty\}$. Hence the result since

$$\{\langle M \rangle_\infty < \infty\} = \bigcup_{K \geq 1} \{\tau_K = \infty\}.$$

B. Note that

$$E[M_{n+1}^2 \mid \mathcal{F}_n] = M_n^2 + \langle M \rangle_{n+1} - \langle M \rangle_n.$$

Define

$$V_n = M_n^2, \quad \gamma_n = \langle M \rangle_{n+1} - \langle M \rangle_n, \quad a_n = \langle M \rangle_{n+1}^2.$$

The result then follows from Part 3 of Corollary 13.3.30 (observe that there exists a k_0 such that $a_k \geq 1$ for $k \geq k_0$ and $\sum_{k=k_0}^{\infty} \gamma_k / a_k = \sum_{k=k_0}^{\infty} (\langle M \rangle_{k+1} - \langle M \rangle_k) / \langle M \rangle_{k+1}^2 \leq \int_1^\infty x^{-2} \, dx < \infty$) which says, in particular, that $\sqrt{V_{n+1}/a_n} = M_{n+1}/\langle M \rangle_{n+1}$ converges to 0. $\qquad \square$

Remark 13.3.37 We do not have in general $\{\langle M \rangle_\infty < \infty\} = \{\{M_n\}_{n \geq 0} \text{ converges}\}$.

The following is a *conditioned version of the Borel–Cantelli lemma*. Note that, in this form, we have a necessary and sufficient condition.

Corollary 13.3.38 Let $\{\mathcal{F}_n\}_{n \geq 1}$ be a filtration and let $\{A_n\}_{n \geq 1}$ be a sequence of events such that $A_n \in \mathcal{F}_n$ $(n \geq 1)$. Then

$$\left\{ \sum_{n \geq 1} P(A_n \mid \mathcal{F}_{n-1}) = \infty \right\} \equiv \left\{ \sum_{n \geq 1} 1_{A_n} = \infty \right\}.$$

Proof. Define $\{M_n\}_{n \geq 0}$ by $M_0 := 0$ and for $n \geq 1$,

$$M_n := \sum_{k=1}^{n} (1_{A_k} - P(A_k \mid \mathcal{F}_{k-1})).$$

This is a square-integrable $\mathcal{F}_n$-martingale, with bracket process

$$\langle M \rangle_n = \sum_{k=1}^{n} P(A_k \mid \mathcal{F}_{k-1})(1 - P(A_k \mid \mathcal{F}_{k-1})).$$

In particular,

$$\langle M \rangle_n \le \sum_{k=1}^{n} P(A_k \,|\, \mathcal{F}_{k-1}).$$

A. Suppose that $\sum_{k=1}^{\infty} P(A_k \,|\, \mathcal{F}_{k-1}) < \infty$. Then, by the above inequality, $\langle M \rangle_\infty < \infty$, and therefore, by Part A of Theorem 13.3.36, M_n converges. Since by hypothesis, $\sum_{k=1}^{\infty} P(A_k \,|\, \mathcal{F}_{k-1}) < \infty$, this implies that $\sum_{k=1}^{\infty} 1_{A_k} < \infty$.

B. Suppose that $\sum_{k=1}^{\infty} P(A_k \,|\, \mathcal{F}_{k-1}) = \infty$ and $\langle M \rangle_\infty < \infty$. Then M_n converges to a finite random variable and therefore

$$\frac{M_n}{\sum_{k=1}^{n} P(A_k \,|\, \mathcal{F}_{k-1})} = \frac{\sum_{k=1}^{n} 1_{A_k}}{\sum_{k=1}^{n} P(A_k \,|\, \mathcal{F}_{k-1})} - 1 \to 0.$$

C. Suppose that $\sum_{k=1}^{\infty} P(A_k \,|\, \mathcal{F}_{k-1}) = \infty$ and $\langle M \rangle_\infty = \infty$. Then $\frac{M_n}{\langle M \rangle_n} \to 0$ and a fortiori,

$$\frac{M_n}{\sum_{k=1}^{n} P(A_k \,|\, \mathcal{F}_{k-1})} \to 0,$$

that is,

$$\frac{\sum_{k=1}^{n} 1_{A_k}}{\sum_{k=1}^{n} P(A_k \,|\, \mathcal{F}_{k-1})} \to 1.$$

$\square$

The Robbins–Monro algorithm

Consider an input-output relationship $u \in \mathbb{R} \to y \in \mathbb{R}$ of the form

$$x = g(u, \varepsilon)$$

where ε is a random variable, and let

$$\Phi(u) := \mathrm{E}[g(u, \varepsilon)].$$

We wish to determine u^* such that $\Phi(u^*) = \alpha$, where α is given.

Remark 13.3.39 This is a dosage problem: u is the dose and $\Phi(u)$ is the (average) effect produced by this dose; u^* is the dose realizing the desired effect α.

Φ is assumed non-decreasing, but is otherwise unknown. In order to determine u^*, one makes a series of experiments. Experiment $n \ge 0$ associates with the input U_n (an experimental dose) the output

$$X_{n+1} = g(U_n, \varepsilon_{n+1}),$$

where $\{\varepsilon_n\}_{n \ge 1}$ is IID. The input U_n is a function of the previous experimental results $X_1, \ldots, X_n$, and therefore

$$\mathrm{E}[X_{n+1} \,|\, \mathcal{F}_n] = \Phi(U_n),$$

where $\mathcal{F}_n = \sigma(X_1, \ldots, X_n)$. We want to choose U_n as a function of $X_1, \ldots, X_n$ that converges almost surely to u^*. The following strategy is reasonable: reduce the dose if $X_{n+1} > \alpha$, augment it otherwise. This remark has led to the Robbins–Monro algorithm:

$$U_{n+1} = U_n - \gamma_n(X_{n+1} - \alpha), \quad n \geq 0, \tag{13.44}$$

where $\gamma_n \geq 0$ for all $n \geq 0$.

The question is: Under what conditions does $\lim_{n\uparrow\infty} U_n = u^*$?

We shall need the following deterministic lemma:[3]

Lemma 13.3.40 *Let* $f : \mathbb{R} \to \mathbb{R}$ *be a continuous function such that for some* x^* *and some* $\alpha \in \mathbb{R}$, $f(x^*) = \alpha$ *and, for all* $x \neq x^*$,

$$(f(x) - \alpha)(x - x^*) < 0$$

and

$$|f(x)| \leq K(1 + |x|)$$

for some constant $K > 0$. *Let* $\{\gamma_n\}_{n\geq0}$ *be a non-increasing non-negative deterministic sequence such that*

$$\sum_{n\geq0} \gamma_n = \infty,$$

and let $\{\varepsilon_n\}_{n\geq1}$ *be a deterministic sequence such that*

$$\sum_{n\geq0} \gamma_n\varepsilon_{n+1} \text{ converges}.$$

Then, the sequence $\{x_n\}_{n\geq0}$ *defined by*

$$x_{n+1} = x_n + \gamma_n \left(f(x_n) - \alpha + \varepsilon_{n+1}\right), \quad n \geq 0,$$

converges to x^* *for any initial condition* x_0.

Let $\{X_n\}_{n\geq0}$ and $\{Y_n\}_{n\geq0}$ be sequences of square-integrable random vectors of dimension d adapted to some filtration $\{\mathcal{F}_n\}_{n\geq0}$. Let $\{\gamma_n\}_{n\geq0}$ be a non-increasing sequence of non-negative random variables such that

$$\lim_{n\uparrow\infty} \gamma_n = 0,$$

and $\gamma_0 \leq C < \infty$ for some deterministic constant C. Suppose that

$$X_{n+1} = X_n + \gamma_n Y_{n+1}, \quad n \geq 0,$$

and that, moreover, for all $n \geq 0$,

$$\mathrm{E}[Y_{n+1} \mid \mathcal{F}_n] = f(X_n), \qquad \mathrm{E}[|Y_{n+1} - f(X_n)|^2 \mid \mathcal{F}_n] = \sigma^2(X_n),$$

where the function $f : \mathbb{R} \to \mathbb{R}$ is continuous, such that for some $x^* \in \mathbb{R}^d$,

$$f(x^*) = 0$$

and for all $x \in \mathbb{R}$ such that $x \neq x^*$,

$$f(x) \times (x - x^*) < 0.$$

---

[3][Duflo, 1997], Proposition 1.2.3.

Theorem 13.3.41 *Suppose in addition that*

$$|f(x)| \leq K(1 + |x|)$$

for some constant $K > 0$, and

$$\sum_{n \geq 0} \gamma_n = \infty \text{ and } \sum_{n \geq 0} \gamma_n \sigma^2(X_n) < \infty.$$

Then, $\lim_{n \uparrow \infty} X_n = x^*$.

Proof. As

$$X_{n+1} = X_n + \gamma_n f(X_n) + \gamma_n(Y_{n+1} - f(X_n)) \quad (n \geq 0),$$

the process $\{M_n\}_{n \geq 0}$ defined by $M_0 := 0$ and

$$M_n := \sum_{k=1}^{n} \gamma_{k-1}(Y_k - f(X_{k-1})) \quad (n \geq 1)$$

is a square-integrable $\mathcal{F}_n$-martingale and

$$\langle M \rangle_n = \sum_{k=1}^{n} \gamma_{k-1}^2 \sigma^2(X_{k-1}).$$

In particular, since $\langle M \rangle_\infty < \infty$ by hypothesis, $\{M_n\}_{n \geq 0}$ converges to a finite limit. The result then follows from Lemma 13.3.40 with $\varepsilon_{n+1} = Y_{n+1} - f(X_n)$.   □

EXAMPLE 13.3.42: BACK TO THE DOSAGE PROBLEM. Consider the algorithm (13.44). We apply Theorem 13.3.41 with $f = \Phi - \alpha$. The conditions guaranteeing that $\lim_n U_n = u^*$ are therefore, besides $\Phi(u^*) = \alpha$ and Φ continuous,

$$(\Phi(u) - \alpha)(u - u^*) \text{ for all } u \neq u^*,$$

$$\sum_{n \geq 0} \gamma_n = \infty, \sum_{n \geq 0} \gamma_n^2 < \infty$$

and, for some $K < \infty$,

$$E[g(u, \varepsilon)]^2 \leq K(1 + |u|^2).$$

13.4   Continuous-time Martingales

The definition of a martingale in continuous time is similar to the one in discrete time and we shall see that most of the results in discrete-time find counterparts in continuous-time.

Let $\{\mathcal{F}_t\}_{t \geq 0}$ be a history (or filtration) on $\mathbb{R}_+$.

Definition 13.4.1 *A complex stochastic process $\{Y(t)\}_{t\geq 0}$ such that for all $t \in \mathbb{R}_+$*

(i) $Y(t)$ is $\mathcal{F}_t$–measurable, and

(ii) $E[|Y(t)|] < \infty$,

is called a $(P, \mathcal{F}_t)$-martingale (resp., submartingale, supermartingale*) if for all $s, t \in \mathbb{R}_+$ such that $s \leq t$,*

$$E[Y(t) \mid \mathcal{F}_s] = Y(s) \qquad (resp., \; \geq Y(s), \leq Y(s)). \tag{13.45}$$

When $Y(t) \geq 0$ for all $t \in \mathbb{R}_+$, the integrability condition is not required.

EXAMPLE 13.4.2: COMPENSATED COUNTING PROCESS. The counting process $\{N(t)\}_{t\geq 0}$ of Example 5.1.5 is such that

$$Y(t) := N(t) - \lambda t \quad (t \geq 0)$$

is an $\mathcal{F}_t^Y$-martingale (Exercise 13.5.24).

This result admits a converse:

Theorem 13.4.3 ([4]) *Let N be a simple locally finite point process on $\mathbb{R}_+$ such that for some filtration $\{\mathcal{F}_t\}_{t\geq 0}$ the stochastic process*

$$M(t) := N(t) - \lambda t \quad (t \in \mathbb{R}_+)$$

is an $\mathcal{F}_t$-martingale. Then N is a homogeneous Poisson process with intensity λ, and for any interval $(a, b] \in \mathbb{R}_+$, $N((a, b])$ is independent of $\mathcal{F}_a$.

Proof. In view of Theorem 7.1.8, it suffices to show that for all $T > 0$, and for all non-negative bounded real-valued stochastic processes $\{Z(t)\}_{t\geq 0}$ with left-continuous trajectories and adapted to $\{\mathcal{F}_t\}_{t\geq 0}$,

$$E\left[\int_{(0,T]} Z(t)\, N(\mathrm{d}t)\right] = E\left[\int_0^T Z(t)\lambda\, \mathrm{d}t\right]. \tag{$\star$}$$

The proof then is along the same lines as that of Theorem 7.1.8. Equality $(\star)$ is true for

$$Z(t, \omega) := 1_A(\omega)\, 1_{(a,b]}(t)$$

for any interval $(a, b] \subset \mathbb{R}_+$ and any $A \in \mathcal{F}_a$, since in this case, $(\star)$ reads

$$E\left[1_A N((a, b])\right] = E\left[1_A(b - a)\right],$$

that is, since A is arbitrary in $\mathcal{F}_a$,

$$E\left[N((a, b]) \mid \mathcal{F}_a\right] = (b - a),$$

which is the martingale hypothesis. The extension to non-negative bounded real-valued stochastic processes $\{Z(t)\}_{t\geq 0}$ with left-continuous trajectories is then done as above via the approximation (7.5). □

---

[4][Watanabe, 1964].

Theorem 13.4.4 *A supermartingale with constant mean is a martingale.*

Proof. This follows from the fact that two integrable random variables X and Y such that $X \leq Y$ and $\mathrm{E}[X] = \mathrm{E}[Y]$ are almost surely equal. $\square$

Definition 13.4.5 *A complex stochastic process $\{Y(t)\}_{t \geq 0}$ is called a $(P, \mathcal{F}_t)$-local martingale (resp., local submartingale, local supermartingale) if there exists a non-decreasing sequence of $\mathcal{F}_t$-stopping times $\{\tau_n\}_{n \geq 1}$ (the localizing sequence) such that*

(a) $\lim_{n \uparrow \infty} \tau_n = \infty$, *and*

(b) for all $n \geq 1$, $\{Y(t \wedge \tau_n)\}_{t \geq 0}$ is a $(P, \mathcal{F}_t)$-martingale (resp., supermartingale, submartingale).

Theorem 13.4.6 *A non-negative local martingale is a supermartingale.*

Proof. By Fatou's lemma,

$$\mathrm{E}\left[M(t) \mid \mathcal{F}_s\right] = \mathrm{E}\left[\lim_n M(t \wedge T_n) \mid \mathcal{F}_s\right]$$
$$\leq \liminf_n \mathrm{E}\left[M(t \wedge T_n) \mid \mathcal{F}_s\right] = \liminf_n M(s \wedge T_n) = M(s).$$

$\square$

The following characterization of the martingale property can be viewed as a kind of converse of Doob's optional sampling theorem. It will be referred to as Komatsu's lemma.

Lemma 13.4.7 *Let $\{\mathcal{F}_t\}_{t \geq 0}$ be a history. A real-valued $\mathcal{F}_t$-progressive stochastic process $\{X(t)\}_{t \geq 0}$ with the property that, for all bounded $\mathcal{F}_t$-stopping times T such that $X(T)$ is integrable, $\mathrm{E}[X(T)] = \mathrm{E}[X(0)]$, is an $\mathcal{F}_t$-martingale.*

Proof. Exercise 13.5.26.

$\square$

13.4.1   From Discrete Time to Continuous Time

Many among the results given for discrete-time martingales extend easily to the continuous-time right-continuous martingales.

For the extension of Kolmogorov and Doob's inequalities to right-continuous martingales, it suffices to observe that for a right-continuous process $\{X(t)\}_{t \geq 0}$, $\sup_{t \in \mathbb{R}} |X(t)| = \sup_{t \in \mathbb{Q}} |X(t)|$, where $\mathbb{Q}$ is the set of rational numbers.

Theorem 13.4.8 (Kolmogorov's inequality) *Let $\{Y(t)\}_{t \geq 0}$ be a right-continuous $\mathcal{F}_t$-submartingale. Then, for all $\lambda \in \mathbb{R}$ and all $a \in \mathbb{R}_+$,*

$$\lambda P\left(\sup_{0 \leq t \leq a} Y(t) > \lambda\right) \leq \mathrm{E}\left[Y(a) 1_{\{\sup_{0 \leq t \leq a} Y(t) > \lambda\}}\right]. \tag{13.46}$$

In particular, if $M\{(t)\}_{t \geq 0}$ is a right-continuous $\mathcal{F}_t$-martingale, then, for all $p \geq 1$, all $\lambda \in \mathbb{R}$ and all $a \in \mathbb{R}_+$,

$$\lambda^p P\left(\sup_{0 \leq t \leq a} |M(t)| > \lambda\right) \leq \mathrm{E}[|M(a)|^p]. \tag{13.47}$$

Theorem 13.4.9 (Doob's inequality) *Let* $\{M(t)\}_{n\geq 0}$ *be a right-continuous $\mathcal{F}_t$-mart-ingale. For all $p > 1$ and all $a \in \mathbb{R}_+$,*

$$\| M(a) \|_p \leq \| \sup_{0\leq t\leq a} |M(t)| \|_p \leq q \| M(a) \|_p, \qquad (13.48)$$

where q is defined by $\frac{1}{p} + \frac{1}{q} = 1$.

For the martingale convergence theorem, it suffices to show that the upcrossing inequality holds true. This is done in the proof of the following extension to continuous time of the discrete-time result. As above, one takes advantage of the right-continuity assumption.

Theorem 13.4.10 *Let* $\{Y(t)\}_{t\geq 0}$ *be a right-continuous $\mathcal{F}_t$-submartingale, L_1-bounded, that is, such that*

$$\sup_{t\geq 0} E[|Y(t)|] < \infty. \qquad (13.49)$$

Then $\{Y(t)\}_{t\geq 0}$ converges P-a.s. as $t \uparrow \infty$ to an integrable random variable $Y(\infty)$.

Proof. Let $D_n := \left\{\frac{k}{2^n}\right\}_{k\in\mathbb{N}}$ and $D := \cup_{n\in\mathbb{N}} D_n$. For given $0 \leq a < b$, let $\nu_n([a,b], K)$ and $\nu([a,b], K)$ be the number of upcrossings of $[a,b]$ respectively by $\{Y(t)\}_{t\in\Delta_n\cap[0,K]}$ and $\{Y(t)\}_{t\in\Delta\cap[0,K]}$. The upcrossing inequality for discrete-time submartingales give $(b-a)E[\nu_n([a,b], K)] \leq E[(Y(K) - a)^+]$, and therefore, passing to the limit as $n \uparrow \infty$, $(b-a)E[\nu([a,b], K)] \leq E[(Y(K) - a)^+]$. By the right-continuity assumption, $\nu([a,b]) = \sup_{K\in\mathbb{N}} \nu([a,b], K)$ and therefore, $(b-a)E[\nu([a,b])] \leq \sup_{K\in\mathbb{N}} E[(Y(K) - a)^+] < \infty$. The rest of the proof is then as in Theorem 13.3.2. $\qquad\square$

The continuous-time extensions of Theorems 13.3.8 and 13.3.14, and of Corollary 13.3.18, follow from Theorem 13.4.10 in the same way as their original discrete-time counterparts follow from Theorem 13.3.2. We leave to the reader the task of formulating these extensions.

The next result is the extension of Theorem 13.3.10 to continuous time, and its proof is left for the reader.

Theorem 13.4.11 *Let* $\{Y(t)\}_{t\geq 0}$ *be a right-continuous $\mathcal{F}_t$-submartingale, uniformly integrable. Then $\{Y(t)\}_{t\geq 0}$ converges a.s. and in L^1 to an integrable random variable denoted by $Y(\infty)$ and $E[Y(\infty) \,|\, \mathcal{F}_t] \geq Y(t)$.*

The above theorems required only a slight adaptation of their discrete-time versions. For the results that are stated in terms of convergence in a metric space, the adaptation to continuous time is even more immediate, and is based on the following lemma of analysis.

Lemma 13.4.12 *Let (E, d) be a complete metric space. A family $\{x_t\}_{t\geq 0}$ of elements of E converges to some element $x \in E$ as $t \uparrow \infty$ if and only if for any non-decreasing sequence of times $\{t_n\}_{n\geq 1}$ increasing to ∞ as $n \uparrow \infty$, the sequence $\{x_{t_n}\}_{n\geq 1}$ converges to x as $n \uparrow \infty$.*

Therefore any statement of convergence as $t \uparrow \infty$ that can be expressed in terms of convergence in a complete metric space can be obtained from the discrete-time version. This is the case for convergence in $L^p_{\mathbb{C}}(P)$ $(p \geq 1)$ and also for convergence in probability (which can indeed be expressed in terms of convergence in a complete metric space, see Theorem 4.2.4).

We now proceed to the statement and proof of Doob's optional sampling theorem in continuous time, which requires a little more work and the use of the reverse martingale convergence theorem.

Theorem 13.4.13 *Let $\{Y(t)\}_{t \geq 0}$ be a uniformly integrable right-continuous $\mathcal{F}_t$-submartingale, and let S and T be $\mathcal{F}_t$-stopping times such that $S \leq T$. Then*

$$E\left[Y(T) \mid \mathcal{F}_S\right] \geq (resp., =)Y(S).$$

Proof. For any $\mathcal{F}_t$-stopping time τ, let $\tau(n)$ be the approximation of the stopping time τ given in Theorem 5.3.13. Recall that this is an $\mathcal{F}_t$-stopping-time *decreasing* to τ as $n \uparrow \infty$. Now, fix n and let $\mathcal{G}^{(n)}_k := \mathcal{F}_{\frac{k}{2^n}}$ $(k \in \mathbb{N})$. Observe that $\tau(n)$ is a $\mathcal{G}^{(n)}_k$-stopping time. From Doob's optional sampling theorem applied to the $\mathcal{G}^{(n)}_k$-submartingale $\{Y(\frac{k}{2^n})\}_{k \geq 0}$,

$$E\left[Y(T(n)) \mid \mathcal{F}_{S(n)}\right] \geq Y(S(n)).$$

In particular, for all $A \in \mathcal{F}_S \subseteq \mathcal{F}_{S(n)} \subseteq \mathcal{F}_{T(n)}$ (by (v) of Theorem 5.3.19),

$$E\left[1_A Y(T(n))\right] \geq E\left[1_A Y(S(n))\right]. \qquad \star$$

Let $Z_{-n} := Y(T(n))$ and $\mathcal{A}_{-n} := \mathcal{F}_{T(n)}$ $(n \geq 0)$. By the reverse martingale convergence theorem (Theorem 13.3.14) applied to the submartingale $\{Z_n\}_{n \leq 0}$ adapted to the filtration $\{\mathcal{A}_n\}_{n \leq 0}$, the latter converges to $X(T)$ almost surely (right-continuity hypothesis) and also in L_1 (Theorem 13.3.10). A similar statement holds for S and therefore we can pass to the limit in ($\star$) to obtain

$$E\left[1_A Y(T)\right] \geq (resp. =)E\left[1_A Y(S)\right].$$

$\square$

EXAMPLE 13.4.14: THE RISK MODEL, TAKE 4. The *probability of ruin* corresponding to an initial capital u is

$$\Psi(u) := P(u + X(t) < 0 \text{ for some } t > 0). \tag{13.50}$$

It is a simple exercise (Exercise 13.5.25) to show that

$$E\left[e^{-rX(t)}\right] = e^{tg(r)},$$

where

$$g(r) = \lambda h(r) - rc = \lambda \left(\int_0^\infty e^{rv}\,\mathrm{d}G(v) - 1\right) - rc.$$

For any $u \in \mathbb{R}_+$,

$$M_u(t) := \frac{e^{-r(u + X(t))}}{e^{tg(r)}} \qquad (t \in \mathbb{R}_+)$$

is an $\mathcal{F}_t^X$-martingale. Indeed, for $0 \le s \le t < \infty$,

$$\begin{aligned}
\mathrm{E}\left[M_u(t) \mid \mathcal{F}_s^X\right] &= \mathrm{E}\left[\frac{e^{-r(u+X(t))}}{e^{tg(r)}} \mid \mathcal{F}_s^X\right] \\
&= \mathrm{E}\left[\frac{e^{-r(u+X(s))}}{e^{sg(r)}} \frac{e^{-r(X(t)-X(s))}}{e^{(t-s)g(r)}} \mid \mathcal{F}_s^X\right] \\
&= M_u(s)\mathrm{E}\left[\frac{e^{-r(X(t)-X(s))}}{e^{(t-s)g(r)}} \mid \mathcal{F}_s^X\right] = M_u(s).
\end{aligned}$$

For all $u \ge 0$,
$$T_u := \inf\{t \ge 0\,;\; u + X(t) < 0\}$$

(with the usual convention that the infimum of an empty set is infinite) is an $\mathcal{F}_t^X$-stopping time. For any $t_0 < \infty$, since $T_u \wedge t_0$ is a bounded stopping time and since $\{M_u(t)\}_{t\ge 0}$ is a (positive) martingale, we may apply Doob's optional stopping theorem:

$$\begin{aligned}
e^{-ru} = M_u(0) &= \mathrm{E}\left[M_u(T_u \wedge t_0)\right] \\
&= \mathrm{E}\left[M_u(T_u \wedge t_0) \mid T_u \wedge t_0 < t_0\right] P(T_u \wedge t_0 < t_0) \\
&\quad + \mathrm{E}\left[M_u(T_u \wedge t_0) \mid T_u \wedge t_0 \ge t_0\right] P(T_u \wedge t_0 \ge t_0) \\
&\ge \mathrm{E}\left[M_u(T_u \wedge t_0) \mid T_u \wedge t_0 < t_0\right] P(T_u \wedge t_0 < t_0) \\
&= \mathrm{E}\left[M_u(T_u) \mid T_u < t_0\right] P(T_u < t_0).
\end{aligned}$$

But $u + X(T_u) \le 0$ on $\{T_u < \infty\}$, and therefore

$$\begin{aligned}
P(T_u < t_0) &\le \frac{e^{-ru}}{\mathrm{E}\left[M_u(T_u) \mid T_u < t_0\right]} \\
&\le \frac{e^{-ru}}{\mathrm{E}\left[e^{-T_u g(r)} \mid T_u < t_0\right]} \le e^{-ru} \sup_{0 \le t \le t_0} e^{tg(r)}.
\end{aligned}$$

Letting $t_0 \to \infty$,
$$\Psi(u) \le e^{-ru} \sup_{t \ge 0} e^{tg(r)}. \tag{$\star$}$$

We choose, under the assumption that $\sup_{t\ge 0} e^{tg(r)} < \infty$, the r maximizing the right-hand side of $(\star)$, that is
$$R = \sup\{r\,;\; g(r) \le 0\},$$

i.e. the positive solution of
$$h(r) = \frac{cr}{\lambda}.$$

This is the celebrated *Lundberg's inequality*: $\Psi(u) \le e^{-Ru}$.

Theorem 13.4.15 *Let $\{Y(t)\}_{t\ge 0}$ be an $\mathcal{F}_t$-martingale and let T be $\mathcal{F}_t$-stopping time. Then $\{Y(t \wedge T)\}_{t\ge 0}$ is an $\mathcal{F}_t$-martingale.*

Proof. First suppose $\{Y(t)\}_{t\ge 0}$ is uniformly integrable. By Theorem 13.4.11, $\mathrm{E}[Y(\infty) \mid \mathcal{F}_{t\vee T}] = Y(t \vee T)$, and therefore

$$\begin{aligned}
\mathrm{E}[Y(\infty) - Y(T) \mid \mathcal{F}_{t\vee T}] &= Y(t \vee T) - Y(T) \\
&= 1_{\{T \le t\}}(Y(t) - Y(t \wedge T)) = Y(t) - Y(t \wedge T).
\end{aligned}$$

This variable is $\mathcal{F}_t$-measurable and therefore

$$E[Y(\infty) - Y(T) \mid \mathcal{F}_t] = Y(t) - Y(t \wedge T)$$

that is, since $E[Y(\infty) \mid \mathcal{F}_t] = Y(t)$,

$$E[Y(T) \mid \mathcal{F}_t] = Y(t \wedge T).$$

We now get rid of the uniform integrability assumption. For any $a \geq 0$, the $\mathcal{F}_{t \wedge a}$-martingale $\{Y(t \wedge a)\}_{t \geq 0}$ is uniformly integrable (Theorem 4.2.13) and therefore, by Theorem 13.4.13, for $t \leq a$,

$$E[Y(T \wedge a) \mid \mathcal{F}_t] = Y(t \wedge T \wedge a) = Y(t \wedge T).$$

$\square$

Predictable Quadratic Variation Processes

Definition 13.4.16 *Let $\{\mathcal{F}_t\}_{t \geq 0}$ be a history. Let $\{M(t)\}_{t \geq 0}$ be a local $\mathcal{F}_t$-martingale. A non-decreasing $\mathcal{F}_t$-predictable stochastic process $\{\langle M \rangle(t)\}_{t \geq 0}$ such that $M(t)^2 - \langle M \rangle(t)$ is a local $\mathcal{F}_t$-martingale is called the* predictable quadratic variation process *of the local martingale $\{M(t)\}_{t \geq 0}$.*

The following result of martingale theory is quoted without proof.

Theorem 13.4.17 *Let $\{\mathcal{F}_t\}_{t \geq 0}$ be a history. Let $\{M(t)\}_{t \geq 0}$ be a local square-integrable $\mathcal{F}_t$-martingale with quadratic variation process $\{\langle M \rangle(t)\}_{t \geq 0}$.*

a. If $\langle M \rangle(\infty) < \infty$, then $M(t)$ converges to a finite limit as $t \uparrow \infty$.

b. If $\langle M \rangle(\infty) = \infty$, then

$$\lim_{t \uparrow \infty} \frac{M(t)}{\langle M \rangle(t)} = 0.$$

13.4.2   The Banach Space $\mathcal{M}^p$

Definition 13.4.18 *Let $p \geq 1$. An $\mathcal{F}_t$-martingale $\{M(t)\}_{t \geq 0}$ is called p-integrable if*

$$\sup_{t \geq 0} E[|M(t)|^p] < \infty.$$

It is called p-integrable on the finite interval $[0, a]$ if $E[|M(a)|^2] < \infty$.

Remark 13.4.19 Condition (13.4.18) implies that this martingale is uniformly integrable (Theorem 4.2.14).

Note that when $p \geq 1$, $E[|M(a)|^p] < \infty$ implies $\sup_{t \in [0,a]} E[|M(t)|^p] < \infty$ since $\{|M(t)|^p\}_{t \geq 0}$ is then an $\mathcal{F}_t$-submartingale.

For $p \geq 1$, let $\mathcal{M}^p([0,1])$ be the collection of p-integrable $\mathcal{F}_t$-martingales over $[0,1]$. We shall not distinguish between versions, that is to say, an element $\mathcal{M}^p$ is an equivalence class for the equivalence $M \sim M'$ defined by $M(t) = M'(t)$ P-a.s. for all $t \in [0,1]$.

Theorem 13.4.20 *For $p \geq 1$, $\mathcal{M}^p([0,1])$ is a Banach space for the norm*

$$||M||_p = E[|M(1)|^p]. \tag{13.51}$$

Proof. First, we verify that (13.51) defines a norm. Only the fact that $||M||_p = 0$ implies $M = 0$ (that is, $P(M(t) = 0) = 1$ for all $t \in [0,1]$) is perhaps not obvious. By Jensen's inequality for conditional expectations, for all $t \in [0,1]$,

$$E[|M(t)|^p] = E[|E[M(1) \mid \mathcal{F}_t]|^p] \leq E[E[|M(1)|^p \mid \mathcal{F}_t]] = E[|M(1)|^p] = 0$$

which implies in particular that $P(M(t) = 0) = 1$ for all $t \in [0,1]$.

Let now $\{M_n\}_{n \geq 1}$ be a Cauchy sequence of $\mathcal{M}^p([0,1])$, that is,

$$\lim_{k,n \uparrow \infty} E[|M_n(1) - M_k(1)|^p] = 0 \,.$$

By the same Jensen-type argument as above, for all $t \in [0,1]$,

$$\lim_{k,n \uparrow \infty} E[|M_n(t) - M_k(t)|^p] = 0 \,.$$

Therefore, for all $t \in [0,1]$, there exists a limit in $L^p_{\mathbb{R}}(P, \mathcal{F}_t)$ of the sequence of random variables $\{M_n(t)\}_{n \geq 1}$ that we call $M(t)$.

It remains to show that the process $\{M(t)\}_{n \geq 0}$ so defined is an $\mathcal{F}_t$-martingale, that is, for all $[a, b] \subset [0,1]$ and all $A \in \mathcal{F}_a$

$$E[1_A M(b)] = E[1_A M(a)].$$

Using the assumption that for all $n \geq 1$, $\{M_n(t)\}_{t \geq 0}$ is an $\mathcal{F}_t$-martingale (and therefore

$$E[1_A M_n(b)] = E[1_A M_n(a)])$$

and the fact that for all $t \in [0,1]$, $M_n(t)$ tends to $M(t)$ in $L^p_{\mathbb{R}}(P)$, we have that $\lim_{n \uparrow \infty} E[1_A M_n(a)] = E[1_A M(a)]$ and $\lim_{n \uparrow \infty} E[1_A M_n(b)] = E[1_A M(b)]$. $\square$

13.4.3 Time Scaling

In this subsection, we define changes of the time scale "adapted" to a given history.

Definition 13.4.21 *Let $\{\mathcal{F}_t\}_{t \geq 0}$ be a history. A process $A = \{A(t)\}_{t \geq 0}$ is called a standard non-decreasing stochastic process if it has non-decreasing right-continuous trajectories $t \mapsto A(t, \omega)$ and if moreover $A(0, \omega) \equiv 0$. A standard non-decreasing process $\{T(t)\}_{t \geq 0}$ is called an $\mathcal{F}_t$-change of time if, for all $t \geq 0$, $T(t)$ is an $\mathcal{F}_t$-stopping time.*

Theorem 13.4.22 *Let $\{\mathcal{F}_t\}_{t \geq 0}$ be a right-continuous history and let $\{T(t)\}_{t \geq 0}$ be an $\mathcal{F}_t$-change of time. Then, the family $\{\mathcal{F}_{T(t)}\}_{t \geq 0}$ is a right-continuous history. Moreover, if the stochastic process $\{X(t)\}_{t \geq 0}$ is $\mathcal{F}_t$-progressive, then the stochastic process $\{Y(t)\}_{t \geq 0}$ defined by $Y(t) = X(T(t))1_{\{T(t) < \infty\}}$ is $\mathcal{F}_{T_t}$-adapted.*

Proof. The right-continuity of $\{\mathcal{F}_{T(t)}\}_{t \geq 0}$ follows from the fact that since $\{\mathcal{F}_t\}_{t \geq 0}$ is right-continuous, $\mathcal{F}_T = \cap_{n \geq 1} \mathcal{F}_{T_n}$ for any non-increasing sequence of $\mathcal{F}_t$-stopping times $\{T_n\}_{n \geq 1}$ with limit T. The last assertion of the theorem follows from the fact that if X is $\mathcal{F}_t$-progressive and if T is an $\mathcal{F}_t$-stopping time, then $X(T)$ and T are $\mathcal{F}_T$-measurable. $\square$

Theorem 13.4.23 Let $A = \{A(t)\}_{t\geq 0}$ be a standard non-decreasing stochastic process Associate to it its inverse process $C = \{C(t)\}_{t\geq 0}$ defined by

$$C(t,\omega) = \inf\{s; A(s,\omega) > t\}.$$

If A is adapted to the right-continuous history $\{\mathcal{F}_t\}_{t\geq 0}$, C is an $\mathcal{F}_t$-change of time.

Proof. Indeed, for all $s \geq 0$, all $a \geq 0$,

$$\{C(t) < a\} = \cup_{n\geq 1}\{A(a - \frac{1}{n}) > t\} \in \vee_{n\geq 1}\mathcal{F}_{a-\frac{1}{n}} = \mathcal{F}_{a-} \subseteq \mathcal{F}_a.$$

$\square$

Theorem 13.4.24 Let $X = \{X(t)\}_{t\geq 0}$ and $Y = \{Y(t)\}_{t\geq 0}$ be two stochastic processes adapted to the right-continuous history $\{\mathcal{F}_t\}_{t\geq 0}$, and such that for any $\mathcal{F}_t$-stopping time S,

$$\mathrm{E}\left[X(S)1_{\{S<\infty\}}\right] = \mathrm{E}\left[Y(S)1_{\{S<\infty\}}\right].$$

Then, for any standard non-decreasing $\mathcal{F}_t$-adapted stochastic process $A = \{A(t)\}_{t\geq 0}$ and any $\mathcal{F}_t$-stopping time T,

$$\mathrm{E}\left[\int_0^T X(s)\,\mathrm{d}A(s)\right] = \mathrm{E}\left[\int_0^T Y(s)\,\mathrm{d}A(s)\right].$$

Proof. First case: $T \equiv \infty$. By Theorem B.8.1 and Fubini,

$$\mathrm{E}\left[\int_0^\infty X(s)\,\mathrm{d}A(s)\right] = \mathrm{E}\left[\int_0^\infty X(C(s))\,\mathrm{d}s\right] = \int_0^\infty \mathrm{E}\left[X(C(s))\right]\,\mathrm{d}s$$

and similarly for the process Y. This suffices to prove the theorem in this case. General case: Replace A by B defined by $B(t) = A(t \wedge T)$.

$\square$

The main result of this subsection is the following "integration by parts" formula.

Theorem 13.4.25 Let $A = \{A(t)\}_{t\geq 0}$ be a standard non-decreasing stochastic process adapted to the right-continuous history $\{\mathcal{F}_t\}_{t\geq 0}$, and let $\{M(t)\}_{t\geq 0}$ be a right-continuous uniformly integrable $\mathcal{F}_t$-martingale. Then, for any $\mathcal{F}_t$-stopping time T,

$$\mathrm{E}\left[\int_0^T M(t)\,\mathrm{d}A(t)\right] = \mathrm{E}\left[M(T)A(T)\right]. \tag{13.52}$$

Proof. Recall that under the conditions of the theorem, there exists an integrable variable $M(\infty)$ such that $\lim_{t\uparrow\infty} M(t) = M(\infty)$ almost surely and for all $t \geq 0$, $M(t) = \mathrm{E}\left[M(\infty)|\mathcal{F}_t\right]$. This gives a meaning to $M(T)$ when $T = \infty$. Define the processes X and Y by $X(t) = M(t)1_{\{t\leq T\}}$, $Y(t) = M(T)1_{\{t\leq T\}}$. By Doob's optional sampling theorem,

$$\mathrm{E}\left[X(S)1_{\{S<\infty\}}\right] = \mathrm{E}\left[Y(S)1_{\{S<\infty\}}\right]$$

for any $\mathcal{F}_t$-stopping time. Therefore, Theorem 13.4.24 applies, yielding the announced result.

$\square$

Chapter VII of [Shiryaev, 1984, 1996] for discrete-time martingales. For continuous-time martingales, see the more advanced [Durrett, 1996].

13.5 Exercises

Exercise 13.5.1. DISCOUNTED PRODUCT
Let $\{X_n\}_{n \geq 1}$ be a sequence of independent integrable random variables with a common mean $m \neq 0$. Show that

$$Y_n := m^{-n} X_1 X_2 \cdots X_n \quad (n \geq 1)$$

is an $\mathcal{F}_n^X$-martingale.

Exercise 13.5.2. THE LÉVY MARTINGALE
Let $\{X_n\}_{n \geq 0}$ be an HMC with state space E and transition matrix $\mathbf{P}$, and let $f : E \to \mathbb{R}$ be a bounded function. Show that the process

$$M_n^f = f(X_n) - f(X_0) - \sum_{k=0}^{n-1} (\mathbf{P} - I) f(X_k)$$

is an $\mathcal{F}_n^X$-martingale.

Exercise 13.5.3. A MARTINGALE
Let $\{X_n\}_{n \geq 0}$ be an HMC with state space E and transition matrix $\mathbf{P}$. Let $f : \mathbb{N} \times E \to \mathbb{R}$ be a function such that for all $n \geq 0$ and $i \in E$,

$$E_i[|f(n, X_n)|] < \infty$$

and

$$\sum_{j \in E} p_{ij} f(n+1, j) = f(n, i).$$

Show that

$$M_n = f(n, X_n)$$

defines a $\mathcal{F}_n^X$-martingale $\{M_n\}_{n \geq 0}$.

Exercise 13.5.4. MARTINGALE CHARACTERIZATION OF AN HMC
Let $\{X_n\}_{n \geq 0}$ be a stochastic process with values in the countable space E. It is not assumed to be an HMC. Let $\mathbf{P}$ be some transition matrix on E. Prove that if for all bounded $f : E \to \mathbb{R}$, $\{M_n^f\}_{n \geq 0}$ defined in Exercise 13.5.2 is a martingale with respect to $\{X_n\}_{n \geq 0}$, then $\{X_n\}_{n \geq 0}$ is an HMC with transition matrix $\mathbf{P}$.

Exercise 13.5.5. PROBABILITY OF HIT
Let $\{X_n\}_{n \geq 0}$ be an HMC with state space E and let B be a closed subset of states, that is,

$$i \in B \Rightarrow \sum_{j \in B} p_{ij} = 1.$$

Let T be the hitting time of B and define for $i \in E$,

$$h(i) = P_i(T < \infty).$$

Show that $\{h(X_n)\}_{n \geq 0}$ is an $\mathcal{F}_n^X$-martingale.

Exercise 13.5.6. A MARTINGALE REPRESENTATION THEOREM
Let $\{X_n\}_{n \geq 0}$ be a sequence of $\{0, 1\}$-valued random variables and let $\lambda_{n-1} := \mathrm{E}\left[X_n \mid \mathcal{F}_{n-}^X\right]$ $(n \geq 0)$, where $\mathcal{F}_{-1} := \{\varnothing, \Omega\}$. Show that any $\mathcal{F}_n^X$-martingale $\{M_n\}_{n \geq 0}$ is of the form

$$M_n = M_0 + \sum_{j=0}^{n} H_j(X_j - \lambda_{j-1}),$$

where $\{H_n\}_{n \geq 0}$ is an $\mathcal{F}_n^X$-predictable sequence.

Exercise 13.5.7. A MARTINGALE BUILT ON A PERMUTATION
Let $a_1, \ldots, a_k \in \mathbb{R}$ be such that $\sum_{j=1}^{k} a_j = 0$. Let π be a completely random permutation of $\{1, \ldots, k\}$, that is, $P(\pi = \pi_0) = \frac{1}{k!}$ for all permutations π_0 of $\{1, \ldots, k\}$. Let $\mathcal{F}_n := \sigma(\pi(1), \ldots, \pi(n))$ $(1 \leq n \leq k)$ and

$$X_n := \frac{k}{k-n} \sum_{j=1}^{n} a_{\pi(j)}.$$

Show that $\{X_n\}_{1 \leq n \leq k}$ is an $\mathcal{F}_n$-martingale.

Exercise 13.5.8. RUINED AGAIN
Show that the function $h(i) = \left(\frac{1-p}{p}\right)^i$ is harmonic for the nonsymmetric random walk on $\mathbb{Z}$ (with $p_{i,i+1} = p, p_{i,i-1} = 1 - p$, where $p \in (0, 1)$ and $p \neq \frac{1}{2}$). Apply the optional sampling theorem to obtain the ruin probability in Example 6.1.10.

Exercise 13.5.9. MEAN HITTING TIME VIA MARTINGALES
Let $\{X_n\}_{n \geq 0}$ be a symmetric random walk on $\mathbb{Z}$. Show that $\{X_n\}_{n \geq 0}$ and $\{X_n^2 - n\}_{n \geq 0}$ are martingales with respect to $\{X_n\}_{n \geq 0}$. Deduce from this the mean hitting time of $\{-a, b\}$, where a and b are positive integers.

Exercise 13.5.10. ABSORPTION PROBABILITY
Consider the HMC $\{X_n\}_{n \geq 0}$ with state space $E = \{0, 1, \ldots, m\}$ and transition probabilities

$$p_{ij} = \binom{m}{j} \left(\frac{i}{m}\right)^j \left(1 - \frac{i}{m}\right)^{m-j}.$$

In particular, 0 and m are absorbing states.

(a) Show that $\{X_n\}_{n \geq 0}$ is a martingale.

(b) Compute the probability of absorption by state 0.

Exercise 13.5.11. UPCROSSINGS
Let $\{M_n\}_{n \geq 0}$ be an $\mathcal{F}_n$-martingale. Let $a, b \in \mathbb{R}$ with $a < b$, and let ν_n be the number of upcrossings of $[a, b]$ before ($\leq$) time n. For $k \geq 1$, let A_k be the event that there are exactly $k - 1$ upcrossings of $[a, b]$ before ($\leq$) time n. Show that

$$(b - a)P(\nu_n \geq k) \leq \mathrm{E}\left[(a - M_n)1_{A_k}\right].$$

Exercise 13.5.12. REPAIR SHOP, TAKE 4
Examples 6.1.7, 6.2.3 and 6.2.14 continued. This HMC satisfies the recurrence equation

$$X_{n+1} = (X_n - 1)^+ + Z_{n+1} \quad (n \geq 0),$$

where $\{Z_n\}_{n\geq 1}$ is an IID sequence independent of the initial state X_0. Recall that in terms of the probability distribution $P(Z_1 = k) = a_k, \ (k \geq 0)$, its transition matrix is

$$\mathbf{P} = \begin{pmatrix} a_0 & a_1 & a_2 & a_3 & \cdots \\ a_0 & a_1 & a_2 & a_3 & \cdots \\ 0 & a_0 & a_1 & a_2 & \cdots \\ 0 & 0 & a_0 & a_1 & \cdots \\ \vdots & \vdots & \vdots & \vdots & \end{pmatrix}.$$

Assume this chain is irreducible, that is, $P(Z_1 = 0) < 1$ and $P(Z_1 \geq 2) > 0$. Also suppose that $E[Z_1] > 1$. Apply Corollary 13.3.27 with a function h of the form $h(j) = 1 - \zeta^j$ to prove that the chain is transient.

Exercise 13.5.13. $E[X \mid \mathcal{F}_\tau] = X(\tau)$
Let τ be a stopping time for the filtration $\{\mathcal{F}_n\}_{n\geq 1}$. Let $X_n := E[X \mid \mathcal{F}_n] \ (n \geq 1)$ where X is an integrable random variable. Prove that $E[X \mid \mathcal{F}_\tau] = X(\tau)$.

Exercise 13.5.14. MARTINGALE BOUNDED BY AN INTEGRABLE RANDOM VARIABLE
Let $\{X_n\}_{n\geq 1}$ be an $\mathcal{F}_n$-martingale and let Z be an integrable random variable such that $X_n \leq Z \ (n \geq 1)$. Prove that $\{X_n\}_{n\geq 1}$ converges almost surely.

Exercise 13.5.15. KRICKEBERG'S DECOMPOSITION
Prove that an $\mathcal{F}_n$-martingale $\{M_n\}_{n\geq 0}$ such that $\sup_{n\geq 0} E[|M_n|] < \infty$ is the difference of two non-negative $\mathcal{F}_n$-martingales. (Hint: Doob's decomposition applied to $|M_n|$.)

Exercise 13.5.16. POLYA'S URN
At time 0 an urn contains exactly one black ball and one white ball. At time $n \geq 0$, a ball is drawn at random and then at time $n + 1$ this ball is put back into the urn together with another ball of the same color. In particular, there are at time n exactly $n + 2$ balls in the urn. Let B_n be the number of black balls in the urn. Let $X_n := \frac{B_n}{n+2}$ be the proportion of black balls at time n. Show that $\{X_n\}_{n\geq 0}$ is a martingale and that the ratio of the number of black balls to the number of white balls converges.

Exercise 13.5.17. RECORDS
Let $\{X_n\}_{n\geq 1}$ be an IID sequence of random variables with a common cumulative distribution F that is continuous. For $1 \leq i \leq n$, let $Y_i := 1$ if and only if $X_i = \max(X_1, \ldots, X_i)$. We shall admit that X_i is uniformly distributed on $\{1, \ldots, i\}$ and that $\{Y_i\}_{1\leq i\leq n}$ is IID. Let $Z_n := \sum_{i=1}^n 1_{\{Y_i=1\}}$ (the number of times a record is broken, that is, the number of i's such that $X_i > \max(X_1, \ldots, X_{i-1})$). Prove that $\frac{Z_n}{\ln n} \to 1$ almost surely.

Exercise 13.5.18. A MAXIMAL INEQUALITY
Let $\{X_n\}_{n\geq 0}$ be a centered square-integrable martingale. Let $\lambda > 0$. Prove the following inequality:

$$P\left(\max_{0\leq k\leq n} X_k > \lambda \right) \leq \frac{E[X_n^2]}{E[X_n^2] + \lambda^2}.$$

Hint: With $c > 0$, work with the sequence $\{(X_n + c)^2\}_{n \geq 0}$ and then select an appropriate c.

Exercise 13.5.19. AN EXTENSION OF HOEFFDING'S INEQUALITY
Let M be a real $\mathcal{F}_n^X$-martingale such that, for some sequence $d_1, d_2, \ldots$ of real numbers,

$$P(B_n \leq M_n - M_{n-1} \leq B_n + d_n) = 1, \quad n \geq 1,$$

where for each $n \geq 1$, B_n is a function of X_0^{n-1}. Prove that, for all $x \geq 0$,

$$P(|M_n - M_0| \geq x) \leq 2 \exp\left(-2x^2 \Big/ \sum_{i=1}^{n} d_i^2\right).$$

Exercise 13.5.20. THE DERIVATIVE OF A LIPSCHITZ CONTINUOUS FUNCTION
Let $f : [0, 1) \to \mathbb{R}$ satisfy a Lipschitz condition, that is,

$$|f(x) - f(y)| \leq M|x - y| \quad (x, y \in [0, 1)),$$

where $M < \infty$. Let $\Omega = [0, 1)$, $\mathcal{F} = \mathcal{B}([0, 1))$ and let P be the Lebesgue measure on $[0, 1)$. Let for all $n \geq 1$

$$\xi_n(\omega) := \sum_{k=1}^{2^n} 1_{\{[(k-1)2^{-n}, k2^{-n})\}}(\omega)$$

and

$$\mathcal{F}_n = \sigma(\xi_k ; 1 \leq k \leq n).$$

(i) Show that $\mathcal{F}_n = \sigma(\xi_n)$ and $\vee_n \mathcal{F}_n = \mathcal{B}([0, 1))$.

(ii) Let

$$X_n := \frac{f(\xi_n + 2^{-n}) - f(\xi_n)}{2^{-n}}.$$

Show that $\{X_n\}_{n \geq 1}$ is a uniformly integrable $\mathcal{F}_n$-martingale.

(iii) Show that there exists a measurable function $g : [0, 1) \to \mathbb{R}$ such that $X_n \to g$ P-almost surely and that $X_n = \mathrm{E}[g \mid \mathcal{F}_n]$.

(iv) Show that for all $n \geq 1$ and all k $(1 \leq k \leq 2^n)$

$$f(k2^{-n}) - f(0) = \int_0^{k2^{-n}} g(x)\,\mathrm{d}x$$

and deduce from this that

$$f(x) - f(0) = \int_0^x g(y)\,\mathrm{d}y \quad (x \in [0, 1)).$$

Exercise 13.5.21. A NON-UNIFORMLY INTEGRABLE MARTINGALE
Let $\{X_n\}_{n \geq 0}$ be a sequence of IID random variables such that $P(X_n = 0) = P(X_n = 2) = \frac{1}{2}$ $(n \geq 0)$. Define

$$Z_n := \prod_{j=1}^{n} X_j \quad (n \geq 0).$$

Show that $\{Z_n\}_{n\geq 0}$ is an $\mathcal{F}_n^X$-martingale and prove that it is not uniformly integrable.

Exercise 13.5.22. THE BALLOT PROBLEM VIA MARTINGALES
This exercise proposes an alternative proof for the ballot problem of Example 1.2.13.
Let $k := a + b$ and let D_n be the difference between the number of votes for A and the number of votes for B at time $n \geq 1$. Prove that

$$X_n = \frac{D_{k-n}}{k-n} \quad (1 \leq n \leq k)$$

is a martingale. Deduce from this that the probability that A leads throughout the voting process is $(a - b)/(a + b)$. Hint: $\tau := \inf\{n \,;\, X_n = 0\} \wedge (k - 1)$.

Exercise 13.5.23. A VOTING MODEL
Let $G = (V, \mathcal{E})$ be a finite graph. Each vertex v shelters a random variable $X_n(v)$ representing the opinion (0 or 1) at time n of the voter located at this vertex. At each time n, an edge $\langle v, w \rangle$ is chosen at random, and one of the two vertices, again chosen at random (say v), reconsiders his opinion passing from $X_n(v)$ to $X_{n+1}(v) = X_n(w)$. The initial opinions at time 0 are given. Let Z_n be the total number of votes for 1 at time n. Show that $\{Z_n\}_{n\geq 1}$ is a martingale that converges in finite random time to a random variable Z_∞ taking the values 0 or $|V|$, the probability that all opinions are eventually 1 being equal to the initial proportion of 1's.

Exercise 13.5.24. THE FUNDAMENTAL MARTINGALE OF AN HPP
Prove that for the counting process $\{N(t)\}_{t\geq 0}$ of Example 5.1.5, $\{N(t) - \lambda t\}_{t\geq 0}$ is an $\mathcal{F}_t^Y$-martingale.

Exercise 13.5.25. COMPOUND POISSON PROCESSES
Let N be an HPP on $\mathbb{R}_+$ with intensity λ and point sequence $\{T_n\}_{n\geq 1}$. Let $\{Z_n\}_{n\geq 1}$ be an IID real-valued sequence independent of N, with common CDF F. Define for all $t \geq 0$,

$$Y(t) = \sum_{n\geq 1} Z_n 1_{(0,t]}(T_n) \,.$$

(The process $\{Y(t)\}_{t\geq 0}$ is called a *compound Poisson process*.) Show that

$$\mathrm{E}\left[e^{-rY(t)}\right] = e^{\lambda t(1-h(r))},$$

where $h(r) := \mathrm{E}\left[e^{-rZ_1}\right] = \int_0^\infty e^{-rx}\,\mathrm{d}F(x)$.

Exercise 13.5.26. KOMATSU'S LEMMA
Prove the following: A right-continuous real stochastic process $\{X(t)\}_{t\geq 0}$ adapted to the filtration $\{\mathcal{F}_t\}_{t\geq 0}$ is an $\mathcal{F}_t$-martingale if and only if for all bounded $\mathcal{F}_t$-stopping times τ,

$$\mathrm{E}\left[X(\tau)\right] = \mathrm{E}\left[X(0)\right] \,.$$

Hint: Show that for all $0 \leq a \leq b$ and all $A \in \mathcal{F}_a$,

$$\tau := a1_A + b1_{\overline{A}}$$

defines an $\mathcal{F}_t$-stopping time.

Exercise 13.5.27. 0 IS AN ABSORBING STATE FOR NON-NEGATIVE MARTINGALES
Prove that for a non-negative martingale $\{M(t)\}_{t\geq}$, $\{M(s) = 0\} \subseteq \{M(t) = 0\}$ whenever
$0 \leq s < t$.

Exercise 13.5.28. AVOIDING 0
Let $\{M(t)\}_{t\geq 0}$ be a right-continuous martingale.

A. Let $\tau := \inf\{t \geq 0\,;\, M(t) = 0\}$. Prove that $M(\tau) = 0$ on $\{\tau < \infty\}$.

B. Prove that if $P(M(T) > 0) = 1$ for some $T \geq 0$, then $P(M(t) > 0$ for all $t \leq T) =$
    1. (Hint: use the stopping times T and $T \wedge \tau$.)

Exercise 13.5.29. p-INTEGRABLE MARTINGALES
If $\{M(t)\}_{t\geq 0}$ is right-continuous and p-integrable, then

$$\sup_{T \in \mathcal{T}} E[|M(T)|^p] < \infty,$$

where $\mathcal{T}$ is the collection of all finite $\mathcal{F}_t$-stopping times.

Exercise 13.5.30. LOCAL MARTINGALES
Prove that if $\{M(t)\}_{t\geq 0}$ is a right-continuous local $\mathcal{F}_t$-martingale such that

$$E\left[\sup_{0\leq s\leq t} |M(s)|\right] < \infty \quad (t \geq 0),$$

it is in fact a martingale.

Chapter 14

A Glimpse at Itô's Stochastic Calculus

The Itô integral is an extension of the Wiener integral to a class of non-deterministic integrands. It is the basic tool of the Itô stochastic calculus, of which this chapter is a brief introduction.

14.1 The Itô Integral

14.1.1 Construction

The Itô integral will be constructed via an isometric extension analogous to that used for the construction of the Wiener–Doob integral.

Let $\mathcal{A}(\mathbb{R}_+)$ be the collection of $\mathcal{F}_t$-progressive complex-valued stochastic processes $\varphi = \{\varphi(t)\}_{t\geq 0}$ such that

$$\mathrm{E}\left[\int_{\mathbb{R}_+} |\varphi(t)|^2 \, \mathrm{d}t\right] < \infty.$$

We view $\mathcal{A}(\mathbb{R}_+)$ as a complex Hilbert space with inner product

$$\langle \varphi_1, \varphi_2 \rangle_{\mathcal{A}(\mathbb{R}_+)} := \mathrm{E}\left[\int_{\mathbb{R}_+} \varphi_1(t)\varphi_2(t)^\star \, \mathrm{d}t\right].$$

(To be more precise, an element of $\mathcal{A}(\mathbb{R}_+)$ is an equivalence class of such processes with respect to the equivalence relation

$$\varphi \sim \varphi' \text{ if and only if } \varphi(t,\omega) = \varphi'(t,\omega), \quad P(\mathrm{d}\omega) \times \mathrm{d}t \text{ a.e.})$$

For $T \geq 0$, let $\mathcal{A}([0,T])$ be the collection of $\mathcal{F}_t$-progressive complex-valued stochastic processes φ such that

$$\mathrm{E}\left[\int_0^T |\varphi(t)|^2 \, \mathrm{d}t\right] < \infty,$$

and let $\mathcal{A}_{loc}$ be the collection of $\mathcal{F}_t$-progressive complex-valued stochastic processes φ such that $\varphi \in \mathcal{A}([0,T])$ for all $T \geq 0$.

Finally, let $\mathcal{B}_{loc}$ be the collection of $\mathcal{F}_t$-progressive complex-valued stochastic processes φ such that P-a.s.

549

© Springer Nature Switzerland AG 2020
P. Brémaud, *Probability Theory and Stochastic Processes*, Universitext,
https://doi.org/10.1007/978-3-030-40183-2_14

$$\int_0^t |\varphi(s)|^2 \mathrm{d}s < \infty \quad \text{P-a.s.} \quad (t \geq 0).$$

(14.1)

Definition 14.1.1 *A real stochastic processes* $\varphi := \{\varphi(t)\}_{t\geq 0}$ *of the form*

$$\varphi(t,\omega) = \sum_{i=1}^{K-1} Z_i(\omega) 1_{(t_i, t_{i+1}]}(t),$$

(14.2)

where $K \in \mathbb{N}_+$, $0 \leq t_1 < t_2 < \cdots < t_K < \infty$ *and where* Z_i $(1 \leq i \leq K)$ *is a complex square-integrable* $\mathcal{F}_{t_i}$-*measurable random variable, is called an* elementary $\mathcal{F}_t$-predictable *process.*

Such stochastic processes are in $\mathcal{A}(\mathbb{R}_+)$.

Lemma 14.1.2 *The vector subspace* $\mathcal{G}$ *of* $\mathcal{A}(\mathbb{R}_+)$ *consisting of the elementary* $\mathcal{F}_t$-*predictable processes is dense in* $\mathcal{A}(\mathbb{R}_+)$.

Proof. First, consider the operators P_n $(n \geq 1)$ acting on the functions $f \in L^2_{\mathbb{C}}(\mathbb{R}_+)$ as follows:

$$[P_n f](t) := n \sum_{i=1}^{n^2} \left(\int_{(i-1)/n}^{i/n} f(s)\, \mathrm{d}s \right) 1_{(i/n, (i+1)/n]}(t).$$

By Schwarz's inequality, for all $t \in (i/n, (i+1)/n]$,

$$\begin{aligned}
|[P_n f](t)|^2 &= \left(\int_{(i-1)/n}^{i/n} n f(s)\, \mathrm{d}s \right)^2 \\
&\leq \int_{(i-1)/n}^{i/n} n^2\, \mathrm{d}s \times \int_{(i-1)/n}^{i/n} f(s)^2\, \mathrm{d}s \\
&= n \int_{(i-1)/n}^{i/n} |f(s)|^2\, \mathrm{d}s,
\end{aligned}$$

and therefore

$$\int_{\mathbb{R}_+} |[P_n f](t)|^2\, \mathrm{d}t \leq \int_{\mathbb{R}_+} |f(t)|^2\, \mathrm{d}t.$$

$(\star)$

Also

$$P_n f \to f \text{ in } L^2_{\mathbb{R}}(\mathbb{R}_+)$$

$(\star\star)$

for all functions $f \in \mathcal{C}^0_c$ (continuous with compact support), and therefore for all $f \in L^2_{\mathbb{R}}(\mathbb{R}_+)$ by density of $\mathcal{C}^0_c$ in $L^2_{\mathbb{R}}(\mathbb{R}_+)$.

Let now φ be in $\mathcal{A}(\mathbb{R}_+)$. For fixed ω, $[P_n \varphi](\cdot, \omega)$ is the function obtained by applying P_n to the function $t \to \varphi(t, \omega)$. By $(\star)$,

$$\mathrm{E}\left[\int_{\mathbb{R}_+} |[P_n \varphi](t)|^2\, \mathrm{d}t \right] \leq \mathrm{E}\left[\int_{\mathbb{R}_+} |\varphi(t)|^2\, \mathrm{d}t \right] < \infty \quad a.s.,$$

and therefore the function $t \to [P_n \varphi](t, \omega)$ is in $L^2_{\mathbb{C}}(\mathbb{R}_+)$ for P-almost all ω. The stochastic process $\{P_n \varphi(t)\}_{t\geq 0}$ is in $\mathcal{G}$ (note that by Theorem 5.3.9, $\int_{(i-1)/n}^{i/n} \varphi(s)\, \mathrm{d}s$ is $\mathcal{F}_{i/n}$-measurable since $\{\varphi(t)\}_{t\geq 0}$ is $\mathcal{F}_t$-progressive).

As $n \uparrow \infty$, $\{[P_n\varphi](t)\}_{t\geq 0}$ converges in $\mathcal{A}(\mathbb{R}_+)$ to $\{\varphi(t)\}_{t\geq 0}$. In fact, by $(\star\star)$,

$$\int_{\mathbb{R}_+} |[P_n\varphi(\cdot,\omega)](t) - \varphi(t,\omega)|^2 \, dt \to 0$$

and therefore

$$\|P_n\varphi - \varphi\|^2_{\mathcal{A}(\mathbb{R}_+)} = \mathrm{E}\left[\int_{\mathbb{R}_+} |[P_n\varphi(\cdot,\omega)](t) - \varphi(t,\omega)|^2 \, dt\right] \to 0$$

(by dominated convergence since, by $(\star\star)$ and the triangle inequality,

$$\int_{\mathbb{R}_+} |[P_n\varphi(\cdot,\omega)](t) - \varphi(t,\omega)|^2 \, dt \leq \left(\|[P_n\varphi(\cdot,\omega)]\|_{L^2_{\mathbb{C}}(\mathbb{R}_+)} + \|\varphi(\cdot,\omega)\|_{L^2_{\mathbb{C}}(\mathbb{R}_+)}\right)^2$$

$$\leq 4\|\varphi(\cdot,\omega)\|^2_{L^2_{\mathbb{C}}(\mathbb{R}_+)}$$

and $\mathrm{E}\left[\varphi(\cdot,\omega)\|^2_{L^2_{\mathbb{C}}(\mathbb{R}_+)}\right] = \|\varphi\|^2_{\mathcal{A}(\mathbb{R}_+)} < \infty$). $\qquad\square$

Let $L^2_{0,\mathbb{C}}(P)$ be the Hilbert subspace of $L^2_{\mathbb{C}}(P)$ consisting of the complex centered square-integrable variables. Define the mapping $I : \mathcal{G} \to L^2_{0,\mathbb{C}}(P)$ by

$$I(\varphi) := \sum_{i=1}^{K-1} Z_i\left(W(t_{i+1}) - W(t_i)\right). \tag{14.3}$$

One verifies that for all $\varphi \in \mathcal{G}$, $\mathrm{E}[I(\varphi)] = 0$. Also, for all $\varphi_1, \varphi_2 \in \mathcal{G}$,

$$\mathrm{E}[I(\varphi_1)I(\varphi_2)^\star] = \langle\varphi_1, \varphi_2\rangle_{\mathcal{A}(\mathbb{R}_+)}.$$

Proof. By polarization, it suffices to treat the case $\varphi_1 = \varphi_2 = \varphi$ and to write

$$\mathrm{E}\left[I(\varphi)^2\right] = \sum_{i=1}^{K-1} \mathrm{E}\left[|Z_i|^2(W(t_{i+1}) - W(t_i))^2\right]$$

$$+ 2\sum_{i<\ell;i,\ell=1}^{K-1} \mathrm{E}\left[Z_i Z_\ell^\star(W(t_{i+1}) - W(t_i))(W(t_{\ell+1}) - W(t_\ell))\right]$$

$$= \sum_{i=1}^{K-1} \mathrm{E}\left[|Z_i|^2\right]\mathrm{E}\left[(W(t_{i+1}) - W(t_i))^2\right]$$

$$+ 2\sum_{i<\ell;i,\ell=1}^{K-1} \mathrm{E}\left[Z_i(W(t_{i+1}) - W(t_i))Z_\ell^\star\right]\mathrm{E}\left[(W(t_{\ell+1}) - W(t_\ell))\right]$$

$$= \sum_{i=1}^{K-1} \mathrm{E}\left[|Z_i|^2\right]\mathrm{E}\left[(W(t_{i+1}) - W(t_i))^2\right]$$

$$= \sum_{i=1}^{K-1} \mathrm{E}\left[|Z_i|^2\right](t_{i+1} - t_i) = \mathrm{E}\left[\int_{\mathbb{R}_+} |\varphi(t)|^2 \, dt\right],$$

where it was observed that $W(t_{\ell+1}) - W(t_\ell)$ $(i < \ell)$ is independent of Z_i, $W(t_{i+1}) - W(t_i)$ and Z_ℓ, since all these variables are $\mathcal{F}_{t_\ell}$-measurable. $\qquad\square$

The mapping I is therefore a linear isometry from $\mathcal{G}$ into $L^2_{0,\mathbb{C}}(P)$. Since $\mathcal{G}$ is a dense subset of $\mathcal{A}(\mathbb{R}_+)$, it can be uniquely extended to a linear isometry (Theorem C.3.2), still denoted by I, from $\mathcal{A}(\mathbb{R}_+)$ into $L^2_{0,\mathbb{C}}(P)$. For $\varphi \in \mathcal{A}(\mathbb{R}_+)$, the square-integrable random variable $I(\varphi)$ will be denoted by

$$\int_{\mathbb{R}_+} \varphi(t)\,\mathrm{d}W(t)$$

and called the *Itô integral* of $\{\varphi(t)\}_{t \geq 0}$ with respect to the $\mathcal{F}_t$-Wiener process $\{W(t)\}_{t \in [0,1]}$. Since $I(\varphi)$ is in $L^2_{0,\mathbb{C}}(P)$,

$$\mathrm{E}\left[\int_{\mathbb{R}_+} \varphi(t)\,\mathrm{d}W(t)\right] = 0. \tag{14.4}$$

The isometry is expressed by

$$\mathrm{E}\left[\left(\int_{\mathbb{R}_+} \varphi(t)\,\mathrm{d}W(t)\right)\left(\int_{\mathbb{R}_+} \psi(t)\,\mathrm{d}W(t)\right)^*\right] = \mathrm{E}\left[\int_{\mathbb{R}_+} \varphi(t)\psi(t)^*\,\mathrm{d}t\right], \tag{14.5}$$

for all φ, ψ in $\mathcal{A}(\mathbb{R}_+)$.

Remark 14.1.3 If $\varphi \in \mathcal{A}(\mathbb{R}_+)$ is moreover continuous, then

$$\int_{\mathbb{R}_+} \varphi(t)\,\mathrm{d}W(t) = \lim_{n \uparrow \infty} \sum_{i=1}^{n^2} \varphi\left(\frac{i}{n}\right)\left(W\left(\frac{i+1}{n}\right) - W\left(\frac{i}{n}\right)\right) \text{ in } L^2_{\mathbb{C}}(P)$$

(Exercise 14.4.8).

14.1.2   Properties of the Itô Integral Process

For $\varphi \in \mathcal{A}_{loc}$ and $t \in \mathbb{R}_+$, let

$$\int_0^t \varphi(s)\,\mathrm{d}W(s) := \int_{\mathbb{R}_+} 1_{\{s \leq t\}}\varphi(s)\,\mathrm{d}W(s).$$

Theorem 14.1.4 *For $\varphi \in \mathcal{A}_{loc}$, the stochastic process $\{\int_0^t \varphi(s)\,\mathrm{d}W(s)\}_{t \geq 0}$ is a square-integrable $\mathcal{F}_t$-martingale.*

Proof. It is enough to show this for $\varphi \in \mathcal{A}(\mathbb{R}_+)$. Let $\{\varphi_n\}_{n \geq 1}$ be a sequence in $\mathcal{G}$ approximating φ in $\mathcal{A}(\mathbb{R}_+)$. One verifies that for all $n \geq 1$, $\int_0^t \varphi_n(s)\,\mathrm{d}W(s)$ $(t \geq 0)$ is an $\mathcal{F}_t$-square-integrable martingale. Therefore for all $0 \leq s \leq t$ and all $A \in \mathcal{F}_s$,

$$\mathrm{E}\left[1_A \int_s^t \varphi_n(u)\,\mathrm{d}W(u)\right] = 0,$$

and therefore, passing to the limit in $L^2_{\mathbb{C}}(P)$, by continuity of the inner product,

$$\mathrm{E}\left[1_A \int_s^t \varphi(u)\,\mathrm{d}W(u)\right] = 0.$$

This being true for all $A \in \mathcal{F}_s$, $\mathrm{E}\left[\int_s^t \varphi(u)\,\mathrm{d}W(u) \mid \mathcal{F}_s\right] = 0$. $\qquad \square$

When $\varphi \in \mathcal{A}(\mathbb{R}_+)$, by (14.5),

$$\sup_{t \geq 0} \mathrm{E}\left[\left|\int_0^t \varphi(s)\,dW(s)\right|^2\right] = \mathrm{E}\left[\int_{\mathbb{R}_+} |\varphi(t)|^2\,dt\right] < \infty$$

and therefore (Theorem 4.2.14), $\{\int_0^t \varphi(s)\,dW(s)\}_{t \geq 0}$ is a uniformly integrable martingale.

Theorem 14.1.5 *Let $\{W(t)\}_{t \geq 0}$ be an $\mathcal{F}_t$-Wiener process and let $\varphi \in \mathcal{A}(\mathbb{R}_+)$. Let $M(t) := \int_0^t \varphi(s)\,dW(s)$ and $\langle M \rangle(t) := \int_0^t \varphi(s)^2\,ds$ $(t \geq 0)$. The stochastic process $\{M(t)^2 - \langle M \rangle(t)\}_{t \geq 0}$ is an $\mathcal{F}_t$-martingale.*

Proof. Since

$$\mathrm{E}\left[M(t)^2 - M(s)^2 \mid \mathcal{F}_s\right] = \mathrm{E}\left[(M(t) - M(s))^2 \mid \mathcal{F}_s\right],$$

it suffices to show that

$$\mathrm{E}\left[(M(t) - M(s))^2 \mid \mathcal{F}_s\right] := \mathrm{E}\left[\int_s^t \varphi(u)\,dW(u))^2 \mid \mathcal{F}_s\right] = \mathrm{E}\left[\int_s^t \varphi(u)^2\,du) \mid \mathcal{F}_s\right].$$

Convergence in L^2 implies convergence in L^1 of the squares. Therefore, with the obvious notation (see Definition 14.1.1 and the construction of the Itô integral), by continuity in L^1 of conditional expectation,

$$\mathrm{E}\left[\int_s^t \varphi(u)\,dW(u))^2 \mid \mathcal{F}_s\right] = \mathrm{E}\left[\lim_{n \uparrow \infty}\left(\sum_{i=1}^n Z_i \Delta W_i\right)^2 \mid \mathcal{F}_s\right]$$

$$= \lim_{n \uparrow \infty} \mathrm{E}\left[\left(\sum_{i=1}^n Z_i \Delta W_i\right)^2 \mid \mathcal{F}_s\right]$$

$$= \lim_{n \uparrow \infty}\left(\sum_i \mathrm{E}\left[\mathrm{E}\left[Z_i^2 (\Delta W_i)^2 \mid \mathcal{F}_{t_i}\right] \mid \mathcal{F}_s\right] + 2\sum_{i<j} \mathrm{E}\left[\mathrm{E}\left[Z_i Z_j \Delta W_i \Delta W_j \mid \mathcal{F}_{t_j}\right] \mid \mathcal{F}_s\right]\right)$$

$$= \lim_{n \uparrow \infty} \sum_i \mathrm{E}\left[\mathrm{E}\left[Z_i^2 (\Delta W_i)^2 \mid \mathcal{F}_{t_i}\right] \mid \mathcal{F}_s\right]$$

$$= \lim_{n \uparrow \infty} \sum_i \mathrm{E}\left[Z_i^2 (t_i - t_{i-1}) \mid \mathcal{F}_s\right] = \mathrm{E}\left[\int_s^t \varphi(u)^2\,du \mid \mathcal{F}_s\right].$$

$\square$

The Itô integrals are defined *globally* over $(\Omega, \mathcal{F}, P)$, as limits in quadratic mean and *not trajectorywise* (ω by ω). If they were defined trajectorywise, then for any random time τ we would have

$$\int_0^{\tau(\omega)} \varphi(s, \omega)\,dW(s, \omega) = \int_0^t \mathbf{1}_{\{s \leq \tau(\omega)\}} \varphi(s, \omega)\,dW(s, \omega) \quad (\omega \in \Omega),$$

that is to say, for each $\omega \in \Omega$, the value at $\tau(\omega)$ of the process $X(t) := \int_0^t \varphi(s)\,dW(s)$ is equal to the value at t of the process $\int_0^t \mathbf{1}_{\{s \leq \tau\}} \varphi(s)\,dW(s)$ $(t \geq 0)$. In fact this is true, although not obvious, when τ is an $\mathcal{F}_t$-stopping time:

Theorem 14.1.6 *Let* $\{W(t)\}_{t\geq 0}$ *be an* $\mathcal{F}_t$-*Wiener process and let* $\varphi \in \mathcal{A}(\mathbb{R}_+)$. *Let* τ *be a bounded* $\mathcal{F}_t$-*stopping time. Then*

$$X(\tau) = \int_{\mathbb{R}_+} 1_{\{s\leq\tau\}}\varphi(s)\,\mathrm{d}W(s).$$

Proof. Since

$$\sup_{t\geq 0} \mathrm{E}[|X(t)|^2] = \sup_{t\geq 0}\mathrm{E}\left[\int_0^t \varphi(s)^2\,\mathrm{d}s\right] \leq \mathrm{E}\left[\int_0^\infty \varphi(s)^2\,\mathrm{d}s\right] < \infty,$$

the martingale $\{X(t)\}_{t\geq 0}$ is uniformly integrable (Theorem 4.2.14). In particular, by optional sampling (Corollary 13.2.8),

$$X(\tau) = \mathrm{E}\left[X(\infty)\,|\,\mathcal{F}_\tau\right] = \mathrm{E}\left[\int_0^\infty \varphi(s)\,\mathrm{d}W(s)\,|\,\mathcal{F}_\tau\right].$$

We have to show that for all $A \in \mathcal{F}_\tau$,

$$\mathrm{E}\left[1_A \int_0^\infty \varphi(s)\,\mathrm{d}W(s)\right] = \mathrm{E}\left[1_A \int_0^\infty \varphi(s)\,1_{\{s\leq\tau\}}\,\mathrm{d}W(s)\right]$$

or, equivalently, that

$$\mathrm{E}\left[\int_0^\infty 1_A 1_{(\tau,\infty)}(s)\,\varphi(s)\,\mathrm{d}W(s)\right] = 0.$$

But the process $\{\varphi(t)1_A 1_{(\tau,\infty)}(t)\}_{t\geq 0}$ is in $\mathcal{A}(\mathbb{R}_+)$ and the result then follows by (14.4). $\square$

Theorem 14.1.7 *The stochastic process* $\{\int_0^t \varphi(s)\,\mathrm{d}W(s)\}_{t\geq 0}$, *where* $\varphi \in \mathcal{A}(\mathbb{R}_+)$, *is continuous in the quadratic mean and admits a continuous version.*

Proof. The continuity in the quadratic mean follows from

$$\mathrm{E}\left[\left|\int_0^b \varphi(s)\,\mathrm{d}W(s) - \int_0^a \varphi(s)\,\mathrm{d}W(s)\right|^2\right] = \mathrm{E}\left[\left|\int_a^b \varphi(s)\,\mathrm{d}W(s)\right|^2\right]$$
$$= \mathrm{E}\left[\int_a^b \varphi(s)^2\,\mathrm{d}s\right],$$

a quantity which tends to 0 as either $b \downarrow a$ or $a \uparrow b$. The stochastic process

$$M_n(t) := \int_0^t [P_n\varphi](s)\,\mathrm{d}W(s) \quad (t \geq 0)$$

is a continuous $\mathcal{F}_t$-martingale and in particular, by Kolmogorov's inequality (Theorem 13.47), for all $T > 0$ and all $a > 0$,

$$P\left(\sup_{t\in[0,T]} |M_n(t) - M_m(t)| > a\right) \leq \frac{1}{a^2}\mathrm{E}\left[\int_0^T \left([P_n\varphi](s) - [P_m\varphi](s)\right)^2\,\mathrm{d}t\right],$$

a quantity that tends to 0 as $n, m \to \infty$. We can therefore find a sequence $\{n_k\}_{k\geq 1}$ strictly increasing to ∞ such that

$$P\left(\sup_{t\in[0,T]} |M_{n_k}(t) - M_{n_{k-1}}(t)| > 2^{-k}\right) \le 2^{-k},$$

so that, by the Borel–Cantelli lemma,

$$P\left(\sup_{t\in[0,T]} |M_{n_k}(t) - M_{n_{k-1}}(t)| > 2^{-k} \text{ i.o.}\right) = 0.$$

Therefore for P-almost all ω, there exists a finite integer $K(\omega)$ such that for $k > K(\omega)$,

$$\sup_{t\in[0,T]} |M_{n_k}(t) - M_{n_{k-1}}(t)| \le 2^{-k}.$$

This implies that for P-almost all ω the function $t \mapsto M_{n_k}(t,\omega)$ converges uniformly on $[0,T]$ to a function $t \mapsto M_\infty(t,\omega)$ which is continuous (as a uniform limit of continuous functions). Since $M_n(t) \to \int_0^t \varphi(s)\,dW(s)$ in quadratic mean and $M_n(t) \to M_\infty(t)$ P-a.s., both limits are P-a.s. equal, and therefore the stochastic process $\{M_\infty(t)\}_{t\ge 0}$ is a continuous version of $\{\int_0^t \varphi(s)\,dW(s)\}_{t\ge 0}$. □

14.1.3 Itô's Integrals Defined as Limits in Probability

It is possible to define the integral $\int_0^t \varphi(s)\,dW(s)$ when φ is only in $\mathcal{B}_{loc}$, not necessarily in $\mathcal{A}_{loc}$. For this, define for each $n \ge 1$ the $\mathcal{F}_t$-stopping time

$$T_n := \inf\left\{t; \int_0^t |\varphi(s)|^2\,ds \ge n\right\}$$

with the usual convention $\inf \varnothing = \infty$. Clearly, P-a.s., $T_n \uparrow \infty$ and $\int_0^{T_n} |\varphi(s)|^2\,ds \le n$. In particular, the stochastic process

$$\varphi_n(t) := \varphi(t)\,1_{\{t\le T_n\}} \quad (t \ge 0) \tag{14.6}$$

is in $\mathcal{A}(\mathbb{R}_+)$ and $I_n(t) := \int_{\mathbb{R}_+} \varphi_n(s)\,dW(s)$ is therefore well defined. For any n, m and for any $\varepsilon > 0$,

$$P\left(\left|\int_0^t \varphi_n(s)\,dW(s) - \int_0^t \varphi_m(s)\,dW(s)\right| \ge \varepsilon\right)$$
$$\le P\left(\int_0^t |\varphi(s)|^2\,ds \ge min(n,m)\right) \to 0 \text{ as } n, m \uparrow \infty.$$

By Cauchy's criterion of convergence in probability (Theorem 4.2.3), for each $t \ge 0$, $I_n(t)$ converges in probability to some random variable denoted $I(\varphi, t)$.

If $\varphi \in \mathcal{A}(\mathbb{R}_+)$, $\lim_{n\uparrow\infty} I_n(t) = \int_0^t \varphi(s)\,dW(s)$ in $L^2_{\mathbb{C}}(P)$ and therefore also in probability. Therefore, in this case $I(\varphi) = \int_0^t \varphi(s)\,dW(s)$. Therefore, for $\varphi \in \mathcal{B}_{loc}$,

$$\int_0^t \varphi(s)\,dW(s) := \lim_{n\uparrow\infty} \int_0^t \varphi_n(s)\,dW(s),$$

where the limit is in probability, is an extension of the definition of the Itô integral from integrands in $\mathcal{A}(\mathbb{R}_+)$ to integrands in $\mathcal{B}_{loc}$.

The following result is a direct consequence of Theorems 14.1.7 and 14.1.6 (Exercise 14.4.6).

Theorem 14.1.8 *Let* $\{W(t)\}_{t\geq0}$ *be an* $\mathcal{F}_t$-*Wiener process and let* $\varphi \in \mathcal{B}_{loc}$. *The process* $\{X(t)\}_{t\geq0} := \{\int_0^t \varphi(s)\,dW(s)\}_{t\geq0}$ *is then an* $\mathcal{F}_t$-*local martingale which admits a continuous version. Moreover, for any* $\mathcal{F}_t$-*stopping time* τ,

$$X(t \wedge \tau) = \int_0^t \varphi(s) 1_{\{s\leq\tau\}}\,dW(s) \quad (t \geq 0).$$
(14.7)

14.2 Itô's Differential Formula

14.2.1 Elementary Form

The ordinary rules of calculus do not apply to functions of Brownian motion, as the following example shows:

EXAMPLE 14.2.1: SQUARED BROWNIAN MOTION. Let $\{W(t)\}_{t\geq0}$ be an $\mathcal{F}_t$-Wiener process. With $t_i := \frac{it}{n}$,

$$W(t)^2 = \sum_{i=1}^n (W(t_i)^2 - W(t_{i-1})^2)$$

$$= 2\sum_{i=1}^n W(t_{i-1})(W(t_i) - W(t_{i-1})) + \sum_{i=1}^n ((W(t_i) - W(t_{i-1}))^2 := A_n + B_n\,.$$

As $n \uparrow \infty$, using Remark 14.1.3, A_n converges in $L^2(P)$ to $2\int_0^t W(s)\,dW(s)$, whereas B_n converges to t (Theorem 11.2.10). Therefore

$$W(t)^2 = 2\int_0^t W(s)\,dW(s) + t\,.$$

If the trajectories of the Wiener process were of bounded variation, integration by parts would give the (wrong) formula $W(t)^2 = 2\int_0^t W(s)\,dW(s)$.

Functions of Brownian Motion

Let $\mathcal{C}_b^2$ denote the collection of functions $F : \mathbb{R} \to \mathbb{C}$ that are twice continuously differentiable and such that F, F' and F'' are bounded.

Theorem 14.2.2 *Let* $\{W(t)\}_{t\geq0}$ *be a standard* $\mathcal{F}_t$-*Wiener process and let* $F \in \mathcal{C}_b^2$. *Then*

$$F(W(t)) = F(W(0)) + \int_0^t F'(W(s))\,dW(s) + \frac{1}{2}\int_0^t F''(W(s))\,ds\,.$$
(14.8)

Proof. It suffices to treat the case of real-valued functions F. Let $t_i := \frac{it}{n}$. Taylor's formula at the second order gives

$$F(W(t)) - F(W(0)) = \sum_{i=1}^n (F(W(t_i)) - F(W(t_{i-1})))$$

$$= \sum_{i=1}^n F'(W(t_i))(W(t_i) - W(t_{i-1})) + \frac{1}{2}\sum_{i=1}^n F''(\xi_i)(W(t_i) - W(t_{i-1}))^2$$

$$= \sum_{i=1}^n F'(W(t_i))(W(t_i) - W(t_{i-1})) + \frac{1}{2}\sum_{i=1}^n F''(W(\theta_i))(W(t_i) - W(t_{i-1}))^2,$$

where $\xi_i \in [W(t_{i-1}), W(t_i)]$ (and therefore, since the trajectories of the Brownian motion are continuous, $\xi_i = W(\theta_i)$ for some $\theta_i = \theta_i(n, \omega) \in (t_{i-1}, t_i)$. The first term on the right-hand side tends in $L^2(P)$ to $\int_0^t F''(W(s)) \, dW(s)$ (Remark 14.1.3). Let now

$$A_n := \sum_{i=1}^n F''(W(\theta_i))(W(t_i) - W(t_{i-1}))^2,$$

$$B_n := \sum_{i=1}^n F''(W(t_{i-1}))(W(t_i) - W(t_{i-1}))^2,$$

$$C_n := \sum_{i=1}^n F''(W(t_{i-1}))(t_i - t_{i-1}).$$

By Schwarz's inequality, dominated convergence and Theorem 11.2.10,

$$\mathrm{E}\left[|A_n - B_n|\right] \leq \mathrm{E}\left[\sup_i |F''(W(\theta_i)) - F''(W(t_{i-1}))| \times \sum_i (W(t_i) - W(t_{i-1}))^2\right]$$

$$\leq \left(\mathrm{E}\left[\sup_i |F''(W(\theta_i)) - F''(W(t_i))|^2\right] \times \mathrm{E}\left[\left(\sum_i (W(t_i) - W(t_{i-1}))^2\right)^2\right]\right)^{\frac{1}{2}}$$

$$\to (0 \times t^2)^{\frac{1}{2}} = 0.$$

Now,

$$\mathrm{E}\left[|B_n - C_n|^2\right] \leq (\sup F'')^2 \times \sum_i \mathrm{E}\left[|(W(t_i) - W(t_{i-1}))^2 - (t_i - t_{i-1})|^2\right]$$

$$= (\sup F'')^2 \times 2 \sum_i (t_i - t_{i-1})^2 \to 0.$$

Finally, $C_n \to \int_0^t F''(W(s)) \, dW(s)$ in $L^2(P)$, and consequently in $L^1(P)$. Therefore, for all t, the announced equality holds in $L^1(P)$ and consequently P-almost surely. Since the stochastic processes in both sides of the equality are continuous, we have that P-almost surely, this equality holds for all $t \in \mathbb{R}_+$.   □

Remark 14.2.3 Theorem 14.2.2 remains true if we only suppose that $\{W(t)\}_{t\geq 0}$ is a real continuous $\mathcal{F}_t$-martingale such that $\{W(t)^2 - t\}_{t\geq 0}$ is also an $\mathcal{F}_t$-martingale. We shall not prove this, although the proof is very close to the one given in the special case.

_____

EXAMPLE 14.2.4: ITÔ'S RULE FOR EXPONENTIALS. Let $F(x, t) = e^x$ and $X(t) = \int_0^t \varphi(s) \, dW(s) - \frac{1}{2}\int_0^t \varphi(s)^2 ds$ where $\varphi \in \mathcal{B}_{loc}$. Application of rule (14.11) yields

$$L(t) := \exp\left(\int_0^t \varphi(s) \, dW(s) - \frac{1}{2}\int_0^t \varphi(s)^2 ds\right) = 1 + \int_0^t L(s)\varphi(s) \, dW(s). \qquad (14.9)$$

_____

Lévy's Characterization of Brownian Motion

Theorem 14.2.5 *A real continuous $\mathcal{F}_t$-adapted stochastic process $\{W(t)\}_{t \geq 0}$ that is an $\mathcal{F}_t$-martingale and such that $\{W(t)^2 - t\}_{t \geq 0}$ is also an $\mathcal{F}_t$-martingale is a standard $\mathcal{F}_t$-Wiener process.*

Proof. Applying Itô's differentiation rule[1] with $F(x) = e^{iux}$ $(u \in \mathbb{R})$,

$$e^{iuW(t)} - e^{iuW(s)} = iu \int_s^t e^{iuW(z)} dW(z) - \frac{1}{2} u^2 \int_s^t e^{iuW(z)} dz \quad (0 \leq s \leq t).$$

Now, since $E[\int_0^t | e^{iuW(s)} |^2 ds] = t < \infty$, it follows that $\int_0^t e^{iuW(s)} dW(s)$ is an $\mathcal{F}_t$-martingale and therefore, for all $A \in \mathcal{F}_s$,

$$\int_A (e^{iuW(t)} - e^{iuW(s)}) \, dP = -\frac{1}{2} u^2 \int_A \int_s^t e^{iuW(z)} \, dz \, dP. \tag{14.10}$$

Dividing both sides of the above equation by $e^{iuW(s)}$ and applying Fubini's theorem to the right-hand side,

$$\int_A e^{iu(W(t) - W(s))} \, dP = P(A) - \frac{1}{2} u^2 \int_s^t \int_A e^{iu(W(z) - W(s))} \, dP \, dz,$$

and therefore

$$\int_A e^{iu(W(t) - W(s))} dP = P(A) e^{-\frac{1}{2} \frac{u^2}{t-s}}. \tag{$\star$}$$

This equality is valid for all $0 \leq s \leq t$, all $u \in \mathbb{R}$ and all $A \in \mathcal{F}_s$. With $A = \Omega$,

$$E[e^{iu(W(t) - W(s))}] = e^{-(1/2)/u^2/(t-s)},$$

that is, $W(t) - W(s)$ is a centered Gaussian variable with unit variance. Equation $(\star)$ then reads

$$E[1_A e^{iu(W(t) - W(s))}] = P(A) E[e^{iu(W(t) - W(s))}] \quad (A \in \mathcal{F}_s),$$

from which it follows that $W(t) - W(s)$ is independent of $\mathcal{F}_s$. □

14.2.2   Some Extensions

Theorem 14.2.2 can be extended in several directions that do not involve new ideas. The proofs, using arguments very similar to those in the proof of Theorem 14.2.2 are therefore omitted.

Theorem 14.2.2 dealt with functions of the Brownian motion. We now consider functions of an Itô process, whose definition follows.

Definition 14.2.6 *A stochastic process of the form*

$$X(t) := X(0) + \int_0^t f(s) \, ds + \int_0^t \varphi(s) \, dW(s) \quad (t \geq 0), \tag{$\star$}$$

where $X(0)$ is an $\mathcal{F}_0$-measurable random variable and $\{\varphi(t)\}_{t \geq 0}$ and $\{f(t)\}_{t \geq 0}$ are $\mathcal{F}_t$-progressively measurable stochastic processes in $\mathcal{A}_{loc}$ or $\mathcal{B}_{loc}$ is called an Itô process.

---

[1]See Remark 14.2.3.

Theorem 14.2.7 *Suppose that* $\varphi, f \in \mathcal{A}_{loc}$ *and* $X(0)$ *is square integrable. Then, for all functions* $F \in \mathcal{C}^2$,

$$F(X(t) = F(X(0)) + \int_0^t F'(X(s))\varphi(s)\,dW(s)$$
$$+ \int_0^t F'(X(s))f(s)\,ds + \frac{1}{2}\int_0^t F''(X(s))\varphi(s)^2\,ds\,.$$

With the notation, $dX(s) := \varphi(s)\,dW(s) + \psi(s)\,ds$, this formula can written as

$$F(X(t)) = F(X(0)) + \int_0^t F'(X(s))\,dX(s) + \frac{1}{2}\int_0^t F''(X(s))\varphi(s)^2\,ds\,.$$

Therefore, with

$$\langle X \rangle(t) := \int_0^t \varphi(s)^2\,ds$$

(the bracket process of the martingale part of the Itô process),

$$F(X(t) = F(X(0)) + \int_0^t F'(X(s))\,dX(s) + \frac{1}{2}\int_0^t F''(X(s))\,d\langle X \rangle(s)$$

or, in differential form,

$$dF(X(t)) = F'(X(t))\,dX(t) + F''(X(t))\,d\langle X \rangle(t)\,.$$

EXAMPLE 14.2.8: THE GEOMETRIC BROWNIAN MOTION.
Let $Z(t) := \exp\{\sigma W(t) + \mu t\}$ $(t \geq 0)$. By Itô's differential rule,

$$Z(t) = \sigma \int_0^t Z(s)\,dW(s) + \left(\mu + \frac{\sigma^2}{2}\right)\int_0^t Z(s)\,ds\,.$$

In particular, with $\mu = -\frac{\sigma^2}{2}$, the process

$$Z(t) := \exp\left\{\sigma W(t) - \frac{\sigma^2}{2}t\right\} \quad (t \geq 0)$$

is a martingale, called the *geometric Brownian motion*

Let $\{W(t)\}_{t \geq 0}$ be an $\mathcal{F}_t$-Wiener process. Let $\{\varphi(t)\}_{t \geq 0}$ and $\{f(t)\}_{t \geq 0}$ be in $\mathcal{B}_{loc}$. Let $\{X(t)\}_{t \geq 0}$ be an Itô process.

Theorem 14.2.9 *Let* $F : (x, t) \in \mathbb{R}^2 \to F(x, t) \in \mathbb{C}$ *be twice continuously differentiable in the first variable* x *and once continuously differentiable in the second variable* t. *Then*

$$F(X(t), t) = F(X(0), 0) + \int_0^t \frac{\partial F}{\partial t}(X(s), s)\,ds + \int_0^t \frac{\partial F}{\partial x}(X(s), s)\,dX(s)$$
$$+ \frac{1}{2}\int_0^t \frac{\partial^2 F}{\partial x^2}(X(s), s)\,\varphi(s)^2\,ds, \tag{14.11}$$

where

$$\int_0^t \frac{\partial F}{\partial x}(X(s), s)\,dX(s) := \int_0^t \frac{\partial F}{\partial x}(X(s), s)\,\varphi(s)\,dW(s) + \int_0^t \frac{\partial F}{\partial x}(X(s), s)\,f(s)\,ds\,.$$
$$\tag{14.12}$$

A Finite Number of Discontinuities

The Itô differentiation rule remains valid in situations where the functions F, F' and F'' are $\mathcal{C}^2$ in the x argument only on $\mathbb{R}\backslash E$, where E is a finite set $E = \{x_1, \ldots, x_k\}$.

The method of proof is the same as the following one in the simple case of a function of the Brownian motion.

Theorem 14.2.10 *Let $F : \mathbb{R} \to \mathbb{R}$ be $\mathcal{C}^1$ on $\mathbb{R}$ and $\mathcal{C}^2$ on $\mathbb{R}\backslash E$, where $E = \{x_1, \ldots, x_k\}$. If F'' remains bounded on a neighborhood of these points, the Itô formula applies:*

$$F(W(t)) = F(W(0)) + \int_0^t F'(W(s))\, \mathrm{d}W(s) + \frac{1}{2}\int_0^t F''(W(s))\, \mathrm{d}s. \qquad (14.13)$$

Concerning the last (Lebesgue) integral, note that the Lebesgue measure of $\{t\,;\, W(t, \omega) \in E\}$ is almost surely null.

Proof. We first show the existence of a sequence of functions F_n that are $\mathcal{C}^2$ in $\mathbb{R}$ with the following properties:

(i) $F_n \to F$ and $F'_n \to F'$ uniformly in $\mathbb{R}$, and

(ii) for all $x \notin E$, $F''_n(x) \to F''(x)$ and F''_n is uniformly bounded on a neighborhood of E.

Let α be some function in $\mathcal{C}^\infty$ with a compact support, equal to 1 in a neighborhood of E. In particular, $(1 - \alpha)F$ is in $\mathcal{C}^2$ and the Itô formula applies for this function. It remains to show that it also applies to αF, or more generally to F as before, but with compact support. For such a function, consider the approximation

$$F_n := F * \varphi_n$$

where $\varphi_n(x) := n\varphi(nx)$ and φ is $\mathcal{C}^\infty$ with compact support and such that $0 \leq \varphi \leq 1$. Such a function satisfies requirements (i) and (ii) above, and therefore the Itô formula applies:

$$F_n(W(t)) = F_n(W(0)) + \int_0^t F'_n(W(s))\, \mathrm{d}W(s) + \frac{1}{2}\int_0^t F''_n(W(s))\, \mathrm{d}s.$$

The terms of this equality converge as $n \uparrow \infty$ to the corresponding terms of (14.13), the first two terms by uniform convergence of the F_n's, the third by uniform convergence of the F'_n's and the isometry formula. For the third term, observe that, by Schwarz's inequality,

$$\mathrm{E}\left[\left(\int_0^t F_n(W(s))\, \mathrm{d}s - \int_0^t F''_n(W(s))\, \mathrm{d}s\right)^2\right]$$

$$\leq t \int_0^t \mathrm{E}\left[|F_n(W(s)) - F_n(W(s))|^2\right]\, \mathrm{d}s,$$

a quantity that tends to 0 by dominated convergence. $\qquad \square$

The Vectorial Differentiation Rule

Let $C^{1,2}(\mathbb{R}_+ \times \mathbb{R}^d)$ denote the collection of functions $F : \mathbb{R}_+ \times \mathbb{R}^d \to \mathbb{R}$ that are once continuously differentiable in the first coordinate and twice continuously differentiable in the second coordinate. Let $\{W(t)\}_{t\geq 0}$ be a k-dimensional standard Wiener process. Let

$$\varphi(t) := \{\varphi_{i,j}(t)\}_{1\leq i\leq d, 1\leq j\leq k} \quad (t \geq 0)$$

be a real $d \times k$-matrix valued stochastic process such that for all $1 \leq i \leq d$ and all $1 \leq j \leq k$ the process $\{\varphi_{i,j}(t)\}_{t\geq 0}$ is $\mathcal{F}_t^W$-adapted and in $\mathcal{B}_{loc}$. Denote by $\int_0^t \varphi(s)\,dW(s)$ $(t \geq 0)$ the d-dimensional stochastic process whose i-th component is $\sum_{j=1}^k \int_0^t \varphi_{i,j}(s)\,dW_j(s)$ $(t \geq 0)$. Let

$$\psi(t) := \{\psi_i(t)\}_{1\leq i\leq d} \quad (t \geq 0)$$

be a real d-dimensional stochastic process such that for all $1 \leq i \leq d$ the process $\{\psi_i(t)\}_{t\geq 0}$ is $\mathcal{F}_t$-adapted and in $\mathcal{B}_{loc}$. Define the d-dimensional Itô process $\{X(t)\}_{t\geq 0}$ by

$$X(t) := X(0) + \int_0^t \varphi(s)\,dW(s) + \int_0^t \psi(s)\,ds,$$

where $X(0)$ is a vector of integrable random variables. Then:

Theorem 14.2.11 *Under the above conditions, for $F : \mathbb{R}_+ \times \mathbb{R}^d \to \mathbb{R}$ once continuously differentiable in the first coordinate and twice continuously differentiable in the second coordinate, we have the formula*

$$F(t, X(t)) = F(0, X(0)) + \int_0^t \frac{\partial}{\partial s} F(s, X(s))\,ds + \sum_{i=1}^d \int_0^t \frac{\partial}{\partial x_i} F(s, X(s))\psi_i(s)\,ds$$

$$+ \sum_{i=1}^d \int_0^t \frac{\partial}{\partial x_i} F(s, X(s)) \sum_{j=1}^k \varphi_{i,j}(s)\,dW_j(s)$$

$$+ \frac{1}{2} \int_0^t \sum_{i,\ell} \frac{\partial^2}{\partial x_i \partial x_\ell} F(s, X(s)) \left(\sum_{j=1}^k \varphi_{i,j}(s)\varphi_{\ell,j}(s) \right) ds.$$

EXAMPLE 14.2.12: AN INTEGRATION BY PARTS FORMULA. Let for $i = 1, 2$

$$X_i(t) = X_i(0) + \int_0^t \varphi_i(s)\,dW(s) + \int_0^t \psi_i(s)\,ds \quad (t \geq 0)$$

where the φ_i's and ψ_i's satisfy the conditions of Definition 14.2.6. Then

$$X_1(t)X_2(t) = X_1(0)X_2(0) + \int_0^t X_1(s)\,dX_2(s) + \int_0^t X_2(s)\,dX_1(s) + \int_0^t \varphi_1(s)\varphi_2(s)\,ds.$$

14.3   Selected Applications

14.3.1   Square-integrable Brownian Functionals

Theorem 14.3.1 *Let $\{W(t)\}_{t\in[0,1]}$ be an $\mathcal{F}_t^W$-Wiener process, and let $\{m(t)\}_{t\in[0,1]}$ be a real-valued $\mathcal{F}_t^W$-square-integrable martingale on $[0,1]$. Then there exists a real stochastic process $\{\varphi(t)\}_{t\in[0,1]} \in \mathcal{A}([0,1])$ such that*

$$m(t) = m(0) + \int_0^t \varphi(s)\, \mathrm{d}W(s) \quad (t \in [0,1]).\tag{14.14}$$

Lemma 14.3.2 *Let $\{W(t)\}_{t\in[0,1]}$ be an $\mathcal{F}_t^W$-Wiener process.*

A. The collection of random variables

$$\mathcal{H} := \{e^X \,;\, X \in \mathcal{U}\}$$

where $\mathcal{U}$ is the collection of random variables

$$\sum_{j=1}^k a_j W(t_j) \quad (k \in \mathbb{N}, 0 \le t_1 < \cdots < t_k \le 1, a_1, \ldots, a_k \in \mathbb{R})$$

is total in the Hilbert space $L_{\mathbb{R}}^2(\mathcal{F}_1^W, P)$.

B. The collection of random variables that are linear combinations of elements of

$$\mathcal{K} := \left\{ \overline{M}^f(1) = e^{\int_0^1 f(s)\, \mathrm{d}W(s) - \frac{1}{2}\int_0^t f(s)^2\, \mathrm{d}s} \,;\, f : \mathbb{R} \to \mathbb{R} \text{ measurable and bounded} \right\}$$

is dense in the Hilbert space $L_{\mathbb{R}}^2(\mathcal{F}_1^W, P)$.

Proof. Part B is an immediate consequence of Part A. For the proof of A, observe that $\mathcal{H} \subset L_{\mathbb{R}}^2(\mathcal{F}_1^W, P)$ and that $1 \in \mathcal{H}$. We have to prove that if $Y \in L_{\mathbb{R}}^2(\mathcal{F}_1^W, P)$ is such that $\mathrm{E}\left[Ye^X\right] = 0$ for all $X \in \mathcal{U}$, then $P(Y = 0) = 1$. As $1 \in \mathcal{H}$, $\mathrm{E}[Y] = 0$. Multiplying Y by a constant if necessary, we may suppose that $\mathrm{E}\left[|Y|\right] = 2$ and in particular, since $\mathrm{E}[Y] = 0$, $\mathrm{E}[Y^+] = 1$ and $\mathrm{E}[Y^-] = 1$. Therefore $Q_+ := Y^+ P$ and $Q_- := Y^- P$ are probability measures. Since $\mathrm{E}\left[Ye^X\right] = 0$ for all $X \in \mathcal{U}$, we have that $\mathrm{E}\left[Y^+ e^X\right] = \mathrm{E}\left[Y^- e^X\right]$ or, equivalently, $\mathrm{E}_{Q_+}\left[e^X\right] = \mathrm{E}_{Q_-}\left[e^X\right]$. Therefore the Laplace transforms of the vectors of the type $(W(t_1), \ldots, W(t_k))$ are the same under Q_+ and Q_-. This implies in particular that Q_+ and Q_- agree on $\mathcal{F}_1^W$. Therefore $\mathrm{E}\left[1_{\{Y^+ > Y^-\}}(Y^+ - Y^-)\right] = 0$ and $\mathrm{E}\left[1_{\{Y^+ < Y^-\}}(Y^+ - Y^-)\right] = 0$, which implies that $P(Y = 0) = 1$. □

Proof. (of Theorem 14.3.1) Let $f : \mathbb{R} \to \mathbb{R}$ be bounded and measurable, and define

$$M^f(t) = \exp\left\{ \int_0^t f(s)\, \mathrm{d}W(s) - \frac{1}{2}\int_0^t f(s)^2 \mathrm{d}s \right\}.\tag{14.15}$$

By Itô's differentiation rule,

$$M^f(t) = 1 + \int_0^t M^f(s) f(s)\, \mathrm{d}W(s).$$

For each $n \ge 1$, let T_n be the $\mathcal{F}_t$-stopping time defined by

$$T_n = \begin{cases} \inf\{t \in (0,1] \mid M^f(t) \ge n\} & \text{if } \{\dots\} \ne \varnothing, \\ 1 & \text{otherwise.} \end{cases}$$

We have

$$M^f(t \wedge T_n) = 1 + \int_0^{t \wedge T_n} M^f(s)f(s)\,\mathrm{d}W(s) = 1 + \int_0^t X(s)f(s)\mathbf{1}\{s \le T_n\}\,\mathrm{d}W(s).$$

Since f is bounded,

$$\mathrm{E}\left[\int_0^t \mid M^f(s)f(s)\mathbf{1}_{\{s \le T_n\}} \mid^2 \,\mathrm{d}W(s)\right] < \infty,$$

and therefore, by isometry,

$$\mathrm{E}[\mid M^f(t \wedge T_n - 1 \mid^2] = \mathrm{E}\left[\int_0^t \mid M^f(s)f(s)\mathbf{1}_{\{s \le T_n\}} \mid^2 \,\mathrm{d}s\right] < \infty.$$

In particular,

$$\mathrm{E}[\mid M^f(t \wedge T_n) \mid^2] \le 1 + \int_0^t \mathrm{E}[\mid M^f(s \wedge T_n) \mid^2]f(s)^2 \mathrm{d}s,$$

so that by Gronwall's lemma,

$$\mathrm{E}[\mid M^f(t \wedge T_n) \mid^2] \le \exp\left\{\int_0^t f(s)^2 \mathrm{d}s\right\}.$$

A fortiori,

$$\mathrm{E}[\mid M^f(t)\mathbf{1}_{\{t \le T_n\}} \mid^2] \le \exp\left\{\int_0^t f(s)^2 \mathrm{d}s\right\},$$

and therefore, by monotone convergence,

$$\mathrm{E}[\mid M^f(t) \mid^2] \le \exp\left\{\int_0^t f(s)^2 \mathrm{d}s\right\} < \infty$$

and

$$\mathrm{E}\left[\int_0^t \mid X(s)f(s) \mid^2 \,\mathrm{d}s\right] < \infty \quad (t \in [0,1]),$$

which implies that $\{M^f(t)\}_{t \in [0,1]}$ is a $(P, \mathcal{F}_t)$-square-integrable martingale.

We are now ready for the proof. Let $L_0^2(\mathcal{F}_1^W, P)$ denote the set of square-integrable random variables of $(\Omega, \mathcal{F}_1^W, P)$ that have mean 0. Let $\{m(t)\}_{t \in [0,1]}$ be a $(P, \mathcal{F}_t^W)$-square-integrable martingale of mean 0. The random variable $m(1)$ is in $L_0^2(\mathcal{F}_1^W, P)$, and by Part B of Lemma 14.3.2, it is a limit in quadratic mean of finite linear combinations of random variables of the form

$$M^f(1) - 1 = \exp\left\{\int_0^1 f(s)\,\mathrm{d}W(s) - \frac{1}{2}\int_0^1 f(s)^2 \mathrm{d}s\right\} - 1,$$

where f is deterministic bounded and measurable. Now, by Itô's differentiation rule,

$$M^f(1) - 1 = \int_0^1 M^f(s)f(s)\,\mathrm{d}W(s).$$

Therefore $m(1)$ is the limit in quadratic mean of elements of the form

$$\int_0^1 \varphi_n(s)\,\mathrm{d}W(s)\,,$$

where $\{\varphi_n(t)\}_{t\in[0,1]}$ is in $\mathcal{A}([0,1])$. Therefore

$$m(1) = \int_0^1 \varphi(s)\,\mathrm{d}W(s)\,,$$

for some $\{\varphi(t)\}_{t\in[0,1]}$ in $\mathcal{A}([0,1])$. Now, $m(t) = \mathrm{E}[m(1) \mid \mathcal{F}_t^W]$. The process $m'(t) := \int_0^t \varphi(s)\,\mathrm{d}W(s)$ $(t \in [0,1])$ is a $(P, \mathcal{F}_t^W)$-square-integrable martingale. In particular, $m'(t) = \mathrm{E}[m'(1) \mid \mathcal{F}_t^W]$. But $m(1) = m'(1)$, P-a.s., and therefore $P(m(t) = m'(t)) = 1$ for all $t \geq 0$. $\qquad\square$

14.3.2 Girsanov's Theorem

This is a generalization of the Cameron–Martin formula (Theorem 11.3.11).

Theorem 14.3.3 *Let $\{X(t)\}_{t\geq0}$ be a $(P, \mathcal{F}_t)$-Wiener process and let $\{\varphi(t)\}_{t\geq0}$ be in $\mathcal{B}_{loc}$. Let*

$$L(t) := L_\varphi(t) := \exp\left\{\int_0^t \varphi(s)\mathrm{d}X(s) - \frac{1}{2}\int_0^t \varphi(s)^2\mathrm{d}s\right\} \quad (t \geq 0)\,.$$

If

$$\mathrm{E}_P[L(T)] = 1\,, \tag{14.16}$$

the formula $\frac{\mathrm{d}\widetilde{P}}{\mathrm{d}P} = L(T)$ defines a probability $\widetilde{P}$ on $(\Omega, \mathcal{F})$ such that

$$\widetilde{W}(t) := X(t) - \int_0^t \varphi(s)\mathrm{d}s \quad (t \in [0,T]) \tag{14.17}$$

is a $(\widetilde{P}, \mathcal{F}_t)$-Wiener process.

The technical lemma below gives (14.16) under additional conditions.

Lemma 14.3.4

A. *Let φ be a real $\mathcal{F}_t$-progressively measurable process such that*

$$\int_0^t |\varphi(t)|^2\,\mathrm{d}t \leq C \quad (t \in \mathbb{R}_+) \tag{14.18}$$

for some $C < \infty$. Then for all $t \in \mathbb{R}_+$, $L(t)$ is square integrable and $\mathrm{E}[L(t)] = 1$.

B. *Let φ be a complex $\mathcal{F}_t$-progressively measurable process satisfying (14.18) for some $C < \infty$. Then for all $t \geq 0$, $|L(t)|$ is square-integrable and $\mathrm{E}[|L(t)|] = 1$.*

Proof. A. (i) We begin with the additional hypothesis that $|\varphi|$ is bounded by some finite number K. Itô's differentiation rule gives

$$L(t) = 1 + \int_0^t L(s)\varphi(s)\,\mathrm{d}X(s),$$

and from the inequality $(a+b)^2 \le 2(a^2 + b^2)$, we have that

$$E\left[L(t)^2\right] \le 2 + 2E\left[\left(\int_0^t L(s)\,\varphi(s)\,\mathrm{d}X(s)\right)^2\right]$$

$$= 2 + 2E\left[\int_0^t L(s)^2\,\varphi(s)^2\,\mathrm{d}s\right] \le 2 + K^2 \int_0^t E\left[L(s)^2\right]\mathrm{d}s.$$

Therefore, by Gronwall's lemma (Lemma B.6.1),

$$E\left[L(t)^2\right] \le 2e^{2K^2 t}$$

and therefore $\{L(t)\varphi(t)\}_{t\ge 0} \in \mathcal{A}_{loc}$, so that $E\left[\int_0^t L(s)\varphi(s)\,\mathrm{d}W(s)\right] = 0$ and then $E\left[L(t)\right] = 1$.

(ii) Back to hypothesis (14.18), let

$$\varphi_n(t) := \varphi(t)1_{\{|\varphi(t)|\le n\}}.$$

Then, by Schwarz's inequality,

$$E\left[L_{\varphi_n}(t)^2\right] = E\left[\exp\left\{2\int_0^t \varphi_n(s)\,\mathrm{d}X(s) - \int_0^t \varphi_n(s)^2\,\mathrm{d}s\right\}\right]$$

$$= E\left[\exp\left\{\int_0^t 2\varphi_n(s)\,\mathrm{d}X(s) - 4\int_0^t \varphi_n(s)^2\,\mathrm{d}s\right\} \times \exp\left\{\int_0^t 3\varphi_n(s)^2\,\mathrm{d}s\right\}\right]$$

$$\le E\left[\exp\left\{\int_0^t 4\varphi_n(s)\,\mathrm{d}X(s) - \frac{1}{2}\int_0^t (4\varphi_n(s))^2\,\mathrm{d}s\right\}\right]^{\frac{1}{2}} \times e^{3C}$$

$$\le E\left[\exp\left\{\int_0^t 4\varphi_n(s)\,\mathrm{d}X(s) - \frac{1}{2}\int_0^t (4\varphi_n(s))^2\,\mathrm{d}s\right\}\right]^{\frac{1}{2}} e^{3C} = 1 \times e^{3C}.$$

Therefore $\sup_n E\left[L_{\varphi_n}(t)^2\right] < \infty$, which implies that the sequence $\{L_{\varphi_n}(t)^2\}_{n\ge 1}$ is uniformly integrable (Theorem 4.2.14). Moreover, this sequence converges in probability to $L(t)$ since $\int_0^t \varphi_n(s)^2\,\mathrm{d}s \uparrow \int_0^t \varphi(s)^2\,\mathrm{d}s$ (monotone convergence) and $\int_0^t \varphi_n(s)\,\mathrm{d}X(s) \to \int_0^t \varphi(s)\,\mathrm{d}X(s)$ in L^2. Therefore (Theorem 4.2.16),

$$E\left[L(t)\right] = \lim_n E\left[L_{\varphi_n}(t)\right] = \lim_n 1 = 1.$$

B. Let $\varphi = \alpha + i\beta$. We have

$$E\left[|L(t)|^2\right] = E\left[\left|\exp\int_0^t (\alpha(s) + i\beta(s))\,\mathrm{d}X(s) - \frac{1}{2}\int_0^t (\alpha(s) + i\beta(s))^2\,\mathrm{d}s\right|^2\right]$$

$$= E\left[L_\alpha(t)^2\right]\exp\left\{\frac{1}{2}\int_0^t \alpha(s)\,\mathrm{d}s\right\} \le E\left[L_\alpha(t)^2\right] \times e^C < \infty.$$

Also:

$$\int_0^t \mathrm{E}\left[|L(s)|^2 |\varphi(s)|^2\right] \mathrm{d}s \le \mathrm{e}^C \int_0^t \mathrm{E}\left[L_\alpha(s)^2 |\varphi(s)|^2\right] \mathrm{d}s$$

$$\le \mathrm{e}^C \int_0^t \mathrm{E}\left[\mathrm{E}\left[L_\alpha(t)^2 \mid \mathcal{F}_s\right]\right] |\varphi(s)|^2 \, \mathrm{d}s$$

$$= \mathrm{e}^C \int_0^t \mathrm{E}\left[L_\alpha(t)^2 |\varphi(s)|^2\right] \mathrm{d}s$$

$$= \mathrm{e}^C \mathrm{E}\left[L_\alpha(t)^2\right] \int_0^t |\varphi(s)|^2 \, \mathrm{d}s$$

$$\le C\mathrm{e}^C \mathrm{E}\left[L_\alpha(t)^2\right] < \infty.$$

Therefore $\mathrm{E}\left[\int_0^t L(s)\varphi(s)\,\mathrm{d}X(s)\right] = 0$ and then, from

$$L(t) = 1 + \int_0^t L(s)\varphi(s)\,\mathrm{d}W(s)\,,$$

we conclude that $\mathrm{E}\left[L(t)\right] = 1$.

$\square$

Proof. (of Girsanov's theorem) In order to prove (14.17), it suffices to show that for all $0 < t_1 < \cdots < t_p = T$ and all $u_1, \cdots, u_p$ in $\mathbb{R}$,

$$\mathrm{E}\left[L(T) \exp\left\{ i \sum_{j=1}^p u_j X(t_j) \right\}\right] = \exp\left\{ -\frac{1}{2} \sum_{j,k=1}^p u_j u_k (t_j \wedge t_k) \right\}. \tag{14.19}$$

We do this with the additional assumption that $\int_0^T \varphi(s)^2 \, \mathrm{d}s < C < \infty$ almost surely. Then the complex stochastic process

$$\psi(t) := \varphi(t) + i \sum_{j=1}^p u_j 1_{[0,t_j]}(t)$$

satisfies a constraint of the type

$$\mathrm{E}\left[\int_0^T |\psi(t)|^2\right] < C' > \infty.$$

Therefore, by Lemma 14.3.4, $[L_\psi(T)] = 1$ or, equivalently,

$$\mathrm{E}\left[L(T) \exp\left\{ i \sum_{j=1}^p u_j X(t_j) \right\}\right] \times \exp\left\{ \frac{1}{2} \sum_{j,k=1}^p u_j u_k (t_j \wedge t_k) \right\} = 1\,, \tag{$\star$}$$

that is, (14.19).

It remains to get rid of the additional assumption. For this, we introduce the processes

$$\varphi_n(t) := \varphi(t) 1_{[0,\tau_n]}(t)$$

where

$$\tau_n := \inf\{t \,;\, \int_0^t \varphi(s)^2 \, \mathrm{d}s \ge n\}.$$

By Lemma 14.3.4, $\mathrm{E}\,[L_{\varphi_n}(T)] = 1 = \mathrm{E}\,[L_\varphi(T)]$ (the last equality is a hypothesis of the theorem). Also $L_{\varphi_n}(T) \to L_\varphi(T)$. Since moreover $L_{\varphi_n}(T) \geq 0$, the conditions of application of Scheffé's lemma (Lemma 4.4.24) are satisfied, and therefore

$$L_{\varphi_n}(T) \to L_\varphi(T) \text{ in } L^1 . \tag{$\dagger$}$$

But $(\star)$ is true for φ_n, that is,

$$\mathrm{E}\left[L_{\varphi_n}(T)\exp\left\{i\sum_{j=1}^p u_j X(t_j)\right\}\right] \times \exp\left\{\frac{1}{2}\sum_{j,k=1}^p u_j u_k(t_j \wedge t_k)\right\} = 1 .$$

Letting $X_n(t) := W(t) - \int_0^t \varphi_n(s)\,\mathrm{d}s$,

$$L_{\varphi_n}(T)\exp\left\{i\sum_{j=1}^p u_j X_n(t_j)\right\} - L_\varphi(T)\exp\left\{i\sum_{j=1}^p u_j X(t_j)\right\}$$

$$= (L_{\varphi_n}(T) - L_\varphi(T))\exp\left\{i\sum_{j=1}^p u_j X_n(t_j)\right\}$$

$$+ L_\varphi(T)\left(\exp\left\{i\sum_{j=1}^p u_j X_n(t_j)\right\} - \exp\left\{i\sum_{j=1}^p u_j X(t_j)\right\}\right) .$$

Using $(\dagger)$ and the fact that $\exp\left\{i\sum_{j=1}^p u_j X_n(t_j)\right\}$ is uniformly bounded and converges, we obtain by dominated convergence that

$$L_{\varphi_n}(T)\exp\left\{i\sum_{j=1}^p u_j X_n(t_j)\right\} \to L_\varphi(T)\exp\left\{i\sum_{j=1}^p u_j X_n(t_j)\right\} \text{ in } L^1 ,$$

which gives $(\star)$. $\qquad\square$

The following result is a sufficient condition for the hypothesis (14.16) to hold.

Lemma 14.3.5 *If in Theorem 14.3.3* $\{\varphi(t)\}_{t\in[0,1]}$ *is bounded, then (14.16) holds. Moreover,* $\{L(t)\}_{t\in[0,1]}$ *thereof is a* $(P,\mathcal{F}_t)$-*square-integrable martingale.*

Proof. For each $n \geq 1$, let S_n be the $\mathcal{F}_t$-stopping time defined by

$$S_n = \begin{cases} \inf\left\{t \mid \int_0^t \varphi(s)^2\,\mathrm{d}s + L(t) \geq n\right\} & \text{if } \{\dots\} \neq \emptyset, \\ +\infty & \text{otherwise}. \end{cases} \tag{14.20}$$

Then $\{L(t \wedge S_n)\}_{t\in[0,1]}$ is a square-integrable $\mathcal{F}_t$-martingale and, by isometry,

$$\mathrm{E}[\mid L(t \wedge S_n) - 1\mid^2] = \mathrm{E}\left[\int_0^t \mid L(s)\varphi(s)\mid^2 1_{\{s\leq S_n\}}\,\mathrm{d}s\right] .$$

In particular,

$$\mathrm{E}[\mid L(t \wedge S_n)\mid^2] \leq 1 + \int_0^t \mathrm{E}[\mid L(s)\varphi(s)\mid^2 1_{\{s\leq S_n\}}]\mathrm{d}s$$

$$\leq 1 + \sup(\varphi)\int_0^t \mathrm{E}[\mid L(s \wedge S_n)\mid^2]\mathrm{d}s .$$

The rest follows by Gronwall's lemma (Theorem B.6.1). $\qquad\square$

Theorem 14.3.6 *If* $\mathrm{E}\,[L(T)] = 1$, *then* $\{L(t)\}_{t\in[0,T]}$ *is an* $\mathcal{F}_t$-*martingale.*

Proof. Itô's differentiation rule applied to $F(x) = \mathrm{e}^x$ yields (14.9). Since $L(t)$ and $\int_0^t \varphi(s)^2 \mathrm{d}s$ are finite continuous processes, $S_n \uparrow \infty$, P-a.s. Also

$$\int_0^t \mid L(s)\varphi(s)1_{\{s\leq S_n\}}\mid^2 \mathrm{d}s \leq n^2,$$

and therefore $\{L(t \wedge S_n)\}_{t\geq 0}$ is a $(P, \mathcal{F}_t)$-square-integrable martingale, and therefore $\{L(t)\}_{t\geq 0}$ is a $(P, \mathcal{F}_t)$-local martingale. Being non-negative, it is also a supermartingale, and a $(P, \mathcal{F}_t)$-martingale on $[0, T]$ when (14.16) is satisfied. □

The Strong Markov Property of Brownian Motion

This result (Theorem 11.2.1) was admitted in Chapter 11. The following proof is based on stochastic calculus, more precisely, on the following characterization:

Lemma 14.3.7 *For a continuous real stochastic process* $\{Y(t)\}_{t\geq 0}$ *adapted to the history* $\{\mathcal{F}_t\}_{t\geq 0}$ *to be an* $\mathcal{F}_t$-*Brownian motion, it is necessary and sufficient that for all* $\lambda \in \mathbb{R}$, *the process*

$$\mathrm{e}^{\lambda Y(t)-\frac{1}{2}\lambda^2 t} \quad (t \geq 0)$$

be an $\mathcal{F}_t$-*martingale. In this case, it is independent of* $\mathcal{F}_0$.

Proof. The necessity results from an elementary computation on Gaussian variables. For the sufficiency, observe that the martingale condition implies that for all intervals $[a, b] \subset \mathbb{R}$,

$$\mathrm{E}\left[\mathrm{e}^{\lambda(Y(b)-Y(a))} \mid \mathcal{F}_a\right] = \mathrm{e}^{\frac{1}{2}\lambda^2(b-a)}.$$

By taking expectations it follows that $Y(b) - Y(a)$ is a centered Gaussian variable of variance $(b - a)$ and then that it is independent of $\mathcal{F}_a$, which implies the independence property of the increments as well as their independence from $\mathcal{F}_0$. □

We now prove Theorem 11.2.1.

Proof. The stochastic process

$$\varphi(t) := 1_A 1_{(\tau+a,\tau+b]}(t) \quad (t \geq 0),$$

where $A \in \mathcal{F}_{\tau+a}$, is $\mathcal{F}_t$-progressively measurable, and therefore by Lemma 14.3.4:

$$\mathrm{E}\left[\exp\left\{\lambda 1_A(W(\tau + b) - W(\tau + a)) - \frac{1}{2}\lambda^2 1_A(b - a)\right\}\right] = P(A)$$

or, equivalently,

$$\mathrm{E}\left[\exp\left\{\lambda(W(\tau + b) - W(\tau + a)) - \frac{1}{2}\lambda^2(b - a)\right\}1_A\right] = 1.$$

Therefore, since A is arbitrary in $\mathcal{F}_{\tau+a}$,

$$\mathrm{E}\left[\exp\left\{\lambda(W(\tau + b) - W(\tau + a))\right\} \mid \mathcal{F}_{\tau+a}\right] = \exp\left\{\frac{1}{2}\lambda^2(b - a)\right\}$$

rom which we deduce as in Lemma 14.3.7 that $W(\tau+b)-W(\tau+a)$ is a centered Gaussian variable with variance $b-a$ independent of $\mathcal{F}_{\tau+a}$. In particular, it has independent increments and therefore $\{W(\tau+t)\}_{t\geq 0}$ is a Wiener process. Moreover, still by Lemma 14.3.7, this process is independent of $\mathcal{F}_\tau$. $\qquad\square$

Remark 14.3.8 The above proof is similar to that of the strong Markov property of Poisson processes (Theorem 7.1.9).

14.3.3 Stochastic Differential Equations

This is a very brief introduction to a vast subject. Let $\{W(t)\}_{t\geq 0}$ be a standard $\mathcal{F}_t$-Brownian motion. We are going to discuss the existence and unicity of a measurable $\mathcal{F}_t$-adapted stochastic process $\{X(t)\}_{t\geq 0}$ such that almost surely

$$X(t) = X(0) + \int_0^t b(X(s))\,\mathrm{d}s + \int_0^t \sigma(X(s))\,\mathrm{d}W(s) \quad (t\geq 0), \tag{14.21}$$

where b and σ are measurable functions such that almost surely

$$\int_0^t \left(|b(X(s))|^2 + |\sigma(X(s))|^2\right)\,\mathrm{d}s < \infty \quad (t\geq 0). \tag{14.22}$$

One then calls $\{X(t)\}_{t\geq 0}$ the solution of the *stochastic differential equation* (14.21).

Condition (14.22) guarantees in particular that the integrand of the Itô integral of (14.21) is in $\mathcal{A}_{loc}$.

Theorem 14.3.9 *If $X(0) \in L^2_{\mathbb{R}}(P)$ and if for some $K < \infty$*

$$|b(x) - b(y)| + |\sigma(x) - \sigma(y)| \leq K\,|x - y|\,, \tag{14.23}$$

there exists a unique solution of (14.21) satisfying condition (14.22).

Proof. A. Uniqueness. If $\{X(t)\}_{t\geq 0}$ and $\{Y(t)\}_{t\geq 0}$ are two solutions,

$$X(t) - Y(t) = \int_0^t (b(X(s)) - b(Y(s)))\,\mathrm{d}s + \int_0^t (\sigma(X(s)) - \sigma(Y(s)))\,\mathrm{d}W(s)$$

and therefore, taking into account the inequality $(a+b)^2 \leq 2(a^2 + b^2)$, Schwarz's inequality, the property of isometry of Itô's integrals and the Lipschitz condition (14.23),

$$\mathrm{E}\left[(X(t) - Y(t))^2\right] \leq 2\mathrm{E}\left[\left(\int_0^t (b(X(s)) - b(Y(s)))\,\mathrm{d}s\right)^2\right]$$

$$+ 2\mathrm{E}\left[\left(\int_0^t (\sigma(X(s)) - \sigma(Y(s)))\,\mathrm{d}W(s)\right)^2\right]$$

$$\leq 2t\mathrm{E}\left[\int_0^t (b(X(s)) - b(Y(s)))^2\,\mathrm{d}s\right]$$

$$+ 2\mathrm{E}\left[\int_0^t (\sigma(X(s)) - \sigma(Y(s)))^2\,\mathrm{d}s\right]$$

$$\leq 2(t+1)K^2\mathrm{E}\left[\int_0^t |X(s) - Y(s)|^2\,\mathrm{d}s\right]$$

$$\leq 2(T+1)K^2\mathrm{E}\left[\int_0^T |X(s) - Y(s)|^2\,\mathrm{d}s\right] \quad (t\leq T).$$

By Gronwall's lemma, for all $t \in [0,T]$, $\mathrm{E}\left[(X(t) - Y(t))^2\right] = 0$, and therefore $P(X(t) = Y(t)) = 1$.

B. Existence. Define recursively the stochastic processes $\{X_n(t)\}_{t \geq 0}$ $(n \geq 0)$ by $X_0(t) := X(0)$ $(t \geq 0)$ and for $n \geq 0$ by

$$X_{n+1}(t) := X(0) + \int_0^t b(X_n(s))\,\mathrm{d}s + \int_0^t \sigma(X_n(s))\,\mathrm{d}W(s). \tag{$\star$}$$

With arguments similar to those of Part A,

$$\mathrm{E}\left[(X_{n+1}(t) - X_n(t))^2\right] \leq 2(T+1)K^2 \mathrm{E}\left[\int_0^T |X_{n+1}(s) - X_n(s)|^2\,\mathrm{d}s\right] \quad (t \leq T).$$

Letting $C_T := 2(T+1)K^2$ and $a := \max_{t \leq T} \mathrm{E}\left[(X_1(t) - X(0))^2\right]$ (a finite quantity, less than a constant times $T^3 \mathrm{E}\left[X(0)^2\right]$), one checks by recurrence that

$$\mathrm{E}\left[(X_{n+1}(t) - X_n(t))^2\right] \leq a\frac{C_T^n t^{n-1}}{(n-1)!}.$$

Therefore

$$\|X_{n+1} - X_n\|_{\mathcal{A}[0,T]} \leq a\frac{(C_T T)^n}{n!}$$

and then

$$\|X_{n+\ell} - X_n\|_{\mathcal{A}[0,T]} \leq a^{\frac{1}{2}} \sum_{k \geq n} \left(\frac{(C_T T)^k}{k!}\right)^{\frac{1}{2}},$$

a quantity that tends to 0 as $n \to \infty$. This shows that $\{X_n(t)\}_{t \geq 0}$ $(n \geq 0)$ is a Cauchy sequence of the Hilbert space $\mathcal{A}[0,T]$ and therefore converges in this space to some $\{X(t)\}_{t \geq 0}$. The Lipschitz condition (14.23) allows us to pass to the limit in $(\star)$ to obtain (14.21). The property (14.22) is easily verified. $\qquad\square$

EXAMPLE 14.3.10:   AN EXPLICIT SOLUTION. It can be checked (Exercise 14.4.15) that the differential equation

$$\mathrm{d}X(t) = \sqrt{1 + X(t)^2}\,\mathrm{d}W(t) + \left[\sqrt{1 + X(t)^2} + \frac{1}{2}X(t)\right]\mathrm{d}t$$

admits the stochastic process

$$X(t) := \sinh W(t) + \sinh^{-1} W(t) + t \quad (t \geq 0)$$

as a solution. Theorem 14.3.9 then guarantees its unicity.

Strong and Weak Solutions

So far we have considered strong solutions. This means that the problem was posed in terms of a preexisting Wiener process and given initial state, and that the solution took the general form

$$X(t) = F\left(t, X(0), \{W(s)\}_{0 \leq s \leq t}\right).$$

A weak solution associated with the parameters (functions) b and σ and a probability distribution π on $\mathbb{R}$ consists of a probability space on which are given

1. a filtration $\{\mathcal{F}_t\}_{t\geq 0}$,

2. a standard $\mathcal{F}_t$-Wiener process $\{W_t\}_{t\geq 0}$,

3. a random variable $X(0)$ with a given distribution π and independent of the above Wiener process, and finally

4. an $\mathcal{F}_t$-progressive stochastic process $\{X_t\}_{t\geq 0}$ such that

$$X(t) = X(0) + \int_0^t b(X(s))\,\mathrm{d}s + \int_0^t \sigma(X(s))\,\mathrm{d}W(s).$$

In this definition, the Wiener process is part of the solution.

EXAMPLE 14.3.11: TANAKA'S STOCHASTIC DIFFERENTIAL EQUATION. In the equation

$$X(t) = \int_0^t \mathrm{sgn}(X(s))\,\mathrm{d}W(s) \quad (t \geq 0),$$

where $\mathrm{sgn}(x) = +1$ if $x \geq 0$ and $\mathrm{sgn}(x) = -1$ if $x < 0$, the Lipschitz conditions of Theorem 14.3.9 are not satisfied. We shall give rough arguments showing that there exist a solution that cannot be a strong solution.

First note that if there exists a solution, it is a Brownian motion. To see this it suffices to show that for all $\lambda > 0$, the process

$$\exp\{\lambda X(t) - \frac{1}{2}\lambda^2 t\} \quad (t \geq 0)$$

is a martingale (Lemma 14.3.7) (Exercise 14.4.17). Therefore we have unicity in law of the solution. Note that it is the best we can do concerning unicity since $\{-X(t)\}_{t\geq 0}$ is another solution.

By the same arguments as above, for any solution $\{X(t)\}_{t\geq 0}$ (which is a Brownian motion), the process $\{W(t)\}_{t\geq 0}$ defined by

$$W(t) := \int_0^t \mathrm{sgn}(X(s))\,\mathrm{d}X(s) \quad (t \geq 0) \tag{$\star$}$$

is a Brownian motion. By differentiation, $\mathrm{d}W(t) = \mathrm{sgn}(X(t))\,\mathrm{d}X(t)$, and therefore, since $\mathrm{sgn}(x)^{-1} = \mathrm{sgn}(x)$, $\mathrm{d}X(t) = \mathrm{sgn}(X(t))\,\mathrm{d}W(t)$. We have therefore obtained a solution.

This solution cannot be a strong solution. Indeed, from $(\star)$, we deduce that $W(t)$ is $\mathcal{F}_t^{|X|}$-measurable. If $\{X(t)\}_{t\geq 0}$ were a strong solution, it would be $\mathcal{F}_t^W$-adapted, that is, $\mathcal{F}_t^{|X|}$-adapted. But $\mathcal{F}_t^X$ contains more information than $\mathcal{F}_t^{|X|}$!

14.3.4 The Dirichlet Problem

This subsection gives a simple example of the interaction between the theory of stochastic differential equations and that of partial differential equations.

Definition 14.3.12 *Let* $u : \mathbb{R}^d \to \mathbb{R}$ *be a function of class* $\mathcal{C}^2$ *(twice differentiable with continuous derivatives) on an open set* $O \subseteq \mathbb{R}^d$. *Its* Laplacian, *defined in* O, *is the function*

$$\triangle u(x) := \sum_{i=1}^d \frac{\partial^2}{\partial x_i^2} u(x).$$

Definition 14.3.13 *The function $u : \mathbb{R}^d \to \mathbb{R}$ of class $\mathcal{C}^2$ on a domain (open and connected set) $D \subseteq \mathbb{R}^d$ is said to be* harmonic on D *if*

$$\triangle u(x) = 0 \quad (x \in D).$$

EXAMPLE 14.3.14: EXAMPLES OF HARMONIC FUNCTIONS. In dimension 2: $(x_1, x_2) \to \ln(x_1^2 + x_2^2)$, and $(x_1, x_2) \to e^{x_1}$ are harmonic on $D = \mathbb{R}^2$. In dimension 3: $x \to |x|^{2-d}$ is harmonic on $D = \mathbb{R}^3 \backslash \{0\}$.

Let $F : \mathbb{R}^d \to \mathbb{R}$ be of class $\mathcal{C}^2$. In the following $\{W(t)\}_{t \geq 0}$ will represent the Brownian motion starting from $a \in \mathbb{R}^d$ (that is, $W(0) = a$). Itô's formula gives, denoting by ∇f the gradient of a function f,

$$F(W(t)) = F(a) + \int_0^t \nabla F(W(s)) \, \mathrm{d}W(s) + \int_0^t \triangle F(W(s)) \, \mathrm{d}s, \qquad (14.24)$$

where

$$M(t) := F(a) + \int_0^t \nabla F(W(s)) \, \mathrm{d}W(s) \quad (t \geq 0)$$

is an $\mathcal{F}_t^W$-local martingale since $\int_0^t (\nabla F(W(s)))^2 \, \mathrm{d}s < \infty \ (t \geq 0)$.

Theorem 14.3.15 *Let $u : \mathbb{R}^d \to \mathbb{R}$ be harmonic on the domain $D \subseteq \mathbb{R}^d$, and let $G \subset D$ be an open set whose closure $\operatorname{clos} G \subset D$. Let*

$$\tau_G := \inf\{t \geq 0 \, ; W(t) \notin G\}$$

be the entrance time of the Brownian motion in $\overline{G}$. Then

$$u(W(t \wedge \tau_G)) - u(a) \quad (t \geq 0)$$

is a centered $\mathcal{F}_t^W$-martingale.

Proof. Let F be a function in $\mathcal{C}^2(\mathbb{R}^d)$ whose restriction to G is u, for instance

$$F(x) := ([(1_{G_{2\delta}} * \alpha] \times u)(x) 1_D(x),$$

where $G_{2\delta}$ is the 2δ-neighborhood of G, $4\delta = d(G, \overline{D})$ and α is a $\mathcal{C}^\infty$ function of integral 1 on $\mathbb{R}^d$ and null outside $\mathcal{B}(0, \delta)$. Then, by (14.24),

$$F(W(t \wedge \tau_G)) = F(0) + \int_0^{t \wedge \tau_G} \nabla F(W(s)) \, \mathrm{d}W(s) + \int_0^{t\tau_G} \triangle F(W(s)) \, \mathrm{d}s,$$

and since $F(x) = u(x)$ is harmonic on G,

$$u(W(t \wedge \tau_G)) = F(0) + \int_0^{t\tau_G} \nabla F(W(s)) \, \mathrm{d}W(s) \quad (t \geq 0),$$

a square-integrable $\mathcal{F}_t^W$-martingale (not just a local martingale, since ∇F is bounded on $\operatorname{clos} G$). $\qquad \square$

Definition 14.3.16 *Let $D \subset \mathbb{R}^d$ be a bounded domain, and let $f : \partial D \to \mathbb{R}$ be a continuous function. The* Dirichlet problem (D, f) *consists in finding a function u that is harmonic on D and equal to f on δD.*

Let D^ε be the ε-interior of D. From Theorem 14.3.15 with $G = D^\varepsilon$ and ε small enough,

$$u(x) = E_x \left[u(W(t \wedge \tau_{D^\varepsilon})) \right] \quad (x \in D),$$

where the notation E_x denotes expectation given that the initial position of the Brownian motion is x. Since D is bounded, $P(\tau_{D^\varepsilon} < \infty) = 1$, and therefore, letting $t \to \infty$,

$$u(x) = E_x \left[u(W(\tau_{D^\varepsilon})) \right]$$

by dominated convergence. Let now $\varepsilon \to 0$. Since $\tau_D < \infty$ and $W(\tau_{D^\varepsilon}) \to W(\tau_D)$,

$$u(x) = E_x \left[u(W(\tau_D)) \right]$$

by dominated convergence. By the boundary condition,

$$u(x) = E_x \left[f(W(\tau_D)) \right]. \tag{14.25}$$

Therefore the solution to the Dirichlet problem is unique, if it exists. We shall not prove existence, which can be obtained by analytical as well as probabilistic arguments. We shall be content with the fact that the solution has a probabilistic interpretation, given by (14.25).

Complementary reading

[Kuo, 2006] and [Baldi, 2018]. The latter has a chapter on mathematical finance and a large collection of corrected exercises. [Oksendal, 1995] has many examples in diverse areas.

14.4 Exercises

Exercise 14.4.1. AN $\mathcal{F}_t$-MARTINGALE
Let $\{W(t)\}_{t \geq 0}$ be an $\mathcal{F}_t$-Wiener process and let $\varphi, \psi \in \mathcal{A}(\mathbb{R}_+)$. Let $M(t) := \int_0^t \varphi(s) \, dW(s)$ and $N(t) := \int_0^t \psi(s) \, dW(s)$. Show that $M(t)N(t) - \int_0^t \varphi(s)\psi(s) \, ds$ $(t \geq 0)$ is an $\mathcal{F}_t$-martingale.

Exercise 14.4.2. PROOF OF THE REFLECTION PRINCIPLE
Prove Theorem 11.2.2 using Itô calculus (Hint: see the proof of the strong Markov property of the Brownian motion given in Subsection 14.3.2.)

Exercise 14.4.3. ITÔ INTEGRALS AS LEBESGUE INTEGRALS, I
Let $\{W(t)\}_{t \geq 0}$ be a standard Brownian motion, and $0 \leq a < b$. Prove that

$$\int_a^b W(s)^3 \, dW(s) = \frac{1}{4} \left(W(b)^4 - W(a)^4 \right) - \frac{3}{2} \int_a^b W(s)^2 \, ds.$$

Exercise 14.4.4. ITÔ INTEGRALS AS LEBESGUE INTEGRALS, II
Let $\{W(t)\}_{t \geq 0}$ be a standard Brownian motion, and $0 \leq a < b$. Prove that

$$\int_a^b e^{W(s)} \, dW(s) = e^{W(b)} - e^{W(a)} - \frac{1}{2} \int_a^b e^{W(s)} \, ds.$$

Exercise 14.4.5. BROWNIAN MOTION ON THE CIRCLE
Let $\{W(t)\}_{t \geq 0}$ be a standard Brownian motion. Let

$$V(t) := \begin{pmatrix} \cos W(t) \\ \sin W(t) \end{pmatrix}.$$

Show that

$$dV(t) = \begin{pmatrix} 0 & -1 \\ 1 & 0 \end{pmatrix} V(t)\, dW(t) - \frac{1}{2} V(t)\, dt \quad \text{with } V(0) = \begin{pmatrix} 1 \\ 0 \end{pmatrix}.$$

Exercise 14.4.6. PROOF OF THEOREM 14.1.8
Give the details of the proof of Theorem 14.1.8.

Exercise 14.4.7. THE AREA UNDER A BROWNIAN MOTION
Let $\{W(t)\}_{t \geq 0}$ be a standard Brownian motion. Show that

$$\int_0^t W(s)\, dW(s) - \frac{1}{3} W(t)^3$$

is an $\mathcal{F}_t^W$-martingale. Use this to compute the expected value of the area of the set

$$\{(t, x)\,;\, t \in [0, T_{a,b}], x \in [0, |W(t)|]\},$$

where $T_{a,b} := \inf\{t \geq 0\,;\, W(t) \in \{-b, a\}\}$ and $a > 0, b > 0$.

Exercise 14.4.8. CONTINUOUS INTEGRANDS FOR THE ITÔ INTEGRAL
Prove the statement of Remark 14.1.3.

Exercise 14.4.9. A MARTINGALE
Let $\{W(t)\}_{t \geq 0}$ be a standard Brownian motion.

(i) Show that the stochastic process

$$Y(t) := tW(t) - \int_0^t W(s)\, dW(s) \quad (t \geq 0)$$

is a martingale.

(ii) Show that for all $u \in [a, b] \subset \mathbb{R}$, $Y(b) - Y(a)$ is orthogonal (in $L_{\mathbb{R}}^2(P)$) to $H(W(s); s \in [0, a])$.

Exercise 14.4.10. THE N-TH POWER OF A BROWNIAN MOTION
Let $\{W(t)\}_{t \geq 0}$ be a standard Brownian motion. Show that

$$W(t)^n - \frac{n(n-1)}{2} \int_0^t W(s)^{n-2}\, ds \quad (t \geq 0)$$

is a martingale.

Exercise 14.4.11. THE PRODUCT OF INDEPENDENT BROWNIAN MOTIONS
Let $\{W_1(t)\}_{t\geq 0}$ and $\{W_2(t)\}_{t\geq 0}$ be independent standard Brownian motions. Is the claim

$$W_1(t)W_2(t) = \int_0^t W_1(s)\, dW_2(s) + \int_0^t W_2(s)\, dW_1(s)$$

resulting from a naive application of the formula of integration by parts true?

Exercise 14.4.12. A DIFFERENTIAL EQUATION FOR THE BROWNIAN BRIDGE
Using the results of Exercise 11.5.13, show that the Brownian bridge thereof satisfies the
following equation:

$$Z(t) = -\int_0^t \frac{1}{1-s} Z(s)\, ds + W(t).$$

Exercise 14.4.13. BROWNIAN MOTION ON THE CIRCLE
Let $\{W(t)\}_{t\geq 0}$ be a standard Brownian motion. Show that the vector process $V(t) = (\cos W(t), \sin W(t))^T$ $(t \geq 0)$ satisfies a stochastic differential equation of the form

$$dV(t) = AV(t)\, dW(t) - \frac{1}{2}V(t)\, dt$$

with initial condition $V(0) = (1,0)^T$, where A is a matrix to be identified.

Exercise 14.4.14. A MOTION ON THE CONE
Let $\{W_1(t)\}_{t\geq 0}$ and $\{W_2(t)\}_{t\geq 0}$ be two independent standard Brownian motions. Show
that the vector process $V(t) = (e^{W_1(t)}\cos(W_2(t)), e^{W_1(t)}\sin(W_2(t)), e^{W_1(t)})^T$ $(t \geq 0)$ satisfies a stochastic differential equation of the form

$$dV(t) = AV(t)\, dW_1(t) + BV(t)\, dW_2(t) + CV(t)\, dt$$

with initial condition $V(0) = (1,0,1)^T$, where A, B and C are matrices to be identified.

Exercise 14.4.15. A STOCHASTIC DIFFERENTIAL EQUATION
Prove that the stochastic process

$$X(t) := \sinh W(t) + \sinh^{-1} W(t) + t \quad (t \geq 0)$$

is a solution of the differential equation

$$dX(t) = \sqrt{1 + X(t)^2}\, dW(t) + \left[\sqrt{1 + X(t)^2} + \frac{1}{2}X(t)\right] dt.$$

Exercise 14.4.16. THE VASICEK MODEL
Consider the stochastic differential equation

$$dX(t) = (-bX(t) + c)\, dt + \sigma W(t).$$

Prove that the unique solution with initial state $X(0)$ is given by

$$X(t) = \frac{c}{b} + \left(X(0) - \frac{c}{b}\right)e^{-bt} + \sigma \int_0^t e^{-b(t-s)}\, dW(s).$$

Exercise 14.4.17. TANAKA'S DIFFERENTIAL EQUATION
Prove that any solution of the stochastic differential equation

$$X(t) = \int_0^t \operatorname{sgn}(X(s)) \, dW(s) \quad (t \geq 0)$$

is such that for all $\lambda > 0$, the process

$$\exp\left\{ \lambda X(t) - \frac{1}{2}\lambda^2 t \right\} \quad (t \geq 0)$$

is a martingale.

Exercise 14.4.18. ITÔ INTEGRALS AS RIEMANN INTEGRALS
Let $f : \mathbb{R} \to \mathbb{R}$ be a continuous function with continuous derivative f'. Let $F(x) :=$ $\int_0^x f(t) \, dt$. Show that

$$\int_0^t f(W(s)) \, dW(s) = F(W(t)) - F(0) - \frac{1}{2} \int_0^t f'(W(s)) \, ds.$$

Use this result to express the following Itô integrals in terms of Riemann integrals:

$$\int_0^t W(s) e^{W(s)} \, dW(s),$$

$$\int_0^t \frac{1}{1 + W(s)^2} \, dW(s),$$

$$\int_0^t e^{W(s) - \frac{1}{2}s} \, dW(s),$$

$$\int_0^t \frac{W(s)}{1 + W(s)^2} \, dW(s).$$

Chapter 15

Point Processes with a Stochastic Intensity

Let $\{\mathcal{F}_t\}_{t\in\mathbb{R}}$ be some history of a simple locally finite point process N on $\mathbb{R}$ (that is, a non-decreasing family of σ-fields such that for all $t \in \mathbb{R}$ and all $a \leq b \leq t$, $N((a,b])$ is $\mathcal{F}_t$-measurable). If it holds that for all $t \in \mathbb{R}$,

$$\lim_{h\downarrow 0} \frac{1}{h} \, \mathrm{E}[N((t,t+h])|\mathcal{F}_t] = \lambda(t) \quad P\text{-a.s.} , \qquad (\star)$$

for some non-negative locally integrable $\mathcal{F}_t$-adapted stochastic process $\{\lambda(t)\}_{t\in\mathbb{R}}$, the latter is called a stochastic $\mathcal{F}_t$-intensity of N.

This local definition of intensity is advantageously replaced by a global definition not involving a limiting derivative-type procedure and is more amenable to rigorous analysis. It opens a connection with the rich theory of martingales and offers among other things a unified view of stochastic systems driven by point processes.

This point of view will reveal a striking analogy with the contents of Chapter 14, the first instance of which is found in Paul Lévy's martingale characterization of Brownian motion (Theorem 14.2.5) and in Watanabe's martingale characterization of the standard Poisson process. The proof of Theorem 13.4.3 is a first example of the point process "stochastic calculus".

15.1 Stochastic Intensity

15.1.1 The Martingale Definition

For a Poisson process N on the real line with locally integrable intensity function $\lambda(t)$, it holds that for all intervals $[c,d] \subset \mathbb{R}$,

$$\mathrm{E}\left[N((c,d]) \mid \mathcal{F}_c^N\right] = \int_c^d \lambda(s)\,\mathrm{d}s$$

or, equivalently since the right-hand side is a deterministic quantity,

$$\mathrm{E}\left[N((c,d]) \mid \mathcal{F}_c^N\right] = \mathrm{E}\left[\int_c^d \lambda(s)\,\mathrm{d}s \mid \mathcal{F}_c^N\right].$$

This motivates the following definition of stochastic intensity.

© Springer Nature Switzerland AG 2020
P. Brémaud, *Probability Theory and Stochastic Processes*, Universitext,
https://doi.org/10.1007/978-3-030-40183-2_15

Definition 15.1.1 *Let N be a simple locally finite point process on $\mathbb{R}$, let $\{\mathcal{F}_t\}_{t\in\mathbb{R}}$ b a history of N and let $\{\lambda(t)\}_{t\in\mathbb{R}}$ be a non-negative a.s. locally integrable real-valued $\mathcal{F}_t$ progressively measurable stochastic process. If for all $a \in \mathbb{R}$ and all intervals $(c,d] \subset (a, \infty)$,*

$$E[N((c \wedge T_n^{(a)}, d \wedge T_n^{(a)})) \mid \mathcal{F}_c] = E\left[\int_{c \wedge T_n^{(a)}}^{d \wedge T_n^{(a)}} \lambda(s)ds \mid \mathcal{F}_c\right], \tag{15.1}$$

where $\{T_n^{(a)}\}_{n\geq 1}$ is a non-decreasing sequence of $\mathcal{F}_t$-stopping times such that $T_n^{(a)} > a$, $\lim_{n\uparrow\infty} T_n^{(a)} = \infty$ and $E\left[N((a, T_n^{(a)}))\right] < \infty$, N is then said to admit the stochastic $(P, \mathcal{F}_t)$-intensity $\{\lambda(t)\}_{t\in\mathbb{R}}$.

The connection between stochastic intensity (Definition 15.1.1) and martingales is the following:

$$M(t) := N((0, t]) - \int_0^t \lambda(s)\,ds$$

is a local $(P, \mathcal{F}_t)$-martingale (Exercise 15.4.1), called the *fundamental (local) $\mathcal{F}_t$-martingale* of the point process N. When the choice of probability P is clear from the context, one says "the $\mathcal{F}_t$-intensity" instead of "the $(P, \mathcal{F}_t)$-intensity".

Remark 15.1.2 The requirement of $\mathcal{F}_t$-progressiveness of the stochastic intensity guarantees that the integrated intensity process $\{\int_0^t \lambda(s)\,ds\}_{t\in\mathbb{R}}$ is measurable and $\mathcal{F}_t$-adapted (Exercise 5.4.2).

Remark 15.1.3 When considering point processes on the positive half-line, the intervention of the a's is superfluous, and it suffices to require that (15.1) holds for $a = 0$. However, the slightly more complicated definition given above is needed to handle point processes on the whole real line, especially stationary point processes.

Remark 15.1.4 The reason why requirement (15.1) cannot be replaced by the simpler one,

$$E\left[N((c, d]) \mid \mathcal{F}_c\right] = E\left[\int_c^d \lambda(s)\,ds \mid \mathcal{F}_c\right], \tag{$\dagger$}$$

not involving the stopping times $T_n^{(a)}$, is that it may occur that both sides of ($\dagger$) are infinite, in which case the information contained in ($\dagger$) is nil. This happens for instance when N is a homogeneous Cox process whose random intensity Λ has an infinite expectation (see Example 15.1.5 below).

EXAMPLE 15.1.5: POISSON AND COX PROCESSES. Let N be a Cox process on $\mathbb{R}_+$ with conditional intensity measure ν with respect to $\mathcal{G} \supseteq \mathcal{F}^\nu$ (see Definition 8.2.5) and suppose that $\nu(C) := \int_C \lambda(s)\,ds$ $(C \in \mathcal{B}(\mathbb{R}_+))$, where $\{\lambda(t)\}_{t\geq 0}$ is a locally integrable non-negative process. Then N admits this process as an $\mathcal{F}_t$-intensity, where $\mathcal{F}_t := \mathcal{F}_t^N \vee \mathcal{G}$ $(t \geq 0)$ (Exercise 15.4.4).

Theorem 15.1.11 below gives a formula that can be considered both a refinement of Campbell's formula and an extension of the smoothing formula for HPPs (Theorem 7.1.7). The notion of predictable process will be needed.

Definition 15.1.6 *Let* $\mathbf{T} = \mathbb{R}$ *or* $\mathbb{R}+$. *Let* $\{\mathcal{F}_t\}_{t \in \mathbf{T}}$ *be a history. The* predictable σ-*field* $\mathcal{P}(\mathcal{F})$ *on* $\mathbf{T} \times \Omega$ *is the* σ-*field generated by the collection of sets*

$$(a, b] \times A \qquad ([a, b] \subset \mathbf{T}, \, A \in \mathcal{F}_a), \tag{15.2}$$

to which one must add, in the case $\mathbf{T} = \mathbb{R}_+$, *the sets* $\{0\} \times A$ $(A \in \mathcal{F}_0)$. *A stochastic process* $\{X(t)\}_{t \in \mathbf{T}}$ *taking its values in a measurable space* $(E, \mathcal{E})$ *is called an* $\mathcal{F}_t$-*predictable process if the mapping* $(t, \omega) \to X(t, \omega)$ *is* $\mathcal{P}(\mathcal{F})$-*measurable.*

For short, one then says: $\{X(t)\}_{t \in \mathbf{T}}$ is in $\mathcal{P}(\mathcal{F})$.

Definition 15.1.7 *Let* $\mathbf{T} = \mathbb{R}$ *or* $\mathbb{R}+$ *and let* $\{\mathcal{F}_t\}_{t \in \mathbf{T}}$ *be a history. Let* $(K, \mathcal{K})$ *be some measurable space. Let* $H : (\mathbf{T} \times \Omega \times K, \mathcal{P}(\mathcal{F}) \otimes \mathcal{K}) \to (\overline{\mathbb{R}}, \mathcal{B}(\overline{\mathbb{R}}))$. *One then says that* $\{H(t, z)\}_{t \in \mathbf{T}, z \in K}$ *is an* $\mathcal{F}_t$-*predictable stochastic process indexed by* K.

Remark 15.1.8 An $\mathcal{F}_t$-predictable process is $\mathcal{F}_t$-progressive (Exercise 15.4.6).

EXAMPLE 15.1.9: LEFT-CONTINUITY AND PREDICTABILITY. A complex-valued stochastic process $\{X(t)\}_{t \in \mathbb{R}}$ adapted to $\{\mathcal{F}_t\}_{t \in \mathbb{R}}$ and with *left-continuous trajectories* is $\mathcal{F}_t$-predictable. In fact, by left-continuity, $X(t, \omega) = \lim_{n \uparrow \infty} X_n(t, \omega)$, where

$$X_n(t, \omega) := \sum_{k=-n2^n}^{+n2^n} X(k2^{-n}, \omega) 1_{(k2^{-n}, (k+1)2^{-n}]}(t),$$

and since $X(k2^{-n})$ is $\mathcal{F}_{k2^{-n}}$-measurable, $(t, \omega) \to X_n(t, \omega)$ is $\mathcal{P}(\mathcal{F})$-measurable.

EXAMPLE 15.1.10: ANOTHER TYPICAL $\mathcal{F}_t$-PREDICTABLE PROCESS. Let S and τ be two $\mathcal{F}_t$-stopping times such that $S \leq \tau$, and let $\varphi : \mathbb{R}_+ \times \mathbb{R} \to \mathbb{R}$ be a measurable function. Then

$$X(t, \omega) = \varphi(S(\omega), t) 1_{\{S(\omega) < t \leq \tau(\omega)\}}$$

defines an $\mathcal{F}_t$-predictable process (Exercise 15.4.6).

Theorem 15.1.11 *Let* N *be a simple point process* N *on* $\mathbb{R}$ *and let* $\{\mathcal{F}_t\}_{t \in \mathbb{R}}$ *be a history of* N. *Let* $\{\lambda(t)\}_{t \in \mathbb{R}}$ *be a non-negative locally integrable* $\mathcal{F}_t$-*progressively measurable stochastic process.*

(i) If for all non-negative left-continuous $\mathcal{F}_t$-*adapted processes* $\{H(t)\}_{t \in \mathbb{R}}$

$$\mathrm{E}\left[\int_{\mathbb{R}} H(t) \, N(\mathrm{d}t)\right] = \mathrm{E}\left[\int_{\mathbb{R}} H(t) \lambda(t) \, \mathrm{d}t\right], \tag{15.3}$$

then $\{\lambda(t)\}_{t \in \mathbb{R}}$ *is an* $\mathcal{F}_t$-*intensity of* N.

(ii) If $\{\lambda(t)\}_{t \in \mathbb{R}}$ *is an* $\mathcal{F}_t$-*intensity of* N, *(15.3) holds true for all non-negative* $\mathcal{F}_t$-*predictable processes* $\{H(t)\}_{t \in \mathbb{R}}$.

Proof. (i) Let $a \in \mathbb{R}$ and let

$$T_n^{(a)} := \inf \left\{ t \geq a \,;\, \int_a^t \lambda(s)\mathrm{d}s \geq n \right\}.$$

Since $\{\lambda(t)\}_{t \in \mathbb{R}}$ is locally integrable, $\lim_{n \uparrow \infty} T_n^{(a)} = +\infty$. Also, for each $n \geq 1$, $T_n^{(a)}$ is an $\mathcal{F}_t$-stopping time. Let for $[c,d] \subseteq [a, +\infty)$ and $C \in \mathcal{F}_c$

$$H(t, \omega) := 1_C(\omega) 1_{(c,d]}(t) 1_{(a, T_n^{(a)}(\omega)]}(t).$$

This defines a left-continuous $\mathcal{F}_t$-adapted stochastic process. Inserting this process into (15.3) gives

$$\mathrm{E}[1_C N(c \wedge T_n^{(a)}, d \wedge T_n^{(a)}]] = \mathrm{E}\left[1_C \int_{c \wedge T_n^{(a)}}^{d \wedge T_n^{(a)}} \lambda(s)\mathrm{d}s \right].$$

Since C is arbitrary in $\mathcal{F}_c$, this is equivalent to (15.1).

(ii) Let $a \in \mathbb{R}$ and let $T_n^{(a)}$ be as in Definition 15.1.1. Define for each $n \geq 1$ the point process N_n by

$$N_n(C) := N(C \cap (a, T_n^{(a)}])$$

and let $\mathcal{H}$ be the collection of non-negative $\mathcal{F}_t$-predictable processes $\{H(t)\}_{t \in \mathbb{R}}$ such that

$$\mathrm{E}\left[\int_{[a, +\infty)} H(t) N_n(\mathrm{d}t) \right] = \mathrm{E}\left[\int_a^{T_n^{(a)}} H(t)\lambda(t)\mathrm{d}t \right]. \tag{15.4}$$

$\mathcal{H}$ is a d-system of functions (Definition 2.1.26) containing the constant 1 as well as the functions of the form 1_D $(D \in \mathcal{S})$ where

$$\mathcal{S} := \{ C \times (c,d]; a < c < d < +\infty, C \in \mathcal{F}_c \}.$$

Moreover, $\mathcal{S}$ is a π-system generating $\mathcal{P}(\mathcal{F})$. Therefore, by Dynkin's theorem (Theorem 2.1.25), $\mathcal{H}$ contains all the non-negative $\mathcal{F}_t$-predictable processes. Letting $n \uparrow +\infty$ in (15.4) and then $a \downarrow -\infty$ gives the result by monotone convergence. $\qquad \square$

Remark 15.1.12 The smoothing formula (15.3) says that the "non-smooth" measure $P(\mathrm{d}\omega) N(\omega, \mathrm{d}t)$ and the "smooth" measure $P(\mathrm{d}\omega)\lambda(t, \omega)\,\mathrm{d}t$ agree on $\mathcal{P}(\mathcal{F})$.

EXAMPLE 15.1.13: TRANSITION TIMES OF A CONTINUOUS-TIME HMC, TAKE 1. Refer to the construction in Section 7.2.3 of a continuous-time HMC $\{X(t)\}_{t \geq 0}$ with countable state space E and whose infinitesimal generator $\mathbf{A} = \{q_{ij}\}_{i,j \in E}$ is stable and conservative, and suppose that this HMC is regular (no accumulation point of the transition times). This construction is based on a family N_{ij} $(i \neq j)$ of independent homogeneous Poisson processes with respective intensities q_{ij} $(i \neq j)$ and the transition times from state i to state j form a point process $\widetilde{N}_{ij}$ given by

$$\widetilde{N}_{ij}(t) = \int_{(0,t]} Z_i(s-)\, N_{ij}(\mathrm{d}s),$$

where $Z_i(t) := 1_{\{X(t)=i\}}$. By Theorem 7.1.7, for all non-negative left-continuous $\mathcal{F}_t^X$-adapted processes $\{H(t)\}_{t \in \mathbb{R}}$

$$E\left[\int_{\mathbb{R}} H(t)\tilde{N}_{ij}(\mathrm{d}t)\right] = E\left[\int_{\mathbb{R}} H(t)Z_i(t-)\,N_{ij}(\mathrm{d}t)\right]$$

$$= E\left[\int_{\mathbb{R}} H(t)Z_i(t-)\,q_{ij}\,\mathrm{d}t\right] = E\left[\int_{\mathbb{R}} H(t)Z_i(t)\,q_{ij}\,\mathrm{d}t\right].$$

Therefore, by (i) of Theorem 15.1.11, $\tilde{N}_{ij}$ admits the $\mathcal{F}_t^X$-intensity $\tilde{\lambda}_{ij}(t) = Z_i(t)q_{ij}$.

EXAMPLE 15.1.14: THE HAWKES BRANCHING POINT PROCESS. This point process N counts the arrivals of individuals in a population, by immigration or by birth. The immigration process is a homogeneous Poisson process of intensity $\nu > 0$. Each individual, immigrant or native, arriving or born at time s gives birth to descendants, the birth times forming a Poisson process of intensity $h(t-s)$ where $h(t) = 0$ if $t \le 0$ and ≥ 0 otherwise. The different lineages and the immigration flow are assumed independent. Then (Exercise 15.4.19), N is a simple point process if $\int_{\mathbb{R}_+} h(t)\,\mathrm{d}t < 1$ that admits the $\mathcal{F}_t^N$-intensity

$$\lambda(t) = \nu + \int_{(0,t]} h(t-s)\,N(\mathrm{d}s).$$

EXAMPLE 15.1.15: STOCHASTIC FAILURE RATE. For a point process N on $\mathbb{R}$ and any $t \in \mathbb{R}$, let $N_{-\infty}^t \cup \varnothing_t^\infty$ denote the point process coinciding with N on $(-\infty, t]$ and without points on $(t, +\infty)$. Let N be a simple point process on $\mathbb{R}$ with $\mathcal{F}_t^N$-intensity $\{\lambda(t)\}_{t\in\mathbb{R}}$ of the form

$$\lambda(t,\omega) = v(t, (N_{-\infty}^t \cup \varnothing_t^\infty)(\omega)),$$

where $v : \mathbb{R} \times M(\mathbb{R}) \to \mathbb{R}_+$ is measurable with respect to the σ-fields $\mathcal{B}(\mathbb{R}) \otimes \mathcal{M}(\mathbb{R})$ and $\mathcal{B}(\mathbb{R}_+)$. Then, for all $[a,b] \subset \mathbb{R}$,

$$P(N((a,b]) = 0 \,|\, \mathcal{F}_a) = \exp\left\{-\int_a^b v(s, N_{-\infty}^a \cup \varnothing_a^\infty)\,\mathrm{d}s\right\}.$$

Proof. Let $Z(t) := 1_{\{N((a,t])=0\}}$ for $t \ge a$. Then,

$$Z(t) = 1 - \int_{(a,t]} Z(s-)N(\mathrm{d}s).$$

For all $A \in \mathcal{F}_a^N$, by Theorem 15.1.11 (since $(t,\omega) \to 1_A(\omega)Z(t-,\omega)$ is in $\mathcal{P}(\mathcal{F}_\cdot^N)$),

$$
\begin{aligned}
E[1_A Z(t)] &= P(A) - E\left[\int_{(a,t]} 1_A Z(s-)\lambda(s)\,\mathrm{d}s\right] \\
&= P(A) - E\left[\int_{(a,t]} 1_A Z(s)\lambda(s)\,\mathrm{d}s\right] \\
&= P(A) - E\left[\int_a^t 1_A Z(s)v(s, N_{-\infty}^s \cup \varnothing_s^\infty)\,\mathrm{d}s\right] \\
&= P(A) - E\left[\int_a^t 1_A Z(s)v(s, N_{-\infty}^a \cup \varnothing_a^\infty)\,\mathrm{d}s\right] \\
&= P(A) - E\left[1_A \int_a^t E[Z(s)\,|\,\mathcal{F}_a]v(s, N_{-\infty}^a \cup \varnothing_a^\infty)\,\mathrm{d}s\right].
\end{aligned}
$$

Therefore

$$E[Z(t) \mid \mathcal{F}_a] = 1 - \int_a^t E[Z(s) \mid \mathcal{F}_a] v(s, N_{-\infty}^a \cup \varnothing_a^\infty) \, ds$$

and finally

$$E[Z(t) \mid \mathcal{F}_a] = \exp\left\{ -\int_a^t v(s, N_{-\infty}^a \cup \varnothing_a^\infty) \, ds \right\}.$$

□

The same proof, this time with $Z(t) := 1_{\{N((T_n, T_n + t]) = 0\}}$ $(t > 0)$, is easily extended (Exercise 15.4.8) to prove that on $\{T_n < \infty\}$ and for all $h > 0$,

$$P(N((T_n, T_n + h]) = 0 \mid \mathcal{F}_{T_n}) = \exp\left\{ -\int_{T_n}^{T_n + h} v(s, N_{-\infty}^{T_n} \cup \varnothing_{T_n}^\infty) \, ds \right\},$$

that is

$$P(T_{n+1} - T_n > h \mid \mathcal{F}_{T_n}^N) = \exp\left\{ -\int_{T_n}^{T_n + h} v(s, N_{-\infty}^{T_n} \cup \varnothing_{T_n}^\infty) \, ds \right\}.$$

This gives the interpretation of stochastic intensity as a stochastic failure rate.

The following is yet another avatar of the Borel–Cantelli lemma that can be considered the continuous time version of Theorem 13.3.38.

Theorem 15.1.16 *Let N be a simple locally finite point process on $\mathbb{R}_+$ with $\mathcal{F}_t$-intensity $\{\lambda(t)\}_{t \geq 0}$. Then*

$$N(\infty) < \infty \iff \int_0^\infty \lambda(s) \, ds < \infty \qquad \text{P-a.s.}$$

Proof. We first give a partial result with an elementary proof:

$$P(N(0, \infty) = \infty) = 1 \iff P\left(\int_0^\infty \lambda(s) \, ds = \infty \right) = 1.$$

The complete proof will be given soon after this.

A. Suppose that $N(0, \infty) = \infty$ a.s. and that it is not true that $\int_0^\infty \lambda(s) \, ds = \infty$ a.s. Then, there exists a $c < \infty$ such that $P\left(\int_0^\infty \lambda(s) \, ds \leq c \right) > 0$. By the smoothing formula,

$$E\left[\int_0^\infty 1_{\{\int_0^s \lambda(u)\, du \leq c\}} N(ds) \right] = E\left[\int_0^\infty 1_{\{\int_0^s \lambda(u)\, du \leq c\}} \lambda(s) \, ds \right].$$

The right-hand side is bounded above by c, whereas the left-hand side is bounded below by

$$E\left[1_{\{\int_0^\infty \lambda(u)\, du \leq c\}} N(0, \infty) \right] = \infty,$$

a contradiction. Therefore $\int_0^\infty \lambda(s) \, ds = \infty$ a.s.

B. Suppose that $\int_0^\infty \lambda(s) \, ds = \infty$ a.s. and that it is not true that $N(0, \infty) = \infty$ a.s. Then there exists a $c < \infty$ such that $P(N(0, \infty) \leq c) > 0$. By the smoothing formula (watch the parentheses for left-continuity),

$$E\left[\int_0^\infty 1_{\{N((0,s)) \leq c\}} N(ds) \right] = E\left[\int_0^\infty 1_{\{N((0,s)) \leq c\}} \lambda(s) \, ds \right].$$

The left-hand side is bounded by $c + 1$, whereas the right-hand side is bounded below by

$$\mathrm{E}\left[1_{\{N(0,\infty)\le c\}} \int_0^\infty \lambda(u)\,\mathrm{d}u\right] = \infty\,,$$

a contradiction. Therefore $N(0,\infty) = \infty$ a.s. $\qquad\square$

The complete proof uses Theorem 13.4.17 after the following observation (Exercise 15.4.13):

Theorem 15.1.17 *Let N be a simple locally finite point process on $\mathbb{R}_+$ with (locally integrable) $\mathcal{F}_t$-intensity $\{\lambda(t)\}_{t\ge 0}$. Define for $t \ge 0$, $M(t) := N(t) - \int_0^t \lambda(s)\,\mathrm{d}s$. Then*

$$M(t)^2 - \int_0^t \lambda(s)\,\mathrm{d}s \quad (t \ge 0)$$

is a local $\mathcal{F}_t$-martingale. In other words, the bracket process of the martingale $\{M(t)\}_{t\ge 0}$ is

$$\langle M\rangle(t) = \int_0^t \lambda(s)\,\mathrm{d}s\,.$$

We may now apply Theorem 13.4.17 which we recall for convenience:

a. If $\langle M\rangle(\infty) < \infty$, then $M(t)$ converges to a finite limit as $t \uparrow \infty$.

b. If $\langle M\rangle(\infty) = \infty$, then

$$\lim_{t\uparrow\infty} \frac{M(t)}{\langle M\rangle(t)} = 0\,.$$

That is, in the particular case of point processes with a stochastic intensity,

a. If $\int_0^\infty \lambda(s)\,\mathrm{d}s < \infty$, $\lim_{t\uparrow\infty}\left(N(t) - \int_0^t \lambda(s)\,\mathrm{d}s\right) < \infty$, and therefore $N(\infty) = \infty$.

b. If $\int_0^\infty \lambda(s)\,\mathrm{d}s = \infty$, $\lim_{t\uparrow\infty} \frac{N(t) - \int_0^t \lambda(s)\,\mathrm{d}s}{\int_0^t \lambda(s)\,\mathrm{d}s} = \infty$. That is

$$\lim_{t\uparrow\infty} \frac{N(t)}{\int_0^t \lambda(s)\,\mathrm{d}s} = \infty$$

and therefore since $\int_0^\infty \lambda(s)\,\mathrm{d}s = \infty$, $N(\infty) = \infty$.

Remark 15.1.18 The second instance of the Poisson–Wiener analogy, this time between the Brownian motion stochastic calculus and the Poisson process stochastic calculus, is perhaps more interesting. The definition of the stochastic $\mathcal{F}_t$-intensity $\{\lambda(t)\}_{t\ge 0}$ of a point process on the positive half-line is that

$$N(t) = \int_0^t \lambda(s)\,\mathrm{d}s + M(t)\,, \qquad\qquad (\star)$$

where $\{M(t)\}_{t\ge 0}$ is a (local) $\mathcal{F}_t$-martingale. From an informational point of view, the point process N is a corrupted version of the "integrated signal" process $\{\int_0^t \lambda(s)\,\mathrm{d}s\}_{t\ge 0}$, as it is obtained from the latter by addition of an "integrated noise" $\{M(t)\}_{t\ge 0}$, that is, a centered and uncorrelated[1] process. This point of view is particularly interesting

---

[1]At least if this martingale is square-integrable.

in a number of applications, in the analysis of spike trains in nervous fibers, in photon based communication systems or in the analysis and detection of any phenomenon whose manifestation is a sequence of events whose rate depends on the "intensity" of the said phenomenon.

In the Brownian motion calculus, the analogy of N is the (Itô) process

$$X(t) = \int_0^t \varphi(s)\,ds + W(t),\qquad\qquad (\star\star)$$

where the Brownian motion $\{W(t)\}_{t\geq 0}$ is indeed a centered and uncorrelated stochastic process, actually the prototype of "integrated white noise".

For the reader familiar with signal analysis, the model $(\star\star)$ corresponds to the technique of amplitude modulation, whereas the model $(\star)$ recalls the technique of amplitude modulation.

15.1.2   Stochastic Intensity Kernels

The notion of stochastic intensity will now be extended to point processes on $\mathbb{R} \times K$ where $(K, \mathcal{K})$ is some measurable space. The main role is attributed to the first ("time") coordinate axis $\mathbb{R}$.

Let $(\Omega, \mathcal{F}, P)$ be a probability space, and let $(K, \mathcal{K})$ be a measurable space. Let $\lambda : \mathbb{R} \times \Omega \times \mathcal{K} \to \overline{\mathbb{R}}_+$ be a mapping such that

- for all $\omega \in \Omega$, all $t \in \mathbb{R}$, $\lambda(\omega, t, \cdot)$ is a sigma-finite measure on $(K, \mathcal{K})$, and
- for all $L \in \mathcal{K}$, $(t, \omega) \to \lambda(t, \omega, L)$ is measurable.

Then $\lambda(\cdot, \cdot)$ is called a *stochastic kernel* from $(\mathbb{R} \times \Omega, \mathcal{B}(\mathbb{R}) \otimes \mathcal{F})$ to $(K, \mathcal{K})$. If moreover, for some history $\{\mathcal{F}_t\}_{t \in \mathbb{R}}$ and for all $L \in \mathcal{K}$, the stochastic process $\{\lambda(t, L)\}_{t \in \mathbb{R}}$ is $\mathcal{F}_t$-adapted (resp. $\mathcal{F}_t$-progressive, $\mathcal{F}_t$-predictable), the stochastic kernel λ is called $\mathcal{F}_t$-adapted (resp. $\mathcal{F}_t$-progressive, $\mathcal{F}_t$-predictable). If there exists a sequence $\{L_k\}_{k\geq 1}$ of sets of $\mathcal{K}$ increasing to K and such that for all bounded intervals $[a, b]$, and all $k \geq 1$, $\int_a^b \lambda(s, L_k)ds < \infty$ a.s., the kernel λ is called *locally integrable*.

Let now $\overline{N}$ be a point process on $\mathbb{R} \times K$. Define for all $t \in \mathbb{R}$

$$\mathcal{F}_t^{\overline{N}} := \sigma\{\overline{N}(A)\,;\, A \in \mathcal{B}(\mathbb{R}) \otimes \mathcal{K}, A \subseteq (-\infty, t] \times K\}.\qquad (15.5)$$

The family $\{\mathcal{F}_t^{\overline{N}}\}_{t \in \mathbb{R}}$ is the *internal history of* $\overline{N}$ and any history $\{\mathcal{F}_t\}_{t \in \mathbb{R}}$ such that $\mathcal{F}_t^{\overline{N}} \subseteq \mathcal{F}_t$ for all $t \in \mathbb{R}$ is called a history of $\overline{N}$. We shall also use the notation $N^L(C) := \overline{N}(C \times L)$.

Definition 15.1.19 *A point process* $\overline{N}$ *on* $\mathbb{R} \times K$ *is called* simple and locally finite *if there exists a sequence* $\{L_k\}_{k\geq 1}$ *in* $\mathcal{K}$ *increasing to* K *and such that for all* $k \geq 1$, N^{L_k} *is a simple locally finite point process on* $\mathbb{R}$.

Definition 15.1.20 *Let* $\overline{N}$ *be a simple locally finite point process on* $\mathbb{R} \times K$ *and let* $\{\mathcal{F}_t\}_{t \in \mathbb{R}}$ *be a history of* $\overline{N}$. *Let* $\lambda(\cdot, \cdot)$ *be an* $\mathcal{F}_t$-adapted locally integrable stochastic kernel *from* $(\mathbb{R} \times \Omega, \mathcal{B}(\mathbb{R}) \otimes \mathcal{F})$ *to* $(K, \mathcal{K})$. *The point process* $\overline{N}$ *is said to admit the stochastic* $(P, \mathcal{F}_t)$-*intensity kernel* $\lambda(\cdot, \cdot)$ *if for all* $[a, b] \subset \mathbb{R}$, *all* $L \in \mathcal{K}$ *and all* $k \geq 1$, *the point process* $N^{L \cap L_k}$ *admits the stochastic* $(P, \mathcal{F}_t)$-*intensity* $\{\lambda(t, L \cap L_k)\}_{t \in \mathbb{R}}$.

We shall also write "the $\mathcal{F}_t$-intensity kernel $\lambda(t, dz)$" instead of "the $\mathcal{F}_t$-intensity kernel $\lambda(\cdot, \cdot)$".

Remark 15.1.21 Note that the sequences $\{L_k\}_{k \geq 1}$ of the definition of a locally integrable kernel and of Definitions 15.1.19 and 15.1.20 can always be taken to be identical without loss of generality.

The proof of the next result is an easy adaptation of that of Theorem 15.1.11.

Theorem 15.1.22 *Let $\overline{N}$ be a simple locally finite point process on $\mathbb{R} \times K$. It admits the stochastic $\mathcal{F}_t$-intensity kernel $\lambda(\cdot, \cdot)$ if and only if*

$$\mathrm{E}\left[\int_{\mathbb{R} \times K} \int H(t, z) \overline{N}(dt \times dz)\right] = \mathrm{E}\left[\int \int_{\mathbb{R} \times K} H(t, z) \lambda(t, dz) dt\right] \qquad (15.6)$$

for all non-negative $\mathcal{P}(\mathcal{F}.) \otimes \mathcal{K}$-measurable functions $H : \mathbb{R} \times \Omega \times K \to \mathbb{R}$. For the "if" part, (15.6) is required only for all non-negative $\mathcal{B}(\mathbb{R}) \otimes \mathcal{F} \otimes \mathcal{K}$-measurable functions that are adapted to $\{\mathcal{F}_t\}_{t \in \mathbb{R}}$ and left-continuous in the t-variable.

An immediate corollary concerns *complex-valued* integrands:

Corollary 15.1.23 *Let $\overline{N}$ be as in Theorem 15.1.22, and let $H \in \mathcal{P}(\mathcal{F}.) \otimes \mathcal{K}$ be a complex-valued function such that at least one of the following statements is true:*

$$\mathrm{E}\left[\int_{\mathbb{R} \times K} \int |H(t, z)| \overline{N}(dt \times dz)\right] < \infty,$$

$$\mathrm{E}\left[\int_{\mathbb{R}} \int_K |H(t, z)| \lambda(t, dz) dt\right] < \infty.$$

Then the other statement is also true, $\int_{\mathbb{R}} \int_K H(t, z) \overline{N}(dt \times dz)$ and $\int_{\mathbb{R}} \int_K H(t, z) \lambda(t, dz) dt$ are well defined and finite quantities, and equality (15.6) holds true.

The Case of Marked Point Processes

So far, the projection on $\mathbb{R}$ of $\overline{N}$ need not be a locally finite point process, that is, $N(C) := \overline{N}(C \times K)$ may be infinite for some bounded Borel set C (for instance, if $\overline{N}$ is an HPP of intensity 1). The case of a marked point process is different from this point of view, but it is nevertheless a particular case to which the general results apply. Recall the following definition:

Definition 15.1.24 *Let N be a simple locally finite point process on $\mathbb{R}$, called the base, with time sequence $\{T_n\}_{n \in \mathbb{Z}}$. Let $Z := \{Z_n\}_{n \in \mathbb{Z}}$ be a random sequence with values in some measurable space $(K, \mathcal{K})$. We then say that (N, Z) is a simple and locally finite point process with marks in K and define the lifted point process N_Z on $\mathbb{R} \times K$ by*

$$N_Z(A) := \sum_{n \in \mathbb{Z}} 1_A((T_n, Z_n)) \quad (A \in \mathcal{B} \otimes \mathcal{K}).$$

(Remember the convention according to which the sum extends only to those indices n for which $|T_n| < \infty$.) Replacing $\overline{N}$ by N_Z in (15.5), we obtain the definition of $\mathcal{F}_t^{N_Z}$, which we shall also write as $\mathcal{F}_t^{N,Z}$.

Definition 15.1.25 *Let (N, Z) be a simple and locally finite marked point process with marks in K and with associated lifted point process N_Z on $\mathbb{R} \times K$. Let $\{\mathcal{F}_t\}_{t \in \mathbb{R}}$ be a history of N_Z. An $\mathcal{F}_t$-adapted stochastic kernel from $(\mathbb{R} \times \Omega, \mathcal{B}(\mathbb{R}) \otimes \mathcal{F})$ to $(K, \mathcal{K})$ is called a $(P, \mathcal{F}_t)$-intensity kernel of (N, Z) if it is a $(P, \mathcal{F}_t)$-intensity kernel of N_Z.*

Remark 15.1.26 In this case, there is no need for the sequence $\{L_k\}_{k \geq 1}$ in Definition 15.1.20 since for each $C \in \mathcal{B}$ and for each $L \in \mathcal{K}$, the point process N_Z^L on $\mathbb{R}$ defined by

$$N_Z^L(C) = N_Z(C \times L) = \sum_{n \in \mathbb{Z}} 1_C(T_n) 1_L(Z_n)$$

is simple and locally finite.

Let (N, Z) be a simple and locally finite marked point process with marks in K, with associated lifted point process N_Z on $\mathbb{R} \times K$, and suppose that it admits the $\mathcal{F}_t$-stochastic kernel $\lambda(t, \mathrm{d}z)$. In particular, the stochastic process

$$\lambda(t) := \lambda(t, K) \quad (t \geq 0)$$

is the $\mathcal{F}_t$-intensity of N. Let now

$$\Phi(t, C) := \frac{\lambda(t, C)}{\lambda(t, K)},$$

with the convention $\frac{0}{0} = 0$, so that

$$\lambda(t, \mathrm{d}z) = \lambda(t)\Phi(t, \mathrm{d}z).$$

The marked point process (N, Z) is then said to have the *local $\mathcal{F}_t$-characteristics* $(\lambda(t), \Phi(t, \mathrm{d}z))$. If $\mathcal{F}_t = \mathcal{F}_t^{N,Z}$ for all $t \geq 0$, the interpretation of Φ is that on $\{T_n < \infty\}$,

$$\Phi(T_n, L) = P(Z_n \in L \mid \mathcal{F}_{T_n-})$$

(Theorem 15.1.43) where we recall that $\mathcal{F}_{T_n-} = \sigma(T_k, Z_k \, ; \, k \leq n - 1) \vee \sigma(T_n)$.

EXAMPLE 15.1.27: POISSON WITH INDEPENDENT IID MARKS. If the base point process N is a Poisson process with (deterministic) intensity $\nu(t)$, and if the sequence of marks is IID and independent of N, (N, Z) admits the $\mathcal{F}_t^{N,Z}$-intensity kernel $\nu(t)Q(\mathrm{d}z)$ or, in other words, the local $\mathcal{F}_t^{N,Z}$-characteristics $(\nu(t)\,\mathrm{d}tQ(\mathrm{d}z))$, where Q is the common distribution of the marks (Exercise 15.4.7).

EXAMPLE 15.1.28: MULTIVARIATE POINT PROCESSES, TAKE 1. The definitions and results above will now be specialized to multivariate point processes by letting $K = E$, a denumerable set. For each $i \in E$, define a point process N_i on $\mathbb{R}$ by $N_i(C) = \overline{N}(C \times \{i\})$ for all $C \in \mathcal{B}(\mathbb{R})$, and let $\lambda_i(t) = \lambda(t, \{i\})$. This defines a *multivariate point process* $\{N_i\}_{i \in E}$, each N_i having the $\mathcal{F}_t$-intensity $\{\lambda_i(t)\}_{t \geq 0}$ where $\{\mathcal{F}_t\}_{t \geq 0}$ is a common history of the N_i's.

EXAMPLE 15.1.29: TRANSITION TIMES OF A CONTINUOUS-TIME HMC, TAKE 2. The HMC of Example 15.1.13 can be viewed as a marked point process $(\widetilde{N}, \widetilde{Z})$ where $\widetilde{N}$ counts all the transitions and $\widetilde{Z}_n := X(\widetilde{T}_n)$. Let $\widetilde{N}_{ij}$ be the point process counting the transitions from state i to j. Recall the notation $Z_i(t) = 1_{\{X(t)=i\}}$. The base point process $\widetilde{N}$ admits the $\mathcal{F}_t^N$-intensity

$$\widetilde{\lambda}(t) = \sum_{i,j\in E\,;\,i\neq j} \widetilde{\lambda}_{ij}(t) = \sum_{i\in E}\sum_{j\,,\,j\neq i} q_{ij}Z_i(t) = \sum_{i\in E} q_i Z_i(t) = q_{X(t)}\,.$$

Let $\widetilde{\lambda}_i(t)$ be the $\mathcal{F}_t^X$-intensity of the point process $\widetilde{N}_i$ counting the transitions bringing the HMC into state i. As $\widetilde{N}_i = \sum_{k\in E\,;\,k\neq i}\widetilde{N}_{ki}$,

$$\widetilde{\lambda}_i(t) = \sum_{k\in E\,;\,k\neq i} Z_k(t)\, q_{ki} = q_{X(t),i}\,.$$

EXAMPLE 15.1.30: STRESS RELEASE EARTHQUAKE MODEL The original appellation of this model is "Self-correcting point process".[2] Its stochastic $\mathcal{F}_t$-intensity is

$$\lambda(t) := e^{X(0)+ct-\sum_{i=1}^{N((0,t))} Z_i},$$

where $X(0)$ is a given random variable, $c > 0$, and $\{Z_n\}_{n\geq 1}$ is an IID sequence of non-negative real random variables. Here

$$\mathcal{F}_t := \sigma(X(0)) \vee \mathcal{F}_t^{(N,Z)},$$

where (N, Z) is the marked point process with base N and mark sequence $\{Z_n\}_{n\geq 1}$.

The following corollary just rephrases Theorem 15.1.22.

Theorem 15.1.31 *Let (N, Z) be a simple marked point process on $\mathbb{R}$ with marks in K and admitting the $\mathcal{F}_t$-stochastic kernel $\lambda(t, \mathrm{d}z)$. Then*

$$\mathrm{E}\left[\sum_{n\in\mathbb{Z}} H(T_n, Z_n)\right] = \mathrm{E}\left[\int_{\mathbb{R}}\int_K H(t, z)\lambda(t, \mathrm{d}z)\right]$$

for all non-negative $\mathcal{P}(\mathcal{F}.) \otimes \mathcal{K}$-measurable functions $H : \mathbb{R} \times \Omega \times K \to \mathbb{R}$.

EXAMPLE 15.1.32: MEAN BUSY PERIOD IN AN $M/GI/1/\infty$ QUEUE. Let the sequence $\{(T_n, \sigma_n)\}_{n\in\mathbb{N}}$ be an M/GI input flow. This means that $\{T_n\}_{n\in\mathbb{N}\setminus\{0\}}$ is the event-times sequence of an HPP A with intensity λ, $T_0 := 0$, and $\{\sigma_n\}_{n\in\mathbb{N}}$ is an IID sequence of non-negative random variables with common CDF G and independent of A. In queueing theory, A counts the arrival times in a "system" (called the queue) and σ_n is the service required by the customer arriving at time T_n. Let $W(t)$ be the workload at time t, that is, the total amount of service remaining to be done at time t, assuming that service is provided at unit rate as long as there is some work to be done. More explicitly, for $t \geq 0$,

---

[2][Isham and Westcott, 1979].

$$W(t) = \sigma_0 + \sum_{n \geq 1} \sigma_n 1_{(0,t]}(T_n) - \int_0^t 1_{\{W(s)>0\}} \, ds \,.$$

Let R_1 be the first strictly positive time at which the system is empty (∞ if the system never empties).

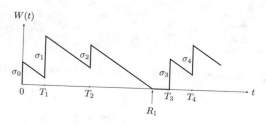

Clearly R_1 is an $\mathcal{F}_t^{\overline{N}}$-stopping time, where $\overline{N}$ is the point process on $\mathbb{R}_+ \times \mathbb{R}_+$ with point sequence $\{(T_n, \sigma_n)\}_{n \in \mathbb{N}}$. For all $M > 0$,

$$R_1 \wedge M \leq \sigma_0 + \sum_{k \geq 1} \sigma_k 1_{(0,R_1 \wedge M]}(T_k) = \sigma_0 + \int_0^\infty \int_E \sigma 1_{(0,R_1 \wedge M]}(t) \overline{N}(dt \times d\sigma) \,. \quad (15.7)$$

The mapping

$$(t, \omega, \sigma) \to H(t, \omega, \sigma) = \sigma 1_{(0,R_1(\omega) \wedge M]}(t),$$

is in $\mathcal{P}(\mathcal{F}^{\overline{N}}) \otimes \mathcal{B}(\mathbb{R}_+)$ (noting that it is left-continuous in the t-argument). Since the $\mathcal{F}_t^{\overline{N}}$-intensity kernel of $\overline{N}$ is $\lambda G(dz)$, we obtain from (15.7) and Theorem 15.1.31

$$\mathrm{E}[R_1 \wedge M] \leq \mathrm{E}[\sigma_0] + \mathrm{E}\left[\int_0^\infty \int_{\mathbb{R}_+} \sigma 1_{(0,R_1 \wedge M]}(t) \lambda G(d\sigma) \, dt\right]$$
$$= \mathrm{E}[\sigma_0] + \lambda \mathrm{E}[\sigma_0] \mathrm{E}[R_1 \wedge M]. \quad (15.8)$$

In particular, $\mathrm{E}[R_1 \wedge M](1 - \lambda \mathrm{E}[\sigma_0]) \leq \mathrm{E}[\sigma_0]$, and therefore, if $\lambda \mathrm{E}[\sigma_0] < 1$,

$$\mathrm{E}[R_1 \wedge M] \leq \frac{\mathrm{E}[\sigma_0]}{1 - \lambda \mathrm{E}[\sigma_0]}$$

for all $M > 0$, and therefore $\mathrm{E}[R_1] \leq \frac{\mathrm{E}[\sigma_0]}{1 - \lambda \mathrm{E}[\sigma_0]} < \infty$. Reproducing the calculation with R_1 replacing $R_1 \wedge M$, we have, since R_1 is almost surely finite, the *equality*

$$\mathrm{E}[R_1] = \frac{\mathrm{E}[\sigma_0]}{1 - \lambda \mathrm{E}[\sigma_0]} \,.$$

In summary: $\lambda \mathrm{E}[\sigma_0] < 1$ is a sufficient condition for $\mathrm{E}[R_1]$ to be finite in an $M/GI/1/\infty$ queue. It is also a necessary condition if $\mathrm{E}[\sigma_0] > 0$, because when R_1 is finite,

$$R_1 = \sigma_0 + \sum_{k \geq 1} \sigma_k 1_{(0,R_1]}(T_k)$$

and therefore $\mathrm{E}[R_1] = \mathrm{E}[\sigma_0] + \lambda \mathrm{E}[\sigma_0] \mathrm{E}[R_1]$, that is $\mathrm{E}[R_1](1 - \lambda \mathrm{E}[\sigma_0]) = \mathrm{E}[\sigma_0]$, which implies that $1 - \lambda \mathrm{E}[\sigma_0] > 0$.

Let (N, Z) be a simple locally finite marked point process on $\mathbb{R}_+$ with marks in the measurable space $(K, \mathcal{K})$ and let $\{\mathcal{F}_t\}_{t \geq 0}$ be a history of (N, Z). Let (N, Z) admit the $\mathcal{F}_t$-intensity kernel $\lambda(t, dz)$. In the sequel, the following notation will be used

$$\int_{(0,t] \times K} H(s, z) \, M_Z(ds \times dz)$$
$$:= \int_{(0,t] \times K} H(s, z) \, N_Z(ds \times dz) - \int_{(0,t] \times K} H(s, z) \, \lambda(s, dz) \, ds \, ,$$

provided the right-hand side is well defined. Therefore, formally,

$$M_Z(ds \times dz) := N_Z(ds \times dz) - \lambda(s, dz) \, ds \, .$$

Stochastic Integrals and Martingales

Theorem 15.1.33 *Let $\{H(t, z)\}_{t \geq 0}$ be an $\mathcal{F}_t$-predictable real-valued stochastic process indexed by K such that for all $t \geq 0$,*

$$\mathrm{E}\left[\int_{(0,t] \times K} |H(s, z)| \lambda(s, dz) \, ds\right] < \infty. \tag{15.9}$$

Then the stochastic process

$$M(t) := \int_{(0,t] \times K} H(s, z) \, M_Z(ds \times dz) \tag{15.10}$$

is a well-defined centered $\mathcal{F}_t$-martingale.

Proof. Condition (15.9) is equivalent to $\mathrm{E}\left[\int_{(0,t] \times K} |H(s, z)| \, N_Z(ds \times dz)\right] < \infty$ for all $t \geq 0$ (Theorem 15.1.31). Therefore, for all $t \geq 0$, $\int_{(0,t] \times K} |H(s, z)| \lambda(s, dz) \, ds < \infty$ and $\int_{(0,t] \times K} |H(s, z)| \, N_Z(ds \times dz) < \infty$, P-a.s. In particular, $\{M(t)\}_{t \geq 0}$ is P-a.s. a well-defined and finite stochastic process. For all $a, b \in \mathbb{R}_+$ ($0 \leq a \leq b$) and for all $A \in \mathcal{F}_a$,

$$\mathrm{E}\left[1_A(M(b) - M(a)\right] = \mathrm{E}\left[\int_{\mathbb{R}_+ \times K} H'(t, z) \, M_Z(ds \times dz)\right] \, ,$$

where $H'(t, z) := H(t, z) 1_A 1_{(a,b]}(t)$ defines an $\mathcal{F}_t$-predictable real-valued stochastic process indexed by K. By Theorem 15.1.31,

$$\mathrm{E}\left[\int_{\mathbb{R}_+ \times K} H'(t, z) \, N_Z(ds \times dz)\right] = \mathrm{E}\left[\int_{\mathbb{R}_+ \times K} H'(t, z) \, \lambda(s, dz) \, ds\right]$$

and therefore, for all $A \in \mathcal{F}_a$, $\mathrm{E}\left[1_A(M(b) - M(a)\right] = 0$. $\square$

Corollary 15.1.34 *Replacing assumption (15.9) of Theorem 15.1.33 by the condition that P-almost surely*

$$\int_{(0,t] \times K} |H(s, z)| \lambda(s, dz) \, ds < \infty \text{ for all } t \geq 0, \tag{15.11}$$

the stochastic process $\{M(t)\}_{t \geq 0}$ defined by (15.10) is then a local $\mathcal{F}_t$-martingale.

Proof. The random time

$$S_n := \inf\{t > 0 \,;\, \int_{(0,t] \times K} |H(s,z)| \lambda(s, \mathrm{d}z) \, \mathrm{d}s \geq n\},$$

with the usual convention $\inf \varnothing = +\infty$, is for each $n \geq 1$ an $\mathcal{F}_t$-stopping time, and $\lim_{n \uparrow \infty} S_n = \infty$. Moreover, $H(t,z) 1_{\{t \leq S_n\}}$ satisfies condition (15.9). Therefore by Theorem 15.1.33, $\{M(t \wedge S_n)\}_{t \geq 0}$ is a well-defined $\mathcal{F}_t$-martingale. $\qquad \square$

Theorem 15.1.35 *Let H be an $\mathcal{F}_t$-predictable real-valued stochastic process indexed by K such that for all $t \geq 0$, P-a.s.*

$$\mathrm{E}\left[\int_{(0,t] \times K} |H(s,z)|^2 \lambda(s, \mathrm{d}z) \, \mathrm{d}s\right] < \infty. \tag{15.12}$$

Then the stochastic process

$$M(t) := \int_{(0,t] \times K} H(s,z) \, M_Z(\mathrm{d}s \times \mathrm{d}z)$$

is well defined and a square-integrable $\mathcal{F}_t$-martingale. Moreover,

$$\mathrm{E}\left[M(t)^2\right] = \mathrm{E}\left[\int_{(0,t] \times K} |H(s,z)|^2 \lambda(s, \mathrm{d}z) \, \mathrm{d}s\right]. \tag{15.13}$$

Proof. Let T_n be the n-th event time of the base point process. The proof that $M(t)$ is well defined follows from Theorem 15.1.33. In fact, observing that $\mathrm{E}\left[\int_{(0,T_n]} \lambda(s) \, \mathrm{d}s\right] = \mathrm{E}\left[N(0,T_n]\right] = n$,

$$\mathrm{E}\left[\int_{(0,t \wedge T_n] \times K} |H(s,z)| \lambda(s, \mathrm{d}z) \, \mathrm{d}s\right] \leq \mathrm{E}\left[\int_{(0,t \wedge T_n] \times K} (1 + |H(s,z)|^2) \lambda(s, \mathrm{d}z) \, \mathrm{d}s\right]$$

$$= \mathrm{E}\left[N(0,T_n]\right] + \mathrm{E}\left[\int_{(0,t \wedge T_n] \times K} |H(s,z)|^2 \lambda(s, \mathrm{d}z) \, \mathrm{d}s\right] < \infty$$

$$= n + \mathrm{E}\left[\int_{(0,t \wedge T_n] \times K} |H(s,z)|^2 \lambda(s, \mathrm{d}z) \, \mathrm{d}s\right] < \infty.$$

Therefore $\{M(t \wedge T_n)\}_{t \geq 0}$ is well defined, and so is $\{M(t)\}_{t \geq 0}$ since $\lim_{n \uparrow \infty} T_n = \infty$.

We now turn to the proof of (15.13). By the product rule of Stieltjes–Lebesgue calculus,

$$M(t)^2 = \int_{(0,t]} M(t-) \, \mathrm{d}M(t) + \int_{(0,t] \times K} H(s,z)^2 N_Z(\mathrm{d}s \times \mathrm{d}z).$$

Since

$$m(t) := \int_{(0,t]} M(t-) \, \mathrm{d}M(t) = \int_{(0,t] \times K} M(s-) H(s,z) \, M_Z(\mathrm{d}s \times \mathrm{d}z)$$

is a local $\mathcal{F}_t$-martingale with respect to the localizing stopping times

$$V_n := \inf\left\{t \geq 0 \,;\, |M(t-)| + \int_{(0,t] \times K} |H(s,z)| \lambda(s, \mathrm{d}z) \, \mathrm{d}s \geq n\right\} \wedge T_n,$$

we have that

$$E\left[M(t \wedge V_n)^2\right] = E\left[\int_{(0,t \wedge V_n] \times K} H(s,z)^2 \, N_Z(\mathrm{d}s \times \mathrm{d}z)\right]$$

$$= E\left[\int_{(0,t \wedge V_n] \times K} H(s,z)^2 \lambda(s,\mathrm{d}z) \, \mathrm{d}s\right],$$

from which (15.13) follows, if we can show that $\lim_{n \uparrow \infty} E\left[M(t \wedge V_n)^2\right] = E\left[M(t)^2\right]$. This will be the case because (as will soon be proved) if $M(t \wedge V_n)$ converges in $L^2_{\mathbb{C}}(P)$ to some limit. This limit is necessarily $M(t)$, the almost sure limit of $M(t \wedge V_n)$. The $L^2_{\mathbb{C}}(P)$-convergence of $M(t \wedge V_n)$ to be proved follows from the Cauchy criterion since, by a computation similar to the one above, with $m \geq n$,

$$E\left[(M(t \wedge V_m) - M(t \wedge V_n))^2\right]$$

$$= E\left[m(t \wedge V_m) - m(t \wedge V_n)\right] + E\left[\int_{(t \wedge V_n, t \wedge V_m] \times K} H(s,z)^2 \lambda(s,\mathrm{d}z) \, \mathrm{d}s\right]$$

$$= E\left[\int_{(t \wedge V_n, t \wedge V_m] \times K} H(s,z)^2 \lambda(s,\mathrm{d}z) \, \mathrm{d}s\right],$$

a quantity that vanishes as $m, n \uparrow \infty$. □

Corollary 15.1.36 *If assumption (15.15) of Theorem 15.4.11 is replaced by*

$$\int_{(0,t] \times K} |H(s,z)|^2 \lambda(s,\mathrm{d}z) \, \mathrm{d}s < \infty \quad P\text{-}a.s., (t \geq 0), \tag{15.14}$$

the stochastic process

$$M(t) := \int_{(0,t] \times K} H(s,z) \, M_Z(\mathrm{d}s \times \mathrm{d}z) \quad (t \geq 0)$$

is a square-integrable local $\mathcal{F}_t$-martingale.

Proof. This follows from Theorem 15.4.11 in the same manner as Corollary 15.1.34 followed from Theorem 15.1.33. □

15.1.3 Martingales as Stochastic Integrals

Let (N, Z) be a marked point process on $[0, 1]$ with marks in the measurable space $(K, \mathcal{K})$ such that the associated lifted point process N_Z is Poisson with intensity measure

$$\nu(\mathrm{d}t \times \mathrm{d}z) := \lambda(t,z) \, \mathrm{d}t \, Q(\mathrm{d}z),$$

where Q is a probability measure on the measurable space $(K, \mathcal{K})$ of the marks. We suppose that $\nu([0, 1] \times K) < \infty$.

Theorem 15.1.37 *Let $\{m(t)\}_{t\in[0,1]}$ be a real-valued centered $\mathcal{F}_t^W$-square-integrable martingale on $[0,1]$. Then*

$$M(t) = \int_{[0,t]\times K} C(s,z)\, M_Z(\mathrm{d}s \times \mathrm{d}z),$$

where C is an $\mathcal{F}_t$-predictable real-valued stochastic process indexed by K such that for all $t \geq 0$, P-a.s.

$$\mathrm{E}\left[\int_{(0,t]\times K} |C(s,z)|^2 \lambda(s,\mathrm{d}z)\,\mathrm{d}s\right] < \infty. \tag{15.15}$$

Lemma 15.1.38

A. *The collection of random variables*

$$\mathcal{H} := \{e^X ; X \in \mathcal{U}\}$$

where $\mathcal{U}$ is the collection of random variables

$$\sum_{j=1}^k a_j N_Z(C_j \times L_j) \quad (k \in \mathbb{N},\, C_j \in \mathcal{B}([0,1]),\, L_j \in K, a_1,\dots,a_k \in \mathbb{R})$$

is total in the Hilbert space $L_{\mathbb{R}}^2(\mathcal{F}_1^{(N,Z)}, P)$.

B. *The collection of random variables that are linear combinations of elements of*

$$\mathcal{K}_0 := \left\{ M_f(1) := \exp\left\{\int_{[0,1]\times K} f(s,z)\, N_Z(\mathrm{d}s \times \mathrm{d}z) + \int_{[0,t]\times K} \left(e^{f(s,z)} - 1\right) \lambda(s,z)\, Q(\mathrm{d}z)\right\}\right\},$$

where $f : [0,1] \times K \to \mathbb{R}$ is a measurable function such that

$$\int_{[0,1]\times K} \left(e^{f(s,z)} - 1\right)^2 \lambda(t,z)\,\mathrm{d}t\, Q(\mathrm{d}z) < \infty,$$

is dense in the Hilbert space $L_{\mathbb{R}}^2(\mathcal{F}_1^{(N,Z)}, P)$.

The proof is an immediate adaptation of Lemma 14.3.2.

The proof of Theorem 15.1.37 is in turn an easy adaptation of the proof of Theorem 14.3.1 and is based on the following lemma:

Lemma 15.1.39 *Let $f : [0,1] \times K \to \mathbb{R}$ be a measurable function such that*

$$\int_{[0,1]\times K} \left(e^{f(s,z)} - 1\right)^2 \lambda(t,z)\,\mathrm{d}t\, Q(\mathrm{d}z) < \infty.$$

(This is the case if f is non-positive or bounded.) Let for $t \in [0,1]$

$$M_f(t) := \exp\left\{\int_{[0,t]\times K} f(s,z)\, N_Z(\mathrm{d}s \times \mathrm{d}z) + \int_{[0,t]\times K} \left(e^{f(s,z)} - 1\right) \lambda(s,z)\, Q(\mathrm{d}z)\right\}.$$

Under the above conditions,

(a):

$$M_f(t) = 1 + \int_{[0,t] \times K} M_f(s-) \left(e^{f(s,z)} - 1 \right) M_Z(\mathrm{d}s \times \mathrm{d}z), \qquad (\star)$$

(b): $\{M_f(t)\}_{t \in [0,1]}$ *is a square integrable martingale, and*

(c): $\mathrm{E}\left[M_f(t)^2 - 1 \right] = \mathrm{E}\left[\int_{[0,t] \times K} M_f(s)^2 \left(e^{f(s,z)} - 1 \right)^2 \lambda(s,z) Q(\mathrm{d}z) \right]$ $(t \in [0,1])$.

Proof. (a): Observe that at an event-time t of the base point process with corresponding mark $z \in K$,

$$M_f(t) - M_f(t-) = M_f(t-) \left(e^{f(t,z)} - 1 \right),$$

at a time t between two event times

$$\frac{\mathrm{d}M_f(t)}{\mathrm{d}t} = M_f(t) \int_K \left(e^{f(t,z)} - 1 \right) \lambda(t,z) Q(\mathrm{d}z).$$

(b): It suffices to show, in view of Theorem 15.4.11, that

$$\mathrm{E}\left[\int_0^1 \int_K M_f(s)^2 \left(e^{f(s,z)} - 1 \right)^2 \lambda(s,z) Q(\mathrm{d}z) \, \mathrm{d}s \right] < \infty. \qquad (15.16)$$

This is true when $M_f(t)$ is replaced by $M_f(t \wedge S_n)$, where

$$S_n := \inf\{t \geq 0 \,;\, M_f(t-) \geq n\}.$$

In particular,

$$\mathrm{E}\left[(M_f(t \wedge S_n) - 1)^2 \right] = \mathrm{E}\left[\int_0^t \int_K M_f(s)^2 1_{\{s \leq S_n\}} \left(e^{f(s,z)} - 1 \right)^2 \lambda(s,z) Q(\mathrm{d}z) \, \mathrm{d}s \right]. \qquad (15.17)$$

Now,

$$\mathrm{E}\left[(M_f(t \wedge S_n) - 1)^2 \right] = \mathrm{E}\left[M_f(t \wedge S_n)^2 \right] - 1$$

and moreover

$$\mathrm{E}\left[M_f(t \wedge S_n)^2 \right] \geq \mathrm{E}\left[M_f(t)^2 1_{\{t \leq S_n\}} \right].$$

Therefore

$$\mathrm{E}\left[M_f(t)^2 1_{\{t \leq S_n\}} \right] \leq 1 + \mathrm{E}\left[\int_0^t \int_K M_f(s)^2 1_{\{s \leq S_n\}} \left(e^{f(s,z)} - 1 \right)^2 \lambda(s,z) Q(\mathrm{d}z) \right]$$

$$= 1 + \int_0^t \mathrm{E}\left[M_f(s)^2 1_{\{s \leq S_n\}} \right] \int_K \left(e^{f(s,z)} - 1 \right)^2 \lambda(s,z) Q(\mathrm{d}z) \, \mathrm{d}s.$$

By Gronwall's lemma, for all $t \geq 0$,

$$\mathrm{E}\left[M_f(t)^2 1_{\{t \leq S_n\}} \right] \leq \exp\left(\int_0^t \int_K \left(e^{f(s,z)} - 1 \right)^2 \lambda(s,z) Q(\mathrm{d}z) \, \mathrm{d}s \right) < \infty.$$

Since $\lim_{n \uparrow \infty} S_n = \infty$, by monotone convergence

$$\mathrm{E}\left[M_f(t)^2 \right] \leq \exp\left(\int_0^1 \int_K \left(e^{f(s,z)} - 1 \right)^2 \lambda(s,z) Q(\mathrm{d}z) \right) := C < \infty,$$

and therefore

$$E\left[\int_0^1\int_K M_f(s)^2\left(e^{f(s,z)}-1\right)^2\lambda(s,z)\,Q(\mathrm{d}z)\right]$$
$$\leq C\int_{[0,1]\times K}\left(e^{f(s,z)}-1\right)^2\lambda(t,z)\,\mathrm{d}t\,Q(\mathrm{d}z),$$

a finite quantity by hypothesis. Therefore (15.16) is proved.

(c): Start from (15.17) and let $n\uparrow\infty$ to obtain

$$E\left[(M_f(t))^2\right]=1+E\left[\int_0^t\int_K M_f(s)^2\left(e^{f(s,z)}-1\right)^2\lambda(s,z)\,Q(\mathrm{d}z)\,\mathrm{d}s\right].$$

This is a differential equation in $E\left[M_f(t)^2\right]$ whose solution gives (c). $\qquad\square$

Remark 15.1.40 The running analogy with Brownian motion stochastic calculus is perhaps more evident in the case of a standard Poisson process N on $\mathbb{R}_+$. In this case, the results of Theorem 15.1.37 specialize as follows. Any centered square-integrable $\mathcal{F}_t^N$-martingale on $[0,1]$ is of the form

$$m(t)=\int_{(0,t]}C(s)\,\mathrm{d}M(s),$$

where $\{C(t)\}_{t\in[0,1]}$ is an $\mathcal{F}_t^N$-predictable process such that $\int_0^1 H(s)^2\,\mathrm{d}s<\infty$.

15.1.4 The Regenerative Form of the Stochastic Intensity

Example 15.1.15 has shown that the notion of stochastic intensity is a generalization of that of hazard rate. Such a result can be extended to marked point processes, still in the special case where the history for which the stochastic intensity (kernel) is defined is the internal history "plus a prehistory".

A few results on the structure of point process histories will be needed. These are intuitive results whose technical proofs have been omitted.

Let $\{T_n\}_{n\geq 1}$ be a simple point process on $\mathbb{R}_+$, that is, a non-decreasing sequence of *positive* random variables possibly taking the value $+\infty$, and strictly increasing on $\mathbb{R}_+$ ($T_n<\infty\Rightarrow T_n<T_{n+1}$). In particular, it may be a finite point process (if $T_n=\infty$ for some $n\in\mathbb{N}$) and it need not be locally finite ($T_\infty:=\lim_{n\uparrow\infty}$ may be finite). Set $T_0\equiv 0$. Let $\{Z_n\}_{n\geq 1}$ be a sequence of random variables taking their values in the measurable space $(K,\mathcal{K})$. The sequence $\{(T_n,Z_n)\}_{n\geq 1}$ is called a marked point process on $\mathbb{R}_+$ with marks in K.

For each $L\in\mathcal{K}$, define the (simple) point process N^L on $\mathbb{R}_+$ by

$$N^L(C):=\sum_{n\geq 1}1_C(T_n)\,1_L(Z_n)\qquad(C\in\mathcal{B}(\mathbb{R}_+)).$$

Note that $N^L(\{0\})=0$. Define the (simple) point process N_Z on $\mathbb{R}_+\times K$ by

$$N_Z(C\times L):=N^L(C)\qquad(C\in\mathcal{B}(\mathbb{R}_+),L\in\mathcal{K}).$$

Define the internal history $\{\mathcal{F}_t^{N,Z}\}_{t\geq 0}$ of N_Z by

$$\mathcal{F}_t^{N,Z} := \sigma\left(N_Z(C \times L)\,; \, C \in (0, t], \, L \in \mathcal{K}\right),$$

and the history $\{\mathcal{F}_t\}_{t\geq 0}$ by

$$\mathcal{F}_t = \sigma(Z_0) \vee \mathcal{F}_t^{N,Z}, \tag{15.18}$$

where Z_0 is a random element taking values in a measurable space $(L_0, \mathcal{K}_0)$ (for instance, a space of functions). It represents a "prehistory" of the marked point process, in the sense that $\mathcal{F}_0 = \sigma(Z_0)$ contains the information already gathered at time 0 that may influence its future behavior.

It is intuitively clear that

$$\mathcal{F}_{T_n} = \sigma(Z_0) \vee \sigma(T_1, Z_1, \ldots, T_n, Z_n)$$

and

$$\mathcal{F}_{T_n-} = \sigma(Z_0) \vee \sigma(T_1, Z_1, \ldots, T_n).$$

Suppose that for all $n \geq 0$, all $L \in \mathcal{K}$, and all $C \in \mathcal{B}(\mathbb{R}_+)$,

$$P\left(S_{n+1} \in C\,, \, Z_{n+1} \in L \mid \mathcal{F}_{T_n}\right)(\omega) = \int_C g^{(n+1)}(\omega, x, L)\,\mathrm{d}x := G^{(n+1)}(\omega, C, L),$$

where for each $L \in \mathcal{K}$, the mapping $(\omega, x) \to g^{(n+1)}(\omega, x, L)$ is $\mathcal{F}_{T_n} \otimes \mathcal{B}(\mathbb{R}_+)$-measurable, and for each (ω, x), $L \to g^{(n+1)}(\omega, x, L)$ is a σ-finite measure on $(K, \mathcal{K})$. In particular,

$$P\left(S_{n+1} \in C \mid \mathcal{F}_{T_n}\right)(\omega) = \int_C g^{(n+1)}(\omega, x)\,\mathrm{d}x := G^{(n+1)}(\omega, C),$$

where $g^{(n+1)}(\omega, x) = g^{(n+1)}(\omega, x, K))$ and $G^{(n+1)}(\omega, C) = G^{(n+1)}(\omega, C, K)$.

Theorem 15.1.41 *For $L \in \mathcal{K}$ and $t \geq 0$, let*

$$\lambda(t, L) := \sum_{n \geq 0} \frac{g^{(n+1)}(t - T_n, L)}{1 - \int_0^{t-T_n} g^{(n+1)}(x)\,\mathrm{d}x}\, 1_{\{T_n \leq t < T_{n+1}\}}.$$

Then, $\lambda(t, \mathrm{d}z)$ is a stochastic $\mathcal{F}_t$-kernel of the marked point process N_Z.

Theorem 15.1.42 *For each $n \geq 0$, let $G^{(n+1)}(\cdot \times \cdot, \omega)$ be a regular version of the conditional distribution of (S_{n+1}, Z_{n+1}) given $\mathcal{F}_{T_n}$. Then for each $n \geq 0$, for all $C \in \mathcal{B}(\mathbb{R}_+)$, and all $L \in \mathcal{K}$,*

$$G^{(n+1)}(C \times L, \omega) = \int_C g^{(n+1)}(x, L, \omega)\,\mathrm{d}x,$$

for some mapping $(t, \omega) \to g^{(n+1)}(x, L, \omega)$ that is $\mathcal{F}_{T_n} \otimes \mathcal{B}(\mathbb{R}_+)$-measurable, and moreover the $\mathcal{F}_t$-predictable stochastic intensity kernel of the marked point process $\{(T_n, Z_n)\}_{n \geq 1}$ has, for $t \in (T_n, T_{n+1}]$, the form

$$\lambda^{(n)}(t, L, \omega) = \frac{g^{(n+1)}(t - T_n, L, \omega)}{G^{(n+1)}([t - T_n(\omega), \infty]), \omega)}.$$

Let N_Z admit the stochastic $(P, \mathcal{F}_t)$-intensity kernel $\lambda(t, dz)$. Denote by $\{\lambda(t)\}_{t \geq}$ an $\mathcal{F}_t$-predictable stochastic intensity of the basic point process $N := N^K$, and for each $L \in \mathcal{K}$, all $t \geq 0$, define $\Phi(t, L)$ by:

$$\lambda(t, L) = \lambda(t)\Phi(t, L)$$

if $\lambda(t) > 0$, and $\Phi(t, L) = 0$ if $\lambda(t) = 0$. Since $\lambda(T_n) > 0$ P-a.s. on $\{T_n < \infty\}$,

$$\Phi(T_n, L)1_{\{T_n < \infty\}} = \frac{\lambda(T_n, L)}{\lambda(T_n)}1_{\{T_n < \infty\}}.$$

Theorem 15.1.43 *For all $n \geq 1$ and all $L \in \mathcal{K}$,*

$$\Phi(T_n, L) = P(Z_n \in L \mid \mathcal{F}_{T_n-}).$$
$$(15.19)$$

15.2 Transformations of the Stochastic Intensity

15.2.1 Changing the History

The stochastic intensity depends of course on the probability P but also on the history $\{\mathcal{F}_t\}_{t \geq 0}$.

In the definition of the $\mathcal{F}_t$-stochastic intensity $\{\lambda(t)\}_{t \geq 0}$ of the simple point process N, the minimal requirement on the history $\{\mathcal{F}_t\}_{t \geq 0}$ is

$$\mathcal{F}_t \supseteq \mathcal{F}_t^\lambda \vee \mathcal{F}_t^N$$
$$(15.20)$$

because the intensity process and the point process itself are $\mathcal{F}_t$-adapted.

What happens if we modify the history? We first answer the following question: To what extent can we change the history and retain the same stochastic intensity?

Theorem 15.2.1 *Let N be a simple and locally finite point process on $\mathbb{R}$ with $\mathcal{F}_t$-intensity $\{\lambda(t)\}_{t \geq 0}$. Then:*

(i) *If $\{\mathcal{G}_t\}_{t \geq 0}$ is a history such that $\mathcal{G}_\infty$ is independent of $\mathcal{F}_\infty$, $\{\lambda(t)\}_{t \geq 0}$ is an $\mathcal{F}_t \vee \mathcal{G}_t$-intensity of N.*

(ii) *If $\{\widetilde{\mathcal{F}}_t\}_{t \geq 0}$ is a history of N such that $\mathcal{F}_t^\lambda \vee \mathcal{F}_t^N \subseteq \widetilde{\mathcal{F}}_t \subseteq \mathcal{F}_t$, $\{\lambda(t)\}_{t \geq 0}$ is an $\widetilde{\mathcal{F}}_t$-intensity of N.*

Proof. Exercise 15.4.5.

$\square$

We now consider situations where the stochastic intensity is modified by a change of history. Let N have the $\mathcal{F}_t$-intensity $\{\lambda(t)\}_{t \geq 0}$, where the history $\{\mathcal{F}_t\}_{t \geq 0}$ satisfies the minimal requirement (15.20). Suppose that the new history $\{\widetilde{\mathcal{F}}_t\}_{t \geq 0}$ is smaller, that is,

$$\mathcal{F}_t \supseteq \widetilde{\mathcal{F}}_t \supseteq \mathcal{F}_t^N \quad (t \geq 0).$$
$$(15.21)$$

It is possible that $\{\lambda(t)\}_{t \geq 0}$ is not $\widetilde{\mathcal{F}}_t$-adapted and therefore cannot be the $\widetilde{\mathcal{F}}_t$-intensity. Nevertheless, there still exists a stochastic $\widetilde{\mathcal{F}}_t$-intensity, and this intensity is obtained by "projection" of the initial stochastic intensity on the smaller history, in a sense to be described now.

Consider the histories $\{\mathcal{F}_t\}_{t\geq 0}$ and $\{\widetilde{\mathcal{F}}_t\}_{t\geq 0}$ satisfying condition (15.21) for all $t \geq 0$. Let $\{Y(t)\}_{t\geq 0}$ be a non-negative locally integrable measurable process. Let the sigma-finite measures μ_1 and μ_2 on $(\mathbb{R} \times \Omega, \mathcal{P}(\mathcal{F}.))$ be defined respectively by:

$$\mu_1(H) := \mathrm{E}\left[\int_{\mathbb{R}} H(t)Y(t)\mathrm{d}t\right] \text{ and } \mu_2(H) := \mathrm{E}\left[\int_{\mathbb{R}} H(t)\mathrm{d}t\right]$$

for all non-negative $H : \mathbb{R} \times \Omega$ that are $\mathcal{P}(\widetilde{\mathcal{F}})$-measurable. Note that μ_2 is the product measure $P \times \ell$ on $(\Omega \times \mathbb{R}, \mathcal{P}(\mathcal{F}.))$. Clearly $\mu_1 \ll \mu_2$, and therefore there exists a Radon–Nikodým (RND) derivative $\widetilde{Y}(t,\omega) = \frac{\mathrm{d}\mu_1}{\mathrm{d}\mu_2}(t,\omega)$ that is $\mathcal{P}(\widetilde{\mathcal{F}}.)$-measurable (and therefore defines an $\widetilde{\mathcal{F}}_t$-predictable process $\{\widetilde{Y}(t)\}_{t\geq 0}$) such that

$$\mu_1(H) = \mu_2(\widetilde{Y}H) = \mathrm{E}\left[\int_{\mathbb{R}} H(t)\widetilde{Y}(t)\,\mathrm{d}t\right].$$

Moreover, this RND is μ_2-unique, that is to say, if there exists another such RND, say $\overline{Y}$, then

$$\widetilde{Y}(t,\omega) = \overline{Y}(t,\omega), \qquad P(\mathrm{d}\omega) \times \mathrm{d}t \quad \text{a.e.} \tag{15.22}$$

Definition 15.2.2 *The above stochastic process $\{\widetilde{Y}(t)\}_{t\geq 0}$ is called the* predictable projection *of $\{Y(t)\}_{t\geq 0}$ on $\{\widetilde{\mathcal{F}}_t\}_{t\geq 0}$, or the $\widetilde{\mathcal{F}}_t$-predictable projection of $\{Y(t)\}_{t\geq 0}$.*

Theorem 15.2.3 *Let the simple locally finite point process N on $\mathbb{R}$ have the stochastic $\mathcal{F}_t$-intensity $\{\lambda(t)\}_{t\geq 0}$. Let $\{\widetilde{\mathcal{F}}_t\}_{t\geq 0}$ be another history, satisfying condition (15.21). Then N has the stochastic $\widetilde{\mathcal{F}}_t$-intensity $\{\widetilde{\lambda}(t)\}_{t\geq 0}$, the $\widetilde{\mathcal{F}}_t$-predictable projection of $\{\lambda(t)\}_{t\geq 0}$.*

Proof. Let $\{H(t)\}_{t\geq 0}$ be a non-negative process that is $\widetilde{\mathcal{F}}_t$-predictable. It is *a fortiori* an $\mathcal{F}_t$-predictable process, and therefore

$$\mathrm{E}\left[\int_{\mathbb{R}} H(t)\,N(\mathrm{d}t)\right] = \mathrm{E}\left[\int_{\mathbb{R}} H(t)\,\lambda(t)\,\mathrm{d}t\right] = \mathrm{E}\left[\int_{\mathbb{R}} H(t)\,\widetilde{\lambda}(t)\,\mathrm{d}t\right].$$

$\square$

For a given history $\{\mathcal{F}_t\}_{t\geq 0}$ and a given probability P, there exist many versions of the $(P, \mathcal{F}_t)$-intensity $(\lambda(t))_{t\geq 0}$. Such an intensity can always been chosen to be $\mathcal{F}_t$-predictable, since if it is not, one can replace it by its $\mathcal{F}_t$-predictable projection, which exists and is essentially unique in a sense to be described below.

Theorem 15.2.4

(a) *The $\mathcal{F}_t$-predictable version is $P(\mathrm{d}\omega)\,\mathrm{d}t$-unique and therefore $P(\mathrm{d}\omega)\,N(\omega,\mathrm{d}t)$-uniqu In particular, if $\{\lambda'(t)\}_{t\geq 0}$ is another $\mathcal{F}_t$-predictable version, then $P(\lambda'(T_n) = \lambda(T_n)\,(n \geq 1)) = 1$.*

(b) *If $\{\lambda(t)\}_{t\geq 0}$ is a predictable version of the $\mathcal{F}_t$-intensity of the locally finite simple point process N, then*

$$\lambda(t,\omega) > 0 \quad \lambda(t,\omega)P(\mathrm{d}\omega)\mathrm{d}t \text{ and } N(\mathrm{d}t,\omega)P(\mathrm{d}\omega)\text{-}a.e. \tag{15.23}$$

In particular,

$$\lambda(T_n) > 0 \quad (n \geq 1), \qquad P\text{-}a.s. \tag{15.24}$$

Proof. (a) follows from the uniqueness property (15.27). For (b), note that $H(t) = 1_{\{\lambda(t)=0\}}$ is an $\mathcal{F}_t$-predictable process. Inserting this into the smoothing formula, we obtain

$$E\left[\sum_{n\geq 0} 1_{\{\lambda(T_n)=0\}}\right] = E\left[\int_{\mathbb{R}_+} 1_{\{\lambda(t)=0\}}\lambda(t)\,\mathrm{d}t\right] = 0,$$

which implies (15.23).

$\square$

In particular, if the locally finite simple marked point process (N, Z) has the $\mathcal{F}_t$-predictable stochastic intensity kernel $\lambda(t)\Phi(t, \mathrm{d}z)$, then, for all $L \in \mathcal{K}$,

$$\lambda(T_n)\Phi(T_n, L) > 0 \text{ on } \{T_n < \infty\}.$$

The above results extend straightforwardly to marked point processes and their stochastic intensity kernels as follows.

Let the simple locally finite marked point process (N, Z) have the $\mathcal{F}_t$-intensity kernel $\lambda(t)\Phi(t, \mathrm{d}z)$, where

$$\Phi(t, \mathrm{d}z) = \lambda(t)\mu(t, z)Q(t, \mathrm{d}z) \tag{15.25}$$

for some $\mathcal{F}_t^{N,Z}$-predictable kernel $Q(t, \mathrm{d}z)$. Let $\{\widetilde{\mathcal{F}}_t\}_{t\geq 0}$ be a history such that

$$\mathcal{F}_t \supseteq \widetilde{\mathcal{F}}_t \supseteq \mathcal{F}_t^{N,Z} \quad (t \geq 0). \tag{15.26}$$

It is possible that $\lambda(t)\Phi(t, \mathrm{d}z)$ is not $\widetilde{\mathcal{F}}_t$-adapted and therefore cannot be the $\widetilde{\mathcal{F}}_t$-intensity kernel. Nevertheless, there still exists a stochastic $\widetilde{\mathcal{F}}_t$-intensity kernel, and it is obtained by "projection" of the initial stochastic intensity kernel on the smaller history, in a sense to be made precise now.

Recall the terminology: if the mapping $Y : (t, \omega, z) \rightarrow Y(t, \omega, z) \in \mathbb{R}$ is $\mathcal{B}(\mathbb{R}_+)\otimes\mathcal{F}\otimes\mathcal{K}$, one says that $Y(t, z)$ is a measurable process indexed by K. It is said to be $\mathcal{F}_t$-adapted (*resp. $\mathcal{F}_t$-predictable*) if moreover for all $z \in K$, the stochastic process $\{Y(t, z)\}_{t\geq 0}$ is $\mathcal{F}_t$-adapted (*resp. $\mathcal{F}_t$-predictable*).

Let the histories $\{\mathcal{F}_t\}_{t\geq 0}$ and $\{\widetilde{\mathcal{F}}_t\}_{t\geq 0}$ satisfy condition (15.26). Let $\{Y(t, z)\}_{t\geq 0}$ be a non-negative measurable process indexed by K. Let the sigma-finite measures μ_1 and μ_2 on $(\mathbb{R} \times \Omega \times K, \mathcal{P}(\mathcal{F}) \otimes \mathcal{K})$ be defined respectively by:

$$\mu_1(H) := \mathrm{E}\left[\int_\mathbb{R}\int_K H(t,z)Y(t,z)\,Q(t,\mathrm{d}z)\,\mathrm{d}t\right]$$

and

$$\mu_2(H) := \mathrm{E}\left[\int_\mathbb{R}\int_K H(t,z)\,Q(t,\mathrm{d}z)\,\mathrm{d}t\right]$$

for all non-negative mappings $H : \mathbb{R}\times\Omega\times K$ that are $\mathcal{P}(\widetilde{F})\otimes\mathcal{K}$-measurable. Note that μ_2 is the product measure $P(\mathrm{d}\omega)\times Q(t,\omega,\mathrm{d}z)\,\mathrm{d}t$ on $(\Omega\times\mathbb{R}\times K, \mathcal{P}(\mathcal{F}.)\otimes\mathcal{K})$. Clearly $\mu_1 \ll \mu_2$, and therefore there exists a Radon–Nikodým (RND) derivative $\widetilde{Y}(t,\omega,z) = \frac{\mathrm{d}\mu_1}{\mathrm{d}\mu_2}(t,\omega,z)$ that is $\mathcal{P}(\widetilde{F}.)\otimes\mathcal{K}$-measurable and therefore defines an $\widetilde{\mathcal{F}}_t$–predictable process indexed by K, $\widetilde{Y}(t,z)$, such that

$$\mu_1(H) = \mu_2(\widetilde{Y}H) = \mathrm{E}\left[\int_\mathbb{R}\int_K H(t,z)\widetilde{Y}(t,z)\,Q(t,\mathrm{d}z)\,\mathrm{d}t\right].$$

Moreover, this RND is μ_2-unique, that is to say, if there exists another such RND, say $\overline{Y}$, then

$$\widetilde{Y}(t,\omega,z) = \overline{Y}(t,\omega,z)\,, \qquad P(\mathrm{d}\omega)Q(t,\omega,\mathrm{d}z)\mathrm{d}t \quad \text{a.e.} \tag{15.27}$$

Definition 15.2.5 *The above stochastic process $\widetilde{Y}(t,z)$ indexed by K is called the pre-dictable projection of $Y(t,z)$ on $\{\widetilde{\mathcal{F}}_t\}_{t\geq0}$, or the $\widetilde{\mathcal{F}}_t$-predictable projection of $Y(t,z)$.*

Theorem 15.2.6 *Let the simple locally finite marked point process (N,Z) on $\mathbb{R}_+$ have the stochastic $\mathcal{F}_t$-intensity kernel (15.25) for some $\mathcal{F}_t^{N,Z}$-predictable kernel $Q(t,\mathrm{d}z)$. Let $\{\widetilde{\mathcal{F}}_t\}_{t\geq0}$ be another history satisfying condition (15.26). Then (N,Z) has the stochastic $\widetilde{\mathcal{F}}_t$-intensity kernel $\widetilde{\lambda}(t)\widetilde{h}(t,z)Q(t,\mathrm{d}z)$ where $\{\widetilde{\lambda}(t)\}_{t\geq0}$ is the $\widetilde{\mathcal{F}}_t$-predictable projection of $\{\lambda(t)\}_{t\geq0}$ and $\widetilde{h}(t,z)$ is the $\widetilde{\mathcal{F}}_t$-predictable projection of $h(t,z)$.*

Proof. Let $H(t,z)$ be a non-negative $\widetilde{\mathcal{F}}_t$-predictable indexed stochastic process. It is *a fortiori* an $\mathcal{F}_t$-predictable indexed stochastic process, and therefore

$$\mathrm{E}\left[\int_\mathbb{R}\int_K H(t,z))\,N(\mathrm{d}t\times\mathrm{d}z)\right] = \mathrm{E}\left[\int_\mathbb{R}\int_K H(t,z))\,\lambda(t)h(t,z)\,Q(t,\mathrm{d}z)\,\mathrm{d}t\right]$$
$$= \mathrm{E}\left[\int_\mathbb{R}\int_K H(t,z))\,\widetilde{v}(t,z)\,Q(t,\mathrm{d}z)\,\mathrm{d}t\right],$$

where $\widetilde{v}(t,z)$ is the $\widetilde{\mathcal{F}}_t$-predictable projection of $\lambda(t)h(t,z)$. Let now

$$\widetilde{h}(t,\omega,z) := \frac{\widetilde{v}(t,\omega,z)}{\widetilde{\lambda}(t,\omega)}\,,$$

a quantity that is $P(\mathrm{d}\omega)\times N(\omega,\mathrm{d}t\times\mathrm{d}z)$- and $P(\mathrm{d}\omega)\times\lambda(t,\omega)Q(t,\omega,\mathrm{d}z)\mathrm{d}t$-well defined in view of Theorem 15.2.4. We have that for all non-negative $\widetilde{\mathcal{F}}_t$-predictable indexed process H,

$$\mathrm{E}\left[\int_\mathbb{R}\int_K H(t,z)\,N(\mathrm{d}t\times\mathrm{d}z)\right] = \mathrm{E}\left[\int_\mathbb{R}\int_K H(t,z)\widetilde{\lambda}(t)\frac{\widetilde{v}(t,z)}{\widetilde{\lambda}(t)}\,Q(t,\mathrm{d}z)\,\mathrm{d}t\right]$$
$$= \mathrm{E}\left[\int_\mathbb{R}\int_K H(t,z)\lambda(t)\,h(t,z)\,Q(t,\mathrm{d}z)\,\mathrm{d}t\right].$$

Equivalently,

$$
\mathrm{E}\left[\int_{\mathbb{R}}\left(\int_K H(t,z)\,\frac{\widetilde{v}(t,z)}{\widetilde{\lambda}(t)}\,Q(t,\mathrm{d}z)\right)\widetilde{\lambda}(t)\,\mathrm{d}t\right]
$$
$$
= \mathrm{E}\left[\int_{\mathbb{R}}\left(\int_K H(t,z)\,h(t,z)\,Q(t,\mathrm{d}z)\right)\lambda(t)\,\mathrm{d}t\right].
$$

Now

$$
\mathrm{E}\left[\int_{\mathbb{R}}\left(\int_K H(t,z)\,h(t,z)\,Q(t,\mathrm{d}z)\right)\lambda(t)\,\mathrm{d}t\right]
$$
$$
= \mathrm{E}\left[\int_{\mathbb{R}}\left(\int_K H(t,z)\,h(t,z)\,Q(t,\mathrm{d}z)\right)N(\mathrm{d}t)\right]
$$
$$
= \mathrm{E}\left[\int_{\mathbb{R}}\left(\int_K H(t,z)\,h(t,z)\,Q(t,\mathrm{d}z)\right)\widetilde{\lambda}(t)\,\mathrm{d}t\right],
$$

and therefore

$$
\mathrm{E}\left[\int_{\mathbb{R}}\left(\int_K H(t,z)\,\frac{\widetilde{v}(t,z)}{\widetilde{\lambda}(t)}\,Q(t,\mathrm{d}z)\right)\widetilde{\lambda}(t)\,\mathrm{d}t\right]
$$
$$
= \mathrm{E}\left[\int_{\mathbb{R}}\left(\int_K H(t,z)\,h(t,z)\,Q(t,\mathrm{d}z)\right)\widetilde{\lambda}(t)\,\mathrm{d}t\right].
$$

Replacing $H(t,z)$ by $\dfrac{H(t,z)}{\widetilde{\lambda}(t)}$,

$$
\mathrm{E}\left[\int_{\mathbb{R}}\left(\int_K H(t,z)\,\frac{\widetilde{v}(t,z)}{\widetilde{\lambda}(t)}\,Q(t,\mathrm{d}z)\right)\mathrm{d}t\right]
$$
$$
= \mathrm{E}\left[\int_{\mathbb{R}}\left(\int_K H(t,z)\,h(t,z)\,Q(t,\mathrm{d}z)\right)\mathrm{d}t\right],
$$

which shows that $\dfrac{\widetilde{v}(t,z)}{\widetilde{\lambda}(t)} = \widetilde{h}(t,z)$. $\qquad\square$

15.2.2 Absolutely Continuous Change of Probability

We now consider changes of intensity entailed by an absolutely continuous change of probability measure. This subsection is of special interest in statistics where the concept of likelihood ratio is of central importance, in particular in hypothesis testing. It is a sweeping generalization of the results of Section 8.3.3.

Let (N, Z) be a simple and locally finite point process on $\mathbb{R}_+$ with marks in K and associated lifted process N_Z on $\mathbb{R}_+ \times K$. Let $\{\mathcal{F}_t\}_{t\geq0}$ be a history of N_Z and suppose that N_Z admits the $(P, \mathcal{F}_t)$-local characteristics $(\lambda(t), \Phi(t, \mathrm{d}z))$.

Let $\{\mu(t)\}_{t\geq0}$ be a non-negative $\mathcal{F}_t$-predictable process and let $\{h(t,z)\}_{t\geq0,z\in K}$ be a non-negative $\mathcal{F}_t$-predictable K-indexed stochastic process, such that for all $t \geq 0$

$$
P\left(\int_0^t \lambda(s)\mu(s)\,\mathrm{d}s < \infty\right) = 1 \tag{15.28}
$$

and

$$P\left(\int_K h(t,z)\Phi(t,dz)\,dz = 1\right) = 1. \tag{15.29}$$

Define for each $t \geq 0$

$$L(t) := L(0)\left(\prod_{T_n \in (0,t]} \mu(T_n)h(T_n, Z_n)\right) \times \cdots$$

$$\exp\left(-\int_{(0,t]}\int_K (\mu(s)h(s,z)-1)\lambda(s)\Phi(s,dz)\,ds\right), \tag{15.30}$$

where $L(0)$ is a non-negative $\mathcal{F}_0$-measurable random variable such that $E[L(0)] = 1$.

Theorem 15.2.7 *Under the above conditions,*

(1) $\{L(t)\}_{t\geq 0}$ *is a non-negative $(P, \mathcal{F}_t)$-local martingale. If, moreover, $E[L(t)] = 1$ for all $t \geq 0$, it is a non-negative $(P, \mathcal{F}_t)$-martingale.*

(2) *If $E[L(T)] = 1$ for some $T > 0$, and if we define the probability Q by the Radon–Nikodým derivative process*

$$\frac{dQ}{dP} = L(T) \tag{15.31}$$

the marked point N_Z admits the $(Q, \mathcal{F}_t)$-local characteristics $(\mu(t)\lambda(t), h(t,z)\Phi(t,dz))$ on $[0, T]$.

Proof.

(1) By the exponential rule of Stieltjes–Lebesgue calculus,

$$L(t) = L(0) + \int_{(0,t]}\int_K (\mu(s)h(s,z)-1)L(s-)M_Z(ds \times dz),$$

where $M_Z(ds \times dz) := N_Z(ds \times dz) - \lambda(s)\Phi(s,dz)\,ds$. Let for $n \geq 1$,

$$S_n = \inf\left(t\,; L(t-) + \int_0^t \mu(s)\lambda(s)\,ds \geq n\right). \tag{15.32}$$

Then, by Theorem 15.1.31, $\{L(t \wedge S_n)\}_{t\geq 0}$ is a $(P, \mathcal{F}_t)$ martingale, and since under conditions (15.37) and (15.38), $P(\lim_{n\uparrow\infty} S_n = \infty) = 1$, $\{L(t)\}_{t\geq 0}$ is a $(P, \mathcal{F}_t)$-local martingale. Being non-negative, it is also a $(P, \mathcal{F}_t)$-supermartingale. But a supermartingale with constant mean is a martingale.

(2) We have to prove that for any non-negative $\mathcal{F}_t$-predictable K-indexed stochastic process $\{H(t,z)\}_{t\geq 0, z\in K}$ and all $t \in [0, T]$,

$$E_Q\left[\int_{(0,t]}\int_K H(s,z)N_Z(ds \times dz)\right] = E_Q\left[\int_{(0,t]}\int_K H(s,z)\mu(s)\lambda(s)h(s,z)\Phi(s,dz)\,ds\right].$$

This is done through the following sequence of equalities (with appropriate justifications at the end)

$$E_Q\left[\int_{(0,t]}\int_K H(s,z)N_Z(\mathrm{d}s\times\mathrm{d}z)\right]$$

$$= E\left[L(t)\int_{(0,t]}\int_K H(s,z)N_Z(\mathrm{d}s\times\mathrm{d}z)\right]$$

$$= E\left[\int_{(0,t]}\int_K L(s)H(s,z)N_Z(\mathrm{d}s\times\mathrm{d}z)\right]$$

$$= E\left[\int_{(0,t]}\int_K L(s-)H(s,z)\mu(s)h(s,z)N_Z(\mathrm{d}s\times\mathrm{d}z)\right]$$

$$= E\left[\int_{(0,t]}\int_K L(s-)H(s,z)\mu(s)h(s,z)\lambda(s)\Phi(s,\mathrm{d}z)\,\mathrm{d}s\right]$$

$$= E\left[\int_{(0,t]}\int_K L(s)H(s,z)\mu(s)h(s,z)\lambda(s)\Phi(s,\mathrm{d}z)\,\mathrm{d}s\right]$$

$$= E\left[L(t)\int_{(0,t]}\int_K H(s,z)\mu(s)h(s,z)\lambda(s)\Phi(s,\mathrm{d}z)\,\mathrm{d}s\right]$$

$$= E_Q\left[\int_{(0,t]}\int_K H(s,z)\mu(s)\lambda(s)h(s,z)\Phi(s,\mathrm{d}z)\,\mathrm{d}s\right].$$

The first equality follows from (15.31) and the fact that for a nonnegative $\mathcal{F}_t$-measurable random variable $V(t)$, $E_Q[V(t)] = E_P[L(t)V(t)]$. The second equality follows from Theorem 13.4.25, the third one from the observation that at a point $T_n < \infty$ of N, $L(T_n) = L(T_n-)\mu(T_n)h(T_n, Z_n)$. The third equality uses the smoothing theorem, the fifth is by Theorem 13.4.25 and the last one uses (15.31). $\qquad\square$

Remark 15.2.8 The main condition to verify when using Theorem 15.2.7 is $E_P[L(T)] = 1$. A general method to do this consists in finding some $\gamma > 1$ such that for the sequence of stopping times $\{S_n\}_{n\geq 1}$ defined by (15.32), $\sup_{n\geq 1} E_P[L(T\wedge S_n)^\gamma] < \infty$. This implies that the sequence $\{L(T\wedge S_n)\}_{n\geq 1}$ is uniformly integrable, and therefore

$$\lim_{n\uparrow\infty} E[L(T\wedge S_n)] = E\left[\lim_{n\uparrow\infty} L(T\wedge S_n)\right].$$

But, by Part (1) of Theorem 15.2.7, $E[L(T\wedge S_n)] = 1$ and $S_n \uparrow \infty$. Therefore $E[L(T)] = 1$.

EXAMPLE 15.2.9: THE LIKELIHOOD RATIO FOR A SIMPLE POINT PROCESS ON A FINITE INTERVAL. Let N be under probability P an $\mathcal{F}_t$-Poisson process of intensity 1. Let $\{\lambda(t)\}_{t\geq 0}$ be a non-negative *bounded* $\mathcal{F}_t$-predictable process. Define for all $t \geq 0$

$$L(t) = L(0)\left(\prod_{n\geq 1}\lambda(T_n)\right)\exp\left(\int_0^t(\lambda(s)-1)\,\mathrm{d}s\right),$$

where $L(0)$ is a non-negative square integrable random variable such that $E[L(0)^2] < \infty$. By the exponential formula of Stieltjes–Lebesgue calculus,

$$L(t) = L(0) + \int_{(0,t]} L(s-)(\lambda(s) - 1)\, (N(\mathrm{d}s) - \mathrm{d}s)\,.$$

By the product rule of Stieltjes–Lebesgue calculus

$$L(t)^2 = L(0)^2 + 2\int_{(0,t]} L(s-)\,\mathrm{d}L(s) + \sum_{s \le t} L(s-)\Delta L(s)$$

$$= L(0)^2 + 2\int_{(0,t]} L(s-)\,\mathrm{d}L(s) + \int_{(0,t]} L(s-)^2(\lambda(s) - 1)\, N(\mathrm{d}s)$$

$$= L(0)^2 + 2\int_{(0,t]} L(s-)\,\mathrm{d}L(s) + \int_{(0,t]} L(s-)^2(\lambda(s) - 1)\,(N(\mathrm{d}s) - \mathrm{d}s) \qquad (15.33)$$

$$+ \int_{(0,t]} L(s)^2\,(\lambda(s) - 1)\,\mathrm{d}s \qquad (15.34)$$

(noting that for Lebesgue-almost all t, $L(t) = L(t-)$). Define for each $n \ge 1$

$$S_n := \inf \left\{ t\,;\, L(t-) + \int_0^t \lambda(s)\,\mathrm{d}s \ge n \right\} \wedge T_n\,,$$

an $\mathcal{F}_t$-stopping time such that $\lim_{n \uparrow \infty} S_n = \infty$. In particular,

$$\int_{(0,t]} L(s-)\,\mathrm{d}L(s) = \int_{(0,t]} L(s-)^2(\lambda(s) - 1)\,(N(\mathrm{d}s) - \mathrm{d}s)$$

is an $\mathcal{F}_t$-local martingale (with localizing sequence $\{S_n\}_{n \ge 1}$) of mean 0. Replacing in (15.34) t by $t \wedge T_n$ and taking expectations,

$$\mathrm{E}\left[L(t \wedge S_n)^2\right] = \mathrm{E}\left[L(0)^2\right] + \int_{(0,t \wedge S_n]} L(s \wedge S_n)^2\,(\lambda(s) - 1)\,\mathrm{d}s\,.$$

In particular, in view of the boundedness assumption on the $\lambda(t)$,

$$\mathrm{E}\left[L(t \wedge S_n)^2\right] \le \mathrm{E}\left[L(0)^2\right] + \mathrm{E}\left[\int_{(0,t]} L(s \wedge S_n)^2\,(\lambda(s) + 1)\,\mathrm{d}s\right]$$

$$= C_1 + \int_{(0,t]} \mathrm{E}\left[L(s \wedge S_n)^2\right](C_2 + 1)\,\mathrm{d}s$$

for some finite positive C_1 and C_2. This implies, by Gronwall's lemma, that

$$\mathrm{E}\left[L(t \wedge S_n)^2\right] \le C_1 \exp\left(\int_0^t (C_2 + 1)\,\mathrm{d}s\right)\,.$$

In particular, for any $T < \infty$, $\sup_{n \ge 1} \mathrm{E}\left[L(T \wedge S_n)^2\right] < \infty$.

EXAMPLE 15.2.10: LIKELIHOOD RATIOS FOR CONTINUOUS-TIME HMCS. Let $\{X(t)\}_{t \ge 0}$ be, under P, a regular continuous-time homogeneous Markov chain with state space E and stable and conservative infinitesimal generator $\{q_{ij}\}_{i,j \in E}$. Let α_{ij} $(i \ne j \in E)$ be non-negative numbers such that for all $i \in E$, $\tilde{q}_i := \sum_{j \ne i \in E} \alpha_{ij} q_{ij} < \infty$. For all $t \ge 0$, let $\mathcal{F}_t := \mathcal{F}_t^X$ and let

$$L(t) := L(0) \left(\prod_{i,j;i \ne j} \alpha_{ij}^{N_{i,j}(t)} \right) \exp\left\{ \int_0^t \sum_{i,j;i \ne j} (\alpha_{ij} - 1)q_{ij}\mathbf{1}_{\{X(s)=i\}}\,\mathrm{d}s \right\}\,.$$

Suppose that $E_P[L(0)] = 1$, that $E_P[L(0)^2] < \infty$ and that $\sum_{j \neq i} \alpha_{ij} q_{ij} < \infty$ ($i \in E$). Then $E_P[L(T)] = 1$ and under probability Q defined by $\frac{dQ}{dP} = L(T)$, the process $\{X(t)\}_{t \geq 0}$ is on the interval $(0, T]$ a regular stable and conservative continuous-time homogeneous Markov chain with state space E and infinitesimal parameters $\tilde{q}_{ij} := \alpha_{ij} q_{ij}$ ($j \neq i$).

Proof. At a discontinuity time t of the chain,

$$L(t) - L(t-) = \sum_{i,j\,;\,j \neq i} L(t-)(\alpha_{ij} - 1)\Delta N_{i,j}(t),$$

whereas for t strictly between two jumps of the chain,

$$\frac{dL(t)}{dt} = L(t) \sum_{i,j\,;\,j \neq i} (\alpha_{ij} - 1)q_{ij}1_{\{X(t)=i\}}.$$

Therefore

$$L(t) = L(0) + \int_{(0,t]} L(s-)\sum_{i \neq j}(\alpha_{ij} - 1)(N_{ij}(ds) - q_{ij}1_{\{X(s)=i\}}ds).$$

By the product rule of Stieltjes–Lebesgue calculus,

$$L(t)^2 = L(0)^2 + 2\int_{(0,t]} L(s-)\,dL(s) + \sum_{s \leq t} L(s-)\Delta L(s)$$

$$= L(0)^2 + 2\int_{(0,t]} L(s-)\,dL(s)$$

$$+ \int_{(0,t]} L(s-)^2 \sum_{i \neq j}(\alpha_{ij} - 1)(N_{ij}(ds) - q_{ij}1_{\{X(s)=i\}}ds)$$

$$+ \int_{(0,t]} \sum_{i \neq j}(\alpha_{ij} - 1)q_{ij}1_{\{X(s)=i\}}\,ds.$$

Using the stopping times of type (15.32), we have that

$$L(t)^2 = L(0)^2 + \text{local martingale} + \int_{(0,t]} \sum_{i \neq j} L(s)^2(\alpha_{ij} - 1)q_{ij}1_{\{X(s)=i\}}\,ds.$$

Then,

$$E\left[L(t \wedge S_n)^2\right] = E\left[L(0)^2\right] + E\left[\int_{(0,t\wedge S_n]} \sum_{i \neq j} L(s \wedge S_n)^2(\alpha_{ij} - 1)q_{ij}1_{\{X(s)=i\}}\,ds\right]$$

and in particular

$$E\left[L(t \wedge S_n)^2\right] \leq E\left[L(0)^2\right] + \int_{(0,t]} E\left[L(s \wedge S_n)^2\right]\sum_{i \neq j}|\alpha_{ij} - 1|q_{ij}ds.$$

Therefore, by Gronwall's lemma,

$$E\left[L(t \wedge S_n)^2\right] \leq E\left[L(0)^2\right]\exp\left\{\sum_i \tilde{q}_i + \sum_i q_it\right\}$$

and finally

$$\sup_{n \geq 1} E\left[L(T \wedge S_n)^2\right] < \infty.$$

□

___

The above two examples can be generalized as follows.

___

EXAMPLE 15.2.11: LIKELIHOOD RATIOS FOR MARKED POINT PROCESSES. Consider the general situation of Theorem 15.2.7. Let (N, Z) be a simple and locally finite point process on $\mathbb{R}_+$ with marks in K and associated lifted process N_Z on $\mathbb{R}_+ \times K$. Let $\{\mathcal{F}_t\}_{t \geq 0}$ be a history of (N, Z) and suppose that (N, Z) admits the $(P, \mathcal{F}_t)$-local characteristics $(\lambda(t), \Phi(t, \mathrm{d}z))$.

Let $\{\mu(t)\}_{t \geq 0}$ be a non-negative $\mathcal{F}_t$-predictable process and let $\{h(t, z)\}_{t \geq 0, z \in K}$ be a non-negative $\mathcal{F}_t$-predictable K-indexed stochastic process, such that for all $t \geq 0$

$$P\left(\int_0^t \lambda(s)\mu(s)\,\mathrm{d}s < \infty\right) = 1 \tag{15.35}$$

and

$$P\left(\int_K h(t, z)\Phi(t, \mathrm{d}z)\,\mathrm{d}z = 1\right) = 1. \tag{15.36}$$

For each $t \geq 0$, define $L(t)$ as in (15.30), with $L(0)$ a non-negative $\mathcal{F}_0$-measurable random variable such that $E[L(0)] = 1$. We suppose in addition that $L(0)$ is square-integrable and that

$$(\mu(t) + 1)h(t, z)\lambda(t) \leq K(t),$$

where $K : \mathbb{R}_+ \to \mathbb{R}_+$ is a deterministic function such that $\int_0^T K(s)\,\mathrm{d}s < \infty$. Then $E[L(T)] = 1$. The proof follows the same lines as the proof in Example 15.2.9 and is left as an exercise.

___

Remark 15.2.12 One need not insist on the analogy with Girsanov's theorem as it is obvious. The proof of Girsanov's result was based on the Itô calculus for Brownian motion. In the case of point processes, the underlying calculus is just the ordinary Stieltjes–Lebesgue calculus.

The Reference Probability Method

Radon–Nikodým derivatives are of course of interest in Statistics (where they are called likelihood ratios), and also in filtering. In the so-called reference probability method, the probability P actually governing the joint statistics of the observation and of the state process is obtained by an absolutely continuous change of probability measure $Q \to P$. This method therefore relies on the Radon–Nikodým results of Subsection 15.2.2. The reference probability Q is chosen such that the observation and the state process are Q-independent and the observation has a simple structure under Q.

The *state process* $\{X(t)\}_{t \geq 0}$ takes its values in some measurable space $(E, \mathcal{E})$ and the observation is a marked point process (N, Z) with time-events sequence $\{T_n\}_{n \geq 1}$ and mark sequence $\{Z_n\}_{n \geq 1}$, the marks taking their values in a measurable space $(K, \mathcal{K})$. Let

$$\mathcal{F}_t := \mathcal{F}_t^{N,Z} \vee \mathcal{F}_\infty^X,$$

where $\{\mathcal{F}_t^X\}_{t\geq 0}$ is the internal history of the state process. The *reference probability* Q is such that (N, Z) is a $(Q, \mathcal{F}_t)$-Poisson process of intensity 1 with independent IID marks of common probability distribution Q. Moreover, under Q, N and $\mathcal{F}_\infty^X$ are independent. Therefore, the stochastic $\mathcal{F}_t$-kernel of the observation under the reference probability Q is $\lambda(t, dz) = 1 \times Q(dz)$.

Let $\{\mu(t)\}_{t\geq 0}$ be a non-negative $\mathcal{F}_t$-predictable process and let $\{h(t, z)\}_{t\geq 0, z\in K}$ be a non-negative $\mathcal{F}_t$-predictable K-indexed stochastic process, such that for all $t \geq 0$

$$Q\left(\int_0^T \mu(s)\,ds < \infty\right) = 1 \tag{15.37}$$

and

$$Q\left(\int_K h(t, z) Q(dz)\,dz = 1\right) = 1. \tag{15.38}$$

Define

$$L(t) := \left(\prod_{T_n \in (0,t]} \mu(T_n) h(T_n, Z_n)\right) \exp\left(-\int_{(0,T]}\int_K (\mu(s)h(s, z) - 1)Q(dz)\,ds\right),$$

and suppose that $E_Q[L(T)] = 1$. Then (Theorem 15.2.7) the marked point (N, Z) admits on $[0, T]$ the $(P, \mathcal{F}_t)$-local characteristics $(\mu(t), h(t, z)Q(dz))$. Moreover, the restrictions of P and Q to $\mathcal{F}_\infty^X$ are the same since $L(0) = 1$. In fact, by hypothesis, $L(0) := \frac{dP_0}{dQ_0} = 1$, that is, P and Q agree on $\mathcal{F}_0 := \mathcal{F}_\infty^X$.

Let $\{Z(t)\}_{t\geq 0}$ be an $\mathcal{F}_t$-adapted real-valued stochastic process.

Lemma 15.2.13 *For all $t \geq 0$,*

$$E_Q\left[L(t)\mid \mathcal{F}_t^N\right] E_P\left[Z(t)\mid \mathcal{F}_t^N\right] = E_Q\left[Z(t)L(t)\mid \mathcal{F}_t^N\right], \qquad Q\text{-a.s.}$$

or, equivalently,

$$E_P\left[Z(t)\mid \mathcal{F}_t^N\right] = \frac{E_Q\left[Z(t)L(t)\mid \mathcal{F}_t^N\right]}{E_Q\left[L(t)\mid \mathcal{F}_t^N\right]}, \qquad P\text{-a.s.}$$

Proof. This is just a rephrasing of Lemma 8.3.12. □

EXAMPLE 15.2.14: ESTIMATING THE RANDOM INTENSITY OF A HOMOGENEOUS COX PROCESS. The above lemma allows us to replace a filtering problem with respect to P by one with respect to Q, which may be a simplification when Q has a simple structure. For instance, if Q is a probability that makes the point process N Poisson with intensity 1, and if Λ is an integrable variable independent, under Q of N, the measure P defined by

$$\frac{dP_t}{dQ_t} = \Lambda^{N(t)} \exp\{(1 - \Lambda)t\}$$

makes N a doubly stochastic process with intensity Λ. Then

$$E\left[\Lambda \mid \mathcal{F}_t^N\right] = \frac{\int_0^\infty \lambda^{N(t)+1} e^{-\lambda t}\,dF(\lambda)}{\int_0^\infty \lambda^{N(t)} e^{-\lambda t}\,dF(\lambda)}.$$

(Exercise 15.4.16.)

15.2.3 Changing the Time Scale

Recall the following elementary result concerning Poisson processes on the line with an intensity. If N is a Poisson process with intensity $\lambda(t)$, defining $\tau(t)$ by $\int_0^{\tau(t)} \lambda(s)\,ds = t$, the point process $\widetilde{N}$ defined by $\widetilde{N}((0,t]) = N((0,\tau(t)])$ is a standard HPP. This result will be extended to the transformation of a point process with given stochastic intensity into a standard HPP.

Let N be a simple locally finite point process on $\mathbb{R}_+$, with the $\mathcal{F}_t$-predictable intensity $\{\lambda(t)\}_{t \geq 0}$, and suppose that $N(0,\infty) = \infty$, P-a.s. or, equivalently (Theorem 15.1.16), $\int_0^\infty \lambda(s)\,ds = \infty$, P-a.s. Define, for each $t \geq 0$, the non-negative random variable $\tau(t)$ by

$$\int_0^{\tau(t)} \lambda(s)\,ds = t. \tag{15.39}$$

For each $t \geq 0$, $\tau(t)$ is well defined since $\int_0^\infty \lambda(s)\,ds = \infty$ (Theorem 15.1.16). For each $t \in \mathbb{R}_+$, $\tau(t)$ is an $\mathcal{F}_t$-stopping time. Indeed, for any $a \in \mathbb{R}$,

$$\{\tau(t) \leq a\} = \left\{ \int_0^a \lambda(s)\,ds \leq t \right\} \in \mathcal{F}_a.$$

Define the simple locally bounded point process $\widetilde{N}$ on $\mathbb{R}_+$ by

$$\widetilde{N}(0,t] := N(0,\tau(t)]. \tag{15.40}$$

Note that $\mathcal{F}_t^{\widetilde{N}} \subseteq \mathcal{F}_{\tau(t)}$, since for all $a \in \mathbb{R}_+$, $\widetilde{N}(0,a] := N(0,\tau(a))$ is $\mathcal{F}_{\tau(a)}^N$-measurable, and therefore $\mathcal{F}_{\tau(a)}$-measurable.

Theorem 15.2.15 $\widetilde{N}$ has $\mathcal{F}_{\tau(t)}$-intensity 1 (and therefore $\mathcal{F}_t^{\widetilde{N}}$-intensity 1).

Proof. Let $[a,b] \in \mathbb{R}$. We must show that

$$\mathrm{E}\left[1_A \widetilde{N}(a,b] \right] = \mathrm{E}\left[1_A (b-a) \right] \quad (A \in \mathcal{F}_{\tau(a)}).$$

But the left-hand side is just

$$\mathrm{E}\left[1_A \int_{\tau(a)}^{\tau(b)} N(dt) \right] = \mathrm{E}\left[1_A \int_0^\infty 1_{(\tau(a),\tau(b)]}(t) N(dt) \right].$$

Since the process $1_A 1_{(\tau(a),\tau(b)]}$ is $\mathcal{F}_t$-predictable (being $\mathcal{F}_t$-adapted and left-continuous), the right-hand side of the above equality is, by the smoothing formula,

$$\mathrm{E}\left[1_A \int_0^\infty 1_{(\tau(a),\tau(b)]}(t)\lambda(t)\,dt \right] = \mathrm{E}\left[1_A \int_{\tau(a)}^{\tau(b)} \lambda(t)\,dt \right] = \mathrm{E}\left[1_A (b-a) \right].$$

$\square$

Remark 15.2.16 By Watanabe's theorem (Theorem 7.1.8), $\widetilde{N}$ is a homogeneous Poisson process of intensity 1. In addition, for all $[a,b] \in \mathbb{R}$, $\widetilde{N}(a,b]$ is independent of $\mathcal{F}_{\tau(a)}$.

Remark 15.2.17 The result of Theorem 15.2.15 may be used to test that a given point process admits a given hypothetical intensity: perform the corresponding change of time and see if the result is a standard Poisson process, by using any available statistical test to assess that a given finite sequence of random variables is IID and exponentially distributed with mean 1.

Remark 15.2.18 The analogous result in the Brownian motion calculus is the one that says roughly that a continuous martingale is a Brownian motion with a different time scale (see, for instance, Section 5.3.2 in [Legall, 2016]).

Cryptology

A question arises naturally: how much information is lost in a change of time scale? Consider for instance the situation of Theorem 15.2.15. Since the change of time transforms the original point process into a homogeneous Poisson process, have we erased all the information previously contained in the stochastic intensity? The answer is: it depends. For instance, suppose that N is a Cox point process with intensity Λ, a non-negative real-valued random variable. In other words, N admits the stochastic $\mathcal{F}_t$-intensity $\lambda(t) \equiv \Lambda$ where $\mathcal{F}_t := \mathcal{F}_t^N \vee \sigma(\Lambda)$. If we perform the time change $\tau(t) = \frac{t}{\Lambda}$, the resulting process $\widetilde{N}$ defined by (15.40) is a standard Poisson process. Moreover, for all $0 \leq c \leq d$, $\widetilde{N}(d) - \widetilde{N}(c)$ is independent of $\mathcal{F}_c = \mathcal{F}_c^N \vee \sigma(\Lambda)$ and in particular of Λ. In this sense, the time change has erased all information concerning Λ, whereas Λ could be recovered from N since, by the strong law of large numbers,

$$\Lambda = \lim_{t\uparrow\infty} \frac{N(t)}{t}. \qquad (\star)$$

In the case of an intrinsic change of time, things are dramatically different. The stochastic $\mathcal{F}_t^N$-intensity of N, $\widehat{\lambda}(t) = \mathrm{E}\left[\Lambda \mid \mathcal{F}_t^N\right]$, is given by (Example 15.2.3):

$$\widehat{\lambda}(t) = \frac{\int_0^\infty \lambda^{N(t)+1} e^{-\lambda t}\, dF(\lambda)}{\int_0^\infty \lambda^{N(t)} e^{-\lambda t}\, dF(\lambda)}.$$

To be even more specific, suppose that $P(\Lambda = a) = P(\Lambda = b) = \frac{1}{2}$ for sone $0 < a < b$, in which case,

$$\widehat{\lambda}(t) = \frac{1 + (b/a)^{N(t)+1} e^{(a-b)t}}{1 + (b/a)^{N(t)} e^{(a-b)t}}.$$

Performing the time change

$$\int_0^{\widehat{\tau}(t)} \widehat{\lambda}(t)\, dt = t,$$

we obtain a point process $\widetilde{N}$ defined by $\widetilde{N}(t) := N(\widehat{\tau}(t))$, which is a standard Poisson process. However, this time, Λ can be entirely recovered from it. In fact, as we now show, N can be reconstructed from $\widetilde{N}$ and then Λ can be obtained by $(\star)$. In fact, if $\widehat{T}_n$ is the n-th point of $\widetilde{N}$, then

$$\int_{\widehat{T}_n}^{\widehat{T}_{n+1}} \widehat{\lambda}(t)\, dt = T_{n+1} - T_n$$

or, more explicitly,

$$\widehat{T}_{n+1} - \widehat{T}_n = f(n, T_{n+1}) - f(n, T_n),$$

where

$$f(n, t) = at - \ln\left(1 + (b/a)^{n+1} e^{(a-b)t}\right).$$

Clearly then, the sequence $\{T_n\}_{n\geq 1}$ can be recovered from $\{\widehat{T}_n\}_{n\geq 1}$.

Remark 15.2.19 An interpretation of the above results in terms of cryptography is the following. If the information is contained in Λ, the intrinsic time change yields a standard Poisson process from which Λ can be extracted only if one knows the "key", that is, the distribution of Λ. (Note however that from a finite trajectory of $\widetilde{N}$ one can only obtain an approximation of Λ. In this sense, secure transmission would be at the price of some unreliability. This unreliability can be controlled at the expense of transmission rate, which is acceptable if one is interested only in storage security.)

Remark 15.2.20 There is an analogous result in the Itô calculus, although this analogy is not a direct one as is the case, for instance, in Girsanov's theorem. We discuss it in very rough terms that will not be further detailed. Consider the "signal + noise" model

$$X(t) = \int_0^t \varphi(s)\, ds + W(t) \quad (t \geq 0),$$

where $\{W(t)\}_{t\geq 0}$ is a standard Wiener process and $\{\varphi(t)\}_{t\geq 0}$ is a locally integrable process (the integrated signal) independent of the Wiener process (the integrated noise). If one denotes by $\{\widehat{\varphi}(t)\}_{t\geq 0}$ a suitable version of the estimated signal process $E\left[\varphi(t) \mid \mathcal{F}_t^X\right]$ $(t \geq 0)$, then

$$\widehat{W}(t) := X(t) - \int_0^t \widehat{\varphi}(s)\, ds \quad (t \geq 0)$$

is a standard Wiener process. A cryptographic interpretation of this result avails in full analogy with the simple Poissonian example of the previous remark.

15.3 Point Processes under a Poisson process

A non-homogeneous Poisson process with (deterministic) intensity function $\lambda(t)$ can be obtained by projecting onto the time axis the points of a homogeneous Poisson process on $\mathbb{R}^2$ of intensity 1 which lie between the curve $y = \lambda(t)$ and the time axis (Exercise 8.5.14). In fact, as we shall see in this section, *any* point process with stochastic $\mathcal{F}_t$-intensity $\{\lambda(t)\}_{t\geq 0}$ not only *can* be obtained in this way (the *direct embedding* theorems) but can always be thought of as having been obtained in this way (the *inverse imbedding* theorems), in general at the cost of an extension of the probability space. The exact formulation and the mathematical details will be given in Subsection 15.3.2. For this the following preliminaries of intrinsic interest are needed.

15.3.1 An Extension of Watanabe's Theorem

The original version of Watanabe's theorem concerns homogeneous Poisson processes on the line. It will be extended, with a proof analogous to the proof of Theorem 7.1.8, to Poisson processes on product spaces of the type $\mathbb{R} \times K$. This new version will play a central role in the proof of the embedding theorems in Subsection 15.3.2. Some notation will be needed for the precise statement of this extension.

Let $\overline{N}$ be a point process on $\mathbb{R} \times K$. The notation $S_t\overline{N}$, where $t \geq 0$, denotes the point process obtained by shifting $\overline{N}$ by t to the left (algebraically). Let $S_t\overline{N}_+$ denote the point process obtained by restricting the shifted process $S_t\overline{N}$ to $\mathbb{R}_+ \times K$. Loosely speaking, $S_t\overline{N}_+$ is the *future* of $\overline{N}$ after time t, and more formally,

$$S_t\overline{N}_+([a,b] \times L) := \overline{N}(\mathbb{R}_+ \cap [a+t, b+t] \times L) \quad ([a,b] \subset \mathbb{R}, L \in \mathcal{K}).$$

One can similarly define $S_t \overline{N}_-$, the restriction of $S_t \overline{N}$ to $\mathbb{R}_- \times K$.

The following version of Watanabe's theorem concerns in particular Cox processes on $\mathbb{R}^2$.

Theorem 15.3.1 *Let $(K, \mathcal{K})$ be a measurable space. Let $\mathcal{G}$ be some σ-field of $\mathcal{F}$. Let $\lambda(t, dz)$ be a locally integrable kernel from $\mathbb{R} \times \Omega$ to $(K, \mathcal{K})$ such that for all $t \geq 0$, $L \in \mathcal{K}$, $\lambda(t, L)$ is $\mathcal{G}$-measurable. Let $\overline{N}$ be a point process on $\mathbb{R} \times K$ that admits the $\mathcal{F}_t$-intensity kernel $\lambda(t, \cdot)$, where*

$$\mathcal{F}_t = \mathcal{H}_t \vee \mathcal{G}$$

and $\{\mathcal{H}_t\}_{t \geq 0}$ is a history of $\overline{N}$. Then $\overline{N}$ is, conditionally on $\mathcal{G}$, a Poisson process with the intensity measure $\overline{\nu}(dt \times dz) = \lambda(t, dz) \times dt$. Furthermore, conditionally on $\mathcal{G}$, for all $t \geq 0$, the future $S_t \overline{N}_+$ of $\overline{N}$ after time t is independent of $\mathcal{H}_t$.

Proof. It suffices to show that for all $a \in \mathbb{R}$, for any finite family of disjoint measurable sets $C_1, \ldots, C_m \subset (a, +\infty) \times K$, and all $t_1, \ldots, t_m \in \mathbb{R}_+$,

$$\mathrm{E}\left[\exp\left(-\sum_{j=1}^{m} t_j \overline{N}(C_j)\right) | \mathcal{F}_a\right] = \prod_{j=1}^{m} \exp\left\{\nu(C_j)(e^{-t_j} - 1)\right\}. \tag{15.41}$$

One may assume that $C_1, \ldots, C_m \subset (a, b] \times L_k$ for some L_k as in Definition 15.1.20. Otherwise, replace the C_j's by $C_j \cap ((a, b] \times L_k)$ and let b and k go to infinity. Denote the above L_k by L.

For all j $(1 \leq j \leq m)$ and all $t \geq 0$, let

$$C_j(t) := C_j \cap \{(-\infty, t] \times K\} \text{ and } C_j^t := \{z \in K; (t, z) \in C_j\}.$$

Define for $t \geq a$

$$Z(t) := \exp\left\{-\sum_{j=1}^{m} t_j N(C_j(t))\right\}. \tag{15.42}$$

In particular,

$$Z(b) = \exp\left\{-\sum_{j=1}^{m} t_j N(C_j)\right\}.$$

Also, since $Z(a) = 1$,

$$Z(t) = 1 + \int_{(a, t]} \int_K Z(s-) \left\{\sum_{j=1}^{m} (e^{-t_j} - 1) 1_{C_j^s}(z)\right\} \overline{N}(ds \times dz). \tag{15.43}$$

For the proof of this equality, observe that any trajectory $t \to Z(t)$ is piecewise constant with discontinuity times that are points of the *simple* point process $N_L(\cdot) := \overline{N}(\cdot \times L)$. Therefore

$$Z(t) = Z(a) + \sum_{s \in (a, t]} (Z(s) - Z(s-)),$$

where $Z(s) \neq Z(s-)$ only if there is a point (s, z) of $\overline{N}$ that belongs to (at most) one of the C_j's. If $(s, z) \in C_j$ and is in $\overline{N}$, $Z(s) = Z(s-)e^{-t_j}$.

Now saying that $(s, z) \in C_j$ is equivalent to saying that $z \in C_j^s$, and therefore, if (s, z) is a point of $\overline{N}$

$$Z(s) - Z(s-) = \sum_{j=1}^{m} Z(s-)(e^{-t_j} - 1)1_{C_j^s}(z),$$

which then gives (15.43) since $Z(a) = 1$.

Let now $A \in \mathcal{F}_a$. We have

$$1_A Z(t) = 1_A + \int\int_{(a,t] \times K} 1_A Z(s-) \left(\sum_{j=1}^{m}(e^{-t_j} - 1)1_{C_j^s}(z) \right) \overline{N}(ds \times dz).$$

The stochastic process indexed by K,

$$H(t, \omega, z) := 1_A(\omega)1_{(a,t]}(t)Z(t_-, \omega) \times \left(\sum_{j=1}^{m}(e^{-t_j} - 1)1_{C_j^t}(z) \right),$$

is $\mathcal{P}(\mathcal{F}.) \otimes \mathcal{K}$ measurable and of constant sign (negative). Therefore,

$$
\begin{aligned}
\mathrm{E}[1_A Z(t)] &= P(A) + \mathrm{E}\left[\int_{(a,t]} \int_K 1_A Z(s-) \left\{ \sum_{j=1}^{m}(e^{-t_j} - 1)1_{C_j^s}(z) \right\} \overline{N}(ds \times dz) \right] \\
&= P(A) + \mathrm{E}\left[\int_a^t \int_K 1_A Z(s) \left\{ \sum_{j=1}^{m}(e^{-t_j} - 1)1_{C_j^s}(z) \right\} \lambda(s, dz)\, ds \right] \\
&= P(A) + \mathrm{E}\left[1_A \int_a^t \int_K Z(s) \left\{ \sum_{j=1}^{m}(e^{-t_j} - 1)1_{C_j^s}(z) \right\} \lambda(s, dz)\, ds \right].
\end{aligned}
$$

Since A is arbitrary in $\mathcal{F}_a$,

$$
\begin{aligned}
\mathrm{E}[Z(t)|\mathcal{F}_a] &= 1 + \mathrm{E}\left[\int_a^t \int_K Z(s) \left\{ \sum_{j=1}^{m}(e^{-t_j} - 1)1_{C_j^s}(z) \right\} \lambda(s, dz)\, ds | \mathcal{F}_a \right] \\
&= 1 + \int_a^t \int_K \mathrm{E}[Z(s)|\mathcal{F}_a] \left\{ \sum_{j=1}^{m}(e^{-t_j} - 1)1_{C_j^s}(z) \right\} \lambda(s, dz)\, s.
\end{aligned}
$$

Therefore

$$\mathrm{E}[Z(t)|\mathcal{F}_a] = \exp\left\{ \sum_{j=1}^{m}(e^{-t_j} - 1)\left\{ \int_a^t \int_K 1_{C_j^s}(z)\lambda(s, dz)ds \right\} \right\}.$$

Letting $t = b$ gives the announced result. $\qquad\square$

15.3.2 Grigelionis' Embedding Theorem

We shall use the following slight extension of the definition of a Poisson process:

Definition 15.3.2 *Let $(K, \mathcal{K})$ be some measurable space. Given a history $\{\mathcal{F}_t\}_{t \in \mathbb{R}}$, the point process $\overline{N}$ on $\mathbb{R} \times K$ is called an $\mathcal{F}_t$-Poisson process if the following conditions are satisfied:*

(i) $\{\mathcal{F}_t\}_{t \in \mathbb{R}}$ is a history of $\overline{N}$ in the sense that

$$\mathcal{F}_t \subseteq \sigma(\overline{N}(D); D \subseteq (-\infty, t] \times K);$$

(ii) $\overline{N}$ is a Poisson process; and

(iii) for any $t \in \mathbb{R}$, $S_t \overline{N}_+$ and $\mathcal{F}_t$ are independent, where $S_t \overline{N}_+$ is defined by

$$S_t \overline{N}_+(C \times L) := \overline{N}\left((C + t) \cap (-\infty, t]\right) \times L\right).$$

Theorem 15.3.3 *Let $(K, \mathcal{K})$ be some measurable space and let Q be some probability measure on it. Let $\overline{N}$ be an $\mathcal{F}_t$-Poisson process on $\mathbb{R} \times K \times \mathbb{R}_+$ with intensity measure $dt \times Q(dz) \times d\sigma$. Let $f : \Omega \times \mathbb{R} \times K \to \mathbb{R}$ be a non-negative function that is $\mathcal{P}(\mathcal{F}.) \otimes \mathcal{K}$-measurable and such that the kernel*

$$\lambda(t, dz) := f(t, z)Q(dz) \tag{15.44}$$

is locally integrable. The marked point process (N, Z) with marks in K defined by

$$N(C \times L) := \int_{\mathbb{R}} \int_K \int_{\mathbb{R}_+} 1_C(t) 1_{(L)}(z) 1_{\{\sigma \le f(t,z)\}} \overline{N}(dt \times dz \times d\sigma) \quad (C \in \mathcal{B}(\mathbb{R}), \in \mathcal{K})$$

admits the $\mathcal{F}_t$-stochastic intensity kernel $\lambda(t, dz)$. $\tag{15.45}$

Proof. We prove the theorem for the unmarked case. The general proof follows exactly the same lines.

Let $\overline{N}$ be a homogeneous $\mathcal{F}_t$-Poisson process on $\mathbb{R} \times \mathbb{R}_+$ with average intensity 1. Let $\{\lambda(t)\}_{t \ge 0}$ be a non-negative locally integrable $\mathcal{F}_t$-predictable stochastic process. Define the point process N on $\mathbb{R}$ by the formula

$$N(C) = \int_C \int_{\mathbb{R}_+} 1_{(0, \lambda(t)]}(z) \overline{N}(dt \times dz) \tag{15.46}$$

for all $C \in \mathcal{B}$. Then, N has the $\mathcal{F}_t$-intensity $\{\lambda(t)\}_{t \ge 0}$.

1. We first show that (15.46) defines a locally finite point process, that is, $N((0, b]) < \infty$ a.s. for all $b \in \mathbb{R}$. Define for all $n \ge 1$

$$\tau_n = \inf\left\{t \ge 0 \; ; \; \int_0^t \lambda(s)ds \ge n\right\}$$

($=\infty$ if $\{\ldots\} = \varnothing$). By the local integrability assumption, $\lim_{n \uparrow \infty} \tau_n = \infty$, a.s. Also τ_n is an $\mathcal{F}_t$-stopping time. By the smoothing formula of Theorem 15.1.22,

$$\mathrm{E}\left[N((0, \tau_n \wedge b])\right] = \mathrm{E}\left[\int_{(0,b]\times\mathbb{R}}\int 1_{(0,\tau_n]}(s)1_{(0,\lambda(s)]}(\sigma)\,\overline{N}(\mathrm{d}s\times\mathrm{d}\sigma)\right]$$

$$= \mathrm{E}\left[\int_{(0,b]\times\mathbb{R}}\int 1_{(0,\tau_n]}(s)1_{(0,\lambda(s)]}(\sigma)\,\mathrm{d}s\,\mathrm{d}\sigma\right]$$

$$= \mathrm{E}\left[\int_0^{\tau_n\wedge b}\lambda(s)\,\mathrm{d}s\right] < \infty.$$

Therefore, a.s., for all $n \geq 1$, $N(0, \tau_n \wedge b] < \infty$.

2. The simplicity is left as an exercise for the reader.

3. In order to prove that N has the $\mathcal{F}_t$-intensity $\{\lambda(t)\}_{t\geq 0}$ it suffices to show that for all $H \in \mathcal{P}(\mathcal{F}.)$, $H \geq 0$,

$$\mathrm{E}\left[\int_{\mathbb{R}} H(t)N(\mathrm{d}t)\right] = \mathrm{E}\left[\int_{\mathbb{R}} H(t)\lambda(t)\mathrm{d}t\right].$$

But the left-hand side of this equality reads

$$\mathrm{E}\left[\int_{\mathbb{R}}\int_{\mathbb{R}_+} H(t)1_{(0,\lambda(t)]}(z)\overline{N}(\mathrm{d}t\times\mathrm{d}z)\right].$$

Since $\overline{N}$ is assumed $\mathcal{F}_t$-Poisson, it admits the $\mathcal{F}_t$-intensity kernel $\lambda(t, \mathrm{d}z) = \mathrm{d}z$. Thus, by Theorem 15.1.22, this is also equal to

$$\mathrm{E}\left[\int_{\mathbb{R}}\int_{\mathbb{R}_+} H(t)1_{(0,\lambda(t)]}(z)\mathrm{d}t\mathrm{d}z\right] = \mathrm{E}\left[\int_{\mathbb{R}} H(t)\lambda(t)\mathrm{d}t\right].$$

$\square$

Theorem 15.3.4 *Let (N, Z) be a locally finite marked point process with marks in the measurable space $(K, \mathcal{K})$ and $\mathcal{F}_t$-stochastic intensity kernel of the form (15.44), where $f : \Omega \times \mathbb{R} \times K \to \mathbb{R}$ is a non-negative function that is $\mathcal{P}(\mathcal{F}.) \otimes \mathcal{K}$-measurable and Q is a probability measure on $(K, \mathcal{K})$. Then, the probability space may be enlarged to accommodate an $\mathcal{F}_t$-Poisson process $\overline{N}$ on $\mathbb{R} \times K \times \mathbb{R}_+$ with intensity measure $\mathrm{d}t \times Q(\mathrm{d}z) \times \mathrm{d}s$ such that (15.45) holds.*

Proof. The result will be proved for the unmarked case, the general case following exactly the same lines. The theorem in this simplified form is as follows. Let N be a simple point process on $\mathbb{R}$ with $\mathcal{F}_t$-predictable intensity $\{\lambda(t)\}_{t\geq 0}$. Then, there exists a homogeneous Poisson process $\overline{N}$ on $\mathbb{R} \times \mathbb{R}_+$ with average intensity 1, such that (15.46) holds. Moreover, this process $\overline{N}$ is an $\mathcal{F}_t \vee \mathcal{F}_t^{\overline{N}}$-Poisson process. As such, for all $a \in \mathbb{R}$, $S_a\overline{N}_+$ is independent of $\mathcal{F}_a$.

Let $\{U_n\}_{n\in\mathbb{Z}}$ be an IID sequence of random variables uniformly distributed on $[0, 1]$, and let $\overline{N}_1$ be a homogeneous Poisson process on $\mathbb{R} \times \mathbb{R}_+$, of intensity 1, such that $\{U_n\}_{n\in\mathbb{Z}}$, $\overline{N}_1$ and $\mathcal{F}_\infty$ are independent. Define $\overline{N}$ by

$$\overline{N}(A) = \int\int_A 1_{(\lambda(t),\infty)}(\sigma)\overline{N}_1(\mathrm{d}t\times\mathrm{d}\sigma) + \sum_{n\in\mathbb{Z}} 1_A((T_n, U_n\lambda(T_n)))$$

for all $A \in \mathcal{B}\otimes\mathcal{B}(\mathbb{R}_+)$. If H is a non-negative function from $\mathbb{R} \times \Omega \times \mathbb{R}_+$ to $\mathbb{R}$,

$$\int_{\mathbb{R}} \int_{\mathbb{R}} H(t,\sigma)\overline{N}(\mathrm{d}t \times \mathrm{d}\sigma) =$$

$$\int_{\mathbb{R}} \int_{\mathbb{R}_+} H(t,\sigma)1_{(\lambda(t),\infty)}(\sigma)\overline{N}_1(\mathrm{d}t \times \mathrm{d}\sigma) + \sum_{n\in\mathbb{Z}} H(T_n, U_n\lambda(T_n)).$$

Denote by N^U the marked point process obtained by attaching the mark U_n to point T_n of N. Let

$$\mathcal{G}_t = \mathcal{F}_t \vee \mathcal{F}_t^{\overline{N}_1} \vee \mathcal{F}_t^{N^U}$$

and suppose that H is $\mathcal{P}(\mathcal{G}_t) \otimes \mathcal{B}(\mathbb{R}_+)$-measurable and non-negative.

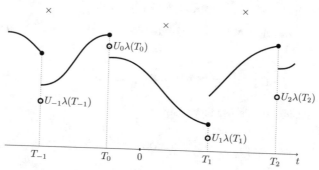

Since $\overline{N}_1$ has the $\mathcal{F}_t^{\overline{N}_1}$-intensity kernel $\lambda_1(t,\mathrm{d}\sigma) = \mathrm{d}\sigma$, the latter is also the $\mathcal{G}_t$-intensity kernel of $\overline{N}_1$ (recall that $\mathcal{F}_\infty^U$ and $\mathcal{F}_\infty$ are independent of $\mathcal{F}_\infty^{\overline{N}_1}$). By the smoothing theorem

$$\mathrm{E}\left[\int_{\mathbb{R}} \int_{\mathbb{R}_+} 1_{(\lambda(t),\infty)}(\sigma)H(t,\sigma)\overline{N}_1(\mathrm{d}t \times \mathrm{d}\sigma)\right] = \mathrm{E}\left[\int_{\mathbb{R}} \int_{\mathbb{R}_+} 1_{(\lambda(t),\infty)}(\sigma)H(t,\sigma)\mathrm{d}t \times \mathrm{d}\sigma\right]$$

for any non-negative $H \in \mathcal{P}(\mathcal{F}.)\otimes\mathcal{B}(\mathbb{R}_+)$. Now, for any non-negative $H \in \mathcal{P}(\mathcal{F}.)\otimes\mathcal{B}(\mathbb{R}_+)$,

$$\mathrm{E}\left[\sum_{n\in\mathbb{Z}} H(T_n, U_n\lambda(T_n))\right] = \mathrm{E}\left[\int_{\mathbb{R}} \int_{[0,1]} H(t, u\lambda(t))N^U(\mathrm{d}t \times \mathrm{d}u)\right]$$

$$= \mathrm{E}\left[\int_{\mathbb{R}} \int_0^1 H(t, u\lambda(t))\,\mathrm{d}t\,\mathrm{d}u\right].$$

(In fact, N^U admits the $\mathcal{F}_t \vee \mathcal{F}_t^{N^U}$-intensity kernel $\lambda(t)1_{[0,1]}(u)\,\mathrm{d}u$. This is also a $\mathcal{G}_t$-stochastic kernel for N^U since $\overline{N}_1$ is independent of N^U and $\mathcal{F}_\infty$. The map $(t,\omega,u) \to H(t, u\lambda(t))$ is $\mathcal{P}(\mathcal{G}_t) \otimes \mathcal{B}(\mathbb{R})$-measurable in view of the $\mathcal{F}_t$-predictability of $\{\lambda(t)\}_{t\geq 0}$, and of the measurability assumptions on H. This justifies the above use of the smoothing theorem.) This term is also equal to

$$\mathrm{E}\left[\int_{\mathbb{R}} \int_{\mathbb{R}_+} H(t,\sigma)1_{(0,\lambda(t)]}(\sigma)\mathrm{d}t\mathrm{d}\sigma\right],$$

by the change of variables $\sigma = u\lambda(t)$. Therefore

$$\mathrm{E}\left[\int_{\mathbb{R}} \int_{\mathbb{R}_+} H(t,\sigma)\overline{N}(\mathrm{d}t \times \mathrm{d}\sigma)\right]$$

$$= \mathrm{E}\left[\int_{\mathbb{R}} \int_{\mathbb{R}_+} H(t,\sigma)1_{(\lambda(t),\infty)}(\sigma)\mathrm{d}t\mathrm{d}\sigma\right] + \mathrm{E}\left[\int_{\mathbb{R}\times\mathbb{R}_+} \int H(t,\sigma)1_{(0,\lambda(t)]}(\sigma)\mathrm{d}t\mathrm{d}\sigma\right]$$

$$= \mathrm{E}\left[\int_{\mathbb{R}} \int_{\mathbb{R}_+} H(t,\sigma)\mathrm{d}t\mathrm{d}\sigma\right]$$

or all non-negative $H \in \mathcal{P}(\mathcal{G}) \otimes \mathcal{B}_+$. Therefore, by Theorem 15.3.1, $\overline{N}$ is a homogeneous $\mathcal{J}_t$-Poisson process on $(\mathbb{R} \times \mathbb{R}_+, \mathcal{B}(\mathbb{R}) \otimes \mathcal{B}(\mathbb{R}_+))$ with intensity 1. It is *a fortiori* an $\mathcal{F}_t \vee \mathcal{F}_t^{\overline{N}}$-Poisson process. $\square$

Variants of the Embedding Theorems

The following results are of the same kind as the previous ones, but they assume boundedness of the intensity kernel. We state them in the unmarked case. Their proofs are left as an exercise (Exercise 15.4.18).

Theorem 15.3.5 (Direct embedding) *Let $\widetilde{N}$ be an* HPP *on $\mathbb{R}$ with intensity $\widetilde{\lambda}$ and let $\{U_n\}_{n \in \mathbb{Z}}$ be an* IID *sequence of marks, uniformly distributed on $[0,1]$. Let $\widetilde{N}^U$ be the associated lifted point process on $\mathbb{R} \times [0,1]$. Let $\{\mathcal{G}_t\}_{t \geq 0}$ be a history independent of $\widetilde{N}_U$, and let for all $t \geq 0$,*

$$\mathcal{F}_t := \mathcal{G}_t \vee \mathcal{F}_t^{\widetilde{N}_U}.$$

Let $\{\lambda(t)\}_{t \geq 0}$ be a non-negative $\mathcal{F}_t$-predictable process bounded by $\widetilde{\lambda} < \infty$, and define a point process N on $(\mathbb{R}, \mathcal{B}(\mathbb{R}))$ by

$$N(C) = \sum_{n \in \mathbb{Z}} 1_C(\widetilde{T}_n) 1_{\left\{0 \leq U_n \leq \frac{\lambda(\widetilde{T}_n)}{\widetilde{\lambda}}\right\}} \qquad (C \in \mathcal{B}(\mathbb{R})). \qquad (15.47)$$

Then N admits the $\mathcal{F}_t$-intensity $\{\lambda(t)\}_{t \geq 0}$.

Theorem 15.3.6 (Inverse embedding) *Let N be a simple point process on $(\mathbb{R}, \mathcal{B}(\mathbb{R}))$ with $\mathcal{F}_t$-intensity $\{\lambda(t)\}_{t \geq 0}$ (assumed $\mathcal{F}_t$-predictable, without loss of generality). Suppose that there exists a constant $\widetilde{\lambda} < \infty$ such that P-a.s.*

$$\lambda(t, \omega) \leq \widetilde{\lambda} \qquad (t \geq 0).$$

Then, there exists a compound Poisson process $(\widetilde{N}, U)$ on $(\mathbb{R}, \mathcal{B}(\mathbb{R}))$ with marks in $([0,1], \mathcal{B}([0,1]))$, characteristics $(\widetilde{\lambda}, Q)$ where Q is the uniform distribution on $[0,1]$, and such that (15.47) holds.

Remark 15.3.7 Grigelionis' construction is of particular importance for coupling point processes. Usually one starts with a point process N_1 from which one constructs $\overline{N}$ via the inverse embedding theorem, and then using this same $\overline{N}$, one constructs another point process N'.

EXAMPLE 15.3.8: LEWIS–SHEDLER–OGATA SIMULATION ALGORITHMS. An immediate application of practical importance of the direct embedding theorems is to the simulation of point processes with a stochastic intensity. We start with a simple example of the methodology. Suppose that one wishes to simulate a point process on the positive half-line with a stochastic intensity of the form

$$\lambda(t, \omega) = v(t, N_{[0,t)}(\omega)),$$

where $v : \mathbb{R}_+ \times M_p(\mathbb{R}_+) \to \mathbb{R}_+$ is measurable with respect to $\mathcal{B}(\mathbb{R}_+) \otimes \mathcal{M}_p(\mathbb{R}_+)$ and $\mathcal{B}(\mathbb{R}_+)$ and bounded, say, by $K < \infty$. For this we can use Theorem 15.3.6, which says that it suffices to thin a homogeneous Poisson process $\widetilde{N}$ on $\mathbb{R}_+$ with intensity K, examining its points sequentially, keeping a point $\widetilde{T}_n$ of it if and only if $U_n \leq \frac{v(\widetilde{T}_n, N_{[0,\widetilde{T}_n)})}{K}$, where $\{U_n\}_{n \geq 1}$ is an IID sequence of random variables uniformly distributed on $[0, 1$ and independent of $\tilde{N}$.

This method is easily adapted to the case of point processes with an $\mathcal{F}_t^N \vee \mathcal{G}_0$-intensity of the form

$$\lambda(t, \omega) = v(t, X(t-, \omega), N_{[0,t)}(\omega)),$$ $(\star)$

where

(i) $\{X(t)\}_{t \geq 0}$ is some CORLOL stochastic process with values in some measurable space $(K, \mathcal{K})$ and $\mathcal{G}_0 = \mathcal{F}_\infty^X := \vee_{t \geq 0} \mathcal{F}_t^X$,

(ii) $v : \mathbb{R}_+ \times M_p(\mathbb{R}_+) \to \mathbb{R}_+$ is measurable with respect to $\mathcal{B}(\mathbb{R}_+) \otimes \mathcal{K} \otimes M_p(\mathbb{R}_+)$ and $\mathcal{B}(\mathbb{R}_+)$, and is bounded by K,

(In other words, the point process we seek to simulate is a semi-Cox point process.)

(iii) $\{X(t)\}_{t \geq 0}$ is independent of $\tilde{N}$, and

(iv) we suppose that we have at our disposition at any time t the value $X(t)$.

A point $\tilde{T}_n$ of $\tilde{N}$ is kept if and only if $U_n \leq \frac{v(\tilde{T}_n, X(T_n-), N_{(0,\tilde{T}_n)})}{K}$.

We can make another step towards generalization assuming an $\mathcal{F}_t^N \vee \mathcal{F}_t^X$-intensity of the form $(\star)$. But this time we must add the condition that, denoting by T_n the n-th point of N, one can construct $X(t)$ from T_n based on the knowledge of $N_{[0,t)}$. One example is when the process $\{X(t)\}_{t \geq 0}$ is of the form

$$X(t) = X(0) + \int_0^t \varphi(N(t-)) \, dt + \int_0^t \sigma(N(t-)) \, dW(t),$$

where $\{W(t)\}_{t \geq 0}$ is a Wiener process independent of $\tilde{N}$.

Other examples, concerning for instance mutually exciting point processes, are left to the imagination of the reader. The relaxation of the hypothesis of boundedness of the stochastic intensity is not a problem since one can always vary the intensity of the Poisson process $\tilde{N}$ when needed.

Complementary reading

[Last and Brandt, 1995] is an advanced text on the martingale theory of point processes on the line, with or without a stochastic intensity. See also [Brémaud, 2020].

15.4 Exercises

Exercise 15.4.1. THE FUNDAMENTAL (LOCAL) MARTINGALE
Show that Definition 15.1.1 implies that for all $a \in \mathbb{R}$, $M_a(t) := N(a, t] - \int_a^t \lambda(s) \, ds$ $(t \geq a)$ is an $\mathcal{F}_t$-local martingale.

Exercise 15.4.2. CONNECTING TO THE INTUITIVE DEFINITION
Let the simple and locally finite point process N on $\mathbb{R}$ have the $\mathcal{F}_t$-intensity $\{\lambda(t)\}_{t \geq 0}$, and suppose that $t \mapsto \lambda(t, \omega)$ is for all $\omega \in \Omega$ a right-continuous function, and that the function $(t, \omega) \mapsto \lambda(t, \omega)$ is uniformly bounded.

Show that

$$\lim_{h \downarrow 0} \frac{1}{h} E[N(t, t+h)|\mathcal{F}_t] = \lambda(t) , \qquad P\text{-a.s.}$$

Exercise 15.4.3. THE TRAFFIC EQUATIONS
Show that in equilibrium, the traffic equations of the Jackson network (see Chapter 9) receive the following interpretation:

$$\lambda_i = E[A_i(0,1]], \quad \lambda_i r_{ij} = E[A_{i,j}(0,1]],$$

where A_i is the point process counting the arrivals (external and internal) into station i, and $A_{i,j}$ is the point process counting the transfers from station i to station j.

Exercise 15.4.4. COX PROCESSES
Let N be a doubly stochastic Poisson process (Cox process) with respect to the σ-field $\mathcal{G}$, with locally integrable stochastic intensity $\{\lambda(t)\}_{t \geq 0}$. Remember that this means that $\mathcal{G} \supseteq \mathcal{F}_\infty^\lambda := \sigma(\lambda(s), s \in \mathbb{R})$ and that whenever $C_1, \ldots, C_K$ are bounded disjoint measurable subsets of $\mathbb{R}$,

$$E\left[\exp\left\{i \sum_{j=1}^K u_j N(C_j)\right\} \Big| \mathcal{G}\right] = \exp\left\{\sum_{j=1}^K (e^{iu_j} - 1) \int_{C_j} \lambda(s)\mathrm{d}s\right\}$$

for all $u_1, \ldots, u_K \in \mathbb{R}$. Show that N admits the $\mathcal{F}_t$-intensity $\{\lambda(t)\}_{t \geq 0}$, where $\mathcal{F}_t = \mathcal{G} \vee \mathcal{F}_t^N$.

Exercise 15.4.5. OTHER HISTORIES
Let N be a simple and locally finite point process on $\mathbb{R}$ with $\mathcal{F}_t$-intensity $\{\lambda(t)\}_{t \geq 0}$. Prove the following:

(i) If $\{\mathcal{G}_t\}_{t \geq 0}$ is a history such that $\mathcal{G}_\infty$ is independent of $\mathcal{F}_\infty$, then $\{\lambda(t)\}_{t \geq 0}$ is an $\mathcal{F}_t \vee \mathcal{G}_t$-intensity of N.

(ii) If $\{\widetilde{\mathcal{F}}_t\}_{t \geq 0}$ is a history of N such that $\mathcal{F}_t^\lambda \vee \mathcal{F}_t^N \subseteq \widetilde{\mathcal{F}}_t \subseteq \mathcal{F}_t$, then $\{\lambda(t)\}_{t \geq 0}$ is an $\widetilde{\mathcal{F}}_t$-intensity of N.

Exercise 15.4.6. ABOUT PREDICTABILITY

(a) Show that a deterministic measurable process is $\mathcal{F}_t$-predictable, and that an $\mathcal{F}_t$-predictable process is $\mathcal{F}_t$-progressively measurable, and in particular measurable and $\mathcal{F}_t$-adapted.

(b) Let S and τ be two $\mathcal{F}_t$-stopping times such that $S \leq \tau$, and let $\varphi : \mathbb{R}_+ \times \mathbb{R} \to \mathbb{R}$ be a measurable function. Then

$$X(t, \omega) = \varphi(S(\omega), t)\mathbf{1}_{\{S(\omega) < t \leq \tau(\omega)\}}$$

defines an $\mathcal{F}_t$-predictable process.

Exercise 15.4.7. POISSON WITH INDEPENDENT IID MARKS
Let (N, Z) be a marked point process on $\mathbb{R}$ with marks in $(K, \mathcal{K})$. Prove that if the base point process N is a simple locally finite Poisson process with (deterministic) intensity $\lambda(t)$ and if the sequence of marks is IID and independent of N, then (N, Z) admits the $\mathcal{F}_t^{N,Z}$-intensity kernel $\lambda(t)Q(dz)$, where Q is the common distribution of the marks.

Exercise 15.4.8. STOCHASTIC FAILURE RATE
Prove the last statement in Example 15.1.15.

Exercise 15.4.9. WATANABE'S THEOREM FOR CONDITIONAL POISSON PROCESSES
Let N be a simple point process on $\mathbb{R}_+$, and let $\{\mathcal{F}_t\}_{t\geq 0}$ be a history of the form

$$\mathcal{F}_t = \mathcal{F}_t^N \vee \mathcal{G}, \tag{15.48}$$

for some σ-field $\mathcal{G}$. Let $\{\lambda(t)\}_{t\geq 0}$ be a measurable stochastic process, locally integrable and such that the random variable $\lambda(t)$ is $\mathcal{G}$-measurable for all t. Suppose that $\mathrm{E}[N((a,b]) \mid \mathcal{F}_a] = \int_a^b \lambda(s)\,ds$ for all $0 \leq a \leq b$. Show that

$$\mathrm{E}[e^{iuN((a,b])} \mid \mathcal{G} \vee \mathcal{F}_a^N] = \exp\left((e^{iu} - 1)\int_a^b \lambda(t)dt\right).$$

Exercise 15.4.10. A MULTIVARIATE WATANABE'S THEOREM
Let N_j $(j \geq 1)$ be locally finite simple point processes without common points and let $\mathcal{F}_t$ be of the form (15.48) with

$$\mathcal{F}_t^N = \bigvee_{j=1}^\infty \mathcal{F}_t^{N_j}$$

$(N := (N_1, N_2, \ldots))$. For all $j \geq 1$, let $\{\lambda_j(t)\}_{t\geq 0}$ be a $\mathcal{G}$-measurable locally integrable stochastic process and suppose that for all $(a, b] \subseteq \mathbb{R}$,

$$\mathrm{E}[N_j(a, b] \mid \mathcal{F}_a] = \int_a^b \lambda_j(s)\,ds.$$

Show that N_j is a $\mathcal{G}$-conditional Poisson process with $\mathcal{G}$-conditional associated measure $C \rightarrow \mathrm{E}[N_j(C)|\mathcal{G}] = \int_C \lambda_j(t)dt$ and that the N_j's are independent conditionally with respect to $\mathcal{G}$.

Exercise 15.4.11. SQUARE-INTEGRABLE STOCHASTIC INTEGRALS

A. Let H be an $\mathcal{F}_t$-predictable real-valued stochastic process indexed by K such that for all $t \geq 0$, P-a.s.

$$\mathrm{E}\left[\int_{(0,t]\times K} |H(s,z)|^2 \lambda(s, dz)\,ds\right] < \infty.$$

Prove that the stochastic process

$$M(t) := \int_{(0,t]\times K} H(s,z)\, M_Z(ds \times dz)$$

is well defined and a square-integrable $\mathcal{F}_t$-martingale, and that moreover

$$\mathrm{E}\left[M(t)^2\right] = \mathrm{E}\left[\int_{(0,t]\times K} |H(s,z)|^2 \lambda(s, dz)\,ds\right].$$

B. If the assumption of A. is replaced by

$$\int_{(0,t]\times K} |H(s,z)|^2 \lambda(s,\mathrm{d}z)\,\mathrm{d}s < \infty \quad P\text{-a.s.}, (t \geq 0),$$

prove that the stochastic process

$$M(t) := \int_{(0,t]\times K} H(s,z)\, M_Z(\mathrm{d}s \times \mathrm{d}z) \quad (t \geq 0)$$

is a square-integrable local $\mathcal{F}_t$-martingale.

Exercise 15.4.12. STRONG MARKOV PROPERTY OF HPPS
In the setting of Exercise 15.4.10 with possibly $K = \infty$, let T be a finite $\mathcal{F}_t$-stopping time and let for all $t \geq 0$

$$\mathcal{F}'_t := \mathcal{F}^N_{T+t}.$$

Define the point processes N'_j ($1 \leq j \leq K$) by

$$N'_j(a,b] = N_j(a+T,\ b+T].$$

In particular $\mathcal{F}'_t = \mathcal{G}' \vee \mathcal{F}^{N'}_t$, where $\mathcal{G}' = \mathcal{G} \vee \mathcal{F}^N_T$. Prove that for all j, N'_j admits the $\mathcal{F}'_t$-intensity $\{\lambda_j(t+T)\}_{t\geq 0}$.

Comment: An interesting special case is when for all j ($1 \leq j \leq K$), $\lambda_j(t) \equiv \lambda_j$ for all t, that is, the original point processes N_j are Poisson processes with average intensities λ_j, mutually independent and independent of $\mathcal{G}$. In this case, the delayed point processes $S_T N_j$ (defined by $S_T N_j(C) = N_j(C + \tau)$) are independent Poisson processes with intensities λ_j, and they are independent of $\mathcal{G}$ and of $\mathcal{F}^N_T = \vee^K_{j=1}\mathcal{F}^{N_j}_T$. The latter independence property is the strong Markov property of multivariate Poisson processes. This was proved by other means in Theorem 7.1.5.

Exercise 15.4.13. $M(t)^2 - \int_0^t \lambda(s)\,\mathrm{d}s$
Let N be a simple locally finite point process on $\mathbb{R}_+$ with (locally integrable) $\mathcal{F}_t$-intensity $\{\lambda(t)\}_{t\geq 0}$. Let for $t \geq 0$, $M(t) := N(t) - \int_0^t \lambda(s)\,\mathrm{d}s$. Prove that $\{M(t)^2 - \int_0^t \lambda(s)\,\mathrm{d}s\}_{t\geq 0}$ is an $\mathcal{F}_t$-local martingale.

Exercise 15.4.14. POISSON WITH INDEPENDENT IID MARKS
Show that if $\overline{N}$ is a Poisson process on $\mathbb{R} \times K$ with intensity measure

$$\overline{\nu}(\mathrm{d}t \times \mathrm{d}z) = \lambda(t)\mathrm{d}t \times \nu(\mathrm{d}z),$$

where ν is a sigma-finite measure on $(K,\mathcal{K})$ and $t \to \lambda(t)$ is a deterministic non-negative locally integrable function, then $\overline{N}$ is a locally finite simple point process with stochastic $\mathcal{F}^{\overline{N}}_t$-intensity kernel

$$\lambda(t,\mathrm{d}z) = \lambda(t)\nu(\mathrm{d}z).$$

Exercise 15.4.15. IID MARKS INDEPENDENT OF THE BASIC POINT PROCESS
Let (N, Z) be a simple and locally finite marked point process with marks in K, with associated (lifted) point process N_Z on $\mathbb{R} \times K$. Suppose that N has $\mathcal{F}_t$-intensity $\{\lambda(t)\}_{t\geq 0}$ and that the mark sequence $\{Z_n\}_{n\in\mathbb{Z}}$ is IID, with common distribution $P(Z_1 \in C) = Q(C)$, and independent of $\mathcal{F}_\infty$.

Show that N_Z has $(\mathcal{F}_t^{N_Z} \vee \mathcal{F}_t)$-intensity kernel

$$\lambda(t, dz) = \lambda(t)Q(dz).$$

Exercise 15.4.16. ESTIMATING THE INTENSITY OF A HOMOGENEOUS COX PROCESS
Prove the statement of Example 15.2.3.

Exercise 15.4.17. LIKELIHOOD RATIOS FOR MARKED POINT PROCESSES
Prove the assertion of Example 15.2.11.

Exercise 15.4.18. ANOTHER TYPE OF EMBEDDING
Prove the following results:

Theorem 1: direct embedding. Let $\widetilde{N}$ be an HPP on $\mathbb{R}$ with intensity $\widetilde{\lambda}$, and let $\{U_n\}_{n \in \mathbb{Z}}$ be an IID sequence of marks, uniformly distributed on $[0, 1]$. Let $\widetilde{N}^U$ be the associated lifted point process on $\mathbb{R} \times [0, 1]$. Let $\{\mathcal{G}_t\}_{t \geq 0}$ be a history independent of $\widetilde{N}_U$, and define for all $t \geq 0$,

$$\mathcal{F}_t = \mathcal{G}_t \vee \mathcal{F}_t^{\widetilde{N}_U}.$$

Let $\{\lambda(t)\}_{t \geq 0}$ be a non-negative $\mathcal{F}_t$-predictable process bounded by $\widetilde{\lambda} < \infty$, and define a point process N on $(\mathbb{R}, \mathcal{B})$ by

$$N(C) = \sum_{n \in \mathbb{Z}} 1_C(\widetilde{T}_n) 1_{\{0 \leq \widetilde{\lambda} U_n \leq \lambda(\widetilde{T}_n)\}} \qquad (C \in \mathcal{B}). \tag{$\star$}$$

Then N admits the $\mathcal{F}_t$-intensity $\{\lambda(t)\}_{t \geq 0}$.

Theorem 2: inverse embedding. Let N be a simple point process on $(\mathbb{R}, \mathcal{B}(\mathbb{R}))$ with $\mathcal{F}_t$-intensity $\{\lambda(t)\}_{t \geq 0}$ (assumed $\mathcal{F}_t$-predictable, without loss of generality). Suppose that there exists a constant $\widetilde{\lambda} < \infty$ such that

$$\lambda(t, \omega) \leq \widetilde{\lambda}, \quad \forall t \geq 0, \quad P\text{-a.s.}$$

Then, there exists a compound Poisson process $(\widetilde{N}, U)$ on $(\mathbb{R}, \mathcal{B}(\mathbb{R}))$ with marks in $([0, 1], \mathcal{B}([0, 1]))$, characteristics $(\widetilde{\lambda}, Q)$, where Q is the uniform distribution on $[0, 1]$, and such that $(\star)$ holds.

Exercise 15.4.19. THE HAWKES PROCESS
Prove the statement of Example 15.1.14.

Chapter 16

Ergodic Processes

Historically, ergodicity concerns an issue of interest to physicists, namely, that of finding conditions for the empirical and probabilistic averages of a stationary stochastic process to coincide. A typical example, besides the strong law of large numbers, is that of a positive recurrent aperiodic stationary HMC (Theorem 6.4.1). The results of this chapter will in particular considerably augment the generality of the specific ergodic theorems concerning Markov chains. A notable feature of the presentation given here resides in the connection it establishes with queueing theory, which will be further exploited in Chapter 17.

16.1 Ergodicity and Mixing

16.1.1 Invariant Events and Ergodicity

The modern approach to ergodic theory is in terms of measurable flows. This framework gives powerful results but in a rather abstract form. However, we shall rapidly connect to the more picturesque framework in terms of random sequences and shifts, and examples concerning Markov chains and queueing processes will give to this theory a more familiar aspect.

Definition 16.1.1 *Let $(\Omega, \mathcal{F}, P)$ be a probability space. Let $\theta : (\Omega, \mathcal{F}) \to (\Omega, \mathcal{F})$ be a bijective measurable map with measurable inverse. The family $\{\theta^n\}_{n \in \mathbb{Z}}$ is then called a* discrete (measurable) flow *(on $(\Omega, \mathcal{F})$) and a sequence of random elements $\{Z_n\}_{n \in \mathbb{Z}}$ defined on $(\Omega, \mathcal{F})$ is said to be compatible with this flow (for short: θ-compatible) if for all $n \in \mathbb{Z}$,*

$$Z_n = Z_0 \circ \theta^n \,,$$

that is, for all $\omega \in \Omega$, $Z_n(\omega) = Z_0(\theta^n(\omega))$.

EXAMPLE 16.1.2: THE SHIFT. Let $(E, \mathcal{E})$ be a measurable space and let $(\Omega, \mathcal{F}) := (E^{\mathbb{Z}}, \mathcal{E}^{\otimes \mathbb{Z}})$. Define $\theta : \Omega \to \Omega$ by: $\theta(\omega) = \{x_{n+1}, n \in \mathbb{Z}\}$, where $\omega = \{x_n, n \in \mathbb{Z}\}$. Let $\mathcal{F}$ be the σ-field on Ω generated by the family $\mathcal{C}$ of sets of the form $\{x_n \in C\}$ $(n \in \mathbb{Z}, C \in \mathcal{E})$. Since $\theta^{-1}(\{x_n \in C\}) = \{x_{n-1} \in C\} \in \mathcal{C}$, it follows that θ is measurable. Clearly, it is bijective and its inverse is measurable. Let for all $n \in \mathbb{Z}$ and all $\omega \in E^{\mathbb{Z}}$, $X_n(\omega) := x_n$. The process $\{X_n\}_{n \in \mathbb{Z}}$ is clearly θ-compatible.

© Springer Nature Switzerland AG 2020
P. Brémaud, *Probability Theory and Stochastic Processes*, Universitext,
https://doi.org/10.1007/978-3-030-40183-2_16

Definition 16.1.3 *Let* $(\Omega, \mathcal{F}, P)$ *be a probability space. Let* $\theta : (\Omega, \mathcal{F}) \to (\Omega, \mathcal{F})$ *be bijective measurable map with measurable inverse. The probability* P *is said to be* θ *invariant if* $P \circ \theta^{-1} = P$. *Then* (P, θ) *is called a* stationary framework *(on* $(\Omega, \mathcal{F})$*)*.

The following result is one of the avatars of the celebrated *Poincaré's recurrence theorem.*

Theorem 16.1.4 *Let* (P, θ) *be a stationary framework on* $(\Omega, \mathcal{F})$ *and let* A *be a mea- surable subset of* $\mathcal{F}$ *such that* $P(A) > 0$. *Then for* P-*almost all* $\omega \in A$, $\theta^n \omega \in A$ *for an infinity of indices* $n \geq 1$.

Proof. Consider the measurable set

$$F := \{\omega \in A \, ; \, \theta^n \omega \notin A \text{ for all } n \geq 1\} = A \backslash \cup_{n \geq 1} \{\omega \, ; \, \theta^n \in A\} \, .$$

If $m > n$, $\theta^n F \cap \theta^m F = \varnothing$. (Indeed, if $\theta^{-m} \in F \subseteq A$, we have by definition of F that $\theta^{-n} \omega = \theta^{m-n}(\theta^{-m}\omega) \notin A$.) By the θ-invariance of P, $P(F) = P(\theta^n F)$ for all $n \geq 1$, and therefore, for all $N \geq 1$,

$$P(F) = \frac{1}{M} \sum_{n=1}^{M} P(\theta^n F) = \frac{1}{M} P(\cup_{n=1}^{M} \theta^n F) \leq \frac{1}{M}$$

and therefore $P(F) = 0$ since M is arbitrary. In other words, outside of a negligible set N_1, for every point $\omega \in A$, there exists an n_1 such that $\theta^{n_1} \omega \in A$. By the same argument applied to θ^k ($k \geq 1$), outside of a negligible set N_k, for every point $\omega \in A$, there exists an n_k such that $\theta^{n_k} \omega \in A$. In particular, for all ω outside the negligible set $N := \cup_{k \geq 1} N_k$, there is for all $k \geq 1$ an n_k such that $\theta^{n_k} \omega \in A \cap N_{n_k}$. Therefore for all ω outside N, $\theta^n \omega \in A$ for an infinity of n. $\square$

The following lemma will be useful on several occasions.

Lemma 16.1.5 *Let* (P, θ) *be a stationary framework. Let* Z *be non-negative* P-*a.s. finite random variable such that* $Z - Z \circ \theta \in L^1_{\mathbb{R}}(P)$. *Then* $\mathrm{E}^0[Z - Z \circ \theta] = 0$.

Proof. (The delicate point here is that Z is not assumed integrable.) For any $C > 0$, $|Z \wedge C - (Z \wedge C) \circ \theta| \leq |Z - Z \circ \theta|$. By the θ-invariance of P, $\mathrm{E}[Z \wedge C - (Z \wedge C) \circ \theta] = 0$ and the conclusion follows by dominated convergence, letting $C \uparrow \infty$ in the last equality. $\square$

Let (P, θ) be a stationary framework on $(\Omega, \mathcal{F})$.

Definition 16.1.6 *An event* $A \in \mathcal{F}$ *is called* strictly θ-invariant *if* $\theta^{-1}(A) = A$. *It is called* θ-invariant *if* $P(A \triangle \theta^{-1}(A)) = 0$, *where* $\triangle$ *denotes the symmetric difference.*

Definition 16.1.7 *The discrete flow* $\{\theta^n\}_{n \in \mathbb{Z}}$ *is called* ergodic *(with respect to* P*) if all* θ-*invariant events are* P-*trivial (that is: of probability either 0 or 1). We shall also say:* (P, θ) *is ergodic.*

Observe that for any θ-invariant event A, the event $B = \cap_{n\in\mathbb{Z}} \cup_{k\geq n} \theta^{-k}A$ is strictly θ-invariant and such that $P(A) = P(B)$. Therefore, for any θ-invariant event, there exists a strictly θ-invariant event with the same probability. In particular, the flow is ergodic if and only if all strictly θ-invariant events are trivial.

EXAMPLE 16.1.8: IRRATIONAL TRANSLATIONS ON THE TORUS, TAKE 1. Let $(\Omega, \mathcal{F}) := ((0, 1], \mathcal{B}((0, 1]))$, and let P be the Lebesgue measure on $(0, 1]$. Let d be some real number. Define $\theta : (\Omega, \mathcal{F}) \to (\Omega, \mathcal{F})$ by $\theta(\omega) = \omega + d \mod 1$. We show that (P, θ) is ergodic if and only if d is irrational.

Proof. Clearly, P is θ-invariant. Let $A \in \mathcal{B}((0, 1])$ and consider the Fourier series development of 1_A,

$$1_A(\omega) = \sum_n a_n e^{2i\pi n\omega}, \qquad \ell\text{-almost everywhere,}$$

where $a_n = \int_A e^{-2i\pi n\omega} d\omega$. With $c := e^{2i\pi d}$, we have

$$a_n = \int_{\theta^{-1}(A)} e^{-2i\pi n\theta(\omega)} d\omega = c^{-n} \int_{\theta^{-1}(A)} e^{-2i\pi n\omega} d\omega.$$

Therefore the n-th Fourier coefficient of $1_{\theta^{-1}(A)}$ is $c^n a_n$. In particular,

$$1_{\theta^{-1}(A)}(\omega) = \sum_n c^n a_n e^{2i\pi n\omega}, \qquad \ell\text{-almost everywhere.}$$

Therefore, A is strictly θ-invariant if and only if $c^n a_n = a_n$ for all n. Now, when d is irrational, c is not a root of unity and then necessarily $a_n = 0$ for all $n \neq 0$, that is, 1_A is a.s. a constant $(= a_0)$, necessarily 0 or 1: A is a trivial set.

The fact that (P, θ) is *not* ergodic if d is rational is left as an exercise (Exercise 16.4.1). □

Theorem 16.1.9 *If (P, θ) is ergodic and if $A \in \mathcal{F}$ satisfies either $A \subseteq \theta^{-1}A$ or $\theta^{-1}A \subseteq A$, then A is trivial.*

In other words, if (P, θ) is ergodic, an event that is either contracted or expanded by θ is necessarily trivial.

Proof. Since P is θ-invariant, for all $A \in \mathcal{F}$,

$$P(A - A \cap \theta^{-1}A) = P(A) - P(A \cap \theta^{-1}A)$$
$$= P(\theta^{-1}A) - P(A \cap \theta^{-1}A) = P(\theta^{-1}A - A \cap \theta^{-1}A)$$

and therefore

$$P(A \triangle \theta^{-1}A) = 2P(A \cap \overline{\theta^{-1}A}) = 2P(\theta^{-1}A \cap \overline{A}).$$

Therefore A is θ-invariant if and only if at least one (and then both) of $P(A \cap \overline{\theta^{-1}A})$ and $P(\theta^{-1}A \cap \overline{A})$ is null. In particular, if $A \subseteq \theta^{-1}A$, then $P(A \cap \overline{\theta^{-1}A}) = 0$ and therefore A is θ-invariant, and therefore trivial. □

The main result of the current chapter is *Birkhoff's pointwise ergodic theorem*:

Theorem 16.1.10 *If (P, θ) is ergodic, then for all $f \in L^1_\mathbb{R}(P)$,*

$$\lim_{N \uparrow \infty} \frac{1}{N} \sum_1^N (f \circ \theta^n) = \mathrm{E}^0[f], \quad P\text{-a.s.} \tag{16.1}$$

The proof will be given in Section 16.2.

Remark 16.1.11 Theorem 16.1.10 entails a considerable improvement on the ergodic theorem for irreducible positive recurrent aperiodic HMCs which are indeed ergodic (actually, mixing) in the sense given to this word in the present chapter.[1] At the price of transferring this chain to the canonical space of random sequences equipped with the canonical shift we have that for any non-negative measurable function $g : E^\mathbb{N} \to \mathbb{R}$,[2]

$$\lim_{N \uparrow \infty} \frac{1}{N} \sum_{k=1}^N (g(X_k, X_{k+1}, \ldots)) = \mathrm{E}_\pi[g(X_0, X_1, \ldots)], \quad P_\mu\text{-a.s.}$$

for any initial distribution of the chain, and where E_π denotes expectation with respect to the initial distribution π, the stationary distribution. Note that there is a slight difference with the ergodic theorem in that the initial distribution may be different from π. We may take this liberty because an irreducible positive recurrent aperiodic chain starting with an arbitrary distribution eventually couples with a stationary chain.

16.1.2   Mixing

We now introduce a particular form of ergodicity.

Definition 16.1.12 *The discrete flow $\{\theta^n\}_{n \in \mathbb{Z}}$ is called P-mixing if for all events $A, B \in \mathcal{F}$,*

$$\lim_{n \uparrow \infty} P(A \cap \theta^{-n} B) = P(A)P(B). \tag{16.2}$$

We shall also say: (P, θ) is mixing.

Mixing is a property of "forgetfulness of the initial conditions" since condition (16.2) is equivalent to

$$\lim_n P(\theta^{-n} B | A) = P(B).$$

Theorem 16.1.13 *If (16.2) holds for all $A, B \in \mathcal{A}$, where $\mathcal{A}$ is an algebra generating $\mathcal{F}$, then (P, θ) is mixing.*

---

[1]There is an unfortunate tradition that reserves the term "ergodic" for an HMC that is irreducible positive recurrent *and aperiodic*. Such a chain is in fact more than ergodic, since it is mixing.

[2]For instance $g(X_0, X_1, \ldots)$ could be the number of consecutive visits of the chain to a given state i without visiting another given state j in between.

Proof. To any fixed $A, B \in \mathcal{F}$ and any $\varepsilon > 0$, one can associate $A', B' \in \mathcal{A}$ such that $A \triangle A'$ and $B \triangle B'$ have probabilities less that ε (Lemma 4.3.8). The same is true of $(\theta^{-n}A) \triangle (\theta^{-n}A') = \theta^{-n}(A \triangle A')$ and of $(\theta^{-n}B) \triangle (\theta^{-n}B')$. In particular, $(A \cap \theta^{-n}B) \triangle (A' \cap \theta^{-n}B')$ has probability less than 2ε, and

$$P(A' \cap \theta^{-n}B') - 2\varepsilon \le P(A \cap \theta^{-n}B) \le P(A' \cap \theta^{-n}B') + 2\varepsilon .$$

Taking the lim sup and the lim inf, and then letting $\varepsilon \downarrow 0$ yields the result. $\qquad\square$

EXAMPLE 16.1.14: MIXING HOMOGENEOUS MARKOV CHAINS. This example continues Example 16.1.2, to which the reader is referred for the notation. The set E is now assumed countable. Let $\mathbf{P}$ be a transition matrix indexed by E that is irreducible positive recurrent, with (unique) stationary distribution π. Let P be the unique probability measure on $(E^{\mathbb{Z}}, \mathcal{E}^{\oplus \mathbb{Z}})$ that makes $\{X_n\}_{n \in \mathbb{Z}}$ a stationary HMC with transition matrix $\mathbf{P}$. Suppose moreover that $\mathbf{P}$ is aperiodic. Then (P, θ) is mixing. Indeed, it suffices, in view of Theorem 16.1.13, to verify (16.2) for sets A and B of the form

$$A = \{X_{k_1} = i_1, \dots, X_{k_p} = i_p\}, \quad B = \{X_{\ell_1} = j_1, \dots, X_{\ell_q} = j_q\},$$

where $k_1 < \cdots < k_p$ and $\ell_1 < \cdots < \ell_q$. This is true since, for $n > k_p$,

$$P(A \cap \theta^{-n}B) = \left(\pi(i_1)p_{i_1,i_2}(k_2 - k_1) \cdots p_{i_{p-1},i_p}(k_p - k_{p-1})\right) \times \left(p_{i_p,j_1}(n + \ell_1 - k_p)p_{j_1,j_2}(\ell_2 - \ell_1) \cdots p_{j_{q-1},j_q}(\ell_q - \ell_{q-1})\right),$$

and $\lim_{n \uparrow \infty} p_{i_p,j_1}(n + \ell_1 - k_p) = \pi(j_1)$, so that

$$\lim_{n \uparrow \infty} p_{i_p,j_1}(n + \ell_1 - k_p) \cdots p_{j_{q-1},j_q}(\ell_q - \ell_{q-1})$$
$$= \pi(j_1)p_{j_1,j_2}(\ell_2 - \ell_1) \cdots p_{j_{q-1},j_q}(\ell_q - \ell_{q-1}) = P(B).$$

If A is strictly invariant for the mixing flow θ, then $P(A) = P(A)^2$, so that $P(A)$ is 0 or 1. Therefore:

Theorem 16.1.15 *A mixing flow is ergodic.*

However, an ergodic flow is not necessarily mixing:

EXAMPLE 16.1.16: IRRATIONAL TRANSLATIONS ON THE TORUS, TAKE 2. The flow of Example 16.1.8 is not mixing even if d is irrational. To see this, take $A = B = (0, \frac{1}{2}]$. Since the set $\{nd \bmod 1 ; n \ge 1\}$ is dense in $(0, 1]$ when d is irrational (this is the celebrated Weyl's equidistribution theorem), $\theta^{-n}A$ and B arbitrarily nearly coincide for an infinite number of indices n. Therefore (16.2) cannot hold.

Theorem 16.1.17 *For any algebra $\mathcal{A}$ generating $\mathcal{F}$, (P, θ) is ergodic if and only if for all $A, B \in \mathcal{A}$,*

$$\lim_{n \uparrow \infty} \frac{1}{n} \sum_{k=1}^{n} P(A \cap \theta^{-k}(B)) = P(A)P(B) . \tag{16.3}$$

Proof. If (P, θ) is ergodic then, by the ergodic theorem, for all $A, B \in \mathcal{F}$

$$\lim_{n\uparrow\infty} \frac{1}{n} \sum_{k=1}^{n} 1_A(\omega) 1_B(\theta^k(\omega)) = 1_A(\omega) P(B),$$

and therefore, taking expectations, (16.3) follows. Conversely, if (16.3) is true for all $A, B \in \mathcal{F}$, then taking an invariant set $A = B$, we obtain that $P(A) = P(A)^2$, and therefore A has probability 0 or 1.

The fact that we can restrict A and B to be in $\mathcal{A}$ is proved in the same way as in Theorem 16.1.13.

$\square$

EXAMPLE 16.1.18: ERGODIC BUT NOT MIXING HMC. The setting is as in Example 16.1.14, except that we do not assume aperiodicity. If the period is ≥ 2, the shift is not mixing any more. To see this let C_0 and C_1 be two consecutive cyclic classes of the chain. Then it is not true that $\lim_{n\uparrow\infty} P(X_0 = i, X_n = j) = \pi(i)\pi(j)$ when $i \in C_0$ and $j \in C_1$. However, with a proof similar to that of Example 16.1.14, one can prove ergodicity using Theorem 16.1.17 and the fact that for a positive recurrent HMC

$$\lim_{n\uparrow\infty} \frac{1}{n} \sum_{k=1}^{n} p_{ij}(k) = \pi(j).$$

(Exercise 16.4.3.)

The Stochastic Process Point of View

In applications, one speaks in terms of a stochastic process rather than flows. The connection between the two points of view is made via canonical spaces, as follows.

To any stochastic process $\{X_n\}_{n\in\mathbb{Z}}$ taking values in the measurable space $(E, \mathcal{E})$ and defined on the probability space $(\Omega, \mathcal{F}, P)$, one can associate a canonical version, by transporting the process on the canonical measurable space $(E^{\mathbb{Z}}, \mathcal{E}^{\otimes\mathbb{Z}})$ as explained after the statement of Theorem 5.1.7. Let $\mathcal{P}_X$ denote the probability distribution of $\{X_n\}_{n\in\mathbb{Z}}$ (therefore a probability on the canonical space), and let S denote the shift on the canonical space:

$$S : (x_n, n \in \mathbb{Z}) \to (y_n, n \in \mathbb{Z}) \text{ where } y_n := x_{n+1}.$$

Definition 16.1.19 *The stochastic process $\{X_n\}_{n\in\mathbb{Z}}$ taking values in the measurable space $(E, \mathcal{E})$ is said to be* ergodic *(resp.* mixing*) iff $(\mathcal{P}_X, S)$ is ergodic (resp. mixing).*

Therefore, in discussing ergodicity of a stochastic process, it is best to assume that it is the coordinate process defined on the corresponding canonical space. This is the convention adopted in the sequel. In other words, when speaking of an ergodic stochastic process $\{X_n\}_{n\in\mathbb{Z}}$ taking values in the measurable space $(E, \mathcal{E})$, we implicitly assume that $(\Omega, \mathcal{F}) = (E^{\mathbb{Z}}, \mathcal{E}^{\otimes\mathbb{Z}})$ and that $(P, \theta) = (\mathcal{P}_X, S)$.

Remark 16.1.20 From the above discussion, we see that a way to decide if a given process is ergodic is to see if it can be obtained in the form

$$\{f(\ldots, x_{n-1}, x_n, x_{n+1}, \ldots)\}_{n \in \mathbb{Z}}.$$

The formalization of this is left for the reader. Let $\{X_n\}_{n \in \mathbb{Z}}$ be an ergodic[3] HMC $\{X_n\}_{n \in \mathbb{Z}}$. Then, the process $\{Y_n\}_{n \in \mathbb{Z}}$, where Y_n is the number of times k ($\tau_n \leq k \leq n$) for which $X_k = 0$ and where τ_n is the last time before n where the HMC took the value 1, is ergodic.

16.1.3 The Convex Set of Ergodic Probabilities

The set of ergodic probabilities coincide with the extremal points of the convex set of stationary probabilities. More precisely:

Theorem 16.1.21 (P, θ) *is ergodic if and only if there exists no decomposition*

$$P = \alpha_1 P_1 + \alpha_2 P_2 \text{ with } \alpha_1 + \alpha_2 = 1 \text{ and } \alpha_1 > 0, \ \alpha_2 > 0, \tag{16.4}$$

where P_1 and P_2 are distinct θ-invariant probabilities.

Proof. We need two lemmas.

Lemma 16.1.22 *If (P_1, θ) and (P_2, θ) are both ergodic, then either $P_1 = P_2$ or $P_1 \perp P_2$.*

Proof. If P_1 and P_2 do not coincide, there exists an $A \in \mathcal{F}$ such that $P_1(A) \neq P_2(A)$. In particular, $B_1 \cap B_2 = \varnothing$, where for $i = 1, 2$,

$$B_i = \left\{ \omega; \lim_{n \uparrow \infty} \frac{1}{n} \sum_{k=1}^{n} 1_A \circ \theta^k = P_i(A) \right\}.$$

Also, by ergodicity, $P_1(B_1) = 1$ and $P_2(B_2) = 1$. Therefore $P_1 \perp P_2$. $\qquad \square$

Lemma 16.1.23 (P, θ) *is ergodic if and only if there exists no θ-invariant probability P_1 distinct from P such that $P_1 \ll P$.*

Proof. If (P, θ) is ergodic and $P_1 \ll P$, then (P_1, θ) is ergodic. (In fact, if A is θ-invariant, then, either $P(A) = 0$ or $P(\overline{A}) = 0$, and therefore, by the absolute continuity hypothesis, $P_1(A) = 0$ or $P_1(\overline{A}) = 0$.) By Lemma 16.1.22, only the possibility $P_1 = P$ remains.

Suppose now (P, θ) *not* ergodic. This means there exists a non-trivial θ-invariant set $A \in \mathcal{F}$: $0 < P(A) < 1$. Define $P_1(B) = P(B \mid A)$. In particular, $P_1 \ll P$. Also P_1 is θ-invariant.

---

[3]In the sense of Markov chain theory, that is, irreducible, periodic and positive recurrent.

Indeed:

$$P_1(\theta^{-1}(B)) = \frac{P(\theta^{-1}(B) \cap A)}{P(A)} = \frac{P(\theta^{-1}(B) \cap \theta^{-1}(A))}{P(A)}$$
$$= \frac{P(B \cap A)}{P(A)} = P(B \mid A) = P_1(B).$$

□

We are now ready to prove Theorem 16.1.21. Suppose (P, θ) is ergodic and that

$$P = \alpha_1 P_1 + \alpha_2 P_2,$$

where $\alpha_1 + \alpha_2 = 1$, $\alpha_1 > 0$, $\alpha_2 > 0$, and where P_1 and P_2 are distinct θ-invariant probabilities. In particular, $P_1 \ll P$. Since P_1 and P are distinct, P_1 cannot be ergodic (Lemma 16.1.23).

Suppose (P, θ) is not ergodic. There exists an invariant set $A \in \mathcal{F}$ that is non-trivial: $0 < P(A) < 1$. The decomposition

$$P(B) = P(A)P(B \mid A) + P(\overline{A})P(B \mid \overline{A})$$
$$= \alpha_1 P_1(B) + \alpha_2 P_2(B)$$

is such that $\alpha_1 + \alpha_2 = 1$, $\alpha_1 > 0$, $\alpha_2 > 0$, and P_1 and P_2 are distinct and θ-invariant. □

16.2   A Detour into Queueing Theory

We will provide a proof of Theorem 16.1.10 after taking a detour into queueing theory.

16.2.1   Lindley's Sequence

Let $(\Omega, \mathcal{F}, P)$ be a probability space and let $\theta : (\Omega, \mathcal{F}) \to (\Omega, \mathcal{F})$ be a bijective measurable map with measurable inverse. Suppose that (P, θ) is ergodic.

Let σ and τ be integrable non-negative random variables defined on $(\Omega, \mathcal{F}, P)$. A *Lindley process* associated with these random variables is a stochastic process $\{W_n\}_{n \in \mathbf{T}}$, where $\mathbf{T} = \mathbb{N}$ or $\mathbb{Z}$, satisfying the recursion equation

$$W_{n+1} = (W_n + \sigma_n - \tau_n)^+ \qquad (n \in \mathbf{T}), \tag{16.5}$$

where

$$\sigma_n = \sigma \circ \theta^n, \quad \tau_n = \tau \circ \theta^n.$$

This equation will be interpreted in terms of queueing since this will greatly help our intuition in the forthcoming developments. Define the *event times sequence* $\{T_n\}_{n \in \mathbb{Z}}$, where $T_0 = 0$ and for all $n \in \mathbb{Z}$,

$$T_{n+1} - T_n = \tau_n.$$

We interpret T_n as the arrival time in a queueing system of customer n, and σ_n as the amount of service (in time units) required by this customer. Define

$$\rho = \frac{\mathrm{E}[\sigma]}{\mathrm{E}[\tau]}.$$

f we interpret $E[\tau]^{-1}$ as the rate of arrivals of customers, ρ is the *traffic intensity*, that s, the average amount of work brought into the system per unit of time. (However we shall not need this interpretation.)

Service is provided at unit rate whenever there remains at least one customer. Otherwise there is no further prescription as to service discipline, priorities, and so on. If W_n is the total service remaining to be done *just before customer n arrives* (that is, at time T_n-), then, obviously, the Lindley recurrence (16.5) is satisfied. In this interpretation, the Lindley process is usually called the *workload process*.

16.2.2   Loynes' Equation

When $\mathbf{T} = \mathbb{N}$, the Lindley process is recursively calculable from the initial workload W_0, but in the case $\mathbf{T} = \mathbb{Z}$, we have nowhere to start the recursion. This corresponds to the situation of a queueing system that has been operating from the infinite past. We may expect that under certain circumstances (of course, a good guess is that $\rho < 1$ will do) the workload process has a stationary version. One is therefore led to pose the problem in the following terms: exhibit a *finite* non-negative random variable $\{W(t)\}_{t\in[0,1]}$ such that the Lindley recursion (16.5) is satisfied for $\{W_n := W \circ \theta^{-n}\}_{n\in\mathbb{Z}}$. Equivalently: we try to find a *finite* non-negative random variable $\{W(t)\}_{t\in[0,1]}$ such that

$$W \circ \theta = (W + \sigma - \tau)^+ . \tag{16.6}$$

The above equation is called *Loynes' equation.*

Theorem 16.2.1 ([4]) *If $\rho < 1$, there exists a unique finite non-negative solution* $\{W(t)\}_{t\in[0,1]}$ *of Loynes' equation (16.6).*

Proof. For $n \geq 0$, define M_n to be the workload found by customer 0 assuming that customer $-n$ found an empty queue upon arrival. In particular, $M_0 = 0$. One checks by induction that

$$M_n = \left(\max_{1\leq m\leq n} \sum_{i=1}^{m}(\sigma_{-i} - \tau_{-i}) \right)^+ . \tag{16.7}$$

In particular, M_n is integrable for all $n \in \mathbb{N}$, being smaller than the integrable random variable $\sum_{i=1}^{n} |\sigma_{-i} - \tau_{-i}|$. Furthermore, the sequence $\{M_n\}_{n\geq0}$ satisfies the recurrence relation

$$M_{n+1} \circ \theta = (M_n + \sigma - \tau)^+ \tag{16.8}$$

and (16.7) shows that it is non-decreasing. Denoting by M_∞ the limit

$$M_\infty = \lim_{n\to\infty} \uparrow M_n = \left(\sup_{n\geq1} \sum_{i=1}^{n}(\sigma_{-i} - \tau_{-i}) \right)^+ \tag{16.9}$$

and letting n go to ∞ in (16.8), we see that M_∞ is a non-negative random variable satisfying

$$M_\infty \circ \theta = (M_\infty + \sigma - \tau)^+. \tag{16.10}$$

The random variable M_∞ is often referred to as *Loynes' variable*, while the sequence $\{M_n\}_{n\geq0}$ is called *Loynes' sequence*. The random variable M_∞ can take infinite values. When using the identity $(a - b)^+ = a - a \wedge b$, Equality (16.8) becomes

---

[4][Loynes, 1962].

$$M_{n+1} \circ \theta = M_n - M_n \wedge (\tau - \sigma) \tag{16.11}$$

and therefore, since P is θ-invariant, and $\{M_n\}_{n \geq 1}$ is increasing and integrable,

$$E[M_n \wedge (\tau - \sigma)] = E[M_n - M_{n+1} \circ \theta] = E[M_n - M_{n+1}] \leq 0.$$

It follows by monotone convergence that

$$E[M_\infty \wedge (\tau - \sigma)] \leq 0. \tag{16.12}$$

Equality (16.10) shows that the event $\{M_\infty = \infty\}$ is θ-invariant (recall that σ and τ are finite). Therefore, by ergodicity, $P(M_\infty = \infty)$ is either 0 or 1. In view of (16.12), $P(M_\infty = \infty) = 1$ implies $E[\tau - \sigma] \leq 0$. Therefore, the condition $E[\sigma] < E[\tau]$ implies that $M_\infty < \infty$, P-a.s.

The solution of Loynes' equation that we just gave (that is, M_∞) is the *minimal* non-negative solution. In order to prove this, it suffices to show that $W \leq M_n$ for all $n \geq 0$ (where $\{W(t)\}_{t \in [0,1]}$ is a non-negative solution of Loynes' equation) and then let $n \uparrow \infty$ to obtain $W \geq M_\infty$. This is proved by induction. The first term of the induction is satisfied since $W \geq 0 = M_0$. Now $W \geq M_n$ implies $W \geq M_{n+1}$ (because $M_{n+1} \circ \theta = (M_n + \sigma - \tau)^+ \leq (W + \sigma - \tau)^+ = W \circ \theta$).

It remains to prove uniqueness of a finite solution of (16.6) if $\rho < 1$. Let $\{W(t)\}_{t \in [0,1]}$ be a finite solution, perhaps different from M_∞. We have

$$\sigma - \tau \leq W \circ \theta - W \leq \sigma,$$

and in particular $W \circ \theta - W$ is integrable. Therefore, by Lemma 16.1.5, $E^0[W \circ \theta - W] = 0$.

Since M_∞ is the minimal solution, for any non-negative solution $\{W(t)\}_{t \in [0,1]}$, $\{W = 0\} \subseteq \{W = M_\infty\}$. The latter event is θ-contracting since both $\{W(t)\}_{t \in [0,1]}$ and M_∞ satisfy (16.6). Since (P, θ) is ergodic, we must then have $P(W = M_\infty) = 0$ or 1. It is therefore enough to show that $P(W = 0) > 0$ (which implies $P(W = M_\infty) = 1$, that is, uniqueness). The proof of this follows from the next lemma.

Lemma 16.2.2 *If $P(W = 0) = 0$, for some finite solution $\{W(t)\}_{t \in [0,1]}$ of (16.6), then* $\rho = 1$.

Proof. Indeed, if $W > 0$ P-a.s. or (equivalently) $W \circ \theta > 0$ P-a.s., then $W \circ \theta = W + \sigma - \tau$ P-a.s., and $E[W \circ \theta - W] = 0$ in view of Lemma 16.1.5, and this implies $E[\sigma] = E[\tau]$. □

This completes the proof of Theorem 16.2.1.

□

A partial converse of Theorem 16.2.1 is the following:

Theorem 16.2.3 *If $E[\sigma] > E[\tau]$, (16.6) admits no finite solution.*

Proof. To prove this, it is enough to show that $M_\infty = \infty$, P-a.s., since M_∞ is the minimal non-negative solution of (16.6). This follows from

$$\lim_{n \to \infty} \frac{1}{n} \sum_{i=1}^{n} (\sigma_{-i} - \tau_{-i}) = E[\sigma - \tau] > 0,$$

ince this in turn implies

$$M_\infty = \left(\sup_n \sum_{i=1}^{n} (\sigma_{-i} - \tau_{-i}) \right)^+ = \infty.$$

$\square$

At this stage, we have proved the following: for $\rho > 1$ there is no finite non-negative solution of (16.6), and for $\rho < 1$, M_∞ is the unique non-negative finite solution of (16.6).

In the critical case ($\rho = 1$) the existence of a finite non-negative solution of (16.6) depends on the distribution of the service and inter-arrival sequences. See Exercises 16.4.10 and 16.4.11.

16.3 Birkhoff's Theorem

16.3.1 The Ergodic Case

We can now proceed to the proof of Theorem 16.1.10. It will be given in the equivalent form:

Theorem 16.3.1 *Whenever (P, θ) is ergodic and both σ and τ are non-negative, not identically null, and integrable*

$$\lim_{n \to \infty} \frac{\sum_{i=0}^{n} \sigma \circ \theta^{-i}}{\sum_{i=0}^{n} \tau \circ \theta^{-i}} = \frac{E[\sigma]}{E[\tau]}, \quad \text{P-a.s.}$$

Proof. According to (16.7),

$$\sum_{i=1}^{n} \sigma \circ \theta^{-i} \le \sum_{i=1}^{n} \tau \circ \theta^{-i} + M_n.$$

We know that if $E[\sigma] < E[\tau]$, $M_n \uparrow M_\infty < \infty$ P-a.s. Taking $\sigma = \frac{1}{2}E[\tau] > 0$, it follows that if $E[\tau] > 0$,

$$\lim_{n \to \infty} \sum_{i=1}^{n} \tau \circ \theta^{-i} = \infty, \quad \text{P-a.s.}$$

Therefore, whenever $E[\tau] > 0$ and $E[\sigma] < E^0[\tau]$,

$$\limsup_{n \to \infty} \frac{\sum_{i=0}^{n} \sigma \circ \theta^{-i}}{\sum_{i=0}^{n} \tau \circ \theta^{-i}} \le \lim_{n \to \infty} \frac{M_n}{\sum_{i=0}^{n} \tau \circ \theta^{-i}} + 1 = 1, \quad \text{P-a.s.}$$

If $E[\tau] > 0$, for some integrable σ, take any a such that $aE[\sigma] < E[\tau]$ to obtain from the previous inequality

$$\limsup_{n \to \infty} \frac{\sum_{i=0}^{n} \sigma \circ \theta^{-i}}{\sum_{i=0}^{n} \tau \circ \theta^{-i}} \le \frac{1}{a}$$

and therefore

$$\limsup_{n \to \infty} \frac{\sum_{i=0}^{n} \sigma \circ \theta^{-i}}{\sum_{i=0}^{n} \tau \circ \theta^{-i}} \le \inf \left\{ \frac{1}{a} ; aE[\sigma] < E[\tau] \right\} = \frac{E[\sigma]}{E[\tau]}.$$

Interchanging the roles of σ and τ, we obtain similarly

$$\limsup_{n\to\infty} \frac{\sum_{i=0}^{n} \sigma \circ \theta^{-i}}{\sum_{i=0}^{n} \tau \circ \theta^{-i}} \geq \frac{E[\sigma]}{E[\tau]} .$$

Hence the result.

$\square$

Remark 16.3.2 If a stochastic process is not stationary, the ergodic theorem does not apply directly to obtain almost sure convergence of the empirical means. However if there is convergence of some sort of such a process to stationarity, the convergence of the empirical mean is possible, for instance in the case of an irreducible positive recurrent aperiodic HMC, where convergence in variation is obtained via coupling. See Theorem 6.4.2.

16.3.2   The Non-ergodic Case

"Non-ergodic" refers to the situation where there are nontrivial invariant sets.[5] Define

$$\mathcal{I} = \{A; \, \theta^{-1}(A) = A\}.$$

$\mathcal{I}$ is a σ-field called the *invariant σ-field*.

Definition 16.3.3 *A random variable X is called* invariant *if $X = X \circ \theta$.*

Theorem 16.3.4 *X is invariant if and only if it is $\mathcal{I}$-measurable.*

Proof. Suppose X invariant. Then, for all $a \in \mathbb{R}$, $\{X \leq a\} = \{X \circ \theta \leq a\} = \theta^{-1}\{X \leq a\}$, and therefore $\{X \leq a\} \in \mathcal{I}$. Conversely, suppose $X = 1_A$, where $A \in \mathcal{I}$. Then $X \circ \theta = 1_A \circ \theta = 1_{\theta^{-1}(A)} = 1_A = X$, and therefore indicators of sets in $\mathcal{I}$ are invariant, and so are the weighted sums of such indicator functions, as well as limits of the latter. Since a non-negative $\mathcal{I}$-measurable random variable is a limit of a sequence of weighted sums of indicators of sets in $\mathcal{I}$, it is invariant. For an arbitrary $\mathcal{I}$-measurable random variable, the proof is completed by considering its positive and negative parts as usual. $\square$

Theorem 16.3.5 *Let (P, θ) be a stationary framework. It is ergodic if and only if every invariant real-valued random variable is almost surely a constant.*

Proof. Sufficiency: Let A be invariant. Then $X = 1_A$ is invariant and therefore almost surely a constant, which implies $P(A) = 0$ or 1.

Necessity: Suppose ergodicity and let X be invariant. Then for all $a \in \mathbb{R}$, $P(X \leq a) = 0$ or 1. When a is sufficiently large, this must be 1 because $\lim_{a \uparrow \infty} P(X \leq a) = 1$. Let $a_0 = \inf\{a \in \mathbb{R}; \, P(X \leq a) = 1\}$. Therefore, for all $\varepsilon > 0$, $P(a_0 - \varepsilon < X < a_0 + \varepsilon) = 1$. Let $\varepsilon \downarrow 0$ to obtain $P(X = a_0) = 1$. $\square$

---

[5]Strictly speaking, the results in the ergodic case follow from those in the current subsection. The choice made in the order of treatment is motivated by the facts that "in practice" the ergodic case is the most interesting one for applications and that the proof of the ergodic theorem seized the opportunity of introducing the G/G/1:∞ queue.

Theorem 16.3.6 (6) *Let (P, θ) be a stationary framework and let X be an integrable random variable. Then*

$$\lim_{n \uparrow \infty} \frac{1}{n} \sum_{k=0}^{n-1} X \circ \theta^k = \mathrm{E}[X \,|\, \mathcal{I}] \qquad P\text{-}a.s.$$

The proof rests on *Hopf's lemma*:

Theorem 16.3.7 *Let (P, θ) be ergodic and let X be an integrable random variable. Define $S_k = X + X \circ \theta + \cdots X \circ \theta^{k-1}$ and $M_n = \max(0, S_1, \ldots, S_n)$. Then*

$$\int_{\{M_n > 0\}} X \, dP \geq 0.$$

Proof. For $n \geq k$, $M_n \circ \theta \geq S_k \circ \theta$, and therefore, for $k > 1$,

$$X + M_n \circ \theta \geq X + S_k \circ \theta = S_{k+1}.$$

This is also true for $k = 1$ ($X \geq S_1 - M_n \circ \theta$ because $S_1 = X$ and $M_n \circ \theta \geq 0$). Therefore

$$X \geq \max(S_1, \ldots, S_n) - M_n \circ \theta.$$

In particular,

$$\mathrm{E}\left[X \, 1_{\{M_n > 0\}}\right] \geq \mathrm{E}\left[(\max(S_1, \ldots, S_n) - M_n \circ \theta) \, 1_{\{M_n > 0\}}\right].$$

But $\max(S_1, \ldots, S_n) = M_n$ on $\{M_n > 0\}$, and therefore

$$\mathrm{E}\left[X \, 1_{\{M_n > 0\}}\right] \geq \mathrm{E}\left[(M_n - M_n \circ \theta) \, 1_{\{M_n > 0\}}\right]$$
$$\geq \mathrm{E}\left[M_n - M_n \circ \theta\right] = 0.$$

$\square$

Proof. We can now give the proof of Theorem 16.3.6. We may suppose that $\mathrm{E}[X \,|\, \mathcal{I}] = 0$, otherwise replace X by $X - \mathrm{E}[X \,|\, \mathcal{I}]$. Define $\overline{X} = \limsup_{n \uparrow \infty} \frac{S_n}{n}$. This is an invariant random variable, and therefore the set $C := \{\overline{X} > \varepsilon\}$ is an invariant set for any fixed $\varepsilon > 0$. We show that $P(C) = 0$. Define $X^* = (X - \varepsilon) 1_C$ and let $S_k^* = X^* + \cdots X^* \circ \theta^{k-1}$ and $M_n^* = \max(0, S_1^*, \ldots, S_n^*)$. Then (Hopf's lemma)

$$\int_{\{M_n^* > 0\}} X^* \, dP \geq 0.$$

The sets $H_n = \{M_n^* > 0\} = \{\max(S_1^*, \ldots, S_n^*) > 0\}$, $n \geq 1$, form a non-decreasing sequence whose sequential limit is

$$H := \left\{ \sup_{k \geq 1} S_k^* > 0 \right\} = \left\{ \sup_{k \geq 1} \frac{S_k^*}{k} > 0 \right\} = \left\{ \sup_{k \geq 1} \frac{S_k}{k} > \varepsilon \right\} \cap C.$$

Since $\sup_{k \geq 1} \frac{S_k}{k} \geq \overline{X}$ and $\overline{X} > \varepsilon$, we have that $H = C$. Therefore

---

6[Birkhoff, 1931].

$$\lim_{n\uparrow\infty} \int_{H_n} X^* \, dP = \int_H X^* \, dP = \int_C X^* \, dP$$

by dominated convergence since X^* is integrable ($E[|X^*|] \le E[|X|] + \varepsilon$). Using the fact that C is an invariant event,

$$0 \le \int_C X^* \, dP = \int_C (X - \varepsilon) 1_C) \, dP = \int_C X \, dP - \varepsilon P(C)$$
$$= \int_C E[X \mid \mathcal{I}] \, dP - \varepsilon P(C) = E[X] - \varepsilon P(C) = -\varepsilon P(C).$$

This implies $P(C) = 0$, that is, $P(\overline{X} \le \varepsilon) = 1$. Since $\varepsilon > 0$ is arbitrary, $P(\overline{X} \le 0) = 1$, that is, almost surely

$$\limsup_{n\uparrow\infty} \frac{S_n}{n} \le 0.$$

The same arguments applied with $-X$ instead of X give

$$-\limsup_{n\uparrow\infty} \frac{-S_n}{n} = \liminf_{n\uparrow\infty} \frac{S_n}{n} \ge 0.$$

Therefore $\limsup_{n\uparrow\infty} \frac{S_n}{n} = \liminf_{n\uparrow\infty} \frac{S_n}{n} = 0.$ □

Corollary 16.3.8 *Let (P, θ) be a stationary framework, and let X be an integrable random variable. Then*

$$\lim_{n\uparrow\infty} \frac{1}{n} \sum_{k=0}^{n-1} X \circ \theta^k = E[X \mid \mathcal{I}] \qquad in \ L^1_{\mathbb{C}}(P).$$

Proof. Let for any $K > 0$,

$$X'_K := X 1_{\{|X| \le K\}}, \quad X''_K := X 1_{\{|X| > K\}}.$$

By the pointwise ergodic theorem,

$$\lim_{n\uparrow\infty} \frac{1}{n} \sum_{k=0}^{n-1} X'_K \circ \theta^k = E[X'_K \mid \mathcal{I}],$$

from which it follows by dominated convergence (X'_M is bounded) that

$$\lim_{n\uparrow\infty} E\left[\left|\frac{1}{n} \sum_{k=0}^{n-1} X'_K \circ \theta^k - E[X'_K \mid \mathcal{I}]\right|\right] = 0.$$

Observing that

$$E\left[\left|\frac{1}{n} \sum_{k=0}^{n-1} X''_K \circ \theta^k\right|\right] \le \frac{1}{n} \sum_{k=0}^{n-1} E\left[\left|X''_K \circ \theta^k\right|\right] = E\left[|X''_K|\right]$$

and

$$E\left[|E[X''_K \mid \mathcal{I}]|\right] \le E\left[E\left[|X''_K| \mid \mathcal{I}\right]\right] = E\left[|X''_K|\right],$$

we have that

$$\mathrm{E}\left[\left|\frac{1}{n}\sum_{k=0}^{n-1}X''_K\circ\theta^k-\mathrm{E}[X''_K\mid\mathcal{I}]\right|\right]\le 2\mathrm{E}\left[|X''_K|\right].$$

Therefore

$$\limsup_{n\uparrow\infty}\mathrm{E}\left[\left|\frac{1}{n}\sum_{k=0}^{n-1}X\circ\theta^k-\mathrm{E}[X\mid\mathcal{I}]\right|\right]\le 2\mathrm{E}\left[|X''_K|\right].$$

It then suffices to let K tend to ∞.   □

16.3.3   The Continuous-time Ergodic Theorem

The extension of the discrete-time result begins with the introduction of the notion of measurable flow in continuous time.

EXAMPLE 16.3.9:   SHIFTS ACTING ON FUNCTIONS, TAKE 1. Let Ω be the space of continuous functions $\omega:\mathbb{R}\to\mathbb{R}$, and let $\mathcal{F}$ be the σ-field on Ω generated by the coordinate functions $X(s):\Omega\to\mathbb{R}$, $s\in\mathbb{R}$, where $X(s,\omega)=\omega(s)$. Define for each $t\in\mathbb{R}$ the mapping $\theta_t:\Omega\to\Omega$ by

$$\theta_t(\omega)(s)=\omega(s+t)$$

(θ_t translates a function $\omega\in\Omega$ by $-t$.)

For fixed $t\in\mathbb{R}$, the mapping θ_t of the above example is measurable. However we shall need more measurability.

Definition 16.3.10 *The family $\{\theta_t\}_{t\in\mathbb{R}}$ of measurable maps from the measurable space $(\Omega,\mathcal{F})$ into itself is called a* shift *on $(\Omega,\mathcal{F})$ if:*

(a) θ_t is bijective for all $t\in\mathbb{R}$, and

(b) $\theta_t\circ\theta_s=\theta_{t+s}$ for all $t,s\in\mathbb{R}$.

This shift is called a (measurable) flow *if, in addition,*

(c) $(t,\omega)\to\theta_t(\omega)$ is measurable from $(\mathbb{R}\times\Omega,\mathcal{B}\otimes\mathcal{F})$ to $(\Omega,\mathcal{F})$,

In particular, θ_0 is the identity and $\theta_t^{-1}=\theta_{-t}$. To simplify the notation, we shall write $\theta_t\omega$ instead of $\theta_t(\omega)$.

Definition 16.3.11 *Given a shift as in Definition 16.3.10, a stochastic process $\{Z(t)\}_{t\in\mathbb{R}}$ is called* compatible *with the shift (for short: θ_t-compatible) if for all $t\in\mathbb{R}$,*

$$Z(t)=Z(0)\circ\theta_t,\qquad(16.13)$$

that is, for all $\omega\in\Omega$, $Z(t,\omega)=Z(0,\theta_t\omega)$.

EXAMPLE 16.3.12:   SHIFTS ACTING ON FUNCTIONS, TAKE 2. In Example 16.3.9, the coordinate process is θ_t-compatible.

The stationarity of a compatible process is embodied in the invariance of the underlying probability with respect to the shifts. More precisely:

Definition 16.3.13 *Let* $(\Omega, \mathcal{F}, P)$ *be a probability space and let* $\{\theta_t\}_{t \in \mathbb{R}}$ *be a shift o*
$(\Omega, \mathcal{F})$. *The probability* P *is called* invariant *with respect to this shift if*

$$P \circ \theta_t^{-1} = P \quad (t \in \mathbb{R}). \tag{16.14}$$

We then say: (P, θ_t) *is a* stationary framework *(on* $\mathbb{R}$*).*

The continuous parameter set is now $\mathbb{R}$. We repeat in this setting the definitions
given for discrete-time flows. Let (θ_t, P) be a stationary framework on $(\Omega, \mathcal{F})$.

Definition 16.3.14 *An event* $A \in \mathcal{F}$ *is called* strictly θ_t-invariant *if* $A = \theta_t^{-1} A$ *for all*
$t \in \mathbb{R}$. *It is called* θ_t-invariant *if* $P(A \ \triangle \ \theta_t^{-1} A) = 0$ *for all* $t \in \mathbb{R}$.

By an easy adaptation of the remark following Definition 16.1.7 to continuous-time
flows, we see that in the following definition, "θ_t-invariant" can be replaced by "strictly
θ_t-invariant".

Definition 16.3.15 *The flow* $\{\theta_t\}_{t \in \mathbb{R}}$ *is called* P-ergodic *if all* θ_t-invariant events are
trivial. *One then says:* (P, θ_t) *is* ergodic.

Theorem 16.3.16 *If* (P, θ_t) *is ergodic, then for all* $f \in L^1(P)$,

$$\lim_{T \uparrow \infty} \frac{1}{T} \int_0^T (f \circ \theta_t) \, dt = E[f], \quad P\text{-}a.s. \tag{16.15}$$

Proof. Defining $\theta := \theta_1$, the pair (P, θ) is ergodic. It is enough to prove the theorem for
non-negative $f \in L^1(P)$. In this case, defining $n(T)$ by

$$n(T) \le T < n(T) + 1,$$

we have the bounds

$$\frac{n(T)}{n(T)+1} \frac{1}{n(T)} \int_0^{n(T)} f \circ \theta_t \, dt \le \frac{1}{T} \int_0^T f \circ \theta_t \, dt \le \frac{n(T)+1}{n(T)} \frac{1}{n(T)+1} \int_0^{n(T)+1} f \circ \theta_t \, dt \,.$$

Defining $g := \int_0^1 f \circ \theta_t \, dt$, we have that $(\star)$

$$\int_0^n f \circ \theta_t \, dt = \sum_{k=1}^n g \circ \theta^k$$

and therefore

$$\lim_{n \uparrow \infty} \frac{1}{n} \int_0^n f \circ \theta_t \, dt = E[g] = E\left[\int_0^1 f \circ \theta_t \, dt\right] = E[f].$$

The conclusion then follows from $(\star)$. $\qquad\square$

Theorem 16.3.17 (P, θ_t) *is ergodic if and only if there exists no decomposition*

$$P = \beta_1 P_1 + \beta_2 P_2, \quad \beta_1 + \beta_2 = 1, \ \beta_1 > 0, \quad \beta_2 > 0, \tag{16.16}$$

where P_1 *and* P_2 *are distinct* θ_t-invariant probabilities for all t.

Proof. The proof is analogous to that of Theorem 16.1.21. $\qquad\square$

Complementary reading

Billingsley, 1965] is the classic introduction to ergodic theory and to the theoretical aspects of information theory.

16.4   Exercises

Exercise 16.4.1. $\omega + d \bmod 1$
Let $\Omega = (0, 1]$, and let P be the Lebesgue measure on Ω. Let d be some real number. Define $\theta : (\Omega, \mathcal{F}) \to (\Omega, \mathcal{F})$ by $\theta(\omega) = \omega + d \bmod 1$. Show that (P, θ) is not ergodic if d is rational.

Exercise 16.4.2. θ ERGODIC, θ^2 NOT ERGODIC
Give an example where (P, θ) is ergodic and (P, θ^2) is not ergodic.

Exercise 16.4.3. ERGODIC YET NOT MIXING HMC
Give the details in the proof of Example 16.1.18.

Exercise 16.4.4. $2\omega \bmod 1$
Let $(\Omega, \mathcal{F}) := ([0, 1), \mathcal{B}([0, 1)))$. Let P be the Lebesgue measure on $[0, 1)$. Consider the transformation

$$\theta(\omega) := \begin{cases} 2\omega & \text{if } \omega \in [0, \tfrac{1}{2}), \\ 2\omega - 1 & \text{if } \omega \in [\tfrac{1}{2}, 1). \end{cases}$$

Show that P is θ-invariant and that (P, θ) is mixing. Hint: The intervals of the form $[\tfrac{k}{2^n}, \tfrac{k+1}{2^n})$ generate $\mathcal{B}([0, 1))$.

Exercise 16.4.5. PERIODIC HMC
Show that an irreducible positive recurrent discrete-time HMC with period ≥ 2 cannot be mixing.

Exercise 16.4.6. PRODUCT OF MIXING SHIFTS
For $i = 1, 2$, let $(\Omega_i, \mathcal{F}_i, P_i)$ be a probability space endowed with the measurable shift θ_i such that (P_i, θ_i) is mixing. Define $(\Omega, \mathcal{F}, P)$ to be the product of the above probability spaces. The product shift $\theta := \theta_1 \oplus \theta_2$ is defined in the obvious manner: $\theta((\omega_1, \omega_2)) := (\theta_1(\omega_1), \theta_2(\omega_2))$.

(1) Show that on the product of two probability spaces, each endowed with a mixing shift, the product shift is mixing.

(2) Give a counterexample when "mixing" is replaced by "ergodic" in the previous question.

Exercise 16.4.7. IRRATIONAL TRANSLATIONS ON THE TORUS
Prove that the flow of Example 16.1.8 is not mixing (d rational or irrational).

Exercise 16.4.8. ERGODICITY OF GAUSSIAN PROCESSES
Show that a stationary centered Gaussian sequence $\{X_n\}_{n \geq 0}$ such that $\lim_{n \uparrow \infty} E[X_0 X_n] = 0$ is ergodic.

Exercise 16.4.9. INVARIANT EVENTS OF AN HMC
In Example 16.1.18, identify the invariant events.

Exercise 16.4.10. LOYNES: THE CRITICAL CASE, I
In Loynes' equation, assume that the random variables $\sigma_n - \tau_n$ are centered, IID, and with a positive finite variance. Prove that in this case there exists no finite solution Z of Loynes' equation. Hint: apply the central limit theorem to $\{\sigma_{-n} - \tau_{-n}\}_{n \geq 1}$.

Exercise 16.4.11. LOYNES: THE CRITICAL CASE, II
Show that if $\rho = 1$, and if there is a finite solution, then for any $c \geq 0$, $W = M_\infty + c$ is also a finite solution of (16.6).

Exercise 16.4.12. LINDLEY: RECURRENCE TO ZERO IN THE STABLE CASE
The stability condition $\rho < 1$ is assumed to hold. Let $W = M_\infty$ be the unique non-negative solution of (16.6). Show that there exists an infinity of negative (resp. positive) indices n such that $W \circ \theta^n = 0$.

Chapter 17

Palm Probability

Palm theory (in this chapter: on the line) links two types of stationarity for marked point processes: *time-stationarity* and *event-stationarity*. Two examples will illustrate this.

The first example is the renewal process, for which one distinguishes the time-stationary (necessarily delayed) version from the undelayed version whose distribution is invariant with respect to the shift that translates the first event time to the origin. It was shown in Chapter 10 that there exists a simple relation between the two versions, which are identical except for the distribution of the first event time. In the terminology of Palm theory, the undelayed version is the Palm version of the time-stationary version.

The second example is that of an irreducible positive recurrent continuous-time HMC whose imbedded chain (the chain observed at the transition times) is also positive recurrent. When such a chain is (time-)stationary, it is not true in general that the embedded discrete-time Markov chain is stationary, even when the latter is assumed positive recurrent. However, there is a simple relation between the stationary distribution of the continuous-time chain and the stationary distribution of the imbedded chain. The continuous-time chain starting with the stationary distribution of the imbedded chain is the Palm version of the stationary continuous-time chain. We observe once more that the distribution of the Palm version is invariant with respect to the shift that translates the first event time (transition time) to the origin.

In general, Palm theory on the line is concerned with jointly stationary stochastic processes and point processes, and with the probabilistic situation at event times. It is especially relevant in queueing theory applied to service systems, where there are two distinct points of view, that of the "operator", who is interested in the behavior of a queue at arbitrary times, and that of the "customer", who is generally interested in the situation found upon arrival. The corresponding issues will be treated in Chapter 9.

17.1 Palm Distribution and Palm Probability

17.1.1 Palm Distribution

The story begins with a new look at Campbell's formula for stationary marked point processes. Let N be a simple point process on $\mathbb{R}^m$ with point sequence $\{X_n\}_{n \in \mathbb{N}}$. Let $\{Z_n\}_{n \in \mathbb{N}}$ be a sequence of random variables with values in the measurable space $(K, \mathcal{K})$. Each Z_n is considered as a mark of the corresponding point Z_n. Recall that the point process and its sequence of marks are referred to as "the marked point process (N, Z)",

© Springer Nature Switzerland AG 2020
P. Brémaud, *Probability Theory and Stochastic Processes*, Universitext,
https://doi.org/10.1007/978-3-030-40183-2_17

which can also be represented as a point process N_Z on $\mathbb{R}^m \times K$:

$$N_Z(D) := \sum_{n \in \mathbb{N}} 1_D(X_n, Z_n) \quad (D \in \mathcal{B}(\mathbb{R}^m) \otimes \mathcal{K}).$$

The marked point process (N, Z) is called stationary if for all $x \in \mathbb{R}^m$, the random measure $S_x(N_Z)$ defined by

$$S_x(N_Z)(D) := \sum_{n \in \mathbb{N}} 1_D(X_n + x, Z_n) \quad (D \in \mathcal{B}(\mathbb{R}^m) \otimes \mathcal{K})$$

has the same distribution as N_Z. The intensity of the (stationary) point process N will henceforth be assumed positive and finite:

$$0 < \lambda := \mathrm{E}[N((0, 1]^m)] < \infty. \tag{17.1}$$

Define the σ-finite measure ν_Z on $(\mathbb{R}^m \times K, \mathcal{B}(\mathbb{R}^m) \otimes \mathcal{K})$ by

$$\nu_Z(D) := \mathrm{E}[N_Z(D)] \quad (D \in \mathcal{B}(\mathbb{R}^m) \otimes \mathcal{K}).$$

Recall the notation $C + x := \{y + x; y \in C\}$ $(C \subseteq \mathbb{R}^m, x \in \mathbb{R}^m)$. By stationarity, for all $C \in \mathcal{B}(\mathbb{R}^m)$ and all $L \in \mathcal{K}$,

$$\nu_Z((C + x) \times L) = \nu_Z(C \times L) \quad (x \in \mathbb{R}^m),$$

that is, the measure $C \to \nu_Z(C \times L)$ is for fixed L translation-invariant, and therefore (Theorem 2.1.45) a multiple of the Lebesgue measure ℓ^m on $(\mathbb{R}^m, \mathcal{B}(\mathbb{R}^m))$:

$$\nu_Z(C \times L) = \gamma(L)\ell^m(C),$$

for some $\gamma(L)$. The mapping $L \to \gamma(L)$ is a measure on $(K, \mathcal{K})$ that is finite since $\gamma(K) = \lambda$. In particular, $Q_N^0 := \lambda^{-1}\gamma$ is a probability measure on $(K, \mathcal{K})$ and

$$\nu_Z(C \times L) = \lambda Q_N^0(L)\ell^m(C).$$

Therefore

$$Q_N^0(L) = \frac{\mathrm{E}\left[\sum_{n \in \mathbb{N}} 1_C(X_n) 1_L(Z_n)\right]}{\lambda \ell^m(C)} \tag{17.2}$$

is a probability on $(K, \mathcal{K})$. It is called the *Palm distribution of the marks*.

Theorem 17.1.1 *Let* $f : \mathbb{R}^m \times K \to \mathbb{R}$ *be a non-negative measurable function. Then*

$$\mathrm{E}\left[\int_{\mathbb{R}^m \times K} f(x, z) N_Z(\mathrm{d}x \times \mathrm{d}z)\right] = \lambda \int_{\mathbb{R}^m} \left\{\int_K f(x, z) Q_N^0(\mathrm{d}z)\right\} \mathrm{d}x. \tag{17.3}$$

Proof. Formula (17.3) is true for

$$f(x, z) := 1_C(x) 1_L(z) \quad (C \in \mathcal{B}(\mathbb{R}^m), L \in \mathcal{K})$$

since it then reduces to (17.2). The general case again follows by the usual monotone class argument based on Dynkin's Theorem 2.1.27. $\qquad\square$

Formula (17.3) is the *Palm–Campbell formula* for stationary marked point processes.

17.1.2 Stationary Frameworks

The passage from the Palm distribution of marks to Palm probability will be done in terms of measurable flows on abstract probability spaces.

Measurable Flows

For convenience, we repeat Definition 16.3.10 of Chapter 16.

Definition 17.1.2 *A family $\{\theta_x\}_{x \in \mathbb{R}^m}$ of measurable functions from the measurable space $(\Omega, \mathcal{F})$ into itself is called a* shift *on $(\Omega, \mathcal{F})$ if:*

(a)   *θ_x is bijective for all $x \in \mathbb{R}^m$, and*

(b)   *$\theta_x \circ \theta_y = \theta_{x+y}$ for all $x, y \in \mathbb{R}^m$.*

The shift $\{\theta_x\}_{x \in \mathbb{R}^m}$ on $(\Omega, \mathcal{F})$ is called a measurable flow *if in addition*

(c)   *$(x, \omega) \mapsto \theta_x(\omega)$ is measurable from $\mathcal{B}(\mathbb{R}^m) \otimes \mathcal{F}$ to $\mathcal{F}$.*

EXAMPLE 17.1.3: MEASURABLE FLOW ON A SPACE OF MEASURES. The shift $\{S_x\}_{x \in \mathbb{R}^m}$ acting on $M(\mathbb{R}^m)$, the canonical space of locally finite measures on $\mathbb{R}^m$, and defined by

$$S_x(\mu)(C) := \mu(C - x) \quad (x \in \mathbb{R}^m, C \in \mathcal{B}(\mathbb{R}^m)),$$

is a measurable flow. To prove this, it suffices to show that the mapping

$$(x, \mu) \to (S_x\mu)(f) := \int_{\mathbb{R}^m} f(y - x)\mu(\mathrm{d}x)$$

is measurable whenever $f : \mathbb{R}^m \to \mathbb{R}$ is a non-negative continuous function with compact support. This is the case since $(x, \mu) \mapsto g(x, \mu) := (S_x\mu)(f)$ is continuous in the first argument and measurable in the second. (Indeed, g is the limit as $n \uparrow \infty$ of the measurable functions $\sum_{k \in \mathbb{N}} 1_{C_{k,n}}(x) g(x_{k,n}, \mu)$, where $\{x_{k,n}\}_{k \in \mathbb{N}}$ is an enumeration of the grid $n_{-1}\mathbb{Z}^m$ and $C_{k,n} = x_{k,n} + (-\frac{1}{2n}, \frac{1}{2n}])^m$.)

EXAMPLE 17.1.4:   THE SHIFT ON MARKED POINT PROCESSES. Let $(H, \mathcal{H})$ be some measurable space. One may take $(\Omega, \mathcal{F}) = (M(\mathbb{R}^m \times H), \mathcal{M}(\mathbb{R}^m \times H))$ with $\theta_x = S_x$, where

$$S_x\mu(C \times L) := \mu((C + x) \times L) \quad (C \in \mathcal{B}(\mathbb{R}^m), L \in \mathcal{H}).$$

Compatibility

A central notion is that of compatibility with a flow.

Definition 17.1.5 *Let $\{\theta_x\}_{x \in \mathbb{R}^m}$ be a measurable flow on $(\Omega, \mathcal{F})$. A stochastic process $\{Z(x)\}_{x \in \mathbb{R}^m}$ defined on $(\Omega, \mathcal{F})$ with values in the measurable space $(K, \mathcal{K})$ is called compatible with the flow $\{\theta_x\}_{x \in \mathbb{R}^m}$ (for short: θ_x-compatible) if*

$$Z(x, \omega) = Z(0, \theta_x(\omega)) \quad (\omega \in \Omega, \, x \in \mathbb{R}^m),$$

that is, in shorter notation, $Z(x) = Z(0) \circ \theta_x$.

A random measure N on $\mathbb{R}^m$ is called compatible with the flow $\{\theta_x\}_{x \in \mathbb{R}^m}$ (for short θ_x-compatible) if

$$N(\theta_x(\omega), C) = N(\omega, C + x) \quad (\omega \in \Omega, \, C \in \mathcal{B}(\mathbb{R}^m), \, x \in \mathbb{R}^m),$$

that is, in shorter notation, $N \circ \theta_x = S_x N$, where S_x is the translation operator acting on measures (Example 16.3.9, (ii)).

Note we have three notations for the same object

$$N \circ \theta_x, \; S_x(N), \; N - x.$$

The latter is not to be confused with $N - \varepsilon_x$, which represents $N \backslash \{x\}$ if $x \in N$, and N if $x \notin N$.

Stationary Frameworks

Let $(\Omega, \mathcal{F}, P)$ be a probability space and let $\{\theta_x\}_{x \in \mathbb{R}^m}$ be a measurable flow on $(\Omega, \mathcal{F})$.

Definition 17.1.6 *The probability P is called invariant with respect to the flow $\{\theta_x\}_{x \in \mathbb{R}^m}$ (for short, θ_x-invariant) if for all $x \in \mathbb{R}^m$*

$$P \circ \theta_x^{-1} = P.$$

(P, θ_x) *is then called a* stationary framework *on $\mathbb{R}^m$.*

EXAMPLE 17.1.7: STATIONARY POINT PROCESS. Let (P, θ_x) be a stationary framework on $\mathbb{R}^m$. If the point process N on $\mathbb{R}^m$ is compatible with the shift, it is stationary. Indeed, letting

$$A = \{\omega; \; N(\omega, C_1) = k_1, \ldots, N(\omega, C_m) = k_m\},$$

with $C_1, \ldots, C_m \in \mathcal{B}(\mathbb{R}^m)$, $k_1, \ldots, k_m \in \mathbb{N}$, we have, by definition,

$$\begin{aligned}
\theta_x^{-1} A &= \{\omega; \; \theta_x(\omega) \in A\} \\
&= \{\omega; \; N(\theta_x(\omega), C_1) = k_1, \ldots, N(\theta_x(\omega), C_m) = k_m\} \\
&= \{\omega; \; N(\omega, C_1 + x) = k_1, \ldots, N(\omega, C_m + x) = k_m\}.
\end{aligned}$$

Therefore, since $P \circ \theta_x^{-1} = P$,

$$P(N(C_1) = k_1, \ldots, N(C_m) = k_m) = P(N(C_1 + x) = k_1, \ldots, N(C_m + x) = k_m).$$

In the situation of Example 17.1.7, one sometimes says for short: (N, θ_x, P) is a *stationary point process.*

EXAMPLE 17.1.8: STATIONARY STOCHASTIC PROCESS. Let (P, θ_x) be a stationary framework on $\mathbb{R}^m$. By the same argument as in the example above, a stochastic process $\{Z(x)\}_{x \in \mathbb{R}^m}$ with values in $(K, \mathcal{K})$ that is θ_x-compatible is strictly stationary. For short: (Z, θ_x, P) is a stationary stochastic process. (Here Z stands for $\{Z(x)\}_{x \in \mathbb{R}^m}$.)

17.1.3 Palm Probability and the Campbell–Mecke Formula

Let $((N, Z), \theta_x, P)$ be a *stationary marked point process* on $\mathbb{R}^m$ such that N is simple and with finite positive intensity λ. In fact, the work needed for the definition of Palm probability has already been done in Subsection 17.1.1. It suffices to choose for measurable mark space $(K, \mathcal{K})$ the measurable space $(\Omega, \mathcal{F})$ itself.

For each $n \in \mathbb{Z}$, θ_{X_n} is a random element taking its values in the measurable space $(\Omega, \mathcal{F})$. To see this, write $\theta_x(\omega)$ as $f(x, \omega)$ and remember that the function $(x, \omega) \mapsto f(x, \omega)$ is measurable, and therefore, since the function $\omega \mapsto X_n(\omega)$ is measurable, so is the function $\omega \mapsto f(X_n(\omega), \omega)$. This defines $\theta_{X_n(\omega)}(\omega) := f(X_n(\omega), \omega)$. The sequence $\{\theta_{X_n}\}_{n \in \mathbb{N}}$ is the *universal mark sequence*.

Remark 17.1.9 If $(\Omega, \mathcal{F})$ is the canonical space of point processes on $\mathbb{R}^m$ and the measurable flow is just the shift on this space, the universal mark associated with the point X_n is $N - X_n$, the canonical process N shifted by X_n. In fact, this mark contains as much and no more information than the whole trajectory N!

Take in (17.2) $(K, \mathcal{K}) = (\Omega, \mathcal{F})$ and $Z_n = \theta_{X_n}(\omega)$. Denote in this case Q_N^0 by P_N^0. In particular, P_N^0 is a probability on $(\Omega, \mathcal{F})$. Formula (17.2) then reads for all $C \in \mathcal{B}(\mathbb{R}^m)$ of positive Lebesgue measure

$$P_N^0(A) := \frac{\mathrm{E}\left[\sum_{n \in \mathbb{N}} 1_C(X_n) 1_A \circ \theta_{X_n}\right]}{\lambda \ell^m(C)} \quad (A \in \mathcal{F}). \tag{17.4}$$

The probability P_N^0 defined by (17.4) is called the *Palm probability* associated with P (or, more precisely, with (N, θ_x, P)).

Remark 17.1.10 The definition (17.4) does not depend on the choice of $C \in \mathcal{B}(\mathbb{R}^m)$ of positive Lebesgue measure.

Theorem 17.1.11 ([1]) *Let $v : \mathbb{R}^m \times \Omega \to \mathbb{R}$ be a non-negative measurable function. Then*

$$\mathrm{E}\left[\int_{\mathbb{R}^m} (v(x) \circ \theta_x) \, N(\mathrm{d}x)\right] = \lambda \, \mathrm{E}_N^0 \left[\int_{\mathbb{R}^m} v(x) \, \mathrm{d}x\right]. \tag{17.5}$$

(The left-hand side of the above equality is just $\mathrm{E}\left[\sum_{n \in \mathbb{N}} v(X_n, \theta_{X_n})\right]$.)

Proof. Formula (17.4) is therefore a special case of the announced equality for the choice

$$v(x, \omega) = 1_C(x) 1_A(\omega),$$

from which the general case follows by the usual monotone class argument based on Dynkin's Theorem 2.1.27. □

---

[1] [Mecke, 1967].

Formula (17.5) is the *Campbell–Mecke formula*. It is, as we have seen, a sophisticated avatar of Campbell's formula. It is sometimes used in the alternative equivalent form

$$\mathrm{E}\left[\int_{\mathbb{R}^m} v(x)\, N(\mathrm{d}x)\right] = \lambda \mathrm{E}_N^0\left[\int_{\mathbb{R}^m} (v(x) \circ \theta_{-x})\, \mathrm{d}x\right]. \tag{17.6}$$

EXAMPLE 17.1.12: AN EXPRESSION OF THE RENEWAL FUNCTION. Let (N, θ_t, P) be a stationary simple locally finite point process on $\mathbb{R}$ with finite intensity λ and let P_N^0 be the associated Palm probability. We show that for all $a \geq 0$,

$$\mathrm{E}\left[N((0,a])^2\right] = \int_0^a (2\mathrm{E}_N^0\left[N((-t,0])\right] - 1)\lambda\,\mathrm{d}t\,,$$

and that, in the case of a renewal process,

$$\mathrm{E}\left[N((0,a])^2\right] = \int_0^a (2R(t) - 1)\lambda\,\mathrm{d}t\,,$$

where R is the renewal function.

Proof. From the integration by parts formula (Theorem 2.3.12) (watch the parentheses),

$$N((0,a])^2 = 2\int_{(0,a]} N((0,t])\, N(\mathrm{d}t) + 2\int_{(0,a]} N((0,t))\, N(\mathrm{d}t)$$

$$= 2\int_{(0,a]} N((0,t])\, N(\mathrm{d}t) - N((0,a])\,.$$

Therefore

$$\mathrm{E}\left[N((0,a])^2\right] = 2\mathrm{E}\left[\int_{\mathbb{R}_+} N((0,t])1_{(0,a]}(t)\, N(\mathrm{d}t)\right] + \lambda a\,.$$

In view of (17.6) with $v(t) := N((0,t])1_{(0,a]}(t)$ (and therefore $v(t) \circ \theta_{-t} = N((-t,0])1_{(0,a]}(t)$),

$$\mathrm{E}\left[\int_{\mathbb{R}_+} N((0,t])1_{(0,a]}\, N(\mathrm{d}t)\right] = \mathrm{E}_N^0\left[\int_{\mathbb{R}_+} N((-t,0])1_{(0,a]}(t)\lambda\,\mathrm{d}t\right]$$

$$= \int_{\mathbb{R}_+} \mathrm{E}_N^0\left[N((-t,0])\right] 1_{(0,a]}(t)\lambda\,\mathrm{d}t\,.$$

For a renewal process, observe that $\mathrm{E}_N^0\left[N((-t,0])\right] = \mathrm{E}_N^0\left[N((0,t])\right]$. □

Let $h : \mathbb{R}^m \times M(\mathbb{R}^m) \to \mathbb{R}$ be a non-negative measurable function. Taking $v(x,\omega) := h(x, N(\omega))$ in (17.5), we have

$$\mathrm{E}\left[\sum_{n \in \mathbb{N}} h(X_n, N - X_n)\right] = \lambda \mathrm{E}_N^0\left[\int_{\mathbb{R}^m} h(x, N)\, \mathrm{d}x\right].$$

Specializing this to $h(x, N) := g(x)1_\Gamma(N)$ $(\Gamma \in \mathcal{M}(\mathbb{R}^m))$ gives

$$\mathrm{E}\left[\sum_{n \in \mathbb{N}} g(X_n)1_\Gamma(N - X_n)\right] = \lambda P_N^0(N \in \Gamma)\int_{\mathbb{R}^m} g(x)\, \mathrm{d}x\,.$$

With $g(x) := 1_C(x)$, where $C \in \mathcal{B}(\mathbb{R}^m)$ is of finite positive Lebesgue measure,

$$\mathrm{E}\left[\sum_{n \in \mathbb{N}} 1_C(X_n)1_\Gamma(N - X_n)\right] = \lambda \ell^m(C)P_N^0(N \in \Gamma)\,. \tag{17.7}$$

Remark 17.1.13 A set $\Gamma \in \mathcal{M}(\mathbb{R}^m)$ represents a property that a measure $\mu \in \mathcal{M}(\mathbb{R}^m)$ may or may not possess, and $N - X_n \in \Gamma$ means that the point process seen by an observer placed at the point X_n (that is, precisely, $N - X_n$) possesses this property. For instance, with $\Gamma = \{\mu\,;\,\mu(B(0,a)\backslash\{0\}) = 0\}$, where $B(x,a)$ is the open ball of radius $a \geq 0$ centered at x, $\{N - X_n \in \Gamma\} = \{(N - X_n)(B(0,a)\backslash\{0\}) = 0\}$, that is, $\{N(B(X_n,a)\backslash\{X_n\}) = 0\}$. The sum $\sum_{n \in \mathbb{N}} 1_C(X_n) 1_\Gamma(N - X_n)$ counts the points of N lying in C and whose nearest neighbor is at a distance $\geq a$.

For a general $\Gamma \in \mathcal{M}(\mathbb{R}^m)$, the intensity of the point process

$$N_\Gamma(C) := \mathrm{E}\left[\sum_{n \in \mathbb{N}} 1_C(X_n) 1_\Gamma(N - X_n)\right] \quad (C \in \mathcal{B}(\mathbb{R}^m)) \tag{17.8}$$

has, in view of (17.7), intensity $\lambda_\Gamma = \lambda P_N^0(N \in \Gamma)$.

Theorem 17.1.14 *Under the Palm probability, there is a point at 0 (the origin of $\mathbb{R}^m$), that is,*

$$P_N^0(N(\{0\}) = 1) = 1\,.$$

Proof. With $g(x) = 1_C(x)$ and $\Gamma = \{\mu\,;\,\mu(\{0\}) = 1\}$, Equality (17.7) becomes, since $N - X_n$ always has exactly one point at the origin,

$$\mathrm{E}\left[\sum_{n \in \mathbb{N}} 1_C(X_n)\right] = \lambda \ell^m(C) P_N^0(N(\{0\}) = 1)\,.$$

Noting that the left-hand side is $\lambda \ell^m(C)$, the result is proved. $\qquad\square$

EXAMPLE 17.1.15: SUPERPOSITION OF INDEPENDENT STATIONARY POINT PROCESSES. Recall that S_x is, for any $x \in \mathbb{R}^m$, the translation by x applied to measures $\mu \in M(\mathbb{R}^m)$:

$$S_x(\mu)(C) = \mu(C + x)\,.$$

Let $\mathcal{P}$ be a probability measure on $(M(\mathbb{R}^m), \mathcal{M}(\mathbb{R}^m))$ such that $\mathcal{P} \circ S_x = \mathcal{P}$ for all $x \in \mathbb{R}^m$. Taking N equal to Φ, the identity map of $M(\mathbb{R}^m)$, we obtain a stationary random $(\Phi, S_x, \mathcal{P})$, which is said to be *in canonical form*. Let $(M_i, \mathcal{M}_i, S_x^{(i)}, \Phi_i)$ $(1 \leq i \leq k)$ be replicas of $(M(\mathbb{R}^m), \mathcal{M}(\mathbb{R}^m), S_x, \Phi)$ and let $\mathcal{P}_i$ be a probability on $(M_i, \mathcal{M}_i)$ which is $S_x^{(i)}$-invariant for all $x \in \mathbb{R}^m$. Suppose that for all i $(1 \leq i \leq k)$, Φ_i is $\mathcal{P}_i$-almost surely a simple point process with finite and positive intensity λ_i. Define the product space

$$(\Omega, \mathcal{F}, P) = \left(\prod_{i=1}^k M_i,\ \otimes_{i=1}^k \mathcal{M}_i, \otimes_{i=1}^k \mathcal{P}_i\right)$$

and, for each $x \in \mathbb{R}^m$, define $\theta_x := \otimes_{i=1}^k S_x^{(i)}$, with the meaning that $\theta_x(\omega) = (S_x^{(i)}\mu_i\,;\,1 \leq i \leq k)$, where $\omega = (\mu_i\,;\,1 \leq i \leq k)$. Define

$$N_i(\omega) := \mu_i \quad \text{and} \quad N(\omega) := \sum_{i=1}^k \mu_i\,.$$

Then (N, θ_x, P) is a stationary point process, the *superposition* of the stationary poin processes (N_i, θ_x, P) $(1 \leq i \leq k)$. Denote by $\mathcal{P}_i^0$ the Palm probability associated t $(\Phi_i, \mathcal{P}_i)$. It will be proved below that

$$P_N^0 = \sum_{i=1}^{k} \frac{\lambda_i}{\lambda} \left(\otimes_{j=1}^{i-1} \mathcal{P}_j \right) \otimes \mathcal{P}_i^0 \otimes \left(\otimes_{j=i+1}^{k} \mathcal{P}_j \right), \tag{17.9}$$

where $\lambda = \sum_{i=1}^{k} \lambda_i$.

Remark 17.1.16 The interpretation of (17.9) is the following. With probability $\frac{\lambda_i}{\lambda}$ the point at the origin in the Palm version comes from the i-th point process and the probability distribution of the i-th process is then its Palm probability, whereas the other processes keep their stationary distributions. All the k point processes remain independent.

Proof of (17.9): By definition, for $A = \prod_{i=1}^{k} A_i$, where $A_i \in \mathcal{M}_i$,

$$P_N^0(A) = \frac{1}{\lambda} \mathbb{E} \left[\int_{(0,1]^m} (1_A \circ \theta_x) N(\mathrm{d}x) \right]$$

$$= \frac{1}{\lambda} \int_{M_1} \cdots \int_{M_k} \int_{(0,1]^m} \sum_{j=1}^{k} \left(\prod_{i=1}^{k} 1_{A_i} \circ S_x^{(i)} \right) \Phi_j(\mathrm{d}x) \mathcal{P}_1(\mathrm{d}\mu_1) \ldots \mathcal{P}_k(\mathrm{d}\mu_k)$$

$$= \sum_{j=1}^{k} \frac{1}{\lambda} \int_{M_1} \cdots \int_{M_k} \left\{ \int_{(0,1]^m} \left(\prod_{i=1}^{k} 1_{A_i} \circ S_x^{(i)} \right) \Phi_j(\mathrm{d}x) \right\} \mathcal{P}_1(\mathrm{d}\mu_1) \ldots \mathcal{P}_k(\mathrm{d}\mu_k).$$

But (Fubini and the definition of Palm probability $\mathcal{P}_j^0$)

$$\frac{1}{\lambda_j} \int_{M_1} \cdots \int_{M_k} \left\{ \int_{(0,1]^m} \prod_{i=1}^{k} (1_{A_i} \circ S_x^{(i)}) \Phi_j(\mathrm{d}x) \right\} \mathcal{P}_1(\mathrm{d}\mu_1) \ldots \mathcal{P}_k(\mathrm{d}\mu_k)$$

$$= \mathcal{P}_j^0(A_j) \prod_{i=1,\ i\neq j}^{k} \mathcal{P}_i(A_i),$$

where we have taken into account the $S_x^{(i)}$-invariance of $\mathcal{P}_i$. Therefore

$$P_N^0 \left(\prod_{i=1}^{k} A_i \right) = \sum_{i=1}^{k} \left\{ \frac{\lambda_i}{\lambda} \mathcal{P}_i^0(A_i) \prod_{\substack{1 \leq j \leq k \\ j \neq i}} \mathcal{P}_j(A_j) \right\},$$

which implies (17.9), by Theorem 2.1.42.

Thinning and Conditioning

Let (N, θ_x, P) be a simple stationary point process on $\mathbb{R}^m$ with finite positive intensity. For $U \in \mathcal{F}$, define

$$N_U(\omega, C) = \int_C 1_U(\theta_x(\omega)) N(\omega, \mathrm{d}x) \quad (C \in \mathcal{B}(\mathbb{R}^m)). \tag{17.10}$$

such a point process is therefore obtained by thinning of N, a point $x \in N(\omega)$ being retained if and only if $\theta_x(\omega) \in U$.

EXAMPLE 17.1.17: MARK SELECTION. Let (N, Z, θ_x, P) be a stationary marked point process. Take $U = \{Z_0 \in L\}$ for some $L \in \mathcal{K}$. Then, since $Z_0(\theta_{X_n}(\omega)) = Z_n(\omega)$,

$$N_U(C) = \sum_{n \in \mathbb{N}} 1_L(Z_n) 1_C(X_n).$$

The point process N_U is obtained by thinning N, only retaining the points of X_n with a mark Z_n falling in L.

(N_U, θ_x, P) is obviously a stationary point process and it has a finite intensity (since $N_U \leq N$). If the intensity of λ_U of N_U is positive, its Palm probability is given by

$$P^0_{N_U}(A) = \frac{1}{\lambda_U} E\left[\int_{(0,1]^m} (1_A \circ \theta_x) N_U(dx)\right],$$

where

$$\lambda_U = E\left[\int_{(0,1]^m} (1_U \circ \theta_x) N(dx)\right] = \lambda P^0_N(U).$$

In addition,

$$E\left[\int_{(0,1]^m} (1_A \circ \theta_x) N_U(dx)\right] = E\left[\int_{(0,1]^m} (1_A \circ \theta_x)(1_U \circ \theta_x) N(dx)\right] = \lambda P^0_N(A \cap U).$$

Therefore

$$P^0_{N_U}(A) = \frac{P^0_N(A \cap U)}{P^0_N(U)} = P^0_N(A \mid U).$$

Note that the sequence of marks could take its values in $M(\mathbb{R}^m)$, for instance, $Z_n = N - X_n$. Recall that $N - X_n = S_{X_n}(N)$. Taking $U := \{\omega\,;\, N(\omega) \in \Gamma\}$ where $\Gamma \in \mathcal{M}(\mathbb{R}^m)$, we see that in this case $N_U \equiv N_\Gamma$, where the latter is defined by (17.8).

EXAMPLE 17.1.18: SUPERPOSITION OF POINT PROCESSES. This generalizes Example 17.1.15. Let N_i $(1 \leq i \leq k)$ be point processes on $\mathbb{R}^m$, all compatible with the flow $\{\theta_x\}_{x \in \mathbb{R}^m}$, with positive finite intensities λ_i $(1 \leq i \leq k)$ respectively, but *not necessarily independent*. Call N their superposition. N is *assumed simple*. From Bayes' rule:

$$P^0_N(A) = \sum_{i=1}^{k} P^0_N(N_i(\{0\}) = 1) P^0_N(A \mid N_i(\{0\}) = 1).$$

But

$$P^0_N(N_i(\{0\}) = 1) = \frac{1}{\lambda} E\left[\sum_{n \in \mathbb{N}} 1_{(0,1]^m}(X_n) 1_{\{N_i(\{X_n\})=1\}}\right] = \frac{1}{\lambda} E[N_i(0,1]] = \frac{\lambda_i}{\lambda}.$$

Let $U = \{N_i(\{0\}) = 1\}$. Since we have $N_U = N_i$ (with the notation of (17.10)), we obtain

$$P^0_N(A \mid N_i(\{0\}) = 1) = P^0_{N_i}(A).$$

Therefore

$$P^0_N(A) = \sum_{i=1}^{k} \frac{\lambda_i}{\lambda} P^0_{N_i}(A).$$

17.2   Basic Properties and Formulas

Attention will now be restricted to simple stationary point processes *on the real line*
The notation t instead of x will emphasize the fact that one is working on the real line

17.2.1   Event-time Stationarity

The Palm probability P_N^0 associated with the simple stationary point process (N, θ_t, P)
has, as we saw for the general case in Theorem 17.1.14, its mass concentrated on $\Omega_0 :=$
$\{T_0 = 0\}$. Recall that the sequence of points $\{T_n\}_{n \in \mathbb{Z}}$ is defined in such a way that it is
strictly increasing and such that $T_0 \leq 0 < T_1$.

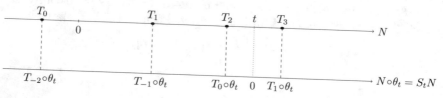

The mapping

$$\theta := \theta_{T_1},$$

defined from Ω_0 into Ω_0, is a bijection, with inverse $\theta^{-1} = \theta_{T_{-1}}$. Also, on Ω_0, $\theta_{T_n} := \theta^n$
for all $n \in \mathbb{Z}^2$. Note that the above is not true on Ω (for instance, the inverse of θ_{T_1} is
not $\theta_{T_{-1}}$; Exercise 17.5.9).

For mappings of the form θ_U with U random, the composition rule (c) of Definition
17.1.2 is no longer valid in that we do not have in general $\theta_U \circ \theta_V = \theta_{U+V}$ when U and
V are random variables. The effect of $\theta_U \circ \theta_V$ on a point process N with sequence of
points $\{T_n\}_{n \in \mathbb{Z}}$ is best understood as follows. One first applies the shift θ_V, obtaining
a point process $N' = \theta_V N$ whose points are of the form $T_n - V$. However the sequence
of these points has to be reindexed to obtain the ordered sequence $\{T'_n\}_{n \in \mathbb{Z}}$ such that
$T'_0 \leq 0 < T'_1$. (For instance, with $U = T_2$, $T'_0 = 0$ and $T'_1 = T_3$.) Once this is done,
one can reiterate the operation with θ_U to obtain a point process N'' whose sequence
of points is $\{T''_n\}_{n \in \mathbb{Z}}$. But beware because this has now shifted N' by $-U(\theta_{V(\omega)})$. For
instance, if $U = T_2$, $V = T_3$, you have to apply to N' the shift of $-T'_3$. Indeed you must
remember that θ_{T_3} means "the shift that moves to 0 the third point strictly to the right
of 0".

The following result is referred to as the "event-time stationarity".

Theorem 17.2.1

$$P_N^0 \text{ is } \theta\text{-invariant.}$$

Proof. First observe that for all $A \in \mathcal{F}$,

$$1_{\theta^{-1}(A)} \circ \theta_{T_n} = 1(\theta_{T_n} \in \theta^{-1}(A)) = 1(\theta_{T_{n+1}} \in A).$$

Formula (17.4) with $A \in \mathcal{F}$ and $C = (0, t]$ yields

---

[2] It is perhaps worthwhile to emphasize the fact that θ_{T_n} is "the shift that moves the n-th
point of a point process to the origin".

$$|P_N^0(A) - P_N^0(\theta^{-1}(A))| \le \frac{1}{\lambda t}\mathrm{E}\left|\sum_{n\in\mathbb{Z}}(1_A\circ\theta_{T_n} - 1_{\theta^{-1}(A)}\circ\theta_{T_n})1_{(0,t]}(T_n)\right|$$

$$= \frac{1}{\lambda t}\mathrm{E}\left|\sum_{n\in\mathbb{Z}}(1_A\circ\theta_{T_n} - 1_A\circ\theta_{T_{n+1}})1_{(0,t]}(T_n)\right| \le \frac{2}{\lambda t}.$$

Letting $t \to \infty$, we obtain $P_N^0(A) = P_N^0(\theta^{-1}(A))$. $\quad\square$

In particular, if $\{Z(t)\}_{t\in\mathbb{R}}$ is compatible with the flow $\{\theta_t\}_{t\in\mathbb{R}}$ and therefore stationary under P, the sequence $\{Z(T_n)\}_{n\in\mathbb{Z}}$ is, under P_N^0, a stationary sequence.

EXAMPLE 17.2.2: PALM–KHINCHIN EQUATIONS. $(^3)$ Let for $k \in \mathbb{N}$ and $t \ge 0$,

$$\varphi_k(t) := P_N^0\left(N((0,t]) = k\right).$$

We have the *Palm–Khinchin equations*:

$$P\left(N((0,t]) > k\right) = \lambda\int_0^t \varphi_k(s)\,\mathrm{d}s.$$

To prove this, observe that

$$1_{N((0,t])>k} = \int_{(0,t]}1_{\{N(s,t]=k\}}\,N(\mathrm{d}s),$$

and deduce from this and $N(s,t] = N(0, t - s]\circ\theta_s$ that

$$1_{\{N((0,t])>k\}} = \int_{\mathbb{R}}1_{(0,t]}(s)\left(1_{\{N(0,t-s]=k\}}\right)\circ\theta_s\,N(\mathrm{d}s).$$

By the Campbell–Mecke formula, the expectation of the right-hand side with respect to P is equal to

$$\lambda\int_{\mathbb{R}}1_{(0,t]}(s)P_N^0\left(N(0,t-s]=k\right)\mathrm{d}s = \lambda\int_0^t P_N^0\left(N(0,t-s]=k\right)\mathrm{d}s$$

$$= \lambda\int_0^t P_N^0\left(N(0,s]=k\right)\mathrm{d}s.$$

Theorem 17.2.3 *Let (N, θ_t, P) be a stationary point process on $\mathbb{R}$ with intensity $0 < \lambda < \infty$ and such that $P(N(\mathbb{R}) = 0) = 0$. Then*

$$\lambda\mathrm{E}_N^0[T_1] = 1.$$

Proof. From the Palm–Khinchin equation with $k = 0$,

$$P(N((0,t]) = 0) = 1 - \lambda\int_0^t \varphi_0(s)\,\mathrm{d}s.$$

But $\varphi_0(s) = P_N^0(N(0,s] = 0) = P(T_1 > s)$, and therefore

$$0 = P(N(\mathbb{R}) = 0) = \lim_{t\uparrow+\infty}P(N((0,t]) = 0)$$

$$= 1 - \lambda\int_0^\infty P_N^0(T_1 > s)\,\mathrm{d}s = 1 - \lambda\mathrm{E}_N^0[T_1].$$

$\quad\square$

---

^{3}[Palm, 1943] for $k = 0$, [Khinchin, 1960] for $k \ge 1$.

17.2.2   Inversion Formulas

How do we pass from the Palm probability to the stationary distribution? The formula that do this are called *inversion formulas*.

Theorem 17.2.4 *Let* (N, P, θ_t) *be a stationary simple point process on* $\mathbb{R}$ *with intensity* $0 < \lambda < \infty$ *and such that* $P(N(\mathbb{R}) = 0) = 0$. *For any non-negative random variable* f,

$$E\left[f\right] = E_N^0\left[\int_0^{T_1}(f \circ \theta_s)\mathrm{d}s\right].$$

One proof, among many others, makes use of the following conservation principle.

The intuition behind the following conservation principle is that in a stationary state, "the smooth variation of a stochastic process is balanced by the variation due to jumps". More precisely:

Theorem 17.2.5 ([4]) *Let* (N, P, θ_t) *be a stationary simple point process on* $\mathbb{R}$ *with intensity* $0 < \lambda < \infty$. *Let* $\{Y(t)\}_{t \in \mathbb{R}}$ *be a real-valued stochastic process, right-continuous with left-hand limits, and let* $\{Y'(t)\}_{t \in \mathbb{R}}$ *be a real-valued stochastic process such that*

$$Y(1) - Y(0) = \int_0^1 Y'(s)\mathrm{d}s + \int_{(0,1]}(Y(s) - Y(s-))N(\mathrm{d}s). \tag{17.11}$$

Suppose that the processes Y *and* Y' *are compatible with the flow. Suppose moreover that*

$$E[|Y'(0)|] < \infty \text{ and } E_N^0[|Y(0) - Y(0-)|] < \infty. \tag{17.12}$$

Then

$$E[Y'(0)] + \lambda E_N^0[Y(0) - Y(0-)] = 0. \tag{17.13}$$

Proof. Observe that $E[\int_0^1 |Y'(s)|\,\mathrm{d}s] = E[|Y'(0)|]$ and $E[\int_{(0,1]} |Y(s) - Y(s-)|\,N(\mathrm{d}s)] = \lambda E_N^0[|Y(0) - Y(0-)|]$. Therefore, condition (17.12) guarantees that $Y(1) - Y(0)$ is P-integrable and by Lemma 16.1.5, $E[Y(1) - Y(0)] = 0$. Equating this to the expectation of the right-hand side of (17.11), we obtain the announced result since $E[\int_0^1 Y'(s)\,\mathrm{d}s] = E[Y'(0)]$ and $E[\int_{(0,1]}(Y(s) - Y(s-))\,N(\mathrm{d}s)] = \lambda E_N^0[Y(0) - Y(0-)]$. □

Remark 17.2.6 $E[Y'(0)]$ is the average rate of increase (per unit time) due to the smooth evolution of the stochastic process Y between jumps, whereas $\lambda E_N^0[Y(0) - Y(0-)]$ is the average rate of increase due to the jumps (at discontinuity times). Therefore Eqn. (17.13) states that the total average rate of increase is null, which is expected since the process is stationary. In this sense, (17.13) is a *conservation equation.*

We now proceed to the proof of Theorem 17.2.4.

Proof. It is enough to prove the result for bounded f. Define

$$Y(t) = \int_t^{T_+(t)}(f \circ \theta_s)\,\mathrm{d}s,$$

---

[4][Miyazawa, 1994].

where $T_+(t) = \inf\{T_n; T_n > t\}$). This process satisfies the conditions of Theorem 17.2.5, with $Y'(0) = -f$ and $Y(0) - Y(0-) = \int_0^{T_1}(f \circ \theta_s)\,ds$ (use the fact that $E_N^0[T_1] = \lambda^{-1} <$ ∞). $\square$

By the θ-invariance of P,

$$E[f] = E_N^0\left[\int_{T_n}^{T_{n+1}}(f \circ \theta_s)\,ds\right], \tag{17.14}$$

and in particular, for $f = 1_A$,

$$P(A) = \int_{-\infty}^{+\infty} P_N^0(T_n < s \le T_{n+1}, \theta_s \in A)\,ds. \tag{17.15}$$

Remark 17.2.7 The inversion formula receives an interesting interpretation when written as

$$E[f] = E_N^0\left[\lambda T_1 \frac{1}{T_1}\int_0^{T_1}(f \circ \theta_t)\,dt\right] = E_N^0\left[\lambda T_1(f \circ \theta_V)\right],$$

where V is a random variable which, "conditionally on everything else", is uniformly distributed on $[0, T_1]$ (for the above to make sense, we must of course enlarge the probability space). This interpretation provides an explicit construction of P from P_N^0, as follows. First construct the probability P_0' by

$$dP_0' = (\lambda T_1)\,dP_N^0.$$

Since $P_N^0(T_0 = 0) = 1$ and P_0' is absolutely continuous with respect to P_N^0, $P_0'(T_0 = 0) = 1$. The stationary probability P is then obtained by placing the origin at random in the interval $[0, T_1]$, that is

$$E[f] = E_0'[f \circ \theta_V].$$

EXAMPLE 17.2.8: MEAN VALUE FORMULAS. Let (N, θ_t, P) be a stationary simple point process on $\mathbb{R}$ with finite positive intensity and associated Palm probability P_N^0. Let $\{Z_t\}_{t \in \mathbb{R}}$ be a θ_t-compatible stochastic process with values in the measurable space $(K, \mathcal{K})$. Here are two particular forms of the inversion formula and of the definition of Palm probability. For all non-negative measurable functions $g : (K, \mathcal{K}) \to (\mathbb{R}, \mathcal{B})$

$$E[g(Z_0)] = \frac{E_N^0\left[\int_0^{T_1} g(Z_t)dt\right]}{E_N^0[T_1]} \tag{17.16}$$

and

$$E_N^0\left[g(Z_0)\right] = \frac{E\left[\sum_{n \in \mathbb{Z}} g(Z_{T_n})1_{\{T_n \in (0,1]\}}\right]}{E\left[\sum_{n \in \mathbb{Z}} 1_{\{T_n \in (0,1]\}}\right]}. \tag{17.17}$$

Backward and Forward Recurrence Times

Theorem 17.2.9 *Let (N, P, θ_t) be a stationary simple point process on $\mathbb{R}$ with intensity $0 < \lambda < \infty$ and such that $P(N(\mathbb{R}) = 0) = 0$, and with associated Palm probability P_N^0. Let F_0 be the cumulative distribution function of T_1 under P_N^0:*

$$F_0(x) = P_N^0(T_1 \leq x).$$

Then, for all $v, w \in \mathbb{R}_+$,

$$P(T_1 > v, -T_0 > w) = \lambda \int_{v+w}^{\infty} (1 - F_0(u)) du. \qquad (17.18)$$

Proof. Taking $A = \{T_1 > v, -T_0 > w\}$ in (17.15) for $n = 0$, where $v, w \in \mathbb{R}_+$, and using the fact that P_N^0-a.s., $-T_0 \circ \theta_t = t$ and $T_1 \circ \theta_t = T_1 - t$ for all $t \in [0, T_1)$, we obtain

$$P(T_1 > v, -T_0 > w) = \lambda E_N^0 \left[\int_0^{T_1} 1_{\{t > v\}} 1_{\{T_1 - t > w\}} \, dt \right]$$

$$= \lambda E_N^0 \left[(T_1 - (v + w))^+ \right].$$

The result then follows from the following computation (valid for any non-negative random variable X and non-negative number a):

$$E[(X - a)^+] = \int_0^{\infty} P((X - a)^+ > x) \, dx$$

$$= \int_0^{\infty} P(X - a > x) \, dx = \int_a^{\infty} P(X > x) \, dx.$$

$\square$

Taking $v = 0$ and $w = 0$, we obtain that $P(T_1 > 0, -T_0 > 0) = 1$ since $\lambda \int_0^{\infty} (1 - F_0(u)) du = \lambda E_N^0 [T_1] = 1$. In particular, we have the following:

Theorem 17.2.10 *Under P, there is no point of N at the time origin.*

Taking now $v = 0$, we obtain

$$P(-T_0 > w) = \lambda \int_w^{\infty} (1 - F_0(u)) \, du.$$

Similarly,

$$P(T_1 > v) = \lambda \int_v^{\infty} (1 - F_0(u)) \, du.$$

Thus $-T_0$ and T_1 are identically distributed under P (but not independent in general). The cumulative distribution function

$$F(x) := \lambda \int_0^x (1 - F_0(u)) du \quad (x \geq 0)$$

is the *excess distribution* of F_0.

17.2.3   The Exchange Formula

Let (N, θ_t, P) and (N', θ_t, P) be two stationary point processes with positive finite intensities λ and λ', respectively. Note that N and N' are *jointly* stationary, in the sense that their stationarity is relative to the same quadruple $(\Omega, \mathcal{F}, P, \theta_t)$. The following result links their Palm distributions.

Theorem 17.2.11 $(^5)$

$$\lambda E_N^0[f] = \lambda' E_{N'}^0 \left[\int_{(0,T_1']} (f \circ \theta_t) N(\mathrm{d}t) \right] \tag{17.19}$$

for all non-negative measurable functions $f : (\Omega, \mathcal{F}) \to (\mathbb{R}, \mathcal{B}(\mathbb{R}))$. *Here* T_n' *is the n-th point of* N'.

Proof. By the monotone convergence theorem, we may assume that f is bounded (say by 1). With such f, associate the function

$$g := \int_{(T_0', T_1']} (f \circ \theta_t) N(\mathrm{d}t).$$

For all $t \in \mathbb{R}_+$, we have

$$\int_{(0,t]} (f \circ \theta_s) \, N(\mathrm{d}s) = \int_{(0,t]} (g \circ \theta_s) \, N'(\mathrm{d}s) + R(t),$$

where, denoting by $T_+'(t)$ the first point of N' strictly after t,

$$R(t) = \int_{(0,T_+'(0)]} (f \circ \theta_s) N(\mathrm{d}s) - \int_{(t,T_+'(t)]} (f \circ \theta_s) N(\mathrm{d}s).$$

For all $\ell > 0$, let

$$f_\ell := f 1_{\{N(T_0', T_1'] \le \ell - 1\}}$$

and

$$g_\ell := \int_{(T_0', T_1']} (f_\ell \circ \theta_t) \, N(\mathrm{d}t).$$

For these functions, each term in $R(t)$ is bounded by ℓ, and therefore the expectations are finite. Moreover, by θ_t-invariance of P, they have the same expectations, so that $E[R(t)] = 0$. We therefore have

$$E \left[\int_{(0,t]} (f_\ell \circ \theta_s) \, N(\mathrm{d}s) \right] = E \left[\int_{(0,t]} (g_\ell \circ \theta_s) \, N'(\mathrm{d}s) \right].$$

This proves (17.19) with $f = f_\ell$, $g = g_\ell$. Letting ℓ go to infinity yields the announced equality. $\qquad\square$

---

^{5}[Neveu, 1968] and [Neveu, 1976].

17.2.4   From Palm to Stationary

Let $\{\theta_t\}_{t\in\mathbb{R}}$ be a measurable flow on $(\Omega,\mathcal{F})$, and let N be a point process compatibl
with this flow. Let P^0 be a probability measure on $(\Omega,\mathcal{F})$ that is

(a) concentrated on $\Omega_0 = \{T_0 = 0\}$, that is, such that $P^0(\Omega_0) = 1$, and

(b) θ_{T_n}-invariant, that is, such that $P^0(\theta_{T_n} \in .) = P^0(.)$ for all $n \in \mathbb{Z}$.

Theorem 17.2.12 ([6]) *Assume moreover that the following three properties are satisfied:*

$$
\begin{array}{lll}
(i) & 0 < E^0[T_1] < \infty, \\
(ii) & P^0[T_1 > 0] = 1, \\
(iii) & E^0[N(0,t_0]] < \infty \text{ for some } t_0 > 0.
\end{array}
\tag{17.20}
$$

*Then P^0 is the Palm probability P_N^0 associated with (N,P), where P is the θ_t-invariant
probability given by*

$$
P(A) = \frac{1}{E^0[T_1]} E^0 \left[\int_0^{T_1} (1_A \circ \theta_t)\, dt \right] \quad (A \in \mathcal{B}(\mathbb{R})).
\tag{17.21}
$$

Proof. We must first show that P is θ_t-invariant for all $t \in \mathbb{R}$, and then that $P_N^0 = P^0$.
On Ω_0, for all $j \in \mathbb{Z}$,

$$
\int_{T_0 \circ \theta_{T_j}}^{T_1 \circ \theta_{T_j}} (1_A \circ \theta_s \circ \theta_{T_j})\, ds = \int_{T_j}^{T_{j+1}} (1_A \circ \theta_s)\, ds,
$$

and therefore, since P^0 is θ_{T_n}-invariant,

$$
P(A) = \frac{1}{n}\frac{1}{E^0[T_1]} E^0 \left[\int_0^{T_n} (1_A \circ \theta_s)\, ds \right].
$$

Also, for all $t \in \mathbb{R}$,

$$
P(\theta_t(A)) = \frac{1}{n}\frac{1}{E^0[T_1]} E^0 \left[\int_0^{T_n} (1_{\theta_t(A)} \circ \theta_s)\, ds \right] = \frac{1}{n}\frac{1}{E^0[T_1]} E^0 \left[\int_{-t}^{T_n - t} (1_A \circ \theta_s)\, ds \right].
$$

Therefore

$$
|P(A) - P(\theta_t(A))| \le \frac{1}{n}\frac{1}{E^0[T_1]} E^0 \left| \int_{T_n - t}^{T_n} (1_A \circ \theta_s)\, ds - \int_{-t}^{0} (1_A \circ \theta_s)\, ds \right| \le \frac{2t}{n E^0[T_1]}.
$$

Letting $n \uparrow \infty$, it follows that $P(A) = P(\theta_t(A))$.

We now prove that the intensity of N is $(E^0[T_1])^{-1}$. From (17.21),

$$
E[f] = \frac{1}{E^0[T_1]} E^0 \left[\int_0^{T_1} (f \circ \theta_t)\, dt \right],
$$

for all non-negative f, and in particular, for all $\varepsilon > 0$,

$$
\frac{E[N(0,\varepsilon]]}{\varepsilon} = \frac{1}{E^0[T_1]} \frac{1}{\varepsilon} E^0 \left[\int_0^{T_1} N(t, t+\varepsilon]\, dt \right].
$$

[6][Ryll-Nardzeweki, 1961], [Slivnyak, 1962].

t follows from assumption (ii) that

$$\lim_{\varepsilon \to 0} \frac{1}{\varepsilon} \int_0^{T_1} N(t, t+\varepsilon] \, dt = 1.$$

Therefore, if we can interchange limit and expectation,

$$\lim_{\varepsilon \to 0} \frac{1}{\varepsilon} E^0 \left[\int_0^{T_1} N(t, t+\varepsilon] \, dt \right] = 1,$$

so that

$$\lim_{\varepsilon \to 0} \frac{E[N(0, \varepsilon]]}{\varepsilon} = \frac{1}{E^0[T_1]}.$$

The interchange is justified by dominated convergence in view of the bound

$$\frac{1}{\varepsilon} \int_0^{T_1} N((t, t+\varepsilon]) dt \le 1 + N((T_1, T_1+\varepsilon])$$

and of assumption (iii).

Since $\frac{E[N(0,\varepsilon]]}{\varepsilon} = \lambda$, we obtain $\lambda = \frac{1}{E^0[T_1]}$ and therefore, in view of (i), $0 < \lambda < \infty$. We may therefore define the Palm probability P_N^0 associated with (N, θ_t, P). It remains to prove that $P^0 = P_N^0$. By the inversion formula, for all $A \in \mathcal{F}$,

$$P(\theta_{T_0} \in A) = \lambda E_N^0 \left[\int_0^{T_1} (1_{\theta_{T_0}^{-1} A} \circ \theta_t) dt \right] = \lambda E_N^0 \left[\int_0^{T_1} 1_A dt \right] = \lambda E_N^0 [T_1 1_A], \qquad (\star)$$

since on Ω^0, $\theta_{T_0} \circ \theta_t$ is the identity for all $t \in (0, T_1]$. From the definition of P in terms of P^0:

$$P(\theta_{T_0} \in A) = \frac{1}{E^0[T_1]} E^0[T_1 1_A].$$

Therefore $E^0[T_1 1_A] = E_N^0[T_1 1_A]$ $(A \in \mathcal{F})$. From this, we obtain that for all non-negative random variables f, $E^0[T_1 f] = E_N^0[T_1 f]$. Since $T_1 > 0$ by construction, we can take $f = T_1^{-1} 1_A$ to obtain $P^0(A) = P_N^0(A)$. $\qquad \square$

EXAMPLE 17.2.13: STATIONARIZATION OF RENEWAL PROCESSES. The result below has already been proved in Chapter 10 on renewal theory. It will be revisited in the light of Palm theory. Suppose that under P_N^0, the inter-event sequence $\{S_n\}_{n \in \mathbb{Z}}$ defined by $S_n = T_{n+1} - T_n$ is IID with finite mean. Then (N, P_N^0) is an *undelayed renewal process* and (N, P) a *stationary renewal process*. The existence of P being granted by the inverse construction, we show that

(a) The distribution of the sequence $S^* = \{S_n\}_{n \in \mathbb{Z} \setminus \{0\}}$ is the same under P and P_N^0.

(b) S_0 and S^* are P-independent.

Proof.

(a) Let $g : (\mathbb{R}^\mathbb{Z}, (\mathcal{B}(\mathbb{R})^\mathbb{Z}) \to (\mathbb{R}, \mathcal{B}(\mathbb{R}))$ be a non-negative measurable function. It suffices to show that

$$E[g(S^*)] = E_N^0[g(S^*)].$$

By the inversion formula

$$\mathrm{E}\left[g(S^*)\right] = \lambda\mathrm{E}_N^0\left[\int_0^{T_1} g(S^*(\theta_u))\mathrm{d}u\right].$$

But if u is in $[0, T_1)$, then $S^*(\theta_u) = S^*$, so that

$$\begin{aligned}
\mathrm{E}\left[g(S^*)\right] &= \lambda\mathrm{E}_N^0\left[\int_0^{T_1} g(S^*)\mathrm{d}u\right] = \lambda\mathrm{E}_N^0\left[T_1 g(S^*)\right]\\
&= \lambda\mathrm{E}_N^0\left[T_1\right]\mathrm{E}_N^0\left[g(S^*)\right] = \mathrm{E}_N^0\left[g(S^*)\right],
\end{aligned}$$

where we have taken into account the independence of $T_1 = S_0$ and S^* under P_N^0.

(b) Similar considerations give

$$\begin{aligned}
\mathrm{E}\left[f(S_0)g(S^*)\right] &= \lambda\mathrm{E}_N^0\left[f(S_0)T_1 g(S^*)\right] = \lambda\mathrm{E}_N^0\left[f(S_0)T_1\right]\mathrm{E}_N^0\left[g(S^*)\right]\\
&= \lambda\mathrm{E}_N^0\left[f(S_0)T_1\right]\mathrm{E}\left[g(S^*)\right] = \mathrm{E}\left[f(S_0)\right]\mathrm{E}\left[g(S^*)\right].
\end{aligned}$$

$\square$

The formula

$$\mathrm{E}[f(S_0)] = \mathrm{E}_N^0[\lambda S_0 f(S_0)]$$

is true for general stationary point processes. It states that on the σ-field generated by S_0, P is absolutely continuous with respect to P_N^0, with Radon–Nikodým derivative λS_0. In particular,

$$P(-T_0 + T_1 \le x) = \int_{[0,x]} \lambda y F_0(\mathrm{d}y),$$

where $F_0(x) = P_N^0(T_1 \le x)$.

EXAMPLE 17.2.14: STATIONARIZATION OF SEMI-MARKOV PROCESSES. The above example can be generalized to semi-Markov processes on a denumerable state space E. Such a stochastic process is constructed as follows. Let $\mathbf{P} = \{p_{ij}\}_{i,j\in E}$ be a stochastic matrix on E, assumed irreducible and positive recurrent. Its unique stationary distribution is denoted by π. For each $i, j \in E$, let G_{ij} be the cumulative distribution function of some strictly positive and proper random variable: thus $G_{ij}(0) = 0$ and $G_{ij}(\infty) = 1$. Denote by m_{ij} the mean

$$m_{ij} = \int_0^\infty t G_{ij}(\mathrm{d}t) = \int_0^\infty (1 - G_{ij}(t))\mathrm{d}t < \infty.$$

Recall at this stage that if U is a random variable uniformly distributed on $[0, 1]$, $G_{ij}^{-1}(U)$ is a random variable with CDF G_{ij} (here G_{ij}^{-1} is the inverse of G_{ij}).

Let $\{X_n\}_{n\in\mathbb{Z}}$ be a stationary Markov chain with transition matrix $\mathbf{P}$, defined on some probability space with a probability $\mathcal{P}^0$, and let $\{U_n\}_{n\in\mathbb{Z}}$, be a sequence of IID random variables, defined on the same space and uniformly distributed on $[0, 1]$. Assume moreover that the sequences $\{U_n\}_{n\in\mathbb{Z}}$ and $\{X_n\}_{n\in\mathbb{Z}}$ are independent under $\mathcal{P}^0$. Define

$$S_n = G_{X_n X_{n+1}}^{-1}(U_n).$$

In particular, conditionally on $X_n = i$ and $X_{n+1} = j$, S_n has the CDF G_{ij}. Moreover, conditionally on the whole sequence $\{X_n\}_{n\in\mathbb{Z}}$, the sequence $\{S_n\}_{n\in\mathbb{Z}}$ forms an independent family of random variables. We can now define a point process N by

$$T_0 = 0, \; T_{n+1} - T_n = S_n \qquad (n \in \mathbb{Z})$$

and the semi-Markov process $\{X(t)\}_{t \in \mathbb{R}}$, by

$$X(t) = X_n, \text{ for } T_n \le t < T_{n+1} \qquad (t \in \mathbb{R}).$$

Observe that there is no explosion (that is, $\lim_{n \to \infty} T_n = \infty$ and $\lim_{n \to -\infty} T_n = -\infty$, almost surely) since $\{X_n\}_{n \in \mathbb{Z}}$ is a recurrent HMC. It will be assumed that

$$E^0[T_1] = \sum_{i \in E} \pi(i) \sum_{j \in E} p_{ij} m_{ij} < \infty$$

and that

$$E^0[N((0, t])] < \infty, \qquad (t \in \mathbb{R}_+).$$

Let $(\Omega, \mathcal{F})$ be the canonical space of continuous functions with limits on the left taking their values in E. Let $\{\theta_t\}_{t \in \mathbb{R}}$ be the usual translation flow acting on functions, and let $P^0 = \mathcal{P}^0 \circ X^{-1}$. On this canonical framework, N is a θ_t-compatible marked point process, the mark of T_n being (X_n, U_n) with $X_n = X(T_n)$. Let $S_n = G_{X_n, X_{n+1}}(U_n)$. Under the above assumptions, all the conditions of the inverse construction are satisfied, so that there exists a θ_t-invariant probability P on $(\Omega, \mathcal{F})$ for which N is a stationary point process and such that $P_N^0 = P^0$.

The probabilistic structure of $\{X(t)\}_{t \in \mathbb{R}}$ under the stationary probability P is the following:

(a) Conditionally on $X_1 = j$, the sequence $S_1, X_2, S_2, X_3, S_3, \ldots$ has the same distribution under P or P_N^0.

(b) Conditionally on $X_0 = i$, the sequence $X_{-1}, S_{-1}, X_{-2}, S_{-2}, \ldots$ has the same distribution under P and P_N^0.

(c) Conditionally on $X_0 = i, X_1 = j, -T_0 > x, T_1 > y$, the sequences in (a) and (b) are independent.

(d) Moreover,

$$P(X_0 = i, \; X_1 = j, \; -T_0 > x, \; T_1 > y) = \lambda \pi(i) p_{ij} \int_{x+y}^{\infty} (1 - G_{ij}(t)) \, dt. \qquad (17.22)$$

Proof. The proof of (a)–(c) is similar to what we saw for renewal processes. For instance, for $n \ge 1$

$$
\begin{aligned}
P(X_n = i, X_{n+1} = j) &= \lambda E_N^0 \left[\int_0^{T_1} 1_{\{X_n \circ \theta_s = i, \; X_{n+1} \circ \theta_s = j\}} \, ds \right] \\
&= \lambda E_N^0 \left[\int_0^{T_1} 1_{\{X_n = i, \; X_{n+1} = j\}} \, ds \right],
\end{aligned}
$$

since $s \in (0, T_1)$ implies $X_n \circ \theta_s = X_n$ for all $n \in \mathbb{Z}$. Therefore

$$
\begin{aligned}
P(X_n = i, \; X_{n+1} = j) &= \lambda E_N^0 \left[T_1 1_{\{X_n = i\}} 1_{\{X_{n+1} = j\}} \right] \\
&= \lambda E_N^0 \left[G_{X_0 X_1}^{-1}(U_0) 1_{\{X_n = i\}} 1_{\{X_{n+1} = j\}} \right] \\
&= \lambda E_N^0 \left[G_{X_0 X_1}^{-1}(U_0) 1_{\{X_n = i\}} \right] p_{ij} \\
&= \lambda E_N^0 \left[T_1 1_{\{X_n = i\}} \right] p_{ij},
\end{aligned}
$$

where we have used the hypothesis $n \geq 1$, guaranteeing that X_{n+1} is conditionall
independent of X_0, X_1, U_0 given $X_n = i$. Similarly (summing the last equality in j)

$$P(X_n = i) = \lambda E_N^0 \left[T_1 1_{\{X_n=i\}} \right]$$

and therefore, for $n \geq 1$,

$$P(X_{n+1} = j | X_n = i) = p_{ij}.$$

More generally, it can be shown with the same type of calculations that $\{X_n\}_{n \in \mathbb{Z}}$ is under
P a Markov chain with transition matrix $\mathbf{P}$. Also, again with the same proof, under P,
$U_1, U_2, \ldots$ are IID random variables uniformly distributed on $[0, 1]$, and independent of
$X_1, X_2, \ldots$, and this proves (a).

As for (b), it suffices to reverse time, and to observe that under P_N^0, the sequence
$\{X_{-n}\}$, $n \in \mathbb{Z}$, is also a stationary and ergodic Markov chain, this time with the
transition matrix $Q = \{q_{ij}\}$, $i, j \in E$, given by

$$q_{ij} = p_{ji} \frac{\pi(j)}{\pi(i)}.$$

The proof of (c) is similar. It remains to give the joint law of X_0, X_1, T_0, T_1 under P. By
the inversion formula, the left-hand side of (17.22) equals

$$\lambda E_N^0 \left[\int_0^{T_1} 1_{\{X_0 \circ \theta_s = i, \, X_1 \circ \theta_s = j\}} 1_{\{-T_0 \circ \theta_s > x, \, T_1 \circ \theta_s > y\}} \, ds \right].$$

But if $s \in (0, T_1)$, $X_0 \circ \theta_s = X_0$, $X_1 \circ \theta_s = X_1$, $-T_0 \circ \theta_s = s$, $T_1 \circ \theta_s = T_1 - s$. Therefore
the last quantity equals

$$\lambda E_N^0 \left[1_{\{X_0=i, \, X_1=j\}} \int_0^{T_1} 1_{\{s>x\}} 1_{\{T_1-s>y\}} \, ds \right]$$

$$= \lambda \pi(i) p_{ij} \, E_N^0 \left[E_N^0 \left[\int_x^\infty 1_{\{T_1 > y+s\}} \, ds \mid X_0 = i, X_1 = j \right] \right]$$

$$= \lambda \pi(i) p_{ij} \int_{x+y}^\infty (1 - G_{ij}(t)) \, dt.$$

Recalling $m_{ij} = \int_0^\infty (1 - G_{ij}(t)) dt$ and setting $x = y = 0$ in (17.22) gives:

$$P(X_0 = i, \, X_1 = j) = \frac{\pi(i) p_{ij} m_{ij}}{\sum_{k \in E} \sum_{l \in E} \pi(k) p_{kl} m_{kl}}$$

and therefore

$$P(-T_0 > x, \, T_1 > y \mid X_0 = i, \, X_1 = j) = \frac{1}{m_{ij}} \int_{x+y} (1 - G_{ij}(t)) dt.$$

$\square$

The G/G/1/∞ Queue in Continuous Time

So far, we have studied the G/G/1/∞ queue at the arrival times or, in other words,
under the Palm probability associated with the arrival process. We now stage all this in
the (time-) stationary framework.

Let $(\Omega, \mathcal{F}, P)$ be a probability space with a measurable flow $\{\theta_t\}_{t \in \mathbb{R}}$, such that (P, θ_t) s ergodic. Let A be a simple point process defined on $(\Omega, \mathcal{F})$, compatible with the flow and with finite average rate:

$$\lambda := \mathrm{E}[A((0,1])] < \infty.$$

A is called the *arrival* (point) process and its n-th point T_n is interpreted as before as the arrival time of customer $\sharp n$ in the service system. Recall the convention $T_n < T_{n+1}$ $(n \in \mathbb{Z})$ and $T_0 \le 0 < T_1$. The *inter-arrival time* between customers $\sharp n$ and $\sharp(n+1)$ is

$$\tau_n := T_{n+1} - T_n.$$

Customer $\sharp n$ carries an amount of (non-negative) required service (or service time) denoted by σ_n, where the sequence $\{\sigma_n\}_{n \in \mathbb{Z}}$ is assumed to be a sequence of marks of the arrival process. Letting P_A^0 be the Palm probability associated with P and A, we define the *traffic intensity*

$$\rho := \lambda \mathrm{E}_A^0[\sigma_0].$$

The sequence $\{(T_n, \sigma_n)\}_{n \in \mathbb{Z}}$ describes the *input* into some queueing system (a G/G *input*, in Kendall's terminology). It is stationary in two distinct (but related) senses: under the Palm probability P_A^0, the sequence $\{(\tau_n, \sigma_n)\}_{n \in \mathbb{Z}}$ is stationary, and under the stationary probability P, the marked point process (A, σ) is stationary.

It was proved in Section 16.2 that if the traffic intensity ρ is strictly smaller than 1, then there exists a unique finite sequence $\{W_n\}_{n \in \mathbb{Z}}$ such that $W_n = W_0 \circ \theta^n$ and that

$$W_{n+1} = (W_n + \sigma_n - \tau_n)^+ \quad (n \in \mathbb{Z}), \quad P_A^0\text{-a.s.} \qquad (\star)$$

By Theorem 17.3.4, $(\star)$ also holds P-a.s. Let $\{W(t)\}_{t \in \mathbb{R}}$ be the (continuous-time) stochastic process defined by

$$W(t) = (W_n + \sigma_n - (t - T_n))^+ \quad (t \in [T_n, T_{n+1})), \qquad (17.23)$$

(the amount of work remaining at time t). Note that $W_n = W(T_n-)$. This defines a θ_t-compatible workload process $\{W(t)\}_{t \in \mathbb{R}}$. Note also that since

$$W_n = 0 \text{ for an infinity of indices } n \in \mathbb{Z}, \quad P_A^0\text{-a.s.},$$

by Theorem 17.3.4 again, this remains true P-a.s.

Recapitulating: Under the stability condition $\rho < 1$, there exists a unique finite non-negative θ_t-compatible process $\{W(t)\}_{t \in \mathbb{R}}$ satisfying equation (17.23). Moreover, there are an infinite number of negative indices n and an infinite number of positive indices n such that $W(T_n-) = 0$.

17.3   Two Interpretations of Palm Probability

This subsection addresses the issue of the "physical" meaning of Palm probability. W
treat in succession the local interpretation and the global (ergodic) interpretation.

17.3.1   The Local Interpretation

Theorem 17.3.1

$$\lim_{t \downarrow 0} \sup_{A \in \mathcal{F}} \ \left| P_N^0(A) - P(\theta_{T_1} \in A \mid T_1 \leq t) \right| = 0. \tag{17.24}$$

In other words, by definition of the distance in variation between two probability measures $(d_V(P_1, P_2) := \sup_{A \in \mathcal{F}} |P_1(A) - P_2(A)|)$

$$\lim_{t \to 0} d_V(P_N^0(\cdot), P(\theta_{T_1} \in \cdot \mid T_1 \leq t)) = 0.$$

In particular for any bounded random variable Z,

$$\lim_{t \downarrow 0} \mathrm{E}\left[Z \circ \theta_{T_1} \mid T_1 \leq t\right] = \mathrm{E}_N^0 \left[Z\right] . \tag{17.25}$$

We prepare the proof with the *Korolyuk–Dobrushin infinitesimal estimates*.

Theorem 17.3.2 *Let (N, θ_t, P) be a simple stationary point process with positive finite intensity λ. Then (Korolyuk's estimate), as $t \downarrow 0$,*

$$P(N((0, t]) > 1) = o(t)$$

and (Dobrushin's estimate)

$$P(N((0, t]) > 0) = \lambda t + o(t).$$

Proof. Let P_N^0 be the associated Palm probability. The inversion formula (17.15), with $n = -1$, gives

$$P(N((0, t]) > 1) = P(T_2 \leq t) = \lambda \int_0^\infty P_N^0(u < -T_{-1}, T_2 \circ \theta_{-u} \leq t)\, \mathrm{d}u.$$

But on Ω_0, $u < -T_{-1}$ implies $T_2 \circ \theta_{-u} = T_1 + u$. Therefore

$$P(N((0, t]) > 1) = \lambda \int_0^t P_N^0(u < -T_{-1}, T_1 \leq t - u)\, \mathrm{d}u$$

$$\leq \lambda \int_0^t P_N^0(T_1 \leq t)\, \mathrm{d}u = \lambda t P_N^0(T_1 \leq t).$$

Since $P_N^0(T_1 > 0) = 1$, $\lim_{t \downarrow 0} P_N^0(T_1 \leq t) = 0$.

For Dobrushin's estimate, we use again the inversion formula (17.15), this time with $n = -1$.

Since $T_1 \circ \theta_{-u} = u$ on $\Omega_0 \cap \{0 < u \le -T_{-1}\}$,

$$
\begin{aligned}
P(N((0,t]) > 0) &= P(T_1 \le t) = \lambda \int_0^\infty P_N^0(u < -T_{-1}, T_1 \circ \theta_{-u} \le t)\,du \\
&= \lambda \int_0^t P_N^0(u < -T_{-1})\,du \\
&= \lambda t - \lambda \int_0^t P_N^0(-T_{-1} \le u)\,du = \lambda t + o(t)\,.
\end{aligned}
$$

$\square$

We now proceed to the proof of Theorem 17.3.1.

Proof. We have to show that

$$
P(\theta_{T_1} \in A \mid T_1 \le t) = \frac{\lambda t}{P(T_1 \le t)}\,\frac{P(T_1 \le t, \theta_{T_1} \in A)}{\lambda t}
$$

tends to $P_N^0(A)$ as $t \downarrow 0$. Taking into account Dobrushin's estimate,

$$
\lim_{t \to 0} \frac{\lambda t}{P(T_1 \le t)} = 1\,,
$$

it suffices to show that, uniformly in A,

$$
\frac{1}{\lambda t}\left| P(T_1 \le t, \theta_{T_1} \in A) - P_N^0(A) \right| \to 0\,.
$$

The inversion formula (17.15) (with $n = -1$) gives

$$
\begin{aligned}
P(T_1 \le t, \theta_{T_1} \in A) &= \lambda \int_0^\infty P_N^0(u < -T_{-1}, T_1 \circ \theta_{-u} \le t, \theta_{T_1} \circ \theta_{-u} \in A)\,du \\
&= \lambda \int_0^t P_N^0(u < -T_{-1}, A)\,du,
\end{aligned}
$$

since $\theta_{T_1} \circ \theta_{-u}$ is the identity and $T_1 \circ \theta_{-u} = u$ on $\Omega_0 \cap \{0 < u < -T_{-1}\}$. Therefore, rewriting

$$
\lambda t P_N^0(A) = \lambda \int_0^t P_N^0(A)\,du,
$$

we obtain

$$
\begin{aligned}
\frac{1}{\lambda t}\left| P(T_1 \le t, \theta_{T_1} \in A) - P_N^0(A) \right| &= \frac{1}{t}\int_0^t P_N^0(u \ge -T_{-1}, A)\,du \\
&\le \frac{1}{t}\int_0^t P_N^0(u \ge -T_{-1})\,du \\
&\le \frac{1}{t}\int_0^t P_N^0(t \ge -T_{-1})\,du = P_N^0(t \ge -T_{-1})\,.
\end{aligned}
$$

But since $P_N^0(-T_{-1} > 0)$, $P_N^0(t \ge -T_{-1}) \to 0$ as $t \to 0$.

$\square$

EXAMPLE 17.3.3: THE INTUITIVE INTERPRETATION. Let $\{X(t)\}_{t\in\mathbb{R}}$ be a θ_t-compatibl
CORLOL stochastic process with values in $\mathbb{R}^m$ and let $f : \mathbb{R}^m \to \mathbb{R}$ be a bounded contin
uous function. Taking $Z := f(X(0-))$ in (17.25), we have that

$$\lim_{h\to 0} \mathrm{E}\left[f(X(T_1-)) \mid N(0,h] \geq 1\right] = \mathrm{E}_N^0\left[f(X(0-))\right].$$

Defining

$$T_+(t) := t + \inf\{h > 0;\ N(t, t+h] = 1\},$$

we have equivalently

$$\lim_{h\to 0} \mathrm{E}\left[f(X(T_+(t)-)) \mid N(t, t+h] \geq 1\right] = \mathrm{E}_N^0\left[f(X(0-))\right].$$

Since the stochastic process $\{Z(t)\}_{t\in\mathbb{R}}$ is CORLOL,

$$\lim_{h\to 0} \mathrm{E}\left[f(X(t)) \mid N(t, t+h] \geq 1\right] = \mathrm{E}_N^0\left[f(X(0-))\right].$$

(A typical use of this arises in queueing theory, when one computes the law of the
number of customers in a stationary system given that some event (departure or arrival)
occurred.)

17.3.2   The Ergodic Interpretation

Let (N, θ_t, P) be a simple locally finite stationary point process on $\mathbb{R}$ with finite positive
intensity λ and let P_N^0 be the Palm probability associated with it. Denote θ_{T_1} by θ.

Theorem 17.3.4

(a) Let $A \in \mathcal{F}$ be an event invariant with respect to the flow $\{\theta_t\}_{t\in\mathbb{R}}$. Then $P(A) = 1$
    if and only if $P_N^0(A) = 1$.

(b) Let $A \in \mathcal{F}$ be a θ-invariant event. Then $P_N^0(A) = 1$ if and only if $P(A) = 1$.

Proof.

(a) Suppose that $P_N^0(A) = 1$. By the inversion formula,

$$\begin{aligned}
P(A) &= \lambda \int_0^\infty P_N^0(u < T_1, \theta_{-u}A)\,du \\
&= \lambda \int_0^\infty P_N^0(u < T_1, A)\,du \quad (\theta_t\text{-invariance of } A) \\
&= \lambda \int_0^\infty P_N^0(u < T_1)\,du \quad (P_N^0(A) = 1) \\
&= \lambda E_0^0[T_1] = 1.
\end{aligned}$$

Conversely, suppose that $P(A) = 1$. By the inversion formula again,

$$1 = P(A) = \lambda \int_0^\infty P_N^0(u < T_1, A)\,du,$$

and therefore, since $1 = \lambda E_N^0[T_1] = \lambda \int_0^\infty P_N^0(u < T_1)\,du$,

$$0 = \lambda \int_0^\infty P_N^0(u < T_1, \overline{A}) \, du = \lambda E_N^0[T_1 1_{\overline{A}}].$$

But $T_1 < \infty$ a.s. since $E_N^0[T_1] = \lambda^{-1} < \infty$. Therefore $E_N^0[T_1 1_{\overline{A}}] = 0$ implies $P_N^0(\overline{A}) = 0$.

(b) If $P(A) = 1$, then from the definition of Palm probability,

$$\begin{aligned}
P_N^0(A) &= \frac{1}{\lambda t} E\left[\sum_n 1_{\theta^{-n}A} 1_{\{0 < T_n \le t\}}\right] \\
&= \frac{1}{\lambda t} E\left[\sum_n 1_A 1_{\{0 < T_n \le t\}}\right] \quad (\theta\text{-invariance of } A) \\
&= \frac{1}{\lambda t} E[1_A N((0, t])] \\
&= \frac{1}{\lambda t} E[N((0, t])] \quad (P(A) = 1) = 1.
\end{aligned}$$

Conversely, assuming that $P_N^0(A) = 1$,

$$\lambda t P_N^0(A) = \lambda t = E[1_A N((0, t])],$$

that is, $E[1_{\overline{A}} N((0, t])] = 0$ for all $t > 0$. Letting $t \uparrow \infty$, we have $E[1_{\overline{A}} N(\mathbb{R}_+)] = 0$, which implies $P(\overline{A}) = 0$ (recall that $P(N(\mathbb{R}_+) = \infty) = 1$). $\qquad\square$

Theorem 17.3.5 (P, θ_t) *is ergodic if and only if* (P_N^0, θ) *is ergodic.*

Proof. Suppose for instance that (P, θ_t) is ergodic and that (P_N^0, θ) is not. Then (Theorem 16.1.21) there exists a decomposition of the type (16.4):

$$P_N^0 = \alpha_1 P_1^0 + \alpha_2 P_2^0$$

where $\alpha_1, \alpha_2 \in (0, 1)$, $\alpha_1 + \alpha_2 = 1$, and P_1^0 and P_1^0 are θ-invariant. Note that $E_1^0[T_1]$ and $E_2^0[T_1]$ are finite since $E_N^0[T_1]$ is finite. Also $P_1^0(T_1 > 0) = P_2^0(T_1 > 0) = 1$ since $P_N^0(T_1 > 0) = 1$. In particular, $E_1^0[T_1]$ and $E_2^0[T_1]$ are positive. The inversion formula gives

$$\begin{aligned}
P(A) = {}&\alpha_1 \frac{E_1^0[T_1]}{E_N^0[T_1]} \left(\frac{1}{E_1^0[T_1]} E_1^0\left[\int_0^{T_1} 1_A \circ \theta_t \, dt\right]\right) \\
&+ \alpha_2 \frac{E_2^0[T_1]}{E_N^0[T_1]} \left(\frac{1}{E_2^0[T_1]} E_2^0\left[\int_0^{T_1} 1_A \circ \theta_t \, dt\right]\right),
\end{aligned}$$

that is

$$P = \beta_1 P_1 + \beta_2 P_2,$$

where $\beta_1 = \alpha_1 \frac{E_1^0[T_1]}{E_N^0[T_1]}$ and $\beta_2 = \alpha_2 \frac{E_2^0[T_1]}{E_N^0[T_1]} \in (0, 1)$ are such that $\beta_1 + \beta_2 = 1$. Moreover, P_1 and P_2, defined by $P_i(A) = \frac{1}{E_i^0[T_1]} E_i^0\left[\int_0^{T_1} 1_A \circ \theta_t \, dt\right]$ $(i = 1, 2)$, are θ_t-invariant for all $t \in \mathbb{R}$. Therefore, by Theorem 16.3.17, P cannot be ergodic, a contradiction.

The proof of the converse part follows the same line of argument (Exercise 17.5.2). $\qquad\square$

The following result is called the *cross-ergodic theorem*. It is essential for queueing theory in that it provides an easy interpretation of the *cross-formulas*, that is, formulas linking P_N^0-means and P-means of operational quantities. See Chapter 9.

Theorem 17.3.6 *Let (N, θ_t, P) be a stationary point process with finite positive inten sity λ, and suppose that (P, θ_t) is ergodic. Let P_N^0 be the associated Palm probability. Le f be in $L^1(P)$. Then*

$$\lim_{T \uparrow \infty} \frac{1}{T} \int_0^T f \circ \theta_t \mathrm{d}t = \mathrm{E}[f], \quad P_N^0\text{-a.s.} \tag{17.26}$$

and

$$\lim_{n \uparrow \infty} \frac{1}{n} \sum_{k=1}^n f \circ \theta_{T_k} = \mathrm{E}_N^0[f], \quad P\text{-a.s.} \tag{17.27}$$

Proof. Since (P, θ_t) is ergodic, the probability of event

$$A := \left\{ \lim_{T \to \infty} \frac{1}{T} \int_0^T f \circ \theta_t \mathrm{d}t = \mathrm{E}[f] \right\}$$

is 1. Moreover, A is θ_t-invariant. Therefore by Lemma 17.3.4, $P_N^0(A) = 1$. This proves (17.26). Similarly, since (P_N^0, θ) is ergodic, for all $f \in L^1(P_N^0)$, the event

$$B := \left\{ \lim_{n \uparrow \infty} \frac{1}{n} \sum_{k=1}^n f \circ \theta_{T_k} = \mathrm{E}_N^0[f] \right\}$$

has probability 1. It is moreover θ-invariant, and therefore $P(B) = 1$ by Lemma 17.3.4. This proves (17.27).

□

17.4   General Principles of Queueing Theory

17.4.1   The PASTA Property

This result belongs to the folklore of queueing theory. It states that if the arrival point process is Poisson, operational characteristics of the system computed just before arrival times and at arbitrary times are the same (the acronym PASTA stands for "Poisson Arrivals See Time Averages"). Here is a precise statement. Let a measurable flow $\{\theta_t\}_{t \in \mathbb{R}}$ and a θ_t-invariant probability P on $(\Omega, \mathcal{F})$ be given. Let A be a θ_t-compatible point process with finite intensity λ (we use the notation A because the PASTA theorem is mostly applied to arrival point processes).

Suppose that for some history $\{\mathcal{F}_t\}_{t \in \mathbb{R}}$, A is an $\mathcal{F}_t$-Poisson process of intensity λ. Let $\{Z(t)\}_{t \in \mathbb{R}}$ be a CORLOL θ_t-compatible $\mathcal{F}_t$-adapted process with values in some topological measurable space $(K, \mathcal{K})$. Let $f : K \to \mathbb{R}$ be a non-negative measurable function. Then, as a particular case of the forthcoming Theorem 17.4.1, we shall see that

$$\mathrm{E}_A^0[f(Z(0-))] = \mathrm{E}[f(Z(0))]. \tag{17.28}$$

Note that, by the cross-ergodic theorem, if (P, θ_t) is ergodic, (17.28) implies that P-almost surely and P-almost surely,

$$\mathrm{E}_A^0[f(Z(0-))] = \lim_{N \to \infty} \frac{1}{N} \sum_{n=1}^N f(Z(T_n-)) = \lim_{T \to \infty} \frac{1}{T} \int_0^T f(Z(s))\mathrm{d}s = \mathrm{E}[f(Z(0))].$$

The following useful extension of the above classical PASTA property is a direct con sequence of the definitions of stochastic intensity and of Palm probability.

Theorem 17.4.1 *Let A, P, $\{\theta_t\}_{t\in\mathbb{R}}$ and $\{\mathcal{F}_t\}_{t\in\mathbb{R}}$ be as above, with the difference that A is not necessarily a Poisson process but has a θ_t-compatible stochastic $\mathcal{F}_t$-intensity $\{\lambda(t)\}_{t\in\mathbb{R}}$. Still assume that its average intensity λ is finite. Let P_A^0 be the Palm probability associated with (A, θ_t, P). Then, for any θ_t-compatible CORLOL stochastic process $\{Z(t)\}_{t\in\mathbb{R}}$ taking its values in some measurable space $(K, \mathcal{K})$, and any non-negative function $f : K \to \mathbb{R}$,*

$$\lambda E_A^0 \left[f(Z(0-)) \right] = E\left[\lambda(0) f(Z(0)) \right]. \tag{17.29}$$

Proof. In the following chain of equalities, the first one is by definition of Palm probability, the second one is the smoothing formula of the stochastic calculus of point processes with a stochastic intensity, the third one makes use of the fact that the discontinuity times of a CORLOL function form a set of null Lebesgue measure and the fifth one uses the stationarity hypothesis.

$$\lambda E_A^0 \left[f(Z(0-)) \right] = E\left[\int_{(0,1]} f(Z(s-)) \, N(ds) \right] = E\left[\int_{(0,1]} f(Z(s-)) \lambda(s) \, ds \right]$$

$$= E\left[\int_{(0,1]} f(Z(s)) \lambda(s) \, ds \right] = \int_{(0,1]} E\left[f(Z(s)) \lambda(s) \right] \, ds$$

$$= \int_{(0,1]} E\left[f(Z(0)) \lambda(0) \right] \, ds = E\left[f(Z(0)) \lambda(0) \right].$$

$\square$

In the Poisson case ($\lambda(t) \equiv \lambda$) we recover the classical PASTA property. However the above generalization has useful applications.

EXAMPLE 17.4.2: THE 'JOB-OBSERVER' PROPERTY. [7] Consider a Gordon–Newell network in stationary regime. The point process A_{ij} counting the transfers from station i to station j admits the $\mathcal{F}_t^X$-intensity

$$\lambda_{ij}(t) = \mu_i r_{ij} 1_{\{X_i(t)>0\}}.$$

Its (average) intensity is therefore $\lambda_{ij} = \mu_i r_{ij} P(X_i(0) > 0)$. By Theorem 17.4.1,

$$\lambda_{ij} P_{A_{ij}}^0 [X(0-) = n] = E[\mu_i r_{ij} 1_{\{X_i(0)>0\}} 1_{\{X(0)=n\}}],$$

Therefore

$$P_{A_{ij}}^0 (X(0-) = n) P(X_i(0) > 0) = \begin{cases} P(X(0) = n) & \text{if } n_i > 0; \\ 0 & \text{otherwise.} \end{cases}$$

In view of the expression (9.21) for the stationary distribution of the Gordon–Newell network,

$$P_{A_{ij}}^0 (X(0-) = n) = \begin{cases} \frac{1}{C} \left(\prod_{l=1,\, l\neq i}^{K} \alpha_l^{n_l} \right) \alpha_i^{n_i-1} & \text{if } n_i > 0; \\ 0 & \text{otherwise,} \end{cases}$$

where C is a constant obtained by normalization:

$$\sum_{n_1+\cdots+n_K=M,\, n_i>0} \left(\prod_{l=1,\, l\neq i}^{K} \alpha_l^{n_l} \right) \alpha_i^{n_i-1} = C.$$

---

[7] [Reiser and Lavenberg, 1980] and [Sevcik and Mitrani, 1981].

After the change of summation variable $n_i \to n_i - 1$, the left-hand side of the above equality becomes

$$\sum_{n_1 + \cdots + n_K = M-1} \left(\prod_{l=1}^{K} \alpha_l^{n_l} \right) = G(K, M-1).$$

Therefore, for n such that $n_i > 0$,

$$P_{A_{ij}}^0 (X(0-) = n) = \frac{1}{G(K, M-1)} \left(\prod_{l=1,\ l\neq i}^{K} \alpha_l^{n_l} \right) \alpha_i^{n_i - 1}.$$

When a customer is transferred from i to j at time t, the situation he sees during his transfer (when he has left i but not yet reached j) for the rest of the network is not $X(t-)$ but $X(t-) - e_i$. Therefore, the state of the network observed by this customer (excluding him) is n with probability $\frac{1}{G(K,M-1)} \prod_{l=1}^{K} \alpha_l^{n_l}$. It is the same as the state of the same network with $M - 1$ customers observed at an arbitrary time by an external observer.

EXAMPLE 17.4.3: INSENSITIVITY OF $M/GI/1/\infty$ LIFO PREEMPTIVE. "Insensitivity" refers to situations where a given average depending on the probability distribution of one or several random variables actually depends on these distributions only through their means. This is the case, for instance, for the $M/GI/\infty$ delay system of Example 8.3.2 in stationary regime, for which the average number of customers in the system depends on the delay only through its mean. There are many other instances of this phenomenon in queueing theory (the distribution of the congestion process of the stationary Erlang queue has this property).[8]

Call $X(t)$ the number of customers in the system at time t in an $M/GI/1/\infty$ LIFO preemptive queue. For fixed $k \geq 1$, denote by N_k the point process counting the arrivals that make the congestion process reach level k:

$$N_k(C) = \sum_{n \in \mathbb{Z}} 1_C(T_n) 1_{\{X(T_n-)=k-1\}}.$$

Let $\{T_n^{(k)}\}$ be the sequence of points N_k, with the usual convention $T_0^{(k)} \leq 0 < T_1^{(k)}$. A customer arriving at time $T_n^{(k)}$ requires the service $\sigma_n^{(k)}$. The following fact is true: $\{\sigma_n^{(k)}\}_{n \in \mathbb{Z}}$ is IID with the same distribution as $\{\sigma_n\}_{n \in \mathbb{Z}}$ (the proof is left for the reader). Because of the LIFO preemptive rule, customer $T_n^{(k)}$ receives all his service when the queue is at level k (see the figure below).

---

[8][Barbour, 1976]; [König and Jansen, 1976].

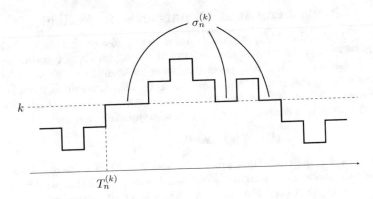

Since k is recurrent (under the stability hypothesis), the law of large numbers gives

$$\lim_{t\to\infty} \frac{\int_0^t 1_{\{X(s)=k\}}\mathrm{d}s}{\int_{(0,t]} 1_{\{X(s-)=k-1\}}N(\mathrm{d}s)} = \mu^{-1} \quad \text{a.s.}$$

By ergodicity,

$$\lim_{t\to\infty} \frac{1}{t}\int_0^t 1_{\{X(s)=k\}}\mathrm{d}s = \pi(k),$$

where π is the stationary distribution of the congestion process. Moreover,

$$\lim_{t\to\infty} \frac{1}{t}\int_{(0,t]} 1_{\{X(s-)=k-1\}}N(\mathrm{d}s) = \lim_{t\to\infty} \frac{N(t)}{t}\frac{1}{N(t)}\int_{(0,t]} 1_{\{X(s-)=k-1\}}N(\mathrm{d}s).$$

But $\lim_{t\to\infty}\frac{N(t)}{t} = \lambda$, and in view of the PASTA property

$$\lim_{t\to\infty} \frac{1}{N(t)}\int_{(0,t]} 1_{\{X(s-)=k-1\}}N(\mathrm{d}s) = \lim_{t\to\infty} \frac{1}{t}\int_0^t 1_{\{X(s)=k-1\}}\mathrm{d}s = \pi(k-1),$$

so that

$$\frac{\pi(k)}{\pi(k-1)} = \rho \quad (k\geq 1).$$

This leads to

$$\pi(k) = (1-\rho)\rho^k \quad (k\geq 0),$$

which shows in particular the insensitivity of the $M/GI/1/\infty$ LIFO preemptive queue, that is, π depends on G only through its mean.

The proof of insensitivity can be extended to the situation where the input process N admits the stochastic $(P,\mathcal{F}_t)$-intensity $\{\lambda_{X(t)}\}_{t\in\mathbb{R}}$, where $\{\mathcal{F}_t\}_{t\in\mathbb{R}}$ is the history recording the past of N at time t and the whole sequence $\{\sigma_n\}_{n\in\mathbb{N}}$. Now, we have to use the extended version of PASTA

$$\lim_{t\to\infty} \frac{1}{N(t)}\int_0^t 1_{\{X(s-)=k-1\}}N(\mathrm{d}s) = \lim_{t\to\infty} \frac{t}{N(t)}\frac{1}{t}\int_0^t 1_{\{X(s)=k-1\}}\lambda_{k-1}\mathrm{d}s = \pi(k-1)\frac{\lambda_{k-1}}{\lambda},$$

which proves insensitivity.

17.4.2   Queue Length at Departures or Arrivals

In the $M/G/1/\infty$ and $G/M/1/\infty$ systems, we were able to compute the stationary queu distributions at departure times and just before an arrival time, respectively. For the $G/G/1/\infty$ queue, there is a very general relation between the stationary distributions at departure times and just before an arrival time. The situation is the following. We have a measurable flow $\{\theta_t\}_{t\in\mathbb{R}}$, two simple point processes A and D without common points and a stochastic process $\{X(t)\}_{t\in\mathbb{R}}$, all θ_t-compatible and such that for all $t \geq 0$

$$X(t) = X(0) + A((0,t]) - D((0,t]) \geq 0.$$

Thus $\{X(t)\}_{t\in\mathbb{R}}$ is the congestion process (queue length) of a queue, A and D being respectively the arrival and departure processes. Also assume that A has a finite intensity $\lambda_A = \lambda$ and that $X(0)$ (and therefore also $X(t)$) has finite expectation. Then, taking expectation in the above evolution equation for the congestion process, we find that $E[A((0,t]] = E[D((0,t])]$ and therefore the intensity of the departure process is $\lambda_D = \lambda$.

Let now $f : \mathbb{N} \to \mathbb{R}$ be a bounded function. For all $t \geq 0$

$$f(X(t)) = f(X(0)) + \int_{(0,t]} \{f(X(s)) - f(X(s-))\}A(ds)$$

$$+ \int_{(0,t]} \{f(X(s)) - f(X(s-))\}D(ds).$$

At an event time of A, $f(X(t)) = f(X(t-)+1)$ and at an event time of D, $f(X(t-)) = f(X(t)+1)$. Therefore

$$f(X(t)) = f(X(0)) + \int_{(0,t]} \{f(X(s-)+1) - f(X(s-))\}A(ds)$$

$$+ \int_{(0,t]} \{f(X(s-)-1) - f(X(s-))\}D(ds).$$

Dividing by λt, taking expectations with respect to P, observing that $E[f(X(t))] = E[f(X(0))]$, we find that

$$E_A^0[f(X(0-)+1) - f(X(0-))] + E_D^0[f(X(0)) - f(X(0)+1)] = 0.$$

With $f(x) = 1_{\{n\}}(x)$, where $n \geq 1$, and letting

$$\pi_{A-}^0(n) := P_A^0(X(0-) = n) , \quad \pi_D^0(n) := P_D^0(X(0) = n),$$

we obtain $\pi_{A-}^0(n-1) - \pi_{A-}^0(n) = \pi_D^0(n) - \pi_D^0(n+1)$. For $n = 0$, observing that $X(0-) \geq 0$ and $X(0-) > 0$ if 0 is a departure time, $\pi_{A-}^0(0) = \pi_D^0(1)$. By summing the last two equalities for all $i \geq 0$:

$$\pi_{A-}^0(i) = \pi_D^0(i+1).$$

In words: the stationary congestion processes observed just before an arrival and just after a departure have the same distribution.

EXAMPLE 17.4.4: THE CASE OF AN $M/GI/1/\infty$ QUEUE. The stationary distribution of the number of customers just after the departure times of a stable $M/G/1/\infty$ queue is equal to the stationary distribution. This is an immediate consequence of the above result and of the PASTA property.

17.4.3 Little's Formula

Let $(\Omega, \mathcal{F}, P)$ be a probability space endowed with a measurable flow $\{\theta_t\}_{t \in \mathbb{R}}$. Assume that (P, θ_t) is ergodic. Consider a point process N compatible with the flow and therefore stationary. Assume its intensity λ finite. The point T_n is interpreted as the arrival time into some "system" or "black box" (say, a service system or a processor) of a "customer" or "job", that remains in the system for the random time W_n. The sequence $\{W_n\}_{n \in \mathbb{Z}}$ is assumed θ_t-compatible (and therefore stationary under the Palm probability P_N^0).

$$\lambda \longrightarrow \boxed{\begin{array}{c} X(t) \\ W_n \end{array}} \longrightarrow \lambda$$

The number $X(t)$ of customers, or jobs, in the system at time t is

$$X(t) = \sum_{n \in \mathbb{Z}} 1_{(-\infty, t]}(T_n) 1_{(t, -\infty)}(T_n + W_n).$$

The process $\{X(t)\}_{t \in \mathbb{R}}$ is by construction θ_t-compatible and therefore stationary. Letting $t = 0$ and taking expectation gives (by the Campbell–Mecke formula and noting that $W_n = W_0 \circ \theta_{T_n}$)

$$E[X(0)] = E\left[\sum_{n \in \mathbb{Z}} 1_{(-\infty, 0]}(T_n) 1_{(0, -\infty)}(T_n + W_0 \circ \theta_{T_n}) \right]$$

$$= \lambda E_N^0 \left[\int_{-\infty}^0 1_{(0, -\infty)}(s + W_0) ds \right]$$

$$= \lambda \int_{-\infty}^0 P_N^0(s + W_0 > 0) ds$$

$$= \lambda \int_0^\infty P_N^0(W_0 > u) du = \lambda E_N^0[W_0].$$

We have just obtained the celebrated *Little's formula*:

$$E[X(0)] = \lambda E_N^0[W_0].$$

If (P, θ_t) is ergodic, the cross-ergodic theorem allows the above equality to be rewritten as

$$E[X(0)] = \lambda \lim_{n \uparrow \infty} \frac{1}{n} \sum_{k=1}^n W_k \qquad P\text{-a.s.}$$

This formula appears in the applied literature in the form $Q = \lambda W$. It is a rather versatile formula, the choice of the "black box" delimiting the system being left to one's imagination.

EXAMPLE 17.4.5: THE EMPTINESS PROBABILITY OF THE $G/G/1/\infty$ QUEUE. Consider the stationary $G/G/1/\infty$ queue of Section 17.2.4 with a service discipline *without preemption* (a service once started cannot be interrupted). Denote by $X(t)$ the number of customers in the whole system (waiting room plus service booth). Take for the black box the service booth, and let W_n be the sojourn time of customer n (time spent in waiting plus time spent in service).

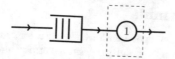

Choose now for the "black box" the service booth. The number of people in the booth is at most 1, actually, equal to $1_{\{X(t)=1\}}$, the expectation of which is $1 - \pi(0)$, where $\pi(0)$ is the stationary probability of finding the system empty at any given time t. The sojourn time of customer n in the booth is σ_n. The θ_t-compatible point process of arrivals in the booth has average intensity λ. By Little's formula

$$\pi(0) = 1 - \rho,$$

where the traffic intensity ρ is given by the formula

$$\rho = \lambda \lim_{n \uparrow \infty} \frac{1}{n} \sum_{k=1}^{n} \sigma_k.$$

This is a universal formula that we have already encountered in various particular cases.

EXAMPLE 17.4.6: A MEAN-VALUE ANALYSIS OF CLOSED JACKSON NETWORKS. Closed Jackson network models arise, for instance, in the situation where an open network is operated with a blocking admission policy: if there are already N customers in the network, the newcomers wait at a gate (in a queue) until some customer is released from the network, at which time one among the blocked customers, if any, is admitted. In the network, there are at most N customers. At "saturation", there is always one customer ready to replace a departing customer, by definition of saturation. Therefore, at saturation, or for all practical purposes near saturation, everything looks as if a departing customer is being immediately recycled.

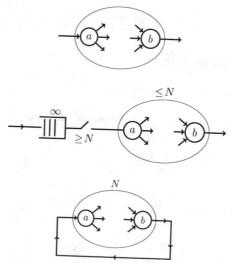

It is important in practice to be able to compute the average number of customers $d(N)$ passing through the entrance point a per unit of time. This quantity is the maximum throughput and is therefore related to the efficiency of the system from the point f view of the operator (who makes money from customers). From the point of view of he quality of service, an important parameter is $W(N)$, the average time spent by a customer between a and b. As a matter of fact, N, $W(N)$, and $d(N)$ are related by

$$d(N)W(N) = N, \qquad (\star)$$

a particular case of Little's formula.

The following algorithm, called *mean value analysis*,[9] makes use of Little's formula and of the job-observer property for computing $d(N)$.

Let v_i denote the average number of visits that a customer entering in a make in station i before leaving the network via b. Let $L_i(N)$ be the stationary average number of customers in station i, let $W_i(N)$ be the customer-average sojourn time in station i and let $\lambda_i(N)$ be the average number of customers served by station i per unit of time. We have

$$\lambda_i(N) = d(N)v_i$$

and from Little's formula

$$L_i(N) = \lambda_i(N)W_i(N) = d(N)v_iW_i(N).$$

Since the total number of customers is N, we have

$$N = \sum_{i=1}^{K} L_i(N) = d(N)\sum_{i=1}^{N} v_iW_i(N).$$

Let $Y_i(N)$ denote the average number of customers in station i seen by a customer arriving at station i. By the job-observer property, $Y_i(N) = L_i(N-1)$. Also, by Wald's identity,

$$W_i(N) = \frac{Y_i(N) + 1}{\mu_i},$$

where μ_i is the service rate at station i. Combining the above relations, we obtain the following recurrence relations:

$$W_i(N) = \frac{1}{\mu_i}(1 + L_i(N-1)),$$

$$T(N) = \frac{N}{\sum_{i=1}^{N} v_iW_i(N)},$$

$$L_i(N) = v_iT(N)W_i(N).$$

One can therefore compute $d(N)$ and then $W(N)$ via $(\star)$, and then choose an operating point N that provides the required balance between the operator's profit and the customer's comfort.

EXAMPLE 17.4.7: THE CHARACTERISTIC FUNCTION OF THE WORKLOAD. Let $\{W(t)\}_{t \in \mathbb{R}}$ be the stationary workload process of a stationary $G/G/1/\infty$ queue (with traffic intensity $\rho < 1$). By looking separately at what happens at the jump times and what happens between the jumps, one can verify the following evolution equation:

---

[9][Reiser and Lavenberg, 1980] and [Sevcik and Mitrani, 1981].

$$e^{iuW(t)} = e^{iuW(0)} - iu \int_0^t e^{iuW(s)} 1_{\{W(s)>0\}} \, ds$$
$$+ \sum_{n\in\mathbb{Z}} \{e^{iuW(T_n)} - e^{iuW(T_n-)}\} 1_{\{0\le T_n\le t\}} \,.$$

Taking expectations and observing that $\mathrm{E}[e^{iuW(t)}] = \mathrm{E}[e^{iuW(0)}]$ and $e^{iuW(s)} 1_{\{W(s)>0\}} = e^{iuW(s)} - 1_{\{X(s)=0\}}$,

$$-iu\mathrm{E}[e^{iuW(0)}] + iuP(X(0) = 0) + \lambda\mathrm{E}_A^0[e^{iuW(0)} - e^{iuW(0-)}] = 0.$$

Suppose that σ_n is independent of $W(T_n-)$ (which is the case in a $GI/GI/1/\infty$ queue, for instance). Then

$$\mathrm{E}_A^0[e^{iuW(0)} - e^{iuW(0-)}] = \mathrm{E}_A^0[e^{iuW(0-)}(e^{iu\sigma_0} - 1)]$$
$$= \mathrm{E}_A^0[e^{iuW(0-)}](\mathrm{E}_A^0[e^{iu\sigma_0}] - 1).$$

In a $GI/GI/1/\infty$ queue, $\mathrm{E}_A^0[e^{iu\sigma_0}] = \mathrm{E}[e^{iu\sigma_0}]$, and therefore, taking into account $P(X(0) = 0) = 1 - \rho$,

$$iu\mathrm{E}[e^{iuW(0)}] = \lambda\mathrm{E}_A^0[e^{iuW(0-)}](\mathrm{E}[e^{iu\sigma_0}] - 1) + iu(1 - \rho).$$

In the special case of a Poisson arrival process $(M/GI/1/\infty)$, the PASTA property gives $\mathrm{E}_A^0[e^{iuW(0-)}] = \mathrm{E}[e^{iuW(0)}]$, and therefore

$$\mathrm{E}[e^{iuW(0)}] = \frac{iu(1 - \rho)}{iu - \lambda(\Psi_\sigma(u) - 1)},$$

where $\Psi_\sigma(u)$ is the characteristic function of σ_0.

Complementary reading

[Baccelli and Brémaud, 2003].

17.5   Exercises

Exercise 17.5.1. A CONTINUOUS-TIME HMC AND ITS EMBEDDED CHAIN
Consider a continuous-time HMC $\{X(t)\}_{t\in\mathbb{R}}$ with state space E that is positive recurrent and stationary. Give an expression in terms of the infinitesimal generator $\mathbf{Q}$ and of the stationary distribution π of the average intensity of the stationary point process counting the transition times of the chain. Assuming this average intensity is finite, give the stationary distribution of the discrete-time HMC $\{X(T_n-)\}_{n\in\mathbb{N}}$ where T_n is the n-th transition time of the chain.

Exercise 17.5.2. ERGODICITY
Give the missing details at the end of the proof of Theorem 17.3.5, that is, prove that if (P_N^0, θ) is ergodic, then (P, θ_t) is ergodic.

Exercise 17.5.3. ERGODICITY AND THE STATIONARY RENEWAL PROCESS
Is a stationary renewal process ergodic? Justify your answer.

Exercise 17.5.4. THE POINT PROCESS WITH EQUISPACED POINTS
Consider the point process with equispaced points (say: $T_{n+1} - T_n = 1$ for all $n \in \mathbb{Z}$). Clearly, for the Palm version, (P_N^0, θ_{T_1}) is mixing. Show that the stationary version (P, θ_t) is not mixing.

Exercise 17.5.5. THE VARIATION DISTANCE BETWEEN THE STATIONARY AND UNDE-LAYED RENEWAL PROCESSES
Show that when the simple locally finite N is under P_N^0 an undelayed renewal point process with an inter-renewal CDF F admitting a density f with respect to the Lebesgue measure,

$$d_V(P, P_N^0) = d_V(F, G),$$

where G is a CDF with density $g(x) = \lambda(1 - F(x))$ (the CDF of the residual inter-renewal distribution). Here P and P_N^0 are restricted to $\mathbb{R}_+$.

Exercise 17.5.6. SELECTED TRANSITIONS OF A STATIONARY HMC
Let (P, θ_t) be a stationary framework and let $\{X(t)\}_{t \in \mathbb{R}}$ be a θ_t-compatible stochastic process with values in the denumerable space E that is moreover an ergodic continuous-time HMC with infinitesimal generator $\mathbf{Q} = \{q_{ij}\}_{i,j \in E}$ and stationary distribution π. Let $N(C)$ be the number of discontinuities of this chain in $C \subseteq E$. This defines a stationary process (N, θ_t, P).

Let now H be a subset of $E \times E - \text{diag}(E \times E)$. Let N_H be the point process counting the H-*transitions* of the chain, that is:

$$N_H(C) := \int_C 1_H(X_{s-}, X_s) N(\mathrm{d}s). \tag{17.30}$$

Then (N_H, θ_t, P) is a stationary point process. In particular, for $H = E \times E - \text{diag}(E \times E)$, we have $N_H = N$, the *basic point process* of the chain, which counts all transitions. Give the expressions of the quantities $P_{N_H}^0(X_{0-} = k)$ and $P_{N_H}^0(X_0 = k)$ in terms of the infinitesimal generator and of the stationary distribution π.

Exercise 17.5.7. MEASURABILITY OF THE SHIFT
Show that, for fixed $x \in \mathbb{R}^m$, the maps θ_x of Example 16.3.9 are measurable.

Exercise 17.5.8. θ_t-COMPATIBILITY
Let $\{\theta_x\}_{t \in \mathbb{R}}$ be a measurable flow and let N be a simple θ_t-compatible point process on $\mathbb{R}$. Let $\{T_n\}_{n \in \mathbb{Z}}$ be its sequence of points. Prove the following: $\{Z_n\}_{n \in \mathbb{Z}}$ is a θ_x-compatible process of marks of N if and only if there exists a stochastic process $\{Z(t)\}_{t \in \mathbb{R}}$ compatible with the flow and such that $Z(T_n) = Z_n$ for all $n \in \mathbb{N}$.

Exercise 17.5.9. θ_{T_1}
Refer to the discussion before Theorem 17.2.1. Give a simple example showing that in general, on Ω, the inverse of θ_{T_1} is *not* $\theta_{T_{-1}}$.

Exercise 17.5.10. THE FIRST POSITIVE EVENT TIME OF A SUPERPOSITION
Take $\mathbb{R}^m = \mathbb{R}$ in Example 17.1.15. Let F_i and F_i^0 be the CDF of the first point of μ under $\mathcal{P}_i$ and $\mathcal{P}_i^0$, respectively. Prove that, under the Palm probability P_N^0, the CDF G^0 of the first point strictly to the right of the origin of the superposition process N is

$$G^0(x) = 1 - \sum_{i=1}^{k} \left(\frac{\lambda_i}{\lambda}(1 - F_i^0(x)) \prod_{\substack{1 \le j \le k \\ j \ne i}} (1 - F_j(x)) \right).$$

Exercise 17.5.11. MOTION-INVARIANT STATIONARY POINT PROCESSES
Let P be a probability on $(M_p(\mathbb{R}^m), \mathcal{M}_p(\mathbb{R}^m))$ that makes of the coordinate process N a motion-invariant stationary point process of finite non-null intensity. Motion invariance means that for any distance-preserving transformation τ of $\mathbb{R}^m$, and any $\Gamma \in \mathcal{M}(\mathbb{R}^m)$, $P(\tau N \in \Gamma) = P(N \in \Gamma)$. Show that the Palm probability P_0^N is invariant under rotation about center 0, that is, for any such rotation r and any $\Gamma \in \mathcal{M}_p(\mathbb{R}^m)$, $P_0^N(rN \in \Gamma) = P_0^N(N \in \Gamma)$.

Exercise 17.5.12. FELLER'S BUS PARADOX
Under the conditions prevailing in Example 17.2.13 with the additional condition that $E_N^0[S_0^2] < \infty$, prove that

$$E[S_0] = \frac{E_N^0[S_0^2]}{E_N^0[S_0]},$$

where $S_0 := T_1 - T_0$. Deduce from this that for S_0 to have the same distribution under P and under P_N^0 it is necessary and sufficient that the interrenewal times be deterministic.

Exercise 17.5.13. THE RELIABILITY POINT PROCESS
Consider a non-delayed renewal process on $\mathbb{R}$ (not just $\mathbb{R}_+$) for which the interrenewal sequence $\{S_n\}_{n \in \mathbb{Z}}$ is of the form $S_n = U_n + V_n$, where $\{U_n\}_{n \in \mathbb{Z}}$ and $\{V_n\}_{n \in \mathbb{Z}}$ are independent IID sequences of random variables with finite mean. Describe the time-stationary version of the point process whose event times are

$$U_0, U_0 + V_0, U_0 + V_0 + U_1, U_0 + V_0 + U_1 + V_1, \ldots$$

Exercise 17.5.14. THE STATIONARY M/G/1/∞ QUEUE
The $M/GI/1/\infty$ queue is a special case of the $G/G/1/\infty$ queue. The service time sequence $\{\sigma_n\}_{n \in \mathbb{Z}}$ is IID with CDF $G(x)$ and mean μ^{-1} and the input process N is Poisson with intensity λ. Let π denote the stationary distribution of the number of customers just after the departure times of a stable M/G/1/∞ queue. Show that the stationary distribution of the congestion process is equal to π: $P(X(0) = i) = \pi(i)$, for all i.

Exercise 17.5.15. TWO EXPRESSIONS OF THE TRAFFIC INTENSITY
Verify the following two expressions of the traffic intensity for the general G/G/1/∞ queue. First in terms of P:

$$\rho = \lambda \lim_{N \to \infty} \frac{1}{N} \sum_{k=1}^{N} \sigma_k, \quad P\text{-a.s.},$$

and then in terms of P_A^0:

$$\rho = \frac{E_A^0[\sigma_0]}{E_A^0[\tau_0]}.$$

Exercise 17.5.16. THE WAITING TIME OF THE $G/G/1/\infty$ QUEUE
In a stationary $G/G/1/\infty$ queue ($\lambda E_A^0[\sigma_0] < 1$), let V_n be the waiting time (spent in the waiting room) of the n-th customer. Give the expression of

$$\lim_{n\uparrow\infty} \frac{1}{n} \sum_{k=1}^{n} V_k$$

in terms of λ, $\pi(0)$ and $E[X(0)]$.

Appendix A

Number Theory and Linear Algebra

A.1   The Greatest Common Divisor

Let $a_1, \ldots, a_k \in \mathbb{N}$ be such that $\max(a_1, \ldots, a_k) > 0$. Their *greatest common divisor* (gcd) is the largest positive integer dividing all of them. It is denoted by $\gcd(a_1, \ldots, a_k)$. Clearly, removing all zero elements does not change the gcd, so that we may assume without loss of generality that all the a_k's are positive.

Let $\{a_n\}_{n \geq 1}$ be a sequence of positive integers. The sequence $\{d_k\}_{k \geq 1}$ defined by $d_k = \gcd(a_1, \ldots, a_k)$ is bounded below by 1 and is non-increasing, and it therefore has a limit $d \geq 1$, a positive integer called the gcd of the sequence $\{a_n\}_{n \geq 1}$. Since the d_k's are integers, the limit is attained after a finite number of steps, and therefore there exists a positive integer k_0 such that $d = \gcd(a_1, \ldots, a_k)$ for all $k \geq k_0$.

Lemma A.1.1 *Let $S \subset \mathbb{Z}$ contain at least one nonzero element and be closed under addition and subtraction. Then S contains a least positive element a, and $S = \{ka \,; k \in \mathbb{Z}\}$.*

Proof. Let $c \in S$, $c \neq 0$. Then $c - c = 0 \in S$. Also $0 - c = -c \in S$. Therefore, S contains at least one positive element. Denote by a the smallest positive element of S. Since S is closed under addition and subtraction, S contains a, $a + a = 2a, \ldots$ and $0 - a = -a, 0 - 2a = -2a, \ldots$, that is, $\{ka \,; k \in \mathbb{Z}\} \subset S$.

Let $c \in S$. Then $c = ka + r$, where $k \in \mathbb{Z}$ and $0 \leq r < a$. Since $r = c - ka \in S$, we cannot have $r > 0$, because this would contradict the definition of a as the smallest positive integer in S. Therefore, $r = 0$, i.e., $c = ka$. Therefore, $S \subset \{ka \,; k \in \mathbb{Z}\}$. □

Lemma A.1.2 *Let $a_1, \ldots, a_k$ be positive integers with greatest common divisor d. There exist $n_1, \ldots, n_k \in \mathbb{Z}$ such that $d = \sum_{i=1}^{k} n_i a_i$.*

Proof. The set $S = \left\{ \sum_{i=1}^{k} n_i a_i \,; n_1, \ldots, n_k \in \mathbb{Z} \right\}$ is closed under addition and subtraction, and therefore, by Lemma A.1.1, $S = \{ka \,; k \in \mathbb{Z}\}$, where $a = \sum_{i=1}^{k} n_i a_i$ is the smallest positive integer in S.

© Springer Nature Switzerland AG 2020
P. Brémaud, *Probability Theory and Stochastic Processes*, Universitext,
https://doi.org/10.1007/978-3-030-40183-2

Since d divides all the a_i's, d divides a, and therefore $0 < d \le a$. Also, each a_i is in A and is therefore a multiple of a, which implies that $a \le \gcd(a_1, \ldots, a_k) = d$. Therefore $d = a$.

Theorem A.1.3 *Let d be the gcd of $A = \{a_n \; ; n \ge 1\}$, a set of positive integers that is closed under addition. Then A contains all but a finite number of the positive multiples of d.*

Proof. We may assume without loss of generality that $d = 1$ (otherwise, divide all the a_n's by d). For some k, $d = 1 = \gcd(a_1, \ldots, a_k)$, and therefore by Lemma A.1.2,

$$1 = \sum_{i=1}^{k} n_i a_i$$

for some $n_1, \ldots, n_k \in \mathbb{Z}$. Separating the positive from the negative terms in the latter equality, we have $1 = M - P$, where M and P are in A.

Let $n \in \mathbb{N}$, $n \ge P(P-1)$. We have $n = aP + r$, where $r \in [0, P-1]$. Necessarily, $a \ge P-1$, otherwise, if $a \le P-2$, then $n = aP + r < P(P-1)$. Using $1 = M - P$, we have that $n = aP + r(M - P) = (a - r)P + rM$. But $a - r \ge 0$. Therefore, n is in A. We have thus shown that any $n \in \mathbb{N}$ sufficiently large, say $n \ge P(P-1)$, is in A. □

A.2 Eigenvalues and Eigenvectors

Let A be a square matrix of dimension $r \times r$, with complex coefficients. If there exists a scalar $\lambda \in \mathbb{C}$ and a column vector $v \in \mathbb{C}^r$, $v \ne 0$, such that

$$Av = \lambda v \qquad (\text{resp., } v^T A = \lambda v^T), \tag{A.1}$$

then v is called a *right-eigenvector* (resp., a *left-eigenvector*) associated with the *eigenvalue* λ. There is no need to distinguish between right and left-eigenvalues because if there exists a left-eigenvector associated with the eigenvalue λ, then there exists a right-eigenvector associated with the same eigenvalue λ. This follows from the facts that the set of eigenvalues of A is exactly the set of roots of the *characteristic equation*

$$\det(\lambda I - A) = 0 \tag{A.2}$$

where I is the $r \times r$ identity matrix, and that

$$\det(\lambda I - A) = \det(\lambda I - A^T).$$

The *algebraic multiplicity* of λ is its multiplicity as a root of the *characteristic polynomial* $\det(\lambda I - A)$.

If $\lambda_1, \cdots, \lambda_k$ are *distinct* eigenvalues corresponding to the right-eigenvectors $v_1, \cdots, v_k$ and the left-eigenvectors $u_1, \cdots, u_k$, then $v_1, \cdots, v_k$ are independent, and so are $u_1, \cdots, u_k$.

Call R_λ (resp. L_λ) the set of right-eigenvectors (resp., left-eigenvectors) associated with the eigenvalue λ, plus the null vector. Both L_λ and R_λ are vector subspaces of $\mathbb{C}^r$, and they have the same dimension, called the *geometric multiplicity* of λ. In particular, the largest number of independent right-eigenvectors (resp., left-eigenvectors) cannot exceed the sum of the geometric multiplicities of the distinct eigenvalues.

The matrix A is called *diagonalizable* if there exists a nonsingular matrix Γ of the same dimension such that

$$\Gamma A \Gamma^{-1} = \Lambda, \qquad (A.3)$$

where

$$\Lambda = \text{diag}(\lambda_1, \cdots, \lambda_r)$$

for some $\lambda_1, \cdots, \lambda_r \in \mathbb{C}$, not necessarily distinct. It follows from (A.3) that with $U = \Gamma^T$, $U^T A = U^T \Lambda$, and with $V = \Gamma^{-1}$, $AV = V\Lambda = \Lambda V$, and therefore $\lambda_1, \cdots, \lambda_r$ are eigenvalues of A, and the ith row of $U^T = \Gamma$ (resp., the ith column of $V = \Gamma^{-1}$) is a left-eigenvector (resp., right-eigenvector) of A associated with the eigenvalue λ_i. Also, $A = V\Lambda U^T$ and therefore

$$A^n = V\Lambda^n U^T. \qquad (A.4)$$

Clearly, if A is diagonalizable, the sum of the geometric multiplicities of A is exactly equal to r. It turns out that the latter is a sufficient condition of diagonalizability of A. Therefore, A is diagonalizable if and only if the sum of the geometric multiplicities of the distinct eigenvalues of A is equal to r.

EXAMPLE A.2.1: DISTINCT EIGENVALUES. By the last result, if the eigenvalues of A are distinct, A is diagonalizable. In this case, the diagonalization process can be described as follows. Let $\lambda_1, \cdots, \lambda_r$ be the r distinct eigenvalues and let $u_1, \cdots, u_r$ and $v_1, \cdots, v_r$ be the associated sequences of left and right-eigenvectors, respectively. As mentioned above, $u_1, \cdots, u_r$ form an independent collection of vectors, and so do $v_1, \cdots, v_r$. Define

$$U = [u_1 \cdots u_r], V = [v_1 \cdots v_r]. \qquad (A.5)$$

Observe that if $i \neq j$, $u_i^T v_j = 0$. Indeed, $\lambda_i u_i^T v_j = u_i^T A v_j = \lambda_j u_i^T v_j$, which implies $(\lambda_i - \lambda_j) u_i^T v_j = 0$, and in turn $u_i^T v_j = 0$, since $\lambda_i \neq \lambda_j$ by hypothesis. Since eigenvectors are determined up to multiplication by a non-null scalar, one can choose them in such a way that $u_i^T v_i = 1$ for all $i \in [1, r]$. Therefore,

$$U^T V = I, \qquad (A.6)$$

where I is the $r \times r$ identity matrix. Also, by definition of U and V,

$$U^T A = \Lambda U^T, AV = \Lambda V. \qquad (A.7)$$

In particular, by (A.6), $A = V\Lambda U^T$. From the last identity and (A.6) again, we obtain for all $n \geq 0$,

$$A^n = V\Lambda^n U^T, \qquad (A.8)$$

that is,

$$A^n = \sum_{i=1}^{r} \lambda_i^n v_i u_i^T. \qquad (A.9)$$

A.3 The Perron–Fröbenius Theorem

We now turn to the more general situation when the eigenvalues may be not distinct. In fact, the result stated below concerns a larger class of square matrices.

Definition A.3.1 *A matrix $A = \{a_{ij}\}_{1 \le i,j \le r}$ with real coefficients is called* nonnegativ *(resp.,* positive*) if all its entries are nonnegative (resp., positive). A nonnegative matri A is called* stochastic *if $\sum_{j=1}^{r} a_{ij} = 1$ for all i, and* substochastic *if $\sum_{j=1}^{r} a_{ij} \le 1$ for al i, with strict inequality for at least one i.*

Nonnegativity (resp., positivity) of A is denoted by $A \ge 0$ (resp., $A > 0$). If A and B are two matrices of the same dimensions with real coefficients, the notation $A \ge B$ (resp., $A > B$) means that $A - B \ge 0$ (resp., $A - B > 0$).

The *communication graph* of a square nonnegative matrix A is the oriented graph with the state space $E = \{1, \dots, r\}$ as its set of vertices and an oriented edge from vertex i to vertex j if and only if $a_{ij} > 0$.

Definition A.3.2 *A nonnegative square matrix A is called* irreducible *(resp.,irreducible* aperiodic*) if it has the same communication graph as an irreducible (resp., irreducible aperiodic) stochastic matrix. It is called* primitive *if there exists an integer k such that $A^k > 0$.*

EXAMPLE A.3.3: A nonnegative matrix is primitive if and only if it is irreducible and aperiodic (exercise).

Theorem A.3.4 *Let A be a nonnegative primitive $r \times r$ matrix. There exists a real eigenvalue λ_1 with algebraic as well as geometric multiplicity one such that $\lambda_1 > 0$, and $\lambda_1 > |\lambda_j|$ for any other eigenvalue λ_j. Moreover, the left eigenvector u_1 and the right eigenvector v_1 associated with λ_1 can be chosen positive and such that $u_1^T v_1 = 1$.*

Let $\lambda_2, \lambda_3, \dots, \lambda_r$ be the eigenvalues of A other than λ_1 ordered in such a way that

$$\lambda_1 > |\lambda_2| \ge \cdots \ge |\lambda_r| \tag{A.10}$$

and if $|\lambda_2| = |\lambda_j|$ for some $j \ge 3$, then $m_2 \ge m_j$, where m_j is the algebraic multiplicity of λ_j. Then

$$A^n = \lambda_1^n v_1 u_1^T + O(n^{m_2 - 1} |\lambda_2|^n), \tag{A.11}$$

where $O(f(n))$ represents a function of n such that there exists $\alpha, \beta \in \mathbb{R}$, $0 < \alpha \le \beta < \infty$, such that $\alpha f(n) \le O(f(n)) \le \beta f(n)$ for all n sufficiently large.

If in addition, A is stochastic (resp., substochastic), then $\lambda_1 = 1$ (resp., $\lambda_1 < 1$).

If A is stochastic but not irreducible, then the algebraic and geometric multiplicities of the eigenvalue 1 are equal to the number of communication classes.

If A is stochastic and irreducible with period $d > 1$, then there are exactly d distinct eigenvalues of modulus 1, namely the dth roots of unity, and all other eigenvalues have modulus strictly less than 1.

For the proof, see [Seneta, 1981].

Appendix B

Analysis

B.1 Infinite Products

Theorem B.1.1 *Let $\{a_n\}_{n\geq 1}$ be a sequence of numbers in the interval $[0,1)$.*

(a) If $\sum_{n=1}^{\infty} a_n < \infty$, then

$$\lim_{n\uparrow\infty} \prod_{k=1}^{n} (1 - a_k) > 0.$$

(b) If $\sum_{n=1}^{\infty} a_n = \infty$, then

$$\lim_{n\uparrow\infty} \prod_{k=1}^{n} (1 - a_k) = 0.$$

Proof. (a): For any numbers $c_1, \ldots, c_n$ in $[0,1)$, it holds that

$$(1 - c_1)(1 - c_2) \cdots (1 - c_n) \geq 1 - c_1 - c_2 - \cdots - c_n$$

(proof by induction). Since $\sum_{n=1}^{\infty} a_n$ converges, there exists an N such that for all $n \geq N$,

$$a_N + \cdots + a_n < \frac{1}{2}.$$

Therefore, defining $\pi(n) = \prod_{k=1}^{n} (1 - a_k)$, we have that for all $n \geq N$,

$$\frac{\pi(n)}{\pi(N-1)} = (1 - a_N) \cdots (1 - a_n) \geq 1 - (a_N + \cdots + a_n) \geq \frac{1}{2}.$$

Therefore, the sequence $\{\pi(n)\}_{n\geq N}$ is non-increasing and bounded from below by $\frac{1}{2}\pi(N-1) > 0$, so that $\lim_{n\uparrow\infty} \pi(n) > 0$.

(b): By the inequality $1 - a \leq e^{-a}$ true when $a \in [0,1)$, we have that $\pi(n) \leq e^{-a_1 - a_2 - \cdots - a_n}$, and therefore, if $\sum_{n=1}^{\infty} a_n = \infty$, $\lim_{n\uparrow\infty} \pi(n) = 0$. $\square$

© Springer Nature Switzerland AG 2020
P. Brémaud, *Probability Theory and Stochastic Processes*, Universitext,
https://doi.org/10.1007/978-3-030-40183-2

B.2 Abel's Theorem

Lemma B.2.1 *Let $\{b_n\}_{n \geq 1}$ and $\{a_n\}_{n \geq 1}$ be two sequences of real numbers such that*

$$b_1 \geq b_2 \geq \cdots \geq b_n \geq 0,$$

and such that for some real numbers m and M, and all $n \geq 1$,

$$m \leq a_1 + \cdots + a_n \leq M.$$

Then, for all $n \geq 1$,

$$b_1 m \leq a_1 b_1 + \cdots + a_n b_n \leq b_1 M. \tag{B.1}$$

Proof. Let $s_n = a_1 + \cdots + a_n$, and use Abel's summation technique to obtain

$$
\begin{aligned}
a_1 b_1 + \cdots + a_n b_n &= b_1 s_1 + b_2 (s_2 - s_1) + \cdots + b_n (s_n - s_{n-1}) \\
&= s_1 [b_1 - b_2] + \cdots + s_{n-1}[b_{n-1} - b_n] + s_n[b_n].
\end{aligned}
$$

The bracketed terms are all non-negative, and therefore replacing each s_i by its lower bound or upper bound yields the result. □

We recall without proof a standard result of calculus.

Lemma B.2.2 *The sum of a uniformly convergent series of continuous functions is a continuous function.*

Theorem B.2.3 *Let $\{a_n\}_{n \geq 1}$ be a sequence of real numbers such that the radius of convergence of the power series $\sum_{n=0}^{\infty} a_n z^n$ is 1. Suppose that the sum $\sum_{n=0}^{\infty} a_n$ is convergent. Then the power series $\sum_{n=0}^{\infty} a_n x^n$ is uniformly convergent in $[0,1]$ and*

$$\lim_{x \uparrow 1} \sum_{n=0}^{\infty} a_n x^n = \sum_{n=0}^{\infty} a_n, \tag{B.2}$$

where $x \uparrow 1$ means that x tends to 1 from below.

Proof. It suffices to prove that $\sum_{n=0}^{\infty} a_n x^n$ is uniformly convergent in $[0,1]$, since (B.2) then follows by Lemma B.2.2. Let $A_n^p := a_n + \cdots + a_p$. By convergence of $\sum_{n=0}^{\infty} a_n$, for all $\epsilon > 0$, there exists an $n_0 \geq 1$ such that $p \geq n \geq n_0$ implies $|A_n^p| \leq \epsilon$, and therefore, since for $x \in [0,1]$, the sequence $\{x^n\}_{n \geq 0}$ is non-increasing, Abel's lemma gives for all $x \in [0,1]$,

$$|a_n x^n + \ldots + a_p x^p| \leq \epsilon x^n \leq \epsilon,$$

from which uniform convergence follows. □

Theorem B.2.4 *Let $\{a_n\}_{n \geq 0}$ be a sequence of non-negative real numbers such that the power series $\sum_{n=0}^{\infty} a_n z^n$ has a radius of convergence 1. If*

$$\lim_{x \uparrow 1} \sum_{n=0}^{\infty} a_n x^n = a \leq \infty, \tag{B.3}$$

then

$$\sum_{n=0}^{\infty} a_n = a. \tag{B.4}$$

Proof. For $x \in [0,1)$, $\sum_{n=0}^{\infty} a_n x^n \leq \sum_{n=0}^{\infty} a_n$ (the a_n are non-negative), and therefore by (B.3), $a \leq \sum_{n=0}^{\infty} a_n$. This proves the result when $a = \infty$.

We now suppose that $a < \infty$. From $\sum_{n=0}^{p} a_n = \lim_{x \uparrow 1} \sum_{n=0}^{p} a_n x^n$, we have that $\sum_{n=0}^{p} a_n \leq a < \infty$. Thus, $\sum_{n=1}^{p} a_n$ is a non-decreasing sequence, converging to some α, $\alpha \leq a < \infty$. By Abel's theorem, $\lim_{x \uparrow 1} \sum_{n=0}^{\infty} a_n x^n = \alpha$, and therefore $\alpha = a$ and $\sum_{n=0}^{\infty} a_n = a$. $\qquad\square$

B.3 Tykhonov's Theorem

Theorem B.3.1 *Let $\{x_n\}_{n \geq 0}$ be a sequence of elements of $[0,1]^{\mathbb{N}}$, that is*

$$x_n = (x_n(0), x_n(1), \ldots),$$

where $x_n(k) \in [0,1]$ for all $k, n \in \mathbb{N}$. There exists a strictly increasing sequence of integers $\{n_l\}_{l \geq 0}$ and an element $x \in [0,1]^{\mathbb{N}}$ such that

$$\lim_{l \uparrow \infty} x_{n_l}(k) = x(k) \tag{B.5}$$

for all $k \in \mathbb{N}$.

Proof. Since the sequence $\{x_n(0)\}_{n \geq 0}$ is contained in the closed interval $[0,1]$, by the Boltzano–Weierstrass theorem, one can extract a subsequence $\{x_{n_0(l)}(0)\}_{l \geq 0}$ such that

$$\lim_{l \uparrow \infty} x_{n_0(l)}(0) = x(0)$$

for some $x(0) \in [0,1]$. In turn, one can extract from $\{x_{n_0(l)}(1)\}_{l \geq 0}$ a subsequence $\{x_{n_1(l)}(1)\}_{l \geq 0}$ such that

$$\lim_{l \uparrow \infty} x_{n_1(l)}(1) = x(1)$$

for some $x(1) \in [0,1]$. Note that

$$\lim_{l \uparrow \infty} x_{n_1(l)}(0) = x(0).$$

Iterating this process, we obtain for all $j \in \mathbb{N}$ a sequence $\left\{x_{n_j(l)}\right\}_{l \geq 0}$ that is a subsequence of each sequence $\{x_{n_0(l)}(1)\}_{l \geq 0}, \ldots, \left\{x_{n_{j-1}(l)}(1)\right\}_{l \geq 0}$ and such that

$$\lim_{l \uparrow \infty} x_{n_j(l)}(k) = x(k)$$

for all $k \leq j$, where $x(1), \ldots, x(j) \in [0,1]$. The diagonal sequence $n_l = n_l(l)$ then establishes (B.5). $\qquad\square$

B.4 Cesàro, Toeplitz and Kronecker's Lemmas

Cesàro's Lemma

Theorem B.4.1 *Let $\{b_n\}_{n \geq 0}$ be a sequence of real numbers such that*

$$\lim_{n \uparrow \infty} b_n = 0.$$

Then

$$\lim_{n \uparrow \infty} \frac{b_1 + \cdots + b_n}{n} = 0.$$

Proof. The sequence $\{b_n\}_{n \geq 0}$ is bounded in absolute value, say by K. For fixed $\epsilon > 0$, there exists an n_0 such that $n > n_0$ implies $|b_n| \leq \epsilon$, and therefore

$$\left| \frac{b_1 + \cdots + b_n}{n} \right| \leq \left| \frac{b_1 + \cdots + b_{n_0}}{n} \right| + \left| \frac{b_{n_0} + \cdots + b_n}{n} \right|$$

$$\leq \frac{n_0 K}{n} + \frac{n - n_0}{n} \epsilon \leq 2\epsilon,$$

if n is sufficiently large.

$\square$

Toeplitz's Lemma

Let $\{x_n\}_{n \geq 1}$ be a sequence of non-negative real numbers converging to a finite limit x. Let $\{a_n\}_{n \geq 1}$ be a sequence of positive real numbers such that for all $n \geq 1$, $b_n := \sum_{j=1}^{n} a_j > 0$ and $\lim_{n \uparrow \infty} b_n = \infty$. Then

$$\lim_{n \uparrow \infty} \frac{1}{b_n} \sum_{j=1}^{n} a_j x_j = x \,.$$

In particular, with $a_n \equiv 1$,

$$\lim_{n \uparrow \infty} \frac{1}{n} \sum_{j=1}^{n} x_j = x \,.$$

Proof. Let $\varepsilon > 0$, and let $n_0 = n_0(\varepsilon)$ be such that $|x_n - x| \leq \frac{\varepsilon}{2}$ if $n \geq n_0$. Take $n_1 > n_0$ such that

$$\frac{1}{b_{n_1}} \sum_{j=1}^{n_0} |x_j - x| < \frac{\varepsilon}{2} \,.$$

Then, for $n > n_1$,

$$\left| \frac{1}{b_n} \sum_{j=1}^{n} a_j x_j - x \right| \leq \frac{1}{b_n} \left| \sum_{j=1}^{n} a_j |x_j - x| \right|$$

$$= \frac{1}{b_n} \sum_{j=1}^{n_0} + \frac{1}{b_n} \sum_{j=n_0+1}^{n}$$

$$\leq \frac{1}{b_{n_1}} \left| \sum_{j=1}^{n_0} a_j |x_j - x| \right| + \frac{1}{b_n} \left| \sum_{j=n_0+1}^{n} a_j |x_j - x| \right|$$

$$\leq \frac{\varepsilon}{2} + \frac{b_n - b_{n_0}}{b_n} \frac{\varepsilon}{2} \leq \varepsilon \,.$$

$\square$

Kronecker's Lemma

Let $\{x_n\}_{n\geq 1}$ be a sequence of real numbers such that $\sum_{j=1} nx_n \to$, and let $\{b_n\}_{n\geq 1}$ be a non-decreasing sequence of positive real numbers such that $\lim_{n\uparrow\infty} b_n = \infty$. Then

$$\lim_{n\uparrow\infty} \frac{1}{b_n} \sum_{j=1}^{n} b_j x_j = 0\,.$$

In particular, with $b_n = n$, $x_n = \frac{y_n}{n}$ and $\sum \frac{y_n}{n} \to$,

$$\lim_{n\uparrow\infty} \frac{y_1 + \cdots y_n}{n} = 0\,.$$

Proof. Let $b_0 := 0$, $S_0 := 0$, $S_n := \sum_{j=1}^{n} x_j$. Then

$$\sum_{j=1}^{n} b_j x_j = \sum_{j=1}^{n} b_j(S_j - S_{j-1}) = b_n S_n - b_0 S_0 - \sum_{j=1}^{n} S_{j-1}(b_j - b_{j-1})$$

and therefore

$$\frac{1}{b_n} \sum_{j=1}^{n} b_j x_j = S_n - \frac{1}{b_n} \sum_{j=1}^{n} S_{j-1} a_j \to 0$$

because, by Toeplitz' lemma, if $S_n \to 0$, then

$$\frac{1}{b_n} \sum_{j=1}^{n} S_{j-1} a_j \to 0\,.$$

$\square$

B.5   Subadditive Functions

Lemma B.5.1 *Let $f : (0,\infty) \to \mathbb{R}$ be a non-negative function such that $\lim_{t\downarrow 0} f(t) = 0$, and such that f is subadditive, that is,*

$$f(t + s) \leq f(t) + f(s)$$

for all $s, t \in (0,\infty)$. Define the (possibly infinite) non-negative real number

$$q := \sup_{t>0} \frac{f(t)}{t}\,.$$

Then

$$\lim_{h\downarrow 0} \frac{f(h)}{h} = q\,.$$

Proof. By definition of q, there exists for each $a \in [0, q)$ a real number $t_0 > 0$ such that $\frac{f(t_0)}{t_0} \geq a$.

To each $t > 0$, there corresponds an $n \in \mathbb{N}$ such that $t_0 = nt + \delta$ with $0 \leq \delta < t$. Since f is subadditive, $f(t_0) \leq nf(t) + f(\delta)$, and therefore

$$a \leq \frac{nt}{t_0} \frac{f(t)}{t} + \frac{f(\delta)}{t_0}\,,$$

which implies

$$a \le \liminf_{t\downarrow 0} \frac{f(t)}{t}.$$

($\delta \to 0$, and therefore $f(\delta) \to 0$, as $t \to 0$; also $\frac{nt}{t_0} \to 1$ as $t \to 0$.) Therefore,

$$a \le \liminf_{t\downarrow 0} \frac{f(t)}{t} \le \limsup_{t\downarrow 0} \frac{f(t)}{t} \le q,$$

from which the result follows, since a can be chosen arbitrarily close to q (arbitrarily large, when $q = \infty$).

$\square$

B.6   Gronwall's Lemma

Theorem B.6.1 *Let* $x : \mathbb{R}^+ \to \mathbb{R}$ *be a positive locally integrable real function such that*

$$x(t) \le a + b \int_0^t x(s)\,\mathrm{d}s \qquad (\star)$$

for some $a \ge 0$ *and* $b > 0$. *Then*

$$x(t) \le ae^{bt}.$$

Proof. Multiplying $(\star)$ by e^{-bt}, we have

$$e^{-bt}x(t) \le ae^{-bt} + be^{-bt}\int_0^t x(s)\,\mathrm{d}s$$

or, equivalently

$$ae^{-bt} \ge e^{-bt}x(t) - be^{-bt}\int_0^t x(s)\,\mathrm{d}s$$

$$= \frac{d}{dt} e^{-bt}\int_0^t x(s)\,\mathrm{d}s.$$

Integrating this inequality:

$$\frac{a}{b}(1 - e^{-bt}) \ge e^{-bt}\int_0^t x(s)\,\mathrm{d}s.$$

Substituting this into $(\star)$:

$$x(t) \le a + be^{-bt}\frac{a}{b}(1 - e^{-bt}) = ae^{bt}.$$

$\square$

B.7   The Abstract Definition of Continuity

Definition B.7.1 *Let* $(X, \mathcal{O})$ *and* $(E, \mathcal{V})$ *be two topological spaces. A function* $f : X \to E$ *is said to be continuous (with respect to these topologies) if*

$$V \in \mathcal{V} \Rightarrow f^{-1}(V) \in \mathcal{O}.$$

The problem consists in connecting this definition to the classical one in metric spaces. In fact:

Theorem B.7.2 *Let (X, d) and (E, ρ) be two metric spaces. A necessary and sufficient condition for a mapping $f : X \to E$ to be continuous according to the definition above (and with respect to the topologies induced by the corresponding metrics) is that for any $\varepsilon > 0$ and any $x \in X$, there exists a $\delta > 0$ such that $d(y, x) \leq \delta$ implies $d(f(y), f(x)) \leq \varepsilon$.*

Lemma B.7.3 *Let $(X, \mathcal{O})$ and $(E, \mathcal{V})$ be two topological spaces. A necessary and sufficient condition for a mapping $f : X \to E$ to be continuous (with respect to these topologies) is that for any $x \in X$ and any open set $V \in \mathcal{V}$ containing $f(x)$, there exists an open set $U \in \mathcal{O}$ containing x and such that $f(U) \subseteq V$.*

Proof. For the necessity, take $U = f^{-1}(V)$. To prove sufficiency, we have to show that if $V \in \mathcal{V}$ then $f^{-1}(V) \in \mathcal{O}$. Take any $x \in f^{-1}(V)$. Since V is an open set containing $f(x)$, there exists, by hypothesis, $U_x \in \mathcal{O}$ containing x such that $f(U_x) \subseteq V$. Clearly $f^{-1}(V) = \cup_{x \in f^{-1}(V)} U_x$, and this is an open set, as a union of open sets. $\square$

Proof. We now turn to the proof of Theorem 2.1.7. Necessity: Take $V = B(f(x), \varepsilon)$. Since $f^{-1}(V)$ is open and $x \in f^{-1}(V)$, there exists a non-empty open ball $B(x, \delta) \subseteq f^{-1}(V)$. Therefore $f(B(x, \delta)) \subseteq V = B(f(x), \varepsilon)$.

For sufficiency, apply Lemma B.7.3: let $x \in X$ and let $V \in \mathcal{V}$ be an open set containing $f(x)$. Since V is an open set, there exists a non-empty open ball $B(f(x), \varepsilon)$ contained in V. There exists an open set U_x (namely $B(x, \delta)$, where δ is as in the statement of the theorem) containing x and such that $f(U_x) \subseteq B(f(x), \varepsilon)$. $\square$

B.8 Change of Time

Theorem B.8.1 *Let $a : \mathbb{R}_+ \to \bar{\mathbb{R}}_+$ be non-decreasing right-continuous and let*

$$c(t) := \inf\{s; a(s) > t\} \quad (t \geq 0).$$

The function $c : \mathbb{R}_+ \to \bar{\mathbb{R}}_+$ is non-decreasing, right-continuous, and $c(t) < \infty$ if and only if $a(+\infty) > t$. Also,

$$a(t) = \inf\{s; c(s) > t\} \quad (t \geq 0).$$

Suppose in addition that the function a takes only finite values and that $a(0) = 0$. Then, for all measurable non-negative functions $f : \mathbb{R}_+ \to \mathbb{R}_+$,

$$\int_0^\infty f(t)\, \mathrm{d}a(t) = \int_0^{a(\infty)} f(c(t))\, \mathrm{d}t = \int_0^\infty 1_{\{c(t) < +\infty\}} f(c(t))\, \mathrm{d}t. \tag{B.6}$$

Proof. Only the change of variable formula is not a classical result. Since there is no mass at 0 for the measure associated with a and the Lebesgue measure, it suffices to prove it for functions f that are indicators of sets of the type $(0, u]$. But in this case, $\int_0^\infty f(t)\, \mathrm{d}a(t) = a(u)$ and $\int_0^{a(\infty)} f(c(t))\, \mathrm{d}t$ is the length of the interval $\{t; c(t) \leq u\}$, that is, $\inf\{t; c(t) > u\}$, which is indeed equal to $a(u)$. $\square$

Corollary B.8.2 *Let* $a : \mathbb{R}_+ \to \mathbb{R}_+$ *be non-decreasing* continuous *function such tha* $a(0) = 0$. *Then, for all measurable functions* $g : \mathbb{R}_+ \to \mathbb{R}_+$

$$\int_0^\infty g((a(t))\,\mathrm{d}a(t) = \int_0^{a(\infty)} g(t)\,\mathrm{d}t. \tag{B.7}$$

Proof. Observe that since a is continuous, $a(c(t)) = t$ when $t \in [0, a(+\infty))$ and apply formula (B.6) with $f = g \circ a$.

$\square$

Appendix C

Hilbert Spaces

C.1 Basic Definitions

Let H be a vector space with scalar field $K = \mathbb{C}$ or $\mathbb{R}$, endowed with a map $(x, y) \in H \times H \to \langle x, y \rangle \in K$ such that for all $x, y, z \in H$ and all $\lambda \in K$,

1. $\langle y, x \rangle = \langle x, y \rangle^*$

2. $\langle \lambda y, x \rangle = \lambda \langle y, x \rangle$

3. $\langle x, y + z \rangle = \langle x, y \rangle + \langle x, z \rangle$

4. $\langle x, x \rangle \geq 0$; and $\langle x, x \rangle = 0$ if and only if $x = 0$.

Then H is called a *pre-Hilbert space* over K and $\langle x, y \rangle$ is called the *inner product* of x and y. For any $x \in E$, define
$$\|x\|^2 = \langle x, x \rangle.$$

The *parallelogram identity*
$$\|x\|^2 + \|y\|^2 = \frac{1}{2}(\|x + y\|^2 + \|x - y\|^2)$$

is obtained by expanding the right-hand side and using the equality
$$\|x + y\|^2 = \|x\|^2 + \|y\|^2 + 2\mathrm{Re}\left\{\langle x, y \rangle\right\}.$$

The *polarization identity*
$$\langle x, y \rangle = \frac{1}{4}\left\{\|x + y\|^2 - \|x - y\|^2 + i\|x + iy\|^2 - i\|x - iy\|^2\right\}$$

is checked by expanding the right-hand side. It shows in particular that two inner products $\langle \cdot, \cdot \rangle_1$ and $\langle \cdot, \cdot \rangle_2$ on E such that $\|\cdot\|_1 = \|\cdot\|_2$ are identical.

C.2 Schwarz's Inequality

Theorem C.2.1 *For all $x, y \in H$,*
$$|\langle x, y \rangle| \leq \|x\| \times \|y\|.$$

Equality occurs if and only if x and y are colinear.

© Springer Nature Switzerland AG 2020
P. Brémaud, *Probability Theory and Stochastic Processes*, Universitext,
https://doi.org/10.1007/978-3-030-40183-2

Proof. Say $K = \mathbb{C}$. If x and y are colinear, that is, $x = \lambda y$ for some $\lambda \in \mathbb{C}$, th
inequality is obviously an equality. If x and y are linearly independent, then for al
$\lambda \in \mathbb{C}$, $x + \lambda y \neq 0$. Therefore

$$0 < \|x + \lambda y\|^2 = \|x\|^2 + |\lambda y|^2\|\lambda y\|^2 + \lambda^*\langle x, y\rangle + \lambda\langle x, y\rangle^*$$
$$= \|x\|^2 + |\lambda|^2\|y\|^2 + 2\mathrm{Re}(\lambda^*\langle x, y\rangle).$$

Take $u \in \mathbb{C}$, $|u| = 1$, such that $u^*\langle x, y\rangle = |\langle x, y\rangle|$. Take any $t \in \mathbb{R}$ and put $\lambda = tu$. Then

$$0 < \|x\|^2 + t^2\|y\|^2 + 2t|\langle x, y\rangle|.$$

This being true for all $t \in \mathbb{R}$, the discriminant of the second degree polynomial in t of
the right-hand side must be strictly negative, that is, $4|\langle x, y\rangle|^2 - 4\|x\|^2 \times \|y\|^2 < 0$. $\square$

Theorem C.2.2 *The mapping $x \to \|x\|$ is a norm on E, that is to say, for all $x, y \in E$,
and all $\alpha \in \mathbb{C}$,*

(a) $\|x\| \geq 0$; and $\|x\| = 0$ if and only if $x = 0$,

(b) $\|\alpha x\| = |\alpha|\,\|x\|$, and

(c) $\|x + y\| \leq \|x\| + \|y\|$ (triangle inequality).

Proof. The proof of (a) and (b) is immediate. For (c) write

$$\|x + y\|^2 = \|x\|^2 + \|y\|^2 + \langle x, y\rangle + \langle y, x\rangle$$

and

$$(\|x\| + \|y\|)^2 = \|x\|^2 + \|y\|^2 + 2\|x\|\|y\|.$$

It therefore suffices to prove

$$\langle x, y\rangle + \langle y, x\rangle = 2\mathrm{Re}(\langle x, y\rangle) \leq 2\,\|x\|\|y\|,$$

which follows from Schwarz's inequality.

$\square$

The norm $\|\cdot\|$ induces a *metric* $d(\cdot, \cdot)$ on H by

$$d(x, y) = \|x - y\|.$$

Recall that a mapping $d : E \times E \to \mathbb{R}_+$ is called a *metric* on E if, for all $x, y, z \in E$,

(a') $d(x, y) \geq 0$; and $d(x, y) = 0$ if and only if $x = y$,

(b') $d(x, y) = d(y, x)$, and

(c') $d(x, y) \geq d(x, z) + d(z, y)$.

The above properties are immediate consequences of (a), (b), and (c) of Theorem C.2.2.
When endowed with a metric, a space H is called a *metric space*.

Definition C.2.3 *A pre-Hilbert space H is called a* Hilbert space *if it is a complete
metric space with respect to the metric d.*

By this, the following is meant: If $\{x_n\}_{n \geq 1}$ is a Cauchy sequence in H, that is, if $\lim_{m,n \uparrow \infty} d(x_m, x_n) = 0$, then there exists an $x \in H$ such that $\lim_{n \uparrow \infty} d(x_n, x) = 0$.

Theorem C.2.4 *Let $\{x_n\}_{n \geq 1}$ and $\{y_n\}_{n \geq 1}$ be sequences in a Hilbert space H that converge to x and y, respectively. Then,*

$$\lim_{m,n \uparrow \infty} \langle x_n, y_m \rangle = \langle x, y \rangle.$$

In other words, the inner product of a Hilbert space is *bicontinuous*. In particular, the norm $x \mapsto \|x\|$ is a continuous function from H to $\mathbb{R}_+$.

Proof. We have for all h_1, h_2 in H,

$$|\langle x + h_1, y + h_2 \rangle - \langle x, y \rangle| = |\langle x, h_2 \rangle + \langle h_1, y \rangle + \langle h_1, h_2 \rangle|.$$

By Schwarz's inequality $|\langle x, h_2 \rangle| \leq \|x\|\|h_2\|$, $|\langle h_1, y \rangle| \leq \|y\|\|h_1\|$, and $|\langle h_1, h_2 \rangle| \leq \|h_1\|\|h_2\|$. Therefore

$$\lim_{\|h_1\|, \|h_2\| \downarrow 0} |\langle x + h_1, y + h_2 \rangle - \langle x, y \rangle| = 0.$$

$\square$

C.3   Isometric Extension

Definition C.3.1 *Let H and K be two Hilbert spaces with inner products denoted by $\langle \cdot, \cdot \rangle_H$ and $\langle \cdot, \cdot \rangle_K$, respectively, and let $\varphi : H \mapsto K$ be a linear mapping such that for all $x, y \in H$*

$$\langle \varphi(x), \varphi(y) \rangle_K = \langle x, y \rangle_H.$$

Then, φ is called a linear isometry *from H into K. If, moreover, φ is from H onto K, then H and K are said to be* isomorphic.

Note that a linear isometry is necessarily injective, since $\varphi(x) = \varphi(y)$ implies $\varphi(x - y) = 0$, and therefore

$$0 = \|\varphi(x - y)\|_K = \|x - y\|_H,$$

which implies $x = y$. In particular, if the linear isometry is *onto*, it is bijective.

Recall that a subset $A \in E$, where (E, d) is a metric space, is said to be *dense* in E if, for all $x \in E$, there exists a sequence $\{x_n\}_{n \geq 1}$ in A converging to x.

Theorem C.3.2 *Let H and K be two Hilbert spaces with inner products denoted by $\langle \cdot, \cdot \rangle_H$ and $\langle \cdot, \cdot \rangle_K$, respectively. Let V be a vector subspace of H that is dense in H, and $\varphi : V \mapsto K$ be a linear isometry from V to K. Then, there exists a unique linear isometry $\widetilde{\varphi} : H \mapsto K$ whose restriction to V is φ.*

Proof. We shall first define $\widetilde{\varphi}(x)$ for $x \in H$. Since V is dense in H, there exists a sequence $\{x_n\}_{n \geq 1}$ in V converging to x. Since φ is isometric,

$$\|\varphi(x_n) - \varphi(x_m)\|_K = \|x_n - x_m\|_H \quad \text{for all } m, n \geq 1.$$

In particular, $\{\varphi(x_n)\}_{n\geq 1}$ is a Cauchy sequence in K and therefore it converges to som element of K, which we denote by $\widetilde{\varphi}(x)$.

The definition of $\widetilde{\varphi}(x)$ is independent of the sequence $\{x_n\}_{n\geq 1}$ converging to x Indeed, for another such sequence $\{y_n\}_{n\geq 1}$,

$$\lim_{n\uparrow\infty}\|\varphi(x_n) - \varphi(y_n)\|_K = \lim_{n\uparrow\infty}\|x_n - y_n\|_H = 0.$$

The mapping $\widetilde{\varphi} : H \mapsto K$ so constructed is clearly an extension of φ (for $x \in V$ one can take for an approximating sequence of x the sequence $\{x_n\}_{n\geq 1}$ such that $x_n \equiv x$).

The mapping $\widetilde{\varphi}$ is linear. Indeed, let $x, y \in H$, $\alpha, \beta \in \mathbb{C}$, and let $\{x_n\}_{n\geq 1}$ and $\{y_n\}_{n\geq 1}$ be two sequences in V converging to x and y, respectively. Then $\{\alpha x_n + \beta y_n\}_{n\geq 1}$ converges to $\alpha x + \beta y$. Therefore

$$\lim_{n\uparrow\infty}\varphi(\alpha x_n + \beta y_n) = \widetilde{\varphi}(\alpha x + \beta y).$$

But

$$\varphi(\alpha x_n + \beta y_n) = \alpha\varphi(x_n) + \beta\varphi(y_n) \to \alpha\widetilde{\varphi}(x) + \beta\widetilde{\varphi}(y)$$

tends to $\widetilde{\varphi}(\alpha x + \beta y) = \alpha\widetilde{\varphi}(x) + \beta\widetilde{\varphi}(y)$.

The mapping $\widetilde{\varphi}$ is isometric since, in view of the bicontinuity of the inner product and of the isometricity of φ, if $\{x_n\}_{n\geq 1}$ and $\{y_n\}_{n\geq 1}$ are two sequences in V converging to x and y, respectively, then

$$\langle\widetilde{\varphi}(x), \widetilde{\varphi}(y)\rangle_K = \lim_{n\uparrow\infty}\langle\varphi(x_n), \varphi(y_n)\rangle_K$$
$$= \lim_{n\uparrow\infty}\langle x_n, y_n\rangle_H = \langle x, y\rangle_H.$$

$\square$

C.4 Orthogonal Projection

Let H be a Hilbert space. A set $G \subseteq H$ is said to be closed in H if every convergent sequence of G has a limit in G.

Theorem C.4.1 *Let $G \subseteq H$ be a vector subspace of the Hilbert space H. Endow G with the inner product which is the restriction to G of the inner product on H. Then, G is a Hilbert space if and only if G is closed in H.*

G is then called a *Hilbert subspace* of H.

Proof. (i) Assume that G is closed. Let $\{x_n\}_{n\in\mathbb{N}}$ be a Cauchy sequence in G. It is *a fortiori* a Cauchy sequence in H, and therefore it converges in H to some x, and this x must be in G, because it is a limit of elements of G and G is closed.

(ii) Assume that G is a Hilbert space with the inner product induced by the inner product of H. In particular, every convergent sequence $\{x_n\}_{n\in\mathbb{N}}$ of elements of G converges to some element of G. Therefore G is closed.

$\square$

Definition C.4.2 *Two elements x, y of the Hilbert space H are said to be orthogonal if $\langle x, y \rangle = 0$. Let G be a Hilbert subspace of the Hilbert space H. The orthogonal complement of G in H, denoted $G^{\perp}$, is defined by*

$$G^{\perp} = \{z \in H : \langle z, x \rangle = 0 \text{ for all } x \in G\}.$$

Clearly, $G^{\perp}$ is a vector space over $\mathbb{C}$. Moreover, it is closed in H since if $\{z_n\}_{n \geq 1}$ is a sequence of elements of $G^{\perp}$ converging to $z \in H$ then, by continuity of the inner product,

$$\langle z, x \rangle = \lim_{n \uparrow \infty} \langle z_n, x \rangle = 0 \quad \text{for all } x \in H.$$

Therefore $G^{\perp}$ is a Hilbert subspace of H.

Observe that a decomposition $x = y + z$, where $y \in G$ and $z \in G^{\perp}$, is necessarily unique. Indeed, let $x = y' + z'$ be another such decomposition. Then, letting $a = y - y'$, $b = z - z'$, we have that $0 = a + b$, where $a \in G$ and $b \in G^{\perp}$. Therefore, in particular, $0 = \langle a, a \rangle + \langle a, b \rangle$. But $\langle a, b \rangle = 0$, and therefore $\langle a, a \rangle = 0$, which implies that $a = 0$. Similarly, $b = 0$.

Theorem C.4.3 *Let G be a Hilbert subspace of H. For all $x \in H$, there exists a unique element $y \in G$ such that $x - y \in G^{\perp}$. Moreover,*

$$\|y - x\| = \inf_{u \in G} \|u - x\|. \tag{C.1}$$

Proof. Let $d(x, G) = \inf_{z \in G} d(x, z)$ and let $\{y_n\}_{n \geq 1}$ be a sequence in G such that

$$d(x, G)^2 \leq d(x, y_n)^2 \leq d(x, G)^2 + \frac{1}{n}. \tag{$\star$}$$

The parallelogram identity gives, for all $m, n \geq 1$,

$$\|y_n - y_m\|^2 = 2(\|x - y_n\|^2 + \|x - y_m\|^2) - 4\|x - \frac{1}{2}(y_m + y_n)\|^2.$$

Since $\frac{1}{2}(y_n + y_m) \in G$,

$$\|x - \frac{1}{2}(y_m + y_n)\|^2 \geq d(x, G)^2,$$

and therefore

$$\|y_n - y_m\|^2 \leq 2\left(\frac{1}{n} + \frac{1}{m}\right).$$

The sequence $\{y_n\}_{n \geq 1}$ is therefore a Cauchy sequence in G and consequently it converges to some $y \in G$ since G is closed. Passing to the limit in $(\star)$ gives (C.1).

Uniqueness of y satisfying (C.1): Let $y' \in G$ be another such element. Then

$$\|x - y'\| = \|x - y\| = d(x, G),$$

and from the parallelogram identity

$$\|y - y'\|^2 = 2\|y - x\|^2 + 2\|y' - x\|^2 - 4\|x - \frac{1}{2}(y + y')\|^2$$

$$= 4d(x, G)^2 - 4\|x - \frac{1}{2}(y + y')\|^2.$$

Since $\frac{1}{2}(y + y') \in G$

$$\|x - \frac{1}{2}(y + y')\|^2 \geq d(x, G)^2,$$

and therefore $\|y - y'\|^2 \leq 0$, which implies $\|y - y'\|^2 = 0$ and therefore $y = y'$.

It now remains to show that $x - y$ is orthogonal to G, that is, $\langle x - y, z \rangle = 0$ for all $z \in G$. Since this is trivially true if $z = 0$, we may assume $z \neq 0$. Because $y + \lambda z \in G$ for all $\lambda \in \mathbb{R}$

$$\|x - (y + \lambda z)\|^2 \geq d(x, G)^2,$$

that is,

$$\|x - y\|^2 + 2\lambda \text{Re}\left\{\langle x - y, z \rangle\right\} + \lambda^2 \|z\|^2 \geq d(x, G)^2.$$

Since $\|x - y\|^2 = d(x, G)^2$, we have

$$- 2\lambda \text{Re}\left\{\langle x - y, z \rangle\right\} + \lambda^2 \|z\|^2 \geq 0 \quad \text{for all } \lambda \in \mathbb{R},$$

which implies $\text{Re}\left\{\langle x - y, z \rangle\right\} = 0$. The same type of calculation with $\lambda \in i\mathbb{R}$ (pure imaginary) leads to $\Im\left\{\langle x - y, z \rangle\right\} = 0$. Therefore $\langle x - y, z \rangle = 0$.

That y is the unique element of G such that $y - x \in G^\perp$ follows from the remark preceding Theorem C.4.3.

$\square$

Definition C.4.4 *The element y in Theorem C.4.3 is called the* orthogonal projection *of x on G and is denoted by $P_G(x)$.*

The projection theorem states, in particular, that for any $x \in G$ there is a unique decomposition

$$x = y + z, \quad y \in G, \ z \in G^\perp,$$

and that $y = P_G(x)$, the (unique) element of G closest to x. Therefore

Theorem C.4.5 *The orthogonal projection $y = P_G(x)$ is characterized by the following two properties:*

(1) $y \in G$;

(2) $\langle y - x, z \rangle = 0$ for all $z \in G$.

This characterization is called the *projection principle*.

Let C be a collection of vectors in the Hilbert space H. The linear span of C, denoted $\text{span}(C)$ is, by definition, the set of all finite linear combinations of vectors of C. This is a vector space. The closure of this vector space, $\overline{\text{span}(C)}$, is called the Hilbert subspace generated by C. By definition, x belongs to this subspace if and only if there exists a sequence of vectors $\{x_n\}_{n \geq 1}$ such that

(i) for all $n \geq 1$, x_n is a finite linear combination of vectors of C, and

(ii) $\lim_{n \uparrow \infty} x_n = x$.

Theorem C.4.6 *An element $\widehat{x} \in H$ is the projection of x onto $G = \overline{\text{span}(C)}$ if and only if*

(α) $\widehat{x} \in G$, *and*

(β) $\langle x - \widehat{x}, z \rangle = 0$ *for all $z \in C$.*

Note that we have to satisfy requirement (β) not for all $z \in G$, but only for all $z \in C$.

We have to show that $\langle x - \widehat{x}, z \rangle = 0$ for all $z \in G$. But $z = \lim_{n \uparrow \infty} z_n$ where $\{z_n\}_{n \geq 1}$ is a sequence of vectors of $\text{span}(C)$ such that $\lim_{n \uparrow \infty} z_n = z$. By hypothesis, for all $n \geq 1$, $\langle x - \widehat{x}, z_n \rangle = 0$. Therefore, by continuity of the inner product,

$$\langle x - \widehat{x}, z \rangle = \lim_{n \uparrow \infty} \langle x - \widehat{x}, z_n \rangle = 0.$$

Theorem C.4.7 *Let G be a Hilbert subspace of the Hilbert space H.*

(α) *The mapping $x \to P_G(x)$ is linear and continuous, and furthermore*

$$\|P_G(x)\| \leq \|x\| \quad \text{for all } x \in H.$$

(β) *If F is a Hilbert subspace of H such that $F \subseteq G$ then $P_F \circ P_G = P_F$. In particular $P_G^2 = P_G$ (P_G is then called idempotent).*

Proof. (α) Let $x_1, x_2 \in H$. They admit the decomposition

$$x_i = P_G(x_i) + w_i \quad (i = 1, 2),$$

where $w_i \in G^\perp$ $(i = 1, 2)$. Therefore

$$x_1 + x_2 = P_G(x_1) + P_G(x_2) + w_1 + w_2$$
$$= P_G(x_1) + P_G(x_2) + w,$$

where $w \in G^\perp$. Now, $x_1 + x_2$ admits a unique decomposition of the type

$$x_1 + x_2 = y + w,$$

where $w \in G^\perp$, $y \in G$, namely: $y = P_G(x_1 + x_2)$. Therefore

$$P_G(x_1 + x_2) = P_G(x_1) + P_G(x_2).$$

One proves similarly that

$$P_G(\alpha x) = \alpha P_G(x) \quad \text{for all } \alpha \in G, \ x \in H.$$

Thus P_G is linear.

From Pythagoras' theorem applied to $x = P_G(x) + w$,

$$\|P_G(x)\| + \|w\|^2 = \|x\|^2,$$

and therefore

$$\|P_G(x)\|^2 \leq \|x\|^2.$$

Hence, P_G is continuous.

(β) The unique decompositions of x on G and $G^\perp$ and of $P_G(x)$ on F and $F^\perp$ are respectively

$$x = P_G(x) + w, \quad P_G(x) = P_F(P_G(x)) + z.$$

From these two equalities, we obtain

$$x = P_F(P_G(x)) + z + w. \tag{$\star$}$$

But $z \in G^\perp$ implies $z \in F^\perp$ since $F \subseteq G$, and therefore $v = z + w \in F^\perp$. On the other hand, $P_F(P_G(x)) \in F$. Therefore $(\star)$ is the unique decomposition of x on F and $F^\perp$; in particular, $P_F(x) = P_F(P_G(x))$.

$\square$

The next result says that the projection operator P_G is "continuous" with respect to G.

Theorem C.4.8

(i) Let $\{G_n\}_{n \geq 1}$ be a non-decreasing sequence of Hilbert subspaces of H. Then, the closure G of $\bigcup_{n \geq 1} G_n$ is a Hilbert subspace of H and for all $x \in H$

$$\lim_{n \uparrow \infty} P_{G_n}(x) = P_G(x).$$

(ii) Let $\{G_n\}$ be a non-increasing sequence of Hilbert subspaces of H. Then $G := \bigcap_{n \geq 1} G_n$ is a Hilbert subspace of H and for all $x \in H$

$$\lim_{n \uparrow \infty} P_{G_n}(x) = P_G(x).$$

Proof. (i) The set $\bigcup_{n \geq 1} G_n$ is evidently a vector subspace of H (however, it is not closed in general). Its closure G is a Hilbert subspace (Theorem C.4.1). To any $y \in G$ one can associate a sequence $\{y_n\}_{n \geq 1}$, where $y_n \in G_n$, and

$$\lim_{n \to \infty} \|y - y_n\| = 0.$$

Take $y = P_G(x)$. By the parallelogram identity

$$
\begin{aligned}
\|P_{G_n}(x) - P_G(x)\|^2 &= \|(x - P_G(x)) - (x - P_{G_n}(x))\|^2 \\
&= 2\|x - P_{G_n}(x)\|^2 + 2\|x - P_G(x)\|^2 \\
&\quad - 4\|x - \tfrac{1}{2}(P_{G_n}(x) + P_G(x))\|.
\end{aligned}
$$

But since $P_{G_n}(x) + P_G(x)$ is a vector in G,

$$\|x - \tfrac{1}{2}(P_{G_n}(x) + P_G(x))\|^2 \geq \|x - P_G(x)\|^2,$$

and therefore

$$
\begin{aligned}
\|P_{G_n}(x) - P_G(x)\|^2 &\leq 2\|x - P_{G_n}(x)\|^2 - 2\|x - P_G(x)\|^2 \\
&\leq 2\|x - y_n\|^2 - 2\|x - P_G(x)\|^2.
\end{aligned}
$$

By the continuity of the norm

$$\lim_{n\uparrow\infty} \|x - y_n\|^2 = \|x - P_G(x)\|^2,$$

and finally

$$\lim_{n\uparrow\infty} \|P_{G_n}(x) - P_G(x)\|^2 = 0.$$

(ii) Devise a direct proof in the spirit of (i) or use the fact that $G^\perp$ is the closure of $\bigcup_{n\geq 1} G_n^\perp$. □

If G_1 and G_2 are orthogonal Hilbert subspaces of the Hilbert space H,

$$G_1 \oplus G_2 := \{z = x_1 + x_2 : x_1 \in G, \, x_2 \in G_2\}$$

is called the *orthogonal sum* of G_1 and G_2.

C.5   Riesz's Representation Theorem

Definition C.5.1 *Let H be a Hilbert space over $\mathbb{C}$ with inner product $(x, y) \in H \rightarrow$ $\langle x, y \rangle$. The linear mapping $f : H \rightarrow \mathbb{C}$ is said to be continuous if there exists an $A \geq 0$ such that*

$$|f(x_1) - f(x_2)| \leq A\|x_1 - x_2\| \quad \text{for all } x_1, x_2 \in H. \tag{C.2}$$

The infimum of the constants A satisfying (C.2) is called the norm *of f.*

Theorem C.5.2 *Let $f : H \mapsto \mathbb{C}$ be a continuous linear form on the Hilbert space H. Then there exists a unique $y \in H$ such that*

$$f(x) = \langle y, x \rangle.$$

Proof. Uniqueness: Let $y, y' \in H$ be such that

$$f(x) = \langle x, y \rangle = \langle x, y' \rangle \quad \text{for all } x \in H.$$

In particular,

$$\langle x, y - y' \rangle = 0 \quad \text{for all } x \in H.$$

The choice $x = y - y'$ leads to $\|y - y'\|^2 = 0$, that is, $y - y'$.

Existence: The kernel $N := \{u \in H : f(u) = 0\}$ of f is a Hilbert subspace of H. Eliminating the trivial case $f \equiv 0$, it is strictly included in H. In particular, $N^\perp$ does not reduces to the singleton $\{0\}$ and therefore, there exists a $z \in N^\perp$, $z \neq 0$. Define y by

$$y = f(z)^* \frac{z}{\|z\|^2}.$$

For all $x \in N$, $\langle x, y \rangle = 0$, and therefore, in particular, $\langle x, y \rangle = f(x)$. Also,

$$\langle z, y \rangle = \langle\langle z, f(z)^* \frac{z}{\|z\|^2}\rangle\rangle = f(z).$$

Therefore the mappings $x \rightarrow f(x)$ and $x \rightarrow \langle x, y \rangle$ coincide on the Hilbert subspace generated by N and z. But this subspace is H itself. In fact, for all $x \in H$,

$$x = \left(x - \frac{f(x)}{f(z)} z\right) + \frac{f(x)}{f(z)} z = u + w,$$

where $u \in N$ and w is colinear with z. □

C.6   Orthonormal expansions

Definition C.6.1 *A sequence $\{e_n\}_{n\geq 0}$ in a Hilbert space H is called an* orthonormal *system of H if it satisfies the conditions*

(α) $\langle e_n, e_k \rangle = 0$ *for all $n \neq k$; and*

(β) $\|e_n\| = 1$ *for all $n \geq 0$.*

If only requirement (α) is met, the sequence is called an orthogonal *system.*

An orthonormal system $\{e_n\}_{n\geq 0}$ is *free* in the sense that a *finite* subset of it is linearly independent. Indeed, taking $(e_1, \ldots, e_k)$ for example, the relation $\sum_{i=1}^{k} \alpha_i e_i = 0$ implies that $\alpha_\ell = \langle e_\ell, \sum_{i=1}^{k} \alpha_i e_i \rangle = 0$ for all $1 \leq \ell \leq k$.

Let $\{e_n\}_{n\geq 0}$ be an orthonormal system of H and let G be the Hilbert subspace of H generated by it.

Theorem C.6.2

(a) *For ay sequence $\{\alpha_n\}_{n\geq 0}$ of complex numbers the series $\sum_{n\geq 0} \alpha_n e_n$ is convergent in H if and only if $\{\alpha_n\}_{n\geq 0} \in \ell_{\mathbb{C}}^2$, in which case*

$$\left\| \sum_{n\geq 0} \alpha_n e_n \right\|^2 = \sum_{n\geq 0} |\alpha_n|^2. \tag{C.3}$$

(b) *For all $x \in H$, we have* Bessel's *inequality:*

$$\sum_{n\geq 0} |\langle x, e_n \rangle|^2 \leq \|x\|^2. \tag{C.4}$$

(c) *For all $x \in H$ the series $\sum_{n\geq 0}\langle x, e_n \rangle e_n$ converges, and*

$$\sum_{n\geq 0} \langle x, e_n \rangle e_n = P_G(x), \tag{C.5}$$

where P_G is the projection on G.

(d) *For all $x, y \in H$ the series $\sum_{n\geq 0}\langle x, e_n \rangle \langle y, e_n \rangle$ is absolutely convergent, and*

$$\sum_{n\geq 0} \langle x, e_n \rangle \langle y, e_n \rangle^* = \langle P_G(x), P_G(y) \rangle. \tag{C.6}$$

Proof. (a) From Pythagoras' theorem we have

$$\left\| \sum_{j=m+1}^{n} \alpha_j e_j \right\|^2 = \sum_{j=m+1}^{n} |\alpha_j|^2,$$

and therefore the sequence $\{\sum_{j=0}^{n} \alpha_j e_j\}_{n\geq 0}$ is a Cauchy sequence in H if and only if $\{\sum_{j=0}^{n} |\alpha_j|^2\}_{n\geq 0}$ is a Cauchy sequence in $\mathbb{R}$. In other words, $\sum_{n\geq 0} \alpha_n e_n$ converges if and only if $\sum_{n\geq 0}|\alpha_n|^2 < \infty$. In this case equality (C.3) follows from the continuity of the norm, by letting n tend to ∞ in the last display.

(b) According to (α) of Theorem C.4.7, $\|x\| \geq \|P_{G_n}(x)\|$, where G_n is the Hilbert subspace spanned by $\{e_1, \ldots, e_n\}$. But

$$P_{G_n}(x) = \sum_{i=0}^{n} \langle x, e_i \rangle e_i,$$

and by Pythagoras' theorem

$$\|P_{G_n}(x)\|^2 = \sum_{i=0}^{n} |\langle x, e_i \rangle|^2.$$

Therefore

$$\|x\|^2 \geq \sum_{i=0}^{n} |\langle x, e_i \rangle|^2,$$

from which Bessel's inequality follows by letting $n \uparrow \infty$.

(c) From (C.4) and (a), the series $\sum_{n\geq 0}\langle x, e_n \rangle e_n$ converges. For any $m \geq 0$ and for all $N \geq m$

$$\langle x - \sum_{n=0}^{N} \langle x, e_n \rangle e_n, e_m \rangle = 0,$$

and therefore, by continuity of the inner product,

$$\langle x - \sum_{n\geq 0} \langle x, e_n \rangle e_n, e_m \rangle = 0 \quad \text{for all } m \geq 0.$$

This implies that $x - \sum_{n\geq 0}\langle x, e_n \rangle e_n$ is orthogonal to G. Also $\sum_{n\geq 0}\langle x, e_n \rangle e_n \in G$. Therefore, by the projection principle,

$$P_G(x) = \sum_{n\geq 0} \langle x, e_n \rangle e_n.$$

(d) By Schwarz's inequality in $\ell_{\mathbb{C}}^2$, for all $N \geq 0$

$$\left(\sum_{n=0}^{N} |\langle x, e_n \rangle \langle y, e_n \rangle^*| \right)^2 \leq \left(\sum_{n=0}^{N} |\langle x, e_n \rangle|^2 \right) \left(\sum_{n=0}^{N} |\langle y, e_n \rangle|^2 \right)$$

$$\leq \|x\|^2 \|y\|^2.$$

Therefore the series $\sum_{n=0}^{\infty} \langle x, e_n \rangle \langle y, e_n \rangle^*$ is absolutely convergent. Also, by an elementary computation,

$$\langle \sum_{n=0}^{N} \langle x, e_n \rangle e_n, \sum_{n=0}^{N} \langle y, e_n \rangle e_n \rangle = \sum_{n=0}^{N} \langle x, e_n \rangle \langle y, e_n \rangle^*.$$

Letting $N \uparrow \infty$ we obtain (C.6) (using (C.5) and the continuity of the inner product). $\square$

Definition C.6.3 *A sequence $\{w_n\}_{n\geq 0}$ in H is said to be total in H if it generates H (that is, the finite linear combinations of the elements of $\{w_n\}_{n\geq 0}$ form a dense subset of H.)*

A sequence $\{w_n\}_{n\geq 0}$ in H is total in H if and only if

$$\langle z, w_n \rangle = 0 \text{ for all } n \geq 0 \Rightarrow z = 0.$$

(†

The proof is left as an exercise.

Theorem C.6.4 *Let* $\{e_n\}_{n\geq 0}$ *be an orthonormal system of* H. *The following properties are equivalent:*

(a) $\{e_n\}_{n\geq 0}$ *is total in* H;

(b) *For all* $x \in H$, *the Plancherel–Parseval identity holds true:*

$$\|x\|^2 = \sum_{n\geq 0} |\langle x, e_n \rangle|^2;$$

(c) *For all* $x \in H$,

$$x = \sum_{n\geq 0} \langle x, e_n \rangle e_n.$$

Proof. (a)$\Rightarrow$(c) According to (c) of Theorem C.6.2,

$$\sum_{n\geq 0} \langle x, e_n \rangle e_n = P_G(x),$$

where G is the Hilbert subspace generated by $\{e_n\}_{n\geq 0}$. Since $\{e_n\}_{n\geq 0}$ is total, it follows by (†) that $G^\perp = \{0\}$, and therefore $P_G(x) = x$.

(c)$\Rightarrow$(b) This follows from (a) of Theorem C.6.2.

(b)$\Rightarrow$(a) From (C.3) and (C.5),

$$\sum_{n\geq 0} |\langle x, e_n \rangle|^2 = \|P_G(x)\|^2,$$

and therefore (b) implies $\|x\|^2 = \|P_G(x)\|^2$. From Pythagoras' theorem

$$\begin{aligned} \|x\|^2 &= \|P_G(x) + x - P_G(x)\|^2 \\ &= \|P_G(x)\|^2 + \|x - P_G(x)\|^2 \\ &= \|x\|^2 + \|x - P_G(x)\|^2, \end{aligned}$$

and therefore $\|x - P_G(x)\|^2 = 0$, which implies $x = P_G(x)$. Since this is true for all $x \in H$ we must have $G \equiv H$, *i.e.* $\{e_n\}_{n\geq 0}$ is total in H. $\square$

A sequence $\{e_n\}_{n\geq 0}$ satisfying one (and hence all) of the conditions of Theorem C.6.4 is called a (denumerable) *Hilbert basis* of H.

Bibliography

Aldous, D., and J.A. Fill, *Reversible Markov Chains and Random Walks on Graphs*, http://www.stat.berkeley.edu/~aldous/RWG/book.html. (2002, 2014).

Alon, N., and J.H. Spencer, *The Probabilistic Method*, Wiley, (1991, 3rd ed. 2010).

Asmussen, S., *Applied Probability and Queues*, Wiley (1987, 2003).

Athreya, K.B., and P.E. Ney, *Branching Processes*, Springer (1972); Dover (2004).

Baccelli, F., and P. Brémaud, *Elements of Queueing Theory*, Springer (1994, 2003).

Baldi, P., *Stochastic Calculus*, Springer (2018).

Berger, M.A., *An Introduction to Probability and Random Processes*, Springer (1993).

Billingsley, P., *Ergodic Theory and Information*, Wiley (1965).

Billingsley, P., *Convergence of Probability Measures*, Wiley (1968).

Billingsley, P., *Probability and Measure*, Wiley (1979, 5th ed. 2012).

Birkhoff, G.D., "Proof of the ergodic theorem", *Proc. Nat. Acad. U.S.A.* **17**, 656–660 (1931).

Blackwell, D., "A renewal theorem", *Duke Mathematical Journal*, **15**, 145–150 (1948).

Bochner, S., *Harmonic Analysis and the Theory of Probability*, Univ. of California Press (1955).

Brémaud, P., *Markov Chains*, Springer (1999), 2nd ed. (2020).

Brémaud, P., *Fourier Analysis and Stochastic Processes* (2014).

Brémaud, P., *Discrete Probability Models and Methods*, Springer (2017).

Brémaud, P., *Point Processes on the Line and in Space*, Springer (2020).

Brockwell, P.J., and R.A. Davis, *Time Series: Theory and Methods*, Springer (1991, second edition).

Burke, P.J., "The output of a queueing system", *Operations Research*, **4**, 6, 699–704 (1956).

Capetanakis, J.I., "Tree algorithm for packet broadcast channels", IEEE Transactions on Information Theory, IT-25, 505–515 (1979).

Çinlar, E., *Introduction to Stochastic Processes*, Prentice-Hall (1975).

© Springer Nature Switzerland AG 2020
P. Brémaud, *Probability Theory and Stochastic Processes*, Universitext,
https://doi.org/10.1007/978-3-030-40183-2

Çinlar, E., *Probability and Stochastics*, Springer (2014).

Cioczek-Georges, R., and B. Mandelbrot, "A class of micropulses and antipersisten fractal Brownian motion", *Stochastic Processes and their Applications*, 60, 1–18, (1995)

Cramér, H., and M.R. Leadbetter, *Stationary and Related Stochastic Processes*, Wiley (1967), Dover (1995).

Daley, D.J., and D. Vere-Jones, *An Introduction to the Theory of Point Processes*, vol. *I*, Springer (1988), 2nd ed. (2003).

Daley, D.J., and D. Vere-Jones, *An Introduction to the Theory of Point Processes*, vol. *II*, Springer (1988), 2nd ed. (2008).

Daley, D.J., and J. Gani, *Epidemic Modelling: An Introduction*, Cambridge University Press(1999).

Dembo, A., and O. Zeitouni, *Large Deviations Techniques and Applications*, Springer (1998, 2nd ed. 2010).

Doeblin, W., "Sur les propriétés asymptotiques de mouvements régis par certains types de chaînes simples", Bulletin Mathématique de la Société Roumaine des Sciences, 39, No. 1, 57–115, No. 2, 3–61 (1937).

Duflo, M., *Random Iterative Models*, Springer (1997).

Durrett, R., *Probability and Examples*, Duxbury Press (1996).

Durrett, R., *Essentials of Stochastic Processes*, Springer (1999, 2012, 2016).

Ethier, S.N. and T.G. Kurtz, *Markov Processes, Characterization and Convergence*, Wiley (1986).

Foss, S., and R.L. Tweedie, "Perfect simulation and backwards coupling", Stochastic Models, 14, 187–203 (1998).

Foster, F.G., "On the Stochastic Matrices Associated with Certain Queuing Processes", The Annals of Mathematical Statistics 24 (3) (1953).

Gordon, W.J., and G.F. Newell, "Closed Queueing Systems with Exponential Servers", *Operations Research*, **15**, 254–265 (1967).

Grandell, J., *Doubly Stochastic Point Processes*, Lecture Notes in Math. 529, Springer, Berlin (1976).

Grandell, J., *Aspects of Risk Theory*, Springer (1991).

Hammersley, J.M., and P. Clifford, "Markov fields on finite graphs and lattices", unpublished manuscript (1968).
http://www.statslab.cam.ac.uk/~grg/books/hammfest/hamm-cliff.pdf.

Harris, T.E., *The Theory of Branching Processes*, Dover, 1989.

Hoeffding, W., "Probability inequalities for sums of bounded random variables", *J. Amer. Statist. Assoc.* 58, 13–30 (1963).

Itô, K., and H.P. McKean, Jr., *Diffusion Processes and their Sample Paths, Classics in Mathematics*, Springer (1974).

Jackson, J.R., "Jobshop-like Queueing Systems", *Management Science*, **10**, 131–142 (1963).

Kallenberg, O., *Foundations of Modern Probability*, Springer (2002).

Kallenberg, O., *Random Measures*, Akademie Verlag (1975); further ed. Academic Press (1983, 2017).

Karatzas, I., and S.E. Shreve, *Brownian Motion and Stochastic Calculus*, Springer (1991, 1998).

Karlin, S., *A First Course in Stochastic Processes*, Academic Press (1966).

Karlin, S., and M.H. Taylor, *A First Course in Stochastic Processes*, Academic Press (1975).

Kelly, F.P., *Reversibility and Stochastic Networks*, Wiley (1979).

Kendall, D.G., "Stochastic Processes Occurring in the Theory of Queues and their Analysis by the Method of the Imbedded Markov Chain", *The Annals of Mathematical Statistics*, **24**, 3, 338–354 (1953).

Kleinrock, L., *Queueing Systems, I, II*, Wiley (1975, 1976).

Klenke, A., *Probability Theory; A Comprehensive Course*, Springer (2008).

Kuo, H.-H., *Introduction to Stochastic Integration*, Springer (2006).

Last, G., and A. Brandt, *Marked Point Process on the Real Line*, Springer (1995).

Last, G., and M. Penrose, *Lectures on the Poisson Process*, Cambridge University Press(2018).

Le Cam, L., "An approximation theorem for the Poisson binomial distribution", Pacific J. Math, **10**, 1181–1197 (1960).

Le Gall, J.-F. *Brownian Motion, Martingales, and Stochastic Calculus*, Springer (2016).

Levin, D.A., Peres, Y., and E.L. Wilmer, *Markov Chains and Mixing Times*, American Mathematical Society (2009).

Liptser, R.S., and A.N., Shiryaev, *Statistics of Random Processes, I General Theory*, Springer (1978, 2001).

Liptser, R.S., and A.N., Shiryaev, *Statistics of Random Processes, II Applications*, Springer (1978).

Little, J.D.C., *A Proof for the Queueing Formula* $L = \lambda W$, Oper. Res. **9**, 383–387 (1961).

Lindvall, T., *Lectures on the Coupling Method*, Wiley (1992).

Loynes, R.M., "The stability of queues with non-independent inter-arrival and service times", Proc. Camb. Phil. Soc., **58**, 444–477 (1962).

Markov, A.A., "Extension of the law of large numbers to dependent events" (in Russian), Bull. Soc. Phys. Math. Kazan, 2, 15, 155–156 (1906).

Mandelbrot, B., and J.W. Van Ness, *Fractional Brownian Motions, Fractional Noises and Applications*, SIAM review, Vol. 10, **4** (1968).

Mazumdar, R., *Performance Modeling, Stochastic Networks, and Statistical Multiplexing* Morgan and Claypool (2010, 2013)

Metropolis, N., M.N. Rosenbluth, A.W. Rosenbluth, A.H. Teller and E. Teller, "Equations of state calculations by fast computing machines", J. Chem. Phys., 21, 1087–92 (1953).

Meyer, P.-A., *Probabilités et Potentiel*, Hermann, Paris (1972).

Meyn, S., and R.L. Tweedie, *Markov Chains and Stochastic Stability*, Cambridge University Press (2009) (first ed. Springer 1993).

Neveu, J., Processus ponctuels, in *École d'été de Saint Flour*, Lect. Notes in Math. **598**, 249–445, Springer (1976).

Norris, J.R., *Markov chains*, Cambridge University Press (1997).

Oksendal, B., *Stochastic Differential Equations*, Springer, (1985), 5th ed. 1995.

Orey, S., *Limit Theorems for Markov Chain Transition Probabilities*, Van Nostrand (1971).

Pakes, A.G., "Some conditions for ergodicity and recurrence of Markov chains", Oper. Res. 17, 1058–1061 (1969).

Pitman, J., "On coupling of Markov chains", Z. für W., 35, 4, 315–322 (1976).

Propp, J.G., and D.B. Wilson, "Exact sampling with coupled Markov chains and applications to statistical mechanics", Rand. Struct. Algs, 9, 223–252 (1996).

Reich, E., "Waiting Times when Queues are in Tandem", *Annals of Mathematical Statistics*, **28**, 768–773. (1957).

Resnick, S., *Adventures in Stochastic Processes*, Birkhäuser (1992).

Revuz, D., *Markov Chains*, North-Holland (1984).

Revuz, D., and M. Yor, *Continuous Martingales and Brownian Motion*, Springer (1991, 1994, 1999).

Rolski, T., H. Schmidli, V. Schmidt and J. Teugels, *Stochastic Processes for Insurance and Finance*, Wiley (1998).

Ross, S., *A Course in Simulation*, Macmillan (1990).

Royden, H.L., *Real Analysis*, Macmillan (1988)

Rudin, W., *Real and Complex Analysis*, McGraw-Hill (1986) .

Seneta, E., *Non-negative Matrices and Markov Chains*, 2nd edition, Springer (1981).

Serfozo, R., *Introduction to Stochastic Networks*, Springer (1999).

Serfozo, R., *Basics of Applied Stochastic Processes*, Springer (2009).

Shiryaev, A.N., *Probability*, Springer (1984, 1996).

Smith, W.L., "Regenerative stochastic processes", Proc. R. Soc. A 232, 6–31 (1955).

Stoica, P., and R. Moses, *Spectral Analysis of Signals*, Prentice-Hall (2005)

Stone, C., "On absolutely continuous components and renewal theory", *Ann. Math. Statist.*, **39**, 271–275 (1963).

Taylor, S.J., *Introduction to Measure and Integration*, Cambridge University Press (1973).

Tsybakov, B.S., and V.A. Mikhailov, "Random multiple packet access: Part-and-try algorithm", Probl. Information Transmission, 16, 4, 305–317, 1980.

Vervaat, W., "Properties of general self-similar processes", *Bull. Int. Statist. Inst.*, 52, 4, 199–216 (1987).

Watanabe, S., "On discontinuous additive functionals and Lévy measures of a Markov process", *Japanese J. Math.* **34**, 53–70 (1964).

Williams, D., *Probability and Martingales*, Cambridge University Press (1991).

Wolff, R., *Stochastic Modeling and the Theory of Queues*, Prentice-Hall, Englewood Cliffs (1989).

Index

© Springer Nature Switzerland AG 2020
P. Brémaud, *Probability Theory and Stochastic Processes*, Universitext,
https://doi.org/10.1007/978-3-030-40183-2

Printed in the United States
By Bookmasters